LASERS, MOLECULES, AND METHODS

ADVANCES IN CHEMICAL PHYSICS

VOLUME LXXIII

EDITORIAL BOARD

LASERS, MOLECULES, AND METHODS

Edited by

JOSEPH O. HIRSCHFELDER

Chemistry Departments,
University of Wisconsin, Madison
and University of California, Santa Barbara

ROBERT E. WYATT

Chemistry Department
University of Texas
Austin, Texas

ROB D. COALSON

Department of Chemistry
University of Pittsburgh
Pittsburgh, Pennsylvania

ADVANCES IN CHEMICAL PHYSICS
VOLUME LXXIII

Series editors

Ilya Prigogine

University of Brussels
Brussels, Belgium
and
University of Texas
Austin, Texas

Stuart A. Rice

Department of Chemistry
and
The James Frank Institute
University of Chicago
Chicago, Illinois

WILEY

AN INTERSCIENCE® PUBLICATION
JOHN WILEY & SONS
NEW YORK · CHICHESTER · BRISBANE · TORONTO · SINGAPORE

Library of Congress Cataloging in Publication Data:

Lasers, molecules, and methods.

　(Advances in chemical physics, ISSN 0065-2385; v. 73)
　Based on papers presented at a symposium held at the
Los Alamos Center for Nonlinear Studies on July 7–11,
1986.

　Includes bibliographies.
　1. Lasers in physics—Congresses.　2. Molecules—
Congresses. I. Hirschfelder, Joseph Oakland, 1911-
II. Wyatt, Robert E. (Robert Eugene)　III. Coalson,
Robert D.　IV. Series.
QD453.A27　vol. 73 [QC685]　541 s [542]　87-25442

ISBN 0-471-62457-8

Printed in the United States of America

10 9 8 7 6 5 4 3 2 1

CONTRIBUTORS

HENK F. ARNOLDUS, Department of Chemistry, State University of New York, Buffalo, New York

J. R. ACKERHALT, Los Alamos National Laboratory, Los Alamos, New Mexico

OSMAN ATABEK, Laboratoire de Photophysique Moleculaire, Université de Paris-Sud, Orsay, France

ANDRÉ D. BANDRAUK, Department of Chemistry, Faculté des Sciences, Université de Sherbrooke, Quebec, Canada

ROB D. COALSON, Department of Chemistry, University of Pittsburgh, Pittsburgh, Pennsylvania

SHIH-I CHU, Department of Chemistry, University of Kansas, Lawrence, Kansas

RICHARD J. COOK, Frank J. Seiler Research Laboratory, U.S. Air Force Academy, Colorado Springs, Colorado

J. D. DOLL, Los Alamos National Laboratory, Los Alamos, New Mexico

J. H. EBERLY, Department of Physics and Astronomy, University of Rochester, Rochester, New York

J. H. FREED, Department of Chemistry, Cornell University, Ithaca, New York

D. L. FREEMAN, Department of Chemistry, University of Rhode Island, Kingston, Rhode Island

PERETZ P. FRIEDMANN, Mechanical, Aerospace and Nuclear Engineering Department, University of California at Los Angeles, Los Angeles, California

T. GEISEL, Universitat Regensburg, Institute fur Theoretische Physik, Regensburg, FRG*.

THOMAS F. GEORGE, Department of Chemistry, State University of New York, Buffalo, New York

M. E. GOGGIN, Los Alamos National Laboratory, Los Alamos, New Mexico

F. T. HIOE, Department of Physics, St. John Fischer College, Rochester, New York

*Present address: Physikalisches Institut, Der Universitat Wurzburg, Wurzburd, FRG.

v

J. O. HIRSCHFELDER, Chemistry Departments, University of Wisconsin, Madison and University of California, Santa Barbara

M. A. HUSSAIN, General Electric Company, Corporate Research and Development, Schenectady, New York

JOHN S. HUTCHINSON, Department of Chemistry and Quantum Institute, Rice University, Houston, Texas

H. J. KIMBLE, Department of Physics, University of Texas at Austin, Austin, Texas

MARY ANN KMETIC, Department of Chemistry, University of Western Ontario, London, Ontario, Canada

WILLIAM J. MEATH, Department of Chemistry, University of Western Ontario, London, Ontario, Canada

P. W. MILONNI, Los Alamos National Laboratory, Los Alamos, New Mexico

RONALD. B. MORGAN, Department of Mathematics, University of Missouri, Columbia, Missouri

SHAUL MUKAMEL, Department of Chemistry, University of Rochester, Rochester, New York

ALI H. NAYFEH, Department of Engineering Science and Mechanics, Virginia Polytechnic Institute and State University, Blacksburg, Virginia

B. NOBLE, Institute of Computational Mathematics, Brunel University, Uxbridge, England

G. RADONS*, Department of Physics and Astronomy, University of Maryland, College Park, Maryland

WILLIAM P. REINHARDT, Department of Chemistry and Laboratory for Research on the Structure of Matter, University of Pennsylvania, Philadelphia, Pennsylvania

J. RUBNER, Physik Department, Technische Universitat, Munchen, FRG

D. J. SCHNEIDER, Department of Chemistry, Cornell University, Ithaca, New York

DAVID S. SCOTT, Intel Scientific Computers, Beaverton, Oregon

JAMES STONE, Rockwell International, Canoga Park, California

R. A. THURAISINGHAM, Department of Chemistry, University of Western Ontario, London, Ontario, Canada

*Present address: Institut für Theoretische Physik, Universität Kiel, Kiel, FRG.

SANDER VAN SMAALEN*, Department of Chemistry, State University of New York, Buffalo, New York

ROBERT E. WYATT, Chemistry Department, University of Texas, Austin, Texas

YI JINGYAN, Department of Chemistry, University of Rochester, Rochester, New York.

*Present address: Laboratory of Inorganic Chemistry, Materials Research Center, State University of Groningen, Nijenborgh, Groningen, The Netherlands.

SERIES INTRODUCTION

Few of us can any longer keep up with the flood of scientific literature, even in specialized subfields. Any attempt to do more and be broadly educated with respect to a large domain of science has the appearance of tilting at windmills. Yet the synthesis of ideas drawn from different subjects into new, powerful, general concepts is as valuable as ever, and the desire to remain educated persists in all scientists. This series, *Advances in Chemical Physics*, is devoted to helping the reader obtain general information about a wide variety of topics in chemical physics, which field we interpret very broadly. Our intent is to have experts present comprehensive analyses of subjects of interest and to encourage the expression of individual points of view. We hope that this approach to the presentation of an overview of a subject will both stimulate new research and serve as a personalized learning text for beginners in a field.

ILYA PRIGOGINE
STUART A. RICE

INTRODUCTION

The systematic investigation of natural phenomena and mathematical systems that are not linear has in recent years emerged as an important and exciting interdisciplinary subject. This "nonlinear science," by blending cleverly conceived computer-based numerical simulations, powerful new methods of mathematical analysis, and experimental and theoretical discoveries in science and engineering, has made remarkable progress toward solving problems that have long been considered simply intractable. Essential to this progress has been the recognition that nonlinear phenomena arising in very different contexts display common features and yield to common methods of analysis. This "universality" of nonlinear phenomena has allowed progress in one discipline to be transferred rapidly to others and confirms the inherently interdisciplinary nature of the subject.

Separately, lasers and small molecules exhibit many important nonlinear effects. Familiar nonlinear phenomena from laser physics include self-focusing, propagation of solitary wave pulses in optical fibers, and resonant transmission in multilevel media. The motion of small molecules, treated at the classical level, can exhibit the deterministically chaotic behavior that has recently been studied extensively in nonlinear dynamical systems. At the quantum level, the behavior of these molecules is related to the exciting but controversial topic of *quantum chaos*. Hence, in studying the interactions of lasers with small molecules, one can hardly be surprised to find nonlinearity playing a crucial role.

This book resulted from a symposium, "Lasers, Molecules, and Methods," held at the Los Alamos Center for Nonlinear Studies on July 7–11, 1986. The conference organizers* applied the approach that has proven successful in many other problems in which nonlinearity dominates. First, the topics presented covered a broad range of apparently disparate but fundamentally related topics from chemical kinetics to stability analysis for helicopter rotor blades. Second, the participants came from a wide variety of disciplines, including aeronautical engineering, experimental and theoretical chemistry and physics, and applied mathematics. Finally, computational, analytic, and experimental techniques were all represented. The interactions resulting from this mix of topics, disciplines, and methodologies proved both novel and stimulating. When Professor Hirschfelder proposed this symposium, we at

---

*The organizing committee was Joseph O. Hirschfelder, Robert E. Wyatt, Jay Ackerhalt, and Rob D. Coalson. In addition, Jay Ackerhalt helped with the local arrangements at Los Alamos.

the Center for Nonlinear Studies were very pleased to join with the National Science Foundation, the Army Research Office, and the Universities of California–Santa Barbara, Wisconsin–Madison, and Texas–Austin in sponsoring it. We hope that these proceedings will serve as a valuable reference for the exciting, rapidly developing, and most emphatically nonlinear subject of laser–molecule interactions.

DAVID K. CAMPBELL

Director,
Los Alamos Center for
Nonlinear Studies
Los Alamos, New Mexico

PREFACE

This book delves deeply and technically into recent advances in both theoretical and experimental research on the wide variety of dynamical and optical phenomena of laser–molecule interactions.*

The very intense laser fields now available are able to tear molecules apart and produce multiphoton ionizations. Furthermore, femtosecond pulses can be used to make a "moving picture" of the motion of the decaying molecular fragments.† Laser research is very exciting and much progress is being made. We hope that our readers will feel some of the excitement that underlies our technical discussions.

The chapters in this book are based upon presentations at the July 1987 "Lasers, Molecules, and Methods" symposium at the Los Alamos Center for Nonlinear Studies. The purpose was to bring together an interdisciplinary group of chemists, physicists, applied mathematicians, and engineers to discuss the most numerically efficient techniques for solving current and anticipated laser–molecule interaction problems. The engineers and applied mathematicians were invited because they have developed a number of techniques that could be useful in solving dynamical problems.

Much of the symposium was devoted to experimental research—the theory suggests experiments and the experiments suggest changes in the theory! Thus, it was important to learn: What are the most experimentally interesting laser problems? How successful have theoreticians been in predicting or explaining the results of these experiments? What techniques did they use and what difficulties did they encounter?

Some of the types of laser–molecule interaction problems discussed in this book are:

- Weak fields and quasi-energy states (Chapter 1)

---

*The National Science Foundation made a thorough study of all chemical applications of lasers. The experimental aspects were summarized by J. I. Steinfeld and M. S. Wrighton, eds., *The Laser Revolution in Energy-Related Research*, a 1976 NSF Report of a workshop held at MIT. Also, T. F. George edited *Theoretical Aspects of Laser Radiation and its Interaction with Atomic and Molecular Systems*, a 1977 National Science Foundation report of a workshop held at Rochester University. These reports gave short discussions of an amazing number of types of laser–molecule interaction problems! They should be brought up to date.

†R. M. Baum, "Subpicosecond Spectroscopy Gives New Data on Reaction Dynamics," *Chem. Eng. News*, p. 36 (January 13, 1986). See also N. F. Scherer, J. L. Knee, D. D. Smith, and A. H. Zewail, "Femtosecond Photofragment Spectroscopy," *J. Phys. Chem.*, **89**, 5141 (1985).

- Very intense field effects (Chapters 17–19)
- Pico- and femtosecond pulses (Chapters 9, 16)
- Classical and quantum chaos in dynamics and spectra (Chapters 21–23)
- Relaxation processes (Chapters 10–12)
- Photochemistry (Chapters 13–15)
- Effects of laser interactions with an adsorbed atom (Chapter 15)
- Fluorescence of single trapped atoms (Chapter 16)
- Effects of adding constant electric fields to lasers (Chapter 8)
- Pathological behavior occurring in dynamical systems (Chapter 3)
- Reduction of quantum noise by squeezed states (Chapter 20)

The overview chapter (Chapter 1) tries to point out relationships between these seemingly different problems and some familiar nonlaser problems.

In all of the chapters, there is a good mix of experimental problems, theory, and methods of solution. Some of the numerically efficient techniques are:

- Multiple-scales method (Chapters 2, 3)
- The *time-slicer* (Chapters 1, 4, 8). In Chapter 4 this is called *the generalized ripple*, and in Chapter 8 it is *the Riemann product integral*
- The Floquet technique (Chapters 5, 6, 10, 17)
- The Lanczos algorithm (Chapters 5, 6, 10)
- The recursive residue generation method (Chapter 5)
- Preconditioning large eigenvalue problems (Chapter 6)
- Monte Carlo methods (Chapter 7)
- The Fokker–Planck approach to modeling dynamics (Chapter 10)
- Time-dependent wavepacket perturbation theory (Chapter 13)

Because of the enthusiastic response of the participants of this symposium, we felt that it would be useful to publish this book. We are grateful for the help of all of the chapter authors and the staff of John Wiley & Sons, which made this possible.

In addition, we want to thank the Los Alamos Center for Nonlinear Studies (CNLS), The National Science Foundation, and The Army Research Office for sponsoring the symposium—and special thanks to David K. Campbell, the Director of CNLS for his encouragement.

J. O. HIRSCHFELDER,
ROBERT E. WYATT,
ROB D. COALSON

CONTENTS

SPECTROSCOPY AND RELAXATION

CHAPTER I

WHERE ARE LASER–MOLECULE INTERACTIONS HEADED?

JOSEPH O. HIRSCHFELDER

Chemistry Departments, University of Wisconsin, Madison, and University of California, Santa Barbara

CONTENTS

I. INTRODUCTION

I have thoroughly enjoyed my 57 years of applying theoretical physics to chemical problems. I was fortunate to get started during the golden age of the 1930s and now I am hoping to participate in the laser age of the 1980s and 1990s.

The 1930s were years of great discoveries in quantum mechanics, but very few people realized their significance. Our present laser age is much more exciting—in *all* research fields, great changes are being made in concepts, experiments, and applications. The lives of everybody are being changed by what we do in our laboratories!

The purpose of this book is to explain the present cutting edge of laser–molecule interaction technology at the working level and where it is headed.

This chapter should be an overview, but since I did not want to paraphrase what is in the other chapters, what I have written is *mainly an overview of what is not in the other chapters*! I have worked very hard trying to learn about some new developments and I would like to tell you what I have learned.

A. Short Discussion of the Chapter Contents

Before getting down to the nitty-gritties, let me explain some of the new or unusual topics discussed in each section of this chapter. Squeezed states are discussed in Section II. These states have stimulated a great deal of interest in reexamining the quantum-dynamical uncertainty principles. Section III considers femtosecond and ultra-intense pulses. The femtosecond pulses are interesting for two reasons: First, they are generated by colliding two picosecond pulses. A very ingenious colliding -pulse ring mode-locking device emits a continuous sequence of femtosecond pulses at regular time intervals. Second, Zewail and others are studying chemical reactions by using one laser to prepare the reagents in a well-defined excited state and using a stream of femtosecond pulses to produce a sequence of spectra that enables the experimenter to watch the progression of a chemical reaction or to determine the manner in which energy is transmitted from one part of a large molecule (e.g., chlorophyll) to another.

The intensity of ultra-intense pulses is so great that they eject a large number of electrons from atoms and produce multicharged ions, split photons into smaller photons, and so on. It is hoped that by studying the effect of ultra-intense pulses, we will get a better understanding of the electron correlations in atoms, about which we know surprisingly little. In atoms or molecules the correlation results from spin–orbit interactions. The importance of electron correlations is illustrated by *superconductivity*, which results from the correlation of Cooper pairs of electrons. Whereas individual electrons are fermions, pairs of electrons are bosons and have a tendency to bunch in momentum space. The ultra-intense-pulse theoretical question is whether the electrons in the outer shell are ejected simultaneously or sequentially. Detailed rigorous quantum-mechanical calculations need to be made in order to have confidence in the answer. However, Bernd Crasemann has derived a statistical model that appears to agree with all of the available experimental data.

I am wondering what other uses there will be for ultra-intense fields. Although their effects are very interesting, they have undesirable side effects and are not specific in the same sense as lower intensity resonances.

Both the experimental and theoretical aspects of nonlinear optics are considered in Section IV. Because of the availability of high-powered lasers, nonlinear optics has become very important. Within the last 10 years, more than 2000 papers have been published about CARS (*coherent anti-Stokes Raman scattering*) and other coherent Raman four-wave mixing processes that involve *hyperpolarizabilities*. There are three reasons why CARS is so popular: Its intensity is 10^5–10^{10} times as intense, its linewidths are narrower, and yet it contains the same information as the usual noncoherent spontaneous Raman spectra. Its disadvantages are that saturation and AC-optical Stark shifts of CARS in gases with laser intensities of only 10^9–10^{11} W/cm^2 have been observed [see Section IV, Ref. 1, p. 367]. Thus, ultra-intense fields probably cannot be used to determine coherent Raman spectra and other nonlinear *four-wave mixing processes*.

There are two kinds of nonlinear equations which occur in laser–molecule problems. The first type occur in CARS the Kerr effect, Josephson junction, and so on, where the nonlinearity is due to the polarization of the medium. Such equations can have soliton solutions. *Solitons* are traveling waves that do not change their size or shape, even after colliding with another soliton. The soliton solutions in *self-induced transparency* produce dramatic deviations from (the otherwise expected) Beer's law. Also, solitons produce surprising electrical properties for conjugate–double-bond polymers such as polyacetylene.

At present the inverse scattering transform (IST) seems to be the only analytical method for determining soliton solutions. Although the IST concept is very general (it has been called the nonlinear equation generalization of the Fourier transform), at present it can be used in only a few model nonlinear equations. For example, only a small perturbation can make the IST unusable. However, a large number of Russians and a few able Americans are trying to improve this situation.

In the second type of equation the nonlinearity is due to relaxations such as collision processes. For such cases, the Gisin techniques can be used to obtain useful solutions. Section IV discusses a very interesting new technique that Gisin, a Swiss mathematician, has devised for solving quantum-mechanical evolution equations that involve both decay and fluctuations. The only limitation to his technique is that the Hamiltonian must be either constant or periodic.

The other general methods for solving coupled sets of nonlinear equations are the multiscalar method of Chapters II and III and the quasi-linearization or periodic shooting of Chapter IV. All three have been widely used by applied mathematicians and engineers and are both accurate and numerically efficient.

Section V discusses quantum electrodynamics (QED), gauge paradoxes, and spin-generalized Poisson brackets. Quantum electrodynamics (QED) is changing. From a calculational (but not a philosophical) standpoint quantized and semiclassical field QED have become reconciled. The price for this correspondence is a modification of semiclassical dynamics. To simulate the fluctuating vacuum field, the Hamiltonian must include the radiation reaction fields of the particles' self-energy. Furthermore, at some initial time the semiclassical electromagnetic fields should have random phase and be treated as operators rather than as functions. Then semiclassical calculations can be made in the Schrödinger, or interaction, picture relatively safely. Thus, it is not necessary to use quantized fields for strong-field laser–molecule interaction problems.

The second part of Section V discusses the quantum-mechanical gauge paradoxes such as the Aharanov–Bohm and Ramsey time delay effects. The thing that they all have in common is that the paradox involves the relative phases of two options. The explanations are obvious: In quantum mechanics, relative phases play an important role in dynamics, whereas the absolute value of the phase has no physical significance.

Few seem to be aware that a classical spin-generalized Poisson bracket exists. First Sudarshan and Mukunda derived it group theoretically. Then Tom Yang and I rederived it on the basis of Breit–Pauli Hamiltonian dynamics.

However, recently Asim Barut has improved on our dynamics. First, he assumed (as we had) that an electron is a real finite-size spherical particle spinning in phase space. Then he succeeded in making a "relativistic classical electron model which exactly reflects all of the properties of the Dirac electron." Thus, he obtained both quantum-mechanical and classical versions of the equation of motion of the electron. Barut claims that the canonical coordinates of the internal motion satisfy (what Barut called) a remarkable Poisson bracket. I hope this result is confirmed.

In Section VI semiclassical and classical techniques are reviewed. There is currently a strong trend toward classical correspondence and time-dependent wavepackets, either the self-consistent variety or those that use angle action variables; the wavepackets can be either in Hilbert or phase space. Benny Gerber and his colleagues express their vibrational state time-dependent self-consistent field (TDSCF) treatments in three forms: quantum mechanical, semiclassical, and classical. (It is amazing that the three versions generally lead to almost the same results.)

With this in mind, I decided to give a detailed Drude's derivation of classical oscillator strengths, etc. In the process of explaining this, I derived a new and completely general solution for the dynamics of a classical oscillator in the presence of an arbitrary time-dependent electric field. (I expect to have fun using this equation!)

Of course, the next step was to consider the semiclassical correspondence and spectral distributions. Then I did a really thorough job explaining dynamical polarizabilities: First I made a strong case for *not* using perturbation theory to calculate polarizabilities; the perturbation series converge slowly, and the continuum (which is usually neglected) can contribute up to 30% of the polarizability of even a ground state. You should always use a variational approach—I give references to upper and lower bounds for dynamical polarizabilities.

Section VII starts with a discussion of when and why we use density matrices and ends with the observation of Sakurai that since density matrices are expressed in the Schrödinger representation, their Liouville evolution equation looks like the Heisenberg equation but has a different sign!

After this food for thought, the different representations and their properties are described. The *interaction representation* is very convenient for many laser–molecule interaction problems since it is the basis of the Dyson series for time development operators. In making canonical transformations, it is important to know about active and passive representations.

The next topic in Section VII is the two techniques for expressing laser–molecule evolution equations in terms of constant Hamiltonians.

1. The time-slicer (TS) technique. Both Bill Meath (Chapter VIII) and Peretz Friedmann (Chapter IV) give very thorough discussions of this technique (although they call it by different names). The TS divides time into small slices so that the Hamiltonian in each slice can be approximated by a constant matrix. In periodic problems, after the time development operator has been calculated for one period, Floquet theory can be used to determine all of the properties of the system for all time. Friedmann has been using the TS for a long time to determine the stability of helicopter rotor blades!

2. Shih-I Chu (Chapter XVII) has done such a brilliant job of describing the standard Floquet technique and generalizing it to apply to quasi-periodic fields with and without arbitrary relaxation that I have to applaud! However, I do show that natural linewidths can be included in the Floquet formulations by changing the Hamiltonian from being Hermitian to having time reversal symmetry. This is easy to do if one replaces the usual Hermitian scalar products by special time-reversal scalar products that Frank Weinhold derived.

The rest of Section VII is devoted to a number of very useful calculation procedures such as;

1. Variational procedures for treating either time-dependent, or periodic, Hamiltonian approximate wavefunctions.
2. Löwdin-type partitioning of Hamiltonian matrices to obtain a smaller dimensional effective Hamiltonian expressed in terms of a Green function. Lippmann–Schwinger type series can then be used to obtain many types of perturbation expansions of the wavefunctions.
3. The general technique (originally derived by Dalgarno) for evaluating all kinds of *sum rules*. It is so simple, it can be so useful, and yet so few people are aware of it that I had to include it in this chapter.

I had expected to include a short discussion of the Kirtman–Certain–Hirschfelder almost-degenerate perturbation formalism. However, I am sure that someone (hopefully me) will soon devise another procedure that can better take advantage of the high-speed computing capabilities of present computers.

Let us end this section with a short discussion of new developments in laser chemistry and noncoherent transient wave functions.

B. Laser Chemistry

There are three approaches to laser chemistry (or reaction kinetics) which we should discuss.

1. *The Dressed-State Energy Surface Technique (DSEST)*

I want to congratulate Tom George and others for developing this relatively simple technique for determining (approximate) rates for all sorts of chemical reactions in the presence of periodic laser fields. The DSEST has played a major role in the development of laser chemistry—a picture of the energy surface is worth a thousand words in explaining laser effects!

When a molecule is placed in a laser field, it becomes distorted, and its discrete states become *dressed*. These dressed states are pseudostationary or quasistationary long-lived transient states.

Mittleman's (1) proof that dressed states have physical significance is that the fluorescent spectrum emitted by the molecule agrees with the theoretically predicted energies and linewidths. If the linewidths are small compared with the energy separation between the states, their rates of decay can often be ignored, and they can be treated as though they were true quantum states.

In laser chemistry, it is fortunate that the near-resonant field strengths are intense enough to saturate but usually they are not intense enough to produce undesirable side effects. Thus, the dressed states can be a good approximation except as follows:

(a) When the states are almost degenerate,
(b) When photons are absorbed or emitted and short-lived transients are produced,
(c) Short-lived transients also occur in the vicinity of almost forbidden transients.

The DSEST comes in all degrees of sophistication. In order to explain the basic approach, let us consider the following derivation of crude semiclassical surfaces for a two-state model of a molecule in a (single-photon) near-resonant periodic field (2).

Nuclear motions are taken to be classical, whereas, the electronic dynamics are quantum mechanical. However, in the determination of the energy surfaces, the Born–Oppenheimer approximation is used. Thus, the unperturbed Schrödinger equation is

$$H^{(0)}(\mathbf{r}, \mathbf{R})\psi_j^{(0)}(\mathbf{r}, \mathbf{R}) = W_j(\mathbf{R})\psi_j^{(0)}(\mathbf{r}, \mathbf{R}),$$

where $H^{(0)}(\mathbf{r}, \mathbf{R})$ is the field-free electronic Hamiltonian for a particular nuclear configuration $\mathbf{R}$.

If the frequency and amplitude of the laser field are ω and $\mathbf{F}_0$, respectively, the detuning is $\Delta(\mathbf{R}) = \frac{1}{2}[W_2 - W_1 - \hbar\omega]$. It follows that the approximate dressed-state wavefunctions for the dressed lower (a) and upper (b) energy states are

$$\Psi_a(\mathbf{r}, \mathbf{R}) = \psi_a^{(0)}(\mathbf{r}, \mathbf{R})\cos[\Theta(\mathbf{R})]\exp[\omega t] - \psi_b^{(0)}(\mathbf{r}, \mathbf{R})\sin[\Theta(\mathbf{R})]$$

and

$$\Psi_b(\mathbf{r}, \mathbf{R}) = \psi_a^{(0)}(\mathbf{r}, \mathbf{R})\sin[\Theta(\mathbf{R})]\exp[\omega t] + \psi_b^{(0)}(\mathbf{r}, \mathbf{R})\cos[\Theta(\mathbf{R})].$$

Here

$$\Theta(\mathbf{R}) = \frac{1}{2}\tan^{-1}\frac{\lambda(\mathbf{R})}{\Delta(\mathbf{R})}$$

and

$$\lambda(\mathbf{R}) = \mathbf{F}_0 \cdot \langle\psi_1|\boldsymbol{\mu}|\psi_2\rangle$$

where $\boldsymbol{\mu}$ is the dipole operator.

It follows that for a nuclear configuration $\mathbf{R}$ the energies of the two dressed states in the rotating-wave approximation are

$$E_a(\mathbf{R}) = \frac{1}{2}[W_1 + W_2 + \hbar\omega] - [\Delta^2 + |\lambda|^2]^{1/2}$$

and

$$E_b(\mathbf{R}) = \frac{1}{2}[W_1 + W_2 + \hbar\omega] + [\Delta^2 + |\lambda|^2]^{1/2}.$$

The energy surfaces $E_a(\mathbf{R})$ and $E_b(\mathbf{R})$ are treated as though they were potential energy surfaces of the molecule in a field-free environment.

It remains to determine the S-matrix elements for a transition from the initial state a to the final state b, where now the a and b include sets of vibrational and rotational as well as electronic states. The quantum-mechanical equations of motion of the molecular system having the potential energy surfaces $E_a(\mathbf{R})$ and $E_b(\mathbf{R})$ are the same as for a similar molecular system in a field-free environment. Thus, standard (field-free) calculational techniques can be used to solve various types of chemical reaction rate problems in the presence of laser fields. In addition, special DSEST perturbation techniques have been developed that simplify the calculations.

I think of Tom George's DSEST, Eyring's theory of absolute reaction rates, and Prigogine's thermodynamics and statistical mechanics of irreversible processes as examples of first-order perturbation responses. They may not be rigorous, but they are useful. All three theories involve long-lived quasi-energy states. Both Eyring (1942) and Prigogine (1950) endowed the quasi-energy states with full sets of thermodynamic properties. It will not be long before we consider the entropy and free energy of ensembles of laser quasi-energy states in addition to the present notion of dressed potential energy surfaces.

2. Exact Time-Developing Wavepacket Solutions

I am amazed that Robert Heather, Xue-Pei Jiang, and Horia Metiu (3–5) have recently succeeded in calculating the (virtually) exact Raman spectrum associated with the photodissociation of the Kulander–Heller (6) model of two excited diabatic states of $H_3{}^+$. In dissociating, the two energy surfaces cross. After the curve crossing, one state dissociates into $H_2{}^+ + H$, while the other dissociates into $H^+ + H + H$. In this model, the nuclear motion occurs on two repulsive two-dimensional potential energy surfaces.

They started (3) with the field-free ground-state wavefunction of the $H_3{}^+$ system and let it evolve in accordance with the Schrödinger equation by expressing the system alternately in the coordinate and momentum representations. [Horia Metiu told me that this use of N-dimensional wavefunctions is more computationally efficient than using N^2-dimensional matrices.] Their reason for calculating the Raman spectrum is that "near-resonant Raman scattering probes the dynamics of photodissociation in the interaction region and generates a large data pool which is sensitive to the dynamic parameters of the model." Coherent anti-Stokes Raman spectra (CARS) has been used to experimentally determine the photofragmentation of many molecules [Sections IV.A and VI.F.1].

The fine details that they "observed" are fantastic:

(a) At first, wavefunction transitions are produced by the absorption of a photon. As a result, the transition probability varies initially as t^2, and a characteristic transient spectrum (with side-bands) is emitted for a time of only 2×10^{-15} s (see section I.C).

(b) Also, in the vicinity of their curve crossing they found that the Raman photons emitted by the two upper energy states interfere. This makes the Raman spectrum sensitive to the existence of the curve crossing and the magnitude of the coupling between the curves (3, 4).

The *fast Fourier transform split-operator technique* (FFTSOT) they use (5) is a generalized version of the procedure developed by Fleck, Morris, and Feit (7) that greatly simplifies the solution of few dimensional dynamical problems that evolve in a short time and are spatially localized. It is very numerically efficient since it converts all matrix operations into Fourier transforms. Whereas multiplication of N-dimensional matrices requires N^2 operations, fast Fourier transforms only require $N \log_e(N)$ operations. The wide range of applicability of the FFTSOT together with its advantages, disadvantages, and optimal conditions for the FFTSOT are discussed in (4).

As in the time-slicer technique, time is divided into M sufficiently small slices, $t_\alpha - \tau < t < t_\alpha$, that within each slice the time variation of the Hamiltonian, $H_\alpha(x)$, can be neglected. Furthermore, the localized space, $0 < x < L$, is spanned by a grid containing N equally spaced points, x_n. This grid serves to define both the N mutually orthogonal states $|x_n\rangle$ in the coordinate representation and also the N conjugate eigenstates $|k_\mu\rangle$ in the momentum space (5, 8). The values of $k_\mu = \hbar p_\mu$ vary in equal intervals from $-N\pi/L$ to $(N-1)\pi/L$. (In order to simplify my derivation, I assume that the space is one-dimensional.)

The Hamiltonian is then expressed as the sum of the kinetic energy $K(k)$ and the potential energy $V_\alpha(x)$. Thus the time development operator for the αth slice is $U_\alpha(\tau) = \exp[-i\tau(K + V_\alpha)/\hbar]$, where K is an operator in momentum space and V_α is an operator in coordinate space. Thus, $\Psi(t_\alpha) = U_\alpha(\tau)\Psi(t_{\alpha-1})$.

Since the coordinates and momenta do not commute, $U_\alpha(\tau)$ is *not equal* to the product $Z_\alpha(\tau) = \exp[-i\tau V_\alpha/\hbar]\exp[-i\tau K/\hbar]$. [Instead, $Z_\alpha(\tau)$ is a first-order approximation to $U_\alpha(\tau)$, which is frequently used in calculations of Feynman path integrals when the second-order term $(\tau/\hbar)[K, V_\alpha]$ is much less than either $K/\hbar$ or $V_\alpha/\hbar$.] However, the Moyal approximation [see Eq. (4.16) of (8)]

$$U_\alpha(\tau) \approx \exp[-i\tau K/2\hbar]\exp[-i\tau V_\alpha/\hbar]\exp[-i\tau K/2\hbar],$$

which is *accurate through the second order*, is used in (3–7).

When the kinetic energy is expressed in the momentum representation,

$$\exp[-i\tau K/\hbar] \quad \text{becomes} \quad \sum_\mu |k_\mu\rangle K(\mu, \tau)\langle k_\mu|$$

where

$$K(\mu, \tau) = \langle k_\mu| \exp[-i\tau K/\hbar]|k_\mu\rangle,$$

and when the potential energy is expressed in the coordinate representation,

$$\exp[-i\tau V_\alpha(x)/\hbar] \quad \text{becomes} \quad \sum_n |x_n\rangle V(\alpha, n)\langle x_n|$$

where

$$V(\alpha, n) = \langle x_n| \exp[-i\tau V_\alpha(x_n)/\hbar]|x_n\rangle.$$

Furthermore, the value of $\Psi(x, t)$ at x_n is $\Psi(x_n, t) = \langle x_n|\Psi(t)\rangle$. [Reference 9, pp. 46–50, discusses coordinate and momentum representations.]

In order to switch from the coordinate to the momentum representation and back again, it is necessary to know the Fourier transform coefficients,

$$\langle k_\mu|x_n\rangle = N^{-1/2}\exp[-ix_n k_\mu] \quad \text{and} \quad \langle x_n|k_\mu\rangle = N^{-1/2}\exp[ix_n k_\mu]$$

Thus, it follows that

$$\langle \Psi(t_\alpha)|x_c\rangle = \sum_{j,\mu,\nu} \langle x_c|k_\mu\rangle K(\mu, \tau/2)\langle k_\mu|x_j\rangle V(\alpha, j)$$
$$\times \langle x_j|k_\nu\rangle K(\nu, \tau/2)\langle k_\nu|x_b\rangle\langle x_b|\Psi(t_{\alpha-1})\rangle,$$

so that four Fourier transforms are required in order to determine $\Psi(x, t_\alpha)$ from $\Psi(x, t_{\alpha-1})$. Thus, it would appear that $4M$ Fourier transforms would be required to determine the wavefunction at a final time $t_f = M\tau$ from $\Psi(x, 0)$. However,

$$\langle \Psi(t_f)|x_c\rangle$$
$$= \sum \langle x_c|k_{\mu_{M+1}}\rangle K(\mu_{M+1}, \tau/2)\langle k_{\mu_{M+1}}|x_{j_M}\rangle V(M, j_M)\langle x_{j_M}|k_{\mu_M}\rangle$$
$$\times K(\mu_M, \tau)\langle k_{\mu_M}|x_{j_{M-1}}\rangle V(M-1, j_{M-1})\langle x_{j_{M-1}}|k_{\mu_{M-1}}\rangle K(\mu_{M-1}, \tau)$$
$$\times \cdots \times V(1, j_1)\langle x_{j_1}|k_{\mu_1}\rangle K(\mu_1, \tau/2)\langle k_{\mu_1}|x_b\rangle\langle x_b|\Psi(0)\rangle$$

where the summation is over all of the N values of each of the $M + 1$ k_μ's and M x_j's. Thus, only $2M + 2$ Fourier transforms are required to determine the wavefunction after M time intervals.

It now takes Metiu et al. only 35 of computing time on a Micro-Vax to determine the Raman spectrum associated with the photodissociation of H_3^+ in spite of the fact that this is a two-state two-dimensional problem with a curve crossing!

Clearly, this technique will be very useful for determining exact solutions of many laser-chemical problems! [Paul L. DeVries (Department of Physics,

Miami University, Oxford, Ohio, personal communication, 1987) has been using a similar technique to calculate the *above-threshold ionization of atoms.* DeVries, too, has been "observing" transient effects, getting very accurate results, and is currently writing a report on this work.]

3. *Applications of Scattering Theory to Laser Chemical Problems*

We are all familiar with collisional scattering (10) and radiative scattering (11). Now we should recognize laser-chemical problems as examples of scattering by both collisions and radiation. From this standpoint, we should be able to use standard semiclassical time-dependent potential scattering to solve chemical dynamic problems in the presence of electromagnetic fields.

There are two steps in the derivation of the scattering formulations. In the first step, the scattering is expressed as a time-developing process [similar to Metiu's (3) calculations of the Raman effect and molecular dynamics as a function of time]. The second (and simplifying) step is to express the scattering in terms of a time-independent continuous flow of ensembles. Usually this is done by making a stationary phase assumption; this would not be applicable in the Metiu Raman problem either in the vicinity of its curve crossing or when transients occurred in the wavefunction. However, the uniform approximation has been used successfully in chemical kinetics in place of the stationary phase approximation to treat curve-crossing problems. I do not know whether the uniform approximation would suffice to treat non-coherent transient wavefunctions, but I am sure that some special technique could be devised.

By using this scattering approach, chemical kinetics with or without a laser field could be expressed as fluid dynamics in phase space!

C. Noncoherent Transient States: Theory and Experiments

Transients occur because nature abhors a sudden change. In mathematics, there is the Fourier series Gibbs phenomenon; in electromagnetic fields, there are surge currents. As field intensities increase and pulse times decrease within the next few years, these transient solutions can become a serious problem in laser–molecule problems.

In 1931, Eugene Wigner predicted that *immediately* after an atom is placed in a *strictly monochromatic* resonant field, its wavefunction would vary linearly with time and therefore its transition probability would vary quadratically with time (12). This was contrary to all existing experimental data and what was then known about chemical kinetics.

Wigner argued that this effect had not been observed because the electromagnetic fields were not sufficiently monochromatic. However, even after the discovery of lasers, this effect was not observed. Indeed, it was not until 1972 that the paradox was first observed (13).

1.   Theory

If an atom in its ground state at $t = 0$ is suddenly placed in a laser field with detuning $\Delta = \frac{1}{2}(\omega - \omega_0)$, it follows from the rotating-wave approximation (but not from Floquet formalism) that its almost-resonant transition probability is (14, 15)

$$P_{\beta \leftarrow \alpha}(t) = \lambda^2 t^2 \sin^2 (\Delta t)/(\Delta t)^2 \qquad (1.1)$$

[Note that Eq. (1.1) is the same as the usual Rabi transition probability except that the detuning Δ occurs in place of the Rabi frequency $q = (\lambda^2 + \Delta^2)^{1/2}$]. Thus, the transition probability *initially* varies quadratically with time. But since $\sin (\Delta t) = 0$ when $t = (\pi/2)\Delta^{-1}$, it is evident that Eq. (1.1) and the quadratic time variation can only apply when $t \ll \Delta^{-1}$ and the radiation is sufficiently coherent.

2.   Experiment

Because this time is so very short, it was necessary for Bill Klemperer [Dyke et al. (13) and Steinfeld (15)] to devise a very ingenious experiment to observe the quadratic and oscillatory behavior. In this experiment, a narrow radio-frequency laser beam passes across a hypersonic atomic beam that has a uniform preselected velocity. Thus, the exposure time for an atom is

$$t = [\text{width of the laser beam}]/[\text{velocity of the laser beam}].$$

In a typical experiment, $t < 10^{-6}$ s. They found that the transition probability as a function of the laser frequency oscillated and was damped by a Lorentzian envelope in almost perfect agreement with Eq. (1.1).

Ordinarily Eq. (1.1) does not apply since $t \gg (\Delta \omega_k)^{-1}$ and the radiation may not be sufficiently coherent. Thus, in order to determine a measurable transition rate, it is usually necessary to average $\Delta \omega_k$ over the linewidth of the radiation. In this case, the transition probability becomes proportional to time in accordance with Fermi's *golden rule* (16), and usual chemical kinetics applies.

References

1. M. H. Mittleman, *Theory of Laser–Atom Interactions*, Plenum, New York, 1982, Chapter 3.
2. T. F. George, I. H. Zimmerman, P. L. DeVries, J.-M. Yuan, K.-S. Lam, J. C. Bellum, H.-W. Lee, M. S. Slutsky, and J.-T. Lin, "Molecular Rate Processes in Intense Laser Radiation," in *Chemical and Biological Applications of Lasers,* Vol. 4. p. 253 C. B. Moore, Ed., Academic, New York, 1979.
3. R. Heather, X.-P. Jiang, and H. Metiu, Use of Raman Spectroscopy to Study Photodissociation Dynamics, *Chem. Phys. Lett.* **142**, 303 (1987).

4. J. Alvarellos and H. Metiu, Evolution of a Wave Function in a Curve Crossing Problem Computed by a Fast Fourier Transform Method, *J. Chem. Phys.* **88**, 4957 (1988).

5. R. Heather and H. Metiu, Calculating Wave Function Evolution by Fast Fourier Transform Methods for Systems with Spatially Extended Wave Functions and Localized Potentials, *J. Chem. Phys.* **86**, 5009 (1987).

6. K. C. Kulander and E. J. Heller, Time Dependent Formulation of Polyatomic Photofragmentation: Application to H_3^+, *J. Chem. Phys.* **69**, 2439 (1978).

7. J. A. Fleck, Jr., J. R. Morris, and M. D. Feit, *Appl. Phys.* **10**, 129 (1967).

8. R. M. Wilcox, Exponential Operators and Parameter Differentiation in Quantum Physics, *J. Math. Phys.* **8**, 962 (1967), Eq. 2.1; R. F. Snider, *J. Math. Phys.* **5**, 1586 (1964).

9. J. J. Sakurai, *Modern Quantum Mechanics*, Benjamin-Cummings, Menlo Park, CA, 1985, p. 181 and pp. 46–50.

10. C. J. Joachim, *Quantum Collision Theory*, North-Holland, Amsterdam, 1975, Chapters 3 and 12.

11. S. Chandrasekhar, *Radiation Transfer*, Oxford University Press, Oxford, 1950.

12. J. O. Hirschfelder, C. F. Curtiss, and R. B. Bird, *Molecular Theory of Gases and Liquids*, Wiley, New York, 1954, 1964, pp. 881–889.

13. T. R. Dyke, G. R. Tomasevich, W. Klemperer, and W. Falconer, Electric Resonance Spectroscopy of Hypersonic Molecular Beams, *J. Chem. Phys.* **57**, 2277 (1972).

14. R. Loudon, *The Quantum Theory of Light*, Oxford University Press, Oxford, 1983.

15. J. I. Steinfeld, *Molecules and Radiation*, MIT Press, Cambridge, MA, 1981. pp. 12–17.

16. P. L. Knight and L. Allen, *Concepts of Quantum Optics*, Pergamon, New York, 1983, pp. 30–33.

II. SQUEEZED STATES

A very important advance in laser technology occurred in 1985 when Richard Slusher and his Bell Laboratory group (1) succeeded in producing a *squeezed-state laser* wavepacket which (in homodyne detection) reduced the quantum noise 4–17% below the *standard quantum limit* (SQL) imposed by the quantum-mechanical uncertainty principle. Subsequently, Jeff Kimble et al. (2) succeeded in reducing the quantum noise by 60%, as Kimble explains in Chapter XX of our present book. Slusher generated his squeezed states by using *four-wave mixing* to create two pairs of phase-conjugated waves; whereas Kimble used *parametric down conversion* in which photons of frequency 2ω were converted to conjugate pairs of frequency ω photons. Since 1970, the possibility of using squeezed states to reduce the quantum noise has been actively discussed and analyzed (3–5). This section is intended to be an introduction to Chapter XX.

To date, the achievements of quantum optics have been based on the measurements of photon correlations. With squeezed states, we seem to be on the verge of being able to measure phase-dependent photon correlation functions.

Squeezed states will have many important (research and practical) applications (3). Since squeezed states increase both the possible signal-to-noise ratio and the useful amplification of signals, they will be useful in many types of interferometric and quantum optics experimentation. Furthermore, they make it possible to increase the number of signals transmitted simultaneously through an optical fiber.

The reduction of quantum noise by squeezed states has stimulated a great deal of interest in uncertainty relations.

1. Han, Kim, and Noz (6) have shown that the time–energy relation *for light waves*, $\Delta t \cdot \Delta \omega \geqslant 1$, is a Lorentz invariant. However, they were not able to derive the corresponding relation for photons because they do not know of any way of mathematically localizing photons, even though they can be somewhat localized in laboratory experiments.

2. Finkel (7) shows how to derive uncertainty relations for statistical measures other than variances by generalizing the *Heisenberg uncertainty principle*,

$$\Delta X \cdot \Delta Y \geqslant \tfrac{1}{2} |\langle \Psi | [X, Y] | \Psi \rangle|.$$

His formulation makes it possible to derive some very interesting relations between angular momentum components.

3. Now that it has become possible to reduce quantum noise, the question arises: Is it possible to measure either distance or momentum with a variance less than the SQL? Measurements with coherent wavepackets can attain the SQL,

$$|\Delta r| = (\tfrac{1}{2}\hbar)^{1/2} = |\Delta p|.$$

Caves (8) and Partovi and Blankenbecler (9) prove that the variance obtained by making successive measurements *cannot* be less than the SQL. However, Yuen (10) has suggested that there might be ways to beat the SQL. The difference is in the proposed way of measuring the variance. According to Eberly (personal communication) an experiment is being set up at Munich to determine which is right.

Dynamic holograms can be made by phase conjugated waves (not necessarily involving squeezed states). Boris Ya Zel'dovich (4) stated:

When a beam of light is reflected by a mirror, the reflected beam is the same as the original beam with the exception that it is time reversed and phase conjugated with respect to the original beam. Traditional *static holography* consists of the following three distinct steps:

1. A hologram is recorded by illuminating a photographic transparency with a two-wave interference pattern resulting from an object and a reference beam.
2. The film is developed.
3. And the hologram is read out with another or the same reference beam.

Four-wave mixing is an example of dynamic holography because all three processes—recording, developing, and reading out—occur simultaneously. The induced variations in the refractive index disappear after the illuminating radiation is turned off and the hologram is continually changing in response to variation in the object beam.

References

1. R. E. Slusher, L. W. Hollberg, B. Yurke, J. R. Klauder, and J. F. Valley, *Phys. Rev. Lett.* **55**, 2409 (1985).
2. Ling-An Wu, H. J. Kimble, J. L. Hall, and Huifa Wu, Generation of Squeezed States by Parametric Down Conversion, *Phys. Rev. Lett.* **57**, 2520 (1986).
3. D. F. Walls, Squeezed States of Light, *Nature* **306**, 141 (1983) gives an excellent review of squeezed states and their possible applications.
4. V. V. Shkunov and B. Ya Zel'dovich, Oprical Phase Conjugation, *Sci. Am.*, January 1985, p. 54.
5. D. M. Pepper, Applications of Optical Phase Conjugation, *Sci. Am.*, January 1986, p. 74.
6. D. Han, Y. S. Kim, and M. E. Noz, *Phys. Rev.* **A35**, 1682 (1987).
7. R. W. Finkel, *Phys. Rev.* **A35**, 1486 (1987).
8. C. M. Caves, Defense of the Standard Limit for Free Mass Position, *Phys. Rev. Lett.* **54**, 2465 (1985).
9. M. H. Partovi and R. Blankenbecler, Quantum Limit for Successive Position Measurements, *Phys. Rev. Lett.* **57**, 2887, 2891 (1986).
10. H. P. Yuen, *Phys. Rev. Lett.* **51**. 719 (1983); **52**, 1730 (1984); **56**, 2176 (1986).

III. SHORT INTENSE PULSES

A. Very Fast Laser Pulses

Very fast laser pulses are becoming very useful in studying detailed mechanisms of molecular energy transfers and chemical kinetics. Every time that a new experimental technique has been developed for generating shorter pulses, it has opened up the possibility of making new applications.

After World War II, the American army left a large number of very powerful antiaircraft search lights in England and declared them surplus. This enabled Professor R. G. W. Norrish to obtain about 20 of them which he thought might be useful for his research at Cambridge on photochemistry. Thus, in 1949 R. G. W. and his student (now Sir) George Porter used batteries of these antiaircraft search lights to produce microsecond photoflashes and make spectroscopic observations of short-lived intermediates in chemical reactions. In contrast, we can *now* explore intermediates that have 10^{-8} times shorter lifetimes.

For a pulse to excite a particular molecular mode, its duration must be comparable with the characteristic time of the mode. I have found Graham Fleming's book *Chemical Applications of Ultrafast Spectroscopy* (1) very useful.

Fleming's (1) Table 1, reproduced here, shows the characteristic time for various molecular processes.

TABLE 1. The Time Range for Molecular Processes

10^{-14}–10^{-13} sec	Collision time in liquids, electronic dephasing, vibrational motions
10^{-14}–10^{-12} sec	Photoionization and Photodissociation
10^{-13}–10^{-12} sec	Solvent relaxation, cage recombination energy transfer in photosynthesis
10^{-13}–10^{-11} sec	Vibrational dephasing and relaxation
10^{-12}–10^{-11} sec	Photosynthesis electron transfer
10^{-12}–10^{-10} sec	Proton transfer
10^{-12}–10^{-9} sec	Molecular rotations
10^{-12}–10^{-8} sec	Photochemical isomerization, electronic relaxation, protein internal motion
10^{-9}–10^{-8} sec	DNA torsional dynamics

Since all of these times are larger than 10^{-15} s, femtosecond pulses should be sufficiently short for molecular or chemical applications. An electron in the nth atomic hydrogen orbit goes around its orbit in $t_n = 0.152n^3$ femtoseconds.

Picosecond pulses are adequate for research on vibrational and rotational motions, whereas femtosecond pulses make it possible to learn a great deal about the motion and coupling of electrons.

Observed spectral linewidth depends upon both dephasing- and population-relaxation rates (which can be quite different). Time-resolved spectroscopy can determine the rates of both kinds of processes. Thus, *the two relaxation constants Γ_1 and Γ_2 in the optical Bloch equation can be experimentally determined.*

It will be vary difficult to decrease the pulse lengths much farther since a femtosecond pulse of visible light (with frequency 10^{14}–10^{15} Hz) can have only one to ten cycles of radiation; similarly, a picosecond infrared pulse (with frequency 10^{12}–10^{14} Hz) can only have one to a hundred cycles.

The properties of *single-cycle pulses* is a topic of considerable curiosity. However, the Rabi frequency depends on the pulse area λT (where T is the pulse length and 2λ is the product of the dipole moment and the field strength) and not on the number of cycles. Therefore, the effects of a single-cycle pulse would be normal except that the transition probabilities it induced would be extremely sensitive to the pulse length.

B. Generation of Femtosecond Pulses

The generation of *femtosecond pulses* (1) is fascinating. A sequence of femtosecond pulses (with a regular interval between them) is generated by making use of *continuous-wave* (cw) *mode-locked organic dye lasers* and *pulse compression*:

(a) *Mode-Locking* consists in placing a shutter in front of the laser cavity and opening it every period $T = 2\pi/\omega = 2L/c$. Here L is the optical distance between two mirrors and c is the velocity of light. The shutter is kept open for the duration of a pulse $\tau = 2\pi/\Delta\omega$. The number of cycles per pulse is $N = T/\tau$. As a result of the shutter, all of the modes that get out have the same phase.

(b) *Colliding-pulse ring mode-locking.* Two pulses collide to provide a standing wave that is much shorter and more intense than the original pulses. The presence of an organic dye such as rhodamine in the space between the two pulses greatly enhances this effect. Amazingly, this device is able to emit a continuous sequence of pulses at regular intervals.

(c) *Pulse compression* makes use of four-stage frequency-doubled Nd–YAG laser amplifier to increase the intensity of the pulse. Although present femtosecond pulses are intense enough to produce saturation, and are quite adequate for most chemical applications, the maximum intensity currently available is many orders of magnitude less than the ultra intensities picosecond pulses can have. Thus, there is considerable research underway to improve their intensity-amplifying process.

C. Femtosecond Moving Pictures

Recently, Ahmed Zewail and others (2–6) experimentally demonstrated how femtosecond pulses can be used to make a "moving picture" of the motion of molecular fragments in chemical dissociation reactions. The data that he obtains can be used to construct *potential energy curves*. Previously, all potential energy curves had been theoretically determined but never observed.

Zewail uses one laser pulse to prepare the reagent in a well-defined rotational–vibrational state of an excited electronic state (which has a repulsive potential energy curve). Then he uses a second laser as a probe pulse to monitor the reaction products at regular time intervals. Thus, he is able to determine the rates at which photofragments are moving apart (2). In this way he studied both the photofragmentation of $H_2O_2 \rightarrow 2OH$ (3, 4) and of $ICN \rightarrow I + CN$ (5). His results agree with those of other types of spectroscopic studies. Furthermore, Zewail has been able to determine the properties of the activated state which occurs in the reactive collision (6) of a hydrogen atom with CO_2 to form HOCO, and which then dissociates into $HO + CO$.

This is truly an exciting new development in chemical dynamics! Indeed, this technique has been so successful that many other people have started to use it.

D. Effects of Varying Laser Intensities

The effects of lasers depend on their strength. For constant fields, strength means *field strength F* expressed in volts per centimeter. Oscillating fields are usually characterized by their cycle-averaged intensity,

$$I = (c/2\pi)|F|^2 = 1.4 \times 10^7 F^2 \qquad (\text{W/cm}^2).$$

The atomic units of field strength and intensity are

$$F_a = \frac{e}{a_0^2} = 5.14 \times 10^9 \text{ V/cm} \quad \text{and} \quad I_a = 7 \times 10^{16} \text{ W/cm}^2.$$

[I think that both Shih-I Chu (Chapter XVII) and Joe Eberly (Chapter XVIII) agree with these numbers.] Note that I_a is between 10^8 and 10^{10} times larger than the laser intensities usually available in a chemistry laboratory. However, picosecond pulses 100 times larger than I_a are now available.

The effects of lasers as a function of intensity can be expressed roughly as in Table 2.

Molecules can no longer stick together if the laser intensity is greater than the mean Coulombic intensity acting upon the wavefunction of the state to be excited. Whereas ultra-intense lasers are required to exceed the mean Coulombic intensity of ground states, even weak lasers exceed the Coulombic field of Rydberg states.

TABLE 2. Optical Resonant Frequency Lasers and Their Effects

Weak fields: $I < 10^6$ W/cm², does not saturate, distorts and polarizes atom

Strong fields: $10^6 < I > 10^8$ W/cm², saturates, bound–bound transitions

Intense fields: $10^8 < I > 10^{12}$ W/cm², many-photon excitations and ionizations more frequent than bound–bound transitions

Very intense fields: $10^{12} < I > 10^{14}$ W/cm², above-threshold ionization, multiply charged ions, harmonic Generation, and fluorescence; low-order perturbation theory applies

Ultra-intense fields: $I > 10^{14}$ W/cm², same effects as very intense fields, but perturbation theory does not apply; whereas picosecond pulses can be ultra-intense; ultra-intense femtosecond pulses have not yet been made

---

Within the next year or two (1988–1989), laser intensities larger than the Compton intensity, $I_{Comp} = 5 \times 10^{19}$ W/cm^2, will become available. Charles Rhodes defines I_{Comp} as the intensity at which nonrelativistically calculated electron quiver velocity is equal to the velocity of light. Although there is no agreement on this definition, everyone agrees I_{Comp} is near 10^{20} W/cm^2. If the intensity exceeded I_{Comp}, the photon–electron scattering cross sections would become very large, and many kinds of relativistic effects would occur. Although the effects of such ultra-intense fields will be interesting and perhaps useful for either nuclear research or *Star Wars*, it is difficult to imagine that they would be useful for studying any molecular properties. As a matter of fact, *laser intensities that are larger than necessary to perform the desired job are likely to produce some unwanted side effects!*

E.   Experimental Effects of Ultra-Intense Short Pulses: ATI, Multiply Charged Ions, and Harmonic Generation

When atoms are irradiated by very strong monochromatic pulses with intensity greater than 10^{12} W/cm^2, three very interesting effects occur: *above-threshold ionization* (ATI), *multiply-charged positive ion formation*, and *harmonic photon generation*. (I hope that my summary will supplement Joe Eberly's excellent discussion of ATI in Chapter XVIII. I greatly admire both his knowledge and his intuition.)

I have just returned from a NATO workshop (7) (Sherbrooke, Quebec, 1987) on the effects of short laser pulses organized by André Bandrauk (author of Chapter XIX). Everyone agreed that the effects of 10^{12}–10^{14} W/cm^2 picosecond pulses can be explained by low-order perturbation theory. However, there was no agreement about the detailed mechanism of shorter, more intense pulses, which we call *ultra-intense* since perturbations cannot explain their effects. Some experts believe that the electrons in the outer atomic shell are ejected sequentially (one at a time); whereas other experts argue that groups of many electrons are ejected simultaneously. I shall present both arguments.

Let us consider the experimental results of Charles Rhodes and his group at the University of Illinois at Chicago using (193-nm) ultraviolet picosecond 10^{15} W/cm^2 pulses (8–11).

(a) *Above-threshold ionization* (ATI). The atom absorbs more than the least number of photons needed to reach the ionization threshold. The photoelectrons produced by ionization gain kinetic energy by absorbing additional photons. Rhodes et al. (8) observed the following numbers of photons absorbed per ion formed: He, 4–6; Ne, 4 and 5; A, 3–6; Kr, 3–5; Xe, 2 and 3.

Many explanations of ATI have been proposed. Both Eberly (Chapter (XVIII) and Ref. 12) and Dulcic (13) have concluded that the bound–

continuum transitions determine the overall ionization rate while the continuum–continuum transitions redistribute the continuum states into ATI peaks! However, there is still an argument as to the role of the pondermotive (quiver) energy in producing ATI (14–17).

(b) *Multiply-charged positive ions* are formed in surprisingly large numbers. Rhodes et al. (8, 9) observed the following multiply-charged positive ions: He^{2+}, Ne^{2+}, Ar^{6+}, Xe^{8+}, Hg^{4+}, and U^{10+}.

Rhodes (8) proposed that the electrons in the inner shells are excited while the loosely bound outer shell is coherently driven by the laser field. Two or more of the outer shell electrons become coupled and move as a unit. Then the excitation energy of the inner shells is transferred to the outer shell electrons, which are thereby simultaneously ejected from the atom by a *reverse Coster–Kronig* type *Auger effect* (9, 10). However, many people do not believe that the experimental evidence for this mechanism is conclusive at the present time (14–21).

(c) *Harmonic photon generation* is produced by the scattering mechanism $N\gamma(\omega) + X \to \gamma(N\omega) + X$ arising from the nonlinear nature of the atomic susceptibility at the high-field strength. Here X can be either an atom or an ion. In the electric dipole approximation, the harmonic order N is restricted to odd integers. Rhodes and co-workers (11) observed the following maximum harmonics: He, 13; Ne, 17; Ar, Kr, 7; and Xe, 9.

F. Theory of Multiphoton Ionization of Atoms: Perturbative, Nonperturbative, and Statistical

1. Lambropoulos and Tang (22–24) have developed a very general procedure for estimating the number of multi charged positive ions formed by a laser pulse. They express the probability of producing a singly charged ion by the absorption of K photons as

$$P_K = 1 - \exp\left(-\sigma_K \int_0^\infty [G(t)]^K \, dt \right),$$

where σ_K is the generalized K-photon ionization cross section (expressed in units of $cm^{2k} \sec^{k-1}$), which can be determined by the lowest nonvanishing order of perturbation theory (24). Furthermore, $G(t)$ is the flux of the pulse (number of photons per cm^2 per sec), which is $6.24 \times 10^{18} I(t)/\hbar\omega$ [where $\hbar\omega$ is expressed in eV and $I(t)$ is in W/cm^2]. They have found that the shape of most experimental pulses is reasonably well fitted by

$$G(t) = G_0 \frac{t^2}{\tau^2} \exp\left(1 - \frac{t^2}{\tau^2} \right)$$

where τ is a scale factor.

Lambropoulos (22–24) shows that in Rhodes' experiments with picosecond ultraviolet 10^{15} W/cm² peak intensity fields, Xe is ionized to Xe⁺ during the first 0.1 ps (in which the field intensity is less than 10^{13} W/cm²) and the Xe⁺ is ionized to Xe²⁺ in less than another 0.1 ps (in which the field is less than 10^{14} W/cm²). These ionization times are typical of all atoms.

Thus, he concludes that (for picosecond pulses) with intensities up to 10^{17} W/cm² the first few stages of ionization proceed sequentially during the first part of the rise to the laser peak intensity during which the intensity is sufficiently low (10^{14} W/cm²) that low-order perturbation theory is applicable. However, Lambropoulos adds that his arguments would probably not apply to femtosecond pulses with the same peak intensity. It is also doubtful whether they will apply to the experiments using 10^{19} W/cm² pulses Rhodes hopes to carry out in 1988.

2. *Electron correlations.* We know surprisingly little about electron couplings. Cooper electron pairs in superconductivity (which behave like a single boson particle with charge $-2e$) are a good example of electron coupling. To a first approximation, a Cooper pair consists of two free electrons in an antisymmetric spin state. Their unusual properties result from the attractive interactions of a Fermi gas (25). It is important to note that although these electrons are correlated in momentum space and *not* in coordinate space (25–26), they can "condense" into a single quantum state in a superconductor.

Chemists have only recently started to realize the importance of electron correlations in atomic and molecular structures (16, 27), which single-configurational wavefunctions cannot represent. Electron correlations produce strong interactions between excited configurations that are otherwise degenerate or nearly degenerate. Strong coupling of these excitations can alter and even change the nature of the spectrum. The electron coupling "affects the behavior of these systems in radiative and collisional processes or in external fields" (16).

The multiconfigurational wavefunctions are eigenfunctions of irreducible representations of angular momentum quasi-spin groups (with seniority) (16, 27). The number of electrons correlated depends upon the dimension of the irreducible representation. A very similar group-theoretic technique has been used since the early days of quantum mechanics for determining the overtones of degenerate molecular vibrational modes (16).

3. *A statistical formulation* of the multiphoton production of multiply-charged ions by strong laser pulses has been derived by Crasemann and co-workers (19, 20) and seems to agree with all the experimental evidence. However, before we can have confidence in its accuracy, we would like to compare it with a really good detailed quantum-mechanical formulation, which does not yet exist.

Crasemann's stochastic approach was based upon research on collisional charge transfer ionization processes (28, 29) such as, $A^{q+} + B \rightarrow A^{(q-k)+} + B^{n+} + (n-k)e$, which assumed that charge-state distributions resulting from collisions (or from laser pulses) depend upon the number of ways in which the final charge state can be realized within the constraints of quantum conditions and conservation laws. The large number of alternative routes leading to multiphoton absorption and multiply-charged ion formation suggested that the process should be treated as a Markovian function of time (i.e., by the most probable path technique). Thus, Crasemann et al maximized the entropy, which was defined in terms of the probability of observing every possible ionic state of every possible charge. In this way they determined two possible ion distribution functions (each having a single adjustable parameter):

(a) *The binomial distribution* corresponds to simultaneous ejection of the electrons. This distribution fits *collisional* charge transfer process data (28).

(b) *The truncated Poisson distribution* is applicable if the multiply-charged ions are formed sequentially by ejection of one electron at a time.

Both distributions fit Rhodes's ultraviolet experimental results quite well, but the truncated Poisson fit the data significantly better than the binomial. Indeed, it appears that in high-intensity ($\geqslant 10^{14}$ W/cm^2) visible (as well as ultraviolet) picosecond pulses, multiply charged ions are predominantly created by a stepwise process.

On the other hand, high intensity *infrared pulses* have a very large number of photons absorbed and favour simultaneous ionization processes such as the tunneling of several electrons along the field axis [in a manner similar to the collisional charge transfer processes (28)]. Thus, the *infrared pulses have binomial charge distributions.*

In general, the more intense the field and the greater the number of photons wandering around inside the atom, the more likely it will be that the multiply-charged ions will be produced by the Coster–Kronig simultaneous electron ejection (binomial distribution function) mechanism.

This statistical formulation of multiply-charged ions should have many interesting application. For example,

(a) Rhodes's forthcoming 10^{19} W/cm^2 pulse experimental data;

(b) X-ray photoionization of K- or L-shell electrons (30), and

(c) nuclear decay Auger processes (either internal conversion of γ-rays or nuclear capture of an electron).

4. *Rhodes response.* In a study by Charles Rhodes and his colleagues (10) they present very convincing experimental evidence that at least two-electron correlations play an important role in

(a) the single-photon ionization of xenon,

(b) X-ray fluoresence spectra and double ionization of xenon, and

(c) the collisional charge transfer with a 400–600 eV electron, $e^- + Xe^{6+} \rightarrow Xe^{9+} + 4e^-$.

Furthermore, they find that the experimental data for ion production of helium and neon agree quite well with the single-electron sequential formulation.

I believe that most of the observations and conclusions that Rhodes et al. make agree with the result which would be expected on the basis of the *statistical formulation*. Thus, it seems that there remains very little to argue about.

5. *Effects of a very intense monochromatic field on a hydrogen atom* have been studied by Shih-I Chu and Cooper (17). They used an extension of the Chu and Reinhardt L_2 non-Hermitian Floquet formalism to determine the multiphoton ionization and ATI of a ground-state hydrogen atom in a very intense $(10^{13} < I < 10^{15} \, \text{W/cm}^2)$ field with frequency 0.5 a.u. $= 2 \times 10^{16} \, \text{s}^{-1}$. To the extent that enough Floquet blocks are included, this procedure gives *exact nonperturbative results* and includes all relevant processes involving various photon numbers. They found that in this intensity regime two- and three-photon ionization rates do not obey the power laws predicted by lowest order perturbation theory. Kulander (31) made a direct integration of the hydrogen-atom time-dependent Schrödinger equation and obtained good agreement with Chu and Cooper's results.

6. *Effects of very intense pulses on helium atoms* have been studied by Kulander (32). He numerically integrated time-dependent Hartree–Fock (TDHF) equations (for intensities up to $10^{15} \, \text{W/cm}^2$) to determine the effects of frequency, intensity, pulse length, and pulse shape on the multiphoton ionization of helium atoms in very intense laser fields. The principal disadvantage of this model is that the instantaneous correlations of the electrons are smeared out or averaged over. This mean-field approximation makes the calculations manageable, but it reduces the accuracy of the results. A good feature of the TDHF is that it gives an indication of the extent to which multiple ionization is a sequential process. Kulander found that the preionization dynamics (PD) (which Lambropoulos says is most important) varies dramatically with wavelength. For very short wavelengths, he had evidence that two-electron collective modes are excited in the PD and produce ATI. Clearly, this work is the beginning of a very interesting series of papers on multiphoton ionizations.

7. *Szöke's theoretical approach.* At present, there is need for rigorous computations of pulse effects to compare with the experimental data.

Fortunately, a number of excellent theoretical research programs are underway. For example, Abraham Szöke (33) is developing a rigorous but manageable theoretical treatment of multiphoton excitations of atoms. The motions of the bound and the free electrons are treated separately in a non perturbative manner, while the ionization is regarded as a perturbation.

By restricting himself to laser pulses that have a narrow spectral width such that $\varepsilon = \Delta\omega/\bar{\omega} \ll 1$, where $\Delta\omega$ is the average spectral width and $\bar{\omega}$ is the average frequency, Szöke is able to make use of the *method of multiple time scales* (34) in which there is a "fast time", $\tau_0 = \int_0^t \omega(t')\,dt'$, which is the "optical phase time", and a "slow time", $\tau_1 = \varepsilon\tau_0$. The Schrödinger equation can then be expressed in terms of these two times and solved by successive expansions in powers of τ_0 and τ_1.

Carrying this out formally, Szöke makes use of Floquet theory, strong-field scattering theory, and many-body diagrammatic perturbation theory. In reading his preprint, I was fascinated by his ingenuity and I am sure that his applications will be very interesting.

As far as I know Szöke's paper is the first laser application of the multiple-scales method. It is a very powerful (and yet quite simple) method for obtaining solutions of physical problems involving widely different time or length scales. Since it is particularly suitable for treating oscillatory problems, Ben Noble and Moeyett Hussain (Chapter II) and Ali Nayfeh (Chapter III) explain the technique and show how it can be applied. The solutions can be as accurate as desired in all degrees of freedom. Thus, in boundary layer problems, the solutions apply to the transition region as well as to the interior and exterior.

G. Nuclear Transitions with Internal Conversion of γ-Rays

These transitions produce multiply charged ions. The γ-ray emitted by the nucleus removes the electrons from the K shell, and the subsequent vacancy cascades (Auger effect) leave the molecule with a very high positive charge (18). For example, when $^{80m}_{35}$Br(4, 4 hr) decays to $^{80}_{35}$Br(18 m), the bromine atom from CH_3Br has been observed (35) to have charges up to $13+$ (with an average of $6+$). From the widths of X-ray fluorescence lines and absorption edges, the time required for all of the Auger electrons to be ejected was estimated to be between 10^{-15} and 10^{-16} s. Since this is much shorter than the vibrational period (10^{-14} s) of methyl bromide, the molecule holds together until after the positive ions have been formed. Then there must be an internal transfer of electrons from the CH_3 to the Br prior to fragmentation that occurs as the result of Coulombic repulsion of redistributed charges.

When an electron in the K or L shell of an atom is captured by a nucleus, as in the case when $^{125}_{53}$I in methyl iodide is transformed into $^{125}_{52}$Te, it gives rise to an Auger cascade (similar to the γ-ray internal conversion cascades). Thus, tellurium ions with charges up to 18 + have been observed (36). These nuclear decay Auger effects are interesting from two standpoints:

(a) In the limit as pulses get shorter (so that the preionization period becomes less important) and as laser frequencies climb into the X-ray range, the radiochemical effects should be the same as the laser pulse effects. Will the mechanism of their multicharged ion formation involve sequential or simultaneous ejection of their outer shell electrons? I would bet on simultaneous ejection.

(b) Radiochemists have learned a great deal about the fragmentation products and chemical changes in *molecules* that result from Auger cascades. Thus, from the radiochemists we can learn what to expect when ultra-intense laser pulses are applied to molecules. In the future, some of the multicharged molecular fragments produced by laser pulses may be useful as, for example, catalysts.

References

1. G. R. Fleming, *Chemical Applications of Ultrafast Spectroscopy*, Oxford University Press, New York, 1986.
2. R. M. Baum, Subpicosecond Spectroscopy Gives New Data on Reaction Dynamics, *Chem. Eng. News*, January 13, 1986, p. 36.
3. N. F. Scherer, F. E. Doany, and A. H. Zewail, *J. Chem. Phys.* **84**, 1932 (1986).
4. N. F. Scherer, J. L. Knee, D. D. Smith, and A. H. Zewail, Femtosecond Photofragment Spectroscopy: The Reaction $ICN \rightarrow I + CN$, *J. Phys. Chem.* **89**, 5141 (1985).
5. M. Dantus, M. J. Rosker, and A. H. Zewail, Real-Time Femtosecond Probing of "Transition States" in Chemical Reactions, *J. Chem. Phys.* **87**, 2395 (1987).
6. N. F. Scherer, L. R. Khundkar, R. B. Bernstein, and A. H. Zewail, Real-Time Picosecond Clocking of the Collision Complex in a Bimolecular Reaction: Birth of OH from $H + CO_2$, *J. Chem. Phys.* **87**, 1451 (1987).
7. A. D. Bandrauk (Ed.), Atomic and Molecular Processes with Short Pulses, (Plenum, New York, 1987).
8. U. Johann, T. S. Luk, H. Egger, and C. K. Rhodes, *Phys. Rev. A* **34**, 1084 (1986).
9. T. S. Luk, U. Johann, H. Egger, H. Pummer, and C. K. Rhodes, Collision-Free Multiple Photon Ionization of Atoms and Molecules at 193 nm, *Phys. Rev. A* **32**, 214 (1985).
10. K. Boyer, H. Jara, T. S. Luk, I. A. McIntyre, A. McPherson, R. Rosman, J. C. Solem, C. K. Rhodes, and A. Szöke, "Role of Many-Electron Motions in Multiphoton Ionization and Excitation," in *International Conference on Multiphoton Processes*, Vol. IV, Eds. S. J. Smith and P. L. Knight, Cambridge Press, London, 1988.
11. A. McPherson, G. Gibson, H. Jara, U. Johann, T. S. Luk, I. A. McIntyre, K. Boyer, and C. K. Rhodes, Studies of Multiphoton Production of Vacuum Ultraviolet Radiation in Rare Gases, *J. Opt. Soc. Am. B* **4**, 595 (1987).
12. Z. Deng and J. H. Eberly, *J. Opt. Soc. Am. B* **2**, 486 (1985); *J. Phys. B* **18**, L287 (1985).
13. A. Dulcic, *Phys. Rev. A* **35**, 1673 (1987).
14. P. R. Freeman, P. H. Bucksbaum, et al., A.T.I. with Picosecond Laser Pulses, *Phys. Rev. Lett.* **59**, 1092 (1987).

15. P. H. Bucksbaum, R. R. Freeman, M. Bashkansky, and T. J. McIlrath, Role of the Pondermotive Potential in A.T.I., *J. Opt. Soc. Am. B* **4**, 760 (1987).

16. D. R. Herrick, New Symmetry Properties of Atoms and Molecules, Adv. Chem. Phys. **52**, 1 (1983).

17. S.-I. Chu and J. Cooper, *Phys. Rev. A* **32**, 2769 (1985).

18. D. Chatterji, *Theory of Auger Transformations*, Academic, New York, 1976.

19. B. Crasemann, Ed., *Atomic Inter-Shell Physics*, Plenum, New York, 1985.

20. X.-D. Mu, T. Åberg, A. Blomberg, and B. Crasemann, Production Multicharged Ions by Strong UV Laser Pulses: Theoretical Evidence for Stepwise Ionization, *Phys. Rev. Lett.* **56**, 1909 (1986).

21. A. L. Robinson, Atoms in Strong Laser Fields Obey the Rules, *Science* **232**, 1193 (1986).

22. P. Lambropoulos and X. Tang, Multiple Excitation and Ionization of Atoms by Strong Lasers, *J. Opt. Soc. Am. B* **4**, 821 (1987); *Phys. Rev. Lett.* **58**, 108 (1987). See also P. Lambropoulos, Generalized Atomic K-Photon Ionization Cross-Sections, *Adv. At. Molec. Phys.* **12**, 87 (1976).

23. P. L. Lambropoulos, Mechanisms for Multiple Ionization of Atoms by Strong Pulsed Lasers, *Phys. Rev. Lett.* **55**, 2141 (1985).

24. S. L. Chin and P. Lambropoulos, Eds., *Multiphoton Ionization of Atoms*, Academic, New York, 1983, see especially, Y. Gontier and M. Trahin, "Theory of Multiphoton Ionization," Chapter 3.

25. C. Kittel, *Quantum Theory of Solids*, Wiley, New York, 1967, p. 153.

26. J. M. Blatt, *Superconductivity*, Academic, New York, 1964, Chapter 3.

27. J. Paldus, "Many-Electron Correlation Problem—Group Theoretic Approach," in *Theoretical Chemistry: Advances and Perspectives*, Vol. 2, H. Eyring and D. J. Henderson, Eds., Academic, New York, 1976.

28. T. Åberg, A. Blomberg, J. Tulkki, and O. Goscinski, Maximum Entropy Theory of Recoil Charge Distribution in Electron-Capture Collisions, *Phys. Rev. Lett.* **52**, 1207 (1984).

29. A. Russek, *Phys. Rev.* **132**, 246 (1963); A. Russek and J. Meli, *Physica* **46**, 222 (1970).

30. T. A. Carlson and R. M. White, Abundances and Recoil Energy Spectra of Fragment Ions Produced by X-ray Interaction with CH_3I, *J. Chem. Phys.* **44**, 4510 (1966).

31. K. C. Kulander, Ionization of H Atoms by Strong Lasers, *Phys. Rev. A.* **35**, 445 (1987).

32. K. C. Kulander, Time Dependent Hartree Fock Theory of Multiphoton Ionization: Helium, *Phys. Rev. A* **36**, 2726 (1987).

33. A. Szöke, Theory of Atoms in Strong, Pulsed Electromagnetic Fields: General Considerations, Lawrence Livermore Nat. Lab. Rep. UCRL-95294 Rev. 1, (1987).

34. C. M. Bender and S. A. Orszag, *Advances in Mathematical Methods for Scientists and Engineers*, McGraw-Hill, New York, 1978.

35. S. Wexler and G. R. Anderson, Dissociation of Methyl Bromide by Isomeric Transition of 4.4-hr Br^{80m}, *J. Chem. Phys.* **33**, 850 (1960).

36. T. A. Carlson and R. M. White, Fragment Ions from CH_3Te^{125} Followed by Decay of CH_3I^{125}, *J. Chem. Phys.* **38**, 2930 (1963).

IV. NONLINEAR OPTICS AND NONLINEAR EQUATIONS

Nonlinear optics has become a very important branch of laser technology because of the availability of high-powered lasers. For example, during the last

10 years, more than 2000 papers have been published about coherent Raman techniques and their applications (1). There are two seemingly different, but related, reasons why the optics and their associated evolution equations should be nonlinear.

1. The non-linearity is proportional to the polarization of the medium, which in turn is a function of the field strength. The solutions for this type of evolution equation can contain *solitons*. Furthermore, it should be possible to use the inverse scattering transformation (IST) technique to solve them.
2. The nonlinearity can result from fluctuations and decay produced by relaxation processes. N. Gisin, a Swiss mathematician, has devised a new technique for solving this type of problem.

Since very few of my colleagues are familiar with either the IST or Gisin techniques, I thought it important to include them in this chapter.

A.   Coherent Anti-Stokes Raman Spectroscopy (CARS)

After the discovery of lasers in 1960 (optically incoherent), spontaneous Raman spectroscopy (SRS) became a practical procedure for determining the concentration of molecular species (from vibrational Raman bands) and the temperature distribution (from band contours) in flames (2). Furthermore, it was found by trial and error that the intensity of Raman spectra could be greatly increased by mixing the radiation of a number of lasers having different frequencies.

It is amazing that only five years after the advent of lasers, Maker and Terhune discovered *coherent anti-Stokes Raman scattering (CARS)*, which is a *four-wave mixing technique* that is much more intense than the SRS, gives the same spectral information as SRS, and has narrower linewidth and excellent spectral resolution (1, 2). Its principal disadvantages are that it is sensitive to laser instabilities and tends to saturate when the laser fields are very intense (2).

In order to avoid confusion, it is important to explain the two meanings of coherence. *Molecular coherence* means that the phase of the induced radiation of a single molecular system is the same as the phase of its incident radiation. In this sense all single-molecule Raman spectra are coherent. Whereas *optical coherence* means that the phase of the induced radiation emitted by each of the N molecules in the system is the same.

Since CARS is optically coherent, it is *superradiant* (Section VII.A), and its intensity is proportional to N^2 (where N is the number of molecules in the system). This is in contrast to the intensity of the optically incoherent SRS, which is proportional to N. Furthermore, since the scattering of the CARS radiation is in the forward direction, the width of the CARS lines is minimal.

CARS has been used a great deal to determine the basic structural, dynamic, and spectroscopic properties of organic and polymeric materials (3) since the hyperpolarizabity γ (Section VI.F) of many of these materials is very large. Indeed, the value of γ for either polyacetylene or polydiacetylene is enormous (on the order of 10^6 a.u. per CH or CH_2 unit) (4) because of their conjugate double bonds.

In addition to the traditional use of CARS for the remote sensing of temperature and species in hostile environments such as plasma, flames, internal combustion engines, and the exhaust from jet engines, CARS spectra of very intense picosecond (or femtosecond) pulses are currently being used to study unusual species and reactive intermediates in laser-induced chemical reactions, photodissociations, and clusters (1).

Thus, the CARS technique could be used to obtain detailed experimental information regarding the photodissociation of $H_3{}^+$, which would be very similar to the theoretical results that Metiu and his colleagues obtained by using a split-operator fast Fourier transform technique to determine the SRS (See Section I.B.2)!

Many theoretical techniques closely related to CARS have been developed that are useful for special applications. These techniques include stimulated Raman gain and loss (SRG and SRL), photoaccoustic Raman spectroscopy (PARS), and Raman-induced Kerr effect spectroscopy (RIKES) (1).

These coherent Raman techniques involve *four-wave mixing* (FWM). A FWM process involves the interaction of three incident laser fields, with wave vectors k_1, k_2, and k_3 and frequencies ω_1, ω_2, and ω_3, together with the emitted field that has the wave vector k_e and frequency ω_e. (The magnitude of a wave vector is $k_j = n_j \omega_j / c$, where n_j is the index of refraction.) Coherent Raman processes are a special type of FWM involving two, rather than three, incoming laser beams since $k_3 = k_1$. Furthermore, $k_e = 2k_1 - k_2$ so that $\omega_e = 2\omega_1 - \omega_2$. The intensities of the various coherent Raman spectra are proportional to the squares of the dynamic hyperpolarizabilities (both real and imaginary parts) corresponding to up to 48 Feynman diagrams. [See Ref. 1 and the review article written by Mukamel (5), author of Chapter XII.] Methods of calculating hyperpolarizabilities are discussed elsewhere (Section VI.F, Refs. 33–40).

B. The Inverse Scattering Transformation

At present, Russian mathematicians and physicists are very much interested in the solution of nonlinear optic problems. Konopelchenko (6) states that the IST method discovered 20 years ago is one of the most powerful tools for solving nonlinear evolution equations of dynamical systems. Indeed, he cites 300 references of physics applications. Recently, Peter Lax in the United States received the 1987 Wolff Award in Mathematics for his matrix operator generalization of the IST formulation. Those nonlinear equations which are integrable by the IST approach have many remarkable properties, such as *soliton* solutions with very surprising scattering behavior and infinite sets of integrals of motion. Solitons are important in hydrodynamics, plasma, the electromagnetic properties of high polymers, and so on.

In the 1970s, there was great interest in using IST to determine (7) the *Kerr Effect, Josephson junction, self-induced-transparency, three- and four-wave mixing, parametric oscillations,* and other nonlinear optical effects produced by short, very intense laser pulses. Most nonlinear optical effects result from a coherent response of a resonant medium to an optical laser pulse (8). For these effects to occur, it is necessary that the pulse length be 10^{-10} seconds or shorter so that the induced polarization in the medium can retain a definite phase relationship with the molecules (8). The amplitude and nonlinearity of these effects increase sharply as the pulse intensity becomes larger and the pulse length becomes shorter.

The best example of the application of the IST approach to nonlinear optics is Kaup's (9) determination [and see Ablowitz and Segur's (7) explanation] of the *exact solution* of the McCall–Hahn *self-induced transparency* effect for an atomic two-state model. Their results agreed with the experimental data much better than previous perturbative treatments. Furthermore, the analytical form of their solution helped to explain the significance of the experimental results. One of the most striking features of self-induced transparency is the dramatic deviations from Beer's law that occur whenever the area of the initial pulse exceeds π. The reason is that the pulse energy is composed of two parts: one part that is transmitted losslessly by solitons and another part that is the continuous spectrum absorbed in accordance with Beer's law. Without IST it would have been virtually impossible to determine the soliton effects.

In order to determine the nonlinear effects, it is necessary to solve the optical Bloch equations together with Maxwell's equations for an assemblage of noninteracting molecules distributed with a uniform density. George L. Lamb, Jr. (8) gives a very detailed explanation of the standard semiclassical procedure for determining the basic equations. Until 1967 it was necessary to obtain approximate solutions of these equations by using perturbative procedures. Fortunately, IST now makes it possible to determine *exact* analytical solutions.

Up to the present time, the electric dipole approximation for the atom–field interaction has sufficed. However, as laser intensities become more intense, it will become necessary to use either the electric quadrupole–magnetic dipole approximation or the semirelativistic Breit–Pauli Hamiltonian.

My object in writing this section is to stimulate new research efforts. In spite of the 1970s successes, interest in using the IST to solve nonlinear laser problems has waned. Most of the "easy" nonlinear optical problems using simple two-state models of the molecular system interacting with an ideal dielectric have already been solved. Now, in order to explain experimental results, it is necessary to use more realistic molecular models and relaxation processes. It will require ingenuity and the ability to improvise new techniques in order to use the IST to solve such complicated laser problems. However, the efforts will certainly be rewarded by the discovery of unexpected properties.

At present, there are many variations of the IST method (6): direct linearization, Riemann–Hilbert, Hirota, dressing, algebraic and algebraic-geometric, τ-function, δ-function, and so on. The best one to use depends on whether your evolution equation is an ordinary or partial differential equation, an integrodifferential equation, or either a differential- or difference–difference equation (6). What they all have in common is that they are used in much the same way as Fourier transforms for linear equations. Probably it will take a number of years before the IST is systematized so that there is no need for all of the variations. It is amazing what it can do and surprising what it cannot do.

The present state of the art is very well summarized in books by Konopelchenko (6), Bullough and Caudrey (10), and Ablowitz and Segur (7). Three of the most important aspects of the IST integrable evolution equations of physical interest are:

(a) Faddeev says (11) that all of the IST integrable evolution equations of physical interest correspond to infinite dimensional Hamiltonian systems where every point in phase space is a dimension. From this standpoint, the IST procedure involves a nonlinear canonical transformation to angle-action type variables. (I keep wondering whether there are any simplifications which occur when the Hamiltonian of a non-linear equation is independent of time, such as the simple form of the time development operator for linear equations.)

(b) The Painlevé test (12) is a useful criterion for identifying IST integrable partial differential equations (with complex variables). In order for an equation to pass this test, poles are the only singularities of its general integral that can exist on arbitrary noncharacteristic (movable) hypersurfaces.

(c) The IST equations may sometimes be modified to accommodate small perturbations provided that they do not introduce any new effects. However, because the perturbation equations may converge (or diverge) very slowly, it was necessary for Hirota (13) to develop a new perturbation procedure, which makes use of Padé approximates and Bäcklund transformations, to derive a special form of bilinear differential equations which can be solved exactly. Once we understand *why* it is so difficult to use perturbation procedures, we will have a much better understanding of IST nonlinear equations! Furthermore, it will become possible to solve nonlinear optics problems with greater accuracy by using relastic molecular models.

1. *The derivation of IST* was motivated (7) by the transformation $v = (\partial \psi / \partial \psi)/\psi$, which converts the nonlinear Ricatti equation

$$\frac{\partial v}{\partial x} + v^2 + u = 0$$

to the linear equation

$$\frac{\partial^2 \psi}{\partial x^2} + u\psi = 0.$$

In 1968, Miura, Gardner, and Kruskal (14) generalized this procedure by starting with the Schrödinger-like equation

$$-\frac{\partial^2 \psi}{\partial x^2} + [u(x,t) - \lambda^2]\psi(x,t) = 0 \tag{4.1}$$

and postulating the associated time evolution equation

$$\frac{\partial \psi}{\partial t} = A(x,t)\psi + B(x,t)\frac{\partial \psi}{\partial x} \tag{4.2}$$

where A and B can be any general scalar functions independent of $\psi(x,t)$.

For example, if $A(x,t) = -\partial u/\partial x$, $B(x,t) = 4\lambda^2 - 2u$ and $\partial \lambda/\partial t = 0$, then since $\partial^3 \psi / \partial x \, \partial t^2 = \partial^3 \psi / \partial t \, \partial x^2$, it follows that the nonlinear *Korteweg-de Vries* (KdV) equation (11, 12, 15) is satisfied,

$$K(u) = \frac{\partial u}{\partial t} - 6u\frac{\partial u}{\partial x} + \frac{\partial^3 u}{\partial x^3} = 0. \tag{4.3}$$

The KdV equation was discovered (9) in 1895 to be the evolution equation of gravity (soliton) waves in shallow water. Since that time, it was found to play an important role in the dynamics of lattice models, plasma, magnetohydrodynamics, and nonlinear optics.

Given a nonlinear partial differential equation, the IST is a procedure for determining the functions $A(x,t)$ and $B(x,t)$ in a linear time evolution equation. Because the choice of $A(x,t)$ and $B(x,t)$ is not unique, the differential equations have an infinite number of constants of motion.

The basic idea of the IST technique is that a *nonlinear* partial differential evolution equation (analogous to the Schrödinger time-dependent equation) can be solved by the IST technique provided that it is possible to transform the variables so as to obtain two *linear* differential equations which have the forms

$$Lq = \lambda q \quad \text{and} \quad \frac{\partial q}{\partial t} = Mq \tag{4.4}$$

where λ is the spectral density corresponding to a matrix operator $L(t, x)$. Also, $M(t, x)$ is a matrix operator associated with time development.

By taking the time derivative of $Lq = \lambda q$ (assuming λ is constant) and making use of $\partial q/\partial t = Mq$, we obtain *Peter Lax's matrix operator form of the IST* (15),

$$\frac{\partial L}{\partial t} = [M, L] = ML - LM. \tag{4.5}$$

Lax found that, for the KdV equation,

$$L = \frac{\partial^2}{\partial x^2} + u(t, x) \quad \text{and} \quad M = 4\frac{\partial^3}{\partial x^3} + 6u\frac{\partial}{\partial x} + 3\frac{\partial u}{\partial x}.$$

Inverse scattering transformations can be considered to be generalizations of Fourier transforms which are used to solve linear equations (6, 7). Their solution procedures are entirely analogous. At each step, the IST only requires the solution of a *linear* equation. In both the nonlinear and linear problems, the form of the evolution equations is characterized by dispersion relations, and both are used in the same manner:

1. *Fourier transforms.* For example, given the dispersion relation $\partial q/\partial t = -\omega \, \partial q/\partial x$, where $\omega(k) = -k^2$ together with the solution $q(x, 0)$, by making a Fourier transform of $q(x, 0)$ we obtain

$$b(k, 0) = \int_{-\infty}^{+\infty} q(x, 0)e^{-ikt}\,dx.$$

Then using the dispersion relation $b(k, t) = b(k, 0)e^{-i\omega t}$ it follows that $q(x, t)$ is the inverse Fourier transform of $b(k, t)$.

2. *The IST technique* proceeds by three similar steps:

$$q(x, 0) \xrightarrow[\text{I}]{} w(\lambda, 0) \xrightarrow[\text{II}]{} w(\lambda, t) \xrightarrow[\text{III}]{} q(x, t).$$

Faddeev (6) says that the first step represents a nonlinear canonical transformation into angle-action type variables. Each of these IST steps requires the solution of a linear equation.

C. Solitons

Solitons were discovered in 1834 by John Scott Russell, a scientist working at the University of Edinburgh, Department of Natural Philosophy, who was watching a boat which was being pulled rapidly by two horses along the narrow Forth-Clyde canal. Russell said (16) that the boat stopped suddenly and the water it had put in motion piled up along its prow. Then the accumulated water left the boat and formed a solitary wave 30 feet long and 1–2 feet high, which travelled 8 miles an hour for miles without changing its form or decreasing its speed.

Solitons and solitary waves are closely related; a solitary wave is a bell-shaped traveling wave that does not change its shape but decreases in size (16, p. 3); whereas a soliton is a solitary wave that preserves its shape and size even after a collision with another soliton (16, p. 6).

The recent interest in the KdV equation started with the discovery that it determines the evolution of solitons. The KdV equation has traveling-wave solutions of the form (15) $u(x,t) = s(x - ct)$ such that s and all of its derivatives approach zero as x approaches $\pm \infty$. For this choice of $u(x,t)$, the KdV equation becomes

$$(s - c)\frac{\partial s}{\partial x} + \frac{\partial^3 s}{\partial x^3} = 0,$$

which, after integrating once, then multiplying by $2\,\partial s/\partial x$, and integrating once more, becomes $- cs^2 + \frac{1}{3}s^3 + (\partial s/\partial x)^2 = 0$. From this relation and the assumption that $s(0) = 0$, it follows that

$$s(x) = 3c \operatorname{sech}^2\left[\tfrac{1}{2}c^{1/2}x\right] = S(x, c). \tag{4.6}$$

thus, for every speed c, the KdV equation has a solution that vanishes when $x = \pm \infty$. Note that $S(x, c)$ is normalized, symmetric in x, decays exponentially as $|x|$ approaches infinity, and its maximum value is $S(0, c) = 3c$. Because of its shape, $S(x, c)$ is called a solitary wave.

As Peter Lax says (15), it is very surprising that all solutions of the KdV equation (which vanish when $x = \pm \infty$) have solitary waves hidden in them. Indeed, there exists an infinite number of discrete positive numbers c_j (called eigenspeeds of u) and sets of phase shifts $\Theta_j^{\pm}$ such that

$$\lim_{t \to \pm \infty} u(x + ct, t) = \begin{cases} S(x - \Theta_j^{\pm}, c_j) & \text{if } c = c_j, \\ 0 & \text{if } c \neq c_j. \end{cases}$$

Of course, it is these eigensolutions that we call Solitons!

Furthermore, the eigenspeeds are the invariant functionals or integrals of the motion that play an important role in the determination of nonlinear optical effects. Thus, $c_j = \int_{-\infty}^{+\infty} F_j(u)\,dx$ and the first three integrals correspond to the conservation of mass, momentum, and energy:

$$c_1 = \int_{-\infty}^{+\infty} u(x,t)\,dx,$$

$$c_2 = \int_{-\infty}^{+\infty} u(x,t)^2\,dx,$$

$$c_3 = \int_{-\infty}^{+\infty} \left[\left(\frac{\partial u}{\partial x}\right)^2 + u^3 \right] dx.$$

According to Konopelchenko (6), c_3 is the "Hamiltonian of the KdV equation." Furthermore, the functionals $F_j(u)$ and the vectors $X_j(u)$ satisfy the conservation laws (Ref. 6, p. 4),

$$\frac{\partial F_j}{\partial t} + \operatorname{div} X_j = 0 \tag{4.7}$$

Ablowitz and Segur (7) and Kaup (9), in their self-induced transparency research, found that one of the IST conservation laws is the time delay. [I was interested to learn this since I showed that *Wigner's time delay* is the same as the *virial theorem* for scattering processes in a central field (17). This is not surprising since Van Kampen (18) has used Noether's fundamental theorem to prove that the virial theorem is an invariant property of all Lagrangian or Hamiltonian systems.] If the virial theorem is one of the conservation laws, it seems likely that all of the conservation laws are *hypervirial theorems*:

1. The *classical hypervirial theorem* (19) states that the Poisson bracket time averaged over its trajectory, $\overline{\{H,W\}}$, vanishes. (Here $\{H,W\}$ is the Poisson bracket and the long bar over it stands for the time averaging.) Furthermore, if $\overline{\{H,W\}} = 0$, then Hamilton's *principle of least action* (20) is satisfied. This means that if the trajectory is perturbed, $\overline{\{H,W\}}$ will continue to vanish through the first order of the perturbation. Here H is the Hamiltonian of the dynamical system and W is an arbitrary analytic function of the coordinates and momenta.

For a system of particles, the usual virial theorem is obtained by taking $W = \sum_j r_j \cdot p_j$, where r_j is the radius vector and p_j is the linear momentum of the jth particle.

2. For quantum systems, the hypervirial theorem is the *Heisenberg*

equation of motion. For time-dependent quantum systems, the hypervirial theorem is (21)

$$\frac{\partial}{\partial t}\langle\Psi|W|\Psi\rangle = \langle\Psi|\frac{\partial W}{\partial t}|\Psi\rangle + i\langle\Psi|[H,W]|\Psi\rangle. \tag{4.8}$$

Furthermore, if Eq. (4.8) is integrated by parts, one obtains an equation having the same form as Eq. (4.7). [The hypervirial theorem for scattering problems (where not all of the quantum states are bound) has a somewhat different form (22)].

It certainly seems reasonable that the conservation laws that hold for nonlinear dynamical systems should also apply to linear systems.

I believe that the *multiscale method* of solving differential equations discussed by Ben Noble and Moeyyet Hussain in Chapter II and by Ali Nayfeh in Chapter III might simplify the solution of some nonlinear soliton containing evolution equations.

D. Gisin's Nonlinear Schrödinger-Like Equations

Gisin (23–26) used the *master equation formalism* to develop nonlinear dissipative Schrödinger-like equations for a large class of problems with constant Hamiltonians that *involve both friction and fluctuations.* Furthermore, he derived surprisingly simple *analytic solutions* for their coherent eigenfunctions!

There are two reasons why I am interested in Gisin's formalism:

(a) Some of his equations can have soliton solutions, and it will be interesting to compare the IST and Gisin treatments of the same evolution equations.

(b) The Gisin formal solutions can be converted to explicit solutions by using either the *fast Fourier tranform split-operator technique* (discussed in Section I.B.2) or else by the new *Chebyshev method* in which the time development operator $U = \exp(-iHdt)$ is expanded in complex Chebyshev polynomials in powers of $-iHdt$ (27, 28).

1. *Relaxation and Dissipative Problems with Constant H*

Gisin (23) considers a system s interacting with a bath b. The (constant) Hamiltonian for $s + b$ is

$$H = H_s + H_b + \lambda H_{sb}. \tag{4.9}$$

After making the usual Markovian approximation, the Schrödinger equation for the evolution of the unnormalized system wavefunction is

$$ih\frac{d\varphi}{dt} = \{H_s + \lambda PH_{sb}P - i\lambda^2(B + iA)\}\varphi(t), \tag{4.10}$$

where the operator P projects the whole space onto the system space and A and B are two formally self-adjoint operators defined by

$$B + iA = \int_0^\infty Pe^{i(H_s + H_b)\tau}H_{sb}[I - P]e^{-i(H_s + H_b)s}[I - P]H_{sb}P\,d\tau. \tag{4.11}$$

Although the functions $\varphi(t)$ can decay to zero, the *coherent state wavefunctions* $\psi(t) = \varphi(t)/|\varphi(t)|$ are normalized to unity. The nonlinear Schrödinger equation for $\psi(t)$ is

$$ih\frac{d\psi}{dt} = H_0\psi(t) + i\lambda^2[\langle\psi(t)|B|\psi(t)\rangle - B]\psi(t), \tag{4.12}$$

where

$$H_0 = H_s + \lambda PH_{sb}P + \lambda^2 A.$$

Furthermore, if both H_0 and B are bounded and B is positive, Eq. (4.12) has the unique formal solution

$$|\psi(t)\rangle = \frac{\exp[-iH_0 t - \lambda^2 Bt]|\psi(0)\rangle}{\{\langle\psi(0)|\exp[iH_0 t - \lambda^2 Bt]\cdot\exp[-iH_0 t - \lambda^2 Bt]|\psi(0)\rangle\}^{1/2}}. \tag{4.13}$$

[Note the close connection between Eq. (4.13) and the corresponding time development relation for linear equations with constant Hamiltonian,

$$|\psi(t)\rangle = U(t,0)|\psi(0)\rangle = \exp(-iHt)|\psi(0)\rangle.$$

This indicates that the time independence of H is essential for Gisin's type of formulation.]

If H_0 and B commute, then

$$\frac{d}{dt}\langle\psi(t)|B|\psi(t)\rangle = -2\lambda^2\langle\psi(t)|B^2|\psi(t)\rangle + 2\lambda^2[\langle\psi(t)|B|\psi(t)\rangle]^2.$$

$$\leqslant 0. \tag{4.14}$$

As t approaches infinity, $\psi(t)$ approaches the normalized projection of $\psi(0)$

on the eigenspace of B, and its eigenvalue becomes the lowest one for which this projection does not vanish.

If, on the other hand, H_0 and B do not commute but λ is sufficiently small, $\psi(\infty)$ is close to the H_0 eigenvector for which $\langle \psi(\infty)|B|\psi(\infty)\rangle$ is minimal.

Gisin (23) has applied this formalism successfully to a number of spin systems in a magnetic field. The quantum mechanical analog of the damped simple harmonic oscillator is his simplest example.

2. Damped Harmonic Oscillator (24)

Allen and Eberly (29) and others have used the classical damped simple harmonic oscillator as a prototype of a two-state atomic system interacting with both a resonant laser field and a relaxation bath. Since the correspondence between classical optics and modern quantum mechanics is very close, it is surprising that it was not until 1981 that anyone succeeded in deriving a quantum-mechanical analog of the equation for the damped simple harmonic oscillator. The difficulty was in knowing how to represent quantum-mechanical friction! However, N. Gisin (24) knew from his master equation formulation and Eq. 4.12) that the nonlinear Schrödinger equation must be

$$i\hbar\frac{\partial\psi}{\partial t} = H_0\psi + ik[\langle\psi|H_0\psi\rangle - H_0]\psi \tag{4.15a}$$

or the equivalent nonlinear Liouville equation

$$i\hbar\frac{d\rho}{dt} = -[\rho, H_0] + i\frac{k}{\hbar}[[\rho, H_0], \rho], \tag{4.15b}$$

where H_0 is the quantum-mechanical simple harmonic oscillator Hamiltonian

$$H_0 = (2m)^{-1}p^2 + (\tfrac{1}{2}m\omega^2)q^2 \tag{4.16}$$

that has the eigenfunctions $|\varphi_n\rangle$ and eigenvalues $\varepsilon_n = (n + \tfrac{1}{2})\hbar\omega$.

If Eq. (4.15) is multiplied by $\langle\psi|H_0|$, it is obvious that

$$\frac{d[\langle\psi|H_0|\psi\rangle]}{dt} = 2k[\langle\psi|H_0|\psi\rangle^2 - \langle\psi|H_0^2|\psi\rangle].$$

From this, it is easy to show that the formal *coherent* solutions to Eq. (4.15) are given by

$$|\psi(t)\rangle = \frac{\exp[-(i+k)H_0 t]|\psi(0)\rangle}{\langle\psi(0)|\exp[-2kHt]\psi(0)\rangle^{1/2}} \tag{4.17}$$

The eigenfunction $|\psi_\alpha(t)\rangle$ can then be expanded as a linear combination of the simple harmonic oscillator functions $|\varphi_n\rangle$ so that $|\psi_\alpha(t)\rangle = \sum_{\alpha,n} C_{\alpha,n}(t)|\varphi_n\rangle$, where

$$C_{\alpha,n}(t) = C_{\alpha,n}(0)\exp\left[-\tfrac{1}{2}(i+k)(2n+1)\omega t\right]N_\alpha(t)^{-1/2} \qquad (4.18)$$

and

$$N_\alpha(t) = \sum_{s=0} |C_{\alpha,s}(0)|^2 \exp\left[-k(2s+1)\omega t\right].$$

Gisin used a Weyl transformation (30) of Eq. (4.15) into semiclassical phase space to prove that Eq. (4.15) agrees (in the limit as $\hbar$ approaches zero) with the classical damped harmonic oscillator equation

$$\frac{d^2q}{dt^2} + 2k\omega\frac{dq}{dt} + (1+k^2)\omega^2 q = 0. \qquad (4.19)$$

The most significant difference between the behavior of this quantum-mechanical model and that of a classical damped oscillator is that the coherent states of the quantum system evolve in semistable limit cycles to the lowest eigenstate of the simple harmonic oscillator. It is surprising that the coherent states remain coherent during their evolution.

3. *Periodic Dissipative Systems*

I believe that periodic dissipative system problems can be solved by first transforming the system into the Shirley (31) and Sambé (32) form of the Floquet representation in which the Hamiltonian is time independent and expressed in terms of infinite dimensional pseudostate space. Then the periodic state problem can be solved by the same master equation formulation Gisin (23) developed for constant Hamiltonian problems.

Instead, Gisin (25) abandoned his master equation formulation and treated the periodic problems as generalized damped harmonic oscillators.

Thus it should be possible to use the Gisin procedure to solve the *Floquet–Liouville Supermatrix* nonlinear optical evolution equations that Shih-I Chu derives in Chapter XVII. This would help to establish the relationship between the IST and the Gisin types of nonlinear integrable differential equations. It would be particularly interesting since many, if not all, nonlinear optical evolution equations have soliton solutions!

References

1. J. W. Nibler and J. J. Yang, Nonlinear Raman Spectroscopy of Gases, *Ann. Rev. Phys. Chem.* **38**, 349 (1987).

2. S. Druet and J.-P. Taran, *Coherent Anti-Stokes Raman Spectroscopy (CARS)*, in *Chemical and Biochemical Applications of Lasers*. Vol. 4, C. B. Moore, Ed., Academic. New York, 1979, p. 187.

3. D. J. Williams, Ed., *Nonlinear Optical Properties of Organic and Polymeric Materials*, A.C.S. Symposium Series 233, American Chemical Society, Washington, DC, 1983.

4. B. Kirtman, Convergence of Finite Chain Approximation for Linear and Nonlinear Polarizabilities of Polyacetylene, *Chem. Phys. Lett.* **143**, 81 (1988).

5. S. Mukamel, Solvation Effects in Four-Wave Mixing and Spontaneous Raman and Fluorescence Lineshapes of Polyatomic Molecules, *Adv. Chem. Phys.* **LXX** (pt. 1), 165 (1987).

6. B. G. Konopelchenko, *Nonlinear Integrable Equations*, Lecture Notes No. 270, Springer-Verlag, Berlin, 1987.

7. M. J. Ablowitz and H. Segur, *Solitons and the Inverse Scattering Transformation*, SIAM, Philadelphia, 1981.

8. G. L. Lamb. Jr, Analytical Description of Ultrashort Optical Pulse Propagation in a Resonant Medium, *Rev. Mod. Phys.* **43**, 99 (1971).

9. D. J. Kaup, Coherent Pulse Propagation: A Comparison of the Complete Solution with the McCall-Hahn Theory and Others, *Phys. Rev. A* **16**, 704 (1977).

10. R. K. Bullough and P. J. Caudrey, Eds., *Topics in Current Physics*, Vol. 17 *Solitons*, Springer, New York, 1980.

11. L. D. Faddeev, "Hamiltonian Interpretation of IST," in *Solitons*, R. K. Bullough and P. J. Caudrey, Springer, New York, 1980, Chapter 11.

12. P. A. Clarkson and C. M. Cosgrove, PainLevé Analysis of the Nonlinear Schrödinger Family of Equations, *J. Phys. A* **20**, 2003 (1987).

13. R. Hirota, "*Direct Methods in Soliton Theory*," in *Solitons*, R. K. Bullough and P. J. Caudry, Springer, New York, 1980, Chapter 5.

14. R. M. Miura, C. S. Gardner, and M. D. Kruskal, *J. Math. Phys.* **9**, 1204 (1968).

15. P. D. Lax, Integrals of Nonlinear Equations of Evolution and Solitary Waves, *Commun. Pure Appl. Math.* **21**, 467 (1968).

16. R. K. Bullough and P. J. Caudrey, "*The Soliton and its History*," in *Solitons*, R. K. Bullough and P. J. Caudrey, Springer, New York, 1980, Chapter 1.

17. J. O. Hirschfelder, Similarity of Wigner's Delay Time to the Virial Theorem for Scattering by a Central Field, *Phys. Rev. A* **19**, 2463 (1979).

18. N. G. Van Kampen, Transformation Groups and the Virial Theorem, *Rep. Math. Phys.* **3**, 235 (1972).

19. J. O. Hirschfelder, The Classical Hypervirial Relations, *Z. Physik. Chem.* **37**, 3/4 (1963).

20. W. Youngrau and S. Mandelstam, *Variational Principles in Dynamics and Quantum Mechanics*, Pitman, New York, 1960.

21. S. T. Epstein, *Variational Methods in Quantum Chemistry*, Academic, New York, 1974, p. 254.

22. P. D. Robinson and J. O. Hirschfelder, Virial Theorem and Generalizations in Scattering Theory, *Phys. Rev.* **129**, 1391 (1963); S. T. Epstein and P. D. Robinson, Hypervirial Theorem for Variational Wave Functions in Scattering Theory, *Phys. Rev.* **129**, 1396 (1963).

23. N. Gisin, Microscopic Derivation of a Class of Nonlinear Dissipative Schrödinger-Like Equations, *Physica* **111A**, 364 (1982).

24. N. Gisin, Simple Nonlinear Dissipative Quantum Evolution Equation. *J. Phys. A* **14**, 2259 (1981).

25. N. Gisin, Dissipative Quantum Dynamics for Systems Period in Time. *Found. Phys.* **13**, 643 (1983).

26. N. Gisin, Generalisation of Wigner's Theorem for Dissipative Systems, *J. Phys. A* **19**, 205 (1986).

27. H. Tal-Ezer and R. Kosloff, Accurate and Efficient Scheme for Propagating the Time Dependent Schrödinger Equation, *J. Chem. Phys.* **81**, 3967 (1984).

28. R. C. Mowrey and D. J. Kouri, Application of Close Coupling Wave Packet Method to Long Lived Resonance States in Molecule–Surface Scattering, *J. Chem. Phys.* **86**, 6140 (1987).

29. L. Allen and J. H. Eberly, *Optical Resonance and Two-Level Atoms*, Wiley, New York, 1975.

30. S. de Groot, *La Transformation de Weyl et la Fonction de Wigner*, Universite de Montreal, Montreal, 1975, p. 41.

31. J. H. Shirley, *Phys. Rev.* **138B**, 979 (1965); unpublished Ph.D. Thesis, California Institute of Technology, 1963.

32. H. Sambé, *Phys. Rev. A* **7**, 2203 (1973).

V. QUANTUM ELECTRODYNAMICS AND CORRESPONDENCE

Quantized field and semiclassical field quantum electrodynamics (QED) have become calculationally (but not philosophically) reconciled.

Thus either formalism can be used to solve most, if not all, QED problems. Which technique is easier and more numerically efficient will depend upon the nature of the problem.

For a number of years it has been realized that there is a close correspondence between classical and quantum-mechanical (nonrelativistic or semirelativistic) dynamical operators that are accurate through all $1/c^2$ or $\alpha^4 mc^2$ fine structural terms. The great changes that are taking place are in our *understanding* of relativistic quantized field QED and the *changes that must be made* in semiclassical QED in order to correspond with it.

A.   Quantum Electrodynamics Is Changing

Although quantized field QED has been extremely successful and useful, it has been difficult to perform its second quantizations and extremely difficult to understand the underlying physics of its fluctuating vacuum field.

Both Edwin T. Jaynes and Asim Barut believe that there are *no QED problems that cannot be solved in principle by semiclassical techniques.* However, this remains to be proven.

As Peter Milonni (co-author of Chapter XXI) explained in his 1976 review article (1), E. T. Jaynes, Timothy Boyer, Jay Ackerhalt, and others (including himself) have discovered semiclassical techniques that make it possible to eliminate many of the differences between the effects of strong classical and quantized fields: Einstein's *A*-coefficient, the Lamb shift, spontaneous emission, Casimir–Polder-delayed van der Waals forces, Dicke's *super-radiance*, and so on.

The effects of the fluctuating vacuum field can be simulated by:

1. Changing the boundary conditions of the classical Maxwell equations so that the homogeneous solution (instead of vanishing identically) is a fluctuating zero-point field of energy density $\frac{1}{2}\hbar\omega$ per mode rather than the usual null source-free solution.

2. The radiation reaction fields of the particles' self-energy, provided that the phase of the reaction fields is taken to be random.

3. This radiation reaction field (like the real zero-point fluctuating field) must satisfy the Lorentz invariance, which requires that the spectral density be proportional to frequency. By letting Planck's constant serve as a multiplicative constant, this condition is satisfied and makes the "theoretical blackbody radiation" agree with experiments.

In 1987 Peter Milonni presented a paper (2) explaining how the calculational differences between the vacuum and reaction field interpretations have been reconciled (but there are still important differences of opinion regarding the physical significance of the vacuum field). He noted:

> Whatever calculations we need can be made in the Schrödinger or interaction picture relatively safely without going into philosophical questions. To emphasize the classical aspects of the vacuum as much as possible, we should choose symmetric ordering of non-commuting atomic and field operators at every stage of a calculation. We have learned to appreciate the important differences in the positive and negative frequency parts of the field in either classical or quantum electrodynamics (and use the symmetric sum of the two). Spontaneous emission is a simple consequence of the fact that oscillating dipoles radiate (however, classical electrodynamics also must be modified as otherwise it would give erroneous results) (2).

As Joe Eberly (author of Chapter XVIII) said (3), "Only in theoretical models do ideal atoms exist which *interact with monochromatic light*. Real atoms experience a fluctuating environment of many perturbing influences, and real lasers can exhibit a variety of fluctuations in phase, frequency, and amplitude!"

In thinking about random fields, it is important to remember that the overall phase of a system has no physical significance *but* the relative phases of different parts of the system can have important observable effects [Section V.D].

Thus in theoretical calculations, it is only at some initial time that we can require that the classical fields have random phases. At all later times, the evolution equations determine their phases.

B. Barut's Semiclassical QED Approach

Asim Barut and co-workers (4–12) have taken advantage of the present state of our knowledge to systematize and simplify the solution of QED problems. His approach is very exiciting and undoubtedly accurate; however, his methods are so novel that they may need small changes (12). Let me summarize.

1. His formulation is based upon the radiation reaction of the particles' self-fields without recourse to a fluctuating vacuum.

2. In previous versions of QED, the self-energy is first neglected and then put into the equations by perturbation theory one photon at a time. Instead, Barut follows the classical procedure of calculating the radiative processes directly from the self-energy of the electrons in the external fields. Then, he uses the Green's functions of the reaction fields to eliminate the self-energies from the coupled Maxwell–Dirac equations, leaving a nonlinear integrodiffer-ential equation to solve. By using Dirac–Coulomb orbitals, there is no need for second quantization.

3. Since QED involves the scattering of radiation, Barut recognized that it is simpler to use an *action* principle rather than solve the equations of motion directly. In other words, he makes a canonical transformation of the system into an angle-action variable representation. His dimensionless action is closely related both to the scattering amplitude in space–time and to the invariant energy-momentum tensor of the matter–field system. This proce-dure is similar to the S-matrix technique used by both physicists and chemists to solve collisional scattering problems.

4. All of Barut's calculations proceed parallel to classical radiation theory. Furthermore, both the Barut and classical radiative terms for non relativistic problems are expressed in closed form.

5. I was delighted to see that Barut's formulation is expressed in phase space because, as Fermi showed, the concept of probability in phase space is virtually the same in both classical and quantum mechanics. This is in contrast to (either the coordinate or momentum) Hilbert space where Van Vleck (13) explained the big differences between classical and quantum mechanical probability concepts! (I will be interested to find out whether Barut's nonrelativistic dynamics is useful for solving laser-molecule interaction problems.)

Barut has derived nonperturbative formulas for determining the energy level and Lamb shifts accurate through all orders of $Z\alpha$ for hydrogenic atoms (9). In order to do this, he uses the Dirac equation together with the (covariant) Möller potential (10) for his (two-particle) equation of motion. In order to solve this equation, he uses relativistic Dirac–Coulomb wavefunctions and renormalizes the system so as to make the sums over the positive and negative energy states finite. (I cannot help wondering whether rotation of the coordinate in the Hamiltonian into the complex plane could be used for this renormalization.)

Barut calculated both the Lamb shift and linewidth (which are the real and complex parts of atomic energy shifts) and he obtained the following spontaneous decay rates of hydrogen and muonium (10):

	Hydrogen	Muonium
$\Gamma(2P \to 1S_{1/2})$	$6.2649 \times 10^8 \, s^{-1}$	$6.2382 \times 10^8 \, s^{-1}$
$\Gamma(2S_{1/2} \to 1S_{1/2})$	$2.4946 \times 10^{-6} \, s^{-1}$	$2.3997 \times 10^{-6} \, s^{-1}$

These values are accurate through terms of order α^5 and agree with the best experimental results.

C. Nonrelativistic Quantized Field QED Applications

Although I have emphasized semiclassical techniques, there are many applications where quantized field procedures are preferable. The treatise by David Craig and T. Thirunamachandran (14) gives a rigorous but readable formal derivation of the quantized field matrix elements for an amazing number of laser–molecule interaction effects: one- and two-photon absorption and emission, Rayleigh and Raman scattering, dispersion forces in a radiation field, radiation-induced chiral discrimination, both linear and nonlinear optical processes such as CARS and laser-induced optical rotation, self-energy, and the Lamb shift. In other words, it is a one-volume encyclopedia! Since it is self-contained and begins with first principles, it should be useful both for beginners and experts. The derivations are based upon the research of Edwin Power (15) and W. P. Healy (16) as well as this author's.

Craig and Thirunamachandran express their wavefunctions in terms of Glauber coherent photon states that correspond to classical electromagnetic fields in the limit that the mean number of photons, $\langle n \rangle$, becomes infinite. [The difference between the transition probability of a Glauber state and its classical counterpart (16) varies as $1/\langle n \rangle$, which for strong fields is negligible.]

I am concerned that number (or Fock) states are frequently used in laser chemistry instead of Glauber states. This certainly could cause errors in the calculations since the number states are not coherent and do not correspond to a classical electric field. Furthermore, in some laser chemistry articles, the quantized field wavefunctions are supposed to evolve, as though they were classical Floquet states with the pseudostate kets $|n\rangle$ being replaced by the photon number kets $|n\rangle$. [Compare Eqs. (1.23) and (2.15) of Ref. 17.] The errors that result may be negligible, but I have never seen a thorough error analysis.

D. Gauge Effects and Paradoxes

In quantum mechanics, the *relative values of the gauge* or phase can play an important role in determining transition probabilities and other observables;

whereas the *absolute value of the gauge* of a wavefunction *has no physical significance*. This is the explanation of the following "paradoxes":

(a) *Ramsey's (18) time delay*, in which the transition probability depends upon the time delay between two pulses.

(b) The well-known *Aharnov–Bohm effect* (19), which has been recently verified (20) for the magnetoresistence of doubly connected geometries in high-mobile GaAs–AlGaAs hetero-structure systems.

(c) The *molecular Aharanov–Bohm effect* (21, 22), which is mathematically similar to the true Aharanov–Bohm effect although it does not involve electromagnetic fields. Whenever two electronic potential energy surfaces experience a conical intersection, an effective vector potential term must be inserted into the nuclear Born–Oppenheimer Hamiltonian in order to make the total (electronic–nuclear) Born–Oppenheimer wavefunction single valued. Calculations of the energy levels of Li_3 show that this effect is quite appreciable.

(d) Recently, Mike Berry's phase rotation of electromagnetic radiation propagating in fiber optics (which is an effect very similar to the Aharanov–Bohm effect) has been experimentally observed (23) and explained as a classical effect (24).

Gauge changes produced by canonical transformations have been studied by Ed Power (25, 26) for the three alternative forms of the Hamiltonian (minimal coupling, multipolar, and generalized space-translation) describing the interaction of radiation with atoms. They *do not* all give the same expectation values for physical observables because of errors (such as omitting the continuum states) in the mathematical model of an atom or the truncation of the approximation after the second order of perturbation. [As Saul Epstein explained in his article (27) entitled "What is H_0?" one can get any answer that one wants in a second-order calculation of energy by making the "proper" choice of the zero-order Hamiltonian!] Power concluded that the multipolar gauge is the simplest to use and the most reliable Hamiltonian for calculating molecular energies.

Tom Yang and I discovered (28) that although the *gauge of the external field potentials is arbitrary* in either the minimal-coupling or Breit–Pauli Hamiltonians, after making a Power–Zienau–Woolley transformation:

(a) The gauge origin is necessarily the center of mass.

(b) The gauge of the *internal fields* associated with the *internal molecular motions* is necessarily *multipolar gauge*.

(c) The gauge of the *external fields* associated with the motion of the center of mass remains arbitrary. (These gauge restrictions probably occur because the Breit–Pauli Hamiltonian and Power–Zienau–Woolley transformations are not covariant.)

E.   Spin-Generalized Classical Poisson Bracket

Very few people seem to be aware of the existence of *spin-generalized Poisson brackets*, although they can be very useful in the solution of semiclassical or classical problems involving spins.

Tom Yang and I were able to generalize the classical Poisson bracket to include spin (29) on the basis of our analysis of *Breit–Pauli Hamiltonian* molecular dynamics (30).

We assumed that a classical electron is a finite-size spinning sphere whose angular momentum (in its rest frame) is its spin and succeeded in making a completely classical derivation of the Breit–Pauli Hamiltonian for molecular systems in the presence of external electromagnetic fields. Furthermore, we proved that (though $1/c^2$ or $\alpha^4 mc^2$ fine structural terms) there is correspondence for all quantum-mechanical dynamic operators. If an electron were not a sphere, it would be necessary to know its angular velocity as well as its spin in order to describe its dynamics. [In 1962, Dirac (31) used a similar classical model of an electron together with the Bohr–Sommerfeld quantization conditions. He found that the rest mass of the first excited state was similar to a muon's.]

If both $f(t, \mathbf{p}, \mathbf{r}, \mathbf{s})$ and $g(t, \mathbf{p}, \mathbf{r}, \mathbf{s})$ are any general functions of time as well as momenta, coordinates, and spins of each of the particles (note that $\mathbf{p}$, $\mathbf{r}$, and $\mathbf{s}$ are vectors), then the *spin-generalized Poisson bracket* is

$$\{f, g\} = \sum_j \left[-\frac{\partial f}{\partial \mathbf{p}_j} \cdot \frac{\partial g}{\partial \mathbf{r}_j} + \frac{\partial f}{\partial \mathbf{r}_j} \cdot \frac{\partial g}{\partial \mathbf{p}_j} - \frac{\partial f}{\partial \mathbf{s}_j} \cdot \left[\mathbf{s}_j \times \frac{\partial g}{\partial \mathbf{s}_j} \right] \right]. \tag{5.1}$$

This generalized Poisson bracket should apply to any system having a Hamiltonian such that

$$\frac{d\mathbf{s}_j}{dt} = -\mathbf{s}_j \times \frac{\partial H}{\partial \mathbf{s}_j}. \tag{5.2}$$

A desirable feature of this Poisson bracket is that

$$\frac{df}{dt} = \frac{\partial f}{\partial t} + \{f, H\}, \tag{5.3}$$

so that we obtained the three *generalized canonical equations*:

$$\frac{d\mathbf{p}_j}{dt} = \{\mathbf{p}_j, H\}, \tag{5.4a}$$

$$\frac{d\mathbf{r}_j}{dt} = \{\mathbf{r}_j, H\}, \tag{5.4b}$$

and

$$\frac{d\mathbf{s}_j}{dt} = \{\mathbf{s}_j, H\}. \tag{5.4c}$$

Previously, Sudarshan and Mukunda (32) had derived the same spin-generalized Poisson bracket by using purely group-theoretic arguments; whereas our derivation is based upon a dynamical approach. The advantage of the group-theoretic approach is its simplicity and elegance, whereas the dynamical approach seems to be required for systems in external electromagnetic fields.

Lakshmanan and Daniel (33) have found the spin-generalized Poisson bracket useful in solving a number of problems including discrete Heisenberg ferromagnetic spin chains.

Recently Asim Barut and Ninozanghi (11) made a very sophisticated study of the dynamics of an electron considered as a real radiating and spinning spherical particle in phase space. Indeed, they made a "relativistic classical electron model which exactly reflects all of the properties of the Dirac electron." For example, their model even undergoes "real *Zitterbewegung*" so that the spin appears as the angular momentum of the *Zitterbewegung*! After separating the center of mass and the relative internal coordinates, they obtained both quantum mechanical and classical versions of the equation of motion of the electron. The canonical coordinates of the internal motion satisfy what Barut calls a remarkable Poisson bracket. [I am anxious to learn more about this approach. I believe, however, that it is closely related to Bill Miller's classical spin matrix research (34).]

References

1. P. W. Milonni, *Semiclassical and Quantum Electrodynamical Approaches in Nonrelativistic Radiation Theory*, North-Holland, Amsterdam, 1976.

2. P. W. Milonni, "Different Ways of Looking at the Electromagnetic Vacuum," in *Proceedings of the Adriatico Research Conference On Vacuum in Nonrelativistic Matter-Radiation Systems*, Trieste, Italy, July 1987.

3. J. H. Eberly, K. Wodkiewicz, and B. W. Shore, Noise in Strong Laser–Atom Interactions: Phase Telegraph Noise, *Phys. Rev. A* **30**, 2381 (1984).

4. A. O. Barut and J. P. Dowling, Quantum Electrodynamics Based on Self-Energy, Without Second Quantization. The Lamb Shift and Long-Range Casimir-Polder Van der Waals Forces Near Boundaries, *Phys. Rev. A* **36**, 2550 (1987).

5. A. O. Barut and J. Kraus, *Found. Phys.* **13** (1983).

6. A. O. Barut and J. F. Van Huele, *Phys. Rev. A* **32**, 3187 (1985).

7. A. O. Barut, "An Approach to Finite Non-Perturbative Quantum Electrodynamics," *Proc. 2nd Int. Symp. Found., Quant. Mech.*, Tokyo, 1986, p. 323.

8. A. O. Barut and N. Ünal, New Approach to Bound-State Quantum Electrodynamics, *Physica* **142A**, 488 (1987).

9. A. O. Barut and Y. I. Salamin, "Relativistic Theory of the Lamb Shift Based on Self Energy" and also "Relativistic Theory of Spontaneous Emission," in *Proceedings of the Adriatico Research Conference On Vacuum in Relativistic Matter-Radiation Systems*, Trieste, Italy, July 1987.

10. A. O. Barut, On the Treatment of Möller and Breit Potentials and the Covariant Two-Body Equation for Positronium and Muonium, *Phys. Script.* **36**, 493 (1987).

11. A. O. Barut, Electron as a Radiating and Spinning Dynamical System and Discrete Internal Quantum System, *Phys. Script.* **35**, 229 (1987); A. O. Barut and N. Zanghi, *Phys. Rev. Lett.* **52**, 2009 (1984).

12. I. Bialynicki-Birula, *Phys. Rev. A* **34**, 3500 (1986). This is a criticism of the procedures used by Barut and Van Huele (6). See also Barut's detailed rebuttal, A. O. Barut, *Phys. Rev. A* **34**, 3502 (1986).

13. J. H. Van Vleck, *Proc. Nat. Acad. Sci.* **14**, 178 (1928).

14. D. P. Craig and T. Thirunamachandran, *Molecular Quantum Electrodynamics*, Academic New York, 1984.

15. E. Power and T. Thirunamachandran, Quantum Electrodynamics with Nonrelativistic Sources, *Phys. Rev. A* **28**, 2649, 2663, 2671 (1983). See also E. A. Power, *Introductory Quantum Electrodynamics*, Longmans, London, 1964.

16. W. P. Healy, *Non-Relativistic Quantum Electrodynamics*, Academic, New York, 1982.

17. P. K. Aravind and J. O. Hirschfelder, Two-State Systems in Semiclassical and Quantized Fields, *J. Phys. Chem.* **55**, 4788 (1984).

18. N. F. Ramsey, *Molecular Beams*, Oxford University Press, London, 1956.

19. J. J. Sakurai, *Advanced Quantum Mechanics*, Addison Wesley, Reading, MA, 1978, p. 17.

20. G. Timp, A. M. Chang, J. E. Cunningham, T. Y. Chang, P. Mankiewich, R. Behringer, and R. E. Howard, Observations of the Aharanov-Bohm Effect for $\omega_c \tau \rangle$, *Phys. Rev. Lett.* **5**, 2814 (1987).

21. C. A. Mead, The Molecular Aharonov-Bohm Effect in Bound States, *Chem. Phys.* **49**, 23 (1980); C. A. Mead and D. G. Truhlar, *J. Chem. Phys.* **70**, 2284 (1979).

22. A. G. Redfield, The theory of Relaxation Processes, *Adv. Magn. Reson.* **1**, 1 (1965).

23. A. Tomita and R. Y. Chiao, Observation of Berry's Topological Phase by Use of Optical Fibers, *Phys. Rev. Lett.* **57**, (1987).

24. J. Segert, Photon Berry's Phase as a Classical Topological Effect, *Phys. Rev. A* **36**, 10 (1987).

25. E. A. Power, "Canonical Transformations," in *Multiphoton Processes*, J. H. Eberly and P. Lambropoulos, Eds., Wiley, New York, 1978, p. 11.

26. E. A. Power and T. Thirunamachandran, On the Nature of the Hamiltonian for Interaction of Radiation with Molecules, *Am. J. Phys.* **46**, 3790 (1978).

27. S. T. Epstein, "What is H_0?" in *Application in Quantum Mechanics*, C. H. Wilcox, Ed. Wiley, New York, 1966, p. 49.

28. K.-H. Yang, J. O. Hirschfelder, and B. R. Johnson, Interaction of Molecules with Electromagnetic Fields: II. Multipole Operators and Dynamics of Molecules with Moving Nuclei in Electromagnetic Fields, *J. Chem. Phys.* **75**, 2321 (1981).

29. K.-H. Yang and J. O. Hirschfelder, Generalizations of Classical Poisson Brackets to Include Spins, *Phys. Rev. A* **22**, 1814 (1980).

30. K.-H. Yang and J. O. Hirschfelder, Interaction of Molecules in Electromagnetic Fields: I. Classical Particles and Fields, *J. Chem. Phys.* **72**, 5863 (1980).

31. P. A. M. Dirac, *Proc. Roy. Soc. (Lond.)* **A264**, 57 (1962); R. G. Nadig, Z. Kunszt, P. Hasenfratz, and J. Kuti, *Ann. Phys.* **116**, 380 (1978).

32. E. C. G. Sudarshan and N. Mukunda, *Classical Dynamics: A Modern Perspective*, Wiley, New York, 1974, Chapters 17 and 18, especially pp. 316–320 and 365–369.

33. M. Lakshmanan and M. Daniel, Classical Models of Electrons and Nuclei and Spin-Generalized Poisson Brackets, *J. Chem. Phys.* **78**, 7505 (1983).

34. H.-D. Meyer and W. H. Miller, Classical Analog for Electronic Degrees of Freedom in Nonadiabatic Collision Processes, *J. Chem. Phys.* **70**, 3214 (1979); and also Classical Model for Electronic Degrees of Freedom: Derivation Via Spin Analogy and Applications to $F^* + H_2 \rightarrow F + H_2$, *J. Chem. Phys.* **71**, 2156 (1979).

VI. SEMICLASSICAL AND CLASSICAL TECHNIQUES

A. Introduction

Thanks to the recent work of Bill Miller (1, 2), Eric Heller (3), Benny Gerber (4), Shih-I Chu (5), and many others, there is a renewed interest in using semiclassical and classical dynamics for solving all kinds of molecular dynamics problems, with or without an external field. Furthermore (as we shall see), there is a close correspondence between Drude's nineteenth-century classical oscillators and quantum-mechanical dressed states.

In the hands of an expert like Bill Miller, semiclassical calculations can have accuracy comparable with quantum mechanical results. (For example, Bob Schrieffer told me that at the Copenhagen symposium honoring the 100th year anniversary of Niels Bohr, he heard that the old Bohr atom energy levels for helium were more accurate than Hartree–Fock calculations.)

John Wheeler (6) made an interesting comparison between semiclassical and quantum-mechanical formalisms. Since the semiclassical method deals with trajectories in phase space, it is easy to understand the physics involved, whereas "*the elegance and apparent simplicity of formal quantum mechanics conceals a great deal of physics!*" In formal quantum mechanics, there is no indication of the behavior in momentum space of a system whose wavefunction is expressed in spatial representation. Therefore, it is difficult to determine the conditions under which a small change in the parameters produces a large change in the properties of the system.

Between 1978 and 1987, Bill Miller and his colleagues published a series of papers (1, 2, 7) showing how classical mechanics can be used to model the electronic (as well as the other) degrees of freedom of a molecular system. This has enabled them to make both classical and semiclassical analogs of general *second-quantized* Hamiltonians for many-electron systems expressed in terms of angle and action variables. Thus, they are establishing the basis for using S-matrix (or angle–action variables in phase space) techniques to solve laser–molecule interaction problems.

Miller's research and McLachlan's time-dependent Hartree–Fock theory (8) must have inspired Benny Gerber, Mark Ratner and their colleagues (4) to develop *quantal, semiclassical, and classical versions* of Dirac's *time-dependent self-consistent field* (TDSCF) (9) approximation, which is a numerically efficient technique of wide applicability. All three versions give good agreement with experimental data for unimolecular dissociation of van der Waal's molecules. One of the most desirable properties of the TDSCF is that it retains the separability of the modes but still allows for energy transfer between the modes. Thus, each mode in quantum mechanics is a quasi-energy state; whereas a mode in classical mechanics is an oscillator.

Recently, Jeff Needels and Shih-I Chu (5) have used Gerber's TDSCF techniques to study infrared laser multiphoton vibrational excitation of SO_2. They found that it was much more efficient than previous methods. Furthermore, their classical and quantum-mechanical calculations of the essential frequency and intensity-dependent dynamical excitation properties were in agreement with the results of experiments. Thus, we can expect that increasing interest will develop in using semiclassical and classical procedures for determining laser interaction effects.

Heller (3) deserves a lot of credit for pioneering semiclassical wavepacket techniques and applying them to a variety of chemical problems including molecular spectroscopy. As Littlejohn noted (10), "Wave-packets are particularly useful because they yield uniform results, require no special attention at caustics, and can be used in any number of dimensions. Because of wave-packet spreading, one cannot simply take the Fourier transform of the time evolution of a wave packet." Instead, Robert Littlejohn (10) has developed a technique for propagating the wavepacket making use of *metaplectic operators* (11) to determine the actions, which serve as his Hamiltonians. Littlejohn noted: "The energy eigenvalues which result are the Einstein–Brillouin–Keller quantized values, and new forms of the eigenfunctions appear. These are free of caustic singularities and represent averages of wave-packets over the invariant torus" (10). [The use of action variables for wavepacket propagation is very closely related both to S-matrix techniques for solving scattering problems and to Barut's formulation of semiclassical QED (see Section V.B). Until Littlejohn solves some physically interesting dynamical problems, it will be difficult to judge the significance of his research.

There are a number of difficulties that the new semiclassical and classical techniques must try to avoid or overcome:

1. First of all, the nonlinear Hamilton–Jacobi equation is generally more difficult to solve than the corresponding Schrödinger equation.

2. In a WBK (Wentzel–Brillouin–Kramers) approximation, if the coordinate does not extend from minus to plus infinity, the solution may not be unique! The question is, should a Langer type of transformation be used? Russell Pack says (12), "For bound vibrational states, *NO*; but for scattering phase shifts, *YES!*" Furthermore, Bill Miller (13) has a simple procedure for estimating WBK errors that occur with or without any of the infinite number of possible Langer-type coordinate transformations.

3. The most serious difficulty is that the *Weyl correspondence rule* (14) prescription for constructing the quantum-mechanical Hermitian operator that corresponds to a given classical function of coordinates and momentum is not always correct if it involves products of noncommuting operators (15). This problem is serious because the Wigner distribution function (WDF) (16) is based upon the Weyl correspondence rule: As Serimas, Javanainen, and Varro (17) explain, "The WDF and related quasiprobability distributions have been used very often because they allow quantum mechanical expectation values to be expressed as phase-space integrals analogous to those of classical mechanics. Classical intuition may then be taken over to quantum mechanics. Thus, semiclassical $\hbar$ or $1/T$ expansions can be carried out remarkably easily" (17).

Thus, there is great interest in improving the WDF and the Weyl rule. For example, Narcowich and O'Connell (18) have determined the necessary and sufficient conditions for the WDF to be accurate (18) and Serimas, Javaanainen, and Varro (17) have developed gauge-invariant Wigner operators and functions. I believe that an improved version of the WDF would be very useful in solving many types of laser–molecule interaction problems such as the intense-field multiphoton ionization problems that Andre' Bandrauk and Osman Atabek discuss in Chapter XIX.

B. The Classical Model

The concept of an *oscillator strength* stems from Drude's nineteenth-century classical mechanical model of the electrical and optical behavior of matter (19–22). In many respects Drude optics corresponds closely with present quantum mechanics.

Atoms are regarded as a set of particles j having a charge e_j and a mass m_j. Each is harmonically and isotropically bound to its individual equilibrium position. Its force constant is k_j so that its natural (angular) frequency of vibration is $\omega_j = k_j/m_j$ and its displacement from its equilibrium position is x_j.

Thus, these charged particles perform forced oscillations when driven by a force such as produced by an electric field with intensity $F(t)$. Furthermore, these oscillations would be damped because each oscillator would radiate electromagnetic energy.

Lorentz (23) worried about the self-consistency problem of accounting for the effect of a single oscillator's own field on its own motion. However, when Einstein showed that electromagnetic fields possess energy, it became clear that the energy of the field–oscillator system is conserved. [G. I. Taylor told me that part of his doctoral thesis (around 1915?) was concerned with showing that it is *energy* and not *entropy* that is conserved!]

From the conservation of energy it follows that the rate of energy loss of an oscillator by radiation is (23)

$$\frac{2}{\tau_{rj}} = 2e_j^2 \omega_j^2 (3m_j c^3)^{-1}. \tag{6.1}$$

The natural lifetime τ_{rj} predicted in this way for an electron oscillating at optical frequencies is on the order of 10^{-7} s, and $1/\tau_{rj} \ll \omega_j$. Thus, the energy loss during a period of oscillation would be very small.

However, in real atomic problems, the natural lifetime of an oscillator can be decreased by the following:

1. power broadening or interactions between oscillators,
2. spontaneous ionization of the atom, and
3. collisions, Doppler effect, and so on.

Thus, the *real natural lifetime* is τ_{0j}, which is less than τ_{rj}. But $1/\tau_{0j}$ usually remains much less than ω_j. [I am surprised that Lorentz and Drude, for example, considered the possibility of their oscillators decaying.]

It follows from the nonrelativistic version of the Lorentz force law that the equation of motion (or change) of the jth particle (or oscillator) is (19–21)

$$\frac{d^2 x_j}{dt^2} + \frac{2}{\tau_{0j}} \frac{dx_j}{dt} + \omega_j^2 x_j = \frac{e_j}{m_j} F(t). \tag{6.2}$$

Here the rate of decay of the dipole amplitude $e_j x_j$ is $1/\tau_{0j}$, and $(e_j/m_j)F(t)$ is the external force.

The *exact solution* of Eq. (6.2) for all values of the parameters and all analytic functions $F(t)$ is given in Kamke's *encyclopedia of solutions of differential equations* (24).

If $\omega_j > 1/\tau_{0j}$ and $(dx_j/dt)_0 = 0$, then

$$x_j(t) = e_j(m_j g_j)^{-1} \int_0^{t-t_0} F(t-y) \exp\left(\frac{-y}{\tau_{0j}}\right) \sin(g_j y)\, dy, \tag{6.3}$$

and if $F(t) = 0$, then

$$x_j(t) = [C_1 \cos(g_j t) + C_2 \sin(g_j t)] \exp\left(\frac{-t}{\tau_{0j}}\right). \tag{6.4}$$

In Eqs. (6.3) and (6.4), $g_j = [\omega_j^2 - (1/\tau_0)^2]^{1/2}$. Furthermore, C_1 and C_2 are the constants of integration required to satisfy the initial values of x_j and dx_j/dt.

Since Eq. (6.3) applies to arbitrary fields, it would be interesting to see what properties this classical model would predict for an atom interacting with either (a) an intense short pulse or (b) a sequence of pulses or (c) a quasiperiodic or aperiodic field.

C. Properties of a Classical Atom in a Periodic Field

Let us follow Fano and Cooper (19) in considering the displacement $x_j(t)$ of a classical oscillator perturbed by an oscillating electric field with the complex intensity $F(t) = F_0 \exp[-i\omega t]$. The solution of Eq. (6.3) for this case is quite complicated. However, after a long time, all of the transients have died out, and the *dipole moment of the oscillator* is

$$e_j x_j(t) = \frac{(e_j^2/m_j)F_0 \exp(-i\omega t)}{\omega_j^2 - \omega^2 - 2i\omega/\tau_{0j}}. \tag{6.5}$$

Thus, the *polarizability of the oscillator,* or the *dipole moment per unit field strength,* is

$$\alpha_j(\omega) = \frac{e_j^2/m_j}{\omega_j^2 - \omega^2 - 2i\omega/\tau_{0j}}. \tag{6.6}$$

I was surprised that in this derivation, Fano and Cooper used the *rotating-wave approximation,* in which the field strength is $F(\omega) = F_0 \exp[-i\omega t]$ rather than the $F(t) = F_0 \cos(\omega t)$ which occurs in experiments. However, it is much simpler and sufficiently accurate for most purposes. [Sometime, I will determine the exact solution of Eq. (6.3) for $F(t) = F_0 \cos(\omega t)$.]

In this classical model of an electron, the effective number of electrons in the atom oscillating with the frequency ω_j is its *oscillator strength* $f_j = e_j^2 m_j/e^2 m$, where e and m are the charge and the mass of a *real physical electron.*

A dilute gas containing N_{at} atoms per unit volume has the *susceptibility*

$$X(\omega) = N_{at}\alpha(\omega) = \frac{N_{at}(e^2/m)\sum_j f_j}{\omega_j^2 - \omega^2 - 2i\omega/\tau_{0j}}. \tag{6.7}$$

Furthermore, if the gas is nonmagnetic, the *susceptibility* $X(\omega)$ is related to the *complex dielectric constant* $\varepsilon(\omega)$ and to the *complex index of refraction* $\eta(\omega) = [n(\omega) + ik(\omega)]$ by the relations

$$\varepsilon(\omega) = 1 + 4\pi X(\omega) = \eta(\omega)^2 = [n(\omega) + ik(\omega)]^2. \qquad (6.8)$$

Furthermore, the *absorption coefficient of electromagnetic radiation* $\lambda(\omega)$ is

$$\lambda(\omega) = 2\frac{\omega}{c}k(\omega) = \frac{2\pi^2 e^2}{mc}N\frac{df}{d\omega}. \qquad (6.9)$$

D. Correspondence of Classical Model

All of the properties of the classical model in a weak laser field correspond closely with their quantum-mechanical counterparts. However, there are conceptual and notational differences:

1. The classical oscillators are associated with the *electrons* in the atom, whereas the quantum-mechanical oscillators are associated with the *states* in the atom. Thus, the number of classical oscillators is equal to the number of electrons in the atom, whereas there are an infinite number of both discrete and continuum quantum states.

2. The quantum-mechanical oscillators are *quasi-energy states* having complex energies $E_j = E_{rj} - iE_{ij}$. The imaginary part of their energy is the reciprocal of their natural lifetime, that is, $E_{ij} = 1/\tau_{0j}$, and E_{ij} is their *linewidth*.

3. Gisin's damped harmonic oscillator and other nonlinear Schrödinger-like equations, which are discussed in Section IV.D, are the quantum-mechanical analog of the classical mechanical Eq. (6.2) and its Eq. (6.4) exact family of solutions.

E. Oscillator Strengths and Spectral Distributions

Fano and Cooper (19, 25) defined the quantum-mechanical oscillator strength for a transition from an atomic state k to a state j as

$$f_{jk} = \frac{2m}{\hbar^2}\omega_{jk}|\langle\psi_j|\mathbf{r}|\psi_k\rangle|^2, \qquad (6.10)$$

where $\omega_{jk} = (E_{jr} - E_{kr})/\hbar$ and E_{jr} and E_{kr} are the real parts of the jth- and kth-state energies.

Inokuti (25) points out that if ω is much larger than all of the energies ω_{jk}, then the Coulombic forces holding the atom together would be negligible compared to the external field forces. In this case the response of the atom would be the same as an assembly of N_{el} free electrons (where N_{el} is the number of electrons in the atom). In order to obtain this limiting behavior, it is necessary that the Thomas–Reiche–Kuhn sum rule apply for all values of ω,

$$\int_j f_{jk} = N_{el} \qquad (6.11)$$

Here $\int_j$ is a sum over all of the discrete states j and an integral over the continuum.

Thus, it follows from both second-order quantum-mechanical perturbation theory and Eqs. (6.7) and (6.10) that the dynamic polarizability of the kth atomic state is

$$\alpha_k(\omega) = \frac{e^2}{m} \int_j \frac{f_{jk}}{\omega_{jk}^2 - \omega^2 - 2i\omega_{jk}/\tau_{0j}}, \qquad (6.12)$$

where $1/\tau_{0j} = (\Gamma_j - \Gamma_k)/\hbar$. Also, Γ_j and Γ_k are the imaginary components of the jth and kth energy levels. Usually, the dynamic polarizability is only determined for the ground state, $k = 0$. However, for laser interaction problems, the dynamic polarizability of excited states is sometimes needed.

Inokuti (25) said that since the resonance energy $\hbar\omega_{jk}$ is generally very large compared to its linewidth $\Gamma_j - \Gamma_k$, "we may use the well-known formula

$$\lim_{y \to +0} \text{ of } \frac{1}{x - iy} = P\left(\frac{1}{x}\right) - i\pi\delta(x), \qquad (6.13)$$

where P denotes the Cauchy principle value and δ is the Dirac delta function." Thus, Eq. (6.12) can be rewritten as

$$\alpha_k(\omega) = \frac{e^2}{m} \left[P \int_j \frac{f_{jk}}{\omega_{jk}^2 - \omega^2} + i\pi \int_j f_{jk}\delta(\omega_{jk}^2 - \omega^2) \right]. \qquad (6.14)$$

The first term in Eq. (6.14) describes the overall dispersion, and the second term gives the overall absorption of radiation in the resonant region without going into a detailed treatment of the anomalous dispersion which occurs."

Inokuti (25) makes the surprising statement that "because of the dominance of continuous spectra in general," the best way to get a comprehensive understanding of the optical properties of any material is to determine the *oscillator strength distribution*

$$\frac{df_k}{dE} = \int_j f_{jk}\delta(E - \hbar\omega_{jk}) \qquad (6.15)$$

as a function of E, the excitation energy. Values of df_k/dE have been experimentally determined for many substances by using a synchroton and other sources of vacuum far-ultraviolet radiation. The experimental data (up to 1982) and their interpretation have been reviewed in (26, 27).

For condensed matter (24), in contrast to gases, it is customary to express the optical properties in terms of the complex dielectric response function

$$\varepsilon(\omega) = 1 + 4\pi N_{mol}\alpha(\omega) = \varepsilon_r(\omega) + i\varepsilon_i(\omega). \tag{6.16}$$

The probability of photoabsorption is proportional to $\omega\varepsilon_i(\omega)$, whereas the probability of energy transfer from fast charged particles is $\omega\varepsilon_i(\omega)/[\varepsilon_r^2(\omega) + \varepsilon_i^2(\omega)]$.

For metals, $\varepsilon_r(\omega)$ is large and negative at $\omega = 0$ and steadily increases with ω; it becomes positive in the far-ultraviolet frequency range and approaches unity as $\omega \to \infty$. The probability of energy transfer from fast charged particles has a sharp maximum where ε_r changes sign.

The value of $\varepsilon_i(\omega)$ is large and positive at $\omega = 0$ and continues to decrease as ω increases, except for peaks at Feshbach resonances where new modes of excitation appear: first at the beginning of interband valence electron excitations, then at the L-shell, and finally at the K-shell thresholds.

For organic materials, $\varepsilon_r(\omega)$ is real for all values of ω and has a minimum when $\hbar\omega$ is less than 20 eV. It is curious that all organic compounds have a maximum probability of energy transfer from charged particles around $\hbar\omega = 20$ eV.

F. Coherent Raman Spectroscopy

The energy $E_n(\omega)$ of the nth dressed state of an atom or molecule in a weak electric field having a frequency ω and a field strength $F(\omega)$ can be expressed (28) as a power series in the field strength $\mathbf{F}(\omega)$:

$$E_n(\omega) = E_n^{(0)}(\omega) - \tfrac{1}{2}\underline{\alpha}_n(\omega):\mathbf{F}(\omega)\mathbf{F}(\omega)$$
$$- \tfrac{1}{24}\underline{\gamma}_n(\omega)::\mathbf{F}(\omega)\mathbf{F}(\omega)\mathbf{F}(\omega)\mathbf{F}(\omega) - \cdots. \tag{6.17}$$

Here the dynamic polarizability is $\underline{\alpha}_n(\omega)$ (a second-order tensor), and $\underline{\gamma}_n(\omega)$ (a fourth-order tensor) is the dynamic hyperpolarizability. If $\omega = 0$, $\underline{\alpha}_n$ and $\underline{\gamma}_n$ are the usual static polarizability and hyperpolarizability, and they do not change significantly if ω is much less than the lowest excitation frequency of the atom. Furthermore, odd powers of $\mathbf{F}(\omega)$ can only appear in Eq. (6.17) if the molecule has permanent multipole moments.

The induced electric dipole moment of the nth state is

$$\mathbf{p}_n(\omega) = -\left[\frac{\partial E_n(\omega)}{\partial \mathbf{F}(\omega)}\right]_{\text{av}}$$

$$= [\underline{\alpha}_n(\omega) \cdot \mathbf{F}(\omega)]_{\text{av}} + \tfrac{1}{6}[\underline{\gamma}_n(\omega) :: \mathbf{F}(\omega)\mathbf{F}(\omega)\mathbf{F}(\omega)]_{\text{av}}, \qquad (6.18)$$

where $[\ldots]_{\text{av}}$ means averaged over a cycle. Strangely, $\underline{\alpha}_n$ is called the first-order polarizability and $\underline{\gamma}_n$ is called the third-order polarizability!

In Chapter VIII, Bill Meath and his colleagues show that the presence of a permanent dipole moment can have a significant effect on the molecular spectrum. Furthermore, if the laser field is augmented by a constant electric field, the effect is greatly increased. A similar (but smaller) effect results from including polarizability.

The ground-state polarizability is always positive, whereas the polarizability of an excited state can be either positive or negative depending upon whether it is more likely to make a transition to a higher or to a lower energy state.

The imaginary component (Im) of the dynamical polarizability of stable states is usually negligible (except in the vicinity of resonance where they peak). The only calculation of Im $[\alpha_k(\omega)]$ that I know about is an article by Manakov and his Russian colleagues (29) on the spectrum of a hydrogenlike atom in a laser emission field (see Bayfield, Ref. 30).

The usual (two-photon resonant and nonresonant) Raman spectra and fluorescence are determined by dynamic polarizabilities. Whereas *four-wave mixing processes*, such as CARS and other coherent Raman effects, are determined by the dynamic hyperpolarizabilities (Section IV.A and Refs. 31 and 32). Both the imaginary and the nonresonant terms in the dynamic hyperpolarizabilities can sometimes be important.

Crude estimates of hyperpolarizabilities may be obtained by considering only the lead term in the corresponding Feynman diagrams. Kirtman (33) used a perturbative technique with a semiempirical INDO Hamiltonian and found that the static hyperpolarizability for either polyacetylene or polydiacetylene is enormous (33) (on the order of 10^6 a.u. per CH or CH_2 unit) because of their conjugate double bonds.

G. Calculation of the Real Part of Polarizabilities and Dynamic Polarizabilities

Most elementary books on laser theory express the real part of the dynamic polarizability of a state k as,

$$\alpha_k(\omega) = \int_j \frac{2|\langle \mathbf{\Psi}_j | e\mathbf{r} | \mathbf{\Psi}_k \rangle|^2 (E_j^{(0)} - E_k^{(0)})}{(E_j^{(0)} - E_k^{(0)})^2 - \hbar^2 \omega^2}. \qquad (6.12')$$

These books give the impression that good values of the polarizability can be obtained by calculating the first few terms in the summation and ignoring the continuum. However, such perturbative procedures are seldom as accurate as simple variational calculations.

The best example I know is the static polarizability of the ground state of a hydrogen atom. As A. David Buckingham explained in a lecture (unpublished) at the University of Wisconsin Theoretical Chemistry Institute, Eq. (6.12′) for the $1s$ hydrogen atom with $\omega = 0$ is (in atomic units)

1. *Second-Order Perturbation Calculations*

$$\alpha = 2 \sum_{n=2}^{\infty} \frac{|\langle 1s| \times |np\rangle|^2}{E_{np}^{(0)} - E_{1s}^{(0)}} + \text{continuum} \tag{6.19}$$

$$= \sum_{n=2}^{\infty} \frac{2^{10} n^9 (n-1)^{2n-6}}{3(n+1)^{2n+6}} + \text{continuum}$$

or

$$\alpha = \underset{65.8\%}{2.960} + \underset{8.9\%}{0.400} + \underset{2.9\%}{0.132} + \cdots + \underset{18.6\%}{\text{continuum}}$$

It is truly amazing that although the rate of transitions to the continuum (spontaneous ionization) is negligible, a large fraction of the polarizability can be due to the continuum states! The reason for this is that the set of discrete wave functions for an unperturbed Hamiltonian is *not* a good basis set for determining accurate values of the dynamic polarizability of either ground-state molecules or excited-state atoms. In general, the continuum fraction of the polarizability is small for atoms or molecules having low-lying energy states, but otherwise, it can be as large as 25 or 30%. Furthermore, the larger the molecule, the smaller is its continuum contribution to the polarizability.

2. *Hylleraas Variational Calculations.* Buckingham determined the polarizability by optimizing parameters in trial first-order wavefunctions $\tilde{\psi}^{(1)}$ in the *Hylleraas variational principle* (28, 34),

$$\alpha(0) \geqslant |\langle \psi^{(0)}|V|\tilde{\psi}^{(1)}\rangle|^2 / \langle \tilde{\psi}^{(1)}|H^{(0)} - E^{(0)}|\tilde{\psi}^{(1)}\rangle. \tag{6.20}$$

For the ground state of a hydrogen atom:

(a) If $\tilde{\psi}^{(1)} = (ax)\exp(-br)$, the optimum values of a and b are $a = -1.310$ and $b = 0.797$ so that the approximate polarizability is $\tilde{\alpha} = 4.475$ a.u., which is 99.4% of the exact value, $\alpha = 4.5$ a.u.

(b) If $\tilde{\psi}^{(1)} = (cxr)\exp(-br)$, the optimum values of b and c are $b = 1.158$ and $c = -1.382$ so that $\tilde{\alpha} = 4.378$ a.u. Although this trial function was not quite as good as the first, the sum of the two trial function is considerably better than either one by itself (as, of course, it must be).

I have very personal reasons for appreciating the superiority of variational to perturbative techniques. When I started my graduate work at Princeton in 1931, Professor Edward Condon suggested that I calculate the values of the parallel and perpendicular polarizabilities of H_2 and H_2^+. The only experimental data at that time were on the depolarization, which provided a measure of $(\alpha_{\parallel} - \alpha_{\perp})^2$ but did not tell whether $\alpha_{\parallel}$ is larger than $\alpha_{\perp}$. I varied two parameters in accordance with the Hylleraas variational principle and made all of my calculations by hand with the aid of a large logarithm table. My results agreed with the depolarization and showed that $\alpha_{\parallel} > \alpha_{\perp}$. Of course, I was proud of my accomplishment. However, at the same time Professor Mrowka in Germany had made a calculation using a truncated perturbation series approximation and found $\alpha_{\perp} \geqslant \alpha_{\parallel}$. Thus, since Mrowka was an established professor and I was only a beginning graduate student, Condon concluded that I was wrong and would not permit me to publish my manuscript (35) until three years later when Mrowka discovered that his method was unsatisfactory. My reason for mentioning this ancient history is to encourage people to use *nonperturbative* techniques for solving laser–molecule problems.

3. *Upper and Lower Bounds for Dynamic Polarizabilities.* There are two difficulties in calculating near-resonant dynamic polarizabilities and hyper-polarizabilities, the almost degeneracy of some of the quasi-energy states and the complex nature of their energies. As Bernie Kirtman (36) points out, almost degenerate perturbation formalisms (37) have been developed to avoid the small denominators in usual Rayleigh–Schrödinger-type treatments (with or without the use of Feynman diagrams). The wavefunctions for four-wave mixing processes such as CARS are generally expressed in terms of Feynman diagrams. Kirtman (36) shows how to use these Feynman diagrams as input into his version of almost degenerate perturbations to calculate the real part of either the polarizabilities or the hyperpolarizabilities. (It should be easy to generalize the almost degenerate perturbation formalism so as to include the imaginary part of the polarizabilities by using the time reversal symmetry technique described in Section VII.G.) Davidson, Engdahl, and Moiseyev (38) have established new bounds to resonant eigenvalues by using the complex coordinate method. They found that the complex analog of the variational theorem gives a stationary condition instead of providing a bound to eigenvalues.

Glover and Weinhold (39) have used the method of Gramian inequalities to derive excellent upper and lower bound variational principles for the determination of dynamic polarizabilities for two-electron atoms with frequencies up to and beyond the first excitation. Their treatment is based upon the Hylleraas (34) [Lower bound of $\alpha(\tilde{\psi})$], the Prager–Hirschfelder (upper bound) (40), and the Rebane (upper and lower bounds). Both the Hylleraas and Prager bounds are only valid at frequencies below the first excitation, whereas the Rebane bounds are valid for larger frequencies. One of the principal difficulties is that the zero-order as well as first-order wavefunctions were not known accurately.

After Glover and Weinhold made their formal derivations, they made extensive numerical calculations of the dynamic polarizability of the $1S$ (ground state) and the $2\,^1S$ and $2\,^3S$ (metastable states) of both He and Li^+ two-electron atoms with frequencies up to the second excitation. The maximum error of these states was on the order of 1% except close to either the first or second excitation. In a previous paper (41), Weinhold calculated the static polarizability with a probable maximum error of 2 parts in 100,000, which agreed with the best experimental value and was almost as precise!

Calculations of the dynamic polarizability of atoms or molecules with more than two electrons is extremely difficult. Thus, Wormer and Rijks (42) had to use extreme care in order to avoid making appreciable errors in their determination of the dynamic polarizability of the ground states of both a neon atom and the dimer Ne_2. Since electron correlations were very important, they used configuration interactions with large basis sets to solve both the zero- and first-order perturbation equations exactly. Solution of the zero-order equation did not require any new techniques. However, to solve the first-order equation, Visser and Wormer (43) developed a procedure "for making ab initio computations of Cauchy moments (moments of negative powers of a multipole oscillator strength distribution) followed by Padé approximant representation of the dynamic polarizabilities".

Thus, it seems that we must rely on experimental data (rather than a priori calculations) to determine good values of the dynamic polarizabilities of most atoms and molecules!

References

1. W. H. Miller and K. A. White, Classical Models for Electronic Degrees of Freedom: The Second-Quantized Many-Electron Hamiltonian, *J. Chem. Phys.* **84**, 5059 (1986).

2. W. H. Miller, A Classical/Semiclassical Theory for Infrared Radiation Interactions with Molecules, *J. Chem. Phys.* **69**, 2188 (1978). See also the comments of P. W. Milonni and Miller's reply, *J. Chem.* **72**, 787 (1980). Also see Jaynes's neoclassical model of an electromagnetic field as a set of classical mechanical harmonic oscillators: M. D. Crisp and E. T. Jaynes, *Phys. Rev.* **179**, 1253 (1969) or W. H. Louisell, *Quantum Statistical Properties of Radiation*, Wiley, New York, 1973, pp. 238–246.

3. E. J. Heller, The Semiclassical Way to Molecular Spectroscopy, *Acc. Chem. Res.* **14**, 368 (1981).

4. R. B. Gerber, V. Buch, and M. A. Ratner, Time-Dependent Self-Consistent Field for Intramolecular Energy Transfer: I. Application to Dissociation of Van der Waals Molecules, *J. Chem. Phys.* **77**, 3022 (1982).

5. J. Needels and S.-I Chu, Time-Dependent Self-Consistent Field Approach to Infrared Laser Multiphoton Excitation, *Chem. Phys. Lett.* **139**, 35 (1987).

6. J. A. Wheeler, in *Studies in Mathematical Physics*, E. H. Lieb, B. Simon, and A. S. Wightman, Eds., Princeton University Press, Princeton, NJ, 1976, p. 351.

7. W. H. Miller et al., *J. Chem. Phys.* **69**, 5163 (1978); **70**, 3177, 3214 (1979); **71**, 2156 (1980); **72**, 2272 (1980); **73**, 3191 (1980); **74**, 6075 (1981); **78**, 6640 (1983); and a series in Chem. Phys. Lett.

8. A. D. McLachlan and M. A. Ball, Time-Dependent Hartree-Fock Theory for Molecules, *Rev. Mod. Phys.* **36**, 844 (1964).

9. P. A. M. Dirac, *Proc. Cambridge Philos. Soc.* **26**, 376 (1930).

10. R. G. Littlejohn, Wave-Packet Evolution and Quantization, *Phys. Rev. Lett.* **56**, (1986); See also his review paper, *Phys. Rep.* **138**, 193 (1986).

11. V. Guillemin and S. Sternberg, *Symplectic Techniques in Physics*, Cambridge University Press, Cambridge, 1984.

12. R. Pack, On Improved WKB: Uniform Asymptotic Quantum Conditions, Dunham Corrections, the Langer Modification and RKR Potentials, *J. Chem. Phys.* **57**, 4612 (1972); The Time Dependent Density Matrix, *Phys. Rev. A* **25**, 1815 (1982).

13. J. E. Adams and W. H. Miller, Nonuniqueness of Langer-Type Transformations, *J. Chem. Phys.* **67**, 5775 (1977).

14. H. Weyl, The Weyl Correspondence Rule, *Zeit. Phys.* **46**, 1 (1927); N. H. McCoy, *Proc. Nat. Acad. Sci.* **18**, 674 (1932).

15. W. H. Miller, The Weyl Correspondence Principle, *J. Chem. Phys.* **54**, 1833 (1974).

16. E. P. Wigner, Wigner Distribution Function, *Phys. Rev.* **40**, 749 (1932).

17. O. T. Serimas, J. Javanainen, and S. Varro, Gauge Independent Wigner Functions: General Formulation, *Phys. Rev. A* **33**, 2913 (1986).

18. F. J. Narcowich and R. F. O'Connell, Necessary and Sufficient Condition for a Phase-Space Function to be a Wigner Distribution, *Phys. Rev. A* **34**, 1 (1986).

19. U. Fano and J. W. Cooper, Spectral Distribution of Atomic Oscillator Strengths, *Rev. Mod. Phys.* **40**, 441 (1968).

20. J. O. Hirschfelder, C. F. Curtiss, and R. B. Bird, *Molecular Theory of Gases and Liquids*, Wiley, New York, 1954, 1964, pp. 881–889.

21. P. K. Drude, *The Theory of Optics*, Longmans, Green, London, 1933.

22. J. S. Levinger, *Nuclear Photodisintegration*, Oxford University Press, Oxford, 1960.

23. L. Allen and J. H. Eberly, *Optical Resonance and Two-Level Atoms*, Wiley, New York, 1975. Note that Section 6.3 and Eq. (6.19) give $R = \Omega^2/\Gamma$, where R is the rate of stimulated emission or absorption, Ω is the Rabi frequency, and Γ is the linewidth.

24. E. Kamke, *Differential Gleichungen, Lösungsmethoden und Lösungen*, 3rd ed., Chelsea, New York, 1948, p. 413, Eqs. 2.36b and 2.35b.

25. M. Inokuti, VUV Absorption and Its Relation to the Effects of Ionizing Corpuscular Radiation, *Photochem. Photobiol.* **44**, 279 (1986).

26. M. Inokuti, J. L. Dehmer, T. Baer, and J. L. Hanson, Oscillator Strength Moments, Stopping

Powers, and Total Inelastic Scattering Cross-Sections of All Atoms Through Strontium, *Phys. Rev. A* **23**, 95 (1981).

27. Schematics of Moments of Dipole Oscillator Strength Distribution for Atoms in First and J. L. Dehmer, M. Inokuti, and R. P. Saxon, Second Rows, *Phys. Rev. A* **12**, 102 (1975); Also see the Addendum, *Phys. Rev. A* **17**, 1229 (1978).

28. J. O. Hirschfelder, W. Byers-Brown, and S. T. Epstein, Recent Developments in Perturbation Theory, *Adv. Quant. Chem.* **1**, 255 (1964); see Section 10 for sum rules.

29. B. A. Zon, N. L. Manakov, and L. P. Rapoport, *Opt. Spectrosc.* **38**, 6 (1975) (translated from Russian); N. L. Manakov, V. D. Ovsyannikov, and L. P. Rapoport, Perturbation Theory for the Quasienergy Spectrum of Atoms in a Strong Monochromatic Field, *Sov. Phys. JETP* **43**, 885 (1976).

30. J. E. Bayfield, Excited Atomic and Molecular States in Strong Electromagnetic Fields, *Phys. Rep.* **51**, 317 (1979).

31. J. W. Nibler and J. J. Yang, Nonlinear Raman Spectroscopy of Gases, *Ann. Rev. Phys. Chem.* **38**, 349 (1987).

32. S. Mukamel, Solvation Effects in Four-Wave Mixing and Spontaneous Raman and Fluorescence Lineshapes of Polyatomic Molecules, *Adv. Chem. Phys.* **LXX** (pt. 1), 165 (1987).

33. B. Kirtman, *Chem. Phys. Lett.* **143**, 81 (1988).

34. S. T. Epstein, *Variational Method in Quantum Chemistry*, Academic, New York, 1974, Appendix C.37.

35. J. O. Hirschfelder, The Polarizability and Related Properties of Molecular Hydrogen and the Diatomic Molecular Ion, *J. Chem. Phys.* **3**, 555 (1935); B. Mwroka, *Zeit. f. Phys.* **76**, 300 (1932).

36. B. Kirtman, Simultaneous Calculation of Several Interacting Electronic States by Generalized Van Vleck Perturbation Theory, *J. Chem. Phys.* **75**, 798 (1981).

37. P. R. Certain and J. O. Hirschfelder, New Partitioning Perturbation Theory, *J. Chem. Phys.* **53**, 2992 (1970).

38. E. R. Davidson, E. Engdahl, and N. Moiseyev, New Bounds to Resonance Eigenvalues, *J. Chem. Phys.* **33**, 2436 (1986). See also N. Moiseyev, P. Froelich, and E. Watkins, Resonances by Complex Coordinate Method with Hermitian Hamiltonian, *J. Chem. Phys.* **80**, 3623 (1984).

39. R. M. Glover and F. Weinhold, Dynamic Polarizabilities of Two-Electron Atoms with Rigorous Upper and Lower Bounds, *J. Chem. Phys.* **65**, 4913 (1976); also Dynamic Polarizabilities of He and Li$^+$ with Rigorous Upper and Lower Bounds, *J. Chem. Phys.* **66**, 185 (1977); Imaginary-Frequency Polarizability and Van der Waals Force Constants of Two-Electron Atoms with Rigorous Upper and Lower Bounds, *J. Chem. Phys.* **66**, 191 (1977).

40. S. Prager and J. O. Hirschfelder, Upper and Lower Bounds for Ground State Second Order Perturbation Energy, *J. Chem. Phys.* **39**, 3289 (1963).

41. F. Weinhold, Mass Polarization and Breit-Pauli Corrections for the Polarizability of ^{4}He, *J. Phys. Chem.* **86**, 1111 (1982).

42. P. E. S. Wormer and W. Rijks, Analysis of Correlation Effects in Molecular Second-Order Time-Dependent Properties: The Dynamic Polarizability of the Neon Atom and the Dispersion Coefficients of Ne$_2$ Dimer, *Phys. Rev. A* **33**, 2928 (1986).

43. F. Visser and P. E. S. Wormer, *Mol. Phys.* **52**, 723 (1984); F. Visser, P. E. S. Wormer, and W. P. J. H. Jacobs, *J. Chem. Phys.* **82**, 3753 (1985).

VII. SOME MATHEMATICAL TECHNIQUES

A. Density Matrices

In 1927 Johnny von Neumann derived density operators (or matrices) ρ to explain the properties of ensembles, which can be quite different from those of the individual molecules. For example, Dicke discovered that an ensemble of N atoms spontaneously emits fluorescent radiation with an intensity that varies as N^2 (instead of N). This phenomena is called superradiance (1). The explanation is that the radiation each of the N atoms emits is proportional to the electric field amplitude, which is proportional to the polarizability of the ensemble, which is proportional to N. For similar reasons, optical nutation and photon echoes have intensities that vary as N^2; whereas the intensity of CARS varies as N^4 because it involves the hyperpolarizability rather than the polarizability.

A Schrödinger equation can be used to describe the evolution of a *single-molecule* wavefunction *unless* there is latitudinal (T_2-type) relaxation involving the transfer of phase. If there is only (T_1-type) relaxation involving the transfer of energy, the Hamiltonian in the Schrödinger equation has time-reversal (rather than Hermitian) symmetry.

For large systems, if the relaxation permits, it may be easier to determine the N-dimensional Schrödinger wavefunction instead of the N^2-dimensional density matrix. Furthermore (as Metiu suggested), it may be desirable to use fast Fourier transforms to express the wavefunction alternately in the coordinate and momentum representations.

Finally, Sakurai (2) has an amusing but thought-provoking remark about the Liouville equation $\partial \underline{\rho}/\partial t = - [\underline{\rho}, H]$:

> This looks like the Heisenberg equation of motion except that *the sign is wrong*! This is not disturbing because $\underline{\rho}$ is not a dynamical observable in the Heisenberg-picture. On the contrary, $\underline{\rho}$ is built up of Schrödinger-picture state kets and bras which evolve in time according to the Schrödinger equation. The classical analogue of the Liouville equation has the same form in which $\rho_{classical}$ is the density of representative points in phase space.

B. Representations

Since Sakurai discusses Heisenberg and Schrödinger representation, I should clarify the definitions of various other representations for some of our readers:

1. *The Schrödinger Representation* (3). Time-dependent *kets*, $|\psi_S(t)\rangle = U(t, t_0)|\psi_S(t_0)\rangle$; time-free *Hamiltonian* H_S, $U(t, t_0) = \exp[-iH_S(t - t_0)/\hbar]$; usually (but not always), *operators* A_S are time independent; time dependence of A_S (if any) is arbitrary.

2. *The Heisenberg representation* (3). Time-independent kets, $|\psi_H\rangle = |\psi_S(t_0)\rangle$; time-dependent Hamiltonian $H_H(t)$,

$$A_H(t) = U^\dagger(t, t_0) A_S(t) U(t, t_0),$$

$$i\hbar \frac{d}{dt} A_H(t) = [A_H(t), H_H(t)] + i\hbar \frac{d}{dt} A_S(t).$$

3. *Interaction Representation* (4). The Hamiltonian is $H_I(t) = H_0 + V_I(t)$, where H_0 is the unperturbed Hamiltonian and $V_I(t)$ is the interaction potential

$$|\psi_I(t)\rangle = \exp\frac{iH_0 t}{\hbar} |\psi_S(t)\rangle,$$

$$i\hbar \frac{d}{dt} |\psi_I(t)\rangle = V_I(t) |\psi_I(t)\rangle,$$

The operators

$$A_I(t) = \exp[iH_0 t/\hbar] A_S \exp[-iH_0 t/\hbar],$$

$$\frac{d}{dt} A_I = \frac{\partial}{\partial t} A_I - i[A_I, H_0]/\hbar.$$

Although the Schrödinger and Heisenberg representations are better known, the interaction representation is more useful for laser–molecule problems. As Blum says in his book (5) on density matrices, the interaction representation is the basis for a rapidly converging perturbation sequence in which the nth order density operator is

$$\rho_I^{(n)}(t) = \rho_I(0) - \frac{i}{\hbar} \int_{t_0}^{t} [V_I(\tau), \rho_I^{(n-1)}(\tau)] \, d\tau.$$

Furthermore, the time development operator can be expressed in terms of the rapidly converging *Dyson series* (2),

$$U_I(t, t_0) = 1 - \frac{i}{\hbar} \int_{t_0}^{t} dt_1 \, V_I(t_1) U_I(t_1, t_0)$$

$$= 1 - \frac{i}{\hbar} \int_{t_0}^{t} dt_1 \, V_I(t_1)$$

$$+ \left(-\frac{i}{\hbar}\right)^2 \int_{t_0}^{t} dt_1 \int_{t_0}^{t_1} dt_2 \, V_I(t_1) V_I(t_2) + \left(-\frac{i}{\hbar}\right)^3 \cdots,$$

where $t \geqslant t_1 \geqslant t_2 \geqslant \cdots$.

Also, you should know about the active and passive representations which are used in transformations such as rotations; otherwise, it can be very confusing to discover that the classical and quantum-mechanical formulations do not appear to correspond, the reason being that they are in different representations.

4. *Active Representation* (4). The *physical system* moves while the coordinates remain unchanged. This is *usual* in quantum-mechanical transformations. If $W = \exp[-iQ]$ is a unitary operator, the transformed wavefunctions and operators are

$$\Psi'(\mathbf{r}, t) = \exp[-iQ]\Psi(\mathbf{r}, t) \quad \text{and} \quad A(\mathbf{r}, \mathbf{p}, t) = WAW^{-1}.$$

If Q commutes with $\partial Q/\partial t$, it is a *gauge transformation*, and the Hamiltonian becomes (4)

$$H' = \exp[-iQ]H\exp[iQ] + \hbar\frac{\partial Q}{\partial t}.$$

However, if Q and $\partial Q/\partial t$ do not commute (6),

$$H'(\mathbf{r}, \mathbf{p}, t) = \exp[-iQ]H\exp[iQ] + \hbar\int_0^1 dg\exp[-igQ]\frac{\partial Q}{\partial t}\exp[igQ]$$

and

$$\frac{\partial}{\partial t}\exp[-iQ] = -i\left[\int_0^1 dg\exp[-igQ]\frac{\partial Q}{\partial t}\exp[igQ]\right]\exp[-iQ].$$

5. *Passive Representation* (4). The physical system remains unchanged while the coordinates are changed. This is the usual classical type of transformation. Thus, the new classical coordinates and momenta are $\mathbf{r}' = W\mathbf{r}W^{-1}$ and $\mathbf{p}' = W\mathbf{p}W^{-1}$. In quantum mechanics, $\mathbf{p}' = -i\hbar\nabla'$. Also in quantum mechanics, the functional form of both the wavefunctions and the Hamiltonian are unchanged, but they become functions of the new coordinates. However, it is very confusing that in quantum mechanics (3) both $\mathbf{p}$ and $\mathbf{p}'$ are designated by $\mathbf{p}$.

C. The Time-Slicer (for Either Pulse or Periodic Fields)

The most general, the simplest, and the most numerically efficient method that I know for determining the time development of an N-level model of a quantum-mechanical system in a classical laser field is what I call the *time-*

slicer (7) because time is divided into a large number of slices. It is the only good *general* procedure for solving pulse problems, and for periodic (or quasi-periodic) field problems, it can be used together with, or else compete with, Floquet techniques.

In each slice, the Hamiltonian $\underline{\mathbf{H}}(t)$ is approximated by a time-independent Hamiltonian $\underline{\mathbf{H}}^{(n)}$. Thus, the time development matrix within this slice is

$$\underline{\mathbf{U}}(t', t_{n-1}) = \exp\left[-\frac{i\mathbf{H}^{(n)}(t' - t_{n-1})}{\hbar} \right], \quad \text{where } t_n > t' > t_{n-1},$$

and the time development matrix after m slices is the product

$$\underline{\mathbf{U}}(t_m, 0) = \underline{\mathbf{U}}(t_m - t_{m-1}) \cdots \underline{\mathbf{U}}(t_2 - t_1)\underline{\mathbf{U}}(t_1 - t_0).$$

All properties of the system can then be determined from the time development matrix.

Thus, the time-slicer can generate dynamical moving pictures on a computer screen! Similar techniques have been applied to all kinds of dynamical and mathematical problems for many years. Some of its virtues are as follows:

1. It is computer ready since it only requires matrix multiplication and diagonalization to obtain eigenfunctions and eigenvalues.
2. Its errors vary as the square of the interval size (8).
3. Since all of the $\underline{\mathbf{H}}^{(n)}$ are independent of each other, the $\underline{\mathbf{U}}^{(n)}(t_n - t_{n-1})$ can be determined efficiently by using parallel arrary processors.
4. Since the Hamiltonian need not be Hermitian, the effects of natural lifetimes and some kinds of relaxation can be determined.
5. It can be used to integrate Liouville or optical Bloch equations as well as the Schrödinger equation.
6. It is applicable to pulses, periodic or aperiodic problems. For periodic systems, the time-slicer need only determine the time development operator during the first period. Thereafter, the Floquet formalism determines all of the properties of the system.

I do not want to go into any more detail since both Peretz Friedmann and Bill Meath (and his co-authors) devote their chapters (IV and VIII, respectively) to this technique. Friedmann calls it the *rectangular ripple* method, whereas Meath calls it the *Riemann product integral* method. Although the name "*Riemann product integral*" has historical significance and "*rectangular ripple*" is appropriate for particular applications, neither title is very descriptive.

In Chapter VIII, Bill Meath makes a very thorough analysis of the time-slicer and its applications to laser—molecule problems. In addition, he discusses the effects of permanent dipole moments on single- and multiphoton molecular spectra. By adding a static electric field to the laser field, he can make these effects quite large. One of the effects is to hinder the rotation of the molecule. I believe that Meath's combined static and electric fields will open up a new and exciting avenue for experimental laser chemistry.

Last summer at a NATO meeting Warren Warren explained an unusual laser—molecule interaction application: In order to make a spectral linewidth as narrow as possible, he used computer experimentation with a time-slicer-like technique to shape his pulse by varying the laser intensity in each slice.

D. The Floquet Form of the Schrödinger Equation

The time-dependent Schrödinger equation can be expressed in matrix form as

$$\underline{\mathscr{H}}(t)\mathbf{\Psi}_k(t) = 0, \tag{7.1}$$

where

$$\underline{\mathscr{H}}(t) = \underline{\mathbf{H}}(t) - i\hbar\underline{\mathbf{I}}\frac{\partial}{\partial t}. \tag{7.2}$$

Here the wavefunction $\mathbf{\Psi}_k(t)$ is a column vector having the jth element, $\Psi_{jk} = \int \varphi_j(\mathbf{r})^* \Psi_k(\mathbf{r},t)\,d\mathbf{r}$, and the Hamiltonian $\underline{\mathbf{H}}(t)$ has the (gj)th element, $\mathbf{H}(t)_{gj} = \int \varphi_g(\mathbf{r})^* \mathbf{H}(\mathbf{r},t)\varphi_j(\mathbf{r})d\mathbf{r}$, where the functions φ_g are any complete set of basis functions spanning coordinate space. Also, $\underline{\mathbf{I}}$ is the unit matrix. Of course, $\mathbf{\Psi}_k(t)$ satisfies boundary conditions as well as Eq. (7.1). We seek the eigenfunctions $\mathbf{X}_k(t)$ and the eigenvalues W_k of the operator $\underline{\mathscr{H}}(t)$,

$$\underline{\mathscr{H}}(t)\mathbf{X}_k(t) = W_k\mathbf{X}_k(t). \tag{7.3}$$

It follows that

$$\mathbf{\Psi}_k(t + t_0) = \mathbf{X}_k(t + t_0)\exp\left[-\frac{iW_k(t + t_0)}{\hbar}\right] = \underline{\mathbf{U}}(t + t_0, t_0)\mathbf{\Psi}_k(t_0). \tag{7.4}$$

In deriving Eq. (7.4), Sambé (3) required that $\mathbf{X}_k(t)$ be an eigenfunction in the coordinate-time Hilbert space of $\underline{\mathscr{H}}(t)$. I think this implies that $\mathbf{X}_k(t)$ must be a *coherent* state but I am not sure. It may be that it is necessary to specify some other special property of $\mathbf{X}_k(t)$ in order for it to have any physical significance. In any case, the only application of Eq. (7.4) that I know is to systems with periodic (or quasi-periodic) Hamiltonians.

E. The Floquet Formalism for Periodic Hamiltonian Systems

Equation (7.3) is the *Floquet equation* for systems with periodic Hamiltonians. In this case, both $\mathscr{H}(t) = \mathscr{H}(t + T)$ and its eigenfunctions are periodic, $X_k(t) = X_k(t + T)$. The subscript k designates the Floquet mode. There are as many modes as there are $\varphi_j(\mathbf{r})$ basis functions in the wavefunction $\Psi(\mathbf{r}, t)$.

Since the $X_k(t)$ are periodic, it follows from Eq. (7.4) that

$$X_k(t_0)\exp\left[-\frac{iW_kT}{\hbar}\right] = \underline{U}(T + t_0, t_0)X_k(t_0). \tag{7.5}$$

Thus, the Floquet functions $X_k(t_0)$ are the eigenfunctions, and the $\exp[-iW_kT/\hbar]$ are the eigenvalues of the time development matrix, $\underline{U}(T + t_0, t_0)$. Furthermore, $X_k(t_0)$ and W_k can be determined by using the time-slicer to calculate the time development matrix through the first period. Once the Floquet functions and pseudo energies have been determined, there is no difficulty in learning the properties of the system for all time.

However, Bob Wyatt [author of Chapter V] prefers the usual Floquet procedure in which both $\mathscr{H}(t)$ and $X(t)$ are expanded in Fourier series composed of linear combinations of the functions $u_n(t) = \exp(in\omega t)$, where n is any positive or negative integer or zero. Thus, the Hamiltonian $\mathscr{H}(t)$ and $X_k(t)$ become implicit (rather than explicit) functions of time, in which case they are called the *Floquet Hamiltonian and functions*, respectively. In this sense, the Floquet equations (7.3) and (7.4) can be expressed (in seemingly time-independent form) in the infinite dimensional Floquet pseudospace. The Floquet equation can then be expressed as a secular equation to determine both the Floquet functions and pseudoenergies W_k. Whether this Floquet technique or the time-slicer is more efficient depends upon the problem.

If the Hamiltonian is both Hermitian and periodic, the W_k are *real* constants (defined to within modulo $2\pi\hbar/T$) and the Floquet modes are well-behaved coherent solutions which Sambé (9) called steady states.

Shih-I Chu (10) has a very thorough presentation of the usual Floquet formalism in Chapter XVII, and he makes a number of ingenious generalizations so that it can be applied to polychromatic fields, above-threshold ionization, and so on. I am particularly interested in his Floquet–Maxwell–Liouville supermatrix formulation for solving nonlinear optical processes since previous techniques have been difficult to apply. Also I am hoping that he will be able to determine the conditions required for the nonlinear processes to have soliton solutions.

F. Non-Hermitian Periodic Hamiltonians

If $\underline{H}(t)$ is periodic but *not Hermitian*, then the $X_k(t)$ are still periodic but the W_k's are complex constants. Peretz Friedmann (in Chapter IV and (11)) is specially interested in non-Hermitian Hamiltonians in connection with his research on the stability of helicopter rotor blades!

If $\text{Im}(W_k) < 0$, the mode grows, whereas, if $\text{Im}(W_k) > 0$, the mode decays. Also, if $H(\mathbf{r}, t)$ is non-Hermitian, chaos or bifurcation can occur.

I think you will find Chapter IV very interesting, easy to read, and directly applicable to laser problems. Friedmann discusses a number of numerically efficient techniques that engineers have developed for solving both linear and nonlinear periodic dynamical problems. For example, he has an unusually accurate form of the fourth-order Runge–Kutta procedure that can be used in connection with the time-slicer. For nonlinear problems he gives both a quasi-linearization technique (which is a generalization of the familiar Newton–Raphson method and has second-order convergence) and a "periodic-shooting" method (I wonder whether the shooting is as interesting as the title). Note that Friedmann uses the following nomenclature:

$$t = \psi, \qquad \Psi = y, \qquad \underline{H}(t) = \frac{A(\psi)}{(i\hbar)},$$

$$U(t, t_0) = \Phi(\psi, \psi_0), \qquad -iW_k = R.$$

Non-Hermitian Hamiltonians can occur in atom–laser interaction problems as a result of spontaneous ionization, spontaneous emission of photons, collision, Doppler broadening, and interactions between the quasi-energy states. Chu shows in Chapter XVII and Ref. 12 how to solve spontaneous ionization problems by rotating the coordinates in the Hamiltonian into the complex plane. Here it will be shown time reversal symmetry can be used to solve many of the other types of non-Hermitian Hamiltonian problems.

G. Decay of Quasi-Energy States: Time Reversal Symmetry

Atoms in electromagnetic fields are best described in terms of their quasi-energy states, which are long lived transients that have complex energies $E_j - i(\frac{1}{2}\Gamma_j)$. Here $E_j = \hbar\omega_j$ is the real part of the energy and $\Gamma_j = \hbar\gamma_j$ is the linewidth. In the Wigner–Weisskopf approximation (13), the mean lifetime of the state is $\tau_j = 1/\gamma_j$. Typically, in the infrared laser fine-structure resonance experiments of Willis Lamb (14) and Silverman and Pipkin (15), γ_j was on the order of $10^{-7} - 10^{-9}$ seconds.

As a result of the decaying states, imaginary diagonal terms are introduced into the otherwise Hermitian Hamiltonian. The resulting Hamiltonian usually has time-reversal or time-inversion symmetry (16–18). [Unsymmetrical pulse problems could be an exception.] The most general definition of time-reversal symmetry is

$$\underline{H}(\mathbf{p}, \mathbf{r}, \mathbf{S}, t, \mathbf{E}, \mathbf{B}) = \underline{H}(-\mathbf{p}, \mathbf{r}, -\mathbf{S}, -t, \mathbf{E}, -\mathbf{B}), \qquad (7.6)$$

where $\mathbf{p}$, $\mathbf{r}$, $\mathbf{S}$, t, $\mathbf{E}$, and $\mathbf{B}$ are the momenta, coordinates, spin, time, electric intensity and magnetic intensity, respectively. The solution to problems having this symmetry can be simplified by making use of the biorthogonal coordinate (19) "t-product" technique that Weinhold (17) and Moiseyev and Certain (18) developed (in connection with the rotation of a Hamiltonian into the complex plane).

In place of using the usual Hermitian scalar products, we use the t-products,

$$\{f|g\} = \int_{\text{space}} f(r, -t)g(r, t)\, dr. \qquad (7.7)$$

[See Ref. 18 for a complete specification of $f(r, -t)$ when the Hamiltonian involves spin]. The advantage of these t-products is that when the Hamiltonian is time reversible, the t-products have the "Hermitian-like" property that $\{Hf|g\} = \{f|Hg\}$. Thus, we use t-products in determining the time development operator $\underline{U}(t, t_0)$ wherever we would normally use Hermitian scalar products. However, since $\underline{U}(t, t_0)^\dagger$ is the *usual* adjoint so that $U_{\alpha\beta}^\dagger = U_{\beta\alpha}^*$, t-products *cannot* be used in deriving the density matrix,

$$\underline{\rho}(t) = \underline{U}(t, t_0)\underline{\rho}(t_0)\underline{U}(t, t_0)^\dagger \qquad (7.8)$$

from the time development operator.

Let us consider a (near) resonant interaction of two transient states α and β that have transitional dipole moment μ_0. If the laser field intensity is $F_0 \cos(\omega t)$, the probability (in the rotating-wave approximation) of a transition from state α to β is

$$P(t) = \exp\left[-\tfrac{1}{2}(\gamma_\alpha + \gamma_\beta)t\right] \frac{\lambda^2}{q_0^2} \sin^2(q_0 t), \qquad (7.9)$$
$$\beta \leftarrow \alpha$$

where $\hbar\lambda = \tfrac{1}{2}\mu_0 F_0$ and the Rabi frequency is $q_0 = [\lambda^2 + \Delta_0^2]^{1/2}$. Notice that the detuning Δ_0 has become complex,

$$\Delta_0 = \tfrac{1}{2}[\omega - (\omega_\beta - \omega_\alpha)] - (i/4)(\gamma_\alpha - \gamma_\beta). \qquad (7.10)$$

The trace of the density matrix is $\mathrm{Tr}(\rho) = \exp[-\frac{1}{2}(\gamma_\alpha + \gamma_\beta)t]$. Furthermore, at resonance $\omega = \omega_\beta - \omega_\alpha$, the Rabi frequency becomes $q_0 = [\lambda^2 - \frac{1}{16}(\gamma_\alpha - \gamma_\beta)^2]^{1/2}$.

Thus, if the field strength is so weak that

$$F_0 < \frac{|\gamma_\alpha - \gamma_\beta|}{2\mu} \quad \text{or} \quad 4\hbar\lambda < |\Gamma_\alpha - \Gamma_\beta|, \tag{7.11}$$

the Rabi frequency becomes imaginary and the transition probability decays without oscillating.

H. Variational Procedures for Approximate Time-Dependent Wavefunctions [Assuming $\underline{H}(t)$ Is Hermitian]

Since perturbations are used very often in determining approximate solutions to laser–molecule interaction problems, it seems useful to discuss the variational conditions that should be satisfied.

The time-dependent Schrödinger equation can be expressed in the form

$$\underline{\mathcal{H}}(t)\Psi(t) = 0 = \left[\underline{H}(t) - i\hbar\underline{I}\frac{\partial}{\partial t} \right]\Psi(t), \tag{7.12}$$

where $\underline{I}$ is the unit matrix.

It is desirable that an approximate (trial) wavefunction $\tilde{\Psi}(t)$ should satisfy three conditions (20):

(a)
$$\langle \tilde{\Psi}(t)|\underline{\mathcal{H}}(t)|\tilde{\Psi}(t)\rangle = 0, \tag{7.13}$$

(the expectation value of $\underline{\mathcal{H}}(t)$ should vanish),

(b)
$$\langle \underline{\mathcal{H}}(t)\tilde{\Psi}(t)|\tilde{\Psi}(t)\rangle = \langle \tilde{\Psi}(t)|\underline{\mathcal{H}}(t)\tilde{\Psi}(t)\rangle, \tag{7.14}$$

(the approximate wavefunction should be in the Hermitian domain of $\underline{\mathcal{H}}(t)$),

(c) Furthermore, $\left\langle i\frac{\partial}{\partial t}\tilde{\Psi}(t)|\tilde{\Psi}(t) \right\rangle = \left\langle \tilde{\Psi}(t)|i\frac{\partial}{\partial t}\tilde{\Psi}(t) \right\rangle,$ (7.15)

where the norm of the wavefunction should be independent of time; otherwise, the trace of $\rho(t)$ would vary with time.

Löwdin and Mukherjee (21) discovered that these conditions can be satisfied by transforming the trial wavefunction $\tilde{\Psi}(t)$ into a new trial function,

$$\bar{\Psi}(t) = \tilde{\Psi}(t)\exp\left\{ -i\int_0^t dt\,[\langle \tilde{\underline{H}}(t)\rangle - \langle \tilde{\underline{f}}(t)\rangle] \right\}, \tag{7.16}$$

where

$$\langle \tilde{\underline{H}}(t) \rangle = \frac{\langle \tilde{\Psi}(t) | \underline{H}(t) | \tilde{\Psi}(t) \rangle}{\langle \tilde{\Psi}(t) | \tilde{\Psi}(t) \rangle}, \tag{7.17}$$

and

$$\langle \tilde{\underline{f}}(t) \rangle = \frac{\langle \tilde{\Psi}(t) | \underline{f} \tilde{\Psi}(t) \rangle}{\langle \tilde{\Psi}(t) | \tilde{\Psi}(t) \rangle}, \tag{7.18}$$

where $\mathbf{f} = i\hbar\underline{\mathbf{I}}(\partial/\partial t)$.

The approximate wavefunctions $\tilde{\Psi}(t)$ have a number of interesting properties (21):

1. Wavefunction $\bar{\Psi}(t) = \bar{\mathbf{X}}(t) \exp[-i\bar{W}(t)]$, where $\bar{W}(t)$ is a *real* function of time that satisfies the equation

$$\frac{\partial \bar{W}}{\partial t} = \langle \bar{\mathbf{X}}(t) | \mathscr{H}(t) | \bar{\mathbf{X}}(t) \rangle. \tag{7.19}$$

2. If $\underline{U}(t)$ is a unitary operator such that $\bar{\Psi}^U(t) = \underline{U}(t)\bar{\Psi}(t)$ can be normalized and considered to be an allowed trial wavefunction, $\bar{\Psi}^U(t)$ statisfies Eqs. (7.13)–(7.15) in which $\underline{H}(t)$ is replaced by

$$\underline{H}^U(t) = i\hbar \frac{d\underline{U}}{dt} \underline{U}^\dagger(t) + \underline{U}(t)\underline{H}(t)\underline{U}^\dagger(t). \tag{7.20}$$

3. If σ is a *real* parameter in $\underline{H}(t)$ such that $\bar{\Psi}(t) + \partial\bar{\Psi}/\partial\sigma$ can be normalized and considered to be an allowable trial wavefunction, the *time-dependent Hellmann–Feynman theorem* (22) is satisfied,

$$\left\langle \bar{\Psi}(t) \left| \frac{\partial \underline{H}}{\partial \sigma} \right| \bar{\Psi}(t) \right\rangle = i\hbar \frac{\partial}{\partial t} \left\langle \bar{\Psi}(t) \left| \frac{\partial \bar{\Psi}}{\partial \sigma} \right. \right\rangle. \tag{7.21}$$

4. If $\underline{U}(t) = \exp[i\underline{G}(t)]$, where $\underline{G}(t)$ is a Hermitian operator such that $i\underline{G}(t)\Psi(t)$ is a possible trial wavefunction, the *time-dependent Hypervirial theorem* or the *generalized Ehrenfest theorem* for $\underline{G}(t)$ is satisfied,

$$\frac{\partial}{\partial t} \langle \bar{\Psi}(t) | \underline{G}(t)\bar{\Psi}(t) \rangle = \left\langle \bar{\Psi}(t) \left| \frac{\partial \underline{G}}{\partial t} \right| \bar{\Psi}(t) \right\rangle$$

$$+ i \langle \bar{\Psi}(t) | [\underline{H}(t), \underline{G}(t)] | \bar{\Psi}(t) \rangle. \tag{7.22}$$

The time-dependent hypervirial theorem has many different applications:

(a) In the unlimited Hartree–Fock approximation, the trial functions consist of one-electron normalized orbitals, and Eq. (7.22) is satisfied by all one-electron $\underline{G}(t)$ operators.
(b) In the spin-unlimited Hartree–Fock, Eq. (7.22) is satisfied by all spin-independent one-electron $\underline{G}(t)$'s.
(c) If $\underline{G}(t)$ is a gauge transformation, Eq. (7.22) implies the conservation of electric charge in either the unlimited or spin-unlimited Hartree–Fock treatments.
(d) If, on the other hand $\underline{H}(t)$ is composed of one-electron operators and $\underline{G}(t) = \underline{H}(t)$, or if $\underline{H}(t)$ contains two-electron terms but $\bar{\Psi}$ is *not* time dependent, Eq. (7.22) gives the *work energy theorem*,

$$\frac{\partial}{\partial t} \langle \bar{\Psi}(t) | \underline{H}(t) | \bar{\Psi}(t) \rangle = \left\langle \bar{\Psi}(t) \left| \frac{\partial \underline{H}(t)}{\partial t} \right| \bar{\Psi}(t) \right\rangle. \tag{7.23}$$

I. Floquet Variational Principles [When $\underline{H}(t)$ Is Periodic]

In this case, $\underline{H}(t + T) = \underline{H}(t)$ and $\bar{\Psi}(t) = \bar{X}(t) \exp[-i \bar{W} t]$, where

$$\bar{X}(t + T) = \bar{X}(t) \quad \text{and} \quad \frac{\partial}{\partial t} \langle \bar{X}(t) | \bar{X}(t) \rangle = 0.$$

Furthermore, $\bar{W}$ is a *real* constant determined by (23)

$$\frac{1}{T} \int_0^T dt \langle \bar{X}(t) | \mathcal{H}(t) - \bar{W} \underline{I} | \bar{X}(t) \rangle = 0, \tag{7.24}$$

which implies the *Heinrich variational principle* (24), $\delta \bar{W} = 0$. Then, as Shih-I Chu says in Chapter XVII, it follows from the Hellmann–Feynman theorem that

$$\frac{1}{T} \int_0^T dt \langle \Psi(t) | \underline{H}(t) | \Psi(t) \rangle = W - \omega \frac{\partial W}{\partial \omega}. \tag{7.25}$$

J. Partitioning of Time-Independent Schrödinger Equation

Partitioning (25, 26) is a good first step in explaining how the Green function, perturbative, or variational procedures are used to solve the Schrödinger equation, $H | \Psi \rangle = E | \Psi \rangle$.

Let the functions $| \varphi_j \rangle$ be a complete orthonormal set of eigenfunctions of the unperturbed Hamiltonian $H^{(0)} = H - V$ (where V is considered to be a small perturbation); then

$$H^{(0)}|\varphi_j\rangle = E_j^{(0)}|\varphi_j\rangle, \quad \sum_j |\varphi_j\rangle\langle\varphi_j| = 1, \quad \text{and} \quad |\Psi\rangle = \sum_j |\varphi_j\rangle C_j. \quad (7.26)$$

Furthermore, the variational principle leads to the linear equations

$$\sum_k [H_{jk} - E\delta_{jk}]C_k = 0, \quad \text{where } H_{jk} = \langle\varphi_j|H|\varphi_k\rangle. \quad (7.27)$$

Thus, the Schrödinger equation can be expressed in matrix form as $\underline{H}\mathbf{C} = E\mathbf{C}$.

If the energy levels E we seek result from the strong interaction of g unperturbed states that are degenerate or almost degenerate, it is desirable to divide (or partition) the complete basis set into two subsets:

(a) composed of the strongly interacting functions

$$|\varphi_{a1}\rangle, \ldots, |\varphi_{ag}\rangle \quad \text{and}$$

(b) composed of the remainder of the basis functions $|\varphi_{bj}\rangle$. If E is nondegenerate, (a) is the single function $\varphi(a)$.

As a result of this partitioning, $\mathbf{C}$ is divided into two parts,

$$\mathbf{C} = \begin{pmatrix} \mathbf{C}_a \\ \mathbf{C}_b \end{pmatrix} \quad \text{so that} \quad \underline{H} = \begin{pmatrix} \underline{H}_{aa} & \underline{H}_{ab} \\ \underline{H}_{ba} & \underline{H}_{bb} \end{pmatrix}. \quad (7.28)$$

Furthermore, the Schrödinger equation is split into two matrix equations,

$$\underline{H}_{aa}\mathbf{C}_a + \underline{H}_{ab}\mathbf{C}_b = E\mathbf{C}_a \quad \text{and} \quad \underline{H}_{ba}\mathbf{C}_a + \underline{H}_{bb}\mathbf{C}_b = E\mathbf{C}_b.$$

By solving the second equation for $\mathbf{C}_b$, we obtain

$$\mathbf{C}_b = (E\underline{I}_{bb} - \underline{H}_{bb})^{-1}\underline{H}_{ba}\mathbf{C}_a$$

where $\underline{I}_{bb}$ is the unit matrix in the b-space. Then the first equation can be expressed as the nonlinear Schrödinger-like equation

$$\bar{\underline{H}}_{aa}\mathbf{C}_a = E\mathbf{C}_a \quad \text{or} \quad \bar{\underline{H}}_{aa}\Psi_a = \underline{E}_a\Psi_a, \quad (7.29)$$

where the *effective Hamiltonian* $\bar{\underline{H}}_{aa}$ is

$$\bar{\underline{H}}_{aa} = \underline{H}_{aa} + \underline{H}_{ab}[E\underline{I}_{bb} - \underline{H}_{bb}]^{-1}\underline{H}_{ba}, \quad (7.30)$$

and $\underline{E}_a$ is a diagonal matrix in a-space. However, this notation is clumsy so we refer to E_a as E.

Here $[E\underline{I}_{bb} - \underline{H}_{bb}]^{-1} = G(E)$ is the usual *Green function or operator* (27). If E is equal to any of the E_{bj}, the Green function does not exist, and Eq. (7.30) is not valid.

(For those readers not familiar with Green functions, this section is intended to be an introduction to both Chapters V and XIX.)

Nonperturbative solutions to Eq. (7.29) can be obtained by using contour integrals to determine the residues of the Green function poles. This is explained in (28, 29).

Perturbative solutions to Eq. (7.29) can be determined either by using the *Lippmann–Schwinger equation* or by successive inversions on the matrix $E\underline{I}_{bb} - \underline{H}_{bb}$. The Lippmann–Schwinger equation is based upon repetitions of the simple operator identity,

$$A^{-1} = B^{-1} + B^{-1}(B - A)A^{-1}. \tag{7.31}$$

For example, if $A^{-1} = G(E)$ and if

$$B^{-1} = (E_0 - H^{(0)})^{-1} = \sum_j \frac{|\varphi_{bj}\rangle\langle\varphi_{bj}|}{E_0 - E_{bj}^{(0)}} = R_0(E_0) \tag{7.32}$$

(where the sum is over all of the b-states and $j \neq 0$), and if $V' = V - (E - E_0)$, we obtain the Lippman–Schwinger equation

$$G(E) = R_0(E_0) + R_0(E_0)V'G(E), \tag{7.33}$$

where the value of E_0 is arbitrary. The operator R_0 is called the *resolvent*. It follows from the Lippmann–Schwinger equation that

$$G(E) = R_0 + R_0 V' R_0 + R_0 V' R_0 V' R_0 + \cdots. \tag{7.33'}$$

It is frequently useful to utilize the nice property of the resolvent,

$$(E_0 - H_{bb}^{(0)})R_0(E_0) = \sum_{bj} |\varphi_{bj}\rangle\langle\varphi_{bj}|$$

$$= 1 - \sum_{ak=1}^{g} |\varphi_{ak}\rangle\langle\varphi_{ak}|. \tag{7.34}$$

If E is not degenerate, the Rayleigh–Schrödinger perturbation series is obtained by substituting Eq. (7.33) into Eq. (7.26). In this nondegenerate case,

$\Psi(E) = [1 + G(E)H]\varphi_0$. [Note that Löwdin (26) calls my $G(E)$ his operator $T(E)$].

Inversion of Matrices

If the matrix $\underline{A}$ is an arbitrary matrix in the partitioned form

$$\underline{A} = \begin{pmatrix} \underline{A}_{11} & \underline{A}_{12} \\ \underline{A}_{21} & \underline{A}_{22} \end{pmatrix}, \quad \text{then (29)}$$

$$\underline{A}^{-1} = \begin{pmatrix} \underline{A}_{11}^{-1} + \underline{N}_{12}\underline{D}_{22}^{-1}\underline{N}_{21} & -\underline{N}_{12}\underline{D}_{22}^{-1} \\ -\underline{D}_{22}^{-1}\underline{N}_{21} & \underline{D}_{22}^{-1} \end{pmatrix} \tag{7.35}$$

where

$$\underline{N}_{12} = \underline{A}_{11}^{-1}\underline{A}_{12}, \qquad \underline{N}_{21} = \underline{A}_{21}\underline{A}_{11}^{-1}, \qquad \underline{D}_{22} = \underline{A}_{22} - \underline{A}_{21}\underline{N}_{12}.$$

If the matrix $\underline{A}_{22}$ is chosen to consist of a single element, $\underline{A}_{12}$ and $\underline{A}_{21}$ are column and row vectors, respectively, and $\underline{D}_{22}$ is a single element. Thus, one can invert any matrix by starting with the upper left corner element and successively adding a row and column to the previously inverted matrix. This corresponds to successively adding a new basis function to the space.

K. General Techniques for Evaluating Sum Rules (30, 31)

All of us are familiar with a few sum rules, but surprisingly few people realize that there is a simple procedure for deriving new sum rules (30, 31). This makes it possible to greatly simplify many problems involving a sum over all the discrete states and an integral over the continuum.

For example, let us consider the large family of functions

$$S_k(\omega) = \sum_j{}' \langle \varphi_0 | W(\omega + E_0^{(0)} - H^{(0)})^k | \varphi_j \rangle \langle \varphi_j | A | \psi \rangle, \tag{7.36}$$

where W and A are operators, ω is any constant, and $|\psi\rangle$ could be any ket including $|\varphi_0\rangle$. The prime on the summation indicates that 0 and all states 0s degenerate with it are excluded from the summation over the discrete states and integration over the continuum.

If k is a positive integer or 0, the sum rule is

$$S_k(\omega) = \langle \varphi_0 | W(\omega + E_0^{(0)} - H^{(0)})^k A | \psi \rangle - \omega^k \sum_{0s} \langle \varphi_0 | W | \varphi_{0s} \rangle \langle \varphi_0 | A | \psi \rangle \tag{7.37}$$

If k is a negative integer, the sum rule is

$$S_k(\omega) = \langle \varphi_0 | F A | \psi \rangle - \sum_{0s} \langle \varphi_0 | F | \varphi_{0s} \rangle \langle \varphi_{0s} | A | \psi \rangle, \tag{7.38}$$

where F is a *real* function satisfying the differential equation

$$[(\omega + E_0^{(0)} - H^{(0)})^{|k|} F - \omega^{|k|} \langle \varphi_0 | F | \varphi_0 \rangle] | \varphi_0 \rangle$$
$$= [W - \langle \varphi_0 | W | \varphi_0 \rangle] | \varphi_0 \rangle. \tag{7.39}$$

It can be useful to study $S_k(\omega)$ both as a function of k and of ω since some of the values of $S_k(\omega)$ may be known empirically; they may help to estimate other values as follows:

If k is positive, $dS_k(\omega)/d\omega = kS_{k-1}(\omega)$.

If k is negative, $d^{|k|}S_{-1}(\omega)/d\omega^{|k|} = (-1)^k |k|! S_{-1-k}(\omega)$.

The sum rules involving oscillator strengths are the most familiar examples,

$$S_k = -2(-1)^k \sum_j{}' (E_0^{(0)} - E_j^{(0)})^k \langle \varphi_0 | \mu | \varphi_j \rangle \langle \varphi_j | \mu | \varphi_0 \rangle. \tag{7.40}$$

Note that $\frac{1}{2}S_{-3}$ is the norm of a first-order perturbation (where $\mu = V$); S_{-2} is the polarizability; S_0 is the Reiche–Thomas–Kuhn sum rule; and so on.

References

1. M. Sargent III, M. O. Scully, and W. E. Lamb, Jr., *Laser Physics*, Addison-Wesley, London, 1974, Appendix G.

2. J. J. Sakurai, *Modern Quantum Mechanics*, Benjamin-Cummings, Menlo Park, 1985, pp. 176, 181.

3. C. Cohen-Tannoudji, B. Diu, and F. Laloë, *Quantum Mechanics*, Vol. 1, Wiley, New York, 1977, p. 312.

4. L. I. Schiff, *Quantum Mechanics*, 3rd ed., McGraw-Hill, New York, 1968, pp. 171, 188.

5. K. Blum, *Density Matrix Theory and Applications*, Plenum, New York, 1981, pp. 55–59.

6. R. M. Wilcox, Exponential Operators and Parameter Differentiation in Quantum Physics, *J. Math. Phys.* **8**, (1967); R. F. Snider, *J. Math. Phys.* **5**, 1586 (1964).

7. J. O. Hirschfelder and R. W. Pyzalski, Simple Approach to Quantum Systems in Arbitrary Fields: The "Time-Slicer", *Phys. Rev. Lett.* **55**, 1244 (1985); J. O. Hirschfelder, Laser and Magnetic Resonance Propagators, *Int. J. Quant. Chem.* **29**, 1139 (1986).

8. C. S. Hsu, On Approximating a General Linear Periodic System, *J. Math. Anal. Appl.* **45**, 234 (1974).

9. H. Sambé, *Phys. Rev. A* **7**, 2203 (1973).

10. S.-I Chu, Dynamics and Symmetries in Intense Field Multiphoton Processes: Floquet Theoretical Approaches, *Adv. Multiphot. Proc. Spectrosc.* **2**, 175 (1986).

11. P. P. Friedmann, Numerical Methods for Determining the Stability and Response of Periodic Systems with Applications to Helicopter Dynamics and Aeroelasticity, *Comp. Math. Appl.* (*Gr. Brit.*) **12A**, 131 (1986).

12. S.-I Chu and W. P. Reinhardt, *Phys. Rev. Lett.* **39**, 1195 (1977); A. Maquet, S.-I. Chu and W. P. Reinhardt, *Phys. Rev. A* **27**, 2946 (1983).

13. V. Weisskopf and E. Wigner, *Z. Phys.* **63**, 54 (1930).

14. W. E. Lamb, *Phys. Rev.* **85**, 259 (1952).

15. M. P. Silverman and F. M. Pipkin, Interaction of a Decaying Atom with a Linearly Polarized Oscillating Field, *J. Phys. B* **5**, 1844 (1972).

16. For the physics and mathematics associated with time reversal see: M. Lax, *Symmetry Principles in Solid State and Molecular Physics*, Wiley, New York, 1974, Chapter 10; "Time Reversal Symmetry in Dissipative Systems," in *Proceedings of the Einstein Centennial Celebration at Ill. Acad. Sci.*, B. Gruber and R. C. Millman, Eds., Plenum, New York, 1980, p. 1.

17. F. Weinhold, *University of Wisconsin Theoretical Chemistry Institute Report* No. 590, January 1978.

18. N. Moiseyev, P. R. Certain and F. Weinhold, Resonance Properties of Complex-Rotated Hamiltonians, *Mol. Phys.* **36**, 1613 (1978).

19. P. M. Morse and H. Feshbach, *Methods of Theoretical Physics*, McGraw-Hill, New York, 1953, p. 884.

20. S. T. Epstein, *Variational Method in Quantum Chemistry*, Academic, New York, 1974, Appendix C.

21. P. O. Löwdin and P. K. Mukherjee, Comments on the Time Dependent Variation Principle, *Chem. Phys. Lett.* **14**, 1 (1972).

22. E. F. Hayes and R. G. Parr, Time Dependent Hellmann-Feynman Theorem, *J. Chem. Phys.* **43**, 1831 (1965).

23. R. H. Young and W. J. Deal, Jr., *J. Math. Phys.* **11**, 3298 (1970).

24. J. Heinrichs, *Phys. Rev.* **172**, 1315 (1968).

25. J. O. Hirschfelder and P.-O. Löwdin, Long Range Interaction of Two 1s-Hydrogen Atoms, *Mol. Phys.* **2**, 229 (1959), Appendices II and III.

26. P.-O. Löwdin, Studies in Perturbation Theory. IV: Solution of Eigenvalue Problems by Projector Operator Formalism, *J. Math. Phys.* **3**, 969 (1962).

27. J. R. Taylor, *Scattering Theory*, Wiley, New York, 1972, Chapter 8. "The Green's and T Operators."

28. N. L. Manakov, V. D. Ovsiannikov and L. P. Rapoport, Atoms in a Laser Field, *Phys. Rep.* **141**, 319 (1985).

29. A. S. Householder, *Principles of Numerical Analysis*, McGraw-Hill, New York, 1953, p. 78.

30. J. O. Hirschfelder, W. Byers-Brown and S. T. Epstein, Recent Developments in Perturbation Theory, *Adv. Quant. Chem.* **1**, 255 (1964), see Section 10 for sum rules.

31. A. Dalgarno, Sum Rules and Atomic Structure, *Rev. Mod. Phys.* **35**, 522 (1963); U. Fano, Spectral Distribution of Atomic Oscillator Strengths, *Rev. Mod. Phys.* **40**, 441 (1968).

Acknowledgment

I want to thank all of the chapter authors of this book and all of the participants of the Laser–Molecule Interaction Symposium at Los Alamos 1986 for the help that they have given me in learning laser technology. Without Dave Campbell, neither the symposium nor this book would have been possible. Also, thanks to André Bandrauk for having invited me to his very stimulating NATO workshop on short intense pulses.

I am grateful to Jay Ackerhalt, P. K. Aravind, Shih-I Chu, Joe Eberly, Tom George, F. T. Hioe, Moeyyet Hussain, E. T. Jaynes, Bernie Kirtman, Bill Meath, Horia Metiu, Peter Milonni, Ben Noble, Russell Pack, and Frank Weinhold for giving me a lot of much-needed advice.

Furthermore, I want to thank M. J. Ablowitz, Gernod Albers, Asim Barut, Bernd Crasemann, P. S. Bucksbaum, Chuck Curtiss, N. Gisin, Mitio Inokuti, Bruce Johnson, P. Lambropoulos, C. Mavroyannis, Hersch Rabitz, Bahaa Saleh, Randy Shirts, and A. Szöke for giving me preprints and reprints of their recent research.

I very much appreciate the help that Bob Wyatt and Rob Coalson have given me in editting this book. Thanks are due to the University Graduate School and Wisconsin Alumni Research Foundation for their financial assistance. From October to April every year I work at the University of California, Santa Barbara, Chemistry Department and Quantum Institute. I am very grateful to their staffs for the use of their facilities and for a great deal of help in my research.

Finally, I want to thank my wife for her good advice, excellent proofreading, and patient understanding of my research efforts. Also, thanks are due to Sheila Aiello for taking care of my secretarial needs. With all of this help, I feel responsible for making this a very useful, exciting chapter—I hope that I have not let you down!

MATHEMATICAL METHODS

CHAPTER II

MULTIPLE SCALING AND A RELATED EXPANSION METHOD, WITH APPLICATIONS

B. NOBLE

Institute of Computational Mathematics,
Brunel University, Uxbridge, England

M. A. HUSSAIN

General Electric Company, Corporate Research and Development,
Schenectady, New York, 12301

CONTENTS

Multiple scaling is a systematic method for obtaining solutions of linear and nonlinear, ordinary and partial differential equations involving a small parameter ε. The physical problem involves widely different time and/or length scales, which are reflected in the explicit introduction of fast and slow variables. This chapter first briefly describes one multiple scaling method and then develops an expansion method based on knowing the qualitative behavior of the solution, often conveniently determined by multiple scaling.

83

The methods are particularly suitable for oscillatory problems. One of our interests is the mean motion (the so-called guiding center) and the envelope.

Section II describes the methods and their application to nonlinear and variable-coefficient equations and the separation of fast and slow components. The examples in this section are relatively simple, to illustrate basic principles. Section III deals with more complicated examples selected from a long and instructive paper by van Kampen (1). We emphasize that we are not advocating the methods in this chapter as better than those of van Kampen. The two are complementary.

The reader who is interested mainly in the examples in Section III need read only Sections II.A, B, referring back to Sections II.C, D, E as necessary; these later sections deal with generalizations of the basic methods, illustrating in particular how difficulties can arise and how they can be overcome.

I. INTRODUCTION

Many phenomena in the physical sciences depend on widely different time and/or length scales, reflected in the occurrence of a small parameter ε in the differential equation describing the system. This phenomenon has led to a large number of so-called singular perturbation techniques for obtaining approximate solutions of such equations in the form

$$x = x_0(t, \varepsilon) + \varepsilon x_1(t, \varepsilon) + \cdots,$$

where the x_i are obtained by solving a sequence of simpler equations, the first involving only x_0, the second x_0 and x_1, and so on.

Multiple scaling is a term applied to a general class of methods that introduces several time scales explicitly into the original equations. In the simplest case, we might introduce $t_k = \varepsilon^k t, k = 0, 1, 2, \ldots$. If we are dealing with ordinary differential equations, the sequence of simpler problems are partial differential equations, but it turns out that these can be solved by ordinary differential equation techniques. We introduce one of several variants of the method by solving a simple example in Section II.A.

This chapter originated in response to a request from Professor J. O. Hirschfelder to give a "methods" lecture at the Los Alamos meeting on the topic of multiple scaling, with particular reference to the type of problems treated in a long, interesting, and instructive paper by van Kampen (1) that contains a large number of concrete examples of physicochemical applications. We found that the multiple-scaling method could reproduce most (if not all) of van Kampen's results as well as illuminate the role of the fast variables that van Kampen deliberately suppresses.

Multiple scaling often provides us, in a comparatively straightforward way, with knowledge about the quantitative form of the solution. Thus, we might find that

$$x \approx A(\varepsilon t)\exp it + \text{c.c.} \qquad (1.1)$$

where A can be complex and c.c. stands for complex conjugate. Here t is the fast variable and εt is the slow variable. Or we might find

$$x \approx B(\varepsilon t) + C(\varepsilon t)\exp(-t), \qquad (1.2)$$

where B and C are real and depend only on the slow variable. It is often instructive to use the insight obtained via multiple scaling (or otherwise) to obtain equations for the slowly varying function A in Eq. 1.1 or B and C in Eq. 1.2. Many expansion methods have been suggested in the literature. Our version depends on knowing the qualitative behavior of the solution, which is often obtained via multiple scaling. An introduction to our method is given at the end of Section II.A.

The example in Section II.A. is very simple. A more complicated example involving a nonlinear damped oscillator with a limit cycle is considered by both multiple scaling and the expansion method in Section II.B.

Sections II.C and II.D treat nonlinear and variable-coefficient equations, combining concrete examples with more general results.

Section II.E. collects miscellaneous comments on the separation of fast and slow variables, including a brief discussion of a reduction of equations to canonical form, which performs the separation of fast and slow variables explicitly.

This concludes Section II, which is concerned with basic principles illustrated by simple examples and covers some simple examples in van Kampen (1). Section III deals with more complicated examples from van Kampen, adequately summarized in the section headings.

An important question is "When do the two methods described in this chapter succeed, and when do they fail?" The first comment is that they are particularly suitable for oscillatory problems, the simplest situation being when the solution is of the form 1.1; but a more complicated example would be a sum of terms such as those in 1.1 and 1.2, including several oscillatory frequencies and/or real exponentials:

$$x = B(\varepsilon t) + C(\varepsilon t)e^{-t} + A(\varepsilon t)e^{it} + \text{c.c.}$$

If the fast terms involve only real exponentials, such as 1.2, it seems the equations are sometimes more amenable to methods that match solutions in different regions, for example, boundary layer techniques, partly because of the kind of equation that arises in that situation. A second comment is that quite often the techniques will fail because they are applied to the wrong form of the equation, or the different time scales are not simply related, or the fast solutions are not simple exponential or trigonometric functions, and so on. Section II not only explains the method, but also shows how some of these causes of failure can be circumvented. This discussion is continued in Section III.F, where we summarize some of the prerequisites for success of the methods, some of the reasons for failure, and some of the advantages and disadvantages of the two methods.

We postpone our remarks on the literature until the bibliographical notes preceding the references. However, we have found that Nayfeh (2), still in print, is a very useful text with a clear description of many of the available methods for singular perturbation problems and many illuminating examples and references to the literature.

II. BASIC PRINCIPLES ILLUSTRATED BY SIMPLE EXAMPLES

Our first objective in Section II is to explain the multiple-scaling method and the related expansion method used throughout the chapter. This is done via two second-order equation examples in Sections II.A and II.B. The next objective is to illustrate the scope of the methods and some of the difficulties that arise by considering nonlinear and variable-coefficient equations in Sections II.C and II.D. Finally various points concerning the separation of fast and slow components not considered earlier are covered in Section II.E.

It is not necessary to read the whole of Section II before reading Section III. Most of Section III can be understood after reading Section II.A, referring back to Section II as necessary. On the other hand, Section II will provide understanding necessary when tackling new problems. Section II.E is particularly important from this point of view.

A.   Introduction to Multiple Scaling and an Expansion Method

To introduce multiple scaling, we consider the linear equation with small damping,

$$\frac{d^2x}{dt^2} + 2\varepsilon\frac{dx}{dt} + x = 0, \qquad x(0) = \alpha, \qquad x'(0) = \beta. \tag{2.1}$$

This equation is commonly used as an introductory example because it illustrates the main points of the method without getting bogged down in excessive algebra. Also, the exact solution is available for comparison, namely,

$$x = e^{-\varepsilon t}\left[\alpha\cos\omega t + \frac{\beta + \varepsilon\alpha}{\omega}\sin\omega t\right], \qquad \omega = (1 - \varepsilon^2)^{1/2}. \qquad (2.2)$$

If we try a straightforward expansion by substituting

$$x = x_0 + \varepsilon x_1 + \varepsilon^2 x_2 + \cdots,$$

equating coefficients of successive powers of ε to 0, we find that the first two terms give

$$x \approx \alpha\cos t + \beta\sin t + \varepsilon a\sin t - \varepsilon t(\alpha\cos t + \beta\sin t). \qquad (2.3)$$

This is obviously a poor approximation to the exact solution 2.2 for large t due to the last term in 2.3, the so-called secular term, which has a multiplier t. Multiple scaling is one method for eliminating secular terms.

We describe one version of the method of multiple scaling. Introduce

$$t_0 = t, \qquad t_1 = \varepsilon t, \ldots, \qquad t_k = \varepsilon^k t, \qquad (2.4)$$

in which k is chosen according to the degree of approximation required, and assume

$$x = x_0(t_0, \ldots, t_k) + \cdots + \varepsilon^k x_k(t_0, \ldots, t_k). \qquad (2.5)$$

We formulate $k + 1$ subproblems for the x_i, $i = 0, \ldots, k$.

We proceed as if the t_i were independent, so that

$$\frac{d}{dt} = \frac{\partial}{\partial t_0} + \varepsilon\frac{\partial}{\partial t_1} + \varepsilon^2\frac{\partial}{\partial t_2} + \cdots. \qquad (2.6)$$

From 2.5 and 2.6,

$$\frac{dx}{dt} = \frac{\partial x_0}{\partial t_0} + \varepsilon\left(\frac{\partial x_1}{\partial t_0} + \frac{\partial x_0}{\partial t_1}\right) + \varepsilon^2\left(\frac{\partial x_2}{\partial t_0} + \frac{\partial x_1}{\partial t_1} + \frac{\partial x_0}{\partial t_2}\right) + \cdots. \qquad (2.7)$$

Similarly,

$$\frac{d^2 x}{dt^2} = \frac{\partial^2 x_0}{\partial t_0^2} + \varepsilon\left(\frac{\partial^2 x_1}{\partial t_0^2} + 2\frac{\partial^2 x_0}{\partial t_0\,\partial t_1}\right)$$

$$+ \varepsilon^2\left(\frac{\partial^2 x_2}{\partial t_0^2} + 2\frac{\partial^2 x_1}{\partial t_0\,\partial t_1} + \frac{\partial^2 x_0}{\partial t_1^2} + 2\frac{\partial^2 x_0}{\partial t_0\,\partial t_2}\right) + \cdots. \qquad (2.8)$$

Substituting 2.7 and 2.8 in 2.1 and equating coefficients of powers of ε to zero, we find equations that can be solved successively for $x_0, x_1, x_2, \ldots$:

$$\frac{\partial^2 x_0}{\partial t_0^2} + x_0 = 0, \tag{2.9}$$

$$\frac{\partial^2 x_1}{\partial t_0^2} + x_1 = -2\frac{\partial^2 x_0}{\partial t_0 \partial t_1} - 2\frac{\partial x_0}{\partial t_0}, \tag{2.10}$$

$$\frac{\partial^2 x_2}{\partial t_0^2} + x_2 = -2\frac{\partial^2 x_1}{\partial t_0 \partial t_1} - \frac{\partial^2 x_0}{\partial t_1^2} - 2\frac{\partial^2 x_0}{\partial t_0 \partial t_2} - 2\frac{\partial x_0}{\partial t_1} - 2\frac{\partial x_1}{\partial t_0}. \tag{2.11}$$

Although, strictly speaking, these are partial differential equations, they can be solved as ordinary differential equations provided we take into account dependence of constants of integration on $t_1, t_2, \ldots$. To obtain a solution accurate to second order in ε, we choose $k = 2$ in 2.4 and 2.5. The first step is then to solve 2.9, which can be done in terms of trigonometric functions or complex exponentials:

$$x_0 = a_0(t_1, t_2)\cos t_0 + b_0(t_1, t_2)\sin t_0 = A_0(t_1, t_2)\exp(it_0) + \text{c.c.} \tag{2.12}$$

Either form can be used, but we proceed with complex quantities because it is easier to deal with one complex unknown and powers of exponentials rather than powers of sines and cosines, which require expansion formulas.

Substituting for x_0 from 2.12 in 2.10, we obtain

$$\frac{\partial^2 x_1}{\partial t_0^2} + x_1 = -2i\left(\frac{\partial A_0}{\partial t_1} + A_0\right)e^{it_0} + \text{c.c.} \tag{2.13}$$

The $\exp(\pm it_0)$ on the right will give rise to secular terms unless we choose A_0 so that

$$\frac{\partial A_0}{\partial t_1} + A_0 = 0.$$

Remembering that $A_0 = A_0(t_1, t_2)$, we see that

$$A_0(t_1, t_2) = A_0(t_2)e^{-t_1}.$$

No confusion should arise from using $A_0(\cdot, \cdot)$ and $A_0(\cdot)$ to represent two different functions.

The right side of 2.13 is then zero, and the general solution of 2.13 is

$$x_1 = A_1(t_1, t_2)e^{it_0} + \text{c.c.} \tag{2.14}$$

We now come to an important point that simplifies the algebra considerably. If we substitute x_0 and x_1 in 2.11, elimination of secular terms leads to the conclusion that $A_1(t_1, t_2) = A_1(t_2)e^{-t_1}$, so that

$$x \approx [A_0(t_2) + \varepsilon A_1(t_2)]e^{-t_1 + it_0}.$$

But the same result could have been obtained by assuming that A_0 in 2.12 is replaced by a function $A(t_1, t_2; \varepsilon)$ that can be expanded in a power series in ε, in particular,

$$A(t_1, t_2; \varepsilon) = A_0(t_1, t_2) + \varepsilon A_1(t_1, t_2),$$

together with $x_1 = 0$ instead of Eq. 2.14. This procedure of disregarding the homogeneous solution at all orders except when it first appears (and then the arbitrary constants depend on ε) is more difficult to justify in general problems. We return to this point at the end of the next section.

Following this method for 2.11, we set

$$x_0 = A(t_2)e^{-t_1 + it_0} + \text{c.c.}, \qquad x_1 = 0, \tag{2.15}$$

on the right side of 2.11. Elimination of secular terms requires that

$$-2i\frac{\partial A(t_2)}{dt_2} + A(t_2) = 0, \qquad A(t_2) = Ae^{-1/2it_2}, \tag{2.16}$$

where A is a complex constant. The right side of 2.11 is then zero, and we again disregard the homogeneous solution so $x_2 = 0$.

Equations 2.15 and 2.16 give

$$x \approx e^{-\varepsilon t}[a\cos(1 - \tfrac{1}{2}\varepsilon^2)t + b\sin(1 - \tfrac{1}{2}\varepsilon^2)t], \tag{2.17}$$

where a, b are real constants. The initial conditions $x(0) = \alpha, x'(0) = \beta$ then give

$$x \approx X = e^{-\varepsilon t}\left(\alpha\cos\omega_0 t + \frac{\beta + \varepsilon\omega}{\omega_0}\sin\omega_0 t\right), \qquad \omega_0 = 1 - \tfrac{1}{2}\varepsilon^2. \tag{2.18}$$

Before examining these approximations in detail, we derive the same result by a different method we have called an "expansion method related to multiple scaling" because we will infer the form of the required expansion from multiple-scaling results. In the simple example under consideration, the first multiple-scaling result 2.12 implies that the dominant part of the solution is of the form

$$x = P(t)e^{it} + \text{c.c.},\qquad(2.19)$$

where $P(t)$ is a slowly varying function of t. In the present context this means that dP/dt and d^2P/dt^2 are of orders ε and ε^2, respectively. Substituting 2.19 in 2.1, we find

$$\frac{d^2P}{dt^2} + 2(i+\varepsilon)\frac{dP}{dt} + 2i\varepsilon P = 0.\qquad(2.20)$$

The dominant terms give

$$\frac{dP_0}{dt} + \varepsilon P_0 = 0,\qquad P_0 = A\varepsilon^{-\varepsilon t},\qquad(2.21)$$

where A is a complex constant.

To obtain the next-order approximation, following the same procedure and guided by 2.21, we set

$$P = P_1 e^{-\varepsilon t}$$

in 2.20. This gives

$$\frac{d^2P_1}{dt^2} + 2i\frac{dP_1}{dt} - \varepsilon^2 P_1 = 0.$$

The dominant terms give

$$\frac{dP_1}{dt} = -\tfrac{1}{2}i\varepsilon^2 P_1,\qquad P_1 = A\exp(-\tfrac{1}{2}i\varepsilon^2 t).$$

Assembling these results,

$$X \approx e^{-\varepsilon t}[A\exp i(1 - \tfrac{1}{2}\varepsilon^2)t + \text{c.c.}],\qquad(2.22)$$

which agrees with 2.17, as, of course, it should.

In complicated problems, particularly when there are several variables of different magnitudes present, it is sometimes awkward to pick out dominant terms in equations. There are several variants of the expansion method designed to make orders of magnitude explicit. One of these is the following. In 2.1, we introduce $\tau = \varepsilon t$:

$$\varepsilon^2 \frac{d^2 x}{d\tau^2} + 2\varepsilon^2 \frac{dx}{d\tau} + x = 0, \qquad x(0) = \alpha, \qquad x'(0) = \frac{\beta}{\varepsilon}. \tag{2.23}$$

Instead of 2.19 we now set

$$x = P(\tau)e^{i\tau/\varepsilon} + \text{c.c.},$$

where $dP/d\tau$ and $d^2 P/d\tau^2$ are of order unity, and

$$\frac{dx}{d\tau} = \left(\frac{dP}{d\tau} + \frac{iP}{\varepsilon} \right)e^{i\tau/\varepsilon} + \text{c.c.},$$

$$\frac{d^2 x}{d\tau^2} = \left(\frac{d^2 P}{d\tau^2} + \frac{2i}{\varepsilon}\frac{dP}{d\tau} - \frac{P}{\varepsilon^2} \right)e^{i\tau/\varepsilon} + \text{c.c.}$$

Substituting in 2.23,

$$\varepsilon \frac{d^2 P}{d\tau^2} + 2(i + \varepsilon)\frac{dP}{d\tau} + 2iP = 0. \tag{2.24}$$

This is equivalent to 2.20, but we now obtain a first approximation by simply setting $\varepsilon = 0$ without the need to examine orders of magnitude of derivatives. A second approximation can be obtained by solving 2.24 approximately in the same way.

An important remark is that the rapidly oscillating component is eliminated in the expansion method; the resulting Eq. 2.24 for the slow component is particularly suitable for numerical work. However, exploiting this lies outside the scope of this chapter. See note at end added in proof.

The reader should not conclude from the simple example considered so far that the expansion method is simpler than multiple scaling, though this is often the case. The purpose of the next section is to solve a nonlinear example that will provide a better comparison. The algebra involved, particularly in higher order approximations, may or may not be heavier in one or another of the methods. In any case, they provide a valuable check on each other, particularly since they may give answers of somewhat different but reconcilable forms.

It is appropriate at this point to make some general comments on accuracy, explaining, in particular, what is meant by saying that an approximation is uniform over an expanding interval of order $1/\varepsilon^k$.

We first note that the crude approximation 2.3 to the exact solution 2.2 of 2.1 is accurate only for $\varepsilon t \ll 1$. Many of van Kampen's approximations are derived on this assumption. The first-order multiple-scaling approximation, derived by assuming $A(t_2) = $ const. in 2.15, gives

$$x \approx X = e^{-\varepsilon t}[\alpha \cos t + (\beta + \varepsilon \alpha) \sin t].$$

From this and the exact result 2.2, we deduce

$$|x - X| \leqslant e^{-\varepsilon t}|A \sin \tfrac{1}{2}\varepsilon^2 t + B\varepsilon^2|.$$

We notice immediately that if $\varepsilon t \leqslant C$, then $|x - X| \leqslant D\varepsilon$, where D is a constant independent of ε, and we say that X is uniformly accurate to order ε *over an expanding interval of order* $1/\varepsilon$. This is the usual kind of estimate possible for multiple scaling. However, because of the presence of the exponential, it is possible to do better in this particular example. Differentiating with respect to t, we find that the bound for $|x - X|$ is a maximum when $\varepsilon t = 1$ approximately. Then $|x - X| \leqslant k\varepsilon$ for some constant k. The approximation is uniformly accurate to order ε for *all* t. For this kind of estimate, it seems to be necessary that damping be present.

The usual estimate in multiple scaling is that if we include time scales up to t_k, the approximation is uniformly accurate to order ε^k over times such that $\varepsilon^k t \leqslant C$, that is, over an expanding interval of length $1/\varepsilon^k$. The reader can show that 2.18 is one order better than this due to the particular nature of the problem. Also, in this problem the inaccuracy is in the phase. The envelope $\exp(-\varepsilon t)$ is completely accurate.

Of course, we were able to obtain these results only because we knew the exact solution. The determination of error estimates when the exact answer is unknown in a technical problem of some difficulty and limited applicability. If we claim that X is an approximate answer to a nonlinear operator equation expressed symbolically as $F(x) = 0$ involving a parameter ε, it is often possible to establish a result of the form

$$|F(x) - F(X)| \leqslant C\varepsilon^k.$$

The difficult step is then to deduce a bound on $|x - X|$. A simple result of the required type is the Gronwall inequality (see Ref. 3, a good starting point for the interested reader).

B. Van der Pol Equation

To understand the potentialities and limitations of the two methods introduced in the last section, we consider an example that is more complicated than the simple linear damped oscillator 2.1. An instructive, related example is the van der Pol equation

$$\frac{d^2x}{dt^2} + x = \varepsilon(1 - x^2)\frac{dx}{dt}. \tag{2.25}$$

This is a nonlinear perturbation of the linear oscillator, but the nature of the right side is such that if x is large, we would expect damping; that is, the amplitude of x will decrease; but if the amplitude of x is small, this amplitude will increase (equivalent to negative ε in 2.1). This indicates that as $t \to \infty$, the solution of 2.25 will have a limiting form consisting of a constant times an oscillatory component.

To obtain a solution via multiple scaling, we follow the procedure in the last section but omit details the reader can supply. The first approximation gives

$$x_0 = A(t_1, t_2)\exp(it_0) + \text{c.c.} \tag{2.26}$$

Then

$$\frac{\partial^2 x_1}{\partial t_0^2} + x_1 = -2\frac{\partial^2 x_0}{\partial t_0 \, \partial t_1} + (1 - x_0^2)\frac{\partial x_0}{\partial t_0}, \tag{2.27}$$

$$\frac{\partial^2 x_2}{\partial t_0^2} + x_2 = -2\frac{\partial^2 x_0}{\partial t_0 \, \partial t_1} - \frac{\partial^2 x_0}{\partial t_1^2} - 2\frac{\partial^2 x_0}{\partial t_0 \, \partial t_2}$$
$$+ (1 - x_0^2)\left(\frac{\partial x_1}{\partial t_0} + \frac{\partial x_0}{\partial t_1}\right) - 2x_0 x_1 \frac{\partial x_0}{\partial t_0}. \tag{2.28}$$

Substituting (2.26) in (2.27), elimination of secular terms requires that

$$\frac{\partial A}{\partial t_1} = \tfrac{1}{2}(1 - A\bar{A})A, \tag{2.29}$$

and the solution of (2.27) is then

$$x_1 = \tfrac{1}{8}iA^3 e^{3it_0} + \text{c.c.}, \tag{2.30}$$

where we omit the homogeneous component as in the last section (we return to this point later).

Substituting 2.26 and 2.30 in 2.28, we find that elimination of secular terms requires

$$\frac{\partial A}{\partial t_2} = \frac{i}{2}\left[\frac{\partial^2 A}{\partial t_1^2} - (1 - 2A\bar{A})\frac{\partial A}{\partial t_1} + A^2\frac{\partial \bar{A}}{\partial t_1} - \tfrac{1}{8}A^3\bar{A}^2\right]. \tag{2.31}$$

To examine the solution of 2.29, we first set

$$A = \tfrac{1}{2}ae^{i\phi} \tag{2.32}$$

in 2.29, which leads to

$$\frac{\partial \phi}{\partial t_1} = 0, \qquad \frac{\partial a}{\partial t_1} = \tfrac{1}{2}(1 - \tfrac{1}{4}a^2)a \tag{2.33}$$

with the solution

$$\phi = \phi(t_2), \qquad a^2 = \frac{4}{1 + c(t_2)e^{-t_1}}. \tag{2.34}$$

We can make considerable progress in solving 2.31 without using 2.32–2.34 by noting that we can deduce from 2.29 that

$$\frac{\partial^2 A}{\partial t_1^2} = \tfrac{1}{2}(1 - 2A\bar{A})\frac{\partial A}{\partial t_1} - \tfrac{1}{2}A^2\frac{\partial \bar{A}}{\partial t_1}.$$

Substituting in 2.31 yields

$$\frac{\partial A}{\partial t_2} = \frac{i}{4}\left[-(1 - 2A\bar{A})\frac{\partial A}{\partial t_1} + A^2\frac{\partial \bar{A}}{\partial t_1} - \tfrac{1}{4}A^3\bar{A}^2 \right].$$

Substituting for $\partial A/\partial t_1$, from 2.30, yields

$$\frac{\partial A}{\partial t_2} = -\frac{i}{8}[1 - 4A\bar{A} + \tfrac{7}{2}(A\bar{A})^2]A.$$

Substituting 2.32 and separating real and imaginary parts gives

$$\frac{\partial a}{\partial t_2} = 0, \qquad \frac{\partial \phi}{\partial t_2} = -\tfrac{1}{8}(1 - a^2 + \tfrac{7}{32}a^4). \tag{2.35}$$

The first equation states that a is independent of t_2 so that $c(t_2)$ in 2.34 is a constant. The second equation then gives

$$\phi = -\tfrac{1}{8}(1 - a^2 + \tfrac{7}{32}a^4)t_2 + \phi_0, \tag{2.36}$$

where ϕ_0 is independent of t_2. Strictly speaking, this contradicts the first equation in 2.33 since 2.34 shows us that a depends on t_1. However, 2.34 shows us that a is a slowly varying function of t, and the conclusion to be drawn is that 2.35 is valid as long as a can be regarded as a constant; that is, as long as t is not too large.

It is possible to avoid the contradiction by generalizing the method of multiple scaling. The difficulty has arisen at the stage of introducing t_2, and it involves t_1, so let us introduce more generally

$$t_2 = \varepsilon F(t_1), \quad \text{where } F(t_1) \to kt_1 \text{ as } t_1 \to 0, \tag{2.37}$$

for some constant k, and F is a function to be determined. This new t_2 is approximately $\varepsilon^2 t$ for some small t, but we have introduced some flexibility as t increases. The only change in the basic equations 2.26–2.28 is that $\partial^2 x_0 / \partial t_0 \, \partial t_2$ in 2.28 is replaced by $F'(t_1)\partial^2 x_0 / \partial t_0 \partial t_1$. This means that an $F'(t_1)$ appears on the left of 2.31, and on introducing 2.32 as before, we find, instead of 2.35,

$$\frac{\partial a}{\partial t_2} = 0, \quad \frac{dF(t_1)}{dt_1}\frac{\partial \phi}{\partial t_2} = -\tfrac{1}{8}(1 - a^2 + \tfrac{7}{32}a^4). \tag{2.38}$$

The first equation again states that a is independent of t_2, but we know that it depends on t_1. We can therefore choose F so that

$$\frac{dF(t_1)}{dt_1} = -\tfrac{1}{8}(1 - a^2 + \tfrac{7}{32}a^4), \quad \frac{\partial \phi}{\partial t_2} = 1, \tag{2.39}$$

and then

$$\phi = t_2 + \phi_0 = \varepsilon F(t_1) + \phi_0,$$

where ϕ_0 is a constant. The first equation in 2.39 can be integrated by rewriting it using 2.33 as

$$\frac{dF}{dt_1} = -\frac{1}{16} - \frac{1}{8}\left(\frac{1}{a} - \frac{7}{4}a\right)\frac{da}{dt_1}.$$

Integration gives

$$F = -\frac{1}{16}t_1 - \frac{1}{8}\left[\log\frac{a}{a_0} - \frac{7}{8}(a^2 - a_0^2)\right],$$

where $a_0 = a(0)$, introduced so that $F \to kt_1$ as $t_1 \to 0$. Summarizing the multiple-scaling result, we have

$$x = \tfrac{1}{2}ae^{i(t+\phi)} + \tfrac{1}{64}\varepsilon ia^3 e^{3i(t+\phi)} + \text{c.c.}, \tag{2.40a}$$

where

$$a = \frac{1}{1 + ce^{-\varepsilon t}}, \qquad \phi = -\frac{\varepsilon^2}{16}t - \frac{\varepsilon}{8}(\log a - \tfrac{7}{8}a^2) + \alpha, \tag{2.40b}$$

where the constants c, α can be determined from the initial conditions.

These results have been obtained by Nayfeh (2, p. 247) by including in 2.30 the solution $B\exp(it) + \text{c.c.}$ of the homogeneous equation corresponding to 2.27. This leads to extra terms in B in 2.31, but the problem is then undetermined, and in this nonlinear case the indeterminacy is not removed by going to higher order. Nayfeh resolves the difficulty by using an ansatz, essentially that B is close to a multiple of A, which led to 2.40 indirectly. We chose to resolve the indeterminancy by assuming that $B = 0$, but this led first to an answer that could be justified only under a restrictive assumption avoided by introducing an arbitrary time scale for t_2. One of the reasons for discussing this example in some detail is to draw the reader's attention to the fact that if the solutions of the homogeneous equations arising at later stages of multiple scaling are included, this introduces extra arbitrary functions, and it may not be obvious how to determine these even with the help of initial conditions on x, which will determine those for the x_i. On the other hand, it is clear from the preceding example that ignoring the homogeneous equation solutions may not be altogether straightforward. One of the reasons for introducing the expansion method is that it gives a valuable crosscheck and an alternative in the case of difficulties.

We next apply the expansion method to the van der Pol example. Equation 2.40 suggests that x is of the form

$$x = Pe^{it} + \varepsilon Q e^{3it} + \cdots, \tag{2.41}$$

where P, Q are slowly varying functions of x. We shall use the method leading to 2.20 rather than the variant associated with 2.23, since the algebra in this latter method is somewhat more laborious in this example and we require only the first approximation. Substituting 2.41 in 2.25, we need the following result, where we retain only terms needed at the next step:

$$x^2 \frac{dx}{dt} = \left[2P\bar{P}\left(\frac{dP}{dt} + iP\right) + P^2\left(\frac{d\bar{P}}{dt} - i\bar{P}\right) + i\varepsilon\bar{P}^2 Q \right]e^{it}$$
$$+ iP^3 e^{3it} + \text{c.c.} + O(\varepsilon^2).$$

Equating the coefficients of exp it and exp $3it$ and retaining terms up to order ε in the first and only the dominant term in the second, we find

$$\frac{d^2P}{dt^2} + 2i\frac{dP}{dt} = \varepsilon(1 - 2P\bar{P})\left(\frac{dP}{dt} + iP\right) - \varepsilon P^2\left(\frac{d\bar{P}}{dt} - i\bar{P}\right) - i\varepsilon^2 \bar{P}^2 Q, \quad (2.42a)$$

$$Q = \tfrac{1}{8}iP^3. \tag{2.42b}$$

The dominant term in (2.42a) gives the first approximation,

$$\frac{dP_0}{dt} = \frac{\varepsilon}{2}(1 - P_0\bar{P}_0)P_0. \tag{2.43}$$

Substituting this in the subdominant terms in 2.42a, approximating d^2P/dt^2 by d^2P_0/dt^2, and using 2.43, we find

$$\frac{dP}{dt} = \frac{\varepsilon}{2}(1 - P\bar{P})P - \frac{i\varepsilon^2}{8}\left[1 - 4P_0\bar{P}_0 + \tfrac{7}{2}(P_0\bar{P}_0)^2\right]P_0.$$

To the same order, we can replace this by

$$\frac{dP}{dt} = \frac{\varepsilon}{2}(1 - P\bar{P})P - \frac{i\varepsilon^2}{8}[1 - 4P\bar{P} + \tfrac{7}{2}(P\bar{P})^2]P.$$

Setting $P = \tfrac{1}{2}be^{i\psi}$ leads to

$$\frac{db}{dt} = \frac{\varepsilon b}{2}(1 - \tfrac{1}{4}b^2) \qquad \frac{d\psi}{dt} = \frac{-\varepsilon^2}{8}(1 - b^2 + \tfrac{7}{32}b^4).$$

These are 2.33 and 2.39, and the solution can be completed as before.

This short treatment has made the expansion method look easier than multiple scaling, but the detailed algebra has been omitted, and we have benefited from doing the multiple-scaling solution first.

C.   Nonlinear Equations (Continued)

To illustrate a number of other points, we first consider the following coupled set of second-degree equations that is a special case of a more general example in Section III.A:

$$\frac{d^2u}{dt^2} = \mu[a(u^2 + v^2) + 2buv], \tag{2.44a}$$

$$\frac{d^2v}{dt^2} = \mu[b(u^2 + v^2) - 2auv] - v, \tag{2.44b}$$

where μ is a small parameter and a, b are constants. The reason for using μ instead of ε will become clear presently.

If we try to solve this system as it stands by multiple scaling, setting $t_k = \mu^k t$, $k = 0, 1, 2$, we find, for the zero-order solution,

$$u_0 = A(t_1, t_2), \qquad v_0 = B(t_1, t_2)e^{it_0} + \text{c.c.}, \tag{2.45}$$

where A is real and B is complex. Elimination of secular terms from the next approximation gives $A^2 + 2B\bar{B} = 0$, which implies $A = 0$, $B = 0$, so that multiple scaling has broken down.

To investigate further, we first set $\mu = 0$ in 2.44, which leads to

$$u = A_0, \qquad v = B_0 e^{it} + \text{c.c.}, \tag{2.46}$$

where A_0, B_0 are constants, which confirms that the dominant part of the solution is given by 2.45, with A, B slowly varying function. This gives a starting point for the expansion method, suggesting that we should insert

$$u = A(t), \qquad v = B(t)e^{it} + \text{c.c.}, \tag{2.47}$$

in 2.44. Separating the fast component in 2.44a, we find that $AB = 0$, so that the expansion method also breaks down, but in a different way.

In a situation such as this, it is often instructive to replace the original problem by a simpler one containing the essential features of the original. In particular, it is often possible to obtain insight by replacing a nonlinear problem with a related linear one that can be solved exactly. In place of 2.44, consider

$$\frac{d^2u}{dt^2} = \mu(cu + dv), \tag{2.48a}$$

$$\frac{d^2v}{dt^2} = \mu(eu + fv) - v. \tag{2.48b}$$

The solutions of this equation are of the form $\exp(\lambda t)$, with

$$\lambda^2 \approx -1, \qquad \lambda^2 \approx \mu(c + f). \tag{2.49}$$

The fast component is oscillatory with $\lambda = \pm i$, but the slow component is real exponential of the form $\exp(\pm \mu^{1/2} kt)$. This is the reason our methods have broken down. We implicitly assumed that the solution depended on integral powers of μ, whereas in fact the solution also depends on $\mu^{1/2}$.

The results in the last paragraph indicate that we should set $\mu = \varepsilon^2$ in 2.44. If we do this and then expand the solution in terms of ε, the zero-order multiple-scaling solution again gives 2.45. The first-order solution is then (ignoring solutions of the homogeneous equation as before)

$$u_1 = 0, \qquad v_1 = 0, \qquad B(t_1, t_2) = B(t_2), \tag{2.50}$$

that is, B does not depend on t_1. The second-order approximation leads to

$$\frac{\partial^2 u_2}{\partial t_0^2} = a(u_0^2 + v_0^2) + 2bu_0 v_0 - \frac{\partial^2 u_0}{\partial t_1^2}, \tag{2.51a}$$

$$\frac{\partial^2 v_2}{\partial t_0^2} + v_2 = b(u_0^2 + v_0^2) - 2au_0 v_0 - 2\frac{\partial^2 v_0}{\partial t_0 \partial t_2}. \tag{2.51b}$$

Elimination of the secular terms gives

$$\frac{\partial^2 A}{\partial t_1^2} = a(A^2 + 2B\bar{B}), \qquad \frac{\partial B}{\partial t_2} = iaAB, \tag{2.52}$$

and the particular solutions of 2.51 give

$$u_2 = -2bABe^{ito} + \cdots,$$

$$v_2 = b(A^2 + 2B\bar{B}) + \cdots,$$

where terms in $\exp(\pm 2it)$ have been omitted. As in the example in Section II.B, the second equation in 2.52 indicates that B depends on t_1, which contradicts 2.50 but is approximately true for small εt. As in the last section, this apparent contradiction can be avoided by introducing, instead of $t_2 = \varepsilon^2 t$, a different $t_2 = \varepsilon F(t_1)$, where $F(t_1) \to kt_1$ as $t \to 0$. The second equation in 2.52 is then replaced by

$$\frac{dF}{dt_1} = A, \qquad \frac{\partial B}{\partial t_2} = iaB,$$

which (omitting details) leads to the results that follow.

The approximation for u, v becomes

$$u = A - 2\varepsilon^2 bABe^{ito} + \text{c.c.}, \tag{2.53a}$$

$$v = \varepsilon^2 b(A^2 + 2B\bar{B}) + Be^{ito} + \text{c.c.}, \qquad (2.53b)$$

where

$$\frac{d^2 A}{dt^2} = \varepsilon^2 a(A^2 + 2B\bar{B}), \qquad \frac{dB}{dt} = i\varepsilon^2 aAB. \qquad (2.54)$$

As we should expect, these differential equations require four initial conditions, two for A and two for the complex unknown B. In a physical problem, these might be derived from the initial values of u, v, and their derivatives.

To obtain the same results by the expansion method, we use the qualitative results obtained earlier to assume that the solution is of the form

$$u = U + P \exp\left(\frac{i\tau}{\varepsilon}\right) + \text{c.c.}, \qquad (2.55a)$$

$$v = V + Q \exp\left(\frac{i\tau}{\varepsilon}\right) + \text{c.c.}, \qquad (2.55b)$$

where, as indicated in Section II.B, it simplifies the problem of extracting dominant terms if we introduce $\tau = \varepsilon t$. In 2.55, unknown functions and their derivatives with respect to τ are all of the same order. Note that we have *not* used the information in 2.53, that P, V are of lower orders of magnitude than U, Q, because it is instructive to see how this appears automatically in the following analysis.

Substituting 2.55 in 2.44 and equating the slowly varying terms and the terms in $\exp(i\tau/\varepsilon)$ but ignoring any terms in $\exp(2i\varepsilon/\tau)$, which would require corresponding extra terms in the assumed form 2.55, we obtain

$$\frac{d^2 U}{d\tau^2} = a(U^2 + V^2 + 2P\bar{P} + 2Q\bar{Q}) + 2b(UV + P\bar{Q} + \bar{P}Q), \qquad (2.56a)$$

$$P = -2\varepsilon^2 b(UQ + PV) + 2i\varepsilon\frac{dP}{d\tau} + \varepsilon^2\frac{d^2 P}{d\tau^2}, \qquad (2.56b)$$

$$V = \varepsilon^2 b(U^2 + V^2 + 2P\bar{P} + 2Q\bar{Q}) - 2\varepsilon^2 a(UV + P\bar{Q} + \bar{P}Q) - \varepsilon^2\frac{d^2 V}{dt^2}, \qquad (2.56c)$$

$$\frac{dQ}{d\tau} = i\varepsilon a(UQ + PV) + \frac{1}{2}i\varepsilon\frac{d^2 Q}{d\tau^2}. \qquad (2.56d)$$

Note that these are well suited for numerical solution, a remark made earlier in connection with another expansion method result.

We see immediately from 2.56b, c that P, V are of order ε^2. This tells us that the dominant terms in 2.56b, c give

$$P = -2\varepsilon^2 bUQ, \qquad V = \varepsilon^2 b(U^2 + 2Q\bar{Q}). \qquad (2.57)$$

Since 2.56d indicates that $dQ/d\tau$ is of order ε, we can neglect the second derivative in 2.56d. The dominant terms in 2.56a, d give

$$\frac{d^2 U}{d\tau^2} = a(U^2 + 2Q\bar{Q}), \qquad \frac{dQ}{d\tau} = i\varepsilon aUQ. \qquad (2.58)$$

Equations 2.55, 2.57, and 2.58 are precisely 2.53, 2.54, $A \equiv U$, $B \equiv Q$.

We next obtain an approximate solution for a problem considered from a completely different point of view in a talk at the Los Alamos meeting (4). An idealized model for the growth of a two-dimensional dendritic crystal in a supercooled liquid leads to the equations

$$\varepsilon^2 \frac{d^3 x}{ds^3} + \frac{dx}{ds} = \cos x, \qquad -\infty < s < \infty, \qquad (2.59a)$$

$$x(0) = 0, \qquad x \to \pm \tfrac{1}{2}\pi \text{ as } s \to \pm\infty. \qquad (2.59b)$$

Here s is a length. We first need to determine the qualitative form of the solution. The fast component has the form $A \exp(\pm is/\varepsilon)$, where A is slow; this can be verified either by substitution into 2.59a and equating dominant terms or by introducing $t = s/\varepsilon$ and then setting $\varepsilon = 0$. This suggests that we can use multiple scaling by defining the time scales as

$$t_0 = \frac{s}{\varepsilon}, \qquad t_1 = s, \qquad t_2 = \varepsilon s, \ldots . \qquad (2.60)$$

To zero order we find

$$\frac{\partial^3 x_0}{\partial t_0^3} + \frac{\partial x_0}{\partial t_0} = 0, \qquad x_0 = A(t_1) + B(t_1)e^{it_0} + \text{c.c.} \qquad (2.61)$$

To satisfy the second condition in 2.59b, it is necessary that $B(t_1) \to 0$ as $t_1 \to \infty$, and this is the question we wish to examine. Because of the $\cos x$ in 2.59a, it is not possible to make further progress with 2.61 for arbitrary B, but it will suffice to assume that B is small. Switching to the expansion method, we substitute

$$x = X + \varepsilon(Pe^{is/\varepsilon} + \text{c.c.})$$

in 2.59a, using

$$\cos x = \cos X - \varepsilon(Pe^{is/\varepsilon} + \text{c.c.}) \sin X + O(\varepsilon^2).$$

Separation of slow and fast components gives

$$\frac{dX}{dt} = \cos X, \qquad X = \sin^{-1}[\tanh(t + D)],$$

$$\frac{dP}{dt} = \tfrac{1}{2}P \sin X, \quad P = c\cosh^{1/2}(t + D),$$

where c is a complex constant and D is real. We see that P is increasing exponentially as $t \to \pm\infty$. If P had been multiplied by a real exponential, one might have been able to localize the effect, but the multiplier is oscillatory and the fast solution is unstable. From a related point of view, if one considers a boundary value problem in a finite region $-S \leqslant s \leqslant S$, the corresponding approximate solution is very sensitive to the boundary conditions. For instance, if the boundary conditions are $x = \pm(1/2\pi - \Delta)$ at $s = \pm S$, resonance occurs for $S/\varepsilon = n\pi$, and there is no solution. We conclude that the idealized model has led to an equation that has unstable solutions, so that the model is physically unrealistic.

We next consider an equation where the fast solution is a real exponential, as opposed to the examples considered so far, which have involved fast oscillatory components. The prototype linear equation is

$$\varepsilon\frac{d^2x}{dt^2} + \frac{dx}{dt} + x = 0, \tag{2.62}$$

the exact solution of which has the form of the sum of constant multiples of $\exp(\lambda t)$ with $\lambda \approx -1/\varepsilon$ and $\lambda \approx -1$; that is, it has the form

$$x = A + Be^{-t/\varepsilon}, \tag{2.63}$$

where A and B are slowly varying functions compared with $\exp(-t/\varepsilon)$. In order to apply multiple scaling, we introduce $t_0 = t/\varepsilon, t_1 = t, t_2 = \varepsilon t$, and so on. Alternatively, it may be convenient to introduce $\tau = t/\varepsilon$ in 2.62:

$$\frac{d^2x}{d\tau^2} + \frac{dx}{d\tau} + \varepsilon x = 0. \tag{2.64}$$

Multiple scaling then uses the standard $t_k = \varepsilon^k\tau$.

The success of multiple scaling in this linear example is somewhat misleading. Nonlinearities can introduce complications. We illustrate by a nonlinear generalization of 2.62 and 2.64. Both forms are used in the following:

$$\varepsilon\frac{d^2x}{dt^2} + \frac{dx}{dt} + x^2 = 0, \qquad \frac{d^2x}{d\tau^2} + \frac{dx}{d\tau} + \varepsilon x^2 = 0. \qquad (2.65)$$

Multiple scaling applied to either form gives ($t_0 = \tau = t/\varepsilon$)

$$x_0 = A(t_1, t_2) + B(t_1, t_2)e^{-t_0},$$

$$\frac{\partial^2 x_1}{\partial t_0^2} + \frac{dx_1}{dt_0} = -x_0^2 - 2\frac{\partial^2 x_0}{\partial t_0 \partial t_1} - \frac{\partial x_0}{\partial t_1}.$$

Eliminating secular terms and omitting the solution of the homogeneous equation, we find

$$\frac{\partial A}{\partial t_1} = -A^2, \qquad A = \frac{1}{t_1 + C(t_2)}, \qquad (2.66a)$$

$$\frac{\partial B}{\partial t_1} = 2AB, \qquad B = k(t_2)[t_1 + C(t_2)]^2, \qquad (2.66b)$$

$$x_1 = -\tfrac{1}{2}B^2 e^{-2t_0}.$$

It might be thought that if we wished to include only t_0 and t_1, so that $C(t_2)$, $k(t_2)$ could be regarded as constants, then from the preceding results,

$$x \approx x_0 + \varepsilon x_1 = \frac{1}{C+t} + kC(C+2t)e^{-t/\varepsilon} - \tfrac{1}{2}\varepsilon k^2 C^4 e^{-2t/\varepsilon}. \qquad (2.67)$$

Note that we have ignored terms of order higher than ε. In particular, the maximum value of $t\exp(-t/\varepsilon)$ is $\varepsilon\exp(-1)$, so this term has been included. This approximation will be compared in what follows with one obtained by means of the expansion method. A straightforward application of the next order of multiple scaling runs into difficulties. If we ignore the solution of the homogeneous equation, we run into the difficulty met earlier that t_2 has to be redefined, but the method used to avoid the difficulty now works only approximately. Part of the difficulty is that C appears in the denominator of A in 2.66a. It is possible to derive the expansion method result 2.70 below by assuming that C in 2.66a is a constant, but then including the solution of the homogeneous equation. However, these devices are somewhat special, and we prefer to pursue this example by the expansion method.

Guided by the multiple-scaling results, we substitute in the first equation in 2.65:

$$x = X + Pe^{-t/\varepsilon} + Qe^{-2t/\varepsilon},$$

where X, P, Q are slow. Separating fast and slow components,

$$\varepsilon\frac{d^2 X}{dt^2} + \frac{dX}{dt} + X^2 = 0, \tag{2.68a}$$

$$\varepsilon\frac{d^2 P}{dt^2} - \frac{dP}{dt} + 2XP = 0, \tag{2.68b}$$

$$2Q + \varepsilon P^2 = 0. \tag{2.68c}$$

The dominant terms in 2.68a give

$$\frac{dX_0}{dt} + X_0^2 = 0, \qquad X_0 = \frac{1}{C+t}. \tag{2.69}$$

The next order can be found by setting $X = X_0 + \varepsilon X_1$, which leads to

$$\frac{dX_1}{dt} + 2\frac{X_1}{C+t} = \frac{-2}{(C+t)^3}, \qquad X_1 = \frac{D - 2\log(C+t)}{(C+t)^2}. \tag{2.70}$$

The dominant terms in 2.68b give

$$\frac{dP_0}{dt} = 2X_0 P_0, \qquad P_0 = k(C+t)^2. \tag{2.71}$$

The expansion result for n given by 2.69–2.71 is correct to first-order, but X_1 is not in the multiple scaling result 2.67. It is necessary to go to the next order of approximation in the multiple scaling method before this first order term can be determined.

One of the points of this example is to emphasize the importance of using two more or less independent methods of solution as a crosscheck.

Problems where the fast solutions are real exponentials, as in the last example, occur commonly in initial-value problems where there is an initial fast transient and in boundary value problems where the fast component occurs in boundary layers and/or sharp changes in the interior of the region. The reader is reminded that for most of these problems, there is a well-developed literature using other methods such as *matched asymptotic expan-*

sions and *composite expansions* (2, 3, 5, 6). Multiple scaling and the expansion method as described in this chapter can be used for such problems, but a detailed discussion would take us too far afield. Our last example illustrated that multiple scaling may be inefficient for this type of problem.

D. Variable-Coefficient Equations

Consider first an example where multiple scaling gives an approximate solution in a straightforward way:

$$\frac{d^2x}{dt^2} + 4\varepsilon \cos^2 t \frac{dx}{dt} + x = 0. \tag{2.72}$$

The zero-order approximation is

$$x_0 = A(t_1)e^{it_0} + \text{c.c.}$$

Using

$$4\cos^2 t_0 \frac{dx_0}{dt_0} = i(2A - \bar{A})e^{it_0} + \text{c.c.} + \text{terms in } \exp(\pm 3it_0),$$

elimination of secular terms in the first-order approximation requires

$$\frac{\partial A}{\partial t_1} = \tfrac{1}{2}(\bar{A} - 2A).$$

This is easily solved by setting $A = a + ib$. We find

$$x_0 = a_0 e^{-1/2\varepsilon t} \cos t + b_0 e^{-3/2\varepsilon t} \sin t, \tag{2.73}$$

where a_0, b_0 are constants. This is an interesting result because simple averaging in the original equation would suggest a damping factor $\exp(-\varepsilon t)$.

Another example where multiple scaling works well is the Mathieu equation

$$\frac{d^2x}{dt^2} + (a + 2\varepsilon \cos t)x = 0. \tag{2.74}$$

For some values of a and ε there are solutions that grow exponentially, whereas others are bounded for all t. A clear account of the use of multiple scaling to investigate the boundaries in the a–ε plane separating stable from unstable solutions is given in Bender and Orszag (6, pp. 560–566).

Difficulties arise when we try to apply multiple scaling to an example such as

$$\frac{d^2x}{dt^2} + (1 + 2\varepsilon t)\frac{dx}{dt} + 2\varepsilon x = 0. \tag{2.75}$$

Intuitively, the difficulty arises because εt is not small when t is large. When multiple scaling is applied, one might obtain a zero-order approximation by setting $\varepsilon = 0$ or by setting $\varepsilon t = t_1$ in parentheses in 2.75 and then proceeding on the assumption that t_0 and t_1 are independent. However, the elimination of secular terms in the subsequent first-order approximations does not lead to satisfactory results. The source of the difficulty is the assumption that suitable time scales are given by $t_0 = t, t_1 = \varepsilon t$. Related situations have occurred earlier that suggest what to do. In the present context, one solution is to introduce

$$t_0 = \frac{1}{\varepsilon}f(t_1), \qquad t_1 = \varepsilon t, \tag{2.76}$$

where f is a function to be determined in such a way that $f(t_1) \to t_1$ as $t_1 \to 0$. We find

$$\frac{dx}{dt} = f'(t_1)\frac{\partial x_0}{\partial t_0} + \varepsilon\left[f'(t_0)\frac{\partial x_1}{\partial t_0} + \frac{\partial x_0}{\partial t_1}\right],$$

$$\frac{d^2x}{dt^2} = (f')^2\frac{\partial^2 x_0}{\partial t_0^2} + \varepsilon\left[(f')^2\frac{\partial^2 x_1}{\partial t_0^2} + 2f'\frac{\partial^2 x_0}{\partial t_0 \partial t_1} + f''(t_1)\frac{\partial x_0}{\partial t_0}\right].$$

Substituting these in the generalization of 2.75,

$$\frac{d^2x}{dt^2} + a(\varepsilon t)\frac{dx}{dt} + \varepsilon b(\varepsilon t)x = 0, \tag{2.77}$$

we find, for the zero-order approximation,

$$f'(t_1)\frac{\partial^2 x_0}{\partial t_0^2} + a(t_1)\frac{\partial x_0}{\partial t_0} = 0.$$

This indicates that we should set

$$\frac{df}{dt_1} = a(t_1), \qquad f(t_1) = \int_0^{t_1} a(\sigma)\,d\sigma.$$

Then

$$x_0 = A(t_1) + B(t_1)e^{-t_0}.$$

Elimination of secular terms in the first-order approximation leads to expressions for A, B given by

$$\frac{dA}{dt_1} = -\frac{b}{a}A, \qquad \frac{dB}{dt_1} = \frac{b}{a}B.$$

We are clearly assuming implicitly that either $a > 0$ or $a < 0$. Introducing

$$I(\varepsilon t) = \int_0^{\varepsilon t} \frac{b(\sigma)}{a(\sigma)} d\sigma,$$

we find

$$x_0 = A_0 \exp\left[-I(\varepsilon t)\right] + B_0 \exp\left[I(\varepsilon t) - \frac{1}{\varepsilon}f(\varepsilon t)\right] \qquad (2.78)$$

where A_0, B_0 are constants.

As an example we can use the preceding result to solve approximately the following boundary value problem in $0 \leqslant x \leqslant 1$:

$$\varepsilon\frac{d^2 X}{dx^2} + a(x)\frac{dX}{dx} + b(x)X = 0, \qquad X(0) = \alpha, \qquad X(1) = \beta.$$

Replacing x by εt and $X(x)$ by $x(t)$, the differential equation becomes 2.77 with boundary conditions $x(0) = \alpha$, $x(1/\varepsilon) = \beta$. To obtain an approximate solution under the assumption $a(x) > 0$, we find from 2.78, neglecting an exponentially small term,

$$A_0 + B_0 = \alpha, \qquad A_0 \exp\left[-I(1)\right] = \beta.$$

Then 2.78 gives

$$X = \beta \exp\left[I(1) - I(t)\right] + \left[\alpha - \beta \exp I(1)\right]\exp\left(-a(0)\frac{x}{\varepsilon}\right). \qquad (2.79)$$

An interesting feature is that it appears automatically that the boundary layer is at $x = 0$, not $x = 1$. If we had assumed $a < 0$, the boundary layer would have appeared at $x = 1$.

We next derive by multiple scaling the familiar WKB approximation for the solution of

$$\frac{d^2x}{dt^2} + \omega^2(\varepsilon t)x = 0, \tag{2.80}$$

where the frequency ω is a slowly varying function of time: As in the last example, a direct application of multiple scaling is unsatisfactory, which suggests that we should define new time scales as in 2.76, where f is to be determined. The zero-order approximation gives

$$[f'(t_1)]^2 \frac{\partial^2 x_0}{\partial t_0^2} + \omega^2(t_1)x_0 = 0, \tag{2.81}$$

which suggests that we set

$$f'(t_1) = \omega(t_1), \qquad f(\varepsilon t) = \int_0^{\varepsilon t} \omega(\sigma)\, d\sigma, \qquad x_0 = A(t_1)e^{it_0} + \text{c.c.} \tag{2.82}$$

In the first-order approximation, elimination of secular terms gives

$$\frac{dA}{dt_1} = -\frac{\omega'}{2\omega}A, \qquad A = \frac{C}{\omega^{1/2}}, \tag{2.83}$$

where C is a complex constant. Hence,

$$x_0 = \frac{C}{\omega^{1/2}(\varepsilon t)} \exp\left(\frac{i}{\varepsilon}\int_0^{\varepsilon t} \omega(\sigma)\, d\sigma\right) + \text{c.c.} \tag{2.84}$$

This result is easily obtained by the expansion method once we realize what we are looking for. Set

$$x = X \exp iS(t) + \text{c.c.},$$

where X is a slow component. Substitute in 2.80:

$$\frac{d^2X}{dt^2} + i\left(2\frac{dS}{dt}\frac{dX}{dt} + X\frac{d^2S}{dt^2}\right) - \left(\frac{dS}{dt}\right)^2 X + \omega^2(\varepsilon t)X = 0. \tag{2.85}$$

To simplify this equation, we can now choose S so that

$$\frac{dS}{dt} = \omega(\varepsilon t), \qquad S = \frac{1}{\varepsilon}\int_0^{\varepsilon t} \omega(\sigma)\, d\sigma.$$

The remaining dominant terms give, on setting $\varepsilon t = \tau$,

$$2\omega(\tau)\frac{dX}{d\tau} + \omega'(\tau)X = 0.$$

These equations are 2.82 and 2.83 and lead to 2.84, as before.

To conclude this section, consider the turning-point problem:

$$\mu^2\frac{d^2x}{d\tau^2} + (1 - \tau)g(\tau)x = 0, \tag{2.86}$$

where $g(\tau) > 0$ is well behaved. The WKB approximation breaks down as $\tau \to 1$. The solution is oscillatory for $\tau < 1$ and real exponential for $\tau > 1$. To examine the behavior near $\tau = 1$, introduce $t = (1 - \tau)/\mu^\nu$, where ν is to be determined:

$$\frac{d^2x}{dt^2} + \mu^{3\nu-2}g(1 - \mu^\nu t)tx = 0.$$

To obtain the correct qualitative behavior near $t = 0$, we balance the two terms by choosing $\nu = \frac{2}{3}$; so we introduce $\varepsilon = \mu^{2/3}$, which gives

$$\frac{d^2x}{dt^2} + g(1 - \varepsilon t)tx = 0. \tag{2.87}$$

The solution for $\varepsilon = 0$ can be expressed in terms of Airy functions that satisfy

$$\frac{d^2F(z)}{dz^2} + zF(z) = 0. \tag{2.88}$$

This indicates that when we introduce the time scales t_0, t_1 defined in 2.76 into 2.87, we should choose $f(t_1)$ to obtain a zero-order approximation of the form (cf. 2.81)

$$[f'(t_1)]^2\frac{\partial^2x_0}{\partial t_0^2} + \frac{1}{\varepsilon}g(1 - \varepsilon t)t_1x_0 + 0.$$

In order to ensure that the solution for x_0 satisfies 2.88, rewrite this using $t_0 = f(t_1)/\varepsilon$:

$$[f'(t_1)]^2\frac{\partial^2x}{\partial t_0^2} + \left(\frac{g(1 - \varepsilon t)t_1}{f(t_1)}\right)t_0x_0 = 0.$$

Choosing f so that

$$f(f')^2 = t_1 g(1 - t_1), \qquad f^{3/2} = \frac{3}{2} \int_0^{\varepsilon t} [\sigma g(1 - \sigma)]^{1/2} \, d\sigma,$$

we obtain the zero-order approximation

$$A(t_1)F_1(t_0) + B(t_1)F_2(t_0),$$

where F_1, F_2 are independent solutions of 2.88. The solution can be completed as in Nayfeh (2, p. 286).

We have included this example to illustrate that multiple scaling can deal with problems where a variable coefficient has to be included in the zero-order solution. The preceding example can also be solved by the expansion method, assuming a solution of the form

$$X(t)F[S(t)],$$

where F is an Airy function, which can be expressed in terms of Bessel functions of order $\frac{1}{3}$, and X, S are to be determined.

E. Separating Fast and Slow Components

As should be clear from previous examples, multiple scaling gives an automatic separation of fast and slow variables "when it works." We use quotes because sometimes multiple scaling gives an obviously wrong solution or a solution that has internal contradictions even though these may be minor. We have seen examples in Sections II.B–II.D where either our choice of ε was wrong, in which case neither multiple scaling nor the expansion method would work, or we were not using the correct variant of multiple scaling. One reason the expansion method is important in this context is because it can be used to confirm multiple-scaling difficulties and/or to suggest how multiple scaling should be modified to deal with them.

Multiple scaling and the variant of the expansion method (in the way that we use it) depend heavily on the assumption that the solution can be expanded as a power series in ε, where the coefficients in the power series are also functions of ε:

$$x = x_0(t, \varepsilon) + \varepsilon x_1(t, \varepsilon) + \cdots. \tag{2.89}$$

We first make the point that even when the form 2.89 is correct, the equation as presented may not be suitable for multiple scaling. An elementary example is

$$\frac{d^2x}{d\tau^2} + \varepsilon^2 \left(2\frac{dx}{d\tau} + x \right) = 0. \tag{2.90}$$

This equation can be solved exactly, and the solution does have an expansion of the form 2.89. The dominant part of the solution is given by $\exp(\pm i\varepsilon\tau)$. Multiple scaling does not produce this directly as the reader can verify. On the other hand, the change of variable $\tau = t/\varepsilon$ reduces 2.90 to the simple linear equation with light damping (2.1) that was used earlier as a prototype example.

As a second simple example that is easily converted into a suitable form, consider the resonance situation:

$$\frac{d^2x}{dt^2} + x = \cos \omega t, \qquad x(0) = \alpha, \; x'(0) = \beta, \tag{2.91}$$

with exact solution

$$x = \alpha \cos t + \beta \sin t + \frac{1}{1-\omega^2}(\cos \omega t - \cos t). \tag{2.92}$$

Resonance occurs when $\omega = 1$. We can then either solve 2.91 directly or set $\omega = 1 + \varepsilon$ in 2.92 and let $\varepsilon \to 0$, obtaining

$$x = \alpha \cos t + (\beta + \tfrac{1}{2}t) \sin t. \tag{2.93}$$

To investigate by multiple scaling what happens for ω near 1, we introduce $\omega = 1 + \varepsilon$ into 2.91,

$$\frac{d^2x}{dt^2} + x = \cos(t + \varepsilon t). \tag{2.94}$$

Multiple scaling fails when applied to this equation as it stands. To see the source of the difficulty, we go back to 2.92, which tells us that the exact solution is of the form $x = \varepsilon^{-1}x_{-1} + x_0 + \cdots$. Hence, it is εx that is a simple power series in ε, so we should replace x in 2.94 by $y = \varepsilon x$:

$$\frac{d^2y}{dt^2} + y = \varepsilon \cos(t + \varepsilon t), \qquad y(0) = \varepsilon\alpha, \quad y'(0) = \varepsilon\beta.$$

Multiple scaling now gives

$$y_0 = Ae^{it_0} + \text{c.c.},$$

$$\frac{\partial^2 y_1}{\partial t_0^2} + y_1 = \tfrac{1}{2}e^{i(t_0 + t_1)} - 2i\frac{\partial A}{\partial t_1}e^{it_0} + \text{c.c.}$$

Eliminating secular terms, that is, terms whose dominant component is fast,

$$\frac{\partial A}{\partial t_0} = -\frac{i}{4}e^{it_1}, \qquad A = -\frac{1}{4}e^{it_1} + B + \text{c.c.}, \qquad y_1 = 0.$$

Determining the complex number B from the initial conditions,

$$x \approx \alpha \cos t + \beta \sin t - \frac{1}{2\varepsilon}[\cos(1 + \varepsilon)t - \cos t],$$

which agrees with 2.92 and 2.93 as $\varepsilon \to 0$.

The moral of these examples is that it may be necessary to rescale independent and/or dependent variables to convert a given equation into a form suitable for multiple scaling.

Small parameters in a given equation are usually identified on physical grounds. The first step is to nondimensionalize so that we can have a basis for comparison. As an example, consider an irreversible exothermic reaction modeled by reaction diffusion equations (7, p. 93):

$$\rho\frac{\partial Y}{\partial t} = \rho D_Y \nabla^2 Y - r(Y, T),$$

$$\rho C\frac{\partial T}{\partial t} = D_T \nabla^2 T - Qr(Y, T),$$

in a region Ω of space, with radiation-type conditions on the boundary of Ω. Here Y is the mass fraction of the reactant, T the temperature, $r(Y, T)$ the rate of reaction, ρ the density, C the specific heat, D_Y a diffusion rate, D_T a thermal conductivity, and Q a heat of reaction. There are also parameters connected with the boundary conditions. The situation is too complicated for us to go into detail (instead, see Ref. 7 and the quoted references).

The first step is to use dimensional analysis to reduce the parameters to comparable constants. The end result, in simplified terms, is that we are interested in situations when the diffusion of Y and T is much less important than chemical reaction diffusion measured by $r(Y, T)$, so we are concerned with the size of two rates, again simplifying the situation:

$$\varepsilon = (\text{diffusion/chemical}) \text{ rate}, \qquad \lambda = (\text{diffusion/conduction}) \text{ rate}.$$

The approximate solution is obtained on the assumption that ε is small. The analysis is carried out by multiple scaling for various λ. It turns out that for small λ there is a stable solution, but as λ increases and passes through a critical value, the perturbations evolve into a stable oscillatory solution. This is an illustration of *bifurcation* of solutions, which is another reason we quote this example.

The next point is that even if the original equations involve a small parameter μ, the solution may not depend on integral powers of μ. In order to obtain a power series of the form 2.89, we may have to introduce a parameter ε that is some function of the given small quantity μ. We have already met one example at the beginning of Section II.C, where we had to introduce $\varepsilon = \mu^{1/2}$, not the μ that appeared in the original equations.

An interesting example is the Born–Oppenheimer approximation (1, p. 145). To simplify the situation, we consider a one-dimensional model of a diatomic molecule, the two nuclei having the same mass (8, p. 75), for which the Schrödinger equation is

$$-\frac{\partial^2 \psi}{\partial x^2} - \mu \frac{\partial^2 \psi}{\partial X^2} + c^2(|x| - X)^2 \psi = \lambda \psi, \tag{2.95}$$

where X is the relative coordinate of the two nuclei, x is the coordinate of an electron relative to the center of mass, and $\mu = m/M$ is the ratio of the mass of an electron to the mass of a nuclei. We should expect that since μ multiplies a second derivative, the solution will depend on $\mu^{1/2}$. However, X has the form $X = \xi + k\zeta$, where ξ is a constant and $k\zeta$ is a small superimposed vibration reflected in the assumption that k is a small parameter multiplying the variable ζ. We then have

$$\mu \frac{\partial^2 \psi}{\partial X^2} = \left(\frac{\mu}{k^2}\right) \frac{\partial^2 \psi}{\partial \zeta^2}. \tag{2.96}$$

A key question is the relation between μ and k. One line of approach, adiabatic approximation, is to seek a solution of the form

$$\psi = \phi(x; \xi)\eta(X), \tag{2.97}$$

where ϕ is obtained by setting $\mu = 0$ and $X = \xi$ (the average value of X) in 2.95. It is then possible to obtain an equation for η on the assumption that

$k = \mu^{1/4} \ll 1$. Most arguments for this seem to be based on energy considerations rather than mathematics per se (8). Also these references need only 2.97 and do not consider the nature of the perturbation expansion.

Suppose we assume $k = \mu^{v}$, where v is to be determined, and we substitute $X = \xi + k\zeta$ in 2.95. Then the potential is a power series in μ^{v}. Also the coefficient of $\partial^2/\partial\zeta^2$ is μ^{1-2v}, and we would expect power series expansions to depend on the square root of this quantity (cf. 1, p. 146). This immediately suggests that in order to equate to zero coefficients of successive powers of a perturbation parameter, we should set $1/2(1 - 2v) = v$ (i.e., $v = 1/4$), resulting in an expansion that is a power series in $\mu^{1/4}$, as in the original Born–Oppenheimer paper (8).

We next consider an example where the definition of ε can be settled on purely mathematical grounds, namely, an equation that at first sight might seem to be innocuous:

$$(1 + \mu)\frac{d^2x}{dt^2} + 2\frac{dx}{dt} + x = 0. \tag{2.98}$$

This equation can be solved exactly, the solutions having the form $\exp(\lambda t)$ with

$$\lambda \approx -1 \pm (-\mu)^{1/2}.$$

Several points arise. If $\mu > 0$, the solutions have an oscillatory component; if $\mu < 0$, they are real exponential. If $\mu = 0$, the equation for λ has a repeated root, and the solutions are of the form

$$x = (a + bt)\exp(-t),$$

and the presence of t in the first parentheses is a danger signal indicating that the character of the solution will change when $\mu \neq 0$.

We deal with this situation by setting

$$x = Xe^{-t}.$$

If $\mu > 0$, set $\mu = \varepsilon^2$ and $\tau = \varepsilon t$; then 2.98 yields

$$(1 + \varepsilon^2)\frac{d^2X}{d\tau^2} - 2\varepsilon\frac{dX}{d\tau} + X = 0. \tag{2.99}$$

If $\mu < 0$, set $\mu = -\varepsilon^2$ and $\tau = \varepsilon t$. Then 2.98 gives

$$(1 - \varepsilon^2)\frac{d^2X}{d\tau^2} + 2\varepsilon\frac{dX}{d\tau} - X = 0. \tag{2.100}$$

These equations can be solved approximately by multiple scaling or the expansion method in a straightforward way.

In a more general context, if $a_i(\mu)$ are real polynomials in μ, and

$$a_0(\mu)\frac{d^2x}{dt^2} + a_1(\mu)\frac{dx}{dt} + a_2(\mu)x = 0,$$

the solutions can be classified into three categories depending on the value of $\Delta(\mu)$, where

$$\Delta(\mu) = a_1^2(\mu) - 4a_0(\mu)a_2(\mu).$$

This is seen by substituting $x = X \exp(\lambda t)$ and solving the resulting quadratic in μ. There are three cases depending on whether $\Delta(\mu)$ is greater than, equal to, or less than zero. The example in the last paragraph is one for which $\Delta(0) = 0$. The solution will bifurcate at $\mu = 0$ and become real exponential or oscillatory, depending on the sign of $\Delta(\mu)$, $\mu \neq 0$.

The solution of the nth-degree equation is similar but more complicated. If the solutions are of the form $X \exp(\lambda t)$, where λ is a root of $P(\lambda, \mu) = 0$, where P is a polynomial in λ, μ, then if λ_0 is a simple root of $P(\lambda, 0)$, the corresponding $\lambda(\mu)$ can be expressed as $\lambda_0 +$ power series in μ. But if λ_0 is of multiplicity m, we have

$$(\lambda - \lambda_0)^m + \mu^r f(\lambda) + \text{higher powers of } \mu = 0,$$

and the expansion of $\lambda(\mu)$ can contain mth roots of μ. The situation is stated precisely in Ref. 9 (p. 65).

We next pursue a related line of inquiry, namely, the writing of equations as a first-order system using matrices. A form that covers many of our examples is

$$\frac{d\mathbf{x}}{dt} = \mathbf{A}\mathbf{x} + \varepsilon\mathbf{P}(\mathbf{x}, \varepsilon), \tag{2.101}$$

where $\mathbf{x}$, $\mathbf{P}$ are $n \times 1$ vectors and $\mathbf{A}$ is a real constant $n \times n$ matrix. It is assumed that the elements of $\mathbf{P}$ can be expanded as power series in ε. Then the nature of the solution when $\varepsilon = 0$ depends on the eigenvalues of $\mathbf{A}$. When the eigenvalues are distinct, there will be n independent eigenvectors, and the solutions can be expanded as a power series in ε for small ε. If $\mathbf{A}$ has repeated eigenvalues, the nature of the expansion will depend on the Jordan canonical form of $\mathbf{A}$. As in the preceding brief discussion of nth-degree equations, the expansion may involve fractional powers of μ. A clear discussion of the relevant perturbation theory can be found in Wilkinson (9).

The situation of most interest in the present context occurs when $\mathbf{A}$ has rank k so that there are $n - k$ zero eigenvalues and the remaining eigenvalues have real parts less than or equal to zero. There will be $n - k$ slow solutions. We could use a similarity transformation to simplify 2.101, obtaining an equation in which $\mathbf{A}$ is replaced by its Jordan canonical form. However, an elementary and practical approach that can be extended to cover some more general nonlinear equations is the following.

The equation $\mathbf{A}^T\mathbf{u} = 0$ has $n - k$ linearly independent solutions:

$$\mathbf{A}^T\mathbf{u}_i = \mathbf{0}, \qquad i = 1, \ldots, n - k. \tag{2.102}$$

Introduce

$$y_i = \mathbf{u}_i^T\mathbf{x}, \qquad i = 1, \ldots, n - k,$$
$$z_i = \mathbf{v}_i^T\mathbf{x}, \qquad i = 1, \ldots, k, \tag{2.103}$$

where the $\mathbf{v}_i$ are any vectors such that the vectors in the set $\mathbf{u}_i$, $\mathbf{v}_i$ are independent. Let $\mathbf{y}$, $\mathbf{z}$ denote the vectors with components y_i, z_i and let $\mathbf{U}$, $\mathbf{V}$ denote matrices whose columns are the $\mathbf{u}_i$, $\mathbf{v}_i$. In this notation, 2.102 and 2.103 are

$$\mathbf{U}^T\mathbf{A} = \mathbf{0}, \qquad \mathbf{y} = \mathbf{U}^T\mathbf{x}, \qquad \mathbf{z} = \mathbf{V}^T\mathbf{x}.$$

On multiplying 2.101 by $\mathbf{U}^T$, $\mathbf{V}^T$ in turn, the resulting equations can be written in the form

$$\frac{d\mathbf{y}}{dt} = \varepsilon\mathbf{g}(\mathbf{y}, \mathbf{z}, \varepsilon), \tag{2.104a}$$

$$\frac{d\mathbf{z}}{dt} = \mathbf{C}\mathbf{y} + \mathbf{D}\mathbf{z} + \varepsilon\mathbf{h}(\mathbf{y}, \mathbf{z}, \varepsilon), \tag{2.104b}$$

which separates the fast and slow variables explicitly.

A similar procedure can sometimes be applied to the more general equation

$$\frac{d\mathbf{x}}{dt} = \mathbf{f}(\mathbf{x}, \varepsilon). \tag{2.105}$$

Suppose we can find R functions $G_r(\mathbf{x})$ such that

$$\frac{d}{dt}G_r(\mathbf{x}) = \sum_{i=1}^{n} \frac{\partial G_r(\mathbf{x})}{\partial x_i} f_i(\mathbf{x}, 0). \tag{2.106}$$

Introduce new variables:

$$y_r = G_r(\mathbf{x}), \qquad r = 1, \dots, R, \qquad z_s = H_s(\mathbf{x}), \qquad s = 1, \dots, n - R,$$

where the $H_s(\mathbf{x})$ are any functions chosen so that the set G_r, H_s is independent, and the $\mathbf{x}$ can be expressed in terms of $\mathbf{y}$ and $\mathbf{z}$. Then

$$\frac{dy_r}{dt} = \sum_{i=1}^{n} \frac{\partial G_r(\mathbf{x})}{\partial x_i} f_i(\mathbf{x}, \varepsilon).$$

If we assume that $\mathbf{f}(\mathbf{x}, \varepsilon)$ can be expanded as a power series in ε, then using 2.106, we see that 2.105 can be transformed to

$$\frac{d\mathbf{y}}{dt} = \varepsilon \mathbf{g}(\mathbf{y}, \mathbf{z}, \varepsilon), \qquad \frac{d\mathbf{z}}{dt} = \mathbf{k}(\mathbf{y}, \mathbf{z}, \varepsilon),$$

which is a generalization of 2.104. This again separates fast and slow components.

As a simple example, consider

$$\frac{dx_1}{dt} = \alpha p(x_1, x_2) + \varepsilon q(x_1, x_2, \varepsilon), \tag{2.107a}$$

$$\frac{dx_2}{dt} = p(x_1, x_2). \tag{2.107b}$$

Set $y = x_1 - \alpha x_2$, $z = x_2$ to obtain

$$\frac{dy}{dt} = \varepsilon q(y + \alpha z, z, \varepsilon), \qquad \frac{dz}{dt} = p(y + \alpha z, z). \tag{2.108}$$

In the remainder of this section we consider the elimination of fast variables, our main objective being to complement the much more detailed discussion in van Kampen (1). Even when we are interested in both fast and slow variables, the expansion method produces equations for slow components that are important if we wish to develop numerical techniques to implement the expansion method.

Consider the following equation, which is a generalization of 2.65:

$$\frac{d^2 x}{dt^2} + \frac{dx}{dt} + \varepsilon F(x) = 0, \tag{2.109}$$

the zero-order multiple scaling approximation being given by

$$x_0 = A(t_1, t_2) + B(t_1, t_2)^{-t_0}$$

and the first-order approximation by

$$\frac{\partial^2 x_1}{\partial t_0^2} + \frac{\partial x_1}{\partial t_0} = -\frac{\partial A}{\partial t_1} - \frac{\partial B}{\partial t_1} - F(A + Be^{-t_0}).$$

If we are interested in only the slow component, we can simply set $B = 0$ to find

$$\frac{\partial A}{\partial t_1} = -F(A). \tag{2.110}$$

Higher order approximations can be obtained similarly, always setting fast components equal to zero. However, there are more efficient methods for finding slow solutions, as we now discuss.

If we are interested in only the slow component, the second derivative in 2.109 will be an order smaller than the first. To make this explicit, we introduce $\tau = \varepsilon t$ and rewrite 2.109 as

$$\frac{dx}{d\tau} + F(x) = -\varepsilon\frac{d^2 x}{d\tau^2}. \tag{2.111}$$

The first approximation is obtained by setting $\varepsilon = 0$, which reproduces 2.110.

Higher approximations can be found in various ways, but we have to be careful. If we wish to obtain a single equation for the slow component of x, we can proceed as follows. From 2.111 we can deduce

$$\frac{d^2 x}{d\tau^2} = -F'\frac{dx}{d\tau} - \varepsilon\frac{d^3 x}{d\tau^3}, \tag{2.112a}$$

$$\frac{d^3 x}{d\tau^3} = -F''\left(\frac{dx}{d\tau}\right)^2 - F'\frac{d^2 x}{d\tau^2} - \varepsilon\frac{d^4 x}{d\tau^4}, \tag{2.112b}$$

where primes denote differentiation with respect to x. If one wishes to stop at this point, one can eliminate $d^2 x/d\tau^2$ and $d^3 x/d\tau^3$ from 2.111 and 2.112a, b to obtain

$$\frac{dx}{d\tau} = -F - \varepsilon FF' - \varepsilon^2[(F')^2 + (FF')] + O(\varepsilon^3). \tag{2.113}$$

The same result can be obtained by solving 2.111 iteratively. An equivalent example is considered in van Kampen (1), but there is a misprint or a slip in the ε^2 term in his (2.15).

The form 2.113 could be used directly in numerical work, but for analytical solutions we will in general still have to solve 2.113 by some approximate method. It may be preferable to set $x = x_0 + \varepsilon x_1 + \varepsilon^2 x_2 + \cdots$ in the original equation 2.111 and then solve in succession:

$$\frac{dx_0}{d\tau} + F(x_0) = 0$$

$$\frac{dx_k}{d\tau} + F'(x_0)x_k = (\cdot), \qquad k = 1, 2, \ldots,$$

where the terms in parentheses can be evaluated from previous results. One advantage of this approach is that at each step after the first, we solve equations involving the same linear operator. If this can be done analytically, this is a good way to go. It is also possible to derive 2.113 via this approach. Details are left to the reader.

Still another approach is to rewrite 2.109 as a system of first-order equations:

$$\frac{dy}{dt} = -\varepsilon F(z), \qquad (2.114a)$$

$$\frac{dz}{dt} = y - z. \qquad (2.114b)$$

If we are interested in only the slow component of z, substitute $\tau = \varepsilon t$ in 2.114, which gives

$$\frac{dy}{d\tau} = -F(z), \qquad \frac{dz}{d\tau} = \frac{y - z}{\varepsilon}. \qquad (2.115)$$

On substituting power series for y and z, expanding $F(z)$, and equating powers of ε,

$$\frac{dy_0}{d\tau} = -F_0, \qquad y_0 = z_0,$$

$$\frac{dy_1}{d\tau} = -F'_0 z_1, \qquad y_1 = z_1 + \frac{dz_0}{d\tau}, \ldots,$$

where the subscript 0 on F denotes that the function is evaluated at $z = z_0$. Eliminating y_0, y_1,

$$\frac{dz_0}{d\tau} + F_0 = 0, \qquad \frac{dz_1}{d\tau} + F_0' z_1 = -F_0 F_0'.$$

It is left to the reader to carry this out to higher orders and recover previous results by this approach.

This method can be generalized to obtain the slow components in

$$\frac{dy}{d\tau} = g(y, z, \varepsilon), \tag{2.116a}$$

$$\frac{dz}{d\tau} = \frac{1}{\varepsilon} h(y, z) + k(y, z, \varepsilon). \tag{2.116b}$$

Assuming that the solutions can be expanded in a power series in ε, the dominant term in 2.116b gives $h(y_0, z_0) = 0$, and we assume that this can be solved in the form $z_0 = \phi(y_0)$. Then the dominant term in 2.116a gives

$$\frac{dy_0}{d\tau} = g(y_0, \phi(y_0), 0),$$

from which y_0 and therefore z_0 can be determined. The method generalizes to systems of equations, where y, z are vectors, see van Kampen (1).

Whether one should use the method just described in connection with 2.114 or methods similar to those in the preceding paragraphs will depend on the specific equations to be solved.

To conclude this section, we repeat that it is not necessary to understand all the points made here in order to understand the examples in Section III. The objective in Section II is not only to explain the basic techniques but also to provide the reader with the understanding necessary to solve new problems that raise difficulties that may not be covered in Section III.

III.   EXAMPLES FROM VAN KAMPEN

The examples in this section were selected to show how the methods in Section II can be applied in more complicated situations. These examples were chosen from van Kampen (1) because his paper is a source of nontrivial problems of direct interest in connection with the Los Alamos meeting. We illustrate that many of his results are obtainable by our methods. We emphasize that although our methods are not necessarily better than van

Kampen's, they are quite different. In particular, van Kampen is concerned mainly with elimination of fast variables, whereas multiple scaling deliberately takes the fast components into account, which gives additional insight into the nature of the solutions even though we may be ultimately interested in only the slow components. The expansion method provides a convenient way of eliminating the fast variables.

A. Two Particles Moving in Two Force Fields, Bound by a Strong Potential

Consider the problem of two particles of unit mass moving in one dimension with positions x, y in force fields $K(x)$, $L(y)$ and bound by a strong potential $(x - y)^2/2\mu$. The equations of motion are (1, p. 92)

$$\frac{d^2x}{dt^2} = K(x) - \frac{1}{\mu}(x - y), \tag{3.1a}$$

$$\frac{d^2y}{dt^2} = L(y) + \frac{1}{\mu}(x - y). \tag{3.1b}$$

A closely related example but with quadratic K, L was considered at the beginning of Section II.C. Guided by this and the discussion of canonical forms in Section II.E, we find that we can simplify the algebra by introducing $u = \frac{1}{2}(x + y)$, $v = \frac{1}{2}(x - y)$, $\mu = \varepsilon^2$, which gives

$$\frac{d^2u}{dt^2} = \frac{1}{2}[K(u + v) + L(u - v)], \tag{3.2a}$$

$$\frac{d^2v}{dt^2} = \frac{1}{2}[K(u + v) - L(u - v)] - \frac{2v}{\varepsilon^2}. \tag{3.2b}$$

Guided by the analysis in Section II.C, we use the expansion method and set (cf. 2.53)

$$u = U + \varepsilon^{2+\lambda}\left[P\exp\left(\sqrt{2}\frac{it}{\varepsilon}\right) + \text{c.c.} \right], \tag{3.3a}$$

$$v = \varepsilon^2 V + \varepsilon^{\lambda}\left[Q\exp\left(\sqrt{2}\frac{it}{\varepsilon}\right) + \text{c.c.} \right]. \tag{3.3b}$$

The reasons for this choice are as follows. Applying the analysis of 2.44 to this more general case, we can show that the slow component in u is of order ε^2 less than that in v and that the fast component in v is of order ε^2 less than that in u.

However, the magnitude of the fast components relative to the slow components is undetermined in general, at least until we insert initial conditions. To make further progress, in order to be able to expand $K(u+v)$ and $L(u-v)$ as power series in U, we introduce the physical assumption that the fast component in v is of small amplitude compared with the slow component in u, so we assume $\lambda > 0$ but leave λ arbitrary to see whether it can be determined in the course of the analysis.

We substitute 3.3 in 3.2 and expand K, L as a Taylor series in ε around U. Then we equate the dominant terms of the slow and fast components. The dominant slow components give

$$\frac{d^2 U}{dt^2} = \tfrac{1}{2}[K(U) + L(U)] + \tfrac{1}{2}\varepsilon^2[K'(U) - L'(U)]V, \tag{3.4a}$$

$$V = \tfrac{1}{4}[K(U) - L(U)]. \tag{3.4b}$$

Equating terms in ε^λ, the dominant fast components give

$$P = -\tfrac{1}{4}[K'(U) - L'(U)]Q,$$

$$\frac{dQ}{dt} = -\frac{i\varepsilon}{4\sqrt{2}}[K'(U) + L'(u)]Q.$$

These results are independent of λ. The most we can say at this point is that the next terms in the Taylor series will give a slow component of order $\varepsilon^{2\lambda}$, which will cause difficulty unless 2λ is an integer.

As in Section II.E, if we were interested in only the slow solution, it could be obtainable by direct iteration or by substitution of power series in 3.2 rather than using the multiple-scaling machinery. It is now convenient to work with μ rather than ε^2. Inserting power series in μ, $u = u_0 + \mu u_1 + \cdots$, $v = v_0 + \mu v_1 + \cdots$, 3.2b gives $v_0 = 0$, so that we can expand $K(u+v)$ and $L(u-v)$ in Taylor series in ε around u_0. This leads to

$$\frac{d^2 u_0}{dt^2} = \tfrac{1}{2}[K(u_0) + L(u_0)], \tag{3.5a}$$

$$v_1 = \tfrac{1}{4}[K(u_0) - L(u_0)], \tag{3.5b}$$

$$\frac{d^2 u_1}{dt^2} = \tfrac{1}{2}(K' + L')u_1 + \tfrac{1}{8}(K' - L')(K - L), \tag{3.5c}$$

where K, K', L, and L' are evaluated at u_0. We can combine 3.5a and 3.5c to give an equation for $w = u_0 + \mu u_1$:

$$\frac{d^2w}{dt^2} = \tfrac{1}{2}[K(w) + L(w)] + \frac{\mu}{8}[K'(w) - L'(w)][K(w) - L(w)] + O(\mu^2). \quad (3.6)$$

The corresponding result in Ref. 1 [p. 94, Eq. (6.18)] has an extra term that seems to be an algebraic slip since his (6.16c), derived from his (6.11c), should have an extra term. The results 3.4 are equivalent to 3.5 and 3.6.

B.   Electron in Electric and Strong Homogeneous Magnetic Fields

The equations for the motion of an electron in the x, y plane in a homogeneous magnetic field of strength ε^{-1} in the z direction and electric fields $F(x, y)$, $G(x, y)$ in the x and y directions are (1, p. 110)

$$\frac{d^2x}{dt^2} = F(x, y) - \frac{1}{\varepsilon}\frac{dy}{dt}, \qquad \frac{d^2y}{dt^2} = G(x, y) + \frac{1}{\varepsilon}\frac{dx}{dt}. \quad (3.7)$$

A heuristic derivation of the equations for the slow motion is obtainable as follows. The dominant terms in 3.7 give

$$\frac{d^2x}{dt^2} = -\frac{1}{\varepsilon}\frac{dy}{dt}, \qquad \frac{d^2y}{dt^2} = \frac{1}{\varepsilon}\frac{dx}{dt},$$

which have solutions

$$x = X_0 + (P_0 e^{it/\varepsilon} + \text{c.c.}), \qquad y = Y_0 - (iP_0 e^{it/\varepsilon} + \text{c.c.}), \quad (3.8)$$

where X_0, Y_0 are real constants and P_0 is a complex constant.

In the actual solution X_0, Y_0 will be slowly varying functions of t [say, $X(t)$, $Y(t)$], where their derivatives are at most of order unity. If their first derivatives are of order unity, 3.7 indicates that their second derivatives are of order ε^{-1}, which cannot be true for the slow components. This observation suggests that their first and second derivatives are of orders ε and ε^2, respectively, so that to a first approximation for the slow components we can neglect the second derivatives in 3.7 and

$$\frac{dY}{dt} = \varepsilon F(X, Y), \qquad \frac{dX}{dt} = -\varepsilon G(X, Y). \quad (3.9)$$

The derivatives of these equations indicate that the second derivatives of X and Y are of second order in ε, which confirms that the second-order derivatives in 3.7 can be neglected in deriving 3.9.

A derivation of 3.9 that provides more information proceeds as follows. We assume that the constants X_0, Y_0, and P_0 in 3.8 can be replaced by slowly varying functions of t. However, the rapid components of x and y may not be as simply related as in 3.8. Also, in order to make progress mathematically in solving 3.7 and guided by one's physical intuition that we have a small oscillation superimposed on a slowly varying "guiding center," we assume that the oscillating components are of smaller order of magnitude than the fast components. So we write ($\mu > 0$)

$$x = X + \varepsilon^\mu (P e^{it/\varepsilon} + \text{c.c.}), \qquad y = Y + \varepsilon^\mu (Q e^{it/\varepsilon} + \text{c.c.}), \qquad (3.10)$$

where X, Y, P, and Q are functions of time of order unity whose derivatives are at most of order unity. Here, $\mu > 0$ is important when 3.10 is substituted in 3.7 and slow and fast parts are separated.

Omitting algebraic details, we substitute 3.9 into 3.7, expand F, G in Taylor series, and separate fast and slow components. The dominant slow terms give

$$\frac{d^2 X}{dt^2} = F(X, Y) - \frac{1}{\varepsilon} \frac{dY}{dt} + \text{terms in } \varepsilon^{2\mu} \qquad (3.11)$$

with a similar second equation. The dominant parts of the equations give 3.9. The terms in $\varepsilon^{2\mu}$ came from the second-order terms in the expansion of $F(x, y)$ and indicate that to obtain a systematic expansion, 2μ must be an (undetermined) integer. The dominant fast terms give

$$\frac{d^2 P}{dt^2} + \frac{2i}{\varepsilon} \frac{dP}{dt} - \frac{P}{\varepsilon^2} = P \frac{\partial F}{\partial x} + Q \frac{\partial F}{\partial y} - \frac{1}{\varepsilon} \left(\frac{dQ}{dt} + \frac{iQ}{\varepsilon} \right),$$

with a similar equation involving G. Both of these equations lead to the conclusion that the dominant parts of P, Q satisfy $P = iQ$, which is in agreement with the crude result 3.8.

As a check, we sketch the derivation of 3.9 by multiple scaling. Following the procedure in Section II.E for using multiple scaling to find only the slow solution, we find, as a first approximation,

$$\frac{\partial X_0}{\partial t_0} = \frac{\partial Y_0}{\partial t_0} = 0, \qquad X_0 = A(t_1), \qquad Y_0 = B(t_1).$$

Since $\partial^2 X_0 / \partial t_0^2 = \partial^2 Y_0 / \partial t_0^2 = 0$, the next approximation gives

$$\frac{\partial Y_1}{\partial t_0} = F(X_0, Y_0) - \frac{dY_0}{dt_1}, \qquad \frac{\partial X_1}{\partial t_0} = - G(X_0, Y_0) - \frac{dX_0}{dt_1}.$$

To eliminate secular terms, the right sides must be zero, which leads to 3.9. The reader will note that we have not used much of the machinery developed earlier but that this machinery would be required for higher approximations.

We make three comments in comparing our treatment with van Kampen's. The first is that we have precisely identified a basic assumption, namely, that the fast component is of lower order of magnitude than the slow. More precisely, our approximation holds for arbitrary μ in 3.10, though the larger the μ, the greater the accuracy of the approximation.

Second, we can say something about accuracy. Our derivation indicated that 3.11 is accurate to $O(\varepsilon^{2\mu})$ and $d^2 X/dt^2$ is $O(\varepsilon^2)$. If $\mu = 1$, then 3.9 is accurate to $O(\varepsilon^3)$. Methods are available to deduce an error estimate of the form X is accurate to $O(\varepsilon^p)$ over an expanding time interval of size $1/\varepsilon^q$, as discussed briefly in Section II.A. This error estimate should lead to a much better result than van Kampen's requirement (1, p. 111) that $\varepsilon t \ll 1$. Finally, we recommend that the reader compare our derivations with the quite different treatment of van Kampen, which involves the introduction of new variables v, z defined by $dx/dt = v \cos z$, $dy/dt = v \sin z$, and elimination of the fast variable z by averaging.

C. An Electron in a Strong Inhomogeneous Magnetic Field

If the magnetic field is $\varepsilon^{-1} B(x, y)$ in the z direction, the electron will move in the x, y plane with equations (1, p. 114)

$$\frac{d^2 x}{dt^2} = -\frac{1}{\varepsilon} B(x, y)\frac{dy}{dt}, \qquad \frac{d^2 y}{dt^2} = \frac{1}{\varepsilon} B(x, y)\frac{dx}{dt}. \qquad (3.12)$$

We note that if we set $t = \varepsilon\tau$, these equations become

$$\frac{d^2 x}{d\tau^2} = - B(x, y)\frac{dy}{d\tau}, \qquad \frac{d^2 y}{d\tau^2} = B(x, y)\frac{dx}{d\tau}. \qquad (3.13)$$

These are *independent* of ε, so x, y are independent of ε. So how can the solution of 3.12 depend on ε? The answer is that we shall make a *physical* assumption not present in the original equations.

The nature of this assumption can be clarified by the following argument. Suppose that $B = \text{const.} = B_0$. Then we can solve 3.12 exactly, and we find, as in the equations leading to 3.8,

$$x = X_0 + \left[P_0 \exp\left(iB_0\frac{t}{\varepsilon} \right) + \text{c.c.} \right], \qquad y = Y_0 - \left[iP_0 \exp\left(iB_0\frac{t}{\varepsilon} \right) + \text{c.c.} \right].$$

$$(3.14)$$

As in Section III.B, we argue that the solution of 3.12 must have qualitatively the same form as 3.14 except that X_0, P_0, and B_0 must be replaced by slowly varying functions of t. In order to make mathematical progress and also to support physical intuition (this is the physical assumption referred to in the last paragraph), we assume that the oscillatory component is an order of magnitude less than the slow component.

The clearest way to make this basic assumption explicit is to work with 3.13, which is independent of ε, and introduce ε by writing

$$x = X + \varepsilon\xi, \qquad y = Y + \varepsilon\eta, \tag{3.15}$$

where X, Y are slow components and ξ, η are fast. It turns out that we do not need to examine the forms of ξ, η until a late stage in the argument. Attempts to solve 3.13 as it stands by either multiple scaling or an expansion method run into troubles of a type already encountered in connection with the WKB method in Section II.D associated with a slowly varying coefficient. One way of overcoming this difficulty in Section II.D was to introduce a new variable, namely, the integral of the slowly varying function $\omega(\varepsilon t)$. A comparison with the present case suggests that we introduce the integral of $B(x, y)$, except that there is the objection that x and y have rapidly oscillating components. However, x, y have the form 3.15 where X, Y are dominant, which leads us to introduce

$$T = \int_0^\tau B(X(\sigma),\ Y(\sigma))\,d\sigma, \qquad \frac{dT}{d\tau} = B(X, Y). \tag{3.16}$$

We change the variable from τ to T in 3.13, setting $dx/d\tau = p$, $dy/d\tau = q$, which leads to

$$\frac{dx}{dT} = \frac{p}{B(X, Y)}, \qquad \frac{dy}{dT} = \frac{q}{B(X, Y)}, \tag{3.17}$$

$$\frac{dp}{dT} = -\frac{B(x, y)}{B(X, Y)}q, \qquad \frac{dq}{dT} = \frac{B(x, y)}{B(X, Y)}p. \tag{3.18}$$

It is clear that we have made the implicit assumption that $B(X, Y) \geqslant \beta > 0$ for some constant β independent of t over relevant times. Using 3.15, we see that

$$\frac{B(x, y)}{B(X, Y)} = 1 + \frac{\varepsilon}{B}\left(\frac{\partial B}{\partial X}\xi + \frac{\partial B}{\partial Y}\eta\right) + \cdots. \tag{3.19}$$

On multiplying the first equation in 3.12 by dx/dt and the second by dy/dt, adding, and integrating, we see that

$$p^2 + q^2 = \left(\frac{dx}{dt}\right)^2 + \left(\frac{dy}{dt}\right)^2 = \text{const.} = v^2. \qquad (3.25)$$

Equation 3.24 together with a similar equation derived from the second equation in 3.18 and the first in 3.17 give, remembering that $dT = B d\tau = (B/\varepsilon) dt$,

$$\frac{dY}{dT} = -\frac{\varepsilon v^2}{2B^2(X, Y)}\frac{\partial B}{\partial X}, \qquad \frac{dX}{dT} = \frac{\varepsilon B^2}{2B^2(X, Y)}\frac{\partial B}{\partial Y},$$

which agrees with van Kampen (1, p. 115).

It is instructive to compare this treatment with that of van Kampen. He starts by deriving 3.25 and makes essential use of this result immediately by setting $p = v \cos z$, $q = v \sin z$, whereas 3.25 is needed only incidentally in our derivation. He also makes an essential use of an ansatz [1, p. 112, Eq. (11.3b)] that is "justified by the result" but is not altogether obvious, though it can be justified by the arguments we used to establish the relative orders of magnitude of fast and slow terms.

D. Two Oscillators with Nonlinear Coupling

Consider two oscillators with nonlinear coupling governed by the Hamiltonian (1, p. 119)

$$H = \tfrac{1}{2}\omega_1(p_1^2 + q_1^2) + \tfrac{1}{2}\omega_2(p_2^2 + q_2^2) + \varepsilon q_1^2 q_2. \qquad (3.26)$$

This leads to the coupled equations, setting $q_1 = x$, $q_2 = y$,

$$\frac{d^2 x}{dt^2} = -\omega_1^2 x - 2\varepsilon\omega_1 xy, \qquad (3.27a)$$

$$\frac{d^2 y}{dt^2} = -\omega_2^2 y - \varepsilon\omega_2 x^2. \qquad (3.27b)$$

Multiple scaling gives, in the usual way,

$$x_0 = A(t_1)\exp(i\omega_1 t_0) + \text{c.c.}, \qquad y_0 = B(t_1)\exp(i\omega_2 t) + \text{c.c.}, \qquad (3.28a)$$

$$\frac{\partial^2 x_1}{\partial t_0^2} + \omega_1^2 x_1 = -2\left[i\omega_1\frac{\partial A}{\partial t_1}\exp(i\omega_1 t_0) + \text{c.c.}\right]$$
$$- 2\omega_1[AB\exp i(\omega_1 + \omega_2)t_0 + \bar{A}B\exp i(\omega_2 - \omega_1)t_0 + \text{c.c.}], \qquad (3.28b)$$

We next assume that

$$p = \varepsilon^\lambda P + \varepsilon\phi, \qquad q = \varepsilon^\lambda Q + \varepsilon\psi, \tag{3.2}$$

where λ is a constant to be determined and P, Q are slow and ϕ, ψ are fast. A the subsequent calculations are approximate. Substituting in 3.18, we obtai

$$\varepsilon^\lambda \frac{dP}{dT} + \varepsilon \frac{d\phi}{dT} = -\left[1 + \frac{\varepsilon}{B}\left(\frac{\partial B}{\partial X}\xi + \frac{\partial B}{\partial Y}\eta\right)\right][\varepsilon^\lambda Q + \varepsilon\psi] \tag{3.21}$$

and a similar second equation. The dominant fast terms give

$$\frac{d\phi}{dT} = -\psi, \qquad \frac{d\psi}{dT} = \phi, \qquad \phi = a\cos T + b\sin T, \qquad \psi = a\sin T - b\cos T.$$
$$\tag{3.22}$$

The dominant fast terms in 3.17 give, remembering that $B(X, Y)$ is slow:

$$\frac{d\xi}{dT} = \frac{\phi}{B(X, Y)}, \qquad \frac{d\eta}{dT} = \frac{\psi}{B(X, Y)}, \qquad \xi \approx \frac{\psi}{B(X, Y)}, \qquad \eta \approx -\frac{\phi}{B(X, Y)}. \tag{3.23}$$

We are now in a position to extract the slow terms from 3.21. The dP/dT on the left is an order of magnitude less than Q on the right. The only other slow term on the right comes from $\xi\psi$, since $\eta\psi$ is fast. Equations 3.22 and 3.23 give

$$\xi\psi = \frac{a^2 + b^2}{2B(X, Y)} + \text{fast terms.}$$

The slow terms in 3.21 then give

$$\varepsilon^\lambda Q + \frac{\varepsilon^2(a^2 + b^2)}{2B^2}\frac{\partial B}{\partial X} = 0.$$

Balancing these terms requires $\lambda = 2$. The slow terms in the second equation 3.17 give

$$\frac{dY}{dT} = \frac{\varepsilon^2 Q}{2B} = -\frac{\varepsilon^2(a^2 + b^2)}{2B^3}\frac{\partial B}{\partial X}. \tag{3.24}$$

We note that

$$\varepsilon^2(a^2 + b^2) = \varepsilon^2(\phi^2 + \psi^2) = p^2 + q^2 + O(\varepsilon^2).$$

with a similar second equation. Inspection of 3.28b indicates that if $\omega_2 \neq \omega_1$, there is a single secular term that gives $\partial A/\partial t_1 = 0$; that is, A is a function of $t_2, t_3, \ldots,$ and the resulting solution is straightforward. However, if $\omega_2 = 2\omega_1$, this solution breaks down due to a term involving $(\omega_2 - 2\omega_1)^{-1}$. From our point of view, the difficulty is reflected in the fact that there are two secular terms on the right of 3.28b if $\omega_2 = 2\omega_1$.

We could solve the problem assuming simply that $\omega_2 = 2\omega_1$ as van Kampen does, but it is instructive to assume that ω_2 is *near* ω_1 using the following standard method (e.g., see 2, p. 263, Eq. 6.2.188):

$$\omega_2 = 2\omega_1 + k\varepsilon,$$

where k is a constant of order unity. To simplify the algebra and without loss of generality, we assume $\omega_1 = 1$, $\omega_2 = 2$, so that 3.28b and the corresponding second equation become

$$\frac{\partial^2 x_1}{\partial t_0^2} + x_1 = -2\left[i\frac{\partial A}{\partial t_1} \exp it_0 + \text{c.c.} \right] - 2[AB \exp i(3t_0 + kt_1)$$
$$+ \bar{A}B \exp i(t_0 + kt_1) + \text{c.c.}], \tag{3.29a}$$

$$\frac{\partial^2 y_1}{\partial t_0^2} + 4y_1 = -2\left[2i\frac{\partial B}{\partial t_1} \exp i(2t_0 + kt_1) + \text{c.c.} \right]$$
$$- 2[2A\bar{A} + A^2 \exp 2it_0 + \text{c.c.}]. \tag{3.29b}$$

The terms $\bar{A}B \exp(ikt_1)$ in 3.29a and $\partial B/\partial t_1 \exp(ikt_1)$ in 3.29b are slowly varying so elimination of secular terms gives

$$\frac{\partial A}{\partial t_1} = i\bar{A}Be^{ikt_1}, \qquad \frac{\partial B}{\partial t_1} = \tfrac{1}{2}iA^2 e^{-ikt_1}, \tag{3.30a}$$

$$x_1 = \tfrac{1}{4}AB \exp i(3t_0 + kt_1) + \text{c.c.}, \qquad y_1 = -A\bar{A}. \tag{3.30b}$$

We note that 3.30a gives

$$A\bar{A} + 2B\bar{B} = \text{const.} \tag{3.31}$$

To derive these results by the expansion method, we start from the equations

$$\frac{d^2x}{dt^2} + x = -2\varepsilon xy, \tag{3.32a}$$

$$\frac{d^2y}{dt^2} + (2 + k\varepsilon)^2 = -\varepsilon(2 + k\varepsilon)x^2. \tag{3.32b}$$

Guided by the form of the multiple-scaling results in 3.28 and 3.30, we set

$$x = \varepsilon^2 X + Pe^{it} + \varepsilon Re^{3it} + \text{c.c.}, \qquad y = \varepsilon Y + Qe^{i(2+k\varepsilon)t} + \text{c.c.}$$

It is left to the reader to show that to a first approximation the expansion method gives results equivalent to 3.30.

To see what is involved in solving 3.30a, we set

$$A = -\frac{i}{2}ae^{i\phi}, \qquad B = -\frac{i}{2}be^{i\psi},$$

where the $-i$ has been introduced to make it easier to compare the results with those of van Kampen. On separating real and imaginary parts,

$$\frac{da}{dt} = -\tfrac{1}{2}\varepsilon ab \cos(2\phi - \psi - \varepsilon kt), \qquad \frac{d\phi}{dt} = \tfrac{1}{2}\varepsilon b \sin(2\phi - \psi - \varepsilon kt),$$

$$\frac{db}{dt} = \tfrac{1}{4}\varepsilon a^2 \cos(2\phi - \psi - \varepsilon kt), \qquad \frac{d\psi}{dt} = \tfrac{1}{4}\varepsilon\frac{a^2}{b} \sin(2\phi - \psi - \varepsilon kt).$$

These correspond to the results at resonance, $k = 0$ (1, p. 120), except that van Kampen has assumed that the times are so short that we can assume $\phi = \phi_0, \psi = \psi_0$, constants equal to the initial phases. In this case we can find a, b explicitly and show that the energy is slowly transferred to the second oscillator at a rate depending on the initial phases.

E. Kramers' Problem in Brownian Motion

The equation governing the Brownian motion of a particle in one dimension, position x, velocity p, under an external force $K(x)$ is

$$\frac{\partial}{\partial p}\left(p\rho + \frac{\partial\rho}{\partial p}\right) = \varepsilon\left\{\frac{\partial\rho}{\partial t} + p\frac{\partial\rho}{\partial x} + K(x)\frac{\partial\rho}{\partial p}\right\}, \tag{3.33}$$

where $\rho = \rho(x, p, t)$ is the probability density at time t. We show that multiple scaling gives a straightforward derivation of the equation for the density averaged over p:

$$\bar{\rho} = \bar{\rho}(x, t) = \int_{-\infty}^{\infty} \rho(x, p, t)\,dp.$$

In the usual way set $\rho = \rho_0 + \varepsilon\rho_1 + \cdots$, where it will be sufficient to assume that the ρ_i are functions of t_0, t_1 only. The dominant term in 3.33 gives

$$\frac{\partial}{\partial p}\left(p\rho_0 + \frac{\partial\rho_0}{\partial p}\right) = 0,$$

$$\rho_0 = e^{-1/2p^2}\int_0^p e^{1/2\xi^2}\,d\xi\,f(x,t_0,t_1) + e^{-1/2p^2}g(x,t_0,t_1),$$

where f and g are arbitrary functions. On physical grounds $\rho_0 \to 0$ as $p \to \pm\infty$, which implies that $f = 0$. From 3.33 the next approximation is given by

$$\frac{\partial}{\partial p}\left(p\rho_1 + \frac{\partial\rho_1}{\partial p}\right) = \frac{\partial\rho_0}{\partial t_0} + p\frac{\partial\rho_0}{\partial x} + K(x)\frac{\partial\rho_0}{\partial p} = e^{-1/2p^2}\left[\frac{\partial g}{\partial t_0} + p\left(\frac{\partial g}{\partial x} - K(x)g\right)\right].$$

(3.34)

For the same reason as $f = 0$ in the preceding, we see that $\partial g/\partial t_0 = 0$, so that $g = g(x,t_1)$. The particular solution of 3.34 is then

$$\rho_1 = pe^{-1/2p^2}\left(-\frac{\partial g}{\partial x} + K(x)g\right),$$

where, as usual, we omit the solution of the homogeneous equation. Since the integral of ρ_1 with respect to p from minus to plus infinity is zero it makes no contribution to $\bar{\rho}$. Note that ρ_0 and ρ_1 have now been expressed in terms of p, $K(x)$, and g.

The next approximation gives

$$\frac{\partial}{\partial p}\left[p\rho_2 + \frac{\partial\rho_2}{\partial p}\right] = \frac{\partial\rho_1}{\partial t_0} + \frac{\partial\rho_0}{\partial t_1} + p\frac{\partial\rho_1}{\partial x} + K(x)\frac{\partial\rho_1}{\partial p} = e^{-1/2p^2}[\cdots].$$

By the same reasoning as the foregoing, the term in parentheses on the right side must be zero so that

$$\frac{\partial g}{\partial t_1} = p^2\left(\frac{\partial^2 g}{\partial x^2} - \frac{\partial}{\partial x}[K(x)g]\right) + (1-p^2)K(x)\left[\frac{\partial g}{\partial x} - K(x)g\right]. \quad (3.35)$$

In order to find $\bar{\rho}_0$, we multiply by $\exp(-1/2p^2)$ and integrate with respect to p from $-\infty$ to $+\infty$ using

$$\int_{-\infty}^{\infty} e^{-1/2p^2}\,dp = \int_{-\infty}^{\infty} p^2 e^{-1/2p^2}\,dp,$$

which leads to the required result:

$$\frac{\partial \bar{\rho}_0}{\partial t} = \varepsilon \left(\frac{\partial^2 \bar{\rho}_0}{\partial x^2} - \frac{\partial}{\partial x} [K(x)\bar{\rho}_0] \right). \tag{3.36}$$

This would seem to be a quite elementary derivation in comparison with that of van Kampen (1, p. 144).

F. Summary and Concluding Remarks

We first summarize various points concerning the approximate solution of singular perturbation problems by multiple scaling and the related expansion method described here.

Favorable conditions for the success of multiple scaling are the following:

  i. The solution of the equations for $\varepsilon = 0$ is a reasonable approximation to the solution for $\varepsilon \neq 0$.

 ii. The solution is a power series of the form $(t_k = \varepsilon^k t)$

$$x = x_0(t_0, t_1, \ldots) + \varepsilon x_1(t_0, t_1, \ldots) + \cdots.$$

iii. The differential equations to be solved successfully for $x_0, x_1, \ldots$ are constant coefficient.

 iv. Suitable qualitative information is known about the solution, for example, to justify the suppression of secular terms.

It is usually obvious when straightforward multiple scaling fails. For instance, the elimination of secular terms may lead to the nonsensical conclusion that the solution is identically zero. A number of examples where multiple scaling breaks down have been examined in Section II, with suggested remedies. Usually the trouble could be spotted and fixed by going through criteria i–iv. The only additional comment made here is that multiple scaling will be troublesome if the character of the solution changes over the range of the independent variable. An example due to Lighthill (2, p. 80) is

$$(t + \varepsilon x)\frac{dx}{dt} + (2 + t)x = 0, \qquad x(1) = e^{-1}.$$

The solution behaves like $(3/\varepsilon)^{2/3}$ near $t = 0$. One would expect trouble for small t due to the occurrence of $t + \varepsilon x$ multiplying the derivative. Another example quoted in Bender and Orszag (6, p. 568) is an equation where the difficulty is due to the difference in the behavior of $\exp(-\varepsilon t)$ for small and large t.

$$\frac{d^2 y}{dt^2} + e^{-\varepsilon t} y = 0, \qquad y(0) = 0, \qquad y'(0) = 1.$$

This can be solved exactly in terms of Bessel functions, and the difficulty is explained by the change in character of $Y_0(z)$, depending on whether z is large or small. Still another example is the van der Pol equation for *large* damping, examined via the equivalent Rayleigh equation where the asymptotic expansion of the period of the limit cycle is given, as in Ref. 6 (p. 477), as

$$T(\varepsilon) = 3 - 2\ln 2 - \alpha\varepsilon^{2/3} + \tfrac{1}{6}\varepsilon\ln\varepsilon + \cdots.$$

The moral is to be alert to the fact that difficulties like these can arise even in apparently innocuous examples.

Success of the expansion method as described in this chapter is connected with the success or failure of multiple scaling. In particular, it is necessary to know the form of the expansion with fast terms exhibited in explicit form (e.g., as exponentials) but in more complicated situations via known functions (e.g., Airy functions in turning-point problems). We, of course, have relied on multiple scaling to provide us with the form of the expansion, but any method for finding the correct form would be acceptable.

We next make some comments comparing the two methods. Multiple scaling automatically separates the fast and slow variables. The expansion method assumes that we know at least the qualitative way in which fast and slow components are represented in the solution. We indicated in Section II.E how multiple scaling could be used to eliminate fast variables, but often this is unnecessarily laborious, and the expansion method is easier.

The algebra in both methods can be tedious, but it can be handled by computer algebra programs like MACSYMA and REDUCE. Multiple scaling requires somewhat more sophisticated facilities for handling partial differentiation and solving differential equations than the expansion method. Our interest in the subject originated from the realization that the algebra could be done by computer, and we hope to return to this subject.

The expansion method would seem to be well suited for numerical solutions of highly oscillatory problems where we are interested in the envelope and the "guiding-center," not the details of the phase. This again lies outside the scope of this chapter.

To relate the two methods described here to others in the literature, we remind the reader that there are two broad classes of problem depending on widely differing time and/or length scales. If the fast solutions are real exponential, the most commonly used methods are described by terms like *boundary layer theory*, *matched asymptotic expansions*, and *composite expansions*. If the fast solutions are oscillatory, the *method of averaging* and *multiple scaling* are widely used. An introduction to all these methods can be found in Nayfeh (2). We have made no mention of the multiple-scaling method that Nayfeh calls the two-variable expansion and only incidentally mention the generalized method. The expansion method described here has connections with some methods in Nayfeh, but this seems difficult to pin down in detail; it is also related to harmonic balance methods in oscillatory problems.

After this chapter was completed, we saw the paper by Rubenfeld (10), which has points in common with our treatment of difficulties in connection with multiple scaling in Section II. Interestingly enough, this led him to introduce what he calls a derivative-expansion method, not to be confused with the multiple-scaling method with exactly the same name in Nayfeh (2), which is the method we use in this chapter. Rubenfeld's expansion method and ours are quite different.

It is fitting to end with the following quotations. The first consists of parts of two sentences from the last paragraph of Rubenfeld's paper (10): "In conclusion we emphasize that the purpose here [in introducing the expansion method] was not to cast doubt upon the usefulness of the multiscale technique," followed by the hope that more applied mathematicians will add his "expansion procedure to their satchel of mathematical methods, with its contents to be used interchangeably and together." The second is from Sherlock Holmes (quoted in 6, p. 484). "When you follow two separate chains of thought, Watson, you will find some point of intersection that should approximate the truth." Multiple scaling and the expansion method involve two separate chains of thought, once the form of the expansion has been determined by multiple scaling or otherwise. A third chain of thought is provided by van Kampen's approach, provided one is interested only in the elimination of fast variables.

BIBLIOGRAPHICAL REMARKS

Apart from van Kampen (1), most of our references do not deal with physical applications. The main reason is that the literature is so enormous and scattered that it would require a separate survey article to do justice to actual and possible applications in physics and chemistry. References to research papers on the problems dealt with in Section III can be found in van Kampen (1).

The object of these notes is to draw the reader's attention to some additional mathematical references particularly relevant to the topics in this chapter.

The review by O'Malley (11) of two books that we reference (12, 13) contains a concise historical review of recent developments in singular perturbation methods. The Kevorkian–Cole book (12) is particularly useful for advanced work in various branches of classical applied mathematics. The Nayfeh book Ref. 13 is a helpful introduction to the more advanced Ref. 2.

We next list a small number of books and articles that cover topics not dealt with in any detail, if at all, in the main text.

Nayfeh (14) is a comprehensive introduction to nonlinear oscillations. Jeffrey and Kawahara (15) and Dodd et al. (16) give an introduction to nonlinear wave theory. Particularly interesting from our point of view is the simultaneous use of multiple scaling in both space and time variables. There is a limited treatment of this in Nayfeh (2). Schuss (17) is a review paper on singular perturbation methods in stochastic differential equations in mathematical physics. Kokotovic (18) is a review paper on applications in control theory. All of these papers contain excellent bibliographies.

Note added in proof: As a sequel to this paper we have shown how the expansion method deals with forcing terms, both initial transients and steady state, and how it can be implemented numerically and by computer symbol manipulation. In this context the expansion method is closely related to harmonic balance; then the expansion (harmonic balance) method can sometimes deal with *large* perturbations, in particular when the perturbation parameters are so large that bifurcation occurs. This work (with J. J. Wu) is being submitted to J. Sound and Vibration.

Acknowledgment

We express our sincere thanks to Joe Hirschfelder, who provided the stimulus and motivation for this chapter, and to Jagdish Chandra, U.S. Army Research Office, for this continued encouragement.

References

1. N. G. van Kampen, "Elimination of Fast Variables," *Phys. Rep.* **124**, 69–160 (1985).
2. A. H. Nayfeh, *Perturbation Methods*, Wiley, New York, 1973.
3. D. R. Smith, *Singular Perturbation Theory*, Cambridge University Press, New York, 1985.
4. M. D. Kruskal and H. Segur, Technical Memo 85-25, Institute of Theoretical Physics, University of California, Santa Barbara, CA 93106.
5. R. E. O'Malley, *Introduction to Singular Perturbation*, Academic, New York, 1973.
6. C. M. Bender and S. A. Orszag, *Advanced Mathematical Methods for Scientists and Engineers*, McGraw-Hill, New York, 1978.
7. A. K. Kapila, *Asymptotic Treatment of Chemically Reacting Systems*, Pitman, Boston, 1983.
8. E. Merzbacher, *Quantum Mechanics*, Wiley, New York, 1970. See also A. S. Davidov, *Quantum Mechanics*, Oxford University Press, New York, 1965. The original paper is M. Born and J. R. Oppenheimer, *Ann. Phys.* **84**, 457 (1927).
9. J. H. Wilkinson, *The Algebraic Eigenvalue Problem*, Cambridge University Press, New York, 1965.
10. L. A. Rubenfeld, "On a Derivative Expansion Technique...," *SIAM Rev.* **20**, 79–105 (1978).

11. R. E. O'Malley, "Book Review," *Bull. Am. Math. Soc.* **7**, 414–410 (1982).

12. J. Kevorkian and J. D. Cole, *Perturbation Methods in Applied Mathematics*, Springer-Verlag, New York, 1981.

13. A. H. Nayfeh, *Introduction to Perturbation Methods*, Wiley, New York, 1981.

14. A. H. Nayfeh and D. T. Mook, *Nonlinear Oscillations*, Wiley, New York, 1980.

15. A. Jeffrey and T. Kawahara, *Asymptotic Methods in Nonlinear Wave Theory*, Pitman, Boston, 1982.

16. R. K. Dodd, J. C. Eilbeck, J. D. Gibbon, and H. C. Morris, *Solitons and Nonlinear Wave Theory*, Academic, New York, 1982.

17. Z. Schuss, *SIAM Rev.* **22**, 119–155 (1980).

18. P. V. Kokotovic, *SIAM Rev.* **26**, 501–549 (1984).

CHAPTER III

APPLICATION OF THE METHOD OF MULTIPLE SCALES TO NONLINEARLY COUPLED OSCILLATORS

ALI H. NAYFEH

Department of Engineering Science and Mechanics, Virginia Polytechnic Institute and State University, Blacksburg, Virginia 24061

CONTENTS

In this chapter, we investigate the response of two-degree-of-freedom systems with quadratic nonlinearities to parametric and external resonant excitations in the presence of two-to-one internal (autoparametric) resonances. We use the method of multiple scales to construct a first-order uniform expansion yielding four first-order nonlinear ordinary differential (averaged) equations governing

137

the modulation of the amplitudes and the phases of the two modes. These equations are used to determine periodic responses and their stability. The autoparametric resonance produces a strong coupling of the modes involved. For some range of parameters, Hopf bifurcations exist. The fixed points of the averaged equations lose their stability when the real part of a complex-conjugate pair of eigenvalues changes sign from negative to positive. In these ranges, steady-state periodic solutions do not exist. Instead, the response consists of amplitude- and phase-modulated motion, and for small damping it may experience period-multiplying bifurcations and chaos.

I. INTRODUCTION

The present chapter is concerned with the use of the method of multiple scales for the determination of the response of nonlinearly coupled oscillators to parametric and external excitations. Parametric excitations result in time-dependent coefficients (parameters) in the governing equations and boundary conditions, and external excitations result in inhomogeneous terms. To describe the method without resorting to lengthy algebra, we restrict the discussion to the determination of uniform first-order expansions of the response of two-degree-of-freedom systems having quadratic nonlinearities to parametric and external resonances in the presence of two-to-one internal (sometimes called autoparametric) resonances; that is, when $\omega_2 \approx 2\omega_1$, where ω_1 and ω_2 are the linear natural frequencies of the system. The following brief discussion of representative examples provides background for the present chapter. For a comprehensive review of the application of the method of multiple scales to single- and multi-degree-or-freedom systems, the reader is referred to the textbooks by Nayfeh (1–3) and Nayfeh and Mook (4). For a comprehensive review of parametric resonances, we refer the reader to the textbooks by Evan-Iwanowski (5), Nayfeh and Mook (4), Ibrahim (6), and Schmidt and Tondl (7).

The present problem describes the forced response of many elastic and dynamic systems, such as ships, robots, elastic pendulums, beams and plates under static loadings, arches, composite plates, shells, and the sloshing of liquids in containers. All lead to systems of coupled, inhomogeneous ordinary differential equations with quadratic nonlinearities and sometimes time-dependent coefficients. The solutions of such systems exhibit particularly complicated long-time behavior when their linear undamped natural frequencies are commensurable (4–7), that is, when these systems possess internal (autoparametric) resonances. As early as 1863, Froude (8) observed that ships have undesirable roll characteristics when the frequency of a small, free oscillation in pitch is twice the frequency of a small, free oscillation in roll, a condition called internal resonance in the recent literature of nonlinear oscillations (4–7). The significance of this frequency ratio cannot be determined from the linearized equations governing the motion of a ship with six degrees of freedom, because the yaw, sway, and roll modes are not coupled with the pitch, heave, and surge modes.

Mettler and Weidenhammer (9), Sethna (10), and Haxton and Barr (11) used the method of averaging to analyze primary resonances of systems governed by equations with quadratic nonlinearities when one linear natural frequency is twice another. Nayfeh, Mook, and Marshall (12) and Mook, Marshall, and Nayfeh (13) used the method of multiple scales to analyze a simple system of two coupled oscillators with quadratic nonlinearities as a model for the coupling of pitch and roll motions of ships. They investigated both primary and secondary resonances. When $\omega_2 \approx 2\omega_1$ and $\Omega \approx \omega_2$, where the ω_n are the linear natural frequencies and Ω is the excitation frequency, they demonstrated the existence of a saturation phenomenon. Moreover, when $\omega_2 \approx 2\omega_1$ and $\Omega \approx \omega_1$, they showed that there are conditions for which stable periodic steady-state motions do not exist. Instead, there exist amplitude- and phase-modulated motions in which the energy is continuously exchanged between the two modes. Both phenomena are discussed later in this chapter.

Later, Yamamoto and co-workers (14, 15) used the method of harmonic balance and analog simulation to investigate the forced response of systems with quadratic and cubic nonlinearities to a harmonic excitation when one frequency is twice the other. They observed amplitude- and phase-modulated steady-state motions in their analog computer simulation both when $\Omega \approx \omega_2$ and $\Omega \approx \omega_1$. Nayfeh and Mook (4) used the method of multiple scales to analyze the response of a beam to a harmonic excitation. They accounted for the interaction of lateral and longitudinal vibrations. Hatwal, Mallik, and Ghosh (16) reported analytical and numerical results for the response of two internally resonant coupled oscillators to a harmonic excitation when the excitation frequency Ω is near the large frequency ω_2. Their results for small amplitudes are equivalent to those of Sethna (10) and Nayfeh, Mook, and Marshall (12). However, their numerical results for sufficiently large amplitudes show that periodic responses are unstable and give way to periodically modulated motions. Hatwal et al. (17) reported experimental data and further numerical results that demonstrate chaotic motions. Haddow, Barr, and Mook (18) conducted an experiment using a two-degree-of-freedom model consisting of two beams and two concentrated masses and observed the saturation phenomenon when $\Omega \approx \omega_2$ and the nonexistence of periodic motions when $\Omega \approx \omega_1$. Miles (19) used the method of averaging to investigate the response of two internally resonant quadratically coupled oscillators to a harmonic excitation. He examined the stability of the analytical solutions and investigated the possible bifurcations. He presented numerical results that demonstrate chaotic and periodically modulated motions when the excitation frequency Ω is near the lower frequency ω_1. Using a model similar to that of Haddow, Barr, and Mook (18), Nayfeh and Zavodney (20) observed amplitude- and phase-modulated motions when $\Omega \approx \omega_1$. Mook et al. (21, 22) used the method of multiple scales to investigate the influence of a two-to-one internal resonance on the response of a system with quadratic nonlinearities to a combination resonance (i.e., $\Omega \approx \omega_2 + \omega_1$) and a subharmonic resonance of the higher mode (i.e., $\Omega \approx 2\omega_2$). They applied the results to an arch and found that the internal resonance significantly reduces the response. Nayfeh and Raouf (23, 24) used the method of multiple scales to investigate the response of a circular cylindrical shell to a harmonic internal pressure when the natural frequency ω_2 of the breathing mode is twice the natural frequency ω_1 of a flexural mode. They demonstrated the saturation phenomenon when $\Omega \approx \omega_2$ as well as the nonexistence of steady-state periodic motions when $\Omega \approx \omega_2$ or $\Omega \approx \omega_1$. Nayfeh (25) used the method of multiple scales to analyze the response of a ship that is restrained to pitch and roll only to a primary excitation. He found conditions for the existence of Hopf bifurcations (periodic motions lose their stability when the real part of a complex conjugate pair of eigenvalues changes sign from negative to positive) when the two frequencies are in the ratio of 2:1 and when $\Omega \approx \omega_2$. Nayfeh and Mook (26) demonstrated the existence of the saturation phenomenon for three-degree-of-freedom systems with quadratic nonlinearities when $\omega_3 \approx \omega_1 + \omega_2$ and $\Omega \approx \omega_3$.

Nayfeh (27) analyzed the response of two-degree-of-freedom systems with quadratic nonlinearites to a principal parametric resonance ($\Omega \approx 2\omega_2$) in the presence of two-to-one internal resonances (i.e., $\omega_2 \approx 2\omega_1$) and found that the second mode may become saturated. He also found multiple stable steady-state solutions leading to the jump phenomenon. Nayfeh (28) also analyzed the response of multi-degree-of-freedom systems with quadratic nonlinearities to a principal parametric resonance in the presence of combination internal resonances. Again, he found that under certain conditions the directly excited mode becomes saturated and that multiple stable steady-state solutions exist. Miles (29, 30) analyzed the response of an internally resonant pendulum and the sloshing of liquid in a circular container when $\omega_2 = 2\omega_1$ and the lower mode is excited by a principal parametric resonance ($\Omega \approx 2\omega_1$). He found that Hopf bifurcations do not exist in this case. Nayfeh (31, 32) used the method of multiple scales to analyze the problems studied by Miles (29, 30) and relaxed his assumption of perfectly tuned internal resonance (i.e., he put $\omega_2 = 2\omega_1 + \varepsilon\sigma_1$). He found Hopf bifurcations and calculated responses with period-multiplying bifurcations, leading to chaos. Nayfeh and Zavodney (33) analyzed the response of internally resonant two-degree-of-freedom systems with quadratic nonlinearities to a combination parametric resonance ($\Omega \approx \omega_2 + \omega_1$ and $\omega_2 \approx 2\omega_1$). They determined steady-state solutions and their stability. They presented numerical results that demonstrate limit-cycle responses and amplitude- and phase-modulated responses.

II. PROBLEM FORMULATION

We consider the nonlinear response of internally resonant two-degree-of-freedom systems with quadratic nonlinearities to harmonic parametric and external excitations. We limit our discussion to first-order expansions for the interesting case of a two-to-one internal (autoparametric) resonance, that is, $\omega_2 \approx 2\omega_1$, where ω_1 and ω_2 are the linear natural frequencies of the system. The autoparametric resonance provides nonlinear coupling of the two oscillators, thereby producing interesting phenomena. We let u_1 and u_2 be two generalized coordinates describing the motion of the system and assume that its kinetic energy can be expressed as a polynomial having quadratic and cubic terms of the form

$$T = \tfrac{1}{2}M_1(1 + m_1u_1 + m_2u_2)\dot{u}_1^2 + \tfrac{1}{2}M_2(1 + m_3u_1 + m_4u_2)\dot{u}_2^2$$
$$+ M_3u_1\dot{u}_1\dot{u}_2 + M_4u_2\dot{u}_1\dot{u}_2 + \cdots \tag{2.1}$$

where the M_n and m_n are constants. Moreover, we assume that the potential energy has the form

$$V = V_1u_1^2 + V_2u_2^2 + V_3u_1^3 + V_4u_1^2u_2 + V_5u_1u_2^2 + V_6u_2^3 + \cdots \tag{2.2}$$

where the V_n are constants. These forms of the kinetic and potential energies yield linearly uncoupled, nonlinearly coupled equations governing the two modes of oscillation. We consider the effect of weak damping described by the dissipation function

$$D = \tfrac{1}{2}d_1\dot{u}_1^2 + \tfrac{1}{2}d_2\dot{u}_2^2 \tag{2.3}$$

where d_1 and d_2 are constants. Using Lagrange's equations and considering simultaneous harmonic parametric and external excitations, we obtain

$$\ddot{u}_1 + \omega_1^2u_1 + \varepsilon[2\mu_1\dot{u}_1 + \delta_1u_1^2 + \delta_2u_1u_2 + \delta_3u_2^2 + \delta_4\dot{u}_1^2$$
$$+ \delta_5\dot{u}_1\dot{u}_2 + \delta_6\dot{u}_2^2 + \delta_7u_1\ddot{u}_1 + \delta_8u_2\ddot{u}_1 + \delta_9u_1\ddot{u}_2 + \delta_{10}u_2\ddot{u}_2 + \cdots$$
$$+ (F_{11}u_1 + F_{12}u_2)\cos\Omega_1t]$$
$$= F_1\cos(\Omega_2t + \tau_1) \tag{2.4}$$

$$\ddot{u}_2 + \omega_2^2u_2 + \varepsilon[2\mu_2\dot{u}_2 + \alpha_1u_1^2 + \alpha_2u_1u_2 + \alpha_3u_2^2 + \alpha_4\dot{u}_1^2$$
$$+ \alpha_5\dot{u}_1\dot{u}_2 + \alpha_6\dot{u}_2^2 + \alpha_7u_1\ddot{u}_1 + \alpha_8u_2\ddot{u}_1 + \alpha_9u_1\ddot{u}_2 + \alpha_{10}u_2\ddot{u}_2 + \cdots$$
$$+ (F_{21}u_1 + F_{22}u_2)\cos(\Omega_1t + \tau)]$$
$$= F_2\cos(\Omega_2t + \tau_2) \tag{2.5}$$

where

$$\{\omega_1^2, \varepsilon[2\mu_1, \delta_1, \delta_2, \delta_3, \delta_4, \delta_5, \delta_6, \delta_7, \delta_8, \delta_9, \delta_{10}]\}$$

$$= \left\{\frac{V_1}{M_1}, \frac{d_1}{M_1}, \frac{3V_3}{M_1}, \frac{2V_4}{M_1}, \frac{V_5}{M_1}, \frac{1}{2}m_1, m_2, \frac{M_4 - (1/2)m_3 M_2}{M_1}, m_1, m_2, \frac{M_3}{M_1}, \frac{M_4}{M_1}\right\}$$

(2.6)

$$\{\omega_2^2, \varepsilon[2\mu_2, \alpha_1, \alpha_2, \alpha_3, \alpha_4, \alpha_5, \alpha_6, \alpha_7, \alpha_8, \alpha_9, \alpha_{10}]\}$$

$$= \left\{\frac{V_2}{M_2}, \frac{d_2}{M_2}, \frac{V_4}{M_2}, \frac{2V_5}{M_2}, \frac{3V_6}{M_2}, \frac{M_3 - (1/2)m_2 M_1}{M_2}, m_3, \frac{1}{2}m_4, \frac{M_3}{M_2}, \frac{M_4}{M_2}, m_3, m_4\right\}$$

(2.7)

The F_n, F_{mn}, Ω_n, and τ_n are constants. The parameter ε is a small dimensionless parameter that is introduced as a bookkeeping device in the following perturbation analysis. The double pendulum treated by Miles (29) is governed by equations that are a special case of Eqs. 2.4 and 2.5. Another special case of Eqs. 2.4 and 2.5 governs the response of a ship constrained to pitch and roll (25) if we put $\delta_1 = \delta_3 = \delta_4 = \delta_6 = \delta_7 = \delta_{10} = \alpha_2 = \alpha_5 = \alpha_8 = \alpha_9 = 0$.

III.　THE CASE OF PARAMETRIC RESONANCE

In this section, the method of multiple scales is used to construct a first-order uniform expansion of the response of Eqs. 2.4 and 2.5 to principal (i.e., $\Omega \approx 2\omega_2$ or $2\omega_1$) and combination (i.e., $\Omega_1 \approx \omega_2 + \omega_1$ or $\omega_2 - \omega_1$) parametric resonances. Four first-order nonlinear ordinary differential equations are derived to describe the modulation of the amplitudes and the phases of the two modes of oscillation. These equations are used to determine the fixed points and their stability. As the amplitude of the excitation or one of the detuning parameters is varied, a fixed point may lose its stability in one of two ways: A real eigenvalue changes sign from negative to positive or the real part of a pair of complex-conjugate eigenvalues changes sign from negative to positive. The first instability corresponds to the jump phenomenon, whereas the second instability corresponds to the extensively studied Hopf bifurcation (34). Based on the Hopf bifurcation theorem (35), the averaged equations are expected to possess limit-cycle solutions, and hence the system is expected to possess amplitude- and phase-modulated responses near the Hopf bifurcation values.

When the higher mode is excited by a principal parametric resonance (i.e., $\Omega_1 \approx 2\omega_2$), the nontrivial steady-state value of its amplitude is a constant that is independent of the excitation amplitude, whereas the amplitude of the lower mode, which is indirectly excited through the internal resonance, increases with the amplitude of the excitation. However, in addition to the jump phenomenon, the steady-state amplitudes exhibit a Hopf bifurcation leading to amplitude- and phase-modulated motions. When the lower mode is excited by a principal parametric resonance ($\Omega_1 \approx 2\omega_1$), the steady-state amplitudes also exhibit both the jump phenomenon and Hopf bifurcations. In some intervals of the parameters, the periodic solutions of the averaged equations, in the latter case, experience period-doubling bifurcations, leading to chaos.

In the case of combination parametric resonance, the effects of detuning both the internal and parametric resonances and varying the phase and magnitude of the second-mode parametric excitation and the initial conditions are investigated. The first-order perturbation solution predicts trivial and nontrivial stable steady-state solutions and illustrates both the quenching and saturation phenomena. In addition to the steady-state solutions, other periodic solutions are predicted by the amplitude and phase modulation equations. These equations predict a transition from constant steady-state nontrivial responses to limit-cycle responses. Some limit cycles are also shown to experience period-doubling bifurcations. The perturbation solutions were verified by numerically integrating the governing differential equations (33).

In this case, $F_n = 0$. Following the method of multiple scales, we assume that an approximation to the solutions of Eqs. 2.4 and 2.5 can be expressed in the form

$$u_n(t; \varepsilon) = u_{n0}(T_0, T_1, T_2, \ldots) + \varepsilon u_{n1}(T_0, T_1, T_2, \ldots) + \varepsilon^2 u_{n2}(T_0, T_1, T_2, \ldots)$$

$$(3.1)$$

where the u_{nm} are functions of the time scales $T_n = \varepsilon^n t$ and do not depend explicitly on ε. Here, $T_0 = t$, a fast scale characterizing motions with the frequencies ω_1, ω_2, and Ω_n, and $T_1 = \varepsilon t$, $T_2 = \varepsilon^2 t$, $T_3 = \varepsilon^3 t, \ldots$, slow scales characterizing the modulations of the amplitudes and phases. Since the original independent variable (scale) is replaced with multiple scales, the derivatives with respect to t become expansions in terms of the partial derivatives with respect to the T_n as

$$\frac{d}{dt} = D_0 + \varepsilon D_1 + \varepsilon^2 D_2 + \cdots \tag{3.2}$$

$$\frac{d^2}{dt^2} = D_0^2 + 2\varepsilon D_0 D_1 + \varepsilon^2 (2D_0 D_2 + D_1^2) + \cdots \tag{3.3}$$

where $D_n = \partial/\partial T_n$. The number of scales needed to form the approximation is determined by the number of terms or the order of the approximation. A first-order expansion requires two scales, a second-order expansion requires three scales, and so on. In this chapter, we restrict ourselves to a first-order expansion and hence use the two scales T_0 and T_1.

Substituting Eqs. 3.1–3.3 into Eqs. 2.4 and 2.5, recalling the fact that $F_n = 0$, and equating like powers of ε, we obtain:

Order ε^0
___

$$D_0^2 u_{10} + \omega_1^2 u_{10} = 0 \tag{3.4}$$

$$D_0^2 u_{20} + \omega_2^2 u_{20} = 0 \tag{3.5}$$

Order ε
___

$$\begin{aligned}
D_0^2 u_{11} + \omega_1^2 u_{11} = {} & -2D_0 D_1 u_{10} - 2\mu_1 D_0 u_{10} - \delta_1 u_{10}^2 - \delta_2 u_{10} u_{20} - \delta_3 u_{20}^2 \\
& - \delta_4 (D_0 u_{10})^2 - \delta_5 (D_0 u_{10})(D_0 u_{20}) - \delta_6 (D_0 u_{20})^2 \\
& - \delta_7 u_{10} D_0^2 u_{10} - \delta_8 u_{20} D_0^2 u_{10} - \delta_9 u_{10} D_0^2 u_{20} \\
& - \delta_{10} u_{20} D_0^2 u_{20} - (F_{11} u_{10} + F_{12} u_{20}) \cos \Omega_1 T_0 \tag{3.6}
\end{aligned}$$

$$\begin{aligned}
D_0^2 u_{21} + \omega_2^2 u_{21} = {} & -2D_0 D_1 u_{20} - 2\mu_2 D_0 u_{20} - \alpha_1 u_{10}^2 - \alpha_2 u_{10} u_{20} - \alpha_3 u_{20}^2 \\
& - \alpha_4 (D_0 u_{10})^2 - \alpha_5 (D_0 u_{10})(D_0 u_{20}) - \alpha_6 (D_0 u_{20})^2 \\
& - \alpha_7 u_{10} D_0^2 u_{10} - \alpha_8 u_{20} D_0^2 u_{10} - \alpha_9 u_{10} D_0^2 u_{20} - \alpha_{10} u_{20} D_0^2 u_{20} \\
& - (F_{21} u_{10} + F_{22} u_{20}) \cos (\Omega_1 T_0 + \tau) \tag{3.7}
\end{aligned}$$

The solutions of Eqs. 3.4 and 3.5 can be expressed in either the form

$$u_{10} = a_1(T_1) \cos [\omega_1 T_0 + \beta_1(T_1)] \tag{3.8}$$

$$u_{20} = a_2(T_1) \cos [\omega_2 T_0 + \beta_2(T_1)] \tag{3.9}$$

or in the form

$$u_{10} = A_1(T_1) e^{i\omega_1 T_0} + \text{c.c.} \tag{3.10}$$

$$u_{20} = A_2(T_1) e^{i\omega_2 T_0} + \text{c.c.} \tag{3.11}$$

where c.c. is the complex conjugate of the preceding terms and $\bar{A}_n$ is the complex conjugate of A_n. Comparing Eqs. 3.8 and 3.9 with Eqs. 3.10 and 3.11, we conclude that the amplitudes a_n and phases β_n are related to the A_n by

$$A_n = \tfrac{1}{2} a_n e^{i\beta_n} \quad \text{for } n = 1, 2 \tag{3.12}$$

The a_n, β_n, and A_n are undetermined functions of T_1 at this level of approximation; they are determined by imposing the solvability conditions at the next level of approximation.

Substituting Eqs. 3.10 and 3.11 into Eqs. 3.6 and 3.7 yields

$$
\begin{aligned}
D_0^2 u_{11} + \omega_1^2 u_{11} = & -2i\omega_1(A_1' + \mu_1 A_1)e^{i\omega_1 T_0} \\
& - [\delta_2 + \delta_5\omega_1\omega_2 - \delta_8\omega_1^2 - \delta_9\omega_2^2]A_2\bar{A}_1 e^{i(\omega_2 - \omega_1)T_0} \\
& + \Lambda_3 A_1^2 e^{2i\omega_1 T_0} + \Lambda_4 A_2^2 e^{2i\omega_2 T_0} + \Lambda_5 A_2 A_1 e^{i(\omega_2 + \omega_1)T_0} \\
& + \Lambda_6 A_1\bar{A} + \Lambda_7 A_2\bar{A}_2 - \tfrac{1}{2}F_{11}A_1 e^{i(\Omega_1 + \omega_1)T_0} \\
& - \tfrac{1}{2}F_{11}\bar{A}_1 e^{i(\Omega_1 - \omega_1)T_0} - \tfrac{1}{2}F_{12}A_2 e^{i(\Omega_1 + \omega_2)T_0} \\
& - \tfrac{1}{2}F_{12}\bar{A}_2 e^{i(\Omega_1 - \omega_2)T_0} + \text{c.c.}
\end{aligned}
\tag{3.13}
$$

$$
\begin{aligned}
D_0^2 u_{21} + \omega_2^2 u_{21} = & -2i\omega_2(A_2' + \mu_2 A_2)e^{i\omega_2 T_0} \\
& - [\alpha_1 - \alpha_4\omega_1^2 - \alpha_7\omega_1^2]A_1^2 e^{2i\omega_1 T_0} + \Lambda_8 A_2^2 e^{2i\omega_2 T_0} \\
& + \Lambda_9 A_2\bar{A}_1 e^{i(\omega_2 - \omega_1)T_0} + \Lambda_{10} A_2 A_1 e^{i(\omega_2 + \omega_1)T_0} + \Lambda_{11} A_1\bar{A}_1 \\
& + \Lambda_{12} A_2\bar{A}_2 - \tfrac{1}{2}F_{21}A_1 e^{i[(\Omega_1 + \omega_1)T_0 + \tau]} \\
& - \tfrac{1}{2}F_{21}\bar{A}_1 e^{i[(\Omega_1 - \omega_1)T_0 + \tau]} - \tfrac{1}{2}F_{22}A_2 e^{i[(\Omega_1 + \omega_2)T_0 + \tau]} \\
& - \tfrac{1}{2}F_{22}\bar{A}_2 e^{i[(\Omega_1 - \omega_2)T_0 + \tau]} + \text{c.c.}
\end{aligned}
\tag{3.14}
$$

where the Λ_n are constants that are functions of the δ_n, α_n, and ω_n. Depending on the functions A_n, a particular solution of Eqs. 3.13 and 3.14 may contain secular terms of the form $T_0 \exp(\pm i\omega_1 T_0)$ and $T_0 \exp(\pm i\omega_2 T_0)$ and, in the case of resonance, small-divisor terms. To first order, in addition to the internal or autoparametric resonance $\omega_2 \approx 2\omega_1$, these resonant conditions are (a) $\Omega_1 \approx 2\omega_1$, principal parametric resonance of the first mode; (b) $\Omega_1 \approx 2\omega_2$, principal parametric resonance of the second mode; and (c) $\Omega_1 \approx \omega_2 \pm \omega_1$, combination parametric resonances of the additive and difference types. These cases are treated in the following three sections.

A. Principal Parametric Resonance of Second Mode

According to the method of multiple scales (1–3), we need to introduce detuning parameters σ_1 and σ_2, which convert the small-devisor terms to secular terms in the scale T_0. In this case, we let

$$
\omega_2 = 2\omega_1 + \varepsilon\sigma_1 \quad \text{and} \quad \Omega_1 = 2\omega_2 + \varepsilon\sigma^2
$$

and write

$$
\omega_2 T_0 = 2\omega_1 T_0 + \sigma_1 T_1 \quad \text{and} \quad \Omega_1 T_0 = 2\omega_2 T_0 + \sigma_2 T_1
\tag{3.15}
$$

Substituting Eqs. 3.15 into Eqs. 3.13 and 3.14 and eliminating the terms that produce secular terms in u_{11} and u_{21}, we obtain the solvability conditions

$$2i(A_1' + \mu_1 A_1) + 4\Lambda_1 A_2 \bar{A}_1 e^{i\sigma_1 T_1} = 0 \tag{3.16}$$

$$2i(A_2' + \mu_2 A_2) + 4\Lambda_2 A_1^2 e^{-i\sigma_1 T_1} + 2g\bar{A}_2 e^{i(\sigma_2 T_1 + \tau)} = 0 \tag{3.17}$$

where $F_{22} = 4\omega_2 g$ and

$$4\omega_1 \Lambda_1 = \delta_2 + \delta_5 \omega_1 \omega_2 - \delta_8 \omega_1^2 - \delta_9 \omega_2^2 \tag{3.18}$$

$$4\omega_2 \Lambda_2 = \alpha_1 - \alpha_4 \omega_1^2 - \alpha_7 \omega_1^2 \tag{3.19}$$

Substituting Eqs. 3.12 into Eqs. 3.16 and 3.17 and separating real and imaginary parts, we obtain

$$a_1' + \mu_1 a_1 + \Lambda_1 a_1 a_2 \sin \gamma_1 = 0 \tag{3.20}$$

$$a_2' + \mu_2 a_2 - \Lambda_2 a_1^2 \sin \gamma_1 + g a_2 \sin \gamma_2 = 0 \tag{3.21}$$

$$-a_1 \beta_1' + \Lambda_1 a_1 a_2 \cos \gamma_1 = 0 \tag{3.22}$$

$$-a_2 \beta_2' + \Lambda_2 a_1^2 \cos \gamma_1 + g a_2 \cos \gamma_2 = 0 \tag{3.23}$$

where

$$\gamma_1 = \beta_2 - 2\beta_1 + \sigma_1 T_1 \quad \text{and} \quad \gamma_2 = \sigma_2 T_1 - 2\beta_2 + \tau \tag{3.24}$$

Periodic solutions of the system correspond to the fixed points (i.e., constant solutions) of Eqs. 3.20–3.24, which in turn correspond to $a_n' = 0$ and $\gamma_n' = 0$. It follows from Eqs. 3.24 that

$$\beta_1' = \tfrac{1}{2}(\sigma_1 + \tfrac{1}{2}\sigma_2) \quad \text{and} \quad \beta_2' = \tfrac{1}{2}\sigma_2 \tag{3.25}$$

Hence, the fixed points of Eqs. 3.20–3.24 are given by

$$\mu_1 a_1 + \Lambda_1 a_1 a_2 \sin \gamma_1 = 0 \tag{3.26}$$

$$\mu_2 a_2 - \Lambda_2 a_1^2 \sin \gamma_1 + g a_2 \sin \gamma_2 = 0 \tag{3.27}$$

$$-\tfrac{1}{2}(\sigma_1 + \tfrac{1}{2}\sigma_2)a_1 + \Lambda_1 a_1 a_2 \cos \gamma_1 = 0 \tag{3.28}$$

$$-\tfrac{1}{2}\sigma_2 a_2 + \Lambda_2 a_1^2 \cos \gamma_1 + g a_2 \cos \gamma_2 = 0 \tag{3.29}$$

There are two possibilities. Either

$$a_1 = a_2 = 0 \tag{3.30}$$

or

$$a_2 = [\mu_1^2 + \tfrac{1}{4}(\sigma_1 + \tfrac{1}{2}\sigma_2)^2]^{1/2}/|\Lambda_1| \qquad (3.31)$$

$$a_1 = \left[-\chi_1 \pm \left(\frac{g^2 a_2^2}{\Lambda_2^2} - \chi_2^2 \right)^{1/2} \right]^{1/2} \qquad (3.32)$$

where

$$\chi_1 = [4\mu_1\mu_2 - \sigma_2(\sigma_1 + \tfrac{1}{2}\sigma_2)]/4\Lambda_1\Lambda_2 \qquad (3.33)$$

$$\chi_2 = [2\mu_1\sigma_2 + \mu_2(\sigma_2 + 2\sigma_1)]/4\Lambda_1\Lambda_2 \qquad (3.34)$$

Therefore, to the first approximation, the response is trivial or is given by

$$u_1 = a_1 \cos(\tfrac{1}{4}\Omega_1 t - \tfrac{1}{2}\gamma_1 - \tfrac{1}{4}\gamma_2 + \tfrac{1}{4}\tau) + \cdots \qquad (3.35)$$

$$u_2 = a_2 \cos(\tfrac{1}{2}\Omega_1 t - \tfrac{1}{2}\gamma_2 + \tfrac{1}{2}\tau) + \cdots \qquad (3.36)$$

where a_1 and a_2 are defined in Eqs. 3.31 and 3.32 and γ_1 and γ_2 can be obtained from Eqs. 3.26 and 3.27.

We note an interesting feature of the response. The amplitude a_2 of the directly excited second mode is either trivial or a constant value given by Eq. 3.31 that is independent of the excitation amplitude g. On the other hand, the amplitude a_1 of the indirectly excited first mode is either trivial or varies with the amplitude of the excitation, as in Eq. 3.32.

Next, we determine the conditions under which Eq. 3.32 has real roots. To this end, we define the two critical values ζ_1 and ζ_2 of g as follows:

$$\zeta_1 = |\chi_2\Lambda_2|/a_2 \quad \text{and} \quad \zeta_2 = [(\chi_1^2 + \chi_2^2)\Lambda_2^2]^{1/2}/a_2 = (\mu_2^2 + \tfrac{1}{4}\sigma_2^2)^{1/2} \qquad (3.37)$$

When $\chi_1 > 0$, there are no real roots when $g < \zeta_2$ and there is one real root when $g \geqslant \zeta_2$. When $\chi_1 < 0$, there are no real roots when $g < \zeta_1$, there is one real root when $g > \zeta_2$, and there are two real roots when $\zeta_1 \leqslant g \leqslant \zeta_2$.

To determine the local stability of the various fixed points and hence the various periodic solutions, we let

$$A_1 = \tfrac{1}{2}(p_1 - iq_1)e^{i[\nu_1 T_1 + (1/4)\tau]} \quad \text{and} \quad A_2 = \tfrac{1}{2}(p_2 - iq_2)e^{i[\nu_2 T_1 + (1/2)\tau]} \qquad (3.38)$$

where the p_n and q_n are real, and

$$\nu_1 = \tfrac{1}{2}(\sigma_1 + \tfrac{1}{2}\sigma_2) \quad \text{and} \quad \nu_2 = \tfrac{1}{2}\sigma_2 \qquad (3.39)$$

in Eqs. 3.16 and 3.17, separate real and imaginary parts, and obtain

$$p_1' + \mu_1 p_1 + \nu_1 q_1 + \Lambda_1(p_2 q_1 - p_1 q_2) = 0 \qquad (3.40)$$

$$q_1' + \mu_1 q_1 - \nu_1 p_1 + \Lambda_1(p_1 p_2 + q_1 q_2) = 0 \tag{3.41}$$

$$p_2' + \mu_2 p_2 + \nu_2 q_2 - 2\Lambda_2 p_1 q_1 + g q_2 = 0 \tag{3.42}$$

$$q_2' + \mu_2 q_2 - \nu_2 p_2 + \Lambda_2(p_1^2 - q_1^2) + g p_2 = 0 \tag{3.43}$$

The stability of a particular fixed point with respect to a perturbation proportional to $\exp(\lambda T_1)$ is determined by the zeros of the characteristic equation

$$
\begin{vmatrix}
\lambda + \mu_1 - \Lambda_1 q_2 & \nu_1 + \Lambda_1 p_2 & \Lambda_1 q_1 & -\Lambda_1 p_1 \\
-\nu_1 + \Lambda_1 p_2 & \lambda + \mu_1 + \Lambda_1 q_2 & \Lambda_1 p_1 & \Lambda_1 q_1 \\
-2\Lambda_2 q_1 & -2\Lambda_r p_1 & \lambda + \mu_2 & \nu_2 + g \\
2\Lambda_2 p_1 & -2\Lambda_2 q_1 & -\nu_2 + g & \lambda + \mu_2
\end{vmatrix} = 0 \tag{3.44}
$$

To analyze the stability of the nontrivial solution, we put $p_n = q_n = 0$ in Eq. 3.44 and obtain

$$[(\lambda + \mu_1)^2 + \nu_1^2][(\lambda + \mu_2)^2 + \nu_2^2 - g^2] = 0 \tag{3.45}$$

or

$$\lambda = -\mu_1 \pm i\nu_1, \quad -\mu_2 \pm (g^2 - \nu_2^2)^{1/2} \tag{3.46}$$

Hence, the necessary and sufficient condition for the stability of the trivial solution is

$$g < \zeta_2 = (\mu_2^2 + \tfrac{1}{4}\sigma_2^2)^{1/2} \tag{3.47}$$

To analyze the stability of the nontrivial solution, we use Eqs. 3.26–3.29, 3.31, and 3.32 in Eq. 3.44 and obtain

$$\lambda^4 + r_1 \lambda^3 + r_2 \lambda^2 + r_3 \lambda + r_4 = 0 \tag{3.48}$$

where

$$r_1 = 2(\mu_1 + \mu_2)$$

$$r_2 = 4\mu_1 \mu_2 + \mu_2^2 + \nu_2^2 - g^2 + 4\Lambda_1 \Lambda_2 a_1^2$$

$$r_3 = 2\mu_1(\mu_2^2 + \nu_2^2 - g^2) + 4(\mu_1 + \mu_2)\Lambda_1 \Lambda_2 a_1^2$$

$$r_4 = 8\Lambda_1 \Lambda_2 a_1^2(\Lambda_1 \Lambda_2 a_1^2 + \mu_1 \mu_2 - \nu_1 \nu_2)$$

According to the Routh–Hurwitz criterion, the necessary and sufficient conditions for all the roots of Eq. 3.48 to possess negative real parts are

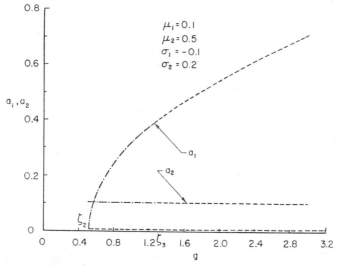

Figure 1. Variation of the amplitudes a_1 and a_2 of the first and second modes with the amplitude g of the excitation for the case of principal parametric resonance of the second mode when there is only one finite-amplitude solution: Solid curves, stable; ———, unstable with at least one eigenvalue being positive; —·—·—, unstable with the real part of a complex-conjugate pair of eigenvalues being positive.

that

$$r_1 r_2 - r_3 > 0 \qquad r_3(r_1 r_2 - r_3) - r_1^2 r_4 > 0 \qquad r_4 > 0 \qquad (3.49)$$

As an example, we let $\Lambda_1 = 1.0$, $\Lambda_2 = 0.5$, $\mu_1 = 0.1$, $\mu_2 = 0.5$, $\sigma_1 = -0.1$, and $\sigma_2 = 0.2$. Figure 1 shows the variation of the steady-state amplitudes a_1 and a_2 with g. In this case, there are two possible solutions: the trivial solution, which is unstable for all values of $g > \zeta_2 \approx 0.5099$, and a nontrivial solution for which $a_2 = 0.1$ for all $g \geqslant \zeta_2$ and a_1 varies with g as in Eq. 3.32. However, the nontrivial solution is unstable for all values of $g \geqslant \zeta_2$ according to the Routh–Hurwitz criterion. Consequently, periodic motions of the type sought (Eqs. 3.35 and 3.36) cannot be realized. As g increases through ζ_2, the nontrivial solution becomes unstable with a pair of complex roots of Eq. 3.48 crossing the imaginary axis into the right half of the complex plane. This stability transition corresponds to the extensively studied problem of Hopf bifurcation (34). Based on the Hopf bifurcation theorem (35), we expect the modulation Eqs. 3.20–3.24 or 3.40–3.43 to exhibit limit cycle oscillations for values of g larger than but close to ζ_2, leading to amplitude- and phase-modulated motions. For $g > \zeta_3 \approx 1.302$, the roots of Eq. 3.48 are real with at

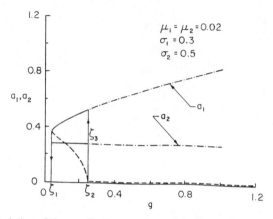

Figure 2. Variation of the amplitudes a_1 and a_2 of the first and second modes with the amplitude g of the excitation for the case of principal parametric resonance of the second mode when there are two finite-amplitude solutions: Solid curves, stable; ———, unstable with at least one eigenvalue being positive; —·—·—, unstable with the real part of a complex-conjugate pair of eigenvalues being positive.

least one of them being positive. Thus, the special points $g = \zeta_2$ and $g = \zeta_3$ are bifurcation points.

Figure 2 shows the variation of the amplitudes a_1 and a_2 with g for $\mu_1 = \mu_2 = 0.02$, $\sigma_1 = 0.3$, and $\sigma_2 = 0.5$. In this case, χ_1 is negative, and hence, Eq. 3.32 has one real root when $g > \zeta_2 \approx 0.2508$ and has two real roots when $\zeta_1 \leqslant g \leqslant \zeta_2$, where $\zeta_1 \approx 0.03808$. The stability analysis shows that the small finite-amplitude solution is unstable, whereas the large finite-amplitude solution is stable for all values of $g \leqslant \zeta_3$, where $\zeta_3 \approx 0.2523$. When $\zeta_3 < g < \zeta_4$, where $\zeta_4 \approx 2.85$, Eq. 3.48 has a pair of complex roots with a positive real part. Again, in this case, steady-state periodic responses do not exist, and for small damping the response may experience period-multiplying bifurcations and chaos. When $g > \zeta_4$, the fixed points are unstable and at least one of the roots of Eq. 3.48 is a positive real quantity. At $g = \zeta_4$, the imaginary part of a complex-conjugate pair with a positive real part vanishes.

Figure 3 shows the frequency response curves for $\sigma_1 = 0$ (i.e., $\omega_2 = 2\omega_1$, a perfectly tuned internal resonance), $g = 0.7$, $\mu_1 = 0.3$, and $\mu_2 = 0.2$. Whereas a_2 is a single-valued function of σ_2, a_1 is a multivalued function of σ_2. Figure 3 also exhibits a jump phenomenon at $\sigma^{(1)} \approx -2.169$, $\sigma^{(2)} \approx -1.342$, $\sigma^{(5)} \approx 1.342$, and $\sigma^{(6)} \approx 2.169$. Moreover, as σ_2 increases beyond $\sigma^{(3)} \approx -1.185$, the nontrivial solution loses its stability with a complex pair of eigenvalues of Eq. 3.48 crossing the imaginary axis into the right-half plane of the complex plane, a Hopf bifurcation. Again, for values of σ_2 larger than but near $\sigma^{(3)}$, the

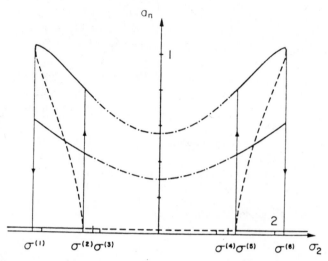

Figure 3. Frequency response curves for the case of principal parametric resonance of the second mode for $\Lambda_1 = 1.0$, $\Lambda_2 = 0.5$, $\mu_1 = 0.3$, $\mu_2 = 0.2$, $\sigma_1 = 0.0$, and $g = 0.7$: Solid curves, stable; ----, unstable with the real part of a complex-conjugate pair of eigenvalues being positive.

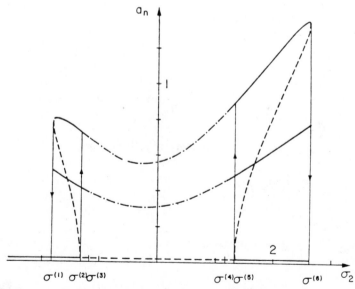

Figure 4. Frequency response curves for the case of principal parametric resonance of the second mode for $\Lambda_1 = 1.0$, $\Lambda_2 = 0.5$, $\mu_1 = 0.3$, $\mu_2 = 0.2$, $\sigma_1 = 0.1$, and $g = 0.7$: Solid curves, stable; ---, unstable with at least one eigenvalue being positive; –·–·–, unstable with the real part of a complex-conjugate pair of eigenvalues being positive.

151

modulation equations are expected to exhibit limit cycle oscillations, producing amplitude- and phase-modulated motions (i.e., doubly periodic motion or motion on a two-torus). Also, as σ_2 decreases below $\sigma^{(4)} \approx 1.185$, the nontrivial solution loses its stability with a complex pair of the eigenvalues of Eq. 3.44 crossing the imaginary axis into the right half of the complex plane, a Hopf bifurcation. Again, when σ_2 is less than but near $\sigma^{(4)}$, the modulation equations are expected to exhibit limit cycle oscillations, yielding amplitude- and phase-modulated motions.

When $\sigma_1 \neq 0$, the frequency response curves lose their symmetry. Figure 4 shows the frequency response curves for $\sigma_1 = 0.1$. In this case, the jumps occur at $\sigma^{(1)} \approx -1.799$, $\sigma^{(2)} \approx -1.342$, $\sigma^{(5)} \approx 1.342$, and $\sigma^{(6)} \approx 2.680$. The Hopf bifurcation points become $\sigma^{(3)} \approx -1.192$ and $\sigma^{(4)} \approx 1.181$.

B. Principal Parametric Resonance of First Mode

In this case, we put

$$\Omega_1 = 2\omega_1 + \varepsilon\sigma_2 \tag{3.50}$$

Then, eliminating the terms that produce secular terms from Eqs. 3.13 and 3.14, we obtain

$$2i(A'_1 + \mu_1 A_1) + 4\Lambda_1 A_2 \bar{A}_1 e^{i\sigma_1 T_1} + 2f\bar{A}_1 e^{i\sigma_2 T_1} = 0 \tag{3.51}$$

$$2i(A'_2 + \mu_2 A_2) + 4\Lambda_2 A_1^2 e^{-i\sigma_1 T_1} = 0 \tag{3.52}$$

where $F_1 = 4\omega_1 f$ and Λ_1 and Λ_2 are defined in Eqs. 3.18 and 3.19. Substituting Eqs. 3.12 into Eqs. 3.51 and 3.52 and separating real and imaginary parts, we obtain

$$a'_1 = -\mu_1 a_1 - \Lambda_1 a_1 a_2 \sin\gamma_1 - fa_1 \sin\gamma_2 \tag{3.53}$$

$$a'_2 = -\mu_2 a_2 + \Lambda_2 a_1^2 \sin\gamma_1 \tag{3.54}$$

$$a_1 \beta'_1 = \Lambda_1 a_1 a_2 \cos\gamma_1 + fa_1 \cos\gamma_2 \tag{3.55}$$

$$a_2 \beta'_2 = \Lambda_2 a_1^2 \cos\gamma_1 \tag{3.56}$$

where

$$\gamma_1 = \sigma_1 T_1 + \beta_2 - 2\beta_1 \quad \text{and} \quad \gamma_2 = \sigma_2 T_1 - 2\beta_1 \tag{3.57}$$

Periodic solutions correspond to the fixed points of Eqs. 3.53–3.57, which in turn correspond to $a'_n = 0$ and $\gamma'_n = 0$. Then, it follows from Eqs. 3.57

that

$$\beta'_1 = v_1 = \tfrac{1}{2}\sigma_2 \quad \text{and} \quad \beta'_2 = v_2 = \sigma_2 - \sigma_1 \tag{3.58}$$

Hence, the fixed points of Eqs. 3.53–3.57 are given by

$$\mu_1 a_1 + \Lambda_1 a_1 a_2 \sin \gamma_1 + f a_1 \cos \gamma_2 = 0 \tag{3.59}$$

$$\mu_2 a_2 - \Lambda_2 a_1^2 \sin \gamma_1 = 0 \tag{3.60}$$

$$v_1 a_1 - \Lambda_1 a_2 a_2 \cos \gamma_1 - f a_1 \cos \gamma_2 = 0 \tag{3.61}$$

$$v_2 a_2 - \Lambda_2 a_1^2 \cos \gamma_1 = 0 \tag{3.62}$$

There are two possibilities: either

$$a_1 = a_2 = 0 \tag{3.63}$$

or

$$a_1^2 = \chi_1 \pm \left[\frac{f^2 [\mu_2^2 + (\sigma_2 - \sigma_1)^2]}{\Lambda_1^2 \Lambda_2^2} - \chi_2^2 \right]^{1/2} \tag{3.64}$$

$$a_2 = |\Lambda_2| a_1^2 [\mu_2^2 + (\sigma_2 - \sigma_1)^2]^{-1/2} \tag{3.65}$$

where

$$\chi_1 = [\tfrac{1}{2}\sigma_2(\sigma_2 - \sigma_1) - \mu_1 \mu_2]/\Lambda_1 \Lambda_2 \tag{3.66}$$

$$\chi_2 = [\tfrac{1}{2}\sigma_2 \mu_2 + \mu_1(\sigma_2 - \sigma_1)]/\Lambda_1 \Lambda_2 \tag{3.67}$$

When $\chi_1 < 0$, Eq. 3.64 has one real root for all $f > \zeta_2$, where

$$\zeta_2 = (\mu_1^2 + v_1^2)^{1/2} \tag{3.68}$$

When $\chi_1 > 0$, Eq. 3.64 has one real root for all $f > \zeta_2$ and two real roots for all $\zeta_1 \leqslant f \leqslant \zeta_2$, where

$$\zeta_1 = \Lambda_1 \Lambda_2 |\chi_2| [\mu_2^2 + (\sigma_2 - \sigma_1)^2]^{-1/2} \tag{3.69}$$

To determine the stability of the fixed points, we substitute Eqs. 3.38, where v_1 and v_2 are defined in Eqs. 3.58, into Eqs. 3.51 and 3.52, separate real and imaginary parts, and obtain

$$p'_1 + \mu_1 p_1 + v_1 q_1 + \Lambda_1(p_2 q_1 - p_1 q_2) + f q_1 = 0 \tag{3.70}$$

$$q_1' + \mu_1 q_1 - v_1 p_1 + \Lambda_1(p_1 p_2 + q_1 q_2) + f p_1 = 0 \qquad (3.71)$$

$$p_2' + \mu_2 p_2 + v_2 q_2 - 2\Lambda_2 p_1 q_1 = 0 \qquad (3.72)$$

$$q_2' + \mu_2 q_2 - v_2 p_2 + \Lambda_2(p_1^2 - q_1^2) = 0 \qquad (3.73)$$

Then, the stability of a particular fixed point with respect to a perturbation proportional to $\exp(\lambda T_1)$ is determined by the zeros of the characteristic equation

$$\begin{vmatrix} \lambda + \mu_1 - \Lambda_1 q_2 & v_1 + \Lambda_1 p_2 + f & \Lambda_1 q_1 & -\Lambda_1 p_1 \\ v_1 + \Lambda_1 p_2 + f & \lambda + \mu_1 + \Lambda_1 q_2 & \Lambda_1 p_1 & \Lambda_1 q_1 \\ -2\Lambda_2 q_1 & -2\Lambda_2 p_1 & \lambda + \mu_2 & v_2 \\ 2\Lambda_2 p_1 & -2\Lambda_2 q_1 & -v_2 & \lambda + \mu_2 \end{vmatrix} = 0 \qquad (3.74)$$

or

$$\lambda_1^2 \lambda_2^2 + v_2^2 \lambda_1^2 + \lambda_2^2 [v_1^2 - (\Lambda_1 p_2 + f)^2 - \Lambda_1^2 q_2^2] + 4\lambda_1 \lambda_2 \Lambda_1 \Lambda_2 a_1^2 + v_1^2 v_2^2$$
$$- v_2^2 [(\Lambda_1 p_2 + f)^2 + \Lambda_1^2 q_2^2] - 4 v_1 v_2 \Lambda_1 \Lambda_2 a_1^2 + 4\Lambda_1^2 \Lambda_2^2 a_1^4$$
$$= 0 \qquad (3.75)$$

where $\lambda_n = \lambda + \mu_n$ and $a_n^2 = p_n^2 + q_n^2$.

To investigate the stability of the trivial solution, we put $p_n = q_n = 0$ in Eq. 3.75 and obtain

$$[(\lambda + \mu_2)^2 + v_2^2][(\lambda + \mu_1)^2 + v_1^2 - f^2] = 0$$

whose solutions are

$$\lambda = -\mu_2 \pm iv_2, \ -\mu_1 \pm (f^2 - v_1^2)^{1/2} \qquad (3.76)$$

Hence, the trivial solution is stable if and only if

$$f \leqslant \zeta_2 = (\mu_1^2 + v_1^2)^{1/2} \qquad (3.77)$$

which is condition 3.68 for the existence of one real root of Eq. 3.64.

To investigate the stability of the nontrivial fixed points, we combine the steady-state Eqs. 3.70 and 3.71 into

$$(\Lambda_1 p_2 + f)^2 + \Lambda_1^2 q_2^2 = \mu_1^2 + v_1^2 \qquad (3.78)$$

Using Eq. 3.78, we rewrite Eq. 3.75 as

$$\lambda^4 + 2(\mu_1 + \mu_2)\lambda^3 + (4\mu_1\mu_2 + \mu_2^2 + \nu_2^2 + 4\Lambda_1\Lambda_2 a_1^2)\lambda^2$$
$$+ [2\mu_1(\mu_2^2 + \nu_2^2) + 4(\mu_1 + \mu_2)\Lambda_1\Lambda_2 a_1^2]\lambda$$
$$+ 4\Lambda_1\Lambda_2 a_1^2(\Lambda_1\Lambda_2 a_1^2 + \mu_1\mu_2 - \nu_1\nu_2)$$
$$= 0 \tag{3.79}$$

Substituting for the δ_n and α_n from Eqs. 2.6 and 2.7 into Eqs. 3.18 and 3.19, one finds that $\Lambda_1\Lambda_2 > 0$. Hence, Eq. 3.79 has real positive roots if and only if

$$\Lambda_1\Lambda_2 a_1^2 + \mu_1\mu_2 - \nu_1\nu_2 < 0 \tag{3.80}$$

which, in conjunction with Eqs. 3.64 and 3.66, implies that the fixed point corresponding to the positive sign in Eq. 3.64 is stable and that corresponding to the negative sign is unstable. Moreover, Eq. 3.79 has a pair of complex-conjugate roots with a positive part unless

$$r_3(r_1 r_2 - r_3) - r_1^2 r_4 > 0 \tag{3.81}$$

where r_1, r_2, r_3, and r_4 are, respectively, the coefficients of $\lambda^3, \lambda^2, \lambda$, and λ^0 in Eq. 3.79.

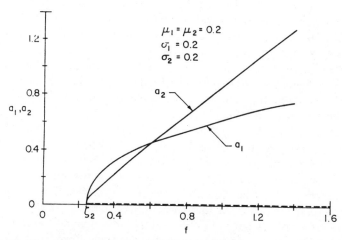

Figure 5. Variation of the amplitudes a_1 and a_2 of the first and second modes with the amplitude f of the excitation for the case of principal parametric resonance of the first mode when there is only one finite-amplitude solution: Solid curves, stable; ———, unstable with at least one eigenvalue being positive.

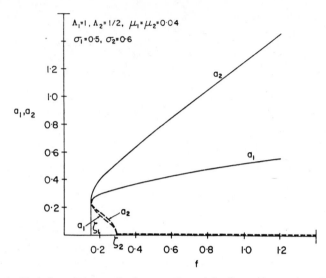

Figure 6. Variation of the amplitudes a_1 and a_2 of the first and second modes with the amplitude f of the excitation for the case of principal parametric resonance of the first mode when there are two finite-amplitude solutions: Solid curves, stable; ---, unstable with at least one eigenvalue being positive.

As an example, we let $\Lambda_1 = 1.0$, $\Lambda_2 = 0.5$, $\mu_1 = \mu_2 = 0.2$, $\sigma_1 = 0.2$, and $\sigma_2 = 0.3$. In this case, $\chi_1 < 0$, and hence Eq. 3.64 has only one solution. Figure 5 shows the variation of a_1 and a_2 with f. As f increases from zero, the response is trivial because $a_1 = a_2 = 0$ is the only possible fixed point, which is stable. As f increases beyond $\zeta_2 = 0.25$, the trivial solution becomes unstable, and Eq. 3.64 has a nontrivial solution, which is stable. Consequently, the response is nontrivial. Figure 5 shows that for $f > 0.7$, the amplitude a_2 of the indirectly excited mode is larger than the amplitude a_1 of the directly excited mode. The critical value $f = \zeta_2$ is a bifurcation point at which one of the real roots of Eq. 3.79 changes sign, and there is a change in the number of fixed points.

Figure 6 shows the variation of a_1 and a_2 with f when $\Lambda_1 = 1.0$, $\Lambda_2 = 0.5$, $\mu_1 = \mu_2 = 0.04$, $\sigma_1 = 0.5$, and $\sigma_2 = 0.6$. In this case, $\chi_1 > 0$, and hence Eq. 3.64 has one stable real solution when $f > \zeta_2 \approx 0.3027$ and two real solutions, the larger of which is stable, when $\zeta_1 \leqslant f \leqslant \zeta_2$, where $\zeta_1 = 0.1486$. Hence, as f increases from zero, the response remains trivial until the bifurcation value $f = \zeta_2$ is reached. As f increases beyond ζ_2, a_1 and a_2 jump up, and the response is nontrivial and the trivial response is unstable. The value $f = \zeta_2$ is a bifurcation point at which there is a change in the number of fixed points. If the amplitude of the excitation is set at a value in the interval (ζ_1, ζ_2), the response is trivial or nontrivial depending on the initial conditions. This is an example of subcritical

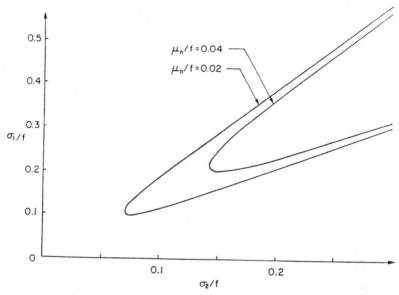

Figure 7. Transition curves across which the real part of a complex-conjugate pair of eigenvalues changes sign (it is positive inside the curves) for $\mu_n/f = 0.02, 0.04$. The case of principal parametric resonance of the first mode.

instability. Again the critical value $f = \zeta_1$ is a bifurcation point at which there is a change in the number of fixed points.

Miles (29, 30) investigated the case of principal parametric resonance of the first mode when $\omega_2 = 2\omega_1$ and found that Hopf bifurcations do not exist in this case. However, the analysis of Miles is for the case of perfectly tuned internal resonance (i.e., $\sigma_1 = 0$). When we allow σ_1 to be different from zero, we find conditions for the existence of Hopf bifurcations (31, 32). When $\mu_n/f = 0.02$ and 0.04, Figure 7 shows two curves corresponding to Hopf bifurcations. As the parameters σ_1/f and σ_2/f are varied to cross the transition curves in Figure 7, the nontrivial solution loses its stability with a complex pair of the roots of Eq. 3.79 crossing the imaginary axis into the right half of the complex plane. These transition curves correspond to the extensively studied problem of Hopf bifurcations (34). Based on the Hopf bifurcation theorem (35), we expect the modulation equations to exhibit limit cycle oscillations for values of σ_1/f and σ_2/f inside but close to the transition curves in Figure 7. Consequently, steady-state periodic motions do not exist. Instead, the motion consists of amplitude- and phase-modulated oscillations; in other words, the motion is doubly periodic (motion on a two-torus). An example is shown in Figure 8 for $\sigma_2/f = 0.16$ and $\sigma_1/f = 0.205$, which shows a projection of the

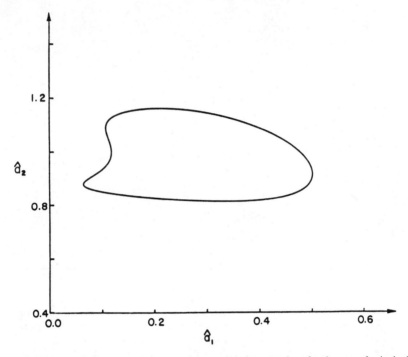

Figure 8. Projection of the trajectory of the modulation equations for the case of principal parametric resonance of the first mode on the $\hat{a}_2$-$\hat{a}_1$ plane, where $\hat{a}_1 = a_1\sqrt{\Lambda_1\Lambda_2}/f$ and $\hat{a}_2 = a_2\Lambda_1/f$, for $\mu_n/f = 0.02$, $\sigma_1/f = 0.205$, and $\sigma_2/f = 0.16$. It exhibits a limit cycle oscillation.

trajectory on the plane $\hat{a}_2 - \hat{a}_1$, where $\hat{a}_2 = a_2\Lambda_1/f$ and $\hat{a}_1 = a_1\sqrt{\Lambda_1\Lambda_2}/f$. However, as σ_1/f decreases, this trajectory loses its stability in a sequence of period-doubling bifurcations, leading to chaos. Figures 9–13 show projections of the trajectories on the plane $\hat{a}_2 - \hat{a}_1$ for σ_1/f of 0.2025, 0.2014, 0.2013, 0.20126, and 0.20125, showing periods 2, 4, 8, 16, and 32, respectively. Decreasing σ_1/f slightly below 0.20125 produces what appears to be a chaotic solution.

C. The Case of Combination Parametric Resonance

For the case of combination resonance of the additive type, we introduce the detuning parameters σ_1 and σ_2 defined according to

$$\omega_2 = 2\omega_1 + \varepsilon\sigma_1 \quad \text{and} \quad \Omega_1 = \omega_2 + \omega_1 + \varepsilon\sigma_2$$

Then, eliminating the terms that produce secular terms from Eqs. 3.13 and

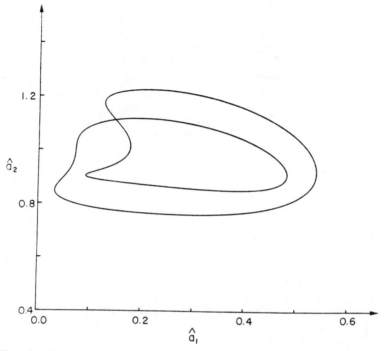

Figure 9. Projection of the trajectory of the modulation equations for the case of principal parametric resonance of the first mode on the $\hat{a}_2$–$\hat{a}_1$ plane, where $\hat{a}_1 = a_1\sqrt{\Lambda_1\Lambda_2}/f$ and $\hat{a}_2 = a_2\Lambda_1/f$, for $\mu_n/f = 0.02$, $\sigma_1/f = 0.2025$, and $\sigma_2/f = 0.16$. It exhibits a period 2 trajectory.

3.14, we obtain

$$2i\omega_1(A_1' + \mu_1 A_1) + 4\omega_1\Lambda_1 A_2\bar{A}_1 e^{i\sigma_1 T_1} + \tfrac{1}{2}F_{12}\bar{A}_2 e^{i\sigma_2 T_1} = 0 \qquad (3.82)$$

$$2i\omega_2(A_2' + \mu_2 A_2) + 4\omega_2\Lambda_2 A_1^2 e^{-i\sigma_1 T_1} + \tfrac{1}{2}F_{21}\bar{A}_1 e^{i(\sigma_2 T_1 + \tau)} = 0 \qquad (3.83)$$

where Λ_1 and Λ_2 are defined in Eqs. 3.18 and 3.19. Substituting the polar form

$$A_1 = \frac{a_1}{8\sqrt{\Lambda_1\Lambda_2}}e^{i\beta_1} \quad \text{and} \quad A_2 = \frac{a_2}{4\Lambda_1}e^{i\beta_2} \qquad (3.84)$$

into Eqs. 3.82 and 3.83 and separating real and imaginary parts, we obtain

$$a_1' = -\mu_1 a_1 - \tfrac{1}{2}a_1 a_2 \sin\gamma_1 - \tfrac{1}{2}f_{12}a_2\sin\gamma_2 \qquad (3.85)$$

$$a_1\beta_1' = \tfrac{1}{2}a_1 a_2 \cos\gamma_1 + \tfrac{1}{2}f_{12}a_2\cos\gamma_2 \qquad (3.86)$$

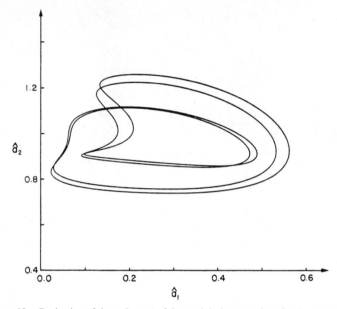

Figure 10. Projection of the trajectory of the modulation equations for the case of principal parametric resonance of the first mode on the $\hat{a}_2$–$\hat{a}_1$ plane, where $\hat{a}_1 = a_1\sqrt{\Lambda_1\Lambda_2}/f$ and $\hat{a}_2 = a_2\Lambda_1/f$, for $\mu_n/f = 0.02$, $\sigma_1/f = 0.2014$, and $\sigma_2/f = 0.16$. It exhibits a period 4 trajectory.

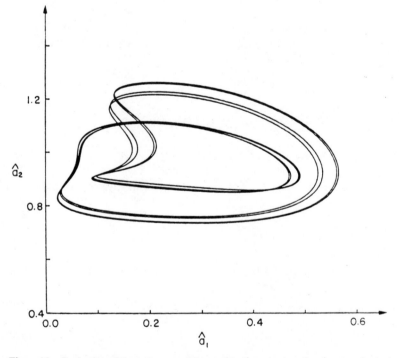

Figure 11. Projection of the trajectory of the modulation equations for the case of principal parametric resonance of the first mode on the $\hat{a}_2$–$\hat{a}_1$ plane, where $\hat{a}_1 = a_1\sqrt{\Lambda_1\Lambda_2}/f$ and $\hat{a}_2 = a_2\Lambda_1/f$, for $\mu_n/f = 0.02$, $\sigma_1/f = 0.2013$, and $\sigma_2/f = 0.16$. It exhibits a period 8 trajectory.

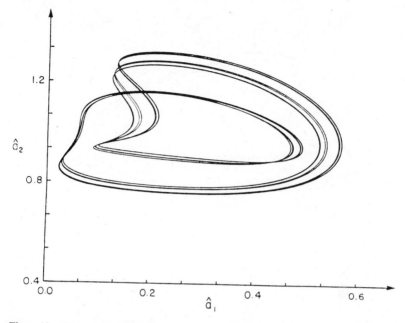

Figure 12. Projection of the trajectory of the modulation equations for the case of principal parametric resonance of the first mode on the $\hat{a}_2$–$\hat{a}_1$ plane, where $\hat{a}_1 = a_1\sqrt{\Lambda_1\Lambda_2}/f$ and $\hat{a}_2 = a_2\Lambda_1/f$, for $\mu_n/f = 0.02$, $\sigma_1/f = 0.20126$, and $\sigma_2/f = 0.16$. It exhibits a period 16 trajectory.

$$a_2' = -\mu_2 a_2 + \tfrac{1}{8}a_1^2 \sin\gamma_1 - \tfrac{1}{4}f_{21}a_1 \sin(\gamma_2 + \tau) \tag{3.87}$$

$$a_2\beta_2' = \tfrac{1}{8}a_1^2 \cos\gamma_1 + \tfrac{1}{4}f_{21}a_1 \cos(\gamma_2 + \tau) \tag{3.88}$$

where

$$f_{12} = \frac{F_{12}}{\omega_1}\sqrt{\frac{\Lambda_2}{\Lambda_1}} \qquad f_{21} = \frac{F_{21}}{2\omega_2}\sqrt{\frac{\Lambda_1}{\Lambda_2}}$$

and

$$\gamma_1 = \beta_2 - 2\beta_1 + \sigma_1 T_1 \qquad \gamma_2 = \sigma_2 T_1 - \beta_2 - \beta_1 \tag{3.89}$$

The scalings in Eqs. 3.84–3.88 are introduced so that we can use the numerical results of Nayfeh and Zavodney (33) directly.

Periodic solutions of Eqs. 2.4 and 2.5 correspond to the fixed points of

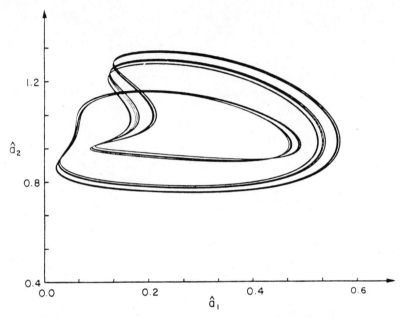

Figure 13. Projection of the trajectory of the modulation equations for the case of principal parametric resonance of the first mode on the $\hat{a}_2$–$\hat{a}_1$ plane, where $\hat{a}_1 = a_1 \sqrt{\Lambda_1 \Lambda_2}/f$ and $\hat{a}_2 = a_2 \Lambda_1/f$, for $\mu_n/f = 0.02$, $\sigma_1/f = 0.20125$, and $\sigma_2/f = 0.16$. It exhibits a period 32 trajectory.

Eqs. 3.85–3.89, which in turn correspond to a_1, a_2, γ_1, and γ_2 being constants. Then, it follows from Eqs. 3.89 that

$$\beta_1' = \tfrac{1}{3}(\sigma_1 + \sigma_2) \quad \text{and} \quad \beta_2' = \tfrac{1}{3}(2\sigma_2 - \sigma_1) \tag{3.90}$$

Consequently, the fixed points of Eqs. 3.85–3.89 are given by

$$\mu_1 a_1 = -\tfrac{1}{2}a_1 a_2 \sin \gamma_1 - \tfrac{1}{2}f_{12}a_2 \sin \gamma_2 \tag{3.91}$$

$$\tfrac{1}{3}(\sigma_1 + \sigma_2)a_1 = \tfrac{1}{2}a_1 a_2 \cos \gamma_1 + \tfrac{1}{2}f_{12}a_2 \cos \gamma_2 \tag{3.92}$$

$$\mu_2 a_2 = \tfrac{1}{8}a_1^2 \sin \gamma_1 - \tfrac{1}{4}f_{21}a_1 \sin (\gamma_2 + \tau) \tag{3.93}$$

$$\tfrac{1}{3}(2\sigma_2 - \sigma_1)a_2 = \tfrac{1}{8}a_1^2 \cos \gamma_1 + \tfrac{1}{4}f_{21}a_1 \cos (\gamma_2 + \tau) \tag{3.94}$$

There are two possibilities. First, $a_1 = a_2 = 0$; second, a_1 and $a_2 \neq 0$ and

Eqs. 3.91–3.94 yield the following solution:

$$a_1 = \frac{2}{\Gamma}[(\mu_1 + \tfrac{1}{2}f_{12}\Gamma \sin \gamma_2)^2 + (\tfrac{1}{3}\sigma_2 + \tfrac{1}{3}\sigma_1 - \tfrac{1}{2}f_{12}\Gamma \cos \gamma_2)^2]^{1/2} \quad (3.95)$$

$$a_2 = \Gamma a_1 \quad (3.96)$$

where

$$\Gamma = [-b \pm (b^2 - c)^{1/2}]^{1/2} \quad (3.97)$$

$$\sin \gamma_2 = d(c_{11}\Gamma^{-1} + c_{12}\Gamma) \qquad \cos \gamma_2 = d(c_{21}\Gamma^{-1} + c_{22}\Gamma)$$

$$b = \frac{2c_{11}c_{12}d^2 + 2c_{21}c_{22}d^2 - 1}{2d^2(c_{12}^2 + c_{22}^2)} \qquad c = \frac{c_{11}^2 + c_{21}^2}{c_{12}^2 + c_{22}^2}$$

if $f_{12} \neq 2f_{21}$ and

$$\Gamma = (-c_{11}/c_{12})^{1/2}$$

$$\gamma_2 = -\tfrac{1}{2}\tau - \arcsin [(\mu_1 + 4\mu_2\Gamma^2)/2\Gamma f_{21} \cos \tfrac{1}{2}\tau]$$

if $f_{12} = 2f_{21}$ provided that $\tau \neq \pm \pi$. When $f_{12} = 2f_{21}$ and $\tau = \pm \pi$, $a_1 = a_2 = 0$ is the only possible solution. The c_{mn} and d are defined in Appendix A. To study the stability of the fixed points, we let

$$A_1 = \frac{1}{8\sqrt{\Lambda_1 \Lambda_2}}(p_1 - iq_1)e^{iv_1 T_1} \quad \text{and} \quad A_2 = \frac{1}{4\Lambda_1}(p_2 - iq_2)e^{iv_2 T_1} \quad (3.98)$$

where

$$v_1 = \tfrac{1}{3}(\sigma_1 + \sigma_2) \quad \text{and} \quad v_2 = \tfrac{1}{3}(2\sigma_2 - \sigma_1) \quad (3.99)$$

in Eqs. 3.82 and 3.83, separate real and imaginary parts, and obtain

$$p_1' + \mu_1 p_1 + v_1 q_1 + \tfrac{1}{2}(p_2 q_1 - p_1 q_2) + \tfrac{1}{2}f_{12}q_2 = 0 \quad (3.100)$$

$$q_1' + \mu_1 q_1 - v_1 p_1 + \tfrac{1}{2}(p_1 p_2 + q_1 q_2) + \tfrac{1}{2}f_{12}p_2 = 0 \quad (3.101)$$

$$p_2' + \mu_2 p_2 + v_2 q_2 - \tfrac{1}{4}p_1 q_1 + \tfrac{1}{4}f_{21}(p_1 \sin \tau + q_1 \cos \tau) = 0 \quad (3.102)$$

$$q_2' + \mu_2 q_2 - v_2 p_2 + \tfrac{1}{8}(p_1^2 - q_1^2) + \tfrac{1}{4}f_{21}(p_1 \cos \tau - q_1 \sin \tau) = 0 \quad (3.103)$$

Then, the stability of a given fixed point to a perturbation proportional to

$\exp(\lambda T_1)$ is determined by the zeros of the characteristic equation

$$
\begin{vmatrix}
\lambda + \mu_1 - \frac{1}{2}q_2 & v_1 + \frac{1}{2}p_2 & \frac{1}{2}q_1 & \frac{1}{2}f_{12} - \frac{1}{2}p_1 \\
-v_1 + \frac{1}{2}p_2 & \lambda + \mu_1 + \frac{1}{2}q_2 & \frac{1}{2}f_{12} + \frac{1}{2}p_1 & \frac{1}{2}q_1 \\
-\frac{1}{4}q_1 + \frac{1}{4}f_{21}\sin\tau & -\frac{1}{4}p_1 + \frac{1}{4}f_{21}\cos\tau & \lambda + \mu_2 & v_2 \\
\frac{1}{4}p_1 + \frac{1}{4}f_{21}\cos\tau & -\frac{1}{4}q_1 - \frac{1}{4}f_{21}\sin\tau & -v_2 & \lambda + \mu_2
\end{vmatrix} = 0
$$

(3.104)

To analyze the stability of a trivial fixed point, we put $p_n = q_n = 0$ in Eq. 3.104 and obtain

$$
\lambda^2 + (\mu_1 + \mu_2 + i\sigma_2)\lambda + \mu_1\mu_2 + i\sigma_2 - \tfrac{1}{8}f_{12}f_{21}e^{-i\tau} = 0 \qquad (3.105)
$$

Hence,

$$
\lambda = -\tfrac{1}{2}(\mu_1 + \mu_2 + i\sigma_2) \pm \tfrac{1}{2}[(\mu_2 - \mu_1)^2 - 2i\sigma_2(\mu_2 - \mu_1)
$$
$$
- \sigma_2^2 + \tfrac{1}{2}f_{12}f_{21}e^{-i\tau}]^{1/2}
$$

(3.106)

Thus, a trivial point is stable if and only if the real parts of both values of λ are less than or equal to zero. The transition curves separating stable from unstable trivial fixed points are given by

$$
(\mu_2 + \mu_1)^2[4\mu_1\mu_2 + \sigma_2^2 - \tfrac{1}{2}f_{12}f_{21}\cos\tau] = [\sigma_2(\mu_2 - \mu_1) + \tfrac{1}{4}f_{12}f_{21}\sin\tau]
$$

(3.107)

To analyze the stability of a nontrivial fixed point, we need to substitute the corresponding values of the p_n and q_n into Eq. 3.104 and check the real parts of its roots. If none of them is greater than zero, the fixed point is stable.

Nayfeh and Zavodney (33) calculated nontrivial solutions and determined the stability of these and of the trivial solutions for various values of the parameters f_{12}, f_{21}, σ_1, σ_2, and τ when $\mu_1 = \mu_1 = 1$, $\delta_1 = 1$, $\delta_2 = 2$, $\delta_3 = 1$, $\alpha_1 = 1$, $\alpha_2 = 2$, $\alpha_3 = 1$, and $\delta_n = \alpha_n = 0$ for $n \geqslant 4$. The perturbation results were verified by numerically integrating the original equations. The results show cases where there are only nontrivial steady-state solutions, only trivial solutions (quenching), only nonsteady-state solutions, and various combinations of the preceding. In addition, the results show the jump phenomenon, modal saturation, Hopf bifurcations, period-doubling bifurcations, and dependence on initial conditions.

Figure 14 shows typical response curves as a function of the detuning parameter σ_1 due to internal resonance. For the parameters chosen, the trivial solutions and the negative branch of Eq. 3.97 are stable. Stable nontrivial fixed points exist only for relatively small values of σ_1 near the origin. When $f_{21} =$

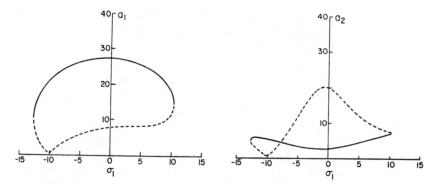

Figure 14. Variation of the amplitudes a_1 and a_2 of the periodic response with the detuning parameter σ_1 of the internal resonance: ——, stable; – – –, unstable; $\tau = 0.2618$ radians, $\sigma_2 = 0$, $f_{12} = 7.00$, $f_{21} = 14.00$. Note that for large values of σ_1, the response of system to parametric resonance is quenched. If $f_{21} = 7.00$, response curves are similar, but the trivial response becomes unstable.

7.00, the response curves are quite similar. However, the trivial response is unstable, so that the solution of Eqs. 3.85–3.89 for any σ_1 at which nontrivial fixed points do not exist is a limit cycle.

Figure 15 shows typical variation of the response curves with the detuning parameter σ_2. For values of τ greater than a critical value τ_c, stable nontrivial fixed points exist only for values of σ_2 away from the origin. However, for small values of τ, the entire negative branch of the nontrivial solution can be stable. These curves and those of Figure 14 illustrate the jump phenomenon. It follows from Figure 15 that if one starts with initial conditions such that the

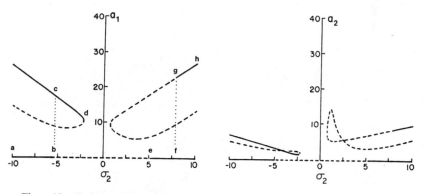

Figure 15. Variation of the amplitudes a_1 and a_2 with the detuning parameter σ_2 due to parametric resonance: ——, stable; – – –, unstable; $\tau = 1.309$ radians, $\sigma_1 = 0$, $f_{12} = 7.00$, $f_{21} = 6.00$.

system response is trivial for $\sigma_2 = -10$ corresponding to point a, and then slowly increases the frequency Ω by increasing σ_2, one eventually reaches the end b of the stable interval for the trivial solution. After this, the amplitude a_1 of the first mode jumps up to the nontrivial value c (corresponding to a periodic response for the system). Increasing σ_2 further causes both a_1 and a_2 to decrease until point d is reached. A further increase in σ_2 causes the response of the system to bifurcate to a torus (corresponding to limit cycles for a_1 and a_2) until point e is reached. The response after point e is now quenched; the only stable steady-state response is trivial. Further increases beyond point f in σ_2 give two possible steady-state solutions for the a_n: trivial and nontrivial; if one is on the trivial response for a given σ_2, disturbances to the system could knock it into a stable periodic motion corresponding to the nontrivial branch $g–h$ and vice-versa. Likewise, if one starts at point h and decreases σ_2, a_1 will jump from g to f, and follow $f–e$. After e, a bifurcation takes place and the a_n execute nonconstant, nontrivial motions of limit cycles (corresponding to a torus response of the system). Decreasing σ_2 beyond point d causes the response amplitudes to stabilize on the fixed point d and then follow the curve $d–c$. Further decreases in σ_2 would cause the system amplitudes to continue on the upper branch unless a disturbance knocks them down to the trivial solution.

The phase τ is a significant parameter in this problem. Figure 16 shows a typical variation of the steady-state amplitudes with τ. For small values of τ, the fixed points corresponding to the negative branch in Eqs. 3.97 are the only stable points. As τ increases, the trivial points become stable, and the nontrivial points become unstable. Figure 16 shows a response curve where an overlap in stable solutions occurs. When this happens, τ_{crit} is given by the maximum value of τ for which stable nontrivial fixed points exist. Then, for

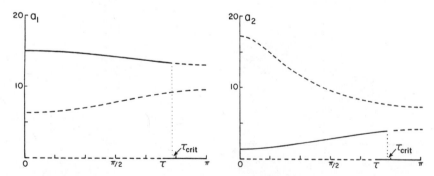

Figure 16. Variation of the amplitudes with the phase angle τ: ——, stable; ———, unstable; $\sigma_1 = \sigma_2 = 0$, $f_{12} = 7.00$, $f_{21} = 7.86$. For $\tau \geqslant \tau_{\mathrm{crit}}$, only the trivial points are stable; therefore the response of the system is quenched.

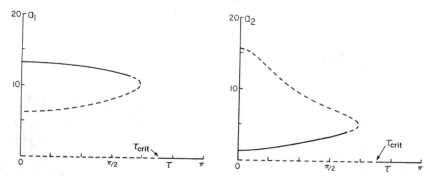

Figure 17. Variation of the response amplitudes with the phase angle τ: ——, stable; ---, unstable; $\sigma_1 = \sigma_2 = 0$, $f_{12} = f_{21} = 7.00$. Note that for $\tau \geqslant \tau_{\text{crit}}$, the response of the system to the parametric resonance is quenched.

$\tau > \tau_{\text{crit}}$, the response of the system is quenched. Decreasing f_{21} causes the two separate branches to merge and form one closed curve, as shown in Figure 17. Since there is no overlapping region in this case, τ_{crit} is given by the trivial stability criterion, and it is determined from Eq. 3.107.

To verify the results in Figure 16, Nayfeh and Zavodney (33) numerically integrated the governing differential Eqs. 2.4 and 2.5 for $\tau = 3.0$. Results are shown in Figure 18 that verify the predictions of the perturbation solution.

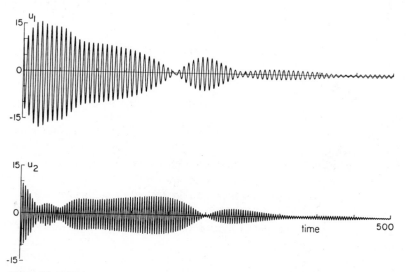

Figure 18. Time history of the response of the system corresponding to the parameters in Figure 16 calculated by numerically integrating governing differential Eqs. 2.4 and 2.5: $\tau = 3.0$ radians, $f_{12} = 7.00$, $f_{21} = 7.86$, $\sigma_1 = \sigma_2 = 0$. Note that the response of the system is quenched.

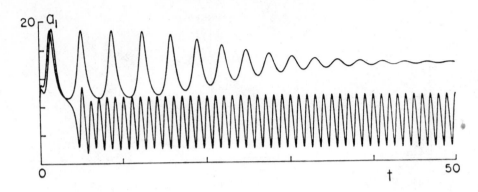

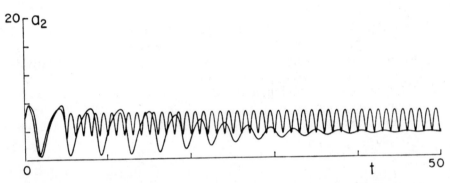

Figure 19. Variation of the u_n with time obtained by numerically integrating the governing differential equations: $\tau = 0.2618$ radians, $\sigma_1 = \sigma_2 = 0$, $f_{12} = 7.00$, $f_{21} = 5.04$, $\varepsilon = 0.01$. Note that the modes of the system achieve constant steady-state amplitudes as predicted by the perturbation solution.

Figure 16 shows that there is only one stable nontrivial fixed point when τ is small. By using different initial conditions (while fixing all other parameters), Nayfeh and Zavodney found that Eqs. 3.85–3.89 also admit a limit cycle solution, as shown in Figure 19. With appropriate initial conditions, the numerical solutions governing Eqs. 2.4 and 2.5 yield a stable, torus response, whose time history is shown in Figure 20.

The addition of a second excitation having the amplitude f_{21} can have a unique effect on the long-time response. Figure 21 shows the variation of the response curves with f_{21}. Figures 22 and 23 show the results of numerically integrating the equations of motion 2.4 and 2.5 with the system parameters corresponding to the two points e and f in Figure 21, again verifying the perturbation results.

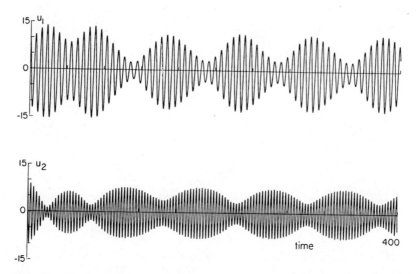

Figure 20. The modulation of the amplitudes of the modes of the system calculated by numerically integrating Eqs. 3.85–3.89, showing both limit cycle and constant steady-state behavior: $\tau = 2.00$ radians, $f_{12} = 7.00$, $f_{21} = 7.86$, $\sigma_1 = \sigma_2 = 0$. The responses correspond to different initial conditions.

In addition to the fixed-point solutions of Eqs. 3.85–3.89, limit cycle solutions have been shown to coexist with the stable fixed-point solutions. What happens as one moves along the branch b–d in Figure 21 is as follows. At b, the fixed-point solution becomes marginally stable because the real part of a complex-conjugate pair of eigenvalues has increased to zero. Increasing

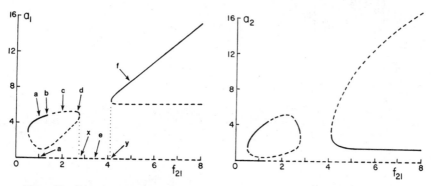

Figure 21. Response of the system corresponding to the parameters in Figure 20. It is calculated by numerically integrating the governing differential Eqs. 2.4 and 2.5 with initial conditions such that the stable response is not a limit cycle (constant amplitude) but a torus.

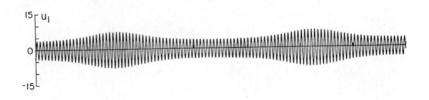

Figure 22. Variation of the steady-state response amplitudes a_1 and a_2 with f_{21} showing three regions of stable nontrivial fixed points: ———, stable; – – –, unstable; $f_{12} = 7.0$, $\tau = 0.2618$ radians, and $\sigma_1 = \sigma_2 = 0$.

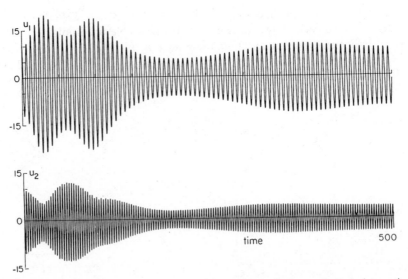

Figure 23. Time histories of the responses of u_1 and u_2 obtained numerically by integrating the governing differential equations: $\tau = 0.2618$ radians, $\sigma_1 = \sigma_2 = 0$, $f_{12} = 7.0$, $f_{21} = 3.30$, and $\varepsilon = 0.01$. Note that the system does not achieve a constant steady-state amplitude in agreement with the perturbation solution.

170

f_{21} causes their real part to become positive (a Hopf bifurcation). The amplitudes a_1 and a_2 begin to oscillate in a limit cycle near the fixed point of the upper previously stable steady-state solution. As f_{21} increases (i.e., as it moves away from the stable fixed point b), the trajectory enlarges; however, the trajectory stays near the upper (now unstable) steady-state solution. As f_{21} approaches the stable fixed point near d, the trajectory size decreases, eventually collapsing to a point, which in this case is the fixed point where stability is again attained, just left of d. Solutions of the averaged equations in the region between points x and y in Figure 21 have limit-cycle trajectories that experience a period-doubling bifurcation. As f_{21} increases from x to y, the period of the limit cycle showly increases. However, at a critical value f_{21}^* of f_{21}, the trajectory undergoes a period-doubling bifurcation. As f_{21} increases to point y, the system bifurcates again because nontrivial fixed points exist, and one of them is stable.

IV. THE CASE OF PRIMARY RESONANCE

In this section, we consider the case of primary resonances. Four first-order nonlinear ordinary differential equations are derived for the modulation of the amplitudes and phases when the excitation frequency is near either the first or the second natural frequency. These equations are used to determine the fixed points and their stability. We demonstrate the saturation phenomenon when the external frequency is near the second natural frequency and demonstrate the existence of a Hopf bifurcation in this case that leads to amplitude- and phase-modulated responses involving both oscillators. We also demonstrate a Hopf bifurcation in the response of internally resonant oscillators when the excitation frequency is near the lower frequency. In this case also the Hopf bifurcation leads to amplitude- and phase-modulated motions. Numerical and experimental results are presented that verify the analytical findings.

In this case, Ω_2 is near either ω_1 or ω_2. To determine a first-order expansion, we need to scale the external excitation so that the effects of resonance, damping, and nonlinearity due to the internal resonance appear in the same order in the same perturbation equation. Consequently, we put $F_n = \varepsilon \hat{F}_n$. Then, substituting Eqs. 3.1–3.3 into Eqs. 2.4 and 2.5, putting the $F_{mn} = 0$, and equating coefficients of like powers of ε on both sides, we obtain:

Order ε^0

$$D_0^2 u_{10} + \omega_1^2 u_{10} = 0 \tag{4.1}$$

$$D_0^2 u_{20} + \omega_2^2 u_{20} = 0 \tag{4.2}$$

Order ε

$$
\begin{aligned}
D_0^2 u_{11} + \omega_1^2 u_{11} = & -2D_0 D_1 u_{10} - 2\mu_1 D_0 u_{10} - \delta_1 u_{10}^2 - \delta_2 u_{10} u_{20} - \delta_3 u_{20}^2 \\
& - \delta_4 (D_0 u_{10})^2 - \delta_5 (D_0 u_{10})(D_0 u_{20}) - \delta_6 (D_0 u_{20})^2 \\
& - \delta_7 u_{10} D_0^2 u_{10} - \delta_8 u_{20} D_0^2 u_{10} - \delta_9 u_{10} D_0^2 u_{20} \\
& - \delta_{10} u_{20} D_0^2 u_{20} + \hat{F}_1 \cos(\Omega_2 T_0 + \tau_1)
\end{aligned}
\tag{4.3}
$$

$$
\begin{aligned}
D_0^2 u_{21} + \omega_2^2 u_{21} = & -2D_0 D_1 u_{20} - 2\mu_2 D_0 u_{20} - \alpha_1 u_{10}^2 - \alpha_2 u_{10} u_{20} - \alpha_3 u_{20}^2 \\
& - \alpha_4 (D_0 u_{10})^2 - \alpha_5 (D_0 u_{10})(D_0 u_{20}) - \alpha_6 (D_0 u_{20})^2 \\
& - \alpha_7 u_{10} D_0^2 u_{10} - \alpha_8 u_{20} D_0^2 u_{10} - \alpha_9 u_{10} D_0^2 u_{20} \\
& - \alpha_{10} u_{20} D_0^2 u_{20} + \hat{F}_2 \cos(\Omega_2 T_0 + \tau_2)
\end{aligned}
\tag{4.4}
$$

The solutions of Eqs. 4.1 and 4.2 are given by Eqs. 3.10 and 3.11. Substituting Eqs. 3.10 and 3.11 into Eqs. 4.3 and 4.4 yields

$$
\begin{aligned}
D_0^2 u_{11} + \omega_1^2 u_{11} = & -2i\omega_1(A_1' + \mu_1 A_1)e^{i\omega_1 T_0} \\
& - [\delta_2 + \delta_5 \omega_1 \omega_2 - \delta_8 \omega_1^2 - \delta_9 \omega_2^2] A_2 \bar{A}_1 e^{i(\omega_2 - \omega_1)T_0} \\
& + \Lambda_3 A_1^2 e^{2i\omega_1 T_0} + \Lambda_4 A_2^2 e^{2i\omega_2 T_0} + \Lambda_5 A_2 A_1 e^{i(\omega_2 + \omega_1)T_0} \\
& + \Lambda_6 A_1 \bar{A}_1 + \Lambda_7 A_2 \bar{A}_2 + \tfrac{1}{2}\hat{F}_1 e^{i(\Omega_2 T_0 + \tau_1)} + \text{c.c.}
\end{aligned}
\tag{4.5}
$$

$$
\begin{aligned}
D_0^2 u_{21} + \omega_2^2 u_{21} = & -2i\omega_2(A_2' + \mu_2 A_2)e^{i\omega_2 T_0} \\
& - [\alpha_1 - \alpha_4 \omega_1^2 - \alpha_7 \omega_1^2] A_1^2 e^{2i\omega_1 T_0} + \Lambda_8 A_2^2 e^{2i\omega_2 T_0} \\
& + \Lambda_9 A_2 \bar{A}_1 e^{i(\omega_2 - \omega_1)T_0} + \Lambda_{10} A_2 A_1 e^{i(\omega_2 + \omega_1)T_0} \\
& + \Lambda_{11} A_1 \bar{A}_1 + \Lambda_{12} A_2 \bar{A}_2 + \tfrac{1}{2}\hat{F}_2 e^{i(\Omega_2 T_0 + \tau_2)} + \text{c.c.}
\end{aligned}
\tag{4.6}
$$

Again, depending on the A_n, a particular solution of Eqs. 4.5 and 4.6 may contain secular terms of the form $T_0 \exp(\pm i\omega_1 T_0)$ and $T_0 \exp(\pm i\omega_2 T_0)$ and, in the case of resonance, small-divisor terms. In addition to the internal or autoparametric resonance $\omega_2 \approx 2\omega_1$, to first order, these resonant conditions are (a) $\Omega_2 \approx \omega_1$, primary resonance of the first mode, and (b) $\Omega_2 \approx \omega_2$, primary resonance of the second mode.

A. Primary Resonance of Second Mode

According to the method of multiple scales, we introduce the detuning parameters σ_1 and σ_2 to convert the small-divisor terms into secular terms and write

$$\Omega_2 T_0 = \omega_2 T_0 + \sigma_2 T_1 \quad \text{and} \quad \omega_2 T_0 = 2\omega_1 T_0 + \sigma_1 T_1 \tag{4.7}$$

Using Eqs. 4.7 and eliminating the terms that produce secular terms from Eqs. 4.5 and 4.6, we obtain

$$2i(A_1' + \mu_1 A_1) + 4\Lambda_1 A_2 \bar{A}_1 e^{i\sigma_1 T_1} = 0 \tag{4.8}$$

$$2i(A_2' + \mu_2 A_2) + 4\Lambda_2 A_1^2 e^{-i\sigma_1 T_1} - f_2 e^{i(\sigma_2 T_1 + \tau_2)} = 0 \tag{4.9}$$

where $\hat{F}_2 = 2\omega_2 f_2$ and the Λ_n are defined in Eqs. 3.18 and 3.19.

Substituting Eqs. 3.12 into Eqs. 4.8 and 4.9 and separating real and imaginary parts, we obtain

$$a_1' = -\mu_1 a_1 - \Lambda_1 a_1 a_2 \sin\gamma_1 \tag{4.10}$$

$$a_2' = -\mu_2 a_2 + \Lambda_2 a_1^2 \sin\gamma_1 + f_2 \sin\gamma_2 \tag{4.11}$$

$$a_1 \beta_1' = \Lambda_1 a_1 a_2 \cos\gamma_1 \tag{4.12}$$

$$a_2 \beta_2' = \Lambda_2 a_1^2 \cos\gamma_1 - f_2 \cos\gamma_2 \tag{4.13}$$

where

$$\gamma_1 = \sigma_1 T_1 + \beta_2 - 2\beta_1 \quad \text{and} \quad \gamma_2 = \sigma_2 T_1 - \beta_2 + \tau_2 \tag{4.14}$$

One can use a scaling of a_1 and a_2 to eliminate Λ_1 and Λ_2 in Eqs. 4.10–4.13. Periodic solutions of Eqs. 2.4 and 2.5 correspond to the fixed points (constant solutions) of Eqs. 4.10–4.14. Thus, they correspond to $a_1' = a_2' = 0$ and $\gamma_1' = \gamma_2' = 0$. It follows from Eqs. 4.14 that

$$\beta_2' = \sigma_2 \quad \text{and} \quad \beta_1' = \tfrac{1}{2}(\sigma_1 + \sigma_2) \tag{4.15}$$

Hence, the fixed points of Eqs. 4.10–4.14 are given by

$$\mu_1 a_1 = -\Lambda_1 a_1 a_2 \sin\gamma_1 \tag{4.16}$$

$$\mu_2 a_2 = \Lambda_2 a_1^2 \sin\gamma_1 + f_2 \sin\gamma_2 \tag{4.17}$$

$$\tfrac{1}{2}(\sigma_1 + \sigma_2)a_1 = \Lambda_1 a_1 a_2 \cos\gamma_1 \tag{4.18}$$

$$\sigma_2 a_2 = \Lambda_2 a_1^2 \cos\gamma_1 - f_2 \cos\gamma_2 \tag{4.19}$$

There are two possible solutions for Eqs. 4.16–4.19. First,

$$a_1 = 0 \quad \text{and} \quad a_2 = \frac{f_2}{\sqrt{\sigma_2^2 + \mu_2^2}} \tag{4.20}$$

and the response is given by

$$u_1 = 0 \quad \text{and} \quad u_2 = a_2 \cos(\Omega_2 t + \tau_2 - \gamma_2) + \cdots \tag{4.21}$$

which is essentially the linear solution. The first mode is unexcited, and the amplitude of the second mode increases linearly with the amplitude f_2 of the excitation. Second,

$$a_1 = (\Lambda_1 \Lambda_2)^{-1/2} [\chi_1 \pm (f_2^2 \Lambda_1^2 - \chi_2^2)^{1/2}]^{1/2} \tag{4.22}$$

$$a_2 = a_2^* = |\Lambda_1|^{-1} [\mu_1^2 + \tfrac{1}{4}(\sigma_1 + \sigma_2)^2]^{1/2} \tag{4.23}$$

where

$$\chi_1 = \tfrac{1}{2}\sigma_2(\sigma_1 + \sigma_2) - \mu_1 \mu_2 \tag{4.24}$$

$$\chi_2 = \sigma_2 \mu_1 + \tfrac{1}{2}\mu_2(\sigma_1 + \sigma_2) \tag{4.25}$$

The response in this case is given by

$$u_1 = a_1 \cos(\tfrac{1}{2}\Omega_2 t + \tfrac{1}{2}\tau_2 - \tfrac{1}{2}\gamma_1 - \tfrac{1}{2}\gamma_2) + \cdots \tag{4.26}$$

$$u_2 = a_2^* \cos(\Omega_2 t + \tau_2 - \gamma_2) + \cdots \tag{4.27}$$

We note a very interesting feature of the second solution; namely, the amplitude a_2 of the directly excited second mode is independent of the amplitude f_2 of the excitation, whereas the amplitude a_1 of the first mode that is not directly excited is a function of the excitation amplitude.

Next, we determine when the roots in Eq. 4.22 are real. To this end, we define two critical values of f_2 as

$$\zeta_1 = |\chi_2/\Lambda_1| \tag{4.28}$$

$$\zeta_2 = [(\chi_1^2 + \chi_2^2)/\Lambda_1^2]^{1/2} = a_2^* \sqrt{\sigma_1^2 + \mu_2^2} \tag{4.29}$$

Clearly, $\zeta_2 \geqslant \zeta_1$. There are two possibilities, depending on whether χ_1 is positive or negative. When $\chi_1 < 0$, there is only one real root of 4.22 when $f_2 \geqslant \zeta_2$. When $\chi_1 > 0$, Eq. 4.22 has two real roots when $\zeta_1 \leqslant f_2 \leqslant \zeta_2$ and one real root when $f_2 > \zeta_2$.

To study the stability of the fixed points and hence the periodic responses, we let

$$A_1 = \tfrac{1}{2}(p_1 - iq_1)e^{i[v_1 T_1 + (1/2)\tau_2]} \qquad A_2 = \tfrac{1}{2}(p_2 - iq_2)e^{i(v_2 T_1 + \tau_2)} \qquad (4.30)$$

where the p_n and q_n are real,

$$v_1 = \tfrac{1}{2}(\sigma_1 + \sigma_2) \quad \text{and} \quad v_2 = \sigma_2 \qquad (4.31)$$

Substituting Eqs. 4.30 and 4.31 into 4.8 and 4.9 and separating real and imaginary parts, we obtain

$$p_1' + \mu_1 p_1 + v_1 q_1 + \Lambda_1(p_2 q_1 - p_1 q_2) = 0 \qquad (4.32)$$

$$q_1' + \mu_1 q_1 - v_1 p_1 + \Lambda_1(p_1 p_2 + q_1 q_2) = 0 \qquad (4.33)$$

$$p_2' + \mu_2 p_2 + v_2 q_2 - 2\Lambda_2 p_1 q_1 = 0 \qquad (4.34)$$

$$q_2' + \mu_2 q_2 - v_2 p_2 + \Lambda_2(p_1^2 - q_1^2) = f_2 \qquad (4.35)$$

The stability of a particular fixed point with respect to a perturbation proportional to $\exp(\lambda T_1)$ is determined by the zeros of

$$
\begin{vmatrix}
\lambda + \mu_1 - \Lambda_1 q_2 & v_1 + \Lambda_1 p_2 & \Lambda_1 q_1 & -\Lambda_1 p_1 \\
-v_1 + \Lambda_1 p_2 & \lambda + \mu_1 + \Lambda_1 q_2 & \Lambda_1 p_1 & \Lambda_1 q_1 \\
-2\Lambda_2 q_1 & -2\Lambda_2 p_1 & \lambda + \mu_2 & v_2 \\
2\Lambda_2 p_1 & -2\Lambda_2 q_1 & -v_2 & \lambda + \mu_2
\end{vmatrix} = 0
$$

or

$$
\lambda^4 + 2(\mu_1 + \mu_2)\lambda^3 + [\mu_1^2 + \mu_2^2 + 4\mu_1\mu_2 + v_1^2 + v_2^2 - \Lambda_1^2 a_2^2 + 4\Lambda_1\Lambda_2 a_1^2]\lambda^2
$$
$$
+ [2\mu_1\mu_2^2 + 2\mu_1^2\mu_2 + 2\mu_1 v_2^2 + 2\mu_2 v_1^2 - 2\mu_2\Lambda_1^2 a_2^2 + 4\Lambda_1\Lambda_2(\mu_1 + \mu_2)a_1^2]\lambda
$$
$$
+ \mu_1^2\mu_2^2 + \mu_1^2 v_2^2 + \mu_2^2 v_1^2 + v_1^2 v_2^2 - \mu_2^2\Lambda_1^2 a_2^2 - v_2^2\Lambda_1^2 a_2^2
$$
$$
+ 4\mu_1\mu_2\Lambda_1\Lambda_2 a_1^2 - 4v_1 v_2\Lambda_1\Lambda_2 a_1^2 + 4\Lambda_1^2\Lambda_2^2 a_1^4 = 0 \qquad (4.36)
$$

where a_1 and a_2 correspond to a specific fixed point.

To investigate the stability of the linear solution given by Eq. 4.20, we put $a_1 = 0$ in Eq. 4.36 and obtain

$$(\lambda^2 + 2\mu_1\lambda + \mu_1^2 + v_1^2 - \Lambda_1^2 a_2^2)(\lambda^2 + 2\mu_2\lambda + \mu_2^2 + v_2^2) = 0$$

Hence,

$$\lambda = -\mu_1 \pm \sqrt{\Lambda_1^2 a_2^2 - v_1^2}, \quad -\mu_2 \pm iv_2 \qquad (4.37)$$

It follows from Eq. 4.37 that the linear solution is stable if and only if $\Lambda_1^2 a_2^2 \leqslant \mu_1^2 + v_1^2$ or $a_2 \leqslant a_2^*$, and hence, $f_2 \leqslant \zeta_2$, according to Eq. 4.29.

To investigate the stability of the nonlinear solution given by Eqs. 4.22 and 4.23, we put $a_2 = a_2^*$ in Eq. 4.36 and obtain

$$\lambda^4 + 2(\mu_1 + \mu_2)\lambda^3 + [\mu_2^2 + 4\mu_1\mu_2 + v_2^2 + 4\Lambda_1\Lambda_2 a_1^2]\lambda^2$$
$$+ [2\mu_1\mu_2^2 + 2\mu_1 v_2^2 + 4\Lambda_1\Lambda_2(\mu_1 + \mu_2)a_1^2]\lambda$$
$$+ 4\Lambda_1\Lambda_2 a_1^2[\Lambda_1\Lambda_2 a_1^2 + \mu_1\mu_2 - v_1 v_2] = 0 \qquad (4.38)$$

The necessary and sufficient conditions that none of the roots of Eq. 4.38 have positive real parts then are

$$\Lambda_1\Lambda_2 a_1^2 + \mu_1\mu_2 - v_1 v_2 > 0 \qquad (4.39)$$

and

$$r_3(r_1 r_2 - r_3) - r_1^2 r_4 > 0 \qquad (4.40)$$

where r_1, r_2, r_3, and r_4 are, respectively, the coefficients of $\lambda^3, \lambda^2, \lambda$, and λ^0 in Eq. 4.38. Substituting for the r_n, we rewrite condition 4.40 as

$$4\mu_1\mu_2(\mu_2^2 + v_2^2)(4\mu_1^2 + 4\mu_1\mu_2 + \mu_2^2 + v_2^2) + 8(\mu_1 + \mu_2)^2 \Lambda_1\Lambda_2 a_1^2$$
$$\cdot(\mu_2^2 + 2\mu_1\mu_2 + 2v_1 v_2 + v_2^2) > 0 \qquad (4.41)$$

Condition 4.39, in conjunction with Eqs. 4.22 and 4.23, implies that the solution corresponding to the positive sign is stable whereas the solution corresponding to the negative sign is unstable. The violation of condition 4.41 would imply the existence of a pair of complex-conjugate roots of Eq. 4.38 with a positive real part. When $v_1 v_2 > 0$, condition 4.41 is satisfied for all values of μ_1, μ_2, and f_2. On the other hand, when $v_1 v_2 < 0$, condition 4.41 may be violated, depending on the values of v_n, μ_n, and f_2.

Figure 24 shows the variation of a_1 and a_2 with f_2 for $\Lambda_1 = 0.5, \Lambda_2 = 0.25$, $\sigma_1 = 0, \sigma_2 = 0, \mu_1 = 0.005$, and $\mu_2 = 0.005$. In this case, $v_1 = 0, v_2 = 0$, and $\chi_1 = -0.000025$. Since $\chi_1 < 0$, Eq. 4.22 has one real root when $f_2 \geqslant \zeta_2$, where $\zeta_2 \approx 0.00005$. Since conditions 4.39 and 4.41 are satisfied, this root is stable. Consequently, when $f_2 \leqslant \zeta_2$, the response is given by Eqs. 4.21; the first mode is not excited because $a_1 = 0$ and the amplitude of the second mode is proportional to f_2. Thus, $f_2 = \zeta_2$ is a bifurcation point at which one of the real roots of 4.36 changes sign and there is a change in the number of fixed points (pitchfork bifurcation). When $f_2 > \zeta_2$, the response is given by Eqs. 4.26 and 4.27 where $a_2 = a_2^* = 0.01 = $ const for all values of f_2 greater than ζ_2 and a_1 is given by Eq. 4.22. Hence, if an experiment is performed by setting $\Omega_2 \approx \omega_2$ and

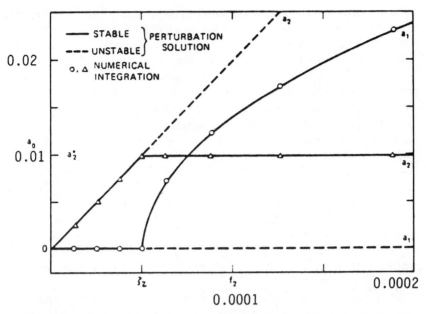

Figure 24. Variation of the steady-state amplitudes a_1 and a_2 with the amplitude of the excitation f_2 when the excitation frequency is near the higher frequency for $\Lambda_1 = 0.5$, $\Lambda_2 = 0.25$, $\sigma_1 = 0.0$, $\sigma_2 = 0.0$, and $\mu_1 = \mu_2 = 0.005$: ———, stable; ———, unstable.

the detunings and damping coefficients are such that $\chi_1 < 0$ and $v_1 v_2 > 0$, one expects the second mode to dominate. This is initially so. But as f_2 increases beyond a critical value ζ_2, a_2 remains constant and equal to a_2^* (i.e., second mode saturates) and the extra energy is spilled over into the first mode. The saturation value a_2^* can be very small if $\sigma_1 + \sigma_2$ and μ_1 are small. The saturation phenomenon was first discovered by Nayfeh, Mook, and Marshall (12) in connection with the pitching and rolling of ships. They found that a ship whose pitch frequency is approximately twice the roll frequency exhibits undesirable roll characteristics, as observed by Froude in 1863. To verify the perturbation result, Nayfeh, Mook, and Marshall integrated a special form of Eqs. 2.4 and 2.5 for a long time and determined the amplitudes of u_1 and u_2. The results of the numerical simulation are in excellent agreement with those predicted by the perturbation solution, as shown in Figure 24.

As a second example, we let $\Lambda_1 = 1.0$, $\Lambda_2 = 0.5$, $\mu_1 = \mu_2 = 0.02$, $\sigma_1 = -0.03$, and $\sigma_2 = 0.12$; then $v_1 = 0.045$, $v_2 = -0.03$, and $\chi_1 = -0.00095$. Hence, Eq. 4.22 has one real root for all values of $f_2 \geqslant \zeta_2$, where $\zeta_2 \approx 0.00178$; condition 4.39 is satisfied for all $f_2 > \zeta_2$; but condition 4.41 is violated for all values of $f_2 \geqslant \zeta_3$, where $\zeta_3 \approx 0.003$. Consequently, the response consists of a pure second mode when $f_2 \leqslant \zeta_2$, combined periodic first and second modes given by Eqs. 4.26 and 4.27 with a_2 being equal to $a_2^* \approx 0.04924$ when $\zeta_2 < f_2 < \zeta_3$, and amplitude- and phase-modulated combined first and second modes when $f_2 \geqslant \zeta_3$. Thus, $f_2 = \zeta_2$ is a bifurcation point at which one of the real roots of Eq. 4.36 changes sign and there is a change in the number of fixed points (pitchfork bifurcation). Moreover, $f_2 = \zeta_3$ is a Hopf bifurcation point, at which the real part of a pair of complex-conjugate roots of Eq. 4.36 changes sign.

Figure 25 shows the variation of a_1 and a_2 with f_2 when $\Lambda_1 = 0.5$, $\Lambda_2 = 0.25$, $\sigma_1 = 0$, $\sigma_2 = 0.025$, and $\mu_1 = \mu_2 = 0.005$. In this case, $v_1 = 0.0125$, $v_2 = 0.025$, and $\chi_1 = 0.0002875$. Hence, Eq. 4.22 has only one real root, which satisfies conditions 4.39 and 4.41 when $f_2 > \zeta_2$, where $\zeta_2 \approx 0.00068$. It has two real roots when $\zeta_1 \leqslant f_2 \leqslant \zeta_2$, where $\zeta_1 \approx 0.00038$. The larger root satisfies conditions 4.39 and 4.41, and hence it is stable, whereas the smaller root satisfies condition 4.41 but violates condition 4.39, and hence it is unstable. Consequently, $f_2 = \zeta_1$ is a nonlocal bifurcation point at which there is a change in the number of fixed points. Moreover, $f_2 = \zeta_2$ is also a bifurcation point at which one of the real roots of Eq. 4.36 changes sign and there is a change in the number of fixed points.

In addition to the saturation phenomenon, Figure 25 exhibits the jump

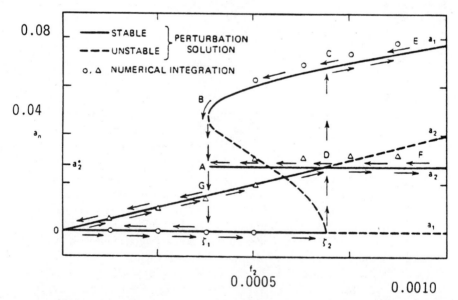

Figure 25. Variation of the steady-state amplitudes a_1 and a_2 with the amplitude of the excitation f_2 when the excitation frequency is near the higher frequency for $\Lambda_1 = 0.5$, $\Lambda_2 = 0.25$, $\sigma_1 = 0.0$, $\sigma_2 = 0.025$, and $\mu_1 = \mu_2 = 0.005$: ——, stable; – – –, unstable.

phenomenon and the sensitivity of the response to the initial conditions. As f_2 increases from zero, the first mode is not excited until f_2 exceeds the critical value ζ_2. As f_2 increases beyond ζ_2, a_1 jumps up from zero to point C and a_2 remains equal to a_2^*, resulting in a large first-mode response. As f_2 increases further, the amplitude a_1 of the first mode increases slowly along the curve BCE whereas the amplitude a_2 of the second mode remains a_2^*. As f_2 decreases from a value corresponding to point E, a_1 decreases slowly along the curve ECB and a_2 remains constant until f_2 reaches the critical value ζ_1. As f_2 decreases beyond ζ_1, a_1 jumps down to zero and a_2 jumps down to point G, resulting in a pure second-mode response. As f_2 decreases further, a_1 remains zero and a_2 decreases linearly with f_2. Again, the results of the numerical solutions of a special form of Eqs. 2.4 and 2.5 obtained by Nayfeh, Mook, and Marshall are in excellent agreement with the predictions of the perturbation solutions, as shown in Figure 25.

Figure 26 shows frequency response curves for the case $\Lambda_1 = 1.0$, $\Lambda_2 = 0.5$, $\mu_1 = \mu_2 = 0.02$, $\sigma_1 = 0.12$, and $f_2 = 0.1$. Again the unstable portions are represented by broken lines. In addition to the jump phenomenon, Figure 26

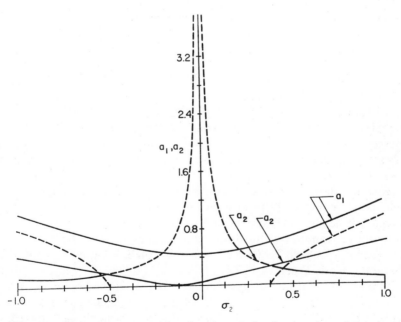

Figure 26. Frequency response curves when the excitation frequency is nearly equal to the higher frequency for $\Lambda_1 = 1.0$, $\Lambda_2 = 0.5$, $\mu_1 = \mu_2 = 0.02$, $\sigma_2 = 0.12$, $f_2 = 0.1$: ———, stable; ----, unstable.

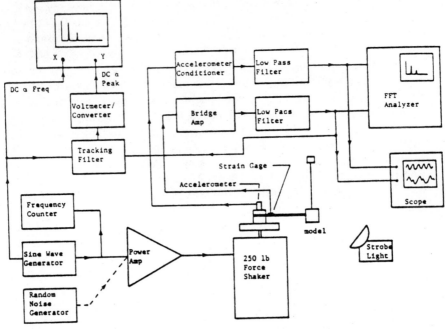

Figure 27.   Instrumentation schematic used in experiments.

shows the interval $-0.047 \leqslant \sigma_2 \leqslant -0.0127$, in which a pair of complex-conjugate roots of Eq. 4.38 have a positive real part, implying the existence of a Hopf bifurcation. In this interval, the response is amplitude- and phase-modulated combined first and second modes. Nayfeh and Raouf (23, 24) presented digital results showing amplitude- and phase-modulated responses of a special form of Eqs. 2.4 and 2.5 and period-multiplying bifurcations of the solutions of the averaged equations.

Haddow, Barr, and Mook (18) used a double-cantilever beam similar to the one shown in Figure 27 to demonstrate the saturation phenomenon and modal interactions. This system can be modeled by two coupled nonlinear differential equations with quadratic nonlinearities of the form in Eqs. 2.4 and 2.5. The linear natural frequencies ω_1 and ω_2 can be adjusted by positioning the upper mass. When $\omega_2 \approx 2\omega_1$, $f_1 = 0$, $f_2 \neq 0$, and $\Omega_2 \approx \omega_2$, the linear solution shows that the steady-state amplitude a_1 of the first mode is zero, whereas the steady-state amplitude a_2 of the second mode increases linearly with f_2, as shown in Figure 24. However, including the nonlinear terms shows that above a threshold value ζ_2 of f_2, the linear solution is unstable, a_2 remains constant (saturates), and the additional energy spills over into the first mode, as shown in Figure 24. Nayfeh, Mook, and Marshall (12) verified the saturation results by numerically integrating the governing equations for a time long enough for a steady-state solution to develop; the results of the numerical simulation are shown in Figure 24. By slightly detuning the excitation frequency from the second natural frequency, one obtains variations of the steady-state amplitudes as shown in Figure 25. When the excitation amplitude f_2 lies in the interval $[\zeta_1, \zeta_2]$, there are three possible steady-state responses. Two of these responses are stable: the trivial response and the larger amplitude response. The response that is attained physically depends on the initial conditions. The instability of the linear solution when $f_2 > \zeta_2$, the saturation phenomenon, and the dependence on the initial conditions have been verified in our facility.

The experiment of Haddow, Barr, and Mook was repeated in our vibration laboratory. The instrumentation schematic is shown in Figure 27. The instrumented model was attached to the shaker table, to which a reference accelerometer was also attached. The length of the upright beam was adjusted so that the second natural frequency was very nearly twice the first natural frequency. The accelerometer signal was conditioned by a constant-current source, and the strain gage was conditioned by a bridge amplifier. These signals were passed through low-pass filters to eliminate the high-frequency noise before they were amplified. An oscilloscope was used at this point to monitor the waveforms. One could readily see the frequency of the response and distorted waveform when displayed side by side with the excitation. The two signals were then analyzed by the a fast Fourier transform (FFT) analyzer. After passing through antialiasing filters, each signal was digitized by an analog-to-digital converter (ADC) for processing. The shaker was driven by its power amplifier, which was driven by a sine wave generator. The frequency was quickly and accurately measured by an independent frequency counter.

To perform the experiment, first we determined approximations to the first and second natural frequencies by exciting the shaker with white noise, using the FFT to compute the frequency-response function (FRF) 30 times, averaging them, and displaying the result. The peaks were located and the frequencies were obtained. The measurements of the natural frequencies were refined by adjusting the frequency of a single-harmonic excitation manually until the amplitude of the response reached its maximum. The transients were allowed to die out by waiting long enough for a steady state to be achieved. The excitation level was adjusted so that the linear solution was stable, and therefore the model responded only at the forcing frequency. The first two natural frequencies were found to be approximately 7.62 and 15.23 Hz.

To demonstrate the instability of the structure when $f_2 > \zeta_2$, we excited the structure near the second natural frequency (15.23 Hz) at a low amplitude. The response consisted of only the second mode, as predicted by linear theory.

When the structure was excited at a larger amplitude, the second mode responded with the amplitude predicted by the linear theory as expected. However, after about 500 cycles, the first mode appeared and began to grow, while simultaneously the second mode diminished. This was readily observed in the FFT of the response and was seen as a distorted waveform on the oscilloscope. Thus, the linear solution was seen to be unstable, and both the first and second modes appeared in the response as predicted by the nonlinear theory and in contrast with the predictions of the linear theory. An interesting point regarding this phenomenon is that the level of excitation was so small that vibration at approximately 15.23 Hz was not even discernible when one's hand was placed on the shaker table. Thus, it was observed that the nonlinearities present were affecting the response dramatically even at extremely low levels of excitation.

To demonstrate the saturation phenomenon, the amplitude of excitation was increased slightly while keeping the excitation frequency fixed at 15.23 Hz. The response, after the transients decayed, showed that the amplitude of the second mode (15.23 Hz) *did not* increase—rather the energy went into the first mode. This was seen on the FFT as an increased peak at the frequency (7.62 Hz) of the first mode. The waveform, originally consisting of the second-mode 15.23-Hz waveform, became distorted as it gradually became dominated by the first mode. Further increases in the amplitude of the excitation continued to increase only the amplitude of the first mode in the steady-state condition. Initially the amplitude of the second mode increased, as observed in the instantaneous FFT of the response; however, as the amplitude of the second mode increased, it pulled up the amplitude of the first mode. This interaction eventually decayed to the steady-state response in which the amplitude of the second mode returned to its saturated level, while the amplitude of the first mode was noticeably higher. The results are documented in a video tape (36).

B. Primary Resonance of First Mode

In this case,

$$\omega_2 = 2\omega_1 + \varepsilon\sigma_1 \quad \text{and} \quad \Omega_2 = \omega_1 + \varepsilon\sigma_2 \tag{4.42}$$

Using Eqs. 4.42 and eliminating the terms that produce secular terms from Eqs. 4.5 and 4.6, we obtain the solvability conditions

$$2i(A_1' + \mu_1 A_1) + 4\Lambda_1 A_2 \bar{A}_1 e^{i\sigma_1 T_1} - f_1 e^{i(\sigma_2 T_1 + \tau_1)} = 0 \tag{4.43}$$

$$2i(A_2' + \mu_2 A_2) + 4\Lambda_2 A_1^2 e^{-i\sigma_1 T_1} = 0 \tag{4.44}$$

where $f_1 = \hat{F}_1/2\omega_1$ and Λ_1 and Λ_2 are defined in Eqs. 3.18 and 3.19.

Substituting Eqs. 3.12 into Eqs. 4.43 and 4.44 and separating real and imaginary parts, we obtain

$$a_1' = -\mu_1 a_1 - \Lambda_1 a_1 a_2 \sin \gamma_1 + f_1 \sin \gamma_2 \tag{4.45}$$

$$a_2' = -\mu_2 a_2 + \Lambda_2 a_1^2 \sin \gamma_1 \tag{4.46}$$

$$a_1 \beta_1' = \Lambda_1 a_1 a_2 \cos \gamma_1 - f_1 \cos \gamma_2 \tag{4.47}$$

$$a_2 \beta_2' = \Lambda_2 a_1^2 \cos \gamma_1 \tag{4.48}$$

$$\gamma_1 = \sigma_1 T_1 + \beta_2 - 2\beta_1 \quad \text{and} \quad \gamma_2 = \sigma_2 T_1 - \beta_1 + \tau_1 \tag{4.49}$$

Again Λ_1 and Λ_2 could be eliminated from Eqs. 4.45–4.48 by a proper scaling of a_1 and a_2.

Periodic solutions correspond to the fixed points of Eqs. 4.45–4.49, which in turn correspond to $a_1' = a_2' = 0$ and $\gamma_1' = \gamma_2' = 0$. It follows from Eqs. 4.49 that

$$\beta_1' = \nu_1 = \sigma_2 \quad \text{and} \quad \beta_2' = \nu_2 = 2\sigma_2 - \sigma_1 \tag{4.50}$$

Then, the fixed points of Eqs. 4.45–4.49 are given by

$$\mu_1 a_1 = -\Lambda_1 a_1 a_2 \sin \gamma_1 + f_1 \sin \gamma_2 \tag{4.51}$$

$$\mu_2 a_2 = \Lambda_2 a_1^2 \sin \gamma_1 \tag{4.52}$$

$$\sigma_2 a_1 = \Lambda_1 a_1 a_2 \cos \gamma_1 - f_1 \cos \gamma_2 \tag{4.53}$$

$$(2\sigma_2 - \sigma_1)a_2 = \Lambda_2 a_1^2 \cos \gamma_1 \tag{4.54}$$

Equations 4.51–4.54 can be manipulated to give

$$\hat{a}_1^6 + 2(\mu_1 \mu_2 - \nu_1 \nu_2)\hat{a}_1^4 + \chi_1^2 \chi_2^2 \hat{a}_1^2 - f_1^2 \chi_2^2 = 0 \tag{4.55}$$

$$\hat{a}_2 = \hat{a}_1^2 / \chi_2 \qquad \hat{a}_1^2 = \Lambda_1 \Lambda_2 a_1^2 \tag{4.56}$$

where

$$\chi_n^2 = \mu_n^2 + \nu_n^2 \tag{4.57}$$

To study the stability of the fixed points, we substitute

$$A_1 = \tfrac{1}{2}(p_1 - iq_1)e^{i(\nu_1 T_1 + \tau_1)} \quad \text{and} \quad A_2 = \tfrac{1}{2}(p_2 - iq_2)e^{i(\nu_2 T_1 + 2\tau_1)}$$

into Eqs. 4.43 and 4.44, separate real and imaginary parts, and obtain

$$p_1' + \mu_1 p_1 + \nu_1 q_1 + \Lambda_1(p_2 q_1 - p_1 q_2) = 0 \tag{4.58}$$

$$q_1' + \mu_1 q_1 - \nu_1 p_1 + \Lambda_1(p_1 p_2 + q_1 q_2) = f_1 \tag{4.59}$$

$$p_2' + \mu_2 p_2 + \nu_2 q_2 - 2\Lambda_2 p_1 q_1 = 0 \tag{4.60}$$

$$q_2' + \mu_2 q_2 - \nu_2 p_2 + \Lambda_2(p_1^2 - q_1^2) = 0 \tag{4.61}$$

Then, the stability of a particular fixed point with respect to a perturbation proportional to $\exp(\lambda T_1)$ is determined by the zeros of the characteristic Eq. 4.36. According to the Routh–Hurwitz criterion, at least one root of Eq. 4.36 has a positive real part, and hence a fixed point given by Eqs. 4.55–4.57 is unstable if all of the following conditions are not satisfied:

$$\chi_1^2 \chi_2^2 - \chi_2^2 \hat{a}_2^2 + 4(\mu_1 \mu_2 - \nu_1 \nu_2 + \hat{a}_1^2)\hat{a}_1^2 > 0 \tag{4.62}$$

$$\mu_2 \chi_2^2 + \mu_1(\chi_1^2 - \hat{a}_2^2) + 2(\mu_1 + \mu_2)(2\mu_1 \mu_2 + \hat{a}_1^2) > 0 \tag{4.63}$$

and

$$\mu_1 \mu_2 (\mu_1 + \mu_2)^4 + 2\mu_1 \mu_2 (\mu_1 + \mu_2)^2 (\nu_1^2 + \nu_2^2) + \mu_1 \mu_2 (\nu_1^2 - \nu_2^2)^2$$
$$- 2\mu_1 \mu_2 [(\mu_1 + \mu_2)^2 + \nu_1^2 - \nu_2^2]\hat{a}_2^2 + 2(\mu_1 + \mu_2)^2 [(\mu_1 + \mu_2)^2 + (\nu_1 + \nu_2)^2]\hat{a}_1^2$$
$$+ \mu_1 \mu_2 \hat{a}_2^4 - 2(\mu_1 + \mu_2)^2 \hat{a}_1^2 \hat{a}_2^2 > 0 \tag{4.64}$$

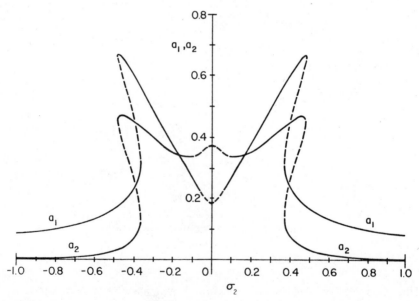

Figure 28. Frequency response curves when the excitation frequency is nearly equal to the lower frequency for $\Lambda_1 = 1.0$, $\Lambda_2 = 0.5$, $\mu_1 = \mu_2 = 0.08$, and $f_1 = 0.08$: ———, stable; ----, unstable.

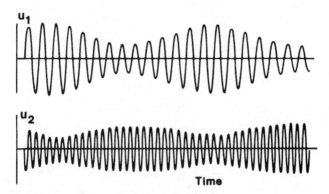

Figure 29.   Numerically calculated amplitude- and phase-modulated motions.

The violation of condition 4.64 would imply the existence of a pair of complex-conjugate values of λ with a positive real part, and hence replacing the inequality by an equality yields the parameters for the Hopf bifurcation.

Figure 28 shows the frequency response curves for $\Lambda_1 = 1.0$, $\Lambda_2 = 0.5$, $\sigma_1 = 0$, $\mu_1 = \mu_2 = 0.08$, and $f_1 = 0.08$. Again, the unstable solutions are represented by broken lines. In addition to the jump phenomenon, Figure 28 shows that there is an interval of σ_2 near zero, namely, $-0.09995 \leqslant \sigma_2 \leqslant 0.09995$, for which there are no stable steady-state periodic solutions. Across this interval, the real part of a pair of complex-conjugate values of λ changes sign, indicating a Hopf bifurcation. Sethna and Bajaj (37) carried out a bifurcation analysis for the latter case and determined the amplitudes of the limit cycles. Hence, the amplitudes a_n and phases β_n of u_1 and u_2 are expected to modulate with time, and consequently, u_1 and u_2 are expected to be aperiodic. Sethna (10) integrated the averaged equations using an analog computer in this interval and found that the a_n and β_n are periodic. A digital solution (12) of a special case of Eqs. 2.4 and 2.5 for a value of σ_2 in this interval is shown in Figure 29; it represents an amplitude- and phase-modulated motion. Similar results were obtained by Yamamoto and Yasuda (14) and Nayfeh and Raouf (23) using analog and digital simulations, respectively. Numerical solutions of the averaged equations show that the limit cycle solutions may undergo period-doubling bifurcations, leading to chaos (19).

To observe these amplitude- and phase-modulated motions, Nayfeh and Zavodney (20) conducted an experiment using the model and setup shown in Figure 27. The lengths of the beams were adjusted so that the linear natural frequencies of the system are $f_1 = 8.03$ Hz and $f_2 = 16.25$ Hz, which are in the ratio of 2:1 (i.e., $f_2 \approx 2f_1$). The mode shapes are shown in Figure 30. Relative displacements of the two masses were measured by strain gages mounted near

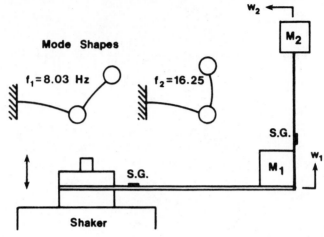

Figure 30. Two-degree-of-freedom model tuned for 2:1 internal resonance and accompanying linear mode shapes.

the bases of the beams. The excitation frequency f was slowly varied up and down between 7.6 and 8.6 Hz while keeping all other parameters constant. The signals were analyzed using an FFT analyzer, and the steady-state amplitudes a_1 and a_2 of the two modes in the periodic cases were determined. They are plotted in Figure 31 as a function of f. The excitation frequencies marking the transition from periodic to aperiodic responses are marked in Figure 31 as the Hopf bifurcation points. The experimental results in this figure are qualitatively in agreement with the theoretically determined frequency response

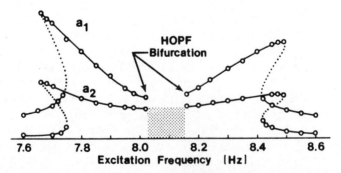

Figure 31. Experimentally determined frequency response curves when the excitation frequency f is near the first natural frequency $f_1 = 8.03$ Hz; the shaded region corresponds to aperiodic long-time motions.

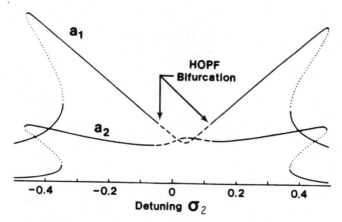

Figure 32. Theoretically determined frequency response curves when the excitation frequency is nearly equal to the lower frequency.

curves in Figure 32. Similar experimental results on a similar model were obtained earlier by Haddow, Barr, and Mook (18).

Representative time traces of the displacement w_1 of mass M_1 are shown in Figure 33 before and after the Hopf bifurcation; this displacement is a linear combination of the generalized coordinates u_1 and u_2 of the two modes. The

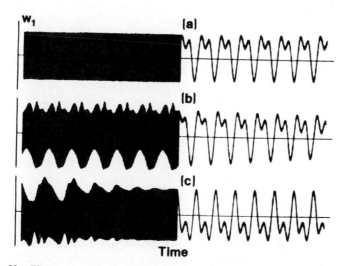

Figure 33. Time traces of the displacement of mass M_1 before and after the bifurcation from periodic to aperiodic responses. The left portion shows the envelope of modulation by compressing time; the right portion shows a sample with an expanded time scale: (a) $f = 8.02$ Hz, (b) $f = 8.03$ Hz, (c) transition from $f = 8.15$ Hz to $f = 8.16$ Hz.

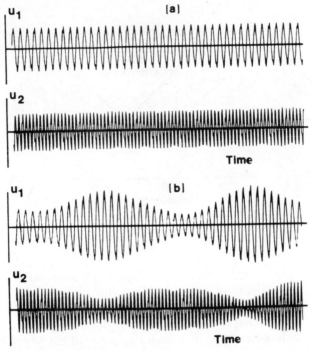

Figure 34. Modal components of the displacement of mass M_1 as a function of time (*a*) before bifurcation ($f = 8.02$ Hz) and (*b*) after the bifurcation ($f = 8.03$ Hz). These traces were filtered from the data of Figure 33.

displacements of the individual modes obtained by filtering the signal are shown in Figure 34. These experimentally determined time traces are qualitatively in agreement with those obtained numerically and shown in Figure 29. The constant-amplitude response before the bifurcation corresponds to a stable fixed point (point attractor) in the projection of the trajectory on the a_2-a_1 plane. These amplitudes were simultaneously monitored on an oscilloscope as the system experienced a bifurcation. The constant-amplitude periodic motion gave rise to an aperiodic motion whose nature changed as the unstable region was penetrated. As the frequency approached the other "Hopf bifurcation point", the aperiodic motion gave way to a constant-amplitude periodic motion. Figure 35 shows a projection of the long-time trajectory on the u_2-u_1 plane. Before the bifurcation, the periodic response is stationary and appears as a figure eight since the period of one mode is exactly twice that of the other mode. Inside the unstable region the figure-eight response is observed to evolve because the two modes are constantly exchanging energy.

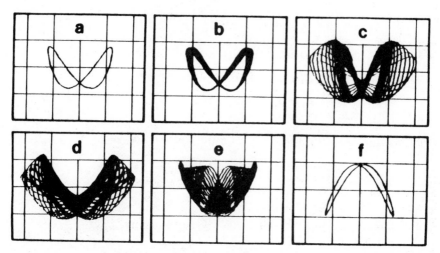

Figure 35. Variation of the displacement u_2 of the second mode with the displacement u_1 of the first mode: (a) periodic motion before the bifurcation, $f = 8.02$ Hz; (b) aperiodic motion after the bifurcation, $f = 8.025$ Hz; (c–e) aperiodic motion in the unstable region, $8.03 < f < 8.15$; and (f) return to periodic motion, $f = 8.16$ Hz. Only portions of a "complete cycle" are shown for the aperiodic responses because the overlapping of a complete cycle obscures the projection of the amplitude- and phase-modulated motion.

V. SECONDARY RESONANCES

In this section, we consider the case of an external excitation that does not lead to primary resonances. Equations are derived for the modulation of the amplitudes and phases for all possible resonances. To first order, these resonances may have one of the following forms: (a) subharmonic resonances of order $\frac{1}{2}$, which lead to modulation equations similar to those that arise from principal parametric resonances; (b) superharmonic resonances of order 2, which lead to modulation equations similar to those that arise from primary resonances; and (c) combination resonances, which lead to modulation equations that arise from combination parametric resonances.

In this case, we take $F_n = O(1)$, $F_{mn} = 0$, and Ω_2 to be away from either ω_1 or ω_2. Substituting Eqs. 3.1 and 3.2 into Eqs. 2.4 and 2.5 and equating coefficients of like powers of ε on both sides, we obtain

Order ε^0

$$D_0^2 u_{10} + \omega_1^2 u_{10} = F_1 \cos(\Omega_2 T_0 + \tau_1) \tag{5.1}$$

$$D_0^2 u_{20} + \omega_2^2 u_{20} = F_2 \cos(\Omega_2 T_0 + \tau_2) \tag{5.2}$$

Order ε

$$D_0^2 u_{11} + \omega_1^2 u_{11} = -2D_0 D_1 u_{10} - 2\mu_1 D_0 u_{10} - \delta_1 u_{10}^2 - \delta_2 u_{10} u_{20} - \delta_3 u_{20}^2$$
$$- \delta_4 (D_0 u_{10})^2 - \delta_5 (D_0 u_{10})(D_0 u_{20}) - \delta_6 (D_0 u_{20})^2$$
$$- \delta_7 u_{10} D_0^2 u_{10} - \delta_8 u_{20} D_0^2 u_{10} - \delta_9 u_{10} D_0^2 u_{20} - \delta_{10} u_{20} D_0^2 u_{20}$$
$$\tag{5.3}$$

$$D_0^2 u_{21} + \omega_2^2 u_{21} = -2D_0 D_1 u_{20} - 2\mu_2 D_0 u_{20} - \alpha_1 u_{10}^2 - \alpha_2 u_{10} u_{20} - \alpha_3 u_{20}^2$$
$$- \alpha_4 (D_0 u_{10})^2 - \alpha_5 (D_0 u_{10})(D_0 u_{20}) - \alpha_6 (D_0 u_{20})^2$$
$$- \alpha_7 u_{10} D_0^2 u_{10} - \alpha_8 u_{20} D_0^2 u_{10} - \alpha_9 u_{10} D_0^2 u_{20} - \alpha_{10} u_{20} D_0^2 u_{20}$$
$$\tag{5.4}$$

The solutions of Eqs. 5.1 and 5.2 can be expressed as

$$u_{10} = A_1(T_1) e^{i\omega_1 T_0} + \Gamma_1 e^{i\Omega_2 T_0} + \text{c.c.} \tag{5.5}$$

$$u_{20} = A_2(T_1) e^{i\omega_2 T_0} + \Gamma_2 e^{i\Omega_2 T_0} + \text{c.c.} \tag{5.6}$$

where

$$\Gamma_n = \tfrac{1}{2} F_n (\omega_n^2 - \Omega_2^2)^{-1} \exp(i\tau_n) \tag{5.7}$$

Substituting Eqs. 5.5 and 5.6 into Eqs. 5.3 and 5.4 yields

$$D_0^2 u_{11} + \omega_1^2 u_{11} = -2i\omega_1 (A_1' + \mu_1 A_1) e^{i\omega_1 T_0} - 4\omega_1 \Lambda_1 A_2 \bar{A}_1 e^{i(\omega_2 - \omega_1) T_0}$$
$$- \xi_{11} e^{2i\Omega_2 T_0} - \xi_{12} \bar{A}_1 e^{i(\Omega_2 - \omega_1) T_0}$$
$$- \xi_{13} \bar{A}_2 e^{i(\Omega_2 - \omega_2) T_0} + \text{c.c.} + \text{NST} \tag{5.8}$$

$$D_0^2 u_{21} + \omega_2^2 u_{21} = -2i\omega_2 (A_2' + \mu_2 A_2) e^{i\omega_2 T_0} - 4\omega_2 \Lambda_2 A_1^2 e^{2i\omega_1 T_0}$$
$$- \xi_{21} e^{2i\Omega_2 T_0} - \xi_{22} \bar{A}_1 e^{i(\Omega_2 - \omega_1) T_0}$$
$$- \xi_{23} \bar{A}_2 e^{i(\Omega_2 - \omega_2) T_0} + \text{c.c.} + \text{NST} \tag{5.9}$$

where Λ_1 and Λ_2 are defined in Eqs. 3.18 and 3.19, NST stands for terms that do not produce secular or small-divisor terms, and the ξ_{mn} are defined in Appendix B. In addition to the internal resonant condition $\omega_2 \approx 2\omega_1$, resonances occur when (a) $\omega_1 \approx 2\Omega_2$, superharmonic resonance of order 2 of the first mode; (b) $2\omega_2 \approx \Omega_2$, subharmonic resonance of order $\frac{1}{2}$ of the second mode; and (c) $\Omega_2 \approx \omega_2 + \omega_1$, combination resonance of the additive type. These cases are considered in the following sections.

A. Superharmonic Resonance of First Mode

In this case, we introduce the detuning parameters σ_1 and σ_2 defined according to

$$\omega_2 = 2\omega_1 + \varepsilon\sigma_1 \quad \text{and} \quad 2\Omega_2 = \omega_1 + \varepsilon\sigma_2 \tag{5.10}$$

Then, eliminating the terms that produce secular terms from Eqs. 5.8 and 5.9 yields

$$2i\omega_1(A_1' + \mu_1 A_1) + 4\omega_1\Lambda_1 A_2\bar{A}_1 e^{i\sigma_1 T_1} + |\xi_{11}|e^{i(\sigma_2 T_1 + \theta_{11})} = 0 \tag{5.11}$$

$$2i(A_2' + \mu_2 A_2) + 4\Lambda_2 A_1^2 e^{-i\sigma_1 T_1} = 0 \tag{5.12}$$

where

$$\xi_{mn} = |\xi_{mn}| \exp(i\theta_{mn}) \tag{5.13}$$

Equations 5.11 and 5.12 have the same form as Eqs. 4.43 and 4.44 obtained for the case of primary resonance of the first mode.

B. Subharmonic Resonance of Second Mode

In this case, we introduce the detuning parameters σ_1 and σ_2 defined according to

$$\omega_2 = 2\omega_1 + \varepsilon\sigma_1 \quad \text{and} \quad \Omega_2 = 2\omega_2 + \varepsilon\sigma_2 \tag{5.14}$$

Then, eliminating the terms that produce secular terms from Eqs. 5.8 and 5.9, we obtain

$$2i(A_1' + \mu_1 A_1) + 4\Lambda_1 A_2\bar{A}_1 e^{i\sigma_1 T_1} = 0 \tag{5.15}$$

$$2i\omega_2(A_2' + \mu_2 A_2) + 4\omega_2\Lambda_2 A_1^2 e^{-i\sigma_1 T_1} + |\xi_{23}|\bar{A}_2 e^{i(\sigma_2 T_1 + \theta_{23})} = 0 \tag{5.16}$$

which have the same form as Eqs. 3.16 and 3.17 obtained for the case of principal parametric resonance of the second mode. Mook, Plaut, and HaQuang (22) used the method of multiple scales to study the influence of an internal resonance on a subharmonic resonance. They obtained numerical results of a special form of Eqs. 2.4 and 2.5 that show a slow continual exchange between the two modes. This type of behavior has been observed in an experiment by Haddow (38) using a model similar to the one in Figure 27. Nayfeh and Zavodney (39) experimentally and theoretically studied this case. They determined conditions for the existence of Hopf bifurcations and presented numerical as well as experimental results showing period-multiplying bifurcations and chaos.

C. Combination Resonance of Additive Type

In this case, we let

$$\omega_2 = 2\omega_1 + \varepsilon\sigma_1 \quad \text{and} \quad \Omega_2 = \omega_2 + \omega_1 + \varepsilon\sigma_2 \tag{5.17}$$

Then, eliminating the terms that produce secular terms from Eqs. 5.8 and 5.9, we obtain

$$2i\omega_1(A_1' + \mu_1 A_1) + 4\omega_1\Lambda_1 A_2\bar{A}_1 e^{i\sigma_1 T_1} + |\xi_{13}|\bar{A}_2 e^{i(\sigma_2 T_1 + \theta_{13})} = 0 \tag{5.18}$$

$$2i\omega_2(A_2' + \mu_2 A_2) + 4\omega_2\Lambda_2 A_1^2 e^{-i\sigma_1 T_1} + |\xi_{22}|\bar{A}_1 e^{i(\sigma_2 T_1 + \theta_{22})} = 0 \tag{5.19}$$

which have the same form as Eqs. 3.82 and 3.83 obtained for the case of combination parametric resonance of the additive type. Mook, HaQuang, and Plaut (21) used the method of multiple scales to determine the influence of an internal resonance on combination resonances. They found that the internal resonance may significantly reduce the response.

VI. INTERACTION OF PARAMETRIC AND EXTERNAL

In this section, we consider the interaction of the parametric and external excitations in Eqs. 2.4 and 2.5. To this end, we let $F_n = O(1)$ with Ω_2 being away from either ω_1 or ω_2. Substituting Eqs. 3.1 and 3.2 into Eqs. 2.4 and 2.5 and equating coefficients of like powers of ε on both sides, we obtain Eqs. 5.1 and 5.2 at $O(\varepsilon^0)$ and Eqs. 3.6 and 3.7 at $O(\varepsilon)$. The solutions of Eqs. 5.1 and 5.2 are given by Eqs. 5.5 and 5.6, which, upon substitution into Eqs. 3.6 and 3.7, yield

$$
\begin{aligned}
D_0^2 u_{11} + \omega_1^2 u_{11} = {} & -2i\omega_1(A_1' + \mu_1 A_1)e^{i\omega_1 T_0} - 4\omega_1\Lambda_1 A_2\bar{A}_1 e^{i(\omega_2 - \omega_1)T_0} \\
& - \xi_{11}e^{2i\Omega_2 T_0} - \xi_{12}\bar{A}_1 e^{i(\Omega_2 - \omega_1)T_0} \\
& - \xi_{13}\bar{A}_2 e^{i(\Omega_2 - \omega_2)T_0} - \tfrac{1}{2}F_{11}A_1 e^{i(\Omega_1 + \omega_1)T_0} \\
& - \tfrac{1}{2}F_{11}\bar{A}_1 e^{i(\Omega_1 - \omega_1)T_0} - \tfrac{1}{2}F_{12}A_2 e^{i(\Omega_1 + \omega_2)T_0} \\
& - \tfrac{1}{2}F_{12}\bar{A}_2 e^{i(\Omega_1 - \omega_2)T_0} - \tfrac{1}{2}(F_{11}\Gamma_1 + F_{12}\Gamma_2)e^{i(\Omega_2 + \Omega_1)T_0} \\
& - \tfrac{1}{2}(F_{11}\Gamma_1 + F_{12}\Gamma_2)e^{i(\Omega_2 - \Omega_1)T_0} + \text{c.c.}
\end{aligned} \tag{6.1}
$$

$$
\begin{aligned}
D_0^2 u_{21} + \omega_2^2 u_{21} = {} & -2i\omega_2(A_2' + \mu_2 A_2)e^{i\omega_2 T_0} - 4\omega_2\Lambda_2 A_1^2 e^{2i\omega_1 T_0} \\
& - \xi_{21}e^{2i\Omega_2 T_0} - \xi_{22}\bar{A}_1 e^{i(\Omega_2 - \omega_1)T_0} \\
& - \xi_{23}\bar{A}_2 e^{i(\Omega_2 - \omega_2)T_0} - \tfrac{1}{2}F_{21}A_1 e^{i[(\Omega_1 + \omega_1)T_0 + \tau]}
\end{aligned}
$$

$$-\tfrac{1}{2}F_{21}\bar{A}_1 e^{i[(\Omega_1-\omega_1)T_0+\tau]} - \tfrac{1}{2}F_{22}A_2 e^{i[(\Omega_1+\omega_2)T_0+\tau]}$$

$$-\tfrac{1}{2}F_{22}\bar{A}_2 e^{i[(\Omega_1-\omega_2)T_0+\tau]} - \tfrac{1}{2}(F_{21}\Gamma_1+F_{22}\Gamma_2)e^{i[(\Omega_2+\Omega_1)T_0+\tau]}$$

$$-\tfrac{1}{2}(F_{21}\Gamma_1+F_{22}\Gamma_2)e^{i[(\Omega_2-\Omega_1)T_0-\tau]} \tag{6.2}$$

In addition to the resonances discussed in Sections III and V, resonances occur when $\omega_1 \approx \Omega_2 \pm \Omega_1$ or $\omega_2 \approx \Omega_2 \pm \Omega_1$, combination resonances of the additive or difference type involving the first or the second mode. We note that more than one resonance can occur simultaneously; that is, resonances can overlap.

A. Combination Resonances

When $\omega_1 \approx \Omega_2 + \Omega_1$ and no other resonances besides the internal resonance occur, we let

$$\omega_2 = 2\omega_1 + \varepsilon\sigma_1 \quad \text{and} \quad \Omega_2+\Omega_1 = \omega_1 + \varepsilon\sigma_2 \tag{6.3}$$

Then, eliminating the terms that lead to secular terms from Eqs. 6.1 and 6.2 yields

$$2i\omega_1(A'_1+\mu_1 A_1) + 4\omega_1\Lambda_1 A_2\bar{A}_1 e^{i\sigma_1 T_1} - f_e e^{i(\sigma_2 T_1+\tau_e)} = 0 \tag{6.4}$$

$$2i(A'_2+\mu_2 A_2) + 4\Lambda_2 A_1^2 e^{-i\sigma_1 T_1} = 0 \tag{6.5}$$

where

$$\tfrac{1}{2}(F_{11}\Gamma_1+F_{12}\Gamma_2) = -f_e e^{i\tau_e} \tag{6.6}$$

Equations 6.4 and 6.5 have the same form as Eqs. 4.43 and 4.44 obtained for the case of primary resonance of the first mode.

When $\omega_1 \approx \Omega_2 - \Omega_1$ and $\omega_2 \approx 2\omega_1$, we let

$$\omega_2 = 2\omega_1 + \varepsilon\sigma_1 \quad \text{and} \quad \Omega_2-\Omega_1 = \omega_1 + \varepsilon\sigma_2 \tag{6.7}$$

Then, eliminating the terms that lead to secular terms from Eqs. 6.1 and 6.2 yields Eqs. 6.4 and 6.5.

When $\Omega_2 \pm \Omega_1 = \omega_2 + \varepsilon\sigma_2$, instead of Eqs. 6.4 and 6.5, we obtain

$$2i(A'_1+\mu_1 A_1) + 4\Lambda_1 A_2\bar{A}_1 e^{i\sigma_1 T_1} = 0 \tag{6.8}$$

$$2i\omega_2(A'_2+\mu_2 A_2) + 4\omega_2\Lambda_2 A_1^2 e^{-i\sigma_1 T_1} - f_e e^{i(\sigma_2 T_1+\tau_e)} = 0 \tag{6.9}$$

$$\tfrac{1}{2}(F_{21}\Gamma_1+F_{22}\Gamma_2) = -f_e e^{i(\tau_e\pm\tau)} \tag{6.10}$$

where the positive and negative signs in Eq. 6.10 correspond to the difference and additive combination resonances, respectively.

B. Overlapping Resonances

One or more of the resonances discussed in Sections III and V and the present section may occur simultaneously. For example,

(a) $\qquad\qquad\qquad \Omega_1 \approx 2\omega_n \qquad$ and $\quad \Omega_2 \approx 2\omega_2$

(b) $\qquad\qquad\qquad \Omega_1 \approx 2\omega_n \qquad$ and $\quad 2\Omega_2 \approx \omega_1$

(c) $\qquad\qquad\qquad \Omega_1 \approx 2\omega_n \qquad$ and $\quad \Omega_2 \approx \omega_2 + \omega_1$

(d) $\qquad\qquad\qquad \Omega_1 \approx \omega_2 + \omega_1 \quad$ and $\quad \Omega_2 \approx \omega_2 + \omega_1$

(e) $\qquad\qquad\qquad \Omega_1 \approx \omega_2 - \omega_1 \quad$ and $\quad \Omega_2 \approx \omega_2 + \omega_1$

might occur simultaneously. When such resonances overlap, they might enhance or reduce their individual influence. For example, Section V.B shows that subharmonic resonances of order $\frac{1}{2}$ are similar to principal parametric resonances. Consequently, their overlapping may enhance or reduce their individual influences, depending on the amplitudes and phases of the excitations. When

$$\Omega_2 = \Omega_1 = 2\omega_2 + \varepsilon\sigma_2 \qquad (6.11)$$

eliminating the terms that lead to secular terms from Eqs. 6.1 and 6.2 yields

$$2i(A_1' + \mu_1 A_1) + 4\Lambda_1 A_2 \bar{A}_1 e^{i\sigma_1 T_1} = 0 \qquad (6.12)$$

$$2i\omega_2(A_2' + \mu_2 A_2) + 4\omega_2 \Lambda_2 A_1^2 e^{-i\sigma_1 T_1} + 2f_e \bar{A}_2 e^{i(\sigma_2 T_1 + \tau_e)} = 0 \qquad (6.13)$$

where

$$f_e e^{i\tau_e} = g e^{i\tau} + \tfrac{1}{2}|\xi_{23}|e^{i\theta_{23}} \qquad (6.14)$$

Thus, if $|f_e| < |g|$, the addition of the subharmonic excitation will result in the reduction of the response to the parametric excitation, and it may even quench it if $|f_e|$ is below the threshold for the excitation of the principal parametric response. Nayfeh (40) demonstrated the quenching of the response of a single-degree-of-freedom system with quadratic and cubic nonlinearities to a principal parametric excitation by the addition of a small subharmonic excitation of order $\frac{1}{2}$ having the proper amplitude and phase.

Acknowledgment

This work was supported by the United States Office of Naval Research under Contract No. N00014-83-K-0184, NR 4322753, the United States Air Force Office of Scientific Research under Grant No. 86-0090, and the National Science Foundation under Grant No. MSM-8521748.

APPENDIX A

$$c_{11} = \mu_1(2f_{21}\cos\tau - f_{12}) - \tfrac{2}{3}(\sigma_2 + \sigma_1)f_{21}\sin\tau \tag{A.1}$$

$$c_{12} = 4\mu_2(2f_{21}\cos\tau - f_{12}) + \tfrac{8}{3}(2\sigma_2 - \sigma_1)f_{21}\sin\tau \tag{A.2}$$

$$c_{21} = \tfrac{1}{3}(\sigma_2 + \sigma_1)(f_{12} + 2f_{21}\cos\tau) + 2\mu_1 f_{21}\sin\tau \tag{A.3}$$

$$c_{22} = -\tfrac{4}{3}(2\sigma_2 - \sigma_1)(f_{12} + 2f_{21}\cos\tau) + 8\mu_2 f_{21}\sin\tau \tag{A.4}$$

$$d = 2(f_{12}^2 - 4f_{21}^2)^{-1} \tag{A.5}$$

APPENDIX B

$$\xi_{11} = (\delta_1 - \delta_4\Omega_2^2 - \delta_7\Omega_2^2)\Gamma_1^2 + (2\delta_2 - \delta_5\Omega_2^2 - \delta_8\Omega_2^2 - \delta_9\Omega_2^2)\Gamma_1\Gamma_2$$
$$+ (\delta_3 - \delta_6\Omega_2^2 - \delta_{10}\Omega_2^2)\Gamma_2^2 \tag{B.1}$$

$$\delta_{12} = (2\delta_1 + 2\delta_4\omega_1\Omega_2 - \delta_7\omega_1^2 - \delta_7\Omega_2^2)\Gamma_1$$
$$+ (\delta_2 + \delta_5\omega_1\Omega_2 - \delta_8\omega_1^2 - \delta_9\Omega_2^2)\Gamma_2 \tag{B.2}$$

$$\xi_{13} = (\delta_2 + \delta_5\omega_2\Omega_2 - \delta_8\Omega_2^2 - \delta_9\omega_2^2)\Gamma_1$$
$$+ (2\delta_3 + 2\delta_6\omega_2\Omega_2 - \delta_{10}\omega_2^2 - \delta_{10}\Omega_2^2)\Gamma_2 \tag{B.3}$$

$$\xi_{21} = (\alpha_1 - \alpha_4\Omega_2^2 - \alpha_7\Omega_2^2)\Gamma_1^2 + (2\alpha_2 - \alpha_5\Omega_2^2 - \alpha_8\Omega_2^2 - \alpha_9\Omega_2^2)\Gamma_1\Gamma_2$$
$$+ (\alpha_3 - \alpha_6\Omega_2^2 - \alpha_{10}\Omega_2^2)\Gamma_2^2 \tag{B.4}$$

$$\xi_{22} = (2\alpha_1 + 2\alpha_4\omega_1\Omega_2 - \alpha_7\Omega_2^2 - \alpha_7\omega_1^2)\Gamma_1$$
$$+ (\alpha_2 + \alpha_5\omega_1\Omega_2 - \alpha_8\omega_1^2 - \alpha_9\Omega_2^2)\Gamma_2 \tag{B.5}$$

$$\xi_{33} = (\alpha_2 + \alpha_5\omega_2\Omega_2 - \alpha_8\Omega_2^2 - \alpha_9\omega_2^2)\Gamma_1$$
$$+ (2\alpha_3 + 2\alpha_6\omega_2\Omega_2 - \alpha_{10}\omega_2^2 - \alpha_{10}\Omega_2^2)\Gamma_2 \tag{B.6}$$

References

1. A. H. Nayfeh, *Perturbation Methods*, Wiley-Interscience, New York, 1973.
2. A. H. Nayfeh, *Introduction to Perturbation Techniques*, Wiley-Interscience, New York, 1981.

3. A. H. Nayfeh, *Problems in Perturbation*, Wiley-Interscience, New York, 1985.
4. A. H. Nayfeh and D. T. Mook, *Nonlinear Oscillations*, Wiley-Interscience, New York, 1979.
5. R. M. Evan-Iwanowski, *Resonance Oscillations in Mechanical Systems*, Elsevier, New York, 1976.
6. R. A. Ibrahim, *Parametric Random Vibration*, Wiley, London, 1985.
7. G. Schmidt and A. Tondl, *Non-Linear Vibrations*, Akademie-Verlag, Berlin, 1986.
8. W. Froude, "Remarks on Mr. Scott-Russell's Paper on Rolling," *Trans. Inst. Nav. Architect.* **4**, 232 (1863).
9. E. Mettler and F. Weidenhammer, *Ing. Archiv* **31**, 421 (1962).
10. P. R. Sethna, *J. Appl. Mechan.* **32**, 576 (1965).
11. R. S. Haxton and A. D. S. Barr, *J. Eng. Ind.* **94**, 119 (1972).
12. A. H. Nayfeh, D. T. Mook, and L. R. Marshall, *J. Hydronaut.* **7**, 145 (1973).
13. D. T. Mook, L. R. Marshall, and A. H. Nayfeh, *J. Hydronaut.* **8**, 32 (1974).
14. T. Yamamoto and K. Yasuda, *Bull. Japan. Soc. Mechan. Eng.* **20**, 168 (1977).
15. T. Yamamoto, K. Yasuda, and I. Nagasaka, *Bull. JSME* **20**, 1093 (1977).
16. H. Hatwal, A. K. Malik, and A. Ghosh, *J. Sound Vibrat.* **81**, 153 (1982).
17. H. Hatwal, A. K. Malik, and A. Ghosh, *J. Appl. Mechan.* **50**, 663 (1983).
18. A. G. Haddow, A. D. S. Barr, and D. T. Mook, *J. Sound Vibrat.* **97**, 451 (1984).
19. J. W. Miles, *Phys. D.* **13**, 247 (1984).
20. A. H. Nayfeh and L. D. Zavodney, *J. Appl. Mechan.* (accepted for publication).
21. D. T. Mook, N. HaQuang, and R. H. Plaut, *J. Sound Vibrat.* **104**, 229 (1986).
22. D. T. Mook, R. H. Plaut, and N. HaQuang, *J. Sound Vibrat.* **102**, 473 (1985).
23. A. H. Nayfeh and R. A. Raouf, *Int. J. Solids Struct.* **23**, 1625 (1987).
24. A. H. Nayfeh and R. A. Raouf, *J. Appl. Mechan.* **54**, 571 (1987).
25. A. H. Nayfeh, *J. Ship Res.* 1988, in press.
26. A. H. Nayfeh and D. T. Mook, *Proceedings of the VIIIth International Conference on Nonlinear Oscillations*, Prague, 1978, Czechoslovak Acad. Sci. p. 511.
27. A. H. Nayfeh, *J. Sound Vibrat.* **88**, 547 (1983).
28. A. H. Nayfeh, *J. Sound Vibrat.* **90**, 237 (1983).
29. J. W. Miles, *J. Appl. Math. Phys.* (*ZAMP*) **36**, 337 (1985).
30. J. W. Miles, *J. Fluid Mechan.* **146**, 285 (1984).
31. A. H. Nayfeh, *J. Sound Vibrat.* **199**, 95 (1987).
32. A. H. Nayfeh, *Phys. Fluids* **30**, 2976 (1987).
33. A. H. Nayfeh and L. D. Zavodney, *J. Sound Vibrat.* **107**, 329 (1986).
34. J. E. Marsden and M. McCracken, *Hopf Bifurcation and its Applications*, Springer-Verlag, New York, 1976.
35. J. K. Hale, *Oscillations in Nonlinear Systems*, McGraw-Hill, New York, 1963.
36. Department of Engineering Science and Mechanics, "Nonlinear Phenomena," videotape, Virginia Polytechnic Institute and State University Blacksburg, VA, 1985.
37. P. R. Sethna and A. K. Bajaj, *J. Appl. Mechan.* **45**, 895 (1978).
38. A. G. Haddow, Theoretical and Experimental Study of Modal Interaction in a Two Degree of Freedom Nonlinear Structure. Ph.D. Thesis, University of Dundee, Scotland, 1983.
39. A. H. Nayfeh and L. D. Zavodney, *J. Appl. Mechan.* (accepted for publication).
40. A. H. Nayfeh, *J. Sound Vibrat.* **96**, 333 (1984).

CHAPTER IV

NUMERICAL TREATMENT OF LINEAR AND NONLINEAR PERIODIC SYSTEMS, WITH APPLICATIONS

PERETZ P. FRIEDMANN

Mechanical, Aerospace and Nuclear Engineering Department, University of California at Los Angeles, Los Angeles, California 90024

CONTENTS

This chapter discusses some recently developed numerical techniques for analyzing the stability, linear response, and nonlinear response of periodic systems, which are governed by ordinary differential equations with periodic coefficients. First, Floquet theory is briefly reviewed, and numerically efficient methods for evaluating the transition matrix at the end of one period are described. Next, the numerical treatment of the linear response problem is discussed. Finally, the numerical solution of the nonlinear response problem is treated using quasi-linearization and periodic shooting. Applications illustrating numerical properties of the methods are presented.

197

I. INTRODUCTION

The equations of motion for a wide class of dynamical systems are mathematically represented by a system of ordinary differential equations with periodic coefficients. These equations can be either linear or nonlinear. Two basic problems associated with these systems are of significant importance: the stability of such systems and their response under various kinds of excitation. In a previous paper [1]* we discussed the efficient numerical treatment of linear periodic systems in considerable detail, with an emphasis on stability problems. In Ref. 1 two numerically efficient methods for dealing with the stability of linear periodic systems were presented and applied to the classical parametrically excited beam problem, a helicopter rotor aeroelastic problem in forward flight, and a ground resonance type of aeromechanical problem. The purpose of this chapter is to discuss some previous work on equations with periodic coefficients and present a number of recently developed numerical methods that are applicable to the treatment of two important problems: (a) response of a linear periodic system under a periodic excitation and (b) response of a nonlinear periodic system under periodic excitation. These two mathematical problems occur frequently in the study of many different branches of science and engineering. However, it should be noted that the author's primary interest and experience with such systems is due to the fact that the efficient numerical treatment of periodic systems plays a central role in helicopter rotor dynamics, aeroelasticity, and structural dynamics [2,3]. A detailed discussion of methods used successfully in problems encountered in helicopter rotor dynamics and aeroelasticity has been presented recently [4]. The emphasis in this chapter is on those particular aspects of these numerical methods that are also widely applicable to other fields and disciplines.

A substantial body of mathematical literature exists on the behavior of ordinary differential equations with periodic coefficients [5–10]. The mathematical interest in this subject is linked to the wide range of applications, in both science and engineering, where periodic systems play an important role. The importance of these applications generated significant research on this topic in various disciplines within science and engineering. Typical examples of such disciplines are

a. control system and network theory [11–13];
b. dynamic stability problems in applied mechanics [14–19] and nonlinear vibrations [20];
c. dynamics and structural dynamics of rotating systems such as helicopter

---

*In this chapter, references are enclosed in brackets—Ed.

rotor blades [1–4, 21], satellite dynamics [22, 23], and rotating shafts [18, 14]; and

d. laser molecule interactions [24–27].

It is not surprising that a considerable communication gap between researchers working in these diverse fields exists and has resulted in a considerable duplication of efforts. The author hopes that this chapter will alleviate this problem by presenting numerical methods that are both general and numerically efficient.

II. FLOQUET THEORY AND THE STABILITY OF LINEAR PERIODIC SYSTEMS

A. Brief Review of Floquet's Theorem and Its Consequences

The stability of systems of linear differential equations with periodic coefficients is governed by Floquet's theorem [5, 8, 12, 13]. Consider a linear periodic homogeneous system

$$\{\dot{y}(\psi) = [A(\psi)]\{y(\psi)\},\tag{1}$$

where $[A(\psi)]$ is a square periodic matrix having a common period T; thus $[A(\psi)] = [A(\psi + T)]$. Based on Floquet's theorem, the transition matrix for the periodic system, represented by Eq. 1, has the form

$$[\Phi(\psi, \psi_0)] = [P(\psi)]^{-1} e^{[R](\psi - \psi_0)} [P(\psi_0)],\tag{2}$$

where $[P(\psi)]$ is also a periodic matrix $[P(\psi + T)] = [P(\psi)]$, $[P(\psi_0)]$ is associated with the initial conditions at ψ_0, and $[R]$ is a constant matrix related to the value of the transition matrix $[\Phi(T, 0)]$ at the end of one period:

$$[\Phi(T, 0)] = e^{[R]T}.\tag{3}$$

Each column of the transition matrix represents a linear independent solution [6] to Eq. 1; thus,

$$[\dot{\Phi}(\psi, \psi_0)] = [A(\psi)][\Phi(\psi, \psi_0)]\tag{4}$$

and $[\Phi(\psi_0, \psi_0)] = [I]$.

In general, two cases can occur: (a) The transition matrix $[\Phi(T, 0)]$ has n distinct eigenvalues, *denoted characteristic multipliers,* which are associated with n independent eigenvectors, and (b) the transition matrix $[\Phi(T,0)]$ has a number of repeated eigenvalues and a certain number of corresponding generalized eigenvectors. A discussion of this case can be found in Ref. 6.

For case (a), where $[\Phi(T,0)]$ has n distinct eigenvalues, these eigenvalues can be related to the n distinct eigenvalues of the matrix $[R]$, which are usually denoted as *characteristic exponents*. From linear algebra, a similarity transformation [28] can be found such that

$$[Q]^{-1}[R][Q] = [\lambda], \tag{5}$$

where the columns of $[Q]$ are the n linearly independent eigenvectors of $[R]$ and $[\lambda]$ is a diagonal matrix containing the eigenvalues of $[R]$. Using the definition of the matrix exponential [28],

$$e^{[R]T} = [I] + \sum_{n=1}^{\infty} \frac{([R]T)^n}{n!}. \tag{6}$$

Combining Eq. 5 and 6 and using Eq. 3,

$$e^{[R]T} = e^{[Q][\lambda][Q]^{-1}T} = [I] + \sum_{n=1}^{\infty} \frac{([Q][\lambda][Q]^{-1}T)^n}{n!}$$

$$= [Q]\left([I] + \sum_{n=1}^{\infty} \frac{([\lambda]T)^n}{n!} \right)[Q]^{-1}$$

$$= [Q]e^{[\lambda]T}[Q]^{-1} = [\Phi(T,0)]. \tag{7}$$

Equation 7 can be also interpreted as a similarity transformation,

$$e^{[\lambda]T} = [Q]^{-1}[\Phi(T,0)][Q] = [\Lambda], \tag{8}$$

where $[\Lambda]$ is a diagonal matrix containing the eigenvalues of the transition matrix at the end of one period, which are usually denoted characteristic multipliers. Equation 8 provides a relation between each characteristic exponent and characteristic multiplier:

$$e^{\lambda_s T} = \Lambda_s, \qquad s = 1, 2, \ldots, n. \tag{9}$$

In general, both λ_s and Λ_s are complex quantities, and thus,

$$\zeta_s = \frac{1}{2T} \ln (\Lambda_{sR}^2 + \Lambda_{sI}^2), \qquad \omega_s = \frac{1}{T} \tan^{-1} \frac{\Lambda_{sI}}{\Lambda_{sR}}. \tag{10}$$

Having provided this background on Floquet theory, the most important consequences of Floquet's theorem can be stated:

1. The stability of the system, Eq. 1, is determined by the matrix $[R]$ or the knowledge of the transition matrix at the end of *one* period, $[\Phi(T,0)]$.

2. Knowledge of the transition matrix over one period determines the solution of the homogeneous system, Eq. 1, everywhere, due to the semigroup [12] or extension property of the transition matrix

$$[\Phi(\psi + pT, 0)] = [\Phi(\psi, 0)](e^{[R]T})^p. \tag{11}$$

3. The stability criteria for the system can be stated in terms of the real part of the characteristic exponent λ_s. The solutions of Eq. 1 are stable; that is, they approach zero as $\psi \to \infty$ if

$$\zeta_s < 0 \quad \text{or} \quad |\Lambda_s| < 1 \quad \text{for } s = 1, 2, \ldots, n.$$

The real part of the characteristic exponent ζ_s is a measure of growth or decay; however, it is difficult to relate it to a physical quantity, such as the damping ratio encountered in constant-coefficient systems, unless one deals with a single-degree-of-freedom system. For single-degree-of-freedom systems, using the energy dissipated per cycle, one can establish an equivalence between a conventional damping coefficient and the real part of the characteristic exponent. The imaginary part of the characteristic exponent ω_s represents a nondimensional frequency. Since $\tan^{-1}$ in Eq. 10 is multivalued, it can be determined only within an integer multiple of the nondimensional period T.

From this discussion it is clear that efficient numerical computation of $[\Phi(T,0)]$ is the basis for the efficient numerical treatment of the stability problem for periodic systems. Furthermore, it will be shown that efficient numerical calculation of transition matrices plays an important role also in the treatment of linear and nonlinear response problems. Numerical methods for evaluating the transition matrix at the end of one period are discussed in the next section.

B.   Numerical Methods for Evaluating $[\Phi(T,0)]$

A detailed discussion of efficient numerical techniques for calculating $[\Phi(T,0)]$ have been presented in Ref. 1. The main advantage of the techniques presented in Ref. 1 is due to the fact that $[\Phi(T,0)]$ is obtained by a single integration pass through Eq. 1, as opposed to n integration passes (for an $n \times n$ system), which was the method employed in some previous studies [29, 30]. Since the publication of Ref. 1, comparative numerical studies [31] have confirmed this conclusion. Two separate schemes were presented in Ref. 1. The first method consists of approximating the periodic matrix $[A(\psi)]$ in Eq. 1 by a series of step functions. This method can be considered to be a generalization of the "rectangular ripple" method [32] to multidimensional systems. Its analytical basis has been provided by Hsu in a series of papers [15–17]. The second method is based on evaluating $[\Phi(T, 0)]$ by using a fourth-order Runge–Kutta scheme to obtain the solution in each step and then cascading the solutions [1]. When carefully programmed, both methods require approximately the same amount of computer time and yield similar accuracy. It will be shown in this chapter that these schemes are important for both stability and response calculations. Therefore, the first method discussed in Ref. 1, which is somewhat more concise and more convenient to implement, is described in what follows.

Each period T is divided in K equal intervals and in each interval the equations are treated as a set of equations with constant coefficients denoting each interval by ψ_k, $k = 0, 1, 2, \ldots, K$, with $0 = \psi_0 < \psi_1 < \cdots < \psi_k = T$. The kth interval (ψ_{k-1}, ψ_k) is denoted by τ_k, and its size is denoted by $\Delta_k = \psi_k - \psi_{k-1}$. In the kth interval the periodic coefficient matrix $[A(\psi)]$ is replaced by a constant matrix $[C_k]$ defined by

$$[C_k] = \frac{1}{\Delta_k} \int_{\psi_{k-1}}^{\psi_k} [A(\xi)] \, d\xi, \qquad \xi \in \tau_k. \tag{12}$$

Thus, the actual system, Eq. 1, is approximated by a constant-coefficient system consisting of a series of step functions:

$$\{\dot{y}_A(\psi; K)\} = [C(\psi, K)]\{y_A(\psi, K)\}, \tag{13}$$

where $\{\dot{y}_A(0, K)\} = \{y(0)\}$, and the subscript A denotes approximate quantities. The theory of linear differential equations with constant coefficients [5] enables one to write the transition matrix for the end of one period, based on Eq. 13, as

$$[\Phi_A(T, K)] = e^{\Delta_K[C_K]} e^{\Delta_{K-1}[C_{K-1}]} \cdots e^{\Delta_1[C_1]}$$

$$= \prod_{k=1}^{K} e^{\Delta_k[C_k]}. \tag{14}$$

With regard to the product sign, it is understood that the order of positioning of the factors is material and the kth factor is to be placed in front of the $(k-1)$th factor. It should be noted that the approximate expression for $[\Phi(T, 0)]$ given by Eq. 14 represents the first level of approximation in this method. It can be shown [17] that when $K \to \infty$, $[\Phi_A(T, K)] \to [\Phi(T, 0)]$.

The efficient numerical evaluation of $[\Phi_A(T, K)]$ is accomplished by using the definition of the matrix exponential for each term in the product given by Eq. 14; thus,

$$e^{\Delta_i[C_i]} = [I] + \sum_{j=1}^{\infty} \frac{(\Delta_i[C_i])^j}{j!}. \tag{15}$$

For sufficiently small time intervals $\Delta_i \to 0$, and the series represented by Eq. 15 converges rapidly; thus, the series can be truncated and represented by a finite number of terms,

$$e^{\Delta_i[C_i]} \cong [I] + \sum_{j=1}^{J} \frac{(\Delta_i[C_i])^j}{j!}, \tag{16}$$

where the error is given by

$$\sum_{j=J+1}^{\infty} \frac{(\Delta_i[C_i])^j}{j!}. \tag{16a}$$

The final approximation for the transition matrix at the end of one period is obtained by combining Eq. 14 and 16,

$$[\Phi_{AA}(T, 0)] = \prod_{i=1}^{K} \left([I] + \sum_{j=1}^{J} \frac{(\Delta_i[C_i])^j}{j!} \right). \tag{17}$$

Equation 16a represents the error in approximating each matrix exponential of the product given by Eq. 14; thus, Eq. 17 represents a second level of approximation inherent in this method, and the double subscript in $[\Phi_{AA}(T, 0)]$ emphasizes this aspect. General asymptotic error estimates for this approximation were obtained in Ref. 17, where it was shown that the error is of order $O(\Delta_i^2)$ for $J \geq 2$. It should be noted, however, that these error estimates were not sharp, and by increasing the number of terms used in approximating the matrix exponential, the accuracy of the method is improved.

Numerical experiments performed in Ref. 1 have shown that the approximation of $[\Phi(T, 0)]$ with $J = 4$, based on Eq. 17, is comparable to a fourth-order Runge–Kutta method in terms of accuracy and computer times. Furthermore, it is evident that Eq. 16 represents an approximation of the matrix exponential by a truncated Taylor series, which represents just one of the many available methods for computing the exponential of a matrix [33, 34].

An alternative procedure for evaluating $[\Phi(T,0)]$ using a fourth-order Runge–Kutta method has been presented in Ref. 1. In a similar manner the transition matrix can be evaluated using a variety of numerical integration procedures that can be implemented in a similar single-pass manner [35], which has been emphasized in this section.

A general procedure for calculating the transition matrix at the end of a period using efficient general-purpose numerical integration procedures, *in a single pass*, is presented next. Let $[D(\psi)]$ be a block diagonal square matrix of size $n^2 \times n^2$ defined as

$$[D(\psi)] = \begin{bmatrix} [A(\psi)] & \cdots & \cdots & 0 \\ \vdots & [A(\psi)] & & \vdots \\ 0 & \cdots & \cdots & [A(\psi)] \end{bmatrix}. \tag{18}$$

Then solve the system

$$\{\dot{z}(\psi)\} = [D(\psi)]\{z(\psi)\} \tag{19}$$

from $\psi = 0$ to $\psi = 2\pi$, with the initial condition vector $\{z(0)\}$ given by

$$\{z(0)\} = \begin{Bmatrix} \mathbf{e}_1 \\ \mathbf{e}_2 \\ \vdots \\ \mathbf{e}_n \end{Bmatrix}, \tag{20}$$

where $\{e_i\}$ is a vector of size n with all its elements equal to 0, except for the ith, which is equal to 1. Let $\{z(T)\}$ be the solution vector of Eq. 19 at the end of a period $\psi = T$ (where usually $T = 2\pi$). Partition $\{Z(T)\}$ into n vectors of size n:

$$\{Z(T)\} = \begin{Bmatrix} \mathbf{z}_1(T) \\ \mathbf{z}_2(T) \\ \vdots \\ \mathbf{z}_n(T) \end{Bmatrix}. \tag{21}$$

The ith column of the transition matrix at the end of the period, $[\Phi(T,0)]$, is then given by $\{z_i(T)\}$, that is,

$$[\Phi(T,0)] = [\{z_1(T)\}|\{z_2(T)\}|\cdots|\{z_n(T)\}]. \tag{22}$$

This algorithm is a single-pass algorithm for obtaining the transition matrix at the end of one period by solving Eq. 1. The implementation of the algorithm does not require the actual coding of $[D(\psi)]$; only the coding of $[A(\psi)]$ is required. This approach was used successfully in Ref. 36. The code used to integrate the equations was DE/STEP, a general-purpose Adams–Bashforth ordinary differential equation solver, which tends to be particularly convenient when the right side of the system equations is expensive to evaluate [37–39]. Another advantage of such a code is its automatic error control.

Recently another approach to calculating the transition matrix at the end of a period using finite elements in the time domain was proposed [40]. This method is based on Hamilton's law of varying action, which is sometimes also denoted as a weak version of Hamilton's principle, which was first discussed by Bailey [41]. Bailey recognized that when the generalized coordinates at a final time t_f are not known, their variation cannot be taken as zero, which introduces as additional term in the classical version of Hamilton's principle. Subsequently Bailey and others [42] considered the direct numerical solution of Hamilton's law of varying action by breaking the time domain into small segments that lead essentially to a finite-element formulation in the time domain. Additional research by Baruch and Ritt [43, 44] has shown that there are ambiguities associated with Hamilton's law of varying action as formulated by Bailey [43] and that the finite-element discretization of this principle does not always lead to numerically stable time domain integration procedures [44, 45] for general dynamic problems. For the case of a periodic system, the periodicity requirement provides additional useful information, and therefore, the method could be implemented successfully for rotor dynamics problems [40] when using cubic interpolation for the finite-element implementation in the time domain [40]. It should be noted, however, that the efficiency and reliability of this method has not been carefully compared to the other numerical methods discussed in this section. However, the method appears to be quite efficient.

III. GENERAL MATHEMATICAL FORM OF PERIODIC NONLINEAR SYSTEMS

Nonlinear periodic systems are frequently encountered in network theory, control system theory, structural dynamics, dynamics, aeroelasticity, and numerous other applications. Numerous examples of such systems can be found in Refs. 2, 4, 14, 20, 46, 47. A convenient description of a coupled system of ordinary differential equations with periodic coefficients can be represented in the following symbolic form:

$$[M]\{\ddot{q}\} + [C(\psi)]\{\dot{q}\} + [K(\psi)]\{q\} = \{F_{NL}(\psi, \mathbf{q}, \dot{\mathbf{q}})\}. \tag{23}$$

In this formulation all nonlinear effects and the excitation are combined in the general vector $\{F_{NL}(\psi, \mathbf{q}, \dot{\mathbf{q}})\}$. The nonlinear terms can be due to numerous effects. However, one frequently distinguishes between two classes of problem: those in which the nonlinearities are weak and those where the nonlinearities are strong. Systems with strong nonlinearities pose major numerical difficulties, and the most effective and general method for dealing with these problems is the use of nonlinear difference equations [46]. Weakly nonlinear periodic systems can be treated using either asymptotic analysis or quasilinearization. The numerical methods presented in this section will be restricted to weakly nonlinear periodic systems, since most of the applications studied by the author belong to this category.

The mathematical and numerical treatment of Eq. 23 is facilitated by rewriting them in first-order state variable form; thus,

$$\{y_1(\psi)\} = \{\dot{q}(\psi)\}, \qquad \{y_2(\psi)\} = \{q(\psi)\}, \tag{24}$$

and Eq. 19 can be rewritten as

$$\{\dot{y}\} = \{Q_{NL}(\psi, \dot{y}, y)\} = \{Z(\psi)\} + [L(\psi)]\{y\} + \{N_1(\psi, y)\} + \{N_2(\psi, y, \dot{y})\}, \tag{25}$$

where

$$[L(\psi)] = \begin{bmatrix} -[M]^{-1}[C(\psi)] & -[M]^{-1}[K(\psi)] \\ [I] & 0 \end{bmatrix} \tag{26}$$

and

$$\{Z(\psi)\} + \{N_1(\psi, y)\} + \{N_2(\psi, y, \dot{y})\} = \begin{Bmatrix} -[M]^{-1}F_{NL}(\psi, y_1, y_2) \\ \{0\} \end{Bmatrix}, \tag{27}$$

with

$$\{y(\psi)\} = \begin{Bmatrix} y_1(\psi) \\ y_2(\psi) \end{Bmatrix}.$$

It should be noted that this transformation doubles the size of the system represented by Eq. 20 when compared to Eq. 19. Since the system is periodic with a period of 2π, one can also assume that the excitation has also a period of 2π. This is usually the case for the applications discussed here. However, note that this assumption introduces no restrictions on the methods discussed here and can be easily generalized to the case when the periodicity of the system and the periodicity of the excitation are different:

$$\{Z(\psi)\} = \{Q(\psi + 2\pi)\}, \tag{28}$$

$$\{L(\psi)\} = [L(\psi + 2\pi)], \tag{29}$$

$${N_1(\psi, \mathbf{y})} = {N_1(\psi + 2\pi, \mathbf{y}, \dot{\mathbf{y}})}, \tag{30}$$

and

$${N_2(\psi, \mathbf{y}, \dot{\mathbf{y}})} = {N_2(\psi + 2\pi, \mathbf{y}, \dot{\mathbf{y}})}. \tag{31}$$

The vector ${Z(\psi)}$ represents a known excitation, the matrix $[L(\psi)]$ contains the periodic coefficients of the linear system, and the vectors ${N_1(\psi, \mathbf{y})}$ and ${N_2(\psi, \mathbf{y}, \mathbf{y})}$ represent the nonlinear terms in the equations, and ${y(\psi)}$ is the state of the system.

The numerical techniques for obtaining the response of both linear and nonlinear periodic systems, discussed in the following sections, will be based on the mathematical form of the equations represented by Eq. 25.

IV. RESPONSE OF LINEAR PERIODIC SYSTEMS

Consider first the linear part of the general nonlinear response problem represented by Eq. 25, which can be written as

$${\dot{y}(\psi)} = {Z(\psi)} + [L(\psi)]{y(\psi)}. \tag{32}$$

Both the system of equations and the excitation for the response problems are periodic, as indicated by Eq. 28 and 29. Furthermore, one can assume without loss of generality that the periodicity of the equations and the excitations is the same and the common period $T = 2\pi$. It is also important to mention that the response problem represented by Eq. 32 plays an important role in the approximation and linearization of nonlinear vibration problems in general and periodic systems in particular. Many examples for this class of problems can be found in Nayfey and Mook's book on nonlinear oscillations [20].

The associated homogeneous system is given by

$${\dot{y}_H(\psi)} = [L(\psi)]{y_H(\psi)}, \tag{33}$$

which is identical to Eq. 1. Urabe [48] has shown that for characteristic multipliers of the corresponding homogeneous system, Eq. 33, that are different from unity (i.e., a stable or unstable homogeneous system according to Floquet theory), there is one, and only one, periodic solution of Eq. 32 given by

$${y(\psi)} = [\Phi(\psi)]\left\{ \int_0^\psi [\Phi(s)]^{-1}{Z(s)}\, ds + ([I] - [\Phi(2\pi)])^{-1} \right.$$
$$\left. \times [\Phi(2\pi)] \int_0^{2\pi} [\Phi(s)]^{-1}{Z(s)}\, ds \right\}. \tag{34}$$

The general solution for any inhomogeneous equation such as Eq. 32, whether periodic or not, can be mathematically written as Eq. 49

$$\{y(\psi)\} = [\Phi(\psi)]\{y(0)\} + [\Phi(\psi)] \int_0^\psi [\Phi(s)]^{-1}\{Z(s)\}\, ds, \tag{35}$$

where, in Eqs. 34 and 35, $[\Phi(\psi)]$ is the transition matrix defined by

$$[\dot\Phi(\psi)] = [L(\psi)][\Phi(\psi)]$$

and $[\Phi(0)] = [I]$.

Note that the condition for the existence of a unique periodic solution, Eq. 34, is that the determinant of $([I] - [\Phi(2\pi)])$ be nonzero. This is satisfied if all the characteristic multipliers of Eq. 33, the Λ_s's, are different from unity, which is equivalent to the condition that the real parts of all characteristic exponents be nonzero (i.e., $\zeta_s \neq 0$). From Floquet theory, if all the ζ_s's < 0, the homogeneous system is asymptotically stable; if any one $\zeta_s > 0$, the homogeneous system becomes asymptotically unstable. In the former case Eq. 34 is the steady-state solution of the linear system Eq. 32. When any one $\zeta_s > 0$, the mathematical expression for $\{y(\psi)\}$, Eq. 34, is still valid; however, the periodic solution of Eq. 32 is of no practical physical significance. Hsu and Cheng (50) have made some interesting comments in this respect.

Comparing Eqs. 34 and 35, it is obvious that Eq. 34 corresponds to the general solution of any inhomogeneous system, Eq. 35, with the initial condition given by

$$\{y(0)\} = ([I] - [\Phi(2\pi)])^{-1}[\Phi(2\pi)] \int_0^{2\pi} [\Phi(s)]^{-1}\{Z(s)\}\, ds. \tag{36}$$

Equation 36 can be simplified by rewriting it in a more convenient manner:

$$\{y(0)\} = ([I] - [\Phi(2\pi)])^{-1} \int_0^{2\pi} [\Phi(2\pi)][\Phi(s)]^{-1}\{Z(s)\}\, ds. \tag{37}$$

Using the semigroup property of the transition matrix [Ref. 12, p. 29] (sometimes denoted the extension property), which establishes a functional equation for the transition matrix

$$[\Phi(\psi,\psi_0)] = [\Phi(\psi,\psi_1)][\Phi(\psi_1,\psi_0)],$$

one can express

$$[\Phi(2\pi,0)] = [\Phi(2\pi,s)][\Phi(s,0)]. \tag{38}$$

Postmultiplying each side of Eq. 38 by $[\Phi(s,0)]^{-1}$ yields

$$[\Phi(2\pi)][\Phi(s,0)]^{-1} = [\Phi(2\pi,s)]. \tag{39}$$

Substitution of Eq. 39 into Eq. 37 yields

$$\{y(0)\} = ([I] - [\Phi(2\pi)])^{-1} \int_0^{2\pi} [\Phi(2\pi,s)]\{Z(s)\}\, ds. \tag{40}$$

Equation 40 is more convenient for the calculation of the initial condition $\{y(0)\}$, Eq. 36, since it does not require the inversion of $[\Phi(s)]$. A very similar solution technique is also described in Ref. 51. The approximation to the transition matrix given, Eq. 17, can be also employed to calculate the value of $[\Phi(2\pi,s)]$ needed in order to evaluate numerically the initial condition $\{y(0)\}$ given by Eq. 40. The value of the integral in Eq. 40 can be evaluated using any conventional integration procedure such as Simpson's rule. This procedure for generating the initial condition is described in what follows.

Divide the period into an even number of equal time steps $2K$, and denote each interval by ψ_k, where $k = 1, 2, \ldots, 2K + 1$; thus, $\psi_1 = 0 < \psi_2 < \psi_3 \cdots < \psi_{2K+1} = 2\pi$.

In the kth interval, the periodic coefficient system matrix is replaced by a constant matrix defined by

$$[C_k] = \frac{1}{\Delta k} \int_{\psi_k}^{\psi_{k+1}} [L(\psi)]\, d\psi. \tag{41}$$

This expression is analogous to Eq. 12. The transition matrix for the interval is given by

$$[\Phi(\psi_{k+1}, \psi_k)] = \exp(\Delta_k [C_k]). \tag{42}$$

The transition matrix from any interior point m to the end of the period can be obtained from the successive multiplication of the transition matrices

$$[\Phi(2\pi, \psi_m)] = \prod_{i=m}^{2K} [\Phi(\psi_{i+1}, \psi_i)] = [\Phi(2\pi, \psi_{2k})] \cdots [\Phi(\psi_{m+1}, \psi_m)]. \tag{43}$$

The order of multiplications in this product is material; thus, the ith term must be postmultiplied by the $(i-1)$th factor.

Applying Simpson's rule to Eq. 40 and combining it with Eq. 43, one obtains a linear system of equations for the initial conditions,

$$([I] - [\Phi(2\pi, 0)])\{y(0)\} = \frac{\Delta}{3} \left\{ [\Phi(2\pi, 0)]\{Z(0)\} + \{Z(2\pi)\} \right.$$

$$+ 4 \sum_{l=2,\text{even}}^{2K} [\Phi(2\pi, \psi_l)]\{Z(\psi_l)\}$$

$$\left. + 2 \sum_{l=3,\text{odd}}^{2K-1} [\Phi(2\pi, \psi_l)]\{Z(\psi_l)\} \right\}, \qquad (44)$$

where $\Delta = 2\pi/2K$.

In Eq. 44 the matrices $[\Phi(2\pi, \psi_l)]$ are evaluated by the successive application of the relations

$$[\Phi(2\pi, \psi_{2K+1} = 2\pi)] = [I], \qquad (45)$$

$$[\Phi(2\pi, \psi_l)] = [\Phi(2\pi, \psi_{l+1})][\exp(\Delta_k[C]_k)]. \qquad (46)$$

The matrix exponentials required for the evaluation of Eq. 46 are obtained from Eq. 16 with $J = 4$.

Taking the initial condition, Eq. 40, the linear system represented by Eq. 25 is integrated numerically using a fourth-order Runge–Kutta scheme with Gill coefficients [1]. The integration is performed with a constant step size that is identical to the step size used in evaluating the transition matrix at the end of one period $[\Phi(T, 0)]$. This means that the use of Eq. 34 is actually, completely bypassed. Convergence of the method is checked by comparing the displacement quantities obtained for the response at the end of the first or second revolution with the initial conditions; that is, compare

$$\{y(0)\} \quad \text{with} \quad \{y(2\pi)\} \quad \text{or} \quad \{y(4\pi)\} \quad \text{or} \quad \{y(6\pi)\}.$$

Depending on the numerical accuracy of the calculations, excellent converged solutions can be obtained with one to three periods.

The procedure described was used successfully in the response calculation of a helicopter rotor blade in forward flight [52], and a slightly modified version of this approach was also effective in obtaining the response of horizontal axis wind turbines [53, 54].

V. RESPONSE OF NONLINEAR PERIODIC SYSTEMS

The solution of the complete nonlinear system represented by Eq. 25 is considered next. For the nonlinear case a number of approaches are possible;

excellent discussions of these methods for the problem encountered in applied mechanics are presented in Refs. 20 and 46. A general mathematical treatment for nonlinear periodic systems based on Galerkin's procedure in the time domain was presented by Urabe [48, 55]. Further refinements of the method proposed by Urabe was considered by a number of authors [56–58]. Nonlinear periodic systems also play an important role in circuit theory. In Refs. 59 and 60 a Newton algorithm for the calculation of the steady-state response of nonlinear periodic systems was presented. When the nonlinearities are *relatively mild*, a number of approaches are available and have been used successfully in a variety of applications [20, 46]. For the case of helicopter rotor blade dynamics, which represents a weakly nonlinear system usually modeled by a complicated system of coupled ordinary differential equations similar in structure to Eq. 25, three approaches have been used (a) quasi-linearization [61], (b) periodic shooting [62], and (c) harmonic balancing [51]. Depending on the implementation, Quasi-linearization and periodic shooting can be equally effective. Harmonic balancing is less convenient to use [51], although some recent developments and extensions of the method [63] indicate that this method is quite useful. It should be noted that harmonic balancing represents an equivalent linearization of a nonlinear system, and thus it places certain mathematical restrictions on this method [64]. An alternative to these methods is a numerical perturbation method, described in Ref. 51, which appears to work only for very mildly nonlinear systems.

The approaches described in this chapter are quasi-linearization and periodic shooting; other methods will be mentioned briefly. Quasi-linearization is essentially a generalized Newton–Raphson method possessing second-order convergence [65]. The quasi-linearization scheme described here is based on an iterative application of the method for obtaining the linear response of periodic systems described in the previous section [61]. Introducing an iteration index k for obtaining the nonlinear periodic response of Eq. 25, one can perform a first-order Taylor series expansion about the kth iteration,

$$\{\dot{y}\}^{k+1} = \{Q_{NL}\}^k + \left[\frac{\partial Q_{NL}}{\partial \mathbf{y}}\right]^k (\{y\}^{k+1} - \{y\}^k) + \left[\frac{\partial Q_{NL}}{\partial \dot{\mathbf{y}}}\right]^k (\{\dot{y}\}^{k+1} - \{\dot{y}\}^k),$$

$$(47)$$

which enables one to write a linear equation in the form

$$\{\dot{y}\}^{k+1} = [A]^k \{y\}^{k+1} + \{f\}^k. \tag{48}$$

Equation 48 is obtained as follows. Express Q_{NL} as given by Eq. 25 and substitute it into Eq. 47. Then $[A]$ and $\{f\}$ in Eq. 48 can be identified as

$$[A]^k = \left([I] - \left[\frac{\partial \mathbf{N}_2}{\partial \dot{\mathbf{y}}} \right]^k \right) \left([L(\psi)] + \left[\frac{\partial \mathbf{N}_1}{\partial \mathbf{y}} \right]^k + \left[\frac{\partial \mathbf{N}_2}{\partial \mathbf{y}} \right]^k \right) \tag{49}$$

and

$$\{f\}^k = \left([I] - \left[\frac{\partial \mathbf{N}_2}{\partial \dot{\mathbf{y}}} \right]^k \right)^{-1} \left\{ \{Z(\psi)\} + \{N_1(\psi, \mathbf{y})\}^k \right.$$

$$\left. + \{N_2(\psi, \mathbf{y}, \dot{\mathbf{y}})\}^k - \left(\left[\frac{\partial \mathbf{N}_1}{\partial \mathbf{y}} \right]^k + \left[\frac{\partial \mathbf{N}_2}{\partial \mathbf{y}} \right]^k \right) \{y\}^k - \left[\frac{\partial \mathbf{N}_2}{\partial \dot{\mathbf{y}}} \right]^k \{\dot{y}\}^k \right\}. \tag{50}$$

To initiate the iterative procedure, the nonlinear terms $\{N_1(\psi, \mathbf{y})\}$ and $\{N_2(\psi, \mathbf{y}, \dot{\mathbf{y}})\}$ are dropped from $\{Q_{\mathrm{NL}}\}$ and the resulting linear system, Eq. 32, is solved for a periodic solution that provides an initial guess; thus $\{y\}^0 = \{y_L\}$, where $\{y_L\}$ is the periodic response of Eq. 32 and is obtained by the method described in the previous section. This method is also used to solve Eq. 48 for $k = 1, 2, \dots$. Thus, quasi-linearization is a sequence of linear iterates based on Eqs. 48–50. For the cases considered [36, 61, 66], the derivatives of $\mathbf{N}_1$ and $\mathbf{N}_2$ were evaluated explicitly. The iterations are terminated when the L_2 norm

$$\|(\{y\}^{k+1} - \{y\}^k)\|_2 < \varepsilon, \tag{51}$$

for all ψ, where ε is a prescribed small parameter. Thus, $k = 1$ corresponds to the linearized response, and $k = 2$ corresponds to the first approximation of the nonlinear periodic response.

The nonlinear periodic shooting method [62] is based on integrating Eq. 25 with an estimated initial condition $\{y(0)\}$ over one period. The integrated solution $\{y(T)\}$ is not periodic and can be ragarded as a first estimate of the periodic solution $\{y_p\}$. The initial conditions can be modified so as to make $\{y(T)\}$ periodic. Applying this procedure in an iterative manner eventually leads to a solution.

The mathematical basis of this method consists of using as initial condition the relation [51, 62]

$$\{y(0)\} = ([I] - [\Phi(2\pi)])^{-1}\{y_E(2\pi)\}, \tag{52}$$

where $\{y_E(2\pi)\}$ is the numerical response of the linear system, Eq. 32, at $\psi = T$ (or 2π) for zero initial conditions. This initial condition, Eq. 52, is identical to Eq. 40, and thus it guarantees a periodic response for the linear system. Since the system represented by Eq. 25 is nonlinear, numerical integration over one period will not yield a periodic solution. Therefore, the method of periodic shooting is based on successive iteration on the initial values in order to satisfy the periodic boundary condition. In each stage of the iteration the full nonlinear equations are solved, whereas the error in the periodic boundary condition is driven to zero. The periodic shooting method described in Ref. 62 is a two-point periodic shooting method for which the algorithm has the following structure:

1. Initial guess for the state vector at the beginning of the period is based on Eq. 52.

2. The nonlinear system is integrated over one period using a suitable numerical integration method.

3. The sensitivity matrix for the state vector at the end of the period is expressed in terms of perturbations in the values of the state vector at the beginning of the period, denoting the sensitivity matrix by $[S]_i$, for the ith iteration:

$$[S]_i = \left[\frac{\partial \{y_f\}}{\partial \{y_0\}} \right]_i, \tag{53}$$

where $\{y_f\}_i$ is the final state vector for the ith iteration and $\{y_0\}$ is the initial state vector for the ith iteration.

4. Using the sensitivity information given by Eq. 53, a revised vector of initial values is calculated, so that the periodic boundary conditions are satisfied. The revised vector of initial conditions for the $(i + 1)$th iteration is given by

$$\{y_0\}_{i+1} = \{y_0\}_i + ([I] - [S]_i)^{-1}(\{y_f\}_i - \{y_0\}_i). \tag{54}$$

5. The nonlinear equations are integrated again over one period using the revised initial condition given by Eq. 54.

6. The periodicity condition is checked within a desired accuracy,

$$|\{y_f\}_{i+1} - \{y_0\}_{i+1}| < \varepsilon. \tag{55}$$

If Eq. 55 is satisfied, the iteration is stopped. If Eq. 55 is violated, one returns to step 3, and the procedure is repeated until convergence is obtained.

A somewhat similar procedure called numerical perturbation is also described in Ref. 51. Comparative numerical studies between quasi-linearization and periodic shooting for rotary wing aeroelastic problems were conducted in Ref. 67. It was shown that the methods are equally efficient in terms of computer times. However, it should be noted that the two-point periodic shooting described in Ref. 67 used a somewhat more efficient method for calculating the sensitivity matrix $[S]_i$, Eq. 53, than the finite-difference method used in previous studies [62]. A careful examination of Eq. 53 shows that infinitesimal changes in the initial values $\{y_0\}_i$ cause infinitesimal changes in the baseline nonlinear solution $\{y\}_i$. Due to the infinitesimal nature of these changes, they are governed by the linearized perturbed system about the baseline solution $\{y\}_i$; furthermore, it can be shown that this perturbed system is homogeneous. Therefore, the sensitivity matrix for the final values with respect to the initial values is simply the transition matrix of the linear perturbed system [67]. This transition matrix can be evaluated using Eq. 17, and thus, the need to use a finite-difference scheme for the calculation of the sensitivity matrix is bypassed. Furthermore, multiple periodic shooting methods were also considered and tested [67]; it was concluded that the success of the shooting as well as the quasi-linearization methods can depend to some extent on the type of problems to which the methods are applied.

Another interesting alternative that is available for the solution of nonlinear periodic systems is the finite-element method in the time domain. The method was initially proposed for nonperiodic systems [41–45] where certain difficulties were encountered with the convergence of the method. The additional periodicity condition present in periodic systems alleviates these problems, as was first recognized by Borri [40]. A limited study of the numerical efficiency of the method, for a helicopter rotor aeroelastic problem, governed by nonlinear ordinary differential equations was conducted in Ref. 68. The method was found to be quite efficient when compared to the other methods described in this section. For the sake of completeness, a brief description of this method is provided.

The equations solved are Eq. 23. The method is based on Hamilton's principle in the weak form, which can be obtained from the principle of virtual work using Eq. 23, premultiplying by virtual displacement $\{\delta q\}^T$ and integrating the resulting equations from an initial time ψ_i to a final time ψ_f:

$$\int_{\psi_i}^{\psi_f} \{\delta q\}^T ([M]\{\ddot{q}\} + [C(\psi)]\{\dot{q}\} + [K(\psi)]\{q\} - \{F_{\mathrm{NL}}(\psi, \mathbf{q}, \dot{\mathbf{q}})\})\, d\psi = 0.$$

$$(56)$$

Integrating by parts, Eq. 56 yields

$$\delta\{q\}^T [M]\{\dot{q}\}\,|_{\psi_i}^{\psi_f} - \int_{\psi_i}^{\psi_f} \{\delta\dot{q}\}^T [M]\{\dot{q}\}\, d\psi + \int_{\psi_i}^{\psi_f} \{\delta q\}^T ([C(\psi)]\{\dot{q}\}$$
$$+ [K(\psi)]\{q\} - \{F_{\mathrm{NL}}(\psi, \mathbf{q}, \dot{\mathbf{q}})\})\, d\psi = 0 \qquad (57)$$

Rearranging Eq. 57 yields

$$\int_{\psi_i}^{\psi_f} \begin{Bmatrix} \delta\{a\} \\ \delta\{\dot{q}\} \end{Bmatrix}^T \begin{Bmatrix} \{F_{NL}(\psi, \mathbf{q}, \dot{\mathbf{q}}) - [C(\psi)]\{\dot{q}\} - [K(\psi)]\{q\}\} \\ [M]\{\dot{q}\} \end{Bmatrix}$$

$$\times d\psi = \begin{Bmatrix} \delta\{q\} \\ \delta\{\dot{q}\} \end{Bmatrix}^T \begin{Bmatrix} [M]\{\dot{q}\} \\ \{0\} \end{Bmatrix} \Big|_{\psi_i}^{\psi_f}. \tag{58}$$

From Eq. 58 one obtains the variational formulation of Hamilton's weak principle:

$$\int_{\psi_i}^{\psi_f} \{\delta y\}^T \{Q\} \, d\psi = \{\delta y\}^T \{b\} \,|_{\psi_i}^{\psi_f}, \tag{59}$$

where

$$\{y\} = \begin{Bmatrix} \mathbf{q}(\psi) \\ \dot{\mathbf{q}}(\psi) \end{Bmatrix}$$

is the same state vector used in Eq. 27, and

$$\delta\{y\} = \begin{Bmatrix} \delta\{q\} \\ \delta\{\dot{q}\} \end{Bmatrix}.$$

Furthermore,

$$\{Q\} = \begin{Bmatrix} \{F_{NL}(\psi, \mathbf{q}, \dot{\mathbf{q}}) - [C(\psi)]\{q\} - [K(\psi)]\{q\}\} \\ [M]\{\dot{q}\} \end{Bmatrix} \tag{60}$$

and

$$\{b\} = \begin{Bmatrix} \{P\} \\ \{0\} \end{Bmatrix},$$

where $\{P\} = [M]\{\dot{q}\}$ is the vector of generalized momenta.

For a periodic response solution one can choose $\psi_i = \psi$ and $\psi_f = \psi + 2\pi$, and for this case the right side of Eq. 59 is zero.

Since $\{Q\}$ is a nonlinear function of $\{q\}$ and $\{\dot{q}\}$, it is useful to take the first-order Taylor series expansion about a given state vector $\{\bar{y}\}$ as a linear approximation. Denote by $\{\bar{Q}\}$ the value of $\{Q\}$ at $\{y\} = \{\bar{y}\}$; then the first-order Taylor series approximation for $\{Q\}$ can be written as

$$\{Q\} = \{\bar{Q}\} + \left(\frac{\partial\{\bar{Q}\}}{\partial\{q\}}\right)\{\Delta q\} + \left(\frac{\partial\{\bar{Q}\}}{\partial\{\dot{q}\}}\right)\{\Delta\dot{q}\}. \tag{61}$$

Evaluating the partial derivatives and substituting into Eq. 59 yields an incremental form of Eq. 59:

$$\int_{\psi_i}^{\psi_f} \{\delta y\}^T \{\bar{Q}\}\, d\psi + \int_{\psi_i}^{\psi_f} \{\delta y\}^T [K_t] \{\Delta y\}\, d\psi = 0, \tag{62}$$

where

$$[K_t] = \left[\begin{array}{c:c} \dfrac{\partial \{F_{NL}(\psi, \mathbf{q}, \dot{\mathbf{q}})\}}{\partial \{q\}} - [K(\psi)] & \dfrac{\partial \{F_{NL}(\psi, \mathbf{q}, \dot{\mathbf{q}})\}}{\partial \{\dot{q}\}} - [C(\psi)] \\ \hdashline \mathbf{0} & [\mathbf{M}] \end{array} \right]. \tag{63}$$

Equation 62 represents the virtual energy of the system, and it can be used as a variational expression for the finite-element discretization in the time domain. The period is discretized into a number of time elements, denoted by N_t. The displacement within each time element is expressed by a suitable interpolation polynomial, where a Hermite (or cubic) interpolation in time seems to be preferable [68]. Denoting the local, nondimensional time coordinate within the element by s, the displacement for any generalized coordinate q_j can be interpolated by

$$q_j(s) = H_{1j}(s)\xi_1 + H_{2j}(s)\xi_2 + H_{3j}(s)\xi_3 + H_{4j}(s)\xi_4. \tag{64}$$

Thus, each element has four nodes, with one degree of freedom at each node, and the state vector can be written as

$$\{y(s)\} = \left\{ \begin{array}{c} \{q(s)\} \\ \{\dot{q}(s)\} \end{array} \right\} = \left\{ \begin{array}{c} [H(s)]\{\xi\} \\ [\dot{H}(s)]\{\xi\} \end{array} \right\}. \tag{65}$$

Substitution of Eq. 65 into Eq. 63 and assembling the element matrices to obtain the global system matrices yields the finite-element equations in the time domain that have a linear form as a result of the first-order Taylor series expansion, Eq. 61:

$$\{\bar{Q}_G\} + [\bar{K}_{tG}]\{\bar{\Delta}\xi_G\} = 0, \tag{66}$$

where the subscript G indicates the assembled global matrices given by

$$\{\bar{Q}_G\} = \sum_{i=1}^{N_t} \int_{\psi_i}^{\psi_{i+1}} \left[\begin{array}{c} [H(s)] \\ [\dot{H}(s)] \end{array} \right]^T \{Q\}\, ds, \tag{67}$$

$$[\bar{K}_{tG}] = \sum_{i=1}^{N_t} \int_{\psi_i}^{\psi_{i+1}} \begin{bmatrix} [H(s)] \\ [\dot{H}(s)] \end{bmatrix}^T [K_t] \begin{bmatrix} [H(s)] \\ [\dot{H}(s)] \end{bmatrix} ds, \tag{68}$$

where $\psi_{i+1} - \psi_i = \Delta\psi_i$ is the time increment for element i.

Clearly Eq. 66 does not represent the complete solution to the nonlinear problem, and therefore, it has to be applied in an iterative manner. The vector $\{\Delta\bar{\xi}_G\}$ represents the incremental global state vector. Applying the periodic boundary condition

$$\{\xi_G(\psi = 0)\} = \{\xi_G(\psi = 2\pi)\} \tag{69}$$

yields the linear equation

$$\{Q_G\} + [K_{tG}]\{\Delta\xi_G\} = 0, \tag{70}$$

which can be solved by any available, efficient, linear equation solver routine. The steady-state response vector $\{\xi_G\}$ is obtained iteratively,

$$\{\xi_G\}^{k+1} = \{\xi_G\}^k + \{\Delta\xi_G\}^{k+1}, \tag{71}$$

where k is the iteration index. The iteration is converged when the L_2 norm

$$\|\{\Delta\xi_G\}^{k+1}\|_2 < \varepsilon, \tag{72}$$

where ε is a suitable chosen error control parameter.

Comparing this method to the quasi-linearization approach described previously in this section indicates that these two methods are based on the concept of linearization and, therefore, are most suitable for weakly nonlinear systems.

VI. APPLICATIONS ILLUSTRATING NUMERICAL ASPECTS OF THE METHODS DISCUSSED

The purpose of this section is to present a few examples where the methods described in this chapter were applied to periodic systems encountered in applied mechanics or engineering. The primary intent is to present information on the numerical aspects, convergence, and efficiency of the methods instead of reproducing numerical results represented in the various references cited.

A. Calculation of Transition Matrices and System Stability for a Beam Example

To illustrate the methods used for evaluating $[\Phi(T, 0)]$ described in Section II.B, the parametric excitation problem of a clamped beam is used

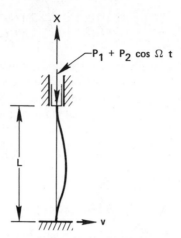

Figure 1. Clamped–clamped column under periodic axial loading; column has uniform mass distribution per unit length *m* and stiffness EI (Ref. 1).

(1, 69, 70). This simple physical system is shown in Figure 1. This example is a very convenient test case for examining numerical schemes for periodic systems. By using a conventional-beam-type finite element, shown in Figure 2, to discretize the spatial dependence of the problem, one can construct homogeneous periodic systems, represented by Eq. 1, in which the system matrix $[A(\psi)]$ can be increased in size in an arbitrary manner by increasing the number of finite elements [1]. In Ref. 1 transition matrices of size 20×20 were computed without any difficulty using the methods described in Section II.B. For larger systems (above 8×8) double precision in the calculation of the transition matrix on IBM machines is usually required. The step size used varied between $60 < K < 120$ per period depending on the size of the system or the number of finite elements used to model the system. For the same step size the computations based on Eq. 17 and the fourth-order Runge–Kutta scheme

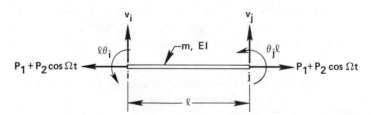

Figure 2. Geometry of beam finite element showing nodal degrees of freedom, v_i, Θ_i, v_j, and Θ_j (Ref. 1).

were equally effective [1]. Considerable information on numerical experimentation involving accuracy, convergence, and computing times can be found in Ref. 1. For this particular example one could obtain approximate closed-form estimates for step size control based on the number of finite elements used for modeling the system.

B. Examples of System Stability and Response Taken from Rotary Wing Aeroelasticity

More complicated examples can be taken from problems of helicopter rotor or wind turbine aeroelastic stability, which are usually denoted by the term *rotary wing aeroelasticity*. The main objective in these problems is to determine the stability and response of a rotating blade under the combined action of fluid dynamic, inertial, and structural loads [2–3]. A typical hingeless rotor blade is shown in Figure 3. The blade has three fundamental degrees of freedom; flap or out of plane of rotation bending, lag or in plane of rotation bending, and torsion or twist of the blade about a spanwise axis. For such blades geometrically nonlinear terms due to moderate deflections are known to be important and have to be incorporated in the blade equations of motion [2]. The resulting equations have the mathematical structure given by Eq. 25, and these equations are mildly nonlinear. A considerable number of studies have indicated that a reliable measure of system stability can be obtained from a linearized system [2, 3]. Denoting by $\{\bar{q}(\psi)\}$ the time-dependent periodic equilibrium of the blade, obtained from solving Eq. 25 using the method described in Section V, the perturbed state about this equilibrium position can be written as

$$\{q(\psi)\} = \{\bar{q}(\psi)\} + \{\Delta q(\psi)\}. \tag{73}$$

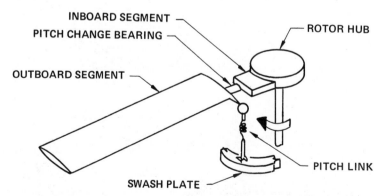

Figure 3. Typical geometry for hingeless, uniform helicopter rotor blade.

Substituting this relation into Eq. 23 and assuming that squares of the perturbation quantities (viz., $\Delta q_i(\psi)\Delta q_j(\psi)$) are small and negligible, one obtains a linearized homogeneous system that can be symbolically written in first-order state variable form as

$$\{\Delta\dot{y}\} = [L(\psi,\bar{y}(\psi))]\{\Delta y\}, \qquad (74)$$

where

$$\{\Delta y\} = \begin{Bmatrix} \Delta\mathbf{q} \\ \Delta\dot{\mathbf{q}} \end{Bmatrix},$$

and $[L(\psi,\bar{y}(\psi))]$ is a coefficient matrix similar to Eq. 26 that depends on both the nondimensional time ψ and the nonlinear periodic equilibrium state of the blade, $\{\bar{y}(\psi)\}$. The matrix $[L(\psi,\bar{y}(\psi))]$ is also a periodic matrix; hence, the stability of the system, Eq. 74, can be obtained from Floquet theory by evaluating the characteristic exponents as described in Section II. This

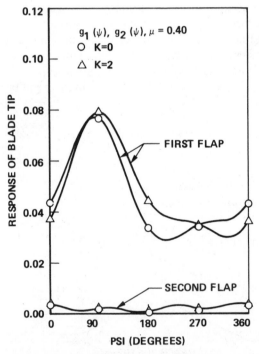

Figure 4. Flap response in first $[g_1(\psi)]$ and second $[g_2(\psi)]$ flapwise modes for a soft-in-plane hingeless rotor blade at advance ratio $\mu = 0.40$.

approach has been used in many recent studies dealing with forward flight [2–4, 36, 61, 68].

The method based on the application of Eq. 40 combined with numerical integration to obtain the response of a linear periodic system was developed initially for the linear periodic response problems of horizontal axis wind turbines [53, 54], where a somewhat more cumbersome version of Eq. 33 (viz., Eq. 37 was used.

Subsequently, this method was applied to the linear and nonlinear response problem of helicopter rotor blades in forward flight in the presence of quasi-steady aerodynamic loads [61] and, more recently, to helicopter rotor blades in forward flight in the presence of unsteady aerodynamic loads [52]. For linear response calculations the step size required for the implementation of the method was similar to the number of steps needed for the calculation of the transition matrix $[\Phi(T,0)]$; thus, $60 < K < 120$ per period.

The quasi-linearization method described in Section V is based on an iterative application of the linear response calculation described in Section IV.

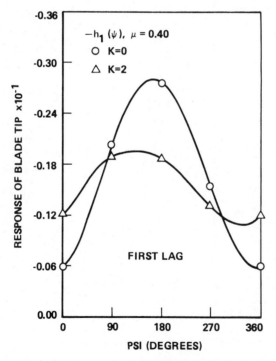

Figure 5. Response in fundamental in-plane or lead–lag mode $[h_1(\psi)]$ for soft-in-plane hingeless rotor blade at advance ratio $\mu = 0.40$ (Ref. 61).

For application to nonlinear periodic system response calculations using Eq. 48, the number of steps that provided good numerical behavior for coupled flap–lag–torsional aeroelastic stability and the response problem in forward flight was approximately $110 < K < 130$ per period [61].

Some typical results using these procedures are presented in Figures 4–6, which illustrate the aeroelastic response of a soft-in-plane hingeless rotor blade at an advance ratio of $\mu = 0.40$, taken from Ref. 61. The aeroelastic analysis is based on two flapwise modes, two lead–lag modes, and two torsional modes. This implies a total of six physical degrees of freedom leading to a 12×1 state vector. The responses in the two flap modes are shown in Figure 4, and the response in the fundamental lead–lag torsional modes are shown in Figures 5 and 6, respectively. For this case the quasi-linearization procedure represented by Eq. 48 converges rapidly. Two curves are shown, one for $k = 0$ and the other for $k = 2$. Recall from the discussion on quasi-linearization that $k = 0$ corresponds to the linear periodic response in which all

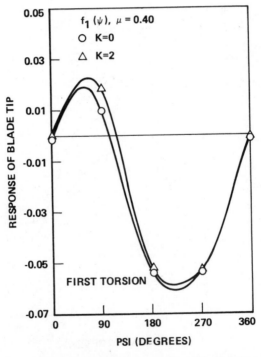

Figure 6. Response in fundamental torsional mode for soft-in-plane hingeless rotor blade at advance ratio $\mu = 0.40$ (Ref. 61).

nonlinear terms are neglected, $k = 1$ represents a linearized response, and $k = 2$ gives the first approximation of the fully nonlinear response. In the figure $k = 2$ represents a fully converged nonlinear response since the values for $k = 3$ are not distinguishable from those for $k = 2$. For the results shown in Figures 4–6, a period was divided into 126 steps, and the computer time required for obtaining a nonlinear response solution was less than 30 sec of CPU time for an IBM 3033 computer.

Another useful attribute of the quasi-linearization technique consists of the information it provides on the relative importance of the nonlinear terms. The nonlinear terms included in Ref. 61 are geometric nonlinearities due to moderate deflections of the blade. Thus, it is evident from Figure 4 that the fundamental flap response is sensitive to the geometric nonlinearities, whereas the second-mode flap response is not. Similarly, from Figure 5 the sensitivity of the fundamental lag response to the nonlinear terms is evident. Figure 6 shows that the fundamental torsional response is relatively insensitive to the nonlinear terms, since the curves for $k = 0$ and $k = 2$ are quite close to each other.

Considerable additional numerical experimentation with quasi-linearization was presented in Ref. 66. In a recent study [67] by Dinyavari, quasi-linearization and periodic shooting were used successfully to obtain the equilibrium position of a hingeless rotor blade in forward flight in the presence of an augmented-state, arbitrary-motion, unsteady aerodynamic model. Reference 67 also contains a number of useful comments pertaining to the numerical implementation of these methods, so as to increase numerical efficiency and reduced storage requirements. Operation counts comparing quasi-linearization with periodic shooting have shown both methods to be equally effective [67].

After establishing the periodic equilibrium position of the blade in forward flight, the aeroelastic stability is determined from the linearized system represented by Eq. 74. A typical stability boundary obtained from such an analysis [61] is presented in Figure 7, which shows the behavior of the real part of the characteristic exponent associated with the fundamental lag mode (or in-plane blade bending degree of freedom) for a stiff-in-plane helicopter rotor blade. The results show an instability in the lag degree of freedom at $\mu \cong 0.39$. A number of details associated with this figure are relevant from a computational point of view. The transition matrix for this case was a 12×12 matrix, which is similar in size to the system treated in Figures 4–6. The step size used was $K = 120$. Furthermore, it is evident from the figure that the quasi-linearization index $k = 0$, 1, 2 associated with the periodic nonlinear equilibrium state about which the system is linearized has a very significant influence on blade stability.

Another important ingredient in the computation of stability boundaries of

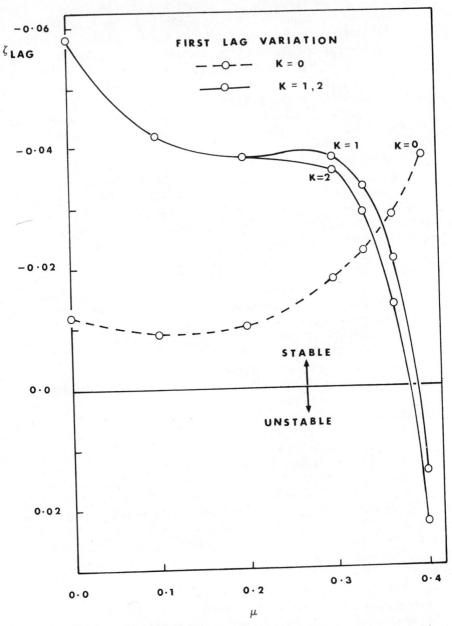

Figure 7. Real part of characteristic exponent for fundamental lag degree of freedom (ζ_{lag}) for stiff-in-plane hingeless rotor blade vs. advance ratio (taken from Ref. 61).

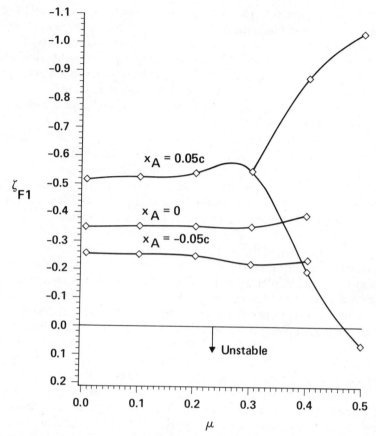

Figure 8. Real part of characteristic exponent for fundamental flap degree of freedom (ζ_{F1}) for soft-in-plane hingeless rotor blade vs. advance ratio μ (taken from Ref. 36).

rotor blades in forward flight is the identification of the characteristic exponents associated with the various degrees of freedom present in the problem. To relate the characteristic exponents to the physical degrees of freedom, both the real and imaginary parts of the characteristic exponents have to be considered. The real part is representative of damping, whereas the imaginary part is related to the frequency of the mode. Both quantities can change significantly with an increase in the advance ratio μ, which is representative of the strength of the parametric excitation present in the system. Furthermore, as mentioned in Section II, since $\tan^{-1}$ in Eq. 10 is multivalued, it can be determined only within an integer multiple of the nondimensional period T. Therefore, the most reliable method for relating the characteristic exponents to the physical degrees of freedom in the system is to start the calculations at $\mu = 0$. For $\mu = 0$ the system has constant coefficients, and the identification based on the real and imaginary part of the eigenvalues is straightforward. Subsequently, the values of μ are incremented in relatively small steps, such as $\Delta\mu \leqslant 0.05$, which result in small changes in ζ_s and ω_s, and thus the characteristic exponents can be identified and traced without any ambiguity.

Another aspect associated with the stability calculation using the linearized system and Floquet theory is evident in Figure 8. This result is based on a coupled flap–lag–torsional analysis of a rotor blade using 12×12 transition matrices. The real part of the characteristic exponent for the fundamental flap degree of freedom, ζ_{F1}, becomes unstable around $\mu = 0.47$ for a particular offset, $x_A = 0.05c$, between the elastic axis and aerodynamic center, which equals 5% of the chord c. The special behavior exhibited by this periodic system is the splitting of the root associated with $x_A = 0.05c$, after which one branch becomes unstable. This behavior is quite typical of periodic systems encountered in helicopter rotor aeroelastic problems and other applications.

It should also be noted that response and stability calculations on rotor blades similar to the ones discussed in the preceding were also obtained in Ref. 68 using finite elements in the time domain. Using six cubic time domain finite elements for a period, good numerical efficiency and accuracy was obtained. The results indicated that the time domain finite-element method produces a fivefold reduction in computing time compared to the computer time obtained by the same authors using a previous method [51], which was less efficient than quasi-linearization. One common aspect of the studies mentioned in this chapter is that they were restricted to systems having 10 physical degrees of freedom or less (i.e., 20×20 transition matrices). Thus, the numerical properties of these methods for larger systems having between 50 and 100 physical degrees remains to be explored.

VII. CONCLUDING REMARKS

This chapter has reviewed a number of numerically efficient techniques for analyzing the stability and response of periodic systems governed by systems of ordinary differential equations with periodic coefficients. The treatment of both linear and nonlinear problems is discussed. The methods discussed are fairly general and robust and should be applicable to numerous problems governed by equations with periodic coefficients. It is also shown that the first-order state variable representation of periodic systems facilitates their numerical treatment for both stability and response type of calculations. The methods presented are numerically efficient and easy to implement on a computer, and the numerical experience gained by applying these methods indicates that in most cases the methods are also numerically reliable. Furthermore, the methods are general and do not have some of the limitations experienced with other methods, such as perturbation methods and harmonic balancing [51]. On the other hand, it should be noted that additional research is needed to establish the reliability of such methods for larger systems, which would be governed by state vectors containing 50–100 state variables.

Acknowledgment

The constructive comments made by Dr. Ron Dinyavari are hereby gratefully acknowledged.

VIII. NOMENCLATURE

$[A(\psi)]$	Periodic matrix
$[C_k]$	Constant matrix approximation of $[A(\psi)]$ over kth iteration
$[C(\psi, k)]$	Step function approximation to $[A(\psi)]$
$[C(\psi)]$	Symbolic matrix representing linear damping effects
$[D(\psi)]$	Block diagonal matrix, Eq. 18
$\{e_i\}$	Initial-condition vector
$\{F_{NL}(\psi, \mathbf{q}, \dot{\mathbf{q}})\}$	Complete nonlinear loading
$H_{ij}(s)$	Hermite interpolation functions
$[H(s)]$	Interpolation matrix in time domain
i	Periodic shooting iteration index
$[I]$	Unit matrix
J	Number of terms used in approximation of matrix exponential
$K, 2K$	Number of intervals into which T is divided
$[K_t]$	Matrix defined in Eq. 63
$[H(s)]$	Symbolic matrix representing linear stiffness effects
k	Quasi-linearization iteration index
$[L(\psi)]$	Linear periodic matrix
$[M]$	Symbolic matrix representing linear inertia effects
N_t	Number of elements in time domain
$\{N_1(\psi, \mathbf{y})\}, \{N_2, \psi, \mathbf{y}, \dot{\mathbf{y}})\}$	Nonlinear parts of systems in state space
$[P(\psi)]$	Periodic matrix used in Floquet theorem
$[Q]$	Constant matrix
$\{Q_{NL}(\psi, \mathbf{y}, \dot{\mathbf{y}})\}$	Nonlinear state vector

$\{Q\}$	Vector defined by Eq. 60
$\{q(\psi)\}$	Generalized coordinate vector, $n \times 1$
$\{\Delta q\}$	Increment in generalized coordinates
$[R]$	Constant matrix used in Floquet theorem
s	Index; also dummy integration variable
$[S]_i$	Sensitivity matrix for periodic shooting, ith iteration
T	Common nondimensional period
t	Time
$\{y\}$	State variable vector, $2n \times 1$
$\{y_A(\psi, k)\}$	Approximate value of $\{y(\psi)\}$
$\{y_0\}$	Value of state vector at beginning of period
$\{y_f\}$	Value of state vector at end of period
$\{y_E\}$	Integrated state at end of period for linear inhomogeneous periodic system with zero initial conditions
$\{Z(\psi)\}$	Excitation in state space
$\{z_i(\psi)\}$	The ith column of transition matrix
ε	Small quantity, error
ζ_s	Real part of characteristic exponent
$[\lambda]$	Diagonal matrix containing characteristic exponents
$[\Lambda]$	Diagonal matrix containing characteristic multipliers
λ_s	Characteristic exponent
Λ_s	Characteristic multiplier
ξ	Dummy variable
$\{\xi\}$	Vector of nodal degrees of freedom
$[\Phi(\psi, \psi_0)]$	Transition matrix at nondimensional time ψ for initial conditions given at ψ_0
$[\Phi_A(T, 0)]$, $[\Phi_{AA}(T, 0)]$	Various approximations of $[\Phi(T, 0)]$ at end of one period
ψ	Nondimensional time coordinate; also azimuth angle of blade $\psi = \Omega t$ measured from straight-aft position
ω_s	Imaginary part of characteristic exponent
Ω	Speed of rotation of rotor

Special Symbols

$[\ ]$	Square matrix
$[\ ]^{-1}$	Inverse
$\Delta_k = \psi_k - \psi_{k-1}$	Step size
$(\ )_R, (\ )_I$	Real and imaginary part of quantity

$(\cdot) = d/d\psi$ Differentiation with respect to
nondimensional time

boldface type   Matrix or vector quantity

References

1. P. Friedmann, C. E. Hammond, and T. Woo. *Int. J. Numer. Meth. Eng.* **11**, 1117 (1977).

2. P. P. Friedmann, *Vertica* **7**, 101 (1983).

3. P. P. Friedmann, *Vertica* **11** (1987).

4. P. P. Friedmann, *Comput. Math. Appl.* **12A**, 131 (1986).

5. E. A. Coddington and N. Levinson, *Theory of Ordinary Differential Equations*, McGraw-Hill, New York, 1955.

6. L. A. Pipes and L. R. Harvill, *Applied Mathematics for Engineers and Physicists*, 3rd ed., McGraw-Hill, New York, 1970.

7. N. P. Erugin, *Linear Systems of Ordinary Differential Equations with Periodic and Quasi-Periodic Coefficients*, Academic, New York, 1966.

8. V. A. Yakubovitch and X. Starzhinskii, *Linear Differential Equations with Periodic Coefficients*, 2 vols., Halsted, Wiley, New York (Israel Program for Scientific Translations), 1975.

9. W. Magnus and S. Winkler, *Hill's Equation*, Wiley, New York, 1966

10. F. M. Arscott, *Periodic Differential Equations*, Pergamon, Oxford, 1964.

11. H. D'Angelo, *Linear Time-Varying Systems: Analysis and Synthesis*, Allen and Bacon, Boston, 1970.

12. R. W. Brocket, *Finite Dimensional Linear Systems*, Wiley, New York, 1970.

13. J. A. Richards, *Analysis of Periodically Time-Varying Systems*, Springer-Verlag, New York, 1983.

14. V. V. Bolotin, *Dynamic Stability of Elastic Systems*, Holden-Day, San Francisco, 1964.

15. C. S. Hsu, *J. Appl. Mech.* **39**, 551 (1972).

16. C. S. Hsu and W. H. Cheng, *J. Appl. Mech.* **40**, 78 (1973).

17. C. S. Hsu, *J. Math. Anal. Appl.* **45**, 234 (1974).

18. F. C. Hu and S. Nemat-Nasser, *AIAA J.* **10**, 30 (1973).

19. A. Schweitzer, *Ing. Archiv.* **46**, 43 (1977).

20. A. F. Nayfeh and D. T. Mook, *Nonlinear Oscillations*, Wiley, New York, 1979.

21. W. Johnson, *Helicopter Theory*, Princeton University Press, Princeton, NJ, 1980.

22. D. L. Mingori, *AIAA J.* **7**, 20 (1969).

23. K. G. Lindh and P. W. Likins, *AIAA J.* **8**, 680 (1970).

24. S. Chu, *Adv. Atom. Mol. Phys.* **21**, 197 (1985).

25. G. F. Thomas and W. J. Meath, *J. Phys. B: Atom. Mol. Phys.* **16**, 951 (1983).

26. R. A. Thuraisingham and W. J. Meath, *Mol. Phys.* **56**, 193 (1985).

27. M. A. Kmetic, R. A. Thuraisingham, and W. J. Meath, *Phys. Rev. A* **33**, 1966 (1986).

28. B. Noble, *Applied Linear Algebra*, Prentice-Hall, Englewood Cliffs, NJ, 1969.

29. M. A. Gockel, *J. Am. Hel. Soc.* **16**, 2 (1972).

30. D. A. Peters and K. H. Hohenemser, *J. Am. Hel. Soc.* **16**, 22 (1971).

31. G. H. Gaonkar, D. S. Prasad, and D. Sastry, *J. Am. Hel. Soc.* **26**, 56 (1981).

32. G. Horvay and S. W. Yuan, *J. Aeronaut. Sci.* **14**, 583 (1947).

33. C. Moler and C. Van Loan, *SIAM Rev.* **20**, 801 (1978).

34. G. J. Bierman, *IEEE. Trans. Autm. Cntrl.*, 228, (1982).

35. C. Von Kerczek, *AIAA J.* 13, 1401 (1975).

36. R. Celi and P. P. Friedmann, *AIAA Paper No. 87-0921CP*, 1987.

37. L. F. Shampine and M. K. Gordon, *Computer Solution of Ordinary Differential Equations: The Initial Value Problem*, Freeman, San Francisco, 1975.

38. L. F. Shampine, H. A. Watts, and S. M. Davenport, *SIAM Rev.* 18, 376 (1976).

39. W. H. Press, B. P. Flannery, S. A. Teukolsky, and W. T. Vetterling, *Numerical Recipes: The Art of Scientific Computing*, Cambridge University Press, Cambridge, 1986, pp. 569–572.

40. M. Borri, *Comp. Math. Appl.* 12A, 149 (1986).

41. C. D. Bailey, *Found. Phys.* 5, 433 (1975).

42. T. E. Simkins, *AIAA J.* 19, 1357 (1981).

43. M. Baruch and R. Riff, *AIAA J.* 20, 687 (1982).

44. R. Riff and M. Baruch, *AIAA J.* 22, 1310 (1984).

45. R. Riff and M. Baruch, *AIAA J.* 22, 1171 (1984).

46. C. S. Hsu, *Adv. Appl. Mech.* 17, 245 (1977).

47. A. M. Samoilenko and N. I. Ronto, *Numerical Analytic Method of Investigating Periodic Solutions*, MIR Publishers, Moscow, 1979.

48. M. Urabe, *Arch. Ratio. Mech. Anal.* 20, 120 (1965).

49. H. K. Wilson, *Ordinary Differential Equations*, Addison-Wesley, Reading, MA, 1971, p. 117.

50. C. S. Hsu and W. H. Cheng, *J. Appl. Mech.* 41, 371 (1974).

51. J. Dugundji and J. H. Wendell, *AIAA J.* 21, 890 (1983).

52. M. A.H. Dinyavari and P. P. Friedmann, *AIAA J.* 24, 1424 (1986).

53. S. B. R. Kottapalli, P. P. Friedmann, and A. Rosen, *AIAA J.* 17, 1381 (1979).

54. W. Warmbrodt and P. Friedmann, *AIAA J.* 18, 1118 (1980).

55. M. Urabe and A. Reiter, *J. Math. Anal. Appl.* 14, 107 (1966).

56. K. R. Schneider, *Computing* 10, 63 (1972).

57. D. Greenspan, *J. Math. Anal. Appl.* 39, 574 (1972).

58. R. Van Dooren, *Numer. Math.* 20, 300 (1973).

59. T. J. Aprille and T. N. Trick, *Proc. IEEE* 60, 108 (1972).

60. T. J. Aprille and T. N. Trick, *IEEE Trans. Cir. Theor.* CT-19, 354 (1972).

61. P. P. Friedmann and S. B. R. Kottapalli, *J. Am. Hel. Soc.* 27, 28 (1982).

62. D. A. Peters and A. P. Izadpanah, *Proceedings of the Thirty-Seventh Annual Forum of the American Helicopter Society*, New Orleans, 1981, pp. 217–226.

63. S. S. Lau, Y. K. Cheung, and S. Y. Wu, *J. Appl. Mech.* 49, 849 (1982).

64. C. A. Borges, L. Cesari, and D. A. Sanchez, *Quart. Appl. Math.*, 457 (1975).

65. S. M. Roberts and J. S. Shipman, *Two Point Boundary Problems: Shooting Methods*, Elsevier, New York, 1972.

66. F. K. Straub and P. P. Friedmann. *Vertica* 5, 75 (1981).

67. M. A. H. Dinyavari, Ph.D. Dissertation, University of California, Los Angeles, 1985.

68. B. Panda and I. Chopra, *Vertica* 11, 187 (1987).

69. T. Iwatsubo, M. Saigo, and Y. Sugijama, *J. Sound Vibrat.* 30, 65 (1973).

70. C. S. Hsu, *J. Appl. Mech.* 30, 367 (1963).

CHAPTER V

THE RECURSIVE RESIDUE GENERATION METHOD

ROBERT E. WYATT

Department of Chemistry and Institute for Theoretical Chemistry, University of Texas, Austin, Texas 78712

CONTENTS

I. INTRODUCTION

A. Recursive Residue Generation Method

The recursive residue generation method RRGM is a recently formulated approach to the computation of transition amplitudes, both time dependent and time independent, in large multi state quantum systems [1, 2].* In studies of the dynamics of large quantum systems (a diverse subject that includes multiphoton excitation, the influence of relaxation processes on spectral lineshapes, interaction of adsorbed particles with surface vibration-electronic states, and collisional energy transfer), the principal limitation has been the inability to obtain eigenvectors of very large Hamiltonian matrices (> 1000 rows or columns), even on Class VI supercomputers. In the RRGM, transition amplitudes are computed one at a time by recursively generating the residues and poles of several Green functions; eigenvectors of large matrices are not computed at all. A novel feature is that the residues can be computed from sets of eigenvalues. Once the residues are known with sufficient accuracy (the calculation can be terminated at various levels of approximation), transition amplitudes are easily computed.

An important step in the RRGM is partial tridiagonalization of the large Hamiltonian matrix, for which the Lanczos algorithm [3–8] is well suited. (A short biography of Lanczos is in the appendix of this chapter.) Over the past decade, a greater appreciation of its powers has arisen through applications to a number of diverse topics in chemistry, physics, and engineering. Some of these applications are mentioned in the next section. In addition, Lanczos recursion and the RRGM have input from a number of subjects in pure and applied mathematics, including Green functions, tridiagonal matrices, continued fractions, orthogonal polynomials, Padé approximants, the moment problem, the algebraic eigenvalue problem, and numerical analysis.

There are several books that should be mentioned because they cover topics closely related to those discussed in this chapter. Volume 35 of *Solid State Physics* [9] is a gold mine of information about recursive approaches to the structure and spectra of disordered solids. The chapters by Heine and by Haydock are highly recommended. In addition, a recent conference proceedings with the appealing title *The Recursion Method* [10] also emphasizes disordered solids, but there are a few chapters on applications to elementary particle and nuclear physics. On the mathematical side, Strang's new undergraduate textbook, *Introduction to Applied Mathematics* [11], contains excellent sections on the Lanczos algorithm and the conjugate gradient and

This work was supported in part by research grants from the National Science Foundation, The Welch Foundation, and Cray Research, Inc.

*In this chapter, references are enclosed in brackets—Ed.

QR algorithms. At a more advanced level, Parlett's book, *The Symmetric Eigenvalue Problem* [4], contains several chapters on how and why Lanczos works in finite precision arithmetic. Cullum and Willoughby's excellent two-volume set, *Lanczos Algorithms for Large Symmetric Eigenvalue Computations* [7], contains or references almost everything known about the Lanczos algorithm. Their philosophy about using Lanczos is similar to Sections II.B–II.E in this chapter.

B. Scope of Lanczos Algorithm: From Elementary Particles to the Atlantic Ocean

The Lanczos recursion algorithm has been used to solve very large eigenproblems in a number of different fields. A partial listing includes the following:

Disordered solids
    Anderson localization [12–15]
    Vibrational spectra [16, 17]
    Surface states [18]
Jahn–Teller effect [19–23]
Infrared spectra, ice [24]
Molecular spectroscopy [25–29]
Multiphoton absorption [30, 31]
Intramolecular dynamics [32, 33]
Photoabsorption/ionization cross sections [34–37]
Spin relaxation [38–42]
Liouville equation/relaxation [43]
Memory functions/relaxation [44]
Wavepacket dynamics [45]

Natural orbital analysis [46]
Percolation on lattices [47]
Thermodynamic properties, error bounds [48]
Reaction rate constants [49, 50]
Molecular scattering amplitudes [51]
Collisional resonances [52]
Electrode modeling [53]
Nuclear shell model [54, 55]
Lattice gauge field theories [56–59]
Structural mechanics
    Dynamic response of buildings [60]
    Offshore platforms [61]
Normal modes, Atlantic Ocean [62]

In addition, the Lanczos algorithm [7] has a number of versions that are appropriate for real symmetric matrices, complex symmetric matrices [39, 41], Hermitian matrices, nonsymmetric matrices [43], and nonorthogonal basis sets [45].

C. Overview

The remainder of this chapter is divided into four major sections. Section II begins with an introduction to transition amplitudes, Green functions, and residues. Then there is a branch into subsections describing two ways to construct the residues: Sections II.B, II.D, and II.E build residues from sets of eigenvalues, and Sections II.F. and II.G are based on use of the QL algorithm to build chain propagators. The remaining sections deal with recursion, Section II.C, and the significant effects of roundoff error, Section II.B, Section II.E.

Section III presents a more detailed view of the recursive procedure.

Sections III.A and III.B are concerned with advantages of the recursive procedure, particularly with how it attempts to build an optimal small subspace within the large space for the process under consideration. That the RRGM works when the Hamiltonian has degenerate eigenvalues is shown by way of an example in Section III.C. When we want a number of transition amplitudes involved in band-to-band transitions, there is a way to significantly reduce the computer time. This topic, residue algebra, is treated in Section III.D. In the final part of this section, different levels of parallelism are described, and reference is made to a dedicated "Lanczos computer" in Section III.F.

Section IV describes another approach to the all-important residues. Moments of the Hamiltonian over the initial state, $\langle 0|H^k|0\rangle$, are not very useful because combinations of them tend to be numerically unstable. However, the moments provide a nice interpretation of the chains (tridiagonal matrices) that are developed recursively. This is the topic of Sections IV.A–IV.C. However, the moments can be analyzed via simple diagrams (Section IV.D) and then recombined to eliminate much of the instability. The definition and use of these linked and irreducible moments is described in Sections IV.E and IV.F.

Section V treats some applications of the RRGM. Correlation functions, power spectra, and multiphoton excitation are treated in Sections V.I and V.II. Finally, some current applications are described in Section V.C.

For those readers who are adventurous enough to try it for themselves, the listing of an RRGM code (based on Sections II.F and II.G), with sample input and output, is available from the author.

II. RECURSIVE PARADIGM

A. The Quantum Transition Amplitude: Importance of Large Residues

The transition amplitude for a system described by a time-independent Hamiltonian H between an initial state $|i\rangle$ (at $t_0 = 0$) and a final state $|f\rangle$ (at time t) is given by

$$\langle f|U(t|0)|i\rangle = \langle f|\exp(-iHt)|i\rangle. \tag{1}$$

Assuming that $|i\rangle$ and $|f\rangle$ belong to the spectrum of a time-independent Hamiltonian H^0, transition between these states is induced by the (time-independent) perturbation $V = H - H^0$, which in some sense makes the original system "unstable." Let us write the eigenproblem for the perturbed system as $H|\alpha\rangle = \varepsilon_\alpha|\alpha\rangle$, $\alpha = 1, 2, \ldots, N$. A discrete nondegenerate spectrum is assumed, together with the usual completeness and orthonormality relations:

$\sum_\alpha |\alpha\rangle\langle\alpha| = 1; \langle\alpha|\beta\rangle = \delta_{\alpha\beta}$. Degeneracy requires a separate discussion; see Section III.C. From Eq. 1, we obtain

$$\langle f | U(t|0) | i \rangle = \sum_{\alpha=1}^{N} \langle f | \alpha \rangle \langle \alpha | i \rangle \exp(-i\varepsilon_\alpha t). \qquad (2)$$

For a relatively small basis of N states, where $N < O(10^3)$, the standard way to compute transition amplitudes is to diagonalize the matrix representation of H in the basis provided by H^0 (e.g., using EISPACK subroutines) so as to obtain *all* eigenvalues and eigenvectors. The matric eigenproblem would be written as $\mathbf{HC} = \mathbf{CE}$ for an orthonormal basis. However, this approach is intractable for larger bases because of the extreme difficulty involved in computing and storing N eigenvectors.

We now make a crucial observation concerning the summation in Eq. 2. Regarding $\langle f | \alpha \rangle$ as the element in row f and column α of the $N \times N$ eigenvector matrix $\mathbf{C}$, the right side of Eq. 2 involves the product of elements in rows i and f with summation over the *columns* of $\mathbf{C}$. In other words, to compute a set of transition amplitudes, the full eigenvectors are not needed. What we require is the *product* of elements in two "eigenrows" (or one row, for the $i \to i$ amplitude). If one of these products, a *residue*, is denoted

$$R_{fi}(\alpha) = \langle f | \alpha \rangle \langle \alpha | i \rangle, \qquad (3)$$

Eq. 2 becomes

$$\langle f | U(t|0) | i \rangle = \sum_{\alpha=1}^{N} R_{fi}(\alpha) \exp(-i\varepsilon_\alpha t). \qquad (4)$$

The residues measure the *"static" capacity* of the eigenstates to convey amplitude during the $i \to f$ transition. *Activation* of each static pathway (through one eigenstate) is provided by the *dynamic factor* $\exp(-i\varepsilon_\alpha t)$.

The set of quantities $R_{fi}(\alpha)$, $\alpha = 1, 2, \ldots, N$, when plotted versus ε_α, define the *"transition envelope."* First, the coefficients $\langle i | \alpha \rangle$ and $\langle f | \alpha \rangle$, when plotted versus ε_α, are frequently largest around the two "unperturbed" energies E_i and E_f. The products $\langle f | \alpha \rangle \langle \alpha | i \rangle$ on the same plot are frequently largest "in-between" E_i and E_f. They indicate how much amplitude leaks from i into f by way of each eigenstate α. Eigenstates oriented in-between $|i\rangle$ and $|f\rangle$ are the best transmitters of the signal that leaves i at $t = 0$ and is received in f at time t. Direct computation of the transition envelope is one goal of the RRGM.

Another comment related to Eq. 4 is that in order to compute a good approximation to the amplitude, up to time t, we only need the largest of the products, say, the largest M of them. The amplitude can then be approximated

by

$$\langle f | U(t|0) | i \rangle = \sum_{\alpha}' R_{fi}(\alpha) \exp(-i\varepsilon_\alpha t), \tag{5}$$

where only M terms are involved in the summation. Stated again, the aim of the RRGM is to directly develop the "most important" contributions to this amplitude.

B.   Building the Green Function: Residues from Eigenvalues

In this section, we will review the original Nauts and Wyatt [1, 2] formulation of the RRGM. An outline of the steps is shown in Figure 1. In the RRGM, direct calculation of the eigenvector coefficients $\langle f | \alpha \rangle$ is avoided. Instead, we focus on the *transition residues* (a real symmetric Hamiltonian matrix and real-valued orthonormal basis functions are assumed):

$$R_{fi}(\alpha) = \langle f | \alpha \rangle \langle \alpha | i \rangle = 1/2[\langle u_0 | \alpha \rangle^2 - \langle v_0 | \alpha \rangle^2] = 1/2[R_u(\alpha) - R_v(\alpha)], \tag{6}$$

where the two orthonormal *transition vectors* associated with the $i \to f$ transition are

$$|u_0\rangle = [|i\rangle + |f\rangle]/\sqrt{2}, \qquad |v_0\rangle = [|i\rangle - |f\rangle]/\sqrt{2}. \tag{7}$$

The quantities $R_u(\alpha)$, $R_v(\alpha)$, and $R_{fi}(\alpha)$ are residues, at pole ε_α, of the diagonal

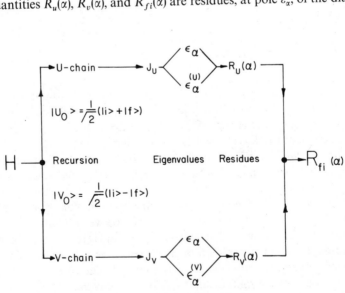

Figure 1.   RRGM flowchart.

and off-diagonal matrix elements of the Green operator $(z - H)^{-1}$, where z does not belong to the spectrum of H; for example,

$$G_u(z) = \langle u_0 | (z - H)^{-1} | u_0 \rangle = \sum_\alpha \frac{\langle u_0 | \alpha \rangle^2}{z - \varepsilon_\alpha} = \sum_\alpha \frac{R_u(\alpha)}{z - \varepsilon_\alpha},$$

$$G_{fi}(z) = \langle f | (z - H)^{-1} | i \rangle = \sum_\alpha \frac{\langle f | \alpha \rangle \langle \alpha | i \rangle}{z - \varepsilon_\alpha} = \sum_\alpha \frac{R_{fi}(\alpha)}{z - \varepsilon_\alpha}. \tag{8}$$

If one of the Green functions, for example, $G_u(z)$, is known, it is a trivial matter to extract the set of residues. For example, at pole ε_α,

$$R_u(\alpha) = \lim_{z \to \varepsilon_\alpha} (z - \varepsilon_\alpha) G_u(z) = \lim_{z \to \varepsilon_\alpha} (z - \varepsilon_\alpha) \langle u_0 | (z - H)^{-1} | u_0 \rangle,$$

$$\alpha = 1, 2, \ldots, N. \tag{9}$$

The problem is to first calculate $G_u(z)$ and $G_v(z)$; residues $\{R_u(\alpha)\}$ and $\{R_v(\alpha)\}$ can then be extracted with Eq. 9, and the time-dependent transition amplitudes can be computed from Eq. 5.

To produce $G_u(z)$, imagine the construction of a new orthonormal basis (the recursion basis), $\{|u_j\rangle; j = 0, 1, \ldots\}$, from the original basis. In this new basis, the equation defining the $N \times N$ Green matrix is

$$(z\mathbf{1} - \mathbf{H})\mathbf{G}(z) = \mathbf{1}, \tag{10}$$

where $\mathbf{G}$ has elements that are the Green functions, $G_{ij}(z) = \langle u_i | (z - H)^{-1} | u_j \rangle$. Each matrix in Eq. 10 can be very large, so direct computation of $\mathbf{G}$ by inverting $(z\mathbf{1} - \mathbf{H})$ is not feasible. However, the key point is that we need *only* the 1, 1 element in the $\mathbf{G}$ matrix; this is the element previously denoted $G_u(z) = \langle u_0 | (z - H)^{-1} | u_0 \rangle$. Now, by the usual procedure to invert matrices,

$$G_u(z) = G_{1,1}(z) = \det [z\mathbf{1} - \mathbf{H}]^r / \det [z\mathbf{1} - \mathbf{H}], \tag{11}$$

where the superscript r denotes the *reduced matrix* obtained by deleting the first row and column from $z\mathbf{1} - \mathbf{H}$. In terms of the eigenvalues of $\mathbf{H}$ and $\mathbf{H}^r$, denoted $\{\varepsilon_\alpha\}$ and $\{\varepsilon_\alpha^{(u)}\}$, respectively, the ratio of determinants becomes

$$G_u(z) = \frac{(z - \varepsilon_1^{(u)}) \cdots (z - \varepsilon_{N-1}^{(u)})}{(z - \varepsilon_1) \cdots (z - \varepsilon_N)}. \tag{12}$$

Note that there are $N - 1$ products in the numerator and N products in the denominator. This result, when combined with Eq. 6, permits the evaluation of residues in terms of eigenvalues (not eigenvectors),

$$R_u(\alpha) = \langle u_0 | \alpha \rangle^2 = \frac{\varepsilon_\alpha - \varepsilon_1^{(u)}}{\varepsilon_\alpha - \varepsilon_1} \cdots \frac{(\varepsilon_\alpha - \varepsilon_{\alpha-1}^{(u)})(\varepsilon_\alpha - \varepsilon_\alpha^{(u)})}{(\varepsilon_\alpha - \varepsilon_{\alpha-1})(\varepsilon_\alpha - \varepsilon_{\alpha+1})} \cdots \frac{\varepsilon_\alpha - \varepsilon_{N-1}^{(u)}}{\varepsilon_\alpha - \varepsilon_N}. \tag{13}$$

There are now $N-1$ terms in this product; each ratio is about unity in magnitude. The significant result is that all residues associated with $G_u(z)$ may be computed from *two* sets of eigenvalues, $\{\varepsilon_\alpha\}$ and $\{\varepsilon_\alpha^{(u)}\}$; no eigenvectors are needed. In a similar way, residues associated with $G_v(z)$ may be computed from $\{\varepsilon_\alpha\}$ and $\{\varepsilon_\alpha^{(v)}\}$.

An $N = 2$ example may clarify these formulas. The defining equation for the Green matrix is

$$\begin{bmatrix} z - H_{11} & H_{12} \\ H_{21} & z - H_{22} \end{bmatrix} \begin{bmatrix} G_{11}(z) & G_{12}(z) \\ G_{21}(z) & G_{22}(z) \end{bmatrix} = \begin{bmatrix} 1 & 0 \\ 0 & 1 \end{bmatrix}. \tag{14}$$

The $(1, 1)$ element of $\mathbf{G}(z)$ is

$$G_u(z) = \frac{z - H_{22}}{(z - H_{11})(z - H_{22}) - H_{12} H_{21}} = \frac{z - \varepsilon_1^r}{(z - \varepsilon_1)(z - \varepsilon_2)}.$$

The two residues are

$$R_u(\varepsilon_1) = \frac{\varepsilon_1 - \varepsilon_1^r}{\varepsilon_1 - \varepsilon_2}, \qquad R_u(\varepsilon_2) = \frac{\varepsilon_2 - \varepsilon_1^r}{\varepsilon_2 - \varepsilon_1}, \tag{15}$$

and they sum to unity: $R_u(\varepsilon_1) + R_u(\varepsilon_2) = 1$.

C. Recursion via Lanczos: Forging Links in the Chain

The main feature that remains to be discussed is how to construct the two recursion bases $\{|u_j\rangle\}$ and $\{|v_j\rangle\}$. For this, we use the powerful Lanczos method [3–7], which builds the new vectors, one at a time, from the old ones in such a way that the Hamiltonian matrix is converted to tridiagonal (Jacobi) form. Initiated with the starter $|u_0\rangle$, M-step Lanczos recursion simultaneously builds elements in the $M \times M$ Jacobi (tridiagonal) matrix $\mathbf{J}$ and the recursion vectors $\{|u_0\rangle, |u_1\rangle, \ldots, |u_{M-1}\rangle\}$, which span the M-dimensional subspace V_M. Obtaining eigenvalues $\{\varepsilon_\alpha\}$ from $\mathbf{J}$ and $\{\varepsilon_\alpha^{(u)}\}$ from $\mathbf{J}^r$ [the $(M-1) \times (M-1)$ reduced Jacobi matrix] is computationally not demanding, as discussed in the next section.

Tridiagonalizing $\mathbf{H}$ is best viewed as a transformation from the original molecular basis $\{|k\rangle, k = 1, 2, \ldots, N\}$ to the recursion basis $\{|u_n\rangle, n =$

$0, 1, \ldots, M - 1$. In this new basis, the only nonzero matrix elements of $\mathbf{J}$ are the diagonal self-energies $\{a_0, a_1, \ldots, a_{M-1}\}$ and the nearest-neighbor coupling (or hopping) energies $\{b_1, b_2, \ldots, b_{M-1}\}$. In effect, we have converted the original problem, with a complex network of interstate couplings, to a one-dimensional disordered chain with M balls $\{a_n\}$ connected by $M - 1$ springs $\{b_n\}$. The chain is disordered in the sense that the diagonal elements all take on different values. We will refer to a_n and b_{n+1} as specifying *one link* in this one-dimensional chain used to portray $\mathbf{J}$ (e.g., see Ref. 12). This one-dimensional chain is the most compact way to represent the problem, aside from the (diagonal) eigenstate representation.

In the Lanczos algorithm, each recursion step, initiated from either $|u_0\rangle$ or $|v_0\rangle$, forges a new link in the chain. Starting with this initial recursion vector $|0\rangle$, we then form the vector $H|0\rangle$; the first self-energy is $a_0 = \langle 0|H|0\rangle$. The first residual vector $\{H|0\rangle - a_0|0\rangle\}$ is then formed; its norm determines the first coupling element in $\mathbf{J}$, $b_1 = \|H|0\rangle - a_0|0\rangle\|^{1/2}$. The next normalized recursion vector is then $|1\rangle = (H|0\rangle - a_0|0\rangle)/b_1$. Now, given the recursion vectors $|n\rangle$ and $|n-1\rangle$, and the previous chain link (a_n, b_n), the next vector is generated from the explicit three-term recurrence relation

$$b_{n+1}|n+1\rangle = H|n\rangle - a_n|n\rangle - b_n|n-1\rangle = |r_n\rangle, \tag{16}$$

where $a_n = \langle n|H|n\rangle$ and b_{n+1} is chosen to normalize the new vector: $b_{n+1}^2 = \langle r_n|r_n\rangle$. By construction, at least in infinite-precision arithmetic, each recursion vector is orthogonal to all previous ones. In order to start the recursion, we choose either $|0\rangle = |u_0\rangle$ or $|0\rangle = |v_0\rangle$.

Having generated the sets of diagonal self-energies and off-diagonal coupling energies from the starter $|0\rangle = |u_0\rangle$, two diagonalizations (using the subroutine TQLRAT from EISPACK) yield eigenvalues of $\mathbf{J}$, denoted $\{\varepsilon_\alpha\}$, and eigenvalues of the reduced Jacobi matrix (in which b_1 is set to zero), denoted $\{\varepsilon_\alpha^{(u)}\}$. From these two sets of eigenvalues, all residues $R_u(\alpha)$ are computed from Eq. 13. Priming the recursion method with the other starting vector $|0\rangle = |v_0\rangle$ then leads to two additional sets of eigenvalues $\{\varepsilon_\alpha\}$ and $\{\varepsilon_\alpha^{(v)}\}$. (Of course, eigenvalues from the two full $\mathbf{J}$ matrices will be identical for the $|u_0\rangle$ or $|v_0\rangle$ starting vectors.) The residues $R_v(\alpha)$ are then computed from an equation analogous to Eq. 13. Finally, the transition residues $R_{fi}(\alpha)$ follow very simply from Eq. 6.

D. Spurious Eigenvalues: Ghosts from Roundoff Error

As recursion proceeds, rounding errors in finite-precision arithmetic lead to a loss of significant figures that produces a gradual (after 30–50 steps) loss of global orthogonality (and sometimes linear independence) in the recursion basis. Numerical experience has shown that this results in multiple copies ("ghosts") of some eigenvalues, usually those on both edges of the range of eigenvalues. In addition, some "extraneous eigenvalues" (i.e., eigenvalues that are poor approximations to any real eigenvalues) are produced, which eventually settle onto actual eigenvalues as M increases. These *spurious eigenvalues*, in the terminology of Cullum and Willoughby [7], must be removed from the eigenvalue lists before computing residues from Eq. 13. This is accomplished in a two-step procedure, as discussed elsewhere [1, 2]; it is an application of the Cullum–Willoughby method. The Cullum–Willoughby comparison test is as follows:

1. Compute eigenvalues for J and J^r. Determine the numerical multiplicities for each eigenvalue.
2. All numerically multiple eigenvalues within each list are "good." All eigenvalues in one list that are also in the other list are "bad" and are deleted from *both* lists.

The validity of the method for our applications was established for $N < 300$ by comparing eigenvalues and residues from direct diagonalization with those from the recursion procedure.

It is important to remove the *multiple* eigenvalues ("ghostbusting") before using Eq. 13 to evaluate residues. Since different numbers of ghost eigenvalues may occur for each eigenvalue of both the full matrix J and the reduced matrix J^r, it is essential to remove multiple copies from both lists $\{\varepsilon_\alpha\}$ and $\{\varepsilon_\alpha^{(u)}\}$. However, single "spurious" eigenvalues that appear in *both* lists lead to canceling contributions to Eq. 13, so they do not have to be removed from either list. A necessary, but not sufficient, check on the residues is the upper bound $R_u(\alpha) < 1$ and the sum rule $\sum_\alpha R_u(\alpha) = 1$. In practice, the sum rule is normally obeyed to within machine precision.

E. How To Cope with Loss of Orthogonality: Ignore It

After Lanczos introduced the recursion algorithm [3] in 1950, it was soon found that the recursion vectors suffered a loss of global orthogonality. Interest in Lanczos recursion quickly waned when Givens and Householder (independently) introduced sequences of plane rotations or elementary reflectors as techniques for reducing a real symmetric matrix to tridiagonal form [4]. The underlying mechanism for the loss of orthogonality was not understood until the work of Paige [6] in the 1970s. In fact, without his penetrating analysis the Lanczos algorithm probably would remain buried. Paige compared several forms of the recursion algorithm with regard to error propagation and stability. It is startling that the orthogonality problem, the appearance of ghost eigenvalues, convergence of some eigenvalues, and roundoff error are all interconnected! In fact, the initial loss of orthogonality of the Lanczos vectors is *caused by* the convergence of some (extreme) eigenvalues [4].

In addition, we are taught that irreducible tridiagonal matrices (no b's are zero) must have *distinct* eigenvalues. However, when the Lanczos algorithm is used to generate the tridiagonal matrix from which eigevalues are found, it is found that one almost always has some numerically multiple eigenvalues (to 12–13 digits, the "machine ε")! This unexpected feature is also connected to roundoff error propagation.

How should we approach the loss-of-orthogonality problem? There are at least several possibilities: total reorthogonalization of all recursion vectors, limited reorthogonalization with respect to converged vectors, and no reorthogonalization: ignore the developing problem. In common with the philosophy of Cullum and Willoughby [7], we will *not* reorthogonalize any vectors—let the pathology develop! In the Nauts–Wyatt version of the RRGM [1, 2], the Cullum–Willoughby identification test was used to eliminate spurious eigenvalues before computing residues from Eq. 13. However, in the later Wyatt–Scott [63] version, we do not eliminate anything. Accurate transition amplitudes are obtained! This will be treated in more detail in the next section.

F. Building the Green Function: The Chain Propagator

In the "original version" of the RRGM, squares of eigenvector coefficients were computed from eigenvalues of the tridiagonal matrix and those of the reduced tridiagonal matrix (one fewer row and column). The eigenvalues and residues were then used to evaluate transition amplitudes and probabilities. As we mentioned earlier, a well-known feature of Lanczos recursion is that spurious eigenvalues may be produced due to the propagation of roundoff error. Comparison of the eigenvalue lists from the original and reduced tridiagonal matrices allows one to eliminate all of these spurious eigenvalues.

In this section, we will review a different route, that of Wyatt and Scott [63], from the tridiagonal matrix to the eigenvalues and residues. The new method *eliminates* the need to compare lists of eigenvalues. All eigenvalues and residues, *even spurious ones*, are used to compute transition amplitudes! In addition, the new method is conceptually and computationally simpler than the original version of the RRGM. Best of all, it works!

Letting $\mathbf{u}_0$ denote the column vector (N elements) representing $|u_0\rangle$, the survival amplitude is simply given by the $(1, 1)$ element of the propagator,

$$\langle 0| \exp(-iHt)|0\rangle = \mathbf{u}_0^t \exp(-iHt)\mathbf{u}_0.$$

As a result, the $i \to f$ transition amplitude in Eq. 3 is computed from the difference between two matrix propagators:

$$A_{if}(t) = 1/2[\mathbf{u}_0^t \exp(-i\mathbf{H}t)\mathbf{u}_0 - \mathbf{v}_0^t \exp(-i\mathbf{H}t)\mathbf{v}_0]. \tag{17}$$

In order to evaluate these two survival amplitudes, we will first focus on the U chain, with recursion vectors $\mathbf{u}_0, \mathbf{u}_1, \ldots$. The set of M vectors $\mathbf{u}_0, \mathbf{u}_1, \ldots, \mathbf{u}_{M-1}$ (each with N rows) form columns in the $M \times M$ matrix $\mathbf{Q}$ that transforms $\mathbf{H}$ into the chain representation

$$\mathbf{Q}^t\mathbf{H}\mathbf{Q} = \mathbf{J}_u. \tag{18}$$

As a result, the survival amplitude is given in terms of the *chain propagator*, $\exp(-i\mathbf{J}_u^t)$:

$$S_u(t) = \mathbf{u}_0^t \mathbf{Q} \exp(-i\mathbf{J}_u^t)\mathbf{Q}^t\mathbf{u}_0 \tag{19}$$

But $\mathbf{u}_0^t\mathbf{Q}$, an M-column row vector, is just the "first" unit vector:

$$\mathbf{u}_0^t\mathbf{Q} = [1, 0, 0, \ldots, 0] = \mathbf{e}_1^t. \tag{20}$$

This arises because $\mathbf{Q}$ is a collection of the M recursion vectors and $\mathbf{u}_i^t\mathbf{u}_j = \delta_{ij}$. As a result, the survival amplitude is the $(1, 1)$ element of the $M \times M$ matrix propagator for the U chain:

$$S_u(t) = (\exp(-i\mathbf{J}_u t))_{1,1}, \tag{21}$$

and the transition amplitude is given by the difference between the $(1, 1)$ elements of the two chain propagators,

$$\langle f | \exp(-i\mathbf{H}t) | i \rangle = 1/2[(\exp(-i\mathbf{J}_u t))_{1,1} - (\exp(-i\mathbf{J}_v t))_{1,1}]. \tag{22}$$

Now, we focus on the matrix eigenproblem involving just one of the tridiagonal matrices (of order M):

$$\mathbf{S}^t\mathbf{J}_u\mathbf{S} = \mathbf{E} = \mathrm{diag}(\varepsilon_1, \varepsilon_2, \ldots, \varepsilon_m), \tag{23}$$

Then the survival amplitude is easily calculated:

$$(\exp(-i\mathbf{J}_u t))_{1,1} = \sum_{\alpha=1}^{M} s_{1\alpha}^2 \exp(-i\varepsilon_\alpha t). \tag{24}$$

A very important feature of this equation is that *only the first row of S* is needed.

If we can efficiently evaluate the M eigenvalues and the coefficients in the first row of this $M \times M$ eigenvector matrix, the survival amplitude can be easily computed by performing the summation in Eq. 24.

G. QL Algorithm for Residues

The traditional technique for computing all of the eigenvalues of a symmetric matrix is to use a finite sequence of orthogonal similarity transforms to tridiagonalize the matrix followed by use of the QL algorithm. The QL algorithm uses a theoretically infinite sequence of orthogonal similarity transformations *that preserves the tridiagonal form* and converges to a diagonal matrix. The resulting eigenvector matrix is the product of all of the transformation matrices. If the eigenvectors are desired, these transformations can be accumulated as they are applied to the matrix. The QL algorithm costs $O(M^2)$ operations to compute the eigenvalues but also costs $O(M^3)$ if *all* of the eigenvectors are computed. Strang provides a fine introduction to the closely related QR algorithm (11).

In our context, the matrix J is already tridiagonal, and only the first row of the eigenvector matrix (S in Eq. 23) is needed. Fortunately, it is possible to obtain the first row of the eigenvector matrix for $O(M^2)$ operations. The subroutine IMTQL2 from the EISPACK library is recommended, but in order to understand the basic idea, we will consider the simplest version of the algorithm. In order to initiate the QL algorithm, we first factor J (now relabeled J_1) into an orthogonal matrix and a lower triangular matrix:

$$J_1 = Q_1 L_1 \rightarrow Q_1^t J_1 = L_1. \tag{25}$$

When these factors are multiplied in **reverse order**, the next Jacobi matrix in the sequence is formed:

$$J_2 = L_1 Q_1 \rightarrow Q_1^t J_1 Q_1. \tag{26}$$

The process is continued by "doggedly iterating" [4]. At step k, we have

$$J_k = Q_k L_k,$$

so the next tridiagonal matrix is

$$
\begin{aligned}
J_{k+1} = L_k Q_k &= Q_k^t J_k Q_k \\
&= (Q_k^t \cdots Q_2^t Q_1^t) J_1 (Q_1 Q_2 \cdots Q_k) \\
&= S_k^t J_1 S_k.
\end{aligned} \tag{27}
$$

There are several significant features about this iterative process:

1. The tridiagonal form is preserved in the sequence of $\mathbf{J}$ matrices.
2. The sequence $\mathbf{J}_1, \mathbf{J}_2, \ldots, \mathbf{J}_k$ converges to a diagonal (eigenvalue) metrix

$$\mathbf{J}_1 \rightarrow \mathbf{J}_2 \rightarrow \cdots \rightarrow \mathbf{J}_k = \text{diag}[\varepsilon_1, \varepsilon_2, \ldots, \varepsilon_m]. \tag{28}$$

3. As $\mathbf{J}_k$ becomes diagonal, $\mathbf{S}_k$ simultaneously converges to the eigenvector matrix of $\mathbf{J}$:

$$\mathbf{S}_k^t \mathbf{J}_k \mathbf{S}_k = \text{diag}(\varepsilon_1, \ldots, \varepsilon_m). \tag{29}$$

A simple application of the use of this algorithm is provided in Ref. 63.

In the preceding section, we emphasized that only the top row $(s_{11}, s_{12}, \ldots, s_{1m})$ of the eigenvector matrix of $\mathbf{J}$ is needed in order to compute survival amplitudes. The QL algorithm, as normally implemented, computes *all* eigenvectors of $\mathbf{J}$; the converged $M \times M$ matrix $\mathbf{S}_k$ is generated. However, the algorithm may be modified to produce only the first row of $\mathbf{J}$. The key observation is the fact that each new transformation is applied *on the right* of the current approximation to the eigenvector matrix. That is, if $\mathbf{Q}_1, \mathbf{Q}_2, \ldots, \mathbf{Q}_k$ are the transformations used so far, the approximation to the eigenvector matrix is

$$\mathbf{S}_k = \mathbf{Q}_1 \mathbf{Q}_2 \cdots \mathbf{Q}_k, \tag{30}$$

with the $\mathbf{Q}$'s multiplied in that order: $\mathbf{S}_{\text{new}} = \mathbf{S}_{\text{old}} \times \mathbf{Q}_{\text{new}}$. Formally, the first row of the eigenvector is

$$\mathbf{e}_1^t \mathbf{S}_k = \mathbf{e}_1^t \mathbf{Q}_1 \mathbf{Q}_2 \cdots \mathbf{Q}_k, \tag{31}$$

where $\mathbf{e}_1^t = (1, 0, 0, \ldots, 0)$. Thus, to obtain the desired row, it is only necessary to update a single vector by operating on the right by each transformation as it is generated.

III. UNDERSTANDING AND DOING RECURSION

A. Why Recur?

Several important advantages of the recursion method are as follows (see also the discussion by Heine [18]):

1. The structure of the RRGM incorporates as much of the physics as possible by developing controlled approximations to the matrix ele-

ments ("observables") of interest. Single matrix elements, or groups of them (band $1 \rightarrow$ band 2 transitions), are computed.

2. We are not overwhelmed by thousands of eigenvalues and eigenvectors, which contain far too much information for the problem at hand.

3. There is (almost) unlimited freedom in choosing the starting vector, $|0\rangle$, to suit the problem at hand. Whereas **H** describes the model and specifies the law of motion through the propagator, $\exp(-iHt)$, the starting vector and the recursion vectors developed from it form the appropriate chain, which contains the physical essence of the problem.

4. The Hamiltonian matrix **H** enters only through the matrix–vector products $H|\text{old}\rangle = |\text{new}\rangle$; **H** is not modified by the computations. Depending on the problem and its size, **H** may be formed "on the fly," or in other cases, for example, just the nonzero elements may be in fast storage. In many cases, **H** is sparse, and only the nonzero elements are needed. In addition, storage space for just a few vectors are required for the Lanczos recursion.

5. For many EISPACK routines designed for relatively small eigen-problems $[N < O(10^3)]$, the CPU time scales as $O(N^3)$. With Lanczos, the variation is $O(N^2)$, which leads to orders-of-magnitude reductions in CPU time in favorable cases. It is for these reasons that recursion methods offer a promising way to solve very large problems.

B. Designer Subspaces: Concentrating the Dynamics

Moro and Freed, in connection with their use of the Lanczos algorithm to model ESR spectra [39–41], mentioned that the Lanczos method can be viewed in two ways: (a) as a *numerical algorithm* to tridiagonalize symmetric matrices and (b) "a theoretical method that can concisely extract the relevant information from a general description of physical systems." With regard to the latter view, they emphasized the important role played by the optimal reduced space (ORS). This is the subspace *with the least dimension*, for a specified $N \times N$ matrix **H**, that generates results (in our case, transition probabilities) correct to within a fixed accuracy. Another way of viewing the ORS is that it is spanned by the *smallest number* of eigenstates of the full Hamiltonian needed to construct the starting state $|0\rangle$ to within some accuracy.

As M increases, the Lanczos algorithm constructs subspaces V_M (spanned by the recursion vectors) that progressively approximate the ORS. As recursion proceeds, each chain link generates a more distant environment of the transition of interest. (A diagrammatic analysis of this feature will be presented in Section IV). As a result, the eigenvalues and largest residues that are most important for the $i \rightarrow f$ transition are generated quickly for small values of M; refinement of these values, along with the generation of small residues and their eigenvalues, occurs as M increases. This feature has been demonstrated elsewhere for multiphoton excitation [1, 2]; for applications to solid-state physics [12], and in the previously mentioned calculation of ESR lineshapes [39].

It is important to appreciate that the physics developed through $\mathbf{J}$ by the M recursion vectors is *not equivalent* to carefully selecting M of the *original N* basis functions in order to generate an $M \times M$ block of the full Hamiltonian matrix. In order to clarify this, consider an example with $N = 6$. Assume the Hamiltonian matrix

$$\mathbf{H} = \begin{bmatrix} 0 & 1 & 1 & 0 & 0 & 0 \\ 1 & 2 & 0 & 1 & 0 & 0 \\ 1 & 0 & 3 & 0 & 1 & 0 \\ 0 & 1 & 0 & 4 & 0 & 1 \\ 0 & 0 & 1 & 0 & 5 & 0 \\ 0 & 0 & 0 & 1 & 0 & 6 \end{bmatrix} \tag{32}$$

and the starting vector

$$\mathbf{u}_0 = [1, 0, 0, 0, 0, 0]^{tr}. \tag{33}$$

Then the first three Krylov vectors $(\mathbf{K}_n = \mathbf{H}^n \mathbf{u}_0)$ are

$$\mathbf{K}_0 = [1, 0, 0, 0, 0, 0]^{tr},$$
$$\mathbf{K}_1 = [0, 1, 1, 0, 0, 0]^{tr},$$
$$\mathbf{K}_2 = [2, 2, 3, 1, 1, 1]^{tr}. \tag{34}$$

The Krylov vectors span a three-dimensional space, but they are not orthonormal. Using the Gram–Schmidt orthonormalization algorithm, we obtain the Lanczos vectors:

$$\mathbf{u}_0 = [1, 0, 0, 0, 0, 0]^{tr,}$$
$$\mathbf{u}_1 = (1/\sqrt{2}) [0, 1, 1, 0, 0, 0]^{tr},$$
$$\mathbf{u}_2 = (\sqrt{2}/\sqrt{7}) [0, -1/2, 1/2, 1, 1, 1]^{tr}, \tag{35}$$

where $\mathbf{u}_i^{tr} \mathbf{u}_j = \delta_{ij}$. Note that the two-dimensional space spanned by $\mathbf{u}_0$, $\mathbf{u}_1$ contains three of the original basis states, whereas the three-dimensional space spanned by $\mathbf{u}_0$, $\mathbf{u}_1$, and $\mathbf{u}_2$ "feels" the full six-dimensional space. Through its action on $\mathbf{u}_0$, the Hamiltonian is designing its M-dimensional subspace within the full N-space; since each Lanczos vector is a linear combination of the N basis vectors, in no way is this equivalent to a subspace spanned by only M of the original basis vectors. The Hamiltonian specifies the model, and the question posed by $\mathbf{u}_0$ is answered by developing the ORS.

C. How to Cope with Degeneracy: Use the Same Formulas

When the Hamiltonian has degenerate eigenvalues, the RRGM automatically yields the total "strength" or residue associated with each of the degenerate energy *levels*. Nauts and Chapuisat elegantly proved this to be the case [64, 65]. We will take a simpler route and demonstrate this feature by way of one example. Consider the following Hamiltonian matrix for a three-state problem:

$$\mathbf{H} = \begin{bmatrix} 0 & 1 & 1 \\ 1 & 0 & 1 \\ 1 & 1 & 0 \end{bmatrix}. \tag{36}$$

The eigenvector matrix is

$$\mathbf{C} = \begin{bmatrix} 1/\sqrt{2} & 1/\sqrt{6} & 1/\sqrt{3} \\ -1/\sqrt{2} & 1/\sqrt{6} & 1/\sqrt{3} \\ 0 & -2/\sqrt{6} & 1/\sqrt{3} \end{bmatrix}, \tag{37}$$

where the columns correspond to the eigenvalues $\varepsilon_1 = -1$, $\varepsilon_2 = -1$, and $\varepsilon_3 = +2$. Using these eigenvalues and eigenvectors, we can find the survival amplitude in state number 1 (for example):

$$\begin{aligned}
\langle 1|\exp(-iHt)|1\rangle &= \langle 1|\alpha_1\rangle^2 \exp(-i\varepsilon_1 t) + \langle 1|\alpha_2\rangle^2 \exp(-i\varepsilon_2 t) \\
&\quad + \langle 1|\alpha_3\rangle^2 \exp(-i\varepsilon_3 t) \\
&= [(1/\sqrt{2})^2 + (1/\sqrt{6})^2] \exp(+it) \\
&\quad + (1/\sqrt{3})^2 \exp(-2it) \\
&= 2/3 \exp(+it) + 1/3 \exp(-2it).
\end{aligned} \tag{38}$$

Note that the total residue for the degenerate eigenvalue is $\frac{2}{3}$.

Turning now to the RRGM, we initiate the recursion with the vector

$$|0\rangle = \begin{bmatrix} 1 \\ 0 \\ 0 \end{bmatrix} \tag{39}$$

and then generate the next two recursion vectors,

$$|1\rangle = 1/\sqrt{2} \begin{bmatrix} 0 \\ 1 \\ 1 \end{bmatrix}, \qquad |2\rangle = 1/\sqrt{2} \begin{bmatrix} 0 \\ 1 \\ -1 \end{bmatrix}, \tag{40}$$

and the tridiagonal matrix and the reduced tridiagonal matrix,

$$\mathbf{J} = \begin{bmatrix} 0 & \sqrt{2} & 0 \\ \sqrt{2} & 1 & 0 \\ 0 & 0 & -1 \end{bmatrix}, \qquad \mathbf{J}^r = \begin{bmatrix} 1 & 0 \\ 0 & -1 \end{bmatrix}. \tag{41}$$

Since $b_2 = 0$, it was necessary to restart the recursion with vector $|2\rangle$ orthogonal to the two previous vectors. The eigenvalues of $\mathbf{J}$ and $\mathbf{J}^r$ are

$$\varepsilon_1 = -1, \qquad \varepsilon_2 = +2, \qquad \varepsilon_3 = -1,$$
$$\varepsilon_1^r = -1, \qquad \varepsilon_2^r = +1. \tag{42}$$

From these five eigenvalues, the Green function, Eq. 12, is

$$G_{11}(z) = \langle 1|(z-H)^{-1}|1\rangle = \frac{(z+1)(z-1)}{(z+1)(z-2)(z+1)} = \frac{z-1}{(z-2)(z+1)}. \tag{43}$$

The residues associated with the two distinct energy levels are

$$R(-1) = \frac{z-1}{z-2} = \frac{-1-1}{-1-2} = \frac{2}{3},$$
$$R(+2) = \frac{z-1}{z+1} = \frac{2-1}{2+1} = \frac{1}{3}. \tag{44}$$

It is extremely fortunate that the total residue for the degenerate level ($\frac{2}{3}$) is automatically predicted by the residue formula, Eq. 13. Also note that the

$z + 1$ term is canceled out of the numerator and denominator in Eq. 43. Since this factor arose when we restarted the recursion after b_2 was found to be zero, it was *not necessary* to restart the recursion procedure in order to find the third vector, $|2\rangle$, and the value for a_2. If we stop the recursion after just two steps (i.e., until we found a zero value for b_i), we obtain

$$\mathbf{J} = \begin{bmatrix} 0 & \sqrt{2} \\ \sqrt{2} & 1 \end{bmatrix}, \qquad \mathbf{J}^r = [1], \tag{45}$$

with the three eigenvalues $\varepsilon_1 = -1, \varepsilon_2 = +2$, and $\varepsilon_1^r = +1$. The residues are again $\frac{2}{3}$ and $\frac{1}{3}$ for the degenerate level and the nondegenerate level, respectively. In exact arithmetic, we just recur until $b_i = 0$. The size of $\mathbf{J}$ will be the number of energy *levels* (not states). For detailed proofs and other examples, the reader should consult Nauts and Chapuisat [64, 65].

D. Residue Algebra

When we want to calculate a number of transition amplitudes, we can put the whole RRGM within an outer DO loop. A much better way is to take advantage of the residue algebra [46]. Let us assume that we wish to obtain all possible transition amplitudes among a given set of vectors, $\{|i\rangle, i = 1, 2, \ldots, N\}$. Among this set of states, there are N diagonal "transitions" and $N(N-1)/2$ off-diagonal transitions; there are thus a total of $N(N+1)/2$ transitions to compute. Then the transition residues can be obtained formally if we know a set of intermediate amplitudes $\{\langle C|\alpha\rangle\}$:

$$\langle i|\alpha\rangle\langle\alpha|j\rangle = [\langle i|\alpha\rangle\langle\alpha|C\rangle][\langle j|\alpha\rangle\langle\alpha|C\rangle]/[\langle C|\alpha\rangle]^2, \tag{46}$$

where $|C\rangle$ is the normalized *composite vector*,

$$|C\rangle = 1/\sqrt{N}[|1\rangle + |2\rangle + \cdots + |N\rangle]. \tag{47}$$

To compute all $N(N+1)/2$ transition amplitudes, we will divide the calculation into two steps.

1. The survival (diagonal) amplitude of the composite vector is computed by initiating the Lanczos algorithm with the primer

$$(1/\sqrt{N}, 1/\sqrt{N}, \ldots, 1/\sqrt{N}).$$

From this calculation, we obtain the N residues $\langle C|\alpha\rangle^2, \alpha = 1, 2, \ldots, N$.

2. The transition amplitude between each "interesting" state $|i\rangle$ and the composite vector is computed. This in turn requires two recursion

sequences, which are primed with the normalized vectors

$$|u_0\rangle = (|i\rangle + |C\rangle)/[2(1 + S)]^{1/2},$$
$$|v_0\rangle = (|i\rangle - |C\rangle)/[2(1 - S)]^{1/2}, \tag{48}$$

where S is the overlap between the nonorthogonal vectors $|i\rangle$ and $|C\rangle$,

$$S = \langle i|C\rangle = 1/\sqrt{N}. \tag{49}$$

From these two recursion sequences, we generate $2N$ residues, $\{\langle u_0|\alpha\rangle^2$ and $\langle v_0|\alpha\rangle^2;\ \alpha = 1, 2, \ldots, N\}$. The $i \rightarrow C$ transition residues are then generated from the equation

$$\langle i|\alpha\rangle\langle\alpha|C\rangle = 1/2[(1 + S)\langle u_0|\alpha\rangle^2 - (1 - S)\langle v_0|\alpha\rangle^2]. \tag{50}$$

This can be verified by substituting Eq. 48 into the right side of Eq. 50.

At this stage, we have $N + 1$ residues at each energy: $\langle i|\alpha\rangle\langle\alpha|C\rangle$, $i = 1, 2, \ldots, N$, and $\langle C|\alpha\rangle^2$. From these we are able to compute $N(N + 1)/2$ transition residues through Eq. 50. Thus, the residue algebra reduces the computation time by a factor of $N/2$.

E.   The Matrix–Vector Product: Using Transition Vectors

In applications of the Lanczos algorithm, the most time-consuming step is multiplying the Hamiltonian matrix onto an "old" recursion vector. In many cases, $\mathbf{H}$ is a large sparse *structured* matrix. It is structured in the sense that the off-diagonal elements lie in a number of strips located at well-defined increments on either side of the main diagonal. These strips may be fashioned into *transition vectors* whose use greatly speeds up the matrix–vector multiply [66]. In effect, the matrix–vector multiplication is converted to the sum of a *small number* of vector–vector multiplies. This is an optimum strategy for vector/pipelined supercomputers.

Now, consider the effect of multiplying $\mathbf{H}$ onto an old recursion vector to obtain the new vector,

$$\mathbf{U}_{\text{new}} = \mathbf{H}\mathbf{U}_{\text{old}}. \tag{51}$$

Of course, the action of $\mathbf{H}^0$ on $\mathbf{U}_{\text{old}}$ is trivial because the basis vectors are chosen to be eigenvectors of $\mathbf{H}^0$:

$$\mathbf{H}^0|n\rangle = \varepsilon_n|n\rangle. \tag{52}$$

Thus, the difficulty arises in efficiently implementing the operation $\mathbf{V}\mathbf{U}_{\text{old}}$.

The advantages that arise from using transition vectors are easily illustrated by means of a one-dimensional potential. Consider the representation of coordinate q in a basis of five harmonic oscillator functions ($|0\rangle$, $|1\rangle,\ldots,|4\rangle$),

$$
\mathbf{V} = 1/\sqrt{2}
\begin{bmatrix}
0 & \sqrt{1} & 0 & 0 & 0 \\
\sqrt{1} & 0 & \sqrt{2} & 0 & 0 \\
0 & \sqrt{2} & 0 & \sqrt{3} & 0 \\
0 & 0 & \sqrt{3} & 0 & \sqrt{4} \\
0 & 0 & 0 & \sqrt{4} & 0
\end{bmatrix}.
\tag{53}
$$

If $\mathbf{U}_{\text{old}} = [C_1, C_2, C_3, C_4, C_5]^{\text{tr}}$, then by direct multiplication, $\mathbf{qU}_{\text{old}}$ is vector

$$
1/\sqrt{2}
\begin{bmatrix}
\sqrt{1}C_2 \\
\sqrt{1}C_1 + \sqrt{2}C_3 \\
\sqrt{2}C_2 + \sqrt{3}C_4 \\
\sqrt{3}C_3 + \sqrt{4}C_5 \\
\sqrt{4}C_4
\end{bmatrix}.
\tag{54}
$$

In order to emphasize the advantages of vectorization, define two transition vectors: Starting with the (1, 1) element in $\mathbf{V}$, move along the first row or column and then down the strips of nonzero elements paralleling the main diagonal. The two transition vectors are

$$
\mathbf{T}_{(-1)} = 1/\sqrt{2}
\begin{bmatrix}
0 \\
\sqrt{1} \\
\sqrt{2} \\
\sqrt{3} \\
\sqrt{4}
\end{bmatrix},
\qquad
\mathbf{T}_{+1} = 1/\sqrt{2}
\begin{bmatrix}
\sqrt{1} \\
\sqrt{2} \\
\sqrt{3} \\
\sqrt{4} \\
0
\end{bmatrix},
\tag{55}
$$

where $\mathbf{T}_{(-1)}$ comes from the strip "above" the main diagonal.

The subscript labeling each vector is the shift index, which will be used soon. Now, define two new vectors by taking the Schur product of each transition vector with $\mathbf{U}_{\text{old}}$ (if vector $\mathbf{D}$ is the Schur product of vectors $\mathbf{A}$ and $\mathbf{B}$, then $D_i = A_i B_i$):

$$\mathbf{D}_{(-1)} = 1/\sqrt{2} \begin{bmatrix} 0 \\ \sqrt{1}C_2 \\ \sqrt{2}C_3 \\ \sqrt{3}C_4 \\ \sqrt{4}C_5 \end{bmatrix}, \qquad \mathbf{D}_{(+1)} = 1/\sqrt{2} \begin{bmatrix} \sqrt{1}C_1 \\ \sqrt{2}C_2 \\ \sqrt{3}C_3 \\ \sqrt{4}C_4 \\ 0 \end{bmatrix}, \qquad (56)$$

Next, shift the rows by the value of the shift index to obtain the shifted vectors,

$$\mathbf{D}'_{(-1)} = 1/\sqrt{2} \begin{bmatrix} \sqrt{1}C_2 \\ \sqrt{2}C_3 \\ \sqrt{3}C_4 \\ \sqrt{4}C_5 \\ 0 \end{bmatrix}, \qquad \mathbf{D}'_{(+1)} = 1/\sqrt{2} \begin{bmatrix} 0 \\ \sqrt{1}C_1 \\ \sqrt{2}C_2 \\ \sqrt{3}C_3 \\ \sqrt{4}C_4 \end{bmatrix}, \qquad (57)$$

Finally, add the shifted vectors to obtain the final result,

$$\mathbf{VU}_{old} = \mathbf{D}'_{(-1)} + \mathbf{D}'_{(+1)} = 1/\sqrt{2} \begin{bmatrix} \sqrt{1}C_2 \\ \sqrt{1}C_1 + \sqrt{2}C_3 \\ \sqrt{2}C_2 + \sqrt{3}C_4 \\ \sqrt{3}C_3 + \sqrt{4}C_5 \\ \sqrt{4}C_4 \end{bmatrix}. \qquad (58)$$

All multiplication and shift operations explicitly vectorize, so this is a very efficient way to multiply the sparse potential matrix times a vector.

The four steps involved in defining and using transition vectors are as follows:

1. As a preprocessing step, define the transition vectors for all values of the shift index. The shift index is $\pm 1 \times$ (displacement of strip from diagonal), with a -1 for strips in the upper triangle.
2. Obtain the Schur product of each transition vector with $\mathbf{U}_{old}$.
3. Shift the rows in every vector resulting from step 2 by the shift index.
4. Add the vectors resulting from step 3.

For a more general discussion and other examples, the reader is referred to Friesner et al. (66).

F. Toward Parallel Processing

The RRGM code has several levels of parallelism, which one could take advantage of in a multiprocessor environment. Of course, if a number of transition amplitudes are needed, they can be spawned off to separate processors. However, there are several levels of granularity, to use Cray's terminology [67], even when a single transition amplitude is being computed. For example, at the multitasking (subroutine) level, one processor could develop one recursion sequence, say the U chain, while the other processor developed the second (V) chain. The Cray subroutines TSKSTART and TSKWAIT would be used to initiate and synchronize these two tasks. Leforestier (private communication) has recently multitasked an RRGM code over four processors of the 256-MW Cray-2 at L'Ecole Polytechnique in Paris. Also, at the microtasking (DO loop) level, the matrix–vector multiply could be split up on several processors. Exploiting these parallel structures presents new challenges to the developer of efficient software.

In an interesting approach to nuclear shell model calculations, Whitehead et al. [68] proposed and initiated the construction of a dedicated Lanczos computer. Particular emphasis was placed on designing hardware to optimize the time-consuming matrix–vector products that always occur in Lanczos calculations.

IV. UNDERSTANDING CHAINS

A. Generating Operators for Chain States

Since any recursion vector $|k\rangle$ in the chain can be constructed from the source vector $|0\rangle$, a generating operator F_k is defined by [69]

$$|k\rangle = F_k|0\rangle. \tag{59}$$

This operator is naturally a polynomial in the various powers of the Hamiltonian operator, $\{H^m; m = 0, \ldots, k\}$. Since the function $|0\rangle$ serves as a generator of the other functions, all the constructed terms $\{a_n\}$ and $\{b_n\}$ (in the tridiagonal representation of H in the recursion basis) can be expressed in terms of moments of the Hamiltonian over the source vector,

$$\mu_k \equiv \langle 0|H^k|0\rangle. \tag{60}$$

Using this definition, the first two generating operators are given trivially by

$$F_0 = 1, \qquad F_1 = (\mu_2 - \mu_1^2)^{-1/2}(H - \mu_1). \tag{61}$$

The Lanczos recursion relation, Eq. 16, becomes

$$F_n|0\rangle = b_n^{-1}[(H - a_{n-1})F_{n-1} - b_{n-1}F_{n-2}]|0\rangle. \tag{62}$$

If the operators $\{F_k\}$ are restricted to operate only on the source function $|0\rangle$, this explicit relation between functions can be interpreted as a recursion relation for the operators,

$$F_n = b_n^{-1}[(H - a_{n-1})F_{n-1} - b_{n-1}F_{n-2}]. \tag{63}$$

As examples, we calculate F_2 and F_3 ("bottom up" from F_0 and F_1):

$$\begin{aligned}
F_2 &= b_2^{-1}[(H - a_1)F_1 - b_1F_0] \\
&= (b_1b_2)^{-1}[H^2 - (a_0 + a_1)H + (a_0a_1 - b_1^2)], \\
F_3 &= b_3^{-1}[(H - a_2)F_2 - b_2F_1] \\
&= (b_1b_2b_3)^{-1}[H^3 - (a_0 + a_1 + a_2)H^2 \\
&\quad + (a_0a_1 + a_1a_2 + a_0a_2 - b_1^2 - b_2^2)H \\
&\quad - (a_0a_1a_2 - a_0b_2^2 - a_2b_1^2)].
\end{aligned} \tag{64}$$

The chain parameters $\{a_n\}$ and $\{b_n\}$ will be derived later in terms of matrix elements of the operators $\{F_n\}$ evaluated for the source vector. Hence, both the chain elements $\{a_n, b_n\}$ and the coefficients for the polynomial representation of F_k are expressible in terms of the moments $\{\mu_k\}$. This means that the whole physics of the transition event is described in terms of the moments of the full Hamiltonian with respect to the initial and final zero-order states.

B. Chain Parameters from Hamiltonian Moments

Some useful relations that apply to the operator set $\{F_k\}$ are easily obtained [69]:

1. Since F_k is expressed as a linear combination of powers of H, F_k and H commute,

$$[F_k, H] = 0. \tag{65}$$

2. Since the recursion vectors are orthonormal,

$$\langle 0|F_kF_m|0\rangle = \delta_{km}. \tag{66}$$

3. Since a_k denotes the self-energy of chain state $|k\rangle$,

$$a_k = \langle 0|F_k^2 H|0\rangle. \tag{67}$$

4. Since b_k denotes the linking term between $|k-1\rangle$ and $|k\rangle$,

$$b_k = \langle 0|F_{k-1}F_k H|0\rangle. \tag{68}$$

Actually, Eq. 67 is the generating relation for the self-energy a_n in terms of the set of moments $\{\mu_k; k = 1, \ldots, 2n+1\}$ since F_n is a polynomial of order n in H. For example, we calculate a_1:

$$a_1 = \langle 0|F_1^2 H|0\rangle = \langle 0|(H-\mu_1)^2 H|0\rangle/b_1^2$$
$$= (\mu_3 - 2\mu_1\mu_2 + \mu_1^3)/(\mu_2 - \mu_1^2),$$

where we have used (see Eq. 68)

$$b_1 = \langle 0|F_0 F_1 H|0\rangle = (\mu_2 - \mu_1^2)^{1/2}. \tag{69}$$

Noting that b_n is the normalizing factor for recursion vector $|n\rangle$, we obtain

$$b_n^2 = (\langle n-1|H - a_{n-1}\langle n-1| - b_{n-1}\langle n-2|)$$
$$\times (H|n-1\rangle - a_{n-1}|n-1\rangle - b_{n-1}|n-2\rangle)$$
$$= \langle 0|F_{n-1}^2 H^2|0\rangle - a_{n-1}^2 - b_{n-1}^2, \tag{70}$$

where Eqs. 67 and 68 have been used. Since it is assumed that the previous terms a_{n-1} and b_{n-1} are already known, b_n is now expressed in terms of the set of moments $\{\mu_k; k = 1, \ldots, 2n\}$. Consequently, in order to obtain $\mathbf{J}$, the $M \times M$ tridiagonal matrix, one has to calculate $\{\mu_k; k = 1, \ldots, 2M-1\}$. This is the minimal information needed in terms of $|0\rangle$ in order to generate M links in the chain.

C. Interpretation of Chain Elements

The expression for b_n^2, Eq. 70, can be interpreted as follows [69]. The first term, $\langle 0|F_{n-1}^2 H^2|0\rangle$, is the expectation value of H^2 in the chain state $|n-1\rangle$. Since a_{n-1} is the self-energy of this state, b_n^2 reduces to

$$b_n^2 = [\langle n-1|H^2|n-1\rangle - \langle n-1|H|n-1\rangle^2] - b_{n-1}^2 = \sigma_{n-1}^2 - b_{n-1}^2, \tag{71}$$

where σ_{n-1}^2 is the *energy variance* in the state $|n-1\rangle$. Consequently,

$$\sigma_{n-1}^2 = b_n^2 + b_{n-1}^2. \tag{72}$$

The energy variance for the chain state $|n-1\rangle$ is the sum of the squares of the coupling terms of this state to its two neighboring states, $|n-2\rangle$ and $|n\rangle$. This statement asserts that since the state $|n-1\rangle$ is coupled to its nearest neighbors, it is not an eigenstate of the Hamiltonian, and consequently, an energy fluctuation is unavoidable. The expression for b_n^2, Eq. 71, can be slightly changed by noting that

$$b_n^2 = \sigma_{n-1}^2 - (\sigma_{n-2}^2 - b_{n-2}^2) = \cdots = \sigma_{n-1}^2 - \sigma_{n-2}^2 + \sigma_{n-3}^2 - \cdots. \qquad (73)$$

Thus, b_n^2 is the alternating sum of all previous energy fluctuations. This is a consequence of the cumulative construction inherent in the Lanczos procedure. Each coupling term depends not only on the energy variance of the adjacent states, but actually also on the whole set of preceding energy variances.

In addition, we note that b_n^2 is a measure of the transition rate from the state $|n-1\rangle$ to all of the following states $\{|n+k\rangle; k = 0, 1, \ldots\}$ via the adjacent state $|n\rangle$. This can be proved by partitioning the Hilbert space into a primary part consisting of the states $|0\rangle, |1\rangle, \ldots, |n-1\rangle$ and another part containing the states $|n\rangle, |n+1\rangle, \ldots, |N\rangle$. This bisection is achieved by defining two projection operators:

$$P = \sum_{j=0}^{n-1} |j\rangle\langle j|, \qquad (74)$$

$$R = \sum_{j=n}^{N} |j\rangle\langle j|. \qquad (75)$$

The Schrödinger equations for the two subspaces are given by

$$\dot{C}_P = -iH_{PP}C_P - iH_{PR}C_R, \qquad \dot{C}_R = -iH_{RP}C_P - iH_{RR}C_R, \qquad (76)$$

where, for example, $H_{PR} \equiv PHR$, and C_P and C_R are column vectors of the time-dependent expansion coefficients (i.e., C_P contains $\langle 0|\Psi(t)\rangle$, $\langle 1|\Psi(t)\rangle, \ldots, \langle n-1|\Psi(t)\rangle$). As the simplest example, take the P subspace as only the source function $|0\rangle$. Then the Schrödinger equation for $C_0(t)$ reduces, after some manipulations, to

$$\dot{C}_0(t) = -ia_0 C_0(t) - b_1^2 \int_0^t d\tau \exp[-ia_1(t-\tau)]C_0(\tau). \qquad (77)$$

It is clear that the R-space chain states influence C_0 through the hopping term b_1^2.

D. Moment Diagrams

It is useful to construct diagrams to represent the moments μ_k because the chain parameters $\{a_n, b_n\}$ may then be "drawn" directly in terms of these diagrams [70]. First, consider an orthonormal basis $\{|j\rangle\}$ in which the Hamiltonian has diagonal self-energies $\{E_j\}$ and off-diagonal couplings $\{H_{jk}\}$. The kth moment of H over the source vector is given by

$$\mu_k = \langle 0|H^k|0\rangle = \sum_{\{j_p\}} \langle 0|H|j_1\rangle \cdots \langle j_{k-1}|H|0\rangle = \sum_{\text{paths}} \text{(value for path)}. \quad (78)$$

The identity $\sum |j_i\rangle\langle j_i| = 1$ has been inserted $k - 1$ times between the powers of H, *value for path* means the term

$$\langle 0|H|j_1\rangle\langle j_1|H|j_2\rangle \cdots \langle j_{k-1}|H|0\rangle, \quad (79)$$

and the sum over *all* paths refers to summation over the indices $j_1, j_2, \ldots, j_{k-1}$.

To make the diagrams simpler, let $|0\rangle$ represent the ground state and assume that $j = 0, 1, \ldots$. Then, with the source state at the origin, excited states $j = 1, 2, \ldots$ along the ordinate, and powers of H along the abscissa, we draw slant lines to represent off-diagonal couplings $(H_{01}, H_{02}, \ldots)$ and horizontal lines to represent self-energies $(E_0, E_1, \ldots)$. The states that can be reached through *one* application of the Hamiltonian are shown as a column of dots. Another way of viewing this is that we pick out the states produced by one application of H on the starting vector:

$$\begin{bmatrix} E_0 & H_{01} & H_{02} & \cdots \\ H_{10} & E_1 & H_{12} & \cdots \\ H_{20} & H_{21} & E_2 & \cdots \\ \vdots & \vdots & \vdots & \vdots \end{bmatrix} \begin{bmatrix} 1 \\ 0 \\ 0 \\ \vdots \end{bmatrix} = \begin{bmatrix} E_0 \\ H_{10} \\ H_{20} \\ \vdots \end{bmatrix}. \quad (80)$$

The next application of **H** brings in those states directly coupled to states in this *first tier*; these states in the *second tier* have second-order couplings back to the source state, for example, $H_{0j}H_{jk}$. These excursions continue with the application of H to each new tier of states, but the last interaction (the kth power of H) must return all paths to the source state. A moment diagram, μ_5, and one path involving the sequence of states $0 \to 1 \to 1 \to 2 \to 1 \to 0$, are shown in Figure 2.

For the simple case where each state is coupled to its two nearest neighbors,

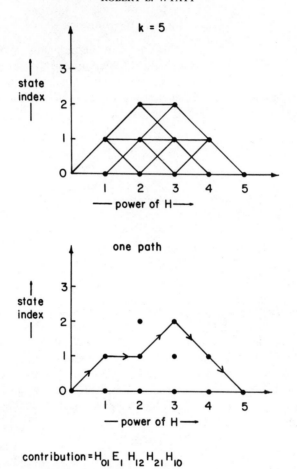

Figure 2. Moment diagram for μ_5 for a system in which each state is coupled to its nearest neighbors. The full-moment diagram is shown, along with one contributing path. Each path must start and end at the source state.

the first several moments are easily written:

$$\mu_0 = 1,$$
$$\mu_1 = E_0, \tag{81}$$
$$\mu_2 = E_0^2 + H_{01}^2 + H_{02}^2,$$
$$\mu_3 = E_0^3 + H_{01}^2 E_1 + H_{02}^2 E_2 + 2E_0 H_{01}^2 + 2E_0 H_{02}^2 + 2H_{01}H_{12}H_{20}.$$

The moment diagrams for $\mu_1, \mu_2,$ and μ_3 shown in Figure 3 involve 1, 3, and 9

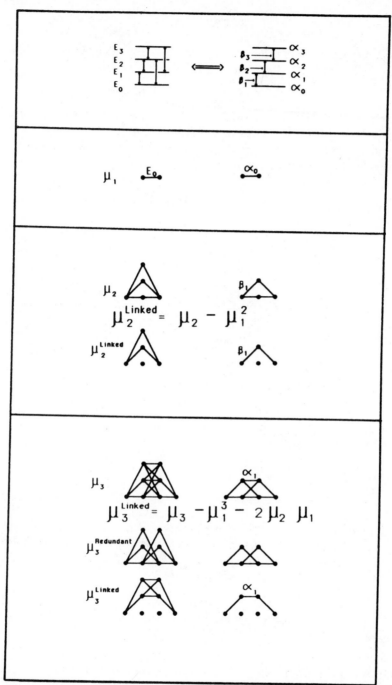

Figure 3. Full and linked moment diagrams μ_k^H and μ_k^J; $k = 1,\ldots,3$ for an anharmonic oscillator. The linked diagrams are the difference between the full diagram and redundant contributions. These redundant paths include the source state as an intermediate node and are subtracted from the full moments. For each moment, the left diagram is in the initial representation and the (equivalent) right diagram is the Jacobi representation.

paths, respectively. For example, one path in μ_3 involves intermediate states 1 and 2 and contributes the term $H_{01}H_{12}H_{20}$.

E. Chain Parameters from Moments

In Section IV.B, we developed expressions for the chain parameters in terms of moments of H over the source state. An important feature arises when these moment expressions are inserted into expressions for the chain parameters [70]. Consider b_1 and a_1 (see Eq. 69):

$$b_1^2 = \mu_2 - \mu_1^2 = \left(\sum_j \langle 0|H|j\rangle\langle j|H|0\rangle\right) - (\langle 0|H|0\rangle)^2$$

$$= \sum_{j>0} H_{0j}H_{j0},$$

$$a_1 = (\mu_3 - 2\mu_1\mu_2 + \mu_1^3)/b_1^2$$

$$= \left\{\sum_{j_1,j_2} \langle 0|H|j_1\rangle\langle j_1|H|j_2\rangle\langle j_2|H|0\rangle \right.$$

$$\left. - 2\langle 0|H|0\rangle\sum_j \langle 0|H|j\rangle\langle j|H|0\rangle + (\langle 0|H|0\rangle)^3\right\}\bigg/ b_1^2$$

$$= \left(\sum_{j,m>0} H_{0j}H_{jm}H_{m0}\right)\bigg/ b_1^2. \tag{82}$$

For the special case where each zero-order state is coupled to its two nearest neighbors, these expressions reduce to

$$b_1^2 = H_{01}^2 + H_{02}^2,$$

$$a_1 = (H_{01}^2 E_1 + H_{02}^2 E_2 + 2H_{01}H_{12}H_{20})/(H_{01}^2 + H_{02}^2). \tag{83}$$

In terms of the diagrams shown in Figure 3, we *leave out all intermediate hops back to the source state*. These condensed moments are called *linked moments*; their diagrams are *linked diagrams*. Several linked moment diagrams, μ_2^{linked} and μ_3^{linked}, are also shown in Figure 3.

We see from Eq. 83 that

$$b_1^2 = \mu_2^{\text{linked}}, \qquad a_1 = \mu_3^{\text{linked}}/\mu_2^{\text{linked}}. \tag{84}$$

All of the higher chain parameters can also be expressed in terms of linked moments [71]. The connection between linked moments and the Lanczos chain parameters was also analyzed by Whitehead and Watt [72]. They also drew a different type of moment diagram to show the contributing paths.

Linked moments emphasize the role of paths that contribute new (relevant) information. For example, by comparing μ_3 and μ_3^{linked}, Eqs. 81 and 83,

$$\mu_3 = \mu_3^{\text{old}} + \mu_3^{\text{linked}}, \tag{85}$$

where

$$\mu_3^{\text{old}} = E_0^3 + 2E_0 H_{01}^2 + 2E_0 H_{02}^2 = 2\mu_1\mu_2 - \mu_1^3.$$

Previously computed "old information" from the lower moments μ_1 and μ_2 is subtracted from μ_3 to obtain new information in the linked moment. In terms of diagrams, five paths from μ_3^{old} are subtracted from the nine paths in μ_3 to leave only four paths in μ_3^{linked}. The spirit is similar to earlier studies of modified and generalized-moment methods [73–76].

The lowest paths of the full moment are removed to obtain the linked moment, which enormously decreases the number of products and additions needed to calculate the relevant contribution (new information) to the tridiagonal chain parameters. As a consequence: (i) a portion of the time consumed by these redundant calculations is saved; (ii) the orthogonality of the high Lanczos states (which include contributions only from the excited zero-order states) is automatically conserved with respect to the lowest Lanczos state $|0\rangle$; and (iii) a significant source of roundoff error is eliminated. *It is important to realize that the "lowest" paths carry a major numerical contribution to the full moment.* For example, for the simplest case of nearest-neighbor couplings in a single mode, where we take all the terms (interactions and diagonal energies) as unity, the 14th full moment is $\mu_{14}^{\text{full}} = 113{,}634$, whereas the linked moment is $\mu_{14}^{\text{linked}} = 15{,}511$. This substantial reduction in the contribution of "old" information to the chain parameters (b_7 in the case of μ_{14}) becomes even more significant when computing the higher chain parameters.

In order to express higher matrix elements of the disordered linear chain in terms of the zero-order energies $\{E_k\}$ and coupling strengths $\{H_{ij}\}$, two properties of the full Hamiltonian (**H**) and the tridiagonal (**J**) representations are exploited. First, these two representations are connected by a similarity transformation and, second, the initial zero-order state in the "H representation" and the source state in the "J representation" are equal. Therefore, moments of the Hamiltonian expressed by the two equivalent representations are equal when calculated with respect to the initial zero-order state and the source state, respectively. Likewise, the diagrams representing these moments are equal. These two types of moments are

$$\mu_k^H = \sum_{\{j_p\}} \langle 0|H|j_1\rangle \langle j_1|H|j_2\rangle \cdots \langle j_{k-1}|H|0\rangle, \tag{86}$$

$$\mu_k^J = \sum_{\{i_p\}} \langle 0|J|i_1\rangle \langle i_1|J|i_2\rangle \cdots \langle i_{k-1}|J|0\rangle,$$

where μ_k^H and μ_k^J are the kth order moments in the H and the J representations, respectively. Each path in the H representation is specified by a set of integers $\{j_k\}$ that determine the intermediate-state indices, and each matrix element in this representation is either an E_k or an H_{jk}. In the J representation, the state $|i\rangle$ is the ith vector generated in the recursion procedure, and each matrix element in this representation is either an a_k or a b_k.

Now, we return to the example where each state in the H representation is coupled to its two nearest neighbors. Of course, in the J representation, each state (recursion vector) is coupled only to its nearest neighbors. The H representation moments μ_k^H ($k = 0, 1, 2, 3$) were given earlier in Eq. 81. In the J representation, we have

$$\mu_1^J = a_0,$$
$$\mu_2^J = a_0^2 + b_1^2,$$
$$\mu_3^J = a_0^3 + 2a_0 b_1^2 + a_1 b_1^2. \tag{87}$$

Equating μ_1^H with μ_1^J gives $a_0 = E_0$. Equating μ_2^H with μ_2^J gives b_1^2 in Eq. 83, and equating μ_3^H with μ_3^J gives a_1 in Eq. 83. Actually, only the linked moments $(\mu_k^H)^{\text{linked}}$ and $(\mu_k^J)^{\text{linked}}$ need to be equated, because the old information cancels from each diagram. For example, for $k = 3$,

$$(\mu_3^H)^{\text{linked}} = H_{01}^2 E_1 + H_{02}^2 E_2 + 2H_{01} H_{12} H_{20},$$
$$(\mu_3^J)^{\text{linked}} = a_1 b_1^2. \tag{88}$$

Equating these two expressions quickly gives the expression for a_1 in Eq. 83.

To summarize this section, in order to obtain the value of the coupling term b_n, the two equivalent linked moments of order $2n$ are equated (as shown in Figure 3 for $2n = 2$). Likewise, to obtain the self-energy a_n, the two equivalent linked moments of order $2n + 1$ are equated (as shown in Figure 3 for $2n + 1 = 3$). Thus, in order to obtain b_n, it is necessary to know $\{b_k, a_k; k = 1, \ldots, n - 1\}$, and to obtain a_n, it is necessary to know $\{b_k, a_k; k = 1, \ldots, n - 1\}$ and b_n.

F.   Chain Parameters from Irreducible Moments

The preceding procedure introduces the notion of linked moments, in which paths refrain from "visiting" the source state in intermediate steps H_{ij} before completing the total interaction.

A question then rises: Is subtraction of lower paths involving intermediate hops back to $|0\rangle$ in moment diagrams the ultimate elimination of redundant contributions to the moments? As can be seen from the level scheme in Figure 4, the chain particle $|1\rangle^J$ is definitely equivalent to the first excited state of the zero-order Hamiltonian $|1\rangle^H$. Consequently, the agruments in

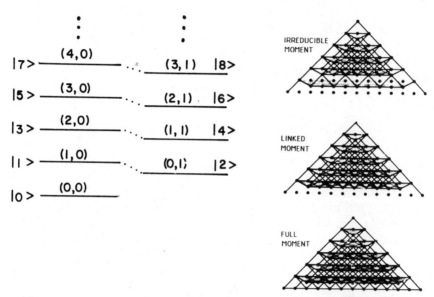

Figure 4. The hierarchy of full, linked, and irreducible moments $\mu_{12}, (\mu_{12})^{\text{linked}}, (\mu_{12})^{\text{ir}}$ for the model system of a radiatively active mode coupled to one background mode. Each dot in the figure represents a state and each bar is Hamiltonian matrix element. The horizontal axis is the number of interactions (powers of H) applied to the source state. The vertical axis (y) is labeled by the state index in the active mode ($y = 0, 1, \ldots, 6$). The axis emerging from the xy plane counts the number of quanta exchanged between the active and the inactive modes ($z = 0, 1$ only). The numerical value of the moment is obtained by summing over all paths linking the left dot with the right dot.

Section IV.E can be repeated to eliminate intermediate visiting of $|1\rangle$ (once it is reached from $|0\rangle$) before finally returning to $|0\rangle$ at the end of the process. By subtracting the corresponding μ_k^J from μ_k^H we get a new difference moment, which is also linked with respect to $|1\rangle$ (i.e., intermediate hops back to $|1\rangle$ are eliminated) and contains the relevant contributions of the next zero-order states to the higher chain states ($|2\rangle, |3\rangle, \ldots$). All the redundant paths are eliminated and need not be numerically calculated. However, there are no further redundant contributions that can be extracted (reduced) from them. In this sense, these *irreducible moments* develop the relevant (essential) part of Lanczos procedure [77, 78].

An example may make this clearer. Using the notation in Figure 4, we easily obtain, by equating the *full moments* for $k = 1, 2, 3$,

$$a_1 = \begin{cases} H_{01} & \text{from } k = 2, \\ E_1 & \text{from } k = 3. \end{cases} \tag{89}$$

Then in order to obtain b_2, we will equate the $k = 4$ moments in the two representations (we include contributions from eight paths in μ_4^H and seven paths in μ_4^J):

$$\mu_4^{full} = E_0^4 + H_{01}^4 + 3H_{01}^2 E_0^2 + H_{01}^2 E_1^2 + H_{01}^2(H_{12}^2 + H_{13}^2)^2$$
$$= a_0^4 + b_1^4 + 3b_1^2 a_0^2 + b_1^2 a_1^2 + b_1^2 b_2^2. \tag{90}$$

Next, substituting E_0 and H_{01} from Eq. 89 into the first part of Eq. 90 yields the linked moments:

$$\mu_4^{linked} = H_{01}^2 E_1^2 + H_{01}^2(H_{12}^2 + H_{13}^2) = b_1^2 a_1^2 + b_1^2 b_2^2. \tag{91}$$

Finally, substituting E_1 from Eq. 89 into the left side of Eq. 91 yields the irreducible moment

$$\mu_4^{ir} = H_{01}^2(H_{12}^2 + H_{13}^2) = b_1^2 b_2^2, \tag{92}$$

from which we obtain

$$b_2 = (H_{12}^2 + H_{13}^2)^{1/2}. \tag{93}$$

Note that eight paths contribute to μ_3^{full} and three contribute to μ_4^{linked} whereas only two contribute to μ_4^{ir} when the moments are computed in the original representation. All of the new information needed to compute b_2 is contained in these latter two paths.

Returning to the "counting problem" involving μ_{14}, we mentioned earlier that $\mu_{14}^{full} = 113,634$. The irreducible moment is only $\mu_{14}^{ir} = 2188$, so that 98.1% of the paths have been discarded because they contain old information!

V. APPLICATIONS OF RRGM

A. Correlation Functions and Power Spectra

In this section, we will demonstrate that the recursive computation of Green function matrix elements in systems with large basis sets $(N > 10^3)$ leads to accurate results for quantum time correlation functions. The controlled nature of the approximations permits a systematic study of convergence and the possibility of producing definitive results [79].

Consider a system described by a time-independent Hamiltonian H. The time autocorrelation function for operator A is given by

$$C(t) = \text{Tr}\{\rho(0)A(0)A(t)\} = \text{Tr}\{\rho(0)A(0)e^{+iHt}A(0)e^{-iHt}\}, \tag{94}$$

where $A(0) \equiv A$ is the operator at time $t = 0$, and $\rho(0)$ is the initial density matrix. For the applications described here, we assume a canonical system, so that $(\beta = 1/kT)$:

$$\rho(0) = e^{-\beta H}/\text{Tr}\{e^{-\beta H}\} = e^{-\beta H}/Q. \tag{95}$$

If we insert a complete set of states, $C(t)$ involves multiple summation [there is a similar expression for $\langle A(t) \rangle$]:

$$C(t) = \frac{1}{Q} \sum_i \sum_j \sum_k \sum_l \sum_m \langle i|e^{-\beta H}|j \rangle$$

$$\times \langle j|e^{+iHt}|k \rangle \langle k|A|l \rangle \langle l|e^{-iHt}|m \rangle \langle m|A|i \rangle. \tag{96}$$

Because of the invariance of the trace to the choice of representation, in all of these equations we are free to use *any convenient* basis set.

For most systems of interest, the sums in Eq. 96 can be truncated for one or more of the following reasons: (a) thermal cutoff, $\langle i|\exp(-\beta H)|j \rangle \ll 1$; (b) time cutoff, $\langle i|\exp(\pm iHt|j \rangle \ll 1$; and (c) range cutoff, $\langle i|A|j \rangle \ll 1$. Thus, $C(t)$ can be computed from a finite set of matrix elements of the operators $e^{-\beta H}$, e^{iHt}, and A.

In the first stage of computation, the RRGM is used to compute accurate values of the thermal and time propagators, $\langle i|\exp(-\beta H)|j \rangle$ and $\langle i|\exp(-iHt)|j \rangle$. This is the most time-consuming part of the calculation. These elements are computed only once and are read into the fast memory as they are needed. For a large class of relevant operators, including various products of coordinates (e.g., $q_i^r q_j^s$), the matrix elements of A are trivial. The final summations are then performed in such a way as to exploit any selection rules inherent in the operator A. Efficient indexing procedures ensure that only nonzero terms are evaluated.

There are two important considerations in designing an RRGM code to calculate statistical averages. First, one would like to minimize the time spent in calculating each Green function, given that a specified level of accuracy in the final averages is desired. Secondly, the set of Green functions that must be computed should be restricted to be as small as possible.

The first decision is a choice of basis set. We adopt a direct-product harmonic oscillator basis because matrix elements of the potential energy are sparse and easily computed if $V(x)$ is a polynomial (it can be fit to one if a numerical representation is all that is available).

Once a basis set has been chosen, a portion of it, the *active space*, is selected for the computation of each matrix element. Because the active space can be easily shifted for each calculation, an optimized space can be constructed for each pair (i, j). This reduces the size of the basis used in each computation and thus reduces time requirements considerably. Accuracy can be checked by systematic increase in the active space dimension, the requirements of which will be dependent on the magnitude of the off-diagonal coupling elements and energy spacings of the basis functions.

Selection of the Green function matrix elements to be calculated involves analysis of the specific Hamiltonian under study. Cutoffs in the total zero-order diagonal energy E_i and in the quantal separation Δn_{ij} of two states are imposed and then systematically increased until convergence is obtained.

Numerical results will be presented for the following system: a field perturbed anharmonic oscillator coupled to a multimode harmonic reservoir. In terms of raising and lowering operators for the anharmonic oscillator $(a^\dagger, a)$ and the reservoir modes $(b_i^\dagger, b_i)$, the Hamiltonian is

$$H = \omega_a[(a^\dagger a) - \chi(a^\dagger a)^2] + V_r[a^\dagger + a] + \sum_{i=1}^{n} \omega_i b_i^\dagger b_i + \sum_{i=1}^{n} V_i[a^\dagger b_i + ab^\dagger_i], \quad (97)$$

where ω_a and ω_i are harmonic energy spacings in the anharmonic and the ith reservoir mode, χ is the anharmonicity parameter, and V_r and V_i are parameters that control the strength of the field and the intramolecular coupling, respectively.

The basis set chosen to represent the Hamiltonian contains products of harmonic oscillator functions for the $n + 1$ modes. If we allow a maximum of p states in each mode, the total basis size is $N = p^{(n+1)}$. In the applications reported here, there are four reservoir modes $(n = 4)$ and $p = 5$. This leads to a basis with the dimension $N = 3125$.

The time correlation function $C(t)$ for the pump mode coordinate x was computed:

$$C(t) = (1/Q)\text{Tr}\{e^{-\beta H}xe^{+iHt}xe^{-iHt}\}. \quad (98)$$

In addition, the power spectrum $I(\omega)$ was obtained from the Fourier transform of $C(t)$:

$$I(\omega) = A\left|\int_0^\infty e^{i\omega t}\Gamma(t)\,dt\right|^2,$$

$$\Gamma(t) = [C(t) - C(\infty)]\exp(-at^2), \quad (99)$$

where A is chosen to scale $I(\omega)$. In addition, $C(\infty)$ is the long-time average of the real part of $C(t)$; it represents a "dc shift" in the motion of the wavepacket due to the anharmonicity in the Hamiltonian. The dc shift is subtracted from $C(t)$ so that very low frequency components will not dominate plots of the

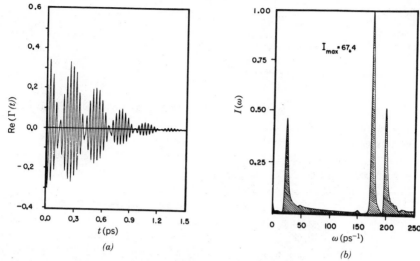

Figure 5. Real part of the time correlation function (a) and the resulting scaled power spectrum (b) for a five-mode system with 3125 basis states for $\beta = 0.002$ cm.

power spectrum. Our interest is in the dynamics relative to C, not in the fact that the correlation function is shifted at long times. Finally, the Gaussian factor $\exp(-at^2)$ defines a "time window" that prevents relatively long times from entering into the calculation of $I(\omega)$.

For the parameters in this Hamiltonian, the states fall into bands separated by about $1000 \, \text{cm}^{-1}$. Thus, in $I(\omega)$ there should be intense lines due to neighboring band transitions around $\omega = (1000 \, \text{cm}^{-1})/(5.3 \, \text{cm}^{-1} \, \text{ps}) = 200 \, \text{ps}^{-1}$. In addition, intraband transitions contribute to lines at much lower frequencies.

Figure 5 shows $\text{Re}[\Gamma(t)]$ and $I(\omega)$ at a relatively high value of β (0.004 cm), which corresponds to $T = 362$ K. The oscillation of $\text{Im}[C(t)]$ about the value zero, large values of $I(\omega)$ near $\omega = 200 \, \text{ps}^{-1}$ due to interband transitions, and less intense values for frequencies below $50 \, \text{ps}^{-1}$ due to intraband transitions are clearly evident.

B. Energy Growth during Multiphoton Excitation

In this section, computational results on the laser-induced excitation of a multimode model will be described [31]. The monochromatic laser (intensity I, frequency ω_0) interacts with an anharmonic pump mode (frequency ω_1, anharmonicity χ) that is coupled to harmonic mode 2, which in turn is coupled to harmonic mode 3, and so on. The total number of molecular modes in this linear chain is m. The Hamiltonian operator for time $t \geq 0$ is [atomic units (au) are assumed]

$$H = \sum_{j=1}^{m} \omega_j \sigma_j^\dagger \sigma_j - \chi(\sigma_1^\dagger \sigma_1)^2 + \beta \sum_{j=1}^{m} (\sigma_j^\dagger \sigma_{j+1} + \sigma_j \sigma_{j+1}^\dagger)$$
$$+ \omega_0 c^\dagger c + \varepsilon(\sigma_1^\dagger + \sigma_1)(c + c^\dagger). \tag{100}$$

The parameter β controls the intermode coupling, and ε is determined by the laser intensity. The laser is "turned off" when $t < 0$. Numerical values for all parameters are $\omega_0 = 1000\,\text{cm}^{-1}$, $\varepsilon = 1.646\,\text{cm}^{-1}$, $\omega_j = 1000\,\text{cm}^{-1}$, $\beta = 0.3\,\text{cm}^{-1}$, and $\chi = 1.0\,\text{cm}^{-1}$. For the studies reported here, the harmonic frequencies for all modes are identical, and the intermode coupling parameter (β) is the same for all pairs of modes.

In addition, c and $c^\dagger$ are lowering and raising operators for the laser mode, whereas σ_j and $\sigma_j^\dagger$ are lowering and raising operators for molecular mode j. For example, if $|n\rangle$ denotes a vibrational state with n quanta ($n = 0, 1, 2, \ldots$), the action of the lowering and raising operators is given by

$$\sigma_j |n\rangle = n^{1/2} |n-1\rangle, \qquad \sigma_j^\dagger |n\rangle = (n+1)^{1/2} |n+1\rangle.$$

Clearly, these operators lower or raise the number of quanta (by -1 or $+1$, respectively) in this mode. The operators c and $c^\dagger$ have a similar effect on a photon number state.

The basis set for the molecule consists of products of harmonic oscillator functions for the various modes. For example, for the four-mode molecule ($m = 4$), the basis consists of products

$$|i\rangle = |n_1\rangle |n_2\rangle |n_3\rangle |n_4\rangle \qquad (n_j = 0, 1, \ldots, n_{\max}).$$

The molecular ground state arises when each oscillator is in its ground state, $|g\rangle = |0\rangle |0\rangle |0\rangle |0\rangle$. For the combined laser molecule system, the basis consists of products of molecular basis functions times photon number states, $\{|i\rangle|N\rangle\}$. For this basis, the photon states are said to dress the molecular states. Although the molecular basis spans a direct-product space, the overall field-dressed molecular basis is not taken as a direct product. The reason is that only dressed states nearly in resonance will be strongly involved in the dynamics. Starting from the dressed ground state $|g\rangle|N\rangle$, whose energy we arbitrarily define as zero, only excited states $|i\rangle|N'\rangle$ whose energies are within the band of energies $[-D, +D]$ are needed to accurately describe the dynamics.

When the laser interacts with the molecule, energy flows through the pump mode into successive modes along the chain. This study emphasizes the rate of energy deposition in the molecule for various values of the intermode coupling parameter (β). In the uncoupled limit when $\beta = 0$, the laser and pump modes act as a coupled two-oscillator system undergoing quasi-periodic energy exchange. At the opposite extreme, when β is large compared to ε, successive background modes down the chain should be quickly excited. For intermediate values of β, the time dependence of the total absorption is of particular interest.

We will now consider the method for calculating the energy absorption in the molecule. If the molecule starts in state $|g\rangle$ at $t = 0$, the molecular energy at time t is

$$\langle E(t) \rangle = \sum_f E_f |A_{fg}(t)|^2, \tag{101}$$

where $A_{fg}(t)$ is the $g \to f$ transition amplitude. The number of photons absorbed from the laser is the ratio of this quantity to the photon energy (ω_0),

$$N(t) = \sum_f \frac{E_f}{\omega_0} |A_{fg}(t)|^2. \tag{102}$$

Another quantity of interest is the average number of quanta absorbed by mode j, $N_j(t)$. In terms of the total wavefunction at time t, this is given by

$$N_j(t) = \langle \Psi(t) | \sigma_j^\dagger \sigma_j | \Psi(t) \rangle, \tag{103}$$

where $(\sigma_j^\dagger \sigma_j)$ is the number operator for oscillator j. This average may also be computed from the amplitudes $A_{fg}(t)$. We will now proceed to examine plots of N_j and N versus t.

In the calculations reported here, the $m = 8$ molecular chain, which has seven background modes, was studied. To generate the molecular basis, we used $n_{max} = 1$ (each mode has a ground state and one excited state), and $256 \times 3 = 768$ field-dressed states were used. Time evolution calculations leading to one of the plots in Figures 6 and 7 (with 500 time steps) required 500 s on the Cray X-MP/24 at the University of Texas System Center for High Performance Computing.

Figure 6 shows $N(t)$ and $N_j(t)$ for all modes when $t \ll 2.5 \times 10^7$ au (1 ps $= 4.3 \times 10^4$ au). Figure 7 is a closeup view (within the dotted box) of the short-time behavior for $t < 9 \times 10^6$ au. The average number of quanta absorbed, $N(t)$, shows linear (average) growth between 4 and 13×10^6 au and then saturates near the value $N = 2.5$ at later times. By $t = 4 \times 10^6$ au, the pump mode damps to a time-averaged value of 0.5, oscillates around this value for 8×10^6 au, and then exhibits a ringing effect between 12 and 16×10^6 au. This ringing effect also appears in $N(t)$, just before saturation, near $t = 15 \times 10^6$ au.

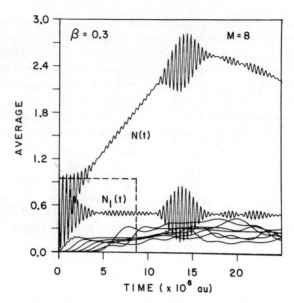

Figure 6. Average number of quanta absorbed, $N(t)$ and average number of quanta in the pump mode, $N_1(t)$, for eight-mode system. Seven background modes are also shown in the lower part. Blowup view within the dotted box is shown in Figure 7.

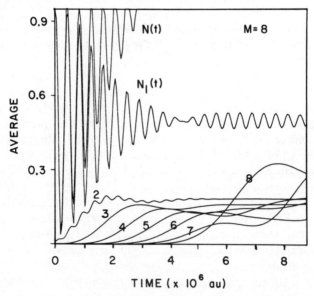

Figure 7. Blowup view within the dotted box in Figure 6 to emphasize time evolution in modes 2–8.

270

Figure 7 more clearly illustrates the growth and saturation phenomena in each of the seven background modes. Each $N_j(t)$ curve is shifted to later times by about $\Delta t = 0.5 \times 10^6$ au, relative to the previous curve. This shift reflects the pulse propagation time from one mode to the next. The excitation curve for the terminal mode, $N_8(t)$, saturates at a higher value (0.35) than the previous six background modes (which saturate between 0.12 and 0.20). After the last mode saturates, the other background modes gradually reach higher levels of excitation (during the time interval 8×10^{-6}–15×10^6 au). After 15×10^6 au, the background modes undergo a complex, irregular sharing of the excitation.

One of the most significant results that is revealed in Figure 7 is the origin of linear growth in the total number of quanta absorbed, $N(t)$. The sequence of displaced growth saturation curves, when added together, yields the linear (on the average) total excitation curve. Further analysis and discussion is presented in Friesner et al. [31].

C. Other Examples

We will briefly mention several other applications of the RRGM.

1. Flux Autocorrelation Functions and Reaction Rate Constants

Another area where time-dependent transition amplitudes enter is in the "direct" computation of Boltzmann averaged quantal rate constants for chemical reactions [49, 50]. In order to set the stage for use of the recursion method, a brief review will be presented of Miller's formulation of the rate constant in terms of the flux autocorrelation function [80]. The exact quantal equilibrium rate constant is given by the Boltzmann average of the reactive flux,

$$k(T)Q = \text{Re}\{\text{Tr}[v\exp(-\beta H)P]\}, \tag{104}$$

where Tr denotes a quantum mechanical trace, $v = \delta(s)p/m$ is the velocity operator, evaluated at the "transition state" ($s = 0$), P is an operator that projects onto states of positive momentum in the distant past (reactants), and Q is the translational partition function per unit length for reactants. In addition to the "momentum" form for P, there is a more useful form, namely, the Heisenberg transform of the positional step function,

$$P = \lim_{t \to \infty} [\exp(iHt)h(s)\exp(-iHt)], \tag{105}$$

where $h(s) = +1$ for $s \geq 0$ and $h(s) = 0$ if $s < 0$. In addition, in order to evaluate matrix elements, we can write

$$P = \lim_{t \to \infty} \int_0^t d\tau(dP/d\tau), \tag{106}$$

where

$$\frac{dP}{d\tau} = (1/2)\exp(iH\tau)[p/m, \delta(s)]_+ \exp(-iH\tau), \tag{107}$$

in which $[\quad]_+$ denotes an anticommutator, and $\delta(s)$ is the delta function. Combining these equations yields

$$kQ = \lim_{t \to \infty} \text{Re} \int_0^t C(\tau)\, d\tau, \tag{108}$$

$$C(\tau) = (1/2m^2)\,\text{Tr}\,\{\delta(s)(\partial/\partial s)U(-t_c^*)[(\partial/\partial s)\delta(s) + \delta(s)\partial/\partial s]U(t_c)\}, \tag{109}$$

in which the propagator for complex time $t_c = \tau - i\beta/2$ is

$$U(t_c) = \exp(-iHt_c). \tag{110}$$

In deriving Eq. 109, we have used the relation

$$\exp(-\beta H)P = \exp(-\beta H/2)P\exp(-\beta H/2).$$

Even though continuum states are preferred for evaluating the trace, there is also the possibility of using a discrete basis [81]. In previous model calculations employing the Eckart potential, a zero-order harmonic oscillator basis was used to evaluate eigenvectors and eigenvalues of H. The trace was then computed in the eigenvector representation. This approach is clearly limited to relatively small problems for which diagonalization of H in the zero-order basis is feasible.

In a discrete real-valued, zero-order orthonormal basis, Eq. 109 becomes

$$C(\tau) = \frac{1}{2m^2} \sum_{i,j,k,l=1}^{N} \phi_i(0)\phi_j'(0)U_{jk}(-t_c^*)$$
$$\times [\phi_k(0)\phi_l'(0) - \phi_l(0)\phi_k'(0)]U_{li}(t_c), \tag{111}$$

where matrix elements of the transition amplitudes are given by

$$U_{li}(t_c) = \langle \phi_1 | \exp(-iHt_c) | \phi_i \rangle. \tag{112}$$

Once again, the RRGM will be used to evaluate these amplitudes [49].

In a recent study [49], the RRGM was used to evaluate the flux autocorrelation function in several discrete zero-order bases for different numbers of recursion steps, M. In calculating $k(T)$ for the symmetric Eckart barrier, both harmonic escillator and particle-in-the-box basis sets were used to evaluate the transition amplitudes in Eq. 112. The flux autocorrelation converged faster in the box basis; see Figure 2 in Ref. 49. The resulting rate constant agreed well with exact calculations based on matrix diagonalization.

2. Intramolecular Energy Transfer

Intramolecular energy transfer from high excited states is another area where time-dependent transition amplitudes play a major role. Two examples will be given. In both cases, the survival amplitude of the initial state $\langle \Psi(0)|\Psi(t)\rangle$ plays a major role in the analysis. Computationally, this is one of the easiest quantities to compute with the RRGM since a single recursion sequence is required.

Marshall and Hutchinson [32] have been concerned with linewidths for overtone relaxation in propargyl alcohol, $H—C\equiv C—CH_2—OH$. Excitation of the $v = 5$ OH overtone leads to a broad band compared to the fourth overtones in other alcohols. With nine degrees of freedom (the CH vibrations on the α-carbon and C—OH torsion are not included), very large basis sets are encountered. For a "small" calculation with about 530 carefully chosen states, the RRGM results could be compared with those from direct diagonalization. The results agreed, but the CPU times on the VAX 11/750 were very different. With direct diagonalization, a large amount of paging is required so that the process is slower than $O(N^3)$ and about 5 h of CPU time are necessary. With the RRGM, no paging is required, and the whole calculation is completed in several minutes! To make sure that the "smart" basis led to correct results, they also ran calculations with up to 9000 basis functions. Fortunately, the results agreed closely with those from the smaller basis.

In a different study of overtone relaxation, Duneczky and Reinhardt [33] analyzed mode-specific energy transfer in a model for n-butane. Starting from the $v = 8$ CH overtone on the terminal carbon, the molecule is expected to undergo fast relaxation through the participation of several tiers of states involving Fermi resonance of the CH stretch with lower frequency modes. Since the frequency for the high-energy CH stretch ($1700\,cm^{-1}$) is about a factor of 2 higher than the HCC bend ($1020\,cm^{-1}$), an important decay pathway should include (stretch, bend) states such as $8, 0 \rightarrow 7, 2 \rightarrow 6, 4$. With the molecule modeled as a CH_3 fragment (three CH Morse oscillators with harmonic HCH and HCC bends) coupled to a C—C—C chain, there are nearly 400 near-resonant states in the first two tiers of states. Using the RRGM, the survival amplitude of the $v = 8$ overtone state and its Fourier transform $I(\omega)$ were computed. With 100 Lanczos recursions, the fast-decay (up to 0.02–0.05 ps) portion of the spectrum was accurately computed when compared to calculations involving fewer recursion steps.

3. Simulation of Electrode Processes

What appears to be a strikingly different application of the RRGM involves simulation by Bard and Friesner of time-dependent electrode processes [53]. Consider the simplest one-dimensional case: The time-dependent concentration of a species at distance x from an electrode is $c(x, t)$. The electrode surface is at $x = 0$, the species undergoes a fast irreversible process at this electrode, and the time-dependent flux of this reactive species at the electrode is the quantity that is to be computed. Let the x axis be discretized into a sequence of N boxes such that the concentration in box j (between x_j and x_{j+1}, with $x_0 = 0$) at time t is $c_j(t)$. Rate equations for the $c_j(t)$ are

$$\dot{c}_1 = Dc_2 - 2Dc_1,$$

$$\dot{c}_2 = Dc_1 - 2Dc_2 + Dc_3,$$

$$\dot{c}_j = Dc_{j-1} - 2Dc_j + Dc_{j+1}. \tag{113}$$

where D is the diffusion coefficient, the $+D$ terms represent gain from neighboring boxes, and the diagonal $-D$ terms represent loss from each box to the two neighboring boxes. There is loss from box 1 to box 0 but no gain in box 1 from box 0 since the species entering box 0, by assumption, reacts quickly and irreversibly. From Eq. 113, the "state vector" $c(t)$, with N components representing concentrations in the various boxes, evolves according to the kinetic equation

$$\dot{c}(t) = Wc(t), \tag{114}$$

where W is the $N \times N$ matrix of diffusion coefficients. Given the vector representing the initial uniform concentration, where $c_j(0)$ has been scaled to unity,

$$I = \begin{bmatrix} 1 \\ 1 \\ 1 \\ \vdots \\ 1 \end{bmatrix}, \tag{115}$$

the time-evolved vector is $\exp(Wt)I$. If F denotes projection into the first box,

$$F = \begin{bmatrix} 1 \\ 0 \\ 0 \\ \vdots \\ 0 \end{bmatrix}, \tag{116}$$

the flux at time t can be computed from to the $I \to F$ matrix element:

$$\text{Flux}(t) = \mathbf{F}^t \exp(\mathbf{W}t)\mathbf{I}. \tag{117}$$

This almost looks like quantum mechanics, so the RRGM now enters in the familiar way: We first define two transition vectors $\mathbf{I} + \mathbf{F}$ and $\mathbf{I} - \mathbf{F}$ and then recursively develop two tridiagonal chains in order to evaluate the two survivals,

$$(\mathbf{I} \pm \mathbf{F})^t \exp(\mathbf{W}t)(\mathbf{I} + \mathbf{F}). \tag{118}$$

In recent computations employing about 1000 boxes, the RRGM fluxes were obtained in about 2% of the CPU time for the standard method. In addition to these simulations of potentiostatic experiments involving a nonvarying electrode potential, it also appears possible to extend the RRGM to cyclic voltammetry where the electrode potential is explicitly time dependent.

Acknowledgments

The RRGM develop through stimulating interactions with my colleagues. I wish to thank Andre Nauts, Claude Leforestier, Richard A. Friesner, David Scott, José E. Castillo, Kent F. Milfeld, Israel Schek, Nimrod Moiseyev, Jean-Philippe Brunet, Xavier Chapuisat, John Hutchinson, and Csilla Duneczky. Access to the Cray X-MP/24 at the University of Texas System Center for High Performance Computing is gratefully acknowledged. This research was supported by the Robert Welch Foundation and the National Science Foundation.

APPENDIX: BIOGRAPHY OF CORNELIUS LANCZOS [82]

Cornelius Lanczos was born on February 2, 1893, in Székesfehérvár, Hungary. Following undergraduate studies at the University of Budapest in 1921, he obtained a Ph.D. in mathematical physics from the University of Szeged. After several years at the University of Freiburg, he worked with Madelung at the University of Frankfurt (1924–1928). During the following year at the University of Berlin, he began a long association with Einstein. In 1931, he moved to the United States and began a 15-year association with Purdue University. It was during this period that he initiated many important studies in numerical analysis. During the latter war years (1943–1944), he was associated with the Bureau of Standards Mathematical Tables Project. After leaving Purdue (1946), he developed an interest in the industrial applications of mathematics through associations with the Boeing Airplane Company, the Bureau of Standards Institute at UCLA, and North American Aviation. Probably his most productive period began in 1954, when he joined the Dublin Institute for Advanced Studies. Author of about 100 scientific papers as well as an accomplished pianist, Lanczos was known as an extremely modest and gentle person. He died on June 24, 1974, during a visit to Budapest.

References

1. A. Nauts and R. E. Wyatt, *Phys. Rev. Lett.* **51**, 2238 (1983).

2. A. Nauts and R. E. Wyatt, *Phys. Rev. A* **30**, 872 (1984).

3. C. Lanczos, *J. Res. Nat. Bur. Stand.* **45**, 255 (1950).

4. B. N. Parlett, *The Symmetric Eigenvalue Problem*, Prentice-Hall, Englewood Cliffs, NJ, 1980, Chapter 13.

5. C. Moler and I. Shavitt, Eds., *Numerical Algorithms in Chemistry: Algebraic Methods*, NRCC, Lawrence Berkeley Laboratories, 1978.

6. C. C. Paige, *J. Inst. Math. Appl.* **10**, 373 (1972); **18**, 341 (1976).

7. J. K. Cullum and R. A. Willoughby, *Lanczos Algorithms for Large Symmetric Eigenvalue Computations*. 2 Vols., Birkhäuser, Boston, 1985.

8. G. H. Golub and C. F. Van Loon, *Matrix Computations*, Johns Hopkins, Baltimore, 1983.

9. H. Ehrenreich, F. Seitz, and D. Turnbull, Eds., *Solid State Physics*, Vol. 35, Academic, New York, 1980.

10. D. G. Pettifor and D. L. Weaire, Eds., *The Recursion Method*, Springer-Verlag, New York, 1985.

11. G. Strang, *Introduction to Applied Mathematics*, Wellesley-Cambridge, Wellesley, 1986.

12. R. Haydock, H. Ehrenreich, F. Seitz, and D. Turnbull, Eds., *Solid State Physics* Vol. 35, Academic, New York, 1980, p. 215.

13. R. Haydock, *Philos. Mag. B* **43**, 203 (1981).

14. S. Kirkpatrick and T. P. Eggarter, *Phys. Rev. B* **6**, 3598 (1972).

15. R. Haydock, *J. Phys. C* **14**, 229 (1981).

16. J. Hafner and G. Punz, *Phys. Rev. B* **30**, (1984).

17. P. E. Meek, *Philos. Mag.* **33**, 897 (1976).

18. V. Heine, in D. G. Pettifor and D. L. Weaire, Eds., *The Recursion Method*, Springer-Verlag, New York, 1985, p. 2.

19. N. Sakamato and S. Muramatsu, *Phys. Rev. B* **17** 868 (1978).

20. S. Muramatsu and N. Sakamoto, *J. Phys. Soc. Japan* **26**, 1273 (1979).

21. S. N. Evangelou, M. C. M. O'Brien, and R. S. Perkins, *J. Phys. C* **13**, 4175 (1980).

22. M. C. M. O'Brien and S. N. Evangelou, *J. Phys. C* **13**, 611 (1980).

23. J. R. Fletcher and D. R. Pooler, *J. Phys. C* **15**, 2695 (1982).

24. M. S. Bergren and S. A. Rice, *J. Chem. Phys.* **77**, 587 (1982).

25. E. Haller, H. Köppel, and L. S. Cederbaum, *J. Mol. Spec.* **111**, 3771 (1985).

26. H. Köppel, L. S. Cederbaum, and W. Domcke, *J. Chem. Phys.* **77**, 2014 (1982).

27. E. Haller and H. Köppel, in J. Cullum and R. A. Willoughby, Eds., *Large Eigenvalue Problems*, North-Holland, Amsterdam, 1986.

28. J. P. Brunet, C. Leforestier, R. A. Friesner, and R. E. Wyatt, to be published.

29. R. A. Friesner and R. E. Wyatt, to be published.

30. J. E. Castillo and R. E. Wyatt, *J. Comp. Phys.* **59**, 120 (1985).

31. R. A. Friesner, J. P. Brunet, R. E. Wyatt, and C. Leforestier, *J. Supercomp. Appl.* **1**, 9 (1987).

32. K. Marshall and J. Hutchinson, *J. Phys. Chem.* **91**, 3219 (1987).

33. C. Duneczky and W. P. Reinhardt, to be published.

34. P. W. Langhoff, in G. H. Diercksen and S. Wilson, Eds., *Methods in Computational Molecular Physics*, Reidel, Dordrecht, 1983, p. 299.

35. M. R. Hermann and P. W. Langhoff, *Int. J. Quant. Chem.* **23**, 135 (1983).

36. M. R. Hermann and P. W. Langhoff, *Phys. Rev. A* **28**, 1957 (1983).

37. M. R. Hermann and P. W. Langhoff, *J. Math Phys.* **24**, 541 (1983).

38. S. Alexander, A. Baram, and Z. Luz, *J. Chem. Phys.* **61**, 992 (1974).

39. G. Moro and J. H. Freed, *J. Phys. Chem.* **84**, 2837 (1980).

40. G. Moro and J. H. Freed, *J. Phys. Chem.* **74**, 3747 (1981).

41. G. Moro and J. H. Freed, *J. Chem. Phys.* **74**, 3747 (1981); **75**, 3157 (1981).

42. A. Baram, *Mol. Phys.* **45**, 309 (1982).

43. W. A. Wassam, Jr., *J. Chem. Phys.* **82**, 3371, 3386 (1985).

44. G. Grosso and G. Pastori Parravicini, in I. Prigogine and S. A. Rice, Eds., *Advances in Chemical Physics*, Vol. LXII, Wiley, New York, 1985, pp. 81, 133.

45. T. J. Park and J. C. Light, *J. Chem. Phys.* **85**, 5871 (1986).

46. N. Moiseyev, R. A. Friesner, and R. E. Wyatt, *J. Chem. Phys.* **85**, 331 (1986).

47. M. Friedrichs and R. A. Friesner, *Chem. Phys. Lett.* **137**, 285 (1987).

48. R. G. Gordon, *J. Math. Phys.* **9**, 655 (1968).

49. R. E. Wyatt, *Chem. Phys. Lett.* **121**, 301 (1985).

50. T. J. Park and J. C. Light, *I. Chem. Phys.* **85**, 5871 (1986); See Section III A.

51. C. Duneczky and R. E. Wyatt, *J. Chem. Phys.* **87**, 4519 (1987).

52. K. F. Milfeld and N. Moiseyev, *Chem. Phys. Lett.* **130**, 145 (1986).

53. M. Friedrichs, R. A. Friesner and A. J. Bard, *J. Electroanal. Chem.*, to be published.

54. H. Nissimov, *Phys. Lett.* **46B**, 1 (1973).

55. R. R. Whitehead, A. Watt, B. J. Cole, and I. Morrison, in M. Baranger and E. Vogt, Eds., *Advances in Nuclear Physics*, Vol. 9, Plenum, New York, 1977, p. 123.

56. A. C. Irving and A. Thomas, *Nucl. Phys. B* **200**, 424 (1982); **215**, 23 (1983).

57. A. C. Irving, in D. G. Pettifor and D. L. Weaire, Eds., *The Recursion Method*, Springer-Verlag, New York, 1985, p. 140.

58. H. Roomany, H. W. Wyld, and L. E. Holloway, *Phys. Rev. D* **21**, 1557 (1980).

59. I. M. Barbour, N. E. Behilil, P. E. Gibbs, G. Schierholz, and Teper, in D. G. Pettifor and D. L. Weaire, Eds., *The Recursive Method*, Springer-Verlag, New York, 1985, p. 149.

60. B. Nour-Omid and R. W. Clough, *Earthquake Eng. and Struct. Dyn.* **12**, 565 (1984).

61. I. M. Smith and E. E. Heshmati, *Earthquake Eng. Struct. Dyn.* **11**, 585 (1983).

62. A. K. Cline, G. H. Golub, and G. W. Platzman, in J. R. Bunch and D. J. Rose, Eds., *Sparse Matrix Computations*, Academic, New York, 1976, p. 409.

63. R. E. Wyatt and D. Scott, in J. Cullum and R. Willoughby, Eds., *Large Eigenvalue Problems*, North-Holland, Amsterdam, 1986, p. 67.

64. A. Nauts, *Chem. Phys. Lett.* **119**, 529 (1985).

65. A. Nauts and X. Chapuisat, *Phys. Rev. A* **32**, 3403 (1985).

66. R. A. Friesner, R. E. Wyatt, C. Hempel, and B. Criner, *J. Comp. Phys.* **64**, 220 (1986).

67. Cray Research Inc., *Multitasking Users Guide*, SN-0222, Cray Research Inc., Mendota Heights, MN, 1984.

68. L. M. MacKinzie, D. Berry, A. M. MacLeod, and R. Whitehead, in D. G. Pettifor and D. L. Weaire, Eds., *The Recursion Method*, Springer-Verlag, New York, 1985, p. 165.

69. I. Schek and R. E. Wyatt, *J. Chem. Phys.* **83**, 3028 (1985).

70. I. Schek and R. E. Wyatt, *J. Chem. Phys.* **83**, 4650 (1985).

71. I. Schek and R. E. Wyatt, *J. Chem. Phys.* **84**, 4497 (1986).

72. R. R. Whitehead and A. Watt, *J. Phys. G* **4**, 835 (1978).

73. P. Lambin and J.-P. Gaspard, *Phys. Rev. B* **26**, 4356 (1982).

74. J.-P. Gaspard and P. Lambin, in D. G. Pettifor and D. L. Wearie, Eds., *The Recursion Method*, Springer-Verlag, New York, 1985, p. 72.

75. J. C. Wheeler, *J. Chem. Phys.* **80**, 472 (1984).

76. J. C. Wheeler and C. Blumstein, *Phys. Rev. B* **6**, 4380 (1972).

77. I. Schek and R. E. Wyatt, *Chem. Phys. Lett.* **129**, 99 (1986).

78. I. Schek and R. E. Wyatt, in Y. Prior, A. Ben-Reuven, and M. Rosenbluh, Eds., *Methods of Laser Spectroscopy*, Plenum, New York, 1986, p. 347.

79. R. A. Friesner and R. E. Wyatt, *J. Chem. Phys.* **82**, 1973 (1985).

80. W. A. Miller, *J. Chem. Phys.* **61**, 1823 (1974).

81. W. H. Miller, S. D. Schwartz, and J. W. Tromp, *J. Chem. Phys.* **79**, 4889 (1983).

82. E. Y. Rodin, Ed., *Computers and Mathematics with Applications: In Memory of Cornelius Lanczos*, Pergamon, New York, 1976.

CHAPTER VI

PRECONDITIONING EIGENVALUE PROBLEMS

RONALD B. MORGAN

Department of Mathematics, The University of Missouri, Columbia, Missouri 65201

DAVID S. SCOTT

Intel Scientific Computers, 15201 NW Greenbrier Parkway, Beaverton, Oregon 97006

CONTENTS

I. INTRODUCTION

There are a number of methods for computing a few eigenvalues of a large, sparse, symmetric matrix (see Ref. 1, Chapters 11–15; and Ref. 2). We will assume that factoring the matrix is impractical due to its size and sparsity structure. Section II of this chapter briefly discusses a few existing methods that are applicable in this situation. Sections III and IV present two new methods that apply the technique of preconditioning (3) to the solution of eigenvalue problems.

Preconditioning was developed for use in solving linear equation problems with the conjugate gradient method (Refs. 4 and 5, Chapters 7 and 8). The conjugate gradient method converges slowly if the distribution of eigenvalues is not favorable. Preconditioning changes the spectrum and improves convergence. In essence, preconditioning means multiplying both sides of a

279

matrix equation with an approximate inverse to the matrix. For example, $Ax = b$ becomes $M^{-1}Ax = M^{-1}b$ (of course, M^{-1} and $M^{-1}A$ are not formed explicitly). The new equation has a matrix $(M^{-1}A)$ with a different and hopefully improved distribution of eigenvalues. The more compressed the eigenvalues are around 1 and away from 0, the faster the convergence. The ultimate preconditioner is $M = A$, but this would require solving a system of linear equations in A, which is the original problem. However, there are techniques for developing approximations to A that are easily factored (6–9). For example, with incomplete factorization (8), M is formed by factoring A, except the same sparsity structure is retained. Whenever a zero element would be changed to a nonzero during the factorization, the nonzero is discarded.

II. THE LANCZOS ALGORITHM AND DAVIDSON'S METHOD

The Lanczos algorithm (1, Chapters 12 and 13) is one of the best methods for finding a few eigenvalues of a large symmetric matrix A. Lanczos uses A only in the form of a matrix–vector product, and it often finds a few eigenvalues quickly. However, the Lanczos algorithm does not always work well. Its effectiveness depends on the distribution of the eigenvalues of A and on the starting vector. This can be understood if Lanczos is viewed as being the Rayleigh–Ritz procedure applied to a Krylov subspace. The Rayleigh–Ritz procedure (1, p. 214) extracts approximate eigenvectors from a given subspace of R^n by reducing to a smaller eigenvalue problem. In the procedure, a matrix is computed with orthonormal columns that span the given subspace. If the columns are computed by the Gram–Schmidt procedure (1, pp. 98, 99), the Rayleigh–Ritz method is expensive for large subspaces.

A Krylov subspace is $Sp\{x, Ax, A^2x, \ldots, A^jx\}$, the span of the sequence of vectors generated by the power method. This subspace is an appropriate choice for the Rayleigh–Ritz procedure. A three-term recurrence formula (1, p. 255) produces orthonormal vectors that span the space. Thus, the orthonormal matrix required by Rayleigh–Ritz can be computed inexpensively. Thus, the Lanczos algorithm is much less expensive than the standard Rayleigh–Ritz. In addition, just as the power method generally produces an approximation to an eigenvector, Krylov subspaces often contain good approximations to several eigenvectors. The space is much more powerful than the sequence and contains approximations much more quickly (see Example 1). A priori bounds can be established for the convergence of Lanczos (1, Chapter 12). The rate of convergence is approximately exponential in the gap ratio (1, p. 251). The gap ratio is a simple indicator of the relative separation of an eigenvalue from the rest of the spectrum. For λ_1, the smallest eigenvalue, the gap ratio is $(\lambda_2 - \lambda_1)/(\lambda_n - \lambda_2)$. Just as the power method has difficulty if the largest eigenvalues are about the same size, the Lanczos algorithm has difficulty if the desired eigenvalues are not well separated from the rest of the spectrum.

It might be possible to correct the convergence problems of Lanczos if the distribution of eigenvalues could be changed. One way to do this is to replace A with the inverted operator $(A - \sigma)^{-1}$. (The notation that will be used for $A - \sigma I$ is $A - \sigma$.) This is shift-and-invert Lanczos (10). The eigenvalues of A near σ will be better separated in the spectrum of the inverted operator.

To implement the inverted operator, it is necessary to solve a large system of linear equations. Factoring is assumed to be too expensive because of the size and structure of the matrix; however, the conjugate gradient method can be used. But like Lanczos the conjugate gradient method is plagued by convergence problems when the distribution of eigenvalues is not favorable. The conjugate gradient method extracts its approximate solution from a Krylov subspace, just as the Lanczos algorithm extracts approximate eigenvectors from a Krylov subspace. Lanczos will generally use the Krylov subspace better than will the conjugate gradient method, since the ultimate goal is to find eigenvalues. Lanczos extracts approximate eigenvectors directly from the subspace. However, if the conjugate gradient method is preconditioned, it has a different and perhaps better subspace than does Lanczos.

Thus, preconditioning can be applied to eigenvalue problems through shift-and-invert Lanczos with the preconditioned conjugate gradient method. Similarly, inverse iteration and the Rayleigh quotient iteration can be implemented with the preconditioned conjugate gradient method [the SYMMLQ (11) variation is needed because the matrices will be indefinite]. This approach was investigated by Szyld (12). However, a direct approach to preconditioning eigenvalue problems would be preferable. Shift-and-invert Lanczos requires very accurate solution of the systems of linear equations, which is expensive. Inverse iteration and the Rayleigh quotient iteration throw away old vectors and find only one eigenvalue at a time. They discard much information.

It is not obvious how eigenvalue problems can be directly preconditioned. Multiplying both sides of the equation $Az = \lambda z$ by an approximate inverse produces a generalized eigenvalue problem that is no easier to solve. However, Davidson's method (13, 14) can be viewed as a method that uses diagonal preconditioning. Davidson's method builds a subspace iteratively through the Rayleigh–Ritz procedure. At each step the Rayleigh–Ritz procedure determines an approximate eigenpair (θ, y) of A. The new vector for the subspace is then $(D - \theta)^{-1}(A - \theta)y$, where D is the diagonal matrix with the same diagonal as A. If the $(D - \theta)^{-1}$ term were eliminated, the method would just be an expensive way of running the Lanczos algorithm. With this small change from Lanczos. Davidson (13) found a big improvement in results. In one example of dimension 372, Davidson's method reduced the norm of the residual vector, $r = Ay - \theta y$, to 10^{-6} in 10 iterations, whereas 28 iterations of Lanczos reduced the residual norm to only 2×10^{-2}.

Davidson (13) derived his method through perturbation analysis. The new vector $(D - \theta)^{-1}(A - \theta)y$ was intended to be a correction to y. But it is also possible to view $(D - \theta)^{-1}$ as an approximate inverse of $A - \theta$. Then we see that the subspace in Davidson's method is generated by the preconditioned matrix $(D - \theta)^{-1}(A - \theta)$. Eventually, θ approaches an eigenvalue λ. Therefore, it is the distribution of the eigenvalues of the matrix $(D - \lambda)^{-1}(A - \lambda)$ that controls the asymptotic convergence of Davidson's method. Here, $(D - \lambda)^{-1}(A - \lambda)$ has an eigenvalue at 0, and the corresponding eigenvector is the same as the eigenvector of A associated with λ (if λ is a multiple eigenvalue, $(D - \lambda)^{-1}(A - \lambda)$ has the same eigenspace as the eigenspace of A). It is the tendency for the nonzero eigenvalues of $(D - \lambda)^{-1}(A - \lambda)$ to be compressed around 1 that makes Davidson's method effective. Then the eigenvalue at 0 stands out in the spectrum.

Example 1. Davidson's method, the Lanczos algorithm, and the power method were compared for computing the smallest eigenvalue of a symmetric matrix A of order 40. Here, A is tridiagonal except that the a_{1n} and a_{n1} positions are nonzero. Also, $a_{ii} = i$ and $a_{i,i+1} = 1$ for all i and $a_{1n} = 1$. The starting vector was $(1, 0.1, 0.1, \ldots, 0.1)^t$ (^{t} denotes transpose). This is a moderately but not exceptionally accurate starting vector. The smallest eigenvalue is 0.238525, the next is 1.78, and the largest is 40.76. Therefore, the gap ratio for the smallest eigenvalue is $(1.78 - 0.2385)/(40.76 - .1.78) = 0.040$. Asymptotically, the preconditioned matrix for Davidson's method approaches $(D - 0.2385)^{-1}(A - 0.2385)$. This matrix has one eigenvalue at zero and the others are between 0.606 and 2. Therefore, the effective asymptotic gap ratio for Davidson's method is $(0.606 - 0)/(2 - 0.606) = 0.43$. As a result, the asymptotic convergence of Davidson's method was much faster than for Lanczos. The smallest Ritz value in Davidson's method attained 12 digits of accuracy at step 13 (when the space was of dimension 13). Lanczos required about 30 steps to reach the same degree of accuracy. As expected, the power method was even slower than the Lanczos (a sequence is used instead of a subspace). The power method with optimal shift 21.37 ($A - 21.37$ was the operator) reached only five digits of accuracy after 100 steps.

III. THE GD METHOD

The diagonal of the matrix A in Example 1 is a fairly good approximation to the matrix, so it is not surprising that diagonal preconditioning improves the spectrum. The molecular energy matrices that Davidson was interested in also tend to have diagonals that are good approximations to the matrices (though Davidson did not state this as a condition for his method). For some problems, Davidson's method is not an improvement over Lanczos. Kalamboukis (15) reported that for certain nondiagonally dominant matrices, Davidson's method converged no faster than Lanczos. For these matrices the spectrum of $(D - \lambda)^{-1}(A - \lambda)$ is no better than that of A.

When the diagonal preconditioning of Davidson's method is not effective, it makes sense to use a better preconditioner. We replace D with M, where M is an approximation to A. This generalization of Davidson's method is referred to as the GD method (14). Unlike the conjugate gradient method, the preconditioner is not required to be positive definite. Also, $M - \theta$ need not be symmetric, though generally we will assume that it is.

An equivalence can be established between the effectiveness of preconditioning for the conjugate gradient method and the effectiveness of preconditioning for the GD method (16, pp. 35–40). The great amount of research on preconditioning linear equations problems can now be applied to eigenvalue problems. If $A - \theta$ is indefinite (i.e., $A - \theta$ has both negative and positive eigenvalues), caution should be taken in applying a preconditioning technique developed for positive definite matrices, such as incomplete factorization (8). Also, there must be caution in the unlikely event that M has exactly the same eigenvalue and eigenvector as the desired eigenvalue and eigenvector of A (16, pp. 53–62).

Example 2. GD was tested on the matrix from Example 1. Here M is chosen to be T, the tridiagonal portion of A, an extremely good approximation. The eigenvalues of $(M - \lambda_1)^{-1}(A - \lambda_1)$ are 0 and 2, and all the others are 1. The gap ratio is 1. The smallest Ritz value converged to 12 digits of accuracy by step 7. But 10 of these digits came in only two steps, from step 5 to step 7. It is even more impressive to compare the improvement in the norm of the residual vector $r = Ay - \theta y$. Measured from the point at which the most rapid convergence began, the norm of the residual vector decreased by 12 orders of magnitude in about 38 steps for Lanczos; in 10 steps for Davidson's method (GD with $M = D$), and in 3 steps for GD with $M = T$.

We have seen that preconditioning can greatly improve the asymptotic rate of convergence (convergence once θ is near λ), but the initial convergence is another matter. When θ is away from λ, the eigenvectors of $(M - \theta)^{-1}(A - \theta)$ are not significant, and generating the subspace with this matrix is not likely to help you find the desired eigenvector of A.

Example 3. Davidson's method was applied again to the matrix in Example 1. This time the starting vector had components chosen randomly on the interval $(-1, 1)$. It took 23 steps for θ to attain one correct digit, and it took 27 steps for

12 correct digits (twice as many steps as with the better starting vector in Example 1).

When the starting vector (or space) is not accurate, the initial θ will probably be far from the desired λ. But if a better estimate of the desired eigenvalue is known, the GD method can be modified. Let σ be the eigenvalue estimate and replace the preconditioner $(M - \theta)^{-1}$ with $(M - \sigma)^{-1}$.

Example 4. This modified GD method was used with $M = D$ and with the random starting vector as in the previous example. Here σ is chosen to be 0.4, which is fairly close to the eigenvalue 0.2385. With $(D - 0.4)^{-1}(A - \theta)$ generating the subspace θ reached one digit of accuracy by step 5 and 12 digits after 12 steps.

Another reason for allowing different shifts for A and M is to avoid the expense of changing the preconditioner at every step. However, it may be best to occasionally change σ to the current value of θ. If σ is never changed, the asymptotic convergence may be slower than for regular GD. An effective criterion for changing σ is given in the literature (16, pp. 66–68).

IV. THE PL ALGORITHM

We will now compare the costs for Lanczos and GD and see that there is a situation for which neither method is cost-effective. We will count the number of multiplications required for a step (iteration) of each method and only consider terms of order n, where A is $n \times n$. Each step of GD has one matrix–vector product using A. This requires mn multiplications, where m is the average number of nonzero elements per row of A. The expense for the rest of the Rayleigh–Ritz procedure is $5jn$ multiplications per step, where j is the dimension of the subspace. This expense increases as the subspace grows in size. It will be called the Rayleigh–Ritz overhead expense. The total cost for one step of the GD method is $mn + 5jn$ multiplications plus the cost for the preconditioner. The simple Lanczos algorithm requires $mn + 5n$ multiplications per step. The expense of Lanczos does not grow as the subspace grows (unless Ritz vectors are formed).

For a particular matrix, two things must be assessed. First, is the preconditioning effective relative to its cost? The preconditioned matrix must have a better distribution of eigenvalues than A, so that the number of steps required is reduced. If the preconditioning is expensive, the number of iterations must be substantially reduced. For example, with incomplete factorization (8), an application of the preconditioner costs about the same as a matrix–vector product (assuming the factorization has already been done). Neglecting the Rayleigh–Ritz overhead expense, the number of iterations would need to be cut at least in half by the preconditioning.

Second, we must consider how much effect the Rayleigh–Ritz overhead cost has. Some matrices are so full that the cost of the matrix–vector product dominates the overhead cost. Other matrices are quite sparse, and the Rayleigh–Ritz overhead is a major factor. When the overhead is too expensive, the GD method can be restarted occasionally. This keeps the overhead down, but it is not always a satisfactory solution because the advantage of having a large subspace is lost.

We conclude that if the preconditioning is not effective (relative to the cost of applying the preconditioner), the Lanczos algorithm is the preferred method. If the preconditioning is effective and the Rayleigh–Ritz overhead expense is negligible, the GD method should be used. However, if the preconditioning is effective but the overhead is large, neither method is good. The Lanczos algorithm is insufficient because too many iterations are required; GD is too expensive per iteration. In this situation a method is needed that has the advantage of a preconditioned spectrum yet has the efficiency of the Lanczos algorithm. Such a method is presented next.

We will look at the matrix $W \equiv L^{-1}(A - \sigma)L^{-t}$, where $M - \sigma = LL^t$ is the Cholesky factorization ($M - \sigma$ must be positive definite). If the eigenvectors of W are multiplied by L^{-t}, they become the eigenvectors of $(M - \sigma)^{-1}(A - \sigma)$. The eigenvectors of $(M - \sigma)^{-1}(A - \sigma)$ are of interest because as σ approaches λ, one of them approaches z (the eigenvector of A associated with λ). As suggested by Scott (17), W can be used in a loop to find improved approximations to an eigenpair of A. Since W is symmetric, its eigenvectors can be computed with the Lanczos algorithm. However, there is no reason to carry the Lanczos computation to full convergence because it is inside another loop.

The algorithm that follows will be referred to as the preconditioned Lanczos (PL) algorithm. A positive definite preconditioner is required so that Cholesky factorization is possible. Note that $M - \sigma$ is replaced by $M(\sigma)$ ($M - \sigma$ with M constant may not always be positive definite).

The PL Algorithm. Given a vector x_0, compute $\sigma_0 = x_0^t A x_0 / x_0^t x_0$, and for $k = 0, 1, 2, \ldots$, do 1–5:

1. Choose $M(\sigma_k)$ to be positive definite and let $L_k L_k^t$ be its Cholesky factorization.

2. Define $W_k \equiv L_k^{-1}(A - \sigma_k)L_k^{-t}$.

3. Run Lanczos with W_k and starting vector $L_k^t x_k$ until the smallest Ritz value is bounded away from zero by the residual norm. Let θ_k be the smallest Ritz value and y_k be the corresponding Ritz vector of unit length.

4. Let $x_{k+1} = L_k^{-t} y_k$

5. Set

$$\sigma_{k+1} = x_{k+1}^t A x_{k+1} / x_{k+1}^t x_{k+1}$$

$$= \sigma_k + \theta_k / x_{k+1}^t x_{k+1}.$$

Here (σ_{k+1}, x_{k+1}) is an approximate eigenpair of A. The PL process should be terminated either when the sequence σ_k settles down or when the residual vector of x_{k+1} has a small enough norm.

$M(\sigma)$ must be positive definite. One way to ensure this is as follows: Choose an M, factor $M - \sigma$ in LDL^t form, take absolute values of the elements of the diagonal matrix D, and let $M(\sigma)$ be $L|D|L^t$. However, if this D has an element near zero, then $M(\sigma)$ will have a large condition number (it will have both large and small eigenvalues). This may distort W and slow convergence. Of course, D can be adjusted if it has a small element. Another approach to ensuring positive definiteness is to simply let $M(\sigma)$ be $M - \eta$, where η is less than the smallest eigenvalue of M.

It is guaranteed that σ_k in the PL algorithm will converge to an eigenvalue of A as long as the preconditioner is implemented in a reasonable manner (16, pp. 86–88). Specifically, $M(\sigma_k)$ and $M(\sigma_k)^{-1}$ should both be bounded in norm. Also, as σ_k converges to ω, $M(\sigma_k)$ should converge to $M(\omega)$.

Let λ be the eigenvalue to which σ_k converges. Here, λ will almost always be the smallest eigenvalue of A (except for a measure zero situation). In addition, the convergence is guaranteed to be at an asymptotically quadratic rate with respect to k (16, pp. 89–93).

Example 5. The PL algorithm was applied to the matrix and starting vector from Example 1. Diagonal preconditioning was used; $M(\sigma) = |D - \sigma|$. As expected, there was quadratic convergence toward λ_1. The term σ_k had 1 correct digit for $k = 4$, 4 correct digits at $k = 5$, and 10 for $k = 6$. The number of Lanczos iterations required for the inside loop increased as k increased. The dimension of the subspace developed by Lanczos was only 2 for $k = 1$ and $k = 2$, but it grew to dimension 9 for $k = 6$. Comparing the total number of matrix–vector products (matrix multiplies with A) required to reach a residual norm below 10^{-10}, GD (Davidson's method) required 17 and PL required 21.

Example 6. Next, PL with tridiagonal preconditioning was tested. Positive definiteness was achieved by the two ways mentioned: first, by taking absolute values after a factorization; second, by letting $M(\sigma_k) = T - 0.0$ for $k = 1, 2$ (after that, $T - \sigma_k$ is positive definite). In the first test, convergence was slow initially. Seventeen matrix–vector products were required to reach a residual norm under 10^{-10}, compared to eight for GD. However, in the second test, where the shift for T was kept outside of the spectrum of T, only nine steps were required.

Example 7. GD and PL were tested on a larger matrix from oil reservoir simulation that was distributed by Sherman (18) for conjugate gradient method tests. The matrix is 1000×1000 and is fairly sparse. It has seven diagonal bands. The bands occur in the 0 (main diagonal), 1, 10, and 100 positions. The matrix is positive definite, with its smallest eigenvalue at 0.00032. However, it is not quite diagonally dominant. The smallest eigenpair of the matrix was computed. The methods were terminated when the residual norm reached 10^{-10}. The elements of the starting vector were selected randomly on the interval (0, 1). Since A is positive definite, an initial estimate of 0.0 was used for the smallest eigenvalue. This estimate was important to the success of the methods. The shift for M was chosen with the schemes given elsewhere (16, pp. 66–68, 105–107). Incomplete factorization was used for the preconditioning [during a factorization, any nonzero elements created in previously zero positions were discarded (8)]. The smallest eigenvalue is not well separated from the others, so the Lanczos algorithm converged very slowly. It required 600 iterations to reach a residual norm of 10^{-10}. To compute eigenvectors from a subspace of dimension 600 would be expensive. In this case GD required only 43 steps, and PL required 80 steps. However, comparing computation time, PL was less than half as expensive as GD.

Eigenvalues other than the smallest one can be computed with PL by choosing a different θ in the Lanczos loop. A deflation scheme is also possible.

V. SUMMARY

The GD and PL methods can greatly reduce the expense of computing eigenvalues of many matrices. Both methods require approximate inverses that can be implemented with reasonable expense. The methods also need either an accurate starting vector (or starting space) or an initial eigenvalue estimate.

The GD method is most useful when the matrix–vector product is expensive. The PL method may be better if the matrix is fairly sparse and only a few extreme eigenvalues are desired.

References

1. B. N. Parlett, *The Symmetric Eigenvalue Problem*, Prentice-Hall, Englewood Cliffs, NJ, 1980.
2. R. K. Nesbet, "Large Matrix Techniques in Quantum Chemistry and Atomic Physics," in I. Duff, Ed., *Sparse Matrices and Their Uses*, Academic, New York, 1981, pp. 161–174.
3. P. Concus, G. H. Golub, and G. Meurant, *SIAM J. Sci. Stat. Comput.* **6**, 220 (1985).

4. A. Cline, "Several Observations on the Use of the Conjugate Gradient Methods," ICASE Report 76-22, NASA Langley Research Center, Hampton, VA, 1976.

5. L. A. Hageman and D. M. Young, *Applied Iterative Methods*, Academic, New York, 1981.

6. O. B. Axelsson, "On Preconditioning and Convergence Acceleration in Sparse Matrix Problems," Technical Report CERN 74-10, CERN, Geneva, Switzerland, 1974.

7. I. Gustafsson, *BIT* **18**, 142 (1978).

8. J. A. Meijerink and H. A. van der Vorst, *Math. Comp.* **31**, 148 (1977).

9. H. L. Stone, *SIAM J. Numer. Anal.* **5**, 530 (1968).

10. T. Ericsson and A. Ruhe, *Math. Comp.* **35**, 1251 (1980).

11. C. C. Paige and M. A. Saunders, *SIAM J. Numer. Anal.* **12**, 617 (1975).

12. D. B. Szyld, "A Two-level Iterative Method for Large Sparse Generalized Eigenvalue Calculations," Ph.D. Thesis, Courant Institute, New York University, 1983.

13. E. R. Davidson, *J. Comp. Phys.* **17**, 87 (1975).

14. R. B. Morgan and D. S. Scott, *SIAM J. Sci. Stat. Comput.* **7**, 817 (1986).

15. T. Z. Kalamboukis, *J. Phys. A: Math. Gen.* **13**, 57 (1980).

16. R. B. Morgan, "Preconditioning Eigenvalue Problems," Ph.D. Thesis, University of Texas, Austin, TX, 1986.

17. D. S. Scott, *SIAM J. Numer. Anal.* **18**, 102 (1981).

18. A. H. Sherman, *Linear Algebra for Reservoir Simulation Comparison Study of Numerical Algorithms*, J. S. Nolen & Associates, Houston, TX, September 1984.

CHAPTER VII

STATIONARY PHASE MONTE CARLO METHODS

J. D. DOLL

*Los Alamos National Laboratory, Mail Stop G-738, Los Alamos,
New Mexico 87545*

D. L. FREEMAN

*Department of Chemistry, University of Rhode Island, Kingston,
Rhode Island 02881*

CONTENTS

Monte Carlo (MC) methods have historically been of use in chemical physics in the numerical study of the equilibrium properties of classical and, more recently, quantum-mechanical many-body systems. The general application of these numerical approaches to time-dependent phenomena, although possible in principle, has been limited in practice to the study of relatively short-time phenomena. The essential numerical difficulty blocking their general use in dynamical problems is that of constructing high-dimensional averages of rapidly oscillatory integrands. The present work explores relatively simple extensions of the ordinary MC procedure that appear to be useful in overcoming such difficulties.

I. INTRODUCTION

Monte Carlo (MC) methods represent an effective numerical approach to the study of a broad class of many-particle problems. These approaches have

289

historically been especially useful within chemical theory in the study of the equilibrium statistical mechanics of condensed-phase classical-mechanical systems (1). More recently, their utility has been extended to the more general area of equilibrium quantum-mechanical calculations (2) where the necessary quantum-mechanical traces or the density matrix elements themselves are reduced to a MC problem through the use of numerical path integration methods (3).

As emphasized by a number of investigators (4–8), time-dependent as well as equilibrium quantum-mechanical information is, in principle, available from such path integral approaches. Miller, for example, has indicated how one might formulate the problem of the calculation of thermal rate constants for chemical reactions in these terms (6). The price paid for this additional dynamical information is the calculation of the density matrix elements for complex rather than real temperatures. Formally, this requirement presents no obstacles, although such formulations are, at present, limited to the generation of relatively short-time information. The bulk of effort in development of such time-dependent methods is currently focused on analytic continuation (5, 6) and basis set techniques (6, 9, 10), although intrinsically approximate classical path methods (11) have also been explored.

In this chapter we develop a different approach to problems in real-time quantum dynamics, an approach based on extensions of traditional MC path integral methods (12). In this work we focus specifically on overcoming the central numerical problem of such approaches, the construction of high-dimensional averages over (possibly) rapidly oscillatory integrands. We show that there exist relatively simple modifications of ordinary MC methods that offer the possibility of a general approach to such numerical tasks. In view of the range of possible applications of such methods, this chapter concentrates attention on the MC methods themselves. Subsequent applications will be reported elsewhere. Section II presents the formal developments, and Section III illustrates these developments with simple numerical examples.

II. FORMAL DEVELOPMENT

We introduce in this section methods designed to handle the generic problem of calculating multidimensional averages of (possibly) rapidly oscillatory integrands. For the present development we consider the specific problem of evaluating numerically averages of the form

$$I(t) = \frac{\int \rho(x) \exp\left[itf(x)\right] dx}{\int \rho(x)\, dx}, \tag{2.1}$$

where $\rho(x)$ is a specified probability distribution, t is a parameter, and $f(x)$ is some function of x. We will be somewhat casual in the following discussions, assuming that neither $\rho(x)$ nor $f(x)$ have any special pathologies. Integrals of the preceding type arise naturally in a variety of contexts, including, as was discussed in the introduction, the treatment of real-time path integral problems. Our primary emphasis here will be in the development of direct MC-based methods for the treatment of such integrals. Approaches based on analytic continuation methods have been discussed elsewhere (3, 5, 6). We draw attention, in this regard, to the comments of Pollock and Ceperley concerning the use of such continuation methods in conjunction with MC-based approaches (13).

We begin by noting the obvious and point out that when the parameter t is "small," the evaluation of $I(t)$ is straightforward. In this case ordinary MC methods (1) are useful, and the average in Eq. 2.1 becomes

$$I(t) = \frac{\sum_{n=1}^{N} \exp\left[itf(x_n)\right]}{N}, \qquad (2.2)$$

where the points (x_n), $n = 1, \ldots, N$, are chosen randomly from the distribution $\rho(x)$. For many purposes this direct MC evaluation of $I(t)$ will prove adequate. Our experience suggests, in fact, that the range of problems for which this direct MC approach is useful may be larger than is generally appreciated.

Historically, progress has been made in the present type of problems by exploiting the separation of length scales that enter the problem as t increases. For large t values the integrand in 2.1 oscillates rapidly as a function of x, except in those regions where $f(x)$ is stationary. The effect of these oscillations is to wipe out contributions to the integral in most regions of configuration space. Such reasoning leads to the familiar "stationary phase" method (14) in which $f(x)$ is replaced by its second-order Taylor expansion about points for which $f'(x) = 0$. Such a replacement, although globally poor, is appropriate since it describes $f(x)$ well in the limited regions that contribute significantly to the integral. For the simple case of a single stationary phase point and a one-dimensional integral, $I(t)$ is given by

$$\lim_{t \to \infty} I(t) = \rho(x_{sp}) \exp\left[itf(x_{sp})\right] \left\{2\pi/\left[-itf''(x_{sp})\right]\right\}^{1/2} \bigg/ \int \rho(x)\, dx, \qquad (2.3)$$

where x_{sp} is the stationary phase point, $f'(x_{sp}) = 0$. The utility of Eq. 2.3 is limited for general multivariable functions, limited in principle because of its approximate asymptotic nature and limited in practice because of difficulties inherent in the stationary phase construction (e.g., the location of stationary phase regions or complexity of the stationary phase problem's analytic structure).

Comparing Eqs. 2.2 and 2.3, we see the difficulty inherent in attempting a direct MC calculation of $I(t)$ when t is large. The stationary phase method utilizes a physical argument rather than an explicit construction to treat or, more precisely, to avoid treating the interference effects present in the original integral. Monte Carlo methods, on the other hand, attempt a brute-force construction of these interference effects. To succeed, a direct application of the MC method must therefore utilize a number of quadrature points that is sufficiently large to assure that interference effects are properly described. If the parameter t is large and the oscillations severe, the number of points required is, in general, impossibly large. The convergence of standard MC estimates is thus being limited by the relatively inefficient numerical process of assembling phase interferences. This is an especially annoying situation since the method is thus working very hard to calculate a vanishingly small contribution to the overall result.

In light of the foregoing discussion, it would appear reasonable to modify the direct MC treatment of such problems in a way that would remove this burden of numerically generating the interference effects. A discussion of one such modification follows (15).

We can exploit the separation of length scales in the present context in a way that is generalizable to large-dimensional problems and in a way that avoids the need for the prior location of stationary phase regions. To motivate the results, consider a thought experiment in which we follow the convergence of a MC estimate of $I(t)$. We begin by estimating $I(t)$ with an N-point sum as in Eq. 2.2. Using the MC points, x_n, we then perform a Wigner–Seitz-like construction (16), filling that portion of configuration space covered by the MC random walk with a set of cells [Voronoi polyhedra (16)] requiring that N be sufficiently large that the variation of $\rho(x)$ (but *not necessarily exp*$[itf(x)]$) over any particular cell is negligible. Since, by assumption, $\rho(x)$ is effectively constant over each cell, we can now improve our direct MC estimate by replacing the representative value of the integrand for cell n used by the direct MC procedure, $\exp[itf(x_n)]$, by the average of $\exp[itf(x)]$ over cell n, $\langle \exp[itf(x)]\rangle_n$. If t is small, $\exp[itf(x)]$ and its average over the cell are essentially the same, and the present procedure reduces to the standard and already viable MC method. If t is large, however, the effect of averaging $\exp[itf(x)]$ over the cell will be to "bundle" together local contributions to the integral building in local phase cancellations in the non-stationary-phase regions. Assuming, for the moment, that these local averages can be generated efficiently, this procedure effectively transfers the task of contructing the phase interferences from the MC method to the local averaging process, thereby improving the convergence characteristics of the statistical procedure. The present method is clearly exact in the limit of large N. Moreover, we anticipate that the procedure will become accurate whenever N is large enough that variation of $\rho(x)$ over the Wigner–Seitz cells becomes negligible.

This approach is completely rigorous but would be somewhat cumbersome to implement in practice, requiring first the construction of numerous Voronoi polyhedra and then the calculation of the associated integrand averages. A more practical procedure is to replace the actual integrand, $\exp[itf(x)]$, with one averaged about the point x, $\langle \exp[itf(x)] \rangle_\varepsilon$. The length scale ε of the average is regarded as a parameter in the calculation. In the limit that ε vanishes, the procedure reduces to ordinary MC and converges rigorously to the exact result. In what follows we argue that accurate results are achieved for finite ε values that are in turn determined by the length scale of the density, $\rho(x)$. More explicitly, if $\delta_\varepsilon(y)$ is a prelimit delta function with a length scale specified by ε,

$$\lim_{\varepsilon \to 0} \delta_\varepsilon(x) = \delta(x), \tag{2.4}$$

then we imagine the local average of the integrand about the point x to be of the form

$$\langle \exp[itf(x)] \rangle_\varepsilon = \int \delta_\varepsilon(y) \exp[itf(x+y)] \, dy. \tag{2.5}$$

The average in Eq. 2.5 "filters" the integrand, removing oscillations on length scales smaller than ε. The present procedure thus amounts to replacing the original integral, $I(t)$, by

$$I_\varepsilon(t) = \int \rho(x) \langle \exp[itf(x)] \rangle_\varepsilon \, dx \Big/ \int \rho(x) \, dx. \tag{2.6}$$

Clearly, $I_\varepsilon(t)$ reduces to $I(t)$ in the limit as $\varepsilon \to 0$. More importantly, however, $I_\varepsilon(t)$ will effectively reach its limit for finite values of ε, values that are small relative to the length scale of variations of $\rho(x)$. To make this point more explicit, we rewrite the original integral in Fourier space, giving

$$I(t) = \int \hat{\rho}(k) \hat{g}(-k) \, dk / 2\pi, \tag{2.7}$$

where the individual transforms are

$$\hat{\rho}(k) = \int dx \exp(ikx) \left[\rho(x) \Big/ \int \rho(x) \, dx \right] \tag{2.8}$$

and

$$\hat{g}(k) = \int dx \exp(ikx) \exp[itf(x)], \tag{2.9}$$

respectively. The natural length scale of $\rho(x)$ limits the range of k values over which the integrand of the Fourier space integral, Eq. 2.7, is significant. That is, if $\rho(x)$ is effectively constant over length scales of order ε, $\hat{\rho}(k)$ will vanish for k values beyond roughly $1/\varepsilon$. The higher order Fourier components of $\hat{g}(k)$ are thus irrelevant to the integral, and we can therefore replace $\hat{g}(k)$ with a functional form that also vanishes outside this k range without modifying the Fourier integral. For example, we can replace $\hat{g}(k)$ by $\hat{g}_\varepsilon(k)$, where $\hat{g}_\varepsilon(k)$ is given by

$$\hat{g}_\varepsilon(k) = \hat{g}(k)\hat{\delta}_\varepsilon(k). \tag{2.10}$$

Here $\hat{\delta}_\varepsilon(k)$ is assumed to be a "bandpass filter" that is unity for k values in the range $(-1/\varepsilon, 1/\varepsilon)$ and tends to zero otherwise. In direct space $\delta_\varepsilon(x)$ is a prelimit delta function of range ε, and Eq. 2.10 thus reduces to 2.5.

A wide variety of explicit forms for the "ε average" in Eq. 2.5 can be obtained depending on the explicit form utilized for the prelimit delta function and depending on the degree of accuracy used in executing the subsequent averaging integrations. The choice of which form is used for any particular application will be, in part, a matter of experience of convenience. Perhaps the simplest of these forms is obtained by writing

$$\langle \exp[itf(x)] \rangle_\varepsilon = \frac{\displaystyle\int_{-\varepsilon}^{\varepsilon} \exp[itf(x+y)]\, dy}{\displaystyle\int_{-\varepsilon}^{\varepsilon} dy}. \tag{2.11}$$

This integral will not be available analytically for arbitratry $f(x)$. Progress can be made, however, by expanding $f(x+y)$ in a Taylor series about the point x. Through first order this expansion gives, in one dimension,

$$\langle \exp[itf(x)] \rangle_\varepsilon = \exp[itf(x)] \sin[\varepsilon tf'(x)]/\varepsilon tf'(x), \tag{2.12}$$

a result easily generalized to higher dimensions. We see that the local averaging process synthesizes the interference effects in the vicinity of x, producing a term that modifies the original integrand in a way that discriminates against non-stationary-phase regions. Expanding $f(x+y)$ through second order and using Eq. 2.5 produces a more involved, but basically similar, result. It is informative to note that these second order results take on a familiar form in one dimension as t becomes large and ε becomes small, namely,

$$\lim_{t \to \infty} \langle \exp[itf(x)] \rangle_\varepsilon = \sum_{sp} \exp[itf(x_{sp})] \delta_\varepsilon(x - x_{sp}) \left(\frac{2\pi}{-itf''} \right)^{1/2}. \quad (2.13)$$

The sum in Eq. 2.13 is over points determined by the requirement that $f'(x_{sp})$ vanish. Replacing the original integrand in Eq. 2.1 with the second order result in Eq. 2.13 produces (essentially) the familiar stationary phase method. Other useful expressions can be obtained by using a Gaussian prelimit form for the delta function. This choice produces a Gaussian "smear" of the integrand,

$$\langle \exp[itf(x)] \rangle_\varepsilon = \frac{\displaystyle\int_{-\infty}^{\infty} dy \exp(-y^2/2\varepsilon^2) \exp[itf(x+y)]}{\displaystyle\int_{-\infty}^{\infty} dy \exp(-y^2/2\varepsilon^2)}. \quad (2.14)$$

First- or second-order gradient expansions are again of use in the actual construction of these Gaussian averages. Expanding $f(x + y)$ through second order, for example, produces, in one dimension (easily generalized),

$$\langle \exp[itf(x)] \rangle_\varepsilon = \exp[itf(x)] \exp[-(t\varepsilon f')^2/2A]/\sqrt{A}, \quad (2.15)$$

where $A = 1 - itf''\varepsilon^2$. Cumulant expansions (17) of the ε averages are also often of use. Explicitly,

$$\langle \exp[itf(x)] \rangle_\varepsilon = \exp \frac{\displaystyle\sum_{m=1}^{\infty} (it)^m \langle\langle f^m \rangle\rangle_\varepsilon}{m!}, \quad (2.16)$$

which, through second order, becomes

$$\langle \exp[itf(x)] \rangle_\varepsilon = \exp(it\langle f \rangle_\varepsilon) \exp(-t^2 \langle\langle f^2 \rangle\rangle_\varepsilon/2). \quad (2.17)$$

The notation $\langle\langle f^m \rangle\rangle$ in Eqs. 2.16 and 2.17 represents the mth order cumulant, as discussed in Ref. 17. In Eq. 2.17 the neighborhood averaging has modified the original integrand, introducing a positive damping factor that discriminates against non-stationary-phase regions. As a practical matter, it is important to note that this positive damping factor can be combined with the original distribution, $\rho(x)$, to give a modified distribution that emphasizes the important regions of configuration space. Such cumulant weighting has the effect of turning the problem of locating the stationary phase regions into a problem in MC sampling. Other filtering strategies have been explored, including those involving filtering in the reciprocal Fourier space. At present no clear-cut winner for the general construction of the local averages has emerged from the variety of results discussed previously. We have found all of the preceding forms to be of use, although different applications do favor particular approaches. We close with the general observation that the introduction of approximations into the construction of the local averages may serve to reduce the upper limit of the ε values that are useful in the present procedure relative to the values anticipated on the basis of $\rho(x)$ length scale estimates.

In summary, it appears that for problems involving the evaluation of integrals of the type indicated in Eq. 2.1, a useful strategy is to replace the original integrand with a suitably "filtered" one. A proper choice of the filter will leave the results unchanged but will generally simplify their calculation. As indicated, such an approach contains the ordinary stationary phase method but is generalizable in ways that familar method is not. In particular, the present results offer the possibility of combining stationary-phase-like ideas with important and useful MC methods. Because of their close association with these familiar approaches, we term the present result the stationary phase Monte Carlo method (SPMC), again emphasizing that the present method is more general than its asymptotic namesake. In particular, the present method is, in principle, always an exact, as opposed to an asymptotic, approach. In closing, we note the close analogy between the present problems and those in the numerical simulation of stochastic differential equations (18), where a separation of time rather than of length scales is frequently the essential simplifying feature.

III. EXAMPLES

We discuss in this section a few illustrative examples of the use of the present methods. These examples are intended to be pedagogical in nature and to indicate the utility and relative simplicity of the approach. Applications dealing with more complex and physically motivated problems will be reported elsewhere.

We begin with the study of a simple integral, one for which the various steps in the SPMC implementation can be carried through analytically. Integrals of the form

$$I(t) = \frac{\int_{-\infty}^{\infty} dx \exp(-x^2/2\sigma^2)\exp(itx^2)}{\int_{-\infty}^{\infty} dx \exp(-x^2/2\sigma^2)} \tag{3.1}$$

arise in the path integral treatment of the density matrix elements of the harmonic oscillator at a complex temperature. Obviously, $I(t)$ is available analytically and is equal to $(1 - 2it\sigma^2)^{-1/2}$. It is interesting, however, to consider the MC and SPMC evaluations of $I(t)$. In the SPMC approach we replace the integrand in 3.1, $\exp(itx^2)$, with a locally averaged result, $\langle \exp(itx^2) \rangle_\varepsilon$, giving

$$I_{\text{SPMC}}(t) = \frac{\displaystyle\int_{-\infty}^{\infty} dx \exp(-x^2/2\sigma^2)\langle \exp(itx^2) \rangle_\varepsilon}{\displaystyle\int_{-\infty}^{\infty} dx \exp(-x^2/2\sigma^2)}. \tag{3.2}$$

For this example we take as the local average in the neighborhood of the point x a Gaussian "smear" (Eq. 2.14),

$$\langle \exp(itx^2) \rangle_\varepsilon = \frac{\displaystyle\int_{-\infty}^{\infty} dy \exp(-y^2/2\varepsilon^2) \exp[it(x + y)^2]}{\displaystyle\int_{-\infty}^{\infty} dy \exp(-y^2/2\varepsilon^2)}$$

$$= \frac{\exp itx^2/(1 - 2it\varepsilon^2)}{\sqrt{1 - 2it\varepsilon^2}}. \tag{3.3}$$

Analytically, $I_{\text{SPMC}}(t)$ is thus

$$I_{\text{SPMC}}(t) = (\sqrt{1 - 2it\sigma^2(1 + \varepsilon^2/\sigma^2)})^{-1}. \tag{3.4}$$

This equation illustrates, for the present example, the nature of the convergence of the SPMC result to the exact value as ε becomes small relative to the length scale of $\rho(x)$. The choice of a numerical value for ε in practical applications is a compromise, small ε values favoring absolute accuracy and larger ε values accelerating the convergence of the MC calculation. For the present example we see from 3.4 that a choice of $\varepsilon/\sigma = 0.1$ will result in relative errors of the SPMC approach (for large values of t) on the order of 1%. If inadequate, this relative error is easily reduced by choosing smaller values of ε. Evaluating 3.1 and 3.2 numerically, we have for the MC and SPMC results, respectively,

$$I_{\text{MC}}(t) = \frac{\displaystyle\sum_{n=1}^{N} \exp(itx_n^2)}{N}, \tag{3.5a}$$

$$I_{SPMC}(t) = \frac{\sum_{n=1}^{N} \langle \exp(itx_n^2) \rangle_\varepsilon}{N}, \tag{3.5b}$$

where the points x_n are chosen randomly from the Gaussian distribution $\exp(-x^2/2\sigma^2)$. Some idea of the effect of the SPMC averaging can be gained by examining Figure 1, where we plot as functions of x the Gaussian probability distribution function ($\sigma = 1$), $\exp(-x^2/2)$, and the real parts of $\exp(itx^2)$ and $\langle \exp(itx^2) \rangle_\varepsilon$ for various values of t. All SPMC averages shown are for $\varepsilon = 0.1$. As is obvious from Figure 1, the SPMC integrand, $\langle \exp(itx^2) \rangle_\varepsilon$, is essentially the same as the original integrand, $\exp(itx^2)$, for small t values but is considerably simpler than the original highly oscillatory integrand for large t values. We anticipate, therefore, that the convergence of the SPMC sum 3.5b will be similar to that for the direct MC result 3.5a for small t, but markedly better than the MC results for large t. Actual numerical results support these expectations. For example, the real parts of the MC and SPMC results for $t = 1$ and $\sigma = 1$ are, respectively, 0.574 ± 0.013 and 0.571 ± 0.012 versus the exact result of 0.569. For $t = 100$ the corresponding results are 0.0541 ± 0.0205 and 0.0476 ± 0.0033 versus the exact result of 0.0501. A fixed number of MC points (in this case 1600) were used in all cases, and the errors reported are one standard deviation estimates of the statistical uncertainties. We see for $t = 1$ that the SPMC and MC methods are equivalent. For $t = 100$, however, the SPMC method represents a significant improvement over the direct MC approach, reducing the statistical error by approximately a factor of 6. Recalling that the overall statistical error in MC methods varies as the square root of the total number of points, the SPMC results thus represent roughly a factor of 40 savings in effort relative to the direct MC approach for $t = 100$. This savings grows as t increases and is over 60,000 by the time t reaches 10^5.

Figure 2 expands our previous comments concerning the range of useful ε values. Shown in Figure 2 is the variation of the SPMC results for the preceding example with respect to ε. All results in Figure 2 are for $t = 10^5$ and were computed with a common number of MC points. The variation of the statistical errors with respect to ε are thus a reflection of the variation of the efficiency of the SPMC method with respect to the choice of ε. Based on our previous discussions (Section II), we would like to pick an ε value as large as possible (to achieve maximum smoothing of the integrand) consistent with the requirement that ε be small relative to the length scale of $\rho(x)$. We anticipate, therefore, that useful ε values will lie in the range of 0.1 or less for the present example. As can be seen, the general trend is that larger ε values produce smaller statistical errors. Explicitly, the ratio of the error for $\varepsilon = 0$ (the direct MC method) and the error for $\varepsilon = 0.1$ is roughly 250, indicating a corresponding savings of effort of approximately 60,000 for the SPMC method relative to

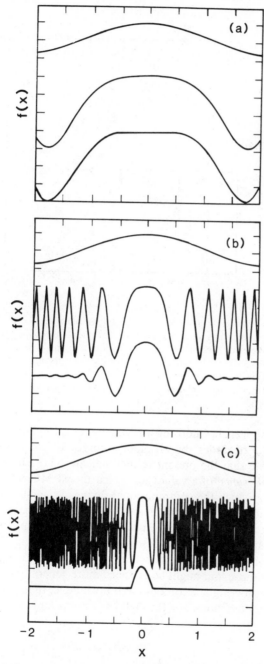

Figure 1. Plots, from top to bottom in each panel, of $\exp(-x^2/2)$ and the real parts of $\exp(itx^2)$ and $\langle \exp(itx^2) \rangle_\varepsilon$ as functions of x for various values of t. The average, $\langle \exp(itx^2) \rangle_\varepsilon$, was computed by means of the gaussian average, Eq. 2.14, with $\varepsilon = 0.1$ in all cases. The three panels correspond, from top to bottom, to t values of 1, 10, and 100.

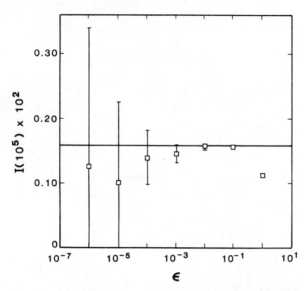

Figure 2. Plots of real parts of SPMC results for the gaussian integral in Section III, Eq. 3.1, for various values of ε. All results were obtained using the same number of MC points (1600) with $t = 10^5$ and $\sigma = 1$. The error bars shown are one-standard-deviation estimates of the statistical uncertainty. Equation 3.3 was utilized to construct the SPMC integrand, $\langle \exp(itx^2) \rangle_\varepsilon$. Although not shown, the $\varepsilon = 0$ results (ordinary MC) are essentially the same as those for $\varepsilon = 10^{-6}$. The solid line represents the exact analytic value of the real part of the integral for $t = 10^5$, 0.158×10^{-2}.

the direct MC approach. Choosing an ε value that is too large relative to the length scale of ρ degrades the absolute accuracy of the results in accord with 3.4. We note that the present results indicate there exists a broad and *easily identifiable range* of ε values for which the SPMC estimate of $I(t)$ is accurate and for which the computational savings is significant.

An important question for the present example is to what extent we have used the answer to get the answer. That is, since we have used the exact result for the Gaussian average in Eq. 3.3 to represent the local average of the integrand, we might question how accurately the computational savings thus obtained reflect those that might be expected more generally. We explore this question by considering a second numerical example, that of evaluating the zero-order Bessel function $J_0(t)$ through its integral representation (19),

$$J_0(t) = \frac{\displaystyle\int_0^\pi d\theta \exp\left[it\cos(\theta)\right]}{\displaystyle\int_0^\pi d\theta}, \qquad (3.6a)$$

and through its SPMC representation,

$$J_{0,\text{SPMC}}(t) = \frac{\int_0^\pi d\theta \, \langle \exp\left[it\cos(\theta)\right] \rangle_\varepsilon}{\int_0^\pi d\theta}. \tag{3.6b}$$

The MC and SPMC estimates of $J_0(t)$ are obtained by generating N θ values randomly from the interval $(0, \pi)$, giving

$$J_{0,\text{MC}}(t) = \sum_{n=1}^N \exp\left[it\cos(\theta_n)\right]/N \tag{3.7a}$$

and

$$J_{0,\text{SPMC}}(t) = \sum_{n=1}^N \langle \exp\left[it\cos(\theta_n)\right] \rangle_\varepsilon /N. \tag{3.7b}$$

We again utilize a Gaussian smear for the ε average in Eq. 3.7b,

$$\langle \exp\left[it\cos(\theta_n)\right] \rangle_\varepsilon = \frac{\int_{-\infty}^\infty dy \exp(-y^2/2\varepsilon^2) \exp\left[it\cos(\theta_n + y)\right]}{\int_{-\infty}^\infty dy \exp(-y^2/2\varepsilon^2)}. \tag{3.8}$$

For the present example we approximate the average in Eq. 3.8 by means of a second-order gradient expression (Eq. 2.15). Figure 3 presents sample numerical results documenting the improvement possible using the SPMC method for the present problem. We plot there the results of SPMC calculations of $J_0(t)$ for a large value of t (in this case, $t = 10^6$) for various values of the averaging width, ε. For each ε value approximately 10,000 MC points were used to estimate $J_0(t)$. It is clear from Figure 3 that there again exists a large and easily identifiable ε range over which the statistical error in the SPMC results is, for the same number of MC points, far smaller than that for the direct MC method ($\varepsilon = 0$) and for which $J_{0,\text{SPMC}}$ approximates J_0 accurately. The savings in effort amounts to approximately a factor of 40,000 for the $\varepsilon = 10^{-2}$ results, for example. From Figure 3 we see that the upper limit of useful ε values for the present example is approximately 10^{-2}, a somewhat smaller value than would have been anticipated based on the general length scale arguments previously presented. It has been our experience that this is a general phenomenon. That is, we have generally found that the introduction of approximations into the local averaging process (gradient expansions,

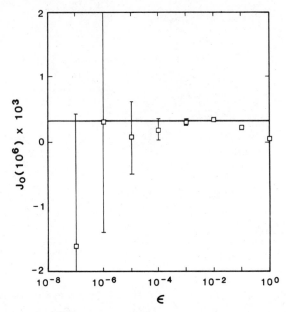

Figure 3. Plots of SPMC estimates of the zero-order Bessel function, $J_0(t)$, for $t = 10^6$ as a function of the averaging width ε. All estimates were calculated with the same number of MC points (10^4) and all used the second-order gradient form of SPMC averaging (Eqs. 3.8 and 2.15). The solid line corresponds to the stationary phase value of $J_0(10^6)$, 0.331×10^{-3}.

cumulant expansions, etc.) reduces slightly the maximum ε values appropriate for use in the SPMC approach.

We close with a simple two-dimensional example chosen to demonstrate that the present approach is not limited in any special way to problems in one dimension. Specifically, we examine the Gaussian average of the form

$$I(t) = \frac{\displaystyle\int_{-\infty}^{\infty} dx\, dy \exp\left[-(x^2 + y^2)/2\right] \exp\left[it(x - y)^2\right]}{\displaystyle\int_{-\infty}^{\infty} dx\, dy \exp\left[-(x^2 + y^2)/2\right]}. \tag{3.9}$$

Obviously, we could exploit the special structure of the present problem, utilizing the fact that the complex exponential depends only on the difference of the two coordinates to reduce the present example to that in Eq. 3.1. We choose, however, to evaluate 3.9 by means of the SPMC method without taking special advantage of this fact. We write the SPMC estimate of $I(t)$

as

$$I(t) = \sum_{n=1}^{N} \langle \exp[it(x - y)^2] \rangle_{n,\varepsilon}/N \qquad (3.10)$$

and use a two-dimensional Gaussian smear of width ε as the local average of the complex exponential in the vincinity of the MC points x_n and y_n. Explicitly, we have

$$\langle \exp[it(x - y)^2] \rangle_{n,\varepsilon} = \exp[it(x_n - y_n)^2/(1 - 4it\varepsilon^2)]/\sqrt{1 - 4it\varepsilon^2}. \qquad (3.11)$$

From 3.11 we see that the exact result for 3.9 is $1/\sqrt{1 - 4it}$. As before, we find that the SPMC method can considerably simplify the numerical evaluation of the integral under study. For example, choosing $\varepsilon = 0.1$, we find for $t = 10^5$ that with 10^4 points the MC and SPMC estimates of the real part of $I(t)$ are 0.845 ± 0.605 and 0.115 ± 0.003, respectively, versus an exact value of 0.112. The savings in effort of the SPMC approach relative to the direct MC method for this example is thus in excess of 60,000, in line with the improvements seen for the one-dimensional results considered. We emphasize again that in implementing the SPMC method for the present example, we have not taken special advantage of the form of 3.9 to reduce the formal two-dimensional integral to an equivalent one-dimensional one.

IV. SUMMARY

We have introduced in the present work modifications of ordinary MC techniques aimed at the evaluation of large-dimensional averages of highly oscillatory integrands. Such problems arise in a variety of contexts, including, for example, the path integral treatment of time-dependent phenomena. In the present approach we have exploited a length scale separation in a manner suggested by an analysis of ordinary stationary phase methods. The resulting SPMC methods appear to offer a general approach to the problem under discussion and, for the numerical examples considered, yield a considerable computational savings relative to conventional statistical methods. We are currently examining the utility of these approaches in real-time finite-temperature path integral problems.

Acknowledgments

We would like to thank Professors W. H. Miller and R. D. Coalson for useful discussions concerning this work. We would also like to thank Professor H. Rabitz for bringing items in Ref. 15 to our attention. Work at the University of Rhode Island was supported in part by grants from Research Corporation, the University of Rhode Island Academic Computer Center, and the

University of Rhode Island Engineering Computer Laboratory. This chapter reflects work performed under the auspices of the U.S. Department of Energy.

References

1. See, for example, J. P. Valleau and S. G. Whittington, "A Guide to Monte Carlo for Statistical Mechanics: Highways," and J. P. Valleau and G. M. Torrie, "A Guide to Monte Carlo for Statistical Mechanics: Byways," in *Modern Theoretical Chemistry*, Vol. 5, B. J. Berne, Ed., Plenum, New York, 1977.

2. For a recent cross section of work in this area see the Proceedings of the Conference on Frontiers of Quantum Monte Carlo, published as *J. Stat. Phys.* **43** (5/6), 729–1244 (1986) and references therein.

3. For a partial discussion of current activity in this area, see B. J. Berne and D. Thirumalai, *Ann. Rev. Phys. Chem.* **37**, 401 (1986).

4. E. C. Behrman, G. A. Jongeward, and P. G. Wolynes, *J. Chem. Phys.* **79**, 6277 (1983); **83**, 668 (1985); E. C. Behrman and P. G. Wolynes, *J. Chem. Phys.* **83**, 5863 (1985).

5. D. Thirumalai and B. J. Berne, *J. Chem. Phys.* **81**, 2512 (1984); **79**, 5029 (1983).

6. W. H. Miller, S. D. Schwartz, and J. W. Tromp, *J. Chem. Phys.* **79**, 4889 (1983).

7. J. D. Doll, *J. Chem. Phys.* **81**, 3536 (1984).

8. R. D. Coalson, *J. Chem. Phys.* **85**, 926 (1986).

9. W. H. Miller, S. D. Schwartz, and J. W. Tromp, *J. Chem. Phys.* **79**, 4889 (1983); K. Yamashita and W. H. Miller, *J. Chem. Phys.* **82**, 5475 (1985).

10. A. Nauts and R. E. Wyatt, *Phys. Rev. Lett.* **51**, 2238 (1983); R. A. Friesner and R. E. Wyatt, *J. Chem. Phys.* **82**, 1973 (1985); R. E. Wyatt, *Chem. Phys. Lett.* **121**, 301 (1985).

11. See, for example, W. H. Miller, *Science* **233**, 171 (1986) and references therein.

12. R. D. Coalson, D. L. Freeman, and J. D. Doll, *J. Chem. Phys.* **85**, 4567 (1986) and references therein.

13. E. L. Pollock and D. M. Ceperley, *Phys. Rev.* **B30**, 2555 (1984).

14. G. G. Stokes, *Mathematical and Physical Papers*, Vol. 2, Cambridge University Press, City, 1883, p. 329; G. F. Carrier, M. Krook, and C. E. Pearson, *Functions of a Complex Variable*, McGraw-Hill, New York, 1966, pp. 272–275.

15. After the present work was completed we learned of similar ideas developed in a different physical context by G. Fisanick-Englot and H. Rabitz, *J. Chem. Phys.* **62**, 2747 (1975), and S. M. Tarr, J. Sampson, and H. Rabitz, *J. Chem. Phys.* **64**, 5291 (1976).

16. See, for example, J. M. Ziman, *Models of Disorder*, Cambridge University Press, 1979, pp. 2–3.

17. R. Kubo, *J. Phys. Soc. Jpn.* **17**, 1100 (1962); N. van Kampen, *Phys. Rep.* **24**, 171 (1976).

18. J. D. Doll and D. L. Freeman, *Science* **234**, 1356 (1986) and references therein.

19. G. N. Watson, *A Treatise on the Theory of Bessel Functions*, Cambridge University Press, 1966, Chapters 2 and 3; M. Abramowitz and L. Stegun, *Handbook of Mathematical Functions*, National Bureau of Standards Applied Mathematics Series, No. 55, U. S. Government Printing Office, Washington, DC, 1966, Eq. (9.1.21).

SPECTROSCOPY AND RELAXATION

CHAPTER VIII

APPLICATIONS OF THE RIEMANN PRODUCT INTEGRAL METHOD TO SPECTROSCOPIC PROBLEMS

WILLIAM J. MEATH, R. A. THURAISINGHAM,
and MARY ANN KMETIC**

Department of Chemistry, University of Western Ontario, London, Ontario,
Canada N6A 5B7

CONTENTS

**The authors also are associated with the Centre for Interdisciplinary Studies in Chemical Physics.

I. INTRODUCTION

In this chapter the Riemann product integral method for solving the time-dependent wave equation is reviewed with a view toward its application to the interaction of multilevel atoms and molecules with applied time-dependent electromagnetic and static electric fields. Various applications of the method are discussed that illustrate the effects of static electric fields, and in particular permanent dipole moments, on the single- and multiphoton spectra of atoms and molecules. Analytic rotating-wave approximations (RWA) for the molecular spectral profiles are used to help interpret the results; these are the molecular analogue of the usual RWA for atoms.

Section II contains a brief discussion of the solution of the time-dependent wave equation for arbitrary time-dependent perturbations using Riemann product integral techniques. The solution of the wave equation is obtained by a numerical algorithm for the evolution operator associated with the time-dependent system, which is based on subdividing the time interval of interest, $[t_0, t]$, into a (sufficient) number of small subintervals. The number of subintervals is chosen so that the time-dependent Hamiltonian is essentially constant over each subinterval, leading to simple results for the subinterval evolution operators; the group property of the evolution operator is then used to cascade the solution from the initial time t_0 to the final time t. The discussion is specialized in Section III to the interaction of atoms and molecules with electromagnetic radiation. Special emphasis is placed on continuous-wave or sinusoidal electromagnetic fields. Here the solution of the time-dependent wave equation over the initial period of the Hamiltonian can be used in an iterative manner to obtain the complete solution for arbitrary times. By recasting this iterative solution into Floquet form, convenient expressions are obtained for the efficient evaluation of both atomic and molecular temporal and steady-state spectra, which again require only the solution of the time-dependent wave equation over the first period of the Hamiltonian. The techniques discussed avoid the use of Floquet secular equations. The Riemann product integral method is particularly efficient for problems involving sinusoidal electromagnetic fields since it need be used only over the first period of the Hamiltonian and seems to be essentially independent of the strengths and frequencies of the fields.

Section IV contains a discussion, with numerical examples, of the effects of permanent dipoles and/or static electric fields on single- and multiphoton resonance profiles and is concerned with situations where the absorbing molecules assume fixed configurations relative to the directions of the applied electromagnetic and static electric fields. A qualitative discussion of the effects of permanent dipoles is given through the use of analytic RWA expressions for the molecular spectral profiles, which are also used to help interpret exact numerical calculations of the spectra. It is shown, for example, that the presence of permanent dipoles can drastically reduce the coupling between the molecule and the electromagnetic field leading to single- and multiphoton resonance profiles that are much narrower than would be expected in the absence of permanent moments. Permanent dipoles can also produce resonance positions to low frequency relative to those predicted for weak electromagnetic fields, in contradistinction to the Block–Siegert shift to the high frequency associated with atoms. Both the presence of permanent dipoles or a static electric field can induce even-, as well as the usual odd-, photon resonances. In the absence of permanent dipoles the resonance positions are always to high frequency relative to those expected in the weak electromagnetic field limit in the absence of a static field; part of this shift corresponds to a Block–Siegert effect and part is due to the static field. In the presence of both permanent dipoles and a static electric field, the resonance positions can be either to low or high frequency relative to the weak-field limit.

Techniques for calculating orientational or rotationally averaged molecular spectra are outlined in Section V. Such rotational averaging is important in applications where the absorbing molecule undergoes free or hindered rotations, for example, gas phase molecules in the absence or presence of an "aligning" field. The resonance profiles for freely rotating molecules are discussed, and in some cases the structure of the absorption spectra associated with fixed-configuration spectra can be lost, whereas in others the width and, especially for multiphoton resonances, the height of the profiles can be reduced by rotational averaging. What is observed depends on the size of the coupling between the permanent dipoles and the electromagnetic field relative to the absorption frequencies. Examples are also given of hindered rotationally averaged spectra, which are functions of temperature, using gas phase molecules in a static electric field and molecules adsorbed on a surface as models. These illustrate, respectively, the sensitivity of the single- and multiphoton molecular resonance profiles to the molecular transition and permanent moments, and their relative orientation, and to the orientation of the absorbing molecule relative to the surface. Both single- and two-photon resonance profiles of adsorbed molecules are considered, at two levels of sophistication, using N_2 and CO adsorbed on graphite as models. Initially the role of the surface is taken to be as a source of a holding potential energy that hinders the rotation of the adsorbed molecule. Then the modification of the electromagnetic field caused by the presence of the surface is taken into account. Broadening mechanisms for the spectral profiles of the adsorbed molecules are also discussed.

Throughout this chapter the interpretation of the single- and multiphoton resonance profiles or absorption spectra, as a function of the molecular and applied field parameters characterizing the problems studied, is facilitated by using two-level ideas and in particular analytic RWA expressions for the resonance profiles and for the molecule– (or atom–) electromagnetic field coupling parameters. Many of the effects discussed would be exceedingly difficult to understand otherwise. For example, the molecular RWA results indicate that the heteronuclear molecule–electromagnetic field coupling can be *reduced* as the strength of the electromagnetic field *increases*, in contradistinction to atoms or homonuclear molecules, a prediction verified by exact model calculations.

Atomic units are used unless indicated otherwise. The atomic units of energy, circular frequency, length, dipole moment, and electric field strength are E_H, $E_H/\hbar$, a_0, ea_0, and $E_H e^{-1} a_0^{-1}$, where E_H is the hartree of energy, a_0 is the Bohr radius, e is the absolute value of the charge of an electron, and $\hbar$ is the reduced Planck constant.

II. SOLUTION OF THE TIME-DEPENDENT WAVE EQUATION USING THE RIEMANN PRODUCT INTEGRAL METHOD

In this section an efficient computational algorithm for the solution of the time-dependent wave equation is discussed with a view toward application to the interaction of multilevel atoms and molecules with applied static and time-dependent electromagnetic fields (Section III). Our approach is based on the Riemann product integral representation (1) of the evolution operator for the system coupled with Frazer's method of mean coefficients (2). The presentation follows that of Thomas and Meath (3); references to related work are given in the sequel.

A. The Time-Dependent Wave Equation and the Evolution Operator

The time-dependent wavefunction for an N-level atom or molecule interacting with a time-dependent electric field $\mathscr{E}(t)$ and a static electric field $\mathscr{E}_s$ can be written as

$$\Psi(\mathbf{r}, t) = \mathbf{\Phi}^T(\mathbf{r})\mathbf{a}(t) \tag{2.1}$$

with

$$\mathbf{a}^\dagger(t)\mathbf{a}(t) = 1 \tag{2.2}$$

where the column vector $\mathbf{a}(t)$ and the row vector $\mathbf{\Phi}^T(\mathbf{r})$ are defined by $(\mathbf{a})_j = a_j(t)$ and $(\mathbf{\Phi}^T(\mathbf{r}))_j = \phi_j(\mathbf{r})$, where ϕ_j is the wavefunction for the jth stationary state of the unperturbed system corresponding to energy E_j. In the semiclassical dipole approximation, and in the Schrödinger representation, the coefficients $a_j(t)$ are obtained by solving, subject to initial conditions $\mathbf{a}(t_0) = \mathbf{A}$, where the

A_j are known constants, the system of coupled first-order differential equations given by

$$\frac{d}{dt}\mathbf{a}(t) = \mathbf{C}(t)\mathbf{a}(t) \tag{2.3}$$

where

$$\mathbf{C}(t) = -i\mathbf{H}(t) = -i[\mathbf{E} - \boldsymbol{\mathscr{E}}(t)\cdot\boldsymbol{\mu} - \boldsymbol{\mathscr{E}}_s\cdot\boldsymbol{\mu}] \tag{2.4}$$

Here $\mathbf{H}(t)$ is the time-dependent Hamiltonian for the interaction of the atom or molecule with the applied fields, the square matrices $\mathbf{E}$ and $\boldsymbol{\mu}$ are defined by $(\mathbf{E})_{ij} = E_j\delta_{ij}$ and $(\boldsymbol{\mu})_{ij} = \boldsymbol{\mu}_{ij} = \langle \phi_i|\boldsymbol{\mu}|\phi_j\rangle$, and $\boldsymbol{\mu}$ is the electric dipole moment operator for the atom or molecule.

In most of this chapter we will be interested in the interaction of electromagnetic radiation with an atom or molecule, sometimes in the presence of a static electric field $\boldsymbol{\mathscr{E}}_s = \hat{e}_s\mathscr{E}_s$, where $\hat{e}_s$ is the unit vector specifying the direction of the static field of magnitude $\mathscr{E}_s$. The time-dependent electric field is taken as

$$\boldsymbol{\mathscr{E}}(t) = \hat{e}f(t)\mathscr{E}_0\cos(\omega t + \delta) \tag{2.5}$$

where $\boldsymbol{\mathscr{E}}_0 = \hat{e}\mathscr{E}_0$, ω, δ, and $\hat{e}$ are the vector amplitude, circular frequency, phase, and direction of polarization, respectively, of a sinusoidal electric field modulated by a possible time-dependent pulse envelope $f(t)$. The method of solution for Eq. 2.3, outlined in what follows is, of course, applicable to any reasonable choice of $\boldsymbol{\mathscr{E}}(t)$.

The solution of Eq. 2.3 can be written in terms of the evolution operator $U(t, t_0)$,

$$\mathbf{a}(t) = \mathbf{U}(t, t_0)\mathbf{a}(t_0) \tag{2.6}$$

where the evolution operator satisfies

$$\frac{d}{dt}\mathbf{U}(t, t_0) = \mathbf{C}(t)\mathbf{U}(t, t_0) \tag{2.7}$$

subject to the initial conditions

$$\mathbf{U}(t_0, t_0) = \mathbf{I} \tag{2.8}$$

with $\mathbf{I}$ being the $N \times N$ unit matrix. A more complete discussion of the properties of the system of coupled differential equations given by Eq. 2.3 and of the evolution operator or, as it is often called, the solution or integral matrix for Eq. 2.3 can be found in various sources (3) (e.g., 4–6). The unitary nature of $U(t, t_0)$ and its group property

$$U(t, t_0) = U(t, t')U(t', t_0) \tag{2.9}$$

are of particular use in the development of the Riemann product integral approach for the construction of the evolution operator.

B. The Riemann Product Integral Representation of the Evolution Operator

To construct the evolution operator $U(t, t_0)$, we begin by subdividing the time interval $[t_0, t]$ into n subintervals by means of the points $t_0 < t_1 < t_2 < t_3 \cdots < t_{n-1} < t_n = t$. Using Eq. 2.9, one can then write

$$U(t, t_0) = U(t, t_{n-1})U(t_{n-1}, t_{n-2}) \cdots U(t_2, t_1)U(t_1, t_0)$$

$$= T \prod_{k=1}^{n} U(t_k, t_{k-1}) \tag{2.10}$$

where the time-ordering operator T arranges the product in chronological order from right to left. Now in each subinterval $[t_{k-1}, t_k]$, of length $\Delta t_k = t_k - t_{k-1}$, choose an arbitrary point t'_k for $k = 1, 2, \ldots, n$. By taking Δt_k sufficiently small, for example, by choosing n sufficiently large, $C(t)$ can be assumed to be essentially constant over the interval $[t_{k-1}, t_k]$ with a value $C(t'_k)$. Equation 2.7 is easy to solve in such circumstances to give

$$U(t_k, t_{k-1}) = \exp[C(t'_k)\Delta t_k]. \tag{2.11}$$

The arbitrary point t'_k in $[t_{k-1}, t_k]$ can always be chosen, via the mean value theorem, such that

$$C(t'_k) = \frac{1}{\Delta t_k} \int_{t_{k-1}}^{t_k} C(t)\, dt = \frac{1}{\Delta t_k} \mathscr{C}(k) \tag{2.12}$$

Using Eqs. 2.12 and 2.11 in Eq. 2.10 then yields

$$U(t, t_0) = T \prod_{k=1}^{n} \exp\left[\int_{t_{k-1}}^{t_k} C(t)\, dt\right] \tag{2.13}$$

This is the operational expression for the Riemann product integral representation of the evolution operator. Following Volterra and Hostinsky (7), the formal definition of the Riemann product integral of $\mathbf{C}(t)$ over $[t_0, t]$ is given by

$$\mathbf{U}(t, t_0) = \lim_{m \to 0} T \prod_{k=1}^{n} \exp[\mathscr{C}(k)] \qquad (2.14)$$

where $m = \max \Delta t_k$ is the length of the longest subinterval $[t_{k-1}, t_k]$. The limit $m \to 0$ implies limit $n \to \infty$.

Thus, to compute the evolution operator using its Riemann product integral representation, the interval $[t_0, t]$ is decomposed into a discrete number of subintervals, the exponential matrix defined by Eqs. 2.11 and 2.12 is evaluated over each subinterval, and the group property of the evolution operator is applied over adjacent subintervals. The actual discretization of $[t_0, t]$ used depends on the problem, on the nature of $\mathbf{C}(t)$ as a function of t, and on the desired accuracy of the solution. Methods for carrying out these calculations have been discussed by Thomas and Meath (3) both for sinusoidal and nonsinusoidal perturbing electromagnetic fields. The approach is ideally suited for the use of array processors. The Riemann product integral representation of $\mathbf{U}(t, t_0)$ is unitary, if $\mathbf{H}(t)$ is self-adjoint, by construction, and so the unitarity of the solution provides no check for the accuracy of the method as applied to a particular problem. This can only be deduced by varying n or m, see Eqs. 2.13 and 2.14.

The properties of Riemann product integrals, and the close analogy of their algebra with that of Riemann integrals, has been reviewed and extended recently by Dollard and Friedman (1, 8), who emphasize their usefulness in constructing solutions to the equations of motion arising in scattering problems. Analogous techniques have been used in applications to stability problems, for example, helicopter rotor dynamics [see Ref. 9, and Chapter IV of this volume]. The evaluation of $\mathbf{U}(t, t_0)$ using Eq. 2.13 appears to have been originally suggested by Frazer in 1931 (see also Ref. 2), who applied the method to second-order differential equations that arise in engineering applications. It is clear that Eq. 2.11 can be obtained by using the first Magnus solution for $\mathbf{U}(t_k, t_{k-1})$ with the neglect of the higher order Magnus terms being justified by choosing n large enough. In this sense the method discussed here was used by Walker and Preston (10) to investigate multiple-photon excitation of the anharmonic oscillator and by Dougherty, Augustin, and Rabitz (11) in their treatment of intermolecular energy transfer in the presence of radiation. It also has features similar in spirit to the work of Wei and Norman (12) and to the stroboscopic method of Minorsky (13), which has been applied, for example, to the study of transitions induced by resonant fields in N-level systems by Askar (14).

The Riemann product integral method yields results for the time-dependent coefficients $a_j(t)$ that correspond to the weighting or amplitude factors for the stationary state wavefunctions in the expansion of the time-dependent wavefunction given by Eq. 2.1. Once these coefficients are known, one can evaluate the probability of finding the atom or molecule in a particular state j either as a function of time or in the steady-state limit. The time-resolved (time-dependent) state amplitude matrix is given by Eq. 2.6 and the time-resolved induced transition probability for excitations to state j for a given N-level system prepared initially with state amplitudes $\mathbf{a}(t_0)$ is given by $|a_j(t)|^2$.

III. THE INTERACTION OF ATOMS AND MOLECULES WITH ELECTROMAGNETIC RADIATION

In the remainder of this chapter we will be concerned with the interaction of an atom or molecule, in the presence or absence of an applied static electric field, with electromagnetic radiation where the time-dependent electric field is given by Eq. 2.5. Apparently, there have been relatively few applications of the Riemann product integration scheme to such problems until quite recently; see, for example, Thomas and Meath (3), and Hirschfelder and Pyzalski (15) and their discussion of the analogous "time-slicer" approach (see also Chapter 1), and what follows.

A. General Considerations

When an atom or molecule interacts with a time-dependent electromagnetic field, given by Eq. 2.5, and a static electric field, the state amplitudes $a_j(t)$ and the transition probabilities $|a_j(t)|^2$ can depend on many variables and/or parameters, for example, the directions $(\hat{e}, \hat{e}_s)$ and magnitudes $(\mathscr{E}_0, \mathscr{E}_s)$ of the applied electromagnetic or static fields, the circular frequency ω, the phase δ, and the pulse envelope $f(t)$, as well as on the choice of initial conditions $\mathbf{a}(t_0)$, the dipole moment matrix $\boldsymbol{\mu}$, the number N of energy levels and their spacing, and, for example, the number n of time steps used in the construction of the Riemann product representation of the evolution operator for the problem. For molecules the results also depend on the relative orientation of the molecule with respect to the applied fields. Clearly, even for small N, *many* calculations are needed for a complete investigation of a given problem, and one therefore needs an efficient method for solving the time-dependent wave equation as a function of the various variables discussed in the preceding sentences. The Riemann product integral approach offers such a method, particularly for the interaction of an atom or molecule with a continuous-wave laser, $f(t) = 1$ in Eq. 2.5; see Section III.B for more details. It avoids many of the problems (3) associated with perturbative and/or Magnus approaches; see also, for example, Pechukas and Light (16), Mehrotra and Boggs (17), and Langhoff, Epstein, and Karplus (18).

The method outlined in Section II.B is capable of yielding essentially exact results for the transition probabilities, bearing in mind that N is finite, the semiclassical electric dipole approximation is used, and we have neglected relaxation or decay processes. The latter can be treated phenomenologically, either through a uniform relaxation mechanism (19, 20) or by introducing appropriate widths (21) for the energy levels being considered; see (22) for more details. The latter amounts to replacing the real diagonal matrix $\mathbf{E}$ in Eq. 2.4 by the complex diagonal matrix $\mathbf{E} - i\gamma$, where γ contains the widths for the relevant energy levels; the method of Section II.B can be used for this problem, bearing in mind that $\mathbf{U}$ is no longer unitary. Decay processes are neglected in this chapter, and so our results will be suitable for use if the effects of the perturbing time-dependent field occur over time intervals that are short relative to the important relaxation times involved in the system.

For nonsinusoidal time-dependent perturbations (e.g., pulsed electromagnetic fields or pulsed lasers), the solution for the evolution operator must be developed directly for all time. This has been discussed (3, 23) using the Riemann product integral method for a Gaussian pulse corresponding to $f(t) = \exp(-t^2/\tau_p^2)$, where τ_p is related to the pulse duration time. Explicit calculations for a two-level system were performed and used to discuss, for example, the dependence of time-resolved and steady-state frequency sweep spectra on the characteristics of the pulse.

For sinusoidal time-dependent perturbations (e.g., sinusoidal electromagnetic fields or continuous-wave lasers), the periodicity of the perturbing field can be used to greatly simplify the solution of Eq. 2.7 and the evaluation of the transition probabilities and their phase and/or long-time (steady-state) averages. This is the case of interest in the rest of this chapter, and so some of the computational aspects related to this problem will be reviewed briefly next.

B. Sinusoidal Electromagnetic Fields

Here $f(t) = 1$ in Eq. 2.5, corresponding to an infinite pulse duration. Since $\mathscr{E}(t)$, and hence $\mathbf{C}(t)$ given by Eq. 2.4, is periodic for all $t \geq t_0$, the evaluation of $\mathbf{U}(t, t_0)$ and the induced transition probabilities for all t requires the knowledge of $\mathbf{U}(t, t_0)$ over only the time interval $[0, 2\pi/\omega]$, where we take $t_0 = 0$ for this continuous-wave problem. In this section only a brief discussion of the working equations for the evaluation of the relevant transition probabilities will be given; complete derivations with various applications can be found, for example, in Shirley (24), Salzman (25), Burrows and Salzman (26), Moloney and Meath (20, 21, 27), Dion and Hirschfelder (28), Chu (29), and references therein. It is relevant to note that the sinusoidal part of Eq. 2.5 corresponds to plane-polarized light; the techniques discussed apply to any periodic perturbation, for example, a molecule interacting with circularly polarized light.

Making use of the periodic properties of $C(t)$, one can show that

$$\mathbf{a}(\theta + 2s\pi) = \mathbf{U}(\theta, 0)[\mathbf{U}(2\pi, 0)]^s \mathbf{c}_0(\delta) \qquad 0 \leqslant \theta \leqslant 2\pi \qquad (3.1)$$

where $s = 0, 1, 2, \ldots$, and $\theta = \omega t + \delta$,

$$\mathbf{c}_0(\delta) = \mathbf{U}^{-1}(\delta, 0)\mathbf{a}(0) \qquad (3.2)$$

and $\mathbf{a}(0) = \mathbf{A}$ is the initial ($t = 0$) value of the state amplitude matrix. It is clear from Eq. 3.1 that the evaluation of the state amplitudes as a function of time requires the knowledge of the evolution operator $\mathbf{U}(\theta, 0)$ only on the θ interval $[0, 2\pi]$; the derivation (21, 27) of Eq. 3.1 is independent of Floquet's theorem (24, 21).

The expression for the state amplitudes given by Eq. 3.1 can be written (21) in Floquet form:

$$\mathbf{a}(\theta + 2s\pi) = \mathbf{Z}(\theta) \exp\left[i\Delta(\theta + 2s\pi)\right]\mathbf{b}_0(\delta) \qquad (3.3)$$

where

$$\mathbf{b}_0(\delta) = \mathbf{S}^{-1}\mathbf{c}_0(\delta) \qquad (3.4)$$

$$\mathbf{Z}(\theta) = \mathbf{U}(\theta, 0)\mathbf{S} \exp(-i\Delta\theta) = \mathbf{Z}(\theta + 2s\pi) \qquad (3.5)$$

and Δ is a diagonal characteristic exponent matrix whose elements are related to the eigenvalues $\lambda_j = \exp(i\Delta_{jj}2\pi)$ of the unitary matrix $\mathbf{U}(2\pi, 0)$ and $\mathbf{S}$ is a matrix whose columns are the orthonormal eigenvectors $\mathbf{S}_j$ of $\mathbf{U}(2\pi, 0)$ corresponding to the λ_j. The Floquet form of the solution for the state amplitudes is very convenient for evaluating the phase and/or long time-averaged transition probabilities for the atom or molecule under consideration. These calculations are carried out using the following expression (21) for the temporal-phase-dependent transition probability $|a_j(t)|^2$,

$$P_j(t) = |a_j(t)|^2 = \sum_{q,s} Z_{jq}(\theta)\beta_{qs}(\theta, \delta)Z_{js}^*(\theta) \qquad (3.6)$$

where $(\Delta_{jj} = \Delta_j)$

$$\beta_{qs}(\theta, \delta) = b_{0q}(\delta)b_{0s}^*(\delta) \exp\left[i(\Delta_q - \Delta_s)\theta\right] \qquad (3.7)$$

The phase-averaged temporal and the phase and long-time-averaged transition probabilites are defined, respectively, by

$$\bar{P}_j(t) = \frac{1}{2\pi} \int_0^{2\pi} |a_j(t)|^2 \, d\delta \qquad (3.8)$$

$$\bar{P}_j = \lim_{\tau \to \infty} \frac{1}{\tau} \int_0^\tau \bar{P}_j(t)\, dt \tag{3.9}$$

The physically meaningful temporal behavior of the transition probability is normally independent of the phase δ and corresponds to the phase-averaged result of Eq. 3.8; in certain cases (see Ref. 30) the phase average can be taken over a domain for δ less than $[0, 2\pi]$. The long-time or steady-state average of this result, Eq. 3.9, often corresponds to the physically observed spectrum of the atom or molecule (31, 32). Using Eqs. 3.6–3.8 yields

$$\bar{P}_j(t) = \sum_{q,s} \exp\left[i(\Delta_q - \Delta_s)\omega t\right] \beta_{qs}^j(\omega t) \tag{3.10}$$

where

$$\beta_{qs}^j(\omega t) = \frac{1}{2\pi} \int_0^{2\pi} Z_{jq}(\omega t + \delta) Z_{js}^*(\omega t + \delta) b_{0q}(\delta) b_{0s}^*(\delta) \exp\left[i(\Delta_q - \Delta_s)\delta\right] d\delta \tag{3.11}$$

and because of the periodicity of $\mathbf{Z}(\theta)$ (see Eq. 3.5),

$$\beta_{qs}^j(\omega t) = \beta_{qs}^j(\omega t + 2s\pi) \tag{3.12}$$

Substituting Eq. 3.10 into Eq. 3.9 and using Eq. 3.12 leads to the following convenient expression for the steady-state induced transition probability (21):

$$\bar{P}_j = \sum_q \frac{1}{2\pi} \int_0^{2\pi} \beta_{qq}^j(\theta')\, d\theta' \qquad \theta' = \omega t \tag{3.13}$$

It is clear from Eqs. 3.1 and 3.10–3.13 that the phase- and time-dependent, the time-dependent phase-averaged, and the phase- and long-time-averaged transition probabilities can all be readily evaluated once the evolution operator is known over the first period of the time-dependent Hamiltonian or $\mathbf{C}(t)$.

Operationally, for a given set of parameters defining the problem, the calculations are carried out using the following procedure:

1. Determine the evolution operator $\mathbf{U}(\theta, 0)$ on the interval $[0, 2\pi]$ in θ.
2. Diagonalize $\mathbf{U}(2\pi, 0)$ in order to obtain the eigenvalues λ_j and eigenvectors $\mathbf{S}_j$ and hence $\mathbf{S}$ and $\mathbf{S}^{-1}$.
3. Use the λ_j to evaluate the characteristic exponents using (21, 32)

$$\Delta_j = \frac{1}{2\pi} \tan^{-1}\left(\text{Im}\,\lambda_j / \text{Re}\,\lambda_j\right) \tag{3.14}$$

4. Determine $b_0(\delta)$ from Eq. 3.4 and $Z(\theta)$ from Eq. 3.5.
5. The various transition probabilities are then obtained as already discussed.

The integrals appearing in the expressions for the averaged transition probabilities have integrands that are normally very well behaved and are readily evaluated by using standard techniques. (3, 20, 21) As well as their use in evaluating transition probabilities, plots of the characteristic exponents as a function of ω can be used to predict resonance positions and widths in frequency sweep spectra and are related to dressed atom or molecule energies. (20, 24, 32, 33)

Clearly, the key step in the procedure is the evaluation of $U(\theta, 0)$ on the first period of the Hamiltonian. This can be done by solving Eq. 2.7 using any convenient method, for example, numerically (25, 26), by matching power series techniques (20, 21, 27, 33), by Floquet secular equation methods (24, 28, 29) (see also chapters V and XVII) or by Riemann product integral methods (as outlined in Section II.B). We have found the latter approach to be much more computationally efficient than most of the others. For example, its efficiency seems to be essentially independent of the strengths of the applied fields and of the values of the frequency. In many applications it is at least a factor of 5 faster than our matching power series method, which is itself quite efficient. The difficulty with the power series method is that as the couplings between the electromagnetic field and the atom or molecule become large, the $[0, 2\pi]$ interval in θ must be broken up into many subintervals to avoid using an excessive number of terms in the power series solution over a given interval; the power series have asymptotic characteristics as the coupling strengths become large. The determination of the power series satisfying appropriate matching conditions at the boundaries of the subintervals is time consuming, and these problems are enhanced by the presence of static electric fields or nonzero diagonal dipole matrix elements in the μ matrix in $C(t)$. The matching power series approach and, in particular, the Riemann product integral method avoid the use of the Floquet secular equation, which can become very large for strong fields (many dressed atom states) and/or large N and quite ill-behaved for very strong coupling strengths.

In what follows most of the examples taken from our work in the literature have been evaluated using the Riemann product integral method, and all of the new work presented has been carried out using this technique.

IV. EXAMPLES SHOWING SOME EFFECTS OF PERMANENT DIPOLE MOMENTS AND STATIC ELECTRIC FIELDS

In this section we discuss some of the features of single- and multiphoton spectra, which depend on the presence of nonzero diagonal dipole matrix elements (*permanent dipole moments*) in the species being probed by the radiation and/or on the interaction of the atom or molecule with a static electric field $\mathscr{E}_s$. Two-level examples are chosen in this section to make interpretations of the spectra relatively straightforward by using rotating-wave approximation (RWA) or Rabi-type expressions for the resonance profiles and other two-level ideas.

A. Effects of Permanent Dipole Moments ($\mathscr{E}_s = 0$)

In this section we take the static field $\mathscr{E}_s = 0$. We begin with a brief review of the usual RWA for atoms and its recent extension to molecules.

1. Rotating-Wave, or Rabi-Type, Approximation

In this approximation, the steady-state transition probability for finding the atom in a higher energy state 2, assuming it was in a lower energy state 1 at $t = 0$, is given by (24, 34)

$$\bar{P}_2 = \frac{|\boldsymbol{\mu} \cdot \mathscr{E}_0|^2}{2[(\Delta E - \omega)^2 + |\boldsymbol{\mu} \cdot \mathscr{E}_0|^2]} \tag{4.1}$$

where $\Delta E = E_2 - E_1$, $\boldsymbol{\mu} = \boldsymbol{\mu}_{12}$ is the transition dipole matrix element connecting states 1 and 2, and this expression corresponds to a one-photon transition. The validity of this result for the spectrum of the atom is most easily discussed by introducing a coupling strength parameter $b = |\boldsymbol{\mu} \cdot \mathscr{E}_0|/\Delta E$. The RWA is reliable for weak coupling problems where $b \ll 1$. Under these conditions the atom–electromagnetic field coupling is given by $|\boldsymbol{\mu} \cdot \mathscr{E}_0|$, which is equal to the half-width at half maximum of the resonance profile ($\bar{P}_2$ vs. ω), and the resonance frequency $\omega_{\text{res}} = \Delta E$; see Figure 1. As the coupling strength increases to $b \lesssim 1$ or $b > 1$, the counter rotating terms neglected in the RWA become more important, and the resonance profile given by Eq. 4.1, which is essentially an on-resonance result, becomes more and more unreliable. The widths of the spectral profiles increase and are not well represented by $2|\boldsymbol{\mu} \cdot \mathscr{E}_0|$, and Bloch–Siegert shifts in the resonance frequencies ($\omega_{\text{res}} > \Delta E$) and dynamic backgrounds in the spectra occur that are absent in the RWA. Similar comments apply for the higher photon transitions (24), where in the weak-field limit $\omega_{\text{res}} \approx \Delta E/N$ for the N-photon transition and for atoms $N = 1, 3, 5, \ldots$, assuming each energy state of the atom involved in the transition has a definite parity [which, e.g., does not apply for the hydrogen atom (33)].

The situation for molecules is considerably different for several reasons. Often $\boldsymbol{\mu}_{ii} \neq 0$, and clearly Eq. 4.1, which holds for atoms where $\boldsymbol{\mu}_{ii} = 0$, will not apply. In addition, molecules, unlike atoms, are not isotropic, and so molecular spectra will depend on the orientation of the molecule relative to the direction of the applied electromagnetic field. Thus, in many applications, calculations of molecular spectra should include averaging over the allowed

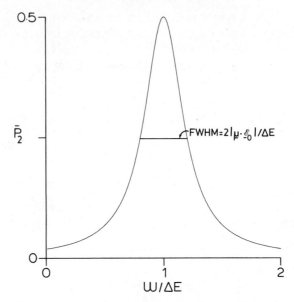

Figure 1. The single-photon RWA resonance profile; $\bar{P}_2$, given by Eq. 4.1, versus $\omega/\Delta E$ for $b = |\mathbf{\mu} \cdot \mathscr{E}_0|/\Delta E = 0.2$. Resonance occurs at $\omega/\Delta E = 1$, where $\bar{P}_2 = 0.5$ and the profile, as a function of $\omega/\Delta E$ has a full width at half maximum (FWHM) of $2b$.

orientations of the molecule relative to the field direction. This can have pronounced effects on the spectra, see Section V.

Some of the features of molecular versus atomic spectra can be explained qualitatively by using an RWA result for two-level resonance profiles, developed by Kmetic and Meath (35), which includes the effects of diagonal dipole matrix elements $\mathbf{\mu}_{ii}$, $i = 1, 2$, as well as those of the transition dipole $\mathbf{\mu} = \mathbf{\mu}_{12}$ in the treatment of the problem. The expression for the steady-state N-photon transition probability is given by

$$\bar{P}_2^{(N)} = \frac{|C(N)|^2}{2[(\Delta E - N\omega)^2 + |C(N)|^2]} \qquad N = 1, 2, 3, \ldots \qquad (4.2)$$

where the coupling between the molecule and the applied electromagnetic field is given by

$$C(N) = \mathbf{\mu} \cdot \mathscr{E}_0 [J_{N-1}(\mathbf{d} \cdot \mathscr{E}_0/\omega) + J_{N+1}(\mathbf{d} \cdot \mathscr{E}_0/\omega)]$$
$$= \mathbf{\mu} \cdot \mathscr{E}_0 (2N)(\mathbf{d} \cdot \mathscr{E}_0/\omega)^{-1} J_N(\mathbf{d} \cdot \mathscr{E}_0/\omega) \qquad (4.3)$$

with

$$\mathbf{d} = \boldsymbol{\mu}_{22} - \boldsymbol{\mu}_{11} \qquad (4.4)$$

and J_k is a Bessel function of integer order k. The usual expression for the coupling between the molecule and the field, $\boldsymbol{\mu} \cdot \mathscr{E}_0$, occurring in the Rabi result for the atom is modulated by the term involving the Bessel functions in Eq. 4.3. The coupling between the molecule and the field is frequency dependent and is an oscillating function of the argument $(\mathbf{d} \cdot \mathscr{E}_0/\omega)$ of the Bessel functions; plots of $C(N)$ versus $(\mathbf{d} \cdot \mathscr{E}_0/\omega)$, for $N = 1, 2, 3$, are given in Figure 2. As $\mathbf{d} \to 0$, $C(N) = \boldsymbol{\mu} \cdot \mathscr{E}_0 \delta_{N,1}$, and the coupling parameter $C(N)$ and the molecular RWA result for $\bar{P}_2^{(N)}$ yield the atomic results. The results of Thomas (36), for $N = 1, 2$, agree with Eq. 4.2.

The effects of permanent dipoles on spectra can be discussed qualitatively for one-photon transitions $(N = 1)$ by comparing the resonance profiles predicted by Eq. 4.2 versus Eq. 4.1; indeed, the result including permanent dipoles can be obtained from the atomic result Eq. 4.1 by replacing the usual coupling term $\boldsymbol{\mu} \cdot \mathscr{E}_0$ by the frequency-dependent coupling parameter $C(1)$, which is generally less than $\boldsymbol{\mu} \cdot \mathscr{E}_0$ (see Figure 2) for $\mathbf{d} \neq 0$. Thus, for a molecule the resonance profile will not be given by the pure Lorentzian result of Figure 1. Rather, this profile will be modified with oscillatory fringes and

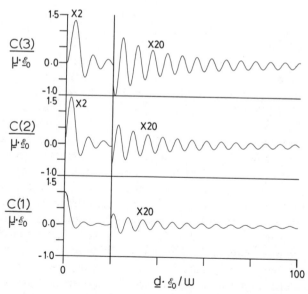

Figure 2. The RWA molecule–electromagnetic field coupling parameter $C(N)$ given by Eq. 4.3. Plotted is $C(N)/(\boldsymbol{\mu} \cdot \mathscr{E}_0)$ for $N = 1, 2, 3$ as a function of $\mathbf{d} \cdot \mathscr{E}_0/\omega$.

asymmetries, with zeros occurring at the zeros of $J_1(\mathbf{d}\cdot\mathscr{E}_0/\omega)$ as a function of ω. The resonance position still occurs at $\omega = \Delta E$ where $\bar{P}_2^{(1)} = 0.5$. For a sufficiently narrow (main) resonance, where $C(1)$ does not vary appreciably over the width of the resonance, the frequency-dependent full width at half maximum (FWHM) of the resonance is given by $2C(1)$, with $\omega = \Delta E$, versus $2\mu\cdot\mathscr{E}_0$ for the atomic case. Hence, another effect of permanent dipoles is to reduce the width of the resonance for $\mathbf{d} \neq 0$ relative to the atomic case; the reduction in width can be appreciable for $(\mathbf{d}\cdot\mathscr{E}_0/\omega)$ sufficiently large and especially if it is near a zero of the Bessel function J_1.

Our general RWA result for the resonance profiles, Eq. 4.2, suggests for $\mathbf{d} \neq 0$ that $\omega_{\text{res}}^N = \Delta E/N$, that the resonance profiles can have oscillatory fringes and asymmetries around the main resonance peaks for all N, and that the widths of the resonances will often be very narrow relative to the $\mathbf{d} = 0$ atomic case. If $\mathbf{d} \neq 0$, at least one of the states involved in the transition is of mixed parity, and, hence, unlike atoms, molecules with permanent dipoles can support even- as well as odd-photon transitions. The FWHM of the N-photon resonance profile is given by

$$(\text{FWHM})^N \approx \frac{2}{N}\{|C(N)|\}_{\omega = \omega_{\text{res}}^N} = 4\mu\cdot\mathscr{E}_0\frac{\Delta E}{N\mathbf{d}\cdot\mathscr{E}_0}J_N(\mathbf{d}\cdot\mathscr{E}_0 N/\Delta E) \qquad (4.5)$$

under the assumption that this result does not vary appreciably with ω across the main N-photon resonance. This suggests that the width of the N-photon resonances decrease as N increases, both through the factor N^{-1} occurring in Eq. 4.5 and since $J_N(\mathbf{d}\cdot\mathscr{E}_0 N/\Delta E)$ often decreases as N increases. While this prediction will often be reliable, it is not rigorous, for example if $\mathbf{d}\cdot\mathscr{E}_0 N/\Delta E$ is near a zero of J_N for N small relative to N large [see the plots of $C(N)$ given in Figure 2].

Some of these qualitative results cannot be deduced from a comparison of Eqs. 4.1 and 4.2 for $N > 2$ or from Eq. 4.2 alone. They arise from this comparison for $N = 1$ and from comparisons of exact calculations for $\mathbf{d} \neq 0$ versus $\mathbf{d} = 0$ problems, interpreted by the use of Eq. 4.2 for all N; also the RWA expression for the $N = 3, 5, \ldots$ resonance profiles for atoms derived by Shirley (24) is very useful in this context. Some of these features were apparently first observed by Thomas and Meath (37) in their investigation of some of the features of spectra that arise from the presence of nonzero diagonal dipole matrix elements and/or near degenerate energy levels using multiphoton vibrational resonances involving HeXe, NeAr, and NeXe as models.

A detailed discussion of Eq. 4.2 and its validity has been given previously (35). For example, it is clear that it is not correct in the limit that $\mathbf{d} \to 0$ for $N = 3$ since a two-level atom will support three-photon transitions while $C(3) \to 0$ in this limit (see Figure 2). The correction terms in the atomic limit arise from higher order perturbation theory treatments of the Floquet secular equation;

the usual RWA result corresponds to the perturbative result through first order only. Also, our result predicts $\omega_{res}^N = \Delta E/N$, whereas exact model calculations (see Section IV.A.2) show that when $\mathbf{d} \neq 0$, shifts of the resonance frequencies to the low-frequency side of the zero field positions ($\omega_{res} = \Delta E/N$) can occur relative to the Bloch–Siegert shifts to high frequency observed in the atomic case. Like all RWA-type approximations, Eq. 4.2 is more reliable as the coupling between the transition dipole and the applied electromagnetic field [here $C(N)$] becomes small. Generally, Eq. 4.2 becomes more reliable as $\boldsymbol{\mu} \cdot \mathscr{E}_0$ decreases and $\mathbf{d} \cdot \mathscr{E}_0$ increases for fixed values of ΔE and bearing in mind that the effects of neighboring states must be considered for many-level problems. Since the RWA result is a "near"-resonance result, it will also break down when the N-photon resonances begin to overlap appreciably.

The expression for the molecule-applied electromagnetic field coupling for N-photon transitions given by Eq. 4.3 is very important. It predicts for $\mathbf{d} \neq 0$, for one-photon transitions, that the molecule–field coupling can be considerably less than the usual atom–field coupling $\boldsymbol{\mu} \cdot \mathscr{E}_0$ as $\mathscr{E}_0$ increases due to the presence of permanent dipoles in molecules. For two-photon transitions, which usually do not exist for atoms, the molecule–field coupling $C(2)$ at first increases and then can oscillate with an overall decrease as $\mathscr{E}_0$ increases (see Figure 2). Similar comments apply for the higher photon transitions, bearing in mind that the RWA results for the molecule–applied field coupling are not reliable as $\mathbf{d} \to 0$ for the odd-photon transitions if $N > 1$. In general, for systems with diagonal dipole matrix elements, the coupling between the system and the applied electromagnetic field can be much less than expected from the usual (atomic) theory. This observation then extends the use of the RWA result to much larger applied fields than one would normally expect. The experimental implications of this are of course relevant; in some cases applied electromagnetic fields with large magnitudes really interact with molecules weakly!

Even with its obvious limitations, the RWA result for the N-photon resonance profile given by Eq. 4.2 is very useful. In particular, it provides an analytical expression that can be used to interpret the effects of permanent dipole moments on spectra as a function of the parameters of the problem. The expression for the molecule–field coupling parameter $C(N)$ plays a crucial role in these interpretations, even when the RWA resonance profile is not a very good representation of the exact results. Some examples that illustrate these points are discussed next.

2. Some Numerical Examples

Here we will consider, explicitly, two examples dealing with two-level molecules with the molecular configurations fixed relative to the applied electromagnetic field such that $\boldsymbol{\mu} \parallel \mathscr{E}_0 \parallel \mathbf{d}$.

To help in the discussion of single- and multiphoton spectra, it is convenient to use three coupling strength parameters:

$$b = |\mathbf{\mu} \cdot \mathscr{E}_0|/\Delta E \tag{4.6}$$

$$\beta(N) = [|C(N)|/\Delta E]_{\omega = \omega_{res}} \tag{4.7}$$

$$\eta(N) = N|\mathbf{d} \cdot \mathscr{E}_0|/\Delta E \tag{4.8}$$

The first of these is the parameter used to discuss the validity of the atomic single-photon resonance profile Eq. 4.1. The second is the N-photon analogue of this that takes the effects of the permanent dipole moments into account when computing the molecule–applied field coupling, and the third represents the coupling of the permanent dipoles with the applied electromagnetic field. The parameter $\eta(N)$ is essentially the argument of the Bessel functions occurring in the molecule–electromagnetic field coupling parameter $C(N)$ evaluated at the approximate N-photon resonance frequency $\omega_{res} \approx \Delta E/N$.

In Figure 3, the exact spectrum (35) for the model two-level system characterized by $b = 6.84$, $\beta(1) = 6.85 \times 10^{-2}$, and $\eta(1) = 27$ is compared with the analogous spectrum with $\mathbf{d} = 0$ $[\eta(1) = 0]$ for circular frequencies as-

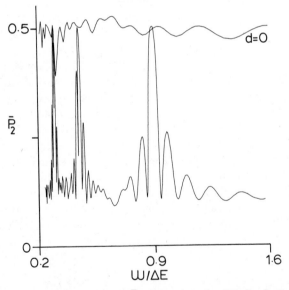

Figure 3. The steady-state spectrum, $\bar{P}_2$ as function of $\omega/\Delta E$, for the two-level model characterized by $\mu = -0.5072$, $d = 2.0$, $\Delta E = 3.706 \times 10^{-5}$, $\mathscr{E}_0 = 5 \times 10^{-4}$, $a_1(0) = 1$, and $\mathbf{\mu} \parallel \mathscr{E}_0 \parallel \mathbf{d}$. The $N = 1, 2$, and 3 zero field resonance positions, $\Delta E/N$, correspond to $\omega/\Delta E = 1, 0.5$, and 0.333, respectively, whereas the exact resonances occur at 0.897, 0.447, and 0.297. The corresponding spectrum, with $d = 0$, is also plotted.

sociated with the one- through three-photon resonances. The spectrum for **d** = 0 is essentially fully saturated as expected for a large coupling strength $b = 6.84$; it is not easy to identify the resonance positions in this "atomic" spectrum. When **d** ≠ 0, the coupling strength parameter $\beta(1)$ is about 100 times smaller than b, resulting in a markedly narrower set of resonance profiles with easily identified resonance positions; the two-photon as well as the odd-photon resonances are clearly present. The oscillatory fringes associated with the main resonances (discussed in Section IV.A.1) are readily seen in Figure 3. It is also clear that the positions of the resonances are shifted to the low-frequency side of the zero field estimates $\Delta E/N$ in contradistinction to the high-frequency shifts expected for the atomic case. Whereas one of the effects of **d** ≠ 0 has been to reduce the dynamic background markedly relative to the atomic spectrum, a background still persists. This, as well as the "negative" Bloch–Siegert shift of the resonance frequencies, is another feature not predicted by our simple RWA result of Eq. 4.2. Overall, the RWA expression for the resonance profiles does reasonably well in predicting the general features of the spectra illustrated in Figure 3 for **d** ≠ 0 but does not account for some of the important quantitative features of the exact calculation (35). There are other examples available (35) for **d** ≠ 0 where the agreement of the RWA with the precise calculations is considerably better than for the cases explicitly discussed here.

Next we consider an example where the RWA gives a less reliable representation of the exact numerical results but, nevertheless, is still useful in interpreting features of the spectrum (38). The molecular parameters chosen are characteristic of a level configuration in substituted aromatic molecules that exhibit intense one-photon transitions (39); the relevant coupling strength parameters are $b = 1.5$, $\beta(1) = 0.22$, and $\eta(1) = 3.25$. The exact results for the steady-state spectrum is shown in Figure 4, where it is compared with the spectrum with the same parameters except for d, which is chosen to be zero $[\eta(1) = 0]$. The effective coupling strength $\beta(1)$ between the molecule and the electromagnetic field for the $d \neq 0$ calculation is much larger for this example than for the corresponding spectrum of Figure 3, whereas the coupling strength b is less than in Figure 3.

When $d = 0$, the dynamic background in the steady-state spectrum is large but does not fully suppress the single- and triple-photon resonances in this "atomic" case. Both these resonances show large Bloch–Siegert shifts to the high-frequency side of the zero field resonance positions. When $d \neq 0$, even as well as odd-photon resonances occur with large shifts in the resonance frequencies to the low-frequency side of the zero field predictions. Since $\beta(1)$ is considerably less than b, two of the effects of the permanent dipoles in the molecular example are to narrow the resonance profiles and to lessen the dynamic background of the spectrum relative to the $d = 0$ case. Some features

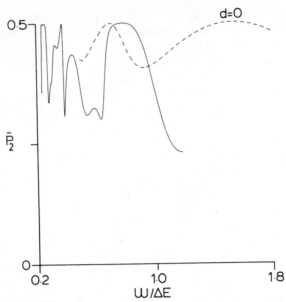

Figure 4. The steady-state spectrum, $\bar{P}_2$ as function of $\omega/\Delta E$, for the two-level molecule characterized by $\mu = 3.0$, $d = 6.50$, $\Delta E = 0.1$, $\mathscr{E}_0 = 5 \times 10^{-2}$, and for $a_1(0) = 1$ and $\mu \parallel \mathscr{E}_0 \parallel \mathbf{d}$. The $N = 1$, 2, and 3 zero field resonance positions correspond to $\omega/\Delta E = 1$, 0.5, and 0.333, respectively, whereas the exact resonances occur at 0.800, 0.377, and 0.257. The corresponding spectrum for $d = 0$ is also plotted, with the $N = 1$, 3 resonance maxima being at $\omega/\Delta E = 1.54$, 0.70, respectively.

of the molecular spectrum can be related to the oscillatory fringes discussed previously, but most of these are suppressed by the remaining relatively high dynamic background in the exact spectral results as compared to zero background in the corresponding RWA calculations.

The molecular calculations discussed in this section are for fixed configurations of the molecule. In many experiments the molecules are free to rotate with respect to the direction of the applied electromagnetic field, and the observed spectra correspond to orientation-averaged, not fixed-configuration, calculations of the steady-state spectrum. When the calculated spectra are properly averaged over the allowed orientations of the molecule relative to the field direction, much of the interesting structure in the spectra due to the presence of permanent dipole moments or diagonal dipole matrix elements tends to average out (30).

To observe this structure, for example, the oscillatory fringes and the large negative resonance shifts discussed previously, experiments must be performed on molecules that are held more or less rigidly in place relative to the direction of the applied electromagnetic field. These types of experiments should be possible by preparing the molecules of interest for the experiment by dissolving them in an orientated liquid crystal media or polymer host. For example, the orientation of the liquid crystal host molecules can be set by applying an electric field, and this causes the molecules to be probed by the electromagnetic field to adopt a similar orientation (40). Work along these lines is currently under investigation.

B. Including the Effects of a Static Electric Field

The Schrödinger time-dependent wave equation, written in the form of Eq. 2.3, is in the representation furnished by the stationary states of the unperturbed Hamiltonian H_0, where $H_0\phi_i = E_i\phi_i$. In the presence of a static electric field $\mathscr{E}_s$, it is often convenient to transform the time-dependent wave equation into the representation that diagonalizes the static part of the Hamiltonian operator $H(t)$, namely, $H_s = H_0 - \boldsymbol{\mu}\cdot\mathscr{E}_s$. This is easy to do by diagonalizing H_s with respect to the $\{\phi_i\}$ to obtain the eigenfunctions and eigenvalues for the new representation and is essentially mandatory if many of the approximate perturbative methods are used to solve to the problem. If exact methods are used, such as that discussed in Section II, the choice of the representation for solving the time-dependent Schrödinger equation is optional. However, from an interpretative point of view, the static Hamiltonian diagonalized represent-ation is always more convenient. Once in this representation, the spectra can be interpreted (20, 33, 41) in an analogous fashion to a spectrum in the absence of a static field, with several obvious additional considerations—for example, the additional static background and the possibility of the mixing of states of different parity caused by the application of the static field.

For a two-level problem the basis functions $\{\phi_1, \phi_2\}$ and $\{\phi_-, \phi_+\}$ that diagonalize H_0 and H_s, respectively, are conveniently denoted as the $(1, 2)$ and the $(-, +)$ representations. Under the influence of the static electric field the stationary state energy levels E_1 and E_2 are shifted to E_- and E_+ by the mixing of the wavefunctions ϕ_1 and ϕ_2 by the field. The energy separation in the $(-, +)$ representation, γ, is given by (42)

$$\gamma = [(\Delta E - \mathbf{d}\cdot\mathscr{E}_s)^2 + 4(\boldsymbol{\mu}\cdot\mathscr{E}_s)^2]^{1/2} = E_+ - E_- \geqslant 0 \qquad (4.9)$$

Clearly, $\gamma > \Delta E$ or $\gamma < \Delta E$ are both possible in general, whereas γ is always greater than ΔE if $\mathbf{d}\cdot\mathscr{E}_s = 0$.

The spectra for the two-level problem are readily interpreted in the $(-, +)$ representation. In the weak electromagnetic field limit the resonances occur at frequencies $\omega_{res}^N \approx \gamma/N$, $N = 1, 2, 3, \ldots$, and both even- and odd-photon transitions are allowed even if $\mathbf{d} = 0$ due to the mixing of the ϕ_1 and ϕ_2 states by $\mathscr{E}_s$; ϕ_- and ϕ_+, in general, are of mixed parity. For $\mathbf{d} = 0, \gamma > \Delta E$ and

$\omega_{res}^N > \Delta E/N$ even for weak electromagnetic fields (small $\mathcal{E}_0$), but this is not a Bloch–Siegert-type shift. For more intense $\mathcal{E}_0$ fields $\omega_{res}^N > \gamma/N$, and this difference contains the Bloch–Siegert shift for the static-field problem. Analogous but more complicated possibilities can arise when $\mathbf{d} \neq 0$ since it is then possible for the shifts in resonance positions due to the electromagnetic field to be either to low or to high frequency relative to γ/N.

The transformation from the $(1, 2)$ to the $(-, +)$ representation, in the context of investigating the effects of permanent dipole moments, has been discussed and applied to obtain analytic exact solutions to some model problems (42). It has also been used to develop RWA-type solutions (38) for the induced transition probabilities and resonance profiles, and expressions for the molecule–electromagnetic field coupling parameters, which contain the effects of both the static electric field and $\mathbf{d} \neq 0$. These analytic results yield Eqs. 4.1–4.3 in the appropriate limits and are useful in interpreting and predicting the effects of static fields on molecular spectra. Comparison of the results obtained from the RWA expressions with two-level exact calculations have been used (38) to help assess the validity of the RWA for these types of problems, and interestingly, the RWA resonance profiles for $\mathcal{E}_s \neq 0$ contain some of the shift of the resonance frequencies to higher frequencies relative to γ/N. The explicit analytic RWA expressions (38) for the resonance profiles are not particularly useful for what follows and so are not given explicitly here.

Consider the two-level problem characterized by $\mathbf{d} = 0$, $\Delta E = 2.15 \times 10^{-7}$, $\gamma = 3.04 \times 10^{-7}$, and $b_y = |\mu \mathcal{E}_0|/\gamma = 0.5$, with $\mathbf{\mu} \| \mathcal{E}_0 \| \mathcal{E}_s$. The more correct RWA coupling strength parameter $\beta_y(1)$ that takes account of the effect of the static electric field in the coupling between the atom and the electromagnetic field is $\beta_y(1) = 0.34 \approx b_y$ for this problem (38). The molecular parameters (43) correspond to the $J = 0 \rightarrow J = 1$ rotational transition in the ground vibrational state of CsI. Figure 5 compares the exact model spectrum for $\mathcal{E}_s \neq 0$ with that corresponding to $\mathcal{E}_s = 0$ over the frequency range corresponding to the one- to three-photon transitions. With $\mathcal{E}_s = 0$ the spectrum consists of only the one- and three-photon transitions, both of which show rather large Bloch–Siegert shifts. For $\mathcal{E}_s \neq 0$, both even- and odd-photon resonances are present. The shifts to high frequency for $\mathcal{E}_s \neq 0$ relative to $\Delta E/N$ are made up of two components, that due to $\mathcal{E}_s$ itself and that due to a Bloch–Siegert-type shift relative to γ/N. In addition to the dynamic background associated with any strong-coupled interaction, the static-field problem has an additional static background (20, 41) given by $\bar{P}_2(\text{stat}) = \frac{1}{2}[1 - (\Delta E/\gamma)^2]$; the background for $\mathcal{E}_s \neq 0$ is considerably greater than for $\mathcal{E}_s = 0$ in the spectra given in Figure 5.

There are other interesting applications (20, 33, 40–42, 44, 45) involving the effects of static electric fields both in the absence or the presence of permanent dipole moments. An example of the later, which also involves performing calculations of orientationally averaged spectra, is discussed in Section V.

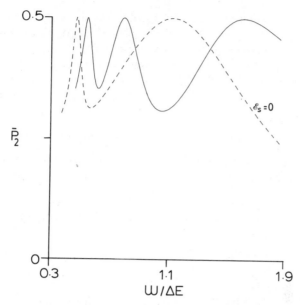

Figure 5. The steady-state spectrum, $\bar{P}_2$ as function of $\omega/\Delta E$, for the two-level molecule characterized by $\mu = 2.6398, d = 0, \Delta E = 2.15 \times 10^{-7}, \mathscr{E}_0 = 5.76 \times 10^{-8}, \mathscr{E}_s = 4.07 \times 10^{-8}$, and for $a_1(0) = 1$ and $\mu \parallel \mathscr{E}_0 \parallel \mathscr{E}_s$. The $N = 1, 2,$ and 3 zero field ($\mathscr{E}_0 \to 0$) resonance positions are at γ/N, which correspond to $\omega/\Delta E = 1.413, 0.707, 0.471$, respectively, whereas exact resonances occur at $\omega/\Delta E = 1.605, 0.791, 0.549$. Corresponding results for $\mathscr{E}_s = 0$ are also plotted with the $N = 1$ and 3 resonance maxima occurring at $\omega/\Delta E = 1.127, 0.472$, respectively, whereas zero field ($\mathscr{E}_0 \to 0$) resonances occur at $\omega/\Delta E = 1, 0.333$.

V. ORIENTATION OR ROTATIONALLY AVERAGED SPECTRA

Many of the calculations in the literature involving (exact) solutions of the time-dependent wave equation have been performed for atoms or for molecules in a fixed orientation relative to the direction of the applied electromagnetic and/or static fields. For molecular problems it is often important to average the transition probabilities with respect to molecular orientations relative to the field directions.

Orientationally or rotationally averaged transition probabilities have been evaluated for perturbation theory results where they can be performed analytically (see, e.g., Refs. 46 and 47). Apparently, little analogous work has been done for the more general case (including orientational averages in exact formalisms for solving the time-dependent wave equation) until very recently (30, 38). This does not include, of course, rotational calculations including the rotational energy levels of molecules explicitly [see, e.g., the Fourier series

Floquet secular equation calculations of infrared multiphoton absorption effects by Ho and Chu (48) and Leasure, Milfeld, and Wyatt (49)].

In this section we will discuss, with examples, both free and hindered rotationally averaged spectra. The former would correspond to gas phase molecules in the absence of static electric fields; the latter corresponds to, for example, gas phase molecules in the presence of static fields and molecules adsorbed on a surface.

A. Scheme for Evaluating Rotational Averages

The rotational or orientational average of the steady-state transition probability defined by Eq. 3.9 is given by (30, 38, 50)

$$(\bar{P}_j)_{\rm rot} = \frac{\displaystyle\int_0^{2\pi}\int_0^{\pi}\int_0^{2\pi} \bar{P}_j(\alpha,\beta,\gamma)\exp\left[-\Delta W/kT\right]\sin\beta\, d\alpha\, d\beta\, d\gamma}{\displaystyle\int_0^{2\pi}\int_0^{\pi}\int_0^{2\pi}\exp\left[-\Delta W/kT\right]\sin\beta\, d\alpha\, d\beta\, d\gamma} \tag{5.1}$$

where α, β, and γ are the Euler angles specifying the rotation of the molecule-fixed coordinate axes with respect to the space-fixed axes and $\bar{P}_j(\alpha,\beta,\gamma)$ is the steady-state resonance profile for the fixed configuration (α,β,γ), which can be evaluated as discussed in Section II. In the absence of a "hindering field" (e.g., static electric field or a surface), all orientations should be weighted equally corresponding to $\Delta W = 0$. In the presence of a hindering field, each orientation is weighted by the appropriate Boltzmann factor $\exp\left[-\Delta W/kT\right]$, where k is the Boltzmann constant, T is the absolute temperature, and the energy difference ΔW depends on the problem (see what follows).

A numerical scheme based on Eq. 5.1 for evaluating the free rotational average of $\bar{P}_j$ has been discussed in considerable detail by Thuraisingham and Meath (30). The modification to incorporate the Boltzmann factor required for hindered rotations has been discussed for molecules in an applied static electric field, with applications to a special case, by Kmetic, Thuraisingham, and Meath (38), and more computational details, including the effects of surfaces, will be available soon (51). Some explicit numerical examples are discussed in what follows.

The need for computational efficiency in determining the transition probabilities, even for problems with a small number of energy levels is readily apparent in the calculation of these rotationally averaged spectra. For example, in two-level calculations with $\mathscr{E}_s \neq 0$ and $\boldsymbol{\mu} \perp \mathbf{d}$, some 300–400 calculations of fixed-orientation steady-state transition probabilities $\bar{P}_j$ are required to evaluate the rotational averaged steady-state result $(\bar{P}_j)_{\rm rot}$ for one value of the frequency. Thus, to evaluate a resonance profile for an N-photon transition requires $\sim 300 \times 50 = 15{,}000$ steady-state fixed-configuration calculations.

B. Applications Involving Free Rotational Averages

The exact fixed-configuration spectrum (30), $\bar{P}_2$ versus ω, for a two-level model system characterized by $\eta(1) = 0.05$ $(d \neq 0)$, $\beta(1) \approx b = 0.1$, and $\mathscr{E}_s = 0$, with $\boldsymbol{\mu} \parallel \mathbf{d} \parallel \mathscr{E}_0$, is illustrated in Figure 6 for frequencies associated with the one- and two-photon resonances that occur at $\omega/\Delta E = 1.003$ and $\omega/\Delta E = 0.507$, respectively, with heights of 0.5. Also included in this figure is the rotationally averaged spectrum (30) $(\bar{P}_2)_{\text{rot}}$ versus ω. Here the resonance peaks have heights slightly less than 0.5 and 0.158, respectively, and occur at $\omega/\Delta E = 1.000$ and $\omega/\Delta E = 0.503$. Further, the one-photon resonance peak and profile are sharpened and narrowed, respectively, and the overall spectral background is decreased by rotionally averaging $\bar{P}_2$.

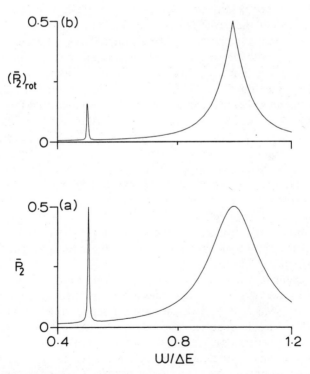

Figure 6. (a) The steady-state spectrum $\bar{P}_2$ corresponding to $\boldsymbol{\mu} \parallel \mathbf{d} \parallel \mathscr{E}_0$ and (b) the rotationally averaged steady-state spectrum $(\bar{P}_2)_{\text{rot}}$, as functions of $\omega/\Delta E$, for two-level model characterized by $\mu = 5.0$, $d = 2.5$, $\boldsymbol{\mu} \parallel \mathbf{d}$, $\Delta E = 0.10$, $\mathscr{E}_s = 0$, $\mathscr{E}_0 = 2 \times 10^{-3}$, and $a_1(0) = 1$.

The differences between the fixed-configuration and rotationally averaged spectra can be qualitatively understood from the fact that the fixed-configuration calculation corresponds to the configuration of maximum molecule–electromagnetic field coupling that contributes to the rotationally averaged spectrum; the minimum coupling is zero corresponding to $\mu \perp \mathscr{E}_0$. Hence, on rotational averaging, b varies between $0 \leqslant b \leqslant 0.1$ so that, on average, the molecule–field coupling is considerably reduced relative to the fixed-configuration case of $b = 0.1$. As discussed in Section IV, as b decreases, resonance profiles become narrower, their backgrounds become lower, and their Bloch–Siegert shifts tend to vanish so that the resonance positions tend to occur at $\omega = \Delta E/N$, in agreement with the trends observed in the comparison between $\bar{P}_2$ and $(\bar{P}_2)_{rot}$ given in the last paragraph. A marked decrease upon rotational averaging, as observed in the height of the two-photon resonance profile, is not observed in the height of the broader one-photon resonance since $\bar{P}_2$ has a value of essentially 0.5 at the one-photon zero field resonance position $\omega = \Delta E$ for this model calculation. The value of $\bar{P}_2$ at the two-photon zero field resonance position, $\omega = \frac{1}{2}\Delta E$, is very small (0.08).

Qualitatively (30) there will be a decrease in a resonance peak height upon rotational averaging for a particular transition, relative to the peak height for the configuration with maximum coupling between μ and $\mathscr{E}_0$, if the value of $\bar{P}_2$ at the zero field resonance position for this fixed configuration is lower than the resonance maximum. Qualitatively the reduction in peak height will depend on the interplay between the Bloch–Siegert shifts and the resonance widths in the $\bar{P}_2$, as a function of coupling strengths, needed to evaluate the rotational average of $\bar{P}_2$. An example (30) of a very marked reduction in peak height is furnished by a two-level system characterized by the parameters of that of Figure 6 except with $\eta(1) = 0$ $(d = 0)$. Here there is no two-photon resonance, and upon rotational averaging, the maximum of the three-photon resonance is reduced from 0.5 to 0.031.

It might seem that spectral widths are always reduced in rotationally averaged spectra relative to those associated with the configuration of maximum $\mu \cdot \mathscr{E}_0$ coupling. This often happens but is not the case in general, as the next example will illustrate.

Figure 7 compares the exact fixed-configuration spectrum $\bar{P}_2$ with $\mathbf{d} \| \mu \| \mathscr{E}_0$ and $\mathscr{E}_s = 0$, for the two-level model molecule characterized by $b = 0.8$, $\beta(1) = 0.02$, and $\eta(1) = 9.6$ $(\mathbf{d} \neq 0)$, with the corresponding rotationally averaged spectrum $(\bar{P}_2)_{rot}$ for frequencies associated with the one- and two-photon transitions. The oscillatory fringes (see Section IV) occurring in the fixed-configuration spectrum are removed by rotational averaging, and contrary to the preceding example, the rotationally averaged resonance profiles are broader than those for the configuration corresponding to the maximum value of $\mu \cdot \mathscr{E}_0$. The narrowness of the peaks and the fact that the positions of the

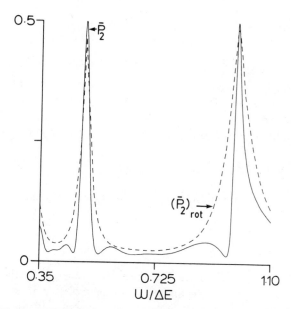

Figure 7. The steady-state spectrum $\bar{P}_2$ with $\boldsymbol{\mu} \parallel \mathbf{d} \parallel \mathscr{E}_0$ and the rotationally averaged profile $(\bar{P}_2)_{\text{rot}}$, as functions of $\omega/\Delta E$, for the two-level system characterized by $\mu = 0.80$, $d = 9.6$, $\boldsymbol{\mu} \parallel \mathbf{d}$, $\Delta E = 5.0 \times 10^{-2}$, $\mathscr{E}_s = 0$, $\mathscr{E}_0 = 5.0 \times 10^{-2}$, and $a_1(0) = 1$. The positions of the one- and two-photon resonances, $\omega/\Delta E = 0.987$, and 0.494, in $\bar{P}_2$ are essentially unchanged by rotationally averaging, but the peak heights have been decreased from 0.5 to 0.485 and 0.466, respectively.

peaks are very close to the zero field limits of 1.0 and 0.5 in spite of the large b value are due to the large values of $\eta(1)$ and $\eta(2)$ in this example, which effectively decrease the molecule–field coupling for all relevant molecular configurations (see also Section IV). A more complete discussion of this example and of others with $\boldsymbol{\mu} \perp \mathbf{d}$ are given in the literature (30).

Even though the interesting oscillatory fringes due to the presence of permanent dipole moments apparently are averaged away in rotationally averaged spectra relative to fixed configuration spectra, the effect of permanent moments is still marked in the orientation-averaged case. The presence of $d \neq 0$ sufficiently large, so that $C(N) \ll b$, markedly reduces the widths and backgrounds associated with large b spectra. For example, the rotational averaged $d \neq 0$ spectrum corresponding to Figure 3 is qualitatively the envelope of the fixed-configuration spectrum, and comparison of this with the atomic ($d = 0$) spectrum given in Figure 3 illustrates this point well.

C. Applications Involving Hindered Rotational Averages

In these applications the molecules are not free to assume random configurations of equal weight with respect to the electromagnetic field direction and

$\Delta W \neq 0$ in Eq. 5.1. The resonance profiles are therefore functions of temperature.

1. Hindered Rotations Caused By Static Electric Field

To better understand the choice of the energy difference ΔW in the Boltzmann factor, consider (38) a two-level molecule characterized by given values of μ, d, and ΔE and with $\mathscr{E}_s \neq 0$. The choice of ΔW appropriate for the initial condition that the molecule is in the lower energy state 1 at time $t = 0$ is given by

$$\Delta W = E_- - E_1 = -\tfrac{1}{2}[\gamma + (\mu_{11} + \mu_{22})\cdot\mathscr{E}_s - (E_1 + E_2)] - E_1 \qquad (5.2)$$

This can be written in a more recognizable form by expanding γ given by Eq. 4.9 to obtain

$$\Delta W = -\mu_{11}\cdot\mathscr{E}_s - (\mu\cdot\mathscr{E}_s)^2/\Delta E + \cdots \qquad (5.3)$$

Thus, the energy difference is the interaction energy in the two-level approximation arising from the interaction of the molecule in state 1 with the applied static field in the absence of the electromagnetic field; the first term on the right side of Eq. 5.3 represents the interaction of the permanent dipole of the molecule with the static field, and the second term is the static-field-induced dipole induction energy.

As a specific example, consider a two-level problem characterized by $\Delta E = 0.1582$, $\mu = 0.31$, $d = 1.18$, and $\mathscr{E}_s = \mathscr{E}_0 = 10^{-3}$ with $\mathscr{E}_s \| \mathscr{E}_0$. If $\mu \perp \mathbf{d}$, the molecular parameters are characteristic (52) of the $S_0 \rightarrow S_2$ transition of nitrobenzene, with state 1 being state S_0. We begin by considering $\mu \| \mathbf{d}$, which is easier to analyze, and will compare this with the $\mu \perp \mathbf{d}$ calculations later.

With $\mu \| \mathbf{d}$ it is easy to show that the rotationally averaged spectra, Eq. 5.1, can be reduced to an integral from -1 to $+1$ over the variable $x = \cos\beta$, where β is the Euler angle between the space-fixed z axis ($\| \mathscr{E}_s$) and the body-fixed z axis ($\| \mu$); for example, $\mu\cdot\mathscr{E}_s = \mu\mathscr{E}_s \cos\beta$ and $\mathbf{d}\cdot\mathscr{E}_s = d\mathscr{E} \cos\beta$. For all β, $\gamma \approx \Delta E$, $\eta(1) < 0.008$, and hence $\beta(1) \approx b$. Thus, the coupling parameter $C(1) \approx \mu\cdot\mathscr{E}_0$ can be used to interpret the coupling between the electromagnetic field and the molecule for the discussion of this example that follows. Plots of exact calculations of the rotationally averaged one-photon resonance profile as a function of ω are given in Figure 8(a) for $T = \infty$ (free rotation) and for $T = 600, 300, 70, 30$, and 20 K.

The behavior of the spectra as a function of T and ω can be predicted and/or explained qualitatively by explicitly considering some typical molecule–field configurations and some of the ideas concerning the interpretation of spectra introduced earlier in this chapter. Three of the possible configurations entering

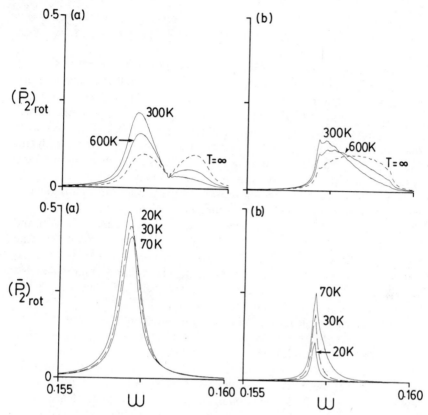

Figure 8. The steady-state rotationally averaged one-photon resonance profile, $(\bar{P}_2)_{rot}$ versus ω, as function of temperature T for (a) $\boldsymbol{\mu} \parallel \mathbf{d}$ (left) and (b) $\boldsymbol{\mu} \perp \mathbf{d}$ (right) for the two-level "nitrobenzene molecule" otherwise characterized by $\mu = 0.31$, $\mu_{11} = 1.574$, $\mu_{22} = 2.754$, $\Delta E = 0.1582$, and with $\mathscr{E}_0 = \mathscr{E}_s = 10^{-3}$, $\mathscr{E}_0 \parallel \mathscr{E}_s$, and $a_1(0) = 1$.

into the evaluation of $(\bar{P}_2)_{rot}$ correspond to (i) $\boldsymbol{\mu}$ and $\mathbf{d}$ aligned with $\mathscr{E}_0$ and $\mathscr{E}_s$ so that $x = 1$ and $\gamma = 0.1570 < \Delta E$ for this attractive configuration, (ii) $\boldsymbol{\mu}$ and $\mathbf{d}$ antialigned with $\mathscr{E}_0$ and $\mathscr{E}_s$ so that $x = -1$ and $\gamma = 0.1594 > \Delta E$ for this repulsive configuration, and (iii) $\boldsymbol{\mu}$ and $\mathbf{d}$ perpendicular to $\mathscr{E}_0$ and $\mathscr{E}_s$ so that $x = 0$, $\gamma = \Delta E$, and $\bar{P}_2 = 0$ for this configuration, which corresponds to zero coupling between the molecule and both the electromagnetic and static fields ($\boldsymbol{\mu} \cdot \mathscr{E}_0 = \boldsymbol{\mu} \cdot \mathscr{E}_s = \mathbf{d} \cdot \mathscr{E}_s = 0$). An analysis of $C(1)$ at $\omega = \gamma$ shows the one-photon coupling parameter has a maximum value of $\sim 3.123 \times 10^{-4}$ when $x = 1$, decreases smoothly to zero at $x = 0$, and then increases in magnitude to $\sim 3.077 \times 10^{-4}$ at $x = -1$; $|C(1)|$ is almost symmetric about $x = 0$, being *slightly* smaller for $x < 0$ than for $x > 0$. For arbitrary configurations of the molecule with respect to the applied fields, the resonance occurs when $\omega = \gamma$ for a particular configuration, and clearly, γ can be greater or less than ΔE as illustrated here and discussed in Section IV.B.

Based on the discussion of the last paragraph, the free rotationally averaged one-photon resonance profile, which corresponds to the high-temperature limit of the Boltzmann-averaged result where $\exp(-\Delta W/kT) = 1$ and therefore to equal weighting to all configurations entering in $(\bar{P}_2)_{\text{rot}}$, should consist of two maxima located at $\omega \approx \gamma$ ($x = 1$) $< \Delta E$ and $\omega = \gamma$ ($x = -1$) $> \Delta E$ with a minimum occurring at $\omega = \Delta E$. This is in qualitative agreement with the calculated result shown in Figure 8(a). For $T \neq \infty$, the Boltzmann factor favors attractive ($\gamma < \Delta E$) configurations versus repulsive ($\gamma > \Delta E$) configurations, and therefore, in the one-photon resonance profile, the high-frequency peak is suppressed and the low-frequency peak is enhanced, relative to the high-temperature limit, as the temperature decreases, in agreement with the exact calculations of Figure 8(a). For sufficiently low temperature the spectrum consists of a single resonance peak of height 0.5 and FWHM $\approx 2C(1)$ located at $\omega = \gamma$ ($x = 1$). The analysis given here is similar to a more detailed presentation based on the $\pi \to \pi^*$ transitions of the pentadienal molecule given by Kmetic, Thuraisingham, and Meath (38).

Now we consider calculations that are again based on the parameters used to generate Figure 8(a) except for taking $\boldsymbol{\mu} \perp \mathbf{d}$. The rotationally averaged one-photon resonance profiles are shown in Figure 8(b) as a function of temperature. A detailed analysis of the spectra is more complicated than for the $\boldsymbol{\mu} \parallel \mathbf{d}$ case since the molecular configurations are characterized by two Euler angles and $\gamma = E_+ - E_-$ is a function of both. However, the basic features can be discussed by taking β to be the angle between $\mathbf{d}$ and the field directions, as before, and by explicitly considering two configurations that are independent of the other angle: (i) $x = \cos \beta = 1$ where $\mathbf{d}$ and $\mathscr{E}_s$ are aligned while $\boldsymbol{\mu} \cdot \mathscr{E}_0 = 0$ and $\boldsymbol{\mu} \cdot \mathscr{E}_s = 0$ so that $\gamma < \Delta E$ and $\bar{P}_2 = 0$ and (ii) $x = -1$ so that $\gamma > \Delta E$ and $\bar{P}_2 = 0$. In addition, for the other possible configurations γ has intermediate values closer to ΔE and $\bar{P}_2 \neq 0$. Hence, one would expect the free rotational ($T = \infty$) averaged spectrum to consist of a single (broad) peak with the maximum located at $\omega = \Delta E$, in agreement with the calculation of Figure 8(b) and in marked contrast to the two-peak spectrum for $\boldsymbol{\mu} \parallel \mathbf{d}$ with the dip at $\omega = \Delta E$ seen in Figure 8(a). The peak in the resonance profile shifts to low frequency relative to $\omega = \Delta E$ as T decreases since in the Boltzmann averaging this favors attractive ($\gamma < \Delta E$) versus repulsive ($\gamma > \Delta E$) configurations. This is illustrated in Figure 8(b), which also shows that the peak heights first increase and then decrease as T decreases, again in marked contrast to the $\boldsymbol{\mu} \parallel \mathbf{d}$ example. For sufficiently low temperature the configuration of maximum attractive alignment ($x = 1$) of $\mathbf{d}$ and $\mathscr{E}_s$ must dominate in the evaluation of $(\bar{P}_2)_{\text{rot}}$, and therefore, the resonance profile becomes centered about $\omega = \gamma$

$(x = 1) < \Delta E$, and its height becomes very small (zero in the hypothetical $T = 0$ limit; $\bar{P}_2 = 0$ for $x = 1$).

The behavior of the single-photon absorption spectra for $\mu \parallel d$ [Figure 8(a)] versus $\mu \perp d$ [Figure 8(b)] as a function of ω and temperature are distinctly different, and such spectra should be useful in deducing molecular properties, including the relative orientation of transition and permanent dipole moments. Investigations of the effects of neighboring molecular levels, of varying the strengths and directions of the applied static and electromagnetic fields, and studies involving higher photon transitions are currently underway for a variety of molecular problems.

2. Hindered Rotations Caused By Molecule–Surface Interactions

In these applications $\mathscr{E}_s = 0$ and the energy difference ΔW in the Boltzmann factor in the expression for the rotationally averaged spectra, Eq. 5.1, is the interaction energy between an adsorbed molecule and the surface. In what follows we will consider model calculations involving N_2 and CO adsorbed over the most favorable site for adsorption on a graphite surface, namely, the center of a carbon hexagon. Lateral interactions are ignored, and so these calculations are most appropriate as models for monolayer and submonolayer coverages (53). The idea is to probe the adsorbed molecule with a continuous-wave light source using the techniques discussed in Section II and the interpretations of the spectra developed in previous sections.

The N_2–graphite and CO–graphite interaction energies are taken to be the atom–atom Lennard–Jones and 6-exp potentials developed by Steele (54) and Talbot, Tildesley, and Steele (53), and by Belak, Kobashi, and Etters (55), respectively. To facilitate the calculations, we use the Fourier series expansions for these molecule–infinite graphite surface potentials. This expansion can be reliably truncated at the first two terms (53, 55); the first depends on the height z between the center of mass of the molecule and the surface plane and on the polar angle θ; the second depends on z, θ, and the azimuthal angle ϕ. The angles θ and ϕ specify the orientation of the molecular bond axis relative to the surface normal and the plane defined by the surface normal and a surface lattice vector, respectively (54).

Initially, model calculations will be discussed in which the role of the surface is taken to be the source of a potential energy that hinders the rotation of the (adsorbed) molecule in the evaluation of the Boltzmann rotationally averaged spectrum through Eq. 5.1. These calculations illustrate how the spectral profiles of the adsorbed molecules can be related to the orientation of the molecule relative to the surface (without considering the complications arising from other effects, which are discussed briefly later). Two types of Boltzmann-averaged steady-state spectra are of interest. One is defined by Eq. 5.1 for fixed z and is denoted $(\bar{P}_j)_{\mathrm{rot}}^z$; the other involves averaging this result with respect to z

and is denoted $(\bar{P}_j)^{z_{av}}_{rot}$. The transition and permanent dipoles and the electronic energies for the various states involved in the electronic transitions studied are taken from McKoy and co-workers (56), Nesbet (57), Huber and Herzberg (58), and Chackerian (59); the transition dipoles and excitation energies are averages for an entire absorption band.

Calculations Neglecting Surface–Electromagnetic Field and Broadening Effects. In the applications that follow the couplings between the electromagnetic field and the adsorbed molecule are weak ($\mu \cdot \mathscr{E}_0/\Delta E < 2 \times 10^{-3}$) so the coupling parameter $\mu \cdot \mathscr{E}_0$ applies. The transition dipoles will always be either parallel or perpendicular to the bond axis **R** of the adsorbed molecule, and the interpretation of the spectra of the adsorbed molecule relative to the orientation of the molecule to the surface can be accomplished by noting that if

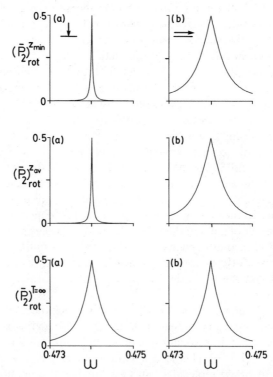

Figure 9. The steady-state rotationally averaged $1 \to 2$ one-photon resonance profiles for N_2 over graphite for $z = z_{min} = 3.3\,\text{Å}$, $(\bar{P}_2)^{z_{min}}_{rot}$, and for z averaging, $(\bar{P}_2)^{z_{av}}_{rot}$. Profile for freely rotating N_2 molecule, $(\bar{P}_2)^{T=\infty}_{rot}$, also given for comparative purposes. The left-hand side of the figure, (a), is for $\mathscr{E}_0$ perpendicular to surface; the right-hand side, (b), is for $\mathscr{E}_0$ parallel to surface. The calculations correspond to $\mu_{12} = 0.5$, $\mu_{12} \parallel \mathbf{R}$, $\Delta E = 0.474$, $\mathscr{E}_0 = 10^{-3}$, $T = 5\,\text{K}$, and $a_1(0) = 1$.

$\mu_{ij} \cdot \mathscr{E}_0 = 0$, the direct $i \to j$ transition is not allowed, whereas if $\mu_{ij} \cdot \mathscr{E}_0 = \mu_{ij} \mathscr{E}_0$, it is maximized for a given $\mathscr{E}_0$.

First consider the transition between the X $^1\Sigma_g^+$ ground state and the C' $^1\Sigma_u^+$ state of N_2, the $1 \to 2$ transition with $\mu_{12} \parallel \mathbf{R}$. The one-photon resonance profile $(\bar{P}_2)_{\text{rot}}^{z = z_{\min}}$ of the adsorbed molecule for $z = z_{\min} = 3.30$ Å, the value of z corresponding to the minimum in the N_2–surface potential over the adsorption site, $(\bar{P}_2)_{\text{rot}}^{z_{\text{av}}}$, and the gas phase free rotationally averaged spectrum, $(\bar{P}_2)_{\text{rot}}^{T = \infty}$, are compared in Figure 9 for $T = 5$ K and $\mathscr{E}_0 = 10^{-3}$. Two cases are considered: $\mathscr{E}_0$ is perpendicular to the surface [Figure 9(a)] and $\mathscr{E}_0$ is parallel to the surface [Figure 9(b)], corresponding to the direction of propagation of the light being parallel or perpendicular to the surface, respectively. The resonances occur at $\omega = E_2 - E_1$ and clearly indicate that the molecule is adsorbed so $\mathbf{R}$ is parallel to the surface at $T = 5$ K, in agreement with the molecular dynamics calculations of Talbot, Tildesley, and Steele (53). The width of the resonance profile for $\mathscr{E}_0$ parallel to the surface is greater than for the free rotationally averaged spectrum and much greater than that for the $\mathscr{E}_0$-perpendicular-to-the-surface spectrum, indicating $\mathbf{R}$ is parallel to the surface since $\mu \parallel \mathbf{R}$. For both directions of $\mathscr{E}_0$ relative to the surface the $z = z_{\min} = 3.3$ Å resonance profiles are essentially identical to the z_{av} spectra.

Next we consider the $1 \to 2$ and the $1 \to 3$ transitions, where state 3 is the B $^1\Pi_u$ electronic state of N_2, in the presence of states $4(A$ $^1\Pi_g)$ and $5(B'$ $^1\Sigma_u^+)$. Here $\mu_{13} \perp \mathbf{R}$ whereas $\mu_{12} \parallel \mathbf{R}$, and we represent the spectrum of the adsorbed molecule by considering the steady-state population of the ground state 1, $\bar{P}_1 = 1 - \sum_{i \neq 1} \bar{P}_i$, as a function of frequency. Figure 10 compares $(\bar{P}_1)_{\text{rot}}^{z}$ as a function of z, $(\bar{P}_1)_{\text{rot}}^{z_{\text{av}}}$, and $(\bar{P}_1)_{\text{rot}}^{T = \infty}$ for values of ω associated with the $1 \to 2$ and $1 \to 3$ single-photon transitions; $\mathscr{E}_0$ is perpendicular to the surface for all calculations and, again, $T = 5$ K and $\mathscr{E}_0 = 10^{-3}$. The resonances occur at $\omega = E_2 - E_1$ and $\omega = E_3 - E_1$. A large depopulation of state 1 corresponds to a strong transition, a small depopulation to a weak transition. Clearly, at $z = 3$ Å, the adsorbed molecule is parallel to the surface since the $1 \to 2$ resonance with $\mu \parallel \mathbf{R}$ is weak, whereas the $1 \to 3$ resonance with $\mu \perp \mathbf{R}$ is relatively much stronger. As z increases, the $1 \to 2$ resonance strengthens whereas the $1 \to 3$ resonance weakens, indicating that the molecular axis $\mathbf{R}$ is, on average, becoming more out of plane with respect to the surface. The rotationally and z-averaged spectrum is very similar to the rotationally averaged spectrum for $z = z_{\min} = 3.3$ Å; in both the molecular axis is essentially parallel to the surface. The free rotationally averaged spectrum, which, of course, is independent of z, is, of all the other cases considered, more similar to, but still quite different than, $(\bar{P}_1)_{\text{rot}}^{z = 5}$.

The nitrogen molecule has no permanent dipole moments, and hence the first multiphoton resonance corresponds to the three-photon transition, and for the reasons discussed in Section V.B, this is essentially eliminated by

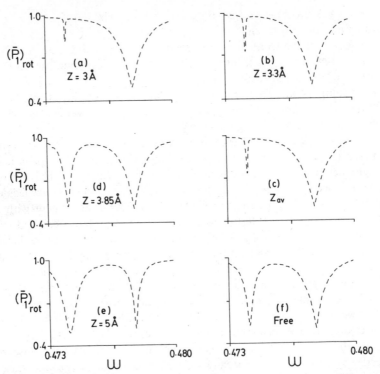

Figure 10. The steady-state rotationally averaged $1 \to 2$ and $1 \to 3$ single-photon spectra, $(\bar{P}_1)_{rot}$, for N_2 over graphite for (a) $z = 3$ Å; (b) $z = 3.3$ Å; (c) z_{av}; (d) $z = 3.85$ Å; (e) $z = 5.0$ Å; (f) free rotation. In all calculations $\mathscr{E}_0$ is perpendicular to surface, $\mathscr{E}_0 = 10^{-3}$, $T = 5$ K, $a_1(0) = 1$, and $\mu_{12} = 0.5$, $\mu_{13} = 1.0$, $\mu_{15} = 1.1$, $\mu_{24} = 0.39$, $\mu_{34} = 1.57$, $\mu_{45} = 0.09$, $\mu_{15} = \mu_{23} = \mu_{25} = \mu_{35} = 0$, $\mu_{12} \| \mu_{15} \| \mu_{34} \| \mathbf{R}$, $\mu_{13} \| \mu_{24} \| \mu_{45} \perp \mathbf{R}$, and $E_1 = 0$, $E_4 = 0.342$, $E_2 = 0.474$, $E_3 = 0.478$, and $E_5 = 0.529$.

rotational averaging. On the other hand, the two-photon transition (for a heteronuclear molecule) has a wider spectral profile and does not so readily average to zero.

As an example, consider the adsorption of CO on graphite and the transitions from the ground $X\,^1\Sigma^+$ electronic state of CO to the $A\,^1\Pi$ and the $C\,^1\Sigma^+$ states, the $1 \to 2$ and $1 \to 3$ transitions, respectively. Here $\mu_{12} \perp \mathbf{R} \| \mathbf{d}_{12} = \mu_{22} - \mu_{11}$ and $\mu_{13} \| \mathbf{R} \| \mathbf{d}_{13} = \mu_{33} - \mu_{11}$.

The single-photon resonance profiles have interpretations analogous to those for N_2 on graphite and so will not be considered further here. Figures 11(a, b), respectively, show $(\bar{P}_3)_{rot}^{z_{av}}$ and $(\bar{P}_2)_{rot}^{z_{av}}$ with $T = 5$ K and $\mathscr{E}_0 = 10^{-3}$ both for $\mathscr{E}_0$ perpendicular to the surface and $\mathscr{E}_0$ parallel to the surface; the corresponding free rotationally averaged spectra are included

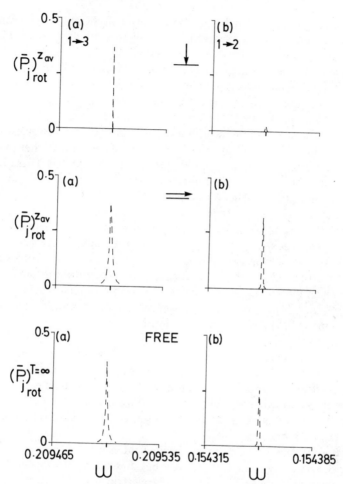

Figure 11. The steady-state rotationally and z-averaged $1 \to 3$ and $1 \to 2$ two-photon resonance profiles for CO over graphite: (a) $(\bar{P}_3)_{\text{rot}}^{z_{av}}$ versus ω and (b) $(\bar{P}_2)_{\text{rot}}^{z_{av}}$ versus ω. Top to bottom: $\mathscr{E}_0$ perpendicular to surface, $\mathscr{E}_0$ parallel to surface, and free rotational average. In all the calculations $T = 5\,\text{K}$, $\mathscr{E}_0 = 10^{-3}$, $a_1(0) = 1$, and $\mu_{13} = 0.760$, $\mu_{33} = 1.77$, $\mu_{11} = 0.044$, $E_3 - E_1 = 0.4190$, $\mu_{12} = 0.560$, $\mu_{22} = 0.8175$, $E_2 - E_1 = 0.3087$, $\mu_{13} \| \mu_{33} \| \mu_{11} \| R$, and $\mu_{12} \perp \mu_{11} \| \mu_{22} \| R$.

for comparative purposes. The two-photon resonances occur at $\omega = \frac{1}{2}(E_3 - E_1)$ and $\omega = \frac{1}{2}(E_2 - E_1)$, respectively.

For the $1 \to 3$ transition, Figure 11(a), the two-photon resonance profile is much broader for $\mathscr{E}_0$ parallel to the surface than for $\mathscr{E}_0$ perpendicular to the surface, indicating a much stronger molecule–electromagnetic field coupling for the former case. The $\mathscr{E}_0$-parallel-to-the-surface profile is slightly broader than the gas phase result. Further, as $\mathscr{E}_0$ is reduced, the height of the resonance peak remains essentially unchanged for $\mathscr{E}_0$ parallel to the surface while it decreases to 0.22 and zero for $\mathscr{E}_0 = 10^{-6}$ and $\mathscr{E}_0 = 10^{-7}$, respectively, if $\mathscr{E}_0$ is perpendicular to the surface. Since $\mathbf{R} \| \boldsymbol{\mu} \| \mathbf{d}$ for this transition, this clearly indicates the molecular axis is parallel to the surface for the conditions of the calculation, in agreement with experimental low-energy electron diffraction (LEED) measurements (60). On the other hand, for the $1 \rightarrow 2$ transition, Figure 11(b), the height of the two-photon resonance profile is essentially zero for $\mathscr{E}_0$ perpendicular to the surface while it is significant, and higher than the gas phase resonance maximum, for $\mathscr{E}_0$ parallel to the surface. These results again indicate the CO molecule is adsorbed, at 5 K, with the molecular axis parallel to the surface. The $\mathscr{E}_0$-perpendicular-to-the-surface resonance is quenched even though $\boldsymbol{\mu} \| \mathscr{E}_0$ since $\mathbf{d} \cdot \mathscr{E}_0 = 0$, that is, effectively, $\mathbf{d} = 0$ and so the two-photon transition is essentially not allowed; the corresponding one-photon resonance is strong. For $\mathscr{E}_0$ parallel to the surface $\mathbf{d} \cdot \mathscr{E}_0 = d\mathscr{E}_0 \neq 0$ and $\boldsymbol{\mu} \cdot \mathscr{E}_0 \neq 0$ as well since the coupling parameter $\boldsymbol{\mu} \cdot \mathscr{E}_0$, for $\boldsymbol{\mu} \perp \mathbf{d}$, involves a term (51), when expressed in terms of Euler angles, that does not vanish when $\theta = 90°$. The width of these very narrow two-photon resonances (note the ω scale in Figure 11) can be increased by increasing the strength of the applied electromagnetic field, as discussed previously.

Discussion Including Surface–Electromagnetic Field and Broadening Effects. The calculations represented in Figures 9–11 do not include two important effects, the modification of the electromagnetic field caused by the presence of the surface and broadening mechanisms for the spectral profiles. These effects can often be understood in terms of perturbations on the initial set of "standard" spectra.

The modification of the electromagnetic field, using the Fresnel equations, has been discussed by several authors (61–63), and here we use the results given by Francis and Ellison (62) for the effective electric field at the position of the molecule in the "adsorbed film." The effective field is given by Eq. 2.5, with $f(t) = 1$ and with the vector amplitude $\mathscr{E}_0$ replaced by $\mathscr{E}_0^{\text{eff}} = \hat{e}_z \mathscr{E}_{0z}^{\text{eff}} + \hat{e}_x \mathscr{E}_{0x}^{\text{eff}}$ for a continuous-wave light source whose polarization vector $\hat{e}$ is parallel to the plane (xz) of incidence (p polarization). The amplitudes $\mathscr{E}_{0z}^{\text{eff}}$ and $\mathscr{E}_{0x}^{\text{eff}}$ are functions of ε, the angle between $\hat{e}$ and the surface normal (angle of incidence $90° - \varepsilon$), and of the optical constants (refractive index, absorption coefficients) of the substrate and the adsorbed molecules. Using the optical constants for graphite and N_2 given in the literature (64, 65), the variation of $\mathscr{E}_{0z}^{\text{eff}}$ and $\mathscr{E}_{0x}^{\text{eff}}$ can be calculated (51) as a function of $0 \leqslant \varepsilon \leqslant 90°$ and ω and is shown in Figure 12 for N_2 on graphite with $\omega = 0.48$ (13.0 eV). Similar variations of these amplitudes with ε occurs for other frequencies in the ultraviolet region of the spectrum and for other molecules on graphite (e.g., CO). In the ultraviolet

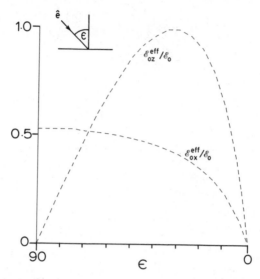

Figure 12. The effective amplitudes $\mathscr{E}_{0x}^{\text{eff}}/\mathscr{E}_0$ and $\mathscr{E}_{0z}^{\text{eff}}/\mathscr{E}_0$ for the electromagnetic field in the presence of N_2 adsorbed on graphite as a function of ε. The frequency of the light is chosen to be $\omega = 0.48$.

and contrary to the case (62, 63) for infrared, $\mathscr{E}_{0x}^{\text{eff}}$ is appreciable for most values of ε and, for example, is not zero for $\varepsilon = 90°$, where it is $0.55\mathscr{E}_0$. Furthermore, there is also a difference in $\mathscr{E}_{0z}^{\text{eff}}$ for the two regions of the spectrum; for example, in the infrared this amplitude is a maximum very close (62, 63) to $\varepsilon = 0°$, whereas in Figure 12 the maximum occurs for a much larger value of $\varepsilon \approx 32°$.

The information represented by Figure 12 imposes certain constraints on the direction of $\hat{e}$ in using electromagnetic fields in the ultraviolet frequency range to probe molecules like N_2 or CO adsorbed on graphite. If $\hat{e}$ is perpendicular to the surface ($\varepsilon = 0°$), there will be essentially no spectrum. However, choosing ε close to zero, one can effectively maximize the z component of the effective electromagnetic field relative to the x component. On the other hand, when $\hat{e}$ is parallel to the surface ($\varepsilon = 90°$), only the x component of the effective electromagnetic field contributes to the interaction with the adsorbed molecule, although the amplitude $\mathscr{E}_{0x}^{\text{eff}}$ is reduced by about a factor of 2 relative to $\mathscr{E}_0$. These observations are illustrated in Figure 13, where the resonance profile for the $X\ ^1\Sigma_g^+ \to C'\ ^1\Sigma_u^+$, $1 \to 2$, transition for N_2 on graphite, calculated (51) including the modification of the electromagnetic field by the surface, is shown for $T = 5\,\text{K}$ and $\mathscr{E}_0 = 1.8 \times 10^{-3}$.

Figures 13(a, b) correspond to the rotationally and z-averaged single-photon resonance profiles for $\varepsilon = 5°$ and $\varepsilon = 90°$, respectively. The amplitude

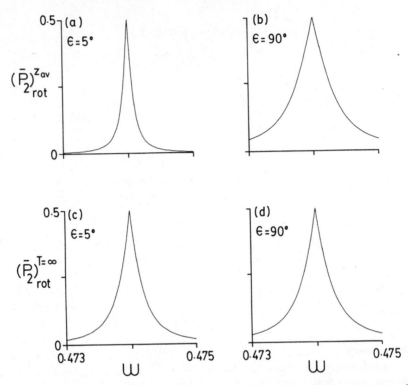

Figure 13. The steady-state rotationally and z-averaged $1 \rightarrow 2$ one-photon resonance profile for N_2 on graphite including the modification of the electromagnetic field caused by the surface: $(\bar{P}_2)^{z\,av}_{rot}$ versus ω for (a) $\varepsilon = 5°$ and (b) $\varepsilon = 90°$. Calculations are characterized as in Figure 9 except that $\mathscr{E}_0 = 1.8 \times 10^{-3}$. The corresponding free rotationally averaged spectra are included (c, d), for comparison.

$\mathscr{E}_0$ is picked in these calculations so that $\mathscr{E}^{eff}_{0x} = 10^{-3}$ for the $\varepsilon = 90°$ case, which yields a resonance profile identical to that given in Figure 9(b), which corresponds to $\varepsilon = 90°$, $\mathscr{E}_0 = 10^{-3}$ and neglect of electromagnetic field–surface coupling. Clearly, for $\varepsilon = 90°$, the effect on the resonance profile of taking the surface–field effect into account for fixed $\mathscr{E}_0$ is simply to reduce the width of the resonance by a factor of ~ 2. The spectrum in Figure 13(a) is much narrower than that in Figure 13(b) since $\varepsilon = 5°$ corresponds to $\mathscr{E}^{eff}_{0z} = 0.72 \times 10^{-3}$ and $\mathscr{E}^{eff}_{0x} = 0.27 \times 10^{-3}$ and only the weaker x component of the electromagnetic field couples with the molecule since $\boldsymbol{\mu} \parallel \mathbf{R}$ is essentially parallel to the surface. Physically, the resonance profile for $\varepsilon = 5°$, including the surface–electromagnetic field coupling effect, can be thought of as an appropriately weighted linear combination of the $\varepsilon = 0°$ and $\varepsilon = 90°$ spectra calculated neglecting this effect. Thus, even though the amplitude of the electromagnetic field used in the calculation of Figure 13(a) is less than that used in Figure 9(a), $\mathscr{E}_0^{\text{eff}} = 0.77 \times 10^{-3}$ versus $\mathscr{E}_0 = 10^{-3}$, the widths of the resonance profiles follow the opposite trend. This is due to the contribution of the component of the electric field parallel to the surface in the calculation of Figure 13(a) that couples strongly with the molecule. Also included in Figure 13 are the free rotationally averaged spectra for $\varepsilon = 5°$ and $\varepsilon = 90°$. The resonance profile for $\varepsilon = 90°$ is slightly broader following the trend in the effective amplitudes of the electromagnetic fields, $\mathscr{E}_0^{\text{eff}} = 0.77 \times 10^{-3}$ versus $\mathscr{E}_0^{\text{eff}} = 10^{-3}$, respectively.

There are several mechanisms that can broaden the spectral profiles of a molecule adsorbed on a surface. The most obvious arises from the fact that absorption bands arising from vibrational fine structure are involved in electronic transitions, and this effect can be enhanced in surface adsorption (66). In addition, important broadening mechanisms arise from energy and/or electron transfer from the adsorbed molecule to the substrate. These effects have been discussed for molecules adsorbed on metals (67, 68). Graphite, unlike a true metal, has a much smaller electron density, and one would expect frequency shifts to be negligible, that is the adsorbed molecule should more or less retain its spectral integrity except for broadening effects, and linewidths due to these effects should be smaller than for metals (69).

The photon-based spectroscopy of adsorbates has been reviewed recently by King (70). In some ways our work is most similar to reflection–absorption infrared spectroscopy (70, 71) since apparently very little analogous ultraviolet work has been done. However, experiments have been carried out involving the optical excitation of rare-gas atoms adsorbed on metals using surface reflection spectroscopy (72). Hopefully, our calculations will be useful in interpreting and/or suggesting future experiments probing molecules adsorbed on surfaces, not only on graphite but also on metals. We emphasize that Figures 9–13 represent models that are partly designed to indicate the capability of computing the spectra of adsorbed molecules. Graphite has been chosen as substrate in these models because of the availability of molecule–graphite potentials for two molecules of particular interest to us. Similar calculations could be performed for other substrates, adsorbates, and frequency ranges, with the appropriate modifications; for example, in the infrared Figure 12 would be modified considerably, even for graphite (71, 73), and for metals different selection rules (71, 74) would have to be considered. Also the various broadening mechanisms can be very dependent on the nature of the surface, the states of the adsorbate being probed, and the frequency of the light involved in the experiment. In addition, surface-heating effects (68) are also dependent on the frequency and the adsorbate material. Each individual case needs careful analysis, and research is continuing along these lines.

If a steady-state resonance profile is not obtainable due to broadening effects, the relevant transitions can be investigated as a function of time by calculating the temporal spectrum for the adsorbed molecule. Lifetimes or widths can be associated with the relevant molecular states and the time-dependent spectra evaluated phenomenologically as a function of the magnitude of the widths, as discussed briefly in Section II. Comparison of the temporal spectra, with and without broadening or decay effects, can be used to help explain the nature of the transitions and the broadening mechanisms.

The discussion of Figures 9–13 has emphasized the use of the spectra of the adsorbed molecule to deduce the orientation of the molecule relative to the surface. Apparently both one- and two-photon transitions can be very specific in this context and complement each other. Of course, such spectra are also important in other contexts, as discussed fully in the literature (67, 68). In this sense, and otherwise, the discussion given here has features in common with the use of other types of spectroscopy (70) not explicitly mentioned previously to probe adsorbed molecules, for example, electron energy loss (68), near edge X-ray absorption fine structure (75), and angle-resolved ultraviolet photo-electron (76) spectroscopies (EELS, NEXAFS, and ARUPS).

VI. CONCLUDING REMARKS

In this chapter we have explicitly considered examples of electric steady-state frequency sweep spectra. The approach discussed in Section II is equally applicable to Stark tuning, Zeeman tuning, and magnetic frequency sweep problems. The time dependence of the transition probabilities can also be readily obtained both with and without decay or broadening effects, and such results are useful in studies of the dynamics of spectral processes. In this regard the time-dependent precursor (35, 36, 38),

$$P_2^{(N)}(t) = \frac{|C(N)|^2}{(\Delta E - N\omega)^2 + |C(N)|^2} \sin^2\left\{\tfrac{1}{2}[(\Delta E - N\omega)^2 + |C(N)|^2]^{1/2}t\right\}$$

of the RWA steady-state resonance profile Eq. 4.2 will play an important role in the interpretation of temporal molecular spectra. It is clear that techniques are available for studying a wide variety of interesting and important problems; see also the other chapters of this volume.

Acknowledgments

The authors would like to thank H. Metiu for very helpful discussions on the role of a surface in broadening spectral profiles of adsorbates. This work was supported by a grant from the Natural Science and Engineering Research Council of Canada and through the award of an NSERC postgraduate scholarship to MAK. Thanks also to J. O. Hirschfelder, J. Ackerhalt, R. Coalson, D.

Campbell, and R. Wyatt and the Center for Nonlinear Studies of the Los Alamos National Laboratory for organizing and supporting the symposium on lasers, molecules, and methods upon which this volume is based.

References

1. J. D. Dollard and C. N. Friedman, *J. Math. Phys.* **18**, 1598 (1977); *J. Funct. Anal.* **28**, 309 (1978).

2. R. A. Frazer, "Disturbed Unsteady Motion and the Numerical Solution of Linear Ordinary Differential Equations," T3179, Ministry of Defence Archives, United Kingdom, 1931; R. A. Frazer, W. J. Duncan, and A. R. Collar, *Elementary Matrices and Some Applications to Dynamics and Differential Equations*, Cambridge University Press, Cambridge, 1965, Sections 7.11–7.15.

3. G. F. Thomas and W. J. Meath, *J. Phys. B.* **16**, 951 (1983).

4. O. Plaat, *Ordinary Differential Equations*, Holden-Day, San Francisco, 1971.

5. N. P. Erugin, *Linear Systems of Ordinary Differential Equations*, Academic, New York, 1966.

6. P. Roman, *Advanced Quantum Theory*, Addison-Wesley, New York, 1965.

7. V. Volterra and B. Hostinsky, *Operations Infinitesimales Lineaires*, Gauthier-Villars, Paris, 1938.

8. J. D. Dollard and C. N. Friedman, *Ann. Phys.* **111**, 251 (1978).

9. P. P. Friedmann, C. E. Hammond, and T. Woo, *Int. J. Numer. Meth. Eng.* **11**, 1117 (1977); P. P. Friedmann, *Comput. Math. Appl.* **12A**, 131 (1986).

10. R. B. Walker and R. K. Preston, *J. Chem. Phys.* **67**, 2017 (1977).

11. E. P. Dougherty, Jr., S. D. Augustin, and H. Rabitz, *J. Chem. Phys.* **74**, 1175 (1981).

12. J. Wei and E. Norman, *Proc. Am. Math. Soc.* **15**, 327 (1964).

13. N. Minorsky, *Nonlinear Oscillations*, Van Nostrand, Amsterdam, 1962.

14. A. Askar, *Phys. Rev. A* **10**, 2395 (1974).

15. J. O. Hirschfelder and R. W. Pyzalsi, *Phys. Rev. Lett.* **55**, 1244 (1985).

16. P. Pechukas and J. C. Light, *J. Chem. Phys.* **44**, 3897 (1966).

17. S. C. Mehrotra and J. E. Boggs, *J. Chem. Phys.* **64**, 2796 (1976).

18. P. W. Langhoff, S. T. Epstein, and M. Karplus, *Rev. Mod. Phys.* **7**, 602 (1972).

19. A. Javan, *Phys. Rev.* **107**, 1579 (1957); K. Shimoda and T. Shimizu, *Nonlinear Spectroscopy of Molecules*, Pergamon, Elmsford, NY, 1972.

20. J. V. Moloney and W. J. Meath, *Mol. Phys.* **35**, 1163 (1978).

21. J. V. Moloney and W. J. Meath, *Mol. Phys.* **31**, 1537 (1976).

22. A. B. Yamashita, S. Nakai, and W. J. Meath, to be published; A. B. Yamashita, Ph.D. Thesis, University of Western Ontario, 1983; M. Quack, *J. Chem. Phys.* **69**, 1282 (1978).

23. G. F. Thomas, *Phys. Rev.* **32**, 1515 (1985).

24. J. H. Shirley, *Phys. Rev. B* **138**, 979 (1965).

25. W. R. Salzman, *Phys. Rev. A* **10**, 461 (1974); **16**, 1552 (1977).

26. M. D. Burrows and W. R. Salzman, *Phys. Rev. A* **15**, 1636 (1977).

27. J. V. Moloney and W. J. Meath, *Mol. Phys.* **30**, 171 (1975).

28. D. R. Dion and J. O. Hirschfelder, *Adv. Chem. Phys.* **35**, 265 (1976).

29. S-I. Chu, *Adv. Atom. Mol. Phys.* **21**, 197 (1985).

30. R. A. Thuraisingham and W. J. Meath, *Mol. Phys.* **56**, 193 (1985).

31. G. Oliver, *Lett. Nuov. C.* **2**, 1075 (1971); *Phys. Rev. A* **15**, 2424 (1975); D. T. Pegg, *J. Phys. B* **6**, 246 (1973); R. Loudon, *Quantum Theory of Light*, Clarendon, New York, 1973, pp. 90–100.

32. J. V. Moloney and W. J. Meath, *Phys. Rev. A* **17**, 1550 (1978) and references therein.

33. J. V. Moloney and W. J. Meath, *J. Phys. B.* **11**, 2641 (1978).

34. M. Sargent III, M. O. Scully, and W. E. Lamb, Jr., *Laser Physics*, Addison-Wesley, Reading, MA, 1974, Chapter 2.

35. M. A. Kmetic and W. J. Meath, *Phys. Lett.* **108A**, 340 (1985).

36. G. F. Thomas, *Phys. Rev. A* **33**, 1033 (1986).

37. G. F. Thomas and W. J. Meath, *Mol. Phys.* **46**, 743 (1982); erratum, *Mol. Phys.* **48**, 649 (1983).

38. M. A. Kmetic, R. A. Thuraisingham, and W. J. Meath, *Phys. Rev. A* **33**, 1688 (1986).

39. W. Liptay, in *Excited States*, Vol. 1, E. C. Lim, Ed., Academic, New York, 1974, p. 198.

40. See, for example, P. Van Pelt and E. E. Havinga, in *Nonlinear Behaviour of Molecules, Atoms and Ions in Electric, Magnetic or Electromagnetic Fields*, L. Neel, Ed., Elsevier, New York, 1979, p. 291; J. Griffiths, *Chem. Brit.* **22**, 997 (1986).

41. A. M. Bonch-Bruevich and V. A. Khodovoi, *Sov. Phys. Usp.* **10**, 637 (1967); A. V. Gaponov, Yu. N. Remkov, N. G. Protopopova, and V. M. Fain, *Opt. Spectros.* **19**, 279 (1965).

42. W. J. Meath and E. A. Power, *Mol. Phys.* **51**, 585 (1984).

43. K. P. Huber and G. Herzberg, *Molecular Spectra and Molecular Structure IV. Constants of Diatomic Molecules*, Van Nostrand-Reinhold, New York, 1978, p. 196.

44. G. F. Thomas, *J. Chem. Phys.* **79**, 4912 (1983); *J. Chem. Phys.* **86**, 71 (1987).

45. H. J. Beyer and K. Kollath, *J. Phys. B* **10**, L5 (1977) and references therein.

46. E. A. Power and T. Thirunamachandran, *J. Chem. Phys.* **60**, 3695 (1974).

47. D. L. Andrews and T. Thirunamachandran, *J. Chem. Phys.* **67**, 5026 (1977).

48. T-S. Ho and S-I. Chu, *J. Chem. Phys.* **79**, 4708 (1983).

49. S. Leasure, K. F. Milfeld, and R. E. Wyatt, *J. Chem. Phys.* **74**, 6197 (1981).

50. K. Yamaoka and E. Charney, *J. Am. Chem. Soc.* **94**, 8693 (1972); W. Liptay and J. Czekalla, *Z. Electrochem.* **65**, 721 (1961).

51. R. A. Thuraisingham and W. J. Meath, *Surf. Sci. and Chem. Phys.*, in press.

52. S. Nagakura, M. Kojima, Y. Morayama, *J. Molec. Spectros.* **13**, 174 (1964) and reference 39, pp. 260–261.

53. J. Talbot, D. J. Tildesley, and W. A. Steele, *Mol. Phys.* **51**, 1331 (1984).

54. W. A. Steele, *J. Phys., Paris* **C4-61**, (1977).

55. J. Belak, K. Kobashi, and R. D. Etters, *Surf. Sci.* **161**, 390 (1985).

56. W. Coughran, J. Rose, T. Shibuya, and V. McKoy, *J. Chem. Phys.* **58**, 2699 (1973); J. Rose, T. Shibuya, and V. McKoy, *J. Chem. Phys.* **58**, 74 (1973); D. L. Yeager and V. McKov, *J. Chem. Phys.* **67**, 2473 (1973).

57. R. K. Nesbet, *J. Chem. Phys.* **43**, 4403 (1965).

58. K. P. Huber and G. Herzberg, *Molecular Spectra and Molecular Structure IV. Constants of Diatomic Molecules*, Van Nostrand-Reinhold, New York, 1978, p. 162.

59. C. Chackerian, *J. Chem. Phys.* **65**, 4228 (1976).

60. H. You and S. C. Fain, *Surf. Sci.* **151**, 361 (1985).

61. T. C. Fry, *J. Opt. Soc. Am.* **22**, 307 (1932).

62. S. A. Francis and A. H. Ellison, *J. Opt. Soc. Am.* **49**, 131 (1959).

63. R. G. Greenler, *J. Chem. Phys.* **44**, 310 (1966).

64. E. A. Taft and H. R. Philipp, *Phys. Rev. A* **138**, 197 (1965).
65. R. C. Weast, Ed. *Handbook of Chemistry and Physics*, 57th ed., CRC, Boca Raton, FL, 1976, p. E224.
66. J E. Demuth and Ph. Avouris, *Phys. Rev. Lett.* **47**, 61 (1981).
67. Ph. Avouris and J. E. Demuth, in *Surface Studies with Lasers*, F. R. Aussenegg, A. Leitner, and M. E. Lippitsch, Eds., Springer-Verlag, New York, 1983, p. 24.
68. S. Nagano, Z-P. Luo, H. Metiu, W. M. Huo, M. A. P. Lima, and V. McKoy, *J. Chem. Phys.* **85**, 6153 (1986).
69. T. Mandel, M. Domke, G. Kaindl, C. Laubschat, M. Prietsch, U. Middelmann, and K. Horn, *Surf. Sci.* **162**, 453 (1985).
70. D. A. King, *Chem. Brit.* **22**, 819 (1986).
71. F. Hoffmann, *Surf. Sci. Rep.* **3**, 107 (1983).
72. J. A. Cunningham, D. K. Greenlaw, and C. P. Flynn, *Phys. Rev. B* **22**, 717 (1980).
73. H. R. Philipp, *Phys. Rev. B* **16**, 2896 (1977).
74. G. W. Rubloff, *Solid State Commun.* **26**, 523 (1978).
75. J. Stöhr and R. Jaeger, *Phys. Rev. B* **26**, 4111 (1982).
76. P. Hofmann, S. R. Bare, N. V. Richardson, and D. A. King, *Solid State Commun.* **42**, 645 (1982).

CHAPTER IX

ANALYTIC SOLUTIONS AND DYNAMIC SYMMETRIES IN LASER-DRIVEN ATOMIC EXCITATIONS

F. T. HIOE

Department of Physics, St. John Fisher College, Rochester, New York 14618

CONTENTS

I. INTRODUCTION

The purpose of this chapter is to present a collection of some analytic results in the problem of laser-driven atomic excitations that have been derived since early 1980 by the author and in some parts by the author in collaboration with C. E. Carroll and J. H. Eberly.

The content can be divided roughly into two parts. The first part, composed of Sections II and III, deals mainly with the coherent transition probabilities of two- and three-level systems driven by lasers. We consider systems in which the frequencies and amplitudes of the driving laser fields are generally dependent on time. Several particular classes of incident laser fields are considered; each consists of an infinite variety of time-dependent amplitude and frequency variations. The results on the final-level occupation probabilities are remarkable in that they are all very simply expressed in terms of elementary functions even though the intermediate steps leading to these simple results involve, in all the cases considered, higher transcendental

351

functions, hypergeometric and generalized hypergeometric functions being two specific examples. These simple analytic solutions are particularly useful for realizing the conditions for the atomic system to be fully excited or to be fully returned to its original state. The latter case is important in relation to the problem of solitary-pulse propagation. Experimentally, the picosecond and femtosecond pulses have many practical applications and are of special interest in studies of the photodissociation of molecules (1). The first implementation of a single laser pulse with continuous amplitude and frequency modulation for optical spectroscopy has recently been reported (2), and it has opened the possibility of a wide variety of coherent spectroscopic applications for which our analytic results may become especially useful.

The second part of this review, composed of Section IV which has several subsections, gives a number of interesting analytic results for a general N-level system when the system possesses a certain dynamic symmetry. A general understanding in terms of group-theoretical concepts and in particular the exploitation of what the author has called the SU (2) type and Gell-Mann type dynamic symmetries not only has unified previously apparently unrelated results, but also has enabled us to predict many new results regarding the existence of constants of motion and the conditions under which, for example, some fraction of the electron population will be trapped in certain levels when the atomic system is coherently excited, and the conditions under which an N-level system can support a multiple solitary pulse propagation.

II. ANALYTIC SOLUTIONS FOR TWO-LEVEL SYSTEMS

The time evolution of a two-level or two-state problem in which the two states are coupled by a possibly time-dependent interaction is governed by the following coupled equations from the time-dependent Schrödinger equation for the probability amplitudes $C_1(t)$ and $C_2(t)$ for the two states $|1\rangle$ and $|2\rangle$ (in units $\hbar = 1$):

$$i\frac{d}{dt}\begin{pmatrix} C_1 \\ C_2 \end{pmatrix} = \begin{pmatrix} H_{11} & H_{12} \\ H_{21} & H_{22} \end{pmatrix}\begin{pmatrix} C_1 \\ C_2 \end{pmatrix}, \qquad (2.1)$$

where $H_{jk} = \langle j|H|k\rangle$ represents the matrix element of the Hamiltonian H.

Beginning with the pioneering work of Rosen and Zener (3), Landau (4), Zener (5), and Rabi (6) some 50 years ago, the problem of finding analytic solutions of Eq. 2.1 for various forms of time-dependent H_{ij} has persisted, notably in the work of Wannier (7), Nikitin (8), Demkov (9), Kaplan (10), Allen and Eberly (11), Crothers (12), Lee and George (13), Bambini and Berman (14), and Bambini and Linberg (15). A considerable unification as well as extension of various analytic results has been obtained by Hioe and Carroll (16–19), whose results are summarized and presented in this section.

By means of a transformation,

$$C_j = a_j \exp\left(-i\int^t H_{jj}\,dt\right), \qquad j = 1, 2. \tag{2.2}$$

Equation 2.1 can be replaced by the following coupled equations with zero diagonal elements:

$$i\frac{d}{dt}\begin{pmatrix} a_1 \\ a_2 \end{pmatrix} = \begin{pmatrix} 0 & -\dfrac{1}{2}\dfrac{dA}{dt}e^{-iB} \\ -\dfrac{1}{2}\dfrac{dA}{dt}e^{iB} & 0 \end{pmatrix}\begin{pmatrix} a_1 \\ a_2 \end{pmatrix}, \tag{2.3}$$

where

$$\frac{dA}{dt}e^{-iB} = -2H_{12}\exp\left(-i\int^t (H_{22} - H_{11})dt\right), \tag{2.4a}$$

and

$$\frac{dA}{dt}e^{iB} = -2H_{21}\exp\left(i\int^t (H_{22} - H_{11})dt\right). \tag{2.4b}$$

In the application to a two-level atom driven by a laser beam, the so-called rotating-wave approximation is usually made in which Eq. 2.1 is replaced by a set of coupled equations of the same form but the matrix elements are now functions that vary slowly with time compared to the optical frequencies. The quantity dA/dt in Eq. 2.3 is the so-called Rabi frequency $\Omega(t)$, which is equal to $2\,dE(t)$, where d is the transition dipole moment and $E(t)$ is the amplitude of the optical frequency electric field (20). We write

$$\Omega(t) = \frac{dA}{dt}, \tag{2.5}$$

where A is the dimensionless pulse area up to time t. The total area of the pulse is

$$\alpha = A(+\infty) - A(-\infty) = \int_{-\infty}^{\infty} \Omega(t)\,dt. \tag{2.6}$$

The quantity dB/dt is the detuning $\Delta(t) \equiv \omega_0 - \omega$, where ω_0 is the transition frequency between the two levels of the atomic system and ω is the frequency of the incident laser field; that is,

$$\Delta(t) = \frac{dB}{dt}. \tag{2.7}$$

Time-dependent detuning might be produced by varying the laser frequency or by varying the differences of atomic energy levels using the Stark or Zeeman shifts. The absolute squares of the solutions of Eq. 2.1 or 2.3, $|a_1|^2$ and $|a_2|^2$, which are equal to $|C_1|^2$ and $|C_2|^2$, respectively, are the occupation probabilities of the respective levels.

Equation 2.3 can also be written in the form of the undamped Bloch equations (20):

$$\frac{d}{dt}\begin{pmatrix} u \\ v \\ w \end{pmatrix} = \begin{pmatrix} 0 & -\Delta(t) & 0 \\ \Delta(t) & 0 & \Omega(t) \\ 0 & -\Omega(t) & 0 \end{pmatrix}\begin{pmatrix} u \\ v \\ w \end{pmatrix}, \tag{2.8}$$

where u, v, and w are the three components of the pseudopolarization in quantum optics or of the magnetization in magnetic resonance that are related to the a_1 and a_2 by

$$u = a_1 a_2^* e^{iB} + \text{c.c.}, \tag{2.9a}$$

$$v = -i a_1 a_2^* e^{iB} + \text{c.c.}, \tag{2.9b}$$

$$w = |a_2|^2 - |a_1|^2. \tag{2.9c}$$

A change of independent variable, which introduces an arbitrary function, will now be introduced. Let z be the new independent variable. It is real when t is real, and dz/dt is nonnegative when t is real. Using 2.5, we rewrite 2.3 as

$$i\frac{d}{dz}\begin{pmatrix} a_1 \\ a_2 \end{pmatrix} = \begin{pmatrix} 0 & -\frac{1}{2}\frac{dA}{dz}e^{-iB} \\ -\frac{1}{2}\frac{dA}{dz}e^{iB} & 0 \end{pmatrix}\begin{pmatrix} a_1 \\ a_2 \end{pmatrix}. \tag{2.10}$$

Elimination of a_2 and a_1 gives

$$\frac{d^2 a_1}{dz^2} + \left[\frac{d}{dz}\left(iB - \ln\frac{dA}{dz}\right)\right]\frac{da_1}{dz} + \frac{1}{4}\left(\frac{dA}{dz}\right)^2 a_1 = 0 \tag{2.11a}$$

and

$$\frac{d^2 a_2}{dz^2} + \left[\frac{d}{dz}\left(-iB - \ln\frac{dA}{dz}\right)\right]\frac{da_2}{dz} + \frac{1}{4}\left(\frac{dA}{dz}\right)^2 a_2 = 0. \tag{2.11b}$$

We shall choose the functions $A(z)$ and $B(z)$ so that a_1 and a_2 can be written in terms of hypergeometric functions. The relation between z and t is arbitrary, which allows an infinite variety of amplitude and frequency modulations.

As we showed in Ref. 19, the requirement of finite area implies, that there are two classes of solutions of Eq. 2.10, each class consisting of three arbitrary parameters α, β, and γ and the arbitrary function dz/dt. We are interested in the final occupation probabilities that are the limits of $|a_1|^2$ and $|a_2|^2$ as $t \to +\infty$ with the initial condition that one of the two states is occupied with unit probability at early times, that is, $|a_1| \to 1$ and $a_2 \to 0$ as $t \to -\infty$.

The first class of solutions is obtained by choosing the functions $A(z)$ and $B(z)$ to be given by

$$\Omega(t) = \frac{dA}{dt} = \frac{\alpha}{\pi}\frac{1}{[z(1-z)]^{1/2}}\frac{dz}{dt} \tag{2.12a}$$

and

$$\Delta(t) = \frac{dB}{dt} = \frac{1}{\pi}\left(\frac{\beta}{z} + \frac{\gamma}{1-z}\right)\frac{dz}{dt}, \tag{2.12b}$$

where α (which satisfies 2.6), β, and γ are arbitrary parameters. The limits of the occupation probabilities as $t \to +\infty$ are

$$|a_1|^2 \to [\cosh(\beta + \gamma) + \cos\Phi]/[2\cosh(\beta)\cosh(\gamma)], \tag{2.13a}$$

and

$$|a_2|^2 \to [\cosh(\beta - \gamma) - \cos\Phi]/[2\cosh(\beta)\cosh(\gamma)], \tag{2.13b}$$

where

$$\Phi = [\alpha^2 - (\beta - \gamma)^2]^{1/2}. \tag{2.14}$$

It can be readily verified that

i. if $\beta = \gamma$, the population completely returns to level 1 at $t = +\infty$ when

$$\Phi = 2n\pi, \qquad n = 1, 2, 3,\ldots \tag{2.15}$$

and

TABLE 1. Explicit Formulas for Detuning Functions Obtained from Seven Simple Pulse-Amplitude Functions

Amplitude Function $\Omega(t)$	Detuning Function $\Delta(t)$, Class I, Eq. 2.12b	Detuning Function $\Delta(t)$, Class II, Eq. 2.17b
$\dfrac{\alpha}{\tau}\operatorname{sech}\dfrac{\pi t}{\tau}$	$\dfrac{1}{\tau}\left(\beta+\gamma-(\beta-\gamma)\tanh\dfrac{\pi t}{\tau}\right)$	$\dfrac{\beta+\gamma-(\beta-\gamma)\sinh(\pi t/\tau)}{\tau\cosh(\pi t/\tau)}$
$\dfrac{\alpha}{\pi^{1/2}\tau}\exp\left(-\dfrac{t^2}{\tau^2}\right)$	$\dfrac{\exp(-t^2/\tau^2)}{\pi^{1/2}\cos[(1/2)\pi\operatorname{erf}(t/\tau)]}$ $\times\left\{\beta+\gamma-(\beta-\gamma)\sin\left[\dfrac{1}{2}\pi\operatorname{erf}\dfrac{t}{\tau}\right]\right\}$	$\dfrac{\exp(-t^2/\tau^2)}{\pi^{1/2}\tau}\left\{\beta+\gamma-(\beta-\gamma)\tan\left[\dfrac{1}{2}\pi\operatorname{erf}\dfrac{t}{\tau}\right]\right\}$
$\dfrac{\alpha\tau}{\pi(t^2+\tau^2)}$	$\dfrac{(\beta+\gamma)(t^2+\tau^2)^{1/2}-(\beta-\gamma)t}{\pi(t^2+\tau^2)}$	$\dfrac{(\beta+\gamma)\tau-(\beta-\gamma)t}{\pi(t^2+\tau^2)}$
$\dfrac{\alpha}{2\tau}$ for $|t|<\tau$ 0 for $|t|\geqslant\tau$	$\dfrac{\beta+\gamma+(\beta-\gamma)\sin(\frac{1}{2}\pi t/\tau)}{2\tau\cos(\frac{1}{2}\pi t/\tau)}$ for $|t|<\tau$	$\dfrac{1}{2\tau}\left[\beta+\gamma+(\beta-\gamma)\tan\dfrac{(1/2)\pi t}{\tau}\right]$ for $|t|<\tau$

$$\frac{\alpha}{2\tau}\exp\left(-\frac{|t|}{\tau}\right)$$

$$\frac{\exp(-|t/\tau|)}{2\tau\sin[(1/2)\pi\exp(-|t/\tau|)]}$$

$$\times\left\{\beta+\gamma-(\beta-\gamma)(\operatorname{sgn}t)\right\}$$

$$\times\cos\left[\tfrac{1}{2}\pi\exp\left(-\frac{|t|}{\tau}\right)\right]$$

$$\frac{\exp(-|t/\tau|)}{2\tau}\left\{\beta+\gamma-(\beta-\gamma)(\operatorname{sgn}t)\cot\left[\tfrac{1}{2}\pi\exp\left(-\frac{|t|}{\tau}\right)\right]\right\}$$

$$\frac{\alpha}{\tau}\exp\left(-\frac{t}{\tau}\right)\ \text{for }t<0$$

$$0\ \text{for }t\geqslant 0$$

$$\frac{\exp(t/\tau)}{\tau\sin[\pi\exp(t/\tau)]}$$

$$\times\left[\beta+\gamma+(\beta-\gamma)\cos\left(\pi\exp-\frac{t}{\tau}\right)\right]$$

$$\text{for }t<0$$

$$\frac{\exp(t/\tau)}{\tau}\left[\beta+\gamma+(\beta-\gamma)\cot\left[\pi\exp-\frac{t}{\tau}\right]\right]\ \text{for }t<0$$

$$0\ \text{for }t\leqslant 0$$

$$\frac{\alpha}{\tau}\exp\left(-\frac{t}{\tau}\right)\ \text{for }t>0$$

$$\frac{\exp(-t/\tau)}{\tau\sin[\pi\exp(-t/\tau)]}$$

$$\times\left[\beta+\gamma-(\beta-\gamma)\cos\left(\pi\exp\left(\frac{t}{\tau}\right)\right)\right]$$

$$\text{for }t>0$$

$$\frac{\exp(-t/\tau)}{\tau}\left\{\beta+\gamma-(\beta-\gamma)\cot\left[\pi\exp\left(\frac{-t}{\tau}\right)\right]\right\}\ \text{for }t>0$$

ii. if $\beta = -\gamma$, the population is completely excited to level 2 at $t = +\infty$ when

$$\Phi = (2n-1)\pi, \qquad n = 1, 2, 3, \ldots . \tag{2.16}$$

The second class of solutions is obtained by choosing the functions $A(z)$ and $B(z)$ to be given by

$$\Omega(t) = \frac{dA}{dt} = \frac{\alpha}{\pi} \frac{1}{z^2 + 1} \frac{dz}{dt} \tag{2.17a}$$

and

$$\Delta(t) = \frac{dB}{dt} = \frac{1}{\pi} \frac{-(\beta - \gamma)z + (\beta + \gamma)}{z^2 + 1} \frac{dz}{dt}, \tag{2.17b}$$

where α (which satisfies 2.6), β, and γ are arbitrary parameters. The limits of the occupation probabilities as $t \to +\infty$ are

$$|a_1|^2 \to [2\cosh(\beta - \gamma)\cos(r)\cos(s) - \cos^2(r) - \cos^2(s)]/\sinh^2(\beta - \gamma) \tag{2.18a}$$

and

$$|a_2|^2 \to [\sinh^2(\beta - \gamma) - 2\cosh(\beta - \gamma)\cos(r)\cos(s)$$
$$+ \cos^2(r) + \cos^2(s)]/\sinh^2(\beta - \gamma), \tag{2.18b}$$

where

$$r = \tfrac{1}{2}\{\alpha^2 + [\beta + \gamma + i(\beta - \gamma)]^2\}^{1/2}, \tag{2.19a}$$

$$s = \tfrac{1}{2}\{\alpha^2 + [\beta + \gamma - i(\beta - \gamma)]^2\}^{1/2}. \tag{2.19b}$$

The case $\beta = -\gamma$ for the second class of solutions (Eqs. 2.17 and 2.18) coincides with the case $\beta = -\gamma$ for the first class of solutions (Eqs. 2.12 and 2.13) for which the occupation probabilities 2.13 and 2.18 can be written more simply as

$$|a_1|^2 \to \operatorname{sech}^2(\tfrac{1}{2}\beta)\cos^2 \tfrac{1}{2}\sqrt{\alpha^2 - 4\beta^2} \tag{2.20a}$$

and

$$|a_2|^2 \to 1 - \operatorname{sech}^2(\tfrac{1}{2}\beta)\cos^2 \tfrac{1}{2}\sqrt{\alpha^2 - 4\beta^2}. \tag{2.20b}$$

The case $\beta = \gamma$ for the second class of solutions will be referred to as the generalized Rabi solution for which the Rabi frequency $\Omega(t)$ and the detuning

TABLE 2. Relations Between z and t Corresponding to Seven Simple Pulse Amplitude Functions

Amplitude Function $\Omega(t)$	Class I, Eq. 2.12 $z(t)$	Class II, Eq. 2.17 $z(t)$
$\dfrac{\alpha}{\tau}\operatorname{sech}\dfrac{\pi t}{\tau}$	$\dfrac{1}{2}\left(1+\tanh\dfrac{\pi t}{\tau}\right)$	$\sinh\dfrac{\pi t}{\tau}$
$\dfrac{\alpha}{\pi^{1/2}\tau}\exp\left(-\dfrac{t^2}{\tau^2}\right)$	$\dfrac{1}{2}\left\{1+\sin\left[\tfrac{1}{2}\pi\operatorname{erf}\dfrac{t}{\tau}\right]\right\}$	$\tan\left(\tfrac{1}{2}\pi\operatorname{erf}\dfrac{t}{\tau}\right)$
$\dfrac{\alpha\tau}{\pi(t^2+\tau^2)}$	$\dfrac{1}{2}\left[1+\dfrac{t}{(t^2+\tau^2)^{1/2}}\right]$	$\dfrac{t}{\tau}$
$\dfrac{\alpha}{2\tau}$ for $|t|<\tau$ 0 for $|t|\geqslant\tau$	$\dfrac{1}{2}\left(1+\sin\dfrac{(1/2)\pi t}{\tau}\right)$ for $|t|<\tau$	$\tan\dfrac{(1/2)\pi t}{\tau}$ for $|t|<\tau$
$\dfrac{\alpha}{2\tau}\exp\left(-\dfrac{|t|}{\tau}\right)$	$\dfrac{1}{2}\left\{1+(\operatorname{sgn}t)\cos\left[\tfrac{1}{2}\pi\exp\left(-\left|\dfrac{t}{\tau}\right|\right)\right]\right\}$	$(\operatorname{sgn}t)\cot\left[\tfrac{1}{2}\pi\exp\left(-\left|\dfrac{t}{\tau}\right|\right)\right]$
$\dfrac{\alpha}{\tau}\exp\left(\dfrac{t}{\tau}\right)$ for $t<0$ 0 for $t\geqslant0$	$\dfrac{1}{2}\left[1-\cos\left(\pi\exp\dfrac{t}{\tau}\right)\right]$ for $t<0$	$-\cot\left(\pi\exp\dfrac{t}{\tau}\right)$ for $t<0$
$\dfrac{\alpha}{\tau}\exp\left(-\dfrac{t}{\tau}\right)$ for $t>0$ 0 for $t\leqslant0$	$\dfrac{1}{2}\left\{1+\cos\left[\pi\exp\left(-\dfrac{t}{\tau}\right)\right]\right\}$ for $t>0$	$\cot\left[\pi\exp\left(-\dfrac{t}{\tau}\right)\right]$ for $t>0$

$\Delta(t)$ have the same time dependence. This case is not covered by the first class of solutions.

For the case in which the detuning $\Delta = dB/dt$ is constant, we use Eq. 2.12 to find the Rabi frequency to be given by

$$\Omega(t) = \frac{(\alpha\Delta/\beta)[z(1-z)]^{1/2}}{1+\gamma/\beta-z} \tag{2.21}$$

and t is related to z by

$$t = (\beta/\pi\Delta)\ln[z(1-z)^{-\gamma/\beta}], \tag{2.22}$$

where we have assumed that β, γ, and Δ have the same sign. The special case given by Eqs. 2.22 and 2.21 was given by Bambini and Berman (14).

For both classes of solutions of the two-state problem, the relation between z and t can be chosen to give an arbitrary pulse amplitude function $\Omega(t)$. Seven simple forms for $\Omega(t)$ and the detuning function $\Delta(t)$ are listed in Table 1. The functions $z(t)$ are given, for each case, in Table 2. The parameters α, β, and γ, but not the shapes of $\Omega(t)$ and $\Delta(t)$, determine the final occupation probabilities given by Eqs. 2.13 and 2.18 for the first and second class of solutions, respectively.

A related result for completely inverting a two-level system by a resonant pulse was given by Robinson (21).

III. ANALYTIC SOLUTIONS FOR THREE-LEVEL SYSTEMS

A. Generalization of Landau–Zener Solution

The Landau–Zener model is widely used in studying the dynamics involving the avoided crossing of two energy levels (22). The probability of a transition between the two levels was found by Landau (4) and Zener (5). When it is applied to quantum optics, the Landau–Zener solution gives the transition probability of a two-level atom when it is driven by a laser beam of constant amplitude and driving frequency that is proportional to the time. The Schrödinger equation for the two-level model can be written as

$$i\frac{d}{dt}\begin{pmatrix} a_1 \\ a_2 \end{pmatrix} = \begin{pmatrix} rt & -\tfrac{1}{2}\Omega \\ -\tfrac{1}{2}\Omega^* & -rt \end{pmatrix}\begin{pmatrix} a_1 \\ a_2 \end{pmatrix}, \tag{3.1.1}$$

where r denotes the rate of sweeping through the avoided crossing and where the Rabi frequency Ω is constant but may be complex. Assuming that state 1 is occupied with unit probability in the limit $t \to -\infty$, the final occupation probabilities as $t \to +\infty$ are

$$|a_1|^2 \to \exp(-4\pi|s|^2) \qquad (3.1.2a)$$

and

$$|a_2|^2 \to 1 - \exp(-4\pi|s|^2), \qquad (3.1.2b)$$

where

$$s = \Omega/4r^{1/2}. \qquad (3.1.3)$$

The occupation probabilities at the intermediate times can be expressed in terms of the confluent hypergeometric, or Whittaker, function.

In the generalization to the three-level Landau–Zener model (23–25), we consider two transitions driven by laser beams of constant amplitude, possibly derived from the same laser, and assume that the detunings of the two laser beams are both proportional to the time. If the frequencies of the two laser beams pass through resonance at very different times, the calculation of transition probabilities can be done by solving a sequence of two two-level problems. Such calculations will not be considered. Thus, we shall assume that the frequencies of the two laser beams pass through resonance at the same time, which we choose to be at $t = 0$ when t goes from $-\infty$ to $+\infty$.

Consider an experiment in which two laser beams drive two transitions in an atom. The driven transitions connect level 1 to level 2 and level 2 to level 3. The 1–3 transition is not driven by the laser field because of the electric dipole selection rules. We assume each laser beam drives only its own transition. The Schrödinger equation for the system has the form

$$i\frac{d}{dt}\begin{pmatrix} a_1 \\ a_2 \\ a_3 \end{pmatrix} = \begin{pmatrix} 2r_1t & -\tfrac{1}{2}\Omega_{12} & 0 \\ -\tfrac{1}{2}\Omega_{12} & 2r_2t & -\tfrac{1}{2}\Omega_{23} \\ 0 & -\tfrac{1}{2}\Omega_{23} & 2r_3t \end{pmatrix}\begin{pmatrix} a_1 \\ a_2 \\ a_3 \end{pmatrix}, \qquad (3.1.4)$$

where a_1, a_2, a_3 are the amplitudes for the three atomic states or atomic levels. The Rabi frequencies are Ω_{12} and Ω_{23}; each of them is twice the product of a transition dipole moment and the corresponding field amplitude. The corresponding detunings are

$$2(r_1 - r_2)t \quad \text{and} \quad 2(r_2 - r_3)t. \qquad (3.1.5)$$

We assume that the amplitude of each frequency field is independent of time, so that Ω_{12} and Ω_{23} are constants.

Eliminating a_2 and a_3 from Eq. 3.1.4, we find the following third-order differential equation for a_1:

$$\frac{d^3 a_1}{dt^3} + 2i(r_1 + r_2 + r_3)t\frac{d^2 a_1}{dt^2}$$

$$+ \{-4(r_1 r_2 + r_2 r_3 + r_3 r_1)t^2 + 2i(2r_1 + r_2) + \tfrac{1}{4}[(\Omega_{12})^2 + (\Omega_{23})^2]\}\frac{da_1}{dt}$$

$$+ \{-8ir_1 r_2 r_3 t^3 - 4r_1(2r_2 + r_3)t + \tfrac{1}{2}i[(\Omega_{12})^2 r_3 + (\Omega_{23})^2 r_1]t\}a_1$$

$$= 0. \tag{3.1.6}$$

Equation 3.1.6 by itself does not reveal any obvious clue as to whether it is some sort of a generalization of the confluent hypergeometric equation satisfied by the Whittaker function, which is known to be the solution of the two-level Landau–Zener model. However, a change of the independent variable from t to $x = t^2$ leads to a differential equation in which all the polynomials appearing as coefficients of the derivatives of a_1 are linear in x. A known method known as the Laplace method (26) will give integral representations of the solutions we have been able to evaluate. It may be remarked that the same method does not seem to be applicable to four or more level models.

Assuming that no two of the rates of sweep, r_1, r_2 and r_3, are equal, the integral representations for the solutions suggest defining two dimensionless parameters

$$p = \frac{(\Omega_{12})^2}{16(r_1 - r_2)} \quad \text{and} \quad q = \frac{(\Omega_{23})^2}{16(r_2 - r_3)}. \tag{3.1.7}$$

The three final occupation probabilities can be expressed in terms of

$$P = \exp(-2\pi|p|) \quad \text{and} \quad Q = \exp(-2\pi|q|), \tag{3.1.8}$$

and they are listed in Table 3 for the various possibilities. It will be noted that the transition probabilities can be discontinuous functions of $r_1 - r_2$ and $r_2 - r_3$. When $r_1 = r_2$ or $r_2 = r_3$, the probability of no transition or the final occupation probability are given in Tables 4 and 5.

Let us represent the triplet of values for the final occupation probabilities of the three levels $(|a_1|^2, |a_2|^2, |a_3|^2)$ by a point inside an equilateral triangle having each altitude equal to unity. The position of the point is determined by the three perpendicular distances of lengths equal to $|a_1|^2$, $|a_2|^2$, and $|a_3|^2$, respectively from the point to the three sides labeled I, II, and III. The three distances representing three probabilities always sum to unity.

TABLE 3. Final Occupation Probabilities for Non resonant Cases[a]

Initial Occupation Probabilities 1, 0, 0	Initial Occupation Probabilities 0, 1, 0	Initial Occupation Probabilities 0, 0, 1
Cases of $r_2 > r_1 > r_3$ and $r_3 > r_1 > r_2$		
$\begin{pmatrix}(1 - Q + PQ)^2 \\ Q(1 - P)(1 + PQ) \\ Q(1 - P)(1 - Q)\end{pmatrix}$	$\begin{matrix}Q(1 - P)(1 + PQ) \\ P^2 Q^2 \\ (1 - Q)(1 + PQ)\end{matrix}$	$\begin{pmatrix}Q(1 - P)(1 - Q) \\ (1 - Q)(1 + PQ) \\ Q^2\end{pmatrix}$
Cases of $r_1 > r_2 > r_3$ and $r_3 > r_2 > r_1$		
$\begin{pmatrix}P^2 \\ (1 - P)(P + Q) \\ (1 - P)(1 - Q)\end{pmatrix}$	$\begin{matrix}(1 - P)(P + Q) \\ (1 - P - Q)^2 \\ (1 - Q)(P + Q)\end{matrix}$	$\begin{pmatrix}(1 - P)(1 - Q) \\ (1 - Q)(P + Q) \\ Q^2\end{pmatrix}$
Cases of $r_1 > r_3 > r_2$ and $r_2 > r_3 > r_1$		
$\begin{pmatrix}P^2 \\ (1 - P)(1 + PQ) \\ P(1 - P)(1 - Q)\end{pmatrix}$	$\begin{matrix}(1 - P)(1 + PQ) \\ P^2 Q^2 \\ P(1 - Q)(1 + PQ)\end{matrix}$	$\begin{pmatrix}P(1 - P)(1 - Q) \\ P(1 - Q)(1 + PQ) \\ (1 - P + PQ)^2\end{pmatrix}$

[a] The initial condition is that one of the three levels is certainly occupied in the limit as $t \to -\infty$. The three possible initial levels correspond to the three columns. The final occupation probabilities for the nth level appear as the nth row in the appropriate matrix. Hence, the probabilities of no transition appear as diagonal matrix elements. P and Q are defined by 3.1.8, and the matrices are symmetric.

TABLE 4. Probability of No Transition for $r_1 = r_2 \neq r_3$ and $r_1 \neq r_2 = r_3$[a]

Resonant Case	Initial Occupation Probabilities for Levels 1, 2, and 3	Probability of No Transition
$r_1 = r_2$	0, 0, 1	$\exp(-4\pi|q|)$
$r_2 = r_3$	1, 0, 0	$\exp(-4\pi|p|)$

[a] Either p or q is defined; see 3.1.7. The initial occupation probabilities must be as in the middle column in order to have definite limits as $t \to -\infty$. Although two occupation probabilities oscillate indefinitely as $t \to +\infty$. The probability of no transition can be calculated.

Figure 1 shows where the point representing the final occupation probabilities of the three levels will lie for various values of p and q for the case $r_1 > r_2 > r_3$, where we have assumed that the initial occupation probabilities of levels 1, 2, 3 are 1, 0, 0. It is seen that a forbidden region is revealed that says that certain values of the three final occupation probabilities cannot be attained for the given initial condition. When the initial occupation probabilities of levels 1, 2, 3 are 0, 1, 0, no forbidden region exists, as shown in Figure 2.

TABLE 5. Final Occupation Probabilities for the Resonant Case in which Sum of the Two Detunings is Always zero[a]

Level	Initial Occupation Probability	Final Occupation Probability
1	0	$\dfrac{(\Omega_{12})^2}{(\Omega_{12})^2 + (\Omega_{23})^2}[1 - \exp(-4\pi|p| - 4\pi|q|)]$
2	1	$\exp(-4\pi|p| - 4\pi|q|)$
3	0	$\dfrac{(\Omega_{23})^2}{(\Omega_{12})^2 + (\Omega_{23})^2}[1 - \exp(-4\pi|p| - 4\pi|q|)]$

[a]Here, $r_1 = r_3 \neq r_2$. Both p and q are defined; see 3.1.7.

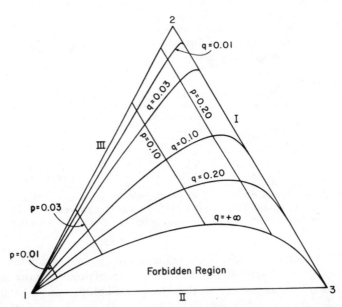

Figure 1. Occupation probabilities at large positive times for atom that starts in level 1 at large negative times. The three occupation probabilities represented by distances inward from the sides of an equilateral triangle having each altitude equal to unity. Lines of constant p and curves of constant q are plotted; the parameters p and q are defined by 3.1.7. If p is small, the probability of occupation of level 1 remains high at all times.

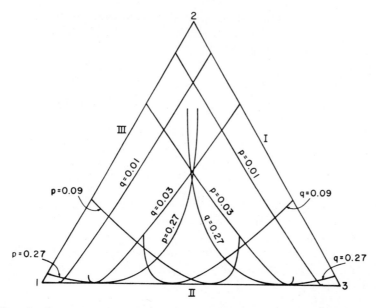

Figure 2. Occupation probabilities at large positive times for atom that starts in level 2 at large negative times. The three occupation probabilities are represented as in Figure 1. Curves of constant p and constant q are plotted. Final occupation probability for level 2 is zero if and only if $\exp(-2\pi p) + \exp(-2\pi q) = 1$; each curve is tangent to the bottom side of the triangle at a point given by this condition. Points near the bottom of the triangle are obtainable from two different pairs of values (p, q) and so are points that are nearly equidistant from sides I and III.

If either of the Rabi frequencies Ω_{12} and Ω_{23} is set equal to zero, our result of course reduces to that of the two-level Landau–Zener model. However, if neither of the Ω_{12} and Ω_{23} is zero, Tables 3–5 show that the results obtained cannot be derived by treating the three-level system as two two-level systems.

B. Class of Amplitude- and Frequency-Modulated Pulses

In this section, we shall consider (27) a three-state atom in which two transitions are driven by laser beams that have amplitude and frequency modulations of the types considered for the two-state model in Section II.
We write the Schrödinger equation as

$$
i\frac{d}{dt}\begin{pmatrix} a_1 \\ a_2 \\ a_3 \end{pmatrix} = \begin{pmatrix} (2\Delta_1 + \Delta_2)/3 & -\tfrac{1}{2}\Omega_1 & 0 \\ -\tfrac{1}{2}\Omega_1 & (-\Delta_1 + \Delta_2)/3 & -\tfrac{1}{2}\Omega_2 \\ 0 & -\tfrac{1}{2}\Omega_2 & -(\Delta_1 + 2\Delta_2)/3 \end{pmatrix}\begin{pmatrix} a_1 \\ a_2 \\ a_3 \end{pmatrix}
$$

$$
(3.2.1)
$$

Here, as before, a_1, a_2, a_3 are the probability amplitudes for the three states and Ω_1, Ω_2 are the Rabi frequencies. If E_1, E_2, and E_3 are the energies of the three states and ω_{12} and ω_{23} are the frequencies of the applied lasers whose signs are chosen so they have the same signs as $E_1 - E_2$ and $E_2 - E_3$, respectively, the signs of the two detunings Δ_1 and Δ_2 follow from their definitions:

$$\Delta_1(t) = E_1 - E_2 - \omega_{12} \tag{3.2.2a}$$

and

$$\Delta_2(t) = E_2 - E_3 - \omega_{23}. \tag{3.2.2b}$$

According to the preceding sign convention for the detunings, $\Delta_1 + \Delta_2 = 0$ corresponds to the two-photon resonance condition, whereas $\Delta_1 = \Delta_2$ will be referred to as the equal-detuning case.

As in the two-state model, we introduce a new independent variable, $z(t)$, and let the Rabi frequencies and detunings assume functional forms of the first class (see Eq. 2.12):

$$\Omega_j(t) = \frac{\alpha_j}{\pi} \frac{1}{[z(1-z)]^{1/2}} \frac{dz}{dt} \tag{3.2.3a}$$

and

$$\Delta_j(t) = \frac{1}{\pi} \left(\frac{\beta_j}{z} + \frac{\gamma_j}{1-z} \right) \frac{dz}{dt}, \qquad j = 1, 2, \tag{3.2.3b}$$

where we have now six real parameters $\alpha_1, \alpha_2, \beta_1, \beta_2, \gamma_1$, and γ_2 and an arbitrary function dz/dt. Elimination of a_2 and a_3 from Eq. 3.2.1 gives a third-order differential equation for a_1. We now wish to express the solution for a_1 in terms of a generalization of the Gauss hypergeometric function $_2F_1(a, b; c; z)$, called the Clausen function $_3F_2(a, b, c; d, e; z)$ (28) which satisfies the following third-order differential equation:

$$\frac{d^3F}{dz^3} + \frac{(3+a+b+c)z - (1+d+e)}{z(z-1)} \frac{d^2F}{dz^2}$$

$$+ \frac{(1+a+b+c+ab+ac+bc)z - de}{z^2(z-1)} \frac{dF}{dz} + \frac{abc}{z^2(z-1)} F = 0. \tag{3.2.4}$$

A useful feature of the Clausen function is that under some special conditions involving the five parameters a, b, c, d, e, the value of the function at $z = 1$ can be expressed in terms of the gamma functions.

The conditions for expressing our solutions in terms of the Clausen function are

$$\beta_1 + \beta_2 = 0 \qquad (3.2.5a)$$

or

$$\gamma_1 + \gamma_2 = 0, \qquad (3.2.5b)$$

respectively. We call these the cases of two-photon resonance at the initial time and at the final time, respectively, since the sum of the two detunings approaches zero as $t \to -\infty$ and $t \to +\infty$, respectively. Condition 3.2.5a or 3.2.5b reduces the number of independent parameters in 3.2.3 to 5. If we further require that the final occupation probabilities be expressible in terms of elementary functions, two more conditions need to be imposed for each of the cases in 3.2.5. These are:

i. For the case $\beta_1 + \beta_2 = 0$,

$$\beta_1 = -\tfrac{1}{2}(\gamma_1 - \gamma_2) \quad \text{and} \quad \alpha_1^2 - \alpha_2^2 = 2(\gamma_1^2 - \gamma_2^2). \qquad (3.2.6a)$$

ii. For the case $\gamma_1 + \gamma_2 = 0$,

$$\gamma_1 = -\tfrac{1}{2}(\beta_1 - \beta_2) \quad \text{and} \quad \alpha_1^2 - \alpha_2^2 = 2(\beta_1^2 - \beta_2^2). \qquad (3.2.6b)$$

Conditions 3.2.5 and 3.2.6 reduce the number of independent parameters in 3.2.3 to 3, which we may assume to be α_1, β_1, and γ_1 for the first pulse and detuning. The parameters α_2, β_2, and γ_2 for the second pulse and detuning must be chosen to satisfy Eqs. 3.2.5 and 3.2.6. We still have an arbitrary function dz/dt.

The final occupation probabilities of the three levels for the two cases 3.2.6a and 3.2.6b for various initial occupation probabilities of the levels are given in Tables 6 and 7, respectively.

Assuming that the atom is initially in state 1, for the case $\beta_1 + \beta_2 = 0$ or two-photon resonance at the initial time, the conditions for complete return to state 1 are

$$\gamma_1 = 0 \quad \text{and} \quad \Phi = 2n\pi, \qquad n = 1, 2, 3, \dots; \qquad (3.2.7a)$$

and the conditions for complete transfer to state 3 are

$$\gamma_2 = 0 \quad \text{and} \quad \Phi = (2n - 1)\pi, \qquad n = 1, 2, 3, \dots, \qquad (3.2.8a)$$

where Φ denotes $\Phi = [\tfrac{1}{4}(\alpha_1^2 + \alpha_2^2) - \tfrac{1}{2}(\gamma_1^2 + \gamma_2^2)]^{1/2}$. For the case $\gamma_1 + \gamma_2 = 0$, or

TABLE 6. Transition Probabilities and Probabilities of No Transition for First Case in Eq. 3.2.6a[a]

Initial Occupation Probabilities	Final Occupation Probabilities
1	$\dfrac{[\cosh(\gamma_1) + \cos\Phi][\cosh(\gamma_2) + \cos\Phi]}{4\cosh(\gamma_1)\cosh[\frac{1}{2}(\gamma_1 + \gamma_2)]\cosh[\frac{1}{2}(\gamma_1 - \gamma_2)]}$
0	$\dfrac{[\cosh(\gamma_1) - \cos\Phi][\cosh(\gamma_2) + \cos\Phi]}{2\cosh(\gamma_1)\cosh(\gamma_2)}$
0	$\dfrac{[\cosh(\gamma_1) - \cos\Phi][\cosh(\gamma_2) - \cos\Phi]}{4\cosh(\gamma_2)\cosh[\frac{1}{2}(\gamma_1 + \gamma_2)]\cosh[\frac{1}{2}(\gamma_1 - \gamma_2)]}$
0	$\dfrac{[\cosh(\gamma_1) + \cos\Phi][\cosh(\gamma_1) - \cos\Phi]}{2\cosh(\gamma_1)\cosh[\frac{1}{2}(\gamma_1 + \gamma_2)]\cosh[\frac{1}{2}(\gamma_1 - \gamma_2)]}$
1	$\dfrac{\cos^2\Phi}{\cosh(\gamma_1)\cosh(\gamma_2)}$
0	$\dfrac{[\cosh(\gamma_2) + \cos\Phi][\cosh(\gamma_2) - \cos\Phi]}{2\cosh(\gamma_2)\cosh[\frac{1}{2}(\gamma_1 + \gamma_2)]\cosh[\frac{1}{2}(\gamma_1 - \gamma_2)]}$
0	$\dfrac{[\cosh(\gamma_1) - \cos\Phi][\cosh(\gamma_2) - \cos\Phi]}{4\cosh(\gamma_1)\cosh[\frac{1}{2}(\gamma_1 + \gamma_2)][\cosh[\frac{1}{2}(\gamma_1 - \gamma_2)]}$
0	$\dfrac{[\cosh(\gamma_1) + \cos\Phi][\cosh(\gamma_2) - \cos\Phi]}{2\cosh(\gamma_1)\cosh(\gamma_2)}$
1	$\dfrac{[\cosh(\gamma_1) + \cos\Phi][\cosh(\gamma_2) + \cos\Phi]}{4\cosh(\gamma_2)\cosh[\frac{1}{2}(\gamma_1 + \gamma_2)]\cosh[\frac{1}{2}(\gamma_1 - \gamma_2)]}$

[a]Here $\Phi = [\frac{1}{2}(\alpha_1^2 + \alpha_2^2) - \frac{1}{2}(\gamma_1^2 + \gamma_2^2)]^{1/2}$, and this angle is real or pure imaginary.

two-photon resonance at the final time, assuming again that the atom is initially in state 1, the conditions for complete return to state 1 are

$$\beta_1 = 0 \quad \text{and} \quad \Phi = 2n\pi, \quad n = 1, 2, 3, \ldots; \qquad (3.2.7b)$$

and the conditions for complete transfer to state 3 are

$$\beta_1 = 0 \quad \text{and} \quad \Phi = (2n - 1)\pi, \quad n = 1, 2, 3, \ldots, \qquad (3.2.8b)$$

where Φ now denotes $\Phi = [\frac{1}{4}(\alpha_1^2 + \alpha_2^2) - \frac{1}{2}(\beta_1^2 + \beta_2^2)]^{1/2}$.

If in Table 6, where we have assumed $\beta_1 + \beta_2 = 0$, we further let $\gamma_1 + \gamma_2 = 0$,

TABLE 7. Transition Probabilities and Probabilities of No Transition for Second Case in Eq. 3.2.6b[a]

Initial Occupation Probabilities	Final Occupation Probabilities
1	$\dfrac{[\cosh(\beta_1) + \cos\Phi][\cosh(\beta_2) + \cos\Phi]}{4\cosh(\beta_1)\cosh[\frac{1}{2}(\beta_1 + \beta_2)]\cosh[\frac{1}{2}(\beta_1 - \beta_2)]}$
0	$\dfrac{[\cosh(\beta_1) + \cos\Phi][\cosh(\beta_1) - \cos\Phi]}{2\cosh(\beta_1)\cosh[\frac{1}{2}(\beta_1 + \beta_2)]\cosh[\frac{1}{2}(\beta_1 - \beta_2)]}$
0	$\dfrac{[\cosh(\beta_1) - \cos\Phi][\cosh(\beta_2) - \cos\Phi]}{4\cosh(\beta_1)\cosh[\frac{1}{2}(\beta_1 + \beta_2)]\cosh[\frac{1}{2}(\beta_1 - \beta_2)]}$
0	$\dfrac{[\cosh(\beta_1) - \cos\Phi][\cosh(\beta_2) + \cos\Phi]}{2\cosh(\beta_1)\cosh(\beta_2)}$
1	$\dfrac{\cos^2\Phi}{\cosh(\beta_1)\cosh(\beta_2)}$
0	$\dfrac{[\cosh(\beta_1) + \cos\Phi][\cosh(\beta_2) - \cos\Phi]}{2\cosh(\beta_1)\cosh(\beta_2)}$
0	$\dfrac{[\cosh(\beta_1) - \cos\Phi][\cosh(\beta_2) - \cos\Phi]}{4\cosh(\beta_2)\cosh[\frac{1}{2}(\beta_1 + \beta_2)]\cosh[\frac{1}{2}(\beta_1 - \beta_2)]}$
0	$\dfrac{[\cosh(\beta_2) + \cos\Phi][\cosh(\beta_2) - \cos\Phi]}{2\cosh(\beta_2)\cosh[\frac{1}{2}(\beta_1 + \beta_2)]\cosh[\frac{1}{2}(\beta_1 - \beta_2)]}$
1	$\dfrac{[\cosh(\beta_1) + \cos\Phi][\cosh(\beta_2) + \cos\Phi]}{4\cosh(\beta_2)\cosh[\frac{1}{2}(\beta_1 + \beta_2)]\cosh[\frac{1}{2}(\beta_1 - \beta_2)]}$

[a]Here $\Phi = [\frac{1}{4}(\alpha_1^2 + \alpha_2^2) - \frac{1}{2}(\beta_1^2 + \beta_2^2)]^{1/2}$; this angle is real or pure imaginary.

or in Table 7, where we have assumed $\gamma_1 + \gamma_2 = 0$, we further let $\beta_1 + \beta_2 = 0$, we are then led to the two-photon resonance condition at all times. Conditions 3.2.6a and 3.2.6b give

$$\alpha_1 = \alpha_2, \qquad \beta_1 + \gamma_1 = 0 \quad \text{and} \quad \beta_2 + \gamma_2 = 0, \qquad (3.2.9)$$

so that the two amplitude functions $\Omega_1(t)$ and $\Omega_2(t)$ are equal, and amplitude functions that are even in t give detuning functions that are odd functions of t. The final occupation probabilities for this case can be obtained from Table 6 or 7.

IV. DYNAMIC SYMMETRIES AND CONSTANTS OF MOTION

A. SU (N) Symmetry

As we proceed from the consideration of the dynamical evolution of two- and three-level atomic systems to that of a general N-level atomic system, we are naturally led to ask not about the occupation probability of a particular level but rather about, what the constants of motion are in the dynamics of the system and what the possibilities are for some fraction of the electron population to be trapped in certain levels. We shall present in this and the following subsections some answers to and some approaches toward answering these questions.

In the place of the time-dependent Schrödinger equation, it is more convenient for our purpose to consider the dynamical evolution of an N-level system in terms of its density matrix $\hat{\rho}(t)$ that satisfies the Liouville equation

$$i\hbar \frac{\partial \hat{\rho}}{\partial t} = [\hat{H}(t), \hat{\rho}(t)], \tag{4.1.1}$$

where the Hamiltonian $\hat{H}(t)$ is generally time dependent. It was first pointed out by Elgin (29) and Hioe and Eberly (30) that for incident laser fields of arbitrary strength, time dependence, and resonance character, the sums of the products of the dynamical variables given by tr $\hat{\rho}(t)^j$, where $j = 1, 2, \ldots$, are constants of motion, that is,

$$\text{tr } \hat{\rho}(t)^j = \text{tr } \hat{\rho}(0)^j = C_j, \quad j = 1, 2, 3, \ldots, \tag{4.1.2}$$

of which N of them are generally independent unless $\hat{\rho}$ is of rank less than N. This can be easily proved by differentiating tr $\hat{\rho}(t)^j$ with respect to the time and using Eq. 4.1.1. Here $\hat{\rho}(0)$ is, in general, assumed to be in a mixed state. The constants of motion are seen to be nonlinear in ρ except for the trivial one when $j = 1$. Although, mathematically, Eq. 4.1.2 is related to a similar equation expressing the conservation of thermodynamic entropy $S = \text{tr } \hat{\rho} \ln \hat{\rho}$ in a reversible process, the conservation laws as expressed by Eq. 4.1.2 have not been previously applied to or exploited in the field of laser physics.

If the Hamiltonian $\hat{H}$ of the N-level system interacting with the laser field is time independent (after the rotating-wave approximation is taken), as in the case when the laser field amplitudes and frequencies are constant, we have the set of linear constants of motion given by (31)

$$\text{tr } [\hat{\rho}(t)\hat{H}^j] = \text{tr } [\hat{\rho}(0)\hat{H}^j] = K_j, \quad j = 0, 1, 2, \ldots, \tag{4.1.3}$$

of which N of them are generally independent. For $j \geq N$, $\hat{H}^j$ can always be expressed in terms of some linear combination of H of lower powers by the Hamilton–Cayley theorem and hence need not be included. Equations 4.1.3 can also be easily proved by differentiation with respect to the time and by the use of Eq. 4.1.1.

The conservation laws 4.1.2 and 4.1.3 hold regardless of the form of the Hamiltonian, and they are consequences of unitarity that will be referred to as the $U(N)$ symmetry.

Just as an N-dimensional vector can be expressed as the sum of N components in suitably chosen basis vectors, the $N \times N$ density matrix $\hat{\rho}(t)$ can be expressed as the sum of N^2 components in any suitably chosen basis operators. The simplest decomposition is clearly to write

$$\hat{\rho}(t) = \sum_{j,k=1}^{N^2} \rho_{jk}(t) \hat{e}_{jk}, \tag{4.1.4}$$

where $\hat{e}_{jk}$ is an $N \times N$ matrix whose (j, k) element is equal to 1 and whose all other elements are equal to 0. Similarly, the $N \times N$ Hamiltonian matrix $\hat{H}(t)$ can be written as

$$\hat{H}(t) = \sum_{j,k=1}^{N^2} H_{jk}(t) \hat{e}_{jk}. \tag{4.1.5}$$

Using the property that $\operatorname{tr}(\hat{e}_{jk}\hat{e}_{k'j'}) = \delta_{jj'}\delta_{kk'}$, we find the inverse relationship

$$\rho_{jk}(t) = \operatorname{tr}\left[\hat{\rho}(t)\hat{e}_{kj}\right], \tag{4.1.6}$$

and

$$H_{jk}(t) = \operatorname{tr}\left[\hat{H}(t)\hat{e}_{kj}\right]. \tag{4.1.7}$$

Instead of using the basis operators $\hat{e}_{jk}$, suppose we now choose any $N^2 - 1$ generators $\hat{s}_j$ of the $SU(N)$ algebra plus the unit operator $\hat{e}$ and represent $\hat{\rho}(t)$ and $\hat{H}(t)$ in terms of these basis operators. Denote the corresponding components of $\hat{\rho}(t)$ and $\hat{H}(t)$ by $\frac{1}{2}S_j(t)$ and $\frac{1}{2}\Gamma_j(t)$, respectively, and write

$$\hat{\rho}(t) = \frac{1}{2}S_0(t)\hat{e} + \frac{1}{2}\sum_{j=1}^{N^2-1} S_j(t)\hat{s}_j, \tag{4.1.8}$$

and

$$\hat{H}(t) = \frac{\hbar}{2}\Gamma_0(t)\hat{e} + \frac{\hbar}{2}\sum_{j=1}^{N^2-1} \Gamma_j(t)\hat{s}_j. \tag{4.1.9}$$

The reason for the factor $\frac{1}{2}$ is that the commonly used $SU(N)$ generators such as the Pauli matrices for the case of SU(2) have the property that

$$\text{tr}\,(\hat{s}_j\hat{s}_k) = 2\delta_{jk}. \tag{4.1.10}$$

The generators of SU(N) algebra are traceless, that is,

$$\text{tr}\,(\hat{s}_j) = 0, \qquad j = 1, 2, \ldots, N^2 - 1. \tag{4.1.11}$$

The components $S_j(t)$ and $\Gamma_j(t)$ are given, from Eqs. 4.1.8–4.1.10, by

$$S_j(t) = \text{tr}\,[\hat{\rho}(t)\hat{s}_j], \tag{4.1.12}$$

$$\hbar\Gamma_j(t) = \text{tr}\,[\hat{H}(t)\hat{s}_j]. \tag{4.1.13}$$

The evolution of the density matrix, instead of the Liouville equation 4.1.1, can be expressed in terms of the evolution of an $(N^2 - 1)$-dimensional real coherence vector $\mathbf{S} = (S_1, S_2, \ldots, S_{N^2-1})$ by

$$\frac{dS_j(t)}{dt} = \sum_{k=1}^{N^2-1} A_{jk}(t)S_k(t), \qquad j = 1, 2, \ldots, N^2 - 1, \tag{4.1.14}$$

where

$$A_{jk} = -\frac{1}{2i\hbar}\,\text{tr}\,(\hat{H}[\hat{s}_j, \hat{s}_k]). \tag{4.1.15}$$

There are infinite number of ways that the SU(N) generators $\hat{s}_j$ can be represented. Given a set of basis operators, one can form another set $\hat{s}'_j$ by an orthogonal transformation or by a unitary transformation $\hat{U}^\dagger\hat{s}_j\hat{U}, j = 1, 2, \ldots,$ where $\hat{U}$ is any $N \times N$ unitary matrix. If the Hamiltonian of the system possesses some special features that, by a suitable choice of the basis operators, result in the matrix $\hat{A}(t)$ in Eq. 4.1.14 to assume a block diagonal form, the system is said to possess a dynamic symmetry of a certain type. When that happens, some superpositions of dynamical variables evolve independently of others, giving rise to the existence of independent sets of constants of motions and the possibility of trapped population.

The question is then to inquire what special features the Hamiltonian $\hat{H}(t)$ must possess for this to happen. Following some clues given in the earlier investigations (32–35), the general answer was given by the author (31, 36–38) and can be stated as follows: If $\hat{H}(t)$ lies entirely in the subspace or subspaces spanned by generators forming a subgroup or subgroups of the SU(N) algebra, the system possesses a certain dynamic symmetry. When $\hat{H}(t)$ lies in the subspace spanned the generators of the O(3) subgroup, the system is said to possess Cook–Shore (34), or SU(2), (39, 37) dynamic symmetry. In that case, the matrix $\hat{A}(t)$ in Eq. 4.1.14 consists of $N-1$ block diagonal matrices of dimensions $3, 5, \ldots, 2N-1$. Another interesting case is when $\hat{H}(t)$ lies in the subspace spanned by the generators of the SU(2) and $N-2$ U(1) subgroups. The system is said to possess what the author (36, 38) has termed Gell-Mann dynamic symmetry. In that case, the matrix $\hat{A}(t)$ in Eq. 4.1.14 consists of N block diagonal matrices: one three-dimensional, one (N^2-N-2)-dimensional, and $N-2$ one-dimensional matrices. The conditions the elements of the Hamiltonian must satisfy to possess these types of symmetries will be discussed in the following sections.

B. SU(2) Symmetry

To reveal the special feature of Cook–Shore, or SU(2), symmetry, we begin by choosing, for the basis operators, the N^2 operators $\hat{T}_q^{(k)}$, $k = 0, 1, 2, \ldots, N-1$ and $q = k, k-1, \ldots, -k$, introduced by Racah (40), which are to be defined in what follows. The value of k specifies the rank of the operator, and the $2k+1$ operators obtained by setting $q = k, k-1, \ldots, -k$, are said to form the components of the tensor operator $\hat{T}^{(k)}$ of rank k. The starting point taken by Racah for these tensor operators is that they should satisfy the commutation relations

$$[\hat{J}_z, \hat{T}_q^{(k)}] = q\hat{T}_q^{(k)},$$

and

$$[\hat{J}_\pm, \hat{T}_q^{(k)}] = [k(k+1) - q(q \pm 1)]^{1/2} \hat{T}_{q\pm1}^{(k)}, \qquad (4.2.1)$$

where the components $\hat{J}_\pm = \hat{J}_x \pm i\hat{J}_y$, $\hat{J}_z$ of the angular momentum operator $\hat{J}$ satisfy the well-known commutation relations

$$[\hat{J}_x, \hat{J}_y] = i\hat{J}_z, \qquad [\hat{J}_y, \hat{J}_z] = i\hat{J}_x, \qquad [\hat{J}_z, \hat{J}_x] = i\hat{J}_y. \qquad (4.2.2)$$

In terms of the projection operator $|m\rangle\langle m'|$, the N^2 operators are given by

$$\hat{T}_q^{(k)} = \sum_{m,m'} (-1)^{J-m'}(2k+1)^{1/2} \begin{pmatrix} J & k & J \\ -m & q & m \end{pmatrix} |m'\rangle\langle m|, \qquad (4.2.3)$$

where

$$m, m' = -J, -J+1, \ldots, J,$$
$$k = 0, 1, 2, \ldots, 2J,$$
$$q = -k, -k+1, \ldots, k,$$
$$J = \tfrac{1}{2}(N-1),$$

and

$$\hat{T}^{(k)}_{-q} = (-1)^q (\hat{T}^{(k)}_q)^\dagger \quad \text{for } q = 1, 2, \ldots, k, \tag{4.2.4}$$

where

$$\begin{pmatrix} j_1 & j_2 & j_3 \\ m_1 & m_2 & m_3 \end{pmatrix}$$

is the $3 - j$ symbol. The set of operators satisfy the following commutation relation:

$$[\hat{T}^{(k_1)}_{q_1}, \hat{T}^{(k_2)}_{q_2}] = \sum_{k_3=0}^{N-1} \sum_{q_3=-k_3}^{k_3} (-1)^{N-q_3-1} [(-1)^{k_1+k_2+k_3} - 1]$$

$$\times [(2k_1+1)(2k_2+1)(2k_3+1)]^{1/2}$$

$$\times \begin{Bmatrix} k_1 & k_2 & k_3 \\ l & l & l \end{Bmatrix} \begin{pmatrix} k_1 & k_2 & k_3 \\ q_1 & q_2 & -q_3 \end{pmatrix} \hat{T}^{(k_3)}_{q_3}, \tag{4.2.5}$$

where $l = \frac{1}{2}(N-1)$ and

$$\begin{Bmatrix} j_1 & j_2 & j_3 \\ l_1 & l_2 & l_3 \end{Bmatrix}$$

is the $6 - j$ symbol. The set of commutation relations 4.2.5 contain 4.2.1 and 4.2.2 as special cases if one notes that the angular mementum operators $\hat{J}_x, \hat{J}_y, \hat{J}_z$ are expressible in terms of the three components $\hat{T}^{(1)}_{-1}, \hat{T}^{(1)}_0, \hat{T}^{(1)}_1$ of the tensor operator of rank 1. It can be shown that $\hat{T}^{(k)}_q$ are orthonormal according to

$$\text{tr}[(\hat{T}^{(k)}_q)^\dagger \hat{T}^{(k')}_{q'}] = \delta(kk')\delta(qq'). \tag{4.2.6}$$

Using the set of $\hat{T}^{(k)}_q$ as our basis operators, we write the density matrix $\hat{\rho}(t)$ as

$$\hat{\rho}(t) = \sum_{k=0}^{N-1} \sum_{q=-k}^{k} T^{(k)}_q(t) \hat{T}^{(k)}_q \tag{4.2.7}$$

where the component $T^{(k)}_q(t)$ of $\hat{\rho}(t)$ along the basis operator $\hat{T}^{(k)}_q$ is given by

$$T^{(k)}_q(t) = \text{tr}[\hat{\rho}(t)(\hat{T}^{(k)}_q)^\dagger]; \tag{4.2.8}$$

and we write the Hamiltonian $\hat{H}(t)$ as

$$\hat{H}(t) = \sum_{k=0}^{N-1} \sum_{q=-k}^{k} \Gamma^{(k)}_q(t) \hat{T}^{(k)}_q, \tag{4.2.9}$$

where

$$\Gamma_q^{(k)}(t) = \text{tr}\,[\hat{H}(t)(\hat{T}_q^{(k)})^\dagger]. \tag{4.2.10}$$

The dynamical evolution of $\hat{\rho}(t)$ can be expressed in terms of the dynamical evolution of $T_q^{(k)}(t)$ as

$$\frac{d}{dt}\,T_q^{(k)}(t) = i \sum_{k',q'} A_{kq,k'q'}(t)\,T_{q'}^{(k')}(t), \tag{4.2.11}$$

where

$$A_{kq,k'q'}(t) = \text{tr}\,\{\hat{H}(t)[(\hat{T}_q^{(k)})^\dagger,\,\hat{T}_{q'}^{(k')}]\}. \tag{4.2.12}$$

The advantage of using the Racah tensors as the basis operators only becomes apparent when the Hamiltonian of the system lies entirely in the subspace spanned by the tensor operator of rank 1 whose components form the $O(3)$ subgroup, that is, if the Hamiltonian can be written as

$$\hat{H}(t) = \sum_{k=0}^{1}\,\sum_{q=-k}^{k}\,\Gamma_q^{(k)}(t)\,\hat{T}_q^{(k)}, \tag{4.2.13}$$

where we have included the unimportant constant term represented by the coefficient of the tensor operator of rank zero which is a unit operator. Equation 4.2.13 can be rewritten in the more familiar form as

$$\hat{H}(t) = c_1(t)\hat{J}_x + c_2(t)\hat{J}_y + c_3(t)\hat{J}_z + d(t), \tag{4.2.14}$$

and the conditions the elements of $\hat{H}(t)$ must satisfy in order that Eq. 4.2.14 or 4.2.13 holds are known as the Cook–Shore conditions (34), which are the following:

i. The N levels are chainwise dipole connected, that is,

$$1\leftrightarrow 2,\,2\leftrightarrow 3,\ldots,N-1\leftrightarrow N. \tag{4.2.15a}$$

ii. The $N-1$ laser fields that drive these transitions (which may come from the same laser) are equally detuned from the respective atomic transitions at all times, namely,

$$\Delta_{12}(t) = \Delta_{23}(t) = \cdots = \Delta_{N-1,N}(t) = \Delta(t). \tag{4.2.15b}$$

iii. The respective Rabi frequencies must satisfy the relations

$$\Omega_j(t) = \sqrt{j(n-j)}\Omega_0(t), \qquad j = 1, 2, \ldots, N-1, \qquad (4.2.15c)$$

where $\Omega_0(t)$ is an arbitrary function of time.

When $\hat{H}(t)$ is given by Eq. 4.2.13 or when the Cook–Shore conditions are satisfied, the N^2 equations of motion, Eq. 4.2.11, separate into N independent sets consisting of $1, 3, 5, \ldots, 2N-1$ coupled equations, respectively. This can be seen from Eqs. 4.2.12, 4.2.13, 4.2.6, and 4.2.5 by noting that (i) the factor $[(-1)^{k_1+k_2+k_3} - 1]$ in Eq. 4.2.5 results in $A_{kq,k'q'} = 0$ unless k and k' are both even or both odd; (ii) because of Eqs. 4.2.13 and 4.2.6, only those terms with $k_3 = 1$ in Eq. 4.2.5 need to be considered in the commutator $[\hat{T}_q^{(k)}, \hat{T}_{q'}^{(k)}]$ for Eq. 4.2.12; and (iii) the triangular condition for the $3-j$ symbol

$$\begin{pmatrix} k_1 & k_2 & k_3 \\ q_1 & q_2 & -q_3 \end{pmatrix}$$

means that if a triangle cannot be formed having sides k_1, k_2, k_3, the $3-j$ symbol vanishes, which when applied to the factor

$$\begin{pmatrix} k & k' & 1 \\ q & q' & -q_3 \end{pmatrix}$$

further requires that $k = k'$ for $A_{kq,k'q'}$ to be nonzero, remembering that k and k' must be both odd or both even from (i). We have thus shown that for each value of k, the set of $T_q^{(k)}(t), q = -k, -k+1, \ldots, k$, forms the basis of a $(2k+1)$-dimensional independent subspace of the dynamical equations of motion. The equations are given by (41)

$$i\frac{d}{dt}T_q^{(k)} = [qc_3(t) + d(t)]T_q^{(k)} + \tfrac{1}{2}[c_1(t) - ic_2(t)][(k-q)(k+q+1)]^{1/2}T_{q+1}^{(k)}$$

$$+ \tfrac{1}{2}[c_1(t) + ic_2(t)][(k+q)(k-q+1)]^{1/2}T_{q-1}^{(k)}. \qquad (4.2.16)$$

In each of the subspaces characterized by the value of $k = 0, 1, 2, \ldots, N-1$, we have (37, 42, 43)

$$[T_0^{(k)}(t)]^2 + 2\sum_{q=1}^{k} (-1)^q T_q^{(k)}(t) T_{-q}^{(k)}(t) = \text{const.} \qquad (4.2.17)$$

The advantage of using the Racah tensors 4.2.3 and the implication of the SU(2) dynamic symmetry 4.2.15 are now clearly demonstrated by Eq. 4.2.17.

The complete solution of Eq. 4.2.16 can also be written (39). Let us denote

$$c_1(t) - ic_2(t) \equiv -\Omega_0(t), \tag{4.2.18a}$$

$$c_3(t) \equiv -\Delta_0(t), \tag{4.2.18b}$$

and we shall ignore the presence of $d(t)$, which only changes the result for $T_q^{(k)}$ by the same phase factor. Let us also denote by $a(t)$ and $-b^*(t)$ the fundamental solution for the amplitude probabilities $C_1(t)$ and $C_2(t)$ for a two-level system that satisfy the equations of motion given by

$$i\frac{d}{dt}\begin{pmatrix} C_1 \\ C_2 \end{pmatrix} = \begin{pmatrix} \frac{1}{2}\Delta_0(t) & -\frac{1}{2}\Omega_0(t) \\ -\frac{1}{2}\Omega_0(t) & -\frac{1}{2}\Delta_0(t) \end{pmatrix}\begin{pmatrix} C_1 \\ C_2 \end{pmatrix} \tag{4.2.19}$$

with the initial condition $|C_1(0)| = 1$, $C_2(0) = 0$, so that the solution of Eq. 4.2.19 for the general initial values of $C_1(0)$ and $C_2(0)$ can be written as

$$\begin{pmatrix} C_1 \\ C_2 \end{pmatrix} = \begin{pmatrix} a(t) & b(t) \\ -b^*(t) & a^*(t) \end{pmatrix}\begin{pmatrix} C_1(0) \\ C_2(0) \end{pmatrix}. \tag{4.2.20}$$

For each value of k in Eq. 4.2.16, the solution for $T_q^{(k)}(t)$ is given by

$$T_q^{(k)} = \sum_{q'=-k}^{k} D_{qq'}^{(k)}(a,b) T_{q'}^{(k)}(0), \qquad q = -k, -k+1, \ldots, k, \tag{4.2.21}$$

where $D_{qq'}^{(k)}(a,b)$ are the matrix elements of the $(2k+1)$-dimensional representation of the $SU(2)$ group. The expression

$$D_{mm'}^{(k)}(a,b) = \sum_{\mu} \frac{[(j-m)!(j+m)!(j-m')!(j+m')!]^{1/2}}{p!q!r!s!} a^p a^{*q} b^r (-b^*)^s \tag{4.2.22}$$

where (i) $p = j - m - \mu$, $q = j + m' - \mu$, $r = \mu$, and $s = m - m' + \mu$ or (ii) $p = m - m' + \mu$, $q = \mu$, $r = j + m' - \mu$, and $s = j + m - \mu$ or (iii) $p = \mu$, $q = m + m' + \mu$, $r = j - m - \mu$, and $s = j - m' - \mu$, where $\mu = 0, 1, 2, \ldots$, can be used, and all will give the same result.

The problem of multiple solitary-pulse propagation through an N-level system, first investigated by Konopnicki, Drummond, and Eberly (44) and subsequently by the author (36, 45, 37), can now be extended and brought to a satisfactory completion with the help of Eq. 4.2.21. Because the time evolution of the atomic system must be consistent with the electric field envelope propagation of the laser pulse given by the reduced Maxwell equation

$$\left(\frac{\partial}{\partial z} + \frac{\partial}{\partial (ct)}\right)\Omega_n(z, t) = -i\frac{2\pi D}{\hbar c}C_n v_n d^2_{n,n+1}\langle \rho_{n,n+1} - \rho_{n+1,n}\rangle, \quad (4.2.23)$$

where the incident electric field propagating in the z direction has $N - 1$ carrier frequencies v_n that are nearly resonant with the successive transition frequencies in a chain of N dipole-connected energy levels, and where $C_n = 1$ if the energy levels increase in energies ($E_{n+1} > E_n$) and $C_n = -1$ if the energy levels decrease in energies ($E_{n+1} < E_n$), $d_{n,n+1}$ is the dipole matrix element, D is the atomic density, and $\langle \ \rangle$ denotes averaging over the Maxwellian velocity distribution of atoms. We find that there are generally $N - 1$ sets of mutually exclusive conditions for supporting the simultaneous solitons (the so-called simulton propagation) through the N-level system satisfying Eqs. 4.2.15. The $N - 1$ mutually exclusive conditions will be numbered by $k_1 = 1, 2, \ldots, N - 1$, and for each value of k_1, the following conditions for the initial population distribution of the energy levels, the energy ordering of the levels, the carrier frequencies, and the dipole moments must be satisfied:

i. The initial population distribution must be given by

$$\rho_{mm}(0) = (-1)^{J-m}\begin{pmatrix} J & J & k_1 \\ m & -m & 0 \end{pmatrix}\delta + (2J + 1)^{-1}, \quad (4.2.24a)$$

where δ, which may be positive or negative, is arbitrary but is restricted by the requirement that $\rho_{mm}(0) \geqslant 0$ for all $m = -J, -J + 1, \ldots, J$.

ii. The ordering of the energy levels, the carrier frequencies, and the dipole moments must satisfy

$$(-1)^{J-m}\frac{C_m v_m d^2_{m,m+1}}{[(J - m)(J + m + 1)]^{1/2}}\begin{pmatrix} J & J & k_1 \\ m & -m-1 & 1 \end{pmatrix} = \text{const.} \quad (4.2.24b)$$

The numbering by $m = -J, -J + 1, \ldots, J$ is related to the numbering by $n = 1, 2, \ldots, N$, by $n = m + J + 1$, and $N = 2J + 1$.

It can be verified that the two different conditions previously given by Konopnicki, Drummond, and Eberly (44) and by Hioe (37, 45) are special cases of $k_1 = 1$ and $k_1 = 2$, respectively. That the conditions for different values of k_1 are mutually exclusive or orthogonal can be appreciated by noting that given the conditions for a particular value of k_1, the evolution of the atomic variables is given by

$$\mathbf{T}^{(k_1)} = \hat{D}^{(k_1)}(a, b)\mathbf{T}^{(k_1)}(0) \quad (4.2.25)$$

and $\mathbf{T}^{(k)} = 0$ at all times for $k \neq k_1$. Assuming that all the coherences are equal to zero initially, that is, $\mathbf{T}_q^{(k_1)}(0) = 0$ for $q \neq 0$, the condition for the atomic variables to return to their original values at, say, $t = +\infty$ is that

$$D_{00}^{(k_1)}(a(+\infty), b(+\infty)) = 1, \tag{4.2.26}$$

or (39)

$$P_{k_1}(x) = 1, \tag{4.2.27}$$

where $P_n(x)$ is the Legendre polynomial of order n, and

$$x = |a(+\infty)|^2 - |b(+\infty)|^2. \tag{4.2.28}$$

This condition can be met by choosing $\Omega_0(t)$ and $\Delta_0(t)$ in Eq. 4.2.19 appropriately, as Section II has shown.

C. Gell-Mann Symmetry

To reveal the special feature of Gell-Mann symmetry (36, 38), we choose, for the basis operators, the $N^2 - 1$ operators, excluding the unit operator, consisting of three types that we call $\hat{A}$, $\hat{B}$, and $\hat{C}$, whose members are $\hat{A}_j$, $j = 1$, 2, 3, $\hat{B}_k$, $k = 1, 2, \ldots, N^2 - N$, and $\hat{C}_l$, $l = 1, 2, \ldots, N - 2$ and that satisfy the following commutation relations:

$$[\hat{A}, \hat{A}] = \hat{A}, \quad [\hat{A}, \hat{B}] = \hat{B}, \quad [\hat{A}, \hat{C}] = 0,$$
$$[\hat{B}, \hat{B}] = \hat{A} + \hat{C}, \quad [\hat{B}, \hat{C}] = \hat{B},$$
$$[\hat{C}, \hat{C}] = 0, \tag{4.3.1}$$

where, for example, $[\hat{A}, \hat{A}] = \hat{A}$ states that the commutator of two different members of type A is equal to a member of type A (multiplied possibly by a constant). The members of type A form a subgroup, the $SU(2)$ subgroup, whose commutation relations among themselves are

$$[\hat{A}_1, \hat{A}_2] = 2i\hat{A}_3, \quad [\hat{A}_2, \hat{A}_3] = 2i\hat{A}_1, \quad [\hat{A}_3, \hat{A}_1] = 2i\hat{A}_2. \tag{4.3.2}$$

Each of the members of type C by itself forms a subgroup, the $U(1)$ subgroup. Members of type C not only commute among themselves but also commute with any member of type A, that is,

$$[\hat{C}_j, \hat{C}_k] = 0, \quad [\hat{C}_j, \hat{A}_k] = 0 \quad \text{for any } j, k. \tag{4.3.3}$$

We shall denote the $N^2 - 1$ basis operators by $\hat{\lambda}_j$, $j = 1, 2, \ldots, N^2 - 1$, with

$j = 1, 2, 3$ for members of type A, $j = n^2 - 1$, $n = 3, 4, 5, \ldots, N$ for members of type C, and the remaining values of j for members of type B, and we choose them so that

$$\mathrm{tr}\,(\hat{\lambda}_j \hat{\lambda}_k) = 2\delta_{jk}, \tag{4.3.4a}$$

and

$$\mathrm{tr}\,(\hat{\lambda}_j) = 0. \tag{4.3.4b}$$

The density matrix $\hat{\rho}(t)$ and the Hamiltonian $\hat{H}(t)$ can be represented, in the space spanned by the basis operators $\hat{\lambda}_j$, by

$$\hat{\rho}(t) = N^{-1}\hat{e} + \frac{1}{2}\sum_{j=1}^{N^2-1} \lambda_j(t)\hat{\lambda}_j, \tag{4.3.5a}$$

where

$$\lambda_j(t) = \mathrm{tr}\,[\hat{\rho}(t)\hat{\lambda}_j]; \tag{4.3.5b}$$

and

$$\hat{H}(t) = \tfrac{1}{2}\hbar\left(\Gamma_0\hat{e} + \sum_{j=1}^{N^2-1} \Gamma_j(t)\hat{\lambda}_j\right), \tag{4.3.6a}$$

where

$$\hbar\Gamma_j(t) = \mathrm{tr}\,[\hat{H}(t)\hat{\lambda}_j]. \tag{4.3.6b}$$

The dynamical evolution of $\hat{\rho}(t)$ can be expressed in terms of the dynamical evolution of $\lambda_j(t)$ as

$$\frac{d\lambda_j(t)}{dt} = \sum_{k=1}^{N^2-1} A_{jk}(t)\lambda_k(t), \qquad j = 1, 2, \ldots, N^2 - 1, \tag{4.3.7}$$

where

$$A_{jk}(t) = \frac{1}{-2i\hbar}\,\mathrm{tr}\,(\hat{H}(t)[\hat{\lambda}_j, \hat{\lambda}_k]). \tag{4.3.8}$$

So far every expression given by Eqs. 4.3.5–4.3.8 is very general. The special feature the Hamiltonian of the system must have in order that the dynamics of the system exhibits the Gell-Mann dynamic symmetry is that the Hamiltonian must lie entirely in the subspaces spanned by members of operators of types A and C, which, as we have already mentioned, form the $SU(2)$ and $U(1)$ subgroups, respectively; namely, the Hamiltonian must be expressible in the form

$$\hat{H}(t) = \sum_{j=1}^{3} a_j(t)\hat{\lambda}_j + \sum_{j=3}^{N} c_{j^2-1}(t)\hat{\lambda}_{j^2-1}, \qquad (4.3.9)$$

remembering our notations for members of types A and C given following Eq. 4.3.3. In Eq. 4.3.9, we have ignored the unimportant constant term accompanying a unit operator. Using Eqs. 4.3.8, 4.3.3, and 4.3.1, it is easy to see that the matrix elements $A_{jk}(t)$ equal zero when j and k belong to members of different types and also when j and k belong to the same or different members of type C. Thus, the dynamical space in Eq. 4.3.7 decomposes into one three-dimensional, one $(N^2 - N - 2)$-dimensional, and $N - 2$ one-dimensional independent subspaces each with its own constants of motion.

An important gap, however, still needs to be filled. That is, since there are infinite number of ways of representing the basis operators $\hat{\lambda}_j$, how do we see that a given Hamiltonian should have the feature given by Eq. 4.3.9? Suppose we choose the simplest representation for $\hat{\lambda}_j$, which we shall denote by $\hat{\lambda}'_j$, in terms of $N \times N$ matrices given by

$$\hat{\lambda}'_1 = \begin{pmatrix} \hat{\sigma}_x & \\ & \hat{0}_{N-2} \end{pmatrix}, \qquad \hat{\lambda}'_2 = \begin{pmatrix} \hat{\sigma}_y & \\ & \hat{0}_{N-2} \end{pmatrix}, \qquad \hat{\lambda}'_3 = \begin{pmatrix} \hat{\sigma}_z & \\ & \hat{0}_{N-2} \end{pmatrix},$$

$$\hat{\lambda}'_{n^2-1} = \frac{1}{[n(n-1)/2]^{1/2}} \begin{pmatrix} \hat{I}_{n-1} & & \\ & -(n-1) & \\ & & \hat{0}_{N-n} \end{pmatrix}, \qquad n = 3, 4, \ldots, N$$

$$(4.3.10a)$$

where $\hat{0}_m$ and $\hat{I}_m$ are $m \times m$ zero and unit matrices, respectively, and $\hat{\sigma}_x$, $\hat{\sigma}_y$, and $\hat{\sigma}_z$ are the 2×2 Pauli matrices

$$\hat{\sigma}_x = \begin{pmatrix} 0 & 1 \\ 1 & 0 \end{pmatrix}, \qquad \hat{\sigma}_y = \begin{pmatrix} 0 & -i \\ i & 0 \end{pmatrix}, \qquad \hat{\sigma}_z = \begin{pmatrix} 1 & 0 \\ 0 & -1 \end{pmatrix}. \qquad (4.3.10b)$$

If we now write

$$\hat{H}'(t) = \sum_{j=1}^{3} a_j(t)\hat{\lambda}'_j + \sum_{j=3}^{N} c_{j^2-1}(t)\hat{\lambda}'_{j^2-1}, \qquad (4.3.11)$$

the question that we posed earlier can be rephrased as follows: For the given Hamiltonian $\hat{H}(t)$, can we find a unitary operator $\hat{U}$ that satisfies $\hat{U}^\dagger \hat{U} = \hat{I}$ such that

$$\hat{U}^\dagger \hat{H}(t) \hat{U} = \hat{H}'(t)? \qquad (4.3.12)$$

If the answer is affirmative, the system possesses the Gell-Mann dynamic symmetry with the consequences we have stated.

A useful approach is to start from Eq. 4.3.11 and try to construct various kinds of unitary operator $\hat{U}$ with elements that appear in the physical Hamiltonian. By applying the unitary transformations

$$\hat{U} \hat{\lambda}'_j \hat{U}^\dagger = \hat{\lambda}_j \qquad (4.3.13)$$

to the right side of Eq. 4.3.11, we can construct many different Hamiltonians whose dynamics possess the Gell-Mann symmetry. The conditions the elements of the Hamiltonian need to satisfy can be quite simply written as a consequence of requiring $\hat{U}$ to satisfy $\hat{U}^\dagger \hat{U} = \hat{I}$.

As an example, a four-level system interacting with a laser field whose Hamiltonian can be expressed in the form

$$\hat{H}(t) = -\hbar \begin{pmatrix} 0 & \alpha_{12} & 0 & \alpha_{14} \\ \alpha_{12} & \Delta & \alpha_{23} & 0 \\ 0 & \alpha_{23} & 0 & \alpha_{34} \\ \alpha_{14} & 0 & \alpha_{34} & \Delta \end{pmatrix}, \qquad (4.3.14)$$

where the α's and Δ's are generally dependent on time, possesses the Gell-Mann dynamic symmetry if the α's have the same time dependence and satisfy the relation

$$\alpha_{12}\alpha_{34} = \alpha_{14}\alpha_{23}. \qquad (4.3.15)$$

The preceding result is obtained by starting from a form of Eq. 4.3.11:

$$\hat{H}'(t) = \tfrac{1}{2}\hbar \left(-2\varepsilon(t)\hat{\lambda}_1 + \Delta(t) \sum_{n=2}^{4} (-1)^n \frac{[n/2]}{\{n(n-2)/2\}^{1/2}} \hat{\lambda}_{n^2-1} \right), \qquad (4.3.16)$$

where $[n/2]$ denotes the integer part of $n/2$. We use the following unitary matrix:

$$\hat{U} = \hat{U}^\dagger = \frac{1}{\varepsilon^{1/2}} \begin{pmatrix} a_1 & 0 & a_3 & 0 \\ 0 & a_2 & 0 & a_4 \\ a_3 & 0 & -a_1 & 0 \\ 0 & a_4 & 0 & -a_2 \end{pmatrix}, \qquad (4.3.17)$$

where the four parameters $a_1, \ldots, a_4$ are required to satisfy

$$a_1^2 + a_3^2 = a_2^2 + a_4^2 \equiv \varepsilon \tag{4.3.18}$$

and are associated with the interaction parameters appearing in the physical Hamiltonian by the relation

$$\alpha_{jk}(t) = a_j(t)a_k(t) = \alpha_{kj}(t). \tag{4.3.19}$$

The characteristic constants of motion for Gell-Mann symmetry in this case are

$$\sum_{j=1}^{3} (\lambda_j(t))^2 = \text{const.}, \qquad \lambda_8(t) = \text{const.}, \qquad \lambda_{15}(t) = \text{const.}, \tag{4.3.20}$$

where the $\lambda_j(t)$ can be expressed in terms of the density matrix elements $\rho_{jk}(t)$ and elements of the Hamiltonian $\alpha_{jk}(t)$ and $\Delta(t)$ from Eqs. 4.3.5b, 4.3.13, 4.3.17–4.3.19, and 4.3.10.

A four-level system with the detunings represented by Eq. 4.3.14 is schematically shown in Figure 3, where the laser-driven transitions $1 \leftrightarrow 2$, $2 \leftrightarrow 3$, $3 \leftrightarrow 4$, and $4 \leftrightarrow 1$ whose strengths are given by the Rabi frequencies $2\alpha_{12}$, $2\alpha_{23}$, $2\alpha_{34}$, and $2\alpha_{14}$ are shown. A consequence of Gell-Mann dynamic symmetry as a result of Eq. 4.3.15 can be illustrated by an example for the case when α_{23} is very large and α_{14} is very small compared to α_{12} and α_{34}. By examining the expressions $\lambda_8(t) = \text{const.}$ and $\lambda_{15}(t) = \text{const.}$, we find that in that case whatever population is initially present in levels 2 and 4 is trapped and remains constant.

A discussion of multiple solitary wave propagation in various subspaces and its analogy with elementary particle physics was given in Refs. 36 and 38. Some interesting work relating to constants of motion and population

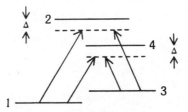

Figure 3.   Schematic representation of a four-level system interacting with a laser field whose Hamiltonian is given by Eq. 4.3.14.

trapping in three- and four-level systems have been given by Radmore and Knight (46), Deng (47), Gottlieb (48), Pegg (49), Buckle, Barnett, Knight, Lauder, and Pegg (50), and Hioe and Eberly (51).

Acknowledgment

This work is partially supported by the Division of Chemical Sciences, U.S. Department of Energy under grant number DE-FG02-84-ER13243. The author is grateful to Professors J. H. Eberly and J. O. Hirschfelder and Dr. C. E. Carroll for helpful discussions.

References

1. N. F. Scherer, J. L. Knee, D. D. Smith, and A. H. Zewail, *J. Phys. Chem.* **89**, 5141 (1985); R. M. Baum, *Chem. Eng. News*, January 13 (1986), p. 36; M. A. Banash and W. S. Warren, *Las. Chem.* **6**, 36 (1986) and references cited.

2. C. P. Lin, J. Bates, J. T. Mayer, and W. S. Warren, "Implementation of a Phase and Amplitude Modulated π Pulse for Coherent Optical Spectroscopy," Preprint, to appear in *J. Chem. Phys.*

3. N. Rosen and C. Zener, *Phys. Rev.* **40**, 502 (1932).

4. L. Landau, *Phys. Z. Sowj.* **2**, 46 (1932).

5. C. Zener, *Proc. Roy. Soc. A* **137**, 696 (1932).

6. I. I. Rabi, *Phys. Rev.* **51**, 652 (1937).

7. G. H. Wannier, *Physics* **1**, 251 (1965).

8. E. E. Nikitin, Opt. Spectrosc. **13**, 431 (1962); *Disc. Faraday Soc.* **33**, 14 (1962).

9. Yu. N. Demkov, *Sov. Phys. JETP* **18**, 138 (1963).

10. A. E. Kaplan, *Sov. Phys. JETP* **38**, 705 (1973); **41**, 409 (1975).

11. L. Allen and J. H. Eberly, *Optical Resonance and Two-Level Atoms*, Wiley, New York, 1975, Section 4.6.

12. D. S. F. Crothers, *J. Phys. B* **11**, 1025 (1978).

13. H. W. Lee and T. F. George, *Phys. Rev. A* **29**, 2509 (1984).

14. A. Bambini and P. R. Berman, *Phys. Rev. A* **23**, 2496 (1981).

15. A. Bambini and M. Linberg, *Phys. Rev. A* **30**, 794 (1984).

16. F. T. Hioe, *Phys. Rev. A* **30**, 2100 (1984).

17. F. T. Hioe and C. E. Carroll, *J. Opt. Soc. Am. B* **2**, 497 (1985).

18. F. T. Hioe and C. E. Carroll, *Phys. Rev. A* **32**, 1541 (1985).

19. C. E. Carroll and F. T. Hioe, *J. Phys. A* **19**, 3579 (1986).

20. L. Allen and J. H. Eberly, *Optical Resonance and Two-Level Atoms*, Wiley, New York, 1975, Chapter 2.

21. E. J. Robinson, *Phys. Rev. A* **31**, 3986 (1985); *J. Phys. B* **18**, L657 (1985); *J. Phys. B* **18**, 3687 (1985).

22. S. Geltman, *Topics in Atomic Collision Theory*, Academic, New York, 1969; Y. Abe and J. Y. Park, *Phys. Rev. C* **28**, 2316 (1983).

23. C. E. Carroll and F. T. Hioe, *J. Opt. Soc. Am. B* **2**, 1355 (1985).

24. C. E. Carroll and F. T. Hioe, *J. Phys. A* **19**, 1151 (1986).

25. C. E. Carroll and F. T. Hioe, *J. Phys. A* **19**, 2061 (1986).

26. J. C. Burkill, *The Theory of Ordinary Differential Equations*, Oliver and Boyd, London, 1962, p. 60.

27. C. E. Carroll and F. T. Hioe, *Phys. Rev.* **36**, 724 (1987), *J. Math. Phys.* **29**, 487 (1988).

28. L. J. Slater, *Generalized Hypergeometric Functions*, Cambridge University Press, 1966; A. Erdélyi, W. Magnus, F. Oberhettinger, and F. G. Tricomi, *Higher Transcendental Functions*, Vol. 1, McGraw-Hill, New York, 1953.

29. J. N. Elgin, *Phys. Lett.* **80A**, 140 (1980).

30. F. T. Hioe and J. H. Eberly, *Phys. Rev. Lett.* **47**, 838 (1981).

31. F. T. Hioe, *Phys. Rev. A* **29**, 3434 (1984).

32. R. G. Brewer and E. L. Hahn, *Phys. Rev. A* **11**, 1641 (1975).

33. H. R. Gray, R. M. Whitley, and C. R. Stroud, Jr., *Opt. Lett.* **3**, 218 (1978).

34. R. J. Cook and B. W. Shore, *Phys. Rev. A* **20**, 539 (1979).

35. F. T. Hioe and J. H. Eberly, *Phys. Rev. A* **25**, 2168 (1982); F. T. Hioe, *Phys. Rev. A* **26**, 1466 (1982).

36. F. T. Hioe, *Phys. Rev. A* **28**, 879 (1983).

37. F. T. Hioe, *Phys. Rev. A* **30**, 3097 (1984).

38. F. T. Hioe, *Phys. Rev. A* **32**, 2824 (1985).

39. F. T. Hioe, *J. Opt. Soc. Am. B* **4**, 1327 (1987). The (3, 1) element for $j = 1$ in eq. (16) of this reference should be $+ b^{*2}$, not $- b^{*2}$.

40. U. Fano and G. Racah, *Irreducible Tensorial Sets*, Academic, New York, 1959; B. R. Judd, *Operator Techniques in Atomic Spectroscopy*, McGraw-Hill, New York, 1963; L. C. Biedenharn and J. D. Louck, *Angular Momentum in Quantum Physics*, Addison-Wesley, Reading, MA, 1981.

41. Equations of this form for the magnetic spin resonance problem were first given by S. Pancharatnam, *Proc. Roy. Soc. Lond. A* **330**, 265 (1972).

42. F. T. Hioe, in *Coherence and Quantum Optics* V, L. Mandel and E. Wolf Eds. Plenum, New York, 1984, p. 965.

43. J. Oreg and S. Goshen, *Phys. Rev. A* **29**, 3205 (1984).

44. M. J. Konopnicki, P. D. Drummond, and J. H. Eberly, *Opt. Commun.* **36**, 313 (1981); M. J. Konopnicki and J. H. Eberly, *Phys. Rev. A* **24**, 2567 (1981).

45. F. T. Hioe, *Phys. Rev. A* **26**, 1466 (1982).

46. R. M. Radmore and P. L. Knight, *J. Phys. B* **15**, 561 (1982).

47. Z. Deng, *Opt. Commun.* **48**, 284 (1983).

48. H. P. W. Gottlieb, *Phys. Rev. A* **26**, 3713 (1982); **32**, 653 (1985).

49. D. T. Pegg, *J. Phys. B* **18**, 415 (1985).

50. S. J. Buckle, S. M. Barnett, P. L. Knight, M. A. Lauder, and D. T. Pegg, *Opt. Acta* **33**, 1129 (1986).

51. F. T. Hioe and J. H. Eberly, *Phys. Rev. A* **29**, 1164 (1984).

Note added in proof: Since the completion of this manuscript, Dr. C. E. Carroll and the author have published three more papers which expanded on some of the work reported here. They can be found in *Phys. Rev. A* **37**, 3000 (1988), *J. Opt. Soc. Am. B* **5**, 859 (1988) and June 1988 issue.

CHAPTER X

SPIN RELAXATION AND MOTIONAL DYNAMICS

D. J. SCHNEIDER and J. H. FREED

Department of Chemistry, Cornell University, Ithaca, New York 14853

CONTENTS

We wish to acknowledge support for this work by NSF grants DMR-86-04200 and CHE 87-03014 and NIH grant GM-25862 and by the Cornell University Materials Science Center.

APPENDICES

I. INTRODUCTION

The study of molecular dynamics in liquids is an active and exciting area in theoretical, computational, and experimental chemical physics. The majority of experimental techniques for studying molecular dynamics in isotropic liquids and liquid crystalline phases involve measuring the response of the system to an externally applied time-dependent perturbation, usually electromagnetic radiation. Examples of such techniques are:

magnetic resonance (electron and nuclear spin resonance),

far-infrared and infrared absorption,

dielectric relaxation,

light scattering (Raman, Rayleigh–Brillouin, etc.),

fluorescence depolarization, and

inelastic neutron scattering.

The raw data from these experiments usually reflect the underlying molecular dynamics in a rather indirect fashion. In many instances it is necessary to explicitly model the response of the system to the perturbing field to extract quantitative information on the molecular motions that modulate the experimentally observed signals. The "inversion" of the spectroscopic data to expose the dynamical information can be a difficult task, especially if the perturbing field is so strong that the system cannot respond linearly to it.

In the weak perturbation limit, linear response theory can be used to simplify the interpretation of time domain experimental data by the connection between the observed signal and an equilibrium-averaged correlation function of the relevant dynamical variables. Likewise, frequency domain measurements in the weak perturbation limit reflect the spectral density of fluctuations in these dynamical variables at equilibrium. These two types of measurements are related by the fact that the spectral density observed in a frequency domain experiment is just the Fourier–Laplace transform of the correlation function obtained in a related time domain experiment. This relationship, a consequence of the fluctuation–dissipation theorem, is exemplified by the well-known equivalence between frequency-domain [continuous-wave (CW)] magnetic resonance spectra and free induction decay signals observed in the time domain. The study of molecular dynamics in liquids by modeling CW electron spin resonance (ESR) spectra in the linear response regime is a central topic of this review.

In more intense fields, where linear response theory is not applicable, the full equations of motion for the system including its coupling to the perturbing field must be solved. This is, in general, a more challenging computational problem, but it can lead to entirely new and informative types of spectroscopic tools to study molecular dynamics. In this review we also include a survey of CW nonlinear methods, such as saturation and double resonance, as well as newer ESR time domain multiple-pulse spin echo methods. These latter methods, when performed in an idealized manner, permit one to separately treat the effects of the intense radiation field and of the molecular dynamics; yet they allow for a great variety of intriguing possibilities in exploring the latter.

The various techniques for studying molecular dynamics in liquids can be roughly divided into two classes depending on the nature of the dynamical variables required to model the spectrum: (i) those that are sensitive to single particle properties and (ii) those that are sensitive to collective motions. For example, the Rayleigh–Brillouin light-scattering spectrum from a pure monatomic fluid such as argon depends on fluctuations in the electric polarizability density over distances on the order of the wavelength of visible light. These fluctuations surely involve the collective motion of many molecules. On the other hand, the ESR spectra of dilute solutions of spin probes in liquids are almost always interpretable in terms of the independent sum of the magnetizations from the individual molecules. The distinction between these two classes can become blurred. For instance, in the study of phase transitions in liquid crystals by ESR, the collective fluctuations in the density or order parameter of the bulk fluid can couple to the spin degrees of freedom and dramatically affect the ESR spectrum.

In addition, the experiments can be classified according to the time scale over which they are most sensitive to fluctuations in the relevant variables. The time scale of a magnetic resonance experiment is determined, in part, by the magnitude of the fluctuations in the spin Hamiltonian produced by the interactions of the spin-bearing molecules with the surrounding solvent and by the spectral resolution. To clarify this statement, we must distinguish two limiting regimes of spin relaxation and two distinct types of relaxation. In the more familiar limit, the motional narrowing limit, one speaks of T_1 and T_2 types of spin relaxation. Now, the effectiveness of the T_1, or spin-lattice, relaxation depends in part on the relative magnitudes of the correlation rate (the inverse of the correlation time) of the dynamical fluctuations and the irradiating frequency. But it also depends on the ratio of the magnitude of the stochastic perturbations of the spin Hamiltonian to the correlation rate. However, the T_2, or transverse, relaxation depends only on the latter ratio. Nevertheless, even when this ratio is small, high-resolution spectroscopy enables the accurate measurement of small contributions to T_2 (e.g., to the CW linewidth). Thus, it is possible to detect very fast processes (correlation rates on the order of $10^{12} \, s^{-1}$), albeit in an indirect manner.

The other limit is the slow-motional limit. The slow-motional regime is where the stochastic perturbations to the spin Hamiltonian are comparable to, or greater than, the correlation rate. In this limit, there is no longer a time scale separation between the characteristic times of molecular motion and spin relaxation, which greatly simplifies the analysis of fast-motional spectra. Instead, the spin degrees of freedom and molecular dynamics become intimately coupled. In this limit, spin relaxation experiments probe the relaxation of these coupled modes and thus provide more direct information on the molecular dynamics. In conventional CWESR experiments with nitroxide spin probes, this limit is usually reached for correlation rates on the order of $10^9 \, s^{-1}$, whereas for nuclear magnetic resonance (NMR), it can be on the order of $10^3 \, s^{-1}$. The characteristic time scale for slow-motional ESR experiments implies that they will be sensitive to motions of spin probes in viscous liquids and/or slowly diffusing spin-labeled macromolecules in solution. On the other hand, the slower time scale of the typical NMR experiment is better suited to the slower dynamics in solid phases. Another feature of the magnetic resonance experiment is that, at least in principle, one can "tune in" the slow-motional regime by properly choosing the value of the magnetic field and the corresponding resonance frequency. This feature allows one to vary the relative magnitudes of some of the relevant terms in the fluctuating part of the spin Hamiltonian to increase the sensitivity of the experiment to a particular time scale of molecular motion. At present, this idea is being used to extend the time scale of the slow-motional ESR experiment to the range 10^{10}–10^{11} s^{-1}, depending on the nature of the spin probe, by working at high fields (90 kG) and high frequencies (250 GHz).

Whereas the analysis of slow-motional spectra holds the potential for greater information on molecular dynamics in liquids, it poses much more complicated computational challenges. In answer to these challenges, there have been a number of major advances in computational methods in recent years. These new computational methods for calculating slow-motional magnetic resonance spectra are the focal point of this review.

With these considerations in mind, the interpretation of dynamical effects on ESR spectra of dilute solutions of spin probes in viscous liquids can proceed in the following manner. Since the experimental observable depends only on the sum of the magnetization due to the individual spins and not on any collective phenomena or interaction between spins on separate particles, it makes sense to base the analysis on an approximate equation of motion for the one-particle spin density matrix. The time evolution of the macroscopic magnetization can then be calculated as the equilibrium average of the contributions due to the individual spins. Moreover, the nature of the interaction of the unpaired electron spin with its environment in a dilute solution is such that the spin Hamiltonian, which drives the time evolution of the density matrix, should depend only on the orientation and/or angular velocity of the spin-bearing molecule with respect to a laboratory-fixed reference frame and on the relative positions of pairs of spin-bearing molecules. This suggests that the positional, orientational, and angular velocity degrees of freedom of the spin probe can be modeled as a stochastic process in which the intricate details of the collisions with other molecules are unimportant.

In this manner, the molecular dynamics is incorporated into the calculation by assuming a specific form of the stochastic modulation of the orientation and position of the spin probe molecules, which can, in general, couple to other degrees of freedom of the fluid. It turns out to be particularly convenient and useful to model the molecular dynamics as a Markov process that modulates the various terms in the spin Hamiltonian. This is not, as we will see, a particularly restrictive approach. One is free to incorporate as many relevant dynamical variables as justified by the experiment into a multidimensional Markov process. However, the greater the number of degrees of freedom that are included, the more challenging is the computational problem. In this review we address the twofold problem presented by this stochastic modeling approach: (i) how to choose appropriate Markovian forms based on the known or presumed molecular physics of the system and (ii) how to solve the resulting equations for the relevant spectral densities or magnetic resonance lineshape. In the latter case, the spin dynamics and the molecular dynamics can be described in a single equation of motion for the spin density matrix, which has been properly generalized to include its dependence on the orientation, angular velocity, and position of the molecule. This equation is usually referred to as the stochastic Liouville equation (SLE). The generalized spin density matrix combines the properties of the usual spin density matrix with those of a classical probability distribution for the dynamical variables incorporated into the Markovian model for the molecular dynamics. The calculation of magnetic resonance spectra from the SLE is the canonical problem dealt with in this review.

After we describe the basic ESR lineshape problem in terms of the SLE in Section II, we review various methods in Section III for reducing the set of coupled partial differential equations represented by the SLE to a tractable set of linear algebraic equations that can be solved on a computer. A survey of methods of solving the SLE based on the classic algorithms of numerical linear algebra are the subject of Section IV. We then focus in Section V on the newer methods based on the Lanczos [1]* and conjugate gradients algorithms [2] that have proven to be extremely powerful for these applications. We describe in some detail their strengths and how they may be employed.

There is a close theoretical connection between these methods and those used by other workers in a variety of fields. The theoretical interrelationships which provide a framework for understanding and justifying these methods are considered in Appendix B.

The modeling of the subclass of stochastic processes that can be described by Fokker–Planck equations is treated in Section VI. It is shown how the inherent symmetry of Fokker–Planck equations, which obey the requirements of detailed balance, allow one to extend the Lanczos and conjugate gradients methods in a particularly simple fashion to the calculation of spectral densities. In fact, what emerges is a very general approach to irreversible processes that obey the preceding restrictions. This general approach can be characterized in the following manner. Whereas the reversible dynamics (both classical and quantum mechanical) is most properly and conveniently analyzed using the familiar unitary or Hilbert space formalism, the inclusion of irreversible terms naturally leads to a formalism involving complex symmetric matrices and complex orthogonal spaces. Since the properties of complex symmetric matrices and nonunitary bilinearly metric spaces such as complex orthogonal spaces are unfamiliar to most readers, these ideas and their more familiar unitary space analogues are summarized in Appendix C because of their importance to the central themes of this review.

A brief survey of the influence of molecular dynamics on nonlinear phenomena and pulsed methods in magnetic resonance is given in Section VII.

---

*In this chapter, references are enclosed in brackets—Ed.

The complex symmetric Lanczos algorithm introduced by Moro and Freed [3] and the closely related complex symmetric conjugate gradients algorithm of Vasavada, Schneider, and Freed [4] arise in a natural fashion in the complex orthogonal space formalism for irreversible processes. A basic message of the present review is the applicability of these methods to computational and theoretical studies of irreversible processes in general and to the study of molecular dynamics in liquids by magnetic resonance in particular.

II. ESR LINESHAPES AND THE STOCHASTIC LIOUVILLE EQUATION

The relationship between the ESR spectral lineshape function, $I(\Delta\omega)$ and the dynamics of motion of a paramagnetic molecule can be expressed in the form [5–12]

$$I(\Delta\omega) = \frac{1}{\pi} \mathrm{Re} \langle v|[i(\Delta\omega \mathbf{I} - \mathbf{L}) + \boldsymbol{\Gamma}]^{-1}|v\rangle, \tag{1}$$

where $\Delta\omega$ is the sweep variable, $\mathbf{L}$ is the Liouville operator associated with the spin Hamiltonian of the probe molecule, and $\boldsymbol{\Gamma}$ is the Markovian operator for the stochastic variables that modulate the magnetic interactions. In most cases, $\boldsymbol{\Gamma}$ is taken to be a Fokker–Planck operator. Also, $|v\rangle$ is the so-called starting vector constructed from the spin transition moment averaged over the equilibrium ensemble. The vectors and operators are defined in the direct product space of the ESR transitions and functions of the stochastic variables. (For typical ESR calculations $\Delta\omega = \omega - \omega_0$, where ω_0 is the Larmor frequency at the center of the spectrum and ω is the angular frequency of the applied radiation field.) Equation 1 is derived from the more general stochastic Liouville equation, which is appropriate for studying the spectrum (cf. Section II.B).

By means of computer calculation of ESR spectra, one may extract information about the dynamics of motion. In particular, spectra in the so-called slow-motional region are sensitive to the form chosen for the diffusion operator $\boldsymbol{\Gamma}$, making it possible to distinguish between different models for the motion by comparison between experimental and theoretical spectra [12–25]. (The slow-motional regime may be defined by the inequality $|\mathbf{L}|/|\boldsymbol{\Gamma}| \geqslant 1$, where $|\mathbf{L}|$ and $|\boldsymbol{\Gamma}|$ are measures of the magnitude of matrix elements of $\mathbf{L}$ and $\boldsymbol{\Gamma}$, respectively.) This application of ESR spectroscopy depends on the efficiency of the algorithm for calculating spectra. As one utilizes more sophisticated models, calculations with matrices that increase geometrically in dimension are required. This need would seem warranted by the development of new ESR techniques that are particularly sensitive to molecular motions [26–35].

Also, there are many applications of the ESR technique, requiring calculation of spectra, to systems of physical or biological relevance with the primary purpose of deriving the relaxation time(s) for the reorientational motion [12, 21–24, 36, 37]. In these applications, simple forms for the diffusion operator are utilized, so the size of the matrix is relatively small. But the use of a very compact algorithm that does not need large memory allows one to use a mini or personal computer directly connected to the ESR spectrometer. Also, one may obtain NMR lineshapes in the slow-motional region, particularly in solids, and these may also be analyzed in terms of Eq. 1 [36–40]. In fact, there continues to be a growing number of applications requiring detailed spectral calculations based on Eq. 1.

Our studies of ESR spectra and the modeling of motional dynamics have made clear that the same algorithms would be applicable to the general class of Fokker–Planck equations, since they may also be represented by operators of the form of those in Eq. 1 due to the existence of both inertial or drift terms and damping terms [3, 4, 41–44]. The calculation of the time correlation functions (or, more precisely, their Fourier–Laplace transforms, which are usually referred to as spectral densities) is also found to proceed from expressions like Eq. 1. Thus, the analysis and discussion in the next several sections will also apply to such cases.

Note that Eq. 1 can be rewritten as

$$I(\Delta\omega) = \frac{1}{\pi} \operatorname{Re} \langle v | u(\Delta\omega) \rangle, \qquad (2)$$

where $|u(\Delta\omega)\rangle$ is the solution of the equation

$$A'(\Delta\omega)|u(\Delta\omega)\rangle = (i\,\Delta\omega\,\mathbf{I} + \mathbf{A})|u(\Delta\omega)\rangle = |v\rangle. \qquad (3)$$

The operator, $\mathbf{A}$ is defined as $\mathbf{A} = \mathbf{\Gamma} - i\mathbf{L}$. The spectrum given by Eq. 3 can be calculated by either solving Eq. 3 for a range of values of $\Delta\omega$ or, alternatively, by diagonalizing $\mathbf{A}$ only once [5–8, 12, 45].

The matrix of the operator $\mathbf{A}$ is in general very large and sparse. Thus, conventional methods [5–8, 12, 45] for solving Eq. 3 by inversion or by diagonalizing $\mathbf{A}$ prove to be too cumbersome (cf. Section IV). One soon runs out of memory even on mainframe computers, and the solution requires prohibitive amounts of computer time. To remedy this situation, the Lanczos algorithm (LA) has been developed for complex-symmetric matrices, since $\mathbf{A}$ is typically of this form or else it can be transformed to this form. It is an efficient method for tridiagonalizing **A** and is particularly well suited to the solution of sets of linear algebraic equations such as Eq. 3, which are characterized by large sparse matrices. We find that it can lead to at least order of magnitude reductions in computation time, and it yields results to the solution of Eq. 1 to a high degree of accuracy [3, 4, 43, 44]. In a more theoretical vein, it was possible to establish the close connection between the LA based on a scheme of projection operators in Hilbert space and the Mori projection scheme in statistical mechanics [43, 44]. Though the emphasis here will be on applications to ESR spectroscopy, the Lanczos methods described in Section V may be regarded as appropriate for a wide range of applications in chemical physics.

There have appeared other reports of computational methods for calculating ESR spectra based on Padé approximants [46] and on the Mori method [47, 48], which may be expected to be formally equivalent to the application of the LA [3, 43, 44] (see also Appendices A and B). This matter has recently been studied in detail by Dammers [49], who finds that whereas all these methods are indeed formally equivalent, the LA is the most stable and efficient from a computational viewpoint. We discuss these matters in more detail in Appendix B.

A. Derivation of the Spectral Lineshape Function in the Linear Response Regime

The slow-motional regime for a tumbling spin probe can be defined operationally as the range of molecular motional rates where a change in motional rate gives rise to an observable effect on the ESR spectra but the spectra cannot be adequately described by a fast-motional theory. The ESR spectra in the fast-motional limit are well understood but are less sensitive to the details of the motions of the spin probes than slow-motional spectra, whereas the rigid limit provides no motional information whatsoever. Thus, one is forced to tackle the problem of the interpretation of the slow-motional spectra.

The breakdown of the fast-motional theories can be traced to their perturbative nature. In the slow-motional regime the dynamics of the spins are strongly coupled to the orientational and/or positional degrees of freedom of the molecule, which render perturbative treatments invalid. To proceed, one must treat both the classical orientational and/or positional degrees of freedom and the quantum-mechanical spin degrees of freedom on a more equal footing. Since solving the exact equations of motion for all the molecules in the sample is obviously an impossible task, some physically reasonable assumptions must be introduced to make the problem tractable.

First, assume that the equation of motion for the density matrix, $\rho(t)$, has the same Hamiltonian $\hat{\mathscr{H}}(t)$ for all members of the ensemble and is given by the quantum-mechanical Liouville equation

$$\frac{\partial \rho}{\partial t} = - i[\hat{\mathscr{H}}(t), \rho], \tag{4}$$

where $\hat{\mathscr{H}}(t)$ is given in angular frequency units.

Now, assume that the time dependence of the spin Hamiltonian $\hat{\mathscr{H}}(t)$ for a spin probe arises from interactions with its environment such that $\hat{\mathscr{H}}(t)$ is fully determined by a complete set of random variables, Ω. Also assume that this time dependence of Ω is described by a stationary Markov process, so that the probability of being in a state Ω_2 at time t_2, if in state Ω_1 at time $t_1 = t_2 - \Delta t$, is independent of the value of Ω at any time earlier than t_1 and depends only on the time difference Δt and not on t_1. A stationary Markov process can be described by a differential equation,

$$\frac{\partial P(\Omega, t)}{\partial t} = - \Gamma(\Omega) P(\Omega, t), \tag{5}$$

where $P(\Omega, t)$ is the probability of the spin probe being in a state Ω at time t.

Since the process is assumed stationary, $\Gamma(\Omega)$ is independent of time. The stochastic evolution operator $\Gamma(\Omega)$ operates only on the random variables Ω and is independent of the spin degrees of freedom and may include such general Markov operators as the diffusion operators given by Fokker–Planck equations and transition rate matrices among discrete states. In most of our examples, Ω will represent Euler angles specifying orientation and $\Gamma(\Omega)$ will be a rotational diffusion or Fokker–Planck operator. It is also assumed that the stochastic process has a unique equilibrium distribution $P_0(\Omega)$ characterized by

$$\Gamma(\Omega) P_0(\Omega) = 0. \tag{6}$$

It can be shown [5, 7, 50] that Eqs. 4–6 lead to the SLE of motion,

$$\begin{aligned}\frac{\partial \rho(\Omega, t)}{\partial t} &= - i[\hat{\mathscr{H}}(\Omega), \rho(\Omega, t)] - \Gamma(\Omega)\rho(\Omega, t) \\ &= - i\mathbf{L}(\Omega)\rho(\Omega, t) - \Gamma(\Omega)\rho(\Omega, t), \end{aligned} \tag{7}$$

where $\rho(\Omega, t)$ is now understood to be the value of the density matrix associated with a particular value of Ω and hence of $\hat{\mathscr{H}}(\Omega)$. Thus, instead of looking at the explicit time dependence of the spin Hamiltonian $\hat{\mathscr{H}}(t)$ involving the interaction with its environment, the spin Hamiltonian is written in terms of random variables Ω, and their modulation (e.g., due to rotational motions) is expressed by the time dependence of Ω.

Equation 7 implicitly neglects the back reaction of the spins on the random variables Ω, so spin relaxation induced by the coupling of the spins to Ω will tend to relax the spins to infinite temperature. This is not a concern for the lineshape problem (cf. Eq. 1). The solution is now well known and will be discussed further in Section VII, where pulsed and nonlinear phenomena are treated, and this matter is important.

The general linear response expression for the imaginary part of the magnetic susceptibility $\chi''(\omega)$ resulting from a very weak linearly polarized microwave field of angular frequency ω being applied to the system [51] is

$$\chi''_{jj}(\omega) = \frac{\omega}{2Nk_BT} \int_0^\infty dt (e^{i\omega t} + e^{-i\omega t}) \operatorname{Tr}\{\mathcal{M}_j(t)\mathcal{M}_j\}, \tag{8}$$

which involves a trace over the macroscopic magnetization operator $\mathcal{M}_j$. In Eq. 8, T is the absolute temperature, k_B Boltzmann's constant, and N is the number of spin eigenstates of the spin probe. Note that M, the macroscopic value of the magnetization is related to $\mathcal{M}$, its associated quantum-mechanical operator, by $M(t) = \operatorname{Tr}\{\rho(t)\mathcal{M}\}$. The oscillating field is taken along the $j = x, y$, or z direction. For our system of noninteracting (or weakly interacting) spin probes with nearly isotropic g value, we have

$$\operatorname{Tr}\{\mathcal{M}_j(t)\mathcal{M}_j\} = \mathcal{N}\gamma_e^2 \operatorname{Tr}\{S_j(t)S_j\} = \mathcal{N}\gamma_e^2 \operatorname{Tr}\{S_jS_j(t)\}, \tag{9}$$

where $\mathcal{N}$ is the number of spins in the sample and γ_e is the magnetogyric ratio of the electron. The spin operator $S_j(t)$ in the Heisenberg representation will obey a SLE equivalent to Eq. 7 given by

$$\frac{\partial S_j(\Omega, t)}{\partial t} = i\mathbf{L}(\Omega)S_j(\Omega, t) - \mathbf{\Gamma}^\dagger(\Omega)\rho(\Omega, t), \tag{10}$$

where the superscript dagger implies the Hermitian adjoint operator. This is required to have the expectation values of the magnetization be identical in the Heisenberg and Schrödinger pictures given by Eqs. 10 and 7, respectively. The SLE in Eq. 10 is subject to the initial condition

$$S_j(\Omega, 0) = P_0(\Omega)S_j. \tag{11}$$

This form will be needed in order to interpret Eq. 9 as a proper equilibrium-averaged correlation function.

Now in Eqs. 8 and 9, the trace over orientational degrees of freedom is replaced by a classical average indicated by an overbar:

$$\overline{S_j S_j(t)} = \overline{S_j S_j(\Omega, t)} = \int d\Omega\, S_j S_j(\Omega, t). \tag{12}$$

In this notation

$$\chi''_{jj}(\pm\omega) = \frac{\mathcal{N}\gamma_e^2\omega}{2Nk_B T}\,\mathrm{Tr}_s\,\overline{\{S_j S_j(\pm\omega, \Omega)\}}, \tag{13}$$

where the trace is only over spin degrees of freedom, $S_j(\pm\omega, \Omega)$ is the Fourier–Laplace transform of $S_j(\Omega, t)$,

$$S_j(\pm\omega, \Omega) \equiv \int_0^\infty dt\, e^{\mp i\omega t} S_j(\Omega, t), \tag{14}$$

and the plus and minus signs are found to correspond to the two counterrotating components of the microwave field, only one of which is important in large static magnetic fields.

From Eqs. 10 and 14 it follows that

$$S_j(\omega, \Omega) = [i(\omega\mathbf{I} - \mathbf{L}) + \mathbf{\Gamma}^\dagger]^{-1} S_j(0, \Omega). \tag{15}$$

Thus,

$$\overline{S_j S_j(\pm\omega, \Omega)} = \int d\Omega\, S_j[i(\omega\mathbf{I} - \mathbf{L} + \mathbf{\Gamma}^\dagger]^{-1} P_0(\Omega) S_j. \tag{16}$$

This expression may be inserted in Eq. 13 to obtain $\chi''_{jj}(\pm\omega)$.

It is convenient at this point to introduce a "symmetrizing" transformation for the stochastic Liouville operator. It is not needed for isotropic liquids with simple models but becomes useful for anisotropic liquids [9] or more sophisticated models [15]. The relevant similarity transformation is

$$\tilde{\mathbf{\Gamma}}(\Omega) = P_0^{-1/2}(\Omega)\mathbf{\Gamma}(\Omega)P_0^{1/2}(\Omega), \tag{17}$$

where here $P_0^{\pm 1/2}(\Omega)$ are regarded as operators. This transformation defines $\tilde{\mathbf{\Gamma}}$ in a form that may be represented by a symmetric matrix that is, in general, complex (cf. Sections VI.B and VI.B.1). The symmetrized diffusion operator $\tilde{\mathbf{\Gamma}}(\Omega)$ will be used in the remainder of this section.

Equation 16 may be rewritten in more symmetric form as

$$\overline{S_j S_j(\pm\omega, \Omega)} = \int d\Omega\, S_j P_0^{1/2}(\Omega)[i(\omega\mathbf{I} - \mathbf{L} + \tilde{\mathbf{\Gamma}}^\dagger]^{-1} P_0^{1/2}(\Omega) S_j. \tag{18}$$

The trace over spin space and averaging over the ensemble in Eq. 18 may be regarded as a scalar product in Liouville space that can be represented as

$$\mathrm{Tr}_s \overline{S_j S_j(\pm\omega,\Omega)} = \langle P_0^{1/2} S_j | [i(\omega\mathbf{I} - \mathbf{L}) + \tilde{\mathbf{\Gamma}}^\dagger]^{-1} | P_0^{1/2} S_j \rangle, \tag{19}$$

which has the form of Eq. 1.

B. The Spin Hamiltonian

The total spin Hamiltonian $\hat{\mathscr{H}}(t)$, expressed in angular frequency units, can be separated into three components,

$$\hat{\mathscr{H}}(t) = \hat{\mathscr{H}}_0 + \hat{\mathscr{H}}_1(\Omega) + \hat{\varepsilon}(t). \tag{20}$$

In the high-field approximation the orientation-independent component $\hat{\mathscr{H}}_0$,

$$\hbar\hat{\mathscr{H}}_0 = \gamma_e B_0 \hat{S}_z - \hbar\sum_i \gamma_i B_0 \hat{I}_z + \hbar\gamma_e \sum_i a_i \hat{S}_z \hat{I}_z, \tag{21}$$

gives the zero-order energy levels and transition frequencies. The orientation-dependent part, $\hat{\mathscr{H}}_1(\Omega)$, can be expressed as the scalar product of two tensors [52]:

$$\hat{\mathscr{H}}_1(\Omega) = \sum_{\mu,i} \sum_{L,M,K} (-1)^K F_{\mu,i}'^{(L,-K)} \mathscr{D}_{KM}^L(\Omega) A_{\mu,i}^{(L,M)}, \tag{22}$$

where the $F_{\mu,i}'^{(L,K)}$ and $A_{\mu,i}^{(L,M)}$ are irreducible tensor components of rank L. The $F_{\mu,i}'^{(L,K)}$ are spatial functions in molecule-fixed coordinates, whereas $A_{\mu,i}^{(L,M)}$ is a spin operator defined in the laboratory axis system. The subscripts μ and i refer to the type of perturbation and to the different nuclei, respectively. The generalized spherical harmonics $\mathscr{D}_{KM}^L(\Omega)$ include the transformation from the molecule-fixed axis system (x',y',z') into the laboratory axis system (x,y,z). For the analysis of most slow-motional ESR spectra of simple free radicals where $S = 1/2$, only second-rank tensors are important, for example, the A- and g-tensors. The calculation of simple types of matrix elements of the Liouville operator derived from $\hat{\mathscr{H}}_0 + \hat{\mathscr{H}}_1$ is summarized in Appendix D.

C. Model Diffusion Operators

When the general method is applied to rotational modulation, Ω can be taken to be the Euler angles for a molecular axis system fixed to a tumbling spin probe with respect to a fixed laboratory axis system. For a molecule undergoing many collisions, causing small random angular reorientations, the resulting isotropic Brownian rotational motion is a Markov process, which can be described by the rotational diffusion equation

$$\frac{\partial P(\Omega, t)}{\partial t} = -R\nabla_\Omega^2 P(\Omega, t), \tag{23}$$

where ∇_Ω^2 is the Laplacian operator on the surface of the unit sphere and R is the rotational diffusion coefficient.

In an isotropic liquid, the equilibrium probability $P_0(\Omega)$ given by Eq. 6 will be equal for all orientations, so that $P_0(\Omega) = 1/8\pi^2$. Here the Markov operator $\Gamma(\Omega)$ for isotropic Brownian rotation is independent of Ω since the liquid is assumed to be isotropic. The operator $\Gamma = R\nabla_\Omega^2$ is of the form of the Hamiltonian for a spherical top; therefore, its orthonormal eigenfunctions are the generalized spherical harmonics

$$\phi_{LMK}(\Omega) = \left(\frac{2L+1}{8\pi^2}\right)^{1/2} \mathscr{D}_{MK}^L(\Omega) \tag{24}$$

with eigenvalues $RL(L+1)$ [53, 54].

Similarly, the Markov operator for axially symmetric Brownian diffusion about a molecule-fixed z axis is formally the Hamiltonian for a symmetric top whose symmetry axis is the z axis. The orthonormal eigenfunctions are again the normalized Wigner rotation matrices with eigenvalues $R_\perp L(L+1) + (R_\parallel - R_\perp)K^2$, where $R_\perp$ and $R_\parallel$ are the rotational diffusion constants about the x and y and about the z axes, respectively [7, 12, 53, 54]. The "quantum numbers" K and M of the Wigner rotation matrices refer to projections along the body-fixed symmetry axis and along a space-fixed axis, respectively. For completely asymmetric Brownian rotation the diffusion constants about the three principal axes are all unequal, and the stochastic operator has more complicated solutions [7, 12, 53, 54].

Some canonical models for rotational reorientation frequently used in ESR spectroscopy are the following

- Brownian rotational diffusion.
- An approximation to free diffusion in which a molecule rotates freely for time τ (i.e., inertial motion with $\tau = I/\beta$, where I is the moment of inertia and β is the friction coefficient) and then reorients instantaneously.
- Jump diffusion in which a molecule has a fixed orientation for a fixed time τ and then "jumps" instantaneously to a new orientation with no inertial effects [13, 55].

For isotropic reorientation, the characteristic relaxation rates (eigenvalues) associated with the generalized spherical harmonic eigenfunctions of rank L

are all degenerate for these three models and are given, respectively, by

$$\tau_L^{-1} = \frac{RL(L+1)}{[1 + R\tau L(L+1)]^{1/2}},$$

$$\tau_L^{-1} = \tau^{-1}\left[1 - \frac{1}{2L+1}\int_0^\pi d\psi\, W(\psi)\frac{\sin(L+1/2)\psi}{\sin(\psi/2)}\right],$$

where $W(\psi)$ is the distribution function for diffusive steps by an angle ψ about an arbitrary axis and is normalized so that

$$\int_0^\pi d\psi\, W(\psi) = 1.$$

More sophisticated Markovian models and the general matter of modeling the molecular dynamics are discussed in Section VI. Some of the simple types of matrix elements of $\tilde{\Gamma}$ that arise in anisotropic liquids are given in Appendix D.

III. OVERVIEW OF DISCRETIZATION METHODS FOR THE SOLUTION OF THE SLE

For typical forms of the diffusion operator, the SLE is a set of coupled partial differential equations (PDEs) governing the time evolution of the orientation-dependent quantum-mechanical spin density matrix subject to specific initial conditions. This set of PDEs can be simplified to a set of coupled linear algebraic equations (LAEs) by Fourier–Laplace transformation of the set of PDEs with respect to time (cf. Eq. 1) followed by discretization of the spatial parts. The discretization is necessary to remove the spatial derivative terms usually present in the diffusion operator. The resulting set of LAEs can then be solved in a variety of ways. The proper choice of method of solution depends on the structure of the matrix and the specific quantity desired.

There are three important techniques to achieve discretization of angular-dependent terms: expansion in a set of basis functions, finite-difference approximation, and finite-element approximation. All of these methods typically give rise to very large sparse matrices characterizing the stochastic Liouville operator $\mathbf{A}$.

The expansion in a set of global basis functions is analogous to the method of variation of constants used in elementary quantum mechanics and the study of differential equations. The set of basis functions used in the expansion is usually not complete in the formal sense since this would imply an infinite set of coupled equations, but it can be made adequate for numerical calculations to any degree of accuracy desired. The most common set of functions to use as basis functions in cases involving rotational diffusion are the generalized spherical harmonics. There are several reasons for this choice:

- They are the eigenfunctions of the three canonical rotational Fokker–Planck operators in isotropic liquids (cf. Section II).
- They, by definition, have well-defined transformation properties with respect to coordinate rotations; thus, powerful group-theoretic techniques can be used to simplify calculations.
- Their eigenvalues have favorable scaling properties with respect to the principal quantum number L.
- They form an orthonormal set, and the infinite set of all generalized spherical harmonics form a basis in which any square integrable function on the unit sphere in four space can be expanded.

Though these functions are the eigenfunctions of the quantum-mechanical rigid rotator Hamiltonian, it is important to realize that these functions are only used to expand the orientation-dependent density matrix in this application and have nothing to do with angular momentum! Nevertheless, all the sophisticated angular momentum coupling and transformation techniques that have been developed in other areas of physics and chemistry can be applied since these functions have well-defined properties under rotations. The properties of the generalized spherical harmonics and applications of angular momentum techniques to density matrix problems are described in detail in various texts [56, 57] and papers [58–64].

For example, the operators $S_j(\Omega, \omega)$ given in Eq. 15 can be expanded as

$$|S_j(\Omega, \omega)\rangle = \sum_{\lambda, m} s_{j, \lambda, m}(\omega)|\lambda, m\rangle, \tag{25}$$

where the ket $|\lambda, m\rangle = |\lambda\rangle|m\rangle$ is a product of spin operators and spatial functions, where λ represents labels for the spin operator basis, m represents the labels associated with the basis of spatial functions, and the expansion coefficients $s_{j, \lambda, m}(\omega)$ are, in general, complex-valued functions. In this notation Eq. 19 becomes

$$\mathrm{Tr_s}\,\overline{S_j S_j(\pm \omega, \Omega)} \propto \mathbf{v}^{\mathrm{tr}}[\mathbf{A}'(\omega)]^{-1}\mathbf{v}, \tag{26}$$

where $\mathbf{A}'(\omega)$ is the matrix of the operator $i(\omega\mathbf{I} - \mathbf{L}) + \tilde{\boldsymbol{\Gamma}}^\dagger$ in the basis $|\lambda, m\rangle$, the superscript tr means transpose, and the elements of the starting vector are

$$v_{\lambda, m} = \langle \lambda, m|P_0^{1/2}S_j\rangle = \frac{1}{N}\mathrm{Tr}\{S_\lambda^\dagger S_j\}\langle m|P_0^{1/2}(\Omega)\rangle. \tag{27}$$

Equation 26 is equivalent to Eq. 1. This discretization procedure can be used to generate a complex symmetric form for $\mathbf{A}'(\omega)$. The evaluation of the components of $\mathbf{v}$ is discussed in Section V.A and in Appendix D.

The finite-difference approximation involves explicit discretization of the spatial variables. By assuming that the radical can only be found at these discrete positions or orientations, it is possible to approximate the Fokker–Planck equation for the particle by a finite-difference equation. This is a very popular technique for solving partial differential equations in many areas of science. The finite-difference approach to solving the SLE was used by Gordon and Messenger for angular variables [45] and by Freed and co-workers for translational problems [65–70], and is still in widespread use [71–76].

The finite-element approach involves approximating the solution of the SLE in a piecewise fashion over finite areas on the unit sphere [77–79] or over finite volumes in Cartesian space [78, 79]. Usually, the solution is well approximated by low-order polynomial functions, and appropriate continuity requirements are enforced along the boundaries of the elements. This matching of elements at the boundaries implies that these functions do not form an orthogonal set, but the solution is not uniquely defined without these conditions. The lack of orthogonality means that the computer solution of the equations generated by applying the finite-element approximation is more difficult by traditional means, as it leads to a generalized eigenvalue problem [80, 81]. The use of finite elements for the SLE and the associated variational problem are discussed in detail by Zientara and Freed [78]. The alternative approach of using global, orthogonal functions instead of piecewise smooth polynomials over small regions is identical to the basis function expansion method already discussed. Derived in this manner, it is known as the global Galërkin variational method [78, 81].

The most prominent exceptions to the pattern of discretization followed by matrix manipulation are the Monte Carlo methods developed by Pedersen [82] and Itzkowitz [83] where the relaxation function is evaluated directly and the spectrum is obtained by Fourier transformation. Though this method seems to be less efficient on conventional computers than the matrix-oriented approaches discussed, the popularity of the Monte Carlo technique in other disciplines has spurred the development of new computer architectures and associated algorithms that should prompt renewed interest in this approach for the CW lineshape problem. The remaining drawback to the Monte Carlo approach, namely, the inaccessibility of the eigenvalue–eigenvector decomposition, makes it inapplicable for spin echo calculations and other applications where these quantities are required.

It is also possible to use numerical integration techniques on the discretized equations of motion to directly evaluate the time evolution, but this has only been attempted when the complete time dependence is required, such as in the detailed investigation of the interaction of the spins with microwave pulses of finite amplitude and duration [84] or when the spin Hamiltonian has a particularly simple form [85].

IV. OVERVIEW OF TRADITIONAL MATRIX METHODS FOR THE SOLUTION OF THE SLE

The algorithms commonly used for the calculation of slow-motional spectra from the SLE fall into two main categories based on efficiency and ability to handle large sparse matrices. First are the traditional algorithms for diagonalizing matrices and solving sets of coupled linear equations. Second, there are the various forms of the LA for tridiagonalization and the related conjugate gradients algorithm for solving sets of linear equations.

The first class of traditional methods is characterized by a variety of difficulties and strengths. These algorithms typically amount to "computational overkill" for the problem at hand. For instance, the Rutishauser–QR diagonalization gives the full set of eigenvectors and eigenvalues though only a small subset are important in the final spectrum. This large computational overhead and their characteristic of modifying the sparsity structure of $\mathbf{A}$ combine to make these algorithms unattractive for present purposes. An important strength of this class of algorithms is their well-characterized stability and reliability.

In contrast, the LA and its kin are much better suited for the efficient calculation of magnetic resonance spectra. They are effective in handling large sparse matrices since they do not modify the original matrix. It is therefore possible to take advantage of the very special sparsity structure of the stochastic Liouville matrix. In addition, all of the quantities calculated in the LA are either used directly in the calculation of the spectrum or are needed in the next recursive step (see Section V.A and Appendix A). In this sense, the LA represents a good approximation to the minimal amount of computation necessary to compute magnetic resonance spectra in the linear response regime.

Before getting into a discussion on the computational aspects of solving the SLE by matrix methods, it is valuable to review the analytic aspects of the problem. In general, the stochastic Liouville matrix $\mathbf{A}$ can always be represented as a complex symmetric matrix (CSM) [3], and $\mathbf{A}$ *cannot* be Hermitian in the presence of relaxation. In addition, there is a band outside of which all matrix elements are identically zero. The class of CSM is quite general, and many of the theorems on diagonalization and related topics do not have simple analogues for non-Hermitian CSMs [86, 87]. For instance, one is not guaranteed that an arbitrary CSM can be diagonalized by a similarity transformation. However, if a CSM is diagonalizable by a similarity transformation, it is diagonalized by a complex orthogonal matrix (COM) $\mathbf{O}$:

$$\mathbf{O}^{tr}\mathbf{A}\mathbf{O} = \Lambda, \tag{28}$$

where Λ is a diagonal matrix containing the eigenvalues λ_i of $\mathbf{A}$. We will *assume* that the stochastic Liouville matrices under consideration are diagonalizable. This is not too drastic an assumption in light of the fact that any square matrix is arbitrarily close to a diagonalizable matrix [87]. The pathologies that can result from nondiagonalizability have been thoroughly studied [86, 87] and only introduce irrelevant complications into the present discussion. The class of COM is also peculiar in many respects. In particular, the magnitude of the elements of a COM is not bounded as is the case for unitary matrices familiar from quantum mechanics. Since A is non-Hermitian, its eigenvalues are not constrained to lie on the real axis. In spite of the lack of the simple behavior characteristic of Hermitian matrices, we can say something useful about the regions in the complex plane that can and cannot contain eigenvalues. Since the diffusion operator is nonnegative (i.e., $\langle\phi|\Gamma|\phi\rangle \geqslant 0$), a well-known theorem from linear algebra states that all the eigenvalues of $\mathbf{A}$ must lie in the closed right half of the complex λ plane [87]. In physical terms, this corresponds to the fact that the relaxation must force the system toward equilibrium. It is very important to note that the localization of eigenvalues to the right-half plane is independent of the dimension of the basis set (i.e., it does not depend on the number of basis functions, finite differences, or finite elements used in the discretization). In addition, the eigenvalues are restricted, by a similar argument, to lie within a band about the real axis, since the spectrum of the Liouville operator is bounded. The width of the band is dependent on the rigid-limit magnetic tensors. Other eigenvalue localization theorems such as Gerschgorin's theorem can be applied, but they do not give rise to transparently useful results except in the fast-motional limit [88], though they do form a basis for some of the traditional diagonalization algorithms. References to several good treatments of the properties of complex orthogonal matrices can be found in the bibliography [86, 87, 89–94].

We will briefly survey the traditional methods for diagonalization and the solution of linear systems of equations from the point of view of calculating magnetic resonance spectra. This is not intended to be a comprehensive or definitive treatment of these topics. For much more information about the theory and usage of the algorithms discussed here, the reader is encouraged to consult standard texts on numerical linear algebra [80, 95, 96] and the references therein.

A. Solving Linear Systems of Equations Using Gaussian Elimination

The most useful classical method of solving the type of linear systems of equations that arise in the calculation of magnetic resonance spectra is

Gaussian elimination with partial pivoting. This method has been used in the past for the calculation of CWESR spectra [13, 97] and is still used in the quantitative analysis of saturation transfer spectra [98, 99] and the calculation of spectral densities in paramagnetic NMR [100].

The idea behind the Gaussian elimination method is to factor the matrix $A'(\Delta\omega)$ in Eq. 3 as a product of a lower triangular matrix L and an upper triangular matrix U, with care taken to arrange the sequence of operations to minimize loss of accuracy from the finite-precision computer arithmetic. In this manner, one solves Eq. 3 by successively solving two simpler triangular systems and never generating the inverse matrix explicitly.

The procedure can be summarized by rewriting Eq. 3 in the form

$$PA'(\Delta\omega)u(\Delta\omega) = LUu(\Delta\omega) = v, \tag{29}$$

where P is a permutation matrix that arises from the sequencing of the operations. The value of factorizing the matrix lies in the fact that the solution of Eq. 3 can be broken down into solving the equation

$$Ly = P^{tr}v \tag{30}$$

for y followed by solving

$$Uu(\Delta\omega) = y \tag{31}$$

for the desired vector $u(\Delta\omega)$. Equations 30 and 31 involve triangular matrices and hence can be solved in $\mathcal{O}(N^2)$ floating-point operations. The LU factorization itself is, however, an $\mathcal{O}(N^3)$ process.

The Gaussian elimination method is quite stable and reliable if the matrix $A'(\Delta\omega)$ is neither singular nor nearly so. It has the definite disadvantage that it requires large amounts of computer time [$\mathcal{O}(N^3)$ operations] and memory (all elements within the bandwidth must be stored). The complete procedure must be carried out for each value of the frequency at which the value $I(\Delta\omega)$ is desired, since the LU factorization is not independent of $\Delta\omega$.

B. Complete Diagonalization Methods

There are two main algorithms for the eigenvalue–eigenvector decomposition of the type of complex symmetric matrices that arise in magnetic resonance problems. The first is a variant of an algorithm devised by Jacobi for the direct diagonalization of real symmetric matrices. The second, more efficient algorithm is a variant of Given's method [101], due to Rutishauser [45, 102], which eliminates some of the drawbacks of the Jacobi algorithm by first tridiagonalizing the matrix and then using the QR iteration to diagonalize the tridiagonal matrix.

The complete diagonalization of a given matrix is usually much more time consuming than solving Eq. 3 by Gaussian elimination with partial pivoting for a single value of $\Delta\omega$. Therefore, it makes sense to use a diagonalization method if $A'(\Delta\omega) = A - i\Delta\omega I$, since in this case the spectrum can be easily computed for any value of $\Delta\omega$. The formula for $I(\Delta\omega)$ can be easily derived by using the diagonalizing transformation (see Eq. 28) to rewrite Eq. 1 as

$$I(\Delta\omega) = \frac{1}{\pi} \operatorname{Re}\left\{ v^{tr}OO^{tr}[A + i\Delta\omega I]^{-1}OO^{tr}v \right\}$$

$$= \frac{1}{\pi} \operatorname{Re}\left\{ (O^{tr}v)^{tr}[A + i\Delta\omega I]^{-1}(O^{tr}v) \right\}. \tag{32}$$

Since $\Lambda + i\Delta\omega I$ is diagonal, this can be collapsed to a simple sum over the eigenvalues of the form

$$I(\Delta\omega) = \frac{1}{\pi} \operatorname{Re}\left\{ \sum_{j=1}^{N} \frac{c_j^2}{\lambda_j + i\Delta\omega} \right\}, \tag{33}$$

where $c_j = (O^{tr}v)_j$ is the projection of the jth eigenvector on the starting vector. The LAEs derived by finite difference or expansion in a set of trial functions typically have this form, though the finite-element method usually gives rise to off-diagonal elements containing ω because of the nonorthogonality of the piecewise polynomial basis functions.

1. Jacobi-Type Methods

The Jacobi algorithm involves the successive zeroing of the largest off-diagonal matrix element by (real) orthogonal similarity transformations. Unfortunately, these successive transformations tend to fill in matrix elements that were initially zero; thus, this algorithm is not too well suited for sparse matrix diagonalization. In spite of the proliferation of nonzero matrix elements, it can be shown that the rotations can be chosen such that the sum of the moduli of the off-diagonal matrix elements is reduced at every stage. The product of these orthogonal matrices is the matrix of eigenvectors while the reduction in the sum of the moduli of the off-diagonal matrix elements causes the original matrix to converge to the desired diagonal matrix of eigenvalues by Gerschgorin's theorem.

The original algorithm, which calls for the selective zeroing of the largest off-diagonal element, requires a complete search of the off-diagonal matrix elements at every stage. This searching is very costly in terms of computer time and is not too productive. In practice, the so-called cyclic Jacobi algorithm is preferred since it does not require searching the off-diagonal matrix elements.

In the cyclic Jacobi algorithm one starts at the diagonal and moves across a row, zeroing each matrix element in turn. When the end of the row is reached, the same procedure is applied to the next row. When the end of the matrix is reached, the entire procedure is repeated until all of the off-diagonal matrix elements are below some small value characteristic of the roundoff error of the computer.

Several authors have discussed the generalization of the cyclic Jacobi algorithm to handle complex symmetric matrices [103–106]. This generalization is rather straightforward.

All variants of the Jacobi algorithm suffer from two major flaws when applied to the large sparse matrices commonplace in magnetic resonance calculations. First, as mentioned previously, matrix elements that were originally zero get replaced by nonzero entries as the algorithm proceeds; thus, these algorithms have storage requirements that scale as the square of the dimension of the matrix to be diagonalized. Second is the fact that the number of Jacobi rotations needed to diagonalize a matrix can be infinite! Because of these drawbacks, the Jacobi methods have been superceded by alternative methods that rely on a reduction to tridiagonal form as an intermediate stage. Nevertheless, there are applications where the Jacobi algorithm is quite useful [80].

2. Tridiagonalization and the QR Algorithm

In most applications it is faster to diagonalize a matrix by first reducing it to tridiagonal form followed by the diagonalization of the resulting tridiagonal matrix than it is to reduce it to diagonal form directly via the Jacobi method. The tridiagonalization method developed by Givens [101] also uses Jacobi rotations to zero off-diagonal matrix elements. The difference between the Jacobi diagonalization and the Givens tridiagonalization algorithms is the sequence in which the off-diagonal matrix elements are annihilated. It can be shown that the Givens tridiagonalization of a symmetric matrix can be accomplished by a finite number of rotations. A variant of the Givens method due to Rutishauser [102] is particularly well suited to the task at hand since it takes advantage of the banded nature of the stochastic Liouville matrix, thereby reducing storage requirements. This tridiagonalization method was popularized in the magnetic resonance community by the work of Gordon and Messenger [45].

The symmetric tridiagonal matrix generated by the Rutishauser algorithm can be diagonalized in several ways. The most common choice, also popularized by Gordon and Messenger [45], is the QR algorithm. Details of the symmetric tridiagonal QR algorithm can be found in standard references on numerical linear algebra [80, 96]. Recently, Cullum and Willoughby [89] have advocated an alternative procedure based on a QL decomposition.

Prior to the introduction of Lanczos-based methods by Moro and Freed [3], the Rutishauser tridiagonalization–QR iteration method was the most efficient method of calculating magnetic resonance spectra. It is, however, a "brute-force" approach in the sense that a large portion of the effort is expended calculating irrelevant eigenvalues whose corresponding eigenvectors have negligible overlap with the starting vector (see Eq. 33). There is no way for the algorithm to differentiate between important and unimportant eigenvalues as the computation proceeds. All eigenvalues and eigenvectors are on an equal footing, and the number of eigenvalues calculated must equal the dimension of the matrix. In contrast, this ability to differentiate between important and unimportant vectors is inherent in the Lanczos-based methods discussed in the next section.

V. LANCZOS AND CONJUGATE GRADIENTS METHODS OF SOLVING THE SLE

The Lanczos algorithm (LA) and the related conjugate gradients algorithm are extremely effective for calculating slow-motional magnetic resonance spectra [3, 4, 20, 71]. In this section we will discuss the basic LA as applied to Hermitian and complex symmetric matrices. More sophisticated variations of the LA involving selective reorthogonalization [80, 96] and the identification of spurious and duplicated eigenvalues [89] have been developed to circumvent known numerical instabilities. The basic algorithms are sufficient for most magnetic resonance calculations.

There are several advantages to employing the LA for sparse matrix problems such as those presented by Eqs. 1 and 3. These advantages include the following:

- Only the nonzero matrix elements need to be stored; although, in general, they may be recomputed as needed to minimize computer memory requirements.
- The original matrix elements are not altered during the operations.
- No new matrix elements are created.
- The entire algorithm can easily be written in less than 100 lines of FORTRAN code.
- The recursive steps or projections on which the algorithm is based are very closely related to the projection methods in statistical mechanics.

These very practical considerations alone are sufficient reason to prefer the Lanczos methods to Gaussian elimination or complete diagonalization methods.

The theoretical advantages of the LA are as important as they are practical to the overall success of the method. The LA provides a convenient conceptual framework for the identification and classification of the important physical features of the lineshape calculation problem through its relationship with the powerful memory function methods of nonequilibrium statistical mechanics [3, 107–112]. The traditional methods for calculating magnetic resonance spectra are based on theorems from linear algebra rather than physical insight. They are, by their very nature, nearly devoid of insight other than that contained in the specification of the terms in the stochastic Liouville matrix and the discretization method.

A. The Lanczos Algorithm

The LA proceeds by recursive projections or steps that produce successively larger tridiagonal matrix approximations to the original matrix. These projections define the so-called Lanczos vectors. If N is the dimension of the matrix and n_s the number of recursive steps needed to converge to an accurate spectrum, then in all cases studied to date $n_s \ll N$. This inequality becomes more dramatic the more complicated the problem. In this sense, the Lanczos projections rapidly seek out, from an initial finite subspace of dimension N that is spanned by the starting basis set of orthonormal vectors $|f_j\rangle$, $j = 1, 2, \ldots, N$, a smaller subspace spanned by the Lanczos vectors (i.e., the basis vectors for the tridiagonalized form of $\mathbf{A}$, which is $\mathbf{T}_n$) written as $|\Phi_k\rangle$, $k = 1, 2, \ldots, n$. When $n = n_s$, these Lanczos vectors are a sufficient basis for accurately representing the spectrum. In this sense the LA constructs subspaces that progressively approximate the "optimal reduced space" for the problem. These subspaces, spanned by the Lanczos vectors, are the Krylov subspaces [80, 89] generated by the span of the vectors $\mathbf{A}^{k-1}|v\rangle$ for $k = 1, 2, \ldots, n$. Thus, the choice of $|v\rangle$ as the starting vector for the LA biases the projections in favor of this optimal reduced space. It is easy to show that this Krylov subspace can only contain eigenvectors of A with a nonzero projection on $|v\rangle$ in exact arithmetic.

Now, consider the recursive steps of the LA. First, identify the starting vector $|v\rangle$ as the first Lanczos vector $|\Phi_1\rangle$ in accordance with the preceding discussion. A Gram–Schmidt orthogonalization on the Krylov sequence $\mathbf{A}^{k-1}|v\rangle$ for $k = 1, 2, \ldots, n$ recursively generates the set of orthornormal Lanczos vectors $|\Phi_k\rangle$ defined as

$$\beta_k|\Phi_{k+1}\rangle = (\mathbf{I} - \mathbf{P}_k)\mathbf{A}|\Phi_k\rangle, \tag{34}$$

where β_k is the normalizing coefficient chosen such that

$$\langle \Phi_{k+1}|\Phi_{k+1}\rangle = 1 \tag{35}$$

and $\mathbf{P}_k$ is the projection operator on the Krylov subspace spanned by the previous Lanczos vectors

$$\mathbf{P}_k = \sum_{j=1}^{k} |\Phi_j\rangle\langle\Phi_j| \qquad k \leqslant n. \tag{36}$$

Equation 34 leads to a three-term recursive relation for generating the $|\Phi_j\rangle$ (cf. Appendix A):

$$\beta_k|\Phi_{k+1}\rangle = (\mathbf{A} - \alpha_k\mathbf{I})|\Phi_k\rangle - \beta_{k-1}|\Phi_{k-1}\rangle, \tag{37}$$

where

$$\alpha_k = \langle\Phi_k|\mathbf{A}|\Phi_k\rangle \tag{38}$$

and

$$\beta_{k-1} = \langle\Phi_k|\mathbf{A}|\Phi_{k-1}\rangle. \tag{39}$$

It may easily be shown that $\mathbf{A}$ has an $n \times n$ tridiagonal approximation $\mathbf{T}_n$ in the basis of Lanczos vectors:

$$\langle\Phi_k|\mathbf{A}|\Phi_j\rangle = 0 \tag{40}$$

if $k \neq j, j \pm 1$, while Eqs. 38 and 39 give the nonzero matrix elements. That is, given the vectors $|\Phi_j\rangle$ in terms of their components $q_{j,k}$ in the original basis set $|f_j\rangle$, $j = 1, 2, \ldots, N$,

$$|\Phi_k\rangle = \sum_{j=1}^{N} q_{j,k}|f_j\rangle, \tag{41}$$

$$q_{j,k} = \langle f_j|\Phi_k\rangle, \tag{42}$$

the column vectors $\mathbf{q}_k$ form the matrix $\mathbf{Q}_n$ with orthonormal columns such that $\mathbf{Q}_n^{\mathrm{tr}}\mathbf{Q}_n = \mathbf{I}_n$ and

$$\mathbf{T}_n = \mathbf{Q}_n^{\mathrm{tr}}\mathbf{A}\mathbf{Q}_n. \tag{43}$$

This is the conventional single-vector LA for real symmetric matrices. The substitution of Hermitian conjugation for transposition in the preceding equations gives the analogous scheme for general Hermitian matrices.

For applications such as the calculation of magnetic resonance spectra and spectral densities associated with Fokker–Planck equations the matrix $\mathbf{A}$ is either complex symmetric or can be transformed to complex symmetric form [3, 44, 43] (cf. Section VI.D). Moro and Freed [3, 43] have shown that one can simplify problems of this type by introducing a generalized norm and scalar product. That is, first consider the general non-Hermitian case. One can introduce a biorthonormal set of functions $|\Phi_j\rangle$ and $|\Phi^{j'}\rangle$ such that

$$\langle \Phi^{j'}|\Phi_j\rangle = \delta_{j',j}, \tag{44}$$

or, alternatively, letting $\mathbf{x}_j$ and $\mathbf{x}^{j'}$ be their column vector representations,

$$(\mathbf{x}^{j'})^\dagger \mathbf{x}_j = \delta_{j',j}. \tag{45}$$

However, for the case of nondefective complex symmetric matrices $\mathbf{A}$, it is possible to let

$$\mathbf{x}^j = \mathbf{x}_j^* \tag{46}$$

such that Eq. 45 becomes

$$\mathbf{x}_{j'}^{\text{tr}}\mathbf{x}_j = \delta_{j',j}. \tag{47}$$

The Lanczos recursion method remains applicable with Eq. 47 defining the generalized scalar product. Note that the left vector is lacking the usual complex conjugation. The important aspects of this generalized scalar product are developed in Appendices A and C.

In general, the time required for the LA tridiagonalization goes approximately as $n_s N(2n_E + 21)$, where n_E is the average number of nonzero matrix elements in a row of $\mathbf{A}$ [3]. This is obviously superior to the traditional methods that require $\mathcal{O}(N^3)$ time since $n_E, n_s \ll N$.

Finally, note that the tridiagonal form of the complex symmetric matrix $\mathbf{T}_n = \mathbf{Q}_n^{\text{tr}} \mathbf{A} \mathbf{Q}_n$ allows the application of very efficient diagonalization methods [45, 89]. The spectrum defined by Eq. 1 can easily be computed from using the eigenvalues of $\mathbf{T}_n$ and the projections of the associated eigenvectors on the starting vector. However, for computing CW spectra, a continued-fraction method [3, 44] can be used directly on the elements of the matrix $\mathbf{T}_n$. That is, since $|v\rangle$ is the first Lanczos vector and the Lanczos vectors are orthogonal in the sense of Eq. 47, the spectrum is given by

$$I(\Delta\omega) = \frac{1}{\pi}[i\,\Delta\omega\,\mathbf{I}_n + \mathbf{T}_n]_{1,1}^{-1}. \tag{48}$$

By examining the structure of the (1, 1) element of the inverse of successively larger principal submatrices of $[i\,\Delta\omega\,\mathbf{I}_n + \mathbf{T}_n]$, it is easy to show that the $I(\Delta\omega)$

can be written in the continued-fraction form (see Appendix B),

$$I(\Delta\omega) = \frac{1}{\pi} \mathrm{Re}\left[\left(i\,\Delta\omega + \alpha_1 - \frac{\beta_1^2}{i\,\Delta\omega + \alpha_2 - \beta_2^2 \cdots}\right)^{-1}\right].$$ (49)

The application of the LA to **A** generates the continued-fraction representation of the spectrum or spectral density. The same result, apart from the identification of $\tilde{\Gamma}$ with the classical Liouville operator, has been derived by H. Mori in the context of the dynamics of systems of interacting particles [107]. As a matter of fact, the same methodology, more specifically the recursive structure of Eq. 34, is the foundation of both the LA and Mori's derivation.

The relation specified by the continued fraction in Eq. 49 is quite general. Analytical calculation of the coefficients α_j and β_j from the explicit operator form of **A** is possible and has been carried out in simple cases [47, 48]. The axially symmetric g-tensor problem is an example of where this type of calculation is practical. This approach quickly leads to extremely complicated formulas that are difficult to handle for the general case. Therefore, numerical implementation of the recursive relation Eq. 37 is essential in calculating enough coefficients of the continued fraction for an accurate approximation of $I(\Delta\omega)$. In practice, one generates the matrix representation of **A** in the $|f_j\rangle$ basis in which the resulting matrix is complex symmetric (see Section VI and Appendix A):

$$\mathbf{A}_{jk} \equiv \langle f_j|\mathbf{A}|f_k\rangle.$$ (50)

From Eq. 37, the standard recursive relation of the LA may be rewritten as

$$\beta_k \mathbf{q}_{k+1} = (\mathbf{A} - \alpha_k \mathbf{I})\mathbf{q}_k - \beta_{k-1}\mathbf{q}_{k-1},$$ (51)

where the column vector $\mathbf{q}_k$ consists of the components $q_{j,k}$ (cf. Eq. 42). The standard computer implementation of the complex symmetric LA [3, 4] can then be used for calculating the coefficients α_j and β_j from which $I(\Delta\omega)$ can be directly calculated using the continued-fraction representation given in Eq. 49.

Finally, we must deal with the calculation of the starting vector $\mathbf{q}_1$. Given Eq. 41 for $|\Phi_1\rangle = |v\rangle$, one can obtain the components $q_{j,1}$ by computing the scalar products $\langle f_j|\Phi_1\rangle$. This direct approach has been used frequently. However, it usually requires numerical integrations that can become unwieldy for several degrees of freedom. An alternative approach is to consider the following expression:

$$\lim_{z\to 0^+} [z\mathbf{I} + \tilde{\Gamma}]\mathbf{q}_0 = \mathbf{c},$$ (52)

where $\mathbf{q}_0$ is the vector representation of $P_0^{1/2}$ and $\mathbf{c}$ is an arbitrary vector with a projection along $\mathbf{q}_0$. This follows from the fact that $P_0^{1/2}$ is the unique stationary solution of $\tilde{\Gamma}$. One can solve this equation using matrix inversion techniques or by using the conjugate gradients algorithm (cf. Section V.B). The complete starting vector $\mathbf{v}$ can be constructed by multiplying the elements of $\mathbf{q}_0$ by the appropriate spin operator terms (cf. Eq. 27).

In discussing the convergence to the correct spectrum of the continued-fraction approximant generated by the LA, it is convenient to use the following definition of the deviation ΔI_n of the true spectral lineshape $I_R(\Delta\omega)$ from that obtained after n iterative steps:

$$\Delta I_n = \int_{-\infty}^{\infty} d\,\Delta\omega |I_n(\Delta\omega) - I_R(\Delta\omega)|, \tag{53}$$

where these spectral lineshape functions have been normalized to unity. It is useful to calculate the "true" converged spectral lineshape $I_R(\Delta\omega)$ using the Rutishauser diagonalization method or some other benchmark procedure whose results are not affected by the peculiar features of the LA. Using this quantity, the sufficient number of steps n_s can be defined as the smallest number n of Lanczos steps that assures an error ΔI_n less than the required

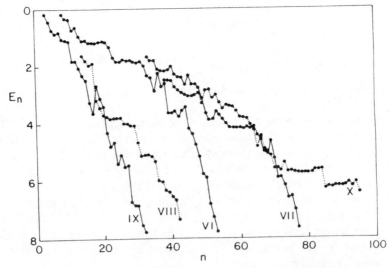

Figure 1.   Behavior of logarithm of error: $E_n = -\log_{10}(\Delta I_n)$ as function of number of steps n in LA for range of cases discussed in Ref. 3 corresponding to simple nitroxide slow-motional spectra.

accuracy for the spectral lineshape. In general, ΔI_n decreases with n, but it may not be strictly monotonic (cf. Figure 1 for some typical trends). For an accuracy $\Delta I_n = 10^{-4}$, n_s is found to be much less than the dimension N of the original matrix except in the extreme motional narrowing limit where N is very small anyway. A significant portion of the computational time saved by using the LA rather than a traditional diagonalization algorithm can be attributed to this phenomenon.

The eigenvalues of $\mathbf{T}_n$ are not strictly relevant to the problem since the spectrum can be calculated directly from the elements of the tridiagonal matrix without resort to diagonalization. Therefore, any approximate form of $\mathbf{T}_n$ is adequate, independent of the effect of the loss of orthogonality of the Lanczos vectors during its generation, if it reproduces the spectral lineshape function well. However, even in this context it is still useful to analyze the eigenvalues associated with $\mathbf{T}_n$ in order to understand how the LA works for this type of problem. The slow-motional ESR spectrum displayed in Figure 2a has a lineshape constructed from a large collection of eigenvalues. In such cases the LA is more efficient in reproducing the overall shape of $I(\Delta\omega)$ than in exactly reproducing the eigenvalues of $\mathbf{A}$. Figure 2b illustrates this fact in displaying the computed eigenvalues of the ESR problem considered in Figure 2a. The dots indicate the exact eigenvalues of the starting matrix $\mathbf{A}$, which has a dimension equal to 42. The triangles represent the 16 eigenvalues of the $n_s = 16$ dimensional tridiagonal matrix whose continued-fraction approximant satisfies $\Delta I_n = 10^{-4}$. From the figure it is clear that there is no simple relation

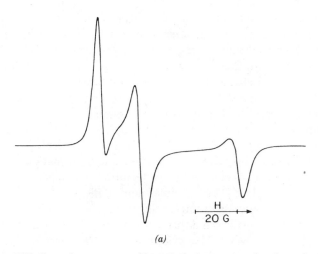

(a)

Figure 2a.   ESR absorption spectrum of hypothetical paramagnetic spin probe. Magnetic and motional parameters the same as case I in Table I of Ref. 3.

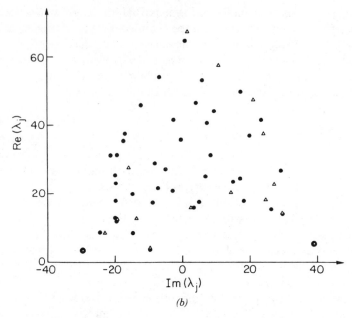

Figure 2b.   Distribution of exact (dots) and approximate (triangles) eigenvalues relative to ESR spectrum displayed in Figure 2a (from Ref. 3).

between overall accuracy in the lineshape function and accuracy in the approximate eigenvalues. Most cannot be simply associated on a one-to-one basis with particular exact eigenvalues. Even when this is possible, the error in the approximate eigenvalues is far greater than the accuracy of 10^{-4} for the full spectrum. In general, the LA generates continued fractions that tend to optimize the overall shape of the spectrum rather than sets of eigenvalues. Although at first such a statement might appear contradictory, it is based on the fact that the spectral density is usually dominated by the eigenvalues of small real part, and the LA is able either to approximate them individually or to provide an "average" to a cluster of eigenvalues sufficient to represent the spectrum. In the case displayed in Figure 2a, one finds that only the first 11 approximate eigenvalues of lowest Re $\{\lambda_j\}$ contribute significantly, whereas of the 42 exact eigenvalues, 23 contribute significantly.

The LA approximates the spectrum surprisingly well in spite of a rather low accuracy for the eigenvalues and their components. We also see that one way to have some insight into this behavior is to look at the results in terms of an approximation of clusters of eigenvalues instead of single eigenvalues. This concept of a cluster is, of course, very qualitative. Moreover, the eigenvalues obtained from the LA cannot be considered independent from one another in the evaluation of their effect on the spectrum. In this sense we refer to the fact that the LA produces an approximation of the optimal reduced space required to represent the spectral function instead of accurately reproducing the eigenvalues. The interpretation in terms of clusters is only a partial and qualitative explanation of this general behavior.

This tendency to approximate the optimal reduced space may be seen as follows. Let us expand the starting vector $|v\rangle$ in terms of the eigenvectors $|\Psi_m\rangle$ of the operator $\mathbf{A}$:

$$|v\rangle = \sum_{m=1}^{N} w_m |\Psi_m\rangle \tag{54}$$

then the lth Krylov vector can be written as

$$|k_l\rangle = \mathbf{A}^{l-1}|v\rangle = \sum_{m=1}^{N} w_m \lambda_m^{l-1} |\Psi_m\rangle. \tag{55}$$

Thus, the n-dimensional subspace spanned by the $|k_l\rangle$, $l = 1, 2, \ldots, n$, cannot contain those eigenvectors $|\Psi_m\rangle$ such that $\langle v|\Psi_m\rangle = w_m = 0$. That is, the Krylov subspace, which is spanned by the Lanczos vectors, is biased to exclude the eigenvectors for which $w_m = 0$ and which therefore cannot contribute to the spectrum.

The fact that the LA only approximates the optimal reduced space is related to what we call the extreme-eigenvalue effect. Increasing the order of the subspace tends to amplify the importance of eigenvectors with larger eigenvalues λ_m but with small w_m. This extreme-eigenvalue effect will be stronger the larger the magnitude of the coefficients w_m for the larger eigenvalues. However, it is the eigenvectors associated with small $|\lambda_m|$ and large projections on the starting vector that are most important in the calculation of the spectrum (cf. Eq. 33). Furthermore, when finite-precision arithmetic is taken into account, a Krylov vector can have a small projection onto eigenvectors that are strictly orthogonal to the starting vector in infinite-precision arithmetic. Since these are frequently eigenvectors with large $|\lambda_m|$, the "extreme-eigenvalue effect" is likely to enhance their role in the subsequent Krylov projections. We see, in Figure 2b, how these effects bring in eigenvalues of large Re $\{\lambda_j\}$ even though they are found to have very small or negligible projections on the starting vector.

Therefore, the combination of the extreme-eigenvalue effect and the particular structure of the matrix generates a departure from ideal behavior of the LA (i.e., the optimal reduced space). This constitutes a negative feature of the application of the LA, but it is inherent in the method, and in spite of this, the algorithm allows us to approximate the matrix in a subspace with dimension much smaller than the initial dimension of the matrix.

This situation can be improved by restricting the initial basis set as closely as possible, so as to exclude those basis elements that contribute appreciably to the eigenvectors with larger $|\lambda_m|$ but are unimportant in the spectrum. This is done in what follows.

Let us, however, first review the disadvantages to the use of the LA for numerical applications. Its main weakness is its loss of orthogonality among the Lanczos vectors it generates from the Krylov vectors by Gram–Schmidt orthogonalization. This is due to accumulation of numerical roundoff errors. As a result, the Lanczos steps can, in practice, be continued beyond the original matrix dimension (i.e., it is possible to have $n > N$). This leads to repeated eigenvalues as well as to spurious eigenvalues (due to introducing Lanczos vectors not contained in the rigorous Krylov subspace defined by A and $|v\rangle$. In general, the ESR spectra (or Fokker–Planck spectral densities) are determined by only a small subset of eigenvectors, in particular, those associated with eigenvalues λ_m with weakest damping (i.e., smallest Re $\{\lambda_m\}$), and approximations to these eigenvalues (or "clusters of eigenvalues") are rapidly generated such that $n_s \ll N$ for convergence to the spectrum, well before roundoff error can accumulate to the point where it can significantly affect the computation. However, roundoff error can become a problem if one works with an unnecessarily large basis set N and/or performs too many Lanczos projections n in the interest of guaranteeing convergence to the correct spectrum.

Another limitation of the LA is the lack of a convenient and objective criterion for determining n_s. One typically calculates the spectrum repeatedly for a sequence of Lanczos steps until convergence is confirmed. This is time consuming, and also it can ultimately lead to substantial accumulation of roundoff error as n becomes large.

Finally, we note the truncation problem: The ESR spectra can be calculated to a good approximation by finite matrices of large enough dimension N; one wishes to truncate the space so as to minimize N consistent with the accurate computation of the spectra. This we referred to as the minimum truncation scheme (MTS). Knowledge of the MTS can greatly speed up calculations. In the past, such knowledge was obtained indirectly by trial-and-error calculation of spectra with different basis sets in order to specify which types of basis vectors are important. This scheme is very time consuming as well as incomplete. In previous practice, one tended to work with sets of basis vectors significantly larger than the MTS, since the latter was not convenient to determine.

We shall now consider an approach that preserves all the advantages of the LA in tridiagonalizing **A** but improves on it by (i) estimating at each recursive step the magnitude of an appropriately chosen residual so that the recursive steps can be terminated when $n = n_s$; (ii) providing a criterion to determine when computer roundoff error has become a problem; and (iii) providing a convenient means of estimating the MTS. It is based on the method of conjugate gradients discussed in the next section.

B. The Conjugate Gradients Algorithm

The conjugate gradient (CG) method of Hestenes and Steifel [2] was originally used for solving equations of form $A|u\rangle = |v\rangle$ with A a real symmetric positive definite (RSPD) matrix. The starting point of the CG method is to consider this equation in the form

$$|r_k\rangle = |v\rangle - A|u_k\rangle, \qquad (56)$$

where $|u_k\rangle$ is the kth approximant to $|u\rangle$, and $|r_k\rangle$ is the residual vector associated with $|u_k\rangle$. The residual vector $|r_k\rangle$ is easily seen to be the vector that gives the negative gradient of the functional $f[u_k] = \langle u_k|A|u_k\rangle - \langle u_k|v\rangle$ provided A is RSPD, so that a minimization of $f[u_k]$ is equivalent to solving $A|u\rangle = |v\rangle$. Equation 56 is solved by successive iterations that do not minimize along the sequence of vectors $|r_k\rangle$, for $k = 1, 2, \ldots, n$, which would be the method of steepest descent, but rather minimize the functional $f[u_k]$ along a set of "conjugate directions" $|p_k\rangle$ for $k = 1, 2, \ldots, n$. This procedure avoids the problem of further minimization steps spoiling the minimization along the previous conjugate direction vectors. The conjugate direction vectors $|p_k\rangle$ are defined by the equations

$$|r_{k+1}\rangle = |r_k\rangle - a_k A|p_k\rangle \qquad (57)$$

and

$$|p_{k+1}\rangle = |r_{k+1}\rangle + b_k|p_k\rangle. \qquad (58)$$

where the a_k and b_k are given by

$$a_k = \frac{\langle r_k|r_k\rangle}{\langle p_k|A|p_k\rangle}, \qquad (59)$$

$$b_k = \frac{\langle r_{k+1}|r_{k+1}\rangle}{\langle r_k|r_k\rangle}. \qquad (60)$$

The residual vectors are easily shown to be mutually orthogonal but not normalized, whereas the set of conjugate direction vectors are "A-conjugate";

that is,

$$\langle p_j | \mathbf{A} | p_k \rangle = 0 \quad \text{if} \quad j \neq k. \tag{61}$$

The vector $|p_k\rangle$ is the closest vector to $|r_k\rangle$ that is $\mathbf{A}$-conjugate to the previous conjugate direction vectors. It is also true that $\langle p_j | r_k \rangle = 0$ for $j = 1, 2, \ldots$, $k - 1$. Also, the $(k + 1)$st approximant to the solution vector $|u_{k+1}\rangle$ is obtained from $|u_k\rangle$ as

$$|u_{k+1}\rangle = |u_k\rangle + a_k |p_k\rangle. \tag{62}$$

These equations permit one to recursively obtain the higher order approximants. That is, let

$$|r_1\rangle = |v\rangle - \mathbf{A}|u_1\rangle \tag{63}$$

and

$$|p_1\rangle = |r_1\rangle, \tag{64}$$

where $|u_1\rangle$ is some initial guess for $|u\rangle$. Then, for $k = 1, 2, \ldots, n - 1$, one calculates, in order, a_k, $|u_{k+1}\rangle$, $|r_{k+1}\rangle$, b_k, and $|p_{k+1}\rangle$ using Eqs. 57–62. At each step the norm of the residual vector $|r_k\rangle$,

$$\|r_k\|^2 = \langle r_k | r_k \rangle, \tag{65}$$

is a measure of the deviation of $|u_k\rangle$ from the true solution $|u\rangle$.

At this stage the LA and CG appear to be very different algorithms; the former tridiagonalizes $\mathbf{A}$, whereas the latter generates a sequence of approximants to the solution of Eq. 3. We consider their equivalence in the following. First we need to consider the applicability of this CG method to complex symmetric matrices. It is not hard to show that for nondefective (i.e., diagonalizable), nonsingular complex symmetric matrices the preceding CG method applies, provided only that we use the generalized norm (see Eq. 47) as in the application of the LA to complex symmetric matrices. However, for the CG method, there is the additional requirement that $\mathbf{A}'(\Delta\omega)$ be nonsingular. Recall from Section IV that the complex symmetric matrix $\mathbf{A}$ will have complex roots: the real parts give the linewidths and the imaginary parts, the resonance frequencies. Since each relevant eigenvalue of $\mathbf{A}$ must, on physical grounds, have a nonzero real part, both $\mathbf{A}$ and $\mathbf{A}'$ will be nonsingular. Nevertheless, it is convenient to replace $i\Delta\omega$ by $i\Delta\omega + T_2'^{-1}$, where $T_2'^{-1}$ is equivalent to an additional linewidth contribution so that one avoids spurious divisions by zero that can occur when

$\langle p_k|\mathbf{A}|p_k\rangle = 0$ in the calculation of a_k (see Eq. 59). Zero divisors of this type can be identified as spurious through the construction of the Lanczos tridiagonal matrix from the various quantities generated by the CG algorithm, as will be shown. The invariance of the Lanczos tridiagonal matrix and the associated continued-fraction approximant to the spectral lineshape function under origin shifts are discussed in Appendix B.

However, it is useful to relax the use of the generalized norm in estimating the error in the approximate solution vector given by Eq. 56, which is used to monitor the convergence of the method. Given $|r_k\rangle$ determined from Eqs. 56 and 57, it is useful to consider two specific forms of the norm, Eq. 65, which we write in terms of its components $y_{j,k}$ of $|r_k\rangle$ in the original basis set $|f_j\rangle$ (cf. Eq. 41):

$$r_{k,\text{ps}}^2 \equiv \left|\sum_j y_{k,j}^2\right|, \tag{66}$$

$$r_{k,\text{H}}^2 \equiv \sum_j |y_{k,j}|^2, \tag{67}$$

whereas letting $|r_{k,\text{true}}\rangle \equiv |v\rangle - \mathbf{A}'(\Delta\omega)|u_k\rangle$ (cf. Eq. 56), at each iterative step we have a third norm:

$$r_{k,\text{true}}^2 = \sum_j |y_{k,j}^{\text{true}}|^2. \tag{68}$$

That is $r_{k,\text{H}}$ is the usual definition of a norm in unitary space, as is $r_{k,\text{true}}$, whereas $r_{k,\text{ps}}$ is the modulus of the generalized or pseudo norm in a complex orthogonal space.

For a complex symmetric matrix, the second and third norms are equal in exact arithmetic and are guaranteed to be real. The first norm is, however, most closely related to the generalized norm required in our CG algorithm for complex symmetric matrices. Taking the absolute value of the generalized norm of $|r_k\rangle$ gives a real value for $r_{k,\text{ps}}^2$, as in Eq. 66, that can be more easily compared to the other two forms of r^2. In practice, it is found that $r_{\text{H}}^2 = r_{\text{true}}^2$ in finite-precision arithmetic until they become on the order of the unit roundoff error [80], whereas r_{ps}^2 is always smaller. Once the value of r_{H}^2 approaches the limit of finite-precision arithmetic, any further iterations fail to improve the quality of the approximate solution vector, and $r_{k,\text{true}}^2$ remains constant as k is increased. In contrast, the values of $r_{k,\text{ps}}^2$ and $r_{k,\text{H}}^2$ continue to decrease and therefore can no longer accurately represent the error in the approximate solution vector. However, these two forms of r^2 are readily available during each step, whereas $r_{k,\text{true}}^2$ requires a substantial amount of extra calculation because the quantity $\mathbf{A}|u_k\rangle$ is not normally part of the CG algorithm. In light of these facts, it is convenient to use $r^2_{k,\mathrm{H}}$ as the criterion for convergence, and it will simply be referred to as r^2. On the other hand, it is advisable to occasionally compute $r^2_{k,\mathrm{true}}$ to check if roundoff error has become a problem.

C. The Equivalence of LA and CG Methods

To make full use of the CG algorithm, its equivalence to the LA must be recognized. First, note that the orthogonal set of vectors $|r_k\rangle$ and the conjugate set $|p_k\rangle$ are contained in the same Krylov subspace generated by $\mathbf{A}$ and $|v\rangle$, and the same is true for the Lanczos vectors [80]. Following Golub and Van Loan [80], an explicit expression for the construction of the Lanczos tridiagonal matrix $\mathbf{T}_n$ by the CG method for a RSPD matrix $\mathbf{A}$ that is valid at each stage of the iteration is

$$\mathbf{T}_k = \mathbf{D}_k^{-1}\mathbf{B}_k^{\mathrm{tr}}\mathscr{A}\mathbf{B}_k\mathbf{D}_k^{-1}, \tag{69}$$

where $\mathscr{A}$ is the diagonal matrix with elements

$$\mathscr{A}_{k,ii} = \langle p_i|\mathbf{A}|p_i\rangle, \qquad i = 1, 2, \ldots, k, \tag{70}$$

and $\mathbf{D}$ is also diagonal with elements

$$D_{k,ii} = \|r_i\| = \rho_i \equiv \left(\sum_j y_{i,j}^2\right)^{1/2}, \qquad i = 1, 2, \ldots, k, \tag{71}$$

whereas $\mathbf{B}_k$ is an upper bidiagonal matrix with elements

$$B_{k,ii} = 1, \qquad B_{k,i,i+1} = -b_i, \qquad i = 1, 2, \ldots, k, \tag{72}$$

with b_i given by Eq. 60. It turns out that the residual vectors are colinear with the Lanczos vectors; more precisely,

$$|\Phi_i\rangle = \pm\rho_i^{-1}|r_i\rangle, \qquad i = 1, 2, \ldots, k. \tag{73}$$

Since the direction of $|r_i\rangle$ and therefore the signs of its components are well defined by Eq. 56, the Lanczos vectors, which are normalized in an arbitrary fashion, bear the sign ambiguity since their direction is not specified by the LA. It follows from Eqs. 69–73 that the matrix elements of $\mathbf{T}_n$ are

$$\alpha_k = \langle p_k|\mathbf{A}|p_k\rangle/\rho_k^2 + (\rho_k^2/\rho_{k-1}^4)\langle p_{k-1}|\mathbf{A}|p_{k-1}\rangle, \tag{74}$$

$$\beta_k = -(\rho_k/\rho_{k-1}^3)\langle p_{k-1}|\mathbf{A}|p_{k-1}\rangle. \tag{75}$$

Thus, the elements α_k and β_k of the Lanczos tridiagonal matrix $\mathbf{T}_n$ are readily obtained from quantities calculated by the CG algorithm for every step. This approach may be used to generate a tridiagonal matrix approximation to $\mathbf{A}$ using the CG algorithm in the same spirit as the LA.

Several points should be made about this equivalence. First, the tridiagonal matrices generated by the LA and CG methods are not equal. The off-diagonal matrix elements can differ in sign because of the sign ambiguity associated with the Lanczos vectors. However, the spectrum is independent of the arbitrary choice of direction of the Lanczos vectors. The change of sign of the off-diagonal matrix elements amounts to performing an equivalence transformation on the continued-fraction approximant to the spectrum (see Appendix B). In addition, one must start with $|r_1\rangle = |v\rangle$, which implies $|u_1\rangle = 0$ from Eq. 63, in order to obtain the correspondence between the Lanczos vectors and the residual vectors given by Eq. 73.

When the complications arising from the application of the CG algorithm to nondefective, nonsingular, complex symmetric matrices are considered, again it is found that this approach is applicable provided that the generalized norm and an origin shift chosen to remove spurious zero divisors are used. This means that ρ_i, as defined in Eq. 71, is a complex quantity.

This equivalence of the tridiagonal matrices generated by the LA and CG methods for the type of complex symmetric matrices $\mathbf{A}$ that arise in the calculation of magnetic resonance spectra has been verified by numerical calculation. It is found that the magnitudes of real and imaginary parts of the matrix elements of $\mathbf{T}_n$ obtained by the LA (Eqs. 38 and 39) and the CG algorithm (Eqs. 74 and 75) agree to at least six significant figures when double-precision arithmetic is used.

In conclusion, the CG method can be applied to complex symmetric matrices $\mathbf{A}$ to give the Lanczos tridiagonal matrix from which spectra may be calculated by the continued-fraction technique. The benefit derived from the very small amount of extra computational work necessary for the CG method as opposed to the LA method is that an objective criterion for the convergence r^2 can be monitored at every step. Finally, note that the basic CG algorithm can be used to directly solve the linear equation problem of Eq. 3 when desired. In fact, this will serve as the basis of our approach to the determination of the MTS.

D. Minimum Truncation Scheme

As discussed previously, it would be highly desirable to have an objective and convenient criterion for selecting the minimum basis set for representing $\mathbf{A}$, which still guarantees convergence of the continued-fraction approximants to the desired accuracy. To this end, the CG method can be used to calculate $|u(\Delta\omega)\rangle$ for different values of the sweep variable, $\Delta\omega$. Since the spectrum is determined by the scalar product $\langle v|u(\Delta\omega)\rangle$ (cf. Eq. 3), the knowledge of the vector $|u(\Delta\omega)\rangle$ in terms of its components $z_j(\Delta\omega) \equiv \langle f_j|u(\Delta\omega)\rangle$ in the original basis set for a representative sample set of sweep positions should provide an accurate assessment of the importance of each basis vector in determining the spectrum.

Consider the jth component z_j. From Eq. 3 and the definition of z_j it follows that

$$z_j(\Delta\omega) = \langle f_j|\mathbf{A}'^{-1}(\Delta\omega)|v\rangle = \sum_m \langle f_j|\psi_m\rangle(i\,\Delta\omega + \lambda_m)^{-1}\langle\psi_m|v\rangle, \quad (76)$$

where, in the last equality of Eq. 76, the eigenvectors $|\psi_m\rangle$ of $\mathbf{A}$ with associated eigenvalues λ_m were introduced. This last expression for $z_j(\Delta\omega)$ in Eq. 76 is a product of three quantities. First, the scalar product $\langle\psi_m|v\rangle$, which is the projection of the mth eigenvector on the transition moment vector, is a measure of the importance of this eigenvector in contributing to the spectrum. Next, $\langle f_j|\psi_m\rangle$ is a measure of the importance of the jth basis vector in contributing to $|\psi_m\rangle$. These two quantities are obviously independent of the sweep variable. Finally, $(i\Delta\omega + \lambda_m)^{-1}$ expresses how the contribution of the mth eigenvector varies across the spectrum. If it resonates far from the frequency of the applied RF field or if it is very broad, this quantity is very small. All these factors are needed to estimate the importance of the basis vector $|f_j\rangle$ to the spectrum. Since they are all included in $z_j(\Delta\omega)$, it can be used as a measure of the importance of $|f_j\rangle$ to the spectrum at the point $\Delta\omega$. Provided that the spectrum is sampled at a sufficient number of values of the sweep variable, the maximum value of $|z_j(\Delta\omega)|$ over the sampled values of the sweep variable can be taken as a measure of the importance of the basis vector $|f_j\rangle$ in determining the spectrum. This may be done by solving Eq. 3 using the CG algorithm and keeping track of the largest value of $|z_j(\Delta\omega)|$ for each basis vector as the sweep variable is varied. In practice, it is useful to monitor a slightly different function that treats the high- and low-amplitude portions of the spectrum more equally.

This approach requires that a basis set larger than the MTS but containing the latter as a subset be utilized initially. Nevertheless, in most applications, wherein calculations are compared to experimental spectra, it is necessary to vary input parameters by small amounts and to repeat the computation many times so the initial efforts at selecting the MTS can often be worthwhile. As problems become larger, it is usually possible to estimate a starting basis set that is not excessively large by extrapolation from an empirical set of rules derived from the MTS obtained from smaller, but closely related problems. This makes the final search for the MTS for the larger problem less time consuming. Our examples below illustrate this.

Given that Eq. 76 is a good criterion for determining the MTS, an alternative way to proceed would be to diagonalize the tridiagonal matrix generated by the LA by the QR transform (see Section IV) but to store the Lanczos and QR transformation matrices and multiply these two matrices together to obtain the projections of the eigenvectors on the various basis vectors and the starting vector. Unfortunately, it would destroy the great efficiency of the LA to keep track of the full transformation matrix. Also, for large matrices ($N \simeq 10^4$ to $N \simeq 10^5$) enormous memory may be required. Consequently, Cullum and Willoughby [89] recommended an inverse iteration procedure to obtain the important eigenvectors once a set of converged eigenvalues have been obtained by the LA. However, as stated previously, the spectra are extremely well approximated long before the actual eigenvalues have converged [3]. Thus, much more effort would be required in implementing the LA in order to achieve accurate enough eigenvalues to construct good eigenvectors by inverse iteration than is required to obtain converged spectra. On the other hand, the basic CG algorithm successfully delivers the needed information to determine the MTS.

The problem of determining the MTS has been studied using the CG method by Vasavada, Schneider, and Freed [4]. Equation 3 is simply solved using the CG algorithm for several values of the sweep variable in the range where the spectrum is expected to be nonzero. It was found that 10–20 values of $\Delta\omega$ are sufficient for slow-motional ESR spectra. In performing the sweep, it is useful to use as the initial vector (cf. Eq. 63) at the mth value of sweep variable, $|u(\Delta\omega)\rangle$ the solution vector from the previous value since this reduced the number of CG steps needed for convergence. This is valid because the Lanczos tridiagonal matrix need not be reconstructed to calculate the amplitude of the spectrum at a single point. In addition, the matrix $\mathbf{A}$ can be preconditioned [80] to enhance the rate of convergence [4] (see also Section V.G). The combined effects of using a good estimate for the solution vector and a preconditioning matrix that minimizes the extreme eigenvalue effect can lead to substantial savings in computational effort [4]. Also, it is sufficient for the present purpose, which is only to estimate the importance of the individual basis vectors, to use weaker convergence criteria for r^2. It was found that a useful measure of the overall importance of the individual basis vectors is given by

$$s_j = \max_{\Delta\omega} \left| \frac{z_j(\Delta\omega)}{\langle v|u(\Delta\omega)\rangle} \right|. \tag{77}$$

This criterion, which involves the normalization of $z_j(\Delta\omega)$ by the amplitude of the spectrum at that point, ensures that the low-amplitude regions of the spectrum are also accurately approximated. This is especially important for two-dimensional electron-spin echo (ESE) spectroscopy, where many of the interesting variations in the contour plots occur in regions of low amplitude.

Based on these studies, it was suggested that a conservative estimate of the MTS consists of those basis vectors for which $s_j > 0.03$ for CWESR spectra and $s_j > 0.0003$ for two-dimensional ESE spectra. Of course, in preliminary calculations, one can use less stringent conditions to obtain rough approximations to the spectra.

The results of these investigations for CWESR and two-dimensional ESE are summarized in Tables 1 and 2, respectively. The initial basis sets used in these calculations were significantly larger than the MTS, as has been the normal procedure for LA calculations. A simple set of truncation rules corresponding to the maximum values of the relevant indices characterizing the basis set that are consistent with our results for the MTS were found. Using this truncation procedure yielded a basis set of dimension N, which is given in the tables. The actual N_{min}, corresponding to the dimension of the MTS for the states that satisfy the s_j criterion for the particular class of calculations, is always smaller than this. Next, a search for patterns involving interrelation-

TABLE 1   Truncation Parameters and MTS for CWESR Spectra[a]

Number	Spin Probe[b]	R	λ	L_{max}^e	L_{max}^o	K_{max}	M_{max}	N	N'	N_{min}
1	TEMPONE	10^7	0	6	3	2	2	42	34	33
2	TEMPONE	10^6	0	14	7	6	2	171	108	100
3	TEMPONE	10^5	0	30	13	10	2	543	285	256
4	TEMPONE	10^4	0	54	15	10	2	990	549	447
5	TEMPONE	10^7	10	10	None	2	2	63	26	26
6	TEMPONE	10^6	5	12	3	2	2	78	54	42
7	TEMPONE	10^6	10	10	None	0	2	33	30	29
8	TEMPONE (90° tilt)	10^7	1	6	3	2	6	288	134	74
9	TEMPONE (90° tilt)	10^7	10	10	9	4	4	822	129	69
10	TEMPONE (90° tilt)	10^6	10	12	11	6	6	1779	533	245
11	CSL	10^6	0	14	7	14	2	231	174	162
12	CSL	10^5	0	30	13	30	2	762	522	474

[a]Symbols: R = rotational diffusion constant (s^{-1}); λ = coefficient of first-order term in expansion of scaled orienting pseudopotential, $-U(\Omega)/k_B T$; L_{max}^e, L_{max}^o = largest even value of L and odd value of L, respectively, for which there exist basis vectors with $s_j > 0.03$, and K_{max}, M_{max} = largest values of K and M for which this occurs; N = dimension of matrix if *all* basis vectors whose indices are less than or equal to L_{max}^e, L_{max}^o, K_{max}, and M_{max}; N' = dimension of basis set derived from lookup table based on specifying M_{min} and M_{max} for every important pair of L and K (for $\psi = 90°$ new selection rules also utilized); and N_{min} = dimension of MTS (number of basis vectors for which $s_j > 0.03$).

[b]Values of g and A tensors for TEMPONE: $g_{xx} = 2.0088$, $g_{yy} = 2.0061$, $g_{zz} = 2.0027$, $A_{xx} = 5.8\,G$, $A_{yy} = 5.8\,G$, $A_{zz} = 30.8\,G$. Values of g and A tensors for CSL: $g_{xx} = 2.0021$, $g_{yy} = 2.0089$, $g_{zz} = 2.0058$, $A_{xx} = 33.44\,G$, $A_{yy} = 5.27\,G$, $A_{zz} = 5.27\,G$. Static magnetic field $B_0 = 3300\,G$, $(\gamma_e T_2')^{-1} = 1.0\,G$.

TABLE 2   Table of Truncation Parameters and MTS for Two-Dimensional ESE Spectra[a]

Number	Spin Probe	R	λ	L_{max}^e	L_{max}^o	K_{max}	M_{max}	N	N'	N_{min}
1	TEMPONE	10^7	0	10	7	6	2	123	94	92
2	TEMPONE	10^6	0	22	17	10	2	429	317	307
3	TEMPONE	10^5	0	44	37	18	2	1485	1010	971
4	TEMPONE	10^4	0	88	71	28	2	4614	2706	2506
5	TEMPONE	10^7	10	16	7	2	2	108	81	76
6	TEMPONE	10^6	5	20	15	8	2	333	223	209
7	TEMPONE	10^6	10	16	11	4	2	168	131	120
8	TEMPONE (90° tilt)	10^7	1	10	7	6	10	1440	752	586
9	TEMPONE (90° tilt)	10^7	10	16	15	6	6	2601	931	607
10	TEMPONE (90° tilt)	10^6	10	20	19	10	12	8196	3804	2835
11	CSL	10^6	0	22	19	22	2	600	503	485
12	CSL	10^5	0	46	37	46	2	2310	1877	1815

[a]All parameters have same meaning as in Table 1, except for s_j, which is taken to be 0.0003.

ships between the different basis set indices that appear in the MTS was made. It was found that a slightly more complicated truncation rule involving the specification of M_{min} and M_{max} for each pair of L and K for which there were basis states satisfying the s_j criterion was very effective. Using a truncation rule of this type in the form of a look-up table, the N-dimensional space could be reduced to N', which is much closer to N_{min} than is N. Such look-up tables are easily implemented to provide useful approximations to the MTS.

E.   Convergence of Lanczos–Conjugate Gradients Projections

The error ΔI_n (Eq. 53) can be used to monitor the convergence of the continued-fraction approximant to the spectral lineshape as a function of the number of Lanczos steps. This is an objective criterion, but it is impractical since it presumes one already has a converged spectrum with which to compare.

Instead, with the CG method the residual r_0^2 (i.e., the values of r^2 calculated for $\Delta\omega = 0$) is readily available at each step and may be used to determine when to terminate the CG algorithm. In general, for a complex symmetric matrix, r_0^2 does not decay monotonically as a function of the number of CG steps n, as it does for an RSPD matrix. Instead, it behaves like a damped oscillator as a function of n; that is, it oscillates, but its local average value decreases as n increases. Typically, CWESR spectra converge very rapidly; performing CG steps until $r_0^2 \simeq 10^{-3}$ to $r_0^2 \simeq 10^{-4}$ is more than adequate. As a conservative criterion, it was suggested that the iterations be terminated when $r_0^2 \simeq 10^{-4}$, even though for most of the spectra a surprisingly low value of $r_0^2 \simeq 10^{-2}$ is sufficient. Here, n_s is taken to be the value of n when r_0^2 first reaches a value of 10^{-4}. For several examples, the continued-fraction approximant derived from the tridiagonal matrix produced by stopping the CG algorithm at $r_0^2 \simeq 10^{-4}$ gives $\Delta I_n \simeq 10^{-7}$ to $\Delta I_n \simeq 10^{-8}$, well within the noise range for experimental spectra.

As the motion slows down, one requires larger basis sets to adequately represent A (i.e., N and N_{min} increase), and also there is an increase in the value of n_s required to achieve the same value of r_0^2. This is true no matter which algorithm is utilized. However, it is observed that $n_s \ll N$ and that n_s increases more slowly than N, so the advantage of the LA or the CG algorithm over traditional ones becomes relatively greater the slower the motions, corresponding to larger N. Some typical results reflecting these features are shown in Table 3. In obtaining the results in Table 3, values of N significantly larger than the MTS have been used, in accordance with previous applications of the LA to the lineshape problem. However, in one case, a smaller value of N that more closely corresponds to the MTS was used. In this case, n_s is only decreased by a factor of 1.34 when N is decreased by a factor of 4.1. This is an example of the general phenomenon of how the LA seeks out an approximation to the "optimal reduced space" already discussed.

TABLE 3   Minimum Number of CG Steps and Associated Residual Values[a]

N	r_0^2	n_s	Case[b]	Number of Matrix Elements[c]
1743	10^{-2}	60	A	32917
	10^{-4}	104		
	10^{-10}	172		
429	10^{-2}	49	A	7701
	10^{-4}	77		
	10^{-10}	128		
3543	10^{-2}	76	B	70399
	10^{-4}	159		
	10^{-10}	315		
7503	10^{-2}	89	C	288085
	10^{-4}	170		
	10^{-10}	326		
8196	10^{-2}	57	D	666965
	10^{-4}	80		
	10^{-10}	143		

[a]Symbols: N = dimension of matrix defined by L_{max}^e, L_{max}^o, K_{max}, and M_{max} and symmetries given in Ref. 20; r_0^2 = residual squared (cf. Eq. 67) calculated at center of spectrum; n_s = number of CG iterations.

[b]Case A: TEMPONE magnetic parameters (cf. Table 1) and $R = 10^6 \text{ s}^{-1}$. Here $N = 429$ corresponds to approximate MTS, cf. Table 2, entry 2, while $N = 1743$ represents typical larger basis set commonly utilized previously in two-dimensional ESE calculations by LA. Case B: TEMPONE magnetic parameters and $R = 10^4 \text{ s}^{-1}$. Case C: CSL magnetic parameters (cf. Table 1) and $R = 10^4 \text{ s}^{-1}$. Case D: TEMPONE magnetic parameters but with strong potential ($\lambda = 10$), $R = 10^6 \text{ s}^{-1}$, and director tilt $\psi = 90°$.

[c]Number of nonzero matrix elements in matrix counting real and imaginary parts separately.

The tendency of the LA to provide a good approximation to the spectrum before it accurately reproduces the important eigenvalues of the original matrix can be understood a little better in terms of the equivalence between the Lanczos and CG algorithms. That is, the Lanczos tridiagonalization of **A** and the solution of Eq. 3 at $\Delta\omega = 0$ using the CG algorithm involve the calculation of an orthonormal basis for the same sequence of Krylov subspaces. Thus, the LA is equivalent to the minimization of the residual to the solution vector by CG at $\Delta\omega = 0$. Furthermore, it is easy to show that the Krylov subspaces generated by $\mathbf{A}'(\Delta\omega) = \mathbf{A} + i\,\Delta\omega\,\mathbf{I}$ must be independent of $\Delta\omega$ [96]. This implies that all of the vectors generated by the CG algorithm, including the approximate solution vector for any value of $\Delta\omega$, can be expressed as linear combinations of the Lanczos vectors obtained from the LA after the same number of steps. In this manner, the equivalence of these two algorithms can be used to extend the approximate *local* solution of the problem (e.g., for $\Delta\omega = 0$) to an approximate *global* solution (i.e., for all values of $\Delta\omega$). The observation that a value of $r_0^2 \simeq 10^{-3}$ to $r_0^2 \simeq 10^{-4}$ at $\Delta\omega = 0$ is sufficient to obtain a converged spectrum for all $\Delta\omega$ is rather empirical but seems to be consistent with our observation that the outer portions (or wings) of the spectrum tend to converge sooner than the central portion for which $\Delta\omega \sim 0$. This observation is a manifestation of the extreme eigenvalue effect (cf. Eq. 55). In addition, the fact that only the projection of the approximate solution vector on the starting vector is important in determining the spectrum can be used to rationalize the relatively large error that can be tolerated in the solution vector (cf. Eq. 3). For the same reason, large errors in the approximate eigenvectors can be tolerated (cf. Eq. 33).

F. Calculating Two-Dimensional ESE Spectra

The CG method has also been used to calculate two-dimensional ESE spectra using the approximate expression of Millhauser and Freed [30] for the two-dimensional ESE signal (cf. Section VII):

$$S(\omega, \omega') \propto \sum_j c_j^2 \frac{T_{2,j}}{1 + \omega^2 T_{2,j}^2} \exp\left(-\frac{(\omega' - \omega_j)^2}{\delta^2} \right), \tag{78}$$

where for the jth "dynamic spin packet" (i.e., the jth eigenvector $|\psi_j\rangle$ of **A** corresponding to the eigenvalue λ_j), $T_{2,j}^{-1} = \mathrm{Re}\{\lambda_j\}$ is its Lorentzian width and $\omega_j = \mathrm{Im}\{\lambda_j\}$ its resonant frequency. Also, the weighting factors are given by

$$c_j^2 = \langle\psi_j|v\rangle^2 \simeq (\mathrm{Re}\,\langle\psi_j|v\rangle)^2, \tag{79}$$

where the approximate equality is valid only in the very slow-motional region where Eq. 78 is approximately valid. The two-dimensional ESE spectrum is inhomogeneously broadened (with respect to the ω' sweep variable) by convolution with a Gaussian distribution of half-width δ. For purposes of testing the computational method, Eq. 78 was utilized with the approximate form for c_j^2 in Eq. 79 even when the motion was too fast for these expressions to accurately represent an experimentally obtainable spectrum.

It is clear from Eq. 78 that it is necessary to obtain estimates of the eigenvalues λ_j that contribute with nonnegligible weight factors. This was done by diagonalizing the tridiagonal matrix $\mathbf{T}_n$ derived from the CG method utilizing standard procedures [3, 45]. The approximate eigenvectors $|\psi_j\rangle$ are then, in principle, obtained in terms of their components $\langle \Phi_k | \psi_j \rangle$ in the Lanczos basis set. However, only the components along $|\Phi_1\rangle = |v\rangle$ are needed, and they form a vector of dimension n_s, which is easily obtained during the procedure [12, 45] (see also Section IV).

Because the two-dimensional ESE spectra require significantly more accurate estimates of the eigenvalues and the weights, the convergence with respect to LA or CG steps occurs only after achieving a residual that is much smaller than what is required for the corresponding CWESR spectrum. In particular, we find that $r_0^2 \simeq 10^{-8}$ to $r_0^2 \simeq 10^{-10}$ is sufficient for two-dimentional ESE spectra, whereas the CW spectra have already converged for $r_0^2 \geq 10^{-4}$. This much more severe requirement for r_0^2 fortunately does not require very many more iterations, as illustrated by the results summarized in Table 3.

In order to study the convergence further, it is possible to introduce, by analogy to Eq. 53, the following definition of the error in the two-dimensional ESE spectrum:

$$\Delta S_n = \int_{-\infty}^{\infty} d\omega \int_{-\infty}^{\infty} d\omega' |S_n(\omega, \omega') - S_R(\omega, \omega')|, \tag{80}$$

where $S_R(\omega, \omega')$ is the normalized "exact two-dimensional ESE spectrum" from a complete diagonalization of $\mathbf{A}$ or some very good approximation to it, and $S_n(\omega, \omega')$ is the normalized approximate spectrum obtained by diagonalizing $\mathbf{T}_n$. The examples studied indicate that $r_0^2 \simeq 10^{-10}$ corresponds to $\Delta S_n \simeq 0.007$ to $\Delta S_n \simeq 0.02$.

Studies of the MTS for two-dimensional ESE spectra (cf. Table 2) showed that it is necessary to retain basis vectors for which $s_j \simeq 0.0006$ to $s_j \simeq 0.0003$, so a cutoff of $s_j = 0.0003$ was recommended for determining the MTS for two-dimensional ESE rather than the $s_j = 0.03$ value recommended for CWESR. It is interesting that a more stringent application of the same criterion (Eq. 77) that was appropriate for CWESR problems is also useful for two-dimensional ESE calculations, which require substantially greater accuracy due to their greater sensitivity to the approximate eigenvalues. This matter is further discussed by Vasavada, Schneider, and Freed [4].

G. Direct Calculation of Spectra and Spectral Densities by Conjugate Gradients

In the preceding section it has been shown how the CG algorithm can be employed to calculate the CWESR spectrum for several values of the sweep variable in order to determine the MTS and the minimum number of CG or Lanczos steps. It is, of course, possible to employ the CG algorithm to directly compute the entire CWESR spectrum from Eq. 3. Preliminary studies show that the calculation of the magnitude of the spectrum at one value of the sweep variable is, on the average, about five times faster than doing enough CG steps to get the entire spectrum from the tridiagonal matrix. This implies that a direct calculation using the CG algorithm for an entire CWESR spectrum using 200 values of the sweep variable would take about 40 times longer than the LA.

One might hope that if the CG calculation is performed for smaller increments of the sweep variable, only a few iterations per sweep position would be required if the solution at the previous sweep position is used as an initial guess for the following point. Numerical experiments show that if the number of sweep positions is increased by a factor of 10 from 20 (for the MTS calculation) to 200, the computer time required to complete the calculation increases by only a factor of 5.5 for termination at $r^2 = 10^{-4}$ and by only a factor of 3.6 for $r^2 = 10^{-2}$. This indicates that one does improve the efficiency of the calculation in this manner, especially if a relatively large r^2 is sufficient. Further improvements in efficiency might also be achieved if the increment in the sweep variable was chosen in an adaptive manner.

Another way to speed up the direct calculation is by preconditioning the $\mathbf{A}'(\Delta\omega)$ matrix, as mentioned previously. Preconditioning is a general device to improve the convergence of an iterative solution to a matrix problem (e.g., Eq. 3). Given an RSPD matrix $\mathbf{A}' = \mathbf{M} - \mathbf{N}$, one finds that the CG method can be accelerated utilizing $\mathbf{M}$ as a preconditioner provided $\mathbf{M}$ is also RSPD [80]. In the case of complex symmetric matrices, if one can let $\mathbf{M} = \mathbf{\Gamma} + T_2^{-1}\mathbf{I}$ and $\mathbf{N} = i(\mathbf{L} - \Delta\omega\mathbf{I})$, $\mathbf{M}$ can be made symmetric and is positive definite. The preconditioned problem one solves is

$$(\mathbf{M}^{-1/2}\mathbf{A}'\mathbf{M}^{-1/2})\mathbf{M}^{1/2}|u\rangle = \mathbf{M}^{-1/2}|v\rangle = \mathbf{A}''|u'\rangle = |v'\rangle. \qquad (81)$$

For isotropic Brownian rotational diffusion, where the eigenvalues of the diffusion operator are proportional to $L(L + 1)$, the stochastic Liouville matrix $\mathbf{A}'$ becomes dominated by the real parts of the diagonal elements for large L. The effect of the preconditioning is to set these real parts equal to unity for all diagonal elements and to scale the other matrix elements accordingly. For more general cases, such as when a restoring potential is needed, it is useful to keep the calculation of M as simple as possible. Taking M to be a diagonal matrix comprised of the diagonal elements of $\Gamma + T_2'^{-1}I$ in these instances seems to work fine. In general, we find that the preconditioned CG algorithm does speed up the convergence of the calculation. Unfortunately, because preconditioning is not a similarity transformation, it cannot be used in a simple way to diagonalize A for purposes of calculating spectra. For an example of where this is possible see Chapter 6.

Based on these results, it is clear that tridiagonalization by the LA is the more efficient method for CWESR, and the direct method does not even apply to two-dimensional ESE using the formula given in Eq. 78. However, there are some cases where the LA method is not as suitable as the direct CG method. These include:

- CWESR of transition metal ions, which require a wide range of sweep of the static magnetic field [17].
- CWESR in the presence of strong saturating radiation fields [97, 99].
- Calculating the effect of finite amplitude–finite length RF pulses on a spin system.

In all cases, the elements of the matrix $A'(\Delta\omega)$ have nontrivial dependence on the sweep variable.

Further enhancement of the speed and efficiency of the direct CG method can be made by acceleration of the convergence by extrapolation methods. We

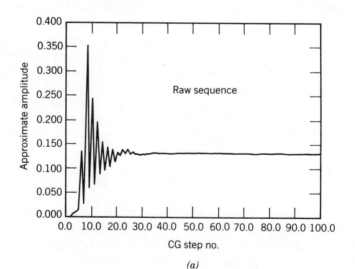

(a)

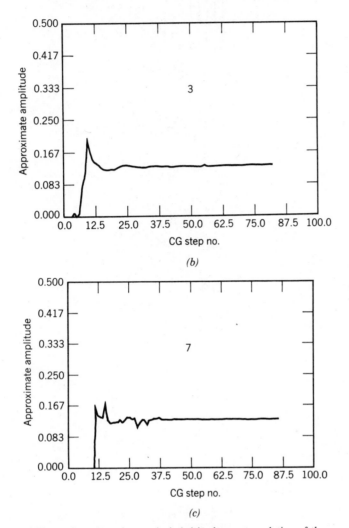

Figure 3. Effects of number of terms included in the ε extrapolation of the approximate values of the amplitude of the spectrum. This calculation is for slow-motional nitroxide spectrum at $\Delta\omega = 0$: (a) raw data; (b) result of three-term ε extrapolation; (c) result of seven-term ε extrapolation.

may expect that later recursive steps merely remove undesirable "transients" in the sequence of approximate values of the spectrum. An effective means of improving the convergence rate of a sequence of this type is the Shanks transformation, or Padé extrapolation, or epsilon expansion. A preliminary study has shown that the Padé extrapolation method can be effective in

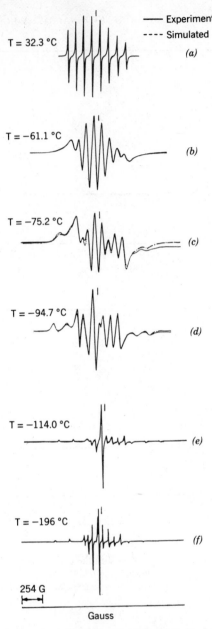

Figure 4. Comparison of experimental and simulated spectra from the fast motional to the rigid limit for VO(acac$_2$(pn)) in toluene. All calculations use a Brownian rotational diffusion model: (a) $\tau_R = 2.06 \times 10^{-11}$ s; (b) $\tau_R = 2.63 \times 10^{-11}$ s; (c) $\tau_R = 5.00 \times 10^{-10}$ s; (d) $\tau_R = 2.25 \times 10^{-10}$ s; (e) $\tau_R = 5.00 \times 10^{-8}$ s; (f) rigid limit. (From Ref. 17. Reprinted with permission from R. F. Campbell and J. H. Freed, *J. Phys. Chem.* **84**, 2668. Copyright 1980 American Chemical Society).

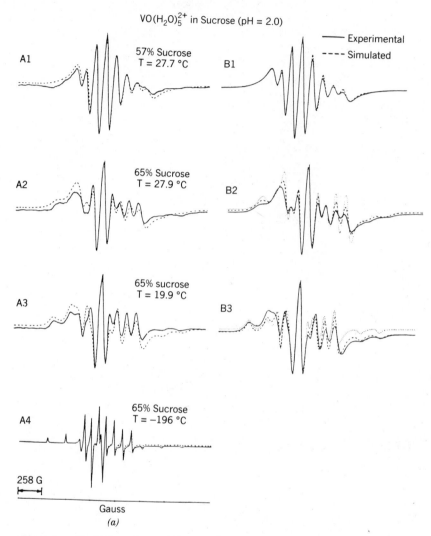

Figure 5a. Model dependence of $VO(H_2O)_5^{2+}$ in sucrose, Series A: Comparison of experiment with moderate-jump diffusion. Series B: Comparison of moderate-jump diffusion (solid lines) with its free (dashed lines) and Brownian (dotted lines) diffusion equivalent. (Moderate jump gave best agreement in all cases.) (A1) $\tau_R^{jump} = 3.4 \times 10^{-10}\,s$; (A2) $\tau_R^{jump} = 6.0 \times 10^{-10}\,s$; (A3) $\tau_R^{jump} = 9.0 \times 10^{-10}\,s$; (A4) rigid limit. (From Ref. 17. Reprinted with permission from R. F. Campbell and J. H. Freed, *J. Phys. Chem.* **84**, 2668. Copyright 1980 American Chemical Society).

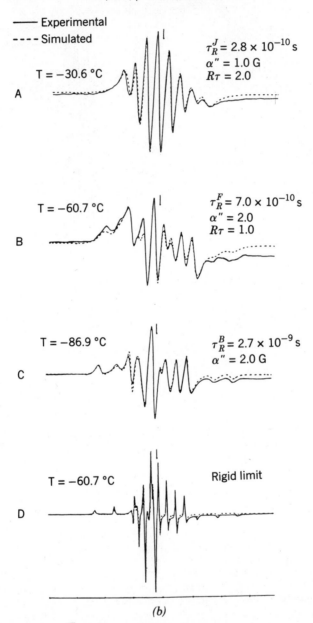

VO(NCS)$_4^{2-}$ in ethylacetate

—— Experimental
- - - - Simulated

A T = −30.6 °C

$\tau_R^J = 2.8 \times 10^{-10}$ s
$\alpha'' = 1.0$ G
$R_T = 2.0$

B T = −60.7 °C

$\tau_R^F = 7.0 \times 10^{-10}$ s
$\alpha'' = 2.0$
$R_T = 1.0$

C T = −86.9 °C

$\tau_R^B = 2.7 \times 10^{-9}$ s
$\alpha'' = 2.0$ G

D T = −60.7 °C

Rigid limit

(b)

Figure 5b. Model dependence of VO(NCS)$_4^{2-}$ in ethyl acetate. Note that A is approximately fit with moderate jump, B with free diffusion, and C with Brownian diffusion. (From Ref. 17.)

436

reducing the amplitude of the rapid variations in the estimated magnitude of the spectrum that occur for the first few CG steps (cf. Figure 3). It is possible that this could be used to reduce the number of CG steps required to give an accurate estimate of the value of the spectrum for a given value of the sweep variable [4]. More details on these techniques and their relationship to the continued-fraction approximants for the spectrum are given in Appendix B. Although the full power of such methods has not been fully explored, they do show substantial promise.

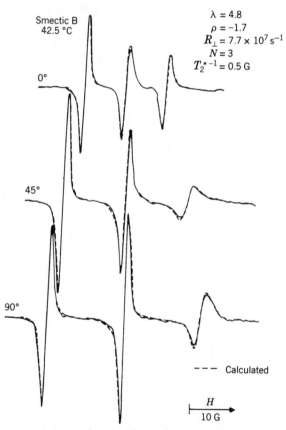

Figure 6. Experimental (solid) and calculated (dashed) ESR spectra of P-probe dissolved in liquid crystal 4O,6 oriented between plates and in smectic B_A phase at 42.5°C. Angle θ between the magnetic field and the plate normal is denoted in the figure. The ordering parameters and rotational rates (given by $R_\perp$, the perpendicular component of the rotational diffusion tensor, and $N = R_\parallel/R_\perp$, the rotational diffusion asymmetry) are on the figures. Both a cylindrically symmetric ordering term, given by $\lambda(5/4\pi)^{1/2} = \varepsilon_0^2/k_B T$, and an asymmetry term of $\rho(5/4\pi)^{1/2} = \varepsilon_2^2/k_B T$, are included (From Ref. 20. Reprinted with permission from E. Meirovitch and J. H. Freed, *J. Phys. Chem.* **88**, 4995. Copyright 1984 American Chemical Society).

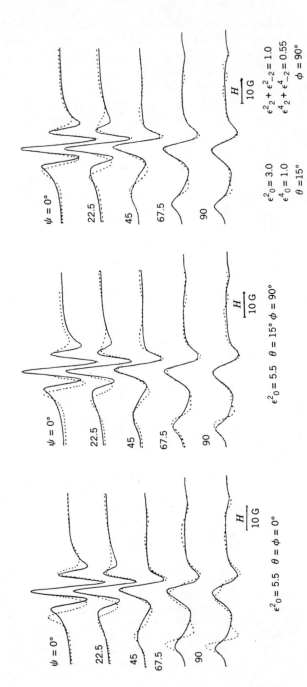

Figure 7. Experimental spectra of homeotropically aligned CSL in liquid crystal S2 at $-8\,°C$ for tilt angle ψ between the liquid crystal director and the static magnetic field (solid lines). Dashed lines: Simulated spectra with anisotropic diffusion in a high ordering potential. $(0, \theta, \phi)$ denote the Euler angles between the magnetic frame and the ordering potential frame. The ε_K^L denote coefficients in the expansion of the scaled ordering potential in spherical harmonics. (From Ref. 21. Reprinted with permission from E. Meirovitch and J. H. Freed, *J. Phys. Chem.* **88**, 4995. Copyright 1984 American Chemical Society).

H. Slow-Motional ESR Spectra: Examples

We have already pointed out that slow-motional spectra provide considerably more information about the microscopic models of rotational dynamics than motionally narrowed spectra. Thus, for example, jump models of rotational reorientation lead to slow-motional spectra that are distinguishable from Brownian reorientational models [12–14]. Whereas nitroxide-type spin probes with ^{14}N nuclear spin of $I = 1$ are commonly studied [12], previous work has shown that VO_2^+ complexes (^{51}V nuclear spin of $I = 7/2$) are more sensitive to the choice of motional model. Also, the latter exhibit slow motions for $\tau_R \simeq 100$ ps, as compared to 1 ns for nitroxides at normal 9-GHz (X-band) frequencies. In particular, the slow-tumbling lineshapes seem to be strongly dependent on the nature of the ligands and of the solvent such that a range of different models had to be used (cf. Figures 4 and 5a, b).

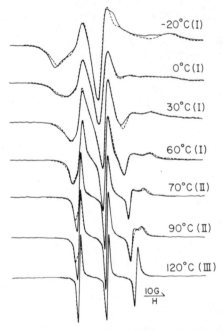

Figure 8. ESR spectra from macroscopically unoriented lipid (DPPC dispersions) with 7 wt % water containing nitroxide end labeled lipid. Solid lines: Experimental spectra. Dashed lines: Calculations based on model of microscopically ordered lipid fragments macroscopically disordered (MOMD model). Phase I: Low-temperature gel phase. Phase II: Liquid crystalline L_α phase. Phase III: New high-temperature phase exhibiting residual microscopic ordering. Typical order parameters: 0.33, 0.18, and 0.02 in Phases I, II, and III, respectively. Typical values of $R_\perp$, perpendicular rotational diffusion coefficient: of 1.0×10^{-8}, 10×10^{-8}, and $20 \times 10^{-8}\,\mathrm{s}^{-1}$, respectively, in these phases. (From Ref. 23. Reprinted with permission from H. Tanaka and J. H. Freed, *J. Phys. Chem.* **88**, 6633. Copyright 1984 American Chemical Society).

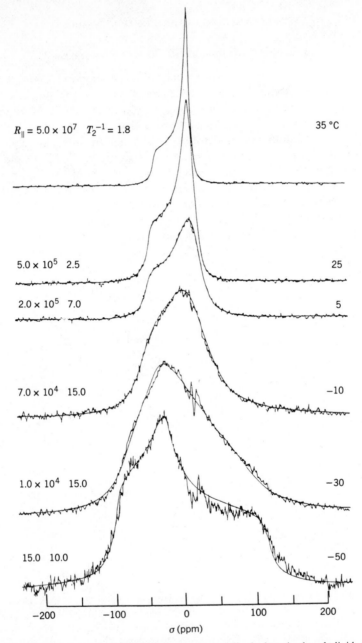

$R_{\parallel} = 5.0 \times 10^{7} \quad T_2^{-1} = 1.8$

35 °C

$5.0 \times 10^{5} \quad 2.5$

25

$2.0 \times 10^{5} \quad 7.0$

5

$7.0 \times 10^{4} \quad 15.0$

−10

$1.0 \times 10^{4} \quad 15.0$

−30

15.0   10.0

−50

−200   −100   0   100   200

σ (ppm)

Figure 9. Experimental ^{31}P NMR spectra from fully hydrated phospholipid (DPPC) bilayers for range of temperatures. Smooth lines: Theoretical calculations showing rotational diffusion coefficient for internal rotation of phosphate moiety in head group. Overall rotation is too slow to affect spectrum. Orientation of "effective" diffusion axis could also be specified. (From Ref. 36. Reprinted with permission from R. F. Campbell, E. Meirovitch, and J. H. Freed, *J. Phys. Chem.* **83**, 525. Copyright 1979 American Chemical Society).

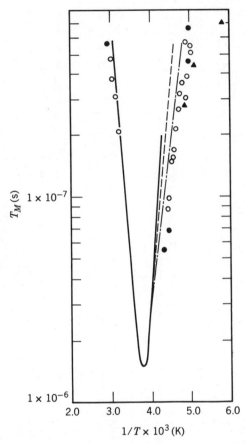

Figure 10. Graph of T_2 (or T_M) versus T^{-1} for PD-TEMPONE in 85% glycerol to 15% water. Circles: conventional T_2 data. Triangles: T_2 from two-dimensional ESE. Different lines represent T_2 in central spectral regime calculated for models of jump diffusion (solid line), free diffusion (dashed line), and Brownian diffusion (dashed–dotted line). Calculations employed values of τ_R extrapolated from motional narrowing regime (From Ref. 30.)

In liquid crystalline phases the most important new feature of molecular rotational dynamics is that it must occur relative to a mean orienting potential. In Figure 6 we show spectra of a nitroxide-labeled liquid crystallike probe molecule in a low-temperature smectic-B phase. The spectrum was fit by a nonspherically symmetric diffusion tensor and a noncylindrically symmetric orienting potential. Orienting potentials in uniaxial liquid crystals may generally be expanded in spherical harmonics, but most experiments provide insufficient information to evaluate beyond the $L = 2$ spherical harmonics. In

TABLE 4 Summary of Rotational Relaxation Mechanisms

Mechanism	Characteristic	Parameters	References
1. Anisotropic diffusion	Unequal reorientation rates about principal axis system fixed in molecule; when reorientation rates are equal, this becomes isotropic Brownian diffusion.	$R_\parallel = (6\tau_{R_\parallel})^{-1}$ rotational diffusion coefficients about molecular symmetry axis; $R_\perp = (6\tau_{R_\perp})^{-1}$ rotational diffusion coefficient about molecular axes perpendicular to symmetry axis; $N = \tau_{R_\perp}/\tau_{R_\parallel}$	5, 12, 13, 16, 149
2. Anisotropic viscosity	Unequal reorientation rates about principal axis system fixed in laboratory (e.g., dc magnetic field axis)	$\hat{R}_\parallel = (6\hat{t}_{R_\parallel})^{-1}$ rotational diffusion coefficient about orienting axis in lab frame; $\hat{R}_\perp = (6\hat{t}_{R})^{-1}$ rotational diffusion coefficient perpendicular to orienting axis in lab frame; $\hat{N} = \hat{t}_{R_\perp}/\hat{t}_{R_\parallel}$	9, 14, 16, 149, 155, 161
3. Fluctuating torques	Anisotropic torques that induce reorientation are themselves relaxing at rate that is not much faster than reorientation of probe molecule	τ_R = relaxation time for rotational diffusion of probe molecule; τ_M = relaxation time for fluctuating torques inducing reorientation of probe molecule; $\varepsilon \equiv (1 + \tau_M/\tau_R)^2$; may be combined with analogues of models 1 (requiring additional specification of $\tau_{M_\parallel}$ and $\tau_{M_\perp}$) or 2 (requiring $\hat{t}_{M_\parallel}$ and $\hat{t}_{M_\perp}$)	14, 16, 41, 153, 161
4. SRLS	Probe molecule relaxes in local potential field, while latter averages out isotropically at significantly slower rate than rate of probe molecule reorientation	$S_l = 1/2 \, (3\cos^2\theta - 1)$, order parameter of probe relative to local anisotropic potential field (i.e., local structure); τ_x = relaxation rate of local structure (may also be combined with analogues of models 1 or 2)	14, 16, 41, 142, 149, 156, 161, 215

5.	Jump diffusion	Molecule reorients by random jumps of arbitrary angle	$R = (\varepsilon^2)_{av}/6\tau$, where τ = time between jumps and $(\varepsilon^2)_{av}$ = mean-square jump angle; can be generalized to include anisotropic diffusion tensors by analogy with models 1 or 2	12–14, 17
6.	Free diffusion	Molecular reorientation is partially due to free gaslike motion that is perturbed by frictional effects of surroundings	τ_R and τ_J, where τ_J = angular momentum relaxation time; they are often related by the Hubbard–Einstein relation: $\tau_R\tau_J = \mathscr{I}/6kT$, $\mathscr{I}$ = moment of intertia; also, $\tau_J^{-1} = \beta$, friction coefficient	12, 15
7.	Director fluctuations	Hydrodynamic effect in which nematic director fluctuates in its orientation with respect to applied magnetic field	$\tau_q^{-1} = Kq^2/\eta$, where τ_q = relaxation time of qth Fourier component of director fluctuation, with K = average elastic constant of liquid crystal and η = viscosity	16, 128
8.	Critical fluctuations	Pretransitional effects near (almost) second-order phase transition; leads to apparent divergences in spin relaxation due to divergences in correlation length and/or relaxation time for order parameter	$\tau_q^{-1} = \omega_\xi(1 + q^2\xi^2)^x$, where, for isotropic–nematic (nematic–smectic) transitions, $\omega_\xi = L/\nu\xi$, with L = force constant for order fluctuations and ν = associated viscosity ($\omega_\xi = \tau_m^{-1}$, with τ_m = relaxation time of smectic clusters), ξ = coherence length of nematic (smectic) order fluctuations and $x = 1(\frac{3}{4})$	128, 141, 150–152
9.	Discrete Jumps	Molecule reorients by discrete jumps among equivalent sites; often this reorientation is about single internal axis	τ_{dJ} = mean time between jumps	40, 213, 214

Reprinted with permission from J. S. Hwang, K. V. S. Rao, and J. H. Freed, *J. Phys. Chem.* **80**, 1970. Copyright 1976 American Chemical Society. (This is an up-dated version).

Figure 7, we illustrate, with an oriented spectrum from a large and rigid nitroxide-labeled molecule, one of the exceptional cases where there is sufficient sensitivity to detect influence of the $L = 4$ spherical harmonics as well.

In spin label studies on oriented lipid multilayers it was possible to discern three transitions, the main [gel to $L_\alpha(1)$ liquid crystalline] transition, a new transition from $L_\alpha(1)$ to a very weakly ordered phase, and finally the transition to an isotropic phase at the highest temperatures. All these phases could be characterized in terms of molecular ordering and dynamics (cf. Figure 8). Thus, the main transition was characterized as primarily a "chain diffusional" transition, whereas the new transition is primarily a "chain orientational" transition. That is, there is a larger relative change in the ordering of the spin probe at the second transition, whereas the diffusion coefficient for the chain labels experiences a more significant relative change at the main transition. These are just a few examples of the kind of information that one can obtain from the slow-motional ESR studies. A related NMR study is shown in Figure 9. As a result of a number of such studies [113], a variety of rotational motional models have been inferred [114]. They are summarized in Table 4.

A clear demonstration of the implications of slow motion may be obtained from ESR spin–echo experiments that directly measure T_2. An echo experiment performed on a nitroxide spin probe in viscous solvent shows that (cf. Figure 10), at high temperatures corresponding to the motional narrowing region, $T_2 \propto \tau_R^{-1}$ (the rotational correlation time), whereas at low temperatures corresponding to the slow-motional regime, $T_2 \propto \tau_R^\alpha$, where $1/2 \leqslant \alpha \leqslant 1$. The value of α depends on the model for reorientation (e.g., $\alpha = 1$ for jump diffusion and $\alpha = 1/2$ for Brownian motion). This leads to a T_2 minimum at the appropriate midrange temperature.

VI. THE FOKKER–PLANCK APPROACH TO MODELING MOLECULAR DYNAMICS

In the Fokker–Planck approach to modeling the molecular dynamics of a complicated many-body system, the exact Newtonian or quantum-mechanical equations of motion for all of the particles in the system are replaced by a much simpler equation only involving the relevant dynamical variables (i.e., those that are relevant for the description of some set of experimental data) and the coupling of these variables to a stochastic thermal bath representing the effects of the neglected variables. The rationale for this procedure is rather well known from a physical point of view for systems at or very near to thermal equilibrium, but the rigorous mathematical justification is quite complicated, especially for systems far from equilibrium [108, 115–121]. The discussion here is mainly directed toward the use of Fokker–Planck operators to model the angular position, angular velocity, and translational degrees of freedom of spin probes in condensed phases and on surfaces for the purpose of interpreting ESR spectra, though the range of applicability of the material presented here extends far beyond this relatively narrow topic.

The reduction of the Fokker–Planck operator for a planar rotator, including both angular position and velocity, to complex symmetric matrix form is carried out in detail as a simple example of the practical use of the theoretical and mathematical concepts presented here.

The correlation functions of interest are of the form

$$g(t) = \overline{f_1^*(t) f_2(0)} \tag{82}$$

$$= \langle f_1 | e^{-\Gamma t} P_0 | f_2 \rangle, \tag{83}$$

where the overbar indicates averaging over the canonical ensemble, and f_1 and f_2 are arbitrary functions of the dynamical variables and external forces. The associated spectral functions are the Fourier–Laplace transforms of the correlation functions

$$\hat{g}(z) = \int_0^\infty e^{-zt} g(t)\, dt \tag{84}$$

$$= \int_0^\infty \langle f_1 | e^{-(z\mathbf{I} + \Gamma)t} P_0 | f_2 \rangle\, dt. \tag{85}$$

The observable spectral densities are given by taking the limit of Eq. 85 as the real part of the complex frequency variable $z = \delta + i\omega$ approaches zero from positive values:

$$\hat{g}(i\omega) = \lim_{\delta \to 0^+} \langle f_1 | [(\delta + i\omega)\mathbf{I} + \Gamma]^{-1} P_0 | f_2 \rangle. \tag{86}$$

This equation is formally equivalent to Eq. 1.

A. General Fokker–Planck Operators

The general multivariate time-independent nonlinear Fokker–Planck operator can be written as [115, 119]

$$\Gamma(q, \lambda) = \sum_{i=1}^N \frac{\partial}{\partial q_i} \left\{ K_i(q, \lambda) - \frac{1}{2} \sum_{j=1}^N \frac{\partial}{\partial q_j} K_{ij}(q, \lambda) \right\}, \tag{87}$$

where q is an N-dimensional real vector of classical stochastic dynamical variables and λ represents the parametric dependence on a set of classical external fields. The term nonlinear here refers to the fact that the real functions $K_i(q, \lambda)$ are not necessarily linear in the dynamical variables and the real functions $K_{ij}(q, \lambda)$ are not constants independent of q. The differential equations for the various probability distributions (Eqs. 88, 89, and 91) are all linear partial differential equations. The set of variables contained in q are assumed to constitute a stationary Markov process. It will also be useful to assume that each of the dynamical variables and external fields have well-defined parity with respect to the operation of (classical) motion reversal (see Section VI.B). The drift coefficients $K_i(q, \lambda)$ and the diffusion coefficients $K_{ij}(q, \lambda) = K_{ji}(q, \lambda)$ do not explicitly depend on time but may depend on the external fields. In modeling the motions of spin probes in liquids and liquid crystals, the external field dependence of the diffusion coefficients can be neglected.

The Fokker–Planck operator determines the time evolution of the conditional probability $P(q'|q; \lambda, t)$ via the forward Kolmogorov equation

$$\frac{dP(q'|q; \lambda, t)}{dt} = -\Gamma(q', \lambda)P(q'|q; \lambda, t) \tag{88}$$

and its adjoint, the backward Kolmogrov equation,

$$\frac{dP(q'|q; \lambda, t)}{dt} = -\Gamma^\dagger(q, \lambda)P(q'|q; \lambda, t), \tag{89}$$

both subject to the same initial condition

$$P(q'|q; \lambda, 0) = \delta(q' - q). \tag{90}$$

It has been assumed here that natural boundary conditions have been enforced to define a unique solution (i.e., enforcing periodicity in angular variables and/or the vanishing of the solution and its derivative for large values of the coordinates and velocities). The conditional probability must satisfy Eqs. 88 and 89 for all positive times.

The Fokker–Planck equation for the time evolution of the probability density,

$$\frac{\partial P(q, t; \lambda)}{\partial t} = -\Gamma(q, \lambda)P(q, t; \lambda), \tag{91}$$

can be derived from Eq. 88 by multiplying both sides by $P(q', t'; \lambda)$ followed by integration over q' and the use of the stationarity properties.

The stationary probability distribution $P_0(q, \lambda)$ is determined by the unique solution to

$$\Gamma(q, \lambda)P_0(q, \lambda) = 0. \tag{92}$$

In most cases discussed in this review, $P_0(q, \lambda)$ is just the equilibrium probability distribution.

B. Time Reversal, Detailed Balance, and Symmetrized Fokker–Planck Operators

The invariance of the fundamental microscopic equations of motion under time reversal is one of the central features of both classical and quantum mechanics. The time reversal invariance of the microscopic equations is closely related to the conditions of detailed balance for transitions between macroscopic states characterized by macroscopic observables. Making the connection between the microscopic dynamics and the macroscopic observables is one of the central problems in statistical mechanics.

A symmetrized form of a general Fokker–Planck operator that satisfies certain detailed balance restrictions is derived in this section [122]. The behavior of the eigenfunctions of the symmetrized Fokker–Planck operator under the combined operation of complex conjugation and classical motion reversal suggest that a set of basis functions that are invariant under this operation will simplify practical calculations. The general scheme for the use of basis sets of this type in forming a finite-dimensional complex symmetric matrix approximation to the symmetrized Fokker–Planck operator for computational work is discussed in Section VI.B.1, and an example of the application of the general method to the planar rotator problem is carried out in Section VI.B.3. The use of the LA for calculating spectral functions is dealt with in Section VI.B.2. The presentation given here regarding detailed balance and its implications relies heavily on the work by van Kampen [115], Graham and Haken [123], Risken [122, 119], Haken [116] and Lax [124, 125].

In the Hamiltonian formulation of classical mechanics the state of a system is described as a point in phase space, and the time evolution is given by Hamilton's equations for the coordinates x_k and momenta p_k,

$$\frac{dx_k}{dt} = \frac{\partial H(x, p)}{\partial p_k}, \tag{93}$$

$$\frac{dp_k}{dt} = -\frac{\partial H(x, p)}{\partial q_k}, \tag{94}$$

for $k = 1, 2, \ldots, N$, where $H(x, p)$ is the Hamiltonian function for the system, which is assumed to be time independent, and is a quadratic function of the momenta. Under these assumptions, the form of Eqs. 93 and 94 is invariant under the substitutions

$$t \to -t, \qquad x \to x, \qquad p \to -p. \tag{95}$$

This is the classical version of time reversal invariance of the microscopic equations of motion for a closed, isolated physical system. This invariance implies that if the system is at the point (x', p') in phase space at time t' and the equations of motion carry this point into a point (x'', p'') at some later time t'', the same equations of motion predict that if the system had started out at the point $(x'', -p'')$ at time t', it would be carried into the point $(x', -p')$ at time t''.

If external forces such as a magnetic field are present, the previous discussion must be modified, since the Hamiltonian is no longer a quadratic function of the momenta. Instead, a more general relation involving the reversal of both the momenta and external fields is needed. In this context it is important to clearly define the operation of motion reversal. The definition of the motion reversal operation used here is the reversal of the momenta of all the particles in the system as well as the currents giving rise to the external fields while leaving the coordinates of the particles in the system unchanged. The macroscopic dynamical variables and fields were previously assumed to have well-defined parity with respect to motion reversal. Let $\tilde{q}$ and $\tilde{\lambda}$ be the sets of motion-reversed variables and fields, respectively. The components of these vectors are given by $\tilde{q}_i = \varepsilon_i q_i$ and $\tilde{\lambda}_j = v_j \lambda_j$, where ε, $v = \pm 1$. It is also convenient to introduce an *antilinear* classical time reversal operator $\mathcal{T}_c$, which is the combined operation of complex conjugation and motion reversal:

$$\mathcal{T}_c[\gamma f(q, \lambda)] = \gamma^* f^*(\tilde{q}, \tilde{\lambda}), \tag{96}$$

where γ is a complex constant and $f(q, \lambda)$ is an arbitrary function of λ. As pointed out by Lax [124], it is unnecessary to include the complex conjugation operation under some circumstances, but it is required here.

The transition to the macroscopic scale can be accomplished by defining a set of macroscopic dynamical variables q that are functions of the microscopic variables x and p followed by a suitable averaging over the equilibrium distribution in phase space. A careful analysis [115] shows that the time reversal symmetry of the microscopic equations leads to the detailed balance relation for the transition rates given by Eqs. 97 and 99.

The definition of detailed balance for a stationary process in the presence of external fields is that the conditional and stationary probabilities satisfy [115, 116]

$$P(q'|q; t, \lambda)P_0(q, \lambda) = P(\tilde{q}|\tilde{q}'; t, \tilde{\lambda})P_0(\tilde{q}', \tilde{\lambda}). \tag{97}$$

An alternative formulation of this definition can be derived by evaluating the derivative with respect to time of both sides of Eq. 97 at $t = 0$ to get a relationship satisfied by the transition probability defined as

$$w(q', q; \lambda) = \frac{dP(q'|q; t, \lambda)}{dt}\bigg|_{t=0} \tag{98}$$

and the stationary distribution. It is

$$w(q', q; \lambda)P_0(q, \lambda) = w(\tilde{q}, \tilde{q}'; \tilde{\lambda})P_0(\tilde{q}', \tilde{\lambda}). \tag{99}$$

It should be noted that these conditions (Eqs. 97 and 99) must be satisfied for all pairs of dynamical variables q' and q independently. More complete derivations of the relationship between macroscopic detailed balance and microscopic time reversal symmetry are given by van Kampen [115] and Lax [124].

It is physically reasonable and can easily be shown that the stationary distribution is invariant under motion reversal,

$$P_0(q, \lambda) = P_0(\tilde{q}, \tilde{\lambda}). \tag{100}$$

The stationary probability distribution can be used to define a real-valued function $\Phi(q, \lambda)$ that can be interpreted as a generalized thermodynamic potential function,

$$P_0(q, \lambda) = Ne^{-\Phi(q,\lambda)}, \tag{101}$$

where N is a normalization factor and the potential function must satisfy $\Phi(q, \lambda) = \Phi(\tilde{q}, \tilde{\lambda})$ since P_0 has this property.

It is now possible to derive a set of conditions that must be satisfied by the drift and diffusion coefficients and the stationary distribution such that the detailed balance conditions given in Eqs. 97 and 99 hold. Since the functions $K_i(q, \lambda)$ are arbitrary well-behaved functions of q and λ, the drift terms in Eq. 87 do not necessarily have simple transformation properties under motion reversal. However, it is straightforward to define linear combinations of the drift coefficients and their motion-reversed counterparts,

$$D_i(q, \lambda) \equiv \tfrac{1}{2}[K_i(q, \lambda) + \varepsilon_i K_i(\tilde{q}, \tilde{\lambda})], \tag{102}$$

$$J_i(q, \lambda) \equiv \tfrac{1}{2}[K_i(q, \lambda) - \varepsilon_i K_i(\tilde{q}, \tilde{\lambda})], \tag{103}$$

to simplify the analysis. The drift terms in Eq. 87 can now be expressed in a form in which the reversible or irreversible nature of the individual terms is made clear.

The insertion of the form of the general Fokker–Planck operator (Eq. 87) into the Fokker–Planck equation (Eq. 91) shows that the Fokker–Planck equation has the form of the divergence of a probability current S,

$$\frac{\partial P(q, t; \lambda)}{\partial t} = - \sum_{i=1}^{N} \frac{\partial S_i(q, t; \lambda)}{\partial q_i}, \tag{104}$$

where the components of S are given by

$$S_i(q, t; \lambda) = \left\{ K_i(q, \lambda) - \frac{1}{2} \frac{\partial}{\partial q_j} K_{ij}(q, \lambda) \right\} P(q, t; \lambda). \tag{105}$$

The components of the probability currents can be broken down into irreversible and reversible parts using the associated definitions for the drift coefficients (Eqs. 102 and 103). The irreversible and reversible components of the probability current are, respectively,

$$S_i^{(+)}(q, t; \lambda) \equiv D_i(q, \lambda) - \frac{\partial}{\partial q_j} [K_{ij}(q, \lambda) P(q, t; \lambda)], \tag{106}$$

$$S_i^{(-)}(q, t; \lambda) \equiv J_i(q, \lambda) P(q, t; \lambda). \tag{107}$$

Equation 92 defining the stationary distribution can now be rewritten as the vanishing of the divergence of the probability current $S_0(q, \lambda)$ associated with $P_0(q, \lambda)$. Inserting $P_0(q, \lambda)$ into Eqs. 106 and 107 give

$$S_{0i}^{(+)}(q, \lambda) = D_i(q, \lambda) P_0(q, \lambda) - \frac{\partial}{\partial q_j} [K_{i,j}(q, \lambda) P_0(q, \lambda)]. \tag{108}$$

$$S_{0i}^{(-)}(q, \lambda) = J_i(q, \lambda) P_0(q, \lambda), \tag{109}$$

Requiring that the divergence of S_0 vanish and using the known symmetry of $P_0(q, \lambda)$ with respect to motion reversal gives the following necessary and sufficient conditions for detailed balance to hold [116, 119]:

$$K_{ij}(q, \lambda) = \varepsilon_i \varepsilon_j K_{ij}(\tilde{q}, \tilde{\lambda}), \tag{110}$$

$$\sum_i \frac{\partial}{\partial q_i} [J_i(q, \lambda) P_0(q, \lambda)] = 0, \tag{111}$$

$$D_i(q, \lambda) P_0(q, \lambda) - \sum_j \frac{\partial}{\partial q_j} [K_{i,j}(q, \lambda) P_0(q, \lambda)] = 0. \tag{112}$$

The vanishing of the divergence of the stationary probability current does not mean that the probability current itself must vanish, even in the stationary or equilibrium state.

The definition of the transition probability (Eq. 98) and the forward and backward Kolmogorov equations (Eqs. 88 and 89) at $t = 0$ are very closely related. In fact, inserting the appropriate initial conditions (Eq. 90) and using the definition of the transition probability, it follows that [119, 122]

$$\Gamma(q', \lambda)\delta(q' - q)P_0(q, \lambda) = \Gamma^\dagger(\tilde{q}', \tilde{\lambda})\delta(\tilde{q}' - \tilde{q})P_0(q', \lambda). \tag{113}$$

Note that there are two sets of free variables in this equation, q' and q. Using the properties of the delta function, it follows that $\delta(q' - q) = \delta(\tilde{q}' - \tilde{q})$, and it is possible to replace q for q' in the argument of $P(q', \lambda)$ on the left side to give

$$[\Gamma(q', \lambda)P_0(q', \lambda) - P_0(q', \lambda)\Gamma^\dagger(\tilde{q}', \tilde{\lambda})]\delta(q' - q) = 0. \tag{114}$$

Since Eq. 114 must hold for any arbitrary value of q, it is equivalent to the *operator* equation [116, 122]

$$\Gamma(q', \lambda)P_0(q', \lambda) = P_0(q', \lambda)\Gamma^\dagger(\tilde{q}', \tilde{\lambda}). \tag{115}$$

In this equation, $P_0(q', \lambda)$ is treated as an operator in the same way a potential function is treated as an operator in quantum mechanics. It must be stressed that Eq. 115 is an *operator* equation that must be valid for operating on arbitrary functions. The derivation of Eq. 115 is based on the original presentation of Risken [122] and the review by Haken [116]. For alternative derivations of this important result based on a master equation, see the book by Risken [119] and the paper by Lax [124].

The operator equation 115 can be put into a more symmetric form by pre- and postmultiplication by the operator $P_0^{-1/2}(q', \lambda)$ to give

$$P_0^{-1/2}(q', \lambda)\Gamma(q', \lambda)P_0^{1/2}(q', \lambda) = P_0^{1/2}(q', \lambda)\Gamma^\dagger(\tilde{q}', \tilde{\lambda})P_0^{-1/2}(q', \lambda). \tag{116}$$

The so-called symmetrized Fokker–Planck operator $\tilde{\Gamma}(q, \lambda)$ defined as

$$\tilde{\Gamma}(q', \lambda) = P_0^{-1/2}(q', \lambda)\Gamma(q', \lambda)P_0^{1/2}(q', \lambda) \tag{117}$$

then satisfies

$$\tilde{\Gamma}(q', \lambda) = \tilde{\Gamma}^\dagger(\tilde{q}', \tilde{\lambda}). \tag{118}$$

This symmetrized form of the Fokker–Planck operator will play a central role in the derivation of matrix element selection rules and the calculation of spectral functions using the complex symmetric LA.

The special symmetry of $\tilde{\Gamma}(q, \lambda)$ is also reflected in its eigenvectors. When $\tilde{\Gamma}$ contains both reversible and irreversible terms, the Hermitian and anti-Hermitian parts of $\tilde{\Gamma}$ do not, in general, commute and the eigenvectors of $\tilde{\Gamma}(q, \lambda)$ and $\tilde{\Gamma}^\dagger(q, \lambda)$ are not the same. However, Eq. 118 can be used in conjunction with the usual biorthogonality relations [126–129] to derive a physically important correspondence between members of the two sets of eigenvectors [116, 122].

Let the $\{\Psi_i(q, \lambda)\}$ be the set of eigenfunctions of $\tilde{\Gamma}(q, \lambda)$ with eigenvalues $\{a_i\}$ and the set $\{\Psi^j(q, \lambda)\}$ be the eigenfunctions of $\tilde{\Gamma}^\dagger(q, \lambda)$ with associated eigenvalues $\{b^j\}$,

$$\tilde{\Gamma}(q, \lambda)\Psi_i(q, \lambda) = a_i \Psi_i(q, \lambda), \tag{119}$$

$$\tilde{\Gamma}^\dagger(q, \lambda)\Psi^j(q, \lambda) = b^j \Psi^j(q, \lambda). \tag{120}$$

The biorthogonality relation is easy to derive by examining matrix elements of $\tilde{\Gamma}$ between eigenfunctions Ψ^j and Ψ_i [129]. Using the definition of the adjoint operator and Eqs. 119 and 120, it is clear that

$$\int \Psi^{j*}(q, \lambda)[\tilde{\Gamma}(q, \lambda)\Psi_i(q, \lambda)]\, dq = a_i \int \Psi^{j*}(q, \lambda)\Psi_i(q, \lambda)\, dq \tag{121}$$

$$= \int [\tilde{\Gamma}^\dagger(q, \lambda)\Psi^j(q, \lambda)]^* \Psi_i(q, \lambda)\, dq \tag{122}$$

$$= b^{j*} \int \Psi^{j*}(q, \lambda)\Psi_i(q, \lambda)\, dq. \tag{123}$$

Taking the difference between Eqs. 121 and 123 gives

$$(a_i - b^{j*}) \int \Psi^{j*}(q, \lambda)\Psi_i(q, \lambda)\, dq = 0, \tag{124}$$

which implies that

$$\int \Psi^{j*}(q, \lambda)\Psi_i(q, \lambda)\, dq = 0 \tag{125}$$

if $a_i \neq b^{j*}$. Carrying out the same steps with the adjoints of Eqs. 119 and 120 shows that the sets of eigenvalues must be complex conjugates of one another; that is $a_i = b^{j*}$ for some pair of indices i, j. A relabeling of the eigenfunctions and eigenvalues can then be carried out such that the usual form of the biorthogonality relation

$$\int \Psi^{j*}(q, \lambda)\Psi_i(q, \lambda)\,dq = \delta_{ij} \tag{126}$$

is valid. If any of the eigenvalues are degenerate and the sets of eigenfunctions are complete, an argument similar to the one used in quantum mechanics can be used here to show that the eigenfunctions corresponding to degenerate eigenvalues can also be chosen such that they are orthogonal.

The derivation of the general biorthogonality relation (Eq. 126) does not take advantage of the known symmetry of $\tilde{\Gamma}(q, \lambda)$ under motion reversal (Eq. 118). The implications of this symmetry are made most evident by looking at the motion-reversed counterpart of Eq. 120,

$$\begin{aligned}
\tilde{\Gamma}^\dagger(\tilde{q}, \tilde{\lambda})\Psi^j(\tilde{q}, \tilde{\lambda}) &= \tilde{\Gamma}(q, \lambda)\Psi^j(\tilde{q}, \tilde{\lambda}) \\
&= b^j \Psi^j(\tilde{q}, \tilde{\lambda}) \\
&= a_j^* \Psi^j(\tilde{q}, \tilde{\lambda}).
\end{aligned} \tag{127}$$

Equations 119 and 127, and the fact that $\tilde{\Gamma}(q, \lambda)$ is invariant under complex conjugation imply that

$$\Psi_i^*(\tilde{q}, \tilde{\lambda}) = \Psi^i(q, \lambda); \tag{128}$$

that is, biorthogonal partners are simply related by the classical time reversal operator. Inserting this result into Eq. 126 shows that an eigenfunction of $\tilde{\Gamma}(q, \lambda)$ must be orthogonal to the motion reversed counterparts of all the other eigenfunctions,

$$\int \Psi_j(\tilde{q}, \tilde{\lambda})\Psi_i(q, \lambda)\,dq = \delta_{ij}. \tag{129}$$

If the eigenfunctions $\{\Psi_i(q, \lambda)\}$ are expanded in a complete orthonormal set of basis functions $\{\phi_i(q, \lambda)\}$ that are invariant under the classical time reversal operation,

$$\phi_i(q, \lambda) = \mathscr{T}_c \phi_i(q, \lambda) = \phi_i^*(\tilde{q}, \tilde{\lambda}), \tag{130}$$

then Eq. 129 can be reexpressed in terms of the components of the eigenfunctions in this basis,

$$\sum_k c_i^k c_j^k = \delta_{ij}, \tag{131}$$

where $c_i^k = \int \phi_k^*(q, \lambda) \Psi_i(q, \lambda) \, dq$. Thus, the eigenvectors are not orthogonal in the usual unitary or Hilbert space sense. Instead, vectors orthogonal in the sense of Eq. 131 are called rectanormal rather than orthonormal [90] (see Appendices A and C). Basis functions obeying Eq. 130 can be constructed from an arbitrary complete orthonormal set by projecting the linear combinations that are even or odd with respect to motion reversal and multiplying the latter by i.

In writing Eq. 131 we have assumed for simplicity that any external fields exhibit motion reversal (i.e. $\tilde{\lambda}_j = \lambda_j$), and/or $\tilde{\Gamma}(q, \lambda)$ (cf. Eq. 118) is invariant with respect to $\lambda \to \tilde{\lambda}$. More generally, as Lax [125] has shown, a generalized motion reversal operator may be introduced (e.g. by multiplying by a two-fold rotation in space) so that the field (e.g. a magnetic field) is invariant to this generalized motion reversal. We assume below that the condition(s) necessary for Eq. 131 to be valid are fulfilled.

A matrix $\mathbf{C}$ constructed from the components of the eigenvectors ($C_{ij} = c_i^j$) in a basis of this type must be a complex orthogonal matrix, $\mathbf{C}^{tr}\mathbf{C} = \mathbf{I}$, by Eq. 131. This matrix, by its construction, is the transformation matrix that diagonalizes the matrix $\mathbf{A}$ of the operator $\tilde{\Gamma}$,

$$\mathbf{C}^{tr}\mathbf{A}\mathbf{C} = \text{diag}(a_1, a_2, \dots). \tag{132}$$

Since $\mathbf{A}$ is diagonalized by a complex orthogonal transformation, it must be a complex symmetric matrix in the basis $\{\phi_i(q, \lambda)\}$. It is extremely important to realize the fact that $\mathbf{A}$ is complex and symmetric. This is a consequence of the following:

- Symmetry of $\tilde{\Gamma}(q, \lambda)$ (Eq. 118).
- Choice of basis set adapted to symmetry of $\tilde{\Gamma}$ (Eq. 130).
- Invariance of operators $\tilde{\Gamma}(q, \lambda)$ and $\tilde{\Gamma}^\dagger(q, \lambda)$ under complex conjugation (Eqs. 87, 101, and 117).

There may also be bases that do not satisfy Eq. 130 that also render $\mathbf{A}$ complex symmetric. This is usually traceable to additional symmetries under which $\tilde{\Gamma}(q, \lambda)$ is invariant. An example of this phenomenon is given in Section VI.B.3. Thus, the validity of Eq. 130 is a sufficient but not necessary condition for the matrix of $\tilde{\Gamma}$ to be complex symmetric. The properties of complex symmetric matrices, complex orthogonal matrices, and rectanormal vectors are studied in greater detail in Appendices A and C.

In concluding, we may mention that, in general, any matrix representation of Γ may be transformed to complex symmetric form $\mathbf{A}$ (representing $\tilde{\Gamma}$) by a similarity transformation $\mathbf{S}$, which can be written in "polar form" by $\mathbf{S} = \mathbf{U}\mathbf{H}$, where $\mathbf{U}$ is a unitary operator and $\mathbf{H}$ is a positive definite Hermitian operator [87]. As a result of the above symmetries, it is immediately possible to identify

$\mathbf{H}$ as $P_0^{-1/2}$ and $\mathbf{U}$ as the transformation from an arbitrary orthonormal basis set to one obeying Eq. 130. Then the special properties of complex *orthogonal* spaces will apply to $\mathbf{A}$ and $\mathbf{C}$. Note, however, that the matrix forming $\mathbf{A}$ and the scalar products forming $\mathbf{C}$ are computed in the context of the appropriate *unitary* (or Hilbert) space in which the operators $\boldsymbol{\Gamma}$ and $\tilde{\boldsymbol{\Gamma}}$ are defined (e.g., Eq. 149 or Eq. 279 below). This duality will be important in the analysis of time correlation functions (cf. Eq. 133) given in Section 6.B.2.

1. Reduction of General Fokker–Planck Operators to Complex Symmetric Matrix Form

The reduction of a general multivariate Fokker–Planck operator to a finite dimensional complex symmetric matrix form is a prerequisite for the implementation of the complex symmetric LA for the calculation of classical correlation functions [3, 43]. A finite dimensional matrix approximation to the symmetrized Fokker–Planck operator is easily obtained by the method of expansion in a finite set of basis functions that are invariant under the classical time reversal operation as outlined in Section VI.B. This discretization method is very convenient because it allows one to choose basis functions for the expansion in which the matrix of the symmetrized Fokker–Planck operator is automatically complex symmetric.

It is useful to make a few comments on the nature of the approximations made in this discretization process. The first comment is related to the necessity of identifying a physically relevant time scale for the process of interest, and the second is related to the close relationship between the method of moments in Hilbert space and the LA.

First, it is important to keep in mind the spirit of approximations initially used in constructing the Fokker–Planck operator. In particular, only the relevant dynamical variables with *long* relaxation times are included in the set q (see also Section VI.C). The remaining variables, whose relaxation times are much shorter, are not explicitly included in the Fokker–Planck description of the stochastic process. Thus, a physically reasonable time scale must be identified to even write down an appropriate Fokker–Planck operator. The mathematical nature of the general Fokker–Planck operator is such that it may have an infinite number of eigenvalues and eigenfunctions. Furthermore, the real parts of the eigenvalues, which correspond to the relaxation rates of the corresponding eigenfunctions, may be arbitrarily large. The inclusion of these rapidly decaying modes is somewhat unsatisfactory since the dynamics of these modes may be affected by the time dependence of some of the dynamical variables that were neglected in the construction of the Fokker–Planck operator. Thus, a finite dimensional approximation to the Fokker–Planck operator may be obtained by only using the projection of the Fokker–Planck operator on the subspace spanned by the slowly relaxing eigenfunctions. The approximation derived in this manner is consistent with the spirit of the physical approximations needed to use a Fokker–Planck operator to model the dynamics.

The difficulty in implementing this projection scheme is that if the eigenvalues and eigenfunctions of the full Fokker–Planck operator are known, the problem has already been solved! The resolution of this tautology is to use physical arguments to identify a set of basis functions whose span should contain the subspace of slow modes. This process is often simplified by considering a closely related problem whose solution is known analytically, the use of asymptotic approximations, or considering the Krylov vectors generated by $\tilde{\Gamma}$ and $|P_0^{1/2}f_2\rangle$ or $\tilde{\Gamma}^\dagger$ and $|P_0^{1/2}f_1\rangle$ (cf. Eq. 137a below). Once a set of basis functions has been chosen, it is easy to generate a finite dimensional approximation to $\tilde{\Gamma}$ by simply evaluating matrix elements.

2. Calculation of Classical Time Correlation Functions and Spectral Densities with the Complex Symmetric Lanczos Algorithm

In order to utilize the complex symmetric LA to calculate spectral functions, it is necessary to generate a finite dimensional complex symmetric matrix approximation to the symmetrized Fokker–Planck operator $\tilde{\Gamma}$ (see Eq. 117). This reduction is discussed in Sections VI.B and VI.B.1. The general forms of the correlation function (Eq. 83) and spectral function (Eq. 85) can be reexpressed in terms of the symmetrized Fokker–Planck operator (Eq. 117),

$$g(t) = \langle P_0^{1/2}f_1|e^{-\tilde{\Gamma}t}|P_0^{1/2}f_2\rangle, \tag{133}$$

$$\hat{g}(z) = \langle P_0^{1/2}f_1|[z\mathbf{I}+\tilde{\Gamma}]^{-1}|P_0^{1/2}f_2\rangle. \tag{134}$$

The complex symmetric LA is applicable in cases where the components of $|P_0^{1/2}f_1\rangle$ are the complex conjugates of the components of $|P_0^{1/2}f_2\rangle$ in the basis chosen in which the matrix of $\tilde{\Gamma}$ is complex symmetric. This is really not a restriction for the general application of the complex symmetric LA, as will be shown here.

Let $\{\phi_n(q,\lambda), n=1,2,\ldots,N\}$ be a set of orthonormal basis functions that are invariant under the classical time reversal operation and span the subspace of slowly relaxing eigenfunctions of $\tilde{\Gamma}$. The general symmetrized spectral function can be expressed in terms of the expansion in this basis as

$$\hat{g}(z) = \sum_{i,j=1}^{N} \langle P_0^{1/2}f_1|\phi_i\rangle\langle\phi_i|[z\mathbf{I}+\tilde{\Gamma}]^{-1}|\phi_j\rangle\langle\phi_j|P_0^{1/2}f_2\rangle. \tag{135}$$

When rewritten in matrix–vector notation, this reads

$$\hat{g}(z) = \mathbf{u}^\dagger[z\mathbf{I}+\mathbf{A}_N]^{-1}\mathbf{v}, \tag{136}$$

where $u_i = \langle \phi_i | P_0^{1/2} f_1 \rangle$, $v_j = \langle \phi_j | P_0^{1/2} f_2 \rangle$, and $(A_N)_{ij} = \langle \phi_i | \tilde{\Gamma} | \phi_j \rangle$.

Since the functions $f_1(q, \lambda)$ and $f_2(q, \lambda)$ are arbitrary functions, the components of $\mathbf{u}$ and $\mathbf{v}$ are, in general, complex. The simplest case to study is where $f_1(q, \lambda)$ and $f_2(q, \lambda)$ are simply related by the classical time reversal operator. If

$$f_2(q, \lambda) = \alpha \mathscr{T}_c f_1(q, \lambda) \tag{137}$$

for some complex constant α, the vectors $\mathbf{u}$ and $\mathbf{v}$ are linearly dependent. In this case, the sequence of Krylov vectors generated by successive application of $\mathbf{A}_N$ on $\mathbf{v}$ span the same subspace as the Krylov vectors generated by $\mathbf{A}_N^\dagger$ and $\mathbf{u}$. Because of this fact, the complex symmetric LA can be used directly to calculate spectral functions that satisfy Eq. 137. It is important to note that spectral functions corresponding to autocorrelation functions where the components of f_1 are real in the basis chosen fall into this category.

On the other hand, if $f_1(q, \lambda)$ and $f_2(q, \lambda)$ are not related by Eq. 137, $\mathbf{u}$ and $\mathbf{v}$ are linearly independent, the Krylov subspaces generated by $\mathbf{A}_N$ and $\mathbf{v}$ and by $\mathbf{A}_N^\dagger$ and $\mathbf{u}$ are inequivalent, and the basic complex symmetric LA cannot be used without further work. There is a straightforward way to resolve this difficulty. The complex symmetric LA can still be used on spectral functions of this type, since they can be rewritten as a linear combination of three spectral functions for which Eq. 137 does hold [3]; that is,

$$\begin{aligned} \hat{g}(z) = \tfrac{1}{2} \{ &\langle P_0^{1/2}(f_1 + \mathscr{T}_c f_2) | [z\mathbf{I} + \mathbf{A}_N]^{-1} | P_0^{1/2}(\mathscr{T}_c f_1 + f_2) \rangle \\ &- \langle P_0^{1/2} \mathscr{T}_c f_2 | [z\mathbf{I} + \mathbf{A}_N]^{-1} | P_0^{1/2} f_2 \rangle \\ &- \langle P_0^{1/2} f_1 | [z\mathbf{I} + \mathbf{A}_N]^{-1} | P_0^{1/2} \mathscr{T}_c f_1 \rangle \}. \end{aligned} \tag{137a}$$

It is easy to verify that the preceding three spectral functions satisfy the symmetry requirement Eq. 137 and thus can be directly calculated with the complex symmetric LA. In fact if $f_1 = \mathscr{T}_c f_1$ and $f_2 = \mathscr{T}_c f_2$ (e.g. one may choose f_1 and f_2 from the basis functions ϕ_i obeying Eq. 130), then Eq. 137a becomes the sum of three autocorrelation functions, two of which are the autocorrelations of f_1 and f_2.

Alternatively, the biorthogonal LA can be used to directly calculate the spectral function as advocated by Wassam [111, 112]. This approach would be most useful if only the cross-correlation of $f_1(q, \lambda)$ with $f_2(q, \lambda)$ is desired. However, the autocorrelations of $f_1(q, \lambda)$ and $f_2(q, \lambda)$ are usually needed to properly interpret the cross-correlation function. In this spirit, it is also necessary to apply the biorthogonal LA three times to get all the required information. The drawback of using the biorthogonal LA is that it requires twice as many matrix–vector multiplications as the complex symmetric LA and its numerical properties are not well understood. Also, in the relatively simple models studied in detail, it is found that the symmetrized Fokker–Planck operators have simpler matrix element structures. This is not necessarily the case for more complex problems. (A Lanczos algorithm for *real* non-symmetric matrices has recently been described [215], and it can be useful for typical Fokker–Planck operators [41]).

3. Example: The Planar Rotator

The Fokker–Planck operator for the angular position and velocity of a planar rotator in the presence of a potential provides a good illustration of the general reduction method since it is not too complicated but does include most of the possible types of terms that can arise in the general Fokker–Planck operator. This model has been used to gain insight into the problem of including inertial effects in the SLE in the slow-motional regime [15]. In addition, Stillman and Freed have used it as a prototypical Fokker–Planck operator in a study of the stochastic modeling of the non-Markovian many-body features of diffusing molecules [41] by a procedure involving the augmentation of the basis set of dynamical variables. The symmetrization and discretization of the simple planar rotator model is discussed here as an illustration of the practical use of the basic ideas of Sections VI.B and VI.B.1. Augmented Fokker–Planck operators for the planar rotator including several different forms of fluctuating torques are given in Section VI.C after the general discussion on augmented Fokker–Planck equations.

The Fokker–Planck equation for the planar rotator problem is

$$\frac{\partial P(\gamma, \dot{\gamma}, t)}{\partial t} = -\Gamma(\gamma, \dot{\gamma}) P(\gamma, \dot{\gamma}, t), \tag{138}$$

where γ and $\dot{\gamma}$ are, respectively, the angular position and velocity with respect to a fixed laboratory frame. The Fokker–Planck operator itself is the sum of a term that is odd under motion reversal,

$$\Gamma_1(\gamma, \dot{\gamma}) = -\Gamma_1(\gamma, -\dot{\gamma}) \tag{139}$$

$$= \dot{\gamma} \frac{\partial}{\partial \gamma} + \frac{F(\gamma)}{I} \frac{\partial}{\partial \dot{\gamma}}, \tag{140}$$

and a term that is even under motion reversal,

$$\Gamma_2(\gamma, \dot{\gamma}) = \Gamma_2(\gamma, -\dot{\gamma}) \tag{141}$$

$$= \frac{\beta k_B T}{I} \frac{\partial}{\partial \dot{\gamma}} \left(\frac{\partial}{\partial \dot{\gamma}} + \frac{\dot{\gamma}}{I k_B T} \right), \tag{142}$$

where β is a phenomenological friction coefficient, I is the moment of inertia of the rotator, k_B is Boltzmann's constant, and T is the absolute temperature. The force on the rotator $F(\gamma)$ is derived from the potential function $V(\gamma)$,

$$F(\gamma) = -\frac{\partial}{\partial\gamma} V(\gamma). \tag{143}$$

The simplest case to consider is where the restoring force $F(\gamma)$ is not present. In this case, the Fokker–Planck operator

$$\Gamma(\gamma, \dot\gamma) = \dot\gamma \frac{\partial}{\partial\gamma} + \frac{\beta k_B T}{I} \frac{\partial}{\partial\dot\gamma} \left(\frac{\partial}{\partial\dot\gamma} + \frac{\dot\gamma}{I k_B T} \right) \tag{144}$$

together with the Maxwell–Boltzmann equilibrium distribution

$$P_0(\gamma, \dot\gamma) = (2\pi)^{-3/2} (k_B T/I)^{-1/2} \exp\frac{-I\dot\gamma^2}{2k_B T} \tag{145}$$

satisfy the conditions of detailed balance. Applying the symmetrizing transformation to Γ gives

$$\tilde\Gamma(\gamma, \dot\gamma) = \dot\gamma \frac{\partial}{\partial\gamma} + \beta \left(\alpha^{-2} \frac{\partial^2}{\partial\dot\gamma^2} - \tfrac{1}{4}\alpha^2\dot\gamma^2 + \tfrac{1}{2} \right), \tag{146}$$

where $\alpha = \sqrt{I/k_B T}$. The symmetrized Fokker–Planck operator assumes a simpler form when written in terms of the variables $q_1 = \gamma$ and $q_2 = \alpha\dot\gamma$,

$$\tilde\Gamma(q_1, q_2) = \alpha^{-1} q_2 \frac{\partial}{\partial q_1} + \beta \left(\frac{\partial^2}{\partial q_2^2} - \tfrac{1}{4}q_2^2 + \tfrac{1}{2} \right). \tag{147}$$

It should be noted that the second term in Eq. 147, which arises from the irreversible part Γ, has the form of the differential equation for the parabolic cylinder functions [129, 130], or, to within an additive constant, a harmonic oscillator Hamiltonian. This observation, together with the particularly simple dependence on γ, suggest that the direct-product basis functions of complex exponentials and harmonic oscillator wavefunctions,

$$\psi_{lk}(q_1, q_2) = N_k e^{-(1/4)q_2^2 + ilq_1} H_k(q_2/\sqrt{2}), \tag{148}$$

where $k = 0, 1, 2, \ldots, K$, $l = 0, \pm 1, \pm 2, \ldots, \pm L$, $N_k = (2\pi)^{-3/4}(2^k k!)^{-1/2}$, and $H_k(q_2)$ are the Hermite polynomials (57, 129), will be well suited for this problem. This basis is orthonormal but not complete since K and L are finite integers, and it is not invariant under the time reversal operation (see Eq. 130). The matrix elements of $\tilde{\Gamma}$ in this basis are

$$\langle \psi_{lk} | \tilde{\Gamma} | \psi_{mn} \rangle = \int_0^{2\pi} \int_{-\infty}^{\infty} dq_1 \, dq_2 \, \psi_{lk}^*(q_1, q_2) \tilde{\Gamma}(q_1, q_2) \psi_{mn}(q_1, q_2) \qquad (149)$$

$$= \beta n \delta_{k,m} \delta_{l,n} + im\alpha^{-1} \delta_{k,m} [\sqrt{n+1} \, \delta_{l,n+1} + \sqrt{n} \, \delta_{l,n-1}]. \qquad (150)$$

It is easy to verify that the matrix of $\tilde{\Gamma}$ is complex symmetric in this basis, that is, $\langle \psi_{lk} | \tilde{\Gamma} | \psi_{mn} \rangle = \langle \psi_{mn} | \tilde{\Gamma} | \psi_{lk} \rangle$.

In the presence of a potential $[F(\gamma) \neq 0$ cf. Eqs. 140 and 143], the matrix of $\tilde{\Gamma}$ is not necessarily complex symmetric in the basis specified by Eq. 148. A new set of basis functions that satisfy Eq. 130 and span the same subspace can be constructed by taking linear combinations of the old basis functions,

$$\phi_{mnp}(q_1, q_2) = (i)^m N_m ([1 + \delta_{n,0}])^{-1/2} e^{-(1/4)q_2^2} H_m(q_2/\sqrt{2})$$

$$\times \begin{cases} \cos(nq_1) & \text{if } p = 1, \\ \sin(nq_1) & \text{if } p = -1, \end{cases} \qquad (151)$$

where $n = 1, 2, \ldots, N$ if $p = -1$ and $n = 0, 1, 2, \ldots, N$ if $p = 1$. Also, $m = 0, 1, 2, \ldots, M$. The symmetry of the resulting matrix can be verified directly from the integral definition of the matrix elements by using the fact that $F(\gamma)$ is a real function and from the structure of the matrix elements of the position and momentum operators in the harmonic oscillator eigenfunction basis.

This basis is not the only possible choice that satisfies the symmetry condition given in Eq. 130. For instance, the basis derived from $\phi_{mnp}(q_1, q_2)$ by omitting the factor $(i)^m$ for even m is also a valid basis that satisfies the symmetry requirement. In addition, any basis that can be expressed in terms of real linear combinations of the basis $\phi(q_1, q_2)$, (i.e. $A\phi_{mnp} + B\phi_{m'n'p'}$ with A and B real constants) will also satisfy the symmetry requirement. The choice of basis can be adapted to the calculation of a specific spectral function or class of spectral functions by taking advantage of the spatial symmetry group under which $\tilde{\Gamma}$ is invariant. The use of spatial symmetry adapted basis functions may reduce the dimension of the matrix.

We can use the functions given by Eq. 151 to illustrate the correlation functions for which Eqs. 137 and 137a hold, so that auto and cross correlations are readily obtained. That is, we may set $f_1 \propto e^{+(1/4)q_2^2} \phi_{mnp}$ and $f_2 \propto e^{+(1/4)q_2^2} \phi_{m'n'p'}$. In fact, any desired correlation function would be obtained directly or as a linear combination of such correlation functions.

C. Extensions of the Fokker–Plank Approach

Stillman and Freed [41] have outlined an extension of the traditional Fokker–Planck approach that can be used to model the non-Markovian many-body features of diffusing molecules. It introduces, in a transparent manner, the basic physics of the relevant degrees of freedom and their couplings and is not restricted to linear transport laws. This approach can also describe both equilibrium and nonequilibrium dynamics but does require the independent specification of the proper equilibrium or stationary probability distribution for the system by independent means.

In this method, the set of relevant dynamical variables (see sections VI.A and VI.B.1) is augmented with stochastic bath variables assumed to have simple Markovian behavior. The augmented set of variables then represents a multidimensional Markov process that obeys a classical SLE. In general, the SLE does not obey the detailed balance conditions given previously since it ignores the back reaction of the dynamical variables on the bath variables. This is a well-known defect in the SLE approach (see section VII). To proceed, the back reaction(s) are incorported into the model by adding term(s) to the SLE to satisfy the detailed balance conditions. The resulting augmented Fokker–Planck equation (AFPE) describes the relaxation of the system to its stationary state, and under appropriate conditions, it also reduces to a classical Fokker–Planck equation for the initial set of dynamical variables. Augmented Langevin equations (ALE) that automatically satisfy the fluctuation–dissipation relationships may be readily obtained from the AFPE.

The relevant dynamical variables Δ of the system are assumed to obey an equation of motion of the form

$$\frac{d\Delta}{dt} = F(\Delta; \Xi(t), \lambda), \tag{152}$$

where $\Xi(t)$ represents the set of independent stochastic bath variables. Furthermore, the stochastic process for the bath variables is assumed to be stationary and Markovian with an associated master equation,

$$\frac{\partial P(\Xi, t; \lambda)}{\partial t} = -\Gamma_\Xi P(\Xi, t; \lambda). \tag{153}$$

With these assumptions, the SLE for the joint probability distribution function of the augmented set of variables can be written as [131–133]

$$\frac{\partial P(\Delta, \Xi, t; \lambda)}{\partial t} = -[\nabla_\Delta \cdot F(\Delta; \Xi, \lambda) + \Gamma_\Xi] P(\Delta, \Xi, t; \lambda). \tag{154}$$

Here the symbol ∇_Δ represents the divergence with respect to the initial set of dynamical variables Δ. The first term on the right side of Eq. 154 is then the Liouville form of Eq. 152. If the variables Δ and Ξ are merged to form a new set q, Eq. 154 has the form of a generic Fokker–Planck equation (see Eqs. 87 and 91),

$$\frac{\partial P(q, t; \lambda)}{\partial t} = -\Gamma(q, \lambda)P(q, t; \lambda). \tag{155}$$

It must be reemphasized that Eq. 154 is incomplete in the sense that the back reaction of the dynamical variables on the bath has been ignored. Therefore, the stationary solution to Eq. 154 will, in general, only yield the correct Boltzmann distribution in the limit of infinite temperature. Another way of stating this is that the joint probability density in the augmented set of variables does not relax to thermal equilibrium! Clearly this is not satisfactory for present purposes. In order to obtain the physically correct approach to the stationary state, additional terms that have been omitted from Eq. 154 must be included. A sufficient condition for the solution of the SLE to relax to the proper stationary state is that it obey the detailed balance conditions stated in Section VI.B. An AFPE that satisfies the requirements of detailed balance can be obtained by adding appropriate reversible and/or irreversible drift terms to Eq. 155. The correct (or, at least reasonable) form of the terms must be determined from physical considerations. The same reasoning that went into specifying the proper stationary state is also applicable in determining these correction terms.

If desired, it is possible to generate a set of ALE from the AFPE [116]. The ALE for each variable q_l in the augmented set can be written as

$$\frac{dq_l}{dt} = k_l(q, \lambda) + \sum_j g_{lj}(q, \lambda)x_j(t),$$

where the $x_j(t)$ are independent Gaussian δ-correlated random functions of time with zero mean, that is,

$$\langle x_j(t) \rangle = 0,$$

$$\langle x_l(t + \tau)x_j(t) \rangle = \delta_{lj}\delta(\tau).$$

The functions $k_l(q, \lambda)$ and $g_{lj}(q, \lambda)$ are related to the drift and diffusion coefficients by

$$K_l(q, \lambda) = k_l(q, \lambda) + \frac{1}{2}\sum_{k,j}\frac{\partial g_{lj}(q, \lambda)}{\partial q_k}g_{kj}(q, \lambda),$$

$$K_{jl}(q, \lambda) = \sum_i g_{ji}(q, \lambda)g_{il}(q, \lambda).$$

The inversion of these equations to obtain the $g_{ij}(q, \lambda)$ and the $k_l(q, \lambda)$ is discussed in [116]. It is not, in general, unique in the absence of additional constraints [119].

An example of this scheme is the explicit introduction of fluctuating torques into the planar rotator problem. If $N(\gamma)$ and $T(\gamma, \dot{\gamma}, t)$ are the mean field torque and fluctuating torques, respectively, the SLE for the process is

$$\frac{\partial P(\gamma, \dot{\gamma}, t)}{\partial t} = -\left\{ \dot{\gamma} \frac{\partial}{\partial \gamma} + I^{-1} \frac{\partial}{\partial \dot{\gamma}} [N(\gamma) + T(\gamma, \dot{\gamma}, t)] \right\} P(\gamma, \dot{\gamma}, t). \tag{156}$$

One physically plausible model for the fluctuating torque is given by assuming that angular position that minimizes the torque is undergoing a simple diffusion process, but the functional form of the mean square torque is dependent only on the deviation from the minimum. That is, it is reasonable to assume that

$$T(\gamma, \dot{\gamma}, t) = V_0 \sqrt{Ik_B T} f(\gamma - \phi(t)),$$

where $\phi(t)$ is the stochastic variable characterizing the fluctuations in the position of the minimum of the torque, and the associated master equation for a simple diffusion in ϕ is valid:

$$\frac{\partial P(\phi, t)}{\partial t} = -\tau_\phi^{-1} \frac{\partial^2 P(\phi, t)}{\partial t^2}.$$

In addition, the stationary state must be specified. A natural choice is to let the equilibrium state be a Maxwell–Boltzmann distribution with respect to the angular velocity and assume that the mean torque is derivable from a potential function that depends only on $\gamma - \phi(t)$, that is, $N(\gamma) = -dU_N(\gamma - \phi)/d\gamma$ so that the equilibrium state is given by

$$\Phi(\gamma, \dot{\gamma}) = \frac{I\dot{\gamma}^2}{2k_B T} + \frac{U_N(\gamma)}{k_B T}.$$

With these assumptions about the nature of the equilibrium state and the form of the fluctuating torque, Eq. 156 does not satisfy the requirements for detailed balance. In particular, the divergence of the reversible probability current is nonzero (i.e., Eq. 111 is violated). The effects of the back-reaction terms can be included through the addition of a reversible drift term to Eq. 156 of the form $J_\phi = \sqrt{I/k_B T} V_0 \dot{\gamma} g(\gamma - \phi)$, where $f(\gamma - \phi) = -dg/d\gamma$ such that the detailed balance conditions are satisfied.

Another physically interesting choice for the equilibrium distribution is to assume that the system relaxes to the instantaneous value of the fluctuating potential,

$$\Phi(\gamma, \dot{\gamma}) = \frac{I\dot{\gamma}^2}{2k_B T} + \frac{U_N(\gamma)}{k_B T} + \sqrt{Ik_B T}V_0 g(\gamma - \phi).$$

As before, the detailed balance conditions are not met without the addition of further terms. In this case, an irreversible drift term can be added to offset the nonzero irreversible probability current (cf. Eq. 112). This choice of potential is appropriate for cases where the torques relax on a time scale long compared to the angular position such as in the slowly relaxing local structure model used in the interpretation of magnetic resonance spectra of spin probes in liquid crystals and model membranes [16, 134].

The methods developed previously to symmetrize standard Fokker–Planck operators and the use of the LA to calculate spectral functions and spectral densities applies equally well to AFPE. In addition, this method, which relies on the construction of an equation of motion for an augmented set of variables (AFPE or ALE) rather than a generalized Langevin equation (GLE), might prove useful in the area of stochastic molecular dynamics calculations since it is easy to include nonlinear couplings and the coefficients of the ALE are time independent. The absence of the memory kernels characteristic of GLE is due to the fact that the set of relevant variables has been extended to include the effects of these interactions. This can be thought of as a redefinition of the projection operators used to define the memory functions or as imposing some physically relevant, nontrival structure on the bath.

D.   Reduction of the Stochastic Liouville Operator to Complex Symmetric Matrix Form

An analysis of the general problem of the reduction of the stochastic Liouville operator to a complex symmetric matrix form is more complicated than the reduction for the general Fokker–Planck operator discussed previously. The source of the added complication is the coupling of classical and quantum-mechanical degrees of freedom.

From general considerations, Hwang and Freed [135] have shown that spin-dependent force and torque terms must appear in the stochastic Liouville operator. Such terms represent the back reaction of the spin system on the classical orientational, positional, or velocity degrees of freedom. For example, the coupling of a spin system to the SLE for the planar rotator in Eq. 156 would require that the mean field torque $N(\gamma)$ be replaced by $N(\gamma) -$

$(i/2)[d\mathscr{H}_s^+(\gamma)/d\gamma]$, where the new term is the spin-dependent torque due to the angular dependence of the semiclassical spin Hamiltonian $\mathscr{H}_s(\gamma)$, and the plus superscript indicates an anticommutator superoperator. It arises from applying the Poisson bracket with respect to γ in the classical part of the Liouville equation. The anticommutator feature is required to maintain a Hermitian density matrix at all times. Thus, the classical probability distribution must also be replaced by the generalized spin density operator of Eq. 7 and Section VII. The existence of a spin-dependent force term had earlier been inferred by Pedersen and Freed [67] in considering the problem of spin-dependent reactive trajectories of interacting radical pairs. A spin-dependent torque term was also inferred by Vega and Fiat [136]. It is easy to show that within the high-temperature approximation, the equilibrium potential should include the term $\mathscr{H}_s(\Omega)/k_B T$, that is the spin force or spin torque acts as an additional effective potential energy term. It is necessary, but need not be sufficient, to guarantee the relaxation of the spin system to thermal equilibrium [68]. Monchick [137] has more recently obtained similar results, and Wassam and Freed [138, 139] have provided a detailed theory that deals with the inclusion of such terms. Both these more recent works employ the Wigner distribution function to pass from a fully quantum-mechanical treatment to the semiclassical limit.

The preceding considerations show that the spin density matrix of the SLE should relax to a canonical distribution in $\mathscr{H}_s(\Omega)$ as well as to the equilibrium distribution for the molecular degrees of freedom. As long as $|\mathscr{H}_s(\Omega)/k_B T| \ll 1$, it is not necessary to explicitly include these spin force and spin torque terms into the SLE. Instead, the term $\Gamma \rho(\Omega, t)$ can be approximated by $\Gamma_{\mathrm{HT}}[\rho(\Omega, t) - \rho_0(\Omega, t)]$, where Γ_{HT} is the high-temperature limit of Γ that does not include spin force or spin torque terms. This form of the SLE will still be consistent with a suitably extended detailed balance criterion, and after the symmetrization of Γ_{HT}, one can again construct complex symmetric matrix representations of the stochastic Liouville operator. The high-temperature form of the relaxation term in the SLE had been inferred by Freed [8], who found that it was required in order to have the general density matrix theory of the SLE reduce to the correct linear response result in the limit of weak irradiation fields (cf. Section II). This correspondence further required $\rho_0(\Omega, t)$ to be the canonical density matrix associated with the *instantaneous* value of $\mathscr{H}_s(\Omega)$, *including* the effects of the irradiating field. This high-temperature limit of the SLE is the form used in the next section.

We have found that this approach does indeed lead to a complex symmetric form of the stochastic Liouville operator in a careful treatment of the high temperature limiting case involving a generalization of the work of Lax [124, 125]. The previous work by Lynden-Bell [61, 62] and Pyper [59, 60] on the symmetries of the SLE is also useful. It should also be mentioned that a

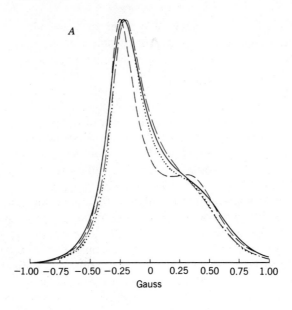

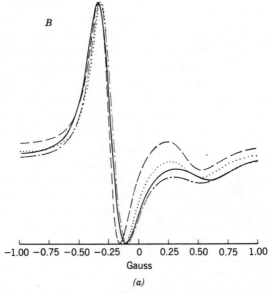

(a)

Figure 11a. Comparison of line shapes for axially symmetric g tensor for different free rotational diffusion models. (A) absorption lineshapes; (B) first-derivative lineshapes. Different rotational diffusion models denoted by dotted lines for Brownian diffusion and solid lines for motion described in full three-dimensional angular momentum space for Brownian particle with damping coefficient $\beta = 4R$; dashed lines: motion described in one-dimensional angular momentum space for Brownian particle with $\beta = 4R$ and $R = 0.13|F|$; dotted-dashed lines: simple free diffusion with $\beta/R = (R\tau)^{-1} = 4$. All have $\tau_R = 1.72 \times 10^{-7}$ s. (From Ref. 15. Reprinted with permission from G. V. Bruno and J. H. Freed, *J. Phys. Chem.* **78**, 935. Copyright 1974 American Chemical Society).

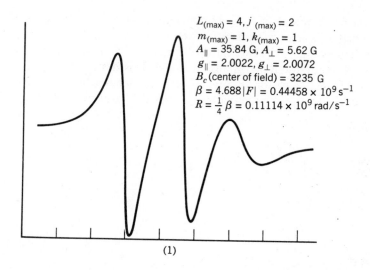

$$L_{(max)} = 4, j_{(max)} = 2$$
$$m_{(max)} = 1, k_{(max)} = 1$$
$$A_{\parallel} = 35.84 \text{ G}, A_{\perp} = 5.62 \text{ G}$$
$$g_{\parallel} = 2.0022, g_{\perp} = 2.0072$$
$$B_c \text{ (center of field)} = 3235 \text{ G}$$
$$\beta = 4.688 |F| = 0.44458 \times 10^9 \text{ s}^{-1}$$
$$R = \tfrac{1}{4} \beta = 0.11114 \times 10^9 \text{ rad/s}^{-1}$$

(1)

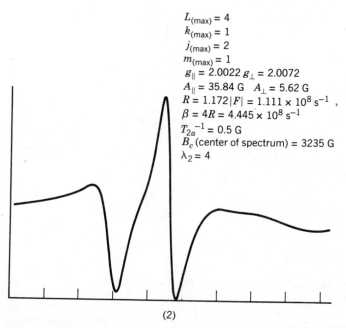

$$L_{(max)} = 4$$
$$k_{(max)} = 1$$
$$j_{(max)} = 2$$
$$m_{(max)} = 1$$
$$g_{\parallel} = 2.0022 \, g_{\perp} = 2.0072$$
$$A_{\parallel} = 35.84 \text{ G} \quad A_{\perp} = 5.62 \text{ G}$$
$$R = 1.172 |F| = 1.111 \times 10^8 \text{ s}^{-1} ,$$
$$\beta = 4R = 4.445 \times 10^8 \text{ s}^{-1}$$
$$T_{2a}^{-1} = 0.5 \text{ G}$$
$$B_c \text{ (center of spectrum)} = 3235 \text{ G}$$
$$\lambda_2 = 4$$

(2)

Figure 11b. Derivative lineshapes for axial nitroxide for motion described in full three-dimensional angular momentum space with $\beta = 4R$ and $\tau_R = 1.5 \times 10^{-9}$ s. Case 1: No ordering potential. Case 2: Potential of $\lambda_0^2 = \tfrac{8}{3}$. (Approximate basis set utilized; cf. Ref. 142.) Separation between x-axis markers is 8.56G.

general prescription for symmetrizing the stochastic Liouville matrix for rotational diffusion problems was given previously [20].

In Figures 11a and 11b we show ESR spectral simulations [15, 140] based on the SLE for rotational diffusion, where Γ is described by a Brownian Fokker–Planck equation in angular and angular momentum space [141]. Related results were obtained with an extended diffusion model [15, 140–142].

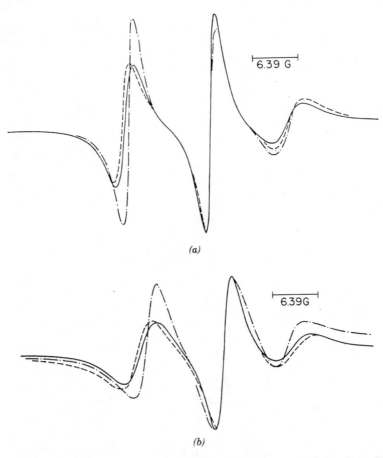

Figure 12. Comparison of experimental and simulated spectra for small nitroxide probe PD-TEMPONE in nematic phase of liquid crystal Phase V. Experimental results at (a) 45 °C, 3450 bars, and (b) 45 °C, 4031 bars, shown as dashed lines. Broken lines are for isotropic Brownian diffusion with (a) 2.25×10^{-9} s; (b) $\tau_R = 4.3 \times 10^{-9}$ s with moderate asymmetric orienting potential; solid lines correspond to calculation based on approximate fluctuating torque model. (From Ref. 114. Reprinted with permission from J. S. Hwang, K. V. S. Rao, and J. H. Freed, *J. Phys. Chem.* **80**, 1790. Copyright 1976 American Chemical Society).

Inertial effects were shown to modify the spectrum for low friction. Methods equivalent to those of this section including the use of a symmetrized diffusion operator were employed [15]. Model calculations based on the fluctuating torque model (cf. Table 4) have also been performed [114] (cf. Figure 12) and are expected to be more relevant in viscous fluids than the inertial models.

VII. NONLINEAR PHENOMENA: SATURATION, DOUBLE RESONANCE, AND SPIN ECHOES

A. The Stochastic Liouville Equation in the Presence of Radiation

In the general case, when linear response theory is no longer valid, we may employ the stochastic Liouville equation of motion for the density matrix $\rho(\Omega, t)$, which is both a spin density operator as well as a classical probability function in the stochastic variables Ω [5, 8, 143–145]:

$$\frac{\partial \rho(\Omega, t)}{\partial t} = [-i\mathbf{L} - \mathbf{\Gamma}(\Omega) - \mathbf{R}][\rho(\Omega, t) - \rho_0(\Omega, t)]. \tag{157}$$

One may recover the ordinary spin density matrix by averaging over Ω:

$$\bar{\rho}(t) = \int d\Omega \, \rho(\Omega, t) = \overline{\rho(\Omega, t)}. \tag{158}$$

The diffusion operator in Eq. 157 has been augmented by the addition of a term $\mathbf{R}$ that is that part of the spin relaxation matrix that is orientation independent [7, 143–145]. The Liouville superoperator may be written as (cf. Eq. 20)

$$\mathbf{L}(\Omega, t) = \mathscr{H}_0^{\times} + \mathscr{H}_1^{\times}(\Omega) + \varepsilon^{\times},$$

where the cross superscript implies the commutator superoperator form, $\mathscr{H}^{\times} \rho \equiv \mathscr{H}\rho - \rho\mathscr{H}$. As before, $\mathscr{H}_0$ is the time- and Ω-independent part of the spin Hamiltonian, $\mathscr{H}_1(\Omega)$ the Ω-dependent part, and $\varepsilon(t)$ the new term due to the interaction with the radiation field:

$$\hbar\varepsilon(t) = \tfrac{1}{2}\hbar\gamma_e B_1 [S_+ e^{-i\omega t} + S_- e^{i\omega t}] \equiv [\varepsilon_+(t) + \varepsilon_-(t)],$$

which is the interaction of the electron spin with a rotating magnetic field $\mathbf{B}_1 = B_1(\hat{i}\cos\omega t + \hat{j}\sin\omega t)$ that defines the x axis of the rotating frame. Relaxation in Eq. 157 is taken toward the instantaneous value of the spin

Hamiltonian [8, 51, 98, 144, 145], that is,

$$\rho_0(\Omega, t) = P_0(\Omega)\frac{\exp[-\hbar\mathcal{H}(\Omega, t)/k_B T]}{\text{Tr}\exp[-\hbar\mathcal{H}(\Omega, t)/k_B T]} \simeq P_0(\Omega)\frac{1}{N}\left[I - \frac{\hbar\mathcal{H}(\Omega, t)}{k_B T}\right],$$
(159)

where N is the total number of spin eigenstates and k_B is Boltzmann's constant. The approximate equality in Eq. 159 is a high-temperature approximation.

At this point it is useful to introduce the operator

$$\chi(\Omega, t) \equiv \rho(\Omega, t) - \rho_0(\Omega, t)$$
(160)

to simplify the form of the following equations. It follows from Eqs. 157 and 160 that

$$\frac{\partial\chi(\Omega, t)}{\partial t} + \frac{\partial\rho_0(\Omega, t)}{\partial t} = [-i\mathbf{L} - \mathbf{\Gamma}]\chi(\Omega, t)$$
(161)

(where $\mathbf{R}$ in Eq. 157 has been omitted for convenience) and from the high-temperature approximate form of Eq. 159:

$$\frac{\partial\rho_0(\Omega, t)}{\partial t} = \frac{i\hbar\omega}{Nk_B T}P_0(\Omega)[\varepsilon_+(t) - \varepsilon_-(t)].$$
(162)

Except for the driving term involving the time derivative of $\rho_0(\Omega, t)$, Eq. 161 is formally similar to Eq. 10 for the spin operator in the Heisenberg representation. By introducing the symmetrized diffusion operator given by Eq. 17 in Section II, Eq. 161 can be converted to

$$\frac{\partial\tilde{\chi}(\Omega, t)}{\partial t} + P_0^{-1/2}(\Omega)\frac{\partial\rho_0(\Omega, t)}{\partial t} = [-i\mathbf{L} - \tilde{\mathbf{\Gamma}}]\tilde{\chi}(\Omega, t),$$
(163)

with

$$\tilde{\chi}(\Omega, t) = P_0^{-1/2}(\Omega)\chi(\Omega, t).$$

Thus, from Eq. 158, one has

$$\bar{\chi}(t) = \int d\Omega\, P_0^{1/2}(\Omega)\tilde{\chi}(\Omega, t) = \overline{P_0^{1/2}(\Omega)\tilde{\chi}(\Omega, t)}.$$

B. General Methods of Solution

The general solution of Eq. 163 is obtained by first expanding $\tilde{\chi}(\Omega, t)$ in a Fourier series in the harmonics of the monochromatic radiation field,

$$\tilde{\chi}(\Omega, t) = \sum_{n=-\infty}^{\infty} \exp\left(in\omega t\right)\tilde{\chi}^{(n)}(\Omega, t). \tag{164}$$

(This may be regarded as a simple application of Floquet theory [129].) The Hermitian nature of ρ and ρ_0 leads to the relation for the matrix elements of $\tilde{\chi}^{(n)}$,

$$\langle a|\tilde{\chi}^{(n)}|b\rangle = \langle b|\tilde{\chi}^{(-n)}|a\rangle^*. \tag{165}$$

In a CW experiment where the steady state is maintained by application of the radiation field at all times, one has

$$\tilde{\chi}^{(n)}(\Omega, t) = \tilde{\chi}_{ss}^{(n)}(\Omega).$$

That is, the Fourier coefficients, which are still spin density operators, are time-independent matrix elements of an ordinary spin operator.

The notation in common use in this field for various types of matrix elements of an ordinary spin operator O is

$$O_\lambda \equiv O_{a^-, b^+}, \tag{166}$$

$$O_{\lambda\pm} \equiv O_{a\pm, b\pm}, \tag{167}$$

$$O_{\lambda_a^\pm, \lambda_b^\pm} \equiv O_{a\pm, b\pm}. \tag{168}$$

In eq. 166 the labeling is such that λ refers to the λth ESR transition between the states a^- and b^+ where the lowercase letters indicate nuclear spin states and the minus and plus refer to the electron spin states $m_s = -\frac{1}{2}$ and $m_s = \frac{1}{2}$. If $a = b$, this is an allowed ESR transition; otherwise, it is a forbidden ESR transition. Equation 167 refers to the diagonal matrix elements, whereas Eq. 168 for $a \neq b$ is a "pseudodiagonal" matrix element [12, 97, 143–148] since it is diagonal with respect to the electron spin but the nuclear spin states are different. These are more properly included with the diagonal matrix elements (in part because of the very small differences in energy between nuclear spin states), and this will be done in what follows. In fact, the summations over $\lambda^\pm$ utilized in the following will imply both types of matrix elements Eqs. 167 and 168.

More generally, it is possible to expand an orientation-dependent operator $O(\Omega)$ in the direct product space of spin operators and the space of square integrable functions of Ω as was done in Eq. 25:

$$|O(\Omega, \omega)\rangle_o = \sum_{\lambda, m} o_{\lambda, m}(\omega)|\lambda, m\rangle,$$

$$|O(\Omega, \omega)\rangle_d = \sum_{\lambda^\pm, m} o_{\lambda^\pm, m}(\omega)|\lambda^\pm, m\rangle. \tag{169}$$

The subscripts o and d refer to the off-diagonal and diagonal subspaces. Then, utilizing Eqs. 162 and 164 for the off-diagonal spin matrix elements of Eq. 163,

$$
\begin{aligned}
\dot{z}^{(n)}_{\lambda,m}(t) = & -i[n\omega - \omega_\lambda]z^{(n)}_{\lambda,m} - \sum_{m'}\langle \lambda,m|\tilde{\Gamma}|\lambda,m'\rangle z^{(n)}_{\lambda,m'}(t) \\
& -i\sum_{\lambda',m'}\langle \lambda,m|\mathscr{H}^{\times}_1(\Omega)|\lambda',m'\rangle z^{(n)}_{\lambda',m'}(t) \\
& -i\sum_{\lambda'\pm,m}\langle \lambda,m|\mathscr{H}^{\times}_1(\Omega)|\lambda'^{\pm},m\rangle \tilde{\chi}^{(n)}_{\lambda'\pm m}(t) \\
& -id_\lambda[\tilde{\chi}(n-1)_{\lambda^+,m} - \tilde{\chi}(n-1)_{\lambda^-,m}] + iq\omega d_\lambda\langle \lambda,m|P^{1/2}_0 S_-\rangle. \quad (170)
\end{aligned}
$$

The off-diagonal matrix elements of $\tilde{\chi}(t)$ in Eq. 170 are denoted $z^{(n)}_{\lambda,m}$, following the established convention [143–145]. Also, the other symbols in Eq. 170 are defined as

$$
\hbar\omega_\lambda = E_{\lambda^+} - E_{\lambda^-}, \quad (171)
$$

where $E_{\lambda^\pm} = (\hbar\mathscr{H}_0)_{\lambda^\pm,\lambda^\pm}$ is the zero-order energy of the $\lambda^\pm$ spin eigenstate of $\mathscr{H}_0$ and the "transition moment" for the λth allowed or forbidden ESR transition d_λ is given by

$$
d_\lambda = \tfrac{1}{2}\omega_1(S_-)_{+,-} = \tfrac{1}{2}\omega_1, \quad (172)
$$

with $\omega_1 = \gamma_e B_1$ and $q = \hbar/Nk_B T$. Note that in the last term on the right in Eq. 170 the matrix element $\langle \lambda,m|P^{1/2}_0 S_-\rangle$ is nonzero only for allowed ESR transitions.

Similarly, the diagonal and pseudodiagonal matrix elements of Eq. 163 are

$$
\begin{aligned}
\dot{\tilde{\chi}}^{(n)}_{\lambda\pm,m}(t) = & -in\omega\tilde{\chi}^{(n)}_{\lambda\pm,m} - \sum_{m'}\langle \lambda^\pm,m|\tilde{\Gamma}|\lambda^\pm,m'\rangle \tilde{\chi}^{(n)}_{\lambda\pm,m'}(t) \\
& -i\sum_{\sigma,m'}\langle \lambda^\pm,m|\mathscr{H}^{\times}_1(\Omega)|\sigma,m'\rangle \tilde{\chi}^{(n)}_{\sigma,m'}(t) \\
& \pm 2d_\lambda[z^{(n+1)}_{\lambda\rightarrow}(t) - z^{(n+1)}_{\lambda\leftarrow}(t)], \quad (173)
\end{aligned}
$$

where the subscript $\lambda^\rightarrow \equiv \lambda$ refers to the O_{a^-,b^+} matrix element as before (cf. Eq. 166), whereas $\lambda^\leftarrow$ refers to the O_{b^-,a^+} matrix element. In the third term on the right in Eq. 173 the summation index σ runs over all $\lambda^\rightarrow$, $\lambda^\leftarrow$, and $\lambda^\pm$. Also note that from Eq. 165, $z^{(n-1)}_{\lambda^\rightarrow}(t) = z^{(-n+1)*}_{\lambda^\leftarrow}(t)$, whereas $\tilde{x}^{(n)}_{\lambda\pm} = \tilde{\chi}^{(-n)*}_{\lambda\pm}$.

The steady-state solutions of Eqs. 170 and 173 are obtained by setting $\dot{z}^{(n)}_{\lambda,m}(t) = \dot{\tilde{\chi}}^{(n)}_{\lambda\pm,m} = 0$ to yield a set of time-independent coupled algebraic equations.

One sees, from Eqs. 170 and 173, that it is only through the effects of the radiation field, where the strength of the interaction with the spins is characterized by ω_1, that the harmonics of the off-diagonal matrix elements $z_{\lambda\rightarrow,m'}^{(n+1)}$ and $z_{\lambda\leftarrow,m''}^{(n-1)}$ are coupled to the harmonics of the diagonal and pseudodiagonal elements $\tilde{\chi}_{\lambda\pm,m}^{(n)}$. An analysis of these equations leads to the result that the extent of coupling depends essentially on the ratio ω_1/ω_λ, which is very small in the presence of large static magnetic fields. Thus, in this case it will be possible to decouple the various harmonics. Next, it will be shown that for high-field saturation cases, it is sufficient to retain only the $z_{\lambda\rightarrow,m'}^{(1)}$, $z_{\lambda\leftarrow,m''}^{(-1)}$ and the $\tilde{\chi}_{\lambda\pm,m}^{(0)}$ terms. The higher harmonics become important in a variety of multiple-resonance schemes [98, 99, 143–145, 149] or experiments done at low static magnetic fields [5, 6].

Consider the power absorbed in a steady-state spectrum given by [51]

$$\mathscr{P} = \omega B_1 \tilde{M}_y = -\tfrac{1}{2}i\omega B_1(M_+ e^{-i\omega t} - M_- e^{i\omega t}), \tag{174}$$

where $\tilde{M}_y$ is the magnetization along the rotating y axis, and the associated operators $\mathscr{M}_\pm$ are given in terms of the electron spin raising and lowering operators $S_\pm$ by

$$\mathscr{M}_\pm(t) = \mathscr{N}\hbar\gamma_e \operatorname{Tr}[\rho(\Omega, t)S_\pm] \tag{175}$$

(cf. Eqs. 8 and 9). By analogy to Eqs. 15 and 16, this can be rewritten as

$$\operatorname{Tr}_s[\bar{\rho}(t)S_\pm] = \operatorname{Tr}_s\overline{\rho(\Omega, t)S_\pm}. \tag{176}$$

Only the terms in $\rho(\Omega, t)$ that contribute to the net time-averaged power absorption via Eqs. 174–176 need be retained. From Eq. 160 we must consider the time evolution of both $\chi(\Omega, t)$ and $\rho_0(\Omega, t)$. We first note that $\rho_0(\Omega, t)$ given by Eq. 159 cannot contribute to the time-averaged absorption since (i) the time-independent terms in $\mathscr{H}_0 + \mathscr{H}_1$ have no components oscillating at $e^{i\omega t}$ needed to cancel the oscillations in Eq. 174, and (ii) the term in $\varepsilon(t)$ that has the needed oscillatory part is found to make contributions to the $M_\pm$ terms in Eq. 174 that are equal in magnitude and opposite in sign. The only terms in the expansion of $\tilde{\chi}(\Omega, t)$ that contribute to the net time-averaged power absorption are the $z_{\lambda\rightarrow,m}^{(1)}$ and $z_{\lambda\leftarrow,m}^{(1)}$ occurring in the form $z_{\lambda,m}^{(1)''} \equiv \operatorname{Im}\{z_{\lambda,m}^{(1)}\} = (2i)^{-1}[z_{\lambda\rightarrow,m}^{(1)} - z_{\lambda\leftarrow,m}^{(-1)}]$. The definition $z_{\lambda,m}^{(1)'} \equiv \operatorname{Re}\{z_{\lambda,m}^{(1)}\}$ will also be needed. Thus, from Eq. 174 the steady-state power absorption is given by

$$\mathscr{P} = 2\mathscr{N}\hbar\omega \sum_{\lambda,m} d_\lambda z_{\lambda,m}^{(1)''}\langle P_0 S_-|\lambda, m\rangle. \tag{177}$$

In the sum over λ only the allowed transitions contribute. In large static magnetic fields only the coupling to $\tilde{\chi}^{(0)}_{\lambda^{\pm},m}$ is required, and these are simply interpreted in terms of deviations from the equilibrium population.

The structure of the coupled differential equations that emerges in the high-field limit from Eqs. 170 and 173 may be written in block matrix form as

$$
\begin{pmatrix}
\dot{z}^{(1)}(t) \\
\dot{z}^{(1)*}(t) \\
\dot{\tilde{\chi}}^{(0)}(t)
\end{pmatrix}
=
\begin{pmatrix}
\mathbf{R} - i\mathbf{K} & \mathbf{0} & i\mathbf{d} \\
\mathbf{0} & \mathbf{R} + i\mathbf{K} & -i\mathbf{d} \\
i\mathbf{d}^{\mathrm{tr}} & -i\mathbf{d}^{\mathrm{tr}} & \mathbf{W}
\end{pmatrix}
\begin{pmatrix}
z^{(1)}(t) \\
z^{(1)*}(t) \\
\tilde{\chi}^{(0)}(t)
\end{pmatrix}
+
\begin{pmatrix}
i\mathbf{Q} \\
-i\mathbf{Q} \\
\mathbf{0}
\end{pmatrix}.
\tag{178}
$$

The vector $\mathbf{z}^{(1)}(t)$ is a vector defined in an $(m \times \lambda)$-dimensional subspace with elements $z^{(1)}_{\lambda,m}$, where $\tilde{\chi}^{(0)}(t)$ is a vector defined in the $(\lambda^{\pm} \times m)$-dimensional subspace with elements $\tilde{\chi}^{(0)}_{\lambda^{\pm},m}(t)$, and $\mathbf{z}^{(1)*}(t)$ is the vector whose elements are the complex conjugates of the elements of $z^{(1)}(t)$.

The relaxation matrix $\mathbf{R}$ in Eq. 178 is the matrix representation of $\boldsymbol{\Gamma} + \mathbf{R}$ from Eq. 157, and the coherence matrix $\mathbf{K}$ is the matrix representation of $\mathbf{L}$ from Eq. 170. By analogy with Section II, let $\mathbf{A}' = \mathbf{R} + i\mathbf{K}$ be the submatrix that governs the dynamics of the off-diagonal density matrix elements, noting, however, that $\mathbf{K}$ can include the effects of multiple quantum coherences, if present [144]. The symmetric matrix $\mathbf{W}$ plays the same role as $\mathbf{A}'$ but for the diagonal and pseudodiagonal density matrix elements [143–146]. The matrix $\mathbf{d}$ is the matrix of transition moments d_{λ} that couples the space of electron spin transitions that contains the vector $z^{(1)}(t)$ to the space of populations and nuclear spin transitions that contains the vector $\chi^{(0)}(t)$, and it is not, in general, square. It has nonzero matrix elements only between electron spin transitions and their associated components in the subspace of diagonal and pseudo-diagonal matrix elements. The vector $\mathbf{Q}$ represents the driving terms in the space of electron spin transitions; its elements are given by the last term on the right side of Eq. 170.

The form of Eq. 178 is sufficiently general to be applicable to multiple-resonance schemes, and the methods for constructing the various matrices are discussed elsewhere [84, 143, 146]. It would, however, be necessary to include higher harmonics than $\mathbf{z}^{(1)}(t)$ and $\chi^{(0)}(t)$. Also note that the terms involving $z^{(0)}_{\lambda,m}$ and $\chi^{(1)}_{\lambda^{\pm},m}$ in Eq. 178 have been neglected. The omission of these terms amounts to neglecting the nonsecular terms from $\mathscr{H}_1^{\times}$; this is readily justified for slow motions in high field [5, 6].

Note that the form of Eq. 178 is of a *matrix* Bloch equation [143, 144], where $\mathbf{z}'$ and $\mathbf{z}''$ are the multidimensional analogues of $\tilde{M}_x$ and $\tilde{M}_y$, respectively, where χ is the analogue of $M_z - M_0$. In the limit of a very weak radiation field where $d_{\lambda} \to 0$, the couplings due to the $\mathbf{d}$ matrices can be neglected in Eq. 178. The neglect of these terms in the weak perturbation limit decouples the off-diagonal space from the diagonal and pseudodiagonal terms.

This allows one to solve just for the latter to obtain the power absorbed (cf. Eq. 174). In fact, in a steady state one has, $\chi_{ss}^{(0)} = 0$, and

$$\mathscr{P} = \tfrac{1}{2}\mathscr{N}\hbar\omega^2 q\omega_1^2 \langle v|[\mathbf{A}'(\omega)]^{-1}|v\rangle \tag{179}$$

from Eqs. 170, 174, and 178 with $|v\rangle$ given by Eqs. 25 and 27. This is seen to be proportional to the expressions for the absorption given by Eqs. 1 and 26. The actual experimentally observed signal is proportional to $\mathscr{P}/\omega_1$ [51].

For finite-amplitude irradiating fields it is necessary to consider the full matrix in Eq. 178, which is of complex symmetric form. Thus, the complex symmetric Lanczos and/or conjugate gradients algorithms can in principle be applied to Eq. 178 also.

C. Steady-State Saturation and Double Resonance

Because of the complexity of Eq. 178, it is beneficial to consider some specific cases where simplifications are possible. For CW saturation [5, 97, 144, 150] and double-resonance schemes such as electron–electron double resonance (ELDOR) [98, 143, 144, 145, 149, 151] or electron–nuclear double resonance (ENDOR) [98, 143, 144], the steady-state solution of Eq. 178 involves setting $\dot{\mathbf{z}}(t) = \dot{\mathbf{z}}^*(t) = \chi(t) = 0$, so the spectrum may be solved either by tridiagonalizing the matrix in Eq. 178 using the LA or by solving the linear algebraic equations directly using the CG algorithm. The latter would appear to have the advantage because it is not possible to remove the sweep variable $\Delta\omega$ from all the diagonal elements. Thus, a Lanczos tridiagonalization would have to be performed for each value of $\Delta\omega$. However, it is possible to rearrange Eq. 178 in a partitioned form to obtain

$$\begin{pmatrix} \mathbf{R} - i\mathbf{K} - \mathbf{S} & \mathbf{S} \\ \mathbf{S} & \mathbf{R} + i\mathbf{K} - \mathbf{S} \end{pmatrix} \begin{pmatrix} \mathbf{z} \\ \mathbf{z}^* \end{pmatrix} = \begin{pmatrix} -i\mathbf{Q} \\ i\mathbf{Q} \end{pmatrix}, \tag{180}$$

where the saturation matrix $\mathbf{S}$ is defined by

$$\mathbf{S} = \mathbf{d}\mathbf{W}^{-1}\mathbf{d}^{\mathrm{tr}}. \tag{181}$$

Equation 180 may now be diagonalized only once to yield the entire spectrum. The details of constructing the symmetric $\mathbf{W}$ matrix and dealing with its singular nature are discussed elsewhere [5, 97, 144, 148]. The solution of Eq. 180 in previous studies has been accomplished by tri-diagonalization using Rutishauser's variant of Givens method (cf. Section IV) followed by diagonalization, but not by the LA. Thus, further work is needed to determine whether the solution of Eq. 180 using the LA or the solution of Eq. 178 (with time derivatives equal to zero) by the CG algorithm is preferred. Methods based on Gaussian elimination (see Section IV) have previously been used to analyze cases involving modulation of the static magnetic field with detection of various harmonics [98, 99].

In Figure 13 we show results of an experimental study on ESR saturation in the slow-motional regime along with the associated calculations.

The general features of the method can be illustrated by considering the simple example of the ESR spectrum of a paramagnetic molecule with an axially symmetric g tensor that is tumbling in an isotropic liquid. In the secular approximation, the orientation-dependent part of the spin Hamiltonian is

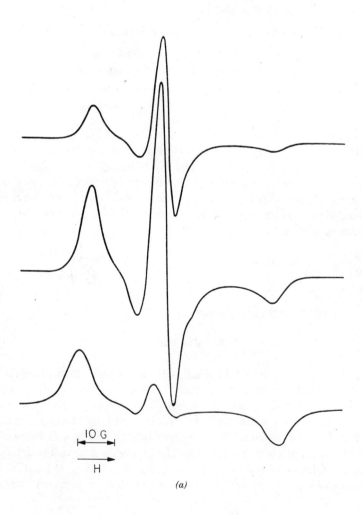

(a)

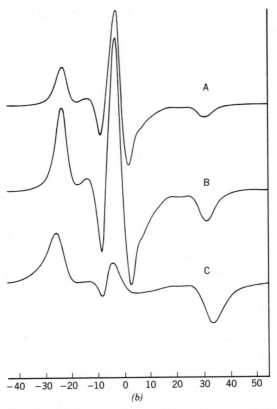

Figure 13. (*a*) Experimental saturation spectra for peroxylamine disulfonate in viscous solvent: "transition moments": (A) 0.025 G; (B) 0.079 G; (C) 0.45 G. (*b*) Simulated spectra for free diffusion with $\tau_R^{\text{free}} = 2.0 \times 10^{-8}$s and appropriate transition moments. Magnetic tensors taken as axially symmetric for simplicity (From Ref. 97b).

simply

$$\mathscr{H}_1(\Omega) = \mathscr{F}\,\mathscr{D}_{00}^2(\Omega)S_z,$$

with

$$\mathscr{F} = \frac{2\beta_e B_0}{3\hbar}(g_\parallel - g_\perp),$$

where $g_\parallel$ and $g_\perp$ are, respectively, the parallel and perpendicular components of the g tensor. It is possible to introduce the orientation-independent contribution to the linewidths, namely the relaxation rates $T_2'^{-1}$ and $T_1'^{-1}$, by including these terms in the operator $\mathbf{R}$. With the inclusion of these terms and the definition

$$\kappa_{L,L'} = [(2L+1)(2L'+1)]^{-1} \begin{pmatrix} L & 2 & L \\ 0 & 0 & 0 \end{pmatrix} \mathscr{F},$$

Eqs. 170 and 173 yields

$$[\Delta\omega - i(T_1'^{-1} + \tau_L^{-1})C_{00}^L(\Delta\omega) - \sum_{L'} \kappa_{L,L'}C_{00}^{L'}(\Delta\omega) + \sqrt{2}\,db_{00}^L(\Delta\omega) = q\omega\,d\delta_{L,0}$$

and   (182)

$$-i(T_1'^{-1} + \tau_L^{-1})b_{00}^L(\Delta\omega) + \sqrt{2}\,d\,\mathrm{Im}\{C_{00}^L(\Delta\omega)\} = 0, \qquad (183)$$

where τ_L^{-1} are the eigenvalues of the isotropic diffusion operator given in Section II. In Eqs. 182 and 183 the standard notation [144] has been used where $z_{\lambda,m}^{(1)} \to C_{00}^L$ and $1/\sqrt{2}(\chi_{\lambda^-,m}^{(0)} - \chi_{\lambda^+,m}^{(0)}) \to b_{00}^L$. The matrix elements of the problem are

$$R_{L,L'} = (-T_2'^{-1} + \tau_L^{-1})\delta_{L,L'},$$

$$K_{L,L'} = \Delta\omega + \kappa_{L,L'},$$

$$\hat{W}_{L,L'} = (T_1'^{-1} + \tau_L^{-1})\delta_{L,L'},$$

$$Q_L = q\omega d\delta_{L,0}.$$

Note that $\mathbf{R}$ and $\hat{\mathbf{W}}$ are diagonal in this representation but $\mathbf{K}$ is not. In this problem $\hat{\mathbf{W}}$ has nonzero elements only in the subspace spanned by the linear combinations b_{00}^L. By the structure of the vector $\mathbf{Q}$, the steady-state absorption lineshape is determined solely by $\mathrm{Im}\{C_{00}^0\}$. The physically interesting feature here may be appreciated by first realizing that b_{00}^0, which represents the ensemble-averaged deviation of the difference in spin population from its equilibrium value due to saturation effects, has a relaxation rate determined by $T_1'^{-1}$ alone. However, the b_{00}^L for $L > 0$, which represent the nonspherically symmetric components of this deviation due to saturation, relax like $T_1'^{-1} + \tau_L^{-1}$. The rotational motion can relax the saturation at one region of the spectrum by transferring the saturated spins to another region of the spectrum.

Another example is the ELDOR signal of the same radical [149]. Here, two radiation fields, a pumping and an observing field, are employed. Therefore, one must label the expansion coefficients with two indices that keep track of the harmonics of each field. Equations 182 and 183 now become

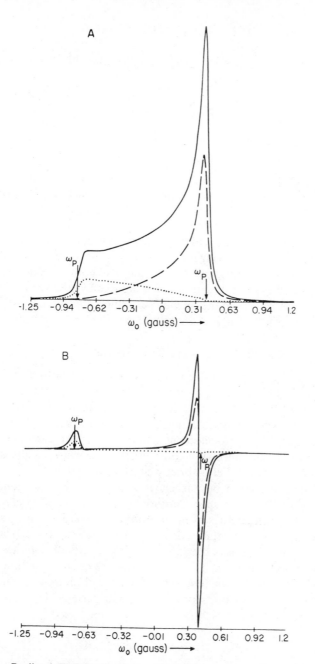

Figure 14. Predicted ELDOR lineshapes for axially symmetric g tensor undergoing isotropic Brownian rotational diffusion. Pump frequency ω_p set at fixed value as shown and observing frequency ω_o is swept. Note how CW saturating effect of pump only partially transmitted to rest of spectrum by rotational motion. Here $\tau_R = 5.75 \times 10^{-6}$ s. $T_1 = 5.7 \times 10^{-6}$ s; (A) absorption; (B) first derivative. Solid line: pure ESR. Dashed and dotted lines: ESR line with pump on. (From Ref. 149.) Courtesy of North-Holland Publ. Co.

$$[\Delta\omega_\alpha - i(T_2'^{-1} + \tau_L^{-1})]C_{00}^L(\alpha) - \sum_{L'} \kappa_{L,L'} C_{00}^{L'}(\alpha) + \sqrt{2}d_\alpha b_{00}^L(\Delta\omega) = q\omega_\alpha d_\alpha \delta_{L,0}$$

and   (184)

$$-i(T_1'^{-1} + \tau_L^{-1})b_{00}^L(\Delta\omega) + \sqrt{2}\sum_\alpha d_\alpha \text{Im}\{C_{00}^L(\alpha)\} = 0, \qquad (185)$$

where $\alpha = o$ or $d = p$, referring to the observing and pumping fields, respectively. Here, $d_\alpha = \omega_{1,\alpha}/2$ and $C_{00}^L(o)$ and $C_{00}^L(p)$ are the $z_{\lambda,m}^{(n_o,n_p)}$ for the $n_o = 1$, $n_p = 0$, and $n_o = 0$ and $n_p = 1$ harmonics, respectively. By setting $\Delta\omega_o$ to be in resonance at positions in the absorption line different from $\Delta\omega_p$, it is possible to monitor the transfer of the saturating effects of the pumping field to the region around the observing field. This absorption of the observing field is proportional to $\text{Im}\{C_{00}^L(o)\}$, which is a function of both $\Delta\omega_o$ and $\Delta\omega_p$. In this simple example, the "saturation transfer" is induced by the rotational diffusion modes with rates τ_L^{-1}, as in the previous example. This "saturation transfer" effect in CW ELDOR is illustrated by the calculations shown in Figure 14.

D. Spin Echoes and Two-Dimensional ESE

For general time domain experiments such as saturation recovery [143, 145, 152] or electron spin echoes [84], one must solve the complete form of Eq. 178 either in the time or frequency domain. However, there are special cases for which the saturation recovery problem can yield analytical forms [143, 144, 145]. Also, electron spin echoes and free induction decays may be greatly simplified by considering somewhat idealized models [26, 28, 30, 33, 35, 146]. Let us briefly consider the latter.

An idealized description of spin echoes is to assume that the radiation field is intense and of sufficiently short duration so that the sole effect of a pulse is a rotation of the density matrix by an appropriate angle. Thus, we assume that $\varepsilon(t)$ dominates the spin Hamiltonian in the rotating frame (i.e., for the equations written in terms of the $z_{\lambda,m}^{(1)}$) during the essentially infinitesimal time that the radiation field is on. Then, in the absence of any radiation, Eqs. 170 and 173 with $d_\lambda = 0$ give the (uncoupled!) time evolution of the off-diagonal and diagonal subspaces during such evolution periods. Thus, for the first evolution period we have

$$\rho_{\text{rf}}(\Omega, t + \tau) = \exp[-i\varepsilon^\times \tau]\rho_{\text{rf}}(\Omega, t), \qquad (186)$$

where the subscript rf implies that only $z_{\lambda,m}^{(1)}$ and $\chi_{\lambda\pm,m}^{(0)}$ are considered in the expansion of $\rho(\Omega, t)$ according to Eqs. 160, 164, and 169. Also, τ is considered as an infinitesimal quantity. The result of this type of idealized pulse is simply a rotation of the density matrix that only affects the electron spins. Whereas for the second period we have

$$\dot{\chi}_{rf}^{(n)}(\Omega, t) = - A_{rf}^{(n)} \chi_{rf}(\Omega, t), \tag{187}$$

where

$$A_{rf}^{(n)} \equiv - i[\mathscr{H}_0^{\times} - n\omega\hat{S}_z^{\times} + \mathscr{H}_1^{\times}(\Omega)] - \Gamma(\Omega)$$

is the rotating frame stochastic Loiuville operator. Thus, for example, a $\pi/2-\tau-\pi-\tau'$ spin echo experiment leads to the following time evolution for $Z^{(1)}(\Omega, t)$:

$$Z^{(1)}(\Omega, \tau + \tau') = \exp[- A_{rf}\tau'] \exp[- A_{rf}^{*}\tau] Z_{\pi/2}^{(1)*}(\Omega, 0^{+}), \tag{188}$$

where the complex conjugation implied by the asterisk in Eq. 188 is more precisely the complex conjugate of all operators linear in the components of electron spin while those operators independent of the electron spin (such as the nuclear Zeeman term) are unaffected, and $Z_{\pi/2}^{(1)}(\Omega, 0^{+})$ gives the value of $Z^{(1)}(\Omega, t)$ resulting from a $\pi/2$ rotation of ρ_0 [28, 146]. In this sequence, one first rotates the equilibrium z magnetization about the rotating x axis into the rotating y axis with the $\pi/2$ pulse. The spins evolve for a time τ, after which the π pulse rotates the y electron spin magnetization into the negative y axis. It then evolves again in the $x-y$ plane for a time τ'. In typical pulsed ESR experiments the echo signal is measured at the time 2τ after the first pulse.

After one introduces expansions of the form of Eq. 169, one obtains from Eq. 187 the matrix differential equation for $z(t)$:

$$\dot{z} = - A'z(t),$$

where the matrix A' is readily obtained from the operator A_{rf}. Also, Eq. 188 yields

$$z(\tau + \tau') = \exp[- A'\tau'] \exp[- A'^{*}\tau] U^{*},$$

where the vector U is obtained from $Z_{\pi/2}(\Omega, 0^{+})$. Finally, the averaged signal may be written as

$$S(\tau + \tau') \propto \text{Re}\{U^{tr} \exp[- A'\tau'] \exp[- A'^{*}\tau] U\}. \tag{189}$$

This expression can be rewritten in terms of eigenvectors and eigenvalues of A' (cf. Eq. [28]) as

$$S(2\tau + t) \propto \text{Re}\{\sum_{l,j} c_{lj} \exp[- (\Lambda_l + \Lambda_j^{*})\tau] \exp[- \Lambda_l t]\}, \tag{190}$$

where $\tau, t > 0$ and $t \equiv \tau' - \tau$. The coefficients c_{lj} are given by

$$c_{lj} = \sum_{k,i,m} U_k O_{kl} O_{il} O_{ij}^* O_{mj}^* U_m^*,\tag{191}$$

where $\mathbf{O}$ is the complex orthogonal matrix that diagonalizes the matrix $\mathbf{A}'$. These eigenvectors are referred to as "dynamic spin packets" [30]. After performing a two-dimensional Fourier transform of Eq. 190 with respect to 2τ and t and using Eq. 192, we recover the general expression for two-dimensional ESE spectra for which Eq. 78 is a special case. The LA may easily be applied, as already discussed in Section V. Equation 191 shows that, in general, one would require the full $\mathbf{O}$ matrix to obtain the signal. This requirement could significantly reduce the power of the LA (but see what follows). However, in the very slow-motional regime where actual experiments of this type are performed, $|\mathscr{H}_1^\times|/|\Gamma| \gg 1$ so that $\mathbf{O}$ must be very nearly real. By setting $\mathbf{O}^* \simeq \mathbf{O}$ in Eq. 191, the approximate form of c_{lj} is

$$c_{lj} \simeq (\mathbf{O}^{\mathrm{tr}}\mathbf{U})_j^2 \delta_{lj} = c_j^2 \delta_{lj}.\tag{192}$$

The calculation of the approximate c_{lj} from Eq. 192 is much simpler than accumulating the full transformation matrix $\mathbf{O}$. The approximate c_{lj} are easily computed by storing only one vector containing the elements $(\mathbf{Q}_D^{\mathrm{tr}}\mathbf{U})$, where the matrix $\mathbf{Q}_D$ is now the matrix that diagonalizes the Lanczos tridiagonal matrix. This procedure is implied in Eq. 78. However, the full matrix representation is needed in Eq. 191. This matter can be studied by first writing $\mathbf{O} \simeq \mathbf{Q}_L\mathbf{Q}_D$, where $\mathbf{Q}_L$ is the matrix of Lanczos vectors that reduces $\mathbf{A}'$ to tridiagonal form ($\mathbf{Q}_L^{\mathrm{tr}}\mathbf{A}'\mathbf{Q}_L = \mathbf{T}_n$). Now Eq. 191 may be approximated as

$$c_{lj} \simeq \sum_{i,r,s} Q_{L,1l} Q_{D,ir} Q_{L,rl} Q_{L,is}^* Q_{D,sj}^* Q_{D,1j}^*,\tag{193}$$

where we have used the fact that $\sum_k U_k Q_{L,kp} = \delta_{p,1}$, since the starting vector $\mathbf{U}$ is just the first Lanczos vector. [Note that the very slow-motional form in Eq. 192 can further be approximated as $c_{lj} \simeq (Q_{D,j1})^2 \delta_{lj}$.] The sum over i in Eq. 193 is over the N-dimensional space spanned by the original basis states, whereas the r and s indices refer to the n_s-dimensional subspace spanned by the Lanczos vectors. The needed Lanczos vectors are seen to be the ones required to adequately represent the relevant eigenvectors and eigenvalues in Eq. 190, and these are the same as those that are important in the approximate form Eq. 192. Thus, it seems reasonable to expect that the MTS required for the approximate solution, discussed in Section V.B and Table 2, would be adequate for the original N-dimensional basis set, whereas the n_s-dimensional

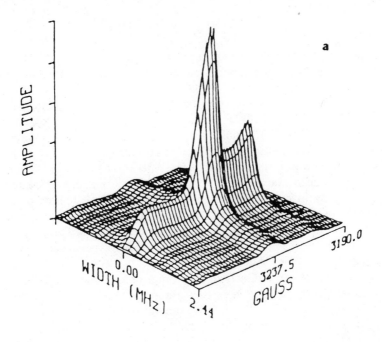

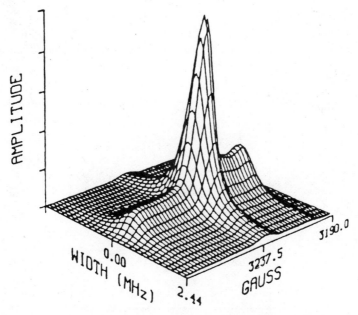

Figure 15. (*a*) Two-dimensional ESE spectra for small nitroxide probe in viscous solvent at −100 °C. (*b*) Simulated spectrum based on theory described in text. (From Ref. 30.)

set of Lanczos vectors from the approximate solution would suffice for the actual calculation of the two-dimensional ESE spectrum. Thus, the number of matrix elements in $\mathbf{Q}_L$ in Eq. 193 would be $N \times n_s$.

Rather than the time-consuming accumulation of the transformation matrices, it should be more efficient to adopt the strategy proposed by Cullum and Willoughby [89]. This would involve applying the LA to obtain $\mathbf{T}_n$ followed by the diagonalization of $\mathbf{T}_n$ by the QR or QL procedure without accumulating the transformation matrix $\mathbf{Q}_D$. After this is complete, the subset of eigenvectors of $\mathbf{T}_n$ corresponding to slowly decaying modes could be reconstructed by inverse iteration [80, 89] and the LA reapplied to $\mathbf{A}'$ to supply the Lanczos vectors required for the transformation of the subset of eigenvectors of $\mathbf{T}_n$ into a set of approximate eigenvectors of $\mathbf{A}'$. This procedure has the advantage that the Lanczos vectors need not be stored; they are regenerated when needed. Alternatively, if it appears that the reapplication of the LA is more time consuming than the storage of the Lanczos vectors, they can be stored. This might often be the case in the class of problems under consideration, where $n_s \ll N$. Another advantage of this approach is that, in general, the number of relevant eigenvectors, n_e, is typically $n_e \ll n_s$, so even less storage and fewer matrix multiplications are required to calculate the c_{lj} by Eq. 193.

The theoretical curve in Figure 10 was obtained from Eq. 190 by setting $t = 0$ corresponding to the echo maximum and then stepping out τ to get the T_2-type decay. Figure 15 illustrates experimental and simulated two-dimensional ESE spectra for a nitroxide in a viscous fluid. The variation of

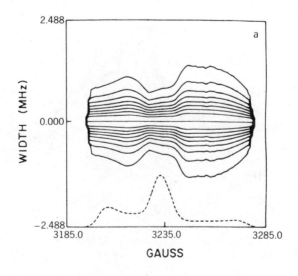

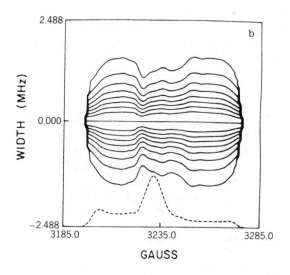

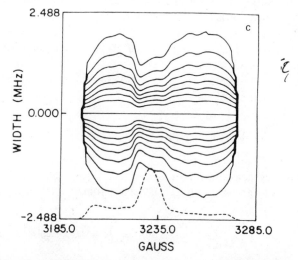

Figure 16. Two-dimensional ESE normalized contour plots. (*a*) Experimental spectrum for same case as Figure 15 but at −75 °C, with signal-to-noise ratio enhancement and dead-time corrections calculated by LPSVD methods. (*b*) Theoretical calculation for a model of isotropic Brownian motion. (*c*) Same as (*b*) but for rotational anisotropy of $N = 2$. Contours normalized to the 0-MHz slice, shown in each figure by dashed lines. (From Ref. 32.)

T_2 across the spectrum is illustrated in the experimental and theoretical contour plots shown in Figure 16. In the very slow-motional regime studied, the CWESR spectra are no longer sensitive to motion, but these two-dimensional ESE spectra remain very sensitive to the motion and to the microscopic model. Note that the two-dimensional spectral resolution and signal-to-noise ratio in the experimental data have been very significantly enhanced by modern techniques of data processing based on linear prediction [32, 153, 154].

Finally, we note that when the approximation of Eq. 192 is not valid and Eq. 191 must be used, the two-dimensional Fourier transform of Eq. 189 with respect to τ and τ' can lead to cross-correlations between the different dynamic spin packets. Analogous types of cross-correlations will show up in the next section.

E. Stimulated Echoes, Magnetization Transfer, and Two-Dimensional Fourier Transform Spectroscopy

We now consider more sophisticated but still idealized pulse sequences. An important pulse sequence is the stimulated echo sequence $\pi/2-\tau_1-\pi/2-T-\pi/2-\tau_2$-echo. In this case, the second $\pi/2$ pulse rotates the y magnetization into alignment along the negative z axis. After evolving for a time T along the z axis, it is returned to the $x-y$ plane by the third $\pi/2$ pulse, and an echo is formed. By stepping out the time T, one can study relaxation of the z· magnetization. We refer to this as a T_1-type, or magnetization transfer, experiment.

This more sophisticated sequence yields the following expression for the desired stimulated echo signal [33, 146]:

$$S(T + \tau_1 + \tau_2) \propto \text{Re}\{\sum_p b_p(\tau_1, \tau_2) \exp[-T/\tau'_p]\}, \tag{194}$$

where τ'_p are the eigenvalues of the $\mathbf{W}$ matrix in Eq. 178 representing relaxation of the diagonal and pseudodiagonal density matrix elements. Also,

$$b_p(\tau_1, \tau_2) = \sum_{l,k,m,n,q,s} (U_q O_{o,qn}) O_{o,mn} O_{d,mp} O_{d,kp} O^*_{o,kj} (O_{o,sj} U_s)^*. \tag{195}$$

In Eq. 195 we distinguish the complex orthogonal transformation that diagonalizes $\mathbf{A}'$ as in Eq. 191 by $\mathbf{O}_o$ and the complex orthogonal transformation that diagonalizes $\mathbf{W}$ by $\mathbf{O}_d$. One might expect $\mathbf{W}$ to be a real symmetric matrix. But because it couples the diagonal and pseudodiagonal matrix elements (which rigorously are off-diagonal density matrix elements), it is, in general, complex and symmetric. However, all complex eigenvalues and eigenvectors occur in complex-conjugate pairs, such that true diagonal density matrix elements, which represent populations, are always real [144–148].

Equation 195 is based on the simplifying assumption that the whole spectrum is irradiated by the idealized pulse. Also it assumes the existence of a "congruence" between the basis sets used to expand the off-diagonal and diagonal subspaces. This congruence, or one-to-one relationship, may be established via the **d** matrix of Eq. 178 in a number of useful cases [144–148]. In order to obtain the full time evolution of the signal, one must separately diagonalize **A** and **W**. Furthermore, there are no known simplifications analogous to Eq. 192 for the two-dimensional ESE case. Thus, in principle, one requires the full transformation matrices O_o and O_d. As already noted in other contexts, this would greatly reduce the speed of the LA, and this procedure has not been used to date on this problem, although the Rutishauser algorithm has been employed [146].

Let us consider what would be involved in employing the LA in such calculation. We start by first comparing Eq. 195 for $b_p(\tau_1, \tau_2)$ versus Eq. 191 for c_{ij}. In the latter the complicating feature was the appearance of the sum $\sum_i O_{o,il} O_{o,ij}^*$, whereas in the former it is the double sum $\sum_{m,k} O_{o,mn} O_{d,mp} O_{d,kp} O_{o,kj}$. The arguments given previously for the efficient calculation of the important subset of eigenvectors of **A′** also apply here, when the vector **U** is used as the starting vector for the LA. One is thus left with the job of diagonalizing **W**. The application of the LA in this case is less well defined given that there is no well-defined time-independent starting vector to use. Given a congruence between the basis sets, it would seem physically reasonable to use the vector U_d that is congruent to **U** as the starting vector in the diagonal subspace. Again, the important subset of eigenvectors of **W** could be constructed by the Cullum–Willoughby procedure. Past experience [34, 35, 146] has shown that only a few of the slowest decaying eigenvectors of **W** are significant. Given that the LA is efficient in generating the eigenvalues of smallest real part (cf. Section V.A), it should not require many Lanczos steps to obtain good approximations to them.

Another possible approach is suggested by rewriting Eqs. 194 and 195 in the form

$$S(\tau_1 + T + \tau_2) \propto \text{Re}\{U^{tr} \exp[-A'\tau_2] V_{\pi/2} \exp[-WT] V_{\pi/2} \exp[-A'\tau_1] U\},$$

(196)

where $V_{\pi/2}$ expresses the effect of the second and third $\pi/2$ pulses. A triple Fourier transform gives

$$S(\omega_1, \omega_2, \omega_T) \propto \text{Re}\{(U[\omega_2 I + A']^{-1}) V_{\pi/2} [\omega_T I + W]^{-1}$$
$$\times V_{\pi/2}([\omega_1 I + A']^{-1} U)\}.$$

(197)

The two expressions in parentheses could, in principle, be calculated using the CG algorithm to solve

$$[\omega_1 I + A']z_1(\omega_i) = U \tag{198}$$

for $z_1(\omega_i)$ for a range of values of ω_1 and ω_2 and to store these vectors. What remains to be calculated is then

$$S(\omega_1, \omega_2, \omega_T) \propto \mathrm{Re}\{z^{\mathrm{tr}}(\omega_2)V_{\pi/2}[\omega_T I + W]^{-1}V_{\pi/2}z(\omega_1)\}. \tag{199}$$

Now, when a simple congruence exists, the effects of $V_{\pi/2}$ are simple and lead to [146] $\chi(\omega_1) = i\sqrt{2}z^*(\omega_1)$ from the second pulse, and $z(\omega_2) = i\sqrt{2}\chi(\omega_2)$ from the third pulse so that Eq. 199 becomes

$$S(\omega_1, \omega_2, \omega_T) \propto \mathrm{Re}\{\chi^{\mathrm{tr}}(\omega_2)[\omega_T I + W]^{-1}\chi(\omega_1)\}. \tag{200}$$

It would probably still be best to diagonalize W for the slowest decaying modes by the LA, as discussed in the preceding.

Equations 199 and 200 suggest that $S(\omega_1, \omega_2, \omega_T)$ can be viewed as a matrix of correlation functions of $\chi(\omega_1)$ with $\chi(\omega_2)$ due to the effects of the relaxation

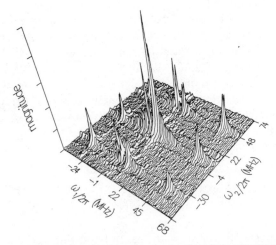

Figure 17. Two-dimensional exchange spectrum of 1.17 mM PD-TEMPONE spin probe in perdeuterated toluene at 21 °C obtained by Fourier transform methods. Cross-peaks are due to Heisenberg spin exchange inducing magnetization transfer between the primary three lines along diagonal. (Spurious peaks, which do not come at the magnetization transfer positions, are due to residual effects of axial peaks). (From Ref. 34.)

processes associated with **W**. Such a correlation matrix (still a function of ω_T) would contain all of the relevant information obtainable by the stimulated echo sequence. A form intermediate between Eqs. 196 and 197, which we write as $S(\omega_1, \omega_2, T)$, can be used to interpret actual two-dimensional exchange experiments that have been performed both by NMR [155] and by ESR [34, 35] (cf. Figure 17). Work is in progress in making the full three-dimensional experiment a practical reality.

We illustrate some experimental and calculated results in Figure 18 for the stimulated echo and inversion recovery sequences, both of which are magnetization transfer experiments. The inversion recovery experiment is based on the following pulse sequence: The first π pulse inverts the z

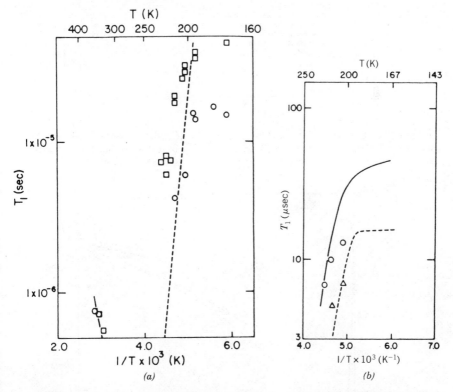

Figure 18. Experimental results and calculations on inversion recovery (IR) and stimulated echo (SE) sequences for small spin probe in viscous solvent. (a) Apparent T_1 versus inverse temperature for IR (squares) and SE (circles) sequences with partial irradiation. Solid line: Measured T_2 for center line. Dashed line: Extrapolated fast-motional τ_R for reference. (b) Comparison of experimental and calculated apparent T_1. Solid line: Experimental IR. Dashed line: Experimental SE. Circles: Calculated IR. Triangles: Calculated SE. (From Ref. 33.) Courtesy of North-Holland Publ. Co.

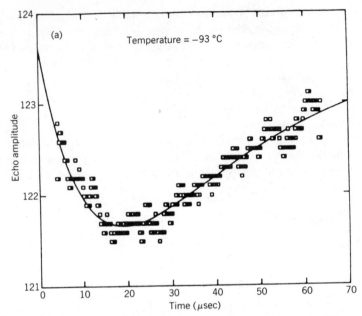

Figure 19a. Experimental stepped-field ELDOR echo amplitude versus time between first π pulse and $\pi/2$ pulse for small spin probe in viscous solvent at $-93\,°C$. Solid line: Best two exponential fit to data. (From Ref. 29.) Courtesy of North-Holland Publ. Co.

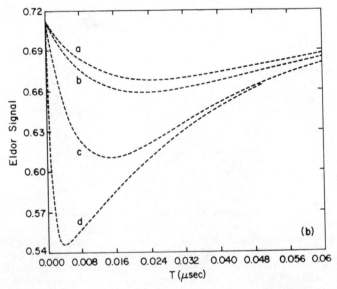

Figure 19b. Calculated ELDOR curves for slowly tumbling nitroxide. In sequence a–d orientation-independent nuclear spin flip rate increased, thereby increasing ELDOR effect. (From Ref. 146.)

magnetization; after a time T its magnitude is probed by forming an echo by tipping the magnetization into the x–y plane with a $\pi/2$ pulse and then refocusing with another π pulse. Closely related to inversion recovery is the echo-ELDOR experiment [29, 156]. It utilizes the same pulse sequence, but the magnetic field is stepped after the first π pulse, so the transfer of inverted magnetization to a different spectral position can be studied as illustrated in Figure 19a and 19b.

The possibilities of a two-dimensional magnetization transfer experiment based on the stimulated echo sequence in studies of motional dynamics are illustrated in Figure 20 for NO_2 adsorbed on a VYCOR surface. The anisotropy of the motion can be immediately discerned from the two-dimensional contour plot.

Another class of experiments probe the z magnetization to study the phenomenon of chemically induced dynamic spin polarization [69]. In this experiment, free radicals are produced with a laser pulse, and the development of the spin polarization with its subsequent decay is monitored by a standard

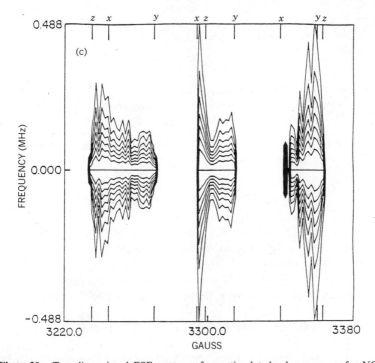

Figure 20. Two-dimensional ESE contours from stimulated echo sequence for NO_2 on vycor at 35 K showing rates of magnetization transfer. It shows relatively rapid rotation about the molecular y axis (i.e., axis parallel to oxygen–oxygen internuclear vector). (From Ref. 33.) Courtesy of North-Holland Publi. Co

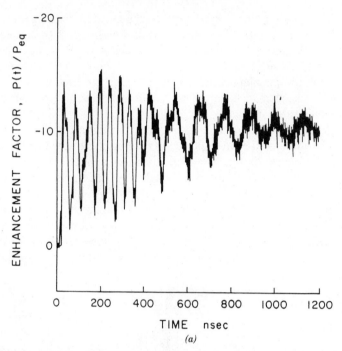

Figure 21a.   Evolution of CIDEP polarization with time from a photoelectron ejected from Rb^- in THF solvent (From Ref. 157. Reprinted with permission from U. Eliav and J. H. Freed, *J. Phys. Chem.* **88**, 1277. Copyright 1984 American Chemical Society).

two-pulse spin echo sequence. In the case of a transient photoelectron generated by the process

$$Rb^+Rb^- \xrightarrow{h\nu} Rb^+ + Rb + e^-$$

in an inert solvent, the polarization is produced with a time constant of 10–50 ns [157]. When the Rb concentration is increased sufficiently, an oscillatory z magnetization is observed (cf. Figures 21a and 21b). It is taken as direct evidence for the radical pair mechanism for the spin polarization, as illustrated by a simple model calculation using the appropriate SLE (cf. Figure 21b) [157].

VIII.   CONCLUSIONS AND FUTURE DIRECTIONS

The underlying theme in this chapter has been the general applicability of the complex symmetric Lanczos and conjugate gradients algorithms to the numerical solution of problems in quantum and classical irreversible statistical mechanics. This theme has been exemplified by specific examples from the

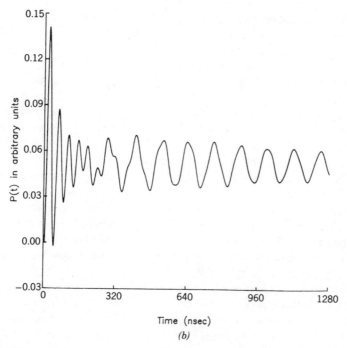

Figure 21b. A predicted oscillatory polarization evolution curve for jump reencounters of the Rb˙ and the e^- with two solvent-dependent species of radical pairs present (From Ref. 157. Reprinted with permission from U. Eliav and J. H. Freed, *J. Phys. Chem.* **88**, 1277. Copyright 1984 American Chemical Society).

study of molecular dynamics in liquids by spin resonance and relaxation. The form of these algorithms has been shown to be appropriate for a wide range of such problems through the examination of the implications of time reversal symmetry and detailed balance. The development of such computational methods, in conjunction with new experimental techniques, should make possible even more powerful methods to study molecular dynamics. Recent work in other areas has further demonstrated the generality and utility of the complex symmetric LA. It has now been used in scattering [158] and optical spectroscopic [159] calculations in conjunction with the complex rotated Hamiltonian method. The properties of the matrices that arise in these studies are closely related to those considered here [126].

We wish to conclude by summarizing some of the ways we believe the application of the LA and CG methods to the class of problems mentioned in the preceding can be improved or enhanced. First, it should be possible to employ a block version [80, 89] of the complex symmetric LA that would be capable of simultaneously calculating the spectral functions corresponding to several auto- and cross-correlation functions. An interesting application of this algorithm would be in cases where the finite-element discretization is used. A block LA might allow one to very efficiently calculate the cross-correlation of dynamical variables in different localized regions of space. In addition, the block LA and CG methods are more easily adapted to take advantage of the architecture of modern supercomputers. The use of nonorthogonal finite-element functions requires a form of the LA designed to treat the generalized eigenvalue problem. Again, this type of algorithm is well known for real matrices. The nonorthogonal LA has only been used in the study of simple problems to date [44, 160, 161], but applications to more complicated magnetic resonance problems might be fruitful.

The CG method was found to be useful in determining the proper minimum basis set for an accurate numerical calculation of spectra. Future work, possibly relying on artificial intelligence techniques, should be directed toward extending the existing database and developing effective methods of accessing this information. Also, further work on preconditioning and sequence convergence acceleration techniques for the direct calculation of spectra using the CG algorithm would be useful.

Given that the underlying mathematical theory for the complex symmetric LA and CG methods involves the theory of bilinearly metric spaces, further study of the properties of these spaces, especially complex orthogonal spaces, could provide further insight into the behavior of these algorithms and into irreversible statistical mechanics in general.

The analysis of the enormous amount of data generated by modern time domain magnetic resonance techniques has become a complex and sophisticated matter. One method that has recently been proven to be very powerful in time domain ESR experiments is based on using linear prediction in conjunction with the singular-value decomposition. We believe that a Lanczos-based singular-value decomposition [89] can be designed to take advantage of the special structure of the matrix generated by the linear prediction modeling of the signal. The development of such an algorithm has the potential of dramatically reducing the computer time and storage requirements for the routine analysis of experimental data using the linear prediction–singular value decomposition method compared to other algorithms that cannot exploit the special structure of the matrix [215].

An important way in which the comparison of theoretical model calculations and experimental data can be improved is by the automation of the process using nonlinear least-squares fitting procedures. Since this process would involve the calculation of large numbers of approximate spectra, a very efficient LA or CG algorithm that can take advantage of a minimum basis set database is required. This is a very complicated effort because the minimum basis set and optimal number of LA or CG steps might vary widely as the nonlinear fitting algorithm searches for the best set of parameters. This is a problem on which members of our group have recently been making some progress utilizing the CG tridiagonalization procedure.

APPENDICES

A. Lanczos Algorithm: A Simple Derivation

A short description of the LA appropriate for matrices of different types is given in this appendix as a reference for readers who are unfamiliar with the algorithm. The derivation given here is slanted toward the generalizations needed for complex symmetric matrices and to simplify the discussions on the connections between the LA method of calculating spectral functions and related methods involving Padé approximants and continued fractions. A convenient starting point in the derivation of the LA is the introduction of a sequence of Krylov vectors generated by successive application of $\mathbf{A}$ on $\mathbf{v}$ (see Section V.A),

$$\mathbf{k}_m = \mathbf{A}^{m-1}\mathbf{v} = \mathbf{A}\mathbf{k}_{m-1}, \qquad m = 1, 2, \ldots. \tag{201}$$

The Lanczos basis is generated by the Gram–Schmidt orthonormalization [80, 87] of the sequence of Krylov vectors in the order of their occurrence.

To start the process, assume that the first member in the orthonormal set, $\mathbf{q}_1$, is parallel to the starting vector $\beta_0\mathbf{q}_1 = \mathbf{k}_1 = \mathbf{v}$. The requirement that the new set of vectors be normalized ($\mathbf{q}_i^{tr}\mathbf{q}_i = 1$) implies $\beta_0 = \|\mathbf{v}\|$.

The second member is given by

$$\beta_1\mathbf{q}_2 = \mathbf{k}_2 - \alpha_1\mathbf{k}_1 = (\mathbf{A} - \alpha_1\mathbf{I})\mathbf{q}_1. \tag{202}$$

The coefficient α_1 is determined by multiplying Eq. 202 on the left by $\mathbf{q}_1^{tr}$, whereas β_1 is chosen to normalize $\mathbf{q}_2$, giving

$$\alpha_1 = \mathbf{q}_1^{tr}\mathbf{A}\mathbf{q}_1, \tag{203}$$

$$\beta_1 = \|(\mathbf{A} - \alpha_1\mathbf{I})\mathbf{q}_1\|. \tag{204}$$

In a similar fashion, the third vector is

$$\beta_2\mathbf{q}_3 = (\mathbf{A} - \alpha_2\mathbf{I})\mathbf{q}_2 - \gamma_1\mathbf{q}_1. \tag{205}$$

Again, γ_1 and α_2 are determined from the orthogonality requirements

$$\gamma_1 = \mathbf{q}_1^{tr}\mathbf{A}\mathbf{q}_2 = \beta_1, \tag{206}$$

$$\alpha_2 = \mathbf{q}_2^{tr} \mathbf{A} \mathbf{q}_2. \tag{207}$$

Inserting these coefficients into Eq. 205 gives a specific case of the general three-term recursion relation that is the heart of the LA,

$$\beta_2 \mathbf{q}_3 = (\mathbf{A} - \alpha_2 \mathbf{I})\mathbf{q}_2 - \beta_1 \mathbf{q}_1. \tag{208}$$

Again, β_2 is chosen such that $\mathbf{q}_3$ is normalized.

The calculation of the fourth vector demonstrates the appearance of the desired recursion relation. In general, $\mathbf{q}_4$ can be expressed as

$$\beta_3 \mathbf{q}_4 = (\mathbf{A} - \alpha_3 \mathbf{I})\mathbf{q}_3 - \gamma_2 \mathbf{q}_2 - \delta_1 \mathbf{q}_1. \tag{209}$$

If Eq. 208 is indeed a prototype of a three-term recursion relation, $\delta_1 = 0$ and $\gamma_2 = \beta_2$. To verify $\delta_1 = 0$, premultiply Eq. 209 by $\mathbf{q}_1^{tr}$ and use the orthogonality property to see that

$$\delta_1 = \mathbf{q}_1^{tr} \mathbf{A} \mathbf{q}_3. \tag{210}$$

Using Eq. 202, this can be rewritten as

$$\delta_1 = (\beta_1 \mathbf{q}_2 + \alpha_1 \mathbf{q}_1)^{tr} \mathbf{q}_3, \tag{211}$$

which vanishes by the orthogonality of the $\mathbf{q}$'s. In addition, premultiplying Eqs. 208 and 209 by $\mathbf{q}_3^{tr}$ gives $\beta_2 = \gamma_2$.

The same behavior is observed for all further vectors. Therefore, the three-term recursion relation

$$\beta_m \mathbf{q}_{m+1} = (\mathbf{A} - \alpha_m \mathbf{I})\mathbf{q}_m - \beta_{m-1}\mathbf{q}_{m-1} \tag{212}$$

can be used to generate all successive $\mathbf{q}$'s so long as the β's are nonzero. The orthonormal basis of $\mathcal{K}_m(\mathbf{A}, \mathbf{v})$ generated in this manner is called the Lanczos basis, and the individual vectors are called Lanczos vectors.

It is informative to rewrite these relations in a matrix form. The transformation matrix $\mathbf{Q}_m$, whose jth column is given by the elements of $\mathbf{q}_j$ for $1 \leqslant j \leqslant m$, transforms the original $N \times N$ matrix $\mathbf{A}$ into an $m \times m$ real symmetric tridiagonal matrix $\mathbf{T}_m$,

$$\mathbf{T}_m = \mathbf{Q}_m^{tr} \mathbf{A} \mathbf{Q}_m. \tag{213}$$

The nonzero matrix elements of $\mathbf{T}_m$ on and above the diagonal are given by

$$(T_m)_{i,i} = \alpha_i, \tag{214}$$

$$(T_m)_{i,i+1} = \beta_i. \tag{215}$$

In addition, the orthonormalization process can be rewritten as a matrix equation relating $\mathbf{Q}_m$ to the matrix of Krylov vectors $\mathbf{K}_m$ and the moments (see Appendix B)

$$\mu_j = \mathbf{v}^{\mathrm{tr}} \mathbf{A}^{j-1} \mathbf{v} \tag{216}$$

using the relations

$$\beta_0 \mathbf{q}_1 = \mathbf{k}_1,$$

$$\beta_1 \mathbf{q}_2 = \mathbf{k}_2 - (\mathbf{k}_2^{\mathrm{tr}} \mathbf{k}_1) \mathbf{k}_1$$

$$= \mathbf{k}_2 - \mu_1 \mathbf{k}_1$$

$$\beta_2 \mathbf{q}_3 = \mathbf{k}_3 - (\mathbf{k}_3^{\mathrm{tr}} \mathbf{k}_2) \mathbf{k}_2 - (\mathbf{k}_3^{\mathrm{tr}} \mathbf{k}_1) \mathbf{k}_1$$

$$= \mathbf{k}_3 - \mu_3 \mathbf{k}_2 - \mu_2 \mathbf{k}_1,$$

and so forth. These relations follow directly from the definitions of the quantities involved. In matrix form these equations read

$$\mathbf{Q}_m \mathbf{B}_m = \mathbf{K}_m \mathbf{U}_m, \tag{217}$$

where $\mathbf{B} = \mathrm{diag}(\beta_0, \beta_1, \beta_2, \dots)$ and the matrix $\mathbf{U}_m$ is the upper triangular matrix

$$\mathbf{U}_m = \begin{bmatrix} 1 & -\mu_1 & -\mu_2 & -\mu_3 & -\mu_4 \cdots \\ 0 & 1 & -\mu_3 & -\mu_4 & -\mu_5 \cdots \\ 0 & 0 & 1 & -\mu_5 & -\mu_6 \cdots \\ 0 & 0 & 0 & 1 & -\mu_7 \cdots \\ 0 & 0 & 0 & 0 & 1 \\ \vdots & & \vdots & \vdots & \ddots \end{bmatrix}. \tag{218}$$

Since $\mathbf{U}_m$ is nonsingular (all its eigenvalues are equal to unity), Eq. 217 can be reexpressed as

$$\mathbf{K}_m = \mathbf{Q}_m(\mathbf{B}_m \mathbf{U}_m^{-1}) = \mathbf{Q}_m \mathbf{R}_m. \tag{219}$$

In this equation $\mathbf{U}_m^{-1}$ and $\mathbf{R}_m$ are upper triangular since $\mathbf{U}_m$ has this property. Therefore, Eq. 219 represents the factorization of the matrix of Krylov vectors into a product of a real matrix with orthonormal columns $\mathbf{Q}_m$ and an upper triangular matrix $\mathbf{R}_m = \mathbf{B}_m \mathbf{U}_m^{-1}$. Furthermore, if $\mathbf{B}_m$ is nonsingular, so is $\mathbf{R}_m$.

Matrix factorizations such as that given by Eq. 219 are of great importance in modern numerical linear algebra. The existence of factorizations of this type for arbitrary real or complex matrices is guaranteed by the following theorem [87].

Theorem 1 (QR Factorization)   *For any given* $N \times M$ *complex matrix* **A** *(* $N \geqslant M$ *), there exists a matrix* **Q** *of the same dimensions with orthonormal columns (* $Q^{\dagger}Q = I_N$ *) and an upper triangular* $M \times M$ *matrix* **R** *such that* $A = QR$. *If* **A** *is real,* **Q** *and* **R** *can also be chosen to be real.*

The Gram–Schmidt orthonormalization procedure can always be used to accomplish this factorization, though there are also many other ways [80].

The extension of the LA to handle Hermitian matrices can be derived in a straightforward manner by replacing the transposition operations with Hermitian conjugation in the previous equations and taking advantage of the symmetry of the usual inner product in a unitary space (see Appendix C). The desired decomposition of the matrix of Krylov vectors is ensured by the QR factorization theorem above. The result of these operations gives the Lanczos recursion relation for a general Hermitian matrix **A**:

$$\beta_m \mathbf{q}_{m+1} = (\mathbf{A} - \alpha_m \mathbf{I})\mathbf{q}_m - \beta_{m-1}\mathbf{q}_{m-1}, \tag{220}$$

where $\alpha_i = \mathbf{q}_i^{\dagger}\mathbf{A}\mathbf{q}_i$ and β_i is again chosen to normalize $\mathbf{q}_{i+1}$. The reduction of **A** to a tridiagonal matrix is given by

$$\mathbf{T}_m = \mathbf{Q}_m^{\dagger}\mathbf{A}\mathbf{Q}_m, \tag{221}$$

where $\mathbf{Q}_m$ satisfies $\mathbf{Q}_m^{\dagger}\mathbf{Q}_m = \mathbf{I}_m$. It is easy to see from the Eq. 220 that an arbitrary Hermitian matrix will still be reduced to a real symmetric tridiagonal form by the general Hermitian LA. It is important to note that the form of the recursion relation is the same as before. The only way the complex nature of the matrix enters is in the definition of the scalar product.

The extension of the LA to handle non-Hermitian, complex symmetric matrices is not so simple. One way to summarize these difficulties is to say that although the QR factorization of the matrix of Krylov vectors generated is natural in the Hermitian case, it is fundamentally inappropriate for the Krylov matrices generated by a complex symmetric matrix. This fact is emphasized by examining Eq. 221 for a case where the Hermitian matrix **A** has no degenerate eigenvalues such that the matrix can be completely reduced to a tridiagonal form [89, 80] of the same dimension by the LA. In this instance, the matrix **Q** becomes unitary and Eq. 221 is a statement that the Hermitian matrix **A** is unitarily similar to Hermitian matrix **T**. However, if two complex symmetric matrices are similar, they must be similar via a complex orthogonal transformation [86, 87]. Thus, a generalization of the LA along different lines must be sought. It turns out that the required generalization amounts to a redefinition of the inner product.

To develop a complex symmetric LA, a different factorization of the matrix of Krylov vectors must be adopted. This amounts to developing an analog of the Gram–Schmidt orthonormalization procedure, which will naturally lead to complex orthogonal matrices rather than unitary matrices. Choudhury and Horn have studied this problem in detail and have shown when such a factorization is possible [90]. In contrast to the Hermitian case, it turns out that there are instances where this generalized Gram–Schmidt procedure fails even where a related factorization of the Krylov matrix exists. In the present context it is the generalization of the Gram–Schmidt procedure that is the more important. The statement of the relevant definitions and theorem and further discussion can be found in Appendix C. Using the result of Theorem 2 in Appendix C, the Gram–Schmidt-like orthogonalization of the Krylov vectors generated by a complex symmetric matrix $\mathbf{A}$ and an arbitrary normalized starting vector depends on the determinants

$$\det(\mathbf{K}_m^{tr}\mathbf{K}_m) \tag{222}$$

being different from zero for all m in question. One of these determinants being zero implies that the associated Lanczos vector is nonzero but has zero norm, and the matrix $\mathbf{Q}_m$ of Lanczos vectors cannot satisfy $\mathbf{Q}_m^{tr}\mathbf{Q}_m = \mathbf{I}_m$. If none of these determinants are nonzero, the complex symmetric LA behaves in the same fashion as the Hermitian LA [80, 89].

Since the columns of $\mathbf{K}_m$ are the Krylov vectors, the product $\mathbf{K}_m^{tr}\mathbf{K}_m$ can be rewritten in terms of the moments. Using the definition of the Krylov vectors (see Eq. 201), it is easy to show that

$$\Delta_m = \det(\mathbf{K}_m^{tr}\mathbf{K}_m) = \det\begin{bmatrix} \mu_0 & \mu_1 & \mu_2 & \mu_3 & \cdots & \mu_m \\ \mu_1 & \mu_2 & \mu_3 & \mu_4 & \cdots & \mu_{m+1} \\ \mu_2 & \mu_3 & \mu_4 & \mu_5 & \cdots & \mu_{m+2} \\ \mu_3 & \mu_4 & \mu_5 & \mu_6 & \cdots & \\ \vdots & \vdots & \vdots & \vdots & \cdots & \vdots \\ \mu_m & \mu_{m+1} & \mu_{m+2} & \cdots & \cdots & \mu_{2m} \end{bmatrix}. \tag{223}$$

In Appendix B determinants of this type will also arise in discussing continued fraction and Padé approximants to the spectrum.

The requirement that the determinants in Eq. 223 be nonzero seems to be the only constructive criterion for the existence of a well-behaved complex symmetric LA in exact arithmetic. A derivation of a similar criterion for Hermitian matrices is given by Householder [162]. The result given here is actually a special case of Householder's analysis of the biorthogonal LA. Other criteria can be found for the existence of a well-behaved complex symmetric LA, such as the requirement that the minimal polynomial of the matrix with respect to the starting vector be a product of distinct linear factors [1]. This type of criterion is not constructive in the sense that if one already knew the minimal polynomial, the problem would already be solved!

B. Continued Fractions, Padé Approximants, and the Lanczos Algorithm

The essence of the Lanczos method of calculating spectral functions is to use the sequence of symmetric tridiagonal matrices generated by the LA to define a sequence of rational, or continued-fraction, approximants to the desired spectral function. The purpose of this appendix is to develop the theory of these approximants and to demonstrate the close connection of this method to related methods involving the calculation of continued fractions and Padé approximants directly from the moment expansion. The focus of this discussion will be on cases where the time evolution of the system is governed by a non-Hermitian complex symmetric matrix $\mathbf{A}_N$, since results similar to those developed here are well known for the Hermitian case.

The generalized spectral density

$$J^{(N)}(z) = \mathbf{u}^\dagger [z\mathbf{I}_N + \mathbf{A}_N]^{-1}\mathbf{v} \tag{224}$$

will serve as a reference point in this discussion. It is assumed that the projections P_N defining the truncated matrix $\mathbf{A}_N = P_N A P_N$ have been chosen such that $J^{(N)}(z)$ is a very good approximation to the "true" spectral function and that the components of $\mathbf{u}$ are the complex conjugates to the components of $\mathbf{v}$ in the orthonormal basis set chosen in which $\mathbf{A}_N$ is complex and symmetric (see Sections VI.B.1 and VI.D) [3]. With these assumptions, $J^{(N)}(z)$ can be expressed as

$$J^{(N)}(z) = \mathbf{v}^{tr}[z\mathbf{I}_N + \mathbf{A}_N]^{-1}\mathbf{v}. \tag{225}$$

1. Continued Fraction Approximants Derived from the Lanczos Algorithm

The rational approximants $J_n^{(N)}(z)$ to $J^{(N)}(z)$ are defined as

$$J_n^{(N)}(z) = (\mathbf{v}^{tr}\mathbf{Q}_n)\mathbf{Q}_n^{tr}[z\mathbf{I}_N + \mathbf{A}_N]^{-1}\mathbf{Q}_n(\mathbf{Q}_n^{tr}\mathbf{v}) \tag{226}$$

$$= (\mathbf{Q}_n^{tr}\mathbf{v})^{tr}[z\mathbf{I}_n + \mathbf{T}_n]^{-1}(\mathbf{Q}_n^{tr}\mathbf{v}). \tag{227}$$

Since the matrix Q_n satisfies $Q_n^{tr} Q_n = I_n$ and the elements in the first column of Q_n are just the components of v, the nth approximant to $J^{(N)}(z)$ is just the $(1, 1)$ matrix element of the inverse of $[zI_n + T_n]$ (see Sections II and II.A):

$$J_n^{(N)}(z) = [zI_n + T_n]_{1,1}^{-1}. \tag{228}$$

A similar result can be derived for the Hermitian case if the complex orthogonal matrix Q_n is replaced with the appropriate unitary matrix

To systematically develop these ideas in a manner that will simplify the discussion of the relationship between the rational approximants, continued fractions, and Padé approximants to the spectral function it is useful to introduce the determinants $D_{l,m}^n$ of the diagonal blocks of $[zI_n + T_n]$,

$$D_{l,m}^n(z) = \det \begin{bmatrix} z + \alpha_l & \beta_l & & & \\ \beta_l & z + \alpha_{l+1} & \beta_{l+1} & & \\ & \beta_{l+1} & z + \alpha_{l+2} & \ddots & \\ & & \ddots & \ddots & \beta_{m-1} \\ & & & \beta_{m-1} & z + \alpha_m \end{bmatrix}, \tag{229}$$

where it is assumed that $m \geq l$.

In terms of these determinants, the application of Cramer's rule to Eq. 228 gives

$$J_n^{(N)}(z) = \frac{D_{2,n}^n(z)}{D_{1,n}^n(z)} \quad \text{for} \quad n \geq 2. \tag{230}$$

It is clear from Eq. 230 that $J_n^{(N)}(z)$ is indeed a rational function since $D_{1,n}^n(z)$ and $D_{2,n}^n$ are polynomials of order at most n and $n - 1$, respectively, in the complex variable z. A recurrence relation for the determinants $D_{1,n+1}^{n+1}(z)$ and $D_{2,n+1}^{n+1}(z)$ can easily be derived by expanding the determinants in question using Laplace's method along the last row or column.

The sequence of rational approximants can be converted to a continued fraction by expanding the determinants in a slightly different fashion. The expansion of $D_{1,n}^n(z)$ about its first row or column gives the result

$$D_{1,n}^n(z) = (z + \alpha_1)D_{2,n}^n(z) - \beta_1^2 D_{3,n}^n(z). \tag{231}$$

Inserting this expansion into the denominator of Eq. 230 and dividing both numerator and denominator by $D_{2,n}^n(z)$ gives

$$J_n^{(N)}(z) = \frac{D_{2,n}^n(z)}{(z + \alpha_1)D_{2,n}^n - \beta_1^2 D_{3,n}^n(z)} \tag{232}$$

$$= \left(z + \alpha_1 - \beta_1^2 \frac{D_{3,n}^n(z)}{D_{2,n}^n(z)} \right)^{-1}. \tag{233}$$

This process can be repeated for the ratio of determinants occurring in the denominator to yield

$$J_n^{(N)}(z) = \left(z + \alpha_1 - \cfrac{\beta_1^2}{z + \alpha_2 - \beta_2^2 \cfrac{D_{4,n}^n(z)}{D_{3,n}^n(z)}} \right)^{-1}. \tag{234}$$

The recursive nature of the expansion of the rational approximant into a continued fraction is now obvious. Using the standard notation for continued fractions, the complete expansion of $J_n^{(N)}(z)$ can be written as

$$J_n^{(N)}(z) = \frac{1}{z + \alpha_1} - \frac{\beta_1^2}{z + \alpha_2} - \frac{\beta_2^2}{z + \alpha_3} - \cdots \frac{\beta_{n-2}^2}{z + \alpha_{n-1}} - \frac{\beta_{n-1}^2}{z + \alpha_n}. \tag{235}$$

A few definitions are in order at this time to establish the terminology and notation for continued fractions used here.

Definition 1 (Elements of Continued Fraction)   *The complex numbers a_m and b_m are the called the elements of the (infinite) continued fraction*

$$\frac{a_1}{b_1} + \frac{a_2}{b_2} + \frac{a_3}{b_3} + \cdots. \tag{236}$$

Definition 2 (Partial Numerators and Denominators)   *The complex numbers a_m and b_m are the mth partial numerator and partial denominator, respectively, of the continued fraction Eq. 236.*

Definition 3 (Convergents)   *The continued fraction*

$$C_m = \frac{a_1}{b_1} + \frac{a_2}{b_2} + \frac{a_3}{b_3} + \cdots \frac{a_m}{b_m} \tag{237}$$

is called the mth convergent of the continued fraction Eq. 236. The mth convergent can be collapsed into a single fraction by simple arithmetic. The numerator A_m and denominator B_m of this fraction are simply called the mth numerator and denominator of Eq. 236. In terms of these quantities, the mth

convergent is just

$$C_m = \frac{A_m}{B_m}.$$ (238)

The approximants $J_n^{(N)}(z)$ are the convergents of the continued fraction Eq. 235.

Some of the most basic and useful results of the theory of continued fractions are the fundamental three-term recurrence relations for the numerators and denominators:

$$A_{m+1} = b_{m+1}A_m + a_{m+1}A_{m-1},$$ (239)

$$B_{m+1} = b_{m+1}B_m + a_{m+1}B_{m-1},$$ (240)

for $m = 0, 1, 2, \ldots$. The value of any convergent is easily calculated using these recurrence relations with the initial conditions

$$A_{-1} = 1, \qquad A_0 = 0,$$ (241)

$$B_{-1} = 0, \qquad B_0 = 1.$$ (242)

These recurrence relations are numerically unstable in most cases but are quite useful for analytical studies.

Definition 4   (Convergence of Continued Fraction)   *A continued fraction such as Eq. 236 is said to converge if the limit of its sequence of convergents*

$$\lim_{m \to \infty} \frac{A_m}{B_m} = \lim_{m \to \infty} C_m$$ (243)

converges to a finite number C and only a finite number of the denominators vanish. The number C is called the value of the continued fraction. A continued fraction that does not satisfy these requirements is said to diverge.

The other basic results from the general theory of continued fractions needed here are those concerned with equivalent continued fractions and equivalence transformations.

Definition 5   (Equivalent Continued Fractions)   *The continued fraction with elements a_m and b_m and convergents C_m is said to be equivalent to another continued fraction with elements a'_m and b'_m and convergents C'_m if $C_m = C'_m$ for all m.*

Definition 6 (Equivalence Transformations) *An equivalence transformation is a mapping of the elements of one continued fraction into an equivalent continued fraction defined by a sequence of nonzero complex constants r_m:*

$$a'_m = r_m r_{m-1} a_m, \qquad m = 0, 1, 2, \ldots, \tag{244}$$

$$b'_m = r_m b_m, \qquad m = 1, 2, 3, \ldots, \tag{245}$$

where $r_0 = 1$.

It is not hard to show that two continued fractions are equivalent if and only if they are related by an equivalence transformation [163].

Continued fractions of the form given in Eq. 235 have been extensively studied by Wall and co-workers and are known as J-fractions [163–167]. The advantage of identifying the continued fractions derived from the tridiagonal matrices generated by the LA as J-fractions is that the general theory of J-fractions can then be used to treat approximants to the spectral functions in both the Hermitian and complex symmetric cases in a uniform fashion.

Of special interest in physical applications is the class of positive definite J-fractions. The class of positive definite J-fractions is characterized by the requirement that the real-valued quadratic form derived from the imaginary parts of the elements of the continued fraction is nonnegative for all n [164].

Definition 7 (Positive Definite J-Fractions) *A continued fraction of the form*

$$\frac{1}{b_1 + \zeta} - \frac{a_1^2}{b_2 + \zeta} - \frac{a_2^2}{b_3 + \zeta} - \cdots \tag{246}$$

is called positive definite J-fraction if, for all M and all real values of y_1, y_2, y_3, \ldots,

$$\sum_{k=1}^{M} \delta_k y_k^2 - 2 \sum_{k=1}^{M-1} \gamma_k y_k y_{k+1} \geqslant 0, \tag{247}$$

where $\delta_k = \mathrm{Im}(b_k)$ and $\gamma_m = \mathrm{Im}(a_k)$.

To simplify the connections to the mathematical literature, it is useful to perform an equivalence transformation on the continued-fraction expansion of the spectral function derived from the Lanczos tridiagonal matrices of the form given in Eq. 235 to yield the alternative form

$$J_n^{(N)}(z) = -iJ_n^{(N)}(\zeta) = \cfrac{1}{b_1 + \zeta -} \cfrac{a_1^2}{b_2 + \zeta -} \cfrac{a_2^2}{b_3 + \zeta -} \cfrac{a_3^2}{b_4 + \zeta -} \cdots \cfrac{a_n^2}{b_n + \zeta},$$

(248)

where $b_m = i\alpha_m$, $a_m = i\beta_m$, and $\zeta = -iz$. The continued fraction on the right side of Eq. 248 is a J-fraction as defined by Wall [164].

The numerator and denominator of a positive definite J-fraction can be found by inserting the elements of the continued fraction Eq. 246 into Eqs. 239 and 240 and the initial conditions in Eqs. 241 and 242:

$$A_n(\zeta) = (b_n + \zeta)A_{n-1} - a_{n-1}^2 A_{n-2}(\zeta),$$

(249)

$$B_n(\zeta) = (b_n + \zeta)B_{n-1} - a_{n-1}^2 B_{n-2}(\zeta),$$

(250)

where a_0 is defined to be unity.

To study the convergence and analytical behavior of infinite J-fractions, it is convenient to introduce the two series

$$X_{n+1}(\zeta) = \frac{A_n(\zeta)}{\displaystyle\prod_{j=1}^{n} a_j},$$

$$Y_{n+1}(\zeta) = \frac{B_n(\zeta)}{\displaystyle\prod_{j=1}^{n} a_j},$$

and the infinite series

$$\sum_{j=1}^{\infty} |X_p(0)|^2,$$

(251)

$$\sum_{j=1}^{\infty} |Y_p(0)|^2.$$

(252)

The polynomials $X_n(\zeta)$ and $Y_n(\zeta)$ each satisfy a three-term recursion relation. If either or both series in Eq. 251 or 252 diverges it is said that the determinate case holds for the continued fraction Eq. 246; otherwise, the indeterminate case holds. It appears that only the indeterminate case is important for the study of spectral functions.

The interest in the class of positive definite J-fractions for which the determinate case holds arises from the following observations [163, 164]:

- The denominators of such continued fractions are nonzero for all complex values of ζ in the open upper half-plane $\text{Im}(\zeta) > 0$.
- They converge uniformly over every closed bounded region in the upper half-plane, and their values are analytic functions of ζ there.
- Explicit truncation error bounds and inclusion regions for the value of an infinite continued fraction can be derived in terms of the elements of the truncated continued fraction.
- The values of the convergents of a continued fraction of this class satisfy $\text{Im}\{C_m(\zeta)\} < 0$ for all ζ such that $\text{Im}\{\zeta\} > 0$.

A comparison of these properties with the requirements that must be satisfied for a spectral function to be physically meaningful [51, 108, 109, 168] shows that finite positive definite J-fractions possess the mathematical features necessary to properly approximate a spectral function. The physical interpretation of the analytic behavior of $J^{(N)}(z)$ in the upper half-plane is that adding a real constant matrix to A_N is equivalent to increasing all the relaxation rates by that same amount, which should not alter the basic mathematical structure of the approximation. A related idea has been used by Dammers [49] as a test of the numerical stability of various methods of calculating magnetic resonance spectra (see what follows).

From the point of view of approximating the true spectral function by the sequence of continued-fraction approximants generated by the LA, the only difference is that the Hermitian LA gives rise to continued-fraction approximants with purely real elements, whereas the complex symmetric LA gives rise to complex elements. The previous discussion on positive definite J-fraction approximants to spectral functions applies equally well to the complex symmetric and Hermitian LA.

2. Spectral Functions as Moment Problems

One systematic way to develop an approximation to $J(z)$ is to use a binomial expansion of Eq. 225:

$$J^{(N)}(z) = \frac{1}{z}\mathbf{v}^{tr}\left[\mathbf{I}_N - \frac{\mathbf{A}_N}{z} + \left(\frac{\mathbf{A}_N}{z}\right)^2 - \left(\frac{\mathbf{A}_N}{z}\right)^3 + \cdots\right]\mathbf{v} \tag{253}$$

$$= \frac{1}{z}\left\{\mathbf{v}^{tr}\mathbf{v} - \mathbf{v}^{tr}\frac{\mathbf{A}_N}{z}\mathbf{v} + \mathbf{v}^{tr}\left(\frac{\mathbf{A}_N}{z}\right)^2\mathbf{v} - \mathbf{v}^{tr}\left(\frac{\mathbf{A}_N}{z}\right)^3\mathbf{v} + \cdots\right\} \tag{254}$$

$$= \frac{1}{z}\sum_{n=0}^{\infty}\left(\frac{-1}{z}\right)^n \mu_{n+1}, \tag{255}$$

where quantities $\mu_{n+1} = \mathbf{v}^{tr} \mathbf{A}_N^n \mathbf{v}$ are known as the moments of $\mathbf{A}$ with respect to $\mathbf{v}$, (cf. Eq. 216). The expansion in Eq. 255 is commonly referred to as the moment expansion. Numerical results using this expansion directly are quite disappointing: The convergence is slow, and the truncated expansion does not have the proper analytic behavior as a function of z.

The generalized, or relaxation, moment expansion Eq. 255 can be used to calculate a continued-fraction approximant with better convergence properties. Several groups have addressed these issues [48, 169, 170] from the point of view of calculating magnetic resonance spectra by developing formulas for the moments μ_n, for $n = 0, 1, 2, \ldots, N$, and using the moments to calculate the elements of a continued fraction of the type of Eq. 248. This seems to be a useful approach for simple problems where tractable formulas for the moments are available. The formulas needed to calculate the moments given by Giordano et al. [48] are extremely complicated even for the simple cases (e.g., where the g and A tensors of the spin probe are axially symmetric in the principal axis frame of the diffusion tensor and nonsecular terms have been omitted from the spin Hamiltonian). Nevertheless, this is useful for such simple problems, especially if analytical formulas are needed. Very often nonaxially symmetric magnetic tensors are required to quantitatively fit experimental data [12]. Experiments have also shown that much more information is available if the spectra of oriented radicals in liquid crystalline phases are studied as a function of the tilt angle between the director and the static field [20, 21]. The lack of generality, together with the fact that extended precision computer arithmetic is often needed to stably compute the elements of the continued fraction [48, 169], suggest that the Lanczos method is more appropriate than the relaxation moment method for the routine analysis of complicated experimental spectra.

An interesting connection exists between the J-fraction approximants of the moment expansion (Eq. 255), the existence of a set of polynomials orthogonal relative to the sequence of moments [164, 171], and the determinants Δ_m (Eq. 223). First, it is necessary to define what is meant by the orthogonality of a set of complex polynomials relative to a complex sequence.

Definition 8 (System of Polynomials Orthogonal Relative to a Sequence)
A sequence of complex polynomials $Q_n(x)$ in the complex variable x,

$$Q_n(x) = \sum_{m=0}^{n} q_m^{(n)} x^m, \qquad n = 0, 1, 2, \ldots, \tag{256}$$

is said to be orthogonal relative to a sequence of complex constants

$\{c_0, c_1, c_2, \ldots\}$ *if*

$$\sum_{i=0}^{m} \sum_{j=0}^{n} q_i^{(m)} q_j^{(n)} c_{i+j} \tag{257}$$

is nonzero if and only if $m = n$.

A necessary and sufficient condition for a set of $M + 1$ such polynomials to exist that are orthogonal relative to the sequence of moments (Eq. 255) is that the determinants Δ_m be nonzero for all $m < M - 1$ [164]. Moreover, it can be shown that these polynomials are identical to the denominator polynomials given by Eq. 250 of the continued fraction given in Eq. 248. Thus, it is evident that the relaxation moment method is equivalent to the Lanczos method.

An alternative approach involving continued fractions is to seek a classical moment expansion of the absorption spectrum or spectral density. There is a vast mathematical literature on classical moment problems and their relationship to the theory of general orthogonal polynomials, continued fractions, and Padé approximants [163, 164, 172, 173]. This viewpoint differs significantly from the LA or relaxation moment approaches, where the continued-fraction expansion of a generalized spectral function is studied. The difference between these approaches lies in the fact that the absorption spectrum or spectral density can be treated as a *real* nonnegative function over the entire range of frequencies, whereas the related generalized spectral function is inherently complex valued.

In the classical moment method, the quantities of interest are the moments of the frequency with respect to the absorption spectrum or spectral density, $I(\omega)$,

$$\langle \omega^n \rangle = \int_{-\infty}^{\infty} \omega^n I(\omega) \, d\omega, \tag{258}$$

and not the moments of the operator $\mathbf{A}_N$, as in Eq. 255.

Gordon [174] has studied the response of a spin system to a pulsed excitation as a classical Hamburger moment problem [172, 173] and has derived truncation error bounds and inclusion regions similar to those alluded to earlier in connection with positive definite J-fractions. This is not surprising given that the solution to the Hamburger moment problem can be expressed as a real, positive definite J-fraction [164]. Lado, Memory, and Parker have used similar techniques in conjunction with memory function ideas [175]. Lonke has also derived related results for the spectral density of excitations in a many-body system of fermions [176]. A nice review of the application of the theory of continued fractions to physical problems involving relaxation is given by Grosso and Pastori Parravicini [170].

3. Padé Approximants to Spectral Functions

Padé approximants [164, 178, 179] have also been used to calculate slow-motional ESR spectra by Dammers, Levine, and Tjon [46, 49]. These calculations involve the determination of the coefficients of a certain sequence of rational approximants to the moment expansion of the spectral function (see Eq. 255). A Padé approximant is a rational approximant that satisfies the restrictions stated in the following definition.

Definition 9 (Padé Approximants) *Let* $A(z) = a_0 + a_1 z + a_2 z^2 + a_3 z^3 + \cdots$
be a given formal power series and $P_L(z)$ *and* $Q_M(z)$ *be polynomials of degree at most* L *and* M, *respectively, where* $Q_M(0) = 1$. *Let*

$$C(L/M) = \det \begin{bmatrix} a_{L-M+1} & a_{L-M+2} & a_{L-M+3} & \cdots & a_L \\ a_{L-M+2} & a_{L-M+3} & a_{L-M+4} & \cdots & a_{L+1} \\ a_{L-M+3} & a_{L-M+4} & a_{L-M+5} & \cdots & a_{L+2} \\ \vdots & \vdots & \vdots & \ddots & \vdots \\ a_L & a_{L+1} & a_{L+2} & \cdots & a_{L+M-1} \end{bmatrix} \neq 0. \qquad (259)$$

If $A(z) - P_L(z)/Q_M(z) = \mathcal{O}(z^{L+M+1})$, *then* $[L/M]_A(z) = P_L(z)/Q_M(z)$ *is called the* L/M *Padé approximant to* $A(z)$.

Since the Padé approximants to the moment expansion might exist for arbitrary L and M, it is necessary to determine which of these approximants have the proper analytical structure as a function of the complex variable z to represent physically reasonable spectral functions. The conclusions given by Dammers from these analyses, together with the examination of model calculations and the connections of this method to Mori's projection operator method, was that the $[M - 1/M]$ Padé approximant to $zJ^{(N)}(z)$ is the proper choice. (See also [3]).

In light of the selection of the class of $[M - 1/M]$ Padé approximants as the preferred set, it is interesting to examine the connections between this class of rational approximants and the continued-fraction approximants from the LA or relaxation moment methods. In particular, the determinant $C(M - 1/M)$ occurring in the definition of the $[M - 1/M]$ Padé approximant (Eq. 259) is identical to the determinant Δ_M (Eq. 223) occurring in the analyses of the LA and the relaxation moment method. Thus, the existence of the class of $[M - 1/M]$ approximants is equivalent to saying that the related LA is well behaved or that the elements of the continued fraction can be calculated from the moments μ_i.

The other way in which Padé approximants have been used in the calculation of magnetic resonance spectra is to accelerate the convergence of the sequence of estimates of the amplitude of a spectrum given by the CG algorithm at a particular field position [4]. Because of the equivalence of the results of the LA and CG algorithm, it is possible to identify the sequence of estimates given by the CG method with the values of the successive convergents of the continued fraction given by the LA evaluated at the same point. Given the set $\{C_m\}$ of CG estimates, the series

$$C_0 + \sum_m (C_{m+1} - C_m) x^m \tag{260}$$

can be constructed. Inserting the eigenvector–eigenvalue decomposition of the matrix $\mathbf{A}_N$ in Eq. 255 for the moment expansion shows that the moment expansion has a form that suggests that a Padé or generalized Shanks resummation technique might accelerate its convergence [130]. It also follows from the preceding discussion that the sequence of values of the convergents of continued-fraction expansion and the series Eq. 260 also have this property. The literature on Padé or generalized Shanks resummation techniques and related topics such as the epsilon algorithm is far too voluminous and technical to be reviewed here, especially since excellent treatments are available elsewhere [130, 179–182].

The general idea of using generalized Shanks transformations to accelerate the convergence of series is equivalent to evaluating a sequence of Padé approximants to the series Eq. 260 at $x = 1$. In practice, it is not necessary to explicitly construct the Padé approximants and evaluate them at $x = 1$. Instead of this cumbersome procedure, the values of the Padé approximants can be calculated very efficiently directly from the coefficients of the series using the scalar epsilon algorithm [180–182].

It is also of some interest that the LA and CG algorithm are closely related to a vector version of the epsilon algorithm [178]. The vector epsilon algorithm can even be used to diagonalize matrices [183]! Matrix-valued continued fractions have also been applied to the solution of Fokker–Planck operators [119] and boundary problems [184].

4. *Summary*

The mathematical thread that ties these various ideas together is the determinants that arise throughout in the discussions about the existence of the various approximations. The determinants Δ_m play a crucial role in:

- The existence of a well-behaved complex symmetric LA (Appendix A, Eqs. 222 and 223);
- The definition of the denominators of the rational approximants and

continued fractions derived from the Lanczos tridiagonal matrix (Eqs. 248 and 250);

- The existence of a J-fraction solution of the generalized or relaxation moment problem (Eq. 257); and
- The definition of the relevant class of Padé approximants to $zJ^{(N)}(z)$ (Eq. 259).

A study of the implications of this connection between these different approaches shows that the LA, relaxation moment problem, and Padé approximant methods are all essentially equivalent from an analytic point of view. The difference between these methods lies in the stability and efficiency of the numerical algorithms required to implement them on a computer.

In his thesis, Dammers [49] compared the Padé approximant [46, 49], relaxation moment [47, 48], and Lanczos methods [3] with respect to the invariance of the calculated spectrum under origin shifts. It is obvious that

$$J(z) = \mathbf{v}[(z - z_0)\mathbf{I} + (\mathbf{A} + z_0\mathbf{I})]^{-1}\mathbf{v}$$

is independent of the value chosen for z_0. However, the exact analytical invariance of the preceding expression is not necessarily reflected by the numerical procedures under comparison. In his study, only the results of the LA were found to be invariant with respect to origin shifts of this type. The other two methods show substantial deviations in the computed value of $J(z)$ as a function of z_0.

Dammers analyzed this phenomenon in terms of the problems associated with the iterative generation of the sequence of moments in the Padé and relaxation moment methods. The LA does not suffer from such problems since the Lanczos projections automatically cancel the effect of the origin shift to the machine precision at every step. These observations are consistent with the well-known problems associated with the numerical stability of calculating high-order Padé approximants [181] and the need for extended precision arithmetic when performing very slow-motional calculations with the relaxation moment method [47, 48]. The use of extended precision arithmetic has never been necessary in any of our calculations with the LA, even for two-dimensional ESE spectra [4, 30, 88] and extremely complicated slow-motional CWESR studies [20, 21]. These studies indicate that the LA is the method of choice for calculating slow-motional ESR spectra.

Whitehead and Watt [185, 186] have also examined this matter from an analytical view point. They found that the rearrangement to the calculation of the elements of the continued fraction from the moments in a manner that eliminates terms that would cancel in exact arithmetic leads to the LA in a natural way. These ideas are also related to the cumulant and moment expansion methods that have previously been applied in analytical studies of magnetic resonance and relaxation [10, 11, 175, 187–189].

The penultimate point of this appendix has been to summarize the various mathematical methods that have been developed to calculate spectral functions. In his pioneering work, Mori [107] presented a general and powerful statistical mechanical methodology to develop the continued fraction form of spectral functions. These different mathematical methods have been shown (e.g. in this appendix) to be analytically equivalent to one another and ultimately are equivalent to the Mori method as well [43, 44, 49, 111, 112]. However, these different mathematical realizations give rise to numerical algorithms which differ significantly in their numerical reliability and stability.

C. Bilinearly Metric Spaces and Relaxation Phenomena

In a very general manner, Moro and Freed [3] showed that calculations of magnetic resonance spectra and related Fokker–Planck forms can be simplified by taking advantage of the fact that the matrix of the operator generating the time evolution of the system can always be put into a complex symmetric form. If A is the non-Hermitian complex symmetric matrix generating the time evolution, and $A^\dagger$ is its complex-conjugate transpose, the following holds:

- The eigenvectors of A are not orthogonal with respect to the usual unitary space inner product.
- The eigenvectors of A and $A^\dagger$ are not identical; instead, they are related by complex conjugation.
- The time evolution of the system is not unitary but has a different structure related to a complex orthogonal transformation.
- The transformation (if it exists) that diagonalizes the matrix that generates the time evolution of the system is also complex orthogonal, not unitary.

It is obvious that there are many aspects of the structure of complex vector spaces equipped with the usual unitary space norm and scalar product that make theoretical and numerical studies quite cumbersome for this class of problems. Two separate approaches have been developed to cope with these difficulties. One approach is to retain the usual unitary space inner-product structure and introduce two basis sets [111, 112]. The basis vectors within each set are not orthogonal to one other but are orthogonal to all but one member of the other set. This type of basis is called biorthogonal. The second approach is to simply redefine the norm and scalar product and retain a single basis set. Though it is not apparent, these two approaches are actually identical for complex symmetric matrices [4, 44]. It is in the spirit of the second approach that this section is directed toward the systematic development of the generalization of the concepts of the norm and scalar product in a finite dimensional complex vector space in which complex symmetric and complex orthogonal matrices play the same respective roles as Hermitian and unitary matrices play in the usual scalar product space. In the latter part of the section some of the peculiar aspects of geometry in this new space will be discussed. All of this material is well known to mathematicians but has not been presented in this fashion to the chemical physics community.

The approach used here is that based on the work of Heuvers [190] on the types of functions that can serve as scalar products in complex vector spaces. In this manner any explicit definition of dual spaces and related matters can be avoided. As emphasized by Znojil [191], it is unnecessary to define a scalar product of any type in order to use the LA. The Lanczos recursion relation given by Eq. 212 can be simply viewed as a linear three-term recursion relation that is well defined by the algebraic properties of the complex vector space in which it is defined. The middle-of-the-road view taken here is valuable in that some notion of geometry can be defined through the introduction of a new type of scalar product that is more natural for the problems at hand. This new geometry is naturally associated with the known mathematical structure of the class of problems under consideration (see Section VI.B).

The concept of linear independence of a set of complex N vectors is strictly a property of the algebraic structure associated with the complex vector space and thus is unrelated to the definitions of orthogonality or normalization of vectors.

Definition 10 (Linear Independence) *A set of vectors* y_i, $y_i \in V \subseteq \mathscr{C}^N$, $i = 1, 2, \ldots, M$, *is called linearly dependent if there exists a corresponding set of constants* $\alpha_i \in \mathscr{C}$, $i = 1, \ldots, M$, *not all of which are zero, such that*

$$\alpha_1 x_1 + \alpha_2 x_2 + \cdots + \alpha_M x_M = 0. \tag{261}$$

A set of vectors that is not linearly dependent is called linearly independent. The linear independence of a set of vectors implies that Eq. 261 can only be satisfied if all the constants are simultaneously equal to zero.

Definition 11 (Bases and Dimension) *A finite set of vectors* $f_i \in V$, $i = 1, \ldots, M$, *is called a basis of* V *if the set* f_i *is linearly independent and if for any* $x \in V$ *there exist constants* $x_i \in \mathscr{C}$, $i = 1, 2, \ldots, M$, *such that*

$$x = \sum_{i=1}^{M} x_i f_i. \tag{262}$$

It can be shown that the maximum number of linearly independent vectors in **V** *is equal to M. This number is independent of the choice of basis and is called the dimension of* **V**.

For reference, it is useful to introduce the concept of bilinear forms as a stepping stone to the usual definitions of norm and inner product. This will enable us to formulate the required generalization of the norm and scalar product of a vector for the class of problems at hand.

Definition 12 (Primitive Bilinear Form) *A complex scalar-valued function* $\rho(\mathbf{x}, \mathbf{y})$ *of two independent vector variables* $\mathbf{x}, \mathbf{y} \in \mathbf{V}$ *is called a primitive bilinear form on* **V** × **V** *if, for all vectors* $\mathbf{z} \in \mathbf{V}$ *and scalars* $\alpha, \beta \in \mathscr{C}$,

$$\rho(\mathbf{x} + \mathbf{z}, \mathbf{y}) = \rho(\mathbf{x}, \mathbf{y}) + \rho(\mathbf{z}, \mathbf{y}), \tag{263}$$

$$\rho(\mathbf{x}, \alpha\mathbf{y} + \beta\mathbf{z}) = \alpha\rho(\mathbf{x}, \mathbf{y}) + \beta\rho(\mathbf{x}, \mathbf{z}). \tag{264}$$

The adjective *primitive* comes from the fact that the usual types of bilinear forms are special cases of these primitive bilinear forms. It is very important to realize that this definition does not specify the relationship of $\rho(\mathbf{x}, \mathbf{y})$ to $\rho(\mathbf{y}, \mathbf{x})$. The reason that this "zero-order" definition is introduced here is that it will serve as a common starting point for the various types of bilinear metrics. The definition of a particular class of bilinear forms involves the specification of the properties of a primitive bilinear form with respect to interchange of the vector variables.

Definition 13 (Classes of Bilinear Forms) *A primitive bilinear form* $\rho(\mathbf{x}, \mathbf{y})$ *on* **V** × **V** *is a symmetric, antisymmetric, or Hermitian bilinear form in* **V** *if, for all vectors* $\mathbf{x}, \mathbf{y} \in \mathbf{V}$,

$$\rho(\mathbf{y}, \mathbf{x}) = \rho(\mathbf{x}, \mathbf{y}), \tag{265}$$

$$\rho(\mathbf{y}, \mathbf{x}) = -\rho(\mathbf{x}, \mathbf{y}), \tag{266}$$

$$\rho(\mathbf{y}, \mathbf{x}) = \rho^*(\mathbf{x}, \mathbf{y}), \tag{267}$$

respectively.

These three types of bilinear forms generate the three essentially unique types of scalar products in complex vector spaces [190].

Definition 14 (Scalar Product) *A scalar product on* **V** *is a bilinear form on* **V** × **V** *that satisfies* $\rho(\mathbf{z}, \mathbf{x}) = 0$ *where* $\mathbf{z} \neq 0$, *for all* **x** *if and only if* $\mathbf{z} = 0$. *A scalar* *product is defined to be symmetric, antisymmetric, or Hermitian if the associated bilinear form has this property.*

In this discussion only symmetric and Hermitian scalar products will be considered. A complex vector space with a positive definite Hermitian scalar product is known as a unitary space, whereas if the scalar product is indefinite, it is known as an indefinite inner-product space. A complex vector space with a symmetric scalar product is called a complex orthogonal, or complex Euclidean, space.

The motivation for introduction of a scalar product in a complex vector space is that it can be used to define a geometry on the space.

Definition 15 (Bilinearly Metric Space) *A complex vector space* $\mathbf{V}$ *with a scalar product is a bilinearly metric space.*

The two most important geometric concepts for the present discussion are those of orthogonality and normalization.

Definition 16 (Orthogonality of Vectors) *Two vectors* $\mathbf{x}, \mathbf{y} \in \mathbf{V}$ *are orthogonal with respect to the scalar product* $\rho(\cdot, \cdot)$ *if* $\rho(\mathbf{x}, \mathbf{y}) = 0$.

Definition 17 (Norms and Normalizable Vectors) *A vector* $\mathbf{x} \in \mathbf{V}$ *is normalizable with respect to the scalar product* $\rho(\cdot, \cdot)$ *if there exists a constant* $\beta \in \mathscr{C}$ *such that*

$$\rho(\beta\mathbf{x}, \beta\mathbf{x}) = 1. \tag{268}$$

The constant β *is called the norm of* $\mathbf{x}$ *if some convention is specified to uniquely determine it. If a vector is nonzero but not normalizable, it will be called nonnormalizable or quasi-null, and it cannot be assigned a norm.*

From this definition it is easy to see that all vectors such that $\rho(\mathbf{x}, \mathbf{x}) \neq 0$ are normalizable. In particular, the usual scalar product of two complex vectors $\mathbf{x}, \mathbf{y} \in \mathscr{C}^N$, $\langle \mathbf{x}, \mathbf{y} \rangle = \sum_{i=1}^{N} x_i^* y_i$ and the vector norm $\|\mathbf{x}\| = \sqrt{\sum_{i=1}^{N} |x|^2}$ conform to the definitions given here of a positive definite Hermitian scalar product.

The adjoint of a matrix with respect to a scalar product is also an important quantity in the study of bilinearly metric spaces.

Definition 18 (Adjoint with Respect to Scalar Product) *The adjoint* $\mathbf{A}^\dagger$ *of a square complex matrix* $\mathbf{A}$ *with respect to a scalar product* $\rho(\mathbf{x}, \mathbf{y})$ *is defined by*

$$\rho(\mathbf{x}, \mathbf{A}\mathbf{y}) = \rho(\mathbf{A}^\dagger\mathbf{x}, \mathbf{y}). \tag{269}$$

A matrix $\mathbf{A}$ *is called self-adjoint with respect to* $\rho(\cdot, \cdot)$ *if* $\mathbf{A} = \mathbf{A}^\dagger$.

If the scalar product is Hermitian, $A^{\ddagger} = A^{\dagger}$; if it is symmetric, $A^{\ddagger} = A^{tr}$. A closely related construct is that of an isometric matrix or transformation in a bilinearly metric space.

Definition 19 (Isometric Transformation)   *An isometric transformation B on a complex vector space* V *with scalar product* $\rho(x, y)$ *is a linear transformation that preserves the scalar product between any two vectors* x, y$\in$V, *that is,*

$$\rho(x, y) = \rho(Bx, By). \tag{270}$$

The matrix of an isometric transformation will also be called isometric.

A table summarizing the features of real Euclidean, unitary, and complex orthogonal spaces is given in the following.

Space	Field	Symmetry of Scalar Product	Isometric Transformation
Euclidean	$\mathscr{R}$	Symmetric	Real orthogonal
Unitary	$\mathscr{C}$	Hermitian	Unitary
Complex orthogonal	$\mathscr{C}$	Symmetric	Complex orthogonal

The geometry in a unitary space is a natural generalization of the geometry of real Euclidean spaces to complex vector spaces (e.g., all nonzero vectors have positive norms, one can define an angle between any two nonzero vectors, the vector norm satisfies the triangle inequality, etc.). The infinite dimensional analogue of a unitary space is a Hilbert space. Despite the wide use of unitary and Hilbert spaces in modern physical theories, there are many applications where these definitions are too restrictive. One way to generalize these ideas that has proven quite useful is to relax the requirement that the inner product be positive definite and work in an indefinite inner-product space rather than a unitary space. The general mathematical theory for both finite and infinite dimensional indefinite inner-product spaces has been worked out (192–198). These ideas have been applied to the study of quantum field theory [199], electrical networks [200], and the stability of systems of differential equations with periodic coefficients [201]. The theory of indefinite inner-product spaces has been tailored to the study of problems where the time evolution of the system is governed by an operator **B** that is self-adjoint with respect to an indefinite Hermitian scalar product. It can be shown that a general complex matrix **B** is similar to its Hermitian conjugate if and only if there exists a nonsingular Hermitian matrix **J** such that $BJ = JB^{\dagger}$ [91, 94]. This Hermitian matrix can then be used to define a Hermitian bilinear form in which **B** is self-adjoint (i.e., $\rho(x, y) = \sum_{i,j=1}^{N} J_{ij} x_i^* y_j$, where $J_{ij} = J_{ij}^*$ and $BJ = JB^{\dagger}$ so that $\rho(x, By) = \rho(Bx, y)$). It is not always possible to find a scalar product with these properties if the time evolution is governed by a complex symmetric matrix. In fact, the existence of the matrix $\mathbf{J}$ with the preceding properties implies that B is similar to a matrix with only real elements [87, 91]. This is certainly not the case for all complex symmetric matrices since any general square complex matrix is similar to a complex symmetric matrix [86, 87]. If the matrix governing the time evolution of the system is not similar to its Hermitian conjugate, the desired Hermitian scalar product does not exist. In practice, since there are no rules or algorithms to derive the matrix $\mathbf{J}$ (if it exists) from the matrix $\mathbf{B}$, the indefinite inner-product formalism cannot be used unless the equation for the time evolution of the system or other physical insight can be used to specify $\mathbf{J}$. It appears that the indefinite inner-product spaces are not the most convenient framework for the study of magnetic resonance lineshapes and related phenomena since the construction of the required indefinite inner product is difficult or impossible.

An alternative to the loosening of the requirement that the (Hermitian) inner product be positive definite is to use a symmetric rather than a Hermitian bilinear form in the definition of the fundamental scalar product. In this manner, it is always possible to construct a scalar product in which the complex symmetric matrix in question is self-adjoint. The simplest choice for a symmetric scalar product is to use the $N \times N$ identity matrix to generate a symmetric bilinear form $\sigma(\mathbf{x}, \mathbf{y}) = (\mathbf{x}, \mathbf{y}) = \sum_{i,j=1}^{N} x_i y_i$ in the same way as is done to construct the usual inner product $\rho(\mathbf{x}, \mathbf{y}) = \langle \mathbf{x}, \mathbf{y} \rangle = \sum_{i=1}^{N} x_i^* y_i$. It is important to note that all complex symmetric matrices are self-adjoint with respect to the scalar product $(\cdot, \cdot)$ the same way that all Hermitian matrices are self-adjoint with respect to $\langle \cdot, \cdot \rangle$.

The geometry in complex orthogonal spaces is much more complicated than that of unitary spaces. The description of the geometry in complex orthogonal spaces given here relies heavily on the work of Choudhury and Horn [90] and Malcev [128]. One of the biggest differences is the fact that the orthogonality of two vectors in a complex orthogonal space does not imply that they are linearly independent. A simple example of this phenomenon is given by the two vectors $\mathbf{u} = (1, i)^{\text{tr}}$ and $\mathbf{v} = \alpha(1, i)$ in $\mathscr{C}^2$ equipped with the symmetric scalar product $(\mathbf{x}, \mathbf{y}) = x_1 y_1 + x_2 y_2$. It is easy to verify that $(\mathbf{u}, \mathbf{v}) = 0$ for all $\alpha \in \mathscr{C}$. Note that when $\alpha = 1$ this example also demonstrates there can be nonzero vectors that are not normalizable. The analogue in a complex orthogonal space of a set of orthonormal vectors in a unitary space is a set of rectanormal vectors (90).

Definition 20 (Rectanormality of a Set of Vectors) *A set of vectors* $\mathbf{y}_i \in V$, *$i = 1, 2, \ldots, M$, is called retanormal with respect to the symmetric scalar product $\sigma(\cdot, \cdot)$ if*

$$\sigma(\mathbf{y}_i, \mathbf{y}_j) = \delta_{i,j} \quad \text{for all } i, j. \tag{271}$$

Obviously a rectanormal set of vectors cannot contain any non-normalizable vectors.

Given the fact that not all nonzero vectors in a complex orthogonal space are normalizable, the existence of a well-defined complex orthogonal space analogue of the unitary space Gram–Schmidt orthonormalization procedure is called into question. Choudhury and Horn have studied this problem and have proven under what conditions a complex orthogonal space Gram–Schmidt rectanormalization procedure can be established. Before stating these conditions, it is useful to define the triangular equivalence of two sets of vectors.

Definition 21 (Triangular Equivalence) *Two sets of vectors* $\mathbf{x}_i$, $i = 1, 2, \ldots, M$, *and* $\mathbf{y}_j$, $j = 1, 2, \ldots, M$, *from an N-dimensional complex orthogonal space* **V** *are called triangularly equivalent if the span of the set* $\mathbf{x}_i$, $i = 1, 2, \ldots, K$, *is identical with the span of* $\mathbf{y}_j$, $j = 1, 2, \ldots, K$, *for all* $K \leqslant M$.

The Gram–Schmidt rectanormalization procedure is the same as the Gram–Schmidt orthonormalization procedure except that the unitary space Hermitian scalar product is replaced everywhere by the symmetric scalar product appropriate for the complex orthogonal space (see Appendix A). The conditions under which this procedure makes sense is the topic of the following theorem by Choudhury and Horn [90]:

Theorem 2 (Existence of Gram–Schmidt Rectanormalization Procedure) *Given a set of linearly independent vectors* $\mathbf{x}_i$, $i = 1, 2, \ldots, M$, *in a complex orthogonal space, define the sequence of matrices* $X_1, X_2, \ldots, X_M$, *where the elements of the jth column of* X_i *are given by the elements of* $\mathbf{x}_j$ *for* $j \leqslant i$. *The set* $\mathbf{x}_i$ *is triangularly equivalent to a rectanormal set* $\mathbf{y}_j$ *(cf. Def. 21) if and only if*

$$\det(X_i^{\mathrm{tr}} X_i) \neq 0 \quad \text{for } 1 \leqslant i \leqslant M. \tag{272}$$

The Gram–Schmidt rectanormalization of the sequence of Krylov vectors generated by a complex symmetric matrix and an arbitrary normalizable starting vector is discussed in Appendix A. The existence of the Gram–Schmidt rectanormalization procedure for the Krylov vectors implies that the complex symmetric LA is well defined. This result was referred to in Appendix A.

In summary, complex orthogonal spaces enjoy the following properties:

- All CSMs are self-adjoint.
- The spectral function can be expressed as a symmetric bilinear form or a sum of symmetric bilinear forms.

- If two CSMs are similar, they are related by an isometric transformation. In particular, the transformation that diagonalizes a diagonalizable CSM is isometric.
- Under suitable restrictions, a Gram–Schmidt rectanormalization procedure can be used to sequentially convert a linearly independent set of vectors into a triangularly equivalent rectanormal set. The conditions under which this procedure is valid are the same as those needed to have a well-defined complex symmetric LA.
- The rectanormality of vectors is closely related to the known properties of the eigenvectors of general Fokker–Planck operators [119] (see Section VI.B).

Thus, complex orthogonal spaces can be used as a similar type of framework for the calculation of magnetic resonance spectra and spectral densities derived from Fokker–Planck forms as unitary spaces provide for quantum-mechanical problems in the absence of relaxation.

D. Matrix Elements in ESR Problems

We consider a spin probe with one hyperfine interaction, described by the spin Hamiltonian (in frequency units) in the high-field approximation

$$\hat{\mathcal{H}}(\Omega) = \omega_0 \hat{S}_z + \gamma_e a \hat{S}_z \hat{I}_z + \sum_{K,M,\mu} (-1)^K F_\mu'^{(2,-K)} \mathscr{D}_{KM}^2(\Omega) A_\mu^{(2,M)}, \quad (273)$$

where the nuclear Zeeman interaction is not taken into account. We assume that the rotational motion is described by the symmetrized Smoluchowski equation:

$$\tilde{\Gamma} = [\mathbf{J} - (1/2k_B T)(\mathbf{J}V)]\mathbf{R}[\mathbf{J} + (1/2k_B T)(\mathbf{J}V)], \quad (274)$$

where $\mathbf{J}$ is the generator of infinitesimal rotations of the molecule, $\mathbf{R}$ is the diffusion tensor, and V is the pseudopotential for oriented phases.

We provide only the simplified expression for the matrix elements of $\tilde{\Gamma} - i\mathbf{L}$ resulting from the following conditions: (i) the contribution of the nonsecular terms to the spectrum is negligible, (ii) the magnetic tensors (g and A) and $\mathbf{R}$ tensor are all diagonal in the same molecular frame, (iii) the diffusion tensor is axially symmetric with respect to the z axis of the molecular frame, and (iv) the orienting pseudopotential is axially symmetric and is given by the relation

$$- V(\Omega)/k_B T = \mathscr{D}_{00}^2(\Omega). \quad (275)$$

The matrix elements of the operator $\tilde{\Gamma} - i\mathbf{L}$ will be calculated in the direct-product space of generalized spherical harmonics and the space of the spin transitions $|m', m''\rangle$, where m' and m'' are the eigenvalues of $\hat{I}_z$ of the initial and final states, respectively, and the electron spin projection indices are implicit,

$$|L, M, K; m', m''\rangle = \left(\frac{2L+1}{8\pi^2}\right)^{1/2} \mathscr{D}_{MK}^L(\Omega)|m', m''\rangle, \qquad (276)$$

where $L = 0, 1, 2, \ldots, L_{max}$, $K = 0, \pm 2, \pm 4, \ldots, \min(L, K_{max})$, and $M = 0, \pm 1, \pm 2, \ldots, \min(L, M_{max})$.

In general, the symmetry of the operator and the definition of the starting vector allow one to work with a reduced basis set. The symmetries in the molecular frame allow one to redefine basis elements as

$$|L, M, K'; m', m''\rangle = [2(1 + \delta_{0,K})]^{-1/2}|L, M, K; m', m''\rangle$$
$$+ (-1)^L|L, M, -K; m', m''\rangle), \qquad (277)$$

with K' nonnegative and even. The symmetry in the laboratory frame allows a further reduction to redefined basis elements:

$$|L, M', K'; p, q\rangle = [2(1 + \delta_{0,M}\delta_{0,p})]^{-1/2}(|L, M, K'; p, q\rangle$$
$$+ (-1)^{L+M}|L, -M, K'; -p, q\rangle), \qquad (278)$$

where the index M' is now nonnegative. The quantum numbers p and q are defined as $p = m' - m''$ and $q = m' + m''$, where $p = 0$ ($p \neq 0$) for an allowed (forbidden) ESR transition. Our calculations were performed specifically for ^{14}N-containing nitroxides with nuclear spin $I = 1$, so $m', m'' = 0, \pm 1$. Thus, $p = 0, \pm 1, \pm 2$ and $q = -Q, -Q+2, \ldots, Q-2, Q$, where $Q = 2 - |p|$. In isotropic fluids and in ordered fluids where the symmetry axis of the potential is colinear with the static magnetic field, there exist symmetries such that only the terms with $p = M'$ need be considered so that $M' = 0, 1, 2$. In the following formulas, the primes on M and K are dropped for convenience.

The matrix elements for the diffusion operator in this basis are given by

$$\langle L_1, M_1, K_1; p_1, q_1|\tilde{\Gamma}|L_2, M_2, K_2; p_2, q_2\rangle$$
$$= \langle p_1, q_1|p_2, q_2\rangle \delta_{K_1,K_2}\delta_{M_1,M_2}R_\perp$$
$$\times \left\{\left[L_1(L_1 + 1) + \left(\frac{R_\parallel}{R_\perp} - 1\right)K_1^2 + \tfrac{3}{10}(\varepsilon_0^2)^2\right]\delta_{L_1,L_2} + (-1)^{M_1+K_1}N_L(L_1, L_2)\right.$$
$$\times \left[3\varepsilon_0^2(1 - \tfrac{1}{14}\varepsilon_0^2)\begin{pmatrix} L_1 & 2 & L_2 \\ K_1 & 0 & -K_2 \end{pmatrix}\begin{pmatrix} L_1 & 2 & L_2 \\ M_1 & 0 & -M_2 \end{pmatrix}\right.$$
$$\left.\left. + \tfrac{18}{35}(\varepsilon_0^2)^2\begin{pmatrix} L_1 & 4 & L_2 \\ K_1 & 0 & -K_2 \end{pmatrix}\begin{pmatrix} L_1 & 4 & L_2 \\ M_1 & 0 & -M_2 \end{pmatrix}\right]\right\}, \qquad (279)$$

where $N_L(L_1, L_2) = [(2L_1 + 1)(2L_2 + 1)]^{1/2}$. The components of the starting vector $|v\rangle$ are given as

$$\langle L, M, K; p, q|v\rangle = \delta_{0,K}\delta_{0,M}\langle p, q|S_-\rangle \int d\Omega \mathscr{D}_{00}^L(\Omega) P_0^{1/2}(\Omega). \quad (280)$$

Finally, the Liouville operator has the following matrix elements:

$$\langle L_1, M_1, K_1; p_1, q_1|L|L_2, M_2, K_2; p_2, q_2 \rangle$$
$$= \langle L_1, M_1, K_1; q_1|L_2, M_2, K_2; q_2 \rangle (\omega_0 + \tfrac{1}{2}aq_1\gamma_e) + (-1)^{M_1}$$
$$\times [(1 + \delta_{2,K_1+K_2})(1 + \delta_{1,M_1+M_2})]^{1/2} N_L(L_1, L_2)$$
$$\times \begin{pmatrix} L_1 & 2 & L_2 \\ K_1 & K_2 - K_1 & K_2 \end{pmatrix}\begin{pmatrix} L_1 & 2 & L_2 \\ M_1 & M_2 - M_1 & M_2 \end{pmatrix}$$
$$\times \sum_\mu F_\mu^{(2,K_2-K_1)} G_\mu(M_1, q_1; M_2, q_2), \quad (281)$$

$$G_g(M_1, q_1; M_2, q_2) = \frac{\sqrt{2/3}\,\omega_0}{\text{Tr}\{g\}}, \quad (282)$$

$$G_A(M_1, q_1; M_2, q_2) = \delta_{M_1,M_2}\delta_{q_1,q_2}q_1\gamma_e/\sqrt{6} + \delta_{1,|M_1-M_2|}\delta_{q_1,q_2\pm1}\gamma_e(M_2 - M_1)$$
$$\times \{I(I+1) - \tfrac{1}{4}(q_1 \pm M_1[M_1 - M_2])(q_2 \pm M_2[M_1 - M_2])\}^{1/2}. \quad (283)$$

We remark that the matrix associated with $\tilde{\Gamma} - iL$ is complex symmetric, both $\tilde{\Gamma}$ and L being real and symmetric.

While it is a sparse matrix, the structure of the matrix elements is clearly complicated. The more general case, wherein the preceding simplifying assumptions are removed, leads to an even more complicated (but still sparse) matrix structure, as is detailed elsewhere [20].

References

1. C. Lanczos, *J. Nat. Bur. Stand.* **45**, 255 (1950).

2. M. R. Hestenes and E. Stiefel, *J. Nat. Bur. Stand.* **49**, 409 (1952).

3. G. Moro and J. H. Freed, *J. Chem. Phys.* **74**, 3757 (1981).

4. K. V. Vasavada, D. J. Schneider, and J. H. Freed, *J. Chem. Phys.* **86**, 647 (1987).

5. J. H. Freed, G. V. Bruno, and C. F. Polnaszek, *J. Phys. Chem.* **75**, 3385 (1971).

6. J. H. Freed, G. V. Bruno, and C. Polnaszek, *J. Chem. Phys.* **55**, 5270 (1971).

7. J. H. Freed, in *Electron-Spin Relaxation in Liquids*, L. T. Muus and P. W. Atkins, Eds., Plenum, New York, 1972.

8. J. H. Freed, *Ann. Rev. Phys. Chem.* **23**, 265 (1972).

9. C. F. Polnaszek, G. V. Bruno, and J. H. Freed, *J. Chem. Phys.* **58**, 3185 (1973).

10. J. H. Freed, *J. Chem. Phys.* **49**, 376 (1968).

11. B. Yoon, J. M. Deutch, and J. H. Freed, *J. Chem. Phys.* **62**, 4687 (1975).

12. J. H. Freed, in *Spin Labeling: Theory and Applications*, Vol. I, L. Berliner, Ed., Academic New York, 1976, Chapter 3.

13. S. A. Goldman, G. V. Bruno, C. F. Polnaszek, and J. H. Freed, *J. Chem. Phys.* **56**, 716 (1972).

14. J. S. Hwang, R. P. Mason, L. P. Hwang, and J. H. Freed, *J. Phys. Chem.* **79**, 489 (1975).

15. G. V. Bruno and J. H. Freed, *J. Phys. Chem.* **78**, 935 (1974).

16. C. F. Polnaszek and J. H. Freed, *J. Phys. Chem.* **79**, 2283 (1975).

17. R. F. Campbell and J. H. Freed, *J. Phys. Chem.* **84**, 2668 (1980).

18. E. Meirovitch and J. H. Freed, *J. Phys. Chem.* **84**, 2459 (1980).

19. M. Shiotani, G. Moro, and J. H. Freed, *J. Chem. Phys.* **74**, 2616 (1981).

20. E. Meirovitch, D. Igner, E. Igner, G. Moro, and J. H. Freed, *J. Chem. Phys.* **77**, 3915 (1982).

21. E. Meirovitch and J. H. Freed, *J. Phys. Chem.* **88**, 4995 (1984).

22. E. Meirovitch, A. Nayeem, and J. H. Freed, *J. Phys. Chem.* **88**, 3454 (1984).

23. H. Tanaka and J. H. Freed, *J. Phys. Chem.* **88**, 6633 (1984).

24. H. Tanaka and J. H. Freed, *J. Phys. Chem.* **89**, 350 (1985).

25. L. Kar, E. Ney-Igner, and J. H. Freed, *Biophys. J.* **48**, 569 (1985).

26. A. E. Stillman, L. J. Schwartz, and J. H. Freed, *J. Chem. Phys.* **73**, 3502 (1980).

27. A. E. Stillman, L. J. Schwartz, and J. H. Freed, *J. Chem. Phys.* **76**, 5658 (1982).

28. L. J. Schwartz, A. E. Stillman, and J. H. Freed, *J. Chem. Phys.* **77**, 5410 (1982).

29. J. P. Hornak and J. H. Freed, *Chem. Phys. Lett.* **101**, 115 (1983).

30. G. L. Millhauser and J. H. Freed, *J. Chem. Phys.* **81**, 37 (1984).

31. L. Kar, G. L. Millhauser, and J. H. Freed, *J. Phys. Chem.* **88**, 3951 (1984).

32. G. L. Millhauser and J. H. Freed, *J. Chem. Phys.* **85**, 63 (1986).

33. L. J. Schwartz, G. L. Millhauser, and J. H. Freed, *Chem. Phys. Lett.* **127**, 60 (1986).

34. J. Gorcester and J. H. Freed, *J. Chem. Phys.* **85**, 5375 (1986); ibid. **88**, 4678 (1988).

35. G. L. Millhauser, J. Gorcester, and J. H. Freed, in *Electron Magnetic Resonance of the Solid State*, J. A. Weil, Ed. Canadian Chemical Society, Ottawa (1987), p. 571.

36. R. F. Campbell, E. Meirovitch, and J. H. Freed, *J. Phys. Chem.* **83**, 525 (1979).

37. O. Pschorn and H. W. Spiess, *J. Magn. Reson.* **39**, 217 (1980).

38. E. Meirovitch and J. H. Freed, *Chem. Phys. Lett.* **64**, 311 (1979).

39. C. F. Polnaszek et al., *Biochemistry* **15**, 954 (1976).

40. S. Alexander, A. Baram, and Z. Luz, *Mol. Phys.* **27**, 441 (1974).

41. A. E. Stillman and J. H. Freed, *J. Chem. Phys.* **72**, 550 (1980).

42. G. Moro and J. H. Freed, *J. Phys. Chem.* **84**, 2837 (1980).

43. G. Moro and J. H. Freed, *J. Chem. Phys.* **75**, 3157 (1981).

44. G. Moro and J. H. Freed, in *Large-Scale Eigenvalue Problems*, Mathematical Studies Series, Vol. 127, J. Cullum and R. Willoughby, Eds., Elsevier, NY 1986.

45. R. G. Gordon and T. Messenger, in *Electron Spin Relaxation in Liquids*, L. T. Muus and P. W. Atkins, Eds., Plenum, New York, 1972, Chapter 13.

46. A. J. Dammers, Y. K. Levine, and J. A. Tjon, *Chem. Phys. Lett.* **88**, 192 (1982).

47. M. Giordano, P. Grigolini, D. Leporini, and P. Martin, *Phys. Rev. A* **28**, 2474 (1983).

48. M. Giordano, P. Grigolini, D. Leporini, and P. Martin, *Adv. Chem. Phys.* **62**, 321 (1985).

49. A. J. Dammers, "Numerical Simulation of Electron Spin Resonance Spectra in the Slow Motion Regime," Ph.D. Thesis, Utrecht, 1985.

50. R. Kubo, *Adv. Chem. Phys.* **16**, 101 (1969).

51. A. Abragam, *Principles of Nuclear Magnetism*, Oxford University Press, New York, 1961.

52. J. H. Freed and G. K. Fraenkel, *J. Chem. Phys.* **39**, 326 (1963).

53. J. H. Freed, *J. Chem. Phys.* **41**, 2077 (1964).

54. L. D. Favro, in *Fluctuation Phenomena in Solids*, R. E. Burgess, Ed., Academic, New York, 1965.

55. P. A. Egelstaff, *J. Chem. Phys.* **53**, 2590 (1970).

56. L. C. Biedenharn and J. D. Louck, *Angular Momentum in Quantum Physics*, Addison-Wesley, London, 1981.

57. A. Messiah, *Quantum Mechanics*, Wiley, New York, 1962.

58. M. E. Rose, *Elementary Theory of Angular Momentum*, Wiley, New York, 1957.

59. N. C. Pyper, *Mol. Phys.* **21**, 1 (1971).

60. N. C. Pyper, *Mol. Phys.* **22**, 433 (1971).

61. R. M. Lynden-Bell, in *Electron Spin Relaxation in Liquids*, L. T. Muus and P. W. Atkins, Eds., Plenum, New York, 1972, Chapter 13A.

62. R. M. Lynden-Bell, *Mol. Phys.* **22**, 837 (1972).

63. A. Omont, *Prog. Quant. Electron.* **5**, 69 (1977).

64. V. Gorini, M. Verri, and E. C. G. Sudarshan, in *Group Theoretical Methods in Physics*, K. Wolf, Ed., Lecture Notes in Physics, Vol. 135, Springer-Verlag, New York, 1980.

65. J. B. Pedersen and J. H. Freed, *J. Chem. Phys.* **57**, 1004 (1972).

66. J. B. Pedersen and J. H. Freed, *J. Chem. Phys.* **58**, 2746 (1973).

67. J. B. Pedersen and J. H. Freed, *J. Chem. Phys.* **59**, 2869 (1973).

68. L. P. Hwang and J. H. Freed, *J. Chem. Phys.* **63**, 4017 (1975).

69. J. H. Freed and J. B. Pedersen, *Adv. Mag. Res.* **8**, 1 (1976).

70. J. H. Freed, in *Chemically Induced Magnetic Polarization*, L. Muus et al., Eds., D. Reidel, Dordrecht-Holland, 1977, Chapter 19.

71. K. Wassmer, E. Ohmes, M. Portugall, H. Ringsdorf, and G. Kothe, *J. Am. Chem. Soc.* **107**, 1511 (1985).

72. K. Müller, P. Meier, and G. Kothe, in *Progress in NMR Spectroscopy* (1986).

73. G. P. Zientara and J. H. Freed, *J. Chem. Phys.* **70**, 1359 (1979).

74. G. P. Zientara and J. H. Freed, *J. Chem. Phys.* **71**, 3861 (1979).

75. G. P. Zientara and J. H. Freed, *J. Phys. Chem.* **83**, 3333 (1979).

76. J. P. Korb, M. Ahadi, G. P. Zientara, and J. H. Freed, *J. Chem. Phys.* **86**, 1125 (1987).

77. A. E. Stillman, G. P. Zientara, and J. H. Freed, *J. Chem. Phys.* **71**, 113 (1979).

78. G. P. Zientara and J. H. Freed, *J. Chem. Phys.* **70**, 2587 (1979).

79. G. P. Zientara and J. H. Freed, *J. Chem. Phys.* **71**, 744 (1979).

80. G. H. Golub and C. Van Loan, *Matrix Computations*, Johns Hopkins University Press, Baltimore, 1983.

81. C. A. J. Fletcher, *Computational Galerkin Methods*, Springer-Verlag, New York, 1984.

82. J. B. Pederson, *J. Chem. Phys.* **57**, 2680 (1972).

83. M. M. Itzkowitz, *J. Chem. Phys.* **46**, 3048 (1967).

84. A. E. Stillman and R. N. Schwartz, in *Time Domain Electron Spin Resonance*, L. Kevan and R. N. Schwartz, Eds., Wiley-Interscience, New York, 1979, Chapter 5.

85. G. P. Zientara and J. H. Freed, *J. Chem. Phys.* **72**, 1285 (1980).

86. F. R. Gantmacher, *The Theory of Matrices*, Chelsea, New York, 1959.

87. R. A. Horn and C. A. Johnson, *Matrix Analysis*, Cambridge University Press, NY 1985.

88. D. J. Schneider, unpublished work, 1986.

89. J. K. Cullum and R. A. Willoughby, *Lanczos Algorithms for Large Sparse Eigenvalue Computations*, Birkhäuser, Boston, 1985.

90. D. Choudhury and R. A. Horn, "An Analog of the Gram–Schmidt Algorithm for Complex Symmetric Forms and Diagonalization of Complex Symmetric Matrices," Technical Report No. 454, Department of Mathematical Sciences, Johns Hopkins University, Baltimore, MD, 1986.

91. O. Taussky, *Lin. Alg. Applic.* **5**, 147 (1972).

92. B. D. Craven, *J. Austral. Math. Soc.* **10**, 341 (1969).

93. M. L. Mehta, *Elements of Matrix Theory*, Hindustan, Delhi, India, 1977.

94. J. W. Duke, *Pacific J. Math.* **31**, 321 (1969).

95. J. H. Wilkinson, *The Algebraic Eigenvalue Problem*, Clarendon, Oxford, 1965.

96. B. N. Parlett, *The Symmetric Eigenvalue Problem*, Prentice-Hall, Englewood Cliffs, NJ, 1980.

97a. S. A. Goldman, G. V. Bruno, and J. H. Freed, *J. Chem. Phys.* **59**, 3071 (1973).

97b. S. A. Goldman, Ph.D. Thesis, Cornell University, (1973).

98. L. A. Dalton and L. R. Dalton, in *Multiple Electron Resonance Spectroscopy*, M. Dorio and J. H. Freed, Eds., Plenum, New York, 1979, Chapter 5.

99. J. S. Hyde and L. R. Dalton, in *Spin Labeling: Theory and Applications*, Vol. II, L. Berliner, Ed., Academic, New York, 1976, Chapter 1.

100. N. Benetis, "Nuclear Spin Relaxation in Paramagnetic Metal Complexes and the Slow Motional Problem for the Electron Spin," Ph.D. Thesis, University of Stockholm, 1984.

101. W. Givens, *Nat. Bur. Stand. Appl. Math. Ser.* **29**, 117 (1952).

102. H. Rutishauser, *Proc. Am. Math. Soc. Symp. Appl. Math.* **15**, 219 (1963).

103. P. Anderson and G. Loizou, *J. Inst. Math. Appl.* **12**, 261 (1973).

104. P. Anderson and G. Loizou, *Numer. Math.* **25**, 347 (1976).

105. P. J. Eberlein, *J. Inst. Math. Appl.* **7**, 377 (1971).

106. J. G. Ruffinatti, *Computing* **15**, 275 (1975).

107. H. Mori, *Progr. Theor. Phys.* **33**, 423 (1965).

108. R. Kubo, M. Toda, and N. Hashitsume, *Statistical Physics II, Nonequilibrium Statistical Mechanics*, Springer Series in Solid-State Science, Vol. 31, Springer-Verlag, New York, 1985.

109. B. J. Berne and R. Pecora, *Dynamic Light Scattering*, Wiley-Interscience, New York, 1976.

110. D. Kivelson and K. Ogan, in *Advances in Magnetic Resonance*, Vol. 7, J. Waugh, Ed., Academic, New York, 1974, Chapter 2.

111. W. A. Wassam, Jr., *J. Chem. Phys.* **82**, 3371 (1985).

112. W. A. Wassam, Jr., *J. Chem. Phys.* **82**, 3386 (1985).

113. J. H. Freed, in *Rotational Dynamics of Small and Macromolecules in Liquids*, T. Dorfmuller and R. Pecora, Eds., Springer-Verlag, New York, (1987), p. 89.

114. J. S. Hwang, K. V. S. Rao, and J. H. Freed, *J. Phys. Chem.* **80**, 1790 (1976).

115. N. G. Van Kampen, *Stochastic Processes in Chemistry and Physics*, North-Holland, Amsterdam, 1981.

116. H. Haken, *Rev. Mod. Phys.* **47**, 67 (1975).

117. H. Haken, *Synergetics*, Springer Series in Synergetics, Vol. 1, 2nd enlarged Ed., Springer-Verlag, New York, 1978.

118. H. Haken, *Advanced Synergetics*, Springer Series in Synergetics, Vol. 20, Springer-Verlag, New York, 1983.

119. H. Risken, *The Fokker–Planck Equation*, Springer Series in Synergetics, Vol. 18, Springer-Verlag, New York, 1984.

120. C. W. Gardiner, *Handbook of Stochastic Methods*, Springer Series in Synergetics, Vol. 13, Springer-Verlag, New York, 1983.

121. P. Hänggi and H. Thomas, *Phys. Rep.* **88**, 207 (1982).

122. H. Risken, *Z. Phy.* **251**, 231 (1972).

123. R. Graham and H. Haken, *Z. Phy.* **243**, 289 (1971).

124. M. Lax, in *Symmetries in Science*, B. Gruber and R. S. Milman, Eds., Plenum, New York, 1980.

125. M. Lax, *Symmetry Principles in Solid State and Molecular Physics*, Wiley, New York, 1974.

126. P. O. Löwdin, *J. Math. Phys.* **24**, 70 (1983).

127. V. V. Voyevodin, *Linear Algebra*, Mir, Moscow, 1983.

128. I. A. Malcev, *Foundations of Linear Algebra*, Freeman, New York, 1963.

129. P. M. Morse and H. Feshbach, *Methods of Mathematical Physics*, 2 Vols., McGraw-Hill, New York, 1953.

130. C. M. Bender and S. A. Orszag, *Advanced Mathematical Methods for Scientists and Engineers*, McGraw-Hill, New York, 1978.

131. N. G. van Kampen, *Phys. Rep.* **24C**, 171 (1976).

132. U. Frish, in *Probabilistic Methods in Applied Mathematics*, A. T. Bharucha Reid, Ed., Academic, New York, 1968.

133. A. Brissaud and U. Frisch, *J. Math. Phys.* **15**, 524 (1974).

134. J. H. Freed, *J. Chem. Phys.* **66**, 4183 (1977).

135. L. P. Hwang and J. H. Freed, *J. Chem. Phys.* **63**, 118 (1975).

136. A. J. Vega and D. Fiat, *J. Chem. Phys.* **60**, 579 (1974).

137. L. Monchick, *J. Chem. Phys.* **74**, 4519 (1981).

138. W. A. Wassam, Jr. and J. H. Freed, *J. Chem. Phys.* **76**, 6133 (1982).

139. W. A. Wassam, Jr. and J. H. Freed, *J. Chem. Phys.* **76**, 6150 (1982).

140. L. P. Hwang and J. H. Freed, unpublished work.

141. M. Fixman and K. Rider, *J. Chem. Phys.* **51**, 2425 (1969).

142. R. G. Gordon, *J. Chem. Phys.* **44**, 1830 (1966).

143. J. H. Freed, *J. Phys. Chem.* **78**, 1155 (1974).

144. J. H. Freed, in *Multiple Electron Resonance Spectroscopy*, M. Dorio and J. H. Freed, Eds., Plenum, New York, 1979, Chapter 3.

145. J. H. Freed, in *Time Domain Electron Spin Resonance*, L. Kevan and R. N. Schwartz, Eds., Wiley-Interscience, New York, 1979, Chapter 2.

146. L. J. Schwartz, "Molecular Reorientation and Time Domain Electron Spin Resonance," Ph.D. Thesis, Cornell University, 1984.

147. S. A. Goldman, G. V. Bruno, and J. H. Freed, *J. Phys. Chem.* **76**, 1858 (1972).

148. G. V. Bruno, "Application of the Stochastic Liouville Method in Calculating ESR Line Shapes in the Slow Tumbling Region and ESR-ELDOR Study of Exchange," Ph.D. Thesis, Cornell University, 1973.

149. G. V. Bruno and J. H. Freed, *Chem. Phys. Lett.* **25**, 328 (1974).

150. J. H. Freed, *J. Chem. Phys.* **43**, 2312 (1965).

151. J. S. Hyde, J. C. W. Chien, and J. H. Freed, *J. Chem. Phys.* **48**, 4211 (1968).

152. M. D. Smigel, L. A. Dalton, L. R. Dalton, and A. L. Kwiram, *Chem. Phys.* **6**, 183 (1974).

153. H. Barkhuijsen, R. de Beer, W. M. M. J. Bovee, and D. van Ormont, *J. Magn. Reson.* **61**, 465 (1985).

154. R. Kumaresan and D. W. Tufts, *IEEE Trans.* **ASSP-30**, 833 (1982).

155. J. Jeener, B. H. Meier, P. Bachmann, and R. R. Ernst, *J. Chem. Phys.* **71**, 11 (1979).

156. S. A. Dzuba, A. G. Maryasov, K. M. Salikhov, and Yu. D. Tsvetkov, *J. Magn. Reson.* **85**, 95 (1984).

157. U. Eliav and J. H. Freed, *J. Phys. Chem.* **88**, 1277 (1984).

158. K. F. Milfield and N. Moiseyev, *Chem. Phys. Lett.* **130**, 145 (1986).

159. E. Haller, H. Köppel, and L. S. Cederbaum, *J. Mol. Spectrosc.* **111**, 377 (1985).

160. G. Moro, *Chem. Phys.* **106**, 89 (1986).

161. G. Moro and P. L. Nordio, *Chem. Phys. Lett.* **96**, 192 (1983).

162. A. S. Householder, *The Theory of Matrices in Numerical Analysis*, Blaisdell, New York, 1964; reprinted by Dover, Mineola, NY, 1975.

163. W. B. Jones and W. J. Thron, *Continued Fractions*, Encyclopedia of Mathematics and Its Applications, Vol. 11, Addison-Wesley, Reading, MA, 1980.

164. H. S. Wall, *Analytical Theory of Continued Fractions*, Van Nostrand, New York, 1948; reprinted by Chelsea, New York, 1967.

165. H. S. Wall and M. Wentzel, *Trans. Am. Math. Soc.* **55**, 373 (1944).

166. H. S. Wall and M. Wentzel, *Duke Math. J.* **11**, 89 (1944).

167. H. S. Wall, *Bull. Am. Math. Soc.* **52**, 671 (1946).

168. I. S. Kac and M. G. Krein, *Am. Math. Soc. Trans.* **103**(2), 1 (1974).

169. P. Giannozzi, G. Grosso, and G. Pastori Parravicini, *Phys. Stat. Sol. B* **128**, 643 (1985).

170. G. Grosso and G. Pastori Parravicini, *Adv. Chem. Phys.* **62**, 81 (1985).

171. N. I. Ahiezer and M. G. Krein, *Some Problems in the Theory of Moments*, Transl. Math. Monographs, Vol. 2, American Mathematical Society, Providence, RI, 1962.

172. N. I. Ahiezer, *The Classical Moment Problem*, Oliver and Boyd, Edinburgh, 1965.

173. J. A. Shohat and J. D. Tamarkin, *The Problem of Moments*, American Mathematical Society, New York, 1943.

174. R. G. Gordon, *J. Math. Phys.* **9**, 1087 (1968).

175. F. Lado, J. D. Parker, and G. W. Memory, *Phys. Rev. B* **4**, 1406 (1971).

176. A. Lonke, *J. Math. Phys.* **12**, 2422 (1971).

177. J. Cullum and R. A. Willoughby in *Large-Scale Eigenvalue Problems*, Mathematical Studies Series, Vol. 127, J. Cullum and R. A. Willoughby, Eds. Elsevier, NY 1986.

178. C. Brezinski, *Padé-Type Approximation and General Orthogonal Polynomials*, International Series of Numerical Mathematics, Vol. 50, Birkhäuser, Boston 1980.

179. G. A. Baker, Jr., *Essentials of Padé Approximants*, Academic, New York, San Francisco, London, 1975.

180. C. Brezinski, *J. Comput. Appl. Math.* **12**, 19 (1985).

181. G. A. Baker, Jr. and P. Graves-Morris, *Padé Approximants*, Encyclopedia of Mathematics and Its Applications, Vols. 13 and 14, Addison-Wesley, Reading, MA, 1983.

182. J. Wimp, *Sequence Transformation Methods and Their Applications*, Academic, New York, 1981.

183. C. Brezinski, *Lin. Alg. Its Applic.* **11**, 7 (1975).

184. F. V. Atkinson, *Discrete and Continuous Boundary Problems*, Academic, New York, 1964.

185. R. R. Whitehead and A. Watt, *J. Phys.* **G4**, 835 (1978).

186. R. R. Whitehead and A. Watt, *J. Phys.* **A14**, 1887 (1981).

187. L. P. Hwang, C. F. Anderson, and H. L. Friedman, *J. Chem. Phys.* **62**, 2098 (1975).

188. J. L. Monroe and H. L. Friedman, *J. Chem. Phys.* **66**, 955 (1977).

189. S. Alexander, A. Baram, and Z. Luz, *J. Chem. Phys.* **61**, 992 (1974).

190. K. J. Heuvers, *Lin. Alg. Its Applic.* **6**, 83 (1973).

191. M. Znojil, *J. Phys. A* **9**, 1 (1976).

192. M. G. Krein, *Am. Math. Soc. Trans.* **93**(2), 103 (1970).

193. M. G. Krein, *Am. Math. Soc. Trans.* **1**(2), 27 (1955).

194. A. Z. Jadczyk, *Rep. Math. Phys.* **2**, 263 (1971).

195. J. Bognár, *Indefinite Inner Product Spaces*, Springer-Verlag, New York, 1974.

196. I. S. Iohvidov, M. G. Krein, and H. Langer, *Introduction to the Spectral Theory of Operators in Spaces with an Indefinite Metric*, Mathematical Research Vol. 9, Akademie-Verlag, Berlin, 1982.

197. I. Gohberg, P. Lancaster, and L. Rodman, *Matrices and Indefinite Scalar Products*, Operator Theory: Advances and Applications, Vol. 8, Birkhäuser, Boston 1983.

198. I. M. Glazman and Ju. I. Ljubič, *Finite-Dimensional Linear Analysis*, MIT Press, Cambridge, MA, 1974.

199. C. Itzykson and J.-B. Zuber, *Quantum Field Theory*, McGraw-Hill, New York, 1980.

200. M. C. Pease III, *Methods of Matrix Algebra*, Mathematics in Science and Engineering, Vol. 16, Academic, New York, 1965.

201. V. A. Yakubovic and V. M. Starzhinskii, *Linear Differential Equations with Periodic Coefficients*, Vols. I and II, Wiley, New York, 1975.

202. G. V. Bruno, J. K. Harrington, and M. P. Eastman, *J. Phys. Chem.* **81**, 1111 (1977).

203. W. J. Lin and J. H. Freed, *J. Phys. Chem.* **83**, 379 (1979).

204. K. V. S. Rao, J. S. Hwang, and J. H. Freed, *Phys. Rev. Lett.* **37**, 515 (1976).

205. S. A. Zager and J. H. Freed, *Chem. Phys. Lett.* **109**, 270 (1984).

206. A. Nayeem, "Electron Spin Relaxation and Molecular Dynamics in Liquid Crystalline Phases," Ph.D. Thesis, Cornell University, 1986.

207. S. A. Zager and J. H. Freed, *J. Chem. Phys.* **77**, 3344, 3360 (1982).

208. G. Moro and P. L. Nordio, *Mol. Cryst. Liq. Crys.* **104**, 361 (1984).

209. G. Moro and P. L. Nordio, *J. Phys. Chem.* **89**, 997 (1985).

210. D. G. Pettifor and D. L. Weaire, Eds., *The Recursion Method and Its Applications*, Springer Series in Solid-State Science, Vol. 58, Springer-Verlag, New York, 1985.

211. I. Efrat and M. Tismenetsky, *IBM J. Res. Develop.* **30**, 184 (1986).

212. E. van der Drift, B. A. C. Rousseeuw, and J. Smidt, *J. Chem. Phys.* **88**, 2275 (1984).

213. Z. Luz and R. Naor, *Mol. Phys.* **46**, 891 (1982).

214. D. Gamliel, Z. Luz, and S. Vega, *J. Chem. Phys.* **85**, 2516 (1986).

215. Added in proof: This has now been successfully implemented.

CHAPTER XI

DENSITY MATRIX METHODS AND RELAXATION PHENOMENA IN LASER-EXCITED POLYATOMIC MOLECULES

JAMES STONE

Rockwell Internationals Rocketdyne Division, Canoga Park, California 91304

CONTENTS

We review and extend the use of the density matrix formalism for modeling multiple-photon excitation and dissociation and spectroscopic linewidths in excited polyatomic molecules. Models for both collisional and intramolecular relaxation are reviewed, and various limiting cases of density matrix solutions are considered where rate equations apply either to the full solution or to a partial time average. New expressions for intramolecular dephasing (T_2) and energy transfer (T_1) rates are derived that take into account the distribution of normal-mode frequencies as well as restrictions on the magnitude of coupling relating to changes in normal-mode quantum numbers. Coupling for molecules

529

bonded by Morse potentials is considered, the relationship of parameters in restricted quantum exchange (RQE) theory pointed out, and generalized models proposed that account for both quantum exchange and symmetry requirements. We extend the generalized master equation (GME) formalism to allow equations for the level populations to be generated in cases where level degeneracies are present. Applications are made to finding time-averaged populations for degenerate ladders and for simple intramolecular coupling models. Finally, we consider the possible development of extensive computational methods based on the GME formalism.

I. INTRODUCTION

In this chapter we will discuss the use of density matrices to model and to do practical computations for infrared laser-induced processes in polyatomic molecules. It has now been over 10 years since the first experimental demonstrations of collisionless multiple-photon dissociation (Ref. 1, 2). Theoretical modeling in this area has focused in turn on efforts to elucidate the mechanism of excitation (3–6) and later to quantitatively model the dynamic process in simple systems (7). The access provided to highly excited vibrational states of the molecule has led to new probes, both experimental (8) and theoretical (9–11), into the question of statisticality or ergodicity. First of interest to chemists during the period when theories for unimolecular dissociation rates were being formulated, this area can now be addressed more directly than before by either multiple-photon or overtone (8) excitation of molecules. The density matrix formalism itself takes advantage of statistical assumptions but can also be used to investigate the basis for these assumptions, in particular, by allowing us to look at the time-averaged behavior of systems.

Many of the other chapters in this volume are directly concerned with modeling multiple-photon excitation dynamics or with studying the dynamics of relaxation in molecules and the associated statistics. In addition, since the molecular multiple-photon experiments, a new class of ultrahigh intensity multiphoton processes in atoms has been demonstrated (12) and appears (13) to have certain features in common with dynamical processes in molecules with some additional complications due to the intensity. The relatively moderate intensity required for molecular excitation ordinarily allows use of the rotating-wave approximation for the interaction between the molecule and field. In the molecular environment, the more difficult problem is that of representing the states of the molecule itself.

Density matrix methods occupy an important niche in the overall scheme of modeling molecular states and their dynamics because they allow nonstationary states to be included. Spectroscopically, zero-order states that are projections of some starting state onto an excited region of the potential surface have been shown (14) in classical dynamics simulations to explain the gross, low-resolution features of observed spectra. These correspond in a density matrix formalism to broadened, finite-lifetime states. Higher resolution features may correspond to succeeding relaxation processes. (The detailed correspondence in this case with the density matrix formalism remains to be explored.) The derivation of Bloch equations for the density matrix (15–17) and the specification of the relaxation parameters that occur in them (18) are a focal point for justifying statistical assumptions inherent in the density matrix treatment and for identifying the physical mechanisms that determine the relaxation rates.

This chapter focuses on two aspects of the density matrix formalism for infrared laser-excited polyatomic molecules. The first is a discussion of the underlying physics, which makes a density matrix description appropriate (16–19) and allows contact to be made with underlying potential surface parameters. An elaboration of density matrix models and their physical interpretation is included in Section II. The calculation of dynamic processes in excited molecules can often be dramatically simplified in the density matrix formalism because of the effective collapsing of many microscopic states into a few significant zero-order states with a finite lifetime. The formulation of the models identifies certain significant relaxation rates that must be calculated before the formalism can actually be applied to anything. In Section III, we review and extend the restricted quantum exchange theory for the relaxation rates and explore its connection with the molecular potential and with symmetry requirements. New expressions for the rates, taking the spread in normal-mode frequencies explicitly into account, are derived and compared with earlier treatments.

The second major aspect is a discussion of methods of treating and analyzing the solutions of the density matrix equations. Specific models differ depending on those physical processes (e.g., collisions or intramolecular interactions) that are being modeled. Approximate (20) or exact (21) equations that can be derived from the density matrix equations often give insight into the physical nature of the solutions as well as suggest simplifications in the process of obtaining solutions. The physical conditions modeled by the density matrix equations often include a weak damping situation where the systems undergo strong Rabi oscillations many times before beginning to approach a steady state. Results in Section II illustrate analytic methods where these oscillations may be averaged over if desired. Then in Section IV we discuss a general *algebraic* reduction scheme that, for an arbitrary degenerate ladder model, results in level population equations for transient or steady-state solutions. These equations extend earlier results where a GME (21, 22) was derived for a simple ladder model. In Section V we speculate on some more advanced applications that would combine diagrammatic and numerical methods to provide a general computational technique.

II. DENSITY MATRIX MODELS FOR POLYATOMIC MOLECULAR STATES

The quantum-mechanical density matrix provides a natural framework in which to describe molecular states that are broadened by various collisional or intramolecular interactions. The broadened states are not eigenstates of the full molecular or larger system Hamiltonian since, clearly, full diagonalization would have eliminated any interaction terms. Yet, in many instances, they are the most natural states in which to describe the system. For collisional broadening, the advantages of considering states of an isolated molecule when possible are obvious. In this case it is possible to experimentally remove or adjust the source of broadening by changing the gas pressure, and the states under the broadened band form a true continuum. In the case of "internal" broadening mechanisms, the application of high-resolution spectroscopy to vibrationally excited molecules (23) can resolve molecular eigenstates in some cases. Thus, the description of a broad band of states as a single broadened state requires more careful justification. Yet the use of a full basis of eigenstates in calculations of multiphoton dissociation dynamics seems out of the question for all but the smallest of molecules. The continued development of models employing broadened states is therefore of great interest.

In the next few paragraphs we will discuss some of the models for collisional relaxation that have been investigated, describe some of the results, and then discuss generalizations that are useful in the treatment of intramolecular interactions.

We shall be specifically concerned with various forms of Bloch equations applied to the problem of the interaction of bound molecules with laser radiation. There are inherently a finite number of states in the analysis, but this number may be extremely large in practice. The Bloch equations allow us to deal with relaxation processes. States with finite lifetimes can be included by introducing appropriate decay constants. The finite-lifetime states may involve many molecular eigenstates in their dynamics. As a result, the use of finite-lifetime states has the potential for allowing the use of many fewer states to describe the dynamics of the molecule. In addition, unimolecular decay into an *unbound* dissociative continuum can be included. While we can go even further and formulate a reduced description involving rate equations, the density matrix formalism is more general because it allows for the possibility of coherent pumping. The use of decay constants as prescribed in a Bloch equation must ultimately be justified through analysis of energy eigenstate and coupling statistics, and we will address these topics in Section II.D

A. Collisional Relaxation

Multilevel Bloch equations describing a simple ladder of radiatively pumped states subject to collisional relaxation can be written in the form

$$\frac{d\rho}{dt} = \frac{i}{\hbar}[\rho, H] + \frac{1}{\tau}(K - 1)\rho \tag{1}$$

where

$$H_{n,n} = \hbar\omega_n \tag{2}$$

and

$$H_{n-1,n} = H_{n,n-1} = -\alpha_n A \sin xt \tag{3}$$

Here the ladder of states is coupled through its transition dipoles to a monochromatic (temporally coherent) light field of amplitude A and frequency χ. Several studies (21, 24–26) of collisional relaxation in coherently driven systems have been carried out using Eq. 1 and collision-induced rates specified by the collision superoperator K (27).

$$K_{nl,mk} = \delta_{nl}\delta_{mk}P_{n \leftarrow m}$$
$$(K\rho)_{n,l} = \delta_{nl}\sum_m P_{n \leftarrow m}\rho_{m,m} \tag{4}$$

In Eq. 1 the physical assumption is that collisions occur randomly (i.e., are Poisson distributed) with frequency $1/\tau$. Each collision completely destroys phase information (setting off-diagonal density matrix elements to zero) and induces transitions depending on what state the molecule is in. The conditional probabilities $P_{n \leftarrow m}$ give the probability that a molecule initially in state m will, after a collision, be found in state n. The complete relaxation is described by the action of the collision superoperator K on the density matrix ρ to produce a new postcollision density matrix $K\rho$ (27).

Recent results (20) obtained using Eqs. 1–4 illustrate the power of the density matrix formalism to simultaneously include both relaxation and coherent driving effects and to analyze the resulting systems. Figure 1 shows the level populations of a four-level system where the natural Rabi transition frequency for transitions between levels is much greater than the collision frequency. The density matrix approaches steady state in a highly oscillatory fashion in this limit, but if we are interested in its averaged approach and not in the details of the oscillation, we can do a "local" time average over the oscillations, such as that accompanying the full solutions in Figure 1. The smooth curves in Figure 1 are not merely "drawn in" visually as an average

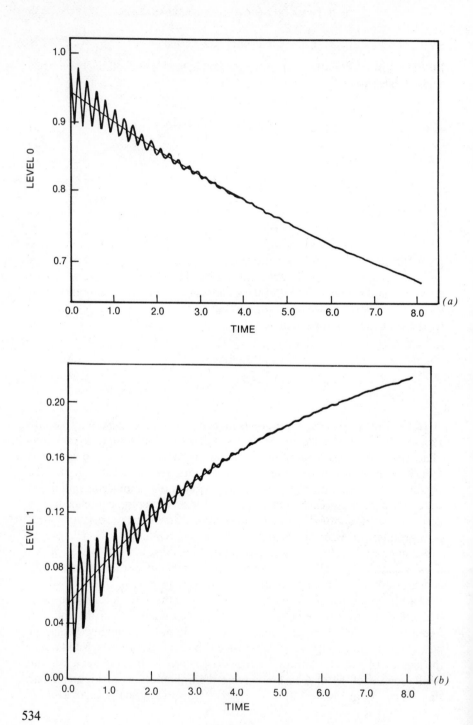

534

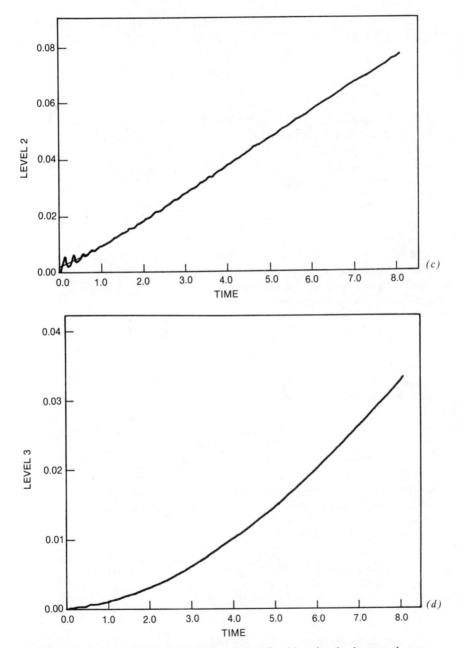

Figure 1. Oscillatory curves show level populations for driven four-level system that are solutions to Eq. 1 for a Landau–Teller collision model ($P_{0 \leftarrow 1} = 0.001, P_{1 \leftarrow 0} = 10^{-6}$, see Ref. 21). These are compared with analytic smoothing approximation given by Eq. 5. Rabi frequencies (in arbitrary time units) are those of a truncated harmonic oscillator with $\Omega_{01} = 0.1$. Level spacing is equal and off resonance by 0.3. Figure from Ref. 20. Used with permission.

over the observed oscillations but were obtained by solving the following rate equation for the averaged level populations:

$$\frac{d\rho_{av}}{dt} = \frac{1}{\tau}(TK - 1)\rho_{av} \tag{5}$$

The initial condition on ρ_{av} is $T\rho(0)$. Here (and in the equation) T is the time-averaged evolution operator for the corresponding collision-free system, defined mathematically as a projection of the full density matrix obtained by solving the Liouville equation (Eq. 1 without the relaxation terms) onto the subspace of zero (Liouville operator) eigenvectors (24),

$$T = \sum_{\mu_i = 0} |\mu_i\rangle\langle\mu_i| \tag{6}$$

The full time evolution operator includes terms for *every* eigenvector, each multiplied by $\exp(\mu_i t)$ (this factor is unity for the zero eigenvalues but an oscillating exponential for all others). The local time average ρ_{av} in the presence of damping can be defined (20) by

$$\rho_{av}(t) = \sum_{\substack{\lambda, \\ \lambda\,\text{real}}} |\lambda\rangle\langle\lambda|\rho(t = 0)e^{\lambda t} \tag{7}$$

Equation 5 correctly predicts this averaged quantity without ever having to deal with the oscillations. A rigorous derivation of this result has been given (20). It can be understood intuitively by considering the effect of collisions during a time interval Δt that is long compared to a Rabi oscillation but short compared to τ. The probability P_1 of a single collision is approximately $\Delta t/\tau$; that of no collisions (P_0) is $1 - \Delta t/\tau$. After a collision, we have, on average, $\rho = K\rho_{av}$; after the system has evolved, $\rho = TK\rho_{av}$, which is the new average ρ for the case where a single collision occurs. When *no* collison occurs, ρ_{av} is unchanged. Weighting these two possibilities by their respective probabilities gives

$$\Delta\rho_{av} = \frac{\Delta t}{\tau}TK\rho_{av} + \left(1 - \frac{\Delta t}{\tau}\right)\rho_{av} - \rho_{av}$$

$$= \frac{1}{\tau}(TK - 1)\rho_{av}\,\Delta t \tag{8}$$

Equation (5) for the time average follows.

At this point we make one final observation on Eq. 5 for future reference. We can make a generalization of a commonly used strong collision model (28) to define a collision superoperator K^i where after a single collision, the system is in a particular state i regardless of its state before the collision. Equation 5, valid in the low pressure-limit, can be used to show that the steady-state solution for this type of collision mechanism has a special interpretation as we go to this limit. From Eq. 5 we see that, since $K^i\rho$ is just the density matrix $\rho_{jk} = \delta_{ji}\delta_{ki}$ regardless of the ρ that K^i operates on, the steady state is just the time-averaged density matrix for a system initially in state i. We will use this observation to help interpret some of our results in Section IV.B.

We note that the kind of damped Rabi oscillation depicted in Figure 1 has been obtained in large-basis-set calculations with molecular eigenstates (29). Also, real eigenvalues obtained from Eq. 5 may be used to help get good starting estimates for eigenvalues of the full density matrix equations. (The remaining eigenvalues not included can be estimated from Schrödinger energy eigenvalue differences.) From the point of view of modeling, however, Eq. 1 is incomplete because it deals only with collisional broadening.

B. Intramolecular Relaxation

A molecular transition can be broadened intramolecularly by the interaction among the various vibrational modes. An especially simple situation arises when only one of these modes is directly responsible for absorption of energy

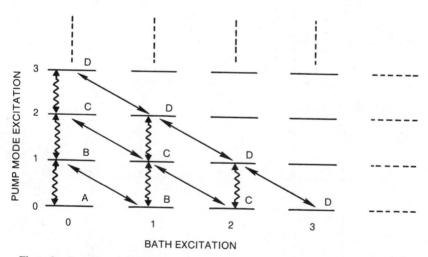

Figure 2. Specification for excited vibrational states in the heat bath feedback model. Total energy in bath is explicitly accounted for. Wavy lines denote transition rates $W_{i,j}$ or dipole coupling specified in Eq. 11 for laser-induced transitions in pump mode. Sloping arrows indicate intramolecular T_1 relaxation between states of same total energy (Figure from Ref. 19. Used with permission).

from the electromagnetic field. Because the states purely in this mode are coupled to other modes (i.e., because they are not eigenstates), the energy relaxes to these other modes, and the states are broadened. The pump mode excitation decreases by this mechanism whereas that of the remaining "heat bath" modes increases as shown in Figure 2. Since further excitation occurs, not in the original ladder of states but in a new one, information about the state of the heat bath is fed back into the calculation. Hence, we call this a "heat bath feedback" (HBFB) model (19).

In contrast, the collision relaxation model of Eq. 1 corresponds to radiative (wavy arrows) and collision-induced relaxations (solid arrows) as shown in Figure 3. The completely parallel nature of the two type of transitions between states as shown reflects the similarity of the assumptions made about energy disposition in the field and in the heat bath; that is, transitions in the molecule due to absorbing or emitting photons cause neligible effects on the field (intense field assumption) and energy exchanged with collision partners in the heat bath causes negligible effect on the heat bath as a whole (large heat bath assumption). Comparison of Figures 2 and 3 points up the fundamental limitation on the ability of Eq. 1 to describe intramolecular relaxation where

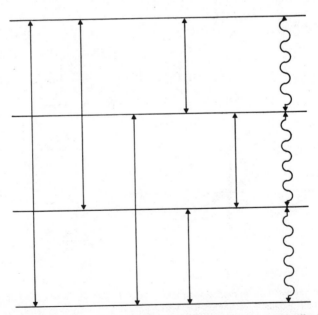

Figure 3. Relaxation and pumping diagram for collisional damping with effectively infinite heat bath. In contrast with HBFB model, radiative (wavy arrow) and collision-induced (solid arrows) are parallel and in competition with each other.

neglecting the effect of added energy on the intramolecular heat bath cannot be justified.

Equations representing the HBFB model (19), that is,

$$\frac{d\rho_{jj}^{(E)}}{dt} = \frac{i}{\hbar}[\rho^{(E)}, H]_{jj} + \sum_{j' \neq j} \left(\frac{1}{T_1}\right)_{j,E \leftarrow j',E'} \rho_{j'j'}^{(E')} - \left(\frac{1}{T_1}\right)_{j',E' \leftarrow j,E} \rho_{jj}^{(E)}$$

$$\frac{d\rho_{jk}^{(E)}}{dt} = \frac{i}{\hbar}[\rho^{(E)}, H]_{jk} - \left(\frac{1}{T_2}\right)_{jk} \rho_{jk}^{(E')} \qquad (9)$$

are distinguished from those for a simple ladder model by the appearance of a new variable E for the amount of energy in the heat bath. The appearance of a new energy E' on the right hand side of Eq. 9 reflects the transitions between different ladders depicted in Figure 2. Relaxation occurs in such a way that the total energy, heat bath plus pump mode, is explicitly conserved [thus, $E' = E + (j - j')\hbar\omega_p$]. This differs from the original concept of the heat bath as an energy sink whose properties are unaffected by energy being added to it.

The importance of using HBFB is clear when considering the process of multiphoton dissociation. Regardless of the reaction *mechanism*, a certain minimum energy will be required for dissociation, and failure to account for energy that has gone into the heat bath makes it impossible to predict when dissociation will occur except for the unlikely case where the mechanism depends only on the energy in the pump mode. In the other extreme case where energy relaxation (T_1) is very rapid and reaction probability depends only on the total energy [RRKM theory (30)], a time scale separation can be made in the HBFB equations, and one can derive (19) a rate equation for the total population on each energy shell irrespective of distribution,

$$\frac{dN_k}{dt} = T_{k \leftarrow k-1} N_{k-1} - T_{k-1 \leftarrow k} N_k + T_{k \leftarrow k+1} N_{k+1} - T_{k+1 \leftarrow k} N_k \qquad (10)$$

where the transition rates

$$T_{k \leftarrow k-1} = \left(\sum_{i=1}^{k} W_{i,i-1}(k-i)\eta_{k'-i}\right)\Bigg/\sum_{i=0}^{k-1} \eta_i$$

$$T_{k-1 \leftarrow k} = \left(\sum_{i=1}^{k} W_{i,i-1}(k-i)\eta_{k-i}\right)\Bigg/\sum_{i=0}^{k} \eta_i \qquad (10a)$$

effectively average rates for individual ladders. Here the η_i's are heat bath densities of states and the $W_{i,i-1}$'s are lifetime-broadened transition rates

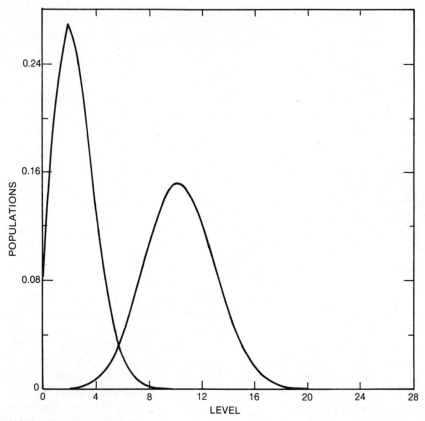

Figure 4. Typical nonmonotonic level population profiles for systems satisfying rate equations of the type shown in Eq. 10, including microscopic reversibility. Results shown use effective density of states factors for a 15-fold degenerate oscillator model and cross sections S_j (defined for each level as $T_{j+1\leftarrow j} - T_{j-1\leftarrow j}$) proportional to $\exp(-0.154j)$. Evolution is followed for two times $3/S_0$ (left curve) and $24/S_0$. (Adapted from Ref. 32. Used with permission).

within each ladder given by (19)

$$W_{i,i-1}(j) = \frac{\alpha_i^2 A^2}{2\hbar^2} \frac{(1/T_2)_{ii-1,j}}{(1/T_2)_{ii-1,j}^2 + (\chi - \omega_{i,j} + \omega_{i-1,j})^2} \tag{11}$$

Here the subscript j on the frequencies reflects a possible energy shift in the pump made from the bath interaction.

An interesting prediction of Eq. 10a is that upward and downward rates are related by ratios of total upper and lower level densities of states (microscopic reversibility.) This general property is characteristic of systems with the type of statistical description we have made here and appears again in connection with intramolecular relaxation rates (see Section III, Eq. 15). The general character of solutions where the rate coefficients have this microscopic reversibility property has been investigated. It was proposed that the solutions are close to thermal distributions (31). This is true in certain instances (32), but more generally the solutions share with thermal distributions the property shown in Figure 4 where, for a particular model (32), the population is plotted as a function of level number for two different times. At a given time, the population is a nonmonotonic function of level, and this is in accord with the level population evolution (33) deduced from experiment. A dependence similar to this has been observed recently in atomic multiphoton processes (12, 34), and theoretical explanations have been advanced (13, 35) to account for it.

C. Generalized Master Equation

A rate equation such as Eq. 10 cannot describe the type of Rabi oscillations shown in Figure 1. Although this form of dynamics may not be present or significant in the quasi-continuum of a moderate-sized polyatomic molecule, there seems little doubt that it is important in small molecules, and it is therefore worthwhile to briefly review in greater detail the way in which the density matrix formalism can be used to handle this type of behavior.

The oscillatory solution in Figure 1 can be found from the full density matrix equations (Eq. 1) by reformulating these as a GME (21, 22),

$$\lambda \rho_{i,i} = \sum_{j \neq i} [P_{i \leftarrow j/\tau} + W_{i,j}(\lambda)\rho_{j,j}] - \rho_{i,i} \sum_{j \neq i} P_{j \leftarrow i/\tau} + W_{j,i}(\lambda) \tag{12}$$

The solution depicted is a sum over solutions corresponding to several eigenvalues λ (where the eigenvalue denotes the time dependence $e^{-\lambda t}$). In a simple ladder model, the radiative transition elements $W_{ij}(\lambda)$ can be obtained by a straightforward algebraic procedure outlined in Refs. (21, 22). We will obtain the generalization of this procedure to a degenerate ladder model in Section IV. There is, however, one point concerning Eq. 12 that is relevant to our present discussion of general modeling questions, that is, that the radiative transition terms in Eq. 11 couple every pair of levels even though the coupling in the Hamiltonian is only between adjacent levels. The GME solution therefore constitutes a reduction of the full density matrix problem, which is distinct from the usual rate equation description. By contrast, the individual radiative transition terms (W) that make up the statistically averaged rates in Eq. 10 couple only between adjacent levels (in a given ladder denoted by the index j).

There are really two major differences between the terms in Eq. 11 and Eq. 12, both related to the fact that rapid T_2 relaxation had to be assumed in order to derive Eq. 11. First, the limit $1/T_2 \to \infty$ in the GME results in nonadjacent $W_{i,j}$'s that are small and can be neglected under the present circumstances (see, however, 25). Second, the relevant GME eigenvalues for Eq. 10 are all close to zero, so that steady-state values of W [i.e., $W(\lambda = 0)$] have been used. Of course, the rates making up the sum in Eq. 10a are just the usual Lorentzian lineshapes expected in this limit, but in general, a full treatment using all λ and nonadjacent W's would be required.

D. Microscopic Energy Levels

An alternative to the density matrix approach for intramolecular dynamics is to solve the time-dependent Schrödinger equation in a large basis of states (29). Without actually adopting this approach, we can take a more fundamental point of view regarding the Bloch equations by noting that the energy eigenstates can be used as a starting point for *deriving* the density matrix formalism. The states in the density matrix formalism are finite-lifetime states, and this choice results in greatly reducing the number of states that have to be explicitly considered. The physical basis for their use can be understood by analyzing those pairs of states carrying optical transition strength.

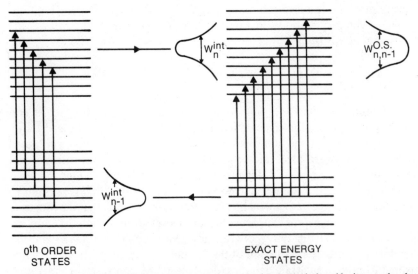

0th ORDER STATES   EXACT ENERGY STATES

Figure 5. Spread of the dipole operator among energy eigenstates induced by intramolecular interactions at upper and lower levels of the transition. Constant length of arrows between zero-order states corresponds to energy of pump mode quantum. Transition arrows from given exact energy state spread over range of energies (Figure from Ref. 18, used with permission).

A schematic picture (18) of the microscopic energy levels and how coupling at both upper and lower quantum levels spreads the transition strength is shown in Figure 5. A homogeneous width results when radiative transitions from a given exact energy state to any of a number of exact energy states as shown at the next level are possible. Contributions to the width from both initial state and final state result because the intramolecular coupling creates a mixture of zero-order states in each energy eigenstate (lower level) and a mixture of energy eigenstates in each zero-order state (upper level). The spacing of the transition in the zero-order states is shown as constant because this just corresponds to the energy of a pump mode quantum—the various states in each zero-order manifold have nearly the same total energy but differ from each other in their heat bath (and possibly pump mode) quantum numbers. Depending on the exact statistics of coupling (18), the transition may have a width that is just a sum of upper and lower level broadenings. This shows how a microscopic picture with many states can be reduced to a few pairs of broadened states, which we can identify with the various possible upper and lower pump mode states included in the HBFB formalism.

The width in transition strength among energy eigenstates is manifested in phase ($\hat{T}_2$) and population (T_1) lifetime effects (36) as well as in the observed low-intensity spectrum. In Section IV, when we consider methods of analyzing the Bloch equation, we will be developing techniques applicable to even more general schemes of broadened states where transition strength arising from degenerate active modes or from more than one infrared active mode can result in radiative coupling to more than one zero-order state at each level. In the meantime (Section III) we will examine more precisely the role of the intramolecular coupling in determining the individual interaction widths in Figure 5 and the various relaxation rates in Eq. 9.

III. RELAXATION RATES AND RESTRICTED INTERMODE COUPLING

When it was first proposed to use the heat bath feedback equations to model actual experimental data, a serious problem became evident. The formulation of the rate equations, Eq. 10, provided a satisfactory picture of the relationship between a rate equation formalism that had been proposed by several groups (6, 33) and a density matrix formalism including heat bath information. Yet the resulting expressions for the pumping rates (Eq. 10a) depended heavily on broadening from unknown T_2 relaxation rates. "First-principle" consider-ations suggested that these rates should be proportional to some coupling factor and an appropriate density of states (the total density of vibrational states?). The coupling factors were unknown, and if one simply assumed that they were constant, a possible one-parameter model for the dependence of linewidth on energy in the molecule simply makes the linewidth proportional to the (total) vibrational density of states. Because of the very strong dependence of this quantity on energy, it was found that no fit to data on SF_6 excitation could be made without introducing an arbitrary cutoff in the linewidths (37).

There had, at the time, been some limited discussions (38) on possible restrictions on the coupling and the effect of these restrictions, if they existed, on the very validity of a mixed-state description of excited polyatomic states (such as the general scheme depicted in Figure 5). General ideas from spectroscopy suggest limited coupling between states with large (normal-mode) quantum number differences. We proposed that this type of restriction might exist (without invalidating the basic picture) and that it would be the dominant mechanism determining the energy dependence of the broadening (39). We proceeded with the analysis of this RQE model and showed, using a vibrational model with degenerate normal-mode frequencies, that the linewidths and relaxation rates would be predicted to *saturate* (18) instead of increasing with energy, as in the simplest model based on the total density of states. This was exactly what seemed to be required for a reasonable model and furthermore was in satisfactory accord with the rough constancy that was being seen in the widths of overtones (40).

According to RQE theory, coupling between nearby energy states is not unrestricted but takes place to a significant degree only between those states that have nearly the same distribution of energy among the various modes. This means that if we calculate the total absolute difference in normal-mode quantum numbers between two states,

$$M(n, n') \equiv \sum_i |n_i - n'_i| \tag{13}$$

the *coupling* between the states is a strongly decreasing function of $M(n, n')$,

$$V_{n,n'} = V_0 \tilde{\alpha}^{M(n,n')} \tag{14}$$

Here $\tilde{\alpha}$ is a dimensionless quantity of magnitude much less than unity and V_0 is an overall scaling factor with the dimensions of energy. Equation 14 is an analytically convenient form having the desired fall off properties. We will discuss some generalizations and the physical interpretation of the parameters in terms of a potential surface in Section III.C.

A complete analysis of the relaxation rates using the degenerate oscillator model and using Eq. 14 for the coupling shows saturation for each of the individual T_1 and T_2 relaxation rates (18), as shown in Figure 6. Saturation occurs for *any* value of α less than 1 (18), albeit at higher and higher energies as 1 is approached. These results and some new results we now discuss are based

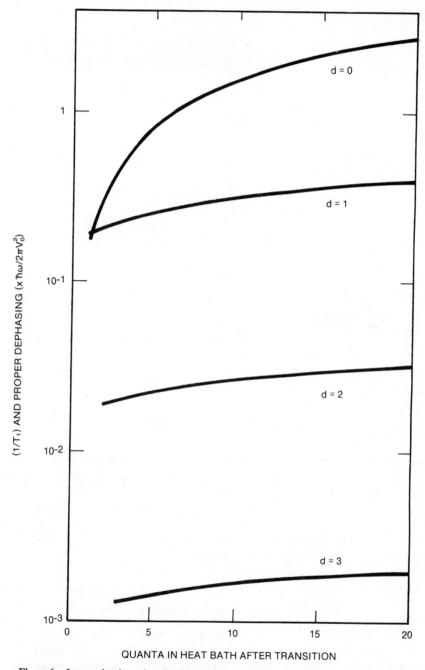

Figure 6. Intramolecular relaxation rates determined for a degenerate oscillator model (Figure from Ref. 18) with 20 modes (frequency ω) as a function of energy in the heat bath. Curves for $d > 0$ correspond to T_1 rates for transfer of d heat quanta into the heat bath. The $d = 0$ curve corresponds to the proper dephasing, which contributes to the total T_2 rate ($\alpha^4 = 0.01$). (Used with permission.)

on a formal derivation (16) of the heat bath feedback equations, Eq. 9, which has resulted in the following formal expressions for relaxation rates in terms of matrix elements:

$$\left(\frac{1}{T_1}\right)_{j,E\leftarrow j',E'} = \frac{2\pi\eta(E)}{\hbar}\gamma^{jj'}(E, E') \tag{15}$$

$$\left(\frac{1}{T_2}\right)_{jk}^{(E)} = \frac{1}{2}[\hat{w}_j(E) + \hat{w}_k(E)]$$

$$+ \frac{1}{2}\sum_{j'\neq j}\left(\frac{1}{T_1}\right)_{j',E'\leftarrow j,E} + \frac{1}{2}\sum_{k'\neq k}\left(\frac{1}{T_1}\right)_{k',E'\leftarrow k,E} \tag{16}$$

where

$$\hat{w}_j(E) = \frac{2\pi\eta(E)}{\hbar}\gamma^{jj}(E, E) \tag{17}$$

$$\gamma^{jj'}(E, E') \equiv \frac{1}{\eta(E)\eta(E')\,\delta E\,\delta E'}\sum_{\substack{\mathbf{n}'_B\in\Omega(E',\delta E')\\ n'_p=j'}}\sum_{\substack{\mathbf{n}_B\in\Omega(E,\delta E)\\ n_p=j}} |V_{\mathbf{nn'}}|^2 \tag{18}$$

Here $\Omega(E, \delta E)$ is the collection of states whose normal-mode quantum numbers satisfy

$$E - \tfrac{1}{2}\delta E \leqslant \sum_{i\in B} n_i\hbar\omega_i \leqslant E + \tfrac{1}{2}\delta E \tag{19}$$

We note that because $\gamma^{jj'}$ is symmetric with respect to exchange of j and j', the T_1 relaxation rates in Eq. 15 exhibit microscopic reversibility. The previous use of these expressions to derive the relaxation rates for a degenerate oscillator model required special ways of assigning the energy intervals that occur in the definitions since the energies there actually occur only at discrete values. Here we derive new expressions for the rates taking into account the actual distribution of molecular frequencies. Density of state rather than discrete combinatorial methods are used so that arbitrary spread in the normal-mode frequencies is allowed for a priori.

A. Analysis of Relaxation Rate Expressions

We will now use the RQE coupling model (Eqs. 13–14) in Eq. 18 to evaluate the relaxation rates for an excited molecule. The energy interval δE in Eq. 19 must be taken small enough so that the variation in the calculated rates would be small for a change of E of order δE but large enough so that the set $\Omega(E, \delta E)$ encompasses numerous states (16). This lower limit restriction on the energy interval has implications for the method that we will use in evaluating the relaxation rates.

Substituting Eq. 14 into Eq. 18, we find

$$\gamma^{jj'}(E, E') = \frac{V_0^2 \alpha^{2|j-j'|}}{\eta(E)\eta(E')} G_\alpha(E, E')$$ (20)

where

$$G_\alpha(E, E') = \frac{1}{\delta E} \frac{1}{\delta E'} \sum_{n' \in \Omega(E', \delta E')} \sum_{n \in \Omega(E, \delta E)} \alpha^{M(n, n')}$$ (21)

and we are making the replacement $\tilde{\alpha}^2 \to \alpha$ for notational simplicity. We can obtain an explicit expression for $G_\alpha(E, E')$ in terms of the energies and heat bath normal-mode frequencies by considering the Laplace transforms of the formal expression in Eq. 21:

$$F(E, \beta') \equiv \int_0^\infty G_\alpha(E, E')e^{-\beta'E'} dE'$$ (22)

and

$$\hat{G}_\alpha(\beta, \beta') = \int_0^\infty F(E, \beta')e^{-\beta E} dE$$ (23)

In the limit of very small δE or $\delta E'$, each subinterval of size δE contains at most a single heat bath state. (For convenience, we are assuming nondegenerate energy levels.) This means that in this limit Eq. 21 is really a weighted sum over δ functions. As already remarked, in the actual evaluation of the relaxation rates, δE must not be allowed to approach this limit. However, for the moment, we introduce the δ-function limit in Eq. 21 and write

$$F(E, \beta') = \frac{1}{\delta E} \sum_{n \in \Omega(E, \delta E)} \int_0^\infty \sum_{n' \in \Omega(E', \delta E')}$$
$$\times \delta(E' - \sum_i n_i' \hbar \omega_i) \alpha^{M(n_i, n_i')} e^{-\beta'E'} dE'$$ (24)

$$F(E, \beta') = \frac{1}{\delta E} \sum_{n \in \Omega(E, \delta E)} \sum_{n'} \alpha^{M(n_i, n_i')} \exp\left(-\beta' \sum_i n_i' \hbar \omega_i\right)$$ (25)

With regard to the δ-function limit, we simply note that there is a natural way of handling the requirement on δE (and in effect avoiding the δ-function limit) once we find the Laplace transforms. The approximate method used to invert the transforms will have the effect of smoothing over unwanted information associated with having taken the δ-function limit here. Note that the sum over n' in Eq. 25 is now over *all* heat bath quantum numbers irrespective of the energy. A second integration can now be performed (also with the δ-function limit for Ω) with the result

$$\hat{G}_\alpha(\beta, \beta') = \sum_n \sum_{n'} \alpha^{M(n_i, n_i')} \exp\left(- \beta \sum n_i \hbar \omega_i - \beta' \sum_i n_i' \hbar \omega_i \right) \tag{26}$$

which can be factored into separate contributions from each of s heat bath modes,

$$\hat{G}_\alpha(\beta, \beta') = \sum_{i=1}^s \sum_{n_i=0}^\infty \sum_{n_i'=0}^\infty e^{-\beta n_i \hbar \omega_i - \beta' n_i' \hbar \omega_i} \alpha^{|n_i - n_i'|} \tag{27}$$

Finally, the sums in Eq. 27 can be expressed in closed form to write $\hat{G}$ in the form

$$\hat{G}_\alpha(\beta, \beta') = f_1(\beta) f_1(\beta') f_2(\beta, \beta) \tag{28}$$

where

$$f_1(\beta) = \prod_{i=1}^s \frac{1}{1 - e^{-\beta \hbar \omega_i} \alpha} \tag{29}$$

$$f_2(\beta, \beta') = \prod_{i=1}^s \frac{e^{(\beta + \beta')\hbar \omega_i/2} - e^{-(\beta + \beta')\hbar \omega_i/2} \alpha^2}{2 \sinh \left[(\beta + \beta')\hbar \omega_i/2 \right]} \tag{30}$$

Now we can obtain an explicit expression for the first Laplace transform $F(E, \beta')$ by applying the Laplace inversion formula (41) to Eq. 28,

$$F(E, \beta') = f_1(\beta') H(E, \beta') \tag{31}$$

where

$$H(E, \beta') = \frac{1}{2\pi i} \int_{c-i\infty}^{c+i\infty} e^{\beta E} f_1(\beta) f_2(\beta, \beta') \, d\beta \tag{32}$$

The contour c must be taken to the right of all poles. The integral can be done in standard fashion by closing the contour at infinity. Since $e^{\beta E}$ vanishes as $\beta \to -\infty$ (E is naturally positive), the part of the integral on the closure vanishes if we make this a semicircle in the left-half plane. As a result, the entire closed contour (c plus the semicircle) encompasses all the poles, and residues at all these poles must be included to obtain the value of the integral.

It is convenient to consider two groups of poles separately, those arising from the f_1 factor and those arising from the f_2 factor. With these two groups we associate the respective terms in the expression

$$H = H_1 + H_2 \tag{33}$$

The contributions to H_1 of interest to us consist of residues at the s distinct real poles of f_1 that occur at the values β_j,

$$\beta_j = \frac{\ln \alpha}{\hbar \omega_j} \tag{34}$$

To each of these corresponds an infinity of *complex* poles having imaginary parts that are multiples of $2\pi i/(\hbar\omega_j)$. The complex poles give rise in the final result to oscillating Fourier components that would serve (in the limit of an infinite series of terms) to produce a δ-function behavior in the energy dependence. In view of the requirement noted in the preceding restricting the energy intervals δE to be large enough to smooth over many states, we specifically avoid including these complex poles and write the contribution from poles of f_1 specifically as a sum just over real residues,

$$\tilde{H}_1(E, \beta') = \sum_{j=1}^{s} P_j \alpha^{E/\hbar\omega_j} \prod_{l=1}^{s} \frac{1 - e^{-\beta'\hbar\omega_l}\alpha^{2 - (\omega_l/\omega_j)}}{1 - e^{-\beta'\hbar\omega_l}\alpha^{-(\omega_l/\omega_j)}} \tag{35}$$

where

$$P_j = \frac{1}{\hbar\omega_j} \prod_{k \neq j} \frac{1}{1 - \alpha^{1 - \omega_k/\omega_j}} \tag{36}$$

and $\tilde{H}$ denotes the approximation to H obtained by inverting $\hat{G}$ subject to the real-pole smoothing approximation.

The real poles of f_2 consist of a single higher order pole at $\beta = -\beta'$, corresponding to the s factors in the denominator of Eq. 30. Here the evaluation requires taking a derivative of order $s - 1$ after multiplying by $(\beta + \beta')^s$. We must differentiate the product of factors that occurs in

$$e^{\beta E}(\beta + \beta')^s f_1 f_2 \tag{37}$$

and this is a straightforward application of Leibnitz's rule provided the singular sinh factors in the denominator of f_2 are grouped together with the $(\beta + \beta')^s$ factors (which go to zero) to form a factor we call g_1. We then refer to the remaining (numerator) factors of f_2 as g_2. With these definitions in hand, we carry out the required operations to obtain the following result:

$$\tilde{H}_2(E, \beta') = \frac{e^{-\beta' E}}{\prod_{i=1}^{s} \hbar\omega_i} \sum_{j+k+l+m=s-1} \frac{E^j}{j!} \left[\sum_{\substack{\{v_i\} \\ \Sigma v_i = k}} \prod_{i=1}^{s} X_{v_i, i}(\beta', \alpha) \right]$$

$$\times \frac{D_l(\{\hbar\omega_i\})}{l!} Y_m(\alpha, \{\hbar\omega_i\}) \tag{38}$$

where the various factors and their correspondence with the factors in Eq. 37 are as follows: The powers of E arise simply from differentiating the exponential in Eq. 37. Beyond this, we have

$$\frac{D_l(\{\hbar\omega_i\})}{l!} = \sum_{\Sigma v_i = l} (\tfrac{1}{2}\hbar\omega_i)^{v_i} \cdots (\tfrac{1}{2}\hbar\omega_s)^{v_s} D_{v_i} \cdots D_{v_s}$$

$$D_{2v} = \frac{2}{(2v)!}(-1)^v (2^{2v-1} - 1) B_{2v}$$

$$D_{2v+1} = 0 \tag{39}$$

which arises from the differentiation of g_1. This factor arises in a similar way in analyzing the ordinary density of states (42) (without any special weighting factors for coupling) and is taken over directly. [The B's are Bernoulli numbers (42).] The factors

$$Y_m(\alpha, \{\hbar\omega_i\}) = \sum_{\substack{v_i \\ \Sigma v_i = m}} \prod_{i=1}^{s} \frac{[(1/2)\hbar\omega_i]^{v_i} - \alpha^2 [-(1/2)\hbar\omega_i]^{v_i}}{v_i!} \tag{40}$$

come from the g_2 factor in Eq. 37, and finally,

$$X_{v,i}(\beta', \alpha) = \frac{(-\hbar\omega_i)^v}{v!} \sum_{m=0}^{v} \mathscr{S}_v^{(m)} \frac{\alpha^m e^{m\beta' \hbar\omega_i} m!}{(1 - e^{\beta' \hbar\omega_i}\alpha)^{m+1}} \tag{41}$$

arises out of successive derivatives of $f_1(\beta)$. Here $\mathscr{S}_v^{(m)}$ is a Stirling number of the second kind (43), and the formula used in Eq. 41 for the higher order derivatives can be proved inductively using the recurrence relation

$$\mathscr{S}_{v+1}^{(m)} = m\mathscr{S}_v^{(m)} + \mathscr{S}_v^{(m-1)} \tag{42}$$

In order to obtain $G_\alpha(E, E')$, we must perform the inverse of the Laplace transform operation indicated in Eq. 22 by carrying out the Laplace inversion integral of $F(E, \beta')$ with respect to β',

$$\int_{c-i\infty}^{c+i\infty} f_1(\beta')[\tilde{H}_1(E, \beta') + \tilde{H}_2(E, \beta')]e^{\beta'E'} d\beta' \tag{43}$$

where we have explicitly used Eqs. 31 and 33 to express $F(E, \beta')$ in terms of the quantities $\tilde{H}_1$ and $\tilde{H}_2$ derived.

The integrand contains the product of f_1 and $H(E, \beta')$, and any poles of either factor would contribute to the integral provided they are enclosed by the contour chosen for the evaluation. It turns out, however, that, considering the exponential factor in Eq. 43, the permissible choice of contour depends on whether the energy E' conjugate to the integration variable β' is greater or less than the energy E. Evidently, the final result cannot depend on whether $E > E'$ or $E < E'$ since the definition of $G_\alpha(E, E')$ in Eq. 21 is manifestly symmetric to the interchange of E and E'. Therefore, let us (at least for the moment) simply *choose* $E' > E$ and evaluate the integral in Eq. 43 accordingly. In this case the asymptotic bahavior in thee terms involving $\tilde{H}_1$ and $\tilde{H}_2$ both allow closure of the contour in the left-half plane. This works out independent of the order of E and E' for the $\tilde{H}_1$ term, and for the $\tilde{H}_2$, we see that the exponential in Eq. 38 gets combined with the exponential in Eq. 43 to form a factor $e^{(\beta'[E'-E])}$ where the energy difference $E' - E$ is positive. Therefore, in evaluating Eq. 43, we choose a common contour for both terms, which, because it is in the left-half plane, surrounds all the poles of the integrand.

The poles in Eq. 43 may come from either the f_1 factor or from $H_1 + H_2$. For the latter, we see from Eqs. 35, 38, and 41 that H_1 and H_2 each have higher order poles at the values $\beta' = -\beta_j$. It can, however, be shown, because of the way these terms arose, that the contributions from these poles must cancel exactly (see Appendix A). The poles of f_1 are thus the *only* poles. These are simple poles (exactly as in Eqs. 35 and 36, and the result is therefore simply

$$\tilde{G}_\alpha(E, E') = \sum_j P_j \alpha^{E'/\hbar\omega_j}[\tilde{H}_1(E, \beta_j) + \tilde{H}_2(E, \beta_j)] \tag{44}$$

We can now use Eq. 34 for β_j in Eq. 44 to obtain the following explicit results:

$$G_\alpha(E, E') = G_\alpha^{(S)}(E, E') + G_\alpha^{(A)}(E^>, E^<) \tag{45}$$

where $G_\alpha^{(S)}(E, E')$ results from the $\tilde{H}_1$ term in Eq. 44 and $G_\alpha^{(A)}(E^>, E^<)$ results from the $\tilde{H}_2$ term. Here $E^> = E'$ and $E^< = E$, and the following explicit expressions apply:

$$G_\alpha^{(S)}(E, E') = \sum_{j=1}^{s} \sum_{k=1}^{s} \alpha^{E/(\hbar\omega_j) + E'/(\hbar\omega_k)} P_j P_k$$

$$\times \prod_{i=1}^{s} \frac{1 - \alpha^{2 - \omega_l(\omega_j^{-1} + \omega_k^{-1})}}{1 - \alpha^{-\omega_l(\omega_l^{-1} + \omega_k^{-1})}} \tag{46}$$

$$G_\alpha^{(A)}(E^>, E^<) = \left[\prod_{i=1}^{s} (\hbar\omega_i) \right]^{-1} \sum_{k=0}^{s-1} \frac{(E^<)^k}{k!} \sum_{l=0}^{s-1-k} \mathscr{D}_l(\alpha)$$

$$\times \sum_{j=1}^{s} \mathscr{P}_{s-1-k-l}^{(j)}(E^> - E^<) \tag{47}$$

where

$$\mathscr{D}_l(\alpha) = \sum_{q=0}^{l} \frac{D_q(\{\hbar\omega_i\})}{q!} \sum_{\substack{v_i \\ \Sigma v_i = l - q}} \prod_{i=1}^{s} \frac{[(1/2)\hbar\omega_i]^{v_i} - \alpha^2 [-(1/2)\hbar\omega_i]^{v_i}}{v_i!} \tag{48}$$

and

$$\mathscr{P}^{(j)}(\Delta E) = C_m^{(j)} \frac{\alpha^{\Delta E/(\hbar\omega_j)}}{\hbar\omega_j} \prod_{k \neq j} \frac{1}{1 - \alpha^{1 - \omega_k/\omega_j}} \tag{49}$$

In the last expression we have used the definition of P_j in Eq. 36. There is also a separate (energy-independent) coefficient $C_m^{(j)}$ for each mode j given by

$$C_m^{(j)} = \sum_{\substack{\{v_i\} \\ \Sigma v_i = m}} \prod_{i=1}^{s} \frac{(-\hbar\omega_i)^{v_i}}{v_i!} \sum_{n=0}^{v_i} \mathscr{S}_{v_i}^{(n)} \frac{n! \alpha^{n(\omega_i + \omega_j)/\omega_j}}{(1 - \alpha^{(\omega_i + \omega_j)/\omega_j})^{n+1}} \tag{49a}$$

In these expressions, $\mathscr{D}$ was derived as a combination of the D_q's and the Y_m's defined in Eq. 40 and $C_m^{(j)}$ involves the X function (see Eq. 41) evaluated for $\beta' = \beta_j$. The first few of these coefficients have been evaluated explicitly in Appendix B. The notation $E^>$, $E^<$ for the two energies introduced in Eq. 47 takes into account the fact that the expression $G_\alpha^{(A)}$ that results from substituting in Eq. 44 is not explicitly symmetric to interchanging the two energies, but in deriving Eq. 44, we had to assume that E' was larger than E. We should comment at this point that if, in evaluating the integral (Eq. 43), we had made the opposite assumption, the contour for the H_2 term would have had to be closed in the right-half plane, and the contribution from this half would then have vanished. (By definition, none of the poles are enclosed since the contour c is to the right of all poles.) But then, in the remaining term, there would be nothing to cancel the poles of H_1. The net result is that we again obtain Eq. 44, but now with the two energies exchanged.

For sufficiently high energies two approximations can be introduced in our expressions for G_α. First, we see that making E and E' large with a fixed difference increases $G_\alpha^{(A)}$ compared with $G_\alpha^{(S)}$, allowing us to drop out $G_\alpha^{(S)}$ altogether. (Actual sample calculations show that $G_\alpha^{(S)}$ can apparently always be neglected even at low energies.) Second, $G_\alpha^{(A)}$ itself consists of a sum involving powers of $E^<$ up to $s-1$. At high energy, we may be able to retain just the highest powers leading to (see Appendix B)

$$
G_\alpha = \left[\prod_{i=1}^{s} (\hbar\omega_i) \right]^{-1} \sum_{j=1}^{s} \frac{\alpha^{\Delta E/(\hbar\omega_j)}}{\hbar\omega_j} \prod_{k \neq j} \frac{1 - \alpha^2}{1 - \alpha^{(\omega_k/\omega_j)+1})(1 - \alpha^{-(\omega_k/\omega_j)+1})}.
$$
$$
\times \left\{ \frac{E^{s-1}}{(s-1)!} + \frac{E^{s-2}}{(s-2)!} \sum_{l=1}^{s} \frac{\hbar\omega}{2^l} \left[\frac{1 + \alpha^2}{1 - \alpha^2} - \frac{2\alpha^{(\omega_l + \omega_j)/\omega_j}}{1 - \alpha^{(\omega_l + \omega_j)/\omega_j}} \right] \right\} \tag{50}
$$

In the following section, we present some numerical results for the evaluation of the full expression for G_α, Eqs. 45–47, and some analysis that pertains to this high-energy limit.

B. Comparison with Earlier Results

The linewidth saturation behavior that we saw in our earlier results (18) (see also Figure 6) should be reflected in how the rates calculated using our final expressions (Eqs. 46 and 47) depend on the bath energies E and E'. In Figure 7, we plot, as a function of initial bath energy E, the proper dephasing $\hat{w}$ and the T_1 relaxation rate for transfer of a single quantum of energy $\hbar\omega$ to the bath. These show the same saturation behavior predicted previously in the degenerate oscillator model. The jagged curves on the same plot show results derived from a previous analysis (44). This analysis also included distributed frequencies, but without the averaging over initial states implied in Eqs. 15 and 17. In constructing Figure 7, we have used our earlier results directly by determining all initial states within successive small energy intervals and calculating their average width. The variation about the average value decreases at high energies, better satisfying the formal requirement in the derivation (16) that there be *no* variation.

The significance of the widths associated with individual (zero-order) initial states can be understood from reference to Figure 5. The nearby energy states *each* interact with the given state so that the profile of its amplitude in the *eigenstates* is spread out from the noninteracting energy. The degree of spread depends on the density and interaction strength of these "final" states. This accounts for one summation in Eq. 18. The other summation is an average over initial states formally required to get the relaxation rates. Figure 7 in

effect compares an explicit numerical averaging of the earlier results with our
present analytic averaging.

   A direct analytic comparison can also be made with earlier linewidth results
for a degenerate oscillator model (18) by taking the results derived here in the
saturation limit (large heat bath energies). The limiting relaxation rates for a

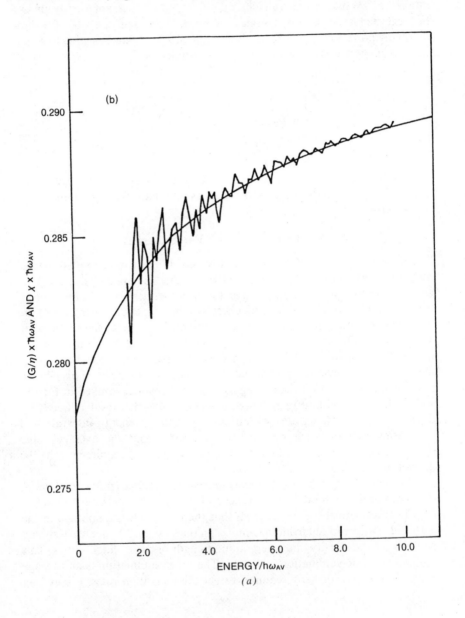

(a)

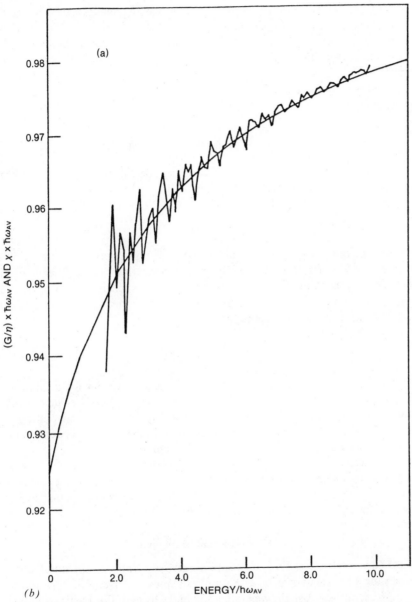

Figure 7. Intramolecular relaxation rates calculated using RQE matrix elements (Eq. 45) and weighted density of states analysis for H_2CO molecular frequencies (smooth curves) compared with explicit averages over initial states for results in Ref. 44. Part (a) shows proper dephasing rates as a function of bath energy and part (b) shows a T_1 rate for transfer of an amount of energy equal to average H_2CO quantum to the entire collection of modes treated as heat bath. In part (a), a correction term taking out unphysical α^0 contributions and discussed in Ref. 44 has been included for both averaged and unaveraged results.

bath of s oscillators with particular normal-mode frequencies is determined by evaluating $G_\alpha(E, E + \Delta E)/\eta(E)$ in the limit of large E. (See Eqs. 15 and 20.) The energy difference ΔE is fixed to a value that indicates the type of process whose limiting rate is being sought; that is, ΔE would be zero for proper dephasing or equal to the energy of one or more pump mode quanta for a T_1 relaxation. Fortunately, this limit is very easy to evaluate because $\eta(E)$ itself is given (42) by an expansion,

$$\eta(E) = \left[\prod_{i=1}^{s} (\hbar\omega_i) \right]^{-1} \sum_{k=0}^{s-1} \frac{E^k}{k!} \frac{D_{s-1-k}(\{\hbar\omega_i\})}{(s-1-k)!} \tag{51}$$

very similar to that in Eq. 47 for G_α. Taking the ratio of the leading order terms in the two expansions leads to the result

$$\chi^\infty(\alpha, \phi, E) = \sum_{j=1}^{s} \frac{\alpha^{(E-E_\phi)/(\hbar\omega_j)}}{\hbar\omega_j} \prod_{i \neq j} \frac{1-\alpha^2}{(1-\alpha^{(\omega_i/\omega_j)+1})(1-\alpha^{-(\omega_i/\omega_j)+1})} \tag{52}$$

However, this expression is identical to a limiting high-energy width expression χ^∞ derived earlier for specific states with large initial excitations (44). The degenerate-frequency limit for these specific-state widths was analyzed elsewhere (44), and this analysis can be extended to prove that if ΔE is equal to an integral number of quanta,

$$\chi^\infty = \sum_{m=d}^{\infty} R_{s,m,d}\alpha^m \tag{53}$$

where the $R_{s,m,d}$'s are combinatorial factors derived earlier for a molecular model with s degenerate normal-mode frequencies. Here d is just $\Delta E/\hbar\omega$. Since Eq. 53 is the expression that appears in Ref. 18 for calculating the limiting linewidths, we see that the distributed frequency results derived here reduce correctly to the earlier degenerate oscillator results.

Actual calculations using Eq. 47 for two molecules with the same number of modes but different normal-mode frequencies show noticeable effects of including the actual frequency distribution. For example, the limiting linewidths for HCOOH and DCOOD differ by about 10%.

C. Generalized Models

There are other restrictions on coupling between states in a real polyatomic molecule besides degree of quantum exchange—most notably symmetry requirements (45). A slight generalization in the form of Eq. 14 consistent with the general requirement to make coupling small when the quantum number differences are large takes the form

$$V_{nn'} = \sum_j \tilde{V}_j \sum_{l=1}^{s} (\alpha_{jl})^{|n_l - n_l'|} \tag{54}$$

where there are now a number of terms j in the potential and each mode l has its own falloff parameter α_{jl} (e.g., see Eq. 56) for each term in the potential. The different terms in the potential could correspond to different physical locations in the molecule—different interatomic bonds, for example—and in a symmetric molecule various terms in Eq. 54 would be related to each other by symmetry operations. Consider, for example, a linear molecule where the potential consists of local Morse potential interactions between adjacent atoms. The matrix elements of the zero-order harmonic states are (46)

$$V_{nn'}^{\text{Morse}} = \tilde{V}_{nn'} + \sum_j D_j \prod_{l=1}^{s} -(\alpha_{jl})^{|n_l - n_l'|} E_{n_l n_l'}(\alpha_{jl})$$

$$- 2D_j \prod_{l=1}^{s} \left(-\frac{\alpha_{jl}}{2}\right)^{|n_l - n_l'|} E_{n_l n_l'} \frac{\alpha_{jl}}{2} \tag{55}$$

where D_j is the dissociation energy of the jth bond,

$$\alpha_{jl} = C_{lj}\left(\frac{2\hbar\omega_l}{D_j}\right)^{1/2} \tag{56}$$

and

$$E_{nm}(\alpha) = E_{mn}(\alpha) = e^{\alpha^2/4}\left(2^{m-n}\frac{m!}{n!}\right)^{1/2} L_m^{(n-m)}\left(-\frac{\alpha^2}{2}\right) \qquad n \geq m \tag{57}$$

Here C_{lj} is the expansion coefficient for the lth normal mode in the jth bond, a quantity of order unity or less. (The contribution $V_{nn'}$ to the potential connects states differing by two quanta within a *given* mode with no other changes, and the net effect is statistically unimportant.) There are two important things to notice about α_{jl}. First, whenever the energy of the normal-mode quantum is small compared with the bond dissociation energy—the most common, if not quite universal, case—α_{jl} is small and serves as the required order parameter in the generalized RQE expression (Eq. 54). [In this context the $E_{nm}(\alpha)$ factors contribute, in addition to the leading order α^{n-m} factors, α-independent numerical factors that give rise to a small *energy* dependence in the effective α values.] Second, contribution from the expansion coefficient factor automatically brings in the effects of molecular symmetry. For example, consider the linear molecule shown schematically in Figure 8. By choosing periodic boundary conditions and making all atoms and all bonds equivalent (47), we impose particularly straightforward symmetry requirements on the normal modes. The normal modes for bond stretches along the chain are completely determined by the symmetry, and the coefficients of the lth normal mode can

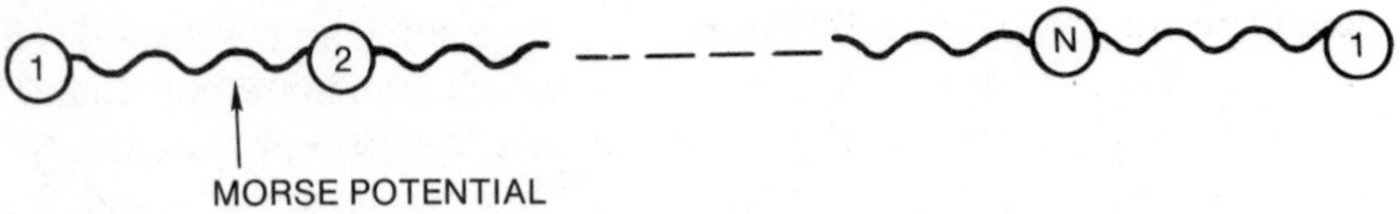

JAMES STONE

MORSE POTENTIAL

Figure 8. Linear molecule consisting of N identical atoms with Morse interactions "Periodic" bonding of atom 1 to atom N also allows for N identical bonds.

be written,

$$\left(\frac{1}{N}\right)^{1/2} \exp\frac{2\pi ijl}{N+1} \tag{58}$$

Clearly, when combinations of these factors are summed over the bond index j in Eq. 54, the terms can cancel to give a vanishing matrix element, as may be required by symmetry. Thus, the generalized RQE expression, Eq. 54, is sufficiently flexible to include both symmetry restrictions and quantum exchange restrictions. Finally, we note that for simple bond potentials, the correspondence between Eqs. 54 and 55 serves to identify both the RQE V_0 and the α parameters.

IV. DEGENERATE LADDER MODELS

The GME formalism has proved to be useful in analyzing the steady-state (21) and transient (22) behavior of driven multilevel systems. Up until now, however, this technique has only been applied to simple ladder systems. Even in a system as complex as a polyatomic molecule with intramolecular interactions, simple ladder models can often be applied. [This includes the heat bath feedback model, e.g., because even though there are multiple ladders (see Figure 2), there is no *radiative* coupling from one ladder to another, only relaxation terms.] However, even in the absence of detailed models to describe more general situations, we can anticipate that models with degenerate-level structure will play an important role.

A detailed model for radiative pumping and the rates of subsequent intramolecular relaxation in a polyatomic molecule requires only simple ladder models when the transition strength for the radiative transitions arises from a single infrared active pump mode. In general, we expect more complicated models to arise when there are multiple active modes fairly close in energy. (Modes that degenerate because of symmetry form an important subcase.) Also, if the intramolecular coupling is relatively weak, the effective density of states may be low enough that the bands of states formed as a result of the interaction may be better described by several instead of a single broadened state. (The first transfer of energy to the heat bath will always have a relatively small effective density of states.) An analysis of the available states in SF_6 using ideas about the coupling closely related to those expounded in Section III (48) shows an unexpectedly low effective density of states, for example. A general description covering all these cases would allow there to be *several* states at each level accessible by radiative transitions, each with its own broadening and relaxation rates.

A four-level system showing the type of generalized structure we have in mind is shown in Figure 9. At each successive level there are several states that may be accessed radiatively from one or more of the states at the preceding level. A formal description of a state now requires both a primary index to specify which level and a secondary degeneracy index. A density matrix

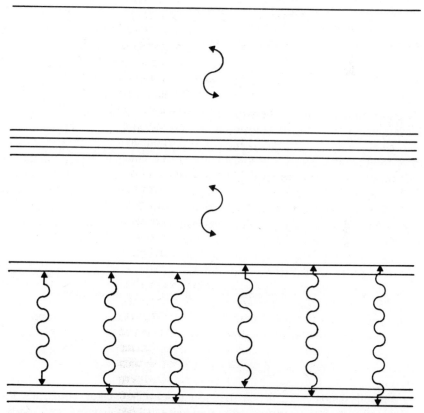

Figure 9. Schematic diagram of possible transitions in a ladder model with several states at each level (a "degenerate-ladder" model). Transitions explicitly indicated between first two levels. Formalism allows each state to have its own widths and relaxation rates.

equation incorporating this type of level structure is again described by Eq. 1,

$$\frac{d\rho}{dt} = \frac{i}{\hbar}[\rho, H] + \frac{1}{\tau}(K-1)\rho \tag{1}$$

Similar to the nondegenerate case, except that now we have

$$H_{n,\xi;n,\xi} = \hbar\omega_{n\xi} \tag{59}$$

$$H_{n-1,\xi;n,\eta} = H_{n,\eta;n-1,\xi} = \alpha_n^{\xi\eta} A \sin \chi t \tag{60}$$

Here K is a relaxation operator that could refer to a detailed model of intramolecular relaxation for cases such as those already discussed. However, since such models remain to be developed, for definiteness, here we use a collisional relaxation model for K based on a model of energy-changing (T_1) and phase-changing (T_2) collisions previously considered in Ref. 21, generalized to take into account the more complex level structure

$$(Kp)_{n\xi,l\phi} = \delta_{nl}\delta_{\xi\phi}\sum_{m,\eta} P_{(n,\xi)\leftarrow(m,\eta)}\rho_{m,\eta;m,\eta} \tag{61}$$

To make the meaning of the various expressions here and in what follows clearer, we are following the convention that Greek letters are used to specify all degeneracy indices. Here $1/\tau$ is the collision frequency, and $P_{(n,\xi)\leftarrow(m,\eta)}$ is the probability that during a collision, a molecule initially in level m in the state η will make a transition to level n in the state ξ. It is convenient to refer to collisions in which any such transitions occur as T_1-type collisions even if the transition is between two sublevels of the same main level. *All* collisions according to Eqs. 1 and 61 result in dephasing by setting off-diagonal density matrix elements to zero.

A. Generalized Master Equation for Degenerate Systems

In preparation for the GME analysis, we go to the rotating-wave approximation and make the transformation

$$\rho_{n,\xi;m,\eta} = C_{n,\xi;m,\eta}e^{i(m-n)\chi t} \tag{62}$$

on the off-diagonal density matrix elements. The vector equations for $dC_{n,\xi;m,\eta}/dt$ are then a matrix equation with a time-independent matrix coefficient. Thus, there are eigensolutions with the time dependence $e^{-\lambda t}$, and the following sets of equations are obtained for these eigensolutions of the density matrix:

$$\lambda C_{n,\xi;n,\xi} + R_{n\xi} + \sum_{\eta'} \beta_n^{\eta'\eta} C_{n,\xi;n-1,\eta'} - \sum_{\eta'} \beta_{n+1}^{\eta\eta'} C_{n,\xi;n+1,\eta'}$$

$$+ \sum_{\xi'} \beta_n^{\xi'\xi} C_{n-1,\xi';n\eta} - \sum_{\xi'} \beta_{n+1}^{\xi\xi'} C_{n+1,\xi';n\eta}$$

$$= 0 \tag{63}$$

$$- C_{n\xi,mn} \left[\frac{1}{\tau} - \lambda + i(m-n)\chi + i(\omega_{n\xi} - \omega_{mn}) \right] + \sum_{\eta'} \beta_m^{\eta'\eta} C_{n,\xi;m-1,\eta'}$$

$$- \sum_{\eta'} \beta_{m+1}^{\eta\eta'} C_{n,\xi;m+1,\eta'} + \sum_{\xi'} \beta_n^{\xi'\xi} C_{n-1,\xi';mn} - \sum_{\xi'} \beta_{n+1}^{\xi\xi'} C_{n+1,\xi';mn}$$

$$= 0 \tag{64}$$

In the level population equations (Eq. 63) we have referred to the collisional transition terms by a single compact matrix R defined by

$$R_{n\xi} = \frac{1}{\tau} \sum_{(m,\eta) \neq (n\xi)} P_{(n,\xi) \leftarrow (m,\eta)} C_{m,\eta;m,\eta}$$

$$- C_{n\xi,n\xi} \frac{1}{\tau} \sum_{(m,\eta) \neq (n,\xi)} P_{(m,\eta) \leftarrow (n,\xi)} \tag{65}$$

The coupling terms $\beta_m^{\eta\eta'}$ are the Rabi frequencies

$$\beta_m^{\eta\eta'} = \frac{\alpha_m^{\eta\eta'} A}{2\hbar} \tag{66}$$

where the index m refers to the upper level of the (radiative) transition. (Since coupling between *adjacent* levels is assumed, the lower level is then $m - 1$.) The two superindices then refer, respectively, to the state within the lower level (η) and the state within the upper level (η'). We have been able to write Eqs. 63 and 64 in a general form not depending on whether either of the levels is the ground (0) or highest ($N - 1$) level by defining β_0 and β_N (independent of any degeneracy index) to be zero.

Our algebraic reduction of the full set of density matrix equations to equations involving just the level populations will exploit the fact that coupling is only present between adjacent levels. The first immediate consequence of this choice of coupling is that the $(i, \xi; j, \eta)$ element of the Liouville operator $[, H]$ in Eq. 1 includes, besides $c_{(i,\xi;j\eta)}$ itself, only those density matrix elements $c_{(i',\xi';j';\eta')}$ where $i' = i \pm 1$ (with $j' = j$) or $j' = j \pm 1$ (with $i' = i$). Furthermore, if we consider the level differences $k = i - j$ and $k' =$

$i' - j'$, we find that the coupling is between density matrix elements where k and k' differ by 1. It is therefore useful to define a new notation for the density matrix elements, diagonal and off-diagonal, where those elements having the same level *difference* are grouped together in distinct vectors,

$$(q^k)_i^{\xi\eta} = C_{i-1,\xi;N-k+i-1,\eta} \tag{67}$$

Here we have taken the level difference to be $N - k$, which results for the nondegenerate case in k being equal to the number of elements in the vector (N level populations where the difference is zero but just 1 where it is $N - 1$

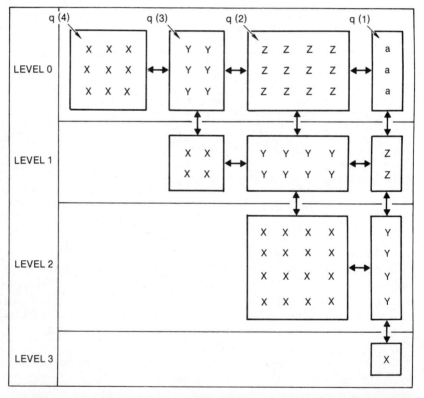

Figure 10. Organization of degenerate-ladder density matrix elements for derivation of generalized master equation. Matrix elements are grouped together into a vector $q^{(i)}$ according to the number of levels separating them, irrespective of state specification within a level. Rabi-frequency-type couplings appear between individual density matrix elements in the blocks joined by arrows. Equivalently, elements of $q^{(i)}$ couple to elements of $q^{(i+1)}$ and $q^{(i-1)}$.

since this difference only occurs between the two extreme levels). The grouping is more involved for the degenerate case but entirely analogous as illustrated in Figure 10 for a four-level system where the respective degeneracies of the levels starting with the ground level are 3, 2, 4, and 1. Each letter x, y, z, or a represents the position of a density matrix element where the level difference $j - i \geqslant 0$. The set of elements where the primary level difference is zero includes but is not limited to the diagonal elements or level populations since $c_{(i,\xi;j,\eta)}$ requires both $j = i$ and $\xi = \eta$ to be diagonal. Nevertheless, all elements with $j = i$ are given the same symbol, x, in Figure 10 and are all considered to be part of the vector q^4. Elements within a given block of x's all have the same level i corresponding to the index pair i, i, but only those elements along the diagonal of each block qualify as diagonal elements of the density matrix. Similarly, for example, q^3 collects together those elements where $j - i = 1$ (y in Figure 10, blocked together according to the particular values of i and j). In what follows it will be convenient to distinguish the off-diagonal elements such as those in q^3 where the primary levels differ as "totally off-diagonal" elements and refer to the elements of q^4 collectively as "primarily diagonal."

The leading term in Eq. 64 involves diagonal terms from the Hamiltonian (energy differences between particular states), effective compensating field energy differences resulting from the time dependence of the transformation to the rotating frame in Eq. 62, a damping constant $1/\tau$, and the λ coming from the assumed exponential time dependence. Together these form a quantity that is in most respects similar to a simple two-level resonance denominator,

$$(Q^k)_i^{\xi\eta} = \frac{1}{\tau} - \lambda + i(N - k)\chi + i(\omega_{i-1,\xi} - \omega_{i-1+N-k,\eta}) \tag{68}$$

However, in the multilevel system, the effect of a single resonance pair is not isolated but enters into the determination of the response of the system as a whole, as in the coupled equations, Eq. 64. With the definitions given in the preceding for the vector components q^k, we can now rewrite these coupled equations in the form

$$(Q^k)_i^{\xi\eta}(q^k)_i^{\xi\eta} = (r^k)_i^{\xi\eta} + (s^k)_i^{\xi\eta} \tag{69}$$

The coupling to components of the neighboring vectors q^{k-1} and q^{k+1} is included here in separate terms written as components of vectors s^k and r^k, where

$$(r^k)_i^{\xi\eta} = \sum_{\eta'} \beta_{N+i-k-1}^{\eta'\eta}(q^{k+1})_i^{\xi\eta'} - \sum_{\xi'} \beta_i^{\xi\xi'}(q^{k+1})_{i+1}^{\xi'\eta} \tag{70}$$

and

$$(s_i^k)^{\xi\eta} = \sum_{\xi'} \beta_{i-1}^{\xi'\xi}(q^{k-1})_{i-1}^{\xi'\eta} - \sum_{\eta'} \beta_{N-k+i}^{\eta\eta'}(q^{k-1})_i^{\xi\eta'} \tag{71}$$

For $k = 1$, the level index i in Eq. 69 only takes on the value 1. In this case the s^k term is not present since q^0 is undefined and the β's in both terms of Eq. 71 (β_0 and β_N, respectively) are zero. Thus, Eq. 69 for $k = 1$ is just a linear relationship between the vectors q^1 and r^1 expressed by the matrix equation

$$(r^1)_i^{\xi\eta} = \sum_{juv} A_{i(\xi\eta),j(uv)}^{(1)}(q^1)_j^{uv} \tag{72}$$

where $A^{(1)}$ is the diagonal matrix defined by

$$A_{i(\xi\eta),i(\xi',\eta')}^{(1)} = \delta_{\xi\xi'}\delta_{\eta\eta'}(Q^1)_i^{\xi\eta} \tag{73}$$

This matrix is the first of a set of matrices used to reduce the problem. The Q in Eq. 73 contains energy differences between the *extreme* levels 0 and $N - 1$ (see Eq. 69). The index i (and j in Eq. 72) only takes the value 1 since there is only one pair of levels separated by $N - 1$. We will find a *general* prescription that gives a matrix $A^{(k)}$ relating the q and r vectors,

$$(r^k)_i^{\xi\eta} = \sum_{juv} A_{i(\xi\eta),j(uv)}^{(k)}(q^k)_j^{uv} \tag{74}$$

The ultimate usefulness of this is that for $k = N - 1$ we will be able to invert A and use the result to eliminate all of the so-called totally off-diagonal density matrix elements from the equations for the level populations.

The process of obtaining the higher A matrices is recursive. Assuming that $A^{(k-1)}$ is known, an expression for q^{k-1} in terms of r^{k-1} is

$$(q^{k-1})_i^{\xi\eta} = \sum_{juv} M_{i(\xi\eta),j(uv)}^{(k-1)}(r^{k-1})_j^{uv} \tag{75}$$

where $M^{(k-1)}$ is the matrix inverse of $A^{(k-1)}$. Now consider Eq. 69 for the next group of off-diagonal elements q^k. The following sequence of substitutions is used to replace s^k on the right side: We first use Eq. 71 to replace s^k by q^{k-1} in Eq. 69. We then use Eq. 75 to replace $q^{(k-1)}$ in the resulting expression by r^{k-1}. Finally, we use Eq. 70 (with k replaced by $k - 1$) to replace r^{k-1} by q^k. The final result is that the quantity $(s^k)_i^{\xi\eta}$ is replaced by

$$\sum_{j(\xi'\eta')} B_{i(\xi\eta),j(\xi'\eta')}(q^k)_j^{\xi'\eta'} \tag{76}$$

where

$$B_{i(\xi\eta),j(\xi'\eta)}^{(k)} = \sum_u \sum_v \beta_{N-k+i}^{\eta u} M_{i(\xi u),j(\xi'v)}^{(k-1)} \beta_{N-k+j}^{\eta'v}$$
$$- \sum_u \sum_v \beta_{N-k+i}^{\eta u} M_{i(\xi u),j-1(v\eta')}^{(k-1)} \beta_{j-1}^{v\xi'}$$
$$- \sum_u \sum_v \beta_{i-1}^{u\xi} M_{i-1(u\eta),j(\xi'v)}^{(u-1)} \beta_{N+j-k}^{\eta'v}$$
$$+ \sum_u \sum_v \beta_{i-1}^{u\xi} M_{i-1(u\eta),j-1(v\eta')} \beta_{j-1}^{v\xi'} \qquad (77)$$

Thus, as we had hoped, r^k can now be expressed in terms of q^k using the matrix $A^{(k)}$ defined by

$$A_{i(\xi\eta),j(\xi',\eta')}^{(k)} = \delta_{ij}\delta_{\xi\xi'}\delta_{\eta\eta'}(Q^k)_i^{\xi\eta} + B_{i(\xi\eta),j(\xi'\eta')}^{(k)} \qquad (78)$$

This formula, together with the inversion of the matrix $A^{(k-1)}$ to obtain $M^{(k-1)}$, provides the necessary recursive steps to produce the required quantities up through $A^{(N-1)}$ and its inverse, $M^{(N-1)}$. For the simple nondegenerate ladder case, these steps reduce to those specified earlier, where a single term appears for each sum over degeneracy indices.

For the nondegenerate cases, the final step (21) in obtaining a GME was to use $M^{(N-1)}$ to eliminate those off-diagonal elements occurring in the equations for the diagonal density matrix elements (the $q^{(N-1)}$'s and their conjugate elements just below the diagonal). In the present case, a similar set of algebraic manipulations can be applied to the full set of $q^{(N)}$'s. However, the resulting closed set of equations now includes both diagonal elements, which in the vector notation are

$$n_{i-1,\xi} = (q^N)_i^{\xi\xi} \qquad (79)$$

and off-diagonal elements between different states at the same level,

$$c_{i-1,\xi;i-1,v} = (q^N)_i^{\xi v} \qquad (80)$$

The first step in obtaining these *partially* reduced equations is, as in the nondegenerate case, to eliminate (totally) off-diagonal elements whose level indices differ by ± 1. The required conjugacy relations for elements below the diagonal are

$$c_{i-1,\xi;i-2,\xi'} = [c_{i-2,\xi';i-1,\xi}]^* \qquad (81)$$

In vector notation,

$$(q^{(N+1)})_i^{\xi\eta} = (q^{(N-1)})_{i-1}^{\eta\xi} \qquad i = 2, N \qquad (82)$$

Finally, substituting Eq. 82 into Eq. 70 and comparing with Eq. 71, we have

$$(r^N)_i^{\xi\eta} = (s^N)_i^{\eta\xi*} \tag{83}$$

Now we must consider Eq. 63 for the diagonal density matrix elements and Eq. 69 with $k = N$ for the subset of off-diagonal elements in Eq. 80, that is, those so-called primarily diagonal elements that are off-diagonal but only between degenerate states at the same level. Using Eq. 83, we find

$$\lambda(q^N)_{n+1}^{\xi\xi} + R_{\eta\xi} + (s^N)_{n+1}^{\xi\xi} + [(s^N)_{n+1}^{\xi\xi}]^* \tag{84}$$

$$(Q^N)_i^{\xi v}(q^N)_i^{\xi v} = (s^N)_{n+1}^{\xi v} + [(s^N)_{n+1}^{\xi v}]^* \tag{85}$$

The sequence of substitutions leading to Eq. 76 apply here also to eliminate S^N in favor of q^N, and the result is

$$-\lambda(\mathbf{q}^N)^D = \mathbf{R} + Z^{DD}(\mathbf{q}^N)^D + Z^{DO}(\mathbf{q}^N)^O \tag{86}$$

$$Z^{OO}(\mathbf{q}^N)^O = Z^{OD}(\mathbf{q}^N)^D \tag{87}$$

Here the superscripts D and O on the matrix Z and the vector $\mathbf{q}^N$ refer, respectively, to the submatrices or subvectors where the corresponding indices are diagonal or off-diagonal in the degeneracy, and Z is the matrix defined by

$$Z_{i(\xi\eta),j(\xi'\eta')} = \delta_{ij}\delta_{\xi\xi'}\delta_{\eta\eta'}(Q_i^N)^{\xi\eta}(1 - \delta_{\xi\eta}) + B_{i(\xi\eta),j(\xi'\eta')}^{(N)} + B_{i(\eta\xi),j(\eta'\xi')}^{(N)*} \tag{88}$$

Finally, Eq. 87 is solved for $(\mathbf{q}^N)^O$, and the result is substituted into Eq. 86 to give

$$\lambda\rho_{i\xi,i\xi} = \sum_{\substack{j \neq i \\ \eta \neq \xi}}\left[\frac{P_{(i,\xi)\leftarrow(j,\eta)}}{\tau} + W_{i\xi,j\eta}(\lambda)\right]\rho_{j\eta,j\eta} - \rho_{i\xi,i\xi}\sum_{\substack{j \neq i \\ \eta \neq \xi}}\frac{P_{(j,\eta)\leftarrow(i,\xi)}}{\tau} + W_{j\eta,i\xi}(\lambda) \tag{89}$$

where

$$W = -Z^{DD} + Z^{DO}[Z^{OO}]^{-1}Z^{OD} \tag{90}$$

The expressions for the radiative transition terms W_{ij} taken from this formal matrix representation are the generalization of our previous GME analysis for a degenerate level structure. For a simple ladder, Eq. 90 gives our earlier results exactly (21, 22) since in this case there is no off-diagonal degeneracy index space and the contribution from the second term, involving Z^{OO}, is absent, and we are left with the result in Eq. 88.

The relationship among the level populations defined by Eq. 89 defines either a steady-state solution ($\lambda = 0$) or a time-dependent component of a transient solution. Since the relationship only involves the diagonal elements of the density matrix, once we can write Eq. 89 explicitly—including values for all the W's—the problem of finding a solution is much simpler than solving Eqs. 63 and 64 directly. The hard part is finding the W values. For the nondegenerate ladder case, the most difficult step in obtaining the steady-state W's [or $W(\lambda)$ for a known value of λ] is constructing and inverting an $(N-1) \times (N-1)$ matrix, a substantially easier operation than the original $N^2 \times N^2$ problem. For degenerate ladders, the amount of work saved depends on the total number of states and their distribution among levels. Some examples are considered in the following section.

B.   Quantum Time Averages in Degenerate Systems

We have incorporated the steps in the recurrence procedure into a computer program that can be used to find steady-state or transient components for arbitrarily defined degenerate ladders, subject only to the storage and precision limitations of the machine used. As an illustration of the results that can be obtained using this methodology, we consider some specially defined steady-state solutions that correspond approximately to quantum-mechanical time-averaged solutions in degenerate ladder systems. The connection with the time average is obtained by using strong collision models k^i of the type discussed in Section II.A, following Eq. 8 and allowing the collision frequency in the calculation to become small. The steady state is found to approach a low-pressure limit within the accuracy of the calculation.

The first example is a four-level system with a twofold degeneracy in each of its levels. Radiative coupling is allowed between every pair of states at adjacent levels, as illustrated in Figure 11, where the arrows show the allowed transitions and give an indication of their relative strength for four different systems. In each case one of the two ground states, the one on the left, is the "initial" state, meaning the state that collisions force the system into. All transitions are assumed to be on resonance, but systems A, B, C, and D differ in the strength of the transitions leading between a lower level in the left (right) ladder and an upper level in the right (left) ladder, with a dotted arrow indicating a transition that is weak compared to the "intraladder" transition strength. The cross-hatched areas within each box give an indication of the fraction of the total population found in the state represented. The calculations in part (*a*) were done at relatively high pressure, whereas those in (*b*) were lower pressure and can be assumed to be at or near the limiting values.

In system A, where all transitions are of equal strength, degenerate states in each pair above the ground level are completely interchangeable and must therefore, on the basis of symmetry, have the same populations. This is, of course, observed to be the case. Not surprisingly, in *all* systems, when the pressure is high, the collisionally forced state has by far the largest population. But it also has the highest single population even in the low-pressure limit, and this is really an example of a statistical requirement (49) that in the time average the population of the initial state cannot be *less* than it would be if the systems were *randomly* distributed (in this case one-eighth of the population in each state). There are differences in the efficacy with which the various excited states become populated, depending on pressure and the transition strength model. This could be interesting from the point of view of selective excitation models, but we do not purse this point as the results are intended mainly to be illustrative.

A second model that is of interest relates to intramolecular coupling. A single "ground" or initial state is connected to a second level, which may have many states, which we assume to be spaced out in energy. This two-level system may be thought of as representing a spectroscopic "doorway" state coupled to a set of nearby background states. In our calculations we have included only a very small number of "background" states. We have spaced

(a)

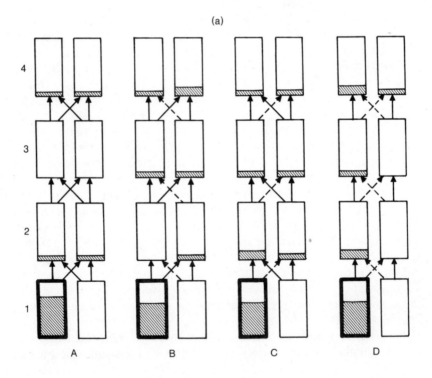

(b)

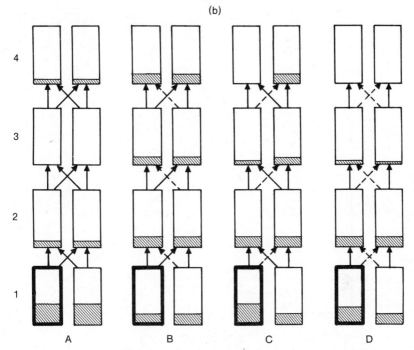

Figure 11. Level populations in various four-level systems with two states at each level. We formally identify the states as belonging to two different ladders with weak or strong cross-couplings between the ladders as indicated in the different models A–D. Each state is indicated by an appropriately placed box, whose steady-state population corresponds to the fraction of cross-hatched area. The "initial" state (see discussion) corresponds in each case to the heavily outlined box in the corner. The basic Rabi frequency corresponds to 1.0 in dimensionless units. We treat a "high-pressure" case [part (a)] where $1/\tau = \beta$ and a "low-pressure" case [part (b)] where $1/\tau = 0.25\beta$. The "weak" couplings (indicated by dotted arrows) use a reduced Rabi frequency of 0.3β.

them with equal energy increments and varied the size of this increment relative to the coupling strength. In Figure 12, we show the populations of the various states, this time just in the low pressure (time averaging) limit, in the second level ordered according to their energy, with the central state being isoenergetic with the initial state. In part (a) the coupling is small enough so that edge effects are unimportant, and the results are substantially equivalent to what we would obtain with an infinite number of states [the Bixon–Jortner model (50)]. Even with only five background states, the basic features of a familiar type of spectroscopic lineshape are beginning to emerge, with the spread in the time-averaged population exhibiting a central peak and wings. Now, however, as the coupling is increased [parts (b) and (c)],

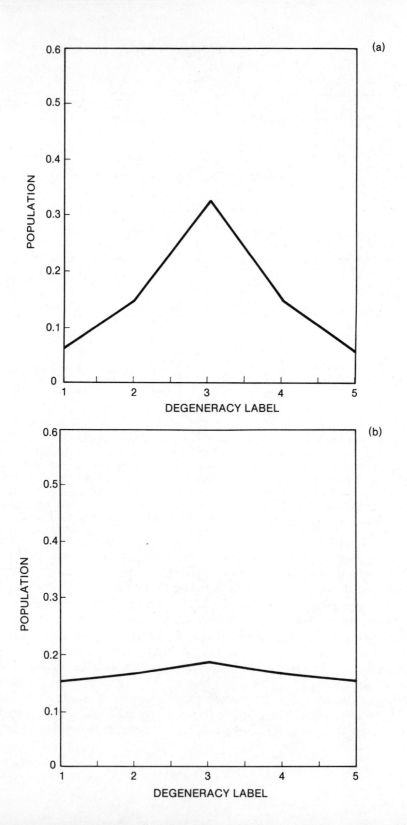

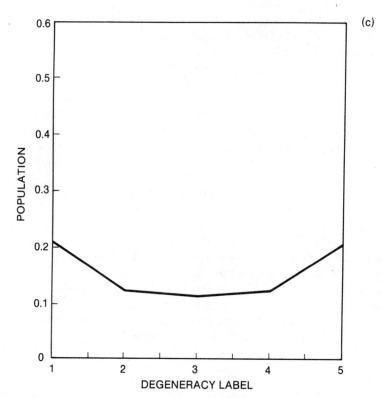

Figure 12. Level populations in five background states for energy separations equal to (*a*) 2.5, (*b*) 1.5, and (*c*) 1.0 in units of the coupling strength (assumed equal from the initial state to each background state).

this feature is lost. The linewidth requires an energy width V^2/Δ in which to "express" itself, and if this is not available, the population "piles up" around the edges.

We comment briefly on the computational effort required to implement the GME formalism for the two cases considered here. In the two-fold degenerate ladder problem, the primary level indices for the $A^{(k)}$ and $M^{(k)}$ matrices go over the same values as in a simple ladder problem $(1, \ldots, k)$, but the total matrix size is multiplied by the product of the degeneracy at each lower level by the degeneracy of the appropriate upper level (i.e., the one $N - k$ levels above the given lower level). Since the degeneracies in this problem are all 2, it follows that $A^{(1)}$ is a 4×4 matrix, $A^{(2)}$ an 8×8, and so on. Then at the last step, the elimination of the last group of off-diagonal elements requires the inversion of a $2N \times 2N$ matrix (involving the primarily diagonal off-diagonals between states at the same level but mixing the state from the left ladder to the right ladder and vice versa). Thus, the largest matrix size is $4(N - 1) \times 4(N - 1)$ or $2N \times 2N$, whichever is larger.

From the point of view of the GME formalism, the second (intramolecular coupling) example is just a degenerate two-level system. Hence, there are no recurrence steps required to obtain the $M^{(N-1)}$ matrix elements required in the *final* step. However, for g background states, the final step itself requires the elimination of $g(g - 1)$ off-diagonal elements (which remain because they are *primarily* diagonal) so that the amount of computational savings realizable compared with the original $(g + 1)^2 \times (g + 1)^2$ problem is small in this case.

V. GENERAL COMMENTS AND CONCLUSIONS

The field of intramolecular dynamics (especially laser-induced dynamics) is gradually moving from a conceptual emphasis to the practical problem of representing the dynamics and its experimental consequences in an exact a manner as possible. Much of the work of the 1986 conference on lasers, molecules, and lasers described in this book shows this orientation but is, for the most part, focused on the use of the molecular eigenstates to do extremely large basis set calculations. The rapidly increasing computational power available increases the likelihood of success with these approaches, at least for molecules of reasonable sizes. Nevertheless, as the limits of these methods are approached, density matrix methods present a possible alternative.

The discussion of models in Section II showed in a general way how the use of a density matrix formalism makes effective use of a relatively small number of important zero-order states to dynamically model many exact energy eigenstates. For this to be of any practical use, we must have a way of incorporating specific information about a molecule into density matrix equations. The heat bath feedback equations, Eq. 9, provide a natural framework in which to treat many molecules. The specific relaxation rates that appear there can all be calculated using the restricted quantum exchange results derived here (Eqs. 15–20 and 47), including specific information concerning the normal-mode frequencies of the molecule. In addition, it appears promising from the discussions in Section III.C that more detailed information from the potential surfaces will tie the anharmonic or overall strength parameters more or less directly to the V_0 and α parameters in the RQE theory. We have considered the Morse potential here. However, detailed studies of the matrix elements for various other generic forms of potentials that are used to model molecular potentials (51) should be valuable.

A significant problem that arises if the density matrix formalism is adopted, either to model internal relaxation processes or to include such external sources of damping as collisions, is that for a system having N states, there are N^2 elements of the density matrix to evaluate. As a result, if the potential advantage of the density matrix methods is to be realized, there is a great advantage in reducing the initial problem of finding N^2 elements to a smaller one involving just the *diagonal* elements of the density matrix. The GME formalism provides a general technique for doing just that, and the extensions of the formalism developed here greatly increase its range of applicability.

In practice, since the reduction scheme cannot in most instances lead to an actual closed rate equation (i.e., we find equations involving λ times the elements of ρ rather than equations for $d\rho/dt$), we must account in general for an eigenvalue dependence in our radiative transition coefficients $W(\lambda)$ (e.g., Eq. 12). There can be up to N^2 distinct eigenvalues, and these must be self-consistent in the sense that the level population equations that are obtained using $W(\lambda)$ as a parameter must yield λ as one of its eigenvalues. Iteration to self-consistency is a possible strategy, but this has the disadvantage that, if done entirely with numerical methods, the reduction scheme must be repeated at each iteration and for every eigenvalue.

Numerical implementation of the reduction scheme is not, however, the only possibility. For the simple ladder case, diagrammatic methods have been devised (22) that lead to analytic expressions for $W(\lambda)$ as a *function* of λ. A computer program designed to carry out the diagrammatic analysis need only be run a single time to create a database of appropriately represented analytic expressions. Iteration on λ then would involve substitution of successive iterates for λ into these analytic expressions.

Success of the iteration scheme lies to a large degree in being able to make reasonable initial guesses for λ, and some of the results presented in Section II are relevant to this problem. For large damping, where a rate equation is essentially exact, we would find that all *relevant* eigenvalues are sufficiently close to zero that $W(\lambda = 0)$ is a sufficiently good approximation for $W(\lambda)$, and iteration may not even be required. On the other hand, we may often be closer to the opposite (weak damping) limit, in which case N small (and real) eigenvalues giving the average time dependence will be given by the eigenvalues of Eq. 5. The remaining eigenvalues would be complex but would have only small real parts and would lie close to pure imaginary values given by the various $N^2 - N$ energy differences. The energies here are those of the system with all radiative couplings included but no damping. They can therefore be found just by solving the $N \times N$ Schrödinger equation in the limited basis of zero-order states. Thus, estimates for *all* the eigenvalues can be obtained, and we are therefore in a position to find all components of the solution.

We close with several remarks specifically concerned with the degenerate level GME formalism given here for the first time. First, we note that in view of the suggested incorporation of diagrammatic methods as part of a general computational scheme, the extension of these methods to the degenerate case would be worthwhile. The existence of this extension to the formalism gives another option in developing computational methods making it unnecessary to abandon density matrix approaches (or specific implementations of them involving the GME) just because of complications in the physics. These complications in themselves would be a worthwhile subject of investigation (and rates appearing in a formalism involving several active modes and a smaller heat bath could be obtained using the results of Section III) but have been left for future investigation.

The actual applications of the degenerate GME formalism have involved the use of special steady-state solutions to investigate the time-averaged statistics of some simple systems. Some possible extensions along these lines are suggested here. In Section IV.B, we have investigated an intramolecular coupling model that, in the limit of many states, is equivalent to the Bixon–Jortner model. Figure 13 shows some coupling schemes that could be investigated and are based on restricted quantum exchange (or, more precisely, *single* quantum exchange) in two- and three-degree-of-freedom

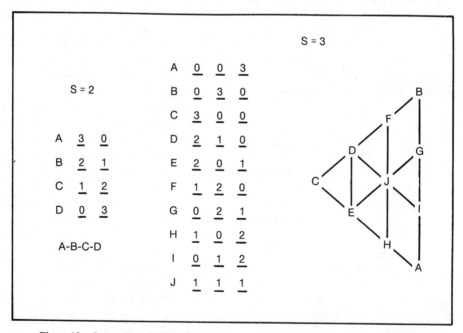

Figure 13. Some generalized intramolecular coupling schemes whose statistical behavior could be investigated using degenerate-ladder models and the methods presented in Section IV. Schemes depicted correspond to single quantum exchange for degenerate $s = 2$ and $s = 3$ oscillators.

systems. In each case, we illustrate with an energy shell having the energy of three (degenerate) quanta and enumerate the possible states (A–D or A–J). The coupling for $s = 2$ just gives a single ladder model, but with three degrees of freedom, we now have a degenerate ladder, with state A constituting one level, D and E the next, and so on. Coupling *within* each level is also present, and because it *is* within one level, this would be straightforward to include in the formalism by adding appropriate terms to quantities occurring in the recurrence scheme. The systematic investigation would give important insight into the effect of the details of coupling on the process of intramolecular relaxation.

APPENDIX A: CANCELLATION OF POLES IN EQ. 43

We show that $H_1(E, \beta') + H_2(E, \beta')$ has no poles at $\beta' = -\beta_j$ even though F_1 and F_2 each have poles there. The basic reason is as follows: The $-\beta_j$ poles arise because in our original inversion problem leading to $F(E, \beta)$, there were poles at $\beta = \beta_j$ and at $\beta = -\beta'$. As β', considered purely as a parameter of the problem approaches $-\beta_j$, poles that were originally separate come together and become higher order than at general points in the parameter space. Hence, expressions valid for $\beta' \neq -\beta_j$ become indeterminate at $\beta' = -\beta_j$, and the contributions from $\beta = \beta_j$ (i.e., F_1) and $\beta = -\beta'(F_2)$ taken separately become singular. However, we can show that the limit of the indeterminate form is finite and that no new poles can arise in this way.

From Eq. 19 we consider the integrand $e^{\beta E} f_1(\beta) f_2(\beta, \beta')$ and define the function

$$g_j(\beta, \beta') = e^{\beta E} f_1(\beta) f_2(\beta, \beta')(\beta + \beta')^s(\beta - \beta_j)$$

We see that because of its definition, g_j is an analytic function of β in a neighborhood of $-\beta'$ and of β_j up to the nearest other β_i ($i \neq j$). Therefore, g_j is suitable for investigating what happens as $-\beta'$ comes close to β_j. The residue at $-\beta'$ is

$$R(-\beta') = \frac{1}{(s-1)!} \left(\frac{d^{s-1}}{d\beta^{s-1}} \frac{g_j(\beta, \beta')}{(\beta - \beta_j)} \right)_{\beta = -\beta'}$$

We can use Leibnitz's rule to rewrite this in the form

$$R(-\beta') = -\frac{1}{(\beta_j + \beta')^s} \sum_{p=0}^{s-1} \frac{1}{p!} \left[\frac{\partial^p g_j(\beta, \beta')}{\partial \beta^p} \right]_{\beta = -\beta'} (\beta_j + \beta')^p$$

This obviously becomes singular as β' goes to $-\beta_j$. Let us consider, however, the other residue. At β_j, we have

$$R(\beta_j) = \frac{g_j(\beta_j, \beta')}{(\beta_j + \beta)^s}$$

However, since g_j is analytic in a neighborhood of $\beta = -\beta'$, which will include β_j, we can expand $g_j(\beta_j, \beta')$ in a power series about $\beta = -\beta'$ to obtain

$$R(\beta_j) = \frac{1}{(\beta_j + \beta')^s} \sum_{p=0}^{\infty} \frac{1}{p!} \left[\frac{\partial^p g_j(\beta, \beta')}{\partial \beta^p} \right]_{\beta = -\beta'} (\beta_j + \beta')^p$$

Now if we add the two residues, we find

$$R(-\beta') + R(\beta_j) = \sum_{p=s}^{\infty} \frac{1}{p!} \left[\frac{\partial^p g_j(\beta, \beta')}{\partial \beta^p} \right]_{\beta = -\beta'} (\beta_j + \beta')^{p-s}$$

Since the sum is restricted to $p \geqslant s$, the combined residues considered as a function of β' do not have a pole at $\beta' = -\beta_j$, despite poles in the individual terms.

APPENDIX B: EXPRESSIONS FOR HIGH ENERGY EXPANSION

We give here the first few of the C and $\mathscr{D}$ coefficients defined in Eqs. 48 and 49:

$$C_0^{(j)} = \prod_{i=1}^{s} \frac{1}{1 - \alpha^{(\omega_i/\omega_j)+1}}$$

$$C_1^{(j)} = -\prod_{i=1}^{s} \frac{1}{1 - \alpha^{(\omega_i/\omega_j)+1}} \sum_{k=1}^{s} \frac{\hbar\omega_k \alpha^{(\omega_k + \omega_j)/\omega_j}}{1 - \alpha^{(\omega_k + \omega_j)/\omega_j}}$$

$$\mathscr{D}_0(\alpha) = (1 - \alpha^2)^s$$

$$\mathscr{D}_1(\alpha) = \tfrac{1}{2}(1 - \alpha^2)^{s-1}(1 + \alpha) \sum_{k=1}^{s} \hbar\omega_k$$

In a high energy expansion, the first terms then take the form

$$G_\alpha = \left[\prod_{i=1}^{s} (\hbar\omega_i) \right]^{-1} \left(\frac{E^{s-1}}{(s-1)!} \mathscr{D}_0(\alpha) \sum_{j=1}^{s} \mathscr{P}_0(j)(E^> - E^<) \right.$$

$$+ \frac{E^{s-2}}{(s-2)!} \left[\mathscr{D}_0(\alpha) \sum_{j=1}^{s} \mathscr{P}_1(j)(E^> - E^<) \right.$$

$$\left. \left. + \mathscr{D}_1(\alpha) \sum_{j=1}^{s} \mathscr{P}_0^{(j)}(E^> - E^<) \right] \right)$$

Acknowledgments

The author owes much to long-standing and stimulating working relationships with Drs. Everett Thiele and Myron Goodman, first at Johns Hopkins University and then at the University of Southern California. Useful ideas have come from collaborations with Drs. David Dows and Ken Kay and with the research groups at Exxon Research and Engineering Company and the National Bureau of Standards. Finally, much is owed to personnel at Rockwell for assistance in preparation of the manuscript and especially to Dr. Jeff Goldstone for reading and making helpful comments on the material covered.

References

1. J. L. Lyman, R. J. Jensen, J. Rink, C. P. Robinson, and S. D. Rockwood, *Appl. Phys. Lett.* **27**, 87 (1975).

2. R. V. Ambartzumian, Yu. A. Gorokov, V. S. Letokhov, and G. N. Makarov, *JETP Lett.* **21**, 171 (1975).

3. M. F. Goodman, J. Stone, and D. A. Dows, *J. Chem. Phys.* **65**, 5052 (1976).

4. M. Quack, *J. Chem. Phys.* **69**, 1282 (1978).

5. D. M. Larsen and N. Bloembergen, *Opt. Commun.* **37**, 11 (1981).

6. J. G. Black, E. Yablonovitch, N. Bloembergen, and S. Mukamel, *Phys. Rev. Lett.* **38**, 1131 (1977).

7. A. Nauts and R. E. Wyatt, *Phys. Rev. Lett.* **51**, 2238 (1983).

8. K. V. Reddy, D. F. Heller, and M. J. Berry, *J. Chem. Phys.* **76**, 2814 (1982); E. Abramson, R. W. Field, D. Imre, K. K. Innes, and J. L. Kinsey, *J. Chem. Phys.* **80**, 2298 (1984).

9. D. W. Noid, M. L. Koszykowski, and R. A. Marcus, *J. Chem. Phys.* **71**, 2864 (1979).

10. E. Thiele and J. Stone, *J. Chem. Phys.* **80**, 5187 (1987).

11. E. Thiele, M. F. Goodman, and J. Stone, *J. Chem. Phys.* **82**, 2598 (1985).

12. A. L'Huiller, L. A. Lompre, G. Mainfray, and C. Manus, *Phys. Rev. Lett.* **48**, 1814 (1982).

13. Z. Deng and J. H. Eberly, *Phys. Rev. Lett.* **53**, 1810 (1984); *J. Opt. Soc. Am. B* **2**, 486 (1985).

14. E. J. Heller, *Accts. Chem. Res.* **14**, 368 (1981).

15. F. Bloch, *Phys. Rev.* **70**, 460 (1946).

16. K. G. Kay, *J. Chem. Phys.* **75**, 1691 (1981).

17. K. G. Kay, J. Stone, E. Thiele, and M. F. Goodman, *Chem. Phys. Lett.* **82**, 539 (1981).

18. J. Stone, E. Thiele, and M. F. Goodman, *J. Chem. Phys.* **75**, 1712 (1981).

19. J. Stone and M. F. Goodman, *J. Chem. Phys.* **71**, 408 (1979).

20. J. Stone, *Phys. Rev. A* **30**, 2517 (1984).

21. J. Stone, E. Thiele, and M. F. Goodman, *J. Chem. Phys.* **59**, 2909 (1973).

22. J. Stone and M. F. Goodman, *Phys. Rev. A* **18**, 2618 (1978).

23. E. Borsella, R. Fantoni, A. Giandini-Guidonia, and C. D. Cantrell, *Chem. Phys. Lett.* **87**, 284 (1982).

24. J. Stone, *Phys. Rev. A* **26**, 1157 (1982).

25. J. Stone and M. F. Goodman, *Phys. Rev. A* **14**, 380 (1976).

26. J. Stone, E. Thiele, and M. F. Goodman, *J. Chem. Phys.* **63**, 2936 (1975).

27. M. F. Goodman and E. Thiele, *Phys. Rev. A* **5**, 1355 (1972).

28. R. Karplus and J. Schwinger, *Phys. Rev.* **73**, 1020 (1948).

29. R. E. Wyatt, private communication.

30. W. Forst, *Theory of Unimolecular Reactions*, Academic, New York, 1973.

31. J. G. Black, P. Kolodner, M. J. Shultz, E. Yablonovitch, and N. Bloembergen, *Phys. Rev. A* **19**, 704 (1979).

32. E. Thiele, J. Stone, and M. F. Goodman, *Chem. Phys. Lett.* **66**, 457 (1979).

33. E. R. Grant, P. A. Schulz, Aa. S. Sudbo, Y. R. Shen, and Y. T. Lee, *Phys. Rev. Lett.* **40**, 115 (1978).

34. T. S. Luk, H. Pummer, K. Boyer, M. Shahidi, H. Egger, and C. K. Rhodes, *Phys. Rev. Lett.* **51**, 110 (1983); K. Boyer, H. Egger, T. S. Luk, H. Pummer, and C. K. Rhodes, *J. Opt. Soc. Am. B* **1**, 3 (1984); T. S. Luk, H. Pummer, and C. K. Rhodes, *Phys. Rev. A* **32**, 214 (1985).

35. M. Crance, *J. Phys. B* **18**, L155 (1985); X.-D. Mu, T. Aberg, A. Blomberg, and B. Crasemann, *Phys. Rev. Lett.* **56**, 1909 (1986).

36. M. F. Goodman, J. Stone, and E. Thiele, in *Multiple-Photon Excitation and Dissociation of Polyatomic Molecules*, C. D. Cantrell, ed., Springer-Verlag, Berlin, 1986, p. 159.

37. J. A. Horsley, J. Stone, M. F. Goodman, and D. A. Dows, *Chem. Phys. Lett.* **66**, 461 (1979).

38. A. Kaldor, M. F. Goodman, D. A. Dows, and R. Thomas, private communication.

39. J. Stone, E. Thiele, and M. F. Goodman, *Chem. Phys. Lett.* **71**, 171 (1980).

40. R. G. Bray and M. J. Berry, *J. Chem. Phys.* **71**, 4909 (1979).

41. G. A. Korn and T. M. Korn, *Mathematical Handbook for Scientists and Engineers*, McGraw-Hill, New York, 1961, p. 217.

42. E. Thiele, *J. Chem. Phys.* **39**, 3258 (1963); P. C. Haarhoff, *Mol. Phys.* **6**, 337 (1963).

43. M. Abramowitz and I. A. Stegun, *Handbook of Mathematical Functions*, National Bureau of Standards, 1964, p. 824.

44. J. Stone and M. F. Goodman, *J. Phys. Chem.* **89**, 1250 (1985).

45. S. M. Lederman, J. H. Runnels, and R. A. Marcus, *J. Phys. Chem.* **87**, 4364 (1983).

46. K. G. Kay and J. Stone, unpublished work.

47. D. A. Dows, private communication.

48. H. W. Galbraith and J. R. Ackerhalt, *Chem. Phys. Lett.* **84**, 458 (1981).

49. E. J. Heller and M. J. Davis, *J. Phys. Chem.* **86**, 2218 (1982).

50. M. Bixon and J. Jortner, *J. Chem. Phys.* **48**, 715 (1968).

51. J. N. Murrell, S. Carter, S. C. Farentos, P. Huxley, and A. J. C. Varandas, *Molecular Potential Energy Functions*, Wiley, New York, 1984; S. Carter, Comp. Phys. Rep. **3**, 209 (1986).

CHAPTER XII

REDUCED EQUATIONS OF MOTION FOR MOLECULAR LINESHAPES AND SEMICLASSICAL DYNAMICS IN LIOUVILLE SPACE

SHAUL MUKAMEL* and YI JING YAN

Department of Chemistry, University of Rochester, Rochester, New York, 14627

CONTENTS

Reduced equations of motion (REM) for semiclassical dynamics in Liouville space, which may be used in the calculation of intramolecular and intermolecular dynamics, correlation functions, and electronic lineshapes, are developed. The method is based on constructing a Gaussian propagator whose equations of motion are obtained by requiring that the first two moments of the coordinates and momenta have the exact time evolution. The present method can be applied for the time evolution of mixed states in phase space and may therefore be particularly useful for molecular processes in condensed phases. The absorption lineshapes and Raman excitation profiles of model anharmonic molecules at finite temperatures are calculated as a demonstration.

*Camille and Henry Dreyfus Teacher–Scholar.

I. INTRODUCTION

Much of the current effort in semiclassical molecular dynamics is focused on the development of efficient methods for the numerical solution of the Schrödinger equation for interacting many-body systems [1–18]. The calculation of molecular dynamical processes such as energy transfer, vibrational relaxation, scattering processes, and spectral lineshapes [19, 20] depends crucially on the development of suitable approximate methods for propagating the molecular wavefunction. Some of the methods commonly used include the path integral formalism [4–8] grid methods [9] and wavepacket dynamics [12–18]. Performing dynamical calculations in Liouville space [21] using the density matrix (instead of the wavefunction) offers a major theoretical challenge. The connection between classical and quantum mechanics is much more transparent in phase space, particularly if we make use of the Wigner representation [22–24], which allows for a systematic expansion of the dynamics in powers of $\hbar$. However, this formal elegance carries a price tag. The dimensionality of phase space is twice that of the configuration space. When the wavefunction can be represented by n coefficients corresponding to some basis set, the density matrix will require n^2 such coefficients. This often poses a significant difficulty. On the other hand, the development of a *reduced description* of molecular processes in which we follow explicitly only a few chosen degrees of freedom can only be achieved using the density matrix [20, 21] since an ensemble-averaged wavefunction (unlike the density matrix) has no physical significance. The future of molecular dynamics in large polyatomic molecules and in condensed phases depends, therefore, on the extension of the semiclassical methods to the evolution of the density matrix in Liouville space. We have recently proposed a new type of self-consistent semiclassical REM in Liouville space [25–27] based on projection operator techniques of nonequilibrium statistical mechanics [28–31]. To lowest order, the equations provide a time-dependent mean field approximation [32, 33], which can be systematically improved. Since this approach is formulated in terms of the density matrix in Liouville space [21, 34], it can be applied to pure states as well as to mixed states. In that respect, it should be particularly useful for molecular dynamics in condensed phases (spectral lineshapes of molecules in solution and on solid surfaces, desorption, vibrational relaxation, etc.). In this chapter we review this Liouville space propagation scheme and discuss possible applications to vibrational dynamics of anharmonic molecules and to electronic spectroscopy; absorption and Raman and fluorescence spectroscopy of isolated and solvated polyatomic molecules.

In Section II, we consider the dynamics in Liouville space and define the propagators relevant for a general time evolution and for the calculation of correlation functions. In Section III, we consider a molecule with two electronic states and develop a semiclassical propagator for the calculation of electronic absorption. This is done using projection operators in Liouville space. In Section IV, we consider the evolution of the density matrix on a single potential surface. This is a special case of the formalism developed in Section III. The time-dependent self-consistent field (TDSCF) (Hartree) approximation is developed and analyzed. In Section V, we discuss and summarize our results, and in Section VI, we present numerical calculations of absorption lineshapes and Raman excitation profiles of model anharmonic systems obtained using the present Liouville space propagation scheme.

II. TIME EVOLUTION IN LIOUVILLE SPACE

We consider a quantum system characterized by N coordinates x_j, $j = 1, \ldots, N$, their conjugate momenta $\hat{p}_j = -i\hbar\,\partial/\partial x_j$, and masses m_j. Its Hamiltonian is

$$H = \tfrac{1}{2}\sum_{j,k}\Omega_{jk}\hat{p}_j\hat{p}_k + V(x_1, x_2, \ldots, x_N), \tag{1}$$

where V is the interaction potential and Ω_{jk} is a mass matrix. For Cartesian coordinates we have $\Omega_{jk} = 1/m_j\delta_{jk}$. We shall hereafter introduce a vector notation and define the N-dimensional vectors $\underline{\mathbf{x}}$ and $\hat{\underline{\mathbf{p}}}$ with components x_j and $\hat{p}_j$, respectively, and the $N \times N$ matrix Ω with matrix elements Ω_{jk}. A matrix is represented by a boldface letter, whereas a vector is represented by a boldface letter with an underbar. Equation 1 then assumes the form

$$H = \tfrac{1}{2}\hat{\underline{\mathbf{p}}}^T\Omega\hat{\underline{\mathbf{p}}} + V(\underline{\mathbf{x}}). \tag{2}$$

Here $\hat{\underline{\mathbf{p}}}^T$ is the transpose of $\hat{\underline{\mathbf{p}}}$. Our goal is to develop a semiclassical self-consistent procedure for the approximate solution of the Liouville equation

$$\frac{d\rho}{dt} = -iL\rho \equiv -\frac{i}{\hbar}[H, \rho]. \tag{3a}$$

Here L is the Liouville operator, and $\rho(\mathbf{x}, \mathbf{x}'; t)$ is the system density matrix. We further define the density matrix in the Wigner representation ρ_w [22–24]:

$$\rho_w(\underline{\mathbf{q}}, \underline{\mathbf{p}}; t) \equiv \frac{1}{(\pi\hbar)^N}\int_{-\infty}^{\infty}\rho(\underline{\mathbf{q}} + \underline{\mathbf{s}}, \underline{\mathbf{q}} - \underline{\mathbf{s}}; t)\exp\left(-\frac{2i\underline{\mathbf{p}}\underline{\mathbf{s}}}{\hbar}\right)d\underline{\mathbf{s}}. \tag{3b}$$

The inverse transform of Eq. 3b is

$$\rho(\underline{\mathbf{q}} + \underline{\mathbf{s}}, \underline{\mathbf{q}} - \underline{\mathbf{s}}; t) = \int_{-\infty}^{\infty}\rho_w(\underline{\mathbf{q}}, \underline{\mathbf{p}}; t)\exp\left(\frac{2i\underline{\mathbf{p}}\underline{\mathbf{s}}}{\hbar}\right)d\underline{\mathbf{p}}. \tag{3c}$$

In Eqs. 3 we have used the substitutions

$$\underline{q} = \tfrac{1}{2}(\underline{x} + \underline{x}'),\tag{4a}$$

$$\underline{s} = \tfrac{1}{2}(\underline{x} - \underline{x}').\tag{4b}$$

The density matrix is normalized as

$$\iint d\underline{x}\, d\underline{x}'\, \rho(\underline{x}, \underline{x}'; t) = \iint \rho_w(\underline{q}, \underline{p}; t)\, d\underline{q}\, d\underline{p} = 1.\tag{5}$$

The equation of motion for the density matrix in the Wigner representation is

$$\frac{d\rho_w}{dt} = -iL_w\rho_w,\tag{6a}$$

where the Liouville operator in the Wigner representation is

$$iL_w\rho_w = \frac{2}{\hbar}H_c \sin\left[\frac{\hbar}{2}\left(\frac{\overleftarrow{\partial}}{\partial \underline{p}}\frac{\overrightarrow{\partial}}{\partial \underline{q}} - \frac{\overleftarrow{\partial}}{\partial \underline{q}}\frac{\overrightarrow{\partial}}{\partial \underline{p}}\right)\right]\rho_w.\tag{6b}$$

Here H_c is the classical Hamiltonian, and the arrows indicate the direction of operation of the derivative. Note that when Eq. 6 is expanded in powers of $\hbar$, the leading (zero-order) term is the classical Liouville equation.

The simplest dynamical problem we may be interested in, is the calculation of the expectation value of some dynamical operator A:

$$\langle A(t) \rangle = \mathrm{Tr}\,[A\rho(t)].\tag{7}$$

Equation 7 can be written in the coordinate representation

$$\langle A(t) \rangle = \iint d\underline{x}\, d\underline{x}'\, A(\underline{x}', \underline{x})\rho(\underline{x}, \underline{x}'; t)\tag{8a}$$

or in the Wigner phase space representation

$$\langle A(t) \rangle = \iint d\underline{p}\, d\underline{q}\, A_w(\underline{p}, \underline{q})\rho_w(\underline{p}, \underline{q}; t).\tag{8b}$$

Since Eqs. 8a and 8b simply correspond to different representations, we shall introduce hereafter a unified notation and recast them in the form

$$\langle A(t) \rangle = \int d\underline{\Gamma}\, A(\underline{\Gamma})\rho(\underline{\Gamma}; t).\tag{9}$$

Equation 9 should be understood as follows: We either express A and ρ in the coordinate representation, and then $d\Gamma = d\underline{x}\,d\underline{x}'$, or A and ρ are given in the Wigner representation, and then $d\Gamma = d\underline{p}\,d\underline{q}$. The evaluation of $\langle A(t)\rangle$ then reduces to solving the Liouville equation (Eq. 3a or 6) for $\rho(\underline{\Gamma};t)$ and then performing the integration (Eq. 9).

Correlation functions constitute another type of dynamical quantities whose calculation is of considerable interest. Experimental (e.g., spectroscopic) observables can usually be expressed in terms of appropriate correlation functions, and their calculation is a key step in the interpretation of these experiments. Given two dynamical operators A and B and the equilibrium density matrix of the system ρ_{eq}, we define the two time correlation function [35]

$$\langle A(t)B(0)\rangle \equiv \mathrm{Tr}\,[A\exp(-iHt/\hbar)B\rho_{eq}\exp(iHt/\hbar)], \tag{10a}$$

or making use of the Liouville operator,

$$\langle A(t)B(0)\rangle \equiv \mathrm{Tr}\,[A\exp(-iLt)B\rho_{eq}], \tag{10b}$$

with

$$\rho_{eq} = \frac{\exp(-H/kT)}{\mathrm{Tr}\exp(-H/kT)}. \tag{11}$$

Equation 10 can be rewritten in the form

$$\langle A(t)B(0)\rangle = \int d\Gamma\, A(\Gamma)\bar{\rho}(\underline{\Gamma};t), \tag{12a}$$

where $\bar{\rho}(\underline{\Gamma};t)$ is given by

$$\bar{\rho}(\underline{\Gamma};t) = \exp(-iLt)B\rho_{eq}. \tag{12b}$$

Here $\bar{\rho}(\underline{\Gamma};t)$ can be obtained from the solution of the Liouville equation (Eq. 3 or 6) with the initial condition

$$\bar{\rho}(\underline{\Gamma};0) = B\rho_{eq}. \tag{13}$$

Note that $\bar{\rho}$, which satisfies the Liouville equation (Eq. 3a) with the initial condition Eq. 13, is not the density matrix of the system. It should be rather interpreted as a *generating function* that allows the calculation of any two time correlation function using Eqs. 12. In the next section, we shall develop a semiclassical procedure for solving the Liouville equation and calculating $\rho(\underline{\Gamma}; t)$ or $\bar{\rho}(\underline{\Gamma}; t)$.

III. REDUCED EQUATIONS OF MOTION FOR MOLECULAR ELECTRONIC SPECTROSCOPY

We shall now develop a semiclassical procedure for the efficient calculation of molecular electronic spectra. We consider a polyatomic molecule with N vibrational degrees of freedom and two electronic states: a ground state $|g\rangle$ and an electronically excited state $|e\rangle$. The molecular Hamiltonian is given by [20]

$$H = |g\rangle H_g \langle g| + |e\rangle(\omega_{eq} + H_e)\langle e|, \tag{14}$$

where

$$H_g = \tfrac{1}{2}\hat{\mathbf{p}}^T \mathbf{\Omega}\hat{\mathbf{p}} + V_g(\underline{\mathbf{x}}). \tag{15b}$$

$$H_e = \tfrac{1}{2}\hat{\mathbf{p}}^T \mathbf{\Omega}\hat{\mathbf{p}} + V_e(\underline{\mathbf{x}}). \tag{15b}$$

Here V_g and V_e are the adiabatic potentials for the ground and the excited states, respectively, and ω_{eg} is the electronic energy gap between their minima. The absorption lineshape is given by

$$I(\omega_L) = \int dt \exp\left[i(\omega_L - \omega_{eg})t\right]\langle \mu(t)\mu(0)\rangle, \tag{16}$$

where ω_L is the photon frequency, $\mu(\underline{\mathbf{x}})$ is the dipole operator, and the dipole correlation function is given by

$$\langle \mu(t)\mu(0)\rangle = \text{Tr}\left[\mu \exp(-iH_e t/\hbar)\mu\rho_{eq}\exp(iH_g t/\hbar)\right] \tag{17}$$

with

$$\rho_{eq} = \frac{\exp(-H_g/kT)}{\text{Tr}\exp(-H_g/kT)}. \tag{18}$$

Equation 17 can be rewritten in the form

$$\langle \mu(t)\mu(0)\rangle = \int d\underline{\Gamma}\, \mu(\underline{\Gamma})\bar{\rho}(\underline{\Gamma}; t), \tag{19}$$

where $\bar{\rho}(\underline{\Gamma}; t)$ satisfies the equation

$$\frac{d\tilde{\rho}}{dt} = -\frac{i}{\hbar}[H_e\tilde{\rho} - \tilde{\rho}H_g] \tag{20a}$$

with the initial condition

$$\tilde{\rho}(\underline{\Gamma};0) = \mu\rho_{eq}. \tag{20b}$$

We note that $\tilde{\rho}(\underline{\Gamma};t)$ [similar to $\tilde{\rho}(\underline{\Gamma};t)$] is not a density matrix but rather a generating function for electronic spectra. The solution of the Liouville equation (Eq. 20) has been the subject of numerous studies in the context of spectral line broadening [20, 21, 36–38]. It bears some resemblance to the dynamics of nonadiabatic transitions (curve crossing) [39]. The ordinary time evolution in Liouville space considered in Section II, is a special case of the present formulation constructed for electronic spectroscopy, since Eq. 20a reduces to Eq. 3a when we set $H_e = H_g$. In this section, we shall develop reduced equations of motion (REM) toward the approximate semiclassical solution of Eq. 20a.

Our REM will be derived as follows. We start with a set of dynamical operators whose expectation values are believed to be relevant for the dynamics. In the present reduced description, we choose to consider the following operators: x_j, $\hat{p}_j$, $x_j x_k$, $\hat{p}_j \hat{p}_k$, and $\hat{p}_j x_k$, $j, k = 1, 2, \ldots, N$. These $M = 2N^2 + 3N$ operators constitute the complete set of linear and bilinear products of x_j and $\hat{p}_j$. Let us denote these operators by A_α, $\alpha = 1, \ldots, M$. We shall further introduce the "expectation values" σ_α of these operators with respect to the generating function $\tilde{\rho}(\underline{\Gamma};t)$, that is,

$$\sigma_\alpha(t) = \langle A_\alpha \rangle = \int d\underline{\Gamma} \, A_\alpha(\underline{\Gamma})\tilde{\rho}(\underline{\Gamma};t) \Big/ \int d\underline{\Gamma} \, \tilde{\rho}(\underline{\Gamma};t), \qquad \alpha = 1, 2, \ldots, M. \tag{21a}$$

Note that $\tilde{\rho}(\underline{\Gamma};t)$ is not a density matrix; therefore, the σ_α are not really expectation values. They are, however, M parameters characterizing $\tilde{\rho}(\underline{\Gamma};t)$, and they represent the complete set of first and second "moments" of the coordinates and momenta. It should further be noted that the trace of $\tilde{\rho}(\underline{\Gamma};t)$ is not conserved when $H_e \neq H_g$, and its variation with time carries a valuable dynamical information. We shall therefore introduce another operator $A_0 = 1$, the unit operator, and define

$$\sigma_0(t) = \int d\underline{\Gamma} \, A_0 \tilde{\rho}(\underline{\Gamma};t). \tag{21b}$$

To simplify the notation, we shall rearrange our operators in a matrix form. To that end, we introduce the following quantities:

$$A_0 = 1,$$

$$\underline{A}_1 = \underline{x},$$

$$\underline{A}_2 = \hat{\underline{p}},$$

$$A_3 = (\underline{x} - \underline{x}^0) \cdot (\underline{x} - \underline{x}^0)^T,$$

$$A_4 = (\hat{\underline{p}} - \underline{p}^0) \cdot (\hat{\underline{p}} - \underline{p}^0)^T,$$

$$A_5 = (\underline{x} - \underline{x}^0) \cdot (\hat{\underline{p}} - \underline{p}^0)^T - \tfrac{1}{2} i\hbar \mathbf{I}, \tag{22a}$$

with

$$\underline{x}^0 = \int d\underline{\Gamma} \, \underline{x} \tilde{\rho}(\underline{\Gamma}; t) \Big/ \int d\underline{\Gamma} \, \tilde{\rho}(\underline{\Gamma}; t),$$

$$\underline{p}^0 = \int d\underline{\Gamma} \, \hat{\underline{p}} \tilde{\rho}(\underline{\Gamma}; t) \Big/ \int d\underline{\Gamma} \, \tilde{\rho}(\underline{\Gamma}; t). \tag{22b}$$

Here A_0 is a number representing the unit operator in phase space; $\underline{A}_1$ and $\underline{A}_2$ are N-component vectors each containing N operators; whereas A_3, A_4, and A_5 are $N \times N$ matrices each containing N^2 operators. The matrix $\mathbf{I}$ is the unit matrix. Since A_3 and A_4 are symmetric matrices, Eqs. 22 contain altogether $M + 1$ distinct operators, where $M = 2N^2 + 3N$. Similarly, we shall rearrange σ_α in a matrix form analogous to A_α. Here σ_0 will be a number; $\boldsymbol{\sigma}_1$ and $\boldsymbol{\sigma}_2$ will be N-component vectors; and $\boldsymbol{\sigma}_3$, $\boldsymbol{\sigma}_4$, and $\boldsymbol{\sigma}_5$ are $N \times N$ matrices. We have thus identified $M + 1$ relevant dynamical variables (Eq. 22) and $M + 1$ parameters σ_α related to the expectation values of these operators. The REM will now be derived by introducing a reduced propagator $\sigma(\underline{\Gamma}; t)$:

$$\sigma(\underline{\Gamma}; t) \cong \tilde{\rho}(\underline{\Gamma}; t). \tag{23}$$

We shall choose the following form for $\sigma(\underline{p}, \underline{q}; t)$ in the Wigner representation:

$$\sigma(\underline{p}, \underline{q}; t) = \frac{\sigma_0(t)}{(2\pi)^N [\det \mathbf{W}(t)]^{1/2}}$$

$$\times \exp \left\{ -\tfrac{1}{2} [\underline{q}^T - \boldsymbol{\sigma}_1^T(t), \underline{p}^T - \boldsymbol{\sigma}_2^T(t)] \mathbf{W}^{-1}(t) \begin{bmatrix} \underline{q} - \boldsymbol{\sigma}_1(t) \\ \underline{p} - \boldsymbol{\sigma}_2(t) \end{bmatrix} \right\} \tag{24}$$

with

$$\mathbf{W}(t) = \begin{bmatrix} \boldsymbol{\sigma}_3(t) & \boldsymbol{\sigma}_5(t) \\ \boldsymbol{\sigma}_5^T(t) & \boldsymbol{\sigma}_4(t) \end{bmatrix}. \tag{25}$$

Here $\det \mathbf{W}(t)$ denotes the determinant of the $\mathbf{W}(t)$ matrix, and $\mathbf{A}^T$ denotes the transpose of $\mathbf{A}$.

Equation 24 implies that $\sigma(\underline{\Gamma}; t)$ is Gaussian at all times. If initially $\sigma(\underline{\Gamma}; 0)$ or $\tilde{\rho}(\underline{\Gamma}; 0)$ (Eq. 20b) is not a Gaussian, we can always represent it as a superposition of Gaussians. In the Appendix we give $\sigma(\underline{x}, \underline{x}'; t)$ in the coordinate representation. Equation 24 depends on $M + 1$ time-dependent parameters, which consist of a number σ_0, two N-dimensional vectors $\boldsymbol{\sigma}_1$ and $\boldsymbol{\sigma}_2$, and three $N \times N$ matrices $\boldsymbol{\sigma}_3, \boldsymbol{\sigma}_4$ and $\boldsymbol{\sigma}_5$. These time-dependent parameters may be uniquely determined by requiring the expectation values of our operators A_α (Eqs. 22) evaluated using the exact ($\tilde{\rho}$) and the approximate (σ) propagators to be the same:

$$\sigma_0(t) = \int d\underline{\Gamma} \, \tilde{\rho}(\underline{\Gamma}; t) = \int d\underline{\Gamma} \, \sigma(\underline{\Gamma}; t), \qquad (26a)$$

$$\sigma_\alpha(t) = \int d\underline{\Gamma} \, A_\alpha(\underline{\Gamma}) \tilde{\rho}(\underline{\Gamma}; t) \bigg/ \int d\underline{\Gamma} \, \tilde{\rho}(\underline{\Gamma}; t)$$

$$= \int d\underline{\Gamma} \, A_\alpha(\underline{\Gamma}) \sigma(\underline{\Gamma}; t) \bigg/ \int d\underline{\Gamma} \, \sigma(\underline{\Gamma}; t), \qquad \alpha = 1, 2, \ldots, M. \qquad (26b)$$

We further define the "overlap" $(M + 1) \times (M + 1)$ matrix with matrix elements

$$S_{\alpha\beta}(t) = \langle\langle A_\alpha | \sigma(\underline{\Gamma}; t) A_\beta \rangle\rangle \equiv \text{Tr}\,[A_\alpha^\dagger \sigma(\underline{\Gamma}; t) A_\beta], \qquad \alpha, \beta = 0, \ldots, M, \qquad (27a)$$

and introduce the Liouville space projection operator

$$P(t) = \sum_{\alpha,\beta=0}^{M} |\sigma(t) A_\alpha \rangle\rangle [S(t)]_{\alpha\beta}^{-1} \langle\langle A_\beta| \qquad (27b)$$

and the complementary projection

$$Q(t) = 1 - P(t). \qquad (27c)$$

We are using Liouville space notation [21, 34], whereby an ordinary operator A is represented by a ket $|A\rangle\rangle$ and the scalar product of two operators is defined by $\langle\langle A|B \rangle\rangle \equiv \text{Tr}\,(A^\dagger B)$. Here $\langle\langle A|L|B \rangle\rangle \equiv \text{Tr}\,(A^\dagger L B)$ is a Liouville space "matrix element." The properties of the projection operator $P(t)$ have been discussed previously [25–27]. In order to simplify the notation, we introduce the auxiliary quantities $\tilde{\sigma}_\alpha$, $\alpha = 0, 1, \ldots, M$, defined by

$$\tilde{\sigma}_0 = \sigma_0, \qquad (28a)$$

$$\tilde{\sigma}_\alpha = \sigma_\alpha \sigma_0. \qquad (28b)$$

We further introduce the generalized Liouville operator $\tilde{L}$ by its action on an arbitrary dynamical variable A:

$$\tilde{L}A \equiv \frac{1}{\hbar}(H_e A - A H_g).\qquad(29)$$

Using the assumption that at some initial time t_0, $\tilde{\rho}(\underline{\Gamma}; t_0) = \sigma(\underline{\Gamma}; t_0)$, we have derived the following exact REM for $\tilde{\sigma}_\alpha$:

$$\dot{\tilde{\sigma}}_\alpha(t) = -i\langle\langle A_\alpha|\tilde{L}|\sigma(t)\rangle\rangle - \int_{t_0}^t ds\,\langle\langle A_\alpha|\tilde{L}K(t,s)Q(s)\tilde{L}|\sigma(s)\rangle\rangle,$$

$$\alpha = 0,\ldots,M,\quad(30a)$$

where

$$K(t,s) = \exp_+\left[-i\int_s^t d\tau\, Q(\tau)\tilde{L}\right],\qquad(30b)$$

and $\exp_+$ denotes the time-ordered exponential

$$\exp_+\left[-i\int_s^t d\tau\, Q(\tau)\tilde{L}\right] \equiv 1 - i\int_s^t d\tau\, Q(\tau)\tilde{L}$$

$$+ (-i)^2\int_s^t d\tau_1 \int_s^{\tau_1} d\tau_2\, Q(\tau_1)\tilde{L}Q(\tau_2)\tilde{L} + \cdots.\qquad(30c)$$

The first term on the right side of the REM Eq. 30a represents the mean field evolution which is exact if $\tilde{\rho}(\underline{\Gamma}; t) = \sigma(\underline{\Gamma}; t)$ for all times, whereas the second term, the fluctuation kernel, arises from the fact that, in general, $\tilde{\rho}(\underline{\Gamma}; t) \neq \sigma(\underline{\Gamma}; t)$ for $t > t_0$, and it corrects the time evolution of $\sigma(\underline{\Gamma}; t)$. In this REM the time derivative of $\tilde{\sigma}_\alpha$ at time t depends on the values of $\tilde{\sigma}_\beta$ at all previous times $t_0 < s < t$. The derivation of the REM (Eqs. 30) is given elsewhere [25–27].

The mean field approximation is obtained by retaining only the first term on the right side of Eq. 30a;

$$\dot{\tilde{\sigma}}_\alpha(t) = -i\,\mathrm{Tr}\,[A_\alpha^\dagger\tilde{L}\sigma(\underline{\Gamma}; t)].\qquad(31)$$

We have evaluated Eq. 31 explicitly and obtained a closed set of equations of motion for $\tilde{\sigma}_\alpha$, $\alpha = 0, 1, \ldots, M$. The REM for σ_α may then be obtained making use of Eqs. 28. Using the matrix notation introduced in Eq. 22, the REM assume the form [27]

$$\dot{\sigma}_0(t) = -\frac{i}{\hbar}U_0\sigma_0(t),$$

$$\dot{\underline{\sigma}}_1(t) = \Omega\underline{\sigma}_2 - \frac{i}{\hbar}\sigma_3\underline{U}_1,$$

$$\dot{\underline{\sigma}}_2(t) = -\mathbf{V}_1 - \frac{i}{\hbar}\sigma_5^T\underline{U}_1,$$

$$\dot{\sigma}_3(t) = \sigma_5\Omega + \Omega\sigma_5^T - \frac{i}{\hbar}\sigma_3 U_2\sigma_3,$$

$$\dot{\sigma}_4(t) = -\mathbf{V}_2\sigma_5 - \sigma_5^T\mathbf{V}_2 - \frac{i}{\hbar}(\sigma_5^T U_2\sigma_5 - \tfrac{1}{4}\hbar^2 U_2),$$

$$\dot{\sigma}_5(t) = -\sigma_3\mathbf{V}_2 + \Omega\sigma_4 - \frac{i}{\hbar}\sigma_3 U_2\sigma_5, \tag{32}$$

where

$$V = \tfrac{1}{2}(V_g + V_e), \tag{33a}$$

$$\mathbf{V}_1 = \left\langle \frac{\partial V}{\partial \underline{\mathbf{x}}} \right\rangle, \tag{33b}$$

$$\mathbf{V}_2 = \left\langle \frac{\partial^2 V}{\partial \underline{\mathbf{x}}\cdot\partial \underline{\mathbf{x}}^T} \right\rangle, \tag{33c}$$

and

$$U = V_e - V_g, \tag{34a}$$

$$U_0 = \langle U \rangle, \tag{34b}$$

$$\underline{U}_1 = \left\langle \frac{\partial U}{\partial \underline{\mathbf{x}}} \right\rangle, \tag{34c}$$

$$U_2 = \left\langle \frac{\partial^2 U}{\partial \underline{\mathbf{x}}\cdot\partial \underline{\mathbf{x}}^T} \right\rangle. \tag{34d}$$

The angular brackets denote

$$\langle R \rangle = \int d\underline{\Gamma}\, R(\underline{\Gamma})\sigma(\underline{\Gamma};t) \bigg/ \int d\underline{\Gamma}\, \sigma(\underline{\Gamma};t). \tag{35}$$

Here $\sigma(\underline{\Gamma};t)$ may be used for the calculation of Raman excitation profiles as well. Consider a Raman process in which the molecule starts at the vibronic state $|a\rangle$ and ends in the vibronic state $|c\rangle$ (both belonging to the ground-state manifold) by absorbing an ω_L photon and emitting an ω_S photon such that $\omega_L - \omega_S = \omega_{ca}$. The Raman excitation profile is the intensity of this Raman transition versus the incident frequency ω_L. It is given by [21, 26, 27]

$$Q_{ca}(\omega_L) = \left| \int_0^\infty dt \exp[i(\omega_L - \omega_{eg})t] G_{ca}(t) \right|^2, \qquad (36a)$$

with

$$G_{ca}(t) = \text{Tr}[D_f \exp(-iH_e t/\hbar) D_i \exp(iH_g t/\hbar)]. \qquad (36b)$$

Here

$$D_i = \mu |a\rangle\langle a|, \qquad (36c)$$

and

$$D_f = |a\rangle\langle c| \mu, \qquad (36d)$$

Equation 36b can be written as

$$G_{ca}(t) = \int d\underline{\Gamma} \, D_f(\underline{\Gamma}) \tilde{\rho}(\underline{\Gamma}; t), \qquad (37a)$$

where $\tilde{\rho}(\underline{\Gamma}; t)$ satisfies Eq. 20a with the initial condition

$$\tilde{\rho}(\underline{\Gamma}; 0) = D_i. \qquad (37b)$$

Further applications of the present REM (Eqs. 32) can be made to fluorescence spectroscopy and to nonlinear optical lineshapes, such as four-wave mixing [21, 27].

IV. SEMICLASSICAL CALCULATION OF THE DENSITY MATRIX AND THE TIME-DEPENDENT SELF-CONSISTENT FIELD (HARTREE) APPROXIMATION

In Section III, we developed REM suitable for the calculation of molecular electronic lineshapes. In that case $\tilde{\rho}(\underline{\Gamma}; t)$ is not a density matrix. It obeys Eq. 20a, in which H_e acts on $\tilde{\rho}$ from the left, and H_g acts from the right. The diagonal elements $\tilde{\rho}(\underline{x}, \underline{x}; t)$ are complex, and its normalization $\sigma_0(t)$ (Eq. 26a) changes with time. We shall now return to the ordinary Liouville space evolution as introduced in Section II and consider the time evolution of the density matrix. Equation 3a is a special case of Eq. 20a obtained by taking $H_e = H_g$. Equation 24 thus provides an approximate solution for the density matrix (Eq. 3b). By setting $U \equiv H_e - H_g = 0$, the REM (Eqs. 32) assume the form

$$\dot{\underline{\sigma}}_1 = \Omega \underline{\sigma}_2,$$

$$\dot{\underline{\sigma}}_2 = -\underline{V}_1,$$

$$\dot{\sigma}_3 = \sigma_5 \Omega + \Omega \sigma_5^T,$$

$$\dot{\sigma}_4 = -V_2 \sigma_5 - \sigma_5^T V_2,$$

$$\dot{\sigma}_5 = -\sigma_3 V_2 + \Omega \sigma_4. \tag{38}$$

Since the normalization of the density matrix is conserved $[\sigma_0(t) = 1]$, we omit σ_0 and do not consider it in this section as a dynamical variable. These equations provide a general semiclassical mean field theory for molecular dynamics. They focus on the complete set of $M = 2N^2 + 3N$ linear and bilinear products of the coordinates and momenta. Such a choice may be useful for few-body problems (small N) but is impractical for macroscopic systems due to the large number of variables involved. A simplified and commonly used procedure is the TDSCF, or Hartree, approximation. In this case we consider only $5N$ expectation values corresponding to $\underline{A}_1$ and $\underline{A}_2$ (denoted σ_{1j} and σ_{2j}, respectively, $j = 1, \ldots, N$) and the diagonal elements of $A_3, A_4,$ and A_5 (denoted $\sigma_{3j}, \sigma_{4j},$ and σ_{5j}, respectively, $j = 1, \ldots, N$). We thus consider only single-particle operators and do not consider explicitly correlations among particles.

Similarly, the approximate density matrix $\sigma(\underline{\Gamma}; t)$ is taken to be in the form of a product of single-particle density matrices

$$\sigma(\underline{\Gamma}; t) = \prod_{j=1}^{N} \sigma_j(\Gamma_j; t), \tag{39}$$

where in the Wigner (phase space) representation,

$$\sigma_j(q_j, p_j; t) = \frac{1}{2\pi \sqrt{\sigma_{3j}\sigma_{4j} - \sigma_{5j}^2}}$$

$$\times \exp\left\{ -\frac{1}{2(\sigma_{3j}\sigma_{4j} - \sigma_{5j}^2)} [\sigma_{4j}(q_j - \sigma_{1j})^2 + \sigma_{3j}(p_j - \sigma_{2j})^2 \right.$$

$$\left. - 2\sigma_{5j}(q_j - \sigma_{1j})(p_j - \sigma_{2j})] \right\}, \tag{40a}$$

and in the coordinate representation,

$$\sigma_j(x_j, x_j'; t) = \frac{1}{\sqrt{2\pi\sigma_{3j}}} \exp\left(-\frac{\sigma_{1j}^2}{2\sigma_{3j}}\right)$$

$$\times \exp\left\{ -\frac{1}{2\sigma_{3j}} \left[\tfrac{1}{4}(x_j + x_j')^2 + \frac{1}{\hbar^2}(\sigma_{3j}\sigma_{4j} - \sigma_{5j}^2)(x_j - x_j')^2 \right. \right.$$

$$+ \frac{\sigma_{5j}}{i\hbar}(x_j + x'_j)(x_j - x'_j) - \sigma_{ij}(x_j + x'_j)$$

$$- \frac{2}{i\hbar}(\sigma_{1j}\sigma_{5j} - \sigma_{2j}\sigma_{3j})(x_j - x'_j)\bigg]\bigg\}. \qquad (40b)$$

In this case Eqs. 38 reduce to the form [26]

$$\dot{\sigma}_{1j} = \sigma_{2j}/m_j, \qquad (41a)$$

$$\dot{\sigma}_{2j} = -\langle V_j(\underline{x})\rangle, \qquad (41b)$$

$$\dot{\sigma}_{3j} = 2\sigma_{5j}/m_j, \qquad (41c)$$

$$\dot{\sigma}_{4j} = -\langle V_{jj}(\underline{x})\rangle\sigma_{5j}, \qquad (41d)$$

$$\dot{\sigma}_{5j} = -\langle V_{jj}(\underline{x})\rangle\sigma_{3j} + \sigma_{4j}/m_j, \qquad (41e)$$

where

$$V_j(\underline{x}) = \frac{\partial V}{\partial x_j}, \qquad (42a)$$

$$V_{jj}(\underline{x}) = \frac{\partial^2 V}{\partial x_j^2}. \qquad (42b)$$

The angular brackets here denote

$$\langle R(\underline{x})\rangle = \int d\underline{x}\, R(\underline{x})\sigma(\underline{x}, \underline{x}; t), \qquad (42c)$$

where

$$\sigma(\underline{x}, \underline{x}; t) = \prod_{j=1}^{N} \frac{1}{\sqrt{2\pi\sigma_{3j}}} \exp\left[-\frac{(x_j - \sigma_{1j})^2}{2\sigma_{3j}}\right]. \qquad (43)$$

Equations 41 constitute the Liouville space TDSCF equations. They provide a cruder approximation for the dynamics than the complete mean field equations 38 since they do not follow explicitly the correlations among particles. Their main advantage over the complete mean field equations 38 is that the number of dynamical variables is $\sim N$ for the TDSCF instead of $\sim N^2$ for the complete mean field equations. This makes the TDSCF particularly useful for large-scale molecular dynamics computations involving many particles. Moreover, in many problems of chemical interest we need to develop a *mixed description* in which some degrees of freedom are treated quantum mechanically and others are treated classically. An example would be the vibrational relaxation of large polyatomic molecules in a solvent [27, 40, 41]. The high-frequency molecular vibrations are quantum mechanical, whereas the solvent degrees of freedom may be treated classically. The TDSCF allows such a mixed description to be developed systematically since all it takes is to keep only the first moments (σ_1 and σ_2) for the classical degrees of freedom and set the other variables (σ_3, σ_4, and σ_5) to zero. At the same time, we may retain all five moments for the other degrees of freedom.

V. DISCUSSION

We shall now analyze the significance of our semicalssical procedure. We first note that Eqs. 38 do not contain $\hbar$. This suggests that they are completely classical. Indeed, the present procedure may be repeated for classical mechanics by replacing L in Eq. 3a with the classical Liouville operator obtained by expanding Eq. 6b to zero order in $\hbar$:

$$-iL\rho = H_c \left[\frac{\overleftarrow{\partial}}{\partial \underline{q}} \frac{\overrightarrow{\partial}}{\partial \underline{p}} - \frac{\overleftarrow{\partial}}{\partial \underline{p}} \frac{\overrightarrow{\partial}}{\partial \underline{q}} \right] \rho = \sum_j \left(\frac{\partial H_c}{\partial q_j} \frac{\partial}{\partial p_j} - \frac{\partial H_c}{\partial p_j} \frac{\partial}{\partial q_j} \right) \rho(\underline{p}, \underline{q}; t). \tag{44}$$

We can then repeat the present derivation step-by-step and obtain Eqs. 38 or 41. *The mean field equations for the moments $\sigma_1, \ldots, \sigma_5$ obtained by using a Gaussian propagator are therefore completely classical.* A major advantage of the present formalism compared with other propagation schemes [12–18] is that Eqs. 38 and 32 are the lowest order of a systematic expansion that may be carried out order by order. The problem of constructing an approximate density matrix for a complicated system using the expectation values of a few dynamical variables is common to many areas of nonequilibrium statistical mechanics [28–31]. A powerful way to achieve that goal is provided by the maximum-entropy formalism [25, 26, 30, 31, 42]. Within this formalism we construct a density matrix $\sigma(\underline{\Gamma}; t)$ that maximizes the entropy subject to the constraints (Eqs. 26). For our chosen set of variables (Eqs. 22), the maximum-entropy distribution is given by [25, 26, 30, 31, 42]

$$\sigma(\underline{\Gamma}; t) = \exp\left(- \sum_{\alpha=0}^{M} \lambda_\alpha A_\alpha \right), \tag{45}$$

where λ_α is a numerical coefficient that may be expressed in terms of σ_α. We have shown [26] that our Gaussian choice (Eq. 24) is identical to the maximum-entropy distribution (Eq. 45). This provides an additional physical insight for our choice (Eq. 24) and connects the present semiclassical procedure with the more general problem of the derivation of REM in nonequilibrium statistical mechanisms. The present equations may be extended by various ways. One possibility is to expand the fluctuation kernel K in Eq. 30a perturbatively. Note that for harmonic systems the mean field equations 32 or 38 are exact. This suggests that an expansion of the kernel K in anharmonicities may be appropriate. Alternatively, we may add more dynamical variables to our chosen set (e.g., x_j^3 and p_j^3) and construct a more elaborate density matrix with more parameters. A natural choice will be to expand the propagator as a Gaussian times a truncated set of Hermite polynomials in p_j and q_j. This will provide an improved reduced description of the system. Once this is done, true quantum propagation effects will enter into the description. We shall now compare our TDSCF equations with the thawed Gaussian (TG) procedure [12], which is widely used in molecular dynamics. We first note that Eqs. 41 may be written as an expansion in the width of the density matrix by expanding the potential around the maximum of the density matrix:

$$\langle V_j(\underline{x}) \rangle = \sum_{n=0}^{\infty} [2^n \sigma_3^{2n}/(2n)!!] V_j^{(2n+1)}(\langle \underline{x} \rangle), \tag{46a}$$

$$\langle V_{jj}(\underline{x}) \rangle = \sum_{n=0}^{\infty} [2^n \sigma_3^{2n}/(2n)!!] V_j^{(2n+2)}(\langle \underline{x} \rangle), \tag{46b}$$

with

$$V_j^{(n)}(\langle \underline{x} \rangle) \equiv \left. \frac{\partial^n V}{\partial x_j^n} \right|_{\underline{x} = \langle \underline{x} \rangle}. \tag{46c}$$

The TG equations may be obtained from our TDSCF equations (Eqs. 41) if the following approximations are made: (i) We take only the zero-order terms in Eqs. 46. Dropping the higher order terms is equivalent to assuming that the density matrix is highly localized and may be approximated by a delta function. We then set

$$\langle V_j(\underline{x}) \rangle = V_j(\langle \underline{x} \rangle), \tag{47a}$$

and

$$\langle V_{jj}(\underline{x}) \rangle = V_{jj}(\langle \underline{x} \rangle). \tag{47b}$$

(ii) We further assume that initially the system is in a pure state with a wavefunction $\Psi(\underline{x}, t)$. For a pure state, our density matrix assumes the form

$$\sigma_j(x_j, x_j'; t) = \psi_j(x_j, t)\psi_j^*(x_j', t). \tag{48}$$

A necessary and sufficient condition for σ_j (Eq. 40) to represent a pure state is

$$\sigma_{5j}^2 = \sigma_{3j}\sigma_{4j} - (\hbar/2)^2. \tag{49}$$

Thus, for a pure state σ_{5j} is uniquely determined, up to a sign, by σ_{3j} and σ_{4j}. Equations 40, 48, and 49 then result in

$$\psi(x_j, t) = (2\pi\sigma_{3j})^{-1/4} \exp\left\{\left(-\frac{1}{4\sigma_{3j}} + \frac{i\sigma_{5j}}{2\hbar\sigma_{3j}}\right)(x_j - \sigma_{1j})^2\right.$$

$$\left. + \frac{i\sigma_{2j}}{\hbar}(x_j - \sigma_{1j}) + \frac{i\gamma_j}{\hbar}\right\}. \tag{50}$$

If we then make the substitutions $\sigma_{1j} = x_t$, $\sigma_{2j} = p_t$, $\sigma_{3j} = \hbar(4\alpha_1)$, and $\sigma_{4j} = \hbar|\alpha_t|^2/\alpha_1$, we obtain the TG equations for x_t, p_t, and α_t [12]. Here α_1 is the imaginary part of α_t and γ is a time-dependent phase factor. We note that the phase information of γ is carried in our REM (Eqs. 32) by the variable σ_0. It should be reiterated that our analysis shows that the mean field equations (Eqs. 38) (and in particular the TG procedure) represent purely classical dynamics. The second-moment bilinear variables (σ_3, σ_4, and σ_5) merely reflect the uncertainty in the initial conditions, but their equations of motion are purely classical since no $\hbar$ appears in Eqs. 38. The appearance of $\hbar$ in Eq. 50 and in the equations of motion of x_t, p_t, and α_t [12] may lead to the conclusion that the propagation is quantum or at least semiclassical. Our analysis shows clearly that this is not the case. Only when the propagator is extended beyond the Guassian form will true quantum propagation effects enter. The REM for $\tilde{\rho}$ (Eqs. 32), unlike Eqs. 38, contain $\hbar$. The appearance of $\hbar$ in Eqs. 32 merely reflects the quantum nature of the electronic two-level system. The nuclear dynamics described by these equations is, however, purely classical, as is evident from our present analysis.

When the density matrix represents a mixed state, Eq. 49 is not satisfied. It can be easily verified from Eqs. 41 that the solution of σ_{5j} is

$$\sigma_{5j}^2(t) = \sigma_{3j}(t)\sigma_{4j}(t) + C, \tag{51}$$

where C is a constant determined by the initial conditions. If $C = -(\hbar/2)^2$, Eq. 49 will be satisfied for all time. Thus, within the mean field approximation (Eq. 29), if the initial single-particle density matrix represents a pure state, it will represent a pure state for all times. Once the fluctuation kernel is included, however, this is no longer the case. Therefore, the fluctuation kernel allows a single-particle pure state to evolve into a mixed state. This is a necessary requirement for a *reduced description*, which should show, for example, how a system relaxes to thermal equilibrium with a thermal bath. For the sake of illustration, let us consider the density matrix for a one-dimensional harmonic system in thermal equilibrium (4):

$$\rho_{eq} = \exp(-\beta H)/\text{Tr}\exp(-\beta H), \tag{52}$$

with $\beta = (kT)^{-1}$, and the Hamiltonian is

$$H(x, \hat{p}) = \hat{p}^2/2m + \tfrac{1}{2}m\omega^2 x^2. \tag{53}$$

In this case we have $\sigma_1 = 0$, $\sigma_2 = 0$, $\sigma_3 = (\hbar/2m\omega)\coth(\tfrac{1}{2}\beta\hbar\omega)$, $\sigma_4 = \tfrac{1}{2}m\omega\hbar\coth(\tfrac{1}{2}\beta\hbar\omega)$, and $\sigma_5 = 0$, and we get

$$\rho_{eq}(x, x') = \frac{1}{\sqrt{2\pi\sigma_3}}\exp\left[-\frac{(x + x')^2}{8\sigma_3} - \frac{\sigma_4(x - x')^2}{2\hbar^2}\right], \tag{54a}$$

and in the Wigner representation,

$$\rho_{eq}(q, p) = \frac{1}{2\pi\sqrt{\sigma_3\sigma_4}}\exp\left[-\frac{q^2}{2\sigma_3} - \frac{p^2}{2\sigma_4}\right]. \tag{54b}$$

The thermal density matrix (Eq. 54) can be represented by our Gaussian form (Eq. 24), but since it corresponds to a mixed state, it cannot be represented by a single wavefunction (Eq. 48). The present phase space REM, based on Eq. 24, may therefore be used to describe the relaxation of a system to thermal equilibrium. Time-dependent self-consistent field equations using pure states were shown to provide useful. approximations for a variety of molecular dynamical problems, including molecular scattering, electronic spectra, the dissociation of clusters, and thermal desorption from surfaces [5–18]. The present Liouville space approach enjoys all these advantages. In addition, it is particularly suitable for dynamics in condensed phases since it may eliminate the necessity of performing tedious thermal averagings. Further applications to fluorescence and four-wave mixing [21, 43] and to solvation dynamics in molecular rate processes [44] are given elsewhere.

Finally, it should be pointed out that the Winger representation is one of several possible prescriptions for converting functions of operators to ordinary functions of parameters. The various representations differ by the choice of time ordering and by the choice of the parameters [45–48]. Although the various representations are all formally exact, they may yield different results when approximations are made. The present formulation may be extended to apply to these alternative representations.

VI. ABSORPTION AND RAMAN EXCITATION PROFILES OF ANHARMONIC MOLECULES

We have applied the present REM (Eqs. 32) toward the calculation of absorption lineshapes and Raman excitation profiles in anharmonic molecules. The absorption spectra were calculated for a two-mode system. We have used two different potentials for the electronically excited state. The first is a Morse potential:

$$V_e(x, x_2) = D_1[1 - \exp(-a_1x_1)]^2 + D_2[1 - \exp(-a_2x_2)]^2. \quad (55a)$$

For this potential the harmonic frequencies (obtained by expanding V_e to quadratic order in x_1 and x_2) are

$$\omega_j' = [2a_j^2 D_j/m_j]^{1/2}, \quad j = 1, 2, \quad (55b)$$

and the anharmonicity parameters [49] are

$$\omega_j' x_e = \frac{a_j}{\sqrt{8m_j D_j}}. \quad (55c)$$

The second model we used for the excited state was a Henon–Heiles potential [50] with cubic anharmonicities:

$$\begin{aligned} V_e(x, x_2) = &\tfrac{1}{2}(m_1\omega_1'^2 x_1^2) + \tfrac{1}{2}(m_2\omega_2'^2 x_2^2) \\ &+ \lambda_1(x_1^2 x_2 - \tfrac{1}{3}x_1^3) + \lambda_2(x_1 x_2^2 - \tfrac{1}{3}x_2^3). \end{aligned} \quad (56a)$$

For this model the anharmonicity parameters are

$$\omega_j' x_e = (\hbar m_j^3 \omega_j'^5)^{-1/2} \lambda_j, \quad j = 1, 2. \quad (56b)$$

The ground-state potential was taken to be harmonic,

$$V_g(x_1, x_2) = \tfrac{1}{2}(\underline{x} - \underline{\Delta})^T \mathbf{g}(\underline{x} - \underline{\Delta}), \quad (57a)$$

with

$$\mathbf{g} = \mathbf{\Omega}^{-1/2}\mathbf{S}^T\boldsymbol{\omega}''^2\mathbf{S}\mathbf{\Omega}^{-1/2}, \quad (57b)$$

where $\mathbf{\Omega}$ and $\boldsymbol{\omega}''$ are diagonal 2×2 matrices with diagonal elements $1/m_j$ and ω_j'', respectively, $j = 1, 2$, and $\mathbf{S}$ represents a Dushinsky rotation matrix [51],

$$\mathbf{S} = \begin{bmatrix} \cos\theta & \sin\theta \\ -\sin\theta & \cos\theta \end{bmatrix}, \quad (58a)$$

with

$$\mathbf{SS}^T = 1. \quad (58b)$$

$\bar{\Delta}_j$ represent the linear displacement of the equilibrium position of mode j

between the ground and the excited states. We further introduce the dimensionless displacement

$$\Delta_j = (m_j \omega_j / \hbar)^{1/2} \bar{\Delta}_j. \tag{59}$$

The following parameters were used in our calculations: $\hbar = 1$, $m_1 = m_2 = 1$, $\sin \theta = 0.25$, and $\omega_1'' = 1000 \, \text{cm}^{-1}$, $\omega_1' = 910 \, \text{cm}^{-1}$, $\Delta_1 = 1$, $\omega_2'' = 300 \, \text{cm}^{-1}$, $\omega_2' = 260 \, \text{cm}^{-1}$, and $\Delta_2 = 1.414$. The system is taken to be initially in thermal equilibrium with temperature T within the ground electronic state. The initial conditions $\sigma_\alpha(0)$ for Eq. 32 are therefore

$$\sigma_0(0) = 1,$$

$$\underline{\sigma}_1(0) = \bar{\underline{\Delta}},$$

$$\underline{\sigma}_2(0) = \underline{0},$$

$$\underline{\sigma}_3(0) = \frac{\hbar}{2} \Omega^{1/2} S^{-1} (1 + 2\bar{n}) \omega''^{-1} S \Omega^{1/2},$$

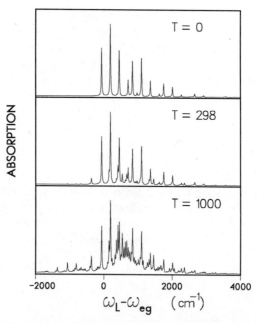

Figure 1. The absorption spectrum of a two-mode harmonic system (Eqs. 57 and 56a with $\lambda_1 = \lambda_2 = 0$) at three temperatures as indicated.

$$\sigma_4(0) = \frac{\hbar}{2}\Omega^{-1/2}S^{-1}(1 + 2\bar{n})\omega''S\Omega^{-1/2},$$

$$\sigma_5(0) = 0, \tag{60}$$

with

$$\bar{n} = [\exp(\beta\hbar\omega'') - 1]^{-1}, \tag{61a}$$

$$\beta = \frac{1}{kT}. \tag{61b}$$

Figure 1 displays the absorption spectrum in the harmonic limit (Eq. 56a) with $\lambda_1 = \lambda_2 = 0$ at three temperatures. Equations 32 were solved with the initial condition (Eqs. 60), and the absorption spectrum was then calculated using Eq. 16 and a standard fast Fourier transform routine. We have further invoked the Condon approximation and set $\mu(\underline{\mathbf{x}}) = 1$. 8192 points were used in the calculation with a resolution of $5\,\mathrm{cm}^{-1}$. The high-temperature calculations require computation time comparable to the zero-temperature calculation.

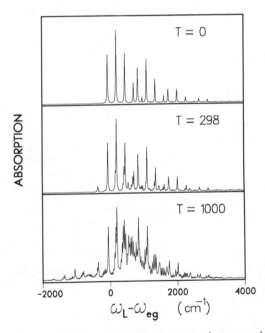

Figure 2. The absorption spectrum of a two-mode system whose ground state is harmonic (Eq. 57) and the excited state is given by a Morse potential (Eq. 55). The anharmonicity parameters are $\omega_1' x_e = \omega_2' x_e = 10^{-4}$.

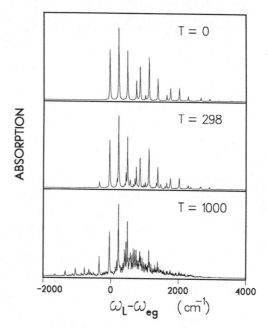

Figure 3. The absorption spectrum of a two-mode system whose ground state is harmonic (Eq. 57) and excited state is given by the cubic potential (Eq. 56). The anharmonicity parameters are $\omega'_1 x_e = \omega'_2 x_e = 0.005$.

This is the main advantage of the present approach compared with calculations using pure states in which the thermal averaging is done at the end and the computation time increases considerably with temperature. Figure 2 displays the absorption spectrum for the Morse potential (Eq. 55) for three temperatures calculated in the same way. The excited-state anharmonicity is $\omega'_1 x_e = \omega'_2 x_e = 10^{-4}$. Figure 3 repeats this calculation for the cubic anharmonic potential (Eq. 56a) with anharmonicity $\omega'_1 x_e = \omega'_2 x_e = 0.005$. A smoothing of the Fourier transform was needed in these calculations. We have also calculated the Raman excitation profiles (Eq. 36) for a single mode whose ground state is harmonic and the excited state is a Morse potential. Figure 4 shows the fundamental 0–1 Raman excitation profile for the single-mode system at $T = 0$ K. We have used $\omega'' = 1050\,\text{cm}^{-1}$, $\omega' = 1000\,\text{cm}^{-1}$, $\omega' x_e = 0.001$, and $\Delta = 1.5$ and a spectral resolution of $20\,\text{cm}^{-1}$. For comparison we show also the exact excitation profile obtained using a sum over states. Figure 5 displays the same calculation on a logarithmic scale (base 10). The good agreement demonstrates the capabilities of the present REM, which should be particularly useful for large systems with many degrees of freedom.

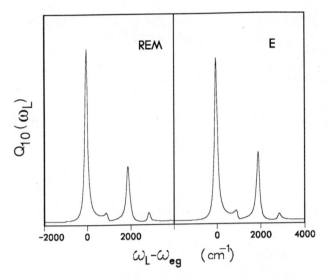

Figure 4. Raman excitation profile (Eq. 36) for a single-mode system with a harmonic ground state and an excited-state Morse potential (Eq. 55 with one mode). Shown is the Raman fundamental 0–1 versus the incident frequency [26]: REM, using the present equations (Eqs. 32); E, exact calculation made by a sum over states.

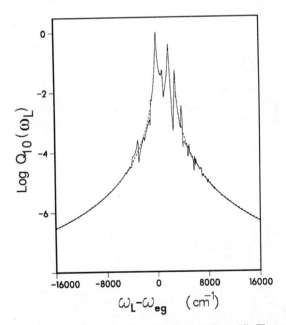

Figure 5. Same as Figure 4 plotted on logarithmic state (base 10). The solid curve is our calculation using the REM and the dashed curve is the exact calculation [26].

APPENDIX: THE GENERATING FUNCTION IN THE COORDINATE REPRESENTATION

Equation 24 may be transformed to the coordinate representation using Eq. 3c. We then get

$$\sigma(\underline{x}, \underline{x}'; t) = C \exp\left[-\underline{\eta}^T \underline{x} - \underline{\delta}^T \underline{x}' - \underline{x}^T \alpha \underline{x} - \underline{x}'^T \beta \underline{x}' - 2\underline{x}^T \gamma \underline{x}'\right],$$

where

$$C = \sigma_0 (2\pi)^{-N/2} [\det \sigma_3]^{-1/2} \exp\left[-\tfrac{1}{2}\underline{\sigma}_1^T \sigma_3^{-1} \underline{\sigma}_1\right],$$

$$\underline{\eta} = \frac{1}{i\hbar}(\underline{\sigma}_2 - \sigma_5^T \sigma_3^{-1} \underline{\sigma}_1) - \tfrac{1}{2}\sigma_3^{-1} \underline{\sigma}_1,$$

$$\underline{\delta} = -\frac{1}{i\hbar}(\underline{\sigma}_2 - \sigma_5^T \sigma_3^{-1} \underline{\sigma}_1) - \tfrac{1}{2}\sigma_3^{-1} \underline{\sigma}_1,$$

$$\alpha = \frac{1}{2\hbar^2}[\sigma_4 - (\sigma_5^T + \tfrac{1}{2}i\hbar I)\sigma_3^{-1}(\sigma_5 + \tfrac{1}{2}i\hbar I)],$$

$$\beta = \frac{1}{2\hbar^2}[\sigma_4 - (\sigma_5^T - \tfrac{1}{2}i\hbar I)\sigma_3^{-1}(\sigma_5 - \tfrac{1}{2}i\hbar I)],$$

$$\gamma = -\frac{1}{2\hbar^2}[\sigma_4 - (\sigma_5^T + \tfrac{1}{2}i\hbar I)\sigma_3^{-1}(\sigma_5 - \tfrac{1}{2}i\hbar I)].$$

Acknowledgments

The support of the National Science Foundation, the Office of Naval Research, the U.S. Army Research Office, and the donors of the Petroleum Research Fund, administered by the American Chemical Society, is gratefully acknowledged.

References

1. W. H. Miller, *Adv. Chem. Phys.* **25**, 69 (1974); **30**, 77 (1975).
2. D. W. Noid, M. L. Koszykowski, and R. A. Marcus, *Ann. Rev. Phys. Chem.* **32**, 267 (1981).
3. P. Pechukas, *Phys. Rev.* **181**, 166, 174 (1969).
4. R. P. Feynman and A. R. Hibbs, *Quantum Mechanics and Path Integrals*, McGraw-Hill, New York, 1965; R. P. Feynman, *Statistical Mechanics A Set of Lectures*, Benjamin Cummings, Reading, PA, 1982.
5. D. Chandler and P. G. Wolynes, *J. Chem. Phys.* **74**, 4078 (1981); D. Chandler, Y. Singh, and D. M. Richardson, *J. Chem. Phys.* **81**, 1975 (1984); A. L. Nichols, D. Chandler, Y. Singh, and D. M. Richardson, *J. Chem. Phys.* **81**, 5109 (1984).
6. C. D. Jonah, C. Romero, and A. Rahman, *Chem. Phys. Lett.* **123**, 209 (1986).

7. J. D. Doll, *J. Chem. Phys.* **81**, 3536 (1984); J. D. Doll, R. D. Coalson, and D. L. Freeman, *Phys. Rev. Lett.* **55**, 1 (1985); J. D. Doll and D. L. Freeman, *Science* **234**, 1356 (1986).

8. R. D. Coalson, *J. Chem. Phys.* **85**, 926 (1986); R. D. Coalson, D. L. Freeman, and J. D. Doll, *J. Chem. Phys.* **85**, 4567 (1986).

9. D. Kosloff and R. Kosloff, *J. Comp. Phys.* **52**, 35 (1983); D. Kosloff and R. Kosloff, *J. Chem. Phys.* **79**, 1823 (1983); R. Kosloff and C. Cerjan, *J. Chem. Phys.* **81**, 3722 (1984).

10. S. A. Adelman and J. D. Doll, *J. Chem. Phys.* **63**, 4908 (1975).

11. J. C. Tully, *Acc. Chem. Res.* **14**, 188 (1981).

12. E. J. Heller, *J. Chem. Phys.* **62**, 1544 (1975); *Acct. Chem. Res.* **14**, 368 (1981).

13. R. D. Coalson and M. Karplus, *Chem. Phys. Lett.* **90**, 301 (1982); *J. Chem. Phys.* **79**, 6150 (1983).

14. R. F. Grote and A. E. DePristo, *Surf. Sci.* **131**, 491 (1983); D. C. Clary and A. E. DePristo, *J. Chem. Phys.* **81**, 5167 (1984).

15. R. T. Skodje and D. G. Truhlar, *J. Chem. Phys.* **80**, 3123 (1984).

16. D. Thirumalai and B. J. Berne, *J. Chem. Phys.* **79**, 5029 (1983); D. Thirumalai, E. J. Bruskin, and B. J. Berne, *J. Chem. Phys.* **79**, 5063 (1983).

17. K. Singer and W. Smith, *Mol. Phys.* **57**, 761 (1986).

18. R. Kosloff, *J. Chem. Phys.* **92**, 2087 (1988).

19. See papers in *Farad. Disc. Chem. Soc.* **75** (1983); A. H. Zewail, ed., *Photochemistry and Photobiology*, Hardwood, New York, 1983.

20. S. Mukamel, *J. Phys. Chem.* **89**, 1077 (1985); *J. Chem. Phys.* **77**, 173 (1982).

21. S. Mukamel, *Phys. Rep.* **93**, 1 (1982); S. Mukamel and R. F. Loring, *J. Opt. Soc. Am. B* **3**, 595 (1986); J. Sue, Y. J. Yan, and S. Mukamel, *J. Chem. Phys.* **85**, 462 (1986); Y. J. Yan and S. Mukamel, *J. Chem. Phys.* **86**, 6085 (1987).

22. M. Hillery, R. F. O'Connel, M. O. Scully, and E. P. Wigner, *Phys. Rep.* **106**, 121 (1984).

23. K. Imre, E. Ozizmir, M. Rosenbaum, and P. E. Zweifel, *J. Math. Phys.* **8**, 1097 (1967).

24. H. Mori, I. Oppenheim, and J. Ross, in *Studies in Statistical Mechanics*, Vol. 1, J. deBoer and G. E. Unlenbeck, eds., North-Holland, Amsterdam, 1962; J. T. Hynes, J. M. Deutch, C. H. Wang, and I. Oppenheim, *J. Chem. Phys.* **48**, 3085 (1968).

25. S. Mukamel, *J. Phys. Chem.* **88**, 3185 (1984).

26. (a) J. Grad, Y. J. Yan, and S. Mukamel, *Chem. Phys. Lett.* **134**, 291 (1987); (b) J. Grad, Y. J. Yan, A. Haque, and S. Mukamel, *J. Chem. Phys.* **86**, 3441 (1987).

27. Y. J. Yan and S. Mukamel, *J. Chem. Phys.* **88**, 5735 (1988).

28. L. Onsager and S. Machlup, *Phys. Rev.* **91**, 1505 (1953).

29. R. Zwanzig, *Supp. Prog. Theo. Phys.* **64**, 74 (1978).

30. E. T. Jaynes, *Phys. Rev.* **106**, 620 (1957); **108**, 171 (1957).

31. B. Robertson, *Phys. Rev.* **144**, 151 (1966); **160**, 175 (1967).

32. D. J. Thouless, *The Quantum Mechanics of Many-Body Systems*, Academic, New York, 1961.

33. R. B. Gerber, V. Buch, and M. Ratner, *J. Chem. Phys.* **77**, 3022 (1982); V. Buch, R. B. Gerber, and M. A. Ratner, *Chem. Phys. Lett.* **101**, 44 (1983); G. C. Schatz, V. Buch, M. A. Ratner, and R. B. Gerber, *J. Chem. Phys.* **79**, 1808 (1983).

34. S. Abe and S. Mukamel, *J. Chem. Phys.* **79**, 5457 (1983).

35. D. Forster, *Hydrodynamic Fluctuations, Broken Symmetry, and Correlation Functions*, Benjamin, New York, 1975.

36. R. G. Breene, *Theories of Spectral Lineshape*, Wiley, New York, 1981.

37. S. Mukamel, S. Abe, Y. J. Yan, and R. Islampour, *J. Phys. Chem.* **89**, 201 (1985).

38. R. D. Coalson, *J. Chem. Phys.* **83**, 688 (1985).

39. See, e.g., J. C. Tully, in *Dynamics of Molecular Collisions in Modern Theoretical Chemistry*, Vol. 2, W. H. Miller, ed., Plenum, New York, 1976.

40. A. Sellmeier, P. O. J. Scherer, and W. Kaiser, *Chem. Phys. Lett.* **105**, 140 (1984); N. H. Gottfried, A. Sellmeier, and W. Kaiser, *Chem. Phys. Lett.* **111**, 326 (1984).

41. E. R. Henry, W. A. Eaton, and R. M. Hochstrasser, *Proc. Nat. Acad. Sci. USA* **83**, 8982 (1986).

42. R. D. Levine and M. Tribus, eds., *The Maximum Entropy Formalism*, MIT Press, Cambridge, MA, 1979.

43. Z. Deng and S. Mukamel, *J. Chem. Phys.* **85**, 1738 (1986); S. Mukamel, *Adv. Chem. Phys.* **70**, 165 (1988). R. F. Loring, Y. J. Yan, and S. Mukamel, *J. Chem. Phys.* **87**, 5840 (1987).

44. M. Sparpaglione and S. Mukamel, *J. Chem. Phys.* **88**, 3263, 4300 (1988).

45. R. M. Wilcox, *J. Math. Phys.* **8**, 962 (1967).

46. W. H. Louisell, *Quantum Statistical Properties of Radiation*, Wiley, New York, 1973.

47. P. D. Drummond and C. W. Gardiner, *J. Phys. A* **13**, 2353 (1980).

48. W. H. Miller, *J. Chem. Phys.* **61**, 1823 (1974).

49. G. Herzberg, *Molecular Spectra and Molecular Structure*, Van Nostrand, New York, 1966.

50. M. Henon and C. Heiles, *Astron. J.* **69**, 73 (1964).

51. Y. J. Yan and S. Mukamel, *J. Chem. Phys.* **85**, 5908 (1986).

CHAPTER XIII

TIME-DEPENDENT WAVEPACKET APPROACH TO OPTICAL SPECTROSCOPY INVOLVING NONADIABATICALLY COUPLED POTENTIAL SURFACES

ROB D. COALSON

Department of Chemistry, University of Pittsburgh, Pittsburgh, Pennsylvania 15260

CONTENTS

INTRODUCTION

In recent years there has been a great deal of interest in nonadiabatic quantum dynamics, or nuclear motion in situations where a single Born–Oppenheimer potential is insufficient (1, 3). Under such circumstances, two or more potential surfaces must be considered and a coupling mechanism invoked in order to provide the possibility of transfer of probability amplitude between surfaces. This class of problems has a venerable history, extending back to early interest in the quantum theory of atomic collisions. The famous Landau–Zener–Stueckelberg (1) analyses of collisionally induced curve crossing, in which

605

there is a possibility of altering the electronic angular momentum state of colliding atoms, has long served as a model for picturing and computing transition rates associated with curve-crossing processes. However, the Landau–Zener–Stueckelberg analysis is both approximate and restricted to one (radial) dimension, which precludes its application to many interesting cases of nonadiabatic dynamics in polyatomic or other many-body (e.g., condensed-phase) systems. Such a shortcoming is lamentable, particularly in the face of an ever-increasing body of high-quality experimental information. Among the most precise measurements of nonadiabatic dynamical behavior currently possible are those obtained via excited-state optical spectroscopic techniques (2). The primary aim of this chapter is to discuss concepts and techniques of time-dependent quantum mechanics that show some promise for understanding optical spectroscopy of nonadiabatically coupled excited-state dynamics of multimode molecular systems (3).

The basic strategy employed in what follows separates into two stages. In the first stage the appropriate frequency domain formula for the optically induced process under consideration is transcribed into the time domain. This means that the burden of computing energy eigenvalues and eigenfunctions of the appropriate (coupled multisurface) molecular Hamiltonian is eliminated. It is replaced by the necessity of propagating multicomponent wavepackets according to the time-dependent Schrödinger equation. In this chapter we trace through the time-dependent formulations of electronic absorption, resonance Raman, and photofragmentation spectroscopy in some detail.

The second basic step is then to develop methods for computing the required multisurface evolutions. One method we have investigated at some length (4, 5) involves a time-dependent perturbation theory expansion in the nonadiabatic coupling strength, with localized (Gaussian) wavepacket propagation techniques (6) employed to compute motion on the zero-order potential curves. Because this approach appears well suited for application to certain many-body problems, considerable attention is focused on it in this exposition, which is organized as follows.

In Section I we present the coupled potential surface model for nonadiabatically coupled excited-state quantum mechanics, which forms the cornerstone for subsequent developments. It is then a simple matter to record in Section II the frequency domain formulas for the cross sections pertaining to one-photon electronic absorption and (two-photon) Raman scattering. With these in hand, transcription into the time domain is carried out in Section III. The resultant formalism hinges on coupled surface wavepacket propagation. In Section IV we discuss the theory and application of two wave mechanical techniques for obtaining such dynamics. After a brief review of coordinate grid integration methods in Section IV.A, we turn our attention in Section IV.B to the subject of wavepacket perturbation theory (WPT). The appealing feature of this approach is that multiple surface dynamics can be determined as a superposition of many single-surface wavepacket trajectories. It is possible in favorable circumstances to obtain each of the latter quantities using standard localized Gaussian wavepacket techniques.

Section V summarizes some recent applications of the time-dependent WPT approach to the computation of nonadiabatic electronic absorption and Raman spectra. Section VI is then concerned with adaptation of the ideas presented in Sections I–V to the process of photo*dissociation* involving nonadiabatically coupled potential surfaces. The new feature here is the direct detection of probabilities for the system to make transitions onto each of the zeroth-order excited surfaces. This feature can be incorporated in a natural way into the time-dependent formalism of Section III. We discuss how to do so and also review some numerical results that are natural photofragmentation analogs of the total absorption and Raman cross-sectional numerics discussed in Section V. Section VII summarizes our current perspective on nonadiabatically coupled excited-state spectroscopy and suggests some avenues for further research.

I. ELEMENTS OF NONADIABATIC QUANTUM MECHANICS

The simplest model of electronic absorption invokes the existence of a manifold of Born–Oppenheimer potential surfaces all well separated in energy, as depicted schematically in Figure 1. These potential functions specify effective force fields for the nuclear motion associated with a given electronic state. Because of the large energy separation between electronic states, a

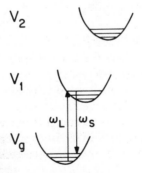

Figure 1. Schematic depiction of radiatively induced interaction between ground-state potential surface V_g and excited surface V_1. If the next lowest lying excited surface, V_2, lies sufficiently far above V_1 in energy, it does not participate in dynamical events initiated by absorption or emission of radiation at frequencies on resonance with transitions between V_g and V_1.

system prepared in state i is assumed to remain in that state, with its nuclei moving in the potential V_i unless perturbed by a high-energy disturbance. Such a disturbance can be caused by an optical or ultraviolet photon of frequency ω_L. As indicated in Figure 1, absorption of this photon can induce a transition of the molecular system from the electronic state in which it was initially prepared (usually the ground electronic state) to a low-lying excited state labeled 1. Within the standard Born–Oppenheimer picture (3a), higher lying excited states are assumed to play no role in the absorption dynamics just described, again due to the large energy gap presumed to exist between these states and excited state 1. More complicated dynamics, such as subsequent emission of a photon of frequency ω_S, can result from the interaction of states g and 1. However, as in the one-photon absorption event, excited states 2, 3,... are not involved.

Although the model depicted in Figure 1 has proven of central importance in understanding molecular spectra, it is by no means universally applicable. In particular, it often happens that the excited states 1 and 2 (and in general others as well) are *not* well separated in energy. In this case the resultant nuclear dynamics in the energy regime co-occupied by electronic states 1 and 2 is not describable by a single potential function. The nuclear motion must instead be described by a coupled-surface Hamiltonian of the form

$$\hat{h}_e = \begin{bmatrix} \hat{h}_1 & g(\hat{x}) \\ g(\hat{x}) & \hat{h}_2 \end{bmatrix}. \tag{1.1}$$

In Eq. 1.1, $\hat{h}_{1,2} = \hat{p}^2/2 + V_{1,2}(\hat{x})$, where $\hat{x}$ and $\hat{p}$ are position and linear momentum operators, respectively, and $V_{1,2}$ are potential functions governing zeroth-order motion. Motion on the surfaces $V_{1,2}$ is then coupled by a non-adiabatic coupling function $g(x)$(7). In this section we use one-dimensional notation since the generalization to D Cartesian dimensions is straightforward. We also set all masses and $\hbar$ to 1 without loss of generality; this can always be accomplished through a judicious choice of units. Finally, we specialize throughout to the case of two coupled Born–Oppenheimer surfaces. Extension to the N-surface case is again straightforward once the two-surface case is understood.

Concomitant with the adoption of the 2×2 matrix operator $\hat{h}_e$, appropriate state vectors must be introduced. These take the two-component form

$$\begin{bmatrix} |\varphi_1(x)\rangle \\ |\varphi_2(x)\rangle \end{bmatrix}, \tag{1.2}$$

where $\langle x|\varphi_{1,2}(x)\rangle$ is the probability amplitude for the system to be located at nuclear coordinate x *and* on the surface $V_{1,2}$. The eigenstates of the Hamiltonian 1.1 are the two-component vectors that satisfy

$$
\begin{bmatrix} \hat{h}_1 & g(\hat{x}) \\ g(\hat{x}) & \hat{h}_2 \end{bmatrix} \begin{bmatrix} |\varphi_1^{(j)}(x)\rangle \\ |\varphi_2^{(j)}(x)\rangle \end{bmatrix} = E_e^{(j)} \begin{bmatrix} |\varphi_1^{(j)}(x)\rangle \\ |\varphi_2^{(j)}(x)\rangle \end{bmatrix},
\tag{1.3}
$$

where the index j labels the eigenstate and the corresponding energy eigenvalue, $E_e^{(j)}$.

Optical spectroscopy probes nuclear motion on the coupled excited surfaces by inducing transitions to or through various energy eigenstates of $\hat{h}_e$. Thus, these objects and their associated eigenvalues comprise the essential ingredients in the microscopic theory of coupled excited-state spectroscopy. (Curiously, the techniques that constitute the central focus of this review *avoid* explicit computation of the energy eigenstates and eigenvalues in question. Nevertheless, formal frequency domain recipes serve as a point of departure for deducing appropriate time-dependent approaches. Thus, we endeavor to establish the former first.)

It frequently proves convenient to describe the content of Eqs. 1.1–1.3 using full Dirac notation, that is, by introducing "electronic state vectors" that behave like spin state vectors in the quantum theory of spin. These vectors have no nuclear coordinate dependence but obey the inner-product relations

$$
\langle e_j | e_k \rangle = \delta_{j,k}, \qquad j, k = 1, 2.
$$

The complete state vector is then built out of product electronic and nuclear factors,

$$
|\Psi\rangle = |\varphi_1(x)\rangle |e_1\rangle + |\varphi_2(x)\rangle |e_2\rangle.
\tag{1.4}
$$

In this notation the coupled time-independent Schrödinger equation for the molecular eigenstates of $\hat{h}_e$ reads simply

$$
\hat{h}_e | \Psi_e^{(j)} \rangle = E_e^{(j)} | \Psi_e^{(j)} \rangle,
\tag{1.5a}
$$

with

$$
\hat{h}_e = \hat{h}_1 |e_1\rangle\langle e_1| + \hat{h}_2 |e_2\rangle\langle e_2| + g(\hat{x})\{|e_1\rangle\langle e_2| + |e_2\rangle\langle e_1|\},
\tag{1.5b}
$$

and

$$
| \Psi_e^{(j)} \rangle = |\varphi_1^{(j)}(x)\rangle |e_1\rangle + |\varphi_2^{(j)}(x)\rangle |e_2\rangle.
\tag{1.5c}
$$

More pragmatic details concerning numerical computation of the eigenstates and eigenvalues associated with the nonadiabatically coupled Hamiltonian $\hat{h}_e$

are attended to at the end of the next section, where they are needed to completely specify the frequency domain prescriptions for electronic absorption and Raman spectroscopy in the presence of nonadiabatic coupling. Note finally that the time-dependent Schrödinger equation has not yet made an appearance in the discussion. In the present work we will utilize it only in a transcriptive sense, that is, as a tool for looking at old physics (interaction of light with molecules) in a new way. This will become clear in Section III.

II. FREQUENCY DOMAIN FORMULAS FOR NONADIABATIC ELECTRONIC ABSORPTION AND RAMAN SPECTRA

In the previous section some basic elements of the mathematical structure of nonadiabatic quantum mechanics were reviewed. With these concepts in place, it is straightforward to specialize the Franck–Condon principle (8) for electronic absorption to the case where two nonadiabatically coupled excited surfaces are relevant. The resultant formula (3a, 4) has the same appearance as the one familiar from the single-excited-surface case, namely,

$$\sigma_i(\omega_L) = \sum_j \langle \Psi_g^{(i)} | \hat{\mu}_R | \Psi_e^{(j)} \rangle \langle \Psi_e^{(j)} | \hat{\mu}_R | \Psi_g^{(i)} \rangle \delta(\omega_L - [E_e^{(j)} - E_g^{(i)}]). \quad (2.1)$$

(A multiplicative factor of ω_L has been suppressed from the right side of this equation, since it generally varies slowly over the width of the absorption profile.) Thus, the spectrum is given by a set of "sticks" whose position is consistent with energy conservation and whose height or intensity depends on the Franck–Condon overlap $|\langle \Psi_e^{(j)} | \hat{\mu}_R | \Psi_g^{(i)} \rangle|^2$. The additional complications introduced by the presence of coupled excited states in the frequency regime of interest are contained in the two-state nature of the state vectors and operators appearing in 2.1. The term $|\Psi_e^{(j)}\rangle$ has been introduced in the preceding section. The relevant initial state on the ground surface, $|\Psi_g^{(i)}\rangle$, also has a direct-product form,

$$|\Psi_g^{(i)}\rangle = |\varphi_g^{(i)}(x)\rangle |g\rangle, \quad (2.2)$$

with $|\varphi_g^{(i)}(x)\rangle$ the ith spatial eigenstate of the ground-state surface (corresponding to energy eigenvalue $E_g^{(i)}$) and $|g\rangle$ the ground-electronic-state vector. The latter is analogous to the excited-electronic-state vectors $|e_{1,2}\rangle$. In particular, it has no nuclear coordinate dependence and obeys the inner-product relations $\langle e_g | e_{1,2}\rangle = 0$, $\langle e_g | e_g \rangle = 1$.

The only part of Eq. 2.1 that remains to be specified is the radiative coupling operator $\hat{\mu}_R$. This effects transitions between ground and excited surface rovibrational manifolds induced by interaction between the radiation field and the (instantaneous dipole moment of the) molecule. For present purposes, the relevant form of this operator is

$$\hat{\mu}_R = \sum_{k=1}^{2} \hat{\mu}_k [|g\rangle\langle e_k| + |e_k\rangle\langle g|], \tag{2.3}$$

where $\mu_k(\hat{x})$ is an operator on spatial coordinates; its scale provides a measure of the strength of the radiative coupling between V_g and V_k.

The analogous adaptation of the Kramers–Heisenberg–Dirac (KHD) formula for resonance Raman scattering (9) to the case where there are multiple coupled excited states in the resonance region follows in a similar fashion. The result is that the Raman excitation profile (REP) for the intensity of the transition from $|\varphi_g^{(i)}(x)\rangle$ to $|\varphi_g^{(f)}(x)\rangle$ as a function of incident light frequency is given by

$$I_{fi}(\omega_L) = |A_{fi}(\omega_L)|^2, \tag{2.4}$$

where A_{fi} is the KHD amplitude (4)

$$A_{fi}(\omega_L) = \sum_j \frac{\langle \Psi_g^{(f)}|\hat{\mu}_R|\Psi_e^{(j)}\rangle\langle \Psi_e^{(j)}|\hat{\mu}_R|\Psi_g^{(i)}\rangle}{\omega_L - [E_e^{(j)} - E_g^{(i)}] + i\Gamma}, \tag{2.5}$$

and Γ is a phenomenological lifetime factor. [As in the absorption case, a slowly varying frequency factor (18d) has been suppressed from the right side of 2.4.] Equation 2.5 expresses the fact that the transition from internal states $|\varphi_g^{(i)}(x)\rangle$ to $|\varphi_g^{(f)}(x)\rangle$ on the ground-state surface is mediated by the electronically excited states $|\Psi_e^{(j)}\rangle$. It is helpful to visualize the process as one in which the molecule initially prepared in $\varphi_g^{(i)}$ on the ground-state surface is promoted to the excited state h_e by absorption of a photon of frequency ω_L. It then falls back into $\varphi_g^{(f)}$ of the ground-state surface by emitting a photon of frequency ω_S such that $\omega_L - \omega_S = E_g^{(f)} - E_g^{(i)}$. The dependence, as a function of incident light frequency, of the intensity of the line in the scattered light spectrum that corresponds to the net $i \to f$ ground-state transition is given by Eqs. 2.4 and 2.5. This picture is quite standard (6) except for the complication introduced by the presence of multiple nonadiabatically coupled excited surfaces in the model under study. The necessary formal alterations are incorporated, as already indicated, by writing down the well-known expressions relevant in the single excited-surface case, then converting all state vectors to two-state objects, and modifying the transition moment operator to act on them appropriately. Having thus established recipes for the computation of both electronic absorption and Raman spectra, it is appropriate next to consider some implementational details.

One way to evaluate frequency domain expressions for absorption (Eq. 2.1) and Raman (Eqs. 2.4 and 2.5) spectra is to compute the excited-state eigenfunctions and eigenvalues $|\Psi_e^{(j)}\rangle$ and $E_e^{(j)}$ directly. For bound excited surfaces this may be accomplished by diagonalizing $\hat{h}_e$ in a discrete basis. In the interest of completeness we outline the basic procedure next.

One begins by expanding (3a, 4) the spatial components $|\varphi_{1,2}^{(j)}(x)\rangle$ of $|\Psi_e^{(j)}\rangle$ (cf. Eq. 1.5c) in terms of basis functions

$$|\varphi_l^{(j)}(x)\rangle \cong \sum_{k=1}^N a_{kl}^{(j)}|\chi_{kl}(x)\rangle, \quad l=1,2, \qquad (2.6)$$

where the χ's are known basis functions and the a's are superposition coefficients. Different sets of basis functions may be used to expand φ_1 and φ_2 if this proves convenient (hence the l subscript on χ). Furthermore, the basis functions χ_l may be assumed without loss of generality to be orthonormal (i.e., $\langle \chi_{kl}|\chi_{k'l}\rangle = \delta_{k,k'}$), since if they are not naturally orthonormal [e.g., in the case of all-Gaussian basis sets (10)], they can easily be orthogonalized in a preliminary step. The approximate nature of the equality in 2.6 is a consequence of truncating the basis at a finite size N. In this basis the matrix eigenvalue–eigenvector problem associated with Eq. 1.5 takes the standard form

$$\mathbf{h}_e\mathbf{a} = E_e\mathbf{a}. \qquad (2.7)$$

In Eq. 2.7 $\mathbf{h}_e$ is the $2N \times 2N$ matrix

$$\mathbf{h}_e = \begin{bmatrix} \mathbf{h}_1 & \mathbf{g} \\ \mathbf{g}^\dagger & \mathbf{h}_2 \end{bmatrix}, \qquad (2.8)$$

where $\mathbf{h}_1$, $\mathbf{h}_2$, and $\mathbf{g}$ are $N \times N$ matrices whose elements are given by

$$(\mathbf{h}_l)_{jk} = \langle \chi_{jl}|\hat{h}_l|\chi_{kl}\rangle, \quad l=1,2, \qquad (2.9a)$$

and

$$(\mathbf{g})_{jk} = \langle \chi_{j1}|\hat{g}|\chi_{k2}\rangle. \qquad (2.9b)$$

When $\mathbf{h}_e$ is diagonalized, the $2N$ resultant eigenvalues are approximations to energy eigenvalues $E_e^{(j)}$ of $\hat{h}_e$, and the associated eigenvectors, $\mathbf{a}^{(j)}$, specify approximations to the eigenstates $|\Psi_e^{(j)}\rangle$. More precisely, the top N entries in the $2N$-dimensional vector $\mathbf{a}^{(j)}$ contain the superposition coefficients $a_{k1}^{(j)}$, $k=1-N$, that determine the spatial characteristics of $|\varphi_1^{(j)}(x)\rangle$. The bottom half of $\mathbf{a}^{(j)}$ contains the analogous information about $|\varphi_2^{(j)}(x)\rangle$. Once the eigenvectors and eigenvalues of $\hat{h}_e$ have been obtained, evaluation of the desired absorption or

Raman spectra then proceeds through computation of the overlaps $\langle \Psi_e^{(j)} | \hat{\mu}_R | \Psi_g^{(k)} \rangle$, where $k = i$ for absorption and $k = i, f$ for Raman scattering. These are easily synthesized once the excited eigenstates $| \Psi_e^{(j)} \rangle$ have been decomposed into a known basis set (3a, 4).

III. TIME-DEPENDENT FORMULATION OF NONADIABATIC ELECTRONIC AND RAMAN SPECTROSCOPY

The frequency domain procedure just described is in principle applicable to all problems in bound coupled excited-state spectroscopy. Questions pertaining to the modifications necessary in order to treat situations where one or both excited surfaces have *dissociative* channels remain to be addressed (cf. Section VI). However, it is not clear, even in the bound excited-state case, that the problem of computing optical spectra has been solved in an operational sense for systems of large spatial dimensionality. This concern stems from the observation that, in general, the size of the required basis set grows alarmingly fast with spatial dimensionality. For a coupled D-dimensional system, the only obvious way to span the appropriate configuration space is to use a product basis set, that is, choose the basis functions to be products of D functions, one for each degree of freedom. If n functions are utilized to cover the relevant region of nuclear coordinate space in each spatial dimension, one is confronted with a basis set consisting of n^D functions. In other words, there is an exponential growth in basis set size with spatial dimensionality of the coordinate space. Usually, this limits $D \leqslant 3$ with currently available computers, although some results in $D > 3$ dimensions have recently been reported (11). In an effort to avoid such basis set dilemmas, we turn our attention to alternative prescriptions that avoid an explicit sum-over-states strategy.

Following the standard route (6), we seek time kernels that can be Fourier (or half-Fourier) transformed into the desired frequency domain spectra. In the case of electronic absorption it is easy to show that the Franck–Condon spectrum recorded in Eq. 2.1 can be recast in the following way (3a, 4):

$$\sigma_i(\omega_L) = \frac{\text{Re}}{\pi} \int_0^\infty dt \exp \{ i(\omega_L + E_g^{(i)})t \} C_i(t), \qquad (3.1a)$$

with

$$C_i(t) = \langle \Psi_g^{(i)} | \hat{\mu}_R e^{-i\hat{h}_e t} \hat{\mu}_R | \Psi_g^{(i)} \rangle. \qquad (3.1b)$$

Notice again that this time-dependent prescription is identical in form to the results for the single excited-surface analog (6). And again, the complications embodied by the presence of coupled excited states in the model under study here are contained in the two-state nature of the state vectors and operators appearing in 3.1b. The content of this equation may be extracted in several equivalent ways. If a complete set of excited eigenstates is inserted either before or after the propagator $\exp\{-i\hat{h}_e t\}$, the sum-over-states procedure is easily recovered, since the resultant form for $C_i(t)$ becomes

$$C_i(t) = \sum_j |\langle \Psi_e^{(j)} | \hat{\mu}_R | \Psi_g^{(i)} \rangle|^2 e^{-iE_e^{(j)}t}. \tag{3.2}$$

It is clear from Eq. 3.2 that Fourier transformation of $C_i(t)$ according to 3.1a regenerates the desired frequency domain formula 2.1. However, since we wish to avoid direct synthesis of the spectrum out of Franck–Condon factors, we consider an alternative interpretation of the formal expression 3.1b that involves coupled surface wavepacket propagation. Operation of $\hat{\mu}_R$ on $|\Psi_g^{(i)}\rangle$ generates initial conditions for dynamics on the coupled excited surfaces, namely,

$$|\Psi_0\rangle \equiv \hat{\mu}_R |\Psi_g^{(i)}\rangle = \hat{\mu}_1 |\varphi_g^{(i)}(x)\rangle|e_1\rangle + \hat{\mu}_2 |\varphi_g^{(i)}(x)\rangle|e_2\rangle. \tag{3.3}$$

These initial conditions are then evolved on the coupled excited surfaces until time t according to

$$|\Psi_t\rangle = e^{-i\hat{h}_e t}|\Psi_0\rangle, \tag{3.4}$$

or, equivalently, via the familiar differential equation

$$i\partial_t|\Psi_t\rangle = \hat{h}_e|\Psi_t\rangle. \tag{3.5}$$

With $|\Psi_t\rangle$ in hand (details of the propagation procedure are discussed in the next section), one has only to project onto $|\Psi_0\rangle$ to obtain $C_i(t) = \langle \Psi_0|\Psi_t\rangle$.

Computation of Raman excitation profiles may be organized in a similar manner. The KHD amplitude 2.5 can be written as (4)

$$-iA_{fi}(\omega_L) = \int_0^\infty dt \, \exp\{i(\omega_L + E_g^{(i)})t - \Gamma t\} C_{fi}(t), \tag{3.6a}$$

with

$$C_{fi}(t) = \langle \Psi_g^{(f)} | \hat{\mu}_R e^{-i\hat{h}_e t} \hat{\mu}_R | \Psi_g^{(i)} \rangle. \tag{3.6b}$$

Equivalence of A_{fi} calculated according to 3.6a with formula 2.5 is demonstrated as in the absorption case by inserting a complete set of excited eigenstates. As in the absorption case, the essential computational task becomes the propagation of the initial (two-component) state vector $|\Psi_0\rangle$

defined in the preceding on the coupled excited surfaces $V_{1,2}$ to obtain $|\Psi_t\rangle$ (also previously defined). Once this has been accomplished, the time kernel relevant to the $i \to f$ Raman transition follows from the projection $\langle \Psi_f | \Psi_t \rangle$, with $|\Psi_f\rangle \equiv \hat{\mu}_R |\Psi_g^{(f)}\rangle$. Notice that only one state vector propagation, $|\Psi_t\rangle$, is necessary to obtain the single-photon absorption cross section and *all* the Raman excitation profiles.

IV. WAVE MECHANICAL TECHNIQUES FOR COUPLED-SURFACE PROPAGATIONS

One way to compute $|\Psi_t\rangle$ is to represent the relevant Schrödinger Equation 3.5 in coordinate space. This results in a set of coupled partial differential equations,

$$i\,\partial_t \begin{bmatrix} \varphi_1(x,t) \\ \varphi_2(x,t) \end{bmatrix} = \begin{bmatrix} T_x + V_1(x) & g(x) \\ g(x) & T_x + V_2(x) \end{bmatrix} \begin{bmatrix} \varphi_1(x,t) \\ \varphi_2(x,t) \end{bmatrix}, \qquad (4.1)$$

where T_x is the kinetic energy differential operator, $T_x = -\frac{1}{2}\partial_x^2$. The initial wavepackets of relevance in the present problem are $\varphi_{1,2}(x,0) = \mu_{1,2}(x)\varphi_g^{(i)}(x)$. Since the wave equation 4.1 is first order in time, it provides a direct prescription for updating $\varphi_{1,2}(x,0) \to \varphi_{1,2}(x, \Delta t)$. This procedure can then be iterated. Once $\varphi_{1,2}(x,t)$ has been obtained, the time kernels associated with absorption and Raman scattering are easily determined through wavepacket overlaps,

$$C_{fi}(t) = \int_{-\infty}^{\infty} dx\, \varphi_g^{(f)}(x)[\mu_1(x)\varphi_1(x,t) + \mu_2(x)\varphi_2(x,t)]. \qquad (4.2)$$

[Note that the absorption kernel $C_i(t)$ is identical to the Raman kernel $C_{ii}(t)$.]

A. Spatial Grid Integration

We consider next some specific methods for integrating the Schrödinger wave equation 4.1. An obvious approach is to set up a spatial grid for each wavepacket component, discretize the initial wavepackets on this grid, and update the value of the wavefunction at each grid point via 4.1:

$$\varphi_1(x_j, \Delta t) \cong \varphi_1(x_j, 0) - i\,\Delta t \{[T_x + V_1(x)]\varphi_1(x_j, 0) + g(x_j)\varphi_2(x_j, 0)\}, \quad (4.3)$$

and analogously for $\varphi_2(x_j, \Delta t)$ (by reversing the roles of the subscripts 1 and 2). In practice, it proves efficient to evaluate the action of the kinetic energy operator on $\varphi_{1,2}(x,0)$ using Fast Fourier Transform technology (12). Furthermore, the simple "first-order" (in time) algorithm indicated in Eq. 4.3 can be replaced with more sophisticated "higher order" variations that improve stability and enable convergence using larger time steps (13). Grid integration of partial differential equations has become a very active field in numerical analysis, partly because it has application in many branches of physics [hydrodynamics (14), optics (15), quantum mechanics (12, 13), and so on]. Despite the fact that it has been utilized for studying nonadiabatic quantum dynamics only recently, considerable progress has been made in this direction, particularly if one includes advances in the closely related area of molecular scattering theory (16). A virtue of this methodology is that it can be applied to systems specified by arbitrary potentials and nonadiabatic coupling functions. Its primary disadvantage is that it is dimensionally limited in essentially the same way as basis function approaches. Clearly, the number of grid points required goes up exponentially with spatial dimensionality. It is not possible at present to solve problems involving more than three spatial dimensions using grid methods, and even these require considerable patience or abundant supercomputer time. Therefore, although such technology appears admirably suited for one- and two-dimensional problems, the need to develop approaches with promise in many-body applications remains.

B. Wavepacket Perturbation Theory

One such strategy involves a time-dependent perturbation theory expansion in the nonadiabatic coupling function (4). It then reduces to a superposition of wavepacket evolutions in which each contributing member is obtained via a sequence of *single*-surface propagations. Each of these can be accomplished in favorable circumstances using localized Gaussian wavepacket evolution techniques. The latter can be applied readily in high-dimensional configuration spaces.

To motivate the preceding claims, we begin by partitioning the Hamiltonian into $\hat{h}_e = \hat{h}_0 + \hat{h}'$, with

$$\hat{h}_0 = \hat{h}_1 |e_1\rangle\langle e_1| + \hat{h}_2 |e_2\rangle\langle e_2|, \tag{4.4a}$$

and

$$\hat{h}' = g(\hat{x})\{|e_1\rangle\langle e_2| + |e_2\rangle\langle e_1|\}. \tag{4.4b}$$

Thus, $\hat{h}_0$ accounts for uncoupled motion on the diabatic surfaces V_1 and V_2, whereas the perturbation Hamiltonian $\hat{h}'$ supplies the nonadiabatic interaction between them. It is then straightforward to devise a scheme for computing $|\Psi_t\rangle = \exp\{-i\hat{h}_e t\}|\Psi_0\rangle$ through a time-dependent perturbation theory expansion based on this partitioning. The practical utility of such a scheme depends on two features, namely, simplicity of the zero-order motion (in the present problem, single-surface motion) and smallness of the perturbation.

Nontrivial examples where both conditions obtain are presented here. Before doing so, it is appropriate to present a few more details of the perturbation theory analysis.

Suppose, momentarily, that the initial state $|\Psi_0\rangle$ has nonzero amplitude only on surface 1, that is, $|\Psi_0\rangle = |\varphi_0(x)\rangle|e_1\rangle$. Then it follows directly that $|\Psi_t\rangle \equiv |\varphi_1(x, t)\rangle|e_1\rangle + |\varphi_2(x, t)\rangle|e_2\rangle$, with

$$|\varphi_1(x, t)\rangle = \left\{ e^{-\hat{h}_1 t} - \int_0^t dt' \int_0^{t'} dt'' e^{-i\hat{h}_1(t-t')} \right.$$
$$\left. \times g(\hat{x})e^{-i\hat{h}_2(t'-t'')}g(\hat{x})e^{-i\hat{h}_1 t''} \right\}|\varphi_0(x)\rangle + O(g^4) \qquad (4.5a)$$

$$|\varphi_2(x, t)\rangle = -i \int_0^t dt' \, e^{-i\hat{h}_2(t-t')}g(\hat{x})e^{-i\hat{h}_1 t'}|\varphi_0(x)\rangle + O(g^3). \qquad (4.5b)$$

Hence, in the absence of nonadiabatic coupling, the spatial wavepacket φ_1 propagates unimpeded on V_1, and no amplitude develops on surface 2. When nonadiabatic coupling is turned on, the lowest order corrections to the motion on the two surfaces are given by Eq. 4.5. If either the coupling strength g or the propagation time t becomes sufficiently large, higher order terms in the perturbation expansion must be included. It remains to be verified that substantial effects due to nonadiabatic coupling can be accurately accounted for using corrections obtained from low-order perturbation theory. Affirmation is provided by examples introduced in what follows. Note for the present that since the Schrödinger equation is linear, the general case of spatial amplitude initially on both surfaces can be solved by propagating the component starting on surface 1 in the manner just outlined, then propagating the piece that starts on surface 2 similarly (with the roles of subscripts 1 and 2 reversed), and finally superposing the two results.

To conclude the formal development, it suffices to provide a few details of the operational aspects of wavepacket perturbation theory. Equations 4.5 are implemented by discretizing the indicated time integrals. Thus, for example, in order to compute the lowest order $[O(g)]$ amplitude to find the system at position x on V_2 in the case where $|\Psi_0\rangle = |\varphi_0(x)\rangle|e_1\rangle$, we appeal to Eq. 4.5b, which amounts in practice to a coherent summation of spatial wavepackets,

$$\varphi_1(x, t) \cong -i \sum_{j=1}^N w_j \varphi^{(j)}(x, t_j). \qquad (4.6)$$

In Eq. 4.6, t_j is the jth quadrature point in an N-point quadrature evaluation of the single time integral appearing on the right side of 4.5; w_j is the associated quadrature weight value. Moreover, the wavepacket φ_j is obtained via the following three-step process:

1. Propagate $\varphi_0(x)$ on V_1 from 0 to t_j.
2. Multiply the spatial wavepacket at t_j by $g(x)$.
3. Propagate the wavepacket generated in the preceding step on V_2 from t_j to t.

The procedure for computing the lowest order $[O(g^2)]$ correction due to nonadiabatic coupling effects is obtained by numerically approximating the two-time integral appearing in 4.5a in the analogous way. It should be noted that any number of quadrature formulas may be used to discretize the necessary time integrals. We have found Legendre quadrature (17) to be considerably more efficient in this regard than simple rectangle rule schemes.

V.   WAVEPACKET PERTURBATION THEORY CALCULATIONS OF ELECTRONIC ABSORPTION SPECTRA AND RAMAN EXCITATION PROFILES

In order to explore the feasiblity of wavepacket perturbation theory (WPT), a simple model of nonadiabatic coupling was subjected to extensive numerical study (4). This model consisted of three linearly displaced one-dimensional harmonic oscillator potential wells,

$$V_g = \tfrac{1}{2}m\omega^2(x + \delta_g)^2 - V_g^{(0)},$$
$$V_1 = \tfrac{1}{2}m\omega^2 x^2,$$
$$V_2 = \tfrac{1}{2}m\omega^2(x - \delta_e)^2 + V_e^{(0)}. \tag{5.1}$$

As the notation indicates, V_g functioned as the ground electronic surface and $V_{1,2}$ as the zero-order excited surfaces. These were nonadiabatically coupled by the constant function $g(x) = g$. Further, the ground surface V_g was coupled radiatively to V_1 only, also by a constant function $\mu_1(x) = 1$. Electronic absorption spectra and resonance Raman excitation profiles were computed for the choice of potential parameters $m = \omega = 1$, $\delta_g = \delta_e = 3$, $V_g^{(0)} = -4$, and $V_e^{(0)} = -9$ and for various values of the nonadiabatic coupling strength g. The net model thus specified bears qualitative resemblance to the situation encountered experimentally in methyl iodide photodissociation (18). In that process, the electronic ground state is radiatively coupled by 266-nm light primarily to an excited potential surface (call it V_1) that correlates with electronically excited ($^2P_{1/2}$) iodine atoms as the dissociation product. However, it is observed experimentally that $\sim 25\%$ of the dissociated iodine atoms are in their $(^2P_{3/2})$ electronic ground state. This strongly suggests the presence of a second relevant excited potential surface (18c) not coupled directly by electromagnetic radiation to the ground-state surface but non-radiatively coupled to V_1. Obviously, a harmonic oscillator model cannot represent the full range of photodissociation dynamics taking place on the electronically excited surfaces V_1 and V_2. In particular, the asymptotic separation of fragments is not correctly accounted for. However, for short- and intermediate-time dynamics, which take place in the Franck–Condon region of the nuclear coordinate space (i.e., near the minimum of V_g), a harmonic oscillator model is reasonable. In fact, the primary goal of the investigation under current scrutiny was to study this type of dynamics, as reflected in electronic absorption spectra and all but very high Raman overtones.

It was found that WPT was quite accurate for computing moderate and, in favorable cases, large effects of nonradiative coupling on low-resolution frequency domain spectra (4). [This type of spectrum is observed in absorption and Raman experiments on methyl iodide (18).] According to the adopted model, the appropriate initial wavepacket (φ_0 in Eq. 45) is simply the vibrational ground-state eigenfunction of V_g,

$$\varphi_0(x) = [\omega/\pi]^{1/4} \exp\{-\omega(x + \delta_e)^2/2\}.$$

Because both V_1 and V_2 are harmonic potentials and the initial wavepacket is Gaussian, sequential propagations of the type described in the preceding section are easy to compute analytically. (The time integrals must still be evaluated by numerical quadrature, but each contributing term can be analytically calculated.) Moreover, due to the nature of the invoked radiative coupling, the lowest nonvanishing correction to absorption and Raman spectra is the $O(g^2)$ term in Eq. 4.5a. This correction was computed and compared with exact results. The latter were obtained by basis set diagonalization, as previously discussed. The qualitative effect of introducing non-adiabatic coupling was seen to be *broadening* of electronic absorption spectra and *suppression* of Raman excitation profiles. As regards the accuracy of WPT, effects such as alterations in spectral peak intensity on the order of 20% were quantitatively accounted for using the method and level of implementation just described. Indeed, in higher overtone REPs at larger nonadiabatic coupling strengths, strong suppression effects (e.g., by a factor of ~ 2.5 in the area under the REP curve for the $0 \rightarrow 5$ overtone at $g = 1.5$) were faithfully, if not quite quantitatively, predicted (4).

Wavepacket perturbation theory is most appealing for low-resolution spectroscopy, since such frequency spectra are determined solely by short-time wavepacket dynamics. This feature optimizes both the reliability of low-order time-dependent perturbation theory and the accuracy of efficient localized wavepacket dynamics techniques utilized to compute the numerous single-surface wavepacket propagations that must be coherently superposed in the WPT scheme. [In the current example, it was possible to compute the relevant single-surface propagations exactly and compactly (4). For motion in more general, namely, anharmonic, potentials, this is not the case, and approximate Gaussian wavepacket algorithms must be utilized. The accuracy of these algorithms generally degrades with time.] However, WPT is not inherently a "short-time" technique, as is the case, for example, with certain quasi-classical spectroscopic methods based on the Reflection Principle (19) and related ideas (20). If the nonadiabatic coupling is sufficiently weak, low-order perturbation theory will remain accurate even up to recurrence times (characterized by relevant vibrational frequencies in the excited-surface potentials). This was demonstrated by numerical example in the model under present consideration. Unfortunately, large effects (greater than $\sim 15\%$ in displacement of significant spectral features) could not be reliably accounted for via second-order perturbation theory at the level of resolution characterized by "spectral fingers" (as opposed to sharp lines; the latter require even longer time dynamics to compute) (4). Thus, computation of highly resolved spectra that are strongly perturbed by nonadiabatic effects appears to be a difficult task. Nevertheless, there is a wealth of dynamical information contained in low-resolution spectra [especially from a complete set of structureless Raman excitation profiles (18)], and the ability to implement WPT in many spatial dimensions makes it well suited for obtaining such information in many-body systems.

We conclude our discussion of the WPT approach to nonadiabatic absorption and Raman spectroscopy with two brief notes that serve to elaborate on points made above. The first concerns the ease with which this approach can be applied to systems consisting of many spatial degrees of freedom. An example is provided by the multidimensional version of the model just studied. That is, $V_{g,1,2}$ become D-dimensional harmonic oscillator potentials. For simplicity, consider the case that all three potentials separate in the same coordinate system (i.e., principal axis rotations are excluded). Then all the spectroscopic time kernels introduced in the one-dimensional notation *factor* at all orders of perturbation theory into products of single-mode results, each of which is easy to compute. Thus, the work to compute any WPT corrections to any order in perturbation theory goes *linearly* with spatial dimensionality in this case. On the other hand, when nonadiabatic coupling effects are present, the work involved in computing eigenfunctions and eigenvalues of $\hat{h}_e$, as is required in frequency domain approaches, grows exponentially. This follows from the way the necessary basis set size scales with spatial dimensionality D, namely as $2n^D$, where n is the number of basis functions needed to span each mode. (The factor of 2 is not critical, but it serves to remind that two sets of basis functions, one for the spatial component on each surface, are required. In general, all are coupled together by nonadiabatic effects.)

The second issue that merits further comment here concerns the origin of the observed effects of nonadiabatic coupling on low-resolution absorption and Raman spectra. The time-dependent, and especially wavepacket, approach to computation of molecular spectra provides some insight into this behavior. In particular, for the coupling scheme considered in the preceding, $\mu_1(x) = 1$, $\mu_2(x) = 0$, it is easy to see why low-resolution absorption spectra are broadened as the nonadiabatic coupling strength is increased. The initial effect of nonadiabatic coupling on the dynamics of the relevant spatial wavepackets is clearly to "siphon off" amplitude from the initially populated surface V_1 onto the radiatively dark surface V_2. Recall that the time kernel associated with the electronic absorption frequency spectrum is obtained for the coupling scenario under consideration by overlapping the component of the spatial amplitude that *remains* on surface 1 with the component "placed" on surface 1 at $t = 0$ [i.e., $\varphi_g^{(0)}(x)$ in the present case]. Hence, this kernel, $C_0(t)$, is expected to decay more rapidly with time as the nonadiabatic coupling is increased. As a consequence, the absorption spectrum, given by the Fourier transform of $C_0(t)$ broadens with increasing nonadiabatic coupling strength.

It is instructive to consider a slightly more detailed model that leads to the same qualitative conclusion for absorption and in addition is useful for understanding *suppression* of Raman excitation profiles (REPs) via nonadiabatic coupling effects: In the short period of time over which the wavepacket dynamics on the coupled excited surfaces V_1 and V_2 affects the total absorption cross section, we make the assumption that the *shape* of $\varphi_1(x, t)$, the packet on surface 1, is not significantly distorted from what it would have been in the absence of nonadiabatic coupling. Specifically,

$$\varphi_1(x, t) \cong F(t)\varphi_g^{(0)}(x, t), \tag{5.2}$$

where $\varphi_g^{(0)}(x, t)$ is the propagation of the initial wavepacket $\varphi_g^{(0)}(x)$ on surface 1 [with $g(x) = 0$], and $F(t)$ is a damping factor intended to account for the siphoning effect on φ_1 induced by nonadiabatic coupling. This implies a Raman time kernel

$$C_{f0}(t) = F(t)\langle \varphi_g^{(f)} | \varphi_g^{(0)}(t) \rangle. \tag{5.3}$$

(The absorption kernel C_0 is identical to the Rayleigh Raman kernel C_{00}.) In order to deduce some qualitative ramifications of the model embodied by Eq. 5.2, let us superimpose upon it two further simplifying approximations.

The first of these is that $F(t)$ can be represented by a Gaussian,

$$F(t) \cong \exp(-\tfrac{1}{2}\gamma^2 t^2), \tag{5.4}$$

with γ^2 a parameter that increases monotonically with g^2. Second, we invoke the following form for $\langle \varphi_g^{(f)} | \varphi_g^{(0)}(t) \rangle$:

$$\langle \varphi_g^{(f)} | \varphi_g^{(0)}(t) \rangle \cong B_f t^f \exp\{-i(\omega_0 + \tfrac{1}{2}\omega)t - \tfrac{1}{2}\sigma^2 t^2\}. \tag{5.5}$$

This form for rapidly decaying Raman kernels is well established (21). Indeed, simple and useful (18) relationships between the parameters B_f, ω_0, and σ^2 that appear on the right side of 5.5 and the characteristics of the excited surface in the Franck–Condon region have been deduced (21). For present purposes we need only indicate that $B_0 = 1$. By appealing to Eq. 3.1a, the short-time absorption kernel C_{00} is then seen to correspond to a Gaussian absorption spectrum centered at ω_0 with mean square dispersion σ^2 and area 1.

What is the effect of nonadiabatic coupling (signified by $\gamma^2 > 0$) on the absorption? When the absorption kernel is multiplied by the Gaussian damping function given in 5.4, the associated frequency spectrum remains a Gaussian with total area 1, but is now characterized by dispersion $\sigma^2 + \gamma^2$. Hence, the phenomenon of "absorption broadening" is clearly contained in the model.

We turn our attention next to implications of the model for REPs. The behavior of two important features, namely, the area under the REP and its value on resonance, is easily examined. Since the approximation to the single-excited-surface kernel given in 5.5 is founded on rapid temporal decay (large σ^2), we assume that the phenomenological constant Γ appearing in 3.6a is negligibly small. With this additional assumption, Eqs. 5.3–5.5 generate profiles $I_{f0}(\omega_L)$ (cf. Eqs. 2.4 and 3.6a) with the following properties:

The area under the curve I_{f0},

$$\int_{-\infty}^{\infty} d\omega_L I_{f0}(\omega_L) \propto [\sigma^2 + \gamma^2]^{-(f+1/2)}, \tag{5.6}$$

where the proportionality constant is independent of γ^2. Hence, the ratio of areas for a given profile in the presence or absence of nonadiabatic coupling is $[\sigma^2/(\sigma^2 + \gamma^2)]^{f+1/2}$. The total area under a particular profile curve is therefore reduced by nonadiabatic coupling. Furthermore, the reduction effect becomes more pronounced with increasing f.

It is also possible to analyze the analogous trend in the *value* of I_{f0} on resonance, that is, at $\omega_L = \omega_0$. It is easy to check that

$$I_{f0}(\omega_L) \propto [\sigma^2 + \gamma^2]^{-(f+1)}, \tag{5.7}$$

with the constant of proportionality again independent of γ^2. Thus, for a given f the ratio of on-resonance intensities with or without nonadiabatic coupling is $[\sigma^2/(\sigma^2 + \gamma^2)]^{f+1}$. Again there is a reduction of intensity due to nonadiabatic coupling, and again this reduction becomes stronger for higher overtones. The alterations in peak height and area just described are consistent with the effects seen in the exact numerical results presented in Refs. 4 and 22 and summarized by the term *suppression*.

In the theory of resonance REPs, a quantity of some utility is the ratio of "overtone" $(0 \to 2)$ to "fundamental" $(0 \to 1)$ intensities in a particular mode as a function of incident light frequency (18, 21). For example, in a single-mode, single-excited-surface system for which the rapid time kernel decay approximation given in 5.5 is appropriate, this ratio is predicted to be $\frac{1}{4}\pi$ (21), independent of detailed specific features of the relevant surfaces. Recent measurements on CH_3I and CD_3I (18c) have yielded a value for the overtone–fundamental ratio associated with the carbon–iodine stretch mode (v_3) that is larger than the theoretical prediction. Because there is also evidence of significant nonradiative coupling to a radiatively "dark" excited surface, it has been suggested (18c) that nonadiabatic coupling effects might account for the anomalously large value of the observed ratio. The simple model of nonadiabatic coupling under discussion here does not admit such an enhancement effect. In fact, it predicts the opposite trend, namely (cf. Eq. 5.7) that the on-resonance overtone–fundamental ratio is reduced in the presence of nonadiabatic coupling by the fraction $\sigma^2/(\sigma^2 + \gamma^2)$. Such a trend is supported by exact numerical results presented in Ref. 4 (cf., in particular, Figure 21 of that reference). These calculations provide evidence that some other effect [perhaps spatial dependence of the radiative transition dipole moment operator (18c)] is responsible for the anomalously large overtone–fundamental ratios in CH_3I and its perdeuterated isomer.

VI. NONADIABATIC PHOTODISSOCIATION CROSS SECTIONS

Historically, the theory of quantum dynamics on bound potentials has developed distinctly from its unbound potential counterpart. This is because the theory of quantum dynamics has been developed from a *time-independent* perspective, that is, through solution of the time-independent Schrödinger equation $H\Psi = E\Psi$. When unbound coordinates are present in a molecular system, a number of complications are introduced. In particular, it becomes necessary to deal with a continuous spectrum of energy eigenvalues and their associated eigenfunctions. Moreover, in the multidimensional case there arise multiple degeneracies due to the possibility of attaining a given energy via different partitionings among internal degrees of freedom. This creates both conceptual and computational difficulties. Conceptually, questions arise concerning the correct normalization of continuum eigenfunctions and the specific linear combinations of degenerate eigenfunctions relevant to state-to-state transitions observed under specific experimental conditions. Computationally, since discrete bound-state bases do not have the correct asymptotic boundary conditions, they have been traditionally abandoned in favor of "close-coupled channel" techniques (23) designed specifically to build in these boundary conditions.

Close-coupled schemes have proven exceedingly useful for computing molecular scattering and photodissociation cross sections. Their primary strengths and weaknesses are the same as other basis techniques. They enable exact solution of problems characterized by essentially arbitrary potentials and coupling functions, provided that it is feasible to include enough "channels" to ensure convergence. The number of channels needed scales roughly in the same way as does the number of product basis functions needed for bound-state calculations, that is, exponentially with the number of spatial degrees of freedom.

In an effort to avoid the seemingly omnipresent "basis set dilemma" associated with time-independent approaches to quantum dynamics, wavepacket dynamical formulations of scattering and photodissociation processes have recently been introduced and wavepacket dynamical techniques developed in order to implement them numerically (24). In the case of photodissociation processes, initial emphasis was naturally placed on the simplest experimentally relevant situation, namely, the case involving a single excited surface. Very recently, extension has been made to the case of nonadiabatically coupled excited surfaces (5). Indeed, the formal extension is rather straightforward since it can be accomplished, as has been seen here in the cases of electronic absorption and Raman excitation spectra, by writing down the single-excited-surface energy domain formula in Dirac notation, and then generalizing all state vectors to consist of two-component spinors and all operators to act on these spinors accordingly.

Much of the background necessary in order to present the basic ingredients of time-dependent nonadiabatic photodissociation theory has already been introduced in previous sections. We need to modify the scenario only slightly to represent dissociative features correctly. In the simplest case where there are no internal degrees of freedom, as in collinear photoinduced dissociation of an adsorbed atom from a structureless crystal, all that is required is to redraw V_1 and V_2 so that each approaches an asymptotic value at large internuclear separation y. Concomitantly, the coupling function g must go asymptotically to zero, so that spatial probability amplitude becomes permanently trapped on one surface or the other as the "fragments" separate. This scenario is schematically depicted in Figure 2.

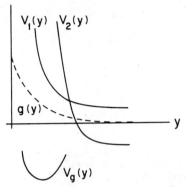

Figure 2. Schematic depiction of potentials V_g, V_1, and V_2 and coupling function g relevant to collinear nonadiabatic photodissociation of adsorbed atom from structureless crystal; the distance from atom to crystal surface is labeled y.

For isolated molecular systems and other many-body problems, degrees of freedom other than the dissociation coordinate are relevant. These "internal" degrees of freedom include vibrational and rotational motion. A number of strategies can be utilized to incorporate the influence of these motions. For example, since the internal degrees of freedom are "quantized" (their energy spectrum is discrete), one way to incorporate them into computational schemes is to "integrate them out." This leads one back to a situation similar to that depicted in Figure 2 except that there are in general *many* potential surfaces for the dissociative motion, and each pair is coupled by a particular "nonadiabatic" coupling function (3*e*).

In certain circumstances, it may be appropriate to retain some subset of the internal degrees of freedom as coordinate space variables to be treated along with the dissociation coordinate using wave mechanical techniques. One such situation is when the dissociation coordinate is coupled to a "bath" of vibrational coordinates. This is illustrated for a prototypical photodesorption

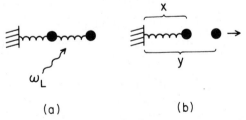

(a)   (b)

Figure 3. Before (*a*) and after (*b*) pictures relevant to two-atom "collinear crystal" undergoing photodetachment by excitation to an excited state in which the end atom is unbounded.

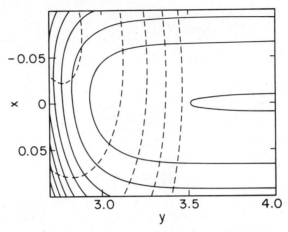

Figure 4.   Typical potential surface V_e relevant to crystal photodetachment is schematized by solid countour lines. The coordinates are labeled as in Figure 3 with $x=0$ in the present figure corresponding to the vibrational equilibrium point of the oscillator fragment shown in Figure 3(b). Here V_e represents an excited-state potential surface accessed on photoabsorption and characterized by repulsion (bond breaking) between atoms x and y. Typical ground-state potential surface V_g also depicted via dashed contour lines for comparison.

process in Figure 3. As can be seen from the figure, the system consists of a dissociating "adatom" located by coordinate y and a bound "crystal" atom located by coordinate x. A typical potential surface relevant to this scenario is schematically depicted in Figure 4. In the case of nonadiabatic coupling effects, there are *two* relevant excited surfaces, each with the qualitative shape indicated for V_e in Figure 4 but differing in detail from each other. Associated with the nonadiabatic interaction of these surfaces is a two-dimensional coupling function g that goes asymptotically to zero at large y. It is this situation on which we concentrate in order to introduce the rudiments of the time-dependent formulation of nonadiabatic photodissociation spectroscopy.

The appropriate Golden Rule formula for the rate of (or "cross section" for) transitions between the ground vibrational eigenstate $\varphi_g^{(i)}$ and the excited-surface fragment state characterized by vibrational quantum number f in the x mode and dissociation onto surface 1 is (3d)

$$\sigma_{f1}(\omega_L) = |\langle \Psi_e^{(f1)}(E)|\hat{\mu}_R|\Psi_g\rangle|^2. \tag{6.1}$$

(As in the total cross-sectional formula 2.1, a factor of ω_L has been suppressed from the right side of Eq. 6.1.) Here E is the energy of the system after it absorbs a photon of frequency ω_L, $E = E_g^{(i)} + \omega_L$. Furthermore, $|\Psi_e^{(f1)}(E)\rangle$ is the eigenstate of $\hat{h}_e$ characterized by energy eigenvalue E and the following asymptotic boundary conditions: It is regular at the origin, and at large values of the dissociation coordinate y has an outgoing (energy-normalized) plane wave component only on surface 1 and in vibrational channel f_1 (in general, it has incoming plane wave components in all energetically accessible vibrational channels on both surfaces). The key ingredient in the recipe is clearly the coupled excited-surface eigenstate $|\Psi_e^{(f_1)}(E)\rangle$. Computation of this quantity (in the coordinate representation) is precisely the goal of close-coupled channel integration methods. We need not pursue the details of close-coupled channel methodology here, since we wish to avoid it by transcribing Eq. 6.1 into the time domain. The result of this exercise is a two-step prescription (5):

1. Propagate the initial two-component wavepacket

$$\begin{bmatrix} \mu_1(\mathbf{x})\varphi_g^{(i)}(\mathbf{x}) \\ \mu_2(\mathbf{x})\varphi_g^{(i)}(\mathbf{x}) \end{bmatrix}$$

on the coupled excited surfaces $V_{1,2}$ according to the time-dependent Schrödinger wave equation 2.1 (exactly as required for computation of total absorption or Raman cross sections).

2. Once the wavepacket components on both surfaces have receded to a sufficiently large value of y that (a) they are trapped on their respective surfaces (i.e., the nonadiabatic coupling has decayed to zero) and (b) the surfaces $V_{1,2}$ separate into a sum of terms, one for each internal mode, then the cross section to dissociate into specific fragment states on surface 1 is obtained by processing φ_1, the wavepacket component *on surface 1* according to the standard procedure employed in the case of single-excited-state photodissociation. This entails a spatial overlap integral of φ_1 with simply prescribed asymptotic fragment eigenstates of V_1. Details of the procedure may be found in many places (24). The analysis of cross sections for dissociation onto V_2 proceeds via analogous processing of the exiting spatial wavepacket on V_2.

The main difficulty encountered in the time-dependent approach lies, as always, in the computation of the coupled-surface wavepacket trajectory indicated in step 1. And, as always, for $D = 1$, $D = 2$, and with considerable effort, $D = 3$, grid integration techniques are appealing because of their reliability and flexibility. However, for higher dimensional systems (e.g., the adatom y attached to a long chain of oscillators that model the crystal lattice), alternative approaches are clearly required. Wavepacket perturbation theory is one possibility, since the essential ingredient of this method, single-surface wavepacket propagations on V_1 and V_2, can be efficiently implemented for adatom-plus-multidimensional-oscillator-bath systems.

In order to explore the feasibility of using WPT for treating nonadiabatic photofragmentation phenomena, the one-dimensional adatom photodissociation model depicted in Figure 2 was studied numerically (5). Suitable exponential repulsive functions were chosen to represent V_1, V_2, and g. Other characteristics of the model were chosen as in the numerical studies of total absorption and Raman spectra discussed in the preceding section. Namely, the initial eigenstate of V_g was assumed Gaussian, as would result from preparation in the vibrational ground state of a harmonic oscillator potential. Further, the radiative coupling invoked was $\mu_1 = 1$ and $\mu_2 = 0$, as in the numerics considered in Section V. Cross sections for the dissociating system to remain on surface 1 as a function of incident light frequency were determined via the two-step prescription previously given, with second-order WPT used to account for the effect of nonadiabatic coupling on φ_1. The procedure was thus similar to the one employed in the total-absorption and Raman spectrum calculations discussed earlier. These sets of calculations did, however, differ in two respects. One minor difference concerns the processing of the dynamics. Rather than synthesizing time kernels, as in the absorption and/or Raman calculations, the exiting wavepacket on surface 1 was projected onto the appropriate asymptotic eigenfunctions, that is, plane waves. A more substantial point of distinction stems from the *an*harmonic nature of dissociative surfaces. The necessary single-surface wavepacket propagations on V_1 and V_2 were obtained via localized Gaussian wavepacket dynamics (24) and were found to be well represented by this approximation. Comparison of WPT with exact results [with the latter obtained via numerical grid integration (12)] produced conclusions similar to those drawn from the REP studies discussed here. Namely, second-order WPT proved rather accurate for describing the siphoning of probability amplitude onto the radiatively "dark" surface. Reduction of the "light" surface cross sections by amounts up to $\sim 50\%$ was quantitatively accounted for via WPT calculations (5).

We close our exposition on nonadiabatic photodissociation theory by mentioning one area in which the time-dependent approach may prove particularly advantageous. This concerns the computation of limited-resolution or "coarse-grained" properties. Particularly when the density of states is large, measurement of individual state-to-state transition probabilities is often not experimentally feasible. Instead, the directly observed quantities represent sums or averages over state-to-state transitions and reflect the constraints of limited experimental resolution. In the phenomenon of nonadiabatic photodissociation, one accessible coarse-grained property is the total cross section for dissociation onto surface 1 (or surface 2, or the ratio of these quantities, termed the *branching ratio*) independent of the final internal states of the fragments (hence summed over all possibilities). It is easy to show (5) that Σ_2, the overall cross section for dissociation onto surface 2, may be computed according to

$$\Sigma_2(\omega_L) = \frac{1}{\pi} \operatorname{Re} \int_0^\infty e^{iEt} \langle \varphi_2(\mathbf{x}, \tau) | \varphi_2(\mathbf{x}, \tau + t) \rangle \, dt. \tag{6.2}$$

In Eq. 6.2 the vector $\mathbf{x}$ includes the dissociation coordinate y and all the bath (internal vibrational) coordinates. Here τ is a time large enough for the packets $\varphi_{1,2}$ to reach sufficiently large values of y that the nonadiabatic coupling has effectively vanished. In words, the operational content of Eq. 6.2 is as follows.

1. Propagate the two-component wavepacket pair (φ_1, φ_2) on the coupled surfaces $V_{1,2}$ according to Eq. 4.1.
2. When, at time τ, these packets have emerged into the region of vanishing nonadiabatic coupling, stop the two-surface propagation: then further propagate the "initial" packet $\varphi_2(x, \tau)$ on V_2 only, recording the overlap kernel $\langle \varphi_2(\mathbf{x}, \tau) | \varphi_2(\mathbf{x}, \tau + t) \rangle$.
3. From the overlap kernel obtained in 2, compute the time transform indicated in Eq. 6.2.

This direct prescription for Σ_2 (which can, of course, be implemented analogously for Σ_1 also) requires substantially less effort than the formally equivalent procedure of summing the probabilities for all state-to-state transitions energetically allowed at E. First, there exists the possibility that $g(\mathbf{x})$ may effectively decay to zero *before* the potential functions V_1 and V_2 have reached their asymptotic behavior. In this case the effort needed to propagate φ_2 onto the asymptotic portion of V_2 is avoided since relatively short-time motion in the interaction region of V_2 suffices to compute the kernel $\langle \varphi_2(\mathbf{x}, \tau) | \varphi_2(\mathbf{x}, \tau + t) \rangle$. Moreover, even if the nonadiabatic coupling does not decay rapidly enough to afford the advantage just described, a second undeniable advantage of the time kernel procedure is that asymptotic overlaps with myriads of final internal vibrational or "phonon" states are avoided.

It is appealing that a correspondence between computational effort and information content in the computed quantities is maintained in a natural way via time-dependent approaches to the theory of quantum-mechanical rate processes. In the present case of nonadiabatic photodissociation, however, the utility of our "shortcut" to computing branching ratios hinges on the efficiency with which the coupled-surface evolution in step 1 can be determined, particularly when many degrees of freedom are involved. One way to proceed is, again, via WPT. To show that this appears feasible for multidimensional systems, we consider the computation of cross sections for absorption onto V_2 when the radiative coupling is given by $\mu_1 = 1$, $\mu_2 = 0$. To $O(g^2)$ this may be expressed as

$$\Sigma_2(\omega_L) = \frac{1}{\pi} \operatorname{Re} \int_0^\infty dt\, e^{iEt} \langle \chi_E | e^{-i\hat{h}_2 t} | \chi_E \rangle + O(g^4). \tag{6.3}$$

with

$$| \chi_E \rangle = g(\hat{\mathbf{x}}) \int_0^\infty dt\, e^{iEt} e^{-i\hat{h}_1 t} | \varphi_g^{(i)} \rangle. \tag{6.4}$$

That is, $\chi_E(\mathbf{x})$ is the spatial function constructed by half-Fourier-transforming the propagation of the initial ground-state surface eigenfunction $\varphi_g^{(i)}$ on the excited surface V_1 and then multiplying the result by the nonadiabatic coupling function g. It is thus more diffuse than $\varphi_g^{(i)}$ but is still of finite spatial extent (due to the decay of g at large y). Consequently, the time kernel utilized in Eq. 6.3 is expected to decay to zero for a sufficiently long time, so that this integral should be well behaved. The spatial function χ_E bears strong resemblance to what has been called the "Raman wavefunction" in applications to multidimensional Raman scattering spectra (involving only one excited surface) (26). In any event, it is clear that the numerical quadratures involved in implementing 6.3 are quadratures in *time* and are not expected to exhibit pronounced dependence on the *spatial* dimensionality of the system.

VII. DISCUSSION AND CONCLUSION

In this chapter, numerous aspects of the theory of optical spectroscopy involving nonadiabatically coupled excited potential surfaces have been developed from a time-dependent perspective. In particular, the processes of electronic absorption, resonance Raman scattering, and photodissociation have been discussed. Computational formalisms that focus on the evolution of well-defined initial states according to the time-dependent Schrödinger equation have previously been established to treat these phenomena when only one excited-state potential surface is relevant (i.e., in the absence of nonadiabatic coupling). A central goal of the present work has been to recount recent generalizations of the single-excited-surface prescriptions that are appropriate when nonadiabatic effects couple dynamics on two or more excited-state surfaces.

An important question arises concerning the utility of energy-domain-to-time-domain transcriptions such as the ones that have occupied us here. Two reasons may be cited: qualitative insight into the content of complicated formulas and computational advantages in certain experimentally interesting situations.

The notion of "physical intuition" is (in our opinion) often strained when discussing quantum-mechanical behavior, which is very far removed from everyday experience in the macroscopic world. In order to avoid this pitfall, we limit ourselves to the following observations concerning conceptual insight provided by time-dependent approaches to molecular spectroscopy. We take as a point of departure the time-honored Golden Rule frequency domain formulas for electronic absorption (Eq. 2.1), Raman scattering (Eq. 2.5), and photodissociation cross sections (Eq. 6.1). The principal restriction on these results is that the intensity of the light source be sufficiently low. In conventional experimental practice, this condition is hard to violate.** If the validity of these Golden Rule formulas is accepted, it follows without any further approximation or assumption that their content can be computed from appropriate time-dependent wavepacket evolution data. In favorable circumstances (e.g., for short-time dynamics on smooth potential surfaces), the qualitative behavior of the wavepacket evolutions is simple [e.g., for single-surface motion, the wavepacket remains nearly Gaussian, and its position and momentum space centers follow classical mechanics (6, 25a). This enables simple correspondences to be made between observed spectral lineshapes and important physical properties of the system such as the shapes of potential surfaces (6, 18, 21)]. When nonadiabatic coupling is present, its effect must be incorporated into any simplistic qualitative dynamical scenarios. An attempt to do this was made in Section V for the case where the primary short-time effect is to siphon off wavepacket amplitude from an initially populated surface via nonadiabatic coupling-induced transfer to an initially unpopulated surface.

We turn next to computational issues. In this work, effort has been made to emphasize the availability of reliable frequency domain techniques (i.e., based on the eigenvalue–eigenvector problem associated with the time-*in*dependent Schrödinger equation) for treating systems of small spatial dimensionality. At the same time, however, it has been pointed out that such techniques become rapidly intractable with increasing spatial dimensionality. It is largely this dimensional limitation that provides the impetus for development of time domain approaches. Although the latter are clearly not a panacea for all problems in quantum dynamics, they have been shown for certain applications to be implementable for many-body systems (27).

Attention has been centered in the current exposition on a particular time domain technique for treating nonadiabatic quantum dynamics, namely, wavepacket perturbation theory. As has been discussed here and elsewhere (5),

---

**With the advent of increasingly powerful light sources, it has recently become possible to study optical spectroscopy "beyond the Golden Rule." See, e.g., the contributions of R. E. Wyatt, A. D. Bandrauk et al., J. H. Eberly, and H. F. Arnoldus et al. in this volume.

it has two primary limitations. First, it is based on a perturbation expansion in the nonadiabatic coupling strength, so that low-order implementations of the formalism are only useful when the nonadiabatic coupling is sufficiently weak. Higher order implementation, although straightforward in principle, becomes rapidly impractical with increasing order. The second problem concerns the ease with which the underlying zero-order motion (i.e. single-surface wavepacket evolutions) can be computed. For quadratic potential functions these evolutions are easy to compute (4). Although the problem of motion on a single quadratic potential is well understood, it is considerably complicated by nonadiabatic interaction of two or more such surfaces. Thus, multidimensional systems that are well represented by such a scheme are good candidates for further exploration via time-dependent approaches of both a perturbative and nonperturbative (22) nature.

When the relevant potentials are *an*harmonic, the problem of obtaining single-surface wavepackets becomes considerably more difficult. Many distinct single-surface trajectories must be determined. The only viable alternative for such a task appears to be localized Gaussian wavepacket propagation techniques (6, 25a). These provide only an approximation to the true dynamics. In general, they are most accurate for short-time motion on smooth potential surfaces. Certainly, caution must be employed in attempting to apply them in the general case.

One merit of developing time domain formalisms of the kind discussed in this work is that numerous computational strategies are thereby suggested. Historically, perturbation expansions have often provided an initial glimpse of behavior slightly removed from well-understood limit behavior. To move far into unknown territory often requires a more radical, nonperturbative approach. A number of possibilities appear worthy of further research. One is the recursive residue generation method (RRGM) of Wyatt and co-workers (11). This technique enables the synthesis of time kernels of the form encountered in the text (e.g., for electronic absorption, cf. Eq. 3.2) without direct diagonalization of the Hamiltonian matrix (in the case of interest here, Eq. 2.8). The rate at which computational labor increases with basis size is thereby reduced considerably. The main novelty introduced by nonadiabatic coupling is that the Hamiltonian matrix 2.8 has a $2N \times 2N$ structure built from four $N \times N$ blocks. This should not substantially impede the implementation of the algorithm.

Another possible route involves the Dirac–Frenkel–McLaughlan (DFM) variational principle (28). Sawada and Metiu (29) have recently shown that this prescription for time-dependent wavepacket evolution can be applied with reasonable efficiency to the problem of nonadiabatically coupled nuclear motion in the event the nuclear wavepacket on each surface is represented by a single Gaussian trial function. The equations of motion for the parameters in the Gaussian trial functions are somewhat more complicated than in the familiar single-surface, single-Gaussian case. Nevertheless, it is very much easier to numerically integrate the closed set of first-order coupled differential equations posed in the two-surface DFM evolution scheme than the exact time-dependent Schrödinger equation. Moreover, DFM is not based on a perturbation expansion in the nonadiabatic strength. It could therefore prove useful for studying medium-to-strong nonadiabatic coupling regimes. The main issue that remains to be resolved is whether two-surface DFM Gaussian approximants give an accurate account of the exact Schrödinger dynamics. In order for this condition to be fulfilled, it is necessary that the exact wavepackets on each surface remain nearly Gaussian over the time interval of interest. Whether they do so, particularly when one or both packets pass through a region where the potential surfaces cross or come very close, has yet to be demonstrated.

Still another alternative time domain route to nonadiabatic quantum dynamics proceeds via the Feynman path integral formulation of quantum mechanics (30). Since this approach has recently been discussed in the context of coupled excited-surface spectroscopy (22), we limit ourselves here to a few brief remarks concerning its strengths and weaknesses. Like the DFM variational principle, path integral techniques are nonperturbative in nature. Hence, they are potentially useful for studying strong coupling situations. Moreover, unlike DFM-guided Gaussian wavepacket dynamics, path integral methods can be implemented without having to adopt uncontrollable assumptions (e.g., that an initially Gaussian wavepacket remain Gaussian as it propagates) by combining the Feynman path integral formalism with Monte Carlo importance sampling techniques. Finally, computational effort increases in a much milder way with spatial dimensionality in the case of Monte Carlo path integral (MCPI) algorithms than it does in the case of basis set methods.

There is, unfortunately, a nontrivial difficulty with MCPI technology when applied to real-time quantum dynamics such as is necessary for the computation of spectroscopic time kernels. In order to obtain the value of a typical time kernel at a specified time, a multitude of oscillating contributions must be summed together. When one attempts to perform this sum via importance sampling, the result is generally characterized by large error bars. Moreover, the error bars become larger for longer time arguments. This feature, combined with the fact that N^2 as many points must be sampled in order to reduce the error bars by a factor of N, has restricted the application of MCPI techniques to date. In particular, they seem well suited for treatment of short-time processes associated with low-resolution, structureless spectra. Examples of this utility specifically in the context of nonadiabatic dynamics have recently been provided (22). For long-time recurrence dynamics, such as is needed to resolve structure in frequency spectra, more efficient sampling procedures are needed. Some progress has recently been reported along these lines in the case of single-surface dynamics (31). Adaptation to the case of coupled multiple-surface motion remains to be accomplished.

In conclusion, we have discussed various spectroscopic processes involving nonadiabatically coupled excited potential surfaces from a time-dependent perspective. Our attempts to do so and to develop associated computational techniques are admittedly primitive. Hopefully, they can provide a stepping stone to more complete understanding of these intriguing and experimentally ubiquitous dynamical scenarios.

Acknowledgments

It is a pleasure to acknowledge the important contributions and encouragement of my co-worker J. L. Kinsey. I also wish to thank M. V. Rama Krishna for helpful comments on the manuscript. This work was supported in part by a Presidential Young Investigator award from the National Science Foundation.

Note Added in Proof: Since this article was prepared, a number of interesting related developments have taken place. Among these are: (1) An application of grid methodology to photodissociation of H_3^+ in a model which involves two nonradiatively coupled excited state surfaces [R. Heather, X.-P. Jiang and H. Metiu, *Chem. Phys. Lett.*, **142**, 303 (1987)]; (2) Successful exploration and extension of the time-dependent self-consistent field (TDSCF) method for computing multidimensional quantum dynamics [R. H. Bisseling et al., *J. Chem. Phys.* **87**, 2760 (1987); J. Kucar, H.-D. Meyer and L. S. Cederbaum, *Chem. Phys. Lett.* **140**, 525 (1987)]. This procedure enables each degree of freedom to be propagated (e.g. via a numerical grid) according to a 1-d Schrödinger equation with an effective potential obtained by averaging over the probability densities associated with the other degrees of freedom. As such it entails a roughly linear growth in computational effort with spatial dimensionality, rather than the exponential growth associated with fully coupled grid methods. To our knowledge, it remains to be demonstrated that TDSCF methods are reliable enough to yield accurate photoabsorption, Raman scattering, or photodissociation cross sections, even for problems in which a single potential surface is involved. Successful extension to multiple surface propagations is expected to be even more difficult. However, since it would greatly reduce the restriction of grid methods to low spatial dimensionalities, such an advance would be welcome. (3) Progress in the development of real time Monte–Carlo Path Integral Techniques has been reported [N. Makri and W. H. Miller, *Chem. Phys. Lett.* **139**, 10 (1987); J. D. Doll, D. L. Freeman, and M. J. Gillan, *Chem. Phys. Lett.* **143**, 277 (1988)]. In particular an algorithmic filtering apparatus has been devised which helps to tame the wild oscillations associated with long-time quantum dynamical propagators. This has only been done for single surface propagations. Extension to the multisurface problems of interest here would be of considerable value.

References

1. L. D. Landau, *Phys. Z. Sowjetunion* **2**, 46 (1932); C. Zener, *Proc. Roy. Soc. Lond. Ser. A* **137**, 696 (1932); E. C. G. Stueckelberg, *Helv. Phys. Acta* **5**, 369 (1932). These results are summarized in

L. D. Landau and E. M. Liftshitz, *Quantum Mechanics (Non-relastic Theory)*, 3rd ed., Pergamon, New York, 1977 Chapter XI, §90.

2. G. ter Horst, D. W. Pratt, and J. Kommandeur, *J. Chem. Phys.* **74**, 3616 (1981); B. J. van der Meer, H. Th. Jonkman, J. Kommandeur, W. L. Meets, and W. A. Majewski, *Chem. Phys. Lett.* **92**, 565 (1982); Y. Matsumoto, L. H. Spangler, and D. W. Pratt, *J. Chem. Phys.* **80**, 5539 (1984).

3. Many others have also explored nonadiabatically coupled excited-state spectroscopy. Recent work that concentrates primarily on bound excited-state motion includes (a) H. Köppel, W. Domcke, and L. S. Cederbaum, *Adv. Chem. Phys.* **57**, 59 (1984); (b) R. L. Whetten, G. S. Ezra, and E. R. Grant, *Ann. Rev. Phys. Chem.* **36**, 277 (1985). (Both reviews summarize the history, mathematical underpinnings, and limitations of the Born–Oppenheimer approximation in some detail.) Recent work on nonadiabatic photodissociation processes includes (c) K. Kodama and A. D. Bandrauk, *Chem. Phys.* **57**, 461 (1981) and A. D. Bandrauk and O. Atabek, *Adv. Chem. Phys.* (this volume); (d) M. Shapiro, *J. Phys. Chem.* **90**, 3644 (1986); (e) J. A. Beswick and M. Glass-Maujean, *Phys. Rev.* **A35**, 3339 (1987). There are also many papers on the semiclassical theory of collisionally induced curve crossing that bear relevance to the problem of multidimensional nonadiabatic spectroscopy; e.g., (f) P. Pechukas, *Phys. Rev.* **181**, 166, 174 (1969); (g) W. H. Miller and T. F. George, *J. Chem. Phys.* **56**, 5637 (1972); (h) J. C. Tuly and R. K. Preston, *J. Chem. Phys.* **55**, 562 (1971); (i) K. Takatsuka and H. Nakamura, *J. Chem. Phys.* **85**, 5779 (1986).

4. R. D. Coalson and J. L. Kinsey, *J. Chem. Phys.* **85**, 4322 (1987).

5. R. D. Coalson, *J. Chem. Phys.*, **86**, 6823 (1987).

6. E. J. Heller, *Accts. Chem. Res.* **14**, 368 (1981).

7. In the present exposition we exclude the possibility that V_1, V_2, or g are momentum dependent. This is not strictly necessary, but it does simplify some of the wavepacket propagation procedures to be discussed in Section IV. The reduction of the full electronic-nuclear motion problem to the problem of nuclear motion on coupled potential surfaces is often performed in such a way that momentum or derivative operators *are* involved in the coupling mechanism. However, by suitable linear transformation, it is possible to transform this (so-called adiabatic) formulation to an equivalent (diabatic) formulation in which only position-dependent potentials and coupling functions appear. Cf. F. T. Smith, *Phys. Rev.* **179**, 111 (1969); also Ref. 3e.

8. J. Franck, *Trans. Farad. Soc.* **21**, 536 (1925); E. U. Condon, *Phys. Rev.* **28**, 1182 (1926); **32**, 858 (1928); G. Herzberg and E. Teller, *Z. Phys. Chem. Abt. B* **21**, 410 (1933).

9. H. A. Kramers and W. Heisenberg, *Z. Phys.* **31**, 681 (1925); P. A. Dirac, *Proc. Roy. Soc. Lond.* **114**, 710 (1927).

10. M. J. Davis and E. J. Heller, *J. Chem. Phys.* **71**, 3383 (1979); I. P. Hamilton and J. C. Light, *J. Chem. Phys.* **84**, 306 (1986).

11. R. E. Wyatt, *Adv. Chem. Phys.* this volume.

12. D. J. Tannor, R. Kosloff, and S. A. Rice, *J. Chem. Phys.* **85**, 5805 (1986).

13. R. Kosloff and D. Kosloff, *J. Chem. Phys.* **79**, 2072 (1983); D. Kosloff and R. Kosloff, *J. Comput. Phys.* **52**, 35 (1983); H. Tal-Ezer and R. Kosloff, *J. Chem. Phys.* **81**, 3967 (1984).

14. W. E. Alley and B. J. Alder, *Phys. Rev. A* **27**, 3158 (1983).

15. M. D. Feit, J. A. Fleck, and L. McCaughan, *J. Opt. Soc. Am.* **73**, 1296 (1983).

16. R. C. Mowrey and D. J. Kouri, *Chem. Phys. Lett.* 119, 285 (1985); *J. Chem. Phys.* **84**, 3535 (1986).

17. M. Abramowitz and I. A. Stegun, eds., *Handbook of Mathematical Functions*, U.S. Department of Commerce, Washington, DC, 1964; W. H. Press, B. P. Flannery, S. A. Teukolsky, and

W. T. Vetterling, *Numerical Recipes: The Art of Scientific Computing*, Cambridge University Press, Cambridge, 1986, Chapter 4.

18. (a) D. Imre. J. L. Kinsey, A. Sinha, and J. Krenos, *J. Phys. Chem.* **88**, 3956 (1984); (b) D. Imre, J. L. Kinsey, R. Field, and D. Katayama, *J. Phys. Chem.* **86**, 2564 (1982); (c) M. O. Hale, G. E. Galica, S. G. Glogover, and J. L. Kinsey, *J. Phys. Chem.* **90**, 4997 (1986); (d) R. L. Sundberg, D. Imre, M. O. Hale, J. L. Kinsey, and R. D. Coalson, *J. Phys. Chem.* **90**, 5001 (1986).

19. S.-Y. Lee, D. J. Tannor, and E. J. Heller, *J. Phys. Chem.* **87**, 2045 (1983).

20. M. Lax, *J. Chem. Phys.* **20**, 1752 (1952).

21. E. J. Heller, R. L. Sundberg, and D. J. Tannor, *J. Phys. Chem.* **86**, 1822 (1982).

22. R. D. Coalson, *J. Chem. Phys.* **86**, 995 (1987).

23. K. C. Kulander and J. C. Light, *J. Chem. Phys.* **73**, 4337 (1980); R. W. Heather and J. C. Light, *J. Chem. Phys.* **78**, 5513 (1983); *J. Chem. Phys.* **79**, 147 (1983); R. Schinke and V. Engel, *J. Chem. Phys.* **83**, 5068 (1985); D. C. Clary, *J. Chem. Phys.* **84**, 4288 (1986); M. Shapiro and R. Bersohn, *Ann. Rev. Phys. Chem.* **33**, 409 (1982).

24. K. C. Kulander and E. J. Heller, *J. Chem. Phys.* **69**, 2439 (1978); S.-Y. Lee and E. J. Heller, *J. Chem. Phys.* **76**, 3035 (1982).

25. (a) E. J. Heller, *J. Chem. Phys.* **62**, 1544 (1975); (b) R. W. Heather and H. Metiu, *Chem. Phys. Lett.* **118**, 558 (1985).

26. R. L. Sundberg and E. J. Heller, *Chem. Phys. Lett.* **93**, 586 (1982).

27. See, e.g., N. Corbin and K. Singer, *Mol. Phys.* **46**, 671 (1982); G. Drolshagen and E. J. Heller, *J. Chem. Phys.* **82**, 226 (1985); S. Mukamel and Y. J. Yan, *Adv. Chem. Phys.*, this volume.

28. E. J. Heller, *J. Chem. Phys.* **64**, 63 (1976).

29. S. Sawada and H. Metiu, *J. Chem. Phys.* **84**, 227, 6293 (1986).

30. R. P. Feynman and A. H. Hibbs, *Quantum Mechanics and Path Integrals*, McGraw-Hill, New York, 1965.

31. J. D. Doll and D. L. Freeman, *Adv. Chem. Phys.* this volume; J. D. Doll, R. D. Coalson and D. L. Freeman, *J. Chem. Phys.*, **87**, 1641 (1987); J. Chang and W. H. Miller ibid., **87**, 1648 (1987).

CHAPTER XIV

LOCAL MODE OVERTONES AND MODE SELECTIVITY

JOHN S. HUTCHINSON

Department of Chemistry and Rice Quantum Institute, Rice University, Houston, Texas 77251

CONTENTS

I. INTRODUCTION

A. Motivation

Our understanding of the dynamics of intramolecular vibrations, vibrational energy transfer, and unimolecular reactivity has been substantially advanced in the past ten years by the study of single-photon overtone excitations of "local" bonds in polyatomic molecules (1–4). The initial experimental observations (5) reported a sequence of vibrational bands at high energy which

637

fit a Morse oscillator energy relationship. These data are naturally interpreted as arising from overtone excitations of light atom–heavy atom (e.g., C–H) bonds behaving as a "diatomiclike" vibration in a polyatomic molecule, thus implying that a large amout of energy (up to 25,000 cm^{-1}) is "deposited" into the local bond during such a transition. However, in polyatomic molecules with more than a few degrees of freedom, these overtone bands most often appear as broad, nearly Lorentzian lineshapes, typically 50–100 cm^{-1} in width. This lifetime broadening implies that the initially localized excitation relaxes, on an ultrafast (fraction of a picosecond) time scale, into the remaining vibrational modes of the molecule. More accurately for the experiments performed to date, the local bond carries the transition dipole moment for these excitations; thus, the vibrational bands appear at energies corresponding closely to the energies of a CH Morse oscillator. However, because the local bond state is significantly mixed with many states in the same energy regime, the oscillator strength of the transition is distributed over many molecular eigenstates. The spectrum is thus heavily congested, leading to the observed broad spectral band.

These different interpretations are physically consistent with one another. Although, as we shall discuss later, experimental studies of these overtone transitions do not prepare truly localized excitations, it is nonetheless clear that (i) with a sufficiently intense and sufficiently short laser pulse, one could prepare a bond-localized excitation, and (ii) such an excitation would be nonstationary, relaxing on a time scale of about 0.1 ps, consistent with an approximately 100-cm^{-1} bandwidth. Studies of high-energy local mode overtone spectra are thus of significant interest for two related reasons. First, the shape and width of these overtone bands, particularly as a function of the molecular environment of the local bond, combined with theoretical analysis has produced a fairly detailed theory of the mechanisms for intramolecular energy transfer involving local bonds at high energy. Second, these (somewhat) mode-selective excitations offer some possibility of altering the reactivity of isolated molecules. In particular, it seems plausible that a localized excitation might demonstrate chemical reactivity (e.g., unimolecular branching ratios) radically different from a thermally excited state at the same energy.

In this chapter, we will examine recent theoretical developments in the analysis and interpretation of both spectroscopic and photochemical studies of local bond overtone vibrations. In particular, in Section II, we will review the classical mechanical theory of nonlinear resonances as has been applied to the relaxation of local bond overtones and will discuss the importance of purely quantum dynamics in enhancing this relaxation in two molecules of experimental interest. In Section III, we discuss the role of resonant energy flow in unimolecular reaction following excitation of a local bond to various overtone states and will illustrate the relative insignificance of mode selectivity in one isomerization reaction. The focus of our study expands in Sections IV and V to include the dynamics of the preparation of the excited state by the laser field. In Section IV, we will see that the excitation process for a bound overtone state competes poorly with intramolecular relaxation, so that the prepared state is actually never truly localized, even when the dipole moment is localized. Finally, we examine in Section V the simultaneous competition of state preparation, intramolecular vibrational relaxation, and unimolecular decay. These latter results are discussed for both long-pulse lasers and ultra-short-pulse lasers.

We first consider a review of the experimental studies of overtone spectroscopy and overtone-induced mode-selective photochemistry, focusing on key results to be discussed in Sections II–V.

B.   Overtone Spectroscopy

As already mentioned, studies of local bond overtones have examined both the spectroscopy of high-energy bound states and the unimolecular photochemistry of high-energy reactive states; we divide our discussion accordingly.

Spectroscopic observation of high overtones is difficult, due primarily to the very low transition moments for such excitations (6). Of course, these transitions would be forbidden if not for vibrational anharmonicity (7), and these anharmonicities appear sufficient to directly excite high overtones only for light atom–heavy atom bonds. Studies have thus focused virtually exclusively on H–X and D–X bonds, where X is carbon, nitrogen, oxygen, or silicon. Even for these molecules, the transition moments typically fall of from that of the fundamental ($v = 1$) excitation by a factor of 6–10 for each additional quanta excited (8); for example, the $v = 6$ overtone is an order of magnitude less intense than that for $v = 5$, which is in turn an order of magnitude less intense than that for $v = 4$, and so on.

For observing such "nearly forbidden" transitions, the method of choice (at least for bound excitations of gas phase molecules) is intracavity laser photoacoustic absorption spectroscopy (8, 9), first applied to gas phase high-overtone spectroscopy by Bray and Berry (5). They observed a sequence of vibrational transitions in benzene up to $19,000 \, \text{cm}^{-1}$ of excitation and further discovered that the frequency of these bands fit a Birge–Sponer relationship (10),

$$\tilde{\nu} = Av + Bv^2 \tag{1}$$

where $\tilde{\nu}$ is the frequency of the v–0 overtone transition. This two-parameter quadratic dependence of frequency with quantum number is consistent with the usual Morse oscillator energies of a diatomic oscillator,

$$E_v = \omega_e(v + \tfrac{1}{2}) - \omega_e x_e(v + \tfrac{1}{2})^2 \tag{2}$$

with $\omega_e = A - B$ and $\omega_e x_e = B$. On this basis, they assigned these bands to local-bond selective excitations and ascribed the associated broad (roughly $100\,\text{cm}^{-1}$) bandwidths to lifetime broadening. Their work was essentially simultaneously analyzed formally by Heller and Mukamel (11) in terms of a bright local bond state providing a "doorway" for the initial excitation with subsequent relaxation into a dense manifold of bath states within the molecule. Underlying this description is the assumption that a sufficient density of states must be present at the excitation energy for the relaxation to proceed rapidly and irreversibly into the bath states.

These early observations have been repeated and have been extended to even higher energies in a wide variety of molecules. Fang et al. (12) have observed CH, OH, and NH overtone spectra in a dazzling variety of molecules and have even discussed the influence of rotational conformation on overtone band positions (13). More recent work by Berry et al. (14, 15) has examined the spectra of acetylenic CH stretch overtones in acetylene and substituted acetylenes, including propyne, trifluropropyne, trimethylpropyne, cyano-acetylene, and virtually all isotopically substituted forms of acetylene. Higher resolution spectra (16), spectra in frozen crystals (17), and the spectra of cooled molecules in supersonic jets (18) have continued to add detail to the now rich body of data available on the vibrations of local bonds at high energy in a variety of molecular environments (19).

Theoretical understanding of these vibrations has progressed accordingly, and we will present our view of intramolecular vibrational relaxation of local bond overtone states in Section II.

C. Overtone-Induced Photochemistry

The possibility of directly probing molecular states at high energy immediately suggests the study of unimolecular reactions induced by overtone excitation (20). Much of the interest in this field lies in the possible observation of mode-selective chemistry (21): the selective excitation of a particular vibrational mode followed by transfer of energy specifically into reactive vibrational modes. For example, Reddy and Berry (22) studied the isomerization of allyl isocyanide to allyl cyanide induced by "bond-selective" excitation of CH bonds in the three different environments in the molecule and observed a significant dependence of reaction rate on the proximity of the CH bond to the CN reactive group.

Other studies in this area have focused on comparisons of reaction rates following overtone excitation to thermal (i.e., non-mode-specific) reaction rates calculated via RRKM (statistical) theory (23). For example, Zare et al. studied the fragmentation of t-butylhydroperoxide by overtone excitation of the OH bond to $v = 5, 6$ and observed some variation of reaction rate from statistical prediction (24). This indicates the presence of O–O bond fragmentation prior to complete equilibration of the OH excitation energy, a plausible result that suggests mode-selective chemistry.

Essentially all other studies of overtone-induced photochemistry have concluded that non-mode-specific vibrational dynamics dominate reaction rates. Jasinski, Frisoli, and Moore (25) initiated isomerizations via CH overtone excitations in cyclobutene, 1-cyclopropylcyclobutene, and 2-methylcyclopentadiene and did not observe any obvious deviations from thermal reaction rates. Crim et al. (18, 26) determined that the overtone-induced decompositions of tetramethyldioxetane and hydrogen peroxide proceed at statistical rates. Kunz and Berry (15, 27) photodissociated formyl fluoride by excitation of various high-energy overtone and combination bands and observed a fairly monotonic increase in reaction rate with increasing energy, indicating the absence of mode-selective events.

The studies listed above are performed either with continuous laser irradiation, typically over many hours, or with pulsed laser excitation using pulses of typically nanosecond length (which we will regard as long pulses). Most recently have come studies from Zewail et al. (28) of picosecond-pulse overtone photochemistry of hydrogen peroxide. Although these studies conclude similarly that mode-selective reactivity does not occur, they nonetheless are crucial in revealing the short-time dynamics that participate in unimolecular reactions.

All of these studies pose substantial theoretical challenges. Specifically, recent theories of resonant energy transfer indicate that local-bond overtones relax (at least initially) along fairly mode-specific pathways; why then are overtone-induced reactions observed to occur nonspecifically? Additionally, we have found (Section IV) that the rate of preparation of the initial overtone state is very much slower than that of either intramolecular relaxation or unimolecular reaction. Given this, what is the nature of the initially prepared state in overtone photochemical studies? As we shall see in Section V, the answer is a subtle function of both the frequency and pulse length of the excitation source. These issues can be addressed only after consideration of the mechanism for overtone relaxation (Section II), reaction (Section III), and state preparation (Section IV).

II.   CLASSICAL AND QUANTUM DYNAMICS OF OVERTONE RELAXATION

A.   Classical Nonlinear Resonances and Vibrational Energy Transfer

Direct spectroscopic observation of high vibrational overtones for local bond modes has been largely restricted to light atom–heavy atom bonds, principally H–X or D–X bonds, where X is carbon, oxygen, nitrogen, or possibly silicon.

These bonds have in common the following important distinctions. First, the light-atom mass obviously gives the H–X bond a high frequency; typically H–X bonds (when relaxed) are the highest frequency modes in a molecule. Applying usual normal-mode concepts, this high frequency tends to isolate the H–X bond as a molecular vibrational mode. Second, the heavier mass (e.g., carbon) in the bond kinetically isolates the H–X vibration. Most of the H–X stretch appears as motion of the light H atom. Furthermore, the H–X stretch is kinetically coupled to the other modes in the molecule (say, a neighboring stretch) via a term inversely proportional to the mass of the heavy atom. As an example of these first two localizing effects, we note that OH (alcoholic) stretches are more localized spectroscopically than are typical CH stretches, due both to the higher OH frequency and to the larger oxygen mass (29, 30).

[Not all HX bonds are good local modes, however. Indeed, an additional criterion for localization of a vibrational mode might be that the bond have a bond angle with its neighbors that is near 90°. This is the case for CH bonds in sp^3 (ethane) or sp^2 (ethene) molecules. However, for acetylenic stretches (14, 15), the CH bond is collinear with the neighboring $C\equiv C$ bond. The bond–bond kinetic coupling, which is proportional to the cosine of the bond angle, is thus much larger for acetylenic CH bonds. These modes are *not* considered local but rather are part of a delocalized "anharmonic normal mode." 31]

Finally and most importantly, observation of a high overtone requires vibrational anharmonicity (otherwise, v–0 transitions are forbidden for $v \neq 1$), and the transition is more intense for larger anharmonicities. Mechanically, HX modes typically have relatively large anharmonic constants, so that observation of overtones is substantially easier for HX bonds than for any other type of molecular mode.

Anharmonicity manifests itself in several physical properties. At the most fundamental level, the potential energy along the vibrational coordinate is not a quadratic function, so that the restoring force of the oscillator is nonlinear. The principal impact of this on vibrational relaxation (beyond making the overtone transition observable) is that the oscillator frequency is a function of the level of vibrational excitation. To illustrate this, we consider the Morse oscillator (32), which has served very effectively to model local mode overtone vibrations. The Morse oscillator vibrational Hamiltonian is

$$H = \frac{p^2}{2\mu} + D(1 - e^{-\alpha x})^2 \tag{3}$$

where μ is the bond reduced mass and D and α are parameters. To find the frequency of the oscillator, we convert to action-angle variables (J, θ) via

$$x = \alpha^{-1} \ln \frac{1 - \rho \sin \theta}{1 - \rho^2}$$

$$p = (2\mu D\rho^2)^{1/2} \frac{(1-\rho^2)^{1/2}\cos\theta}{1-\rho^2\sin\theta} \tag{4}$$

$$\rho = \left(\frac{\Omega J}{D} - \frac{\Omega^2 J^2}{4D^2}\right)^{1/2}$$

where Ω is the harmonic frequency of the anharmonic oscillator (33–35). On substitution, the Hamiltonian becomes independent of the angle coordinate:

$$H = E = \Omega J - \frac{\Omega^2 J^2}{4D} \tag{5}$$

Equation 5 is the exact classical correspondence to the more familiar quantum-mechanical result in Eq. 2. The frequency of the oscillator is just the time derivative of the phase (the angle coordinate); this is easily found from Hamilton's equations (33),

$$\omega = \frac{d\theta}{dt} = \frac{\partial H}{\partial J} = \Omega\left(1 - \frac{\Omega J}{2D}\right) \tag{6}$$

and is clearly not a constant but depends on the energy via the relationship between the energy and the action (Eq. 5). Combining Eqs. 5 and 6 reveals the dependence explicitly,

$$\omega = \Omega\left(1 - \frac{E}{D}\right)^{1/2} \tag{7}$$

from which it is clear that the nonlinear frequency *decreases* monotonically with increasing excitation (34).

Sibert, Hutchinson, Reinhardt, and Hynes (36–40) have shown, by applying the classical mechanical theory of nonlinear resonances of Chirikov (41), Ford (42), and others (34, 35), that this nonlinearity is central to the irreversible relaxation of local bond overtones. Initially, we note that the dependence of the local bond frequency on excitation can create and destroy resonances (in the sense of frequency matches) between vibrational modes. Sibert et al. (36) studied classical energy flow in a model of the stretching vibrations in H_2O consisting of two Morse oscillators kinetically coupled through their momenta p_1 and p_2,

$$H = H_1 + H_2 + G_{12}(\phi)p_1p_2 \tag{8}$$

where G is the stretch–stretch Wilson G matrix element and ϕ is the H_2O bond angle, 104.5°. When the OH bonds are excited with comparable energies (e.g., one bond to $v = 2$ and the other to $v = 3$), then (via Eq. 7) the nonlinear frequencies are close to one another, and the bond modes are in 1:1 resonance. The result, as Sibert et al. (36) show, is quasi-periodic energy exchange between bonds. Specifically, the Hamiltonian in Eq. 8 is converted to action-angle variables, and the 1:1 resonance term is isolated by Fourier expansion of the coupling, yielding the resonance Hamiltonian

$$H = H_1(J_1) + H_2(J_2) - V_0 \cos(\theta_1 - \theta_2) \qquad (9)$$

where

$$V_0 = \frac{4\mu D \cos\phi}{m_0}\left(1 - \frac{E}{D}\right)\frac{1 - (1 - E/2D)^{1/2}}{1 + (1 - E/2D)^{1/2}} \qquad (10)$$

The coupling term in Eq. 9 is significant only when the argument of the cosine is slowly varying in time, that is,

$$\frac{d}{dt}(\theta_1 - \theta_2) = \omega_1 - \omega_2 \approx 0 \qquad (11)$$

which is precisely the condition for a 1:1 nonlinear resonance. Sibert et al. (36) have demonstrated that the simplified resonance Hamiltonian in Eq. 9 reproduces bond–bond energy exchange on resonance very accurately. By contrast, if the vibrational energy is very nonuniformly distributed (e.g., one bond is excited to $v = 5$ and the other is at zero-point energy), then by Eq. 7, the nonlinear frequencies are very different from one another, the argument of the cosine in Eq. 9 varies rapidly in time, and energy exchange is neither expected nor observed classically.

When there are more than two oscillators exchanging energy, the possibility exists that energy transfer out of a locally excited CH bond will be "irreversible" (meaning only that the inevitable recurrence of the initial state occurs at a time that is long enough to be uninteresting). We will illustrate here with a simple model (38) consisting of a hydrogen–carbon Morse oscillator kinetically coupled to a chain of three harmonic CC bonds, each kinetically coupled to its nearest neighbors. The five-atom chain (HCCCC) is frozen in a collinear configuration, so that all four vibrational degrees of freedom are stretches. The CH stretch has Morse parameters $D = 29,400\,\text{cm}^{-1}$ and $\alpha = 1.03a_0^{-1}$ (corresponding to a harmonic frequency of $2860\,\text{cm}^{-1}$ and an anharmonic constant of $69.5\,\text{cm}^{-1}$). The CC bonds have frequency $1850\,\text{cm}^{-1}$. This simple "molecule" serves as an effective model for classical irreversible vibrational relaxation of a CH overtone excitation. In particular, excitation of the CH bond initially to $v = 10$ or above results in classical

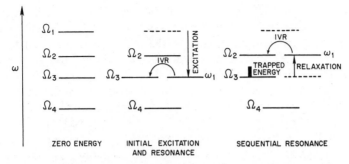

Figure 1. Sequential nonlinear resonance mechanism for relaxation of an initially excited anharmonic mode. The column labeled Zero Energy shows a possible set of harmonic frequencies Ω_j with no excitation. Upon initial excitation, the anharmonic highest frequency mode Ω_1 is excited to a high vibrational level, lowering its frequency, thus bringing it into resonance with Ω_3 and producing energy transfer (IVR). During the IVR, Ω_1 relaxes, raising its frequency into resonance with Ω_2. The energy initially deposited into Ω_3 becomes trapped for long time due to the sequential resonances.

trajectories in which the energy of the CH bond "irreversibly" relaxes. This relaxation may be simply understood as a consequence of "sequential nonlinear resonances," as illustrated in Figure 1. The highest frequency (CH-like) anharmonic normal mode in the chain has a harmonic frequency of $2904 \, \text{cm}^{-1}$. Upon excitation to $v = 10$, the nonlinear frequency is suppressed to $2020 \, \text{cm}^{-1}$, in near 1:1 resonance with one of the lower frequency normal modes of the carbon chain $(1825 \, \text{cm}^{-1})$. As a consequence, energy flows out of the CH-like mode and into the carbon chain mode. However, as this energy flow occurs, the frequency of the CH-mode rises in accord with Eq. 7 until it nearly matches that of the next highest normal mode at $2402 \, \text{cm}^{-1}$. Energy now exchanges between the CH-like mode and this normal mode, and the original resonance is detuned. The result is that *the energy initially deposited into the lower frequency normal mode is trapped!* The classical dynamics give a long-time appearance of irreversible energy transfer due to the nonlinear resonances acting in sequence, and the lifetime of the relaxation process is predicted in this model to be roughly 0.1 ps. This would imply a spectral broadening of the corresponding overtone band of roughly $50 \, \text{cm}^{-1}$, in line with what is observed in many molecules. With very many modes in the large-molecule limit (e.g., benzene), there are generally many such overlapping nonlinear resonances, and irreversible relaxation of CH overtones is expected. Hutchinson et al. (38) demonstrated that this physical picture of irreversibility may be quantified: A resonance Hamiltonian similar to Eq. 9 can be derived that reproduces the overlapping resonances and irreversible relaxation.

Classical nonlinear resonance analysis of overtone vibrational energy relaxation has been successfully applied to a variety of classes of molecules. We note in particular the work of Sibert et al. (40) on benzene and deuterated benzenes. They accurately accounted for the bandwidths observed in the studies of Berry et al. (5, 8) by demonstrating that aryl-CH overtones relax via 2:1 nonlinear resonances between the CH stretch and the local CCH wagging mode. Hutchinson, Hynes, and Reinhardt (39) isolated a similar 2:1 nonlinear resonance in the irreversible relaxation of terminal alkyl CH overtones in n-alkane chains. Of particular interest in the latter case, the CH energy initially is deposited into the two "terminal rocking" normal modes of the CH_3 group. Thus, although the initial excitation does relax rapidly and irreversibly, the energy transfer is extremely mode specific for some time. In both of these cases, nonlinear stretch-to-bend energy transfer dominates the relaxation pathway for the overtone excitation. In Section III, we address the question of whether this mode-specific energy flow can be used to selectively initiate a chemical reaction.

B.　Quantum Correspondence

The classical theory of energy transfer via sequential nonlinear resonances has a direct quantum analogy. Each mode in the classical coupled oscillator system is described by a single-oscillator Hamiltonian. Quantum mechanically, we describe each mode by the set of single-oscillator eigenstates, either harmonic or anharmonic as appropriate. The coupled oscillator system is then represented by the zero-order product states of these single-oscillator eigenstates. For example, for the coupled Morse oscillator system in Eq. 8, the vibrational dynamics are described by the states $|v_1, v_2\rangle = |v_1\rangle|v_2\rangle$, where $|v_j\rangle$ is an eigenstate of the jth oscillator. Energy flow is then represented by a state-to-state transfer of probability of the type $|v_1, v_2\rangle \rightarrow |v_1 - 1, v_2 + 1\rangle$. Hutchinson, Sibert, and Hynes (43) found that this energy transfer should be substantial when the initial and final states in the probability transfer are (i) nearly degenerate and (ii) directly and strongly coupled. Most importantly, *a classical nonlinear resonance between oscillators provides a guarantee that these criteria are met.* Thus, if we observe mode–mode energy transfer classically, we will also observe it quantum mechanically. (As we will discuss in Section II.C, a classical nonlinear resonance is not a *necessary* condition, even though it is sufficient.)

This also provides a clear analogy to the classical relaxation mechanism of sequential nonlinear resonances. Each resonance produces one (or more) state-to-state transfer(s) of probability; as the resonances act in sequence classically, so do the probability transfers. This produces a "tier" structure for relaxation of the initial local overtone state (44, 45). To illustrate, we cite a calculation by Hutchinson, Hynes, and Reinhardt (44) of the many-state quantum dynamics of the relaxation of an initial $v = 5$ excitation of the CH

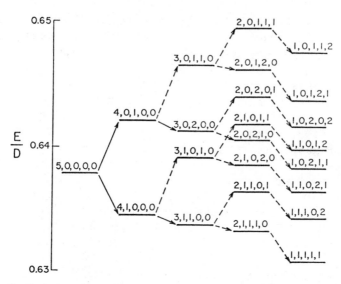

Figure 2. State-to-state probability flow in the quantum relaxation of an initially localized excitation. The initial state $|5, 0, 0, 0\rangle$ corresponds to excitation of a high-frequency nonlinear mode to $v = 5$. This excitation relaxes one quantum at a time through a sequence of tiers of states, in correspondence to a classical sequential nonlinear resonance mechanism, as in Figure 1.

bond in a collinear HCCCC system similar to that discussed in the previous section. The relaxation pathway is illustrated in Figure 2: The initial state $|5, 0, 0, 0\rangle$ in the figure is the fourth overtone of the CH mode with zero-point energy in each of the other three modes. In this particular system, a classical nonlinear resonance of the CH mode with each of the lowest two frequency modes produces a near degeneracy of the state $|5, 0, 0, 0\rangle$ with the bath states $|4, 0, 1, 0\rangle$ and $|4, 1, 0, 0\rangle$. Subsequent transfers occur in sequence, as illustrated, resulting in several tiers of states that participate in relaxing the overtone excitation. The quantum mechanism for relaxation of aryl-CH stretch overtones in benzene was also demonstrated by Sibert et al. (45) to follow a sequential path through several tiers.

C. Quantum Effects in Vibrational Energy Transfer

Due to the success of classical mechanics in describing many features of molecular vibrations, we (29, 31, 43, 44) [and others (46, 47)] have been interested in unveiling the extent and significance of quantum (nonclassical) effects in molecular vibrations. In a particularly interesting study, Davis and Heller (46) observed that classical mechanics often localizes highly vibrationally excited motions. For example in the coupled oscillator system in Eq. 8, an initial excitation of $v = 3$ in one OH bond will remain almost completely localized in that bond; as discussed in Section II.A, the bonds are out of resonance for this energy distribution (36). The quantum analog of this initial classical energy distribution is the initial state $|0\rangle|3\rangle$, which is (in the symmetric H_2O case) exactly degenerate with $|3\rangle|0\rangle$, in which the vibrational energy is in the other bond. Since these states are degenerate, any nonzero coupling will mix them throughly, so that probability flow from $|0\rangle|3\rangle$ to $|3\rangle|0\rangle$ will be complete, as will bond–bond vibrational energy transfer (43). Since this transfer does not occur classically, and since there is no barrier on the vibrational potential energy surface to the classical transfer, Davis and Heller (46) dubbed this quantum effect "dynamic tunnelling."

We have recently proposed a mechanism for this quantum effect (43) based on common intermediate states between the initial and final states in the probability flow. Referring again to the $|0\rangle|3\rangle$ excitation of H_2O, we note that the coupling of this state to $|3\rangle|0\rangle$ is very small in comparison to the coupling to, for example, $|1\rangle|2\rangle$. On the other hand, precisely because the bonds are not in resonance for the $|0\rangle|3\rangle$ energy distribution, the states $|0\rangle|3\rangle$ and $|1\rangle|2\rangle$ differ in energy by several hundred reciprocal centimeters. Nonetheless, we have found that the flow of probability from $|0\rangle|3\rangle$ to $|3\rangle|0\rangle$ *proceeds through* the intermediate states $|1\rangle|2\rangle$ and $|2\rangle|1\rangle$; in particular, the rate of probability flow would be a factor of 20 slower if not for the participation of the intermediate states! This is despite the fact that at no time in the process is there any substantial buildup of probability in the intermediate states. This is illustrated in Figure 3, showing the probability in the initial, final, and intermediate states for the exact quantum dynamics of the energy flow.

The importance of the intermediate states in producing this quantum effect is revealed by perturbation theory (43). The first-order coupling of $|0\rangle|3\rangle$ to $|3\rangle|0\rangle$ produces a splitting of eigenstates too small by the same factor of 20 mentioned in the preceding. In fact, the full eigenstate splitting arises predominantly from higher order terms in the perturbation series, in this case, third order, which incorporates the "indirect" coupling

$$\langle 3,0|H'|2,1\rangle\langle 2,1|H'|1,2\rangle\langle 1,2|H'|0,3\rangle \tag{12}$$

Each matrix element in this product is much larger than the "direct" coupling $\langle 3,0|H'|0,3\rangle$, so that the off-resonance indirect coupling dominates the dynamics. Thus, the transfer of energy between bonds proceeds as

$$|0,3\rangle \rightarrow |1,2\rangle \rightarrow |2,1\rangle \rightarrow |3,0\rangle \tag{13}$$

This indirect pathway is not available classically. The partial transfer of quantum probability from the initial state to the first intermediate state is

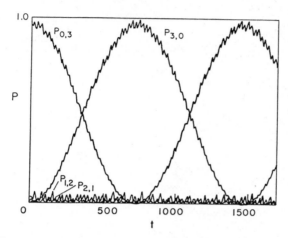

Figure 3. State-to-state probability flow for an initial state $|0, 3\rangle$ of the coupled oscillator system in Eq. 8. Time is in units of OH vibrational periods. Exchange of probability between $|0, 3\rangle$ and $|3, 0\rangle$ is slow but complete and proceeds through the small probabilities in the intermediate states $|1, 2\rangle$ and $|2, 1\rangle$, creating complete exchange of energy between bonds. This energy transfer is classically forbidden.

manifested classically as a very small quasi-periodic transfer of energy between bonds: The classical state tends toward the energy distribution $|0\rangle|3\rangle$ but never attains it and never proceeds beyond it. Thus, this is a purely quantum transfer of energy between bonds.

We have recently shown this indirect coupling mechanism to be significant in the intramolecular vibrations of local overtone excitations in two molecules of spectroscopic interest (29, 31). In the first [cyanoacetylene (linear HCCCN)], Hall and Berry (14) observed broadening of the second, fourth, and fifth overtones of the HC stretch. This is somewhat surprising based on the small number of vibrational modes in this molecule. We have analyzed the spectra (31) in terms of a simple model that incorporates only the stretching degrees of freedom, with normal-mode frequencies of 3489, 2278, 2080, and $863\,\mathrm{cm}^{-1}$. The highest frequency mode (CH-like in character) is strongly anharmonic. Focusing on the second overtone ($v = 3$) of v_1, the classical frequency at this excitation is $3122\,\mathrm{cm}^{-1}$, which closely matches the frequency of a combination mode, $v_2 + v_4$, $3141\,\mathrm{cm}^{-1}$. Thus, we might expect classical energy flow. However, we have determined that the direct coupling of v_1 to $v_2 + v_4$ is very small (48) and is taken to be zero in our model. Classically, the initial state is therefore localized, as is illustrated by an ensemble average of trajectories in Figure 4(a). The quantum excitation is not correspondingly localized [Figure 4(b)] due to the presence of an intermediate pathway: The initial state $|3, 0, 0, 0\rangle$ relaxes (quasi-periodically) by transfer of probability

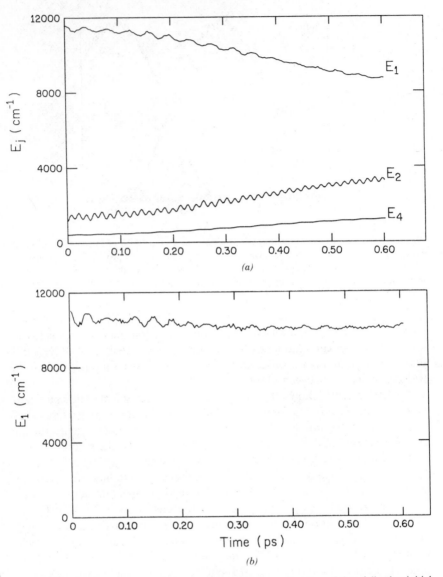

Figure 4. Energy in the v_1 mode of cyanoacetylene as a function of time following initial excitation to $v = 3$. (*a*) Quantum dynamics. Also shown are energies in modes v_2 and v_4, which rise together due to the interaction of the states $|3, 0, 0, 0\rangle$ and $|2, 1, 0, 1\rangle$. (*b*) Classical ensemble average of E_1, which does not relax.

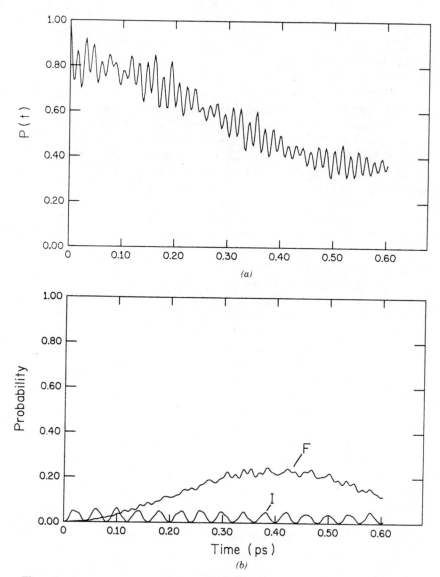

Figure 5. Quantum probability flow in various states following initial excitation of the state $|3, 0, 0, 0\rangle$. (a) Survival probability of the initial state $|3, 0, 0, 0\rangle$. (b) Probability in two bath states $|2, 1, 0, 1\rangle$, labeled F for final, and $|2, 1, 0, 0\rangle$, labeled I for intermediate. Transfer of probability between the nearly degenerate states $|3, 0, 0, 0\rangle$ and $|2, 1, 0, 1\rangle$ occurs through several intermediate states, including $|2, 1, 0, 0\rangle$, accounting for the nonclassical energy transfer in Figure 4.

through common intermediate states $|3,0,0,1\rangle$, $|3,1,0,0\rangle$, $|2,1,0,0\rangle$, and $|2,0,0,1\rangle$ into a nearly degenerate state $|2,1,0,1\rangle$. This indirect flow of probability is illustrated in Figure 5. As in the coupled Morse oscillator case previously discussed, this transfer is due to second-order couplings in the perturbation series and is unavailable classically. The broadening observed spectroscopically for the second CH overtone of cyanoacetylene is thus significantly influenced by purely quantum effects. This effect also contributes to broadening of the fourth CH overtone.

We have observed a similar quantum effect in the fourth OH overtone of propargyl alcohol, $HCCCH_2OH$. Jasinski (30) observed the OH overtone spectra of a wide variety of alcohols at the $v = 5$ level and noted that virtually all of these spectra were relatively narrow compared to the corresponding excitations of CH bonds in typical hydrocarbons. This is reasonably expected, as explained in the opening paragraph of Section II.A. However, Jasinski reported that the fourth OH overtone of propargyl alcohol was roughly $100\,cm^{-1}$ broad, nearly twice that of any other alcohol. To help understand this anomaly, we have studied a simple model of propargyl alcohol, incorporating nine of the vibrational modes (29). In a normal-mode analysis, we find that the CC single-bond stretching mode has an unusually low frequency of $913\,cm^{-1}$, nearly $100\,cm^{-1}$ below that of typical CC single-bond modes, due to its proximity to the neighboring CC triple bond. This anomalous feature produces an accidental nonlinear frequency match between the OH bond excited to $v = 5$ (with a frequency of $2990\,cm^{-1}$) and a combination mode of the CC stretch plus two times the CO stretch ($3009\,cm^{-1}$). It is reasonable to assume, though, that the direct coupling of the OH to this combination mode is very small, both because of the remoteness of the bonds involved and because of the high order of the coupling required (fourth order). If so, this would prevent classical transfer of energy via this resonance. Moreover, we have found that a nonzero direct coupling is *not required* for quantum vibrational energy flow involving the OH bond: An indirect pathway couples the OH overtone state to the combination mode bath state via simple kinetic couplings. Thus, it is plausible, if not likely, that the anomalous fourth-overtone spectrum of propargyl alcohol is accounted for by a nonclassical purely quantum mechanism.

III. UNIMOLECULAR REACTION FOLLOWING MODE-SELECTIVE EXCITATION

The possibility of initiating a chemical reaction with a mode-selective excitation is an intriguing and potentially valuable concept. As discussed in Section I, unimolecular chemical reactions induced by direct excitation of local-mode overtones have been studied frequently in the last ten years (4). The studies of mode–mode energy flow discussed in Section II provide some cause for optimism in this method as well, since the initial stages of the relaxation of high overtones of local-bond excitations are fairly mode specific. For example, Hutchinson, Hynes, and Reinhardt (39) found that a selectively excited local CH overtone in butane relaxed very selectively and very rapidly into the two terminal rocking normal modes of the methyl group. If a similar mode-selective relaxation were to transfer vibrational energy directly to a reaction coordinate, mode-specific nonstatistical enhanced reaction rates could result. For example, Uzer and Hynes (49) originally proposed that resonant stretch-to-bend energy transfer could be used to activate a chemical reaction involving a bending reaction coordinate. An example of such a reaction might be the dissociation of formyl fluoride (27) to HF and CO, in which the reaction coordinate involves HCO bending to carry the hydrogen across to the fluorine atom (49).

To model such a mode-selective reaction, Uzer and Hynes (49) studied the classical dynamics of a simple system in which a Morse oscillator is kinetically coupled to hindered rotor-bending potential. Defining a reaction by a crossing of the hindered rotor barrier, they calculated the probability for reaction as a function of time following excitation of the Morse oscillator to various levels. By construction, a $2:1$ resonance exists in this model between the stretch and bend at the level of $v = 9$ in the stretch, so that energy flow into the reactive bend coordinate should be most pronounced for this excitation. Correspondingly, Uzer and Hynes observed that the probability of reaction is greatest at $v = 9$. Of greatest interest is the fact that *increasing* the total molecular energy by increasing the stretch excitation actually *decreases* the reaction rate, in complete contrast to the predictions of statistical RRKM theory. This nonmonotonic dependence of the reaction rate on molecular energy is a much-sought autograph of mode-selective chemistry. The Uzer–Hynes model, although highly idealized, thus provides a valuable illustration of what conditions might be required in order to observe mode selectivity.

A more realistic model for mode selectivity is the isomerization of hydrogen isocyanide, $HNC \rightarrow HCN$, induced by selective excitation of the HN stretching mode (50). In this case, the reaction coordinate is an internal rotation of the hydrogen about the CN group, a motion that is at zero-order a bend. Thus, resonant energy transfer might be expected to produce mode-selective reaction, as in the Uzer–Hynes model. However, we have found that the addition of one additional mode (the CN stretch) is sufficient to dampen this expectation.

The HN stretching frequency is, at zero energy, $3831 \, cm^{-1}$, many times greater than the frequency of the reactive bend coordinate at $432 \, cm^{-1}$. Thus, the familiar $2:1$ stretch–bend resonance observed in the Uzer–Hynes model plays no role in the transfer of energy into the bend in HNC. Rather, we found

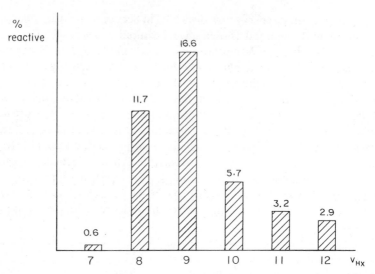

Figure 6. Reaction probability versus initial excitation for the isomerization of HNC to HCN. Reaction probability is calculated as the fraction of classical trajectories that cross over the barrier from HNC to HCN, whether or not they remain trapped, following initial excitation to various vibrational levels of the HN stretching mode. The reaction probability is observed to rise to a maximum and then fall off, a nonmonotonic dependence signaling the possible influence of mode specificity.

that a combination resonance effect substantially enhances the reaction rate. Specifically, at the level of $v = 7$, the nonlinear frequency of the HN stretch is $2546\ cm^{-1}$, in resonance with the combination of the HNC bend and the CN stretch at $2595\ cm^{-1}$. As such, we might expect enhanced transfer of energy into the reaction coordinate at this excitation. In Figure 6, we show the classical reactivity as a function of the level of HN excitation. As hoped, there is evidence of mode selectivity in the nonmonotonic dependence of reactivity on energy; unexpectedly, however, the maximum mode selectivity occurs at $v = 9$, significantly above the location of the combination resonance.

The mechanism for reaction has been worked out in detail by careful examination of classical trajectories (50). First, the reaction is initiated by resonant energy transfer into the CN stretch–HNC bend combination mode. This can be established by "zeroing" the term in the Hamiltonian responsible for the three-mode coupling. In this case, no reactive trajectories are observed, so that in the absence of the combination resonance, energy flow into the reactive coordinate is muted entirely, implicating the combination resonance as the dominant mechanism for energy flow. Nonetheless, the combination resonance is clearly perturbed since the maximum reactivity (Figure 6) does not occur at the resonance center.

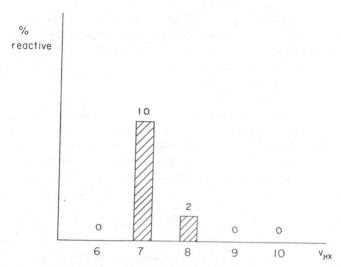

Figure 7. Reaction probability versus initial excitation for the isomerization of HNC to HCN when CN mode has been decoupled from the reaction coordinate. In contrast to Figure 6, these results show marked mode selectivity at single excitation corresponding to the center of a combination resonance.

Although the initial energy transfer occurs via the combination resonance, we have found that this energy flow is insufficient to initiate the isomerization reaction. Thus, following this short-time resonant energy flow, the molecule vibrates with all three degrees of freedom vibrationally "hot." Indeed, the final "kick" of energy into the reaction coordinate necessary for reaction does not come from the HN bond but rather from the CN bond. Moreover, the CN bond acts to drain additional energy from the HN bond at early times. To establish the importance of the CN bond in the reactive dynamics, we can decouple this mode from the reaction coordinate. Very interestingly, the resultant reactive dynamics are *highly* mode selective, as seen in Figure 7. Moreover, the maximum mode selectivity occurs precisely at the resonance center, $v = 7$.

The important result is the enormous increase in complexity of the reaction dynamics introduced by inclusion of the CN bond (compared to the two-degree-of-freedom Uzer–Hynes model). Although energy flow into the reaction coordinate does not occur at all in the absence of the CN mode, this mechanism interferes with the observation of model selectivity in two ways. First, the CN mode depletes some of the initial excitation energy, forcing the maximum mode specificity to a high overtone state, possibly above the range of experimental observation. Second, the initial transfer of energy, which does not induce reaction, simply heats up the molecule, destroying the mode-selective excitation. Thus, *even in a fairly simple system with very few degrees of vibrational freedom, experimental observation of mode selectivity is unlikely.* This, of course, is exactly what has always been found.

Further theoretical study of local-bond overtone-initiated reactivity has been presented by Uzer, Hynes, and Reinhardt (51), who calculated the classical dynamics of dissociation of hydrogen peroxide following excitation of one OH stretch to high overtone levels. Crim et al. (18) and Zewail et al. (28) have studied the gas phase photodissociation of HO_2H by overtone excitation and have observed no evidence of mode-selective enhancement. Uzer et al. (51) posed the question of how the initial OH stretching energy winds up in the OO stretch to dissociate the bond and established a mechanism by which the initial step in the process is a resonant stretch–bend energy transfer from the OH stretch to the OOH bend. Although this energy transfer is rapid, it is not immediately transferred to the OO reaction coordinate. Indeed, the initial energy transfer occurs within 0.5 ps, but the average lifetime of the peroxide molecule is 6 ps. The message is that resonant energy transfer is mode selective but ineffective in inducing unimolecular reaction. Longer time energy flow, sufficient to induce reaction, maintains very little of the initial specificity. This agrees well with the results of our hydrogen isocyanide study and with the majority of experimental photochemical studies. Although overtone-induced mode-selective chemical reactivity remains an elusive target, these studies nonetheless have revealed much of the mechanism of intramolecular vibrational energy transfer required for reaction. The most recent overtone photochemical studies involve picosecond- and subpicosecond-pulse-length excitations (28). We shall see in Section V that this introduces an additional complexity in interpretation of the spectral data in terms of intramolecular energy flow. In particular, preparation of the initial state has not been considered in the theoretical studies discussed in the preceding, and this step competes with vibrational relaxation and unimolecular reaction in the ultrashort-pulse regime. In the next section, we address the question of state preparation.

IV. DYNAMICS OF EXCITATION OF LOCAL-MODE OVERTONES

The studies discussed in the previous two sections assume an initial perfectly localized overtone excitation. Preparation of such a state would have to be essentially instantaneous, however, since these states relax very rapidly, within 100 fs or so. We shall see that, far from being instantaneous, the overtone excitation process is many orders of magnitude slower than intramolecular relaxation (52).

A. Classical Forbiddance of Overtone Excitation

Given the success of classical mechanics in revealing the dynamics of relaxation and reaction following overtone excitation, we pursue a description of the state preparation classically. Gray (53) showed that the classical process of vibrational excitation of a diatomic anharmonic oscillator by a laser may be treated by nonlinear resonance analysis. Specifically, for excitation of the level v of a Morse oscillator, the absorption requires a $v:1$ resonance between the molecular vibration and the laser field oscillator. Gray found that the field–molecule coupling for the $v:1$ transition is proportional to

$$V_{\theta F} = \frac{V_{\text{int}}}{\alpha v}\left(\frac{1+(1-\rho)^2}{\rho}\right)^{-v} \tag{14}$$

where

$$\rho = \left(\frac{E_v}{D}\right)^{1/2} \tag{15}$$

$$V_{\text{int}} = -\left(\frac{\partial \mu}{\partial y}\right)_0 \left(\frac{2\pi E_F}{V_c}\right)^{1/2} \tag{16}$$

Here μ is the dipole moment as a function of the vibrational coordinate y, D and α are parameters of the Morse oscillator, and E_F is the energy of the laser field in a cavity of volume V_c. Because of the inverse exponential dependence of the coupling on v, the field–molecule interaction becomes extremely inefficient for high values of v. As the Morse oscillator (initially at $v=0$) absorbs energy from the field, its frequency decreases according to Eq. 7, thus shifting the oscillator out of resonance with the field. With the very small coupling in Eq. 14, energy absorption is insignificant before the resonance is destroyed; thus, high overtone excitation is a classically forbidden process!

Gray (53) quantified this by calculating the resonance width for the absorption. Defining $p = (I - I_r)/v$, where I_r is the action for the oscillator when it is in resonance with the field and v is the overtone level to which absorption is expected, the resonance width is found to be

$$|p| \leqslant \left(\frac{8D V_{\theta F}}{v^2 \Omega_y^2}\right)^{1/2} \tag{17}$$

where Ω_y is the harmonic frequency and D the dissociation energy of the Morse oscillator. For reasonable values of the laser field energy and for overtones with $v > 2$, the resonance width is sufficiently small that excitation to the overtone energy does not occur.

The classical description of the absorption also estimates poorly the location of the absorption maximum (52–54). Since the molecular oscillator is initially at its zero point, the frequency of the oscillator is near its harmonic value. Thus, for a $v:1$ resonance, the field should be tuned to v times the fundamental. Of course, due to anharmonicity, quantum overtone transitions are observed at a frequency well below v times the fundamental; in essence, the classical oscillator at its ground-state energy does not "know" about its own anharmonicity. Therefore, what little absorption occurs does so at the wrong field frequency.

We have demonstrated these effects in overtone state preparation of a model polyatomic molecule (52). The system models the excitation of a CH stretch kinetically coupled to an adjacent CCH bend via

$$H_{mol} = \frac{P_y^2}{2\mu_{CH}} + D\left(1 - e^{-\alpha y}\right)^2 + \frac{p_x^2}{2m_H(y + y_0)^2} + \tfrac{1}{2}kx^2 \tag{18}$$

where y is the CH stretch mode (with fundamental frequency $3240\,\text{cm}^{-1}$) and x is the bend mode. The system is constructed so that a 2:1 nonlinear resonance exists when the stretch is excited to $v = 4$. The molecular dipole is assumed to be a function of y only, with dipole derivative $d\mu/dy = -0.1732\,D/a_0$ (48), so that the classical Hamiltonian for the interaction of a laser field with the molecule is

$$H = H_{mol} + yV_{int}\cos(\omega_F t + \delta) \tag{19}$$

where V_{int} is given by Eq. 16. We calculated classical trajectories for a range of frequencies in the vicinity of each overtone transition for a relatively high laser intensity of $21.7\,\text{GW/cm}^2$. Calculations were performed with both a linear and a quartic dipole approximation.

We will examine first the results for excitation of the first overtone, $v = 2$. Even for this high laser intensity and for the lowest overtone, very few trajectories absorbed enough energy from the field to reach an excitation concomitant with $v = 2$. This agrees well with the expectations from the preceding discussion. The oscillator detunes from the laser field very quickly as it absorbs energy. Many more trajectories attain an energy equivalent to an excitation of $v = 0.75$, so we will regard, for the sake of analysis, trajectories of this type to be "absorbing trajectories." The percentage of absorbing trajectories is displayed in Figure 8 as a function of laser frequency for both the linear and quartic dipole approximations. Several points can be noted. First, the maximum absorption occurs at $6460\,\text{cm}^{-1}$, which is very close to twice the harmonic frequency of $3240\,\text{cm}^{-1}$ but is significantly away from the quantum transition frequency of $6340\,\text{cm}^{-1}$. Second, the absorption becomes very much

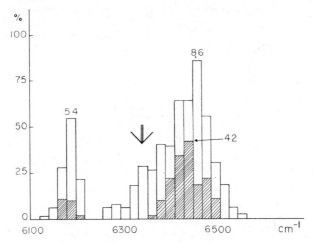

Figure 8. Fraction of absorbing trajectories for the model system in Eq. 18 as a function of laser frequency in the vicinity of the $v = 2$ transition. Shaded results are for a linear approximation to the dipole; open bars are for a quartic approximation to the dipole. The arrow indicates location of the quantum transition in this frequency range.

more probable when a high-order dipole is assumed (54, 55). This is reasonable: By including a term of nth order in the dipole, the oscillator appears to the field to have a frequency component n times as large as its actual frequency. Third, although absorption appears significant at several frequencies, virtually none of these trajectories absorb to the $v = 2$ level. Finally, those trajectories that do reach the $v = 2$ excitation do *not* do so by direct (one-step) absorption into the stretch. This is illustrated in Figure 9, where it is shown that, although the total molecular energy rises substantially, energy flow into the stretch is assisted by its interplay with the bend. The stretch absorbs a small amount of energy but correspondingly detunes from the field. This small amount of energy is then transferred to the bend, thus tuning the stretch back into resonance with the field. Further energy is absorbed from the field, transferred to the bend, and so on. Thus, even though the classical dynamics appear to permit absorption to the $v = 2$ level of the local overtone, the mechanism for this absorption is very different than the quantum dynamics.

The classical dynamics for higher overtones become strikingly uninteresting. For example, upon tuning the laser to frequencies in the vicinity of the $v = 4$ transition, no trajectories absorb energy even to the $v = 0.75$ level. Thus, overtone excitation in a polyatomic molecule is a classically forbidden process, at least for physically plausible laser intensities. Of course, it is clear from experimental observation that overtone excitation is permissible

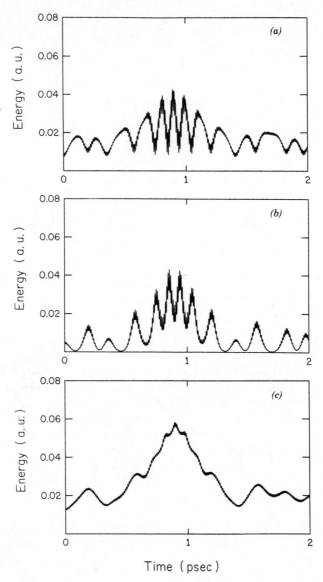

Figure 9. Energy absorption dynamics for a single-absorbing classical trajectory for the model in Eq. 18 for a laser frequency of $6320 \, \mathrm{cm}^{-1}$ (cf. Figure 8). (*a*) Energy in the CH stretching mode. (*b*) Energy in the CCH bending mode. (*c*) Total molecular energy.

quantum mechanically. Preparation of an overtone state occurs through "dynamic tunnelling," an observation first made by Gray (53) and Medvedev (56).

B. Quantum Dynamics of Polyatomic Overtone Excitation

The classical forbiddance of overtone excitation is manifested quantum mechanically, not as a "forbidden" (low-probability) process but as a very slow process. We can get an initial calculation from the Rabi frequency of the excitation. The time τ required for complete transfer of probability out of the ground state and into the excited state is

$$\tau = \pi\hbar/2\Delta \tag{20}$$

where the field–molecule coupling matrix element Δ is

$$\Delta = c V_{int} \langle 0|y|v \rangle \tag{21}$$

and V_{int} is as in Eq. 18. For higher overtones, whose transition moments $\langle 0|y|v \rangle$ are very small, the time scale for absorption increases dramatically. As an example, we consider the $v = 4$ overtone of our model system with a laser intensity of $21.7\,GW/cm^2$. The field–molecule coupling Δ is only $10^{-3}\,cm^{-1}$, so that $\tau = 1.2\,ns$, which is *four orders of magnitude longer* than the time required to transfer energy from stretch to bend. This disparity in time scales results in an excitation delocalized at all times (52).

In an alternative, time-independent view, the small field–molecule coupling forces the excitation to be highly *eigenstate specific*; an eigenstate transition off resonance from the laser frequency by more than the coupling in Eq. 21 will not be significantly excited. As previously noted, this coupling is small, perhaps $10^{-3}\,cm^{-1}$ or less, so that even with a high density of bright states, one and only one eigenstate will be excited by the laser. Since most eigenstates are to some extent delocalized, the excitation is never localized.

We wish to develop these results in more detail (52). Quantum mechanically, we describe the interaction of the field with our model molecular system (Eq. 18) with a conservative, quantized field Hamiltonian,

$$H = H_{mol} + H_f + H_{inf} \tag{22}$$

where H_f is a harmonic oscillator Hamiltonian for the field energy, and

$$H_{int} = iy \left(\frac{\partial \mu}{\partial y}\right)_0 \left(\frac{I}{2\varepsilon_0 c}\right)^{1/2} (|v_f + 1\rangle\langle v_f| + |v_f - 1\rangle\langle v_f|) \tag{23}$$

One Laser Excitation

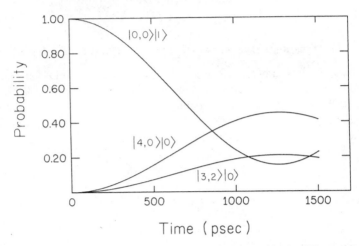

Figure 10. Quantum dynamics of overtone excitation to $v = 4$ in the CH stretching mode. Shown are probabilities in key zero-order states of the model system. The delocalized excitation $|3, 2\rangle$ rises with the same time scale as the localized $|4, 0\rangle$.

where I is laser intensity. As before, the quantum dynamics are projected onto a set of zero-order states, in this case of the form $|v_y\rangle|v_x\rangle|v_f\rangle$. The state $|v_f\rangle$ is an eigenstate of the field Hamiltonian with v_f photons. The initial state in the excitation process is $|0\rangle|0\rangle|v_f\rangle$, and we seek to prepare an excitation of the form $|v\rangle|0\rangle|v_f - 1\rangle$. The resultant quantum dynamics are displayed for $v = 4$ in Figure 10 with two notable features. First, as expected, the time scale for excitation is quite long, on the order of tens of thousands of CH vibrations! Second, the excitation profile of the delocalized state $|3\rangle|2\rangle$ exactly follows that of the localized state $|4\rangle|0\rangle$. This is because we are tuned to a single eigenstate that is a superposition of these zero-order molecular states. If we alternatively tune to the energy of the pure zero-order state $|4\rangle|0\rangle$ in an effort to prepare the purely localized state, there is *no* excitation at all. Thus, it is not possible to prepare a localized state with any significant probability since the time for intramolecular relaxation is so much shorter than the time required to prepare the initial state.

It is illuminating to see how these conclusions vary with the level of overtone excitation and the laser intensity. In Table 1, we calculate the Rabi time for eigenstate excitation (Eqs. 20 and 21) for a variety of such combinations. The clear message is that the time required to prepare a vibrational excitation is longer than the time for intramolecular events, even

TABLE 1   Time Scale (s) for Vibrational Excitation as a Function of Laser Intensity

v	1 kw/cm^2	1 MW/cm^2	1 GW/cm^2	1 TW/cm^2
1	2.4×10^{-8}	7.7×10^{-10}	2.4×10^{-11}	7.7×10^{-13}
2	2.4×10^{-7}	7.6×10^{-9}	2.4×10^{-10}	7.6×10^{-12}
3	1.4×10^{-6}	4.5×10^{-8}	1.4×10^{-9}	4.5×10^{-11}
4	6.5×10^{-6}	2.1×10^{-7}	6.5×10^{-9}	2.1×10^{-10}
5	2.4×10^{-5}	7.7×10^{-7}	2.4×10^{-8}	7.7×10^{-10}
6	8.0×10^{-5}	2.5×10^{-6}	8.0×10^{-8}	2.5×10^{-9}
7	2.3×10^{-4}	7.3×10^{-6}	2.3×10^{-7}	7.3×10^{-9}

for fundamental excitations with reasonably attainable laser intensities.

We will see in the next section conditions that lead to somewhat different results. The results discussed so far apply to purely bound excitations of molecules with not excessively high densities of bright states, prepared with laser pulses long compared to intramolecular time scales. In these case, the only competing processes are state preparation and intramolecular relaxation, and state preparation finishes a poor second.

V.   QUANTUM DYNAMICS OF OVERTONE-INDUCED PHOTODISSOCIATION

As discussed in Section I.C, a particularly informative way to study the role of vibrational energy transfer in chemical reactivity is to photochemically initiate unimolecular reaction by tuning an intense laser to transitions in the vicinity of local-mode high overtones (4). In these cases, the competing processes include unimolecular decay in addition to state preparation and intramolecular energy transfer. If we idealize the mechanism for the overtone-induced photochemical process including these processes, we can picture the photodecay as occurring in three steps:

$$R\text{–}C\text{–}H \xrightarrow{k_e} R\text{–}(C\text{–}H)^* \qquad \text{(local-mode excitation)} \qquad (24a)$$

$$R\text{–}(C\text{–}H)^* \xrightarrow{k_{IVR}} (R\text{–}C\text{–}H)^* \qquad \text{(intramolecular relaxation)} \qquad (24b)$$

$$(R\text{–}C\text{–}H)^* \xrightarrow{k_r} \text{products} \qquad \text{(unimolecular reaction)} \qquad (24c)$$

We have discussed in the preceding sections the time scales associated with each of the steps in this mechanism and the pairwise competition between reaction and relaxation and between excitation and relaxation, establishing that $k_e \ll k_r < k_{IVR}$. In certain cases, therefore, local-mode excitation is the rate-limiting step in the mechanism, in which case the photochemical rate is dominated by the absorption cross section. This will be true for relatively long laser (nanosecond) pulses tuned on resonance to the main absorption feature; we shall see that in this case the initially prepared state is a low-probability steady state of the delocalized excitation (R–C–H)*.

However, if the laser is tuned off resonance from the main spectral transition, then off-resonance oscillations of the excitation become rapid. The initially prepared state is more poorly defined, and the possibility of actually observing intramolecular events arises. Furthermore, if an ultrashort (sub-picosecond) laser pulse is used, state preparation is terminated (though incomplete) on a time comparable to the intramolecular relaxation rate and faster than the unimolecular reaction rate. As a result, the nature of the initially prepared state is a sensitive function of the pulse length and the detuning, and the mechanism and rate of photochemical reaction is affected accordingly. Deconvoluting the photochemical data in such an experiment will require simultaneous treatment of all three steps in the mechanism in Eq. 24. We have recently presented such an analysis of a model unimolecular system that illustrates the competition among these processes. Since the first step in the process is classically forbidden, our approach is purely quantum mechanical throughout.

A. Predissociative Model Molecule

We have studied a system similar to the one discussed in Section IV but modified to give unimolecular decay via vibrational predissociation (57). The model consists of a collinear three-atom system, HCC, where the CH bond (y) is a Morse oscillator and the CC bond (x) is a harmonic oscillator with dissociative behavior. The bonds are coupled via kinetic and potential terms so that the molecular Hamiltonian is

$$H_{mol} = \frac{p_y^2}{2\mu_{HC}} + \frac{p_x^2}{2\mu_{CC}} - \frac{p_x p_y}{m_C} + D(1 - e^{-\alpha x}) + \tfrac{1}{2}kx^2 e^{-\beta x^3}$$
$$+ \tfrac{1}{2}(f_{xy}xy + f_{xxy}x^2 y)\{1 - \tanh[\gamma(x - x_c)]\} \tag{25}$$

The CH bond has a harmonic frequency of $3300\,cm^{-1}$ and an anharmonicity of $65.8\,cm^{-1}$. The CC bond has harmonic frequency of $1407\,cm^{-1}$, and the barrier to dissociation of the CC bond is $6000\,cm^{-1}$. The total potential energy in Eq. 25 is contoured in Figure 11. As in Section IV, the dipole function for the molecule is assumed linear in y, with a dipole derivative of $-0.1732 D/a_0$.

As in Section II and IV, it is useful to discuss quantum vibrational dynamics in terms of a set of zero-order states of the form $|v_y\rangle|v_x\rangle$, where $|v_y\rangle$ is a CH

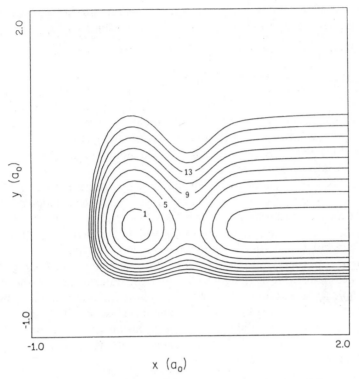

Figure 11. Contour map of the potential energy function for the predissociative system in Eq. 25. Contours are labeled in units of $1000\ cm^{-1}$.

Morse oscillator eigenstate and $|v_x\rangle$ is a quasi-bound eigenstate of the dissociative potential along the CC coordinate. These states have been found numerically using a primitive variational basis of piecewise cubic splines (58). We evaluate matrix elements of the one-dimensional Hamiltonian along x using Gaussian quadrature; eigenstates are calculated by solving the resultant generalized eigenvalue problem with a nonorthogonal basis set. These calculations are very fast and convenient, and convergence of the quasi-bound eigenstates is obtained with a basis of 160 spline functions covering the range $-1a_0 < x < 8a_0$. This establishes a basis for the photochemical dynamics to be discussed.

The model in Eq. 25 has a deliberate mode specificity built into it. Upon excitation of the CH mode to $v = 4$, a 2:1 nonlinear resonance with the CC stretch at zero-point energy is established. As a result, large vibrational energy flow is more facile at this excitation than at either higher or lower energies. This allows us to observe the effects of both resonant and nonresonant vibrational dynamics in the molecular decay.

B. Method of Calculation: Complex Coordinate Interaction Representation

The model system depicted in Figure 11 does not support any truly bound quantum eigenstates; at energies below the reaction threshold (6000 cm^{-1}), potential tunneling permits unimolecular decay, although the lifetime of the ground state is extremely long. At energies above threshold, most states are rapidly dissociative states of the unbound continuum. However, several states can be located in the continuum corresponding to relatively long-lived predissociative resonances (58, 59). Physically, these states possess sufficient energy to escape the molecular well (even without tunneling), but they do not do so immediately for lack of sufficient energy along the reaction coordinate (x). As a simple example, if we excite the CH stretch to 10,000 cm^{-1} without exciting the CC stretch, the molecule will not dissociate until sufficient intramolecular vibrational redistribution has occurred to place over 6000 cm^{-1} of energy into the CC mode. Although vibrational energy flow is fast, lifetimes as long as 100 ps can be found for states well above the reaction threshold. Not coincidentally, these states have large projections onto high overtones of the CH stretch and are thus spectroscopically bright.

To find these predissociative overtone eigenstates, we apply the method of complex coordinates (60). The coordinate x in the dissociative degree of freedom is scaled by a complex phase

$$x \rightarrow xe^{i\theta}, \tag{26}$$

thus destroying the Hermiticity of the vibrational Hamiltonian. Diagonalization of the resultant complex symmetric matrix produces complex energy eigenvalues of the form

$$\tilde{E}_j = E_j - \tfrac{1}{2}i\Gamma_j \tag{27}$$

Where E_j is asssociated with the energy (i.e., transition frequency) of the predissociative resonance. The imaginary part of the complex energy, Γ_j, is the spectral width of the resonance due to a finite lifetime of $1/\Gamma_j$. To see this, we note that the time dependence of an eigenfunction ϕ_j of the Hamiltonian is

$$\Phi_j(t) = \phi_j \exp(-i\tilde{E}_j t) \tag{28}$$

where we have introduced the minor modification that $\tilde{E}_j$ is a complex number. The quantum probability density, calculated as $|\Phi_j(t)|^2$, thus decays

TABLE 2 Predissociative Overtone Eigenstates, Energies, Decay Rates, and Transition
Moments

State	E (cm^{-1})	Γ (cm^{-1})	Transition Moment[a]
Ground state	2,316	—	—
$2v_1 + 2v_2$	11,202	10.8 ± 2	1.9×10^{-4}
$3v_1$	11,390.5	0.088 ± 0.002	2.4×10^{-3}
$3v_1 + 2v_2$	14,084	20.5 ± 2	1.2×10^{-4}
$4v_1$	14,154.5	1.42 ± 0.04	5.2×10^{-4}
$5v_1$	16,785	9.9 ± 0.8	1.3×10^{-4}
$4v_1 + 2v_2$	16,794	68 ± 5	5.0×10^{-5}
$6v_1$	19,359	1.42 ± 0.02	4.6×10^{-5}

[a]Matrix element $\langle 0|y|\phi \rangle$, where $|\phi \rangle$ is the eigenstate and $|0 \rangle$ is the ground eigenstate.

via

$$P(t) = |\Phi_j(t)|^2 = |\phi_j|^2 \exp(-\Gamma_j t) \tag{29}$$

from which it is clear that Γ_j is the rate constant for unimolecular decay of the predissociative eigenstate ϕ_j.

The Hamiltonian in Eq. 25 is thus scaled according to Eq. 26, and eigenstates and complex eigenvalues are found in the zero-order basis, $|v_y \rangle |v_x \rangle$. As noted in the preceding, most of the eigenstates that result have virtually zero lifetime and very little projection onto states in the molecular well. Several predissociative resonances can be identified in the continuum, however, and these are listed in Table 2. The eigenstates in this table are designated by spectroscopic assignments; for example, $3v_1 + 2v_2$ is the quasi-bound eigenstate with principal character in the $|3\rangle|2\rangle$ basis state.

Note that the expected mode specificity of the model is evident in the lifetimes of these quasi-bound states. For example, the state $6v_1$ is much longer lived than the state $5v_1$ at a lower energy by 2574 cm^{-1}, so that the rate of unimolecular reaction actually decreases sharply with increasing energy in this case. This obviously nonstatistical result is due to the 2:1 nonlinear resonance mentioned at the end of Section V.A. Since the $v = 5$ CH excitation is near the resonance center, CH to CC energy flow is enhanced. Quantum mechanically, the zero-order state $|5\rangle|0\rangle$ is near in energy to the states $|4\rangle|2\rangle$ and $|3\rangle|4\rangle$, and the latter of these is very short lived. Thus, the eigenstate $5v_1$, which is dominantly of $|5\rangle|0\rangle$ character, also contains substantial dissociative character and has a short lifetime. By contrast, resonant energy flow does not assist in the reaction of $|6\rangle|0\rangle$, and the corresponding eigenstate $6v_1$ is longer lived.

These states thus clearly reflect the competition between intramolecular energy flow and unimolecular reaction (steps b and c in mechanism Eq. 24). To include the dynamics of state preparation in the discussion, we must incorporate the laser field in the Hamiltonian. As in Section IV, the full field–molecule Hamiltonian is thus given by Eqs. 22, 23, and 25. The laser intensity I is taken to be $200\,GW/cm^2$ for the calculations in this study.

In a photochemical experiment, the state at $t = 0$ (the instant the laser is turned on) is the ground-state eigenfunction dressed with the laser state $|v_f\rangle$. We propagate this initial state with the Hamiltonian in Eq. 22 by adopting the time-dependent interaction representation using a basis of dressed complex energy eigenstates and eigenvalues. The time-dependent quantum state is

$$\Phi(t) = \sum_k \phi_k a_k(t)\exp(-i\tilde{E}_k t) \tag{30}$$

where the states ϕ_k are the dressed molecular eigenfunctions of $H_{mol} + H_f$, and the energies are the complex eigenvalues previously calculated. Inserting this state into the time-dependent Schrödinger equation yields the usual interaction expression for the time-dependent coefficients:

$$\frac{da_k}{dt} = -\frac{i}{\hbar}\sum_l a_l(t)\langle \phi_k | H_{int} | \phi_l \rangle \exp[-i(\tilde{E}_l - \tilde{E}_k^*)t] \tag{31}$$

This equation can now be straightforwardly integrated numerically to give state occupation probabilities and photodissociation yields. However, we add an additional, highly accurate approximation, excluding all states from the calculation that are outside a "window" of $\pm 250\,cm^{-1}$ on either side of the energy of the dressed ground state. Given that field–molecule couplings in this system are on the order of $0.05\,cm^{-1}$, this simplification poses no significant inaccuracy.

This resonant approach significantly reduces computational times in two ways: first, the time of the calculation scales fairly linearly with the number of states included; second, high-frequency components of the dynamics have been eliminated, so that a long time step (100 atomic units, or about 2.4 fs) can be used with confidence. In each calculation, a single laser frequency was assumed. The populations of the ground molecular state, several excited eigenstates, and key zero-order states were calculated from the coefficients in Eq. 32. The normalization $N(t)$ of the quantum state decays with time due to decay of the predissociative states, and the photofragment yield at time t is just $1.0 - N(t)$. To simulate pulsed laser experiments, we simply zero out the laser field after a set pulse time, and the final asymptotic photochemical yield following a pulse of length t is just $1.0 - S(t)$, where $S(t)$ is the survival probability of the ground state following the pulse.

C. Long-Laser-Pulse Photochemical Dynamics

Most of the experiments to date on overtone-induced photochemistry have employed either continuous-wave or long pulsed laser sources. In this range,

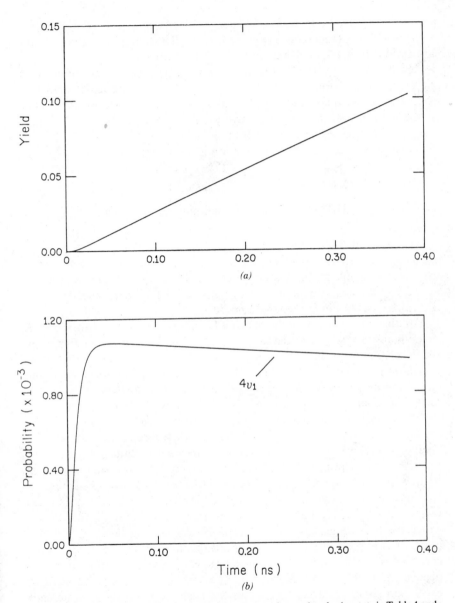

Figure 12. Photodissociation dynamics when the laser is tuned to the $4v_1$ state in Table 1 and is left on continuously during the dynamics. (*a*) Photofragment probability versus time. (*b*) Probability in the predissociative eigenstate $4v_1$.

669

we can categorize our results as either *on resonance* or *off resonance*, determined by whether the laser is tuned to the dominant feature in the absorption spectrum or to one of the darker combination bands.

Results for a typical on-resonance experiment appear in Figure 12, where the laser is tuned to the $4v_1$ state in Table 2. The dynamics are very simple: The population of the quasi-bound eigenstate rises rapidly to a steady-state value of about 10^{-3}, and the photochemical yield rises linearly as a result. These results stem from a simple mechanism in which the second step in Eq. 24 is assumed sufficiently rapid that excitation is delocalized at all times:

$$R\text{--}C\text{--}H \longrightarrow (R\text{--}C\text{--}H)^* \qquad \text{(eigenstate excitation)} \qquad (32a)$$

$$(R\text{--}C\text{--}H)^* \longrightarrow \text{products} \qquad \text{(unimolecular reaction)} \qquad (32b)$$

By a comparison of the time scale for excitation (Table 1) to the lifetime of the excited state (Table 2), it is clear that the excitation step is rate limiting. As usual, then, the intermediate rapidly achieves a steady state. Note that there are no intramolecular dynamics contributing to the photochemical event: A delocalized excitation is prepared and decays.

For different on-resonance excitations (e.g., $5v_1$), the dynamics are the same as in Figure 12, except that the steady-state population is much smaller when the transition moment in Table 2 is smaller; for $5v_1$, the excited-state population is steady at 10^{-6}. Therefore, *the rate of appearance of photofragment is much slower, even though $5v_1$ is a more reactive state than $4v_1$, again due to the rate-limiting excitation step.*

Off resonance, the dynamics are considerably more interesting. We consider next tuning the laser off the main $4v_1$ peak to the adjacent combination band $3v_1 + 2v_2$, a darker but more reactive state. In Figure 13(a), the rate of appearance of photofragment is considerably slower than in Figure 12(a) (although note the difference in time scales). However, in this case, true intramolecular dynamics contribute to the photochemical event. This can be seen in Figures 13(b) and 13(c). As before, the population of the state to which we have tuned rises to a steady state; now, however, appreciable population appears in the off resonance excitation of $4v_1$. This population contributes significantly to the photochemical yield and is moreover time dependent. It is worth noting that, due to the fast off resonance Rabi oscillations, there is a component in the excitation process that actually *competes* with the very fast intramolecular dynamics. Therefore, the "prepared" state in the photochemical experiment is neither localized nor eigenstate specific. It should be possible to observe these intramolecular vibrational dynamics in an experiment that probes photochemistry on a time scale less than about 25 ps.

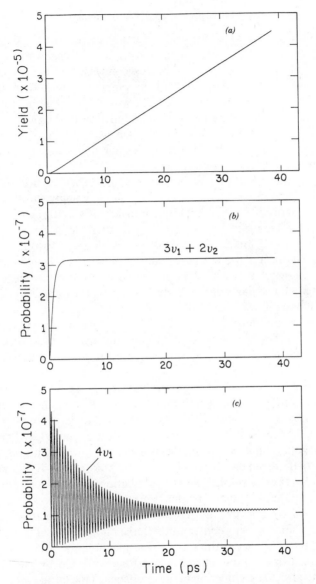

Figure 13. Photodissociation dynamics when the laser is tuned to the state $3v_1 + 2v_2$ in Table 1 and is left on continuously during the dynamics. (a) Photofragment probability. (b) Probability in the predissociative eigenstate $3v_1 + 2v_2$. (c) Probability in the predissociative eigenstate $4v_1$, which is off-resonance from the excitation.

D. Ultrashort-Pulse Photochemistry

In light of the conclusions of the previous section, there should be considerable excitement about the recent experimental work by Zewail and co-workers (28) on overtone photochemistry using subpicosecond laser pulses. As is clear from Figure 13, the excited molecular state at times less than 10 ps is rapidly oscillating in character. For pulse lengths in this time regime, the initially prepared state will therefore be a sensitive function of the pulse length. By probing the photodecay of this initial state as a function of both laser frequency and pulse length, the intramolecular dynamics, for example, as observed in Figure 13(c), should be revealed.

We have examined the photochemical dynamics using subpicosecond pulse lengths for our predissociative model (57). For exceedingly short pulse times (perhaps less than 100 fs), the initial state is in fact very well localized, but with an almost vanishingly small population. For example, if we pump the $4v_1$ transition for 145 fs, the initial state is almost purely $|4\rangle|0\rangle$, but with a population $O(10^{-7})$. If we consider slightly longer pulse lengths, the initial state becomes more delocalized as intramolecular relaxation begins. If the laser is tuned off resonance from the $4v_1$ transition, the population in $|4\rangle|0\rangle$ undergoes rapid off resonance Rabi oscillations; therefore, the state prepared when the pulse ends depends very sensitively on the exact timing of the pulse with respect to these oscillations. As a consequence, we have observed the surprising result that the photochemical yield can *decrease* with increasing pulse length!

These results are illustrated in Figure 14, showing the photochemical yield as a function of laser pulse length both on and off resonance from the $4v_1$ transition. The on resonance yield rises quadratically as a function of pulse length, since the excited state does not reach its steady state in less than 1 ps. By contrast, the off resonance yield increases *nonmonotonically* with pulse length. This is due to the rapid oscillations in the contribution from $4v_1$, as evidenced in Figure 13(c). For example, the Rabi oscillations in $4v_1$ hit a minimum at 485 fs; if the pulse terminates at that time, the contribution of $4v_1$ to the photochemical yield is virtually zero. By contrast, the population in $4v_1$ is a maximum at 240 fs, as is its contribution to the photochemical yield; thus, the yield in Figure 14(b) passes through a local maximum at this time.

If one constructs a photochemical overtone spectrum by measuring the photochemical yield as a function of the frequency of the ultrashort pulse laser excitation, Figure 14 demonstrates that this spectrum will be a sensitive function of the length of the excitation pulse. Therefore, we have calculated the photochemical overtone spectrum as a function of both laser frequency and pulse length. The result for frequencies in the vicinity of the $4v_1$ transition is displayed in Figure 15 as a three-dimensional map. The main central peak

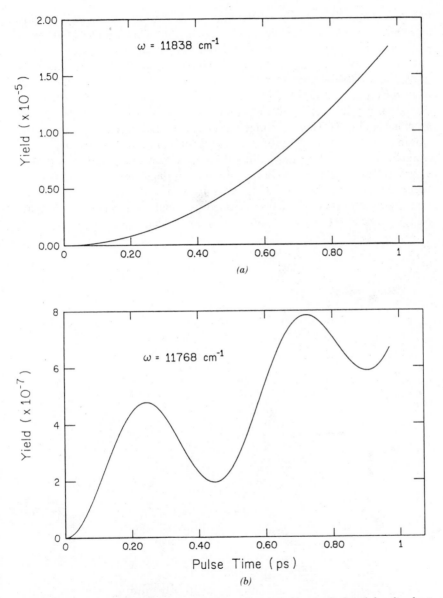

Figure 14. Asymptotic photochemical yield as a function of laser pulse length for ultrashort pulses. The yield is calculated as 1 minus the probability in the ground state when the pulse is terminated. The laser is tuned to (a) $4v_1$ and (b) $3v_1 + 2v_2$.

(which is cropped for clarity of the off resonance features) grows quadratically and monotonically, as we have seen in Figure 14(a). The shoulder to the red of the main peak is the combination band $3v_1 + 2v_2$. Figure 15 clearly reveals the complicated interplay between resonant and off resonant contributions to the photochemical yield as a function of pulse length. The point here is that with ultrashort pulses and with off resonance excitations, the time frame for excitation can interfere with the simple mechanism suggested by eigenstate specificity in Eq. 34. There are three legitimate contenders in the dynamics, and gaining an understanding of the ultrashort photochemistry requires a simultaneous consideration of excitation, relaxation, and reaction. Although experimental construction of a pulse-length-dependent spectrum as in Figure 15 is likely to be arduous, it is clear from the calculations in this chapter that the resultant data is rich in detailed information about intramolecular energy transfer and its contribution to unimolecular reaction.

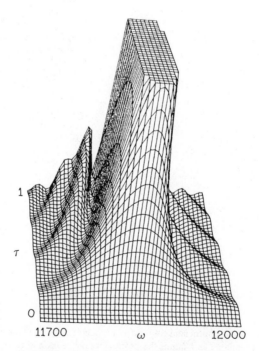

Figure 15. Asymptotic photochemical yield following ultrashort pulses as a function of the pulse length (τ) and frequency (ω) in reciprocal centimeters in the vicinity of the third-overtone state $|4, 0\rangle$. The maximum yield has been cut off at 3×10^{-6} to emphasize the off-resonance features.

VI. CONCLUDING REMARKS

Understanding the role of intramolecular energy transfer in facilitating chemical reactions is one of the fundamental problems of current chemical physics. The real problem for the experimentalist is how to probe these dynamics both before and during the chemical event in addition to the already successful probes of the dynamics post hoc to the reaction. Among the reasons for interest in high-local-mode overtones is the hope that the initial preparation of a high-energy state might be relatively state selective and in favorable cases somewhat mode selective. As we have seen, much can be learned about vibrational dynamics and couplings from a study of the spectra and photochemistry of these states. However, because these excitations are so slow, the most revealing studies should be those in which the length of the excitation is on the order of the intramolecular time scales, that is, subpicosecond. Perhaps unfortunately (perhaps not), the nature of the initial state in the ultrafast pulse excitation is neither mode selective nor state selective, as we have seen. This indicates that considerable theoretical input will be required to interpret experiments in the ultrafast regime. We anticipate that the study of pulse-length-dependent photochemistry will contribute significantly to the understanding of intramolecular energy transfer and unimolecular reactivity.

Acknowledgments

I wish to recognize the enormous contributions and support of my co-workers and students to the work presented here. K. T. Marshall, J. Jiang, P. R. Fleming, and L. G. Spears, Jr. have all provided scientific content, computer assistance, and enlightening discussion. I wish to thank T. A. Holme, in particular, for his diligent studies of overtone excitations and unimolecular reaction, without which many of these problems would not have been undertaken, let alone solved. This work was supported in part by grants from the Robert A. Welch Foundation of Houston, Texas, the Atlantic Richfield Foundation of the Research Corporation, and the National Science Foundation. Acknowledgment is made to the Donors of the Petroleum Research Fund, administered by the American Chemical Society, for partial support of this research.

References

1. B. R. Henry, *Acc. Chem. Res.* **10**, 207 (1977).
2. M. L. Sage and J. Jortner, *Adv. Chem. Phys.* **47**, 293 (1981).
3. L. Halonen and M. S. Child, *Adv. Chem. Phys.* **57**, 1 (1984).
4. F. F. Crim, *Ann. Rev. Phys. Chem.* **35**, 657 (1984).
5. R. G. Bray and M. J. Berry, *J. Chem. Phys.* **71**, 4909 (1979).
6. M. E. Long, R. L. Swofford, and A. C. Albrecht, *Science* **191**, 183 (1976).
7. G. Herzberg, *Infrared and Raman Spectra of Polyatomic Molecules*, Van Nostrand Reinhold, New York, 1945.

8. K. V. Berry, D. F. Heller, and M. J. Berry, *J. Chem. Phys.* **76**, 2814 (1982).

9. G. A. West, J. J. Barrett, D. R. Siebert, and K. V. Reddy, *Rev. Sci. Instrum.* **54**, 797 (1983).

10. R. L. Swofford, M. E. Long, and A. C. Albrecht, *J. Chem. Phys.* **65**, 179 (1976).

11. D. F. Heller and S. Mukamel, *J. Chem. Phys.* **70**, 463 (1979).

12. H. L. Fang and R. L. Swofford, *Chem. Phys. Lett.* **105**, 5 (1984); H. L. Fang, R. L. Swofford, and D. A. C. Compton, *Chem. Phys. Lett.* **108**, 539 (1984); H. L. Fang, D. M. Meister, and R. L. Swofford, *J. Phys. Chem.* **88**, 405 (1984); H. L. Fang and R. L. Swofford, *J. Chem. Phys.* **72**, 6382 (1980).

13. H. L. Fang, D. M. Meister, and R. L. Swofford, *J. Phys. Chem.* **88**, 410 (1984).

14. R. R. Hall, P. R. Fleming, D. T. Halligan, and M. J. Berry, personal communication; R. R. Hall, Ph.D. Dissertation, Rice University, 1984.

15. M. J. Berry, *Proc. Robert A. Welch Found. Conf.* **28**, 133 (1984).

16. G. J. Scherer, K. K. Lehmann, and W. Klemperer, *J. Chem. Phys.* **78**, 2817 (1983).

17. J. W. Perry and A. H. Zewail, *J. Chem. Phys.* **80**, 5333 (1984).

18. L. J. Butler, T. M. Ticich, M. D. Likar, and F. F. Crim, *J. Chem. Phys.* **85**, 2331 (1986); E. S. McGinley and F. F. Crim, *J. Chem. Phys.* **85**, 5741 (1986).

19. J. S. Wong and C. B. Moore, *J. Chem. Phys.* **77**, 603 (1982); M. S. Burberry and A. C. Albrecht, *J. Chem. Phys.* **71**, 4631 (1979); R. A. Bernheim, F. W. Lampe, J. F. O'Keefe, and J. R. Qualey, *Chem. Phys. Lett.* **100**, 45 (1983); J. W. Perry and A. H. Zewail, *J. Phys. Chem.* **85**, 933 (1981).

20. A more detailed review of the experimental work in this area appears in Ref. 4.

21. V. S. Letokhov, *Nature* **305**, 103 (1983); J. I. Steinfeld, in *Laser Induced Chemical Processes*, J. I. Steinfeld, ed., Plenum, New York, 1981, Chapter 4.

22. K. V. Reddy and M. J. Berry, *Chem. Phys. Lett.* **66**, 223 (1979).

23. A review of statistical theories of unimolecular reactions is presented by P. A. Robinson and K. A. Holbrook, *Unimolecular Reactions*, Wiley, New York, 1972.

24. D. W. Chandler, W. E. Farneth, and R. N. Zare, *J. Chem. Phys.* **77**, 4447 (1982); M.-C. Chuang, J. E. Baggott, D. W. Chandler, W. E. Farneth, and R. N. Zare, *Farad. Disc. Chem. Soc.* **75**, 301 (1983).

25. J. M. Jasinski, J. K. Frisoli, and C. B. Moore, *J. Chem. Phys.* **79**, 1312 (1983); *Farad. Disc. Chem. Soc.* **75**, 289 (1983); *J. Phys. Chem.* **87**, 2209 (1983).

26. E. S. McGinley and F. F. Crim, *J. Chem. Phys.* **85**, 5748 (1986); B. D. Cannon and F. F. Crim, *J. Chem. Phys.* **75**, 1752 (1981); T. M. Ticich, T. R. Rizzo, H.-R. Dubal, and F. F. Crim, *J. Chem. Phys.* **84**, 1508 (1986).

27. T. D. Kunz and M. J. Berry, personal communication; T. D. Kunz, Ph.D. Dissertation, Rice University, 1986.

28. N. F. Scherer, F. E. Doany, A. H. Zewail, and J. W. Perry, *J. Chem. Phys.* **84**, 1932 (1986).

29. K. T. Marshall and J. S. Hutchinson, *J. Phys. Chem.*, submitted for publication.

30. J. M. Jasinski, *Chem. Phys. Lett.* **109**, 462 (1984).

31. J. S. Hutchinson, *J. Chem. Phys.* **82**, 22 (1985).

32. P. M. Morse, *Phys. Rev.* **34**, 57 (1929); R. Wallace, *Chem. Phys.* **11**, 189 (1975); M. L. Sage, *Chem. Phys.* **35**, 375 (1978); R. Wallace, *Chem. Phys. Lett.* **37**, 115 (1976).

33. H. Goldstein, *Classical Mechanics*, Addison-Wesley, Reading, MA, 1965, Chapter 9.

34. D. W. Oxtoby and S. A. Rice, *J. Chem. Phys.* **65**, 1676 (1976).

35. C. Jaffé and P. Brumer, *J. Chem. Phys.* **73**, 5646 (1980).

36. E. L. Sibert III, W. P. Reinhardt, and J. T. Hynes, *J. Chem. Phys.* **77**, 3583 (1982).

37. E. L. Sibert III, J. S. Hutchinson, W. P. Reinhardt, and J. T. Hynes, *Int. J. Quant. Chem.* **9**, 375 (1982).

38. J. S. Hutchinson, W. P. Reinhardt, and J. T. Hynes, *J. Chem. Phys.* **79**, 4247 (1983).

39. J. S. Hutchinson, J. T. Hynes, and W. P. Reinhardt, *J. Phys. Chem.* **90**, 3428 (1986).

40. E. L. Sibert III, J. T. Hynes, and W. P. Reinhardt, *J. Chem. Phys.* **81**, 1135 (1984).

41. B. V. Chirikov, *Phys. Rep.* **52**, 263 (1979); F. M. Izrailev and B. V. Chirikov, *Sov. Phys. Dokl.* **11**, 30 (1966).

42. J. Ford, *Adv. Chem. Phys.* **24**, 155 (1973).

43. J. S. Hutchinson, E. L. Sibert III, and J. T. Hynes, *J. Chem. Phys.* **81**, 1314 (1984).

44. J. S. Hutchinson, J. T. Hynes, and W. P. Reinhardt, *Chem. Phys. Lett.* **108**, 353 (1984).

45. E. L. Sibert III, W. P. Reinhardt, and J. T. Hynes, *J. Chem. Phys.* **81**, 1115 (1984); *Chem. Phys. Lett.* **92**, 455 (1982).

46. M. J. Davis and E. J. Heller, *J. Chem. Phys.* **75**, 246 (1981).

47. E. J. Heller, *Chem. Phys. Lett.* **60**, 338 (1979); P. Brumer and M. Shapiro, *Chem. Phys. Lett.* **72**, 528 (1980); K. G. Kay, *J. Chem. Phys.* **72**, 5955 (1980); R. T. Lawton and M. S. Child, *Mol. Phys.* **44**, 709 (1981); J. S. Hutchinson and R. E. Wyatt, *Phys. Rev. A* **73**, 1567 (1981).

48. T. A. Holme and J. S. Hutchinson, *Chem. Phys.* **93**, 419 (1985).

49. T. Uzer and J. T. Hynes, *Chem. Phys. Lett.* **113**, 483 (1985).

50. T. A. Holme and J. S. Hutchinson, *J. Chem. Phys.* **83**, 2860 (1985).

51. T. Uzer, J. T. Hynes, and W. P. Reinhardt, *Chem. Phys. Lett.* **117**, 600 (1985); *J. Chem. Phys.* **85**, 5791 (1986); T. Uzer and J. T. Hynes, in *Stochasticity and Intramolecular Redistribution of Energy*, R. Lefebvre and S. Mukamel, eds., NATO ASI Series, 1986.

52. T. A. Holme and J. S. Hutchinson, *J. Chem. Phys.* **84**, 5455 (1986).

53. S. K. Gray, *Chem. Phys.* **75**, 67 (1983); **83**, 125 (1984).

54. P. S. Dardi and S. K. Gray, *J. Chem. Phys.* **80**, 4738 (1984).

55. K. M. Christoffel and J. W. Bowman, *J. Phys. Chem.* **85**, 2159 (1981).

56. E. S. Medvedev, *J. Mol. Spectrosc.* **114**, 1 (1985).

57. J. S. Hutchinson, *J. Chem. Phys.* **85**, 7087 (1986).

58. R. M. Hedges and W. P. Reinhardt, *J. Chem. Phys.* **78**, 3964 (1983).

59. B. Waite and W. H. Miller, *J. Chem. Phys.* **73**, 3713 (1980).

60. W. P. Reinhardt, *Ann. Rev. Phys. Chem.* **33**, 223 (1982) and references therein.

CHAPTER XV

INTERACTION OF AN ADSORBED ATOM
WITH A LASER

HENK F. ARNOLDUS, SANDER VAN SMAALEN* and THOMAS F.
GEORGE

*Departments of Physics and Astronomy and Chemistry, State University of
New York at Buffalo, Buffalo, New York 14260*

CONTENTS

The irradiation by infrared laser light of an atom adsorbed on the surface of a
harmonic crystal is considered. The dynamic coupling between the atom (its
motion) and the substrate degrees of freedom (phonon field) in the presence of a
confining potential well (van der Waals bond between atom and crystal) gives
rise to thermal relaxation of this adbond configuration. Both the atom and the
substrate are assumed to be transparent, but the bond is allowed to have
nonvanishing dipole moment matrix elements, which couple the external field
to the adsorbate. The equation of motion for the reduced adbond density

Present Address: Laboratory of Inorganic Chemistry, Materials Research Center, State
University of Groningen, Nijenborgh 16, 9747 AG Groningen, The Netherlands.

operator is obtained with reservoir theory, and the relaxation constants are expressed in properties of the crystal. With a similar method, the spectral profile for absorption of weak radiation is derived. Subsequently, the illumination by a strong finite-linewidth laser field that is in close resonance with a single transition of the adbond is examined. The optical Bloch equations in operator form are derived and applied to study the process of the laser heating of the crystal. It is pointed out how this mechanism can be understood as resulting from (multi) photon–phonon conversion reactions which are mediated by the adbond.

I. INTRODUCTION

Applications of tunable laser sources in contemporary experiments in physics and chemistry can roughly be subdivided into two categories. In the first class the laser is used as a diagnostic tool to investigate a dynamical system with spectroscopic methods. Scanning the frequency of a low-power field that is incident on an atom or molecule reveals the resonances of the system as lines in the absorption profile. Interaction of the system with the environment (collisions, spontaneous emission, presence of boundaries, etc.) amounts to relaxation of the molecule, which in turn broadens a spectral line in such a way that the width of the line is proportional to the inverse relaxation rate of the specific transition. More subtle properties of the lineshape, such as the far-wing decay, carry information on the details of the interaction, since the frequency dependence of the absorption is determined by the Fourier–Laplace transform of the time evolution operator of the density operator (1). Therefore, it can be expected that a profound comprehension of dynamical properties can be achieved from an accurate observation of spectral profiles. However, this method requires an elaborate theory that disentangles the contributions of the various mechanisms to the lineshape.

In the second category of experiments, a radiation field is applied in order to modify or even induce a process. An incident photon can supply the necessary activation energy for a reaction that is very unlikely to take place without a field. The advantage of a narrow-band laser is that by tuning the laser frequency, the process can be optimized, in contrast to, for instance, thermal excitation of a species (collisions). Prime examples are laser-induced dissociation and ionization and atomic resonance fluorescence. In more rigorous situations, a strong (pulsed) laser is merely used as a heat gun (melting, vaporization), and the color and polarization of the field are not of crucial importance (2).

In this chapter we consider optical features of atoms which are adsorbed on a crystal. An atom is bounded to the substrate by electromagnetic interaction with the crystal atoms (van der Waals energy), which is effectively described by a potential well. The few (~ 25) vibrational adbond states have transition frequencies (level separations) on the order of 50 cm^{-1} to 500 cm^{-1}, which is in the infrared region of the optical spectrum. Both the atom and the crystal are supposed to be transparent for infrared light, as is, for instance, the case for noble-gas atoms on potassium chloride. The vibrational states, however, exhibit an optical activity in the sense that there are non vanishing dipole matrix elements between the various bound states. Irradiation with an infrared laser will then amount to photon absorption by the bond, and hence, the spectral profile will reflect the properties of the binding potential and the dynamical interaction with the substrate. In this fashion we can study transparent crystals with spectroscopic methods, where the medium for transport of the information is furnished by the adsorbed atoms. This situation is reminiscent of the more familiar problem of collisional redistribution by neutrals in gas phase experiments. Transparent atoms are immersed in a buffer gas, and photons from an incident field are absorbed during collisions only. Accurate measurements of lineshapes then allow the determination of interaction potentials. The only distinction is that collisional redistribution deals with scattering states, whereas the spectroscopy of adsorbates involves bound states. It can be anticipated that the established methods for the gas phase can be converted to suitable surface equivalents, although experiments along this line are still rare (3, 4).

Efforts in the second category of the application of lasers in this system are far more common, and most notably studied is the process of resonant desorption (5–7). An intense infrared laser is tuned into resonance with the transition frequency between a low-lying and a high-lying vibrational state. Absorption of a photon is accompanied by a transition of the adbond to the high-lying state, which is close to the continuum. Then the crystal, regarded as a thermal bath, has only to provide a small amount of energy in order to accomplish the desorption of the atom. Without the driving laser, this process would not occur, since the atomic bond has to be excited to the high-lying state by thermal coupling to the substrate. In this way, one has expected to be able to make clean surfaces without heating (as in thermal desorption) or damaging the materials (8–13). Another idea has been that advantage could be taken of the resonant nature of the process. Different adsorbed species would be desorbed selectively, depending on the laser frequency, which would provide a practical method for isotope separation or less ambitious separation of molecules (12, 13). Although the sketched process will obviously happen if the radiative coupling is sufficiently strong, it does not necessarily mean that the efficiency is very high. Recent experiments (8) show a quantum yield of about 1% (ratio of desorbed atoms to absorbed photons), which conversely implies that almost all radiation is converted to thermal energy of the substrate. Consequently, the solid heats up very fast, and this is precisely what one tries to avoid. Since the thermal coupling is inevitable, the conclusion is that resonant desorption is not a very promising technique. Furthermore, the selectivity has turned out to be very poor, if present at all, which is probably due to a rapid energy exchange between different adsorbates preceding the desorption (14). From a different point of view, however, this mechanism could be applied for laser heating of a transparent crystal. By manipulating the spatial variations of the laser intensity, this process can be used to maintain well-controlled temperature gradients along a gas–solid interface.

II. KINETIC PHONON COUPLING

An atom with mass m is physisorbed on the surface (xy plane) of a crystal, and its motion is assumed to be mainly perpendicular to the surface (z direction). The interaction potential with the solid is subdivided into two parts, V_1 and V_2, where V_1 represents the potential well resulting from the coupling to the closest surface atom, and V_2 is an effective repulsive potential that is brought about by the remainder of the crystal and confines the atom to the region $z \gtrsim 0$. The thermal equilibrium position of the nearest surface atom, with mass M, is taken as the origin of the coordinate system. Then the positions of m and M are denoted by $z\mathbf{e}_z$ and $\mathbf{u}$, respectively, and V_1 depends on $|z\mathbf{e}_z - \mathbf{u}|$ only, whereas V_2 depends on z. Subsequently, it can be asserted that $\mathbf{u}$ (~ 0.1 Å) is much smaller than z ($\gtrsim 1$ Å), which allows a Taylor expansion of V_1 around $\mathbf{u} = 0$. Hence we can write

$$V_1(|z\mathbf{e}_z - \mathbf{u}|) = V_1(z) - \mathbf{u}\cdot\mathbf{e}_z \frac{dV_1}{dz} + \cdots \qquad (2.1)$$

and then omit the ellipsis. In physical terms this means that we neglect multiphonon processes in comparison with single-phonon transitions, as will become clear in due course.

Next we notice that the expansion 2.1 separates the coupling with the lattice vibrations (last term) from the interaction $V_1(z)$ with the static crystal, which equals the interaction for $\mathbf{u} = 0$. A convenient basis set for the adbond wavefunction is therefore provided by the eigenstates of the adbond Hamiltonian H_a, defined as

$$H_a = -\frac{\hbar^2}{2m}\frac{d^2}{dz^2} + V_1(z) + V_2(z), \qquad (2.2)$$

where the first term is the kinetic energy. Once a potential is prescribed, the eigenvalues $\hbar\omega_i$ and the eigenfunctions $|i\rangle$ of the H_a can be evaluated with standard methods. A common choice for the combination $V_1(z) + V_2(z)$ is a

Morse potential (15–19) or a truncated harmonic potential (20). We shall not refer to a specific potential in this chapter but only use the eigenvalue equation

$$H_a |i\rangle = \hbar\omega_i |i\rangle \tag{2.3}$$

and assume the eigenstates to be nondegenerate (as for a Morse potential). Then, in turn, we can represent H_a with respect to its own eigenstates as

$$H_a = \sum_i \hbar\omega_i P_i, \tag{2.4}$$

with $P_i = |i\rangle\langle i|$ the projector on the ith adbond state.

As a model for the crystal we adopt a harmonic lattice representation for which the Hamiltonian reads (21)

$$H_p = \sum_{ks} \hbar\omega(k) a_{ks}^\dagger a_{ks} \tag{2.5}$$

in terms of the annihilation and creation operators for phonons (bosons) in the mode ks. The summation runs over the wave vectors $\mathbf{k}$ and branches s, which are supported by the crystal, and the dispersion relation $\omega = \omega(k)$ is taken to be independent of s and of the direction of $\mathbf{k}$. Coupling between the phonon field and the adbond is brought about by the position operator $\mathbf{u}$ of M, which equals the displacement field at the origin. Explicitly,

$$\mathbf{u} = \sum_{ks} \left(\frac{\hbar v}{2MV\omega(k)} \right)^{1/2} (a_{ks} + a_{-ks}^\dagger) \mathbf{e}_{ks} \tag{2.6}$$

in terms of the volume v of a unit cell and the quantization volume V. Combining everything then gives, for the Hamiltonian of the adbond plus crystal,

$$H = H_a + H_p - RS, \tag{2.7}$$

with

$$R = \mathbf{u} \cdot \mathbf{e}_z, \qquad S = \frac{dV_1}{dz}. \tag{2.8}$$

It will appear in the next section that it is advantageous to work with a density operator formalism, so the equation of motion is

$$i\hbar \frac{d\rho}{dt} = [H, \rho], \qquad \rho^\dagger = \rho, \qquad \mathrm{Tr}\,\rho = 1. \tag{2.9}$$

From Eq. 2.6 we see that the interaction between adbond and crystal, $- RS$, is linear in the creation and annihilation operators, which implies that $- RS$ can only induce transitions between crystal states that differ by not more than one phonon. In other words, by retaining only the linear term in the expansion 2.1, we discard multiphonon processes, which can be justified as long as any set of two adbond states is resonantly coupled by a single-phonon process. Physically, this implies that any level separation $|\omega_i - \omega_j|$ must be smaller than the cutoff frequency of the dispersion relation, which is the Debye frequency ω_D.

III. THERMAL RELAXATION

The full density operator $\rho(t)$ of a single atom and a large crystal has not much significance. Since we are interested in the dynamics of the adbond, we consider the reduced density operator $\rho_0(t)$, defined as

$$\rho_0(t) = \mathrm{Tr}_p\, \rho(t), \tag{3.1}$$

where the trace runs over all phonon states. On the other hand, the crystal is regarded as a large reservoir at temperature T, for which the density matrix equals

$$\bar{\rho}_p = \exp(-\beta H_p)[\mathrm{Tr}_p \exp(-\beta H_p)]^{-1}, \tag{3.2}$$

with $\beta = (kT)^{-1}$. Coupling between the adbond and the heat bath then gives rise to the thermal relaxation of $\rho_0(t)$, and eventually $\rho_0(t)$ reaches a steady state

$$\bar{\rho}_0 = \lim_{t \to \infty} \rho_0(t) \tag{3.3}$$

in which the atomic bond and the crystal are in thermal equilibrium.

It is a standard procedure in relaxation theory to derive an equation for the reduced density operator $\rho_0(t)$. Both projection techniques (22) and a reservoir approach (23, 24) yield the same result, provided that equivalent approximations are made (25). Here we will briefly summarize the derivation of Ref. 24 and adopt a Liouville operator notation, which will allow a concise formulation. A Liouvillian L is related to a Hamiltonian H according to

$$L\rho = \hbar^{-1}[H, \rho], \tag{3.4}$$

which defines the action of L on an arbitrary operator ρ in Hilbert space. Then one regards ρ as a vector in Liouville space and L as a linear operator in that space (26). Consequently, an N-dimensional Hilbert space generates an N^2-dimensional Liouville space. Matrix representations of Liouvillians can be constructed in the very same way as for operators in Hilbert space. A theorem that follows immediately from Eq. 3.4 is

$$\exp(Ls)\rho = \exp(Hs/\hbar)\rho \exp(-Hs/\hbar),\tag{3.5}$$

where, for instance, $s = it$.

Reservoir theory starts with a transformation of the equation of motion 2.9 to the interaction picture, where the coupling $-RS$ is considered as the interaction. With

$$\hat{\rho}(t) = \exp[i(L_a + L_p)t]\rho(t),\tag{3.6}$$

Eq. 2.9 can be written as

$$i\hbar\frac{d}{dt}\hat{\rho}(t) = -W(t)\hat{\rho}(t),\tag{3.7}$$

where $W(t)$ is defined as the Liouvillian

$$W(t)\rho = [(\exp[i(L_a + L_p)t](RS)),\rho]\tag{3.8}$$

for an arbitrary ρ. As the initial condition, we choose

$$\rho(0) = \rho_0(0)\bar{\rho}_p.\tag{3.9}$$

Then the identity

$$\mathrm{Tr}_p(W(t)\rho(0)) = 0\tag{3.10}$$

follows from the fact that $\bar{\rho}_p$ commutes with H_p and from the explicit form of R (Eqs. 2.6 and 2.8), which gives $\mathrm{Tr}_p R\bar{\rho}_p = 0$. If we now integrate Eq. 3.7, take the trace over the phonon states, and use Eq. 3.10, we find

$$\frac{d}{dt}\hat{\rho}_0(t) = -\hbar^{-2}\mathrm{Tr}_p\int_0^t d\tau\, W(t)W(t-\tau)\hat{\rho}(t-\tau),\tag{3.11}$$

which is an exact integral of the equation of motion.

Obviously, the integral in Eq. 3.11 is awkward, and a series of approximations has to be made in order to obtain a manageable expression. The crucial step that has to be made is the factorization

$$\hat{\rho}(t-\tau) \simeq \hat{\rho}_0(t-\tau)\bar{\rho}_p\tag{3.12}$$

in the integrand of Eq. 3.11. Then we substitute the explicit form of $W(t)$, use Eq. 3.5 several times, and transform back to the Schrödinger picture with the inverse relation of Eq. 3.6. Finally, we obtain

$$i\frac{d}{dt}\rho_0(t) = L_a\rho_0(t) - \frac{i}{2\pi}L_S\int_0^t d\tau \exp(-iL_a\tau)L_c(\tau)\rho_0(t-\tau), \qquad (3.13)$$

where L_S and $L_c(\tau)$ are defined by

$$L_S\rho = [S, \rho],$$
$$L_c(\tau)\rho = G(\tau)S\rho - G(\tau)^*\rho S, \qquad (3.14)$$

which involves the reservoir correlation function

$$G(\tau) = 2\pi\hbar^{-2}\mathrm{Tr}_p R\bar{\rho}_p \exp(iL_p\tau)R. \qquad (3.15)$$

Inspection of Eq. 3.13 shows that the coupling to the crystal only enters through the function $G(\tau)$. Approximation 3.12 is sufficient to separate reservoir and adbond operators, and the presence of the solid can be accounted for by a single function $G(\tau)$.

Of paramount importance for the validity of a reservoir approach is that $G(\tau)$ decays to zero (for $\tau \to \infty$) sufficiently quickly. Due to the interference between the many phonon modes, the typical decay time will be on the order of the inverse cutoff frequency of the spectrum, which is ω_D for a crystal. This will become clearer in the next section. Then $L_c(\tau)$ deviates from zero only for $\tau \simeq 0$, and the most crude approximation would be to replace it by an operator that is proportional to a delta function. It is easy, however, to do much better. The net effect of the coupling of the adbond to the crystal is a damping on a time scale that equals the inverse linewidth. Since this is very slow in comparison with ω_D^{-1}, we conclude that in the interaction picture, $\hat{\rho}(t-\tau)$ varies negligibly on a time scale ω_D^{-1}, and therefore, we can replace $\hat{\rho}(t-\tau)$ by $\hat{\rho}(t)$ in Eq. 3.11. This is the Markov approximation, and the equivalent in the Schrödinger picture reads

$$\rho_0(t-\tau) \simeq \exp(iL_a\tau)\rho_0(t), \qquad (3.16)$$

which can be inserted into Eq. 3.13. Then we can replace the upper integration limit by infinity, which yields, for the equation of motion,

$$i\frac{d}{dt}\rho_0(t) = (L_a - i\Gamma)\rho_0(t). \qquad (3.17)$$

Relaxation of the adbond due to the coupling to the phonon reservoir is now completely accounted for by the time-independent Liouvillian

$$\Gamma = \frac{1}{2\pi} L_S \int_0^\infty d\tau \exp(-iL_a\tau) L_c(\tau) \exp(iL_a\tau). \tag{3.18}$$

An expansion of Γ in matrix elements with respect to adbond states will be given in Section V.

If we insert the expression for L_S and $L_c(\tau)$ in Eq. 3.18 and use relation 3.5 for the exponentials, we find that Γ can alternatively be cast in the form

$$\Gamma\rho = L_S(Q\rho - \rho Q^\dagger), \tag{3.19}$$

where the Hilbert space operator Q is defined as

$$Q = \frac{1}{2\pi} \int_0^\infty d\tau \, G(\tau) \exp(-iL_a\tau) S. \tag{3.20}$$

The advantage is that Q only involves a single exponential. Furthermore, we directly find, from the representation 3.19,

$$\text{Tr}_a(\Gamma\rho) = 0, \qquad (\Gamma\rho)^\dagger = \Gamma\rho^\dagger, \tag{3.21}$$

which guarantees the conservation of trace and Hermiticity in the time evolution of $\rho_0(t)$.

IV. RESERVOIR CORRELATION FUNCTION

The properties of the crystal that affect the time evolution of the adbond are all embodied in the correlation function $G(\tau)$. In order to illustrate quantitatively the behavior of $G(\tau)$, we adopt a simple Debye model, which will already reveal the most salient features of a crystal correlation function. In this model, the dispersion relation is taken to be

$$\omega(k) = ckH(\omega_D - ck), \tag{4.1}$$

with c the speed of sound and H the unit step function. Expression 3.15 is easily evaluated with standard techniques (23). It is elucidating to subdivide $G(\tau)$ in a spontaneous and a stimulated part, according to

$$G(\tau) = G(\tau)_{\text{sp}} + G(\tau)_{\text{st}}, \tag{4.2}$$

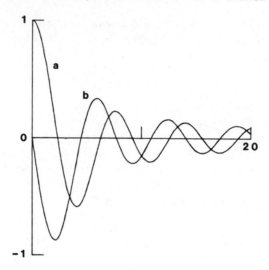

Figure 1. Real (a) and imaginary (b) parts of correlation function $G(\tau)_{sp}$, divided by $\frac{1}{2}\zeta\omega_D^{-1}$ and as a function of $\omega_D\tau$. From Eq. 4.11 it follows that the integral over curve (a) equals zero.

where $G(\tau)_{sp}$ is, by definition, $G(\tau)$ for $T = 0$. Hence, $G(\tau)_{sp}$ is independent of T and accounts for relaxation of the adbond due to the presence of the crystal irrespective of its temperature. We find, for this spontaneous part,

$$G(\tau)_{sp} = \zeta\omega_D^{-1}\frac{(1 + i\omega_D\tau)\exp(-i\omega_D\tau) - 1}{(\omega_D\tau)^2}, \tag{4.3}$$

with the parameter ζ given by

$$\zeta = 3\pi/\hbar M. \tag{4.4}$$

The real and imaginary part of $G(\tau)_{sp}$ are plotted in Figure 1. For the stimulated contribution we obtain

$$G(\tau)_{st} = 2\zeta\omega_D^{-3}\int_0^{\omega_D} d\omega\, n(\omega)\omega\cos(\omega\tau), \tag{4.5}$$

which is plotted in Figure 2 for two values of the temperature. Here $n(\omega)$ is the average number of phonons in mode ω, which is explicitly

$$n(\omega) = [\exp(\beta\hbar\omega) - 1]^{-1}. \tag{4.6}$$

Figures 1 and 2 illustrate that $G(\tau)$ tends to zero indeed on a time scale ω_D^{-1}, although the decay is not exponential.

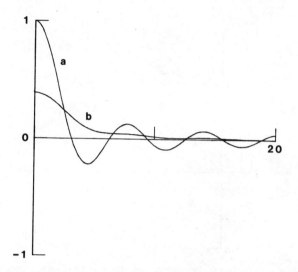

Figure 2. Stimulated part of the reservoir correlation function divided by $2\zeta\omega_D^{-2}kT/\hbar$ and as a function of $\omega_D\tau$. Curve (*a*) corresponds to $\hbar\omega_D = 0.001kT$ (high-temperature limit) and for curve (*b*) we have taken $\hbar\omega_D = 4kT$ (low-temperature region).

From Eq. 3.20 it follows that the relaxation constants (matrix elements of Γ) are determined by the Fourier–Laplace transform of $G(\tau)$,

$$\tilde{G}(\omega) = \frac{1}{\pi} \int_0^\infty d\tau \exp\left(i\omega\tau\right)G(\tau), \tag{4.7}$$

with ω equal to a transition frequency $\omega_i - \omega_j$. From Eqs. 4.3 and 4.5 we readily derive

$$\tilde{G}(\omega)_{sp} = \zeta\omega_D^{-3}\omega H(\omega)H(\omega_D - \omega) - i\pi^{-1}\zeta\omega_D^{-3}\left(\omega_D + \omega \log\left|1 - \frac{\omega_D}{\omega}\right|\right), \tag{4.8}$$

$$\tilde{G}(\omega)_{st} = \zeta\omega_D^{-3}|\omega|n(|\omega|)H(\omega_D - |\omega|)$$
$$+ i\pi^{-1}\zeta\omega_D^{-3}P \int_0^{\omega_D} d\omega'\, n(\omega')\frac{2\omega\omega'}{\omega^2 - \omega'^2}, \tag{4.9}$$

where P is the principal value. These functions are plotted in Figures 3 and 4. For negative values of ω the real part of $\tilde{G}(\omega)_{sp}$ vanishes identically, which implies that this contribution to Γ represents a decaying part. The thermal

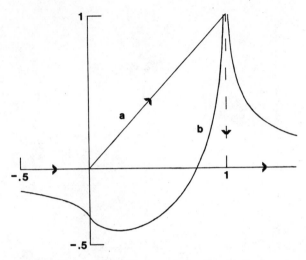

Figure 3. Real (a) and imaginary (b) parts of the Fourier–Laplace transform of the spontaneous part of the correlation function divided by $\zeta\omega_D^{-2}$ and as a function of ω/ω_D. The real part vanishes for $\omega < 0$ and for $\omega > \omega_D$, whereas an imaginary part is present for every ω.

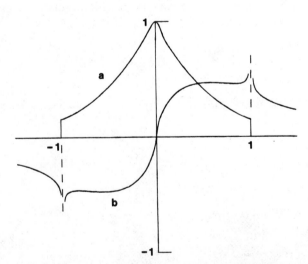

Figure 4. Same as Figure 3 but for the stimulated part. Normalized by $\zeta kT/\hbar\omega_D^3$ and for $\hbar\omega_D = 3kT$. The real part only vanishes for $|\omega| > \omega_D$. Singularities in the imaginary part appear at $\omega = \pm\omega_D$, which is an artefact of the sharp cutoff of the dispersion relation at ω_D. Any smooth decay for $\omega > \omega_D$, but still arbitrarily steep, will result in a finite value of the correlation function at $\omega = \pm\omega_D$.

690

part obeys the relation

$$\tilde{G}(-\omega)_{st} = \tilde{G}(\omega)_{st}^*,$$ (4.10)

which indicates that upward and downward transitions between two levels due to the finite temperature of the crystal have the same rate constant. Hence, the genuine thermal equilibrium distribution is brought about by the spontaneous part of the relaxation operator, which is temperature independent.

Setting ω equal to zero yields immediately

$$\int_0^\infty d\tau\, G(\tau) = \zeta \omega_D^{-2}\left(-i + \frac{\pi kT}{\hbar\omega_D}\right),$$ (4.11)

which can be regarded as a measure of the strength of $G(\tau)$. For $kT \gtrsim \hbar\omega_D$ the stimulated transitions tend to dominate the spontaneous decay.

V. MATRIX ELEMENTS

Solving the equation of motion 3.17 for a particular configuration of adbond states requires an expansion in matrix elements. Taking the (k, l)th matrix element gives

$$\frac{d}{dt}\langle k|\rho_0(t)|l\rangle = -i\langle k|[L_a\rho_0(t)]|l\rangle - \langle k|[\Gamma\rho_0(t)]|l\rangle.$$ (5.1)

It remains to express the right side in matrix elements of $\rho_0(t)$. From the definition 3.4 of L_a and the expression 2.4 of H_a, we easily find, for the first term,

$$\langle k|(L_a\rho)|l\rangle = \Delta_{kl}\langle k|\rho|l\rangle$$ (5.2)

for an arbitrary ρ, where

$$\Delta_{kl} = \omega_k - \omega_l$$ (5.3)

is the level separation between the adbond states $|k\rangle$ and $|l\rangle$.

Slightly more complicated is the evaluation of the relaxation operator Γ. First we notice that the exponential in Eq. 3.20 can be expanded as

$$\exp(-iL_a\tau)\rho = \sum_{kl}\exp(-i\Delta_{kl}\tau)P_k\rho P_l,$$ (5.4)

which yields the representation for Q,

$$Q = \frac{1}{2}\sum_{kl} \tilde{G}(\Delta_{lk})P_k S P_l, \tag{5.5}$$

in terms of the Fourier–Laplace transform of the correlation function. Then we insert Eq. 5.5 into Eq. 3.19 and use the closure relation

$$\sum_i P_i = 1, \tag{5.6}$$

which finally gives

$$\langle k|(\Gamma\rho)|l\rangle = \frac{1}{2}\sum_{mn} (c_{kmmn}\langle n|\rho|l\rangle + c^*_{lmmn}\langle k|\rho|n\rangle)$$
$$- \frac{1}{2}\sum_{mn} (c_{nlkm} + c^*_{mkln})\langle m|\rho|n\rangle \tag{5.7}$$

in terms of the parameters

$$c_{klmn} = \langle k|S|l\rangle\langle m|S|n\rangle \tilde{G}(\Delta_{nm}). \tag{5.8}$$

Equation 5.7 expresses the relaxation operator Γ in matrix elements of the derivative of the potential, $S = dV_1/dz$, and $\tilde{G}(\omega)$, evaluated at the frequencies Δ_{kl}. The combination of Eqs. 5.1, 5.2, and 5.7 turns the equation of motion into a simple set of linear first-order differential equations, which can be solved directly once a potential and an initial state $\rho_0(0)$ are specified.

VI.   ABSORPTION SPECTRUM

A density operator is not directly amenable to observation in an experiment. One way of measuring properties of an adsorbed atom is by probing the system with a weak monochromatic laser. A field with intensity I_L (energy per unit of time that passes through a unit area perpendicular to the direction of propagation) and polarization ε_L is scanned over the resonances of the system, and the power absorption $I(\omega_L)$ (absorbed energy per unit of time) is measured as a function of the laser frequency ω_L. Photons are absorbed from the field by the joint system of atom and crystal and their interaction (although the atom and the crystal separately are both assumed to be transparent), so that the absorption profile exhibits the details of the coupling between the adbond and the phonon field rather than the properties of the adsorbed atom (the potential) alone.

The interaction between the adsorbate and the radiation field is established by a dipole coupling. Since the motion of the atom is restricted to the z direction, the dipole moment operator μ can be written as μe_z, with μ an operator in the Hilbert space spanned by the eigenstates $|k\rangle$ of the adbond Hamiltonian H_a. With E the electric component of the radiation field, we can include the interaction in the equation of motion 2.9 with the substitution

$$H \to H - \mu e_z \cdot E. \tag{6.1}$$

The probe beam is considered to be very weak, so that we can compute the power absorption with the golden rule. From the Appendix of Ref. 1 we copy the formal result

$$I(\omega_L) = I_L \omega_L (\varepsilon_0 \hbar c')^{-1} |e_z \cdot \varepsilon_L|^2 \operatorname{Re} \int_0^\infty d\tau \exp(i\omega_L \tau) \operatorname{Tr} \bar{\rho}[\mu(\tau), \mu], \tag{6.2}$$

where c' is the speed of light and $\bar{\rho}$ is the thermal equilibrium density operator of the entire system but without the interaction $-\mu \cdot E$. The τ dependence of $\mu(\tau)$ is given by $\mu(\tau) = \exp(iL\tau)\mu$. We remark that the commutator gives rise to two terms, where $\mu(\tau)\mu$ $[\mu\mu(\tau)]$ represents stimulated photon absorption (emission). The net absorption is the balance between the loss and gain terms for photons in the laser mode.

Evaluation of expression 6.2 starts with the identity

$$\operatorname{Tr} \bar{\rho}[\mu(\tau), \mu] = \operatorname{Tr} \mu \exp(-iL\tau)[\mu, \bar{\rho}]. \tag{6.3}$$

Then we introduce the quantity

$$D(\tau) = \exp(-iL\tau)[\mu, \bar{\rho}], \tag{6.4}$$

which is an operator in the entire phonon and adbond Hilbert space. With $D_0(\tau) = \operatorname{Tr}_p D(\tau)$, the adbond part of $D(\tau)$, we can write Eq. 6.3 as

$$\operatorname{Tr} \bar{\rho}[\mu(\tau), \mu] = \operatorname{Tr}_a \mu D_0(\tau). \tag{6.5}$$

Then we notice that $D(\tau)$ obeys the differential equation

$$i\frac{d}{d\tau} D(\tau) = LD(\tau), \tag{6.6}$$

with the initial condition

$$D(0) = [\mu, \bar{\rho}_0]\bar{\rho}_p, \tag{6.7}$$

where we assumed that $\bar{\rho} \simeq \bar{\rho}_0 \bar{\rho}_p$. The equation of motion 6.6 is identical to Eq. 2.9, which is sometimes referred to as the quantum regression theorem (27). This implies that the time regression of the dipole correlation function $D(\tau)$ is identical to the time evolution of the density operator $\rho(t)$. This, in turn, is tantamount to the statement that the dynamical properties of the system are reflected in the frequency dependence of the absorption profile, which elucidates the significance of a measurement of $I(\omega_L)$.

From Eq. 6.5 we see that we only have to solve Eq. 6.6 for the reduced correlation function $D_0(\tau)$, which can be done along the very same lines that led to Eq. 3.17. Hence, we can immediately write the solution, which is

$$D_0(\tau) = \exp\left[-i(L_a - i\Gamma)\tau\right][\mu, \bar{\rho}_0]. \qquad (6.8)$$

Substitution into Eq. 6.2 and performing the τ integration then yields

$$I(\omega_L) = I_L \omega_L (\varepsilon_0 \hbar c')^{-1} |e_z \cdot \varepsilon_L|^2 \operatorname{Re} \operatorname{Tr}_a \mu \frac{i}{\omega_L - L_a + i\Gamma} [\mu, \bar{\rho}_0] \qquad (6.9)$$

in terms of the steady-state solution $\bar{\rho}_0$ from Eq. 3.17, which obeys

$$(L_a - i\Gamma)\bar{\rho}_0 = 0, \qquad \operatorname{Tr}_a \bar{\rho}_0 = 1, \qquad \bar{\rho}_0^\dagger = \bar{\rho}_0. \qquad (6.10)$$

From conservation of the trace in the time evolution of $\rho_0(t)$, it follows that

$$\lim_{\tau \to \infty} D_0(\tau) = \bar{\rho}_0 \operatorname{Tr}_a [\mu, \bar{\rho}_0] = 0, \qquad (6.11)$$

which proves that the principal part in the upper integration limit $\tau = \infty$ vanishes. Equation 6.10 is easily solved for any level configuration. Then we insert $\bar{\rho}_0$ into Eq. 6.9, which determines the absorption profile in terms of a matrix inversion.

VII. LINESHAPE

Expression 6.9 represents the complete absorption profile as a function of ω_L. In order to disentangle the contributions from the various adbond resonances, we consider the situation of two levels $|1\rangle$ and $|2\rangle$ with separation $\omega_2 - \omega_1 = \omega_0 > 0$, and we scan the laser over this resonance. The resulting profile $I(\omega_L)$ is then termed the lineshape of this particular transition. With the expansion in matrix elements from Section V, we can immediately write the equation of motion as a linear set of differential equations. With

$\rho_{12} = \langle 1|\rho_0|2\rangle$ and so forth for the other ρ_{ij}, we obtain

$$i\frac{d}{dt}\rho_{22} = -ia_{21}\rho_{22} + ia_{12}\rho_{11}, \tag{7.1}$$

$$i\frac{d}{dt}\rho_{11} = ia_{21}\rho_{22} - ia_{12}\rho_{11}, \tag{7.2}$$

$$i\frac{d}{dt}\rho_{21} = (\omega_0 - i\eta)\rho_{21} + i\eta^* \exp(2i\psi_S)\rho_{12}, \tag{7.3}$$

$$i\frac{d}{dt}\rho_{12} = i\eta \exp(-2i\psi_S)\rho_{21} - (\omega_0 + i\eta^*)\rho_{12}, \tag{7.4}$$

where we introduced the abbreviations for the relaxation parameters

$$a_{kl} = |\langle k|S|l\rangle|^2 \operatorname{Re} \tilde{G}(\Delta_{kl}), \tag{7.5}$$

$$\eta = \tfrac{1}{2}|\langle 2|S|1\rangle|^2 [\tilde{G}(\omega_0) + \tilde{G}(-\omega_0)^*], \tag{7.6}$$

and neglected the couplings due to the small diagonal matrix elements $\langle 1|S|1\rangle$ and $\langle 2|S|2\rangle$. Furthermore, the phase ψ_S of $\langle 2|S|1\rangle$ is defined by

$$\langle 2|S|1\rangle = |\langle 2|S|1\rangle| \exp(i\psi_S). \tag{7.7}$$

Of course, we can choose the phase between $|1\rangle$ and $|2\rangle$ in such a way that $\psi_S = 0$, but it will turn out to be advantageous to postpone the definition of the relative phase.

The steady-state solution of the set 7.1–7.4 is readily found to be

$$\bar{n}_2 = \frac{a_{21}}{a_{21} + a_{12}}, \qquad \bar{n}_1 = \frac{a_{12}}{a_{21} + a_{12}}, \tag{7.8}$$

for the populations $\bar{n}_k = \bar{\rho}_{kk}$, whereas the coherences $\bar{\rho}_{12}$ and $\bar{\rho}_{21}$ vanish. From the set 7.1–7.4 we read the matrix representation

$$\omega_L - L_a + i\Gamma = \begin{bmatrix} \omega_L + ia_{21} & -ia_{12} & 0 & 0 \\ -ia_{21} & \omega_L + ia_{12} & 0 & 0 \\ 0 & 0 & \omega_L - \omega_0 + i\eta & -i\eta^* \exp(2i\psi_S) \\ 0 & 0 & -i\eta \exp(-2i\psi_S) & \omega_L + \omega_0 + i\eta^* \end{bmatrix} \tag{7.9}$$

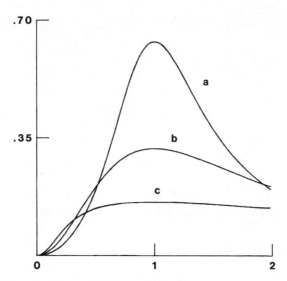

Figure 5. Absorption line as a function of ω_L, and divided by $I_L B(\bar{n}_1 - \bar{n}_2)\hbar$. Frequencies are in units of ω_0. The parameters for these curves are $\mathrm{Im}\,\eta = 0$ and $\mathrm{Re}\,\eta = 0.5$, 1, 2 for a, b, c, respectively. For decreasing $\mathrm{Re}\,\eta$ (approximate linewidth), the profile tends to a Lorentzian with half-width at half maximum equal to $\mathrm{Re}\,\eta$. For $\mathrm{Im}\,\eta = 0$ maximum is always situated at $\omega_L = \omega_0$.

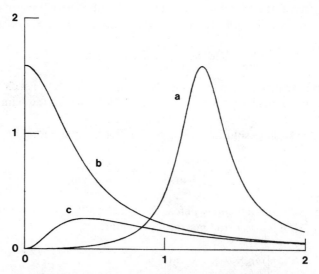

Figure 6. Same as Figure 5 but now with $\mathrm{Re}\,\eta = 0.2$ for all three curves. The imaginary parts of η, the Lamb shift, are $+0.3$, -0.5, -0.6 for a, b and c, respectively. For $\mathrm{Im}\,\eta$ small, the line around $\omega_L = \omega_0$ only shifts over a distance $\mathrm{Im}\,\eta$, but for larger values the lineshape changes dramatically. Top of the curve is always at $\omega_L = \omega_0(|1 + 2\,\mathrm{Im}\,\eta/\omega_0|)^{1/2}$, which is independent of $\mathrm{Re}\,\eta$. For $\mathrm{Im}\,\eta$ small in comparison with ω_0 the top appears at $\omega_L \simeq \omega_0 + \mathrm{Im}\,\eta$.

on the basis $|2\rangle\langle 2|, |1\rangle\langle 1|, |2\rangle\langle 1|$, and $|1\rangle\langle 2|$. The inverse of this matrix is then substituted into Eq. 6.9. Matrix elements of the dipole moment operator are not necessarily real. We write

$$\langle 2|\mu|1\rangle = |\langle 2|\mu|1\rangle|\exp(i\psi_\mu),\tag{7.10}$$

and now we choose the phase of the wavefunction $|1\rangle$ in such a way that $\psi_\mu = \psi_S$. Combining everything then gives, for the absorption line,

$$I(\omega_L) = I_L B(\bar{n}_1 - \bar{n}_2)\hbar\omega_L \frac{1}{\pi} \frac{4\omega_0\omega_L \operatorname{Re}\eta}{(\omega_0^2 - \omega_L^2 + 2\omega_0 \operatorname{Im}\eta)^2 + 4\omega_L^2(\operatorname{Re}\eta)^2},\tag{7.11}$$

where the Einstein coefficient B for stimulated transitions between $|1\rangle$ and $|2\rangle$ is defined as

$$B = \pi(\varepsilon_0\hbar^2 c')^{-1}|\langle 2|\mu|1\rangle \mathbf{e}_z \cdot \mathbf{\varepsilon}_L|^2.\tag{7.12}$$

Due to the fact the $\operatorname{Re}\eta$ and $\operatorname{Im}\eta$ are not necessarily small in comparison with ω_0, the line is not a simple Lorentzian. A few examples of $I(\omega_L)$ are drawn in Figures 5 and 6.

VIII. SECULAR APPROXIMATION

A great simplification arises if the relaxation constants c_{klmn} are small in comparison with the transition frequencies Δ_{kl} of the free evolution of the adbond. In spectral terms this means that the widths and shifts of lines are small in comparison with their central frequency and so $|\eta| \ll \omega_0$ in the notation of the previous section. Then the Liouvillian L_a will dominate the time evolution of $\rho_0(t)$ in Eq. 3.17, and the coupling between eigenvectors of L_a with different eigenvalues can be neglected. Every frequency Δ_{kl} gives rise to a single line, and the time regression of the correlation function no longer couples between different lines. In this approximation, every line evolves in a secular fashion, which explains the origin of the name *secular* for this limiting approximation.

Mathematically, this implies that we can decouple the time evolution of the coherences $\langle k|\rho_0(t)|l\rangle$, $k \neq l$, from the equation for the populations $n_k(t) = \langle k|\rho_0(t)|k\rangle$, since the free evolution of $\langle k|\rho_0(t)|l\rangle$ is proportional to $\exp(-i\Delta_{kl}t)$, whereas $n_k(t)$ evolves with eigenvalue zero ($\Delta_{kk} = 0$). From the matrix representation of Section V it then follows that the coherences decay exponentially to zero and the time evolution of the populations is governed by the master equation

$$\frac{d}{dt}n_k = \sum_l (n_l a_{lk} - n_k a_{kl}),$$
(8.1)

where only the relaxation constants a_{kl} from Eq. 7.5 appear. Equation 8.1 is a simple gain–loss balance for the population of level $|k\rangle$, and we can interpret the term $n_l a_{lk}$ as the rate of transitions from $|l\rangle$ to $|k\rangle$ due to single-phonon emission into the crystal ($\omega_l > \omega_k$) or energy absorption from the solid ($\omega_l < \omega_k$). The set of differential equations 8.1 should be accompanied by the constraint

$$\sum_k n_k = 1,$$
(8.2)

which expresses conservation of trace.

From Eq. 8.1 it follows that the decay constant for level $|k\rangle$ can be written as

$$A_k = \sum_l a_{kl},$$
(8.3)

which turns the master equation into

$$\frac{d}{dt}n_k = -A_k n_k + \sum_l n_l a_{lk}.$$
(8.4)

We notice that in the secular limit, the relaxation constants, which determine the level populations, are real. The coherences, however, still acquire contributions from the imaginary parts of the c_{klmn}'s, which amounts to the line shifts. An advantage of the secular limit is that the imaginary parts can be accounted for by attributing an effective level shift to each frequency ω_k, according to

$$\tilde{\omega}_k = \omega_k + \frac{1}{2}\sum_l \mathrm{Im}\, c_{kllk}.$$
(8.5)

Then we consider this transformation done and suppress the tilde henceforth. For further use, we remark that in the secular limit the relaxation operator can be expressed entirely in projection operators. We obtain

$$\Gamma\rho = \frac{1}{2}\sum_{kl} a_{kl}(P_k\rho + \rho P_k - 2P_l\,\mathrm{Tr}_a\,P_k\rho),$$
(8.6)

as can be checked by inspection.

Obviously, the secular approximation is not suitable for the evaluation of the details of a lineshape, since it turns every line into a Lorentzian. In the remainder of this chapter we shall consider the effect of irradiation with a strong laser. Then the dominant features of this system will be determined by the driving field rather than by the minor details of the coupling to the phonon field. Hence, we can adopt the representation 8.6 for the relaxation operator from here on.

IX. COHERENT EXCITATION

Probing the adsorbate with a weak laser does not alter the dynamics of the system. In a different application of lasers in these configurations, one deliberately tries to modify or affect the behavior of the system by driving a particular transition with a strong, resonant field. The coherent nature of a laser field (well-defined phase in its time evolution) provides essentially a different excitation mechanism than a coupling to a thermal bath, similar to the phonon reservoir. Along with the fact that the laser power can be very high, this then opens the possibility to drive the system away from thermal equilibrium. The presence of the radiation will tend to maintain a certain distribution of the population over the different levels, which has to compete with the thermal relaxation. Therefore, it can be anticipated that it should be feasible to actually change the dynamics of the adbond if the laser is sufficiently intense. As mentioned in the introduction, the first goal has been to enhance the desorption by resonant excitation of the adbond.

Radiative transitions between vibrational adsorbate states have been studied extensively, both in the weak-field case (28–32) and for strong fields (6, 33–35). In this chapter we summarize and extend our own approach (36–38), which is valid for arbitrary laser power and includes the effect of the laser linewidth. We shall again work with a Liouville notation, which allows a concise formulation. Suppose that the laser frequency is in close resonance with a single transition of the adbond only and that it couples a ground state $|g\rangle$ (not necessarily the lowest state of the adbond) with an excited state $|e\rangle$, which are separated by

$$\omega_0 = \omega_e - \omega_g > 0. \tag{9.1}$$

The idea is that $|e\rangle$ is one of the high-lying states in the potential well, which is very unlikely to be populated by thermal excitation. Then this configuration automatically excludes resonant coupling between other levels. Radiative transitions only occur between $|e\rangle$ and $|g\rangle$, but the thermal coupling remains present between all levels.

The electric component of the laser field at the position of the adbond is given by

$$E(t) = E_0 \operatorname{Re} \varepsilon_L \exp\{-i[\omega_L t + \phi(t)]\}, \qquad (9.2)$$

where $\phi(t)$ is a real-valued stochastic process responsible for the broadening of the laser line around its central frequency ω_L. It can be shown (39) that the response of the system is quite insensitive for the stochastic details of $\phi(t)$, provided that we restrict the description to single-mode fields. We shall take $\phi(t)$ as the independent-increment process (40), which covers the more familiar Gaussian white-noise and random-jump processes as special cases. The multiplicative stochastic differential equation for this diffusion process has been solved in the Appendix of Ref. 41, where it turns out that the spectral profile of the laser is a Lorentzian, whose half-width at half maximum is denoted by λ.

The dipole coupling between the adbond and the laser field, Eq. 6.1, now attains the explicit form

$$-\boldsymbol{\mu}\cdot\mathbf{E}(t) = -\tfrac{1}{2}\hbar\Omega|e\rangle\langle g|\exp\{-i[\omega_L t + \phi(t)]\} + \text{Hermitian conjugate}, \qquad (9.3)$$

where the usual rotating-wave approximation (42) has been made and permanent dipole moments have been omitted. The strength of the coupling is determined by the Rabi frequency

$$\Omega = \hbar^{-1}E_0|\langle e|\boldsymbol{\mu}\cdot\varepsilon_L|g\rangle|. \qquad (9.4)$$

Then the equation of motion for the reduced-density operator becomes

$$i\hbar\frac{d}{dt}\rho_0(t) = [H_a - \boldsymbol{\mu}\cdot\mathbf{E}(t), \rho_0(t)] - i\hbar\Gamma\rho_0(t), \qquad (9.5)$$

which now contains the interaction with the external field explicitly.

A convenient way to eliminate the fast oscillations with the optical frequency ω_L from the Hamiltonian is by the introduction of the stochastic unitary transformation (43)

$$\sigma(t) = \exp\{-i[\omega_L t + \phi(t)]L_g\}\rho_0(t), \qquad (9.6)$$

which contains the Liouvillian

$$L_g\rho = [P_g, \rho]. \qquad (9.7)$$

It is easy to check that the transformation only affects the coherences, so that we have

$$\langle k|\sigma(t)|k\rangle = n_k(t) \tag{9.8}$$

for every level $|k\rangle$. With some algebra, we find the transformed equation of motion to be

$$i\frac{d}{dt}\sigma(t) = [L_d + \dot{\phi}(t)L_g - i\Gamma]\sigma(t), \tag{9.9}$$

with $L_d\rho = \hbar^{-1}[H_d, \rho]$ and

$$H_d = H_a + \hbar\omega_L P_g - \tfrac{1}{2}\hbar\Omega(|e\rangle\langle g| + |g\rangle\langle e|) \tag{9.10}$$

the dressed-adbond Hamiltonian. This H_d has the significance of representing the free adbond (H_a), the free evolution of the single laser mode and their dipole coupling. Phonon transitions Γ and phase fluctuations $\dot{\phi}(t)L_g$ couple the eigenstates of H_d, which both give rise to relaxation of $\sigma(t)$.

The appearance of the time derivative $\dot{\phi}(t)$ of the stochastically fluctuating phase turns the equation of motion for $\sigma(t)$ into a multiplicative stochastic differential equation and the density operator $\sigma(t)$ into a stochastic process. Only the average over many realizations of the process $\phi(t)$ can have relevance, and we write

$$\Pi(t) = \{\sigma(t)\}, \tag{9.11}$$

where the brackets $\{\cdots\}$ indicate the average. Then Eq. 9.9 is easily solved for its average, and we obtain the equation of motion for $\Pi(t)$,

$$i\frac{d}{dt}\Pi(t) = (L_d - iW - i\Gamma)\Pi(t), \tag{9.12}$$

where the effective relaxation operator, which accounts for the finite laser linewidth, is given by

$$W = \lambda L_g^2. \tag{9.13}$$

This operator can alternatively be represented as

$$W\rho = \lambda(P_g\rho + \rho P_g - 2P_g Tr_a P_g\rho), \tag{9.14}$$

which is reminiscent of the structure of Γ, Eq. 8.6. In Eq. 9.12 we can incorporate the effect of the laser linewidth by the substitution $a_{gg} \to a_{gg} + 2\lambda$ in the definition of Γ, although in general this is not the correct procedure, as we will see in due course.

X.   DRESSED STATES

Both the laser linewidth and the coupling to the phonon reservoir give rise to a damping of the free evolution of the dressed adbond, which is represented by the Liouvillian L_d in Eq. 9.12. This L_d is the analog of L_a from Eq. 3.17, pertaining to a field-free system. In order to illuminate the physical picture, we diagonalize the Hamiltonian H_d. First we rewrite Eq. 9.10 as

$$H_d = \sum_{i \neq e,g} \hbar\omega_i P_i + \tfrac{1}{2}\hbar(\omega_e + \omega_g + \omega_L)(P_e + P_g)$$
$$- \tfrac{1}{2}\hbar\Delta(P_e - P_g) - \tfrac{1}{2}\hbar\Omega(|e\rangle\langle g| + |g\rangle\langle e|), \tag{10.1}$$

where $\Delta = \omega_L - \omega_0$ is the detuning from resonance. Diagonalization of H_d is now trivial, and we find the eigenvalue equations to be

$$H_d|k\rangle = \hbar\omega_k|k\rangle, \qquad k \neq e,g, \tag{10.2}$$
$$H_d|\pm\rangle = \tfrac{1}{2}\hbar(\omega_e + \omega_g + \omega_L \mp \Omega')|\pm\rangle, \tag{10.3}$$

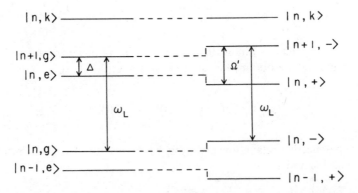

Figure 7.   Diagonalization of the dressed-adbond Hamiltonian. The diagram on the left represents product states of the adbond $|e\rangle, |g\rangle, |k\rangle, \ldots$ and the free evolving laser $|n\rangle$, with $n = 0, 1, 2, \ldots$, the number of photons in the mode. States $|n\rangle$ and $|n + 1\rangle$ are separated by the photon frequency ω_L. In this diagram we have taken $\Delta = \omega_L - \omega_0 > 0$. Since the laser is assumed to be almost on resonance, we have ladder of doublets separated by the detuning Δ. Then the dipole interaction couples the states that form a doublet, but it does not couple two sets of states. A diagonalization in turn gives rise to right diagram, where the separation in the new doublets equals Ω', which is always larger than Δ.

in terms of the generalized Rabi frequency

$$\Omega' = \Delta(1 + \Omega^2/\Delta^2)^{1/2}. \tag{10.4}$$

Hence, the dressed states $|k\rangle$ with $k \neq e, g$ are equal to the adbond states $|k\rangle$, and they have the same eigenvalue, whereas the states $|e\rangle$ and $|g\rangle$ form the linear combination

$$|+\rangle = |g\rangle \sin\tfrac{1}{2}\Theta + |e\rangle \cos\tfrac{1}{2}\Theta, \tag{10.5}$$

$$|-\rangle = |g\rangle \cos\tfrac{1}{2}\Theta - |e\rangle \sin\tfrac{1}{2}\Theta, \tag{10.6}$$

which is parameterized with the angle

$$\Theta = \arctan(\Omega/\Delta). \tag{10.7}$$

The position of the dressed states with respect to the adbond states is illustrated in Figure 7.

XI. STEADY STATE

Solving Eq. 9.12 for a transient regime $[0, t]$ has little significance, since a preparation of a specific initial state $\Pi(0)$ is practically not feasible. Any state $\Pi(0)$, however, will relax to the same long-time solution $\bar{\Pi}$ on a time scale on the order of Γ^{-1}. Therefore, we consider in some more detail the equation

$$(L_d - iW - i\Gamma)\bar{\Pi} = 0. \tag{11.1}$$

Due to the presence of the driving field, the equations for the populations $\bar{\Pi}_k = \langle k|\bar{\Pi}|l\rangle$ will not necessarily decouple anymore from the equations for the coherences between different states. In general, all matrix elements of $\Pi(t)$ couple, but it turns out that in the steady state the coherences for $k \neq l$ vanish provided that $(k, l) \neq (e, g)$. Taking matrix elements of Eq. 11.1 shows that the relation between the populations only involves the coherence between the driven transition $|e\rangle - |g\rangle$, as could be expected. Using $\bar{\Pi}^\dagger = \bar{\Pi}$ gives the important relation between the real and imaginary parts,

$$(A_e + A_g + 2\lambda)\,\mathrm{Re}\,\langle e|\bar{\Pi}|g\rangle = -2\Delta\,\mathrm{Im}\,\langle e|\bar{\Pi}|g\rangle, \tag{11.2}$$

which enables one to combine the equations for $\bar{\Pi}_g$ and $\bar{\Pi}_e$ as

$$a_r(\bar{\Pi}_g - \bar{\Pi}_e) = \Omega\,\mathrm{Im}\,\langle e|\bar{\Pi}|g\rangle \tag{11.3}$$

in terms of the parameter

$$a_r = \tfrac{1}{2}\Omega^2 \frac{(1/2)(A_e + A_g) + \lambda}{[(1/2)(A_e + A_g) + \lambda]^2 + \Delta^2}.$$ (11.4)

With the aid of Eqs. 11.2 and 11.3, we can eliminate the coherence, which finally yields the following set of equations for the populations:

$$\sum_l \bar{\Pi}_l a_{lk} = A_k \bar{\Pi}_k, \qquad k \neq e, g,$$ (11.5)

$$\sum_l \bar{\Pi}_l a_{le} = A_e \bar{\Pi}_e - a_r(\bar{\Pi}_g - \bar{\Pi}_e),$$ (11.6)

$$\sum_l \bar{\Pi}_l a_{lg} = A_g \bar{\Pi}_g + a_r(\bar{\Pi}_g - \bar{\Pi}_e).$$ (11.7)

This set can be considered as a steady-state analog of the master equation 8.1. Inspection of Eqs. 11.5–11.7 shows that we can write the set alternatively as

$$\sum_l \bar{\Pi}_l a'_{lk} = \sum_l \bar{\Pi}_k a'_{kl},$$ (11.8)

where the primed constants are defined as

$$a'_{kl} = \begin{cases} a_{kl}, & k, l \neq e, g \text{ or } g, e, \\ a_{kl} + a_r, & k, l = e, g \text{ or } g, e. \end{cases}$$ (11.9)

In the same fashion as when we identified the quantity $\bar{\Pi}_k a_{kl}$ as the rate of transitions from $|k\rangle$ to $|l\rangle$ due to single-phonon processes, we can now interpret $\bar{\Pi}_e a_r$ and $\bar{\Pi}_g a_r$ as the rates of stimulated radiative transitions from $|e\rangle$ to $|g\rangle$ and from $|g\rangle$ to $|e\rangle$, respectively. We shall see in Section XIV that the optical transitions acquire contributions from both single-photon and multiphoton processes. Notice that the three optical parameters Ω, Δ, and λ only enter the equations for the populations through the combination 11.4 in a_r. Furthermore, the rate constant a_r is linear in the laser power ($\sim \Omega^2$), which implies that the number of radiative transitions increases indefinitely with an increasing incident intensity. Obviously, the net absorbed power by a single adsorbed atom should reach a saturation value in the limit of high irradiances. This will be shown in the next section.

XII. LASER HEATING

Radiative excitations of the adbond occur at a rate $a_r \bar{\Pi}_g$, whereas stimulated emissions of photons in the laser field accompanied by $|e\rangle \to |g\rangle$ transitions happen $a_r \bar{\Pi}_e$ number of times per unit of time. Balancing the rates gives an effective rate of $a_r(\bar{\Pi}_g - \bar{\Pi}_e)$ for the absorption. Every transition corresponds to an effective absorption of a laser photon, so that the power absorption should roughly be equal to $\hbar \omega_L a_r(\bar{\Pi}_g - \bar{\Pi}_e)$. Since the accumulated energy in the vibrational bond must be constant in the steady state, the absorbed power from the laser beam equals the power flow into the crystal. This process of laser heating of the substrate is entirely mediated by adsorbates because the crystal itself was assumed to be transparent.

In Section VI we evaluated the power absorption from a weak monochromatic incident field by applying the golden rule. Obviously, this approach fails for strong, finite-linewidth radiation, so that we must find another way to calculate the power absorption. This is accomplished by considering the work done on the dipole moment by the external field, which is formally given by (42)

$$P(t) = \mathbf{E}(t) \cdot \frac{d}{dt} \langle \boldsymbol{\mu}(t) \rangle. \tag{12.1}$$

Here, $\boldsymbol{\mu}(t) = \exp(iLt)\boldsymbol{\mu}$, and the angle brackets denote the quantum expectation value. Transformation to the Schrödinger picture gives

$$P(t) = \mathrm{Tr}_a[\mathbf{E}(t) \cdot \boldsymbol{\mu}] \frac{d\rho_0}{dt}, \tag{12.2}$$

which only involves the reduced density operator $\rho_0(t)$ of the adbond rather than the Liouvillian L in Eq. 12.1, which pertains to the entire system. With transformation 9.6 we go to the σ representation, which yields

$$P(t) = \hbar\Omega \left\{ \frac{d}{dt} \mathrm{Re} \langle e|\sigma|g \rangle + [\omega_L + \dot{\phi}(t)] \mathrm{Im} \langle e|\sigma|g \rangle \right\}. \tag{12.3}$$

Application of the equation of motion 9.9 for $\sigma(t)$ results in

$$P(t) = -\tfrac{1}{2}\hbar\Omega(A_e + A_g)\,\mathrm{Re}\langle e|\sigma|g\rangle + \hbar\Omega\omega_0\,\mathrm{Im}\langle e|\sigma|g\rangle, \tag{12.4}$$

which eliminates both the time derivative and $\dot{\phi}(t)$ from expression 12.3. Equation 12.4 clearly exhibits that power absorption from an external field is reflected in the appearance of coherences in the system.

The quantity $P(t)$ depends stochastically on time, due to the fluctuations in the laser phase. If we define the steady-state power absorption by

$$I(\omega_L) = \lim_{t \to \infty} \{ P(t) \}, \tag{12.5}$$

we obtain $I(\omega_L)$ from Eq. 12.4 with the substitution $\sigma \to \bar{\Pi}$. Subsequently, we use Eqs. 11.2 and 11.3, which amounts to

$$I(\omega_L) = \hbar \left\{ \omega_L - \frac{2\lambda(\omega_L - \omega_0)}{A_e + A_g + 2\lambda} \right\} a_r (\bar{\Pi}_g - \bar{\Pi}_e). \tag{12.6}$$

For a monochromatic laser this reduces to $I(\omega_L) = \hbar \omega_L a_r (\bar{\Pi}_g - \bar{\Pi}_e)$, as anticipated in the beginning of this section. With increasing laser linewidth λ, the factor in curly brackets tends to ω_0, which reflects that for λ large, the frequency ω_L loses its significance. Then photons are considered to be absorbed in an $|e\rangle \to |g\rangle$ transition, which corresponds to an excitation of the system with energy $\hbar \omega_0$. Furthermore, we remark that the effect of the laser linewidth in Eq. 12.6 cannot be incorporated by the simple substitution $a_{gg} \to a_{gg} + 2\lambda$, as was the case for the equation of motion for $\Pi(t)$.

With Eq. 11.6 we can cast expression 12.6 in the form

$$I(\omega_L) = \hbar \omega_L' \left(A_e \bar{\Pi}_e - \sum_l \bar{\Pi}_l a_{le} \right), \tag{12.7}$$

where ω_L' is an abbreviation for the term in curly brackets in Eq. 12.6. Representation 12.7 of the power absorption only involves phonon relaxation constants and not a_r anymore. From the restriction $0 \leqslant \bar{\Pi}_k \leqslant 1$, which holds for every population, we immediately deduce the upper limit

$$I(\omega_L) \leqslant \hbar \omega_L' A_e \tag{12.8}$$

for the laser heating of the crystal. Although the rate constant a_r for stimulated transitions can become arbitrarily large, the net power absorption exhibits a saturation, where the upper limit is set by the phonon rate constants rather than by optical parameters. This can be understood from the fact that an absorbed photon can only be converted to thermal energy through a phonon transition.

XIII.   TRANSITIONS BETWEEN DRESSED STATES

In Section VIII we adopted the secular approximation for the phonon transition operator Γ with the argument that the time evolution of $\rho_0(t)$ is dominated by the free-evolution Liouvillian L_a of the adbond. This gave rise to the identification of the rate constants a_{kl} for the transitions $|k\rangle \rightarrow |l\rangle$, which occur at a rate $n_k(t)a_{kl}$. For the strongly driven adbond, however, the time evolution of the system is governed by Eq. 9.12 for $\Pi(t)$, which contains the free evolution of the dressed adbond L_d as the dominant part. Hence, it would seem to be more appropriate to define the secular approximation with respect to the dressed states rather than to the field-free states. Essentially, we should start with Eq. 2.9, include the interaction $-\boldsymbol{\mu}\cdot\mathbf{E}(t)$ with the laser field in the Hamiltonian, transform to the σ picture, and incorporate the phonon coupling. This procedure gives an expression for Γ in the presence of a driving laser in which we can subsequently drop the nonsecular terms with respect to dressed states. Elsewhere (36) we executed this laborious scheme, and the results are (almost) identical to the simplified derivation we now give here.

In order to achieve an expression for Γ with respect to dressed states, we merely have to express the projectors P_e and P_g in Eq. 8.6 in dressed-states basis functions $|+\rangle$ and $|-\rangle$. From Eqs. 10.5 and 10.6 we readily find

$$P_e = g_- \hat{P}_+ + g_+ \hat{P}_- - g_0(|+\rangle\langle -| + |-\rangle\langle +|), \tag{13.1}$$

$$P_g = g_+ \hat{P}_+ + g_- \hat{P}_- + g_0(|+\rangle\langle -| + |-\rangle\langle +|), \tag{13.2}$$

where we have introduced the abbreviations

$$g_- = \cos^2 \tfrac{1}{2}\Theta, \qquad g_+ = \sin^2 \tfrac{1}{2}\Theta, \qquad g_0 = \cos\tfrac{1}{2}\Theta \sin\tfrac{1}{2}\Theta, \tag{13.3}$$

and the projectors $\hat{P}_\pm = |\pm\rangle\langle\pm|$ onto the dressed states. Next we substitute the expansions 13.1 and 13.2 into Eq. 8.6 and drop the nonsecular terms. Care should be exercised, however, because a $|+\rangle \rightarrow |+\rangle$ transition between two doublets (Figure 7) is in exact resonance with a $|-\rangle \rightarrow |-\rangle$ transition, and therefore, couplings between these transitions should be retained. Combining everything then results in

$$\Gamma\rho = \frac{1}{2}\sum_{kl} \hat{a}_{kl}(\hat{P}_k\rho + \rho\hat{P}_k - 2\hat{P}_l\,\mathrm{Tr}_a\,\hat{P}_k\rho)$$
$$+ g_0^2(a_{eg} - a_{ee} + a_{ge} - a_{gg})(\hat{P}_+\rho\hat{P}_- + \hat{P}_-\rho\hat{P}_+), \tag{13.4}$$

where the term proportional to g_0^2 comes from the mentioned degeneracy, and $\hat{P}_k$ denotes a projector onto a dressed state. Representation 13.4 gives rise to a master equation with respect to dressed states, which implies that we can interpret the parameters $\hat{a}_{kl}$ as the rate constants for transitions between dressed states.

Of course, the parameters $\hat{a}_{kl}$ can be related to the relaxation constants a_{kl}

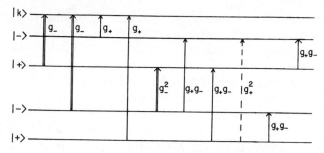

Figure 8. Single-phonon excitations of the laser adbond system. The diagrams with the arrows in the opposite direction (not shown here) correspond to phonon emission into the crystal. The three diagrams with the double arrows persist in the absence of the laser and correspond to radiationless transitions between adbond states. Solid arrows have rate constants proportional to g_+, so they represent single-photon processes. For strong fields the g_+^2 transition appears, which couples the two doublets by a two-photon process. The rightmost diagrams give transitions in a single doublet via a single-photon process. Their rate constants are proportional to the diagonal matrix elements of the derivative of the binding potential, which are small. For zero temperature these transitions vanish identically, since then $a_{ee} = a_{gg} = 0$.

with respect to the adbond states. For $k \neq e, g$ and $l \neq e, g$, we obviously have $\hat{a}_{kl} = a_{kl}$. If one of the states $|k\rangle$, $|l\rangle$ equals a $|+\rangle$ or $|-\rangle$ state, we find

$$\hat{a}_{k\pm} = g_{\mp} a_{ke} + g_{\pm} a_{kg}, \tag{13.5}$$

$$\hat{a}_{\pm k} = g_{\mp} a_{ek} + g_{\pm} a_{gk}, \tag{13.6}$$

in terms of the optical parameters $g_{\pm}$. Transitions between the $|+\rangle$ and $|-\rangle$ states are governed by the rate constants

$$\hat{a}_{\pm\mp} = g_{\pm}^2 a_{ge} + g_{\mp}^2 a_{eg} + g_0^2(a_{ee} + a_{gg}), \tag{13.7}$$

$$\hat{a}_{\pm\pm} = g_0^2(a_{ge} + a_{eg}) + g_{\mp}^2 a_{ee} + g_{\pm}^2 a_{gg}. \tag{13.8}$$

Interaction with the phonon field is now regarded as the occurrence of single-phonon transitions between dressed states. Relaxation constants that connect an adbond state $|k\rangle \neq |e\rangle$ or $|g\rangle$ with a dressed state $|+\rangle$ or $|-\rangle$ acquire a contribution from two distinct processes due to the fact that $|k\rangle$ couples with both the doublets in Figure 7. In order to find which term corresponds to which transitions, we recall the definition 7.5 of the parameter a_{kl}. We notice that the reservoir correlation function $\tilde{G}$ is evaluated at Δ_{kl}, which equals the frequency of the phonon for that particular transition. Therefore, the terms with a_{ek}, a_{ke} represent transitions from and to the upper doublet, respectively, whereas the terms with a_{gk}, a_{kg} describe single-phonon transitions between $|k\rangle$

and the lower states. In a similar way we can interpret the various terms in Eqs. 13.7 and 13.8. In Figure 8 the different processes are indicated by arrows, and the accompanying optical factor specifies the term in Eqs. 13.5–13.8.

XIV. PHOTON–PHONON CONVERSION

Stimulated radiative transitions between adbond states are incorporated in the diagonalization of $H_a - \mathbf{\mu} \cdot \mathbf{E}$, and therefore, only single-phonon transitions persist in a pictorial representation of the various processes with respect to the dressed states. In order to elucidate the mechanism of photon absorption and to establish the relation with the process of laser heating, we consider the limit of (relatively) low laser power. To this end, we first note that $g_\pm$ can be expressed in Ω and Δ according to

$$g_\pm = \tfrac{1}{2}[1 \mp (1 + \Omega^2/\Delta^2)^{-1/2}], \qquad (14.1)$$

which depends only on the optical parameters through the combination Ω^2/Δ^2. For weak fields we then obtain

$$g_- \simeq 1 - \Omega^2/4\Delta^2, \qquad g_+ \simeq \Omega^2/4\Delta^2, \qquad (14.2)$$

which shows that g_- remains present without a radiation field. Rate constants proportional to g_- must consequently correspond to radiationless transitions. On the other hand, the factor g_+ is proportional to the laser power, which implies that every factor g_+ in a $\hat{a}_{kl}$ corresponds to the absorption or emission of a photon. In this fashion we can track down the significance of the optical

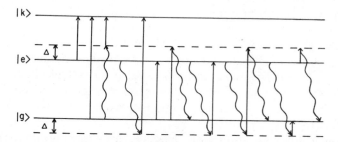

Figure 9. Low-intensity interpretation of diagrams from Figure 8. With respect to the adbond states, a single-phonon process goes together with photon absorptions and emissions in such a way that the resulting diagram is energy conserving. The sequence of processes in a single diagram can only give rise to transitions between the real states $|e\rangle, |g\rangle, |k\rangle, \ldots$, but intermediate states can be virtual in this picture. The diagrams are in the same order as in Figure 8, and for the phonon emission processes we simply reverse the directions of all arrows.

factors in the relaxation constants with respect to dressed states. In Figure 9 we draw the diagrams with respect to the adbond states, and they are in the same order as the corresponding diagrams of Figure 8. All transitions with a rate constant proportional to $g_0^2 = g_- g_+$ are single-photon processes.

Now it should be obvious how the process of laser heating of the crystal can be conceived as a result of many photon–phonon conversion reactions. Every phonon in the diagram corresponds to an energy exchange between the adbond and the crystal, whereas a photon transition amounts to an energy transfer between the laser field and the adbond. A combined photon–phonon diagram therefore represents effectively an exchange of energy between the laser and the crystal, a process mediated by the adsorbate. Summation of the contributions from all diagrams, weighted with the probabilities for the diagrams to occur, then gives the net power absorption by the crystal.

XV. SUMMARY

We have studied theoretically the optical properties of an adsorbed atom in a vibrational bond on a crystal. Coupling of the adbond states with the phonon field of the substrate have been assumed to be brought about by single-phonon transitions only, which is the main approximation in the presented theory. It is straightforward, however, to include higher order processes, especially when the interaction is taken to be a Morse potential. Then all matrix elements can be evaluated analytically, and a system with an arbitrary number of adbond states can be parameterized with the depth, width, and position of the binding potential. We have chosen to restrict the approach to single-phonon transitions, which allows a clear interpretation of the coupling mechanism. Then we can incorporate the interaction with the crystal with a single-phonon field amplitude correlation function $G(\tau)$, which was studied in detail in Section IV. It is the behavior of this crystal response function that determines whether the reservoir approach to thermal relaxation can be justified. In particular, the decay of $G(\tau)$ must be sufficiently fast, in comparison with the adbond relaxation rates, in order to impose a Markovian time evolution on the reduced adbond density operator. This decay time is typically on the order of the inverse Debye frequency, which is reasonably small in comparison with the rate constants for a single-phonon process (approximately one order of magnitude).

Subsequently, the absorption lineshape has been evaluated showing considerable deviation from a Lorentzian in certain cases. The origin of the distortions and the shifts, other than the common Lamb shift, has been tracked down to the presence of nonsecular terms in the time evolution of the density operator, which in turn appear as a consequence of the fact that the damping constants are not necessarily small in comparison with the transition frequencies.

Next, we have considered the irradiation of the adsorbate by an intense nonmonochromatic laser. The laser linewidth is assumed to arise from a stochastically fluctuating diffusive phase of the driving field. Diagonalization of the Hamiltonian is accomplished by a stochastic transformation, which yields the dressed states. These states can be interpreted as the joint eigenstates of the adbond, the single-mode laser, and their interaction. We have analyzed the equation of motion for this system and discussed some properties of the steady state. In particular, the coherence of the driven transition does not disappear in the long-time limit, which indicates that the system is not in thermal equilibrium.

Then the power absorption has been obtained from the work done on the dipole by the external field. The finite bandwidth of the laser gives rise to an effective photon energy $\hbar\omega'_L$, with ω'_L in between ω_0 and ω_L. Then the power absorption can be written as $I(\omega_L) = \hbar\omega'_L a_r(\bar{\Pi}_g - \bar{\Pi}_e)$, with a_r the rate of optical transitions. It has been shown that a_r is linear in the laser intensity, which implies that $\bar{\Pi}_g - \bar{\Pi}_e$ tends to zero in the high-intensity limit, since the absorption rate must remain finite. Stimulated transitions, which occur at a rate a_r (Eq. 11.4) are reduced by an increasing linewidth if the system is driven close to resonance ($\Delta \simeq 0$), whereas far off resonance the linewidth enhances the number of transitions ($a_r \sim \lambda$). This feature is easily understood: In the case $\Delta \simeq 0$ most laser photons are farther off resonance than for monochromatic incident radiation. On the other hand, for a large detuning there is still a considerable amount of photons in close resonance, which have a large probability to be absorbed.

As a last issue, we have derived the rate constants for single-phonon transitions between dressed states. In the low-intensity limit, these processes can be interpreted as single-phonon–multiphoton processes in adbond state diagrams, where "multi" means zero, one, or two. This reveals that photoabsorption is inevitably accompanied by a downward phonon transition, which gives rise to heating of the crystal. Light-induced desorption through resonant excitation of a high-lying state is consequently bound to have an extremely low efficiency.

Acknowledgments

This research was supported by the Air Force Office of Scientific Research (AFSC), United State Air Force, under Contrast F49620-86-C-0009, the Office of Navel Research and the National Science Foundation under Grant CHE-8620274. The United States government is authorized to reproduce and distribute reprints for governmental purposes notwithstanding any copyright notation hereon.

References

1. G. Nienhuis, *Physica* **66**, 245 (1973).

2. R. B. Hall and S. J. Bares, in *Chemistry and Structure at Interfaces*, R. B. Hall and A. B. Elles, eds., VCH Publishers, Deerfield Beach, FL, 1986, pp. 83–149.

3. E. J. Heilweil, M. P. Casassa, R. R. Cavanagh, and J. C. Stephenson, *J. Chem. Phys.* **82**, 5216 (1985).

4. S. Chiang, R. G. Tobin, P. L. Richards, and P. A. Thiel, *Phys. Rev. Lett.* **52**, 648 (1984).

5. M. S. Djidjoev, R. V. Khokhlov, A. V. Kiselev, V. I. Lygin, V. A. Namiot, A. I. Osipov, V. I. Puchenko, and B. I. Provotorov, in *Tunable Lasers and Applications*, A. Mooradian, T. Jaeger, and P. Stokseth, eds., Springer, Berlin, 1976, pp. 100–107.

6. M. S. Slutsky and T. F. George, *Chem. Phys. Lett.* **57**, 474 (1978).

7. M. S. Slutsky and T. F. George, *J. Chem. Phys.* **70**, 1231 (1979).

8. J. Heidberg, H. Stein, and E. Riehl, *Phys. Rev. Lett.* **49**, 666 (1982).

9. T. J. Chuang, *J. Chem. Phys.* **76**, 3828 (1982).

10. T. J. Chuang and H. Seki, *Phys. Rev. Lett.* **49**, 382 (1982).

11. J. Heidberg and I. Hussla, *J. Electron. Spectrosc. Rel. Phen.* **29**, 105 (1983).

12. I. Hussla, H. Seki, T. J. Chuang, Z. W. Gortel, H. J. Kreuzer, and P. Piercy, *Phys. Rev. B.* **32**, 3489 (1985).

13. K. Veeken, P. A. M. van der Heide, L. M. ten Dam, A. R. de Vroomen and J. Reuss, *Surf. Sci.* **166**, 1 (1986).

14. T. J. Chuang, *Surf. Sci. Rep.* **3**, 1 (1983).

15. Z. W. Gortel, H. J. Kreuzer, and R. Teshima, *Phys. Rev. B* **22**, 5655 (1980).

16. S. Efrima, K. F. Freed, C. Jedrzejek, and H. Metiu, *Chem. Phys. Lett.* **74**, 43 (1980).

17. C. Jedrzejek, K. F. Freed, S. Efrima, and H. Metiu, *Chem. Phys. Lett.* **79**, 227 (1981).

18. Z. W. Gortel, H. J. Kreuzer, R. Teshima, and L. A. Turski, *Phys. Rev. B* **24**, 4456 (1981).

19. S. Efrima, C. Jedrzejek, K. F. Freed, E. Hood, and H. Metiu, *J. Chem. Phys.* **79**, 2436 (1983).

20. B. J. Garrison, D. J. Diestler, and S. A. Adelman, *J. Chem. Phys.* **67**, 4317 (1977).

21. A. A. Maradudin, E. W. Montroll, G. H. Weiss, and I. P. Ipatova, *Theory of Lattice Dynamics in the Harmonic Approximation*, Solid State Physics, 2nd ed., Supplement 3, Academic, New York, 1971.

22. R. W. Zwanzig, in *Lectures in Theoretical Physics*, Vol. III, W. E. Britten, B. Downs, and J. Downs, eds., Interscience, New York, 1961, p. 106 ff.

23. W. H. Louisell, *Quantum Statistical Properties of Radiation*, Wiley, New York, 1973, Chapter 6.

24. C. Cohen-Tannoudji, in *Frontiers in Laser Spectroscopy, Proceedings of the Twenty-Seventh Les Houches Summer School*, R. Balian, S. Haroche, and S. Liberman, eds., North-Holland, Amsterdam, 1977.

25. S. van Smaalen and T. F. George, *J. Chem. Phys.* **87**, 5504 (1987).

26. A. Ben-Reuven, *Adv. Chem. Phys.* **33**, 235 (1975).

27. M. Lax, *Phys. Rev.* **172**, 350 (1968).

28. G. Korzeniewski, E. Hood, and H. Metiu, *J. Vac. Sci. Technol.* **20**, 594 (1982).

29. J. Lin, X. Y. Huang, and T. F. George, *Z. Phys. B* **48**, 355 (1982).

30. Z. W. Gortel, H. J. Kreuzer, P. Piercy, and R. Teshima, *Phys. Rev. B* **27**, 5066 (1983).

31. X. Y. Huang, T. F. George, and J. M. Yuan, *J. Opt. Soc. Am. B* **2**, 985 (1985).

32. B. Fain and S. H. Lin, *Physica* **138B**, 63 (1986).

33. J. Lin and T. F. George, *J. Phys. Chem.* **84**, 2957 (1980).

34. J. Lin, X. Y. Huang, and T. F. George, *J. Vac. Sci. Technol. B* **3**, 1525 (1985).

35. A. C. Beri and T. F. George, *J. Vac. Sci. Technol. B* **3**, 1529 (1985).

36. H. F. Arnoldus, S. van Smaalen, and T. F. George, *Phys. Rev. B* **34**, 6902 (1986).
37. S. van Smaalen, H. F. Arnoldus, and T. F. George, *Phys. Rev. B*, **35**, 1142 (1987).
38. H. F. Arnoldus and T. F. George, *J. Opt. Soc. Am. B*, **4**, 195 (1987).
39. H. F. Arnoldus and G. Nienhuis, *J. Phys. B: At. Mol. Phys.* **19**, 873 (1986).
40. N. G. van Kampen, *Stochastic Processes in Physics and Chemistry*, North-Holland, Amsterdam, 1981.
41. H. F. Arnoldus and G. Nienhuis, *J. Phys. B: At. Mol. Phys.* **16**, 2325 (1983).
42. L. Allen and J. H. Eberly, *Optical Resonance and Two-Level Atoms*, Wiley, New York, 1975.
43. G. S. Agarwal, *Phys. Rev. A* **18**, 1490 (1978).

CHAPTER XVI

TELEGRAPHIC ATOMIC FLUORESCENCE

RICHARD J. COOK

*Frank J. Seiler Research Laboratory, United States Air Force Academy,
Colorado Springs, Colorado 80840*

CONTENTS

I. INTRODUCTION

The science of spectroscopy began in the year 1812, when Joseph Fraunhofer first used a slit in conjunction with a prism and discovered the dark lines of the solar spectrum that now bear his name. From that time until about 1980 spectroscopists worked with samples containing large numbers of atoms or molecules. Even in the experiments on photon antibunching (1), where a single-atom resonance fluorescence effect is observed, the integrated fluorescence of many successive atoms is used to test the single-atom theory. Only recently has it become possible to trap a single atom (actually an atomic ion), to cool the atom so as to get rid of Doppler broadening, and to subject the atom to spectroscopic study. This was first achieved in 1980 by Neuhauser, Hohenstatt, Toschek, and Dehmelt (2) using ion-trapping techniques developed earlier by Dehmelt and collaborators (3). Although clouds of neutral sodium atoms have been trapped and cooled by Phillips et al. (4) and Chu et al. (5), the stable trapping of a single neutral atom has yet to be achieved.

We wish to emphasize that the ability to continuously observe the fluorescence of a single atom is important because a number of effects that can be observed in the fluorescence of a single atom (or ion) are totally masked when many atoms contribute to the fluorescence. That is, the fluorescence of a single atom contains information about the atomic dynamics that is not contained in the fluorescence of a many-atom sample. The telegraphic atomic fluorescence considered here is a prime example of this.

715

A second point worth noting is that single-atom fluorescence is not particularly difficult to observe. A strongly driven optical resonance transition radiates on the order of 10^8 photons per second. If a source of this strength is observed with the eye through a $10 \times$ magnifier, the visual brightness of the atomic source is comparable to the brightness of one of the prominent stars of the Big Dipper (constellation Ursa Major) as viewed with the unaided eye.

In the following section we shall discuss, in a qualitative manner, the physics of telegraphic fluorescence and its relation to quantum jumps. This is followed in Section III by a presentation of the theory of ion trapping and cooling, the technology that makes possible the spectroscopy of single atomic ions. In Section IV we present the quantitative theory of telegraphic atomic fluorescence. The chapter concludes in Section V with a review of experiments on telegraphic fluorescence with emphasis on their interpretation as monitors of quantum jumps.

II. PHYSICS OF TELEGRAPHIC FLUORESCENCE

Before attacking the mathematical theory of telegraphic fluorescence, which in some formulations is algebraically cumbersome, we shall give a purely qualitative discussion of the underlying physics. Telegraphic fluorescence is one of those happy phenomena for which the physics is really quite simple and can be understood without mathematics.

One of the reasons telegraphic fluorescence is interesting is that it differs qualitatively from our experience with many-atom samples. We are inclined to think that, for steady illumination, the fluorescent intensity of a spectroscopic sample is some function of the intensities and frequencies of the applied steady fields. All of our experience with many-atom samples indicates that steady illumination leads to steady fluorescence. (Of course, the energy scattered by one or more atoms comes in discrete quanta; but even for a single atom, the rate of photon emission in resonance fluorescence, 10^8 photons per second, is so high that the energy flux may be regarded as constant.) It is therefore somewhat surprising to hear that the fluorescence of a single atom can be intermittent or "telegraphic" when the atom is illuminated with steady laser beams.

To understand how single-atom fluorescence can be intermittent, consider an atom with a strong optical transition, $0 \leftrightarrow 1$, and a weak transition, $0 \leftrightarrow 2$, with a common lower level, as depicted in Figure 1. When we say that the transition $0 \leftrightarrow 1$ is strong, we mean that this is a dipole-allowed optical transition, so that the Einstein spontaneous emission coefficient for level 1, namely A_1, is on the order of 10^8 s^{-1}. When we say that the transition $0 \leftrightarrow 2$ is weak, we have in mind that level 2 is a metastable state with a lifetime of, say, 1 second or an A coefficient $A_2 \approx 1 \text{ s}^{-1}$. Now suppose the atomic electron starts

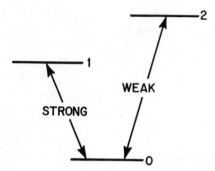

Figure 1. Energy level scheme for observing telegraphic atomic fluorescence.

in level 0. If the strong transition is driven by a strong resonant field and the weak transition is not driven, the electron is quickly promoted to level 1, from which it spontaneously returns to level 0 with the emission of a fluorescence photon. The fluorescence cycle $0 \rightarrow 1 \rightarrow 0$ repeats rapidly, and for a saturating field, the rate of fluorescence (photons per second) is $A_1/2 \sim 10^8$ s^{-1}. If we now apply radiation to the weak transition, the electron is occasionally promoted to level 2, and the strong fluorescence is interrupted because the electron is unavailable for transitions between levels 0 and 1. It is said that the electron is "shelved" in level 2. The fluorescence resumes when the electron returns to level 0 either by spontaneous or stimulated emission. Unless the weak transition is driven very hard, the electron remains in level 2 for a time, on the order of $\tau_2 = 1/A_2 \sim 1$ s, and during this time there is a "dark period" in the atomic fluorescence. Since the weak transitions occur randomly in time, the atomic fluorescence has the form of a random telegraph signal, as illustrated in Figure 2. In this way a single atom illuminated with steady laser beams switches its own fluorescence on and off. The atomic fluorescence is not steady

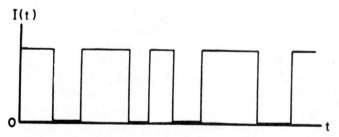

Figure 2. Single-atom fluorescent intensity versus time. Interruptions in the fluorescence are due to excitation of the weak transition $0 \rightarrow 2$.

for steady illumination because the fluorescence depends on the internal dynamics of the atom and not on the illumination alone.

The idea that single-atom fluorescence would abruptly turn off when the atom is shelved in a metastable level was first suggested by Hans Dehmelt (6) as a way to detect a weak transition in single-atom spectroscopy. Dehmelt (7) was interested in the possibility of using trapped and cooled atomic ions as a medium for a more accurate atomic clock. The principal source of inaccuracy of current atomic clocks is transit time broadening of the clock transition. Dehmelt pointed out that this source of error would be eliminated if the beams of current atomic clocks were replaced by stored ions. Each stored ion would continuously interact with the field, and in this way the transient interaction, which is characteristic of atomic beams and gives rise to transit time broadening, is totally eliminated. However, the ion–ion interactions in RF Paul traps cause the ion cloud to heat rapidly. It is therefore highly desirable to use a single atomic ion as the clock medium, since, with only one ion present, there is no ion–ion interaction and the RF heating is negligible. But with only a single ion present, it is difficult to detect the very narrow line profiles one must use for accurate frequency standards. A narrow line implies accurate frequency, but it also implies that the upper state of the active transition is long-lived, so that the natural line breadth is small. Thus, the problem reduces to the detection of very weak lines in a single atomic ion, lines that are too weak to be detected by the direct absorption or scattering of a resonant field. Here the shelving idea provides a way out of this difficulty. Rather than detect the weak line directly, one can infer absorption and emission on the weak transition $0 \leftrightarrow 2$ by its effect on the fluorescence of the strong transition $0 \leftrightarrow 1$, as described in the preceding. Dehmelt emphasized that this is a way to greatly amplify the weak signal one wishes to detect. The absorption of a single photon on the weak transition turns off the emission of 10^8 fluorescence photons per second. Similarly, the emission of a single photon on the weak transition turns on a fluorescent signal of this strength. For this reason Dehmelt refers to his scheme as an "atom amplifier."

It was emphasized by Cook and Kimble (8) that Dehmelt's shelving scheme provides a method for directly observing quantum jumps in a single atom. It is clear from the preceding analysis that the fluorescence is off when the atomic electron is shelved in level 2 and it is on when the electron is not in this level. That is, the excitation and de-excitation of the weak transition is in one-to-one correspondence with the switching on and switching off of the atomic fluorescence. Therefore, the atomic fluorescence emanating from the strong transition directly monitors the quantum jumps on the weak transition, and the detection of the jumps is essentially 100% efficient.

We have said that single-atom fluorescence contains information about the atomic dynamics, specifically the quantum jumps, that is not contained in the fluorescence of a many-atom sample. To see how this information is lost, consider a spectroscopic sample containing n atoms. If the fields radiated by the different atoms are not correlated, the total fluorescent intensity $I(t)$ is the sum of the intensities of the individual atoms, $I(t) = I_1(t) + I_2(t) + \cdots + I_n(t)$. When the sample is driven with two laser beams, in the manner described, each atom produces telegraphic fluorescence, but the switching times are different for the different atoms, and the various fluorescent signals are statistically independent. According to the central-limit theorem, the sum of a large number of statistically independent random variables is a Gaussian distributed random variable with mean equal to the sum of the means, $\bar{I} = n\bar{I}_1$, and variance equal to the sum of the variances, $\sigma_I^2 = n\sigma_1^2$. (Here $\bar{I}_1$ is the mean and σ_1^2 the variance of the single-atom fluorescent intensity.) It follows that the fractional fluctuation of the total fluorescent intensity $(\sigma_1/\bar{I})$ is $n^{-1/2}$ times that of a single atom. So if we have a sample of, say, 10^{20} atoms, the fractional fluctuation of the total fluorescence is smaller by the factor 10^{-10} than for a single atom, a value far too small to be observable. The intensity fluctuations produced by telegraphic fluorescence are observable only for samples containing a very small number of atoms. For one atom the fluorescence assumes one of two possible values: on or off. For two atoms, the fluorescent intensity can take on three values: no atoms on, one atom on and, two atoms on. The one and two atom cases have been demonstrated experimentally by Bergquist et al. (9).

III. ION TRAPPING AND COOLING

In this section we discuss how one traps and cools an atomic ion for the purpose of single-atom spectroscopy. Although it would perhaps be more desirable to trap neutral atoms, it is clear from the start that this is more difficult than the trapping of a charged particle because the electromagnetic forces one can bring to bear on a charged atom are much larger than the dipole forces one must use to manipulate a neutral atom. It is primarily for this reason that atomic ions (and electrons) were successfully trapped some years ago, whereas the trapping of neutral atoms has been demonstrated only recently, and the stable trapping of a single neutral atom has not yet been achieved.

Several methods exist for the trapping of charged particles by means of electric and/or magnetic fields. But if the purpose of the trapping is to perform high-resolution spectroscopy (e.g., for new frequency standards), the methods that make use of magnetic fields, such as Penning traps or magnetic bottles, are not ideal because of the rather large Zeeman distortion of the spectrum one wishes to measure. For this reason, trapping by an electric field alone is desirable. As we shall see, it is possible to trap an ion at a point where the electric field is zero, so that even the relatively small Stark distortion of the spectrum is eliminated.

When considering the possibility of electric field ion trapping, one immediately encounters Earnshaw's theorem (10), which states that there can be no point of stable equilibrium for a charge in an electrostatic field. So, although a charged particle can be stably trapped in a combination of static electric and magnetic fields, as in the Penning trap (11–13), an ion cannot be stably trapped in an electrostatic field alone.

However, it is possible to trap an ion in an oscillating electric field. In a rapidly oscillating field, the motion of the ion can be decomposed into a rapid oscillation of small amplitude at the driving frequency and a slow secular motion. In the following we analyze this motion and find that the secular or guiding center motion is derivable from a time-independent effective potential. The effective potential can have points of stable equilibrium even though the instantaneous electric field has no stable points. It is the effective potential that traps the ion in an RF Paul trap. It should perhaps be noted that with every time-dependent electric field there is an associated magnetic field. But for the low frequencies used in Paul traps, the Zeeman shifts produced by this magnetic field are entirely negligible.

The classical equation of motion for a particle of mass m in a sinusoidal potential

$$V(\mathbf{r}, t) = v(\mathbf{r}) \cos vt \tag{1}$$

with position-dependent amplitude $v(\mathbf{r})$ and frequency v is

$$m \frac{d^2 x^i}{dt^2} = -\frac{\partial V(\mathbf{r}, t)}{\partial x^i}, \tag{2}$$

where x^i, $i = 1, 2, 3$, are the components of the position vector $\mathbf{r}$. We have in mind that the potential V is the potential energy of a charge e in the scalar potential $\phi(\mathbf{r}, t)$ of the electric field [i.e., $V(\mathbf{r}, t) = e\phi(\mathbf{r}, t)$], where the electric field $\mathbf{E} = -\nabla \phi$ is produced by application of a sinusoidal voltage across the trap electrodes.

As already mentioned, the particle motion in a rapidly oscillating field may be decomposed into a component S^i that oscillates rapidly with small amplitude and a component R^i describing the slow secular motion

$$x^i = S^i + R^i. \tag{3}$$

On substituting this into the equation of motion 2, we have

$$m \left(\frac{d^2 S^i}{dt^2} + \frac{d^2 R^i}{dt^2} \right) = -\frac{\partial v(\mathbf{R} + \mathbf{S})}{\partial x^i} \cos vt. \tag{4}$$

Since the displacement $\mathbf{S}$ from the guiding center $\mathbf{R}$ is small, the right side of 4 may be expanded as a Taylor series in the components of **S**. Keeping terms to first order in S^i, we have

$$m\left(\frac{d^2 S^i}{dt^2} + \frac{d^2 R^i}{dt^2}\right) = -\frac{\partial v(\mathbf{R})}{\partial x^i}\cos vt - \frac{\partial^2 v(\mathbf{R})}{\partial x^i \partial x^j}S^j \cos vt. \tag{5}$$

Here we are using the Einstein summation convention according to which a repeated index implies summation on that index. The equality 5 may be satisfied by setting

$$m\frac{d^2 S^i}{dt^2} = -\frac{\partial v(\mathbf{R})}{\partial x^i}\cos vt, \tag{6}$$

and

$$m\frac{d^2 R^i}{dt^2} = -\frac{\partial^2 v(\mathbf{R})}{\partial x^i \partial x^j}S^j \cos vt. \tag{7}$$

Equation 6 states that the oscillatory motion is driven by the force the particle would experience at its guiding center, and 7 accounts for the remainder of Eq. 5. Integration of Eq. 6 twice with respect to time gives

$$S^i = \frac{\partial v(\mathbf{R})}{\partial x^i}\frac{\cos vt}{mv^2}. \tag{8}$$

The terms involving integration constants are secular terms rather than oscillatory ones and accordingly are set to zero. There is no loss of generality in doing this because one can still represent any initial conditions $x^i(0)$ and $\dot{x}^i(0)$ by choosing appropriate initial conditions $R^i(0)$ and $\dot{R}^i(0)$ for the secular motion. Using Eq. 8 in Eq. 7, we get

$$m\frac{d^2 R^i}{dt^2} = -\frac{\partial^2 v(\mathbf{R})}{\partial x^i \partial x^j}\frac{\partial v(\mathbf{R})}{\partial x^j}\frac{\cos^2 vt}{mv^2}. \tag{9}$$

Since the secular motion is slowly varying over the period $T = 2\pi/v$ of the rapidly oscillating field, we can replace the right side of 9 by its time average over this period. The time average of $\cos^2 vt$ is $\frac{1}{2}$. Hence,

$$m\frac{d^2 R^i}{dt^2} = -\frac{1}{2mv^2}\frac{\partial^2 v(\mathbf{R})}{\partial x^i \partial x^j}\frac{\partial v(\mathbf{R})}{\partial x^j}$$

$$= -\frac{\partial}{\partial x^i}\left(\frac{1}{4mv^2}\frac{\partial v(\mathbf{R})}{\partial x^j}\frac{\partial v(\mathbf{R})}{\partial x^j}\right). \tag{10}$$

The final form of this equation indicates that the guiding center motion is derivable from a time-independent effective potential:

$$m\frac{d^2\mathbf{R}}{dt^2} = -\nabla V_{\text{eff}}(\mathbf{R}), \tag{11}$$

$$V_{\text{eff}}(\mathbf{R}) = \frac{\nabla v(\mathbf{R})\cdot\nabla v(\mathbf{R})}{4mv^2}. \tag{12}$$

The important point is that even though the instantaneous potential $V(\mathbf{r}, t)$ has no points of stable equilibrium, the effective potential can have stable equilibrium points. For example, in the Paul trap, one uses the arrangement with two cap electrodes and a ring electrode illustrated in Figure 3 and applies a sinusoidal voltage across the electrodes, as indicated in the figure, to generate the quadrupole potential

$$V(\mathbf{r}, t) = e\phi(\mathbf{r}, t) = e\phi_0(x^2 + y^2 - 2z^2)\cos vt, \tag{13}$$

where ϕ_0 is a constant. As expected from Earnshaw's theorem, this potential has no points of stable equilibrium. On the other hand, the effective potential

$$V_{\text{eff}}(\mathbf{R}) = \frac{e\phi_0(x^2 + y^2 + 4z^2)}{mv^2} \tag{14}$$

does have a point of stable equilibrium at the origin. It is this potential that was used by Neuhauser et al. (2) to trap a single barium ion.

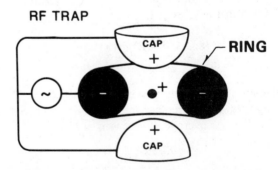

Figure 3.   Paul trap. An oscillating potential difference between the two cap electrodes and the ring electrode produces a sinusoidal quadrupole field near the center of the trap. The effective potential has a point of stable equilibrium at the center where the field strength is zero. The ring electrode is cut away to show the ion (black dot) at center.

The quadrupole trap has another important property. The electric field associated with the quadrupole potential, namely,

$$\mathbf{E} = -2\phi_0(x, y, -2z)\cos vt, \tag{15}$$

is zero at the origin. So, for an atomic particle at the minimum of the effective potential, there is no stark effect to perturb the spectrum. Only when the ion moves off the equilibrium point and experiences a restoring force is the spectrum mildly perturbed.

Of course, the classical theory of ion trapping is only an approximation. A quantum-mechanical derivation of the effective potential was given by Cook, Shankland, and Wells (14). According to this theory, the dominant effect of the rapidly oscillating potential is to add an oscillating phase to the ion wavefunction. The amplitude factor multiplying this phase factor obeys a Schrödinger equation with a time-independent effective potential energy equal to the classical effective potential. Hence, the ion motion in a quadrupole field is that of a three-dimensional quantum-mechanical harmonic oscillator. The general solution of Schrödinger's equation for this case is easily written. Two differences between the classical and quantum results are the discrete energy spectrum and the finite extension of the wavefunction in the quantum case. In other respects the classical and quantum results are remarkably similar. In fact, when the quantum case is formulated in terms of the Wigner function, the dynamics is identical to the classical statistical-mechanical formulation (14).

Ions can be loaded in the trap by electron impact ionization of an atomic beam that passes between the trap electrodes. After the ions are produced, their motion must be damped in some way in order to compensate for the RF heating and to reduce the Doppler spread of spectral lines for high-resolution spectroscopy. The most efficient method for this purpose is laser cooling, a subject we shall now discuss.

Laser cooling is interesting in its own right because it differs from the commonly encountered situation in which the absorption of light heats rather than cools a medium. Two different methods of laser cooling were proposed independently and at about the same time by Hänsch and Schawlow (15) and by Wineland and Dehmelt (16). Both methods make use of the radiation pressure that acts on an atom in near-resonant monochromatic light. Consider a plane traveling wave with propagation vector $\mathbf{k}$ and frequency $\omega = c|\mathbf{k}|$. The photons have momentum $\hbar\mathbf{k}$. When an atom in the wave absorbs a photon, it is promoted from the ground state 0 to the excited state 1; at the same time, it gains the momentum $\hbar\mathbf{k}$ of the photon. From the excited state the atom returns to the ground state by the spontaneous emission of a photon in a random direction. Because the probability for spontaneous emission in a given direction is equal to the probability for emission in the opposite direction, the mean momentum transfer in spontaneous emission is zero. Hence, the net momentum transfer to the atom in absorption followed by spontaneous emission is $\hbar k$, and the rate of these discrete transfers is the rate of spontaneous emission $A_1 P_1$, where A_1 is the Einstein spontaneous emission coefficient for the excited state and P_1 is the probability that this state is occupied. The excited-state probability P_1 is a Lorentzian function of the detuning, $\Delta = (\omega - \mathbf{k} \cdot \mathbf{v}) - \omega_0$, between the atomic transition frequency ω_0 and the Doppler-shifted field frequency $\omega - \mathbf{k} \cdot \mathbf{v}$ (here $\mathbf{v}$ is the velocity of the atom). Specifically,

$$P_1 = \frac{\Omega^2}{4\Delta^2 + A_1^2 + 2\Omega^2}, \tag{16}$$

where Ω is the on-resonance Rabi frequency for the transition ($\Omega = \mu\varepsilon/\hbar$, where μ is the dipole transition moment and $\mathscr{E}$ the electric field amplitude). On collecting terms, we obtain the following expression for the radiation force

$$F = \frac{\hbar k A_1 \Omega^2}{4(\omega - \omega_0 - \mathbf{k} \cdot \mathbf{v})^2 + A_1^2 + 2\Omega^2}. \tag{17}$$

For a weak field ($\Omega \ll A_1$), the power-broadening term $2\Omega^2$ in the denominator is negligible, and the component of the force in direction $\mathbf{k}$ is

$$F = \frac{\hbar k A_1 \Omega^2}{4(\omega - \omega_0 - kv)^2 + A_1^2}, \tag{18}$$

where v is the component of atomic velocity in direction $\mathbf{k}$.

Suppose the field is tuned below the atomic resonance ($\omega < \omega_0$). Then, according to Eq. 18, when the atom moves against the direction of the light, the field is Doppler shifted closer to resonance, and the radiation force increases. When the atom moves in the direction of light propagation, the field is Doppler tuned further from resonance, and the radiation force diminishes. This is the basis of the Hänsch–Schawlow (15) and the Wineland–Dehmelt (16) cooling schemes. Hänsch and Schawlow consider a plane standing wave, $E(x, t) = 2E_0 \cos(kx) \cos(\omega t)$, which is comprised of two counterpropagating running waves, $E_+(x, t) = E_0 \cos(kx - \omega t)$ and $E_- = E_0 \cos(kx + \omega t)$. If the standing wave is tuned below resonance and the atom moves in the positive x direction, the radiation force of the positive-traveling wave diminishes, and that of the negative-traveling wave increases. Therefore, the net force is in the negative-x direction. If the atom moves in the negative-x direction, the other running wave exerts the dominant force, and again the net force opposes the velocity. Thus, the velocity is damped, and the temperature of a collection of atoms decreases. For low-velocity atoms and optimum detuning, the cooling rate on an optical transition can exceed 10^4 K/s, and temperatures on the order of 10^{-3}–10^{-4} K are achievable rather quickly by this technique (17). To cool all the degrees of freedom and not just the motion in the x direction, one needs three superimposed standing waves propagating along the x, y, and z axes. In the experiments of Chu et al. (5) just such an arrangement of standing waves was used to cool and temporarily trap a vapor of neutral sodium atoms. These authors refer to this arrangement of fields as "optical molasses" because of the strong "viscous" force experienced by atoms in it.

For atomic ions, which are already confined in a trap, a single running wave suffices for cooling. Consider the motion of an ion in a quadrupole trap and, in particular, its motion along the x axis. As shown in the preceding, the ion moves in a quadratic effective potential. Hence, the ion oscillates sinusoidally about the origin. Suppose a single running wave tuned below resonance propagates in the x direction. Then, when the ion moves against the radiation, it experiences a strong declerating force because of Doppler tuning closer to resonance. On the other hand, when the ion moves in the direction of the light, it experiences an accelerating force, but this force is quite weak due to Doppler tuning away from resonance on this leg of the oscillation. As a result, the average radiation force over a complete oscillation is dissipative, and consequently, the amplitude of the oscillation decreases. This, in essence, is the cooling scheme of Wineland and Dehmelt (16). A more careful analysis of this scheme shows that the oscillating atom has sidebands on its resonance line, and for optimum cooling, the radiation is tuned to the first sideband below the atomic transition frequency (18, 19). The sideband cooling technique has been used successfully in a number of ion-trapping experiments including ones where a single stored ion is cooled.

IV. THEORY OF TELEGRAPHIC FLUORESCENCE

A qualitative discussion of telegraphic fluorescence was given in Section II. Here we discuss the quantitative theory.

The first detailed treatment of the statistics of intermittent fluorescence was given by Cook and Kimble (8) for the case of incoherent illumination, that is, for the case in which the spectral energy density of the radiation is a slowly varying function of frequency across each atomic line profile. For this case, the excitation of the atom is a rate process, and there exist probabilities per unit time for upward and downward transitions. This makes the mathematical analysis quite simple, and it was for this reason that incoherent excitation was considered first. However, experiments are more likely to be performed using coherent laser excitation. Not long after the paper by Cook and Kimble, there appeared a number of theoretical papers (20–23) which extended the analysis to coherent excitation and pointed out energy level schemes other than the "V" system for which intermittent fluorescence should occur. These papers approached the problem either from the point of view of the density matrix formalism for the three-level system or by studying the correlation functions of the quantized field radiated by the atom. In both of these approaches the formalism can become algebraically cumbersome. For a time there was a controversy about whether the atomic fluorescence would in fact turn on and off abruptly, as described in Section II. More recently a paper by Cohen-Tannoudji and Dalibard (24) presented an approach to telegraphic fluorescence that is both algebraically simple and very clear on the question of "dark periods" in the fluorescence. In this approach one calculates the delay function $D(\tau)$, which is the probability density for the delay τ between fluorescence photons. That is, $D(\tau) d\tau$ is the probability that if a photon is emitted at time t, the *next* photon is emitted in the interval $d\tau$ at time $t + \tau$. In the remainder of this section we develop the delay function formalism by first calculating the delay function for a two-level atom and then extending the analysis to the three-level system.

Consider a two-level atom with transition frequency ω_0 that is driven by a monochromatic field $\mathbf{E} = \hat{\varepsilon}\mathscr{E} \cos \omega t$ of frequency ω and polarization $\hat{\varepsilon}$. For the sake of simplicity, we assume that the atomic dipole transition moment $\boldsymbol{\mu}$ is real and has the same direction as the electric field ($\boldsymbol{\mu} = \hat{\varepsilon}\mu$). Then, in the rotating-wave approximation and in the absence of spontaneous emission, the amplitudes C_0 and C_1 for the atom to be in the lower state or the upper state, respectively, obey the equations

$$\dot{C}_0 = \tfrac{1}{2}i\Omega C_1, \qquad \dot{C}_1 = \tfrac{1}{2}i\Omega C_0 + i\,\Delta C_1, \tag{19}$$

where $\Delta = \omega - \omega_0$ is the detuning frequency between the field frequency ω and the Bohr transition frequency ω_0, and Ω is the on-resonance Rabi frequency.

Equations 19 are most easily derived, as in the preceding, for a classical prescribed electric field. But the same equations correctly describe the interaction of the two-level atom with a quantized single-mode field. In the latter case C_0 is the amplitude for the atom to be in its ground state with n photons in the field, and C_1 is the amplitude to be in the excited state with $n - 1$ photons in the field. For the quantized-field case the Rabi frequency is proportional to the square root of the number of photons initially present in the field, that is, $\Omega = \sqrt{n}\Omega_0$, where Ω_0 is the one-photon Rabi frequency, a quantity determined by the form of the field mode function and the position of the atom in the field. In the absence of any other field modes, the atom can only absorb one photon from the quantized field and return a single photon to the field by stimulated emission. Hence, only two essential states and two amplitudes are required to describe the atom–field interaction.

The dynamics change considerably when spontaneous emission is considered. When all of the quantized electromagnetic modes are included, the excited atom can spontaneously emit photons into any of the modes with frequencies near the atomic transition frequency. These are fluorescence photons. To analyze this process exactly, one must introduce an infinity of amplitudes $C_{0,k\lambda}$. The amplitude $C_{0,k\lambda}$ describes the probability that the atom emits a photon into the mode with wave vector k and polarization λ ($\lambda = 1, 2$) and ends up in the ground state 0. As time proceeds, the driving field mode again excites the atom to state 1, whereas the first fluorescence photon remains in mode $k\lambda$. So we also need amplitudes $C_{1,k\lambda}$ for the atom to absorb a second photon from the driving mode with the first photon in mode $k\lambda$. Then there are amplitudes $C_{0,k\lambda_1 k'\lambda'}$ for the atom to emit a second photon ($k'\lambda'$) and again return to the ground state. Proceeding in this way, one obtains an infinite hierarchy of amplitudes that fully describe resonance fluorescence. These amplitudes are coupled by a hierarchy of equations that are rather cumbersome.

Fortunately, we shall require only the amplitudes C_0 and C_1 for zero photon emission, and the effect of all the higher order amplitudes on these can be expressed quite simply. In the Wigner–Weisskopf approximation, the possibility of spontaneous emission simply adds a damping term $-\frac{1}{2}AC_1$ to the equation for C_1. Equation 19 becomes

$$\dot{C}_0 = \tfrac{1}{2}i\Omega C_1,$$
$$\dot{C}_1 = \tfrac{1}{2}i\Omega C_0 + i\,\Delta C_1 - \tfrac{1}{2}AC_1, \tag{20}$$

where A is the Einstein spontaneous emission coefficient for level 1. Actually, the Wigner–Weisskopf approximation also predicts a shift of the atomic transition frequency, which turns out to be infinite. But, after mass renormalization and an appropriate "cutoff" procedure, the physical shift (Lamb shift) is found to be small and of no particular interest in the present context. Here, we assume that the radiative shift is already included in the measured atomic transition frequency ω_0.

For the following analysis, the interpretation of amplitudes C_0 and C_1 should be kept clearly in mind. The probability that the atom is in the lower (upper) state and has not emitted a fluorescence photon is $P_0 = |C_0|^2 (P_1 = |C_1|^2)$. The probability for the atom to be in its lower (upper) state having emitted one photon into mode $k\lambda$ is $P_{0k\lambda} = |C_{0k\lambda}|^2 (P_{1k\lambda} = |C_{1k\lambda}|^2)$, and the probability for emission of higher numbers of fluorescence photons is the square of higher order amplitudes. In this chapter we shall work exclusively with the amplitudes and probabilities for zero fluorescence photons. If one is not interested in the states of fluorescence photons after they are emitted, the description of resonance fluorescence simplifies considerably. The emission of a photon, or strictly speaking, the detection of the emitted photon, projects the atom into its ground state. Then the atom can once more absorb a photon from the driving mode, emit a fluorescence photon, and again be projected into the ground state. It can be shown that previously emitted fluorescence photons have essentially no effect on the emission of subsequent photons. Hence, each photon emission process, starting from the projection into the ground state, obeys the same equations as for the first photon emission. That is, between photon emissions, the atomic dynamics is described by two amplitudes (C_0 and C_1) that obey Eqs. 20. Immediately after each emission, the amplitudes are reinitiated to the values $C_0 = 1$ and $C_1 = 0$.

To calculate the delay function $D(\tau)$, we proceed as follows. Immediately after emitting a photon at time t, the atom is in its ground state. So the initial conditions for the subsequent time interval are

$$C_0(t) = 1, \qquad C_1(t) = 0. \tag{21}$$

From these initial conditions, the amplitudes $C_0(t + \tau)$ and $C_1(t + \tau)$ for any later time are obtained by solving Eqs. 20. The absolute squares of these amplitudes,

$$P_0 = |C_0(t + \tau)|^2, \qquad P_1 = |C_1(t + \tau)|^2, \tag{22}$$

are the probabilities that the atom is in its ground state or excited state, respectively, and has not emitted the *next* fluorescence photon. Accordingly, the total probability that the atom has not emitted the next photon in time τ is

$$P(\tau) = P_0(\tau) + P_1(\tau)$$
$$= \sum_{i=0}^{1} |C_i(t + \tau)|^2. \tag{23}$$

The rate of change of this probability follows directly from Eqs. 20 and their complex conjugates:

$$\dot{P}(\tau) = -A|C_1(t + \tau)|^2 = -AP_1(\tau). \tag{24}$$

Now the probability per second that the *next* photon is emitted at time $t + \tau$ is the negative of the rate of change of the probability that the atom has not emitted at this time. Hence, the delay function (which is the probability per second that the next photon is emitted at time $t + \tau$) is simply

$$D(\tau) = A|C_1(t + \tau)|^2. \tag{25}$$

The solution to Eqs. 20 with initial conditions 21 is easy to obtain by means of Laplace transforms. The solution is

$$C_0(t+\tau) = \frac{(X_+ + A - i\Delta)e^{X_+\tau/2} - (X_- + A - i\Delta)e^{X_-\tau/2}}{X_+ - X_-} e^{i\Delta\tau/2}, \quad (26a)$$

$$C_1(t+\tau) = \frac{i\Omega(e^{X_+\tau/2} - e^{X_-\tau/2})e^{i\Delta\tau/2}}{X_+ - X_-}, \quad (26b)$$

where

$$X_\pm = -\tfrac{1}{2}A \pm \tfrac{1}{2}\sqrt{A^2 - 4(\Delta^2 + \Omega^2 + iA\Delta)}. \quad (27)$$

From 26b the delay function 25 is found to be

$$D(\tau) = \frac{A\Omega^2[e^{-\gamma_+\tau} + e^{-\gamma_-\tau} - 2e^{-A\tau/2}\cos(v\tau)]}{\{[A^2 - 4(\Delta^2 + \Omega^2)]^2 + 16A^2\Delta^2\}^{1/2}}, \quad (28)$$

where

$$\gamma_\pm = \tfrac{1}{2}\{A \pm \operatorname{Re}[A^2 - 4(\Delta^2 + \Omega^2 + iA\Delta)]^{1/2}\}, \quad (29)$$

$$v = \tfrac{1}{2}\operatorname{Im}[A^2 - 4(\Delta^2 + \Omega^2 + iA\Delta)]^{1/2}. \quad (30)$$

One can see from Eq. 28 that the delay function always tends to zero as $\tau \to 0$. After the atom has emitted, some time is required to excite the atom so that it can again emit. This is the well-known phenomenon of photon antibunching, here appearing in the next-photon delay function.

The case of exact resonance ($\Delta = 0$) is worth examining further. For a "weak" driving field ($\Omega < A/2$), the delay function contains only terms that decay exponentially:

$$D(\tau) = \frac{A\Omega^2(e^{-\gamma_+\tau} + e^{-\gamma_-\tau} - 2e^{-A\tau/2})}{A^2 - 4\Omega^2}, \quad (31)$$

with $\gamma_\pm = \tfrac{1}{2}[A \pm (A^2 - 4\Omega^2)^{1/2}]$. When the driving field is very weak ($\Omega \ll A$), we have $\gamma_+ \approx A$ and $\gamma_- \approx \Omega^2/A$. In this case, the delay function grows from zero to approximately Ω^2/A in a time of order A^{-1}; then it slowly decreases to zero as $(\Omega^2/A)\exp(-\Omega^2\tau/A)$. The small rise and slow decay of the delay function indicates that a weak field is slow to excite the two-level atom. The mean time required for excitation and emission of the next photon is

$$\bar{\tau} = \int_0^\infty \tau D(\tau)\, d\tau$$

$$= \frac{A^2 + 2\Omega^2}{A\Omega^2}. \tag{32}$$

Incidentally, this formula is valid for all values of Ω, not just those less than $A/2$.

For a "strong" driving field ($\Omega > A/2$), the delay function decays exponentially at the rate $A/2$ but with superimposed oscillations at frequency $v = [\Omega^2 - (A/2)^2]^{1/2}$:

$$D(\tau) = \frac{A\Omega^2 e^{-A\tau/2}}{\Omega^2 - (A/2)^2} \sin^2 \frac{\sqrt{\Omega^2 - (A/2)^2}\,\tau}{2}. \tag{33}$$

Note that the next photon cannot be emitted with delays $\tau = 2\pi m / [\Omega^2 - (A/2)^2]^{1/2}$, with m an integer, for the delay function is zero at these times.

The delay function for a two-level atom is difficult to measure. To measure it, one would have to detect each emitted photon with near 100% efficiency to ensure that the *next* photon rather than some later one is registered. Because of limits on photon collection efficiency and detector efficiencies, it appears that a direct measurement of the delay function for a two-level atom is beyond current technology. What can be and has been measured in photon antibunching experiments is the probability that *a* photon, not necessarily the *next* photon, is detected in the interval $d\tau$ at delay τ. This is much less difficult to measure because it is no longer necessary to ensure that no photons are emitted in the interval $[t, t + \tau]$ preceding the interval $d\tau$ under consideration.

Let us consider now the three-level system depicted in Figure 1. When transitions $0 \leftrightarrow 1$ and $0 \leftrightarrow 2$ are driven by Rabi frequencies Ω_1 and Ω_2 and the Einstein A coefficients for levels 1 and 2 are A_1 and A_2, respectively, the equations for the amplitudes C_0, C_1, and C_2 that levels 0, 1, and 2 are occupied and the next photon has not been emitted are

$$\dot{C}_0 = -\tfrac{1}{2}i\Omega_1 C_1 - \tfrac{1}{2}i\Omega_2 C_2, \tag{34a}$$

$$\dot{C}_1 = -\tfrac{1}{2}i\Omega_1 C_0 + (i\Delta_1 - \tfrac{1}{2}A_1)C_1, \tag{34b}$$

$$\dot{C}_2 = -\tfrac{1}{2}i\Omega_2 C_0 + (i\Delta_2 - \tfrac{1}{2}A_2)C_2, \tag{34c}$$

where $\Delta_1 = \omega_1 - \omega_{10}$ and $\Delta_2 = \omega_2 - \omega_{20}$ are the detunings of the laser frequencies ω_1 and ω_2 with respect to the transition frequencies ω_{10} and ω_{20}.

We are assuming there is no spontaneous emission between levels 1 and 2 and that no radiation drives this transition.

Because a photon emitted at time t on either transition projects the system into level 0, the initial conditions for the subsequent time interval are

$$C_0(t) = 1, \qquad C_1(t) = 0, \qquad C_2(t) = 0. \tag{35}$$

From the interpretation of the amplitudes C_0, C_1, and C_2, it follows that the total probability for zero photon emission after delay τ is

$$P(\tau) = |C_0(t + \tau)|^2 + |C_1(t + \tau)|^2 + |C_2(t + \tau)|^2. \tag{36}$$

The negative derivative of this probability is the rate (probability per unit time) of the next-photon emission. Hence, $- \dot{P}(\tau)\, d\tau$ is the probability that the next photon is emitted with delay in the interval $d\tau$ at τ. But from its definition, $D(\tau)\, d\tau$ is this same probability. Hence, as for the two-level atom, $D(\tau) = - \dot{P}(\tau)$, and from the equations of motion 34 and their complex conjugates, we obtain

$$D(\tau) = A_1 |C_1(t + \tau)|^2 + A_2 |C_2(t + \tau)|^2. \tag{37}$$

When we say that the transition $0 \leftrightarrow 1$ is strong and $0 \leftrightarrow 2$ is weak, we mean that the spontaneous and induced transition rates on $0 \leftrightarrow 2$ are much smaller than those on $0 \leftrightarrow 1$:

$$A_2, \Omega_2 \ll A_1, \Omega_1. \tag{38}$$

This being the case, the last term in Eq. 34a is usually much smaller than the first term and may be ignored when the amplitudes C_1 and C_2 are comparable. Equations 34a and 34b then read

$$\dot{C}_0 = -\tfrac{1}{2} i \Omega_1 C_1, \tag{39a}$$

$$\dot{C}_1 = -\tfrac{1}{2} i \Omega_1 C_0 + (i\Delta_1 - \tfrac{1}{2} A_1) C_1. \tag{39b}$$

These equations are formally identical to the two-level Eqs. 20. That the amplitudes to be in levels 0 and 1 should behave approximately as those of a two-level system is expected, since the transition $0 \leftrightarrow 1$ is strongly driven, whereas the slow excitation of the weak transition $0 \leftrightarrow 2$ acts as a small perturbation to this subsystem.

We shall consider only one special case of these equations that is relatively simple and yet illustrates most of the features of the general solution. We take the two driving fields to be exactly resonant with the transitions they drive

$(\Delta_1 = \Delta_2 = 0)$. Then, if the field strength acting on transition $0 \leftrightarrow 1$ is such that $\Omega_1 = A_1/2$, the solution of Eqs. 39 is

$$C_0(t + \tau) = (1 + A_1\tau/4)e^{-A_1\tau/4}, \tag{40a}$$

$$C_1(t + \tau) = -(A_1\tau/4)e^{-A_1\tau/4}. \tag{40b}$$

The remaining equation of the set 34, namely,

$$\dot{C}_2 + \tfrac{1}{2}A_2C_2 = -\tfrac{1}{2}i\Omega_2C_0, \tag{41}$$

shows that the amplitude to be in level 2 is driven by C_0. With initial conditions 35, the solution of Eqs. 41 is

$$C_2(t + \tau) = -\tfrac{1}{2}i\Omega_2 \exp\left(-\tfrac{1}{2}A_2\tau\right) \int_0^\tau \exp\left(\tfrac{1}{2}A_2\tau'\right)C_0(t + \tau')\,d\tau'. \tag{42}$$

From Eq. 40a we see that $C_0(t + \tau')$ is substantial only for $\tau' \lesssim 4/A_1$, and for these values of τ', the exponential in the integrand of Eq. 42 is very close to unity, since $A_2 \ll A_1$. Setting this exponential equal to unity, we have

$$
\begin{aligned}
C_2(t + \tau) &= -\tfrac{1}{2}i\Omega_2 \exp\left(-\tfrac{1}{2}A_2\tau\right) \int_0^\tau C_0(t + \tau')\,d\tau' \\
&= -\frac{4i\Omega_2}{A_1}\left[1 - \left(1 + \frac{A_1\tau}{8}\right)e^{-A_1\tau/4}\right]e^{-A_2\tau/2}, \tag{43}
\end{aligned}
$$

where Eq. 40a was used to obtain the final form. Finally, from Eqs. 40b and 43, the delay function 37 is found to be

$$D(\tau) = D_P(\tau) + D_T(\tau), \tag{44}$$

where

$$D_P(\tau) = \frac{A_1^3\tau^2}{16}e^{-A_1\tau/2}, \tag{45}$$

and

$$D_T(\tau) = \frac{16A_2\Omega_2^2}{A_1^2}\left[1 - \left(1 + \frac{A_1\tau}{8}\right)e^{-A_1\tau/4}\right]^2 e^{-A_2\tau}. \tag{46}$$

This function is sketched in Figure 4. The first term on the right in Eq. 4 corresponds to the large peak in the figure, whereas the second term attaches a long tail to the function. As in the case of the two-level atom, photon

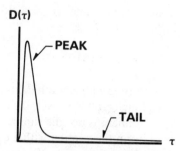

Figure 4. Delay function for three-level atom fluorescence.

antibunching appears as a dip in the function $D(\tau)$ near $\tau = 0$. In general, the area under the peak is near unity. Since the width of this peak is of order $1/A_1$, the large probability associated with the peak means that the chance is very high that the delay between fluorescence photons is of order $1/A_1$. [In fact, in our approximation, the integral of the peak function $D_P(\tau)$ is exactly unity. It would be slightly less in a better approximation.] The long tail $D_T(\tau)$, whose integrated probability is small, describes infrequent long delays between fluorescence photons. For $\tau \gg 1/A_1$, $D_T(\tau)$ is simply

$$D_T(\tau) = \frac{16 A_2 \Omega_2^2}{A_1^2} e^{-A_2\tau}. \tag{47}$$

The total probability in the tail, $(4\Omega_2/A_1)^2$, is very small since $\Omega_2 \ll A_1$.

The solution we have obtained is subject to one more restriction. If Ω_2 is comparable to or larger than A_2, the atomic electron may be driven down from level 2 before it spontaneously decays. This returns the electron to level 0, a process we have neglected by discarding the last term in Eq. 34a. Consequently, our special solution is valid only for $\Omega_2 \ll A_2$. Although we have only looked at a very special case, the qualitative form of the delay function, with a high-probability peak at small τ and a long low-probability tail, is quite general. Other special cases of interest may be found in the literature (20–28).

From the form of the delay function the following picture of the fluorescence emerges. Usually, fluorescence photons are emitted with short delays (of order $1/A_1$), and the high probability for short delays means that long sequences of photons are emitted with only short delays between them. Such a sequence is a "bright period" in the fluorescence. But occasionally a long delay between photons (of order $1/A_2$) occurs, and this is a dark period in the fluorescence. In this way the delay function provides a complete description of telegraphic fluorescence and, in particular, an unequivocal proof that dark periods in the fluorescence exist.

V. EXPERIMENTS

The first experimental demonstration of telegraphic fluorescence was that of Nagourney, Sandberg, and Dehmelt (28). However, the energy level structure of the Ba$^+$ ion used in this experiment is not the simple V system previously discussed. Therefore, we shall begin this section with a discussion of the first experiment on an ion with the V configuration of energy levels, namely, the experiment of Bergquist, Hulet, Itano, and Wineland (9) on the Hg$^+$ ion. We shall then return to the experiments of Nagourney et al and others.

A partial energy level diagram of the mercury ion is given in Figure 5. The resonance transition at 194 nm from the ground state $5d^{10}6s^2\ ^2S_{1/2}$ to the excited state $5d^{10}6p\ ^2P_{1/2}$ is the "strong" transition of the V system. The lifetime of the $^2P_{1/2}$ state (level 1) is 2.3 ns, and the associated A coefficient is $A_1 = 4.3 \times 10^8\ \text{s}^{-1}$. Hence, the saturated rate of fluorescence is $\frac{1}{2}A_1 = 2 \times 10^8\ \text{s}^{-1}$, and in this experiment, the strong transition was driven near saturation by frequency-doubled laser radiation. Actually, the 192-nm radiation was tuned below resonance, since this radiation was used to cool the ion as well as to monitor quantum jumps. The other relevant transition is the 281.5-nm transition from the ground state to the $^2D_{5/2}$ state. This is a "weak" quadrupole transition, the lifetime of the $^2D_{5/2}$ state (level 2) being about 0.1 s. This transition was also driven by frequency-doubled ring dye laser radiation.

A record of the strong fluorescence at 194 nm, with both transitions driven, is shown in Figure 6. Although the fluorescent intensity is somewhat noisy, the abrupt switching on and off of the fluorescence is unmistakable. With the 281.5-nm radiation turned off, the interruptions in the fluorescence become

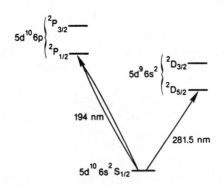

Figure 5.   Energy level diagram for the Hg$^+$ ion.

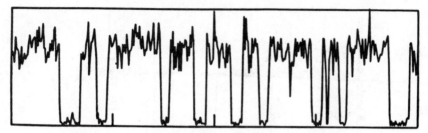

Figure 6. Measured fluorescence intensity for the Hg$^+$ ion with both the 194-nm and the 281.5-nm transitions illuminated.

very infrequent but do not disappear entirely. Presumably, these background events are due to collisions of the Hg$^+$ ion with residual gas in the trap or to infrequent spontaneous emission from the $^2P_{1/2}$ state to the $^2D_{3/2}$ state, which can decay to the $^2D_{5/2}$ state or to the ground state. In any case, the background events do not occur often, and it can be said that this experiment demonstrates the direct observation of quantum jumps on the $^2S_{1/2} \leftrightarrow {}^2D_{5/2}$ transition with near unit efficiency.

Another quantity of interest measured by these authors was the two-time intensity correlation function of the fluorescent intensity. According to the theory of Cook and Kimble (8), this correlation function should show an exponential bump at time differences on the order of the lifetime of level 2. The experimental result was found to be in satisfactory agreement with the theory.

Let us return now to the experiment of Nagourney et al. on the Ba$^+$ ion. A simplified energy level diagram for the barium ion is given in Figure 7. In this experiment, the strong resonance transition at 493.4 nm was driven with a dye laser. The upper level of this transition, $6\,^2P_{1/2}$, has a substantial branching ratio to the metastable $5\,^2D_{3/2}$ state. So, if no additional radiation were applied

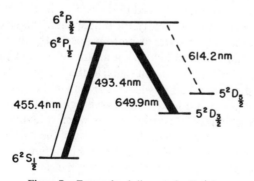

Figure 7. Energy level diagram for Ba$^+$ ion.

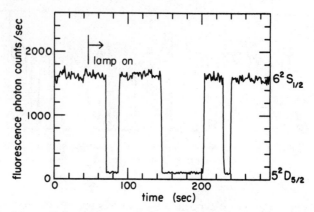

Figure 8. Measured fluorescence intensity for Ba$^+$ ion. The dark periods in the blue flourescence average 30 s duration.

to the atom, the ion would be optically pumped into this metastable state, and the strong fluorescence would cease. To prevent this from occurring, additional laser light is applied to the $5\,^2D_{3/2} \leftrightarrow 6\,^2P_{1/2}$ transition. This returns the atom to the strong transition $6\,^2S_{1/2} \leftrightarrow 6\,^2P_{1/2}$. Whereas the resonance transition in Hg$^+$ is in the ultraviolet and must be observed photoelectrically, the 493.4-nm transition of Ba$^+$ is in the visible (blue) and can be observed with the eye. In Ba$^+$, the shelving, which interrupts the fluorescence, is produced by focusing radiation from a barium hollow cathode lamp onto the ion. The lamp light at 455.4 nm drives the $6\,^2S_{1/2} \rightarrow 6\,^2P_{3/2}$ transition. From the $6\,^2P_{3/2}$ level there is a substantial branching ratio ($\sim 1:3$) into the metastable $5\,^2D_{5/2}$ level. Consequently, after a few excitations to the $6\,^2P_{3/2}$ state, the ion has a high probability of being trapped in the $5\,^2D_{5/2}$ level. The ion remains in this level, with the blue fluorescence switched off, until it spontaneously decays from $5\,^2D_{5/2}$ to the ground state, at which time the fluorescence again turns on.

Notice that when the ion is shelved in the $5\,^2D_{5/2}$ level, it is unaffected by any of the driving fields. Consequently, the time spent in this level is simply the spontaneous lifetime of that state, or more properly, the dwell times in that state are exponentially distributed with a mean value of, in this case, about 30 s. A typical record of the fluorescent signal for Ba$^+$ is given in Figure 8. Here the fluorescence shows considerably less noise than in the case of the Hg$^+$ ion. This is due to the fact that a longer signal-averaging time can be used when looking for interruptions of 30 s duration than when studying dark periods of order 0.1 s. Also, in this experiment the durations of the dark periods (i.e., the dwell times in level $5\,^2D_{5/2}$) were plotted as a histogram and fitted with an exponential to obtain what is probably the best measurement of the natural lifetime of this state, namely, $32 + 5$ s. Other quantum jump experiments on the

Ba$^+$ ion have been reported by Sauter, Neuhauser, Blatt, and Toschek (29). It is remarkable that, in these experiments, one can directly observe the atomic blue fluoresence with dark periods averaging about 30 seconds long.

References

1. H. J. Kimble, M. Dagenais, and L. Mandel, *Phys. Rev. Lett.* **39**, 691 (1977); *Phys. Rev. A* **18**, 201 (1978); M. Dagenais and L. Mandel, *Phys. Rev. A* **18**, 2217 (1978).

2. W. Neuhauser, M. Hohenstatt, P. Toschek, and H. Dehmelt, *Phys. Rev. Lett.* **41**, 233 (1978); *Phys. Rev. A* **22**, 1137 (1980).

3. H. Dehmelt, in *Advances in Laser Spectroscopy*, F. T. Arecchi, F. Strumia, and H. Walther, eds., NATO Advanced Study Institute Series, Vol. 95, Plenum, New York, 1983.

4. W. D. Phillips, J. V. Prodan, and H. J. Metcalf, *J. Opt. Soc. Am.* **B2**, 1751 (1985).

5. S. Chu, L. Holiberg, J. E. Bjorkholm, A. Cable, and A. Ashkin, *Phys. Rev. Lett.* **55**, 48 (1985); S. Chu, J. E. Bjorkholm, A. Ashkin, and A. Cable, *Phys. Rev. Lett.* **57**, 314 (1986).

6. H. Dehmelt, *Bull. Am. Phys. Soc.* **20**, 60 (1975).

7. H. Dehmelt, *IEEE Trans. Instrum. Meas.* **IM-31**, 83 (1982).

8. R. J. Cook and H. J. Kimble, *Phys. Rev. Lett.* **54**, 1023 (1985).

9. J. C. Bergquist, R. G. Hulet, W. M. Itano, and D. J. Wineland, *Phys. Rev. Lett.* **57**, 1699 (1986).

10. J. A. Stratton, *Electromagnetic Theory*, McGraw-Hill, New York, 1941, p.116.

11. D. J. Wineland, R. E. Drullinger, and F. L. Walls, *Phys. Rev. Lett.* **40**, 1639 (1978).

12. R. E. Drullinger, D. J. Wineland, and J. C. Bergquist, *Appl. Phys.* **22**, 365 (1980).

13. D. J. Wineland and W. M. Itano, *Phys. Lett.* **82A**, 75 (1981).

14. R. J. Cook, D. G. Shankland, and A. L. Wells, *Phys. Rev. A* **31**, 564 (1985).

15. T. W. Hänsch and A. L. Schawlow, *Opt. Commun.* **13**, 68 (1975).

16. D. J. Wineland and H. Dehmelt, *Bull. Am. Phys. Soc.* **20**, 637 (1975).

17. R. J. Cook, *Phys. Rev. A* **22**, 1078 (1980).

18. D. J. Wineland, R. E. Drullinger, and F. L. Walls, *Phys. Rev. Lett.* **40**, 1639 (1978).

19. D. J. Wineland and W. M. Itano, *Phys. Rev. A* **20**, 1521 (1979).

20. J. Javanainen, *Phys. Rev. A* **33**, 2121 (1986).

21. D. T. Pegg, R. Loudon, and P. L. Knight, *Phys. Rev. A* **33**, 4085 (1986).

22. F. T. Arecchi, A. Schenzle, R. G. DeVoe, K. Jungman, and R. G. Brewer, *Phys. Rev. A* **33**, 2124 (1986).

23. A Schenzle, R. G. DeVoe, and R. G. Brewer, *Phys. Rev. A* **33**, 2127 (1986).

24. C. Cohen-Tannoudji and J. Dalibard, *Europhys. Lett.* **1**(9), 441 (1986).

25. H. J. Kimble, R. J. Cook, and A. L. Wells, *Phys. Rev. A* **34**, 3190 (1986).

26. A. Schenzle and R. G. Brewer, *Phys. Rev. A* **34**, 3127 (1986).

27. P. Zoller, M. Marte, and D. F. Walls, *Phys. Rev. A* **35**, 198 (1987).

28. W. Nagourney, J. Sandberg, and H. Dehmelt, *Phys. Rev. Lett.* **56**, 2797 (1986).

29. Th. Sauter, W. Neuhauser, R. Blatt, and P. E. Toschek, *Phys. Rev. Lett.* **57**, 1696 (1986).

CHAPTER XVII

GENERALIZED FLOQUET THEORETICAL APPROACHES TO INTENSE-FIELD MULTIPHOTON AND NONLINEAR OPTICAL PROCESSES

SHIH-I CHU

Department of Chemistry, University of Kansas, Lawrence, Kansas 66045

CONTENTS

I. INTRODUCTION

The use of intense laser fields to initiate chemical and physical processes has led to the discovery of many distinctly new nonlinear phenomena such as collisionless multiphoton ionization (MPI) and above-threshold ionization (ATI) of atoms, multiphoton excitation (MPE) and dissociation (MPD) of polyatomic molecules, and so on (1, 2). Although considerable experimental work and theoretical models have been directed toward the understanding of these phenomena, much remains to be explored (1, 2). The difficulties arise from the complexity of the atomic–molecular level structures and the lack of practical and reliable theoretical and numerical methods for ab initio nonperturbative treatments of nonlinear atomic–molecular field interactions of very high order.

Multiphoton transitions may be formally treated in a fully quantum-mechanical or semiclassical fashion. In the former approach, both the system (atom–molecule) and the electromagnetic field are treated quantum mechanically, whereas in the semiclassical approach, the system is described by a time-dependent Schrödinger equation, and the field satisfies the classical Maxwell equations. It is well known that the semiclassical theory (3) gives rise to results that are equivalent to those obtained from the fully quantized theory (4) in strong fields. The semiclassical Hamiltonians usually exhibit sinusoidal time dependence when the field is monochromatic. The use of the Floquet theorem (5) leads to considerable simplification in both the theoretical and the computational analysis of multiphoton processes. Indeed, much recent multiphoton study has been performed based on various extensions of the Floquet theory.

In this chapter, several recent generalizations in Floquet theories and quasi-energy methods for ab initio *nonperturbative* treatments of intense-field multiphoton and nonlinear optical processes will be discussed with primary emphasis on the works completed in the 1983–1986 period in our laboratory. For a general discussion of other previous Floquet approaches, the readers are referred to the two recent Floquet reviews for details (6, 7).

Figure 1 shows an overview of the scope of recent Floquet developments and various processes amenable to generalized Floquet treatments. Not all the subjects listed in the figure will be discussed here. Specifically, I shall address the following new features:

  i. How to handle molecule–field interactions with ultralarge basis. How the artificial intelligence algorithms may be used to preselect the most probable pathways in MPE dynamics.

  ii. How complex coordinate techniques can be invoked to deal with bound–free MPI, ATI, and MPD problems whose Hamiltonians are

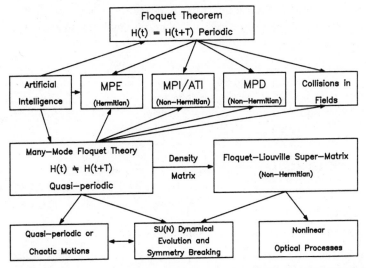

Figure 1. Scope of recent Floquet developments and subjects amenable to generalized Floquet treatments.

non-Hermitian. How the continuum and ionization threshold are shifted in the presence of intense laser fields.

iii. How to extend the Floquet theorem (Hamiltonian periodic in time) to the many-mode Floquet theory (MMFT) to treat multifrequency (multicolor) problems (Hamiltonians nonperiodic or quasi-periodic in time).

iv. How the MMFT can be used to facilitate the study of $SU(N)$ dynamical symmetries and symmetry breaking in multiphoton processes.

v. What is the behavior (quasi-periodic or chaotic) of quantum systems under perturbation by intense polychromatic fields.

vi. How the Floquet theory may be extended to the treatment of the Liouville equation for the time development of a density matrix operator in nonlinear optics [Floquet Liouville supermatrix (FLSM) approach].

vii. How the non-Hermitian FLSM method can be applied to the studies of novel intensity-dependent phenomena in multiphoton resonance fluorescence, nonlinear optical susceptibilities, multiple wave mixings, and so on.

viii. How to formulate the Floquet coupled-states method to probe MPE problems with arbitrary laser pulses.

ix. How to use the Floquet quasi-energy (or dressed-state) diagram to assist the identification of optimum field parameters to achieve population inversion in multilevel systems.

After a brief introduction to the Floquet theorem, the quasi-energy concept, and the conventional time-independent Floquet Hamiltonian method (Section II.A–C), these questions will be analyzed one by one with appropriate extended Floquet techniques, shown in Figure 1, and recent case studies in Sections II–V. This is followed by a conclusion in Section VI.

II. FLOQUET AND GENERALIZED FLOQUET TREATMENTS OF MULTIPHOTON EXCITATION PROCESSES IN PERIODIC FIELDS

A. Floquet Theorem

Consider the general features of the wavefunctions of a quantum system driven by a periodic external field with frequency ω and periodic $\tau(=2\pi/\omega)$. The Schrödinger equation for the system may be written as $(\hbar = 1)$

$$\mathscr{H}(\mathbf{r}, t)\Psi(\mathbf{r}, t) = 0, \tag{1}$$

where

$$\mathscr{H}(\mathbf{r}, t) \equiv H(\mathbf{r}, t) - i\frac{\partial}{\partial t}. \tag{2}$$

The total Hamiltonian $H(\mathbf{r}, t)$ is given by

$$H(\mathbf{r}, t) = H_0(\mathbf{r}) + V(\mathbf{r}, t), \tag{3}$$

where $V(\mathbf{r}, t)$ is the periodic perturbation due to the interaction between the system and the monochromatic field,

$$V(\mathbf{r}, t + \tau) = V(\mathbf{r}, t). \tag{4}$$

The unperturbed Hamiltonian $H_0(\mathbf{r})$ has a complete orthonormal set of eigenfunctions:

$$H_0(\mathbf{r})|\alpha(\mathbf{r})\rangle = E_\alpha^0|\alpha(\mathbf{r})\rangle, \qquad \langle\beta(\mathbf{r})|\alpha(\mathbf{r})\rangle = \delta_{\beta\alpha}. \tag{5}$$

The wavefunction Ψ, called the quasi-energy-state (QES), can be written, according to the Floquet theorem (5), in the form

$$\Psi(\mathbf{r}, t) = \exp(-i\varepsilon t)\Phi(\mathbf{r}, t), \tag{6}$$

where $\Phi(\mathbf{r}, t)$ is periodic in time, that is,

$$\Phi(\mathbf{r}, t + \tau) = \Phi(\mathbf{r}, t), \tag{7}$$

and ε is a real parameter, which is unique up to multiples of $2\pi n/\tau$, called the Floquet characteristic exponent, or the *quasi-energy* (7, 8). The term quasi-energy reflects the formal analogy of the states, Eq. 6, with the Bloch eigenstates in a solid with the quasi-momentum $\mathbf{k}$. Substituting Eq. 6 into Eq. 1, one obtains an eigenvalue equation for the quasi-energy,

$$\mathscr{H}(\mathbf{r}, t)\Phi_\gamma(\mathbf{r}, t) = \varepsilon_\gamma \Phi_\gamma(\mathbf{r}, t), \tag{8}$$

subject to the periodicity condition (7).

For the Hermitian operator $\mathscr{H}(\mathbf{r}, t)$, Sambe (9) introduced the composite Hilbert space $R \oplus T$. The spatial part is spanned by square-integrable (L^2) functions in the configuration space, with the inner product defined by

$$\langle \alpha(\mathbf{r}) | \beta(\mathbf{r}) \rangle \equiv \int \alpha(r)^* \beta(r) \, d\mathbf{r} = \delta_{\alpha\beta}. \tag{9}$$

The temporal part is spanned by the complete orthonormal set of functions $[\exp(im\omega t)]$, where $m = 0, \pm 1, \pm 2, \ldots$ is the Fourier index, and

$$\frac{1}{\tau} \int_0^\tau \exp[i(n - m)\omega t] \, dt = \delta_{nm}. \tag{10}$$

The eigenvectors of $\mathscr{H}$ satisfy the orthonormality condition

$$\langle\langle \Phi_\gamma | \Phi_{\gamma'} \rangle\rangle \equiv \frac{1}{\tau} \int_0^\tau dt \langle \Phi_\gamma(\mathbf{r}, t) | \Phi_{\gamma'}(\mathbf{r}, t) \rangle = \delta_{\gamma\gamma'} \tag{11}$$

and form a complete set in $R \oplus T$:

$$\sum_\gamma |\Phi_\gamma\rangle\langle\Phi_\gamma| = I. \tag{12}$$

B. General Properties of Quasi-Energy States

The QES play a similar role in studying quantum systems in time-periodic fields as the bound states do for the time-independent Hamiltonian. The QES with different quasi-energies ε_γ are mutually orthogonal for each moment of time and form a complete orthonormal set, as indicated in Eqs. 11 and 12.

The quasi-energy eigenvalue equation 8 has the form of the "time-independent" Schrödinger equation in the composite Hilbert space $R \oplus T$. It can be readily shown that all the general quantum-mechanical theorems for the time-independent Schrödinger equation, such as the variational principle, the Hellmann–Feynman, hypervirial, Wigner–Neumann, and other theorems, can be extended also to the QES in periodic fields (7). Thus, for example, the variational form of Eq. 8 can be written as (9)

$$\delta\varepsilon[\Phi] = 0, \qquad \varepsilon[\Phi] \equiv \langle\langle \Phi | \mathscr{H} | \Phi \rangle\rangle / \langle\langle \Phi | \Phi \rangle\rangle. \tag{13}$$

While the energy of the system is not a conserved quantity for an explicitly time-dependent Hamiltonian, it is possible to determine the "mean energy" $\bar{H}_\varepsilon$ of the system in the QES $\Psi_\varepsilon(\mathbf{r}, t)$:

$$\bar{H}_\varepsilon = \frac{1}{\tau} \int_0^\tau \langle \Psi_\varepsilon(\mathbf{r}, t) | H(\mathbf{r}, t) | \Psi_\varepsilon(\mathbf{r}, t) \rangle \, dt$$

$$= \varepsilon + \langle\langle \Phi_\varepsilon(\mathbf{r}, t) | i\frac{\partial}{\partial t} | \Phi_\varepsilon(\mathbf{r}, t) \rangle\rangle. \tag{14}$$

Using the Hellmann–Feynman theorem, it can be shown that (10)

$$\bar{H}_\varepsilon = \varepsilon - \omega \frac{\partial\varepsilon}{\partial\omega}. \tag{15}$$

Other properties of QESs are reviewed in Ref. 7. In the following section, we discuss how the QES eigenvalues and eigenvectors may be determined through the solution of a stationary matrix eigenproblem, a method first introduced by Shirley (3) for a two-level quantum system in a periodic field.

C. Periodic Perturbation as Stationary Problem: Time-Independent Floquet Hamiltonian Method

The QES function $\Psi_\alpha(\mathbf{r}, t)$, Eq. 6, can be expanded in a Fourier series,

$$\Psi_\alpha(\mathbf{r}, t) = \exp(-i\varepsilon_\alpha t) \sum_{n=-\infty}^{\infty} C_\alpha^{(n)}(\mathbf{r}) \exp(-in\omega t). \tag{16}$$

Thus, a QES can be expressed as a superposition of stationary states with energies equal to $\varepsilon_\alpha + n\omega$. The functions $C_\alpha^{(n)}(\mathbf{r})$ of 16 can be further expanded in terms of the orthonormal set of unperturbed eigenfunctions of $H_0(\mathbf{r})$, namely,

$(|\beta(\mathbf{r})\rangle)$,

$$C_\alpha^{(n)}(\mathbf{r}) = \sum_\beta \Phi_{\alpha\beta}^{(n)} |\beta(\mathbf{r})\rangle. \tag{17}$$

Substituting Eqs. 16 and 17 into Eq. 1, we obtain the following system of equations:

$$\sum_n \sum_\beta [\langle \alpha | \hat{H}^{[m-n]} | \beta \rangle - (\varepsilon_\alpha + m\omega)\delta_{mn}\delta_{\alpha\beta}]\Phi_{\alpha\beta}^{(n)} = 0, \tag{18}$$

where

$$\hat{H}^{[q]}(\mathbf{r}) \equiv \frac{\omega}{2\pi} \int_0^{2\pi/\omega} \hat{H}(\mathbf{r}, t)\exp(iq\omega t)\,dt. \tag{19}$$

Note that for a linearly polarized monochromatic field, $V \sim \cos\omega t$, only the matrix elements $\langle \alpha | \hat{H}^{[q]} | \beta \rangle$ with $q = 0$, ± 1 are nonvanishing. The system of Eqs. 18 is similar to the system for a constant (i.e., time-independent) perturbation.

It is convenient to introduce the Floquet state nomenclature (3) $|\alpha n\rangle \equiv |\alpha\rangle|n\rangle$, where α is the system index, and $|n\rangle$ are the Fourier vectors ($n = 0, \pm 1, \pm 2, \ldots$) such that $\langle t|n\rangle = \exp(in\omega t)$. The system of Eqs. 18 can be recast into the form of a matrix eigenvalue equation:

$$\sum_\gamma \sum_k \langle \alpha n | \hat{H}_F | \gamma k \rangle \Phi_{\gamma\beta}^{(k)} = \varepsilon_\beta \Phi_{\alpha\beta}^{(n)}, \tag{20}$$

where $\hat{H}_F$ is the time-*independent* Floquet Hamiltonian whose matrix elements are defined by

$$\langle \alpha n | \hat{H}_F | \beta m \rangle = \hat{H}_{\alpha\beta}^{[n-m]} + n\omega\delta_{\alpha\beta}\delta_{nm}. \tag{21}$$

If follows from Eq. 20 that the quasi-energies are eigenvalues of the secular equation

$$\det|\hat{H}_F - \varepsilon I| = 0. \tag{22}$$

As an example, consider the MPE of the vibrational–rotational states of a diatomic molecule with dipole moment $\boldsymbol{\mu}(\mathbf{r})$ by a monochromatic field with amplitude $\mathbf{E}_0$, frequency ω, and phase ϕ, respectively. In the electric dipole approximation, the interaction potential energy between the quantum system

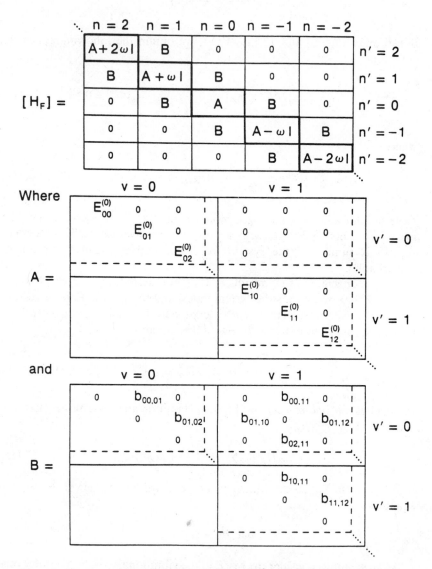

Figure 2. Structure of the exact Floquet Hamiltonian $\hat{H}_F$ in the Floquet state basis $(|vj, n\rangle)$. Hamiltonian is composed of the diagonal Floquet blocks of type A and the off-diagonal blocks of type B. $E_{vj}^{(0)}$ are unperturbed vibrational–rotational energies and $b_{vj,v'j'}$ are electric dipole coupling matrix elements.

and the classical field is given by

$$V(\mathbf{r}, t) = -\boldsymbol{\mu}(\mathbf{r}) \cdot \mathbf{E}_0 \cos(\omega t + \phi). \tag{23}$$

The Floquet matrix $\hat{H}_F$ possesses a block tridiagonal form as shown in Figure 2. The determination of the vibrational–rotational quasi-energies $\varepsilon_{\alpha n}$ and QESs $|\varepsilon_{\alpha n}\rangle$ thus reduces to the solution of a time-independent Floquet matrix eigenproblem. Figure 2 shows that $\hat{H}_F$ has a periodic structure with only the number of ω's in the diagonal elements varying from block to block. This structure endows the quasi-energy eigenvalues and eigenvectors of $\hat{H}_F$ with the following general periodic properties:

$$\varepsilon_{\alpha n} = \varepsilon_{\alpha 0} + n\omega \tag{24}$$

$$\langle \alpha, n + p | \varepsilon_{\beta, m + p} \rangle = \langle \alpha, n | \varepsilon_{\beta m} \rangle. \tag{25}$$

Consider now the transition probability from an initial quantum state $|\alpha\rangle$ to a final quantum state $|\beta\rangle$. The time evolution operator $\hat{U}(t, t_0)$, in matrix form, can be expressed as

$$\hat{U}_{\beta\alpha}(t, t_0) \equiv \langle \beta | \hat{U}(t, t_0) | \alpha \rangle$$

$$= \sum_n \langle \beta n | \exp[-i\hat{H}_F(t - t_0)] | \alpha 0 \rangle \exp(in\omega t). \tag{26}$$

Equation 26 shows that $\hat{U}_{\beta\alpha}(t, t_0)$ can be interpreted as the amplitude that a system initially in the Floquet state $|\alpha 0\rangle$ at time t_0 evolve to the Floquet state $|\beta n\rangle$ by time t according to the time-independent Floquet Hamiltonian $\hat{H}_F$ summed over n with weighting factors $e^{in\omega t}$. The latter interpretation enables one to solve problems involving Hamiltonians periodic in time by methods applicable to time-independent Hamiltonians. The transition probability going from the initial quantum state $|\alpha\rangle$ and a coherent field state to the final quantum state $|\beta\rangle$ summed over all final field states can now be written as

$$P_{\alpha \to \beta}(t, t_0) = |\hat{U}_{\beta\alpha}(t, t_0)|^2$$

$$= \sum_k \sum_m \langle \beta k | \exp[-i\hat{H}_F(t - t_0)] | \alpha 0 \rangle e^{im\omega t_0}$$

$$\times \langle \alpha m | \exp[-i\hat{H}_F(t - t_0)] | \beta k \rangle. \tag{27}$$

The quantity of experimental interest, however, is the transition probability averaged over initial times t_0 (or equivalently averaged over the initial phases of the field seen by the system), keeping the elapsed time $t - t_0$ fixed. This yields

$$P_{\alpha \to \beta}(t - t_0) = \sum_k |\langle \beta k | \exp[-i\hat{H}_F(t - t_0)] | \alpha 0 \rangle|^2. \tag{28}$$

Finally, averaging over $t - t_0$, one obtains the long-time average transition probability

$$\bar{P}_{\alpha \to \beta} = \sum_k \sum_{\gamma l} |\langle \beta k | \varepsilon_{\gamma l} \rangle \langle \varepsilon_{\gamma l} | \alpha 0 \rangle|^2. \tag{29}$$

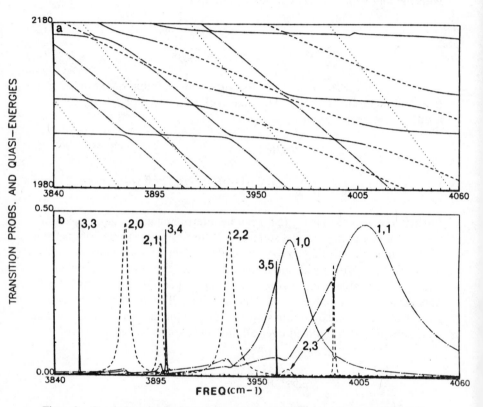

Figure 3. Quasi-energy plots (a) and time-averaged MPE transition probabilities $\bar{P}_{00 \to vj}$ (b) for the HF molecule subject to both the laser ($E_{ac} = 1.0\,\text{TW/cm}^2$) and the dc electric ($E_{dc} = 10^{-4}$ a.u.) fields simultaneously. Dot-dash lines: one-photon peaks. Dashed lines: two-photon peaks. Solid lines: three-photon peaks. Nonlinear effects such as power broadening, dynamical Stark shift, Autler–Townes multiplet splitting, hole burning, and S-hump behaviors are observed and can be correlated with avoided crossing pattern of the quasi-energy levels. (Adapted from Ref. 11).

Much information can be obtained from the plot of the quasi-energy eigenvalues (or the characteristic exponents) of the Floquet Hamiltonian. The main feature of the quasi-energy plot is illustrated in Figure 3 for the case of the HF molecule subject to both the ac and dc fields (11). Nonlinear effects such as power broadening, dynamical Stark effect, Autler–Townes multiplet splitting (5), hole burning, and S-hump behaviors are observed and may be correlated with the avoided crossing patterns in the quasi-energy diagram (Figure 3). Many of the salient features in the spectral lineshapes may be qualitatively understood in terms of an analytical three- or four-level model. The addition of a dc electric field spoils the restriction of the rotational dipole selection rule and induces significant intermixing of the bare molecular rovibrational states. Owing to the greater number of strongly coupled nearby states in the dc field, nonlinear optical effects such as those already mentioned appear at a much lower ac field strength than they would in the absence of the dc field (11). The introduction of an external dc field, therefore, strongly enhances the MPE probabilities and results in a much richer spectrum, in accord with the experimental observations (12).

D. Time Propagator Methods

In addition to the time-independent Floquet Hamiltonian method described in Section II.B, an alternative approach to the study of the multiphoton excitation processes is to determine the time evolution operator $\hat{U}(t, t_0)$ from numerical solution of the Schrödinger equation ($\hbar = 1$)

$$i\frac{\partial \hat{U}(t, t_0)}{\partial t} = \hat{H}(t)\hat{U}(t, t_0), \tag{30}$$

with $\hat{H}(t + \tau) = \hat{H}(t)$ and $\hat{U}(t_0, t_0) = 1$. Numerous numerical methods, such as the Magnus approximation (13), Riemann product integral method (14), time-slicer method (15), rotating-frame transformation method (16), recursive residue generation method (17), and quasi-resonant approximation (18) have been developed for solving $\hat{U}(t, t_0)$. See Ref. 7 for a review of several numerical integration methods.

For a periodic time-dependent perturbation, the time evolution operator $\hat{U}(t, t_0)$ possesses the following important symmetry properties (19):

$$\hat{U}(t + \tau, 0) = \hat{U}(t, 0), \tag{31a}$$

$$\hat{U}(t + \tau, 0) = \hat{U}(t, 0)\hat{U}(\tau, 0), \tag{31b}$$

and

$$\hat{U}(n\tau) = \hat{U}(\tau, 0)^n. \tag{31c}$$

Thus, the time propagator operator over one period, $\hat{U}(\tau, 0)$, provides essentially all the information one needs to determine the long-time behavior of the system. In practice, one truncates the basis set to a finite size, and $\hat{U}(\tau, 0)$ may be diagonlized by some unitary transformation S,

$$S^+ \hat{U}(\tau, 0) S = e^{-iD}, \tag{32}$$

where D is a diagonal matrix. Thus,

$$\hat{U}(\tau, 0) = S e^{-iD} S^+,$$

leading to

$$\hat{U}(n\tau, 0) = \hat{U}(\tau, 0)^n = S e^{-inD} S^+. \tag{33}$$

E. Artificial Intelligence in Multiphoton Dynamics: Most-Probable-Path Approach

Among the many new fields spawned by the proliferation of intense, tunable lasers is that of collisionless multiphoton excitation and dissociation of molecules (1, 2). Experiments have been performed that show that polyatomic molecules (prototypically, SF_6) can be dissociated without collisional assistance by irradiation with intense infrared laser light (2). While the evidence is plentiful and conclusive for medium- and large-size molecules (more than five atoms), the situation is not so clear for smaller polyatomics. Recent studies have conflicted (20, 21) over the detection of dissociation in triatomic molecules, notably, OCS, SO_2, and O_3. It is thus desirable to perform comprehensive theoretical studies of the MPE–MPD dynamics of these sparse intermediate-case molecules.

The *most-probable-path approach* (MPPA) was first introduced by Tietz and Chu (22) in an ab initio study of high-order MPE of SO_2. A brute-force attempt to calculate polyatomic MPE would soon become prohibitive due to the large number of vibrational–rotational states and huge size of the Floquet matrix needed for achieving convergence. For typical 15-photon calculations for SO_2 including only vibrational ladders, for example, a matrix on the order of 10,000 × 10,000 would have to be diagonalized at each frequency and field strength. In any exact Floquet calculation, however, the majority of the molecule–field states are unimportant due to large detuning or very small coupling matrix elements. The MPPA is a practical strategy introduced to determine which molecule–field states are, in fact, important at each step of the multiphoton processes. The procedure is derived from algorithms that utilize *artificial intelligence* to prune the number of choices at each node (photon order) of a decision tree (23). Similar to some minimax game-playing programs, the MPPA examines the possible paths to take at each photon order iteration with the static evaluation function given by Nth-order perturbation theory (this is a breadth-first search). If all paths were followed exhaustively, the problem would be beyond practical solution. In game theory, one answer is to ignore paths that start with very poor moves. The MPPA likewise uses a breadth-limiting heuristic technique and discards any paths for which the Nth-order coupling term is small (with respect to other Nth-order terms).

The MPPA begins by calculating all possible second-order perturbative terms. The N_p largest couplings (where N_p is the number of paths to keep at each step) are chosen as the most probable paths through second order. The initial state and the intermediate states of the chosen paths are marked as important and are used in the final calculations. At each iterative step, the method calculates all possible $(N + 1)$st-order couplings (paths) using only the N_p Nth-order paths saved in the last iteration. The $(N + 1)$st-order couplings are then examined, and the largest N_p are saved for further traversal. The Nth-order states that have now become intermediate to a large $(N + 1)$st-order path are "important" and are marked for later use. By iterating long enough,

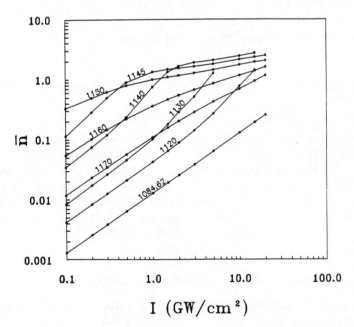

Figure 4. Dependence of the average number of photons absorbed ($\bar{n}$) by SO_2 on laser intensity. The different traces correspond to the indicated excitation frequencies (cm^{-1}) of the laser. Typical infrared MPD of SO_2 requires about forty 9.3-μm photons. The results shown in this figure indicate that MPD of SO_2 cannot be achieved for $I < 20\,GW/cm^2$. (Adapted from Ref. 22).

one can traverse the entire Floquet molecule–field basis space, saving only those states that are important to various ith-order processes. The reduction of the basis set is quite dramatic and leads to many-orders-of-magnitude savings in computer time yet maintains good accuracy in most cases.

Using the MPPA, it is found that (22) collisionless MPD of SO_2 will *not* be achievable at laser intensities under $20\,GW/cm^2$ (Figure 4), in agreement with the recent experimental results of Simpson and Bloembergen (24). The later experiment, however, has extended laser power further up to $300\,GW/cm^2$ and found that appreciable MPD yields begin to occur and that the process is controlled by the laser intensity and not by the laser fluence. Further MPPA study (25) showed that the MPE of SO_2 is primarily a one-ladder pumping phenomenon dominated by the power-broadening effect, and the MPD is likely to occur (though the yield is predicted to be small, $P < 10^{-3}$) at laser intensities above $500\,GW/cm^2$.

The selection of important multiphoton pathways via artificial intelligence algorithms is further exploited recently by Chang and Wyatt in the study of MPE of a spherical top molecule (26) and by Wang and Chu in the study of MPE and three-dimensional quantum diffusion phenomena in Rydberg atoms driven by microwave fields (27). In the latter case, MPPA allows the reduction of an ultralarge Floquet matrix (on the order of millions) to an effective matrix of a manageable size (on the order of several hundreds) (27).

F. Non-Hermitian Floquet Methods for MPI–MPD and Above-Threshold Ionization

The Floquet matrix methods described in previous sections involving the time-independent *Hermitian* Floquet Hamiltonian provide nonperturbative ab initio techniques for the treatment of bound–bound multiphoton transitions. This methods cannot, however, be applied to bound–free transitions such as MPI or MPD processes. One of the major recent extensions of the Floquet theory are the developments of *non-Hermitian* Floquet methods, invoking the use of the method of complex scaling (28) (see Figure 5 for an example) and L^2-continuum discretization techniques (29) for MPI of atoms (30, 31) and MPD of molecules (32).

1. *Non-Hermitian Floquet Hamiltonian Methods for MPI–MPD*

Applying the complex scaling transformation (28), $\mathbf{r} \to \mathbf{r}^{i\alpha}$, to the Schrödinger equation, we obtain, from Eqs. 1–2,

$$i\frac{\partial \Psi(\mathbf{r}e^{i\alpha}, t)}{\partial t} = H(\mathbf{r}e^{i\alpha}, t)\Psi(\mathbf{r}e^{i\alpha}, t), \tag{34}$$

Dilatation transformation $H_Z \longrightarrow H_Z(\alpha)$

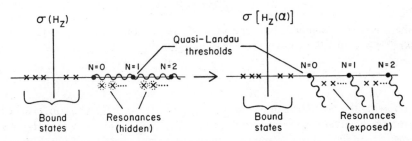

Figure 5. Effect of the complex scaling (or dilatation) transformation, $\mathbf{r} \to \mathbf{r}e^{i\alpha}$, on spectrum, $\sigma(H_z)$, of an atomic Zeeman Hamiltonian H_z. Bound states [the Coulomb series below the first $(N = 0)$ quasi-Landau threshold] are invariant to the transformation while the continua rotate about their respective quasi-Landau thresholds, exposing complex *resonance* Coulomb series (above the $N = 0$ Landau threshold) in appropriate strips of the complex energy plane. These exposed autoionizing resonances may be determined by the use of the complex-coordinate coupled-Landau-channel method (Adapted from Ref. 33).

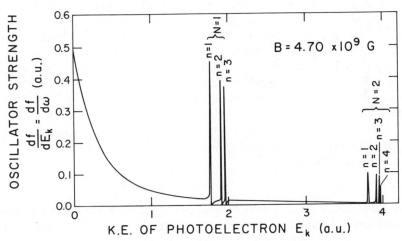

Figure 6. Photoionization oscillator strength $df/d\omega$ of ground state of hydrogen atom $(N = 0, m^\pi = 0^+)$ at the magnetic field strength $B = 4.7 \times 10^9$ G (in astrophysical condensed objects) is given as a function of the kinetic energy of the photoelectron (E_k) near the first $(N = 1)$ and the second $(N = 2)$ excited Landau thresholds. Resonant absorptions due to the presence of the first few autoionizing states are clearly seen. (Adapted from Ref. 34).

where $H(\mathbf{r}e^{i\alpha}, t)$ is now a non-Hermitian periodic Hamiltonian. The wavefunction $\Psi(\mathbf{r}e^{i\alpha}, t)$, can be written according to the Floquet theorem (5),

$$\Psi(\mathbf{r}e^{i\alpha}, t) = \exp(-i\varepsilon t)\Phi(\mathbf{r}e^{i\alpha}, t), \tag{35}$$

where ε is the *complex* quasi-energy, and the periodic function $\Phi(\mathbf{r}e^{i\alpha}, t) = \Phi(\mathbf{r}e^{i\alpha}, t + \tau)$ satisfies the eigenvalue equation

$$\mathscr{H}(\mathbf{r}e^{i\alpha}, t)\Phi(\mathbf{r}e^{i\alpha}, t) = \varepsilon\Phi(\mathbf{r}e^{i\alpha}). \tag{36}$$

Following the procedure described in Section II.C, Eq. 36 can be converted to a matrix eigenvalue equation similar to Eq. 20, except that $\hat{H}_F(\mathbf{r}e^{i\alpha}) \equiv \hat{H}_F(\alpha)$ is now an analytically continued, time-independent *non-Hermitian* Floquet Hamiltonian. The complex scaling transformation distorts the continuous spectrum away from the real axis, exposing the Stark resonances (i.e., complex quasi-energy states) in appropriate higher Riemann sheets and also allowing use of finite variational expansions employing an L^2 basis function chosen from a complete discrete basis. The use of a complete L^2 basis obviates the necessity of explicit introduction of exact atomic or molecular bound and continuum states, thus reducing all computations to those involving finite-dimensional non-Hermitian matrices. The use of complex coordinates not only allows direct calculation of eigenvalue parameters associated with complex dressed states but also completely avoids numerical problems arising from strong coupling between overlapping atomic or molecular continua. The *real* part of the complex quasi-energy (E_R) provides ac Stark-shifted energy, whereas the *imaginary* part ($\Gamma/2$) determines the MPI or MPD rate. The methods have been applied to the study of intensity-dependent MPI of atomic hydrogen in linearly (30) and circularly (31) polarized fields, to atomic autoionization (33) (Figure 5)** and photoionization (34) (Figure 6) in super-strong magnetic fields, to two-photon dissociation of H_2^+ and HD^+ from highly excited vibrational states (32), and to photodissociation of van der Waals molecules (35). For a review of these complex quasi-energy developments and other related MPI–MPD methods, the reader is referred to Refs. 7 and 35 for details.

2. *Above-Threshold Ionization*

A recent advancement in the study of the MPI of atoms is the discovery of an unexpected intense-field phenomenon called *above-threshold ionization*

**The reader is referred to Ref. 87 for a detailed discussion of the nature of quasi-Landau resonances and how the dilatation transformation may be extended to facilitate the determination of these resonances. Reference 88 contains also a good introduction to Landau degeneracy.

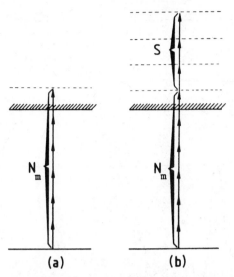

Figure 7. Schematic diagram of above-threshold ionization indicating (a) the minimum number N_m of photons necessary to ionize and (b) the number S of additional photons absorbed above the ionization threshold. The branching ratio to individual continuum depends strongly on laser intensity.

ATI (36, 37). ATI occurs when atoms are irradiated by intense laser fields, and the emitted electron can absorb $N_m + S$ photons, where N_m is the minimum number of photons required to ionize the atom and $S = 0, 1, 2, \ldots$ (Figure 7). Thus, the electron energy spectrum consists of several peaks at $(N_m + S)\hbar\omega - E_g$, where E_g is the ionization potential of the ground state. When the external field strength further increases, the electron peaks broaden and shift, and the slowest electron peaks eventually disappear (37) (see Figure 8 for an example). Various models have been proposed to interpret these phenomena (38). The recent extension of the non-Hermitian Floquet matrix method provides a nonperturbative ab initio technique for the study of the MPI–ATI phenomena (39). It is found that the ionization potential is frequency and intensity dependent and is determined by both the ac Stark shift of the ground state and the continuum threshold *upshift*. The disappearance of the lowest electron energy peaks can be accounted for by the shift of the ionization threshold in intense fields.

The ionization potential in intense fields may be defined as (39)

$$\varepsilon_{\text{th}}(F) = \bar{\varepsilon}_{\text{osc}} + |E_R(F)|, \tag{37}$$

where F is the (peak) field strength, $E_R(F)(<0)$ is the field-dependent

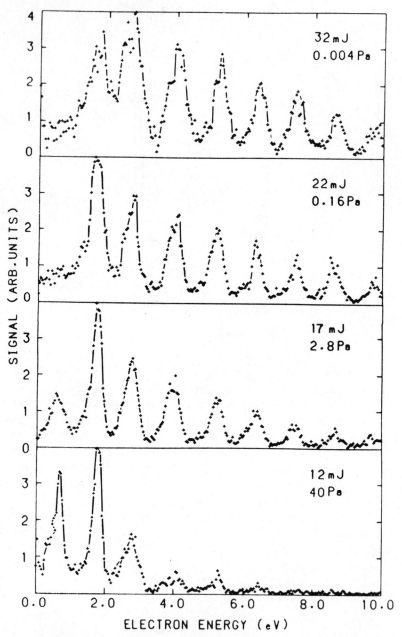

Figure 8. Electron energy spectrum from ionization of xenon gas by the Nd–YAG laser (1.064 μm). The first peak corresponds to 11-photon ionization to $P_{3/2}$ and 12-photon ionization to the $P_{1/2}$ continuum. Note the unexpected rapid disappearance of the slowest photoelectrons when the light intensity is raised (Adapted from Ref. 37).

perturbed ground-state energy obtained from the complex quasi-energy calculation, and

$$\bar{\varepsilon}_{osc} = e^2 F^2 / 4m\omega^2 \tag{38}$$

is the average *quiver* kinetic energy (also known as the *pondermotive potential*) picked up by an electron of mass m and charge e driven sinusoidally by the field. Since in the limit of high quantum numbers, a Rydberg electron becomes a free electron, the continuum threshold is shifted up by an amount equal to $\bar{\varepsilon}_{osc}$. Electrons traversing a laser beam scatter elastically from regions of high

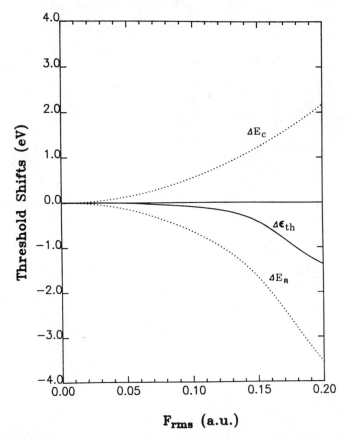

Figure 9. Intensity-dependent threshold shift $\Delta\varepsilon_{th}(F) = \Delta E_C + \Delta E_R$ for $\omega = 0.5$ a.u. ($N_m = 1$). Here $\Delta E_R = E_R(0) - E_R(F)$ is the ac Stark shift of the ground state, while $\Delta E_C = e^2 F^2 / 4m\omega^2 = e^2 F_{rms}^2 / 2m\omega^2$ is the continuum threshold upshift. The rms field strength of $F_{rms} = 1.0$ a.u. corresponds to rms intensity of 7.0×10^{16} W/cm^2.

light intensity by the pondermotive potential. Thus, an electron with energy $\varepsilon_{el}(F)$ less than $\bar{\varepsilon}_{osc}$ cannot escape from the Coulomb potential and is trapped. From Eq. 37, we can define the threshold shift as

$$\Delta\varepsilon_{th}(F) = \varepsilon_{th}(F) - \varepsilon_{th}(F = 0), \tag{39}$$

where $\varepsilon_{th}(F = 0)$ is the field-free ionization threshold.

The total energy of the emitted electron in the field can be written as

$$\varepsilon_{el}(F) = N\hbar\omega + E_R(F) = e^2F^2/4m\omega^2 + P_T^2/2m, \tag{40}$$

where $N = N_m + S$ is the total number of photons absorbed by the electron near the atom. Since a free electron cannot absorb or emit photons after

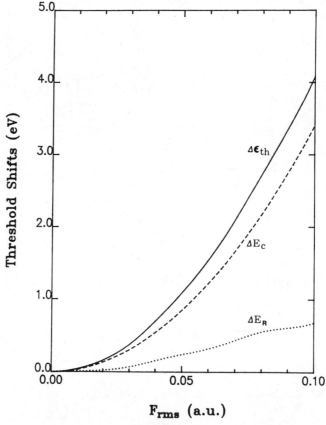

Figure 10. Intensity-dependent threshold shifts for $\omega = 0.2$ a.u. ($N_m = 3$).

leaving the Coulomb field, the electron has an energy ε_{el} that is the same in the laser field as it is at the detector. Thus, the pondermotive potential acts to alter the kinetic energy $(P_T^2/2m)$ of the electron from its value outside the laser field to a lower value inside the laser.

Figures 9 and 10 show typical examples of intensity- and frequency-dependent threshold shifts of atomic hydrogen in intense laser fields, where $\Delta E_R = E_R(F = 0) - E_R(F)$ is the ac Stark shift, $\Delta E_C = e^2 F^2/4m\omega^2$ is the continuum threshold upshift, and $\Delta\varepsilon_{th}(F)$ is the net threshold shift defined by Eq. 39 and is equivalent to the sum of ΔE_R and ΔE_C. Figure 9 shows the threshold shifts typical to the one-photon $(N_m = 1)$ dominant process $(\omega \geqslant 0.5 \, \text{a.u.})$, whereas Figure 10 shows the typical phenomena for the multiphoton $(N_m = 3$ in this case) dominant process $(\omega < 0.5 \, \text{a.u.})$. Note the marked difference between the two cases. For $N_m = 1$ (Figure 9), both the ground state $[E_R(F) > E_R(0)]$ and the continua are upshifted, with the ac Stark shift $|\Delta E_R|$ being greater than ΔE_C. The resulting net threshold shift $\Delta\varepsilon_{th}(F)$ becomes more negative as the field strength increases. Hence, the ionization potential *decreases* with increasing F. On the other hand, for $N_m \geqslant 2$, such as the case $\omega = 0.2 \, \text{a.u.}$ shown in Figure 10, the ground-state energy shifted downward $[E_R(F) < E_R(0)]$, whereas ΔE_C shifts the continuum threshold upward. The result is a large positive net threshold shift, and the ionization potential *increases* rapidly with increasing field strength F. As a general rule, the pondermotive potential ΔE_C becomes more and more important than the ac Stark shift $|\Delta E_R|$ as N_m increases or ω decreases. The consequence is that the ionization potential increases rather rapidly with both F and N_m. The disappearance of the lowest energy electrons in the MPI–ATI experiment of xenon (37) (Figure 8), for example, can be attributed to this threshold shift effect.

III. GENERALIZED MANY-MODE FLOQUET TREATMENTS OF MULTIPHOTON PROCESSES IN POLYCHROMATIC FIELDS

The conventional Floquet theory (3, 5, 6) requires the Hamiltonian to be explicitly periodic in time and is therefore applicable to problems involving only *one* monochromatic radiation field. The desire to develop an exact formalism for treating multifrequency problems is apparent in view of the intense current interest in the study of atomic and molecular processes in polychromatic or multicolor fields. [Examples are multiphoton double resonance, collisions in two laser fields, MPD of polyatomic molecules by two infrared lasers, and multiple wave mixings (1, 2).] Recently an *exact* extension of the one-mode Floquet Hamiltonian formalism (3) to a generalized MMFT has been found (40). This makes it possible to treat, for the first time, the time-dependent problem of any finite-level system exposed to *polychromatic* fields as an equivalent *time-independent* infinite-dimensional eigenvalue problem.

A.   Nonlinear Multiphoton Dynamics in Intense Bichromatic Fields

Without loss of generality, let us consider the interaction of an arbitrary N-level system with two monochromatic radiation fields. In the electric dipole approximation, the Schrödinger equation can be written as ($\hbar = 1$)

$$i\frac{\partial\Psi(\mathbf{r}, t)}{\partial t} = H(\mathbf{r}, t)\Psi(\mathbf{r}, t), \tag{41}$$

where the Hamiltonian $H(\mathbf{r}, t)$ is bichromatic in time,

$$H(\mathbf{r}, t) = H_0(\mathbf{r}) - \boldsymbol{\mu}(\mathbf{r})\cdot[\mathbf{E}_1(t) + \mathbf{E}_2(t)]. \tag{42}$$

Here H_0 and μ are the unperturbed Hamiltonian and the dipole moment of the system, respectively, and $\mathbf{E}_1$ and $\mathbf{E}_2$ are classical fields given by

$$\mathbf{E}_i(t) = \text{Re}\,[E_i\hat{\zeta}_i e^{-i\omega_i t}], \tag{43}$$

where Re signifies the real part of, and E_i, $\hat{\zeta}_i$, and ω_i are, respectively, the electric field amplitude, the polarization vector, and the frequency associated with the ith field. The polarization vector is chosen to be

$$\hat{\zeta} = \hat{z}\,[\text{linear polarization (LP)}]$$

$$= \hat{x} + i\eta\hat{y}\,[\text{circular polarization (CP)}],$$

where $\hat{x}$, $\hat{y}$, $\hat{z}$ are the unit vectors and $\eta = +1(-1)$ corresponds to the left (right) circularly polarized light [LCP (RCP)].

The many-mode Floquet theory (40) allows the *exact* transformation of the bichromatic time-dependent problem, Eq. 41, into an equivalent time-independent infinite-dimensional matrix eigenvalue problem (in Dirac notation):

$$\sum_{\gamma_2}\sum_{k_1}\sum_{k_2}\langle\gamma_1 n_1 n_2|\hat{H}_F|\gamma_2 k_1 k_2\rangle\langle\gamma_2 k_1 k_2|\lambda\rangle = \lambda\langle\gamma_1 n_1 n_2|\lambda\rangle, \tag{44}$$

where

$$\langle\gamma_1 n_1 n_2|\hat{H}_F|\gamma_2 k_1 k_2\rangle = H_{\gamma_1\gamma_2}^{[n_1-k_1,n_2-k_2]}$$
$$+ (n_1\omega_1 + n_2\omega_2)\delta_{\gamma_1\gamma_2}\delta_{n_1 k_1}\delta_{n_2 k_2}, \tag{45}$$

with

$$H_{\gamma_1\gamma_2}^{[n_1,n_2]} = E_{\gamma_1}\delta_{\gamma_1\gamma_2}\delta_{n_10}\delta_{n_20} + \sum_{i=1}^{2} V_{\gamma_1\gamma_2}^{(i)}(\delta_{n_i,1} + \delta_{n_i,-1}), \tag{46}$$

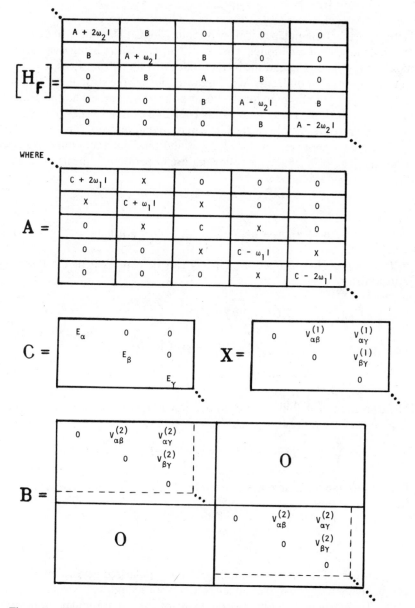

Figure 11. Floquet Hamiltonian for two linearly polarized radiation field problems constructed in a symmetric pattern. Here ω_1 and ω_2 are the two radiation frequencies, E_α, $E_\beta, \ldots$ are unperturbed energies of H_0, and $V_{\alpha\beta}^{(i)}$ $(i = 1, 2)$ is the electric dipole coupling matrix element between unperturbed states $|\alpha\rangle$ and $|\beta\rangle$ by the ith field. Note that the diagonal block A possesses an identical Floquet structure to that of one laser problem (cf. Figure 2). Figure 11 can be generalized to the N-field problem. (Adapted from Ref. 40).

$$V^{(i)}_{\gamma_1\gamma_2} = -\tfrac{1}{2}E_i\langle\gamma_1|\boldsymbol{\mu}\cdot\boldsymbol{\zeta}_i|\gamma_2\rangle. \tag{47}$$

Here $\hat{H}_F$ is the (time-independent) two-mode Floquet Hamiltonian defined in terms of the *generalized* Floquet basis state $|\gamma n_1 n_2\rangle$, with γ being an atomic (or molecular) state of H_0, and the integer index n_i (ranges from $-\infty$ to $+\infty$) a Fourier component of the ith field. Figure 11 depicts the structure of the two-mode Floquet Hamiltonian for the linear polarization case. The components are ordered in such a way that γ runs over unperturbed states (denoted by Greek letters) of H_0 before each change in n_1 and n_1, in turn, runs over before n_2. The quasi-energy eigenvalues $\{\lambda_{\gamma n_1 n_2}\}$ and their corresponding eigenvectors $\{|\lambda_{\gamma n_1 n_2}\rangle\}$ of $\hat{H}_F$ have the following useful periodicity forms:

$$\lambda_{\gamma n_1 n_2} = \lambda_{\gamma 00} + n_1\omega_1 + n_2\omega_2, \tag{48}$$

and

$$\langle\gamma_1, n_1 + q_1, n_2 + q_2|\lambda_{\gamma_2, n_1+q_1, n_2+q_2}\rangle = \langle\gamma_1 n_1 n_2|\lambda_{\gamma_2 n_1 n_2}\rangle. \tag{49}$$

The time evolution operator $\hat{U}(t, t_0)$ can be expressed in the following matrix form:

$$\begin{aligned}
\hat{U}_{\beta\alpha}(t, t_0) &\equiv \langle\beta|\hat{U}(t, t_0)|\alpha\rangle \\
&= \sum_{n_1=-\infty}^{\infty}\sum_{n_2=-\infty}^{\infty}\langle\beta n_1 n_2|\exp[-i\hat{H}_F(t-t_0)]|\alpha 00\rangle \\
&\quad \times \exp[i(n_1\omega_1 + n_2\omega_2)t].
\end{aligned} \tag{50}$$

The transition probability averaged over the initial time t_0 while keeping the elapsed time $t - t_0$ fixed is given by

$$P_{\alpha\rightarrow\beta}(t-t_0) = \sum_{k_1 k_2}|\langle\beta k_1 k_2|\exp[-i\hat{H}_F(t-t_0)]|\alpha 00\rangle|^2. \tag{51}$$

Performing the long-time average over $t - t_0$ gives the time-averaged transition probability

$$\bar{P}_{\alpha\rightarrow\beta} = \sum_{k_1 k_2}\sum_{\gamma l_1 l_2}|\langle\beta k_1 k_2|\lambda_{\gamma l_1 l_2}\rangle\langle\lambda_{\gamma l_1 l_2}|\alpha 00\rangle|^2. \tag{52}$$

In the case of two monochromatic circularly polarized fields, simplification can be made if H_0 possesses either spherical or cylindrical symmetry. Since $[\hat{H}_0, \hat{j}_z] = 0$, where $\hat{j}_z$ is the z component of the total angular momentum operator $\mathbf{j}$, one can write

$$\hat{H}_0|\alpha m\rangle = E^{(0)}_{\alpha m}|\alpha m\rangle, \tag{53}$$

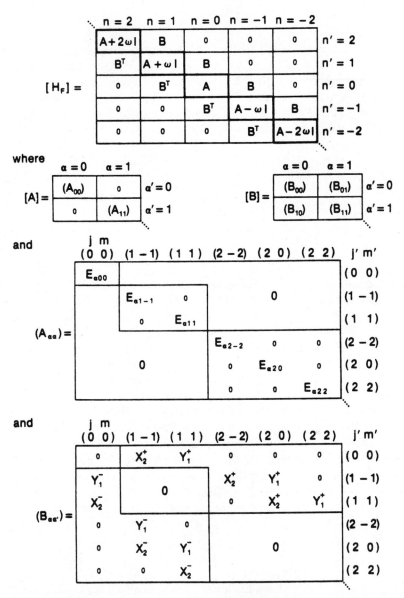

Figure 12. Floquet Hamiltonian for two-rotating-field problems constructed after making rotating-frame transformations. In the diagonal A blocks, $E_{\alpha jm} = E_{\alpha j}^{(0)} + m\Omega$, $\Omega = \frac{1}{2}(\omega_1 + \omega_2)$, and $\omega = \frac{1}{2}(\omega_1 - \omega_2)$. In the off-diagonal B blocks, coupling matrices $X^{\pm}$ and $Y^{\pm}$ are defined as

$$(Y_1^{\pm})_{\alpha jm, \alpha' j'm'} = -\tfrac{1}{2} E_1 \langle \alpha | \mu(r) | \alpha' \rangle \langle jm | \sin\theta e^{-i\phi} | j'm' \rangle \delta_{j',J \pm 1} \delta_{m',m+1}$$

and

$$(X_2^{\pm})_{\alpha jm, \alpha' j'm'} = -\tfrac{1}{2} E_2 \langle \alpha | \mu(r) | \alpha' \rangle \langle jm | \sin\theta e^{+i\phi} | j'm' \rangle \delta_{j',J \pm 1} \delta_{m',m-1}.$$

763

$$\hat{j}_z |\alpha m\rangle = m |\alpha m\rangle, \tag{54}$$

where α denotes the additional quantum number necessary to specify the unperturbed state. By transforming $H(t)$ into a rotating frame using the unitary operator $\hat{S}(t) = \exp(i\hat{j}_z \Omega t)$, the bichromatic problem can be reduced to a monochromatic problem. Thus, for example, the transformed time-dependent Hamiltonian for the two-rotating-field ($\circlearrowright\circlearrowright$) case becomes periodic in time (with period $= 2\pi/\omega$):

$$\tilde{H}(t) = [H_0 - \Omega\hat{j}_z] - [\mu_- E_1 + \mu_+ E_2]e^{i\omega t} - [\mu_+ E_1 + \mu_- E_2]e^{-i\omega t}, \tag{55}$$

where

$$\Omega = \tfrac{1}{2}(\omega_1 + \omega_2), \qquad \omega = \tfrac{1}{2}(\omega_1 - \omega_2),$$

and

$$\mu_\pm = \mu_x \pm i\mu_y = \mu(r) \sin\theta e^{\pm i\phi}.$$

The corresponding Floquet Hamiltonian is shown schematically in Figure 12.

As an example, consider the MPE dynamics in a two-level, or spin $\tfrac{1}{2}$, system ($|\alpha\rangle, |\beta\rangle$, with $E_\beta - E_\alpha = \hbar\omega_0$) driven by bichromatic linearly ($\uparrow\uparrow$) or circularly ($\circlearrowright\circlearrowright$ or $\circlearrowleft\circlearrowleft$) polarized fields with frequencies ω_1, ω_2 and coupling strengths b_1, b_2 ($b_i = V_{\alpha\beta}^{(i)}$ in Eq. 47), respectively. The conservation rule of the total angular momentum of the system plus fields determines the types of multiphoton transitions allowed. We shall use the notation $[n_1, \pm n_2]$ to denote n_1 photons of the first field (ω_1) are absorbed, and n_2 photons of the second field (ω_2) are absorbed ($+$) or emitted ($-$). Thus, for the $\uparrow\uparrow$ case, one can show that the allowed transition types are $[n + 2m + 1, \pm n]$, with $n, m = 0,$ $1, 2, \ldots,$ and the number of MPE pathways for each type is $N_p = (2n + 2m + 1)!/[n!(n + 2m + 1)!]$. On the other hand, only the $[n + 1, -n]$ type is allowed for the circular polarization case, and $N_p = 1$. Further note that for any two-level system, regardless of the number and the polarization of the fields, there is a simple relationship between λ and $\bar{P}$, namely,

$$\bar{P}_{\alpha\to\beta}(\omega_0) = \tfrac{1}{2}\left[1 - 4\left(\frac{\partial\lambda}{\partial\omega_0}\right)^2\right]. \tag{56}$$

Useful analytical expressions for the resonance shift (i.e., the *generalized Bloch–Siegert shift*) and (power-broadening) width of each multiphoton resonance transition may be obtained by invoking some sort of nearly degenerate perturbative treatment (41–43) of the two- (or many-) mode Floquet Hamiltonian $\hat{H}_F$. Thus, for example, the bichromatic resonance shift δ_n [defined as $\delta_n \equiv (n + 1)\omega_1 - n\omega_2 - \omega_0'$, where ω_0' is the resonance frequency such that $\bar{P}_{\alpha \to \beta}(\omega_0') = \frac{1}{2}$] in the linear polarization case can be written as (44)

$$\delta_n^L \cong 2(A + B)/(1 + C + D), \tag{57}$$

where

$$A = |b_1|^2 \left[\frac{1}{n(\omega_1 - \omega_2) + 2\omega_1} \right] + \left[\frac{\theta(n)}{n(\omega_1 - \omega_2)} \right],$$

$$B = |b_2|^2 \left[\frac{1}{(n + 1)(\omega_1 - \omega_2) + 2\omega_2} \right] + \left[\frac{1}{(n + 1)(\omega_1 - \omega_2)} \right],$$

$$C = |b_1|^2 \left[\frac{1}{(n(\omega_1 - \omega_2) + 2\omega_1)^2} \right] + \left[\frac{\theta(n)}{n^2(\omega_1 - \omega_2)^2} \right],$$

$$D = |b_2|^2 \left[\frac{1}{((n + 1)(\omega_1 - \omega_2) + 2\omega_2)^2} \right] + \left[\frac{1}{(n + 1)^2(\omega_1 - \omega_2)^2} \right],$$

while in the circular field case (45)

$$\delta_n^c \cong 2(E + F)/(1 + G + H), \tag{58}$$

where

$$E = |b_1|^2 \left[\frac{\theta(n)}{n(\omega_1 - \omega_2)} \right], \qquad F = |b_2|^2 \left[\frac{1}{(n + 1)(\omega_1 - \omega_2)} \right],$$

$$G = |b_1|^2 \left[\frac{\theta(n)}{n^2(\omega_1 - \omega_2)^2} \right], \qquad H = |b_2|^2 \left[\frac{1}{(n + 1)^2(\omega_1 - \omega_2)^2} \right],$$

and

$$\theta(n) = \begin{cases} 0 & \text{if } n = 0, \\ 1 & \text{otherwise.} \end{cases}$$

Detailed analytical results for resonance shifts, widths, multiphoton transition probabilities, and spectral lineshapes for two- (44, 45) and three-level (46) systems in bichromatic fields are summarized in Ref. 47.

B. SU(N) Dynamical Symmetries and Nonlinear Coherence

It was shown by Feynman, Vernon, and Hellwarth (48) in 1957 that for two-level systems, the description of magnetic and optical resonance phenomena can be greatly simplified by the use of the Bloch spin or pseudospin vector. Extension of the vector model to the N-level system ($N \geq 3$) has recently been made by several groups (49). It is found that the dynamical evolution of N-level nondissipative systems can be expressed in terms of the generalized rotation of an $(N^2 - 1)$-dimensional coherence vector **S** whose property can be analyzed by appealing to SU (N) group symmetry. For example, the time evolution of three-level systems can be described by a coherence vector of constant length rotating in an eight-dimensional space.

The study of SU (N) dynamical evolution of the coherent vector **S** and the symmetry-breaking effects embodied in N-level systems subject to an arbitrary number of monochromatic fields can be greatly facilitated with the use of MMFT (50). Thus, the $(N^2 - 1)$-dimensional coherent vector **S**(t) can be obtained directly from the relation

$$S_j(t) = \text{Tr}\,[\hat{\rho}(t)\hat{s}_j], \qquad j = 1, 2, \ldots, N^2 - 1, \tag{59}$$

where $\hat{s}_j$ are appropriate SU (N) generators, and the density matrix $\hat{\rho}(t)$ is determined by

$$\hat{\rho}(t) = \hat{U}(t, t_0)\hat{\rho}(t_0)\hat{U}^+(t, t_0). \tag{60}$$

Here $\hat{\rho}(t_0)$ is the density matrix at the initial time t_0 (initial conditions) and the time evolution operator $\hat{U}(t, t_0)$ can be determined by means of MMFT and expressed in terms of quasi-energy eigenvalues and eigenvectors, see Eq. 50. Furthermore, the generalized Van Vleck (GVV) nearly degenerate perturbation (42, 43, 50) theory can be extended to the analytical treatment of the time-independent many-mode Floquet Hamiltonian. The general idea behind the MMFT–GVV technique is to block diagonalize the time-independent Floquet Hamiltonian $\hat{H}_F$ (such as Figure 11) so that the coupling between the *model* space and the remainder of the configuration space (called the *external* space) diminishes to a desired order. The perturbed eigenvalues and eigenvectors corresponding to the set of nearly degenerate state chosen can thus be solved approximately by considering only the model space effective Hamiltonian. One important feature of the MMFT–GVV approach is that if the perturbed model space wavefunctions are exact to the nth order, the corresponding quasi-energy eigenvalues in the model space will be accurate to the $(2n + 1)$th order. In that regard, it is interesting to note that the often used rotating-wave approximation (RWA) is merely the lowest order (i.e., $n = 0$) limit, namely, model space wavefunctions correct only to the zero order and eigenvalues accurate to the first order. Furthermore, while the RWA can only deal with sequential one-photon processes, the MMFT–GVV approach is capable of treating both one-photon and multiphoton processes on equal footing. Thus, the MMFT–GVV approach appears to be a natural and powerful extension beyond the conventional RWA limit for the nonperturbative treatment of multiphoton processes in intense polychromatic fields.

The MMFT–GVV method has recently been applied to a unified treatment of both the SU (3) symmetries and symmetry-breaking effects (caused by non-RWA terms) of nondissipative three-level systems at two-photon resonances induced by intense bichromatic fields (50). The MMFT–GVV technique reduces the infinite-dimensional time-independent two-mode Floquet Hamiltonian to a 3×3 (model space) effective Hamiltonian, from which essential analytical properties and vivid geometric motion of the eight-dimensional coherence vector are revealed for the first time (50).

C. Quasi-periodic and Chaotic Motions of Discrete Quantum Systems Under Perturbations by Polychromatic Fields

Under any time-periodic Hamiltonian (i.e., perturbation by a monochromatic field), it is known that a nonresonant bounded quantum system exhibits quasi-periodic behavior and reassembles itself infinitely often in the course of time (51). The behavior of the corresponding quantum systems perturbed by bichromatic or polychromatic fields (where the Hamiltonian is nonperiodic in time) is less clear and has been investigated only recently. In this section, we discuss the behavior of two- and three-level quantum systems driven by intense bichromatic fields.

1. Quasi-periodic Behavior in Nondissipative Systems

As pointed out in Section III.B, the dynamical evolution of any bounded N-level quantum system is governed by the SU (N) dynamical symmetries, in which one can describe the time evolution in terms of a coherence vector $\mathbf{S}(t)$ of constant length rotating in an $(N^2 - 1)$-dimensional space. Using the MMFT, Ho and Chu (50) have recently found that the coherence vector $\mathbf{S}(t)$ displays quasi-periodic behavior when the external fields are not very strong. The behavior of quantum systems in very intense bichromatic fields is somewhat controversial and is discussed in what follows.

Consider a two-level system $(|a\rangle, |b\rangle, E_a < E_b)$ driven by two intense linearly polarized monochromatic fields with amplitudes $(\varepsilon_1, \varepsilon_2)$ and frequencies (ω_1, ω_2), respectively. The time-dependent autocorrelation function for the wavefunction $\Psi(t)$ of the two-level system is defined by

$$C(\tau) \equiv \lim_{T \to \infty} \frac{1}{T} \int_0^T \Psi^*(t)\Psi(t + \tau)\, dt. \qquad (61)$$

If the system exhibits quasi-periodic motion, $C(\tau)$ will be an almost periodic function of the correlation time τ. On the other hand, a rapid decaying autocorrelation function is a signature of quantum chaos (52). In Figure 13, we present the modulus of $C(\tau)$ corresponding to three different field intensities. The physical parameters used are $E_b - E_a = 1.0$, laser frequencies $\omega_1 = 0.85$ and $\omega_2 = \omega_1/\sqrt{15}$, and electric dipole coupling strengths $b_1 = b_2 = 0.1$ (top

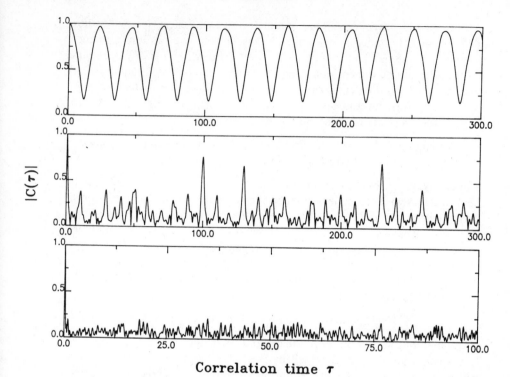

Figure 13. Modulus of the autocorrelation function $C(\tau)$ for a two-level system in the bichromatic fields. From the top to bottom: weak, medium strong, and intense field cases.

figure), 0.5 (middle figure), and 2.0 (bottom figure) arbitrary units. As shown here, the autocorrelation function $|C(\tau)|$ displays a quasi-periodic pattern in the weaker field case (top). For very strong fields (bottom), $|C(\tau)|$ shows a rapid decrease at small τ and then exhibits a noiselike structure at larger τ. Pomeau et al. (53) have studied numerically the similar problem, and they described this phenomenon as "chaotic Rabi flopping." However, a close mathematical study of this noiselike pattern via MMFT reveals that $C(\tau)$ never decays to zero at large correlation times (τ) (as seen in Figure 13), and in fact, it is an almost periodic function of τ regardless of the field strengths (54). Figure 14 shows that the autocorrelation function in strong fields (bottom figure in Figure 13) is reversible in time. This augments the conjecture that for nondissipative bounded quantum systems, the dynamical evolution in intense polychromatic fields is intrinsically quasi-periodic in nature, and no strict quantum stochasticity is possible. More detailed analysis can be found in Ref. 54.

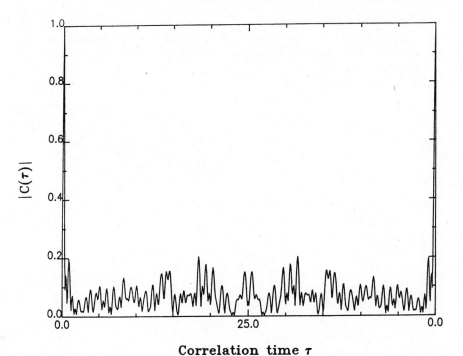

Correlation time τ

Figure 14. Time reversibility of the autocorrelation function (intense-field case in Figure 13).

2. *Non-quasi-periodic Behavior in Dissipative Systems*

Without loss of generality, let us consider now the dynamical behavior of a dissipative three-level system driven by intense bichromatic fields. Figure 15 shows the example of a dissipative Λ-type three-level system whose level 2 (intermediate level) is subject to irreversible decay with population or energy

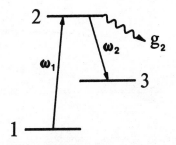

Figure 15. A Λ-type dissipative three-level system. Here g_2 is the population damping constant out of the system from level 2.

damping rate g_2. It is interesting to see how the *dissipative* coherence vector $S(t)$, now a shrinking rotating vector in eight-dimensional space, evolves in time. Under two-photon resonant conditions (Figure 15) and RWA, the coherence vector $S_j(t)$ $(j = 1, 2, \ldots, 8)$ can be solved exactly analytically (55). In the limit $t \to \infty$, we find that

$$S_i(t_\infty) \to 0, \qquad i = 1, 2, \ldots, 7 \tag{62}$$

but

$$S_8(t_\infty) \to 2\beta^2/\sqrt{3}\Omega^2, \tag{63}$$

where $\Omega = (\alpha^2 + \beta^2)^{1/2}$, and α and β are Rabi frequencies given by $\alpha = \frac{1}{2}\langle 1|\boldsymbol{\mu}\cdot\boldsymbol{\varepsilon}_1|2\rangle$ and $\beta = \frac{1}{2}\langle 2|\boldsymbol{\mu}\cdot\boldsymbol{\varepsilon}_2|3\rangle$, respectively. Note that the population scalar $S_0(t)$ $(\equiv \rho_{11} + \rho_{22} + \rho_{33})$ in this case reduces to

$$S_0(t_\infty) \to (\sqrt{3}/2)S_8(t_\infty) = \beta^2/\Omega^2. \tag{64}$$

Thus, at the exact two-photon resonance condition in the RWA limit, the

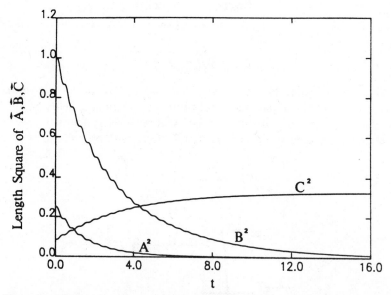

Figure 16. The time evolution of the length square of the subvectors $\mathbf{A} \equiv [S_1(t), S_2(t), S_3(t)]$, $\mathbf{B}(t) \equiv [S_4(t), S_5(t), S_6(t), S_7(t)]$, and $\mathbf{C} \equiv [S_8(t)]$, where $\mathbf{S} = \mathbf{A} + \mathbf{B} + \mathbf{C}$ is the eight-dimensional coherence vector. (Adapted from Ref. 55).

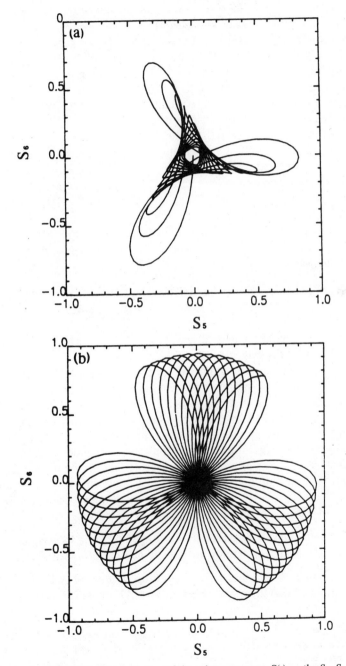

Figure 17. Projection of the trajectory of the coherence vector $S(t)$ on the S_5–S_6 plane for
(a) the dissipative three-level system $(g_2 \neq 0)$ and (b) the corresponding nondissipative system
$(g_2 = 0)$. (Adapted from Ref. 55).

eight-dimensional coherent vector eventually evolves to a one-dimensional scalar, and the population is trapped in the S_8 component only.

If we now relax the RWA limit and perform the exact *non-Hermitian* MMFT analysis (55), we find the coherence vector $S(t)$ first evolves qualitatively similar to the RWA case, namely, the population will be temporarily trapped (quasi-trapped) mainly in the S_8 component (Figure 16). However, at still larger times $(t_\infty \gg 1/g_2)$, the population in the system will eventually completely decay away to the surroundings.

The qualitative difference of the dynamical evolution of dissipative $(g_2 \neq 0)$ and nondissipative $(g_2 = 0)$ systems is illustrated in Figure 17. Shown here are the projections of the trajectories of the coherence vector $S(t)$ onto the S_5-S_6 plane. One sees that the coherence vector exhibits quasi-periodic behavior in the nondissipative case [Figure 17(b)], in accord with the SU(3) dynamical symmetries. On the other hand, for the dissipative case [Figure 17(a)], SU(3) symmetries are broken, and $S(t)$ evolves in a non-quasi-periodic manner from an initially $(N^2 - 1)$-dimensional space $(N = 3$ in this case) to a lower dimensional space.

IV.   FLOQUET LIOUVILLE SUPERMATRIX APPROACHES TO NONLINEAR OPTICAL PROCESSES

The various Floquet approaches discussed in Sections II and III are all based on the solution of the wavefunction (or time propagator) of the Schrödinger equation and are best applied to the studies of intense-field MPE–MPI–MPD processes in which relaxation processes such as spontaneous decays and collisional dampings can be ignored.** However, in nonlinear optical processes such as resonance fluorescence, multiple-wave mixings, and Raman scattering, relaxations often play a significant role (56, 57). The better starting point is to consider the Liouville equation for the time development of the density matrix operator of atoms and molecules, allowing for a more direct introduction of the damping mechanisms. In the following section, we discuss how the Floquet theory can be extended to a formal nonperturbative treatment of the time evolution of the density matrix operator (61).

A.   Floquet Liouville Supermatrix Treatment of Time Development of Density Matrix Operator in Periodic or Polychromatic Fields

The Liouville equation for the time evolution of a set of N-level quantum systems interacting with several coherent linearly polarized monochromatic

---

**However, if only energy or population damping is considered (i.e., the damping operator $\hat{G}$ has only diagonal matrix elements), the nonHermitian MMFT (Ref. 55) based on the Schrödinger equation is still adequate.

fields and undergoing relaxation by Markovian processes (56–58) is ($\hbar = 1$)

$$i\frac{\partial\hat{\rho}(t)}{\partial t} = [\hat{H}(t), \hat{\rho}(t)] + i[\hat{R}, \hat{\rho}(t)], \tag{65}$$

where $\hat{\rho}$ is the density matrix of the system, reduced by an averaging over all irrelevant degrees of freedom acting as a thermal bath, and $\hat{H}(t) = \hat{H}_0 + \hat{V}(t)$. Here $\hat{H}_0$ is the unperturbed Hamiltonian with eigenvalues $\{E_\alpha\}$ and eigenvectors $\{|\alpha\rangle\}$, $\alpha = 0, 1, 2, \ldots, N-1$, and $V(t)$ is the interaction Hamiltonian between the system and the M-mode classical fields given by

$$V(t) = -\sum_{i=1}^{M} \boldsymbol{\mu} \cdot \boldsymbol{\varepsilon}_i \cos(\omega_i t + \phi_i), \tag{66}$$

where $\boldsymbol{\mu}$ is the atomic or molecular dipole moment, ε_i the field amplitude, ω_i the frequency, and ϕ_i the phase of the ith field. The relaxation term $[\hat{R}, \hat{\rho}(t)]$ consists of T_1 (population damping) and T_2 (coherent damping) mechanisms due to the coupling of the system to the thermal bath by radiative decays and collisional relaxations, and so on. More explicitly (58),

$$[\hat{R}, \hat{\rho}]_{\alpha\alpha} = -\Gamma_{\alpha\alpha}\rho_{\alpha\alpha} + \sum_{\beta\neq\alpha} \gamma_{\beta\alpha}\rho_{\beta\beta} \qquad (T_1), \tag{67}$$

$$[\hat{R}, \hat{\rho}]_{\alpha\beta} = -\Gamma_{\alpha\beta}\rho_{\alpha\beta}, \qquad \alpha \neq \beta \qquad (T_2), \tag{68}$$

where the phenomenological damping parameter $\Gamma_{\alpha\alpha}$ describes the population decay, $\Gamma_{\alpha\beta}$ the phase relaxation, and $\gamma_{\beta\alpha}$ the feeding. In the following we shall confine our discussion to closed systems, namely, $\text{Tr}([\hat{R}, \hat{\rho}]) = 0$. Extension to open systems is straightforward.

In the tetradic, or Liouville, space (59, 60) spanned by the basis $\{|\alpha\beta\rangle \equiv |\alpha\rangle\langle\beta|$, where $\alpha, \beta = 0, 1, \ldots, N-1\}$, Eq. 65 may be rewritten as an inhomogeneous supervector equation, $i\,\partial\hat{\rho}(t)/\partial t = \hat{\hat{L}}(t)\hat{\rho}(t) + i\hat{f}$, or in matrix form (61),

$$i\frac{\partial\rho_{\alpha\beta}(t)}{\partial t} = \sum_{\mu\nu} \hat{\hat{L}}_{\alpha\beta,\mu\nu}(t)\rho_{\mu\nu}(t) + i\hat{f}_{\alpha\beta}, \tag{69}$$

where $\hat{\hat{L}}(t)$ is the superoperator or Liouvillian that is nonsingular, whose matrix elements are, assuming $|0\rangle$ is the ground level,

$$\hat{\hat{L}}_{00;\mu\nu}(t) = H_{0\mu}(t)\delta_{0\nu} - H_{\nu 0}(t)\delta_{0\mu} - i(\Gamma_{00} + \gamma_0)\delta_{\mu 0}\delta_{\nu 0}$$
$$- i\left(\sum_{\beta\neq 0}(1 - \delta_{\beta\mu})\gamma_{\beta 0}\right)\delta_{\mu\nu}(1 - \delta_{0\mu}), \tag{70}$$

774 SHIH-I CHU

$$\hat{\hat{L}}_{\alpha\beta;\mu\nu}(t) = H_{\alpha\mu}(t)\delta_{\beta\nu} - H_{\nu\beta}(t)\delta_{\alpha\mu}$$

$$- i(\Gamma_{\mu\nu}\delta_{\alpha\mu}\delta_{\beta\nu} - \gamma_{\mu\beta}\delta_{\alpha\beta}\delta_{\mu\nu}) \qquad (\alpha \neq 0, \beta \neq 0), \qquad (71)$$

and $\hat{f}$ is the source supervector, $f_{\mu\nu} = \gamma_0\delta_{\nu 0}\delta_{\mu 0}$, with $\gamma_0 = \sum_{\beta \neq 0}\gamma_{\beta 0}$. Note that in deriving these equations, use has been made of the relationship $\mathrm{Tr}\,\hat{\rho}(t) = 1$ (closed systems) and the following properties of the Liouville space (60, 61): (i) inner product,

$$\langle \alpha\beta | \mu\nu \rangle \equiv \mathrm{Tr}\,\{ |\beta\rangle\langle\alpha|\mu\rangle\langle\nu| \}, \qquad (72)$$

and (ii) matrix elements of the superoperator $\hat{\hat{O}}$,

$$\langle \alpha\beta | \hat{\hat{O}} | \mu\nu \rangle \equiv \mathrm{Tr}\,\{ |\beta\rangle\langle\alpha| \hat{\hat{O}} |\mu\rangle\langle\nu| \}. \qquad (73)$$

The homogeneous solution of Eq. 69 can be solved most expediently by invoking the MMFT, analogous to solving the Schrödinger equation with Hamiltonian having the same time dependence as that in Eq. 65. The MMFT renders the time-dependent problem into an equivalent time-independent infinite-dimensional supereigenvalue equation (61), namely,

$$\sum_{\sigma\tau}\sum_{\{k\}} \langle \alpha\beta; \{m\} | \hat{\hat{L}}_F | \sigma\tau; \{k\} \rangle \langle \sigma\tau; \{k\} | \Omega_{\mu\nu;\{n\}} \rangle$$

$$= \Omega_{\mu\nu;\{n\}} \langle \alpha\beta; \{m\} | \Omega_{\mu\nu;\{n\}} \rangle, \qquad (74)$$

where $\hat{\hat{L}}_F$ is the *time-independent* many-mode Floquet Liouville superoperator defined in terms of the generalized tetradic Floquet basis $|\alpha\beta; \{m\}\rangle \equiv |\alpha\rangle\langle\beta| \otimes |\{m\}\rangle$, with $\{m\} = m_1, m_2, \ldots, m_M$.

The structure of the Floquet Liouville supermatrix $\hat{\hat{L}}_F$ is illustrated in Figure 18 for the two-level two-mode case. The supereigenvalues $\Omega_{\mu\nu;\{n\}}$ and eigenvectors $|\Omega_{\mu\nu;\{n\}}\rangle$ of $\hat{\hat{L}}_F$ possess the following important properties (61):

(i) $\mathrm{Im}\{\Omega_{\mu\nu;\{n\}}\} < 0$,

(ii) $\Omega_{\mu\nu;\{n+k\}} = \Omega_{\mu\nu;\{n\}} + \sum_{i=1}^{M} k_i\omega_i$, and

(iii) $\langle \alpha\beta; \{m+k\} | \Omega_{\mu\nu;\{n+k\}} \rangle = \langle \alpha\beta; \{m\} | \Omega_{\mu\nu;\{n\}} \rangle$.

Further, it can be shown that in the limit of $\gamma_{\alpha\beta} = \Gamma_{\alpha\beta} = 0$ (i.e., no relaxations), the supereigenvalues Ω and eigenvectors $|\Omega\rangle$ of $\hat{\hat{L}}_F$ are related to the quasi-energy eigenvalues λ and eigenvectors $|\lambda\rangle$ of $\hat{H}_F$, where $\hat{H}_F$ is the Floquet

$$\hat{L}_F = \begin{pmatrix} A+2\omega_2 I & B & 0 & 0 & 0 \\ B^* & A+\omega_2 I & B & 0 & 0 \\ 0 & B^* & A & B & 0 \\ 0 & 0 & B^* & A-\omega_2 I & B \\ 0 & 0 & 0 & B^* & A-2\omega_2 I \end{pmatrix}$$

WHERE

$$A = \begin{pmatrix} C+2\omega_1 I & X & 0 & 0 & 0 \\ X^* & C+\omega_1 I & X & 0 & 0 \\ 0 & X^* & C & X & 0 \\ 0 & 0 & X^* & C-\omega_1 I & X \\ 0 & 0 & 0 & X^* & C-2\omega_1 I \end{pmatrix}$$

$$B = \begin{pmatrix} Y & 0 & 0 & 0 & 0 \\ 0 & Y & 0 & 0 & 0 \\ 0 & 0 & Y & 0 & 0 \\ 0 & 0 & 0 & Y & 0 \\ 0 & 0 & 0 & 0 & Y \end{pmatrix} \qquad C = \begin{pmatrix} -i(\gamma_{ab}+\gamma_{ba}) & 0 & 0 & 0 \\ i\gamma_{ab} & -i\gamma_{ba} & 0 & 0 \\ 0 & 0 & -\omega_{ba}-i\Gamma_{ba} & 0 \\ 0 & 0 & 0 & \omega_{ba}-i\Gamma_{ba} \end{pmatrix}$$

AND

$$X = \begin{pmatrix} 0 & 0 & -V^{(1)}_{ba} & V^{(1)}_{ab} \\ 0 & 0 & V^{(1)}_{ba} & -V^{(1)}_{ab} \\ -V^{(1)}_{ab} & V^{(1)}_{ab} & 0 & 0 \\ V^{(1)}_{ba} & -V^{(1)}_{ba} & 0 & 0 \end{pmatrix} \qquad Y = \begin{pmatrix} 0 & 0 & -V^{(2)}_{ba} & V^{(2)}_{ab} \\ 0 & 0 & V^{(2)}_{ba} & -V^{(2)}_{ab} \\ -V^{(2)}_{ab} & V^{(2)}_{ab} & 0 & 0 \\ V^{(2)}_{ba} & -V^{(2)}_{ba} & 0 & 0 \end{pmatrix}$$

Figure 18. Structure of the Floquet Liouville supermatrix $\hat{\bar{L}}_F$ for the case of two-level systems (with level spacing ω_{ba}) in linearly polarized bichromatic fields. Here ω_1 and ω_2 are the two radiation frequencies; $V^{(i)}_{ab}(i=1,2)$ are the electric dipole couplings; and γ_{ab}, γ_{ba}, and $\Gamma_{ba} \equiv (\gamma_{ab}+\gamma_{ba})/2$ are relaxation parameters. (Adapted from Ref. 61).

Hamiltonian for the nondamping case, by the following relations:

$$\Omega_{\alpha\beta;\{m\}} = \lambda_{\alpha,\{0\}} - \lambda_{\beta,\{0\}} + \sum_{i=1}^{M} m_i \omega_i, \tag{75}$$

$$\langle \mu\nu; \{k\} | \Omega_{\alpha\beta;\{0\}} \rangle = \sum_{\{n\}} \langle \mu, \{n\} | \lambda_{\alpha\{0\}} \rangle \langle \lambda_{\beta\{0\}} | \nu, \{n-k\} \rangle. \tag{76}$$

Thus the supereigenvalues Ω have the physical interpretation as the "difference spectrum" of the quasi-energies.

In terms of the eigenvalues and eigenvectors of the superoperator $\hat{\bar{L}}_F$, the reduced density matrix $\rho(t)$ can be expressed as $\rho(t) = \hat{\bar{U}}(t; t_0)\rho(t_0)$, where $\hat{\bar{U}}$ is the *non-Hermitian* superevolution operator (61) given by, in matrix form,

$$
\begin{aligned}
\hat{\bar{U}}_{\alpha\beta;\mu\nu}(t; t_0) = & \sum_{\{m\}} \langle \alpha\beta; \{m\} | \exp[-i\hat{\bar{L}}_F(t - t_0)] | \mu\nu; \{0\} \rangle \\
& + \gamma_0 \delta_{\mu\nu} \sum_{\sigma\tau} \sum_{\{k\}} \langle \alpha\beta; \{m\} | \Omega_{\sigma\tau;\{k\}} \rangle \langle \Omega^*_{\sigma\tau;\{k\}} | 00; \{0\} \rangle \\
& \times \{1 - \exp[-i\Omega_{\sigma\tau;\{k\}}(t - t_0)]\}/i\Omega_{\sigma\tau;\{k\}} \\
& \times \exp\left(i \sum_{j=1}^{M} m_j \omega_j t\right).
\end{aligned}
\tag{77}
$$

Furthermore, since $\operatorname{Im}\Omega < 0$ for all Ω, the reduced density matrix has a simple form at large times $t \to \infty$,

$$
\begin{aligned}
\rho_{\alpha\beta}(t) \xrightarrow[t \to \infty]{} & \gamma_0 \sum_{\{m\}} \sum_{\sigma\tau} \sum_{\{k\}} (\langle \alpha\beta; \{m\} | \Omega_{\sigma\tau;\{k\}} \rangle \\
& \times \langle \Omega^*_{\sigma\tau;\{k\}} | 00, \{0\} \rangle / i\Omega_{\sigma\tau;\{k\}}) \exp\left(i \sum_{j=1}^{M} m_j \omega_j t\right),
\end{aligned}
\tag{78}
$$

which is *oscillatory* rather than completely *stationary*, as would be the case in the RWA limit.

B. Multiphoton Resonance Fluorescence in Intense Laser Fields

As an example of the usefulness of the FLSM method, we shall first discuss multiple-photon-induced resonance transitions and light scatterings of an ensemble of two-level atoms or molecules illuminated by strong polychromatic fields. While one-photon resonance light scatterings, fluorescence in particular, have been widely discussed (62), especially since the work of Mollow (63), the multiphoton-induced light scatterings have received little attention. In this section, novel nonlinear dynamical features in multiphoton-induced resonance fluorescence spectra will be exploited.

The relevant quantity here is the autocorrelation function (64) of the transition dipole operators $\hat{d}^+ = |b\rangle\langle a|$ and $\hat{d} = |a\rangle\langle b|$, namely, $g(t; t') = \langle \hat{d}^+(t')\hat{d}(t)\rangle$, $t > t'$, in normal order. The two-time average $g(t; t')$ can be evaluated by the use of the quantum regression theorem (65) and expressed as

$$
g(t; t') = \hat{\bar{U}}_{ba;aa}(t; t')\hat{\rho}_{ab}(t') + \hat{\bar{U}}_{ba;ba}(t; t')\hat{\rho}_{bb}(t'),
\tag{79}
$$

where $\hat{\bar{U}}$ is given in Eq. 77. The important feature here is the nonstationary nature of the correlation function $g(t; t')$ at large times t'; that is, $g(t; t')$ depends on both the correlation time $\tau = t - t'$ and the starting time t'. [Note that in the

RWA limit, $g(t; t') \to g(\tau)$, independent of t'.] We can define the time-averaged correlation function as

$$\bar{g}(\tau) = \lim_{T_s \to \infty} \frac{1}{T_s} \int_0^{T_s} dt' \, g(t; t') \qquad (t' \to \infty), \tag{80}$$

involving integrating $g(t; t')$ over all possible times t' while keeping the correlation time $\tau \equiv t - t'$ constant. The power spectrum $\bar{I}(\omega)$ corresponding to the time-averaged correlation function $\bar{g}(\tau)$ can then be evaluated via the Fourier transform:

$$\bar{I}(\omega) \equiv \mathrm{Re}\left[\int_0^\infty d\tau \, e^{i\omega\tau} \bar{g}(\tau) \right]. \tag{81}$$

Here $\bar{I}(\omega)$ can be decomposed into a coherent part $\bar{I}_{\mathrm{coh}}(\omega)$ and an incoherent part $\bar{I}_{\mathrm{inc}}(\omega)$, namely,

$$\begin{aligned}
\bar{I}_{\mathrm{coh}}(\omega) = -\pi\gamma_{ba}^2 \mathrm{Re}\Bigg\{ &\sum_{\{m\}} \Bigg[\sum_{\mu,\nu} \sum_{\{k\}} \langle ba; \{0\} | \Omega_{\mu\nu;\{k\}} \rangle \\
&\times \langle \Omega_{\mu\nu,\{k\}}^* | aa; \{m\} \rangle [\Omega_{\mu\nu;\{k-m\}}]^{-1} \\
&\times \delta\left(\omega - \sum_{i=1}^M m_i \omega_i \right) \Bigg] \Bigg[\sum_{\mu,\nu} \sum_{\{k\}} \langle ab; \{m\} | \Omega_{\mu\nu,\{k\}} \rangle \\
&\times \langle \Omega_{\mu\nu,\{k\}}^* | aa; \{0\} \rangle [\Omega_{\mu\nu;\{k\}}]^{-1} \Bigg] \Bigg\}
\end{aligned} \tag{82}$$

and

$$\bar{I}_{\mathrm{inc}}(\omega) = -\gamma_{ba} \, \mathrm{Re}\left[\sum_{\{m\}} (AB + CD) \right], \tag{83}$$

where

$$A = \sum_{\mu\nu} \sum_{\{k\}} \langle ba; \{0\} | \Omega_{\mu\nu;\{k\}} \rangle \langle \Omega_{\mu\nu;\{k\}}^* | ba; \{m\} \rangle (\Omega_{\mu\nu;\{k\}} - \omega)^{-1},$$

$$B = \sum_{\mu\nu} \sum_{\{k\}} \langle bb; \{m\} | \Omega_{\mu\nu;\{k\}} \rangle \langle \Omega_{\mu\nu;\{k\}}^* | aa; \{0\} \rangle / \Omega_{\mu\nu;\{k\}},$$

$$\begin{aligned}
C = \sum_{\mu\nu} \sum_{\{k\}} &\langle ba; \{0\} | \Omega_{\mu\nu;\{k\}} \rangle \langle \Omega_{\mu\nu;\{k\}}^* | aa; \{m\} \rangle \\
&\times [1 + i\gamma_{ba}(\Omega_{\mu\nu;\{k-m\}})^{-1}](\Omega_{\mu\nu;\{k\}} - \omega)^{-1},
\end{aligned}$$

$$D = \sum_{\mu\nu} \sum_{\{k\}} \langle ab; \{m\} | \Omega_{\mu\nu;\{k\}} \rangle \langle \Omega_{\mu\nu;\{k\}}^* | aa; \{0\} \rangle / \Omega_{\mu\nu;\{k\}}.$$

Equation 82 shows that the coherently scattered light comprises not only
the elastic components (Rayleigh scattering) at $\omega = \omega_1, \omega_2, \ldots, \omega_M$ but also
various harmonic components at $\omega = \Sigma_{i=1}^{M} m_i \omega_i$, where $\Sigma_i m_i$ must be odd
integers. Equation 83 shows that the incoherently scattered light contains
frequency components at the positions defined by the denominator ($\Omega_{\mu\nu;\{k\}}$
$- \omega)^{-1}$, where $\Sigma_{\{k\}} \equiv \Sigma_i k_i$ is an odd integer if $\mu = \nu$ and an even integer if $\mu \neq \nu$.
Thus, there exists in principle an infinite number of components, with the
intensity of each component depending on the degree of mixing of the two
levels by the fields. We emphasize that the expressions Eqs. 82 and 83 are
general results applicable to arbitrary numbers and strengths of the fields.

Resonance fluorescence scattering by atoms or molecules in the presence of
strong laser fields is a delicate nonlinear process at least in two aspects: (i) It is a
cascade process via an infinite number of dressed atomic or molecular states,
and (ii) it requires strong resonance mixings, either by one photon or by
several photons, between unperturbed atomic levels. The strong mixings of
levels produce sidebands due to the ac Stark effect, in addition to those
corresponding to the natural transition frequencies. The resonance flourescence processes of two-level systems driven by a monochromatic laser field of
frequency ω_L are schematically depicted in Figure 19, where each doublet is
characterized by a splitting u between a pair of nearly degenerate quasi-energy
levels, and arrows indicate fluorescence cascade patterns. At each near-
resonance condition $\omega_{ba} \cong (2n + 1)\omega_L$, $n = 0, 1, 2, \ldots,$ the most intense fluorescence light occurs around $\omega = (2n + 1)\omega_L$ and shows a triplet pattern.

There are intimate relationships between the supereigenvalues $\Omega_{\mu\nu;k}$ and the
long-time-averaged population [Figure 20(a) and 20(b)]. Figure 20(a) shows
the strong mixed regions of the two levels caused by one-, three-, and five-
photon resonance transitions, respectively. The splittings and stretches of the
avoided crossing regions reflects the widths of the corresponding lineshapes of
the long-time-averaged excitation spectrum $\bar{\rho}_{bb}$ as a function of ω_L, depicted in
Figure 20(b). Figure 21 shows that the fluorescence power spectra corre-
sponds to the shifted three-photon resonance condition ($\omega_{ba} \sim 3\omega_L$). Note the
strong triplet fluorescence spectra nearby $\omega \cong \omega_L$ and $3\omega_L$ [Figures 21(a) and
21(b)] as well as a much weaker triplet around $\omega \cong 5\omega_L$ [Figure 21(c)].
Particularly interesting is the strongly asymmetric three-peak structure near
$\omega \cong \omega_L$. This asymmetry can be largely attributed to the strong mixings not
only among the resonant, or nearly resonant, unperturbed tetradic Floquet
states (e.g., $|aa;0\rangle$, $|bb;0\rangle$, $|ab; +3\rangle$, and $|ba; -3\rangle$) but also of the
nonresonant states (such as $|ab; +1\rangle$, $|ba; -1\rangle$, etc.). At a much weaker field,
only those nearly resonant states are mixed; thus, intense flourescence light can
only be observed at $3\omega_L$ and possesses a symmetric triple-peak appearance (63). The intense fluorescence light and its asymmetric outlook at $\omega \cong \omega_L$ in
this three-photon resonance case are genuine strong field effects (61).

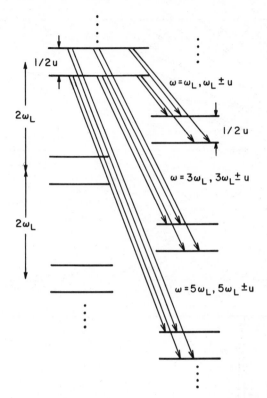

Figure 19. Schematic cascade fluorescence processes of two-level systems driven by a monochromatic field of frequency ω_L. The splitting u of doublets in each column is the splitting of the adjacent quasi-energy levels and is caused by the ac Stark effect and possible detuning $\Delta \equiv \omega_{ba} - (2n + 1)\omega_L$ at nearly resonant conditions. Each column is a collection of quasi-energy levels of the same parity; quasi-energy levels belonging to different columns are of opposite parity. Arrows indicate parts of cascade fluorescence down the infinite number of quasi-energy levels. (Adapted from Ref. 61).

As a general rule concerning the intensity of fluorescence lights, it is found (61) that when the laser field is intense enough to induce multiphoton resonances [say, $\omega_{ba} \cong (2n + 1)\omega_L$], the dominant fluorescence component always occurs at $\omega \cong (2n + 1)\omega_L$. Components in the lower frequency side of the predominant one can have comparable intensities but often exhibit large asymmetry in the three-peak structure, whereas components in the higher frequency side usually decrease rapidly in intensity as the harmonic order increases.

In addition to the time-averaged power spectrum $\bar{I}(\omega)$, another dynamical quantity of interest is the *time-dependent* physical spectrum (66), which can be

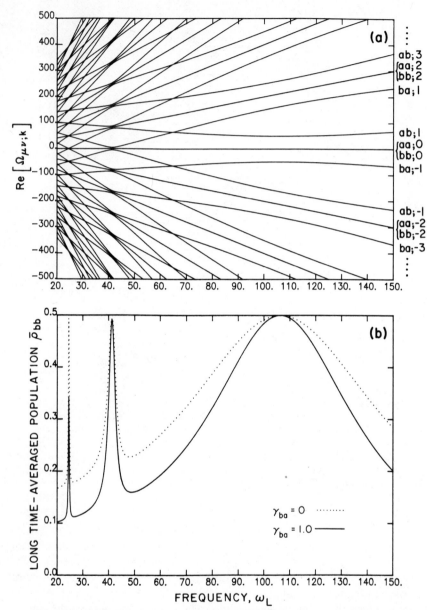

Figure 20. (a) Supereigenvalues (real parts) and (b) long-time-averaged population $\bar{\rho}_{bb}$ for a closed system of two-level atoms driven by an intense monochromatic field of frequency ω_L. Parameters: $\omega_{ba}(\equiv E_b - E_a) = 100.0$; $\gamma_{ba} = 1.0$; $\gamma_{ab} = 0.0$; $|V_{ab}^{(1)}| = 25.0$; $\phi^{(1)} = 0.0$ (arbitrary units). The one-, three-, and five-photon resonances [solid curves in (b)] occur at $\omega_L = 106.335$, 41.295, and 24.525, respectively. Also shown are results of the corresponding nondamping ($\gamma_{ba} = 0.0$) case [dotted curves in (b)] for comparison. (Adapted from Ref. 61).

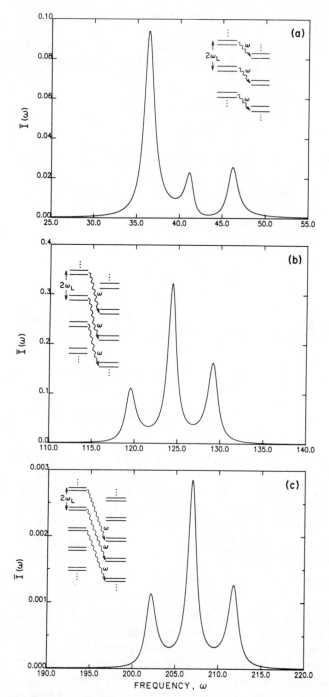

Figure 21. Fluorescence spectra $\bar{I}(\omega)$, Eq. 83, near (a) $\omega \cong \omega_L$, (b) $\omega \cong 3\omega_L$, and (c) $\omega \cong 5\omega_L$ for a system of two-level atoms driven by a monochromatic field; ω_L tuned at the shifted three-photon resonance $\omega_L = 41.295$. Parameters same as in Figure 20. The inset in each figure shows the schematic cascade diagram (not to scale) (cf. Figure 19). (Adapted from Ref. 61).

expressed as a twofold convolution integral over the time and frequency of a filter function and a time-dependent quasi-spectrum that depends on the signal only:

$$\bar{I}(\omega, t; \Gamma) \equiv \int d\omega_0 \int_0^\infty d\tau \, S(\omega_0, \tau; \Gamma) I(\omega - \omega_0, t - \tau). \tag{84}$$

Here $S(\omega_0, \tau; \Gamma)$ is the filter-smoothing function, for example,

$$S(\omega_0, \tau; \Gamma) = (\Gamma^2/2\pi)[(\Gamma/2)^2 + \omega_0^2]^{-1} \exp(-\Gamma\tau), \tag{85}$$

with Γ denoting the filter width and the quasi-spectrum (Page–Lampared spectrum) $I(\nu, t)$ defined as

$$I(\nu, t) = \text{Re}\left[\int_{-\infty}^t dt' \, e^{i\nu(t-t')} g(t; t')\right], \tag{86}$$

with the correlation function given in Eq. 79. Assuming the laser fields are turned on abruptly at time $t = 0$, we find

$$\bar{I}(\omega, t; \Gamma) = \Gamma \, \text{Re}\left[\int_0^t dt_1 \exp[-\Gamma(t - t_1)] \exp(i\omega t_1) \exp(-\Gamma t_1/2)\right.$$
$$\left. \times \int_0^t dt_2 \exp(-i\omega t_2) \exp(\Gamma t_2/2) g(t_1; t_2)\right], \tag{87}$$

which can be carried out analytically. Figure 22 shows the time-dependent physical spectrum $\bar{I}(\omega, t; \Gamma = 0.1)$ at $\omega_L = 41.295$ (shifted three-photon resonance) and $t = 2.5, 5.0, 10.0, 30.0$, respectively. By increasing the time t, we see the three-peak structures, both at $\omega = \omega_L$ and $\omega = 3\omega_L$, gradually grow from zero and converge to the steady-state lineshapes, very close to the ones shown in Figures 21(a) and 21(b), except that each peak now is further broadened by an amount $\Gamma/2$.

C. Intensity-Dependent Generalized Nonlinear Optical Susceptibilities and Multiple-Wave Mixings

The determination of nonlinear optical susceptibilities of atoms and molecules represents a significant area of both experimental and theoretical research in nonlinear optics (56, 57, 67). Calculations of nonlinear susceptibilities $\chi(\omega)$ in a medium with discrete quantum levels are usually performed by means of perturbative methods (56, 67). The perturbative treatment is adequate when both the pump and the probe fields are weak and the corresponding nonlinear

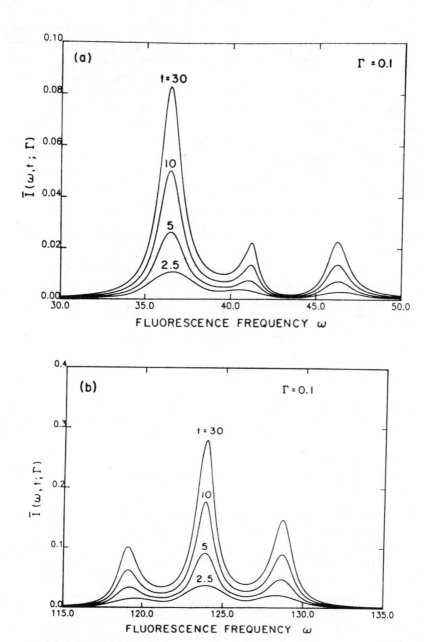

Figure 22. Time-dependent physical spectra $\bar{I}(\omega, t; \Gamma)$, Eq. 87, near (a) $\omega \cong \omega_L$ and (b) $\omega \cong 3\omega_L$ for a system of two-level atoms driven by a monochromatic field; ω_L tuned at the shifted three-photon resonance $\omega_L = 41.295$. The filter width is $\Gamma = 0.1$, assuming the field is abruptly turned on at the time $t = 0$. Parameters same as in Figure 20. (Adapted from Ref. 61).

optical susceptibilities are *independent* of field strengths. However, recent experimental works are often carried out under the conditions that both the pump and the probe fields are strong (68–72). Distinct new features such as subradiative structures, multiphoton absorption peaks, and high-order (up to 27th-order) (68) nonlinear wave mixings have been observed. In most of these nonlinear optical processes, when the fields are intense enough to saturate the transitions, nonlinear susceptibilities become intensity dependent. Nonperturbative response functions are required to explain these intense-field effects (73, 74). In this section, we discuss the extension of the FLSM approach to the study of intensity-dependent $\chi(\omega)$ and multiple wave mixing phenomena, beyond the conventional perturbative approximation and the RWA.

1. *Exact* FLSM *Nonperturbative Treatment*

The nonlinear response of an ensemble of systems to the incident polychromatic fields takes the form of a dielectric polarization density $P(t)$ that acts as a source term in Maxwell's wave equations. The polarization density is related to the expectation value of the dipole moment operator μ and can be calculated from the density matrix $\rho(t)$,

$$\langle P(t) \rangle = N_0 \langle \mu \rangle = N_0 \operatorname{Tr}(\mu\rho(t)), \tag{88}$$

where N_0 is the number density in the ensemble and $\rho(t)$ can be determined by the FLSM method described in Section IV.A.

Without loss of generality, let us consider the response of two-level systems driven by intense M-mode polychromatic fields. The polarization density now takes the form

$$\langle P(t) \rangle = N_0 [\mu_{ba}\rho_{ab}(t) + \mu_{ab}\rho_{ba}(t)], \tag{89}$$

where μ_{ab} is the transition dipole matrix element between the unperturbed atomic states $|a\rangle$ and $|b\rangle$ (assumed to be of opposite parity and $E_a < E_b$).

In the steady state ($t \to \infty$), the polarization density may be expanded as a Fourier series in the incident frequencies (as shown by Eq. 78),

$$\langle P(t) \rangle = \sum_{m_1 m_2 \cdots m_M} P_{m_1 m_2 \cdots m_M}(\omega) e^{-i(m_1\omega_1 + m_2\omega_2 + \cdots + m_M\omega_M)t}, \tag{90}$$

where $P_{m_1 m_2 \cdots m_M}(\omega) \equiv P_{\{m\}}(\omega)$ is the Fourier component at frequency $\omega = m_1\omega_1 + m_2\omega_2 + \cdots + m_M\omega_M$. As an example, consider the two-mode ($M = 2$) case with ω_1 the pump frequency and ω_2 the probe frequency. We have, from Eq. 90,

$$\langle P(t) \rangle = P_{1,0}(\omega_1)e^{-i\omega_1 t} + P_{0,1}(\omega_2)e^{-i\omega_2 t}$$
$$+ P_{2,-1}(2\omega_1 - \omega_2)e^{-i(2\omega_1 - \omega_2)t} + \cdots. \qquad (91)$$

The physical meaning of these terms is as follows: $P_{1,0}(\omega_1)$ and $P_{0,1}(\omega_2)$ give rise to absorption (or amplification) of the pump and probe waves, respectively, while the *mixing* response $P_{2,-1}(2\omega_1 - \omega_2)$ is responsible for generation of an optical wave with frequency $\omega = 2\omega_1 - \omega_2$, and so on. Note that $P_{\{m\}}(\omega)$ is a nonperturbative result. If expanded in terms of a power series of incident fields, $P_{\{m\}}(\omega)$ can be related to the conventional perturbative nonlinear susceptibilities (to infinite order, in principle). For example in the case of bichromatic fields ($M = 2$),

$$P_{2,-1}(2\omega_1 - \omega_2)$$
$$= \chi^{(3)}(-2\omega_1 + \omega_2; \omega_1, \omega_1, -\omega_2)\varepsilon_1^2(\omega_1)\varepsilon_2^*(\omega_2)$$
$$+ \chi^{(5)}(-2\omega_1 + \omega_2; \omega_1, \omega_1, \omega_1, -\omega_1, -\omega_2)\varepsilon_1^3(\omega_1)\varepsilon_1^*(\omega_1)\varepsilon_2^*(\omega_2)$$
$$+ \chi^{(5)}(-2\omega_1 + \omega_2; \omega_1, \omega_1, \omega_2, -\omega_2, -\omega_2)\varepsilon_1^2(\omega_1)|\varepsilon_2(\omega_2)|^2\varepsilon_2^*(\omega_2)$$
$$+ \cdots,$$

where

$$\varepsilon_i(n_i\omega_i) = \begin{cases} \varepsilon_i(\omega_i)^{n_i}, & n_i \geq 0, \\ \varepsilon_i^*(\omega_i)^{|n_i|}, & n_i < 0, \end{cases}$$

is the Fourier transform of the ith optical field at ω_i and $\chi^{(q)}$ is the conventional (intensity-independent) perturbative qth-order optical susceptibility (56, 67). At weak incident fields, the lowest (nonvanishing order susceptibility dominates, and the conventional perturbative approach for $\chi^{(q)}$ is adequate. Thus, if both the pump and probe fields are weak, the generation of a coherent signal at $2\omega_1 - \omega_2$ (four-wave mixing), for example, is described by the third-order ($q = 3$) nonlinear susceptibility $\chi^{(3)}(-2\omega_1 + \omega_2; \omega_1, \omega_1, -\omega_2)$. However, for strong saturating fields, higher order nonlinear susceptibilities can contribute significantly. This leads to the concept of intensity-*dependent* generalized nonlinear optical susceptibility defined by

$$\chi_{\{m\}}(\omega) = P_{\{m\}}(\omega)/[\varepsilon_1(m_1\omega_1)\varepsilon_2(m_2\omega_2)\cdots\varepsilon_M(m_M\omega_M)], \qquad (92)$$

where $\omega = m_1\omega_1 + m_2\omega_2 + \cdots + m_M\omega_M$. In the limit of weak fields, $\chi_{\{m\}}(\omega)$

reduces to the lowest nonvanishing order (intensity-independent) $\chi^{(q)}$, as it should be. Using the results of Eqs. 78, 91, and 92, we obtain the following *nonperturbative* expression for *generalized* nonlinear optical susceptibility (for the two-level M-mode case) in terms of the supereigenvalues and eigenvectors of the Floquet Liouvillian $\hat{L}_F$ (74):

$$\chi_{\{m\}}(\omega = m_1\omega_1 + m_2\omega_2 + \cdots + m_M\omega_M)$$

$$= N_0\gamma_{ba}\left\{\sum_{\sigma\tau}\sum_{\{k\}}(\langle ba; \{m\}|\Omega_{\sigma\tau;\{k\}}\rangle\mu_{ab}\right.$$

$$+ \langle ab; \{m\}|\Omega_{\sigma\tau;\{k\}}\rangle\mu_{ba}) \times \langle\Omega^*_{\sigma\tau;\{k\}}|aa; \{0\}\rangle$$

$$\left. \times [i\Omega_{\sigma\tau;\{k\}}]^{-1}\right\}\bigg/[\varepsilon_1(m_1\omega_1)\varepsilon_2(m_2\omega_2)\cdots\varepsilon(m_M\omega_M)]. \quad (93)$$

2. High-Order Nearly Degenerate Perturbative Treatment (74)

To exploit analytical properties of nonlinear optical processes and to make a connection with commonly used perturbative and RWA approaches, we shall discuss now the extension of Salwen's almost degenerate perturbation theory (41) to the Floquet Liouvillian $\hat{L}_F$. Consider the important class of a system of dipole-allowed two-level atoms (molecules) undergoing $(2|m| + 1)$-photon $[\omega_{ba} \cong (m + 1)\omega_1 - m\omega_2]$ near-resonant transitions in the presence of two intense linearly polarized laser fields characterized by the frequencies (ω_1, ω_2), amplitudes $(\varepsilon_1, \varepsilon_2)$, and initial phases (ϕ_1, ϕ_2), respectively. The two-level $|a\rangle$ and $|b\rangle$ $(E_a < E_b)$ are assumed to be of opposite parity.

In a proper rotating frame (*not* the RWA) defined by the unitary transformation

$$R(t) = \begin{bmatrix} 1 & 0 \\ 0 & e^{i[(m+1)\omega_1 - m\omega_2]t} \end{bmatrix}, \quad (94)$$

the density matrix superoperator $\hat{\rho}(t)$ satisfies approximately the Salwen–Liouville equation, namely,

$$i\frac{\partial\rho(t)}{\partial t} = \hat{\hat{L}}_S\rho(t) + i\hat{f}_S, \quad (95)$$

where $\hat{f}_S$ is the source supervector given by

$$\hat{f}_S = \begin{bmatrix} \gamma_{ba} \\ 0 \\ 0 \\ 0 \end{bmatrix}. \quad (96)$$

When the resonance condition $\omega_{ba} \cong (m+1)\omega_1 - m\omega_2$, m arbitrary integer, is satisfied, the unperturbed tetradic Floquet states $|aa; 00\rangle$, $|bb; 00\rangle$, $|ab; m+1, -m\rangle$, and $|ba; -(m+1), m\rangle$ form a four-dimensional almost degenerate set and span Salwen's "model space." In terms of this model space, the effective Salwen Liouvillian $\hat{\hat{L}}_S$ has the following matrix form:

$$
\hat{\hat{L}}_S = \begin{bmatrix} -i(\gamma_{ab} + \gamma_{ba}) & 0 & -u_a^* & u_a \\ i\gamma_{ab} & -i\gamma_{ba} & -u_b^* & u_b \\ -u_a'^* & -u_b'^* & -(\Delta + \delta) - i\Gamma_{ba} & 0 \\ u_a' & u_b' & 0 & (\Delta + \delta) - i\Gamma_{ba} \end{bmatrix}, \quad (97)
$$

where γ_{ab}, γ_{ba}, and $\Gamma_{ba} (\equiv \Gamma_{ab})$ are the damping constants (cf. Eqs. 67 and 68) due to spontaneous emission and collisions, Δ is the detuning defined by $\Delta = \omega_{ba} - [(m+1)\omega_1 - m\omega_2]$, and δ (bichromatic Bloch–Siegert resonance shift) and u's (power broadening or resonance width parameters) represent intensity-dependent higher order perturbation corrections for the rest of the supermatrix $\hat{\hat{L}}_F$ (called the "external" space).

The steady-state solutions $[d\hat{\rho}(t)/dt = 0]$ for the density matrix in Eq. 95 can be solved readily to give the coherence (i.e., off-diagonal) density matrix elements,

$$
\bar{\rho}_{ba} = -\gamma_{ba}[(\Delta + \delta + i\Gamma_{ba})(\gamma_{ba}u_a' + \gamma_{ab}u_b') + u_b^*(u_b'^*u_a' - u_b'u_a'^*)]/\bar{D}, \quad (98a)
$$

and

$$
\bar{\rho}_{ab} = \bar{\rho}_{ba}^*, \quad (98b)
$$

where

$$
\bar{D} = \gamma_{ba}[(\Delta + \delta)^2 + \Gamma_{ba}^2](\gamma_{ab} + \gamma_{ba}) + 2\operatorname{Re}(z) - 4\operatorname{Im}(u_a'u_b'^*)\operatorname{Im}(u_a u_b^*),
$$

and

$$
z = [\Gamma_{ba} - i(\Delta + \delta)][(\gamma_{ab} + \gamma_{ba})u_b u_b' + \gamma_{ba}u_a u_a' + \gamma_{ab}u_a u_b'].
$$

From Eqs. 92 and 98, the following general analytical expression for intensity-dependent nonlinear optical susceptibility is obtained:

$$
\chi_{m+1, -m}[\omega = (m+1)\omega_1 - m\omega_2]
$$
$$
= -N_0\mu_{ab}\gamma_{ba}\{[(\Delta + \delta) + i\Gamma_{ba}][\gamma_{ba}u_a' + \gamma_{ab}u_b']
$$
$$
+ u_b^*[u_b'^*u_a' - u_b'u_a'^*]\}/(\bar{D}\varepsilon_1[(m+1)\omega_1]\varepsilon_2[-m\omega_2]). \quad (99)
$$

Note that Eqs. 98 and 99 possess the following two distinct features and
advantages over other approaches (56, 67, 73): (i) The intensity-dependent
nature of $\bar{\rho}$ and $\chi(\omega)$ is clearly determined by the two physical parameters δ
and u only and (ii) ρ and $\chi(\omega)$ have the same general functional form as shown
by Eqs. 98 and 99, respectively, regardless of the order $(2|m|+1)$ of

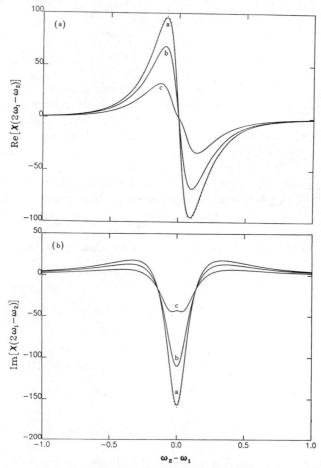

Figure 23. Intensity-dependent nonlinear optical susceptibilities $\chi(\omega)$ corresponding to the
four-wave mixing process, $\omega = 2\omega_1 - \omega_2$, as a function of $\omega_2 - \omega_1$. The pump frequency ω_1 is
fixed at the resonance frequency ($\omega_1 = \omega_{ba} = 100.0$), and the damping parameters used are
$\gamma_{ba} = 0.1$, $\Gamma_{ba} = 5\gamma_{ba}$ (arbitrary units). The probe field strength $|\beta|$ is fixed at 0.01, while the pump
field strength $|\alpha|$ varied. Curves labeled a, b, and c correspond, respectively, to $|\alpha|/\gamma_{ba} = 0.1, 0.5, 1.0$.
The dotted curves are third-order perturbative results, which are intensity-independent. (Adapted
from Ref. 74).

multiphoton processes. Of course, δ and u, depend on m and can be determined via the nearly degenerate perturbative treatment. Analytical expressions for δ and u for various cases can be found in Ref. 74.

Figures 23(a) and 23(b) show the intensity-dependent four-wave mixing nonlinear responses $\chi(\omega = 2\omega_1 - \omega_2)$ as functions of $\omega_2 - \omega_1$ subject to both

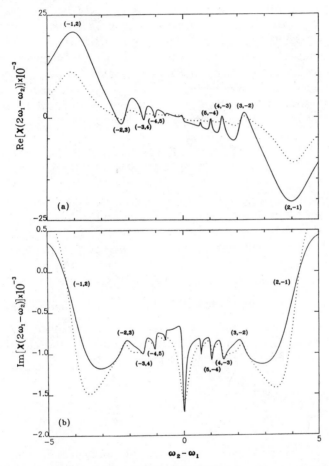

Figure 24. Generalized nonlinear optical susceptibilities $\chi(\omega = 2\omega_1 - \omega_2)$ at intense bichromatic fields are shown as a function of $\omega_2 - \omega_1$. (a) Dispersive responses. (b) Absorptive responses. Solid curves are results for the pure radiative damping case ($\Gamma_{ba} = \frac{1}{2}\gamma_{ba}$); dotted curves include the effects of collisional relaxation ($\Gamma_{ba} = 2\gamma_{ba}$). Parameters used are $\omega_1 = \omega_{ba} = 100.0$, $\gamma_{ba} = 0.1$, $\gamma_{ab} = 0.0$, $|\beta| = 10\gamma_{ba}$, $|\alpha| = 20\gamma_{ba}$, and $\phi_1 = \phi_2 = 0.0$ (arbitrary units). The multiphoton subradiative structures are labeled as (n_1, n_2) corresponding to the contribution from the $n_1\omega_1 + n_2\omega_2$ processes, where n_1 and n_2 are (positive or negative) integers. (Adapted from Ref. 74).

radiative and collisional dampings. The pumping frequency ω_1 is fixed at resonance with the level spacing ω_{ba} ($\omega_1 = \omega_{ba} = 100.0$), the probe field strength $|\beta|$ is fixed at a low value ($|\beta| = 0.01\gamma_{ba}$), and the pump field strength $|\alpha|$ varies from weak to medium strong. First note that in the limit of a weak pump field, $\chi(2\omega_1 - \omega_2)$ approaches the third-order perturbative result (dotted curves), which are intensity independent. However, as the pump field strength $|\alpha|$ increases, significant changes in lineshapes can be seen. In particular, an extra absorption peak (hole) appears at the line center ($\omega_1 = \omega_2 = \omega_{ba}$) at the stronger pump field [Figure 23(b)]. This can be attributed to the contribution from the fifth-order perturbation terms. When both the pump and the probe field intensities are further increased, various higher order contributions will eventually set in, leading to pronounced subharmonic multipeak structures in the generalized nonlinear optical susceptibility, particularly around the line center region [Figures 24(a) and 24(b)].

V. MULTIPHOTON EXCITATION AND POPULATION INVERSION WITH ARBITRARY LASER PULSES

Recently there has been a growing interest in the possibility of preparing multilevel quantum systems in highly excited states by adjusting the laser pulse shape (75), adiabatic sweeping the laser frequency (76, 77), or using trains of pulses (78) with carefully chosen lengths and phases ("composite pulses"). In this section we briefly discuss how the Floquet theory may be generalized to treat the pulse excitation problems (79).

A. Nonadiabatic Coupled Dressed-States Formalism (79)

The time evolution of a nondegenerate N-level system irradiated by a linearly polarized laser pulse field can be described within the electric dipole approximation by the Schrödinger equation

$$i\frac{d|\psi(t)\rangle}{dt} = \{H_0 - \mathbf{\mu} \cdot \mathbf{E}_0(t)\cos[\omega(t)t + \phi(t)]\}|\psi(t)\rangle, \tag{100}$$

where H_0 is the unperturbed Hamiltonian with eigenvalues $E_\alpha^{(0)}$ and eigenvectors $|\alpha\rangle$, $\alpha = 1, 2, \ldots, N$; μ is the electric dipole moment of the system; and $E_0(t)$, $\omega(t)$, and $\phi(t)$ are, respectively, the amplitude, frequency, and phase of the applied laser field. If $E_0(t)$, $\omega(t)$, and $\phi(t)$ change only slightly in any time interval of length $2\pi/\omega(t)$, the pulse or pulse sequence can be viewed in such a way that $E_0(t)$ is the amplitude modulation and $\phi(t)$ is the phase modulation on some carrier wave at frequency $\bar{\omega}(t)$. The discussion that follows will focus on this specific, albeit important, type of pulse excitation.

Collectively, the set of parameters E_0, ω, and ϕ characterizing the laser field shall be abbreviated as $\mathbf{X} = \{\mathbf{E}_0, \omega, \phi\}$. Thus, the total time derivative in Eq. 100 can be written simply as

$$\frac{d}{dt} = \left(\frac{\partial}{\partial t}\right)_{\mathbf{X}} + \dot{\mathbf{X}} \cdot \frac{\partial}{\partial \mathbf{X}}, \tag{101}$$

where

$$\dot{\mathbf{X}} \cdot \frac{\partial}{\partial \mathbf{X}} \equiv \dot{\mathbf{E}}_0 \cdot \frac{\partial}{\partial \mathbf{E}_0} + \dot{\omega} \frac{\partial}{\partial \omega} + \dot{\phi} \frac{\partial}{\partial \phi}. \tag{102}$$

Because $\mathbf{X} = \{\mathbf{E}_0, \omega, \phi\}$ vary only very slowly in time, we can momentarily freeze the parameter $\mathbf{X}$ at a certain value, analogous to Born–Oppenheimer approximation, and establish the parameterized Schrödinger equation,

$$i \left(\frac{\partial}{\partial t}\right)_{\mathbf{X}} |\Phi(t; \mathbf{X})\rangle = \hat{H}(t; \mathbf{X}) |\Phi(t; \mathbf{X})\rangle, \tag{103}$$

where the Hamiltonian $\hat{H}(t; \mathbf{X})$ is given in Eq. 100 with $\mathbf{X}$ fixed at a certain value. The periodic nature of the Hamiltonian $\hat{H}(t; \mathbf{X})$, that is, $\hat{H}(t + T; \mathbf{X}) = \hat{H}(t; \mathbf{X})$ with $T = 2\pi/\omega$, guarantees a complete set of orthonormal quasi-energy state basis $\{|\Phi_\alpha(t; \mathbf{X})\rangle, \alpha = 1, 2, \ldots, N\}$ that possesses the following product form (Floquet theorem):

$$|\Phi_\alpha(t; \mathbf{X})\rangle = \exp[-i\lambda_\alpha(\mathbf{X})t] |\tilde{\Phi}_\alpha(t; \mathbf{X})\rangle, \tag{104}$$

where the quasi-energies λ_α's are real numbers, and the functions $|\tilde{\Phi}_\alpha(t; \mathbf{X})\rangle$ are periodic, that is,

$$|\tilde{\Phi}_\alpha(t + T; \mathbf{X})\rangle = |\tilde{\Phi}_\alpha(t; \mathbf{X})\rangle, \tag{105}$$

and satisfy the orthonormal condition

$$\langle \tilde{\Phi}_\alpha(t; \mathbf{X}) | \tilde{\Phi}_\beta(t; \mathbf{X})\rangle = \langle \Phi_\alpha(t; \mathbf{X}) | \Phi_\beta(t; \mathbf{X})\rangle = \delta_{\alpha\beta}. \tag{106}$$

The quasi-energies $\lambda_\alpha(\mathbf{X})$ and quasi-energy states $|\Phi_\alpha(t; \mathbf{X})\rangle$ of Eq. 103 can be readily solved by the *time-independent* Floquet Hamiltonian method described in Section II.

In general, we can expand the total wavefunction $|\psi(t)\rangle$ of Eq. 100 in terms of $\{|\Phi_\alpha(t; \mathbf{X})\rangle, \alpha = 1, 2, \ldots, N\}$ at each fixed value of $\mathbf{X}$, namely,

$$|\psi(t)\rangle = \sum_{\alpha=1}^{N} a_\alpha(\mathbf{X}) |\Phi_\alpha(t; \mathbf{X})\rangle. \tag{107}$$

Substituting Eq. 107 into Eq. 100 yields a set of coupled equations,

$$i\frac{da_\alpha(\mathbf{X}(t))}{dt} = -i\sum_{\beta=1}^{N} a_\beta(\mathbf{X}(t))\langle\Phi_\alpha(t;\mathbf{X})|\dot{\mathbf{X}}\cdot\frac{\partial}{\partial\mathbf{X}}|\Phi_\beta(t;\mathbf{X})\rangle. \qquad (108)$$

Noting that the changes in $a_\alpha(\mathbf{X}(t))$ and $da_\alpha(\mathbf{X}(t))/dt$ are negligible within a period $T = 2\pi/\omega$, Eq. 108 may be further reduced by averaging it over a duration T to obtain

$$i\frac{da_\alpha(\mathbf{X}(t))}{dt} = -i\sum_{\beta=1}^{N} a_\beta(\mathbf{X}(t))\left\langle\left\langle \Phi_\alpha(t;\mathbf{X})\middle|\dot{\mathbf{X}}\cdot\frac{\partial}{\partial\mathbf{X}}\middle|\Phi_\beta(t;\mathbf{X})\right\rangle\right\rangle. \qquad (109)$$

Here the outer bracket represents a time average, that is,

$$\langle\langle\cdots\rangle\rangle \equiv \frac{1}{T}\int_0^T dt\,\langle\cdots\rangle,$$

so that the coupling matrix elements have no explicit time dependence. Equation 109 shows that transitions between various *adiabatic* quasi-energy states $\{|\Phi_\alpha(t;\mathbf{X})\rangle\}$ are caused by the *nonadiabatic* coupling matrix elements $\langle\langle\Phi_\alpha(t;\mathbf{X})|\dot{\mathbf{X}}\cdot\partial/\partial\mathbf{X}|\Phi_\beta(t;\mathbf{X})\rangle\rangle$ due to the variation of the field quantities $\mathbf{E}_0$, ω, and ϕ in time. The coupling matrix elements can be analytically evaluated using the Hellmann–Feynman theorem (80) and expressed in terms of quasi-energy eigenvalues and eigenvectors (79).

    The coupled dressed-states formalism presented here provides an *exact* nonperturbative treatment, beyond the RWA limit, of the MPE of quantum systems with arbitrary pulse shapes (79). Extension to multiple pulse trains is straightforward (81). We note in passing that similar methods have also recently been developed for the treatment of the following two classes of laser-induced processes is periodic fields.

    (i) *Nonadiabatic theory for high-resolution molecular multiphoton absorption (MPA) spectra* (82). The approach is based on the *adiabatic* separation of *fast* vibrational motion from *slow* rotational motion, incorporating the fact that the infrared laser frequency is close to the frequencies of adjacent vibrational transitions. One thus first solves the quasi-vibrational energy (QVE) states (or, equivalently, the dressed vibrational states) with molecular orientation *fixed*. This reduces the computationally often formidable (vibrational–rotational) Floquet matrix analysis to a manageable scale and, in addition, provides useful physical insights for understanding the nonlinear MPA dynamics. The QVE levels are found to be grouped into distinct energy bands characterized by the infrared frequency, with each band providing an effective potential for molecular rotation. While the interband couplings are totally negligible, the *intraband nonadiabatic angular* couplings are the main driving mechanisms for inducing resonant vibrational–rotational multiphoton transitions.

(ii) *Multicharged ion–atom collisions in intense laser fields: coupled dressed-quasi-molecular-states* (DQMS) *approach* (47, 83). Here we are concerned with nonperturbative treatment of charge transfer processes at *low* collision velocities and *strong* laser intensities. These processes are important in determining the particle densities in Tokamak plasma (84) and have potential usefulness in the development of x-ray lasers (85). As the laser frequency of interest is in the range of quasi-molecular $(A \cdots B)^{+Z}$ electronic energy separations, the laser field oscillates much faster than the motion of the nuclei. It is legitimate to first construct the solutions of the $(A - B)^{+Z}$ + field system, namely, the dressed quasi-molecular electronic states with the internuclear separation $\mathbf{R}$ fixed. The DQMS constructed in this way are *adiabatic*, and their associated quasi-energies (depending parametrically on $\mathbf{R}$) exhibit regions of avoided crossings, where the electronic transition can be induced by the *nondiabatic radial* couplings due to the nuclear movement. By further

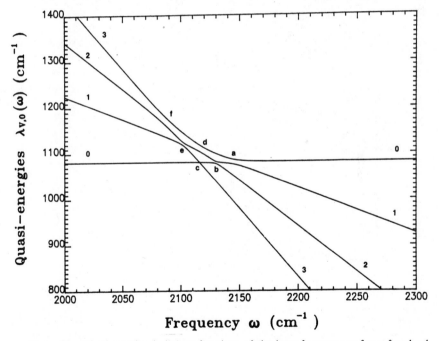

Figure 25. Quasi-energies $\lambda_{v,0}(\omega)$ as functions of the laser frequency ω for a four-level system ($v = 0, 1, 2, 3$ of the CO Morse oscillator) at the laser intensity $50 \, \text{GW/cm}^2$ and laser phase $\phi = 0.0$. (Adapted from Ref. 79).

transforming the *adiabatic* DQMS into an appropriate *diabatic* DQMS representation, defined via the vanishing of the radial couplings already mentioned, one obtains a new set of coupled (diabatic) equations that offer computational advantages.

B. Multiphoton Adiabatic Inversion of Multilevel Systems

In this section we show an example of the application of the coupled dressed-states formalism, Eq. 109, to the study of population inversion (79) among the low-lying vibrational levels of a diatomic molecule induced by a strong laser field of fixed amplitude (E_0) and phase (ϕ) but slowly varying frequency $\omega(t)$. Rotational motion is not considered but can be implemented if needed.

Figures 25 and 26 show the results of the population inversion of the $^{12}C^{16}O$ Morse oscillator from $v = 0$ to $v = 3$ by slowly sweeping the laser frequency ω from 2300 to 2000 cm^{-1}. A linearly scanning rate of the form $\omega(t) = \omega(t=0) - \Lambda t$ is assumed, where Λ is the sweeping rate. The quasi-energies

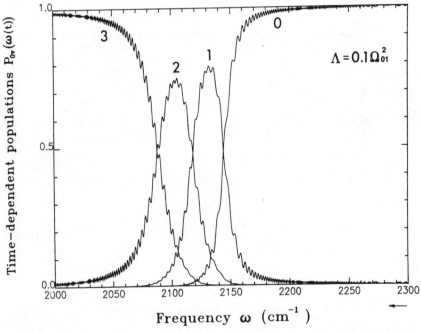

Figure 26. Time-dependent populations $P_{v' \to v}(\omega(t))$, here $v' = 0$, for the four-level system (same as Figure 25) as functions of the laser frequency ω swept at the rate $\Lambda \equiv \dot{\omega}$ equal to $0.1\Omega_{01}^2$, where Ω_{01} is the Rabi frequency. The laser intensity is fixed at 50 GW/cm^2 and the laser phase ϕ is fixed at zero. The arrow, bottom right, indicates the direction of the frequency sweeping. (Adapted from Ref. 79).

$\lambda_{v,0}(\omega)$, $v = 0, 1, 2, 3$, assuming initially the system is at its ground unperturbed vibrational level $(v = 0)$, is depicted in Figure 25. Thus, starting from the adiabatic quasi-energy level connected to $v = 0$ in the extreme right (left) and proceeding to the left (right), one encounters a series of avoided crossings. Population redistribution among the quasi-energy levels can occur at these avoided-crossing points and is induced by the nonadiabatic coupling terms in Eq. 109. Since the narrower are the avoided-crossing regions, the larger will be the nonadiabatic couplings, it is clearly far more favorable to sweep the laser frequency from 2300 to 2000 cm^{-1} than to sweep from the opposite direction. To avoid population redistribution and to preserve adiabaticity (i.e., staying on a single quasi-energy level), at these anticrossing points, the sweeping rate (Λ) has to be sufficiently slow (but faster than the relaxation rates). Figure 26 shows that by sweeping the laser frequency (from right to left) slow enough (where $\Lambda = 0.1\Omega^2$, Ω is the Rabi frequency), one can in fact overcome the bottleneck and achieve nearly 100% population transfer from level 0 (far right) to level 3 (far left). For more complicated systems, there could be many pathways leading from an initial state to a desired final excited state. A Floquet quasi-energy diagram such as Figure 25 is therefore very useful as a guide toward the choice of optimum pathway. In addition, Massey's adiabatic criterion (86) for collision process can be extended here to establish a simple adiabaticity condition for the rate of change of the laser parameter **X** (79).

VI. CONCLUSION

In this chapter, I have discussed several recent Floquet generalizations and their applications to selected multiphoton and nonlinear optical processes, listed in Figure 1. The utilities and advantages of the Floquet methods may be summarized as follows: (i) They provide a simple and consistent physical picture for intensity-dependent nonlinear phenomena in terms of the avoided crossings of a few (real or complex) quasi-energy (or dressed) states. (ii) The methods take into account self-consistently all the intermediate-level shifts and broadenings and multiple coupled continua. (iii) In the case of bound-free MPI–MPD transitions, only bound-state L^2 functions are required, and no asymptotic boundary conditions need to be enforced. (iv) Simplicity exists in the numerical calculations—mainly a (Hermitian or non-Hermitian) matrix or supermatrix eigenvalue problem. (v) The methods are nonperturbative in nature, applicable to arbitrary field strengths, beyond the conventional rotating-wave approximation. It is hoped that the methods described here will provide general and practical nonperturbative techniques for comprehensive and *unified* investigation of single- and multiple-photon, resonant and nonresonant, steady-state and time-dependent phenomena in complex non-linear optical processes in the near future.

Acknowledgments

The work described here was supported by the Alfred P. Sloan Fellowship, J. I. L. A. Visiting Fellowship, Guggenheim Fellowship American Chemical Society–Petroleum Research Fund, and Department of Energy (Division of Chemical Sciences). The author thanks his colleagues and collaborators, particularly Professor William Reinhardt, Professor J. Cooper, Dr. C. Laughlin, Dr. T. S. Ho, Dr. J. V. Tietz, Dr. S. Bhattacharya, Dr. K. K. Datta, and Mr. K. Wang. He is also grateful to Professor J. O. Hirschfelder for constant encouragement of the Floquet developments and his organization of the Los Alamos Symposium on Lasers, Molecules, and Methods.

References

1. Recent reviews can be found, e.g., in *Nonlinear Behavior of Molecules, Atoms, and Ions in Electric, Magnetic or Electromagnetic Fields,* L. Neel, ed., Elsevier, Amsterdam, 1979; *Coherent Nonlinear Optics,* M. S. Feld and V. S. Letokhov, eds., Springer, New York, 1980; *Advances in Chemical Physics,* Vol. 47, (*Photoselective Chemistry*) J. Jortner, R. D. Levine and S. A. Rice, Wiley, New York, 1981; P. A. Schulz, Aa. S. Suadbo, D. J. Krajnovich, H. S. Kwok, Y. R. Shen, and Y. T. Lee, *Ann. Rev. Phys. Chem.* **30**, 311 (1979); V. S. Letokhov, *Nonlinear Laser Chemistry,* Springer, New York, 1983; *Multiphoton Ionization of Atoms,* S. L. Chin and P. Lambropoulos, eds., Academic, New York, 1984; *Advances in Multiphoton Processes and Spectroscopy,* 2 vols. S. H. Lin, ed. World Scientific, Singapore, 1984, 1986.

2. N. Bloembergen and E. Yablonovitch, *Phys. Today* **31**, 23 (1978); A. H. Zewail, V. S. Letokhov, R. N. Zare, R. B. Bernstein, Y. T. Lee, and Y. R. Shen, in a special issue on laser chemistry, *Phys. Today* **33**, 25–59 (1980); T. F. George, *J. Phys. Chem.* **86**, 10 (1982).

3. J. H. Shirley, *Phys. Rev.* **138**, B979 (1965).

4. C. Cohen-Tannoudji and S. Haroche, *J. Phys. (Paris)* **30**, 153 (1969).

5. G. Floquet, *Ann. l'Ecol. Norm. Sup.* **12**, 47 (1883); J. H. Poincaré, *Les Methodes Nouvelles de la Mechanique Celeste,* Vols. I, II, and IV, Paris, 1892, 1893, 1899; S. H. Autler and C. H. Townes, *Phys. Rev.* **100**, 703 (1955).

6. For a review of Floquet developments of two-level systems before 1976, see D. R. Dion and J. O. Hirschfelder, *Adv. Chem. Phys.* **35**, 265 (1976).

7. For a review of more recent developments in Floquet theories (1976–1984), see S. I. Chu, *Adv. At. Mol. Phys.* **21**, 197 (1985).

8. V. I. Ritus, *Sov. Phys.-JETP* **51**, 1544 (1966); Ya. B. Zel'dovich, *Sov. Phys.-JETP* **51**, 1492 (1966).

9. H. Sambe, *Phys. Rev. A* **7**, 2203 (1973).

10. A. G. Fainshtein, N. L. Manakov, and L. P. Rapaport, *J. Phys. B* **11**, 2561 (1978).

11. S. I. Chu, J. V. Tietz, and K. K. Datta, *J. Chem. Phys.* **77**, 2968 (1982). Figure 1 of this paper gives the Floquet matrix structure for the combined (ac + dc) field problem.

12. R. Duperrx and H. van den Bergh, *J. Chem. Phys.* **73**, 585 (1980).

13. See, e.g., W. Magnus, *Comm. Pure Appl. Math.* **7**, 649 (1954); P. Pechukas and J. C. Light, *J. Chem. Phys.* **44**, 3897 (1965); I. Schek, J. Jortner, and M. L. Sage, *Chem. Phys.* **59**, 11 (1981); K. F. Milfeld and R. E. Wyatt, *Phys. Rev. A* **27**, 72 (1983); M. M. Maricq, *J. Chem. Phys.* **85**, 5167 (1986).

14. See, e.g., G. F. Thomas and W. J. Meath, *J. Phys. B* **16**, 951 (1983).

15. J. O. Hirschfelder and R. W. Pyzalski, *Phys. Rev. Lett.* **55**, 1244 (1985).

16. K. B. Whaley and J. C. Light, *Phys. Rev. A* **29**, 1188 (1984).

17. A. Nauts and R. E. Wyatt, *Phys. Rev. Lett.* **51**, 2238 (1983).

18. M. Quack and E. Sutcliffe, *Chem. Phys. Lett.* **105**, 147 (1984).

19. See, e.g., W. R. Salzman, *Phys. Rev. A* **10**, 461 (1974).

20. S. E. Bailkowski and W. A. Guillory, *Chem. Phys. Lett.* **60**, 429 (1979); D. Proch and H. Schröder, *Chem. Phys. Lett.* **61**, 426 (1979).

21. G. L. Wolk, R. E. Weston, and G. W. Flynn, *J. Chem. Phys.* **73**, 1649 (1980); T. B. Simpson and N. Bloembergen, *Opt. Comm.* **37**, 256 (1981).

22. J. V. Tietz and S. I. Chu, *Chem. Phys. Lett.* **101**, 446 (1983).

23. See, e.g., P. H. Winston, *Artificial Intelligence*, Addison-Wesley, Reading, MA, 1979.

24. T. B. Simpson and N. Bloembergen, *Chem. Phys. Lett.* **100**, 325 (1983).

25. J. V. Tietz, Ph.D. Thesis, University of Kansas, 1985.

26. J. Chang and R. E. Wyatt, *J. Chem. Phys.* **85**, 1826 (1986).

27. S. I. Chu and K. Wang, *Phys. Rev. A* (in press).

28. For recent reviews on the method of complex scaling, see (a) W. P. Reinhardt, *Ann. Rev. Phys. Chem.* **33**, 223 (1982); (b) B. R. Junker, *Adv. At. Mol. Phys.* **18**, 208 (1982).

29. See, e.g., H. A. Yamani and W. P. Reinhardt, *Phys. Rev. A* **11**, 1144 (1975).

30. S. I. Chu and W. P. Reinhardt, *Phys. Rev. Lett.* **39**, 1195 (1977); A. Maquet, S. I. Chu, and W. P. Reinhardt, *Phys. Rev. A* **27**, 2946 (1983).

31. S. I. Chu, *Chem. Phys. Lett.* **54**, 367 (1987).

32. S. I. Chu, *J. Chem. Phys.* **75**, 2215 (1981); S. I. Chu, C. Laughlin, and K. K. Datta, *Chem. Phys. Lett.* **98**, 476 (1983); C. Laughlin, K. K. Datta, and S. I. Chu, *J. Chem. Phys.* **85**, 1403 (1986).

33. S. K. Bhattacharya and S. I. Chu, *J. Phys. B* **16**, L471 (1983).

34. S. K. Bhattacharya and S. I. Chu, *J. Phys. B* **18**, L275 (1985); to be published.

35. S. I. Chu, *Int. J. Quantum Chem: Quantum Chem. Symp.* **20**, 129 (1986).

36. See, e.g., P. Agostini, F. Fabre, G. Mainfray, G. Petite, and N. K. Rahman, *Phys. Rev. Lett.* **42**, 1127 (1979); P. Kruit, J. Kimman, and M. J. van der Wiel, *J. Phys. B* **14**, L597 (1981); L. A. Lompre, A. L'Huillier, G. Mainfray, and C. Manus, *J. Opt. Soc. Am. B* **2**, 1906 (1985).

37. P. Kruit, J. Kimman, H. G. Muller, and M. J. van der Wiel, *Phys. Rev. A* **28**, 248 (1983).

38. See, e.g., M. Edwards, L. Pan, and L. Armstrong Jr., *J. Phys. B* **18**, 1927 (1985); Z. Deng and J. H. Eberly, *Phys. Rev. Lett.* **53**, 1810 (1984); M. H. Mittleman, *J. Phys. B* **17**, L351 (1984); M. Crance, *J. Phys. B* **17**, L355 (1984); Y. Gontier, M. Poirier, and M. Trahin, *J. Phys. B* **13**, 1381 (1980); A. Szöke, *J. Phys. B* **18**, L427 (1985); H. G. Muller and A. Tip, *Phys. Rev. A* **30**, 3039 (1984).

39. S. I. Chu and J. Cooper, *Phys. Rev. A* **32**, 2769 (1985).

40. T. S. Ho, S. I. Chu, and J. V. Tietz, *Chem. Phys. Lett.* **96**, 464 (1983).

41. H. Salwen, *Phys. Rev.* **99**, 1274 (1955).

42. B. Kirtman, *J. Chem. Phys.* **49**, 3890 (1968).

43. (a) P. R. Certain and J. O. Hirschfelder, *J. Chem. Phys.* **52**, 5977 (1970); (b) J. O. Hirschfelder, *Chem. Phys. Lett.* **54**, 1 (1978); (c) K. Aravind and J. O. Hirschfelder, *J. Phys. Chem.* **88**, 4788 (1984), where a two-level system in a single monochromatic laser field has been studied in detail using the generalized Van Vleck nearly degenerate perturbation theory.

44. T. S. Ho and S. I. Chu, *J. Phys. B* **17**, 2101 (1984).

45. S. I. Chu and T. S. Ho, *Israel J. Chem.* **24**, 237 (1984).

46. K. Wang, T. S. Ho, and S. I. Chu, *J. Phys. B* **18**, 4539 (1985).

47. S. I. Chu, *Advances in Multiphoton Processes and Spectroscopy*, Vol. 2, World Scientific, Singapore, 1986, pp. 175–238.

48. R. P. Feynman, F. L. Vernon, and R. W. Hellwarth, *J. Appl. Phys.* **28**, 49 (1957).

49. See, e.g., J. Elgin, *Phys. Lett.* **80A**, 140 (1980); F. T. Hioe and J. H. Eberly, *Phys. Rev. A* **25**, 2168 (1982).

50. T. S. Ho and S. I. Chu, *Phys. Rev. A* **31**, 659 (1985).

51. See, e.g., T. Hogg and B. A. Huberman, *Phys. Rev. Lett.* **48**, 711 (1982).

52. See, e.g., M. Shapiro and G. Goelman, *Phys. Rev. Lett.* **53**, 1714 (1984).

53. Y. Pomeau, B. Dorizzi, and B. Grammaticos, *Phys. Rev. Lett.* **56**, 681 (1986).

54. K. Wang and S. I. Chu (to be published).

55. T. S. Ho and S. I. Chu, *Phys. Rev. A* **32**, 377 (1985).

56. Y. R. Shen, *The Principles of Nonlinear Optics*, Wiley, New York, 1984.

57. M. D. Levenson, *Introduction to Nonlinear Laser Spectroscopy*, Academic, New York, 1982.

58. B. Dick and R. M. Hochstrasser, *Chem. Phys.* **75**, 133 (1983).

59. U. Fano, *Phys. Rev.* **131**, 259 (1963).

60. P. O. Löwdin, *Adv. Quant. Chem.* **17**, 285 (1985).

61. T. S. Ho, K. Wang, and S. I. Chu, *Phys. Rev. A* **33**, 1798 (1986).

62. See, e.g., (a) C. Cohen-Tannoudji, in *Frontiers in Laser Spectroscopy*, R. Ballian, S. Haroche, and S. Liberman, eds., Wiley, New York, 1977, p. 103; (b) F. Shuda, C. R. Stroud, Jr., and M. Hercher, *J. Phys. B* **7**, L198 (1973); (c) R. E. Grove, F. Y. Wu, and S. Ezekiel, *Phys. Rev. A* **15**, 227 (1977).

63. B. R. Mollow, *Phys. Rev.* **188**, 1969 (1969).

64. R. Loudon, *The Quantum Theory of Light*, 2nd ed., Oxford University Press, New York 1983.

65. M. Lax, *Phys. Rev.* **129**, 2342 (1963).

66. See, e.g., G. Nienhuis, *J. Phys. B* **16**, 2677 (1983) and references therein.

67. N. Bloembergen, *Nonlinear Optics*, Benjamin, New York, 1965.

68. N. Tan-no, K. Okhawara, and H. Inaba, *Phys. Rev. Lett.* **46**, 1282 (1981).

69. L. Hillman, J. Krasinski, R. W. Boyd, and C. R. Stroud, Jr., *Phys. Rev. Lett.* **52**, 1605 (1984).

70. A. M. Bonch-Bruevich, S. G. Przhibelskii, and N. A. Chigir, *Sov. Phys. JETP* **53**, 285 (1981).

71. R. K. Raj, Q. F. Gao, D. Bloch, and M. Ducloy, *Opt. Commun.* **51**, 117 (1984).

72. G. I. Toptygina and E. E. Fradkin, *Sov. Phys. JETP* **55**, 246 (1982).

73. G. S. Agarwal and N. Nayak, *Phys. Rev. A* **33**, 391 (1986); M. S. Kumar and G. S. Agarwal, *Phys. Rev. A* **33**, 1817 (1986).

74. K. Wang and S. I. Chu, *J. Chem. Phys.* **86**, 3225 (1987).

75. See, e.g., W. S. Warren, *J. Chem. Phys.* **81**, 5437 (1984) and references therein.

76. R. G. Hulet and D. Kleppner, *Phys. Rev. Lett.* **51**, 1430 (1983).

77. G. L. Peterson and C. D. Cantrell, *Phys. Rev. A* **31**, 807 (1985); J. Oreg, F. T. Hioe, and J. H. Eberly, *Phys. Rev. A* **29**, 690 (1984).

78. See, e.g., W. S. Warren and A. H. Zewail, *J. Chem. Phys.* **78**, 2279, 2298 (1983); J. C. Diels and J. Stone, *Phys. Rev. A* **31**, 2397 (1985); M. H. Levitt, D. Suter, and R. R. Ernst, *J. Chem. Phys.* **80**, 3064 (1984).

79. T. S. Ho and S. I. Chu, *Chem. Phys. Lett.* **141**, 315 (1987).

80. See, e.g., C. Cohen-Tannoudji, B. Diu, and F. Laloë, *Quantum Mechanics*, Vol. 2, Wiley, New York, 1977, p. 1192.

81. Y. Huang and S. I. Chu (to be published).

82. S. I. Chu, T. S. Ho, and J. V. Tietz, *Chem. Phys. Lett.* **99**, 422 (1983); T. S. Ho and S. I. Chu, *J. Chem. Phys.* **79**, 4708 (1983).

83. T. S. Ho, S. I. Chu, and C. Laughlin, *J. Chem. Phys.* **81**, 788 (1984); T. S. Ho, C. Laughlin, and S. I. Chu, *Phys. Rev. A* **32**, 122 (1985).

84. See, e.g., J. F. Seely and R. C. Elton, Naval Research Laboratory, Memorandum Report 4317, 1980.

85. R. H. Dixon, J. F. Seely, and R. C. Elton, *Phys. Rev. Lett.* **40**, 122 (1978); W. R. Green, M. D. Wright, J. F. Young, and S. E. Harris, *Phys. Rev. Lett.* **43**, 120 (1979).

86. H. S. W. Massey, *Rep. Prog. Phys.* **12**, 248 (1949).

87. S. I. Chu, *Chem. Phys. Lett.* **58**, 462 (1978).

88. B. R. Johnson, J. O. Hirschfelder, and K.-H. Yang, *Rev. Mod. Phys.* **55**, 109 (1983).

CHAPTER XVIII

QUANTUM OPTICS AT VERY HIGH LASER INTENSITIES: ESSENTIAL STATES AND ABOVE-THRESHOLD IONIZATION

J. H. EBERLY

Department of Physics and Astronomy,* University of Rochester, Rochester, New York 14627

CONTENTS

Workers in quantum optics have long been interested in the interaction of light with matter at high radiation intensities. Most atomic and molecular photoabsorption processes can be characterized as strong or weak according to a simple criterion. This criterion is saturation. If a radiation field is sufficiently intense to put a significant fraction of the initial-state population into a final state, the process is strong whether the light intensity is megawatts or milliwatts. In the simplest case, when the initial state is quasi-stable, the strong-field threshold is reached when the photon flux Φ equals its saturation value $1/\sigma T_1$, where σ is the small-signal absorption cross section for the process and T_1 is the final-state lifetime. It is easy to estimate that saturation can begin at or below $1/\sigma T_1 \simeq 10^{16}$ photons/cm^2-s for optical transitions on resonance, that is, in the miliwatt-per-square-centimeter (mW/cm^2) range of light intensity.

However, if the photon intensity is high enough, bound–bound resonances are insiginificant, and photoionization is the dominant atomic response. Since one optical photon is not able to reach the ionization threshold in most atoms

*Also with the Institute of Optics of the University of Rochester.

and molecules, in almost all cases the ionization process is necessarily multiphoton ionization. Lasers with sufficiently high intensities for ionization-dominant studies have become widely available. In the optical frequency range about $100 \, \text{MW/cm}^2$ is generally high enough.

However, more recent work at intensity levels more than three orders of magnitude higher shows an unexpected variety of new atomic photoprocesses. The surprises offered by these new experimental results are still being debated, and convincing theoretical explanations have not yet appeared. In this sense quantum optics and atomic physics have a common new frontier at ultrahigh intensities. In the space available I will try to describe some of the evidence, give some dimensional estimates and arguments, and present one relatively complete model calculation to try to illuminate this high-intensity frontier.

I. BACKGROUND

Helium and neon are the atoms with the highest ionization potentials, and the ionization of helium and neon requires 19 and 22 photons, respectively, from a Nd–YAG laser beam ($hv \simeq 1.17 \, \text{eV}$, $\lambda \simeq 1.06 \, \mu\text{m}$). Laser intensities available since about 1970 have made the study of such high-order ionization processes possible in a number of laboratories. Of course, lower order multiphoton ionization studies of a wide variety of other atoms and molecules have also been made. A schematic energy level diagram illustrating four-photon ionization is shown in Figure 1.

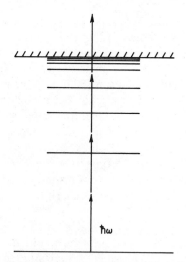

Figure 1. Schematic energy level diagram showing four-photon ionization without intermediate-state resonances.

The principles of perturbation theory mandate that an N-photon ioniz-ation process proceeds at a rate proportional to I^N, or Φ^N, where $I = h\nu\Phi$ is the light intensity (W/cm^2) and Φ is the corresponding photon flux (photons/cm^2-s). Ionization rates are frequently written in terms of a generalized N-photon cross section σ_N as

$$R_N = \sigma_N \Phi^N \tag{1.1}$$

where σ_N then has the dimensions $(cm^2$-s$)^N$ s$^{-1} = cm^{2N}$-s^{N-1}. The principal method of collecting multiphoton ionization data is to count ions. Figure 2 indicates one way that ion count data are commonly presented. The log of the ion signal is plotted as a function of the log of the laser intensity, and the slope of the data points gives N, the multiphoton number, or multiphoton index, according to 1.1.

It might be expected that atomic structure would play an important role in determining the features of the generalized cross section, but experience has tended to indicate otherwise. In Figure 3 we show the ionization cross sections obtained from ion counts for a variety of elements over a range of N from 4 to 22. In the units quoted in the preceding, σ_N is found to obey a "universal" scaling law:

$$\sigma_N \simeq (10^{-34})^N \tag{1.2}$$

In discussing this result, and others to be mentioned, we will assume that the

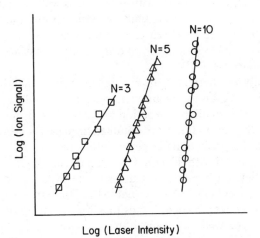

Figure 2. Characteristic method of displaying ion count data. The log–log graph exhibits the multiphoton order of the ionization process as the slope of the line through the data points

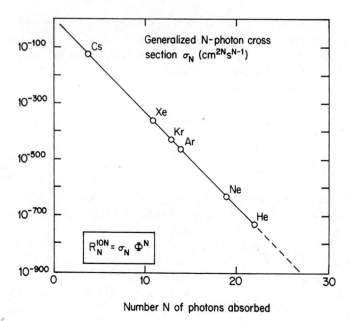

Number N of photons absorbed

Figure 3. Ionization cross sections measured with Nd–YAG laser light (1.17 eV, 1.064 μm) for a variety of elements with multiphoton ionization orders ranging from 4 to 22. [G. Mainfray, private communication.]

term *multiphoton ionization* refers to processes in which $N \geqslant 4$ and $h\nu \geqslant 1$ eV. This restriction is adopted only for the sake of a compact presentation. It excludes important optical work with $N = 2, 3$, as well as ionization studies with infrared and even microwave photons. Multiphoton ionization work of the kind we will include has been carried out since the late 1960s, first by Delone's group in the USSR and very shortly after by the group of Manus and Mainfray in France (1).

One possibility in experiments at very high intensity is that the usual atomic radiation theory may fail to remain valid since it is based on a perturbation series in which the small perturbation parameter is proportional to the intensity. All of the results that were obtained from early experiments were in agreement with the predictions of perturbation theory, despite the extreme intensity levels reached with the lasers used, up to 10^{11} W/cm^2. However, it is easy to argue that no breakdown can really be expected until the atomic electrons sense a laser field strength equal to the Coulomb field strength of the nucleus. This field strength corresponds to a laser intensity of approximately 10^{17} W/cm^2, a value well out of reach until recently.*

*The term "recently" should be understood correctly. This review will appear in print more than two years after the symposium at which it was presented, through no fault of the author or the publisher.

However, in the early 1980s the field took an unexpected turn. Following a suggestion first implemented at Saclay, a number of experimental studies began in which electron detectors were used for the first time. Beginning with a report (2) of a study of cesium from Saclay in 1979, data began to be available on the energy distribution of the ionized photoelectrons. Evidence quickly accumulated to show that a photoelectron is able to absorb additional photons *after* its energy crosses the ionization threshold, that is, after it becomes energetically free. This process is now known as above-threshold ionization (ATI). Of course, the conservation of energy and momentum prevents a truly free electron from absorbing a photon, so the electron's ion must play at least a passive momentum-absorbing role in ATI events.

A subsequent report (3) on a study of xenon from the van der Wiel group at FOM Institute, Amsterdam, greatly enlarged the circle of interest in ATI physics. A sketch of the kind of data presented from FOM, and the original data from Saclay, on ATI electron spectra is shown in Figure 4 in a sequence that covers a range of laser intensities above 10^{11} W/cm^2. The remarkable feature of the FOM data, since confirmed in other laboratories, is that the lowest photopeaks in the electron spectrum, the peaks that correspond to the absorption of the fewest "extra" photons above threshold, become suppressed as the intensity is raised.

At least superficially, the *relative* suppression of the primary photopeak in favor of higher peaks argues that perturbation theory may be inappropriate in ATI photoabsorption. Additional evidence for this conclusion was also contained in the FOM paper (3), which reported the measurement of the multiphoton index N associated with a number of photopeaks. It would naturally be expected that the first peak should show $N = 11$, since 11 photons (from a Nd–YAG laser) are required for ionization. Furthermore, a perturbation-theoretic view of the absorption would require every succeeding peak to show a higher index: $N = 12, 13, \ldots$. However, the reported data indicated $N \simeq 8$ for the first peak and $N \simeq 10$ to $N \simeq 11$ for all higher peaks, completely at variance with any known theory of photoionization.

At the same time, in the early 1980s evidence began to accumulate at both Saclay and the University of Illinois (Chicago) that high-intensity ionization can lead to the ejection of multiple electrons (4) as well as to ATI multiple-photon absorption by a single photoelectron. Multiple-electron production appears to set in at intensities above 10^{13} W/cm^2 and has been observed with Nd–YAG laser photons ($hv \simeq 1$ eV) as well as excimer laser photons ($hv \simeq 6$ eV), whereas ATI effects seem to be more easily produced with the lower energy photons.

New experimental techniques continue to be introduced, and the first informal reports of photon detection have recently appeared from the Rhodes group at the University of Illinois (Chicago). High-order harmonic production (but only of odd harmonics) has apparently been observed. The groups of

Agostini and Petite (Saclay), Lutz (Bielefeld), and Freeman (ATT Bell) have contributed important data on angular distributions (5).

It is becoming clear that not one but several new atomic photoprocesses have been discovered in the past five or six years. It is generally agreed that they are not closely linked, and quite different explanations are likely to be necessary for ATI phenomena compared to multiple-electron production. The evident universality of all of these effects is one of the most striking features of this new domain of atomic physics and quantum optics. It appears possible to observe any or all of the effects mentioned anywhere in the periodic table.

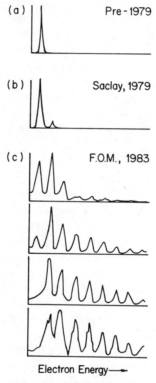

Electron Energy⟶

Figure 4. Artist's sketch of photoelectron energy spectra characteristic of ATI showing the evolution in shape of the peak envelope (from top to bottom of the figure) as the laser intensity increases. The disappearance of the lowest energy photopeak in the last two graphs is presumably not explainable with a traditional perturbative analysis of the absorption process. These schematic spectra are intended to represent (a) the typical spectrum assumed to exist prior to 1979 for $I < 10^{10}$ W/cm^2, (b) the first ATI spectrum recorded at Saclay in 1979 (see Ref. 2) at an intensity $I \simeq 10^{12}$ W/cm^2, (c) series of ATI spectra reported from F.O.M. (see Ref. 3) in 1983 for intensities estimated to range from about $I \simeq 10^{13}$ W/cm^2 (top curve) to about $I \simeq 3 \times 10^{13}$ W/cm^2 (bottom curve).

What is equally remarkable is that evidence for familiar photoprocesses is absent, and the detailed influence of atomic structure on the data is only slight. For example, it was suspected at first that above-threshold photon absorption is due to inverse bremsstrahlung, well-known in laser plasma interactions, and also that ATI photopeaks might be due to the presence of autoionizing states in the atom. The evidence against these interpretations now appears to be conclusive. Also, claims (6) that multiple-electron production may be due to giant dipole-type collective absorption resonances in atoms have been withdrawn.

II. ORDERS OF MAGNITUDE

In thinking about multiphoton ionization some orders of magnitude are important to keep in mind. The Coulomb field of the nucleus provides a benchmark against which laser field strengths can be measured. Following convention, we will use the Coulomb field associated with the first Bohr orbit, $E_c = e/a_0^2$, to estimate the critical laser intensity at which the laser has as much influence on the electron as does the atomic nucleus. If the laser electric field is written $E(t) = \mathscr{E}_0 \exp(-i\omega t) +$ complex conjugate, the cycle-averaged intensity is given by $I = (c/2\pi)|\mathscr{E}_0|^2$. The critical atomic intensity is obtained from this expression by substituting E_c for $\mathscr{E}_0$:

$$I_{at} = ce^2/2\pi a_0^4$$
$$\simeq 10^{17} \text{ W/cm}^2 \tag{2.1}$$

New atomic phenomena will almost certainly be created in processes induced by lasers with $I > I_{at}$.

Another way to orient one's thinking about multiphoton ionization is to compute the number of photons encountered by an individual atomic electron exposed to a laser pulse passing through the electron's orbit. For the sake of illustration we consider Nd–YAG photons and a laser pulse duration of 10 ps and take the area of the orbit to be $(10 \text{ Å})^2$. The results of such estimates are shown in Table 1.

An electron that is nearly ionized is probably not well characterized by a fixed orbital radius or period. Nevertheless, these quantities provide some measure of the role of the ion, which is needed for the electrons to absorb any energy at all. The choice of time durations in the table is of course arbitrary; it is intended to suggest that a meaningful threshold value exists for I in the neighborhood of 10^{12} W/cm^2.

An argument can be constructed as follows. Electrons emit and absorb light only when they are accelerated. Acceleration occurs only during the portion of

TABLE 1

Intensity[a] (W/cm^2)	Photons Encountered	Time Interval
10^6	1	Pulse $\sim 10\,ps$
10^9	$\frac{1}{10}$	Orbit period $\sim 1\,fs$
10^{12}	1	Orbital perigee $\sim 0.01\,fs$
10^{15}	1000	Orbital perigee $\sim 0.01\,fs$
10^{18}	10	Pulse $\sim 1\,ps$

[a]The value 10^{15} is the critical atomic intensity for an electron in an orbit 10 times larger than the first Bohr orbit; at the value 10^{18} the electron is accelerated to relativistic velocity, and its "orbital" area is taken to be a free electron Thomson cross section, $\sigma \simeq 10^{26}\,cm^2$.

the orbit near to the nucleus. We estimate the time for the orbital period at about 1 fs and the fraction of the orbit "near" the nucleus as one-one hundredth of this. The table shows that at $10^{12}\,W/cm^2$ there will be *one photon encountering the electron in this time.* What is the significance of this number? We note that any number greater than 1 will lead to stimulated coherent processes that are *not averaged over the electron's orbit.* In other words, on this argument we expect that $10^{12}\,W/cm^2$ would signal the onset of a regime of absorption characterized by two features: radiative coherence and independence of atomic species.

Perhaps not coincidentally, experimental work agrees with this crude estimate, at least as far as ATI goes. That is, at the 10^{12} level the experimental results first begin to take on the remarkable properties that have attracted so much attention: failure to obey the intensity-scaling relation 1.1 mandated by perturbation theory and apparent universality over the periodic table of elements. That is, universality is what should be expected if photon absorption occurs in a time short compared to an orbital period since the electron requires times comparable with a period to recognize the structure of the atom it resides in. By the same token, coherence of radiative interactions is well known in quantum and nonlinear optics to give rise to electronic response that is not adequately described by perturbation theory—ac Stark effects, saturation, and so on. Most important, one particular message from all ATI experiments is that unexpected new atomic phenomena are created long before the intensity reaches $I_{at} \simeq 10^{17}\,W/cm^2$.

In discussing orders of magnitude, it is necessary to mention *chaos.* This feature of the electron–atom–radiation interaction is dominant at high intensities when the interaction is analyzed classically. The electron in a Coulomb potential subjected to a radiation field is an outstanding example of a periodically driven nonlinear oscillator and exhibits chaotic response if the radiation field is strong enough. Although this aspect of the interaction resists quantization and has not been taken into account in ATI analyses to date, it probably should be. It is strongly suspected that chaos plays a significant role in the theoretical understanding of the microwave ionization experiments of Bayfield (6).

Our semiempirical numerology applies to ATI, but multiple-electron production can be treated in much the same way. Estimates can be constructed that show what photon energy is most efficient in ionizing many electrons in a shell, or in causing the outer electrons to collide with inner shells to create vacancies, or in promoting varieties of many-electron collective action including coupling to nuclear degrees of freedom.

We will not pursue any of these speculations but concentrate completely now on the theory of ATI. In fact, there are currently a number of incomplete theories of ATI, and we will focus on only one of them. This one is based on the postulated existence (at least at asymptotically long times) of a set of "essential states" for the ATI electrons above threshold. This theory may be particularly interesting in the present context because it has antecedents in the chemical physics of nonradiative transitions in polyatomic molecules.

III. INTRODUCTION TO ESSENTIAL STATES OF PHOTOIONIZATION

For orientation we first recall some elementary formulas relating to photo-ionization, which we can temporarily consider to be a one-photon process, as sketched in Figure 5(a). The photoabsorption rate, in lowest order

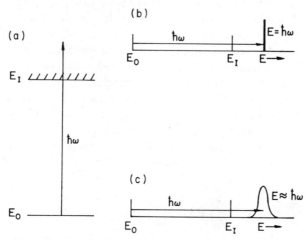

Figure 5. (a) Sketch of energy levels involved in one-photon ionization; (b) corresponding photoelectron energy spectrum according to lowest order perturbation theory; (c) photoelectron energy spectrum, taking finite lifetime of the ground state into account.

perturbation theory, is given by Fermi's golden rule expression (with $\hbar = 1$):

$$R_I = 2\pi |V_{0E}|^2 \rho(E) \tag{3.1}$$

where $|V_{0E}|^2$ is proportional to the light intensity I and the square of an appropriate (usually dipole) atomic matrix element, and ρ denotes the density of continuum states at the energy corresponding to the absorption of one photon. That is, $E - E_0 = \hbar\omega$, where E_0 is the energy of the initial bound state. For our purposes it will be more convenient to measure energy from the initial state, so we will have simply $E = \hbar\omega$. The photoelectron energy spectrum corresponding to this predicted ionization rate is shown in Figure 5(b). In accordance with energy conservation, the energy spectrum is a delta function located at $E = \hbar\omega$.

A slightly more sophisticated view recognizes that the existence of ionization implies a decay in initial-state probability:

$$|C_0(t)|^2 = e^{-R_I t} \tag{3.2}$$

It is immediately evident that the electron energy spectrum shown in Figure 5(b) must be incorrect since the decay of the initial state implies a finite lifetime to the ionization process. The width of the peak in the spectrum must be R_I. Any more particular questions about lineshape are not possible to answer within such a simple theory as this, but an easy extension also provides the lineshape.

A remark about our treatment of the radiation field is in order. We do not quantize the field but assume the form written in the first paragraph of Section II. Despite apparent concern on the part of most photochemists about such an "approximation," there is no significant approximation involved. Quantum-mechanical effects of the radiation field originate in the nonzero nature of the photon operator commutator $[a, a^+] = 1$. This means that quantum radiation effects make a significant contribution only when the strength of all external fields is at the level of one or a few photons. We are discussing in this chapter an entirely different regime of course.

We will work with Schrödinger's equation for the state amplitudes of the initial and final states. We first expand the total electron wavefunction in the unperturbed atomic basis:

$$|\Psi(t)\rangle = C_0(t)|0\rangle + \sum_m \int dE_m C_m(E_m; t)|E_m, m\rangle \tag{3.3}$$

where the states $|E_m, m\rangle$ are the continuum states above the ionization threshold, and the m sum is over all other quantum numbers (e.g., angular momentum) except the energy, which is explicitly indicated by E_m for the mth continuum. By substitution of 3.3 into Schrödinger's equation for the atom, we can obtain the coupled amplitude equations for the C_m.

Now we restrict our attention to a small set of "essential" states, namely, the initial state and those continuum states connected to it via one-photon matrix elements. Normal dipole selection rules for the matrix element mandate that the continuum states differ in orbital angular momentum by one unit from the initial state. For clarity we will adopt a notation that suggests that the gound state (labeled 0) is an S state and the continuum states coupled to it are P states. Then the essential-states equations are

$$\frac{da_0}{dt} = -i \int dE_p \, V_{0p}(E_p) a_p(t) \tag{3.4}$$

$$\frac{da_p}{dt} = -i(E_p - \omega) a_p(t) - i V_{p0}(E_p) a_0(t) \tag{3.5}$$

Note that the absence of a summation on the right side of 3.4 confirms that only states carrying the P label, consistent with the selection rule, make a contribution to the transition. An integral such as $\int dE_p$ obviously runs over all the energies of the p continuum. Later we ignore threshold effects and make the limits infinite in both directions for simplicity. Also, we have made a phase shift in the continuum state amplitudes to obtain the a's from the C's defined in 3.3

$$a_0(t) = C_0(t) \qquad a_p(t) = C_p(E_p; t) e^{i\omega t} \tag{3.6}$$

This involves a kind of rotating-wave approximation. The terms containing factors such as $1 + \exp(\pm 2i\omega t)$ in the interaction have been put equal to 1. This approximation has not been justified rigorously, but it is based on the reasonable expectation that only energies satisfying $E_p \simeq \omega$ will ultimately be important. It ignores the possibility of "diagonal dressing" of the basis states, which was first proposed in different terms by Keldysh and was later explored by Reiss. Recently Lewenstein et al., Becker et al., and Dulcic have applied this procedure specifically to ATI problems. (See Ref. 8 for representative citations.)

The Laplace transforms of the essential-states equations provide a convenient route to their solution. We assume that initially $|\Psi(t)\rangle = a_0 |0\rangle$ and find

$$sA_0 = 1 - i \int dE_p \, V_{0p}(E_p) A_p \tag{3.7}$$

$$sA_p = -i(E_p - \omega) A_p - i V_{p0}(E_p) A_0 \tag{3.8}$$

where the A's are the Laplace transforms of the a's:

$$A_k = \int e^{-st} a_k(t)\, dt \qquad (3.9)$$

The solution for A_0 is

$$A_0 = 1/(s + \gamma + i\delta) \qquad (3.10)$$

where we have made a "pole" approximation by defining

$$\gamma + i\delta = \int dE_p |V_{0p}|^2/[s + i(E_p - \omega)] \qquad (3.11)$$

with the assumption that $\gamma + i\delta$ is independent of s. The validity of this assumption clearly hangs on the E_p dependence of $V_{0p}(E_p)$, and the smoother the dependence the better the validity. Since low-energy continuum matrix elements of one-electron atoms characteristically do not have sharp oscillations, we take the approximation as well justified, particularly in the long-time limit where only very small values of s make any contribution to the Laplace inversions. Of course, the inversion of 3.10 can be determined exactly:

$$a_0(t) = e^{-\gamma_{GR} t} \qquad (3.12)$$

where we have dropped the frequency shift δ, in effect absorbing its value into the definition of the zero of energy and have written γ_{GR} for γ. Here GR means golden rule, in recognition that the limit s–0 of 3.11 reproduces exactly the golden rule result 3.1.

Given 3.12 for a_0, the solution for a_p is equally straightforward:

$$a_p(t) = - iV_{p0} \frac{e^{-i(E_p - \omega)t} - e^{-\gamma_{GR} t}}{\gamma_{GR} - i(E_p - \omega)} \qquad (3.13)$$

In the long-time limit this solution gives the distribution of excited-state probability as a function of electron energy E_p:

$$|a_p(t)|^2 \to \frac{|V_{0p}|^2}{(E_p - \omega)^2 + \gamma_{GR}^2} \qquad (3.14)$$

and this is just what we mean by lineshape. The photoelectron energy spectrum predicted by this formula is indicated in Figure 5(c). Not surprisingly, the lineshape is Lorentz or Breit–Wigner in form, with its full width $2\gamma_{GR}$ equal to the decay rate of the initial state, as anticipated below (3.2).

Our calculation naturally suggests that ionization be interpreted as a decay rather than a scattering process. This is an unusual interpretation to the extent that decay processes are not usually processes tending toward a higher energy state. Nevertheless, it is an interpretation that can be utilized beyond this simple derivation, as will be shown in the next section.

First, however, it will be helpful to broaden the scope of the calculation just outlined so that it applies to multiphoton ionization as well as one-photon ionization. For this purpose we return to the concept of the multiphoton cross section mentioned in Section I. A multiphoton cross section and a multiphoton ionization rate both exist as meaningful concepts only in the absence of intermediate resonances. In the four-photon ionization process shown in Figure 1, for example, there are no resonant intermediate bound states between the initial state and the ionization threshold. This situation is the usual one, since a resonance will be highly improbable unless the laser frequency is adjusted very carefully (which is practically impossible with either Nd–YAG or excimer lasers). The ionization rate can then be written $R_I = \sigma_N \Phi^N$. The only modification needed in our one-photon analysis is thus to interpret the ionization rate $2\gamma_{GR} = R_I$ as given by 1.1 instead of by 3.1.

One effect of this change is to make γ_{GR} a very nonlinear function of laser flux Φ or intensity I:

$$\gamma_{GR} \sim I^N \sim \Phi^N \tag{3.15}$$

This Nth-power dependence may be directly accessible to experimental testing in the future in a way not anticipated previously but suggested by our solution 3.14 for the lineshape. That is, in this section we have shown that by measuring the electron spectral peak width γ_{GR} as well as by measuring the ion rate R_I, one can obtain the value of the multiphoton index N. The present simplified essential-states theory predicts that both measurements will lead to the same value of N. This prediction needs to be tested.

IV.   THE ESSENTIAL-STATES MODEL OF ATI

A point overlooked in traditional ionization theory but recognized by the photochemists Lefevbre and Beswick (9) in a 1972 study of intramolecular nonradiative relaxation is that the first continuum is not the only one. That is, the states labeled by the index p in Section III can , in principle, connect with other continuum states labeled by the index d as well as with the initial state, which, by implication, is an s state. The selection rule for parity allows the orbital angular momentum to increase as well as decrease in both absorption *and* emission.

The relevant equations for the amplitude of the d continuum states, along with the s and p amplitude equations, are the following:

$$\frac{da_0}{dt} = -i \int dE_p \, V_{0p}(E_p) a_p \tag{4.1}$$

$$\frac{da_p}{dt} = -i\Delta_p a_p - iV_{0p} a_0 - i \int dE_d \, V_{pd}(E_p, E_d) a_d \tag{4.2}$$

$$\frac{da_d}{dt} = -i\Delta_d a_d - i \int dE_p \, V_{dp}(E_d, E_p) a_p \tag{4.3}$$

Here Δ denotes a detuning in the continuum: $\Delta_p = E_p - N_p\omega$, where N_p is the number of photons associated with excitation into the p continuum, and so on for all other continua. Of course, one sees that these equations are incomplete; the d continuum should be connected further to the f continuum, in principle, and so on to "higher" continua, but we omit these couplings here. They have been included in earlier papers with Deng (10).

The d states allow further photon absorption by the electron after it has already entered the continuum of positive-energy p states. That is, $|a_p|^2$ can be interpreted as the probability of finding the electron in the first continuum, and $|a_d|^2$ is then obviously the probability that the electron has absorbed another photon and moved into the next continuum. In other words, $|a_d|^2$ represents the first ATI peak. It was Crance and Aymar (11) and Bialynicka-Birula (12) who first applied this kind of model to atomic ionization.

We can attempt to solve Eqs. 4.1–4.3 by the Laplace transform method. Compared to the simpler Eqs. 3.3 and 3.4, one extra assumption is needed here to allow relatively straightforward analytic exact solutions. We assume a factorization of the form $V_{pd} = f_p(E_p)g_d(E_d)$ of the matrix element $V_{pd}(E_p, E_d)$ that connects two continuum states. The algebraic details of the solution are more transparent, however, if the factorization is simplified somewhat, and for our present illustration we will use a factorization of the form

$$V_{pd} = u_p^*(E_p)u_d(E_d) \tag{4.4}$$

The more general factorization, and the ATI formulas it leads to, will be discussed elsewhere (13). For notational simplicity we also write the continuum portion of the bound-free matrix element in factored form by introducing the dummy factor u_0:

$$V_{0p}(E_p) = u_0^* u_p(E_p) \tag{4.5}$$

These factorizations are roughly equivalent to the assumption already implied by the treatment of "normal" ionization in Section III, namely, that the dependence of continuum matrix elements on energy is smooth. The factorization is exactly valid if the matrix elements are extremely smooth, that is, constant functions of energy. Such an extreme-smoothness assumption eliminates principal-value contributions to various continuum integrals in favor of delta function contributions, and we will not rely here on extreme smoothness.

After factoring the continuum integrals in this way, the Laplace transform equations are

$$sA_0 = 1 - iu_0^* \int u_p A_p \tag{4.6a}$$

$$(s + i\Delta_p)A_p = -iV_{p0}A_0 - iu_p^* \int u_d A_d \tag{4.6b}$$

$$(s + i\Delta_d)A_d = -iu_d^* \int u_p A_p \tag{4.6c}$$

In this form it is clear that the integrated products $\int u_m A_m \equiv \int dE_m u_m(E_m)A_m$ are important subsidiary quantities. If we adopt the notation

$$B_m = \int u_m A_m \qquad m = p, d \tag{4.7}$$

we find the equations

$$sA_0 = 1 - iu_0^* B_p \tag{4.8a}$$

$$B_p = -i\kappa_p u_0 A_0 - i\kappa_p B_d \tag{4.8b}$$

$$B_d = -i\kappa_d B_p \tag{4.8c}$$

where we have introduced the new parameter

$$\kappa_m = \int (s + i\Delta_m)^{-1} |u_m|^2 \, dE_m \tag{4.9}$$

Note here that the lack of an extreme-smoothness assumption about the continuum matrix elements means that there will be a principal-part contribution to κ_m, which will be a complex number for that reason.

The solution for A_0 is straightforward and has the same form found in the simpler examples treated in Section III:

$$A_0 = 1/(s + \gamma) \tag{4.10}$$

Thus, for the initial-state amplitude we again, as in Section III, find exponential decay,

$$a_0(t) = e^{-\gamma t} \tag{4.11}$$

but we have solved slightly more general equations here compared with Section III, and the decay rate γ is consequently different:

$$\gamma = \frac{\kappa_p |u_0|^2}{1 + \kappa_p \kappa_d} \tag{4.12}$$

We point out that the numerator is exactly the original expression for γ_{GR} given in 3.11, keeping in mind the factorization 4.5. Also, if $\kappa_d = 0$, $\gamma = \gamma_{GR}$, as should be expected. However, the decay rate as a function of light intensity is now potentially quite different because the $\kappa_p \kappa_d$ term in the denominator of 4.12 is proportional to I. That is, the existence of the d continuum leads to the possibility of a kind of saturation of the ionization rate. This feature has wide implications when we include the infinitely many continua beyond the d continuum (10).

However, we are interested here in the solution of the limited s–p–d model. By combining 4.6 and the solution for A_0, we can easily find the relations

$$
\begin{aligned}
A_p &= \frac{-i}{s + i\Delta_p}\left(\frac{V_{p0}}{s + \gamma} + u_p^* B_d\right) \\
&= \frac{1}{s + i\Delta_p}\frac{1}{s + \gamma}\frac{-iV_{p0}}{1 + Z_{pd}} \\
&= \frac{1}{\gamma - i\Delta_p}\left(\frac{1}{s + i\Delta_p} - \frac{1}{s + \gamma}\right)\frac{-iV_{p0}}{1 + Z_{pd}}
\end{aligned}
\tag{4.13}
$$

where we have introduced a new abbreviation:

$$Z_{pd} = \kappa_p \kappa_d \tag{4.14}$$

The corresponding solution for A_d is similar:

$$A_d = \frac{-1}{\gamma - i\Delta_d}\left(\frac{1}{s + i\Delta_d} - \frac{1}{s + \gamma}\right)\frac{u_d^* \kappa_p u_0}{1 + Z_{pd}} \tag{4.15}$$

Having reduced both A_p and A_d to partial fractions as functions of the Laplace variable s, their inverse Laplace transforms are easily obtained, and so

a_p and a_d are given by constants times the time function $(e^{-i\Delta_m t} - e^{-\gamma t})$, where $m = p, d$. The second of these terms decays to zero, and so the time dependence of the a_m amplitude is asymptotically just $e^{-i\Delta_m t}$.

The probabilities associated with the two continua, namely, $|a_p|^2$ and $|a_d|^2$, thus approach constants that are both Lorentzian functions of energy centered at $\Delta_p = 0$ and at $\Delta_d = 0$, respectively. In particular, the first ATI peak (the second peak above threshold) in this model is given by

$$|a_d(t)|^2 \to \frac{1}{\Delta_d^2 + \gamma^2} \left| \frac{u_d^* \kappa_p u_0}{1 + Z_{pd}} \right|^2 \tag{4.16}$$

We see that the essential-states approach to the coupled continuum amplitudes allows simple analytic expressions to be derived for the time dependence and for the photoelectron energy spectra, including the first ATI peak. Infinite continued fractions govern the magnitude of the decay rate γ as well as the ATI probabilities. Edwards, Pan, and Armstrong (14) and Bialynicka-Birula (12) have analyzed models that are similar in most respects. A related model was put forward earlier by Crance and Aymar (11) but not analyzed in detail. The use of the most general factorization assumption, mentioned above 4.4, causes considerable complication without altering the main conclusions of the essential-states approach. The interesting consequences are already evident in the two-continuum model discussed here. Deng and I have examined the predictions of the infinite-continuum essential-states model in greater detail (10). In the Weisskopf–Wigner limit (integrals such as κ_m are s independent), only some details of intensity dependence change when all higher continua, $f, g, \ldots$, are also included in the calculation.

Figure 6 sketches a special case of the predictions of the essential states for illustration (the continuum matrix elements are chosen to drop off geometrically with higher continuum number in this example). In this special case the Z values associated with higher continuum transitions get progressively smaller. What is strikingly obvious is that the $m + 1$ peak becomes significant only after $Z_{m,m+1}$ exceeds unity. For example, in the figure, we note that the third peak is significant only after $Z_{23} > 1$, which does not occur until frame C. By the same token, the first and second peaks do not grow further and even begin to decrease in height with increasing intensity. This is reminiscent of the "suppression" of low-lying peaks that has been observed experimentally, as shown in Figure 4(c).

The formal saturation properties of the essential-states model are also worth examining. Since we have been assuming that N photons are energetically necessary to reach the ionization threshold, we must take $|V_{0p}|^2$ proportional to I^N. However, the transition matrix elements above the ionization threshold are one-photon in nature, so $|V_{pd}|^2$ and Z_{pd} are both

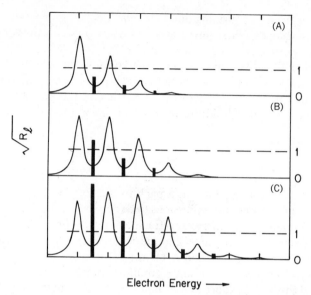

Figure 6. Results of an essential-states calculation of photoelectron energy spectra (see Ref. 10). The heavy vertical bars indicate the values of the Z parameters associated with each continuum–continuum matrix element (recall Eq. 4.14) on the scales indicated at the right side of the graphs. The intensity increases by a factor of 4 from top to middle and from middle to bottom spectra. The Z parameters increase by the same factors. Note that populations in neighboring photopeaks become roughly equalized after the interpeak Z value exceeds unity, but that when $Z < 1$, the population of successive peaks declines monotonically. In this sense Z is an indicator of the onset of nonperturbative behavior.

proportional to I. In view of these intensity dependences, we see that $|u_0|^2$ is proportional to $I^{N-1/2}$. Thus, the probability that the electron reaches the first ATI peak (the d peak here) is proportional to I^{N+1} for low intensities ($Z_{pd} \ll 1$). This is exactly what the principles of perturbation theory suggest: Given N photons to reach the p continuum, it must take $N + 1$ photons to reach the d continuum.

More significantly, we see that as soon as the intensity is high enough so that $Z_{pd} \gg 1$, Z_{pd} will effectively cancel one of the powers of I in the numerator, and for the d peak the ATI peak probability will be proportional to I^N, not I^{N+1}. This is not consistent with a perturbative estimate, of course. The same is true of the decay rate itself, given in 4.12.

Two points deserve special mention. First, we see that there is a natural threshold suggested by the critical value $Z = 1$ at which perturbative physics ceases to be a good guide. The numerical value of intensity that corresponds to $Z = 1$ for a given continuum–continuum transition is difficult to estimate because the continuum wavefunctions are not available in closed form for any element except hydrogen. Studies by Rzażewski and Grobe (15) suggest that this threshold occurs above 10^{15} W/cm^2 in hydrogen. Other estimates have been made by Gontier and Trahin (16) and by Aymar and Crance (17), and somewhat lower values are found, closer to the value 10^{13} W/cm^2 at which experimental ATI effects begin to be evident.

Second, the essential-states model makes a very plain prediction about the linewidths of ATI peaks: All peaks have the same full width 2γ. This is consistent with the uncertainty principle because all of the electron probability originates in the ground state, and the model predicts $1/2\gamma$ to be the lifetime of the initial state. Of course, the uncertainty principle only demands that the rate of growth of all ATI peaks taken together must equal the ground-state decay rate (11). This still permits the ATI linewidths to differ, of course, if other time scales are introduced into the treatment. I have been looking at another artificial but interesting model (perhaps the first exactly solvable one with arbitrary temporal shape of the laser pulse) with Huang and Roso (18) in which pseudo-ATI linewidths differ among themselves.

In the present treatment, the dependence of the ATI widths on γ indicates that the widths should be extremely sensitive to changes in laser intensity if the first bound-free stage of multiphoton ionization (in our discussion the $0-p$ step) involves a large enough value of N. In xenon, for example, eleven Nd–YAG photons are required for ionization, so γ is predicted to be proportional to I^{11}. Merely doubling I then increases γ by three orders of magnitude. Of course, one would perhaps not trust such a prediction until I is already in the neighborhood of the value that makes $Z = 1$, but still it is a prediction that should be tested experimentally.

V. ATOMIC ATI AND MULTIPLE-PHOTON DISSOCIATION OF MOLECULES

The essential-states model has a further attraction in addition to its possible suitability for treatment of the atomic ATI problem. The members of this workshop will understand immediately the analogy between above-threshold ionization in atoms and multiple-photon dissociation (MPD) of polyatomic molecules. MPD was of widespread interest 10 years ago in connection with laser isotope separation. It is not so widely studied anymore for the worst of reasons: There is not much money being thrown at it now. When the energy crisis evaporated, the urgency to understand multiple-photon dissociation went with it.

The MPD data from a wide variety of molecular species are consistent with a picture of photon absorption very close to ATI absorption. In the mid-1970s it was observed that molecules seemed to absorb more photons than they "should" absorb in the dissociation process, and the observed fluence dependence seemed to indicate saturation-like behavior was present. An atom–molecule ATI–MPD analogy can be based on the replacement of the atomic ionization threshold by the threshold of the molecular quasi-continuum.

In the period 1974–1976 a number of workers advanced a rate equation theory of photoabsorption in the quasi-continuum (19). In such a theory the absorption of successive photons can be characterized by rate coefficients. The probabilities of occupation of different quasi-continuum "levels" are then given by

$$\frac{dp_0}{dt} = -R_0(p_0 - P_1) \tag{5.1}$$

$$\frac{dP_1}{dt} = +R_0(p_0 - P_1) - R(P_1 - P_2) \tag{5.2}$$

$$\frac{dP_2}{dt} = +R(P_1 - P_2) - R(P_2 - P_3) \tag{5.3}$$

and so on, for all higher quasi-continuum levels. The lowest level acts as a kind of doorway level, and so the rate coefficient R_0 associated with the 0–1 transition is different from the coefficient R assigned to all the transitions within the quasi-continuum. The set of MPD equations 5.1–5.3 is comparably simple to the essential-states set 4.1–4.3 advanced for the treatment of ATI.

One consequence of Eqs. 5.1–5.3 is the relation

$$\frac{dp_0}{dt} = -\sum_j \frac{dP_j}{dt} \tag{5.4}$$

It is possible to derive the same relation from the ATI equations as well. This is not surprising since 5.4 is merely an expression of probability conservation. Unfortunately, this seems to be as far as the analogy has been developed so far. The ATI equations are fundamental in the sense that they are based on Schrödinger's equation (with rotating-wave and factorization approximations), and the MPD equations are purely empirical. It would be attractive to have an MPD theory also based directly on Schrödinger's equation and molecular matrix elements in the quasi-continuum. It should also be noted that despite some attempts, it has not yet been demonstrated that the ATI equations have an incoherent limit in the form of 5.1–5.3. This by itself is somewhat surprising.

VI. SUMMARY

We have described some of the evidence for new high-power laser phenomena in atomic physics. In particular, I have argued that the universality of ATI is an indication of its fundamental status. Section II contains a dimensional argument showing that the critical field strength associated with ATI may possibly be understood in terms of the onset of stimulated emission due to the presence of many photons inside a reasonable atomic interaction volume. Section III provides a review of photoionization theory utilizing the concept of essential states. From the quantum optics point of view, the ionization process is an induced decay of the atomic ground state, in the course of which the electron probability flows toward the positive-energy continuum states. Our approach to the equations for the Schrödinger state amplitudes, based on Laplace transforms, leads naturally to expressions for the bound-state decay rate, the power-dependent shift of the ionization threshold, and the lineshape of the photoelectron's energy spectrum. In Section IV this approach is generalized to treat multiple continua, leading to various expressions reflecting the existence of continuum–continuum saturation effects. The identification of ATI as a saturation effect is suggested, and the possibility of observing specific saturation phenomena (e.g., linewidths) is mentioned. Section V recalls the rate equation theories of multiple-photon dissociation and identifies points of similarity and difference between ATI and MPD. I suggest that the essential-states approach to MPD would be interesting to pursue.

Acknowledgments

It is a pleasure to acknowledge conversations about the topics discussed here with a great many people, including P. Agostini, L. Armstrong, Jr., J. E. Bayfield, I. Bersons, Z. Bialynicka-Birula, C. Cerjan, M. Crance, Z. Deng, A. Dulcic, F. M. H. Faisal, M. Fedorov, D. Feldmann, Y. Gontier, A. Hazi, D. E. Holmgren, H. Huang, J. Javanainen, R. V. Jensen, K. Kulander, M. Lewenstein, L. A. Lompre, H. O. Lutz, G. Mainfray, C. Manus, M. Mittleman, J. Mostowski, S. Mukamel, L. Pan, G. Petite, L. Roso, K. Rzążewski, Y. R. Shen, B. W. Shore, A. Szöke, and H. B. van den Heuvell.

References

1. References to early multiphoton work are contained in several review articles published over the past decade. A recent review (recall footnote, p. 804) that includes a summary of earlier ones is J. H. Eberly and J. Krasinski, in *Advances in Multiphoton Processes and Spectroscopy*, S. H. Lin, ed. (World Scientific, Singapore, 1985, p. 1. A really recent review of ATI is given in J. H. Eberly and J. Javanainen, *Eur. J. Phys.* (in press, 1988).

2. P. Agostini, F. Fabre, G. Mainfray, G. Petite, and N. K. Rahman, *Phys. Rev. Lett.* **42**, 1127 (1979).

3. P. Kruit, J. Kimman, H. G. Muller , and M. J. Van der Wiel, *Phys. Rev. A* **28**, 248 (1983).

4. A. L'Huillier, L. A. Lomprey, G. Mainfray, and C. Manus, *Phys. Rev. Lett.* **48**, 1814 (1982); T. S. Luk, H. Pummer, K. Boyer, M. Shahidi, H. Eggar, and C. K. Rhodes, *Phys. Rev. Lett.* **51**, 110 (1983).

5. G. Petite, F. Fabre, P. Agostini, M. Crance, and M. Aymar, *Phys. Rev. A* **29**, 2677 (1984); H. J. Humpert, R. Hippler, H. Schwier, and H. O. Lutz, in *Fundamental Processes in Atomic Collision Physics*, H. Kleinpoppen, J. S. Briggs, and H. O. Lutz, eds., Plenum, New York, 1985, p. 649; R. R. Freeman, T. J. McIlrath, P. H. Bucksbaum, and M. Bashkansky, *Phys. Rev. Lett.* **57**, 3156 (1986).

6. C. K. Rhodes, *Science* **229**, 1345 (1985).

7. See, e.g., G. Casati, B. V. Chirikov, and D. L. Shepelyansky, *Phys. Rev. Lett.* **53**, 2525 (1984) and references therein; J. E. Bayfield and L. A. Pinnaduwage, *Phys. Rev. Lett.* **54**, 313 (1985).

8. L. V. Keldysh, *Sov. Phys. JETP* **20**, 1307 (1965); H. R. Reiss, *Phys. Rev. A* **22**, 1786 (1980); M. Lewenstein, J. Mostowski, and M. Trippenbach, *J. Phys. B* **18**, L461 (1985); W. Becker, R. R. Schlicher, and M. O. Scully, *J. Phys. B* **19**, L785 (1986).

9. R. Lefebvre and J. A. Beswick, *J. Mol. Phys.* **23**, 1223 (1972).

10. Z. Deng and J. H. Eberly, *Phys. Rev. Lett.* **53**, 1810 (1984); *J. Opt. Soc. Am. B* **2**, 486 (1985); Z. Deng and J. H. Eberly, *J. Phys. B* **18**, L287 (1985).

11. M. Crance and M. Aymar, *J. Phys. B* **13**, L421 (1980).

12. Z. Bialy nicka-Birula, *J. Phys. B* **17**, 3019 (1984).

13. J. H. Eberly, unpublished. The main expressions appear in capsule form in Z. Deng and J. H. Eberly, *Phys. Rev. Lett.* **58**, 618 (1987).

14. M. Edwards, L. Pan, and L. Armstrong, Jr., *J. Phys. B* **18**, 1927 (1985).

15. K. Rzążewski and R. Grobe, *Phys. Rev. A* **33**, 1855 (1986).

16. Y. Gontier and M. Trahin, *J. Phys. B* **13**, 4383 (1980).

17. M. Aymar and M. Crance, *J. Phys. B* **14**, 3585 (1981).

18. H. Huang, L. Roso, and J. H. Eberly, unpublished.

19. See, e.g., references given in E. R. Grant, P. A. Schulz, Aa. S. Sudbo, M. J. Coggiola, Y. R., Shen and Y. T. Lee, in *Multiphoton Processes*, J. H. Eberly and P. Lambropoulos, eds., Wiley, New York, 1978, p. 359.

CHAPTER XIX

COUPLED-EQUATION METHOD FOR MULTIPHOTON TRANSITIONS IN DIATOMIC MOLECULES: BRIDGING THE WEAK- AND INTENSE-FIELD LIMITS

ANDRÉ D. BANDRAUK

Département de Chimie, Faculté des Sciences, Université de Sherbrooke, Sherbrooke, Quebec, Canada, J1K 2R1

OSMAN ATABEK

Laboratoire de Photophysique Moléculaire, Université de Paris-Sud, Orsay, 91405, France

CONTENTS

I. INTRODUCTION

The advent of intense lasers has led to an increasing interest in the properties of atomic and molecular systems subjected to intense $(I \geqslant 10^8 \text{ W/cm}^2)$ radiation fields where nonlinear effects will predominate (1). A proper theoretical description of molecular processes in such intense fields should therefore be of a nonperturbative nature and should strive as much as possible to include the field in appropriate models. One approach, the dressed-molecule picture, leads to the idea of new adiabatic states created by laser-induced avoided crossings between resonant electronic-field surfaces in molecular systems (2–9). This has led to the prediction of laser-induced resonances that cannot be described by the usual perturbation techniques (6–11). Another approach is to try to develop a complete quantum electrodynamical description of the dressed-molecule concept (12–15). This last method leads to the concept of field-modified adiabatic potential surfaces with appropriate coupled equations for various gauges that are unitary equivalent. In general, it is possible in such theoretical treatments to treat *radiative* and *nonradiative* couplings simultaneously. It will be shown below in a coupled-channel formalism that both radiative and nonradiative couplings are formally equivalent. This enables us in the case of diatomic molecules to make use of semiclassical techniques of the theory of molecular predissociation (16–20) in order to study the presence of laser-induced molecular resonances, that is, bound states that acquire finite lifetimes due to radiative and nonradiative decays.

The necessity of using coupled equations to treat adequately molecular multiphoton processes in the regime of intense fields stems from the fact that molecules are intrinsically multilevel systems; two- or more level descriptions such as used in the dressed-atom picture (1, 21) are no longer adequate. The appearance of continuum nuclear states corresponding to dissociative molecular states requires the proper use of scattering techniques to incorporate finite-lifetime effects arising from multiphoton radiative decays of bound states into continua. We are thus led to study for multiphoton problems the general case of discrete states embedded in continua. These are conveniently treated for all coupling strengths by quantum scattering, or S matrix, techniques (22, 23). Thus, as emphasized by us previously (6, 7), direct photodissociation is the analog of predissociation, where in the former, radiative couplings are equivalent to nonradiative couplings in the latter. In the present work, we shall show that a time-independent collisional treatment of molecules in the presence of an electromagnetic field can lead to a computationally attractive method for accurately calculating general multiphoton lineshapes. This enables us to bridge the weak-field (Franck–Condon) perturbative regime and the strong-field nonperturbative regime. Semiclassical techniques of the theory of predissociation (16, 17, 19, 20) will be shown to be a useful tool to predict the stability of laser-induced resonances as a result of nonlinear radiative interactions between the bound and continuum molecular states. Finally, examples will be given of some numerical simulations of intense-field molecular multiphoton transitions, and a discussion will be presented of future problems to be addressed in order to make the coupled-equations method of multiphoton transitions a viable tool for quantitative predictions in laser photochemistry.

II. TIME-DEPENDENT FORMALISM FOR TRANSITION AMPLITUDES

In this section we consider the general problem of calculating numerically multiphoton cross sections for a molecule. Given that for intense fields, dissociation and ionization are commonly occurring events, the most convenient method of incorporating continua in a computational scheme is via S matrix techniques. Since we shall be considering transitions from initially unperturbed molecules, one must now consider perturbations of the initial and final states by the intense radiative field. Thus, preparation of the initial and final states become now of paramount importance. We shall show that it is possible to define meaningful multiphoton transition probabilities using a generalization of an artificial channel method introduced by Shapiro for the calculation of direct photodissociation cross sections (24). We have developed this general method for previous calculations of direct photodissociation and resonance Raman scattering in strong fields (10, 11). Following the treatment of Bandrauk and Turcotte (20), we shall justify here the coupled-channel method for photodissociation in strong fields by a rigorous time-dependent treatment. This will allow us to illustrate the generality of this method, which we reiterate includes perturbation of the initial state by the field, an aspect neglected in traditional photochemical theories.

The total Hamiltonian for a molecular system submitted to an electromagnetic field can be formally written as

$$H = H_m + H_r + V^r, \tag{1}$$

where H_m is the isolated molecule Hamiltonian, H_r is the free radiation Hamiltonian, and V^r is the radiation matter interaction. The isolated molecule is described by the eigenstates of H_m, which form a complete basis $\{|E_a\rangle\}$. Each molecular state is characterized by its energy E_a and various other quantum indices a (electronic, vibrational, rotational). We further describe the field in the Fock representation (25) by its eigenstates $|n\rangle$ with energies $n\hbar\omega$, where n, ω are, respectively, the number and frequency of the photons. The unperturbed states of the system are eigenstates of $H_0 = H_m + H_r$ and are therefore described by direct products of a molecular and a field state:

$$H_0|E_a, n\rangle = (E_a + n\hbar\omega)|E_a, n\rangle,$$

$$|E_a, n\rangle = |E_a\rangle|n\rangle. \tag{2}$$

We are interested in calculating the temporal evolution of some initial state $|\psi(0)\rangle$ under the influence of the total Hamiltonian H, Eq. [1]. This is formally given by the expressions

$$|\psi(t)\rangle = \exp(-iHt/\hbar)|\psi(0)\rangle,$$

$$= (2i\pi)^{-1} \int dE \exp(-it/\hbar)G(E^+)|\psi(0)\rangle, \qquad [3]$$

where

$$G(E^+) = \lim_{\varepsilon \to 0} (E + i\varepsilon - H)^{-1}. \qquad [4]$$

Expanding in terms of the eigenstates $|a\rangle$ and $|b\rangle$ of H_0, Eq. [2], we can rewrite Eq. [3] as

$$|\psi(t)\rangle = \sum dE_b |b\rangle [\sum I_{ba}^{(t)} \langle a|\psi(0)\rangle], \qquad [5]$$

$$I_{ba}(t) = (2i\pi)^{-1} \int dE \exp(-iEt/\hbar)G_{ba}(E^+). \qquad [6]$$

Therefore, I_{ba} is the time-dependent transition amplitude between the unperturbed states $|E_a, n\rangle$ and $|E_b, n'\rangle$. In Eq. [5], we sum over both bound and continuum states. Using the integral relationship between resolvent G and transition T operators (22, 23),

$$G(E^+) = G_0(E^+) + G_0(E^+)T(E^+)G_0(E^+), \qquad [7]$$

where

$$G_0(E^+) = \lim_{\varepsilon \to 0} (E + i\varepsilon - H_0)^{-1},$$

one can express the general time-dependent amplitude $I_{ba}(t)$ in terms of the time-independent but energy-dependent transition amplitudes $T_{ba}(E^+)$,

$$I_{ba}(t) = (2i\pi)^{-1} \int \frac{dE \exp(-iE^+ t/\hbar) T_{ba}(E^+)}{(E^+ - E_b - n_b\hbar\omega)(E^+ - E_a - n_a\hbar\omega)}. \qquad [8]$$

The transition operator T is obtained from the respective scattering Lippmann–Schwinger equation (22, 23) and is expressible in terms of the resolvent G as

$$T = V + VGV = V + VG_0T. \qquad [9]$$

Thus, calculation of T in terms of the interactions V (radiative and nonradiative) and inserting in Eq. [8] gives the time-dependent transition amplitude. We shall show next that for long-time excitations and (or) fast dissociations (as occurs in intense fields), there is a direct relation between I_{ba} and T_{ba}. The latter is amenable to numerical calculations via coupled-equation techniques for the S or T matrix. We demonstrate this approach in the next section, where we discuss direct photodissociation.

III. PHOTODISSOCIATION IN INTENSE FIELDS

We illustrate in Figure 1 the channels that enter in calculating a direct photodissociation amplitude T_{ac} from some initial molecular bound state $|a\rangle$ to a final molecular continuum state $|c\rangle$. Figure 1(a) corresponds to the weak-field case, whereas Figure 1(b) illustrates the strong-field case. The distinction between the two limits is established in terms of the electronic Rabi frequency ω_{ac}, $\hbar\omega_{ac} = V^r_{ac}$,

$$\omega_{ac}(\text{cm}^{-1}) = \mu_{ac}\langle n|\hat{E}|n \pm 1\rangle = 1.17 \times 10^{-3}\mu_{ac}(\text{a.u.})[I(\text{W/cm}^2)]^{1/2}.$$
$$[10]$$

Here μ_{ac} is the electronic transition moment in atomic units between the electronic states $|a\rangle$ and $|c\rangle$; $\hat{E}$ is the electric field operator, and I is the corresponding electric field intensity. In analogy to the theory of predissociation, the radiative coupling V^r_{ac} is considered *weak* if the corresponding Rabi frequency ω_{ac} is much less than a vibrational frequency ω_v, *intermediate* if it is of the same order as ω_v, and *strong* if it is larger than ω_{ac} (16). It is in the case of both intermediate and strong coupling regimes that perturbative treatments no longer hold. Radiative-induced avoided crossings between the resonant dressed states $|a\rangle$ and $|c\rangle$ [see Figure 1(b)] will occur, giving rise to new states induced by the radiative field. Such states can be termed *adiabatic* dressed states and are best identified by semiclassical techniques (6, 7, 20).

We now introduce artificial channels to transform the physical problem illustrated in Figure 1 into a scattering problem. We illustrate the technique by the appropriate channels in Figure 2. Figure 2(a) illustrates the weak-field direct photodissociation channels. Since the initial state $|a, n\rangle$ is a bound state, an artificial channel $|c_1\rangle$ is coupled weakly to the initial state, enabling one to calculate the S matrix numerically between the artificial continuum $|c_1\rangle$ and the physical continuum $|c\rangle$ in the presence of the bound states of the molecular field potential $|a, n\rangle$. This is related to the transition amplitude by the relation

$$S_{c,c_1}(E) = -2i\pi\exp(i\eta_c)T_{c,c_1}(E)\exp(i\eta_{c_1}),$$
$$[11]$$

where the η's are the pure elastic phase shifts for the two uncoupled

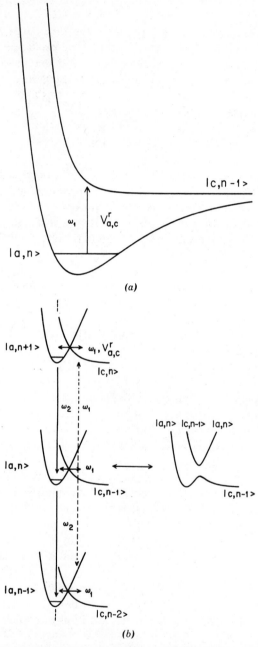

Figure 1. (a) Weak-field single-photon photodissociation with photon occupation numbers ($V_{a,c}^r \neq V_{c,a}^r = 0$). (b) Dressed picture (strong field): ω_1, incident photon, ω_2, emitted photon; $\leftrightarrow$, resonant radiative interaction; $\leftarrow\!\!-\!\!\rightarrow$, virtual radiative interaction ($V_{a,c}^r = V_{c,a}^r \neq 0$, arbitrary strength).

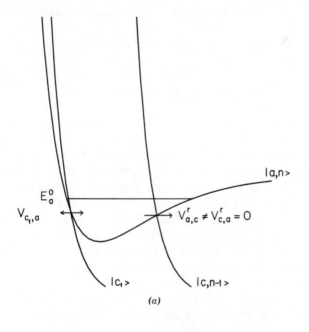

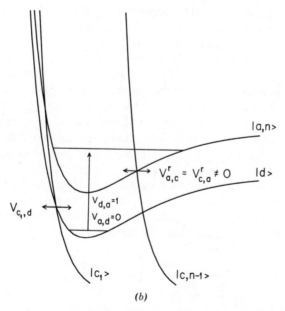

Figure 2. (*a*) Single-photon (weak-field) photodissociation with single entrance channel $|c_1\rangle$. Here $V_{a,c}^r \neq V_{c,a}^r = 0$ gives no radiative corrections to $|a\rangle$. (*b*) Strong-field photodissociation requiring two entrance channels $|c_1\rangle + |d\rangle$. Here $|d\rangle$ is unperturbed (zero-field) initial state. The term $V_{a,c}^r = V_{c,a}^r \neq 0$ now introduces radiative corrections to $|a\rangle$, and $V_{d,a} = 1$, $V_{a,d} = 0$, gives direct overlap between initial state $|d\rangle$ and $|a\rangle$ but no interaction.

829

(unperturbed) continua. Using the second relation expressed in Eq. [9], one readily obtains

$$T_{c,c_1}(E) = V_{c_1,a}(E - E_a + i\Gamma_a/2)^{-1} T_{a,c}.$$ [12]

Thus, the calculation of the S_{c,c_1} or T_{c,c_1} matrix elements between the channels $|c_1\rangle$ and $|c\rangle$ gives access to the direct photodissociation amplitude $T_{a,c}$. Such a procedure has been used in direct photodissociation calculations (24, 26) and has been generalized to calculate photopredissociation (27) and Raman scattering amplitudes (28–30) and other higher order optical amplitudes (31). Implicit in the derivation of Eq. [12] is that the initial state $|a\rangle$ is a well-defined isolated and unperturbed state. Thus, calculations based on Eq. [12] and its generalizations (27–31) are valid only for weak incident fields.

Intense incident fields will produce coherent mixings of the eigenstates of H_0 via the radiative couplings V^r. Superposition of these states as the initially prepared state must be incorporated into the calculation of appropriate amplitudes. Thus, in the case of direct photodissociation, one can calculate via coupled equations the stationary states corresponding to the coupled $|a\rangle$ and $|c\rangle$ manifolds, Figure 2(b), which in actual practice corresponds to the manifold of molecular field states depicted in Figure 1(b). We now extend the artificial channel method further by adding the $|d\rangle$ and $|c_1\rangle$ manifold [Figure 2(b)], where $|d\rangle$ is the true initial unperturbed (zero-field) molecular state, to the total molecular field manifolds [Figure 1(b)], which are radiatively coupled. In practice, $|d\rangle$ is some state $|a, n\rangle$ (i.e., an initial rovibronic level in the presence of n photons) so that the coupling is set as $V_{da} = \langle d|V|a\rangle = 1$, $V_{ad} = 0$, whereas $V_{ac} = V_{ca} = \langle a, n|\mathbf{\mu}\cdot\hat{E}|c, n-1\rangle$. Clearly, using the electronic potential V_d identical to V_a and displacing V_d in energy with respect to V_a ensures that $\langle d|a\rangle = \delta_{a,d}$; that is, the initial condition is set as some $|a\rangle$ state. Finally, $V_{c_1,d} = V_{d,c_1}$ is chosen to be small so as not to perturb the initial state $|d\rangle$ (i.e., negligible widths Γ and shifts ΔE, $\leqslant 10^{-3}$ cm^{-1}).

We now use the formalism of projection operators (32–34) to derive the more general transition amplitudes that take into account coherent mixings of the molecular states by intense fields. We separate the continuum states by the operators P from the bound states by the operators Q and Q',

$$P = \int dE_c |c\rangle\langle c| + \int dE_{c_1} |c_1\rangle\langle c_1|,$$ [13]

$$Q = \sum_a |a\rangle\langle a|, \qquad Q' = Q + \sum_d |d\rangle\langle d|,$$ [14]

such that $P + Q' = 1$, $PQ' = Q'P = PQ = QP = 0$. In terms of these projection operators, the transition operator [9] can be written as (10, 20, 32–34)

$$T = t + t(Q'GQ')t, \qquad\qquad [15]$$

$$t = V + VG_pV, \qquad G_p = P(E^+ - PHP)^{-1}P, \qquad [16]$$

$$Q'GQ' = Q'[E^+ - H_0 - Q'tQ']^{-1}Q'. \qquad [17]$$

The transition amplitude that is to be calculated numerically corresponds to the transition between the artificial channel $|c_1\rangle$ and the physical continuum $|c\rangle$,

$$T_{c,c_1}(E) = \langle c|t|c_1 \rangle + \langle c|t(Q'GQ')t|c_1 \rangle. \qquad [18]$$

Since there is no direct coupling between the two continua, it can be shown that $\langle c|t|c_1 \rangle = \langle a|t|c_1 \rangle = \langle c|t|d \rangle = 0$, so that finally (see Appendix for details)

$$T_{c,c_1}(E) = \sum_d \langle c|tQG|d \rangle \langle d|t|c_1 \rangle, \qquad [19]$$

$$\langle c|tQG|d \rangle = \langle c|tQ(E - H_0 - QtQ)^{-1}|a \rangle \sum_{d,d'} \langle d'|G|d \rangle \langle d|V|c_1 \rangle, \qquad [20]$$

We now expand the projection operator onto the space of biorthogonal eigenstates $|L\rangle$ of the optical potential $(E - H_0 - QtQ)^{-1}$:

$$Q(E - H_0 - QtQ)^{-1}Q = \sum_L |L\rangle \left(E - E_L - \frac{i\Gamma_L}{2} \right)^{-1} |L\rangle,$$

$$E_L = \langle L|H_0|L \rangle + \text{Re} \langle L|QtQ|L \rangle,$$

$$\Gamma_L = -2 \text{Im} \langle L|QtQ|L \rangle. \qquad [21]$$

Furthermore, since $\langle d'|G|d \rangle = \delta_{d,d'}(E - E_{d'} + i\Gamma_{d'}/2)^{-1}$, the numerical transition amplitude T_{c,c_1} factorizes conveniently after using [20] and [21],

$$T_{c,c_1}(E) = T_{c,a}(E) \frac{\langle d|V|c_1 \rangle}{(E - E_d + i\Gamma_d/2)}, \qquad [22]$$

$$T_{c,a}(E) = \sum_L \frac{\langle c|t|L \rangle \langle L|a \rangle}{(E - E_L + i\Gamma_L/2)}. \qquad [23]$$

In obtaining [22], we are assuming the isolated resonance limit, that is, a small linewidth Γ_d that is valid for weak coupling between the two channels $|d\rangle$ and $|c_1 \rangle$. The physical photodissociation amplitude $T_{c,a}(E)$ is thus obtainable from numerical evaluation of the S matrix element $S_{c,c_1}(E)$ for the system illustrated in Figure 2(b):

$$T_{c,a}(E) = i\frac{E - E_d + i\Gamma_d/2}{(2\pi\Gamma_d)^{1/2}} S_{c,c_1}(E)\exp[-i(\eta_c + \eta_{c_1})]. \qquad [24]$$

Returning to the formal definition of $T_{c,a}$, Eq. [23], as opposed to its numerical calculation, Eq. [24], we interpret the amplitude as an overlap of the initial unperturbed state $|a\rangle$ with the complex molecular states $|L\rangle$ dressed by the incident field. These dressed states have finite lifetimes due to the field-induced widths Γ_L. The overlap $\langle a|L\rangle$ corresponds, therefore, to preparation of the initial state $|a\rangle$ into the complex dressed $|L\rangle$ states, as would occur in a sudden excitation. Finally, the system transits into the physical continuum $|c\rangle$ via the transition operator t connecting these dressed states with the physical continuum. The coherence effects of the intense field is thus maintained through the intervention of the complex dressed states $|L\rangle$.

To relate the time-independent photodissociation amplitude $T_{c,a}$ to the time-dependent amplitude $I_{ca}(t)$, we must evaluate $(PGQ)_{ca}$, which gives, in terms of the eigenstates $|L\rangle$ of OGQ,

$$I_{c,a}(t) = (2i\pi)^{-1}\sum_L \int \frac{dE \exp(-iEt/\hbar)t_{cL}\langle L|a\rangle}{(E^+ - E_c - n_c\hbar\omega)(E^+ - E_L + i\Gamma_L/2)}. \qquad [25]$$

Assuming energy-independent matrix elements in the integral [25], the main contribution to the integral comes from the poles at the energies of the continuum $E = E_c + n_c\hbar\omega$ and at the dressed-state energies $E = E_L - i\Gamma_L/2$. These latter poles contribute a factor $\exp(-\Gamma_L t/\hbar)$ that is negligible in the limit $\Gamma_L t/\hbar \gg 1$. Thus, as long as dissociation times and hence lifetimes are shorter than pulse lengths, this will indeed be an excellent approximation. Short radiative lifetimes are attainable in intense fields as a result of fast dissociation times. The final photodissociation amplitude in this limit becomes

$$\lim_{t\to\infty} I_{c,a}(t) = \exp\left[-\frac{i(E_c + n_c\hbar\omega)t}{\hbar}\right]\sum_L \frac{t_{cL}\langle L|E_a, n\rangle}{(E_c + n_c\hbar\omega - E_L + i\Gamma_L/2)}. \qquad [26]$$

The asymptotic function therefore follows from [26] and [23],

$$|\psi(t)\rangle_{\lim t\to\infty} = \int dE_c\, T_{c,a}(E_c)\exp[-i(E_c + n_c\hbar\omega)t/\hbar]|c\rangle, \qquad [27]$$

provided the initial state $|\psi(0)\rangle \equiv |a\rangle$. Thus, the photodissociation probability $P_{a,c}$ for a transition from the initial bound state $|a\rangle$ to the continuum states $|c\rangle$ becomes

$$P_{a,c} = \int dE_{c'} |\langle c'|\psi(t)\rangle|^2 = C \int dE_c |T_{ca}(E_c)|^2, \qquad [28]$$

where C is a constant, and we have used the orthogonality relation $\langle c'|c\rangle = \delta(E_c - E_{c'})$. Equation [28] therefore shows that the amplitude $T_{c,a}(E)$ obtained from the numerical coupled-equations method using the artificial continuum $|c_1\rangle$ coupled to the bound state $|d\rangle = |a\rangle$ that is then coupled to the dressed states is indeed the amplitude needed to calculate the photodissociation probability from the time-dependent formalism.

Invoking the weak coupling limit, one has $|L\rangle \rightarrow |a\rangle$, $\Gamma_L \rightarrow 0$, and $E_L \rightarrow E_a$ so that one obtains (see Appendix) $t = V$, and

$$T_{a,c}(E) = \frac{\langle c|V|a\rangle}{E - E_a + i\Gamma_a/2}. \qquad [29]$$

We therefore conclude that upon calculating the transition amplitude $T_{c_1,c}$ (or S_{c,c_1}), the artificial channel method as depicted in Figure 2(b) furnishes the amplitude $T_{a,c}$, which is the *lineshape* amplitude for the photodissociation of the initial state $|a\rangle$. Expression [23] for this lineshape amplitude (valid for any field strength) reduces to [29], the expected weak-field amplitude. This is to be contrasted to the usual weak-field photodissociation amplitude obtainable from Eq. [12] as a result of the channel configuration in Figure 2(a). The amplitude T_{ac} reduces to V_{ac} in this case, that is, essentially a Franck–Condon factor (24, 26, 29) that gives the Fermi golden rule transition probabilities $\propto |V_{ac}|^2$. Clearly, the coupled-equation method corresponding to Eq. [12] and Figure 2(a) only applies whenever it is justified to neglect the effect of the perturbing field on the initial state, that is, for weak fields or short times in the case of strong fields such that $\Gamma t/\hbar \ll 1$ is respected.

IV. METHOD OF COUPLED EQUATIONS

In order to evaluate the transition amplitudes described, we now specify the states and their couplings. We write the total Hamiltonian 1 as $H = H_m^0 + V^m + H_r^0 + V^r$, where H_m^0 is the field-free molecular Hamiltonian, V_m is an intramolecular *nonradiative* coupling that will depend on the electronic representation used, *diabatic* or *adiabatic* (16–18), H_r^0 is the pure radiation Hamiltonian, and V^r is the *radiative* coupling. As mentioned in Section II, the field is described in the Fock representation by eigenstates $|n\rangle$ with energies $n\hbar\omega$. By writing the total Hamiltonian H [which defines the total resolvent G

(Eq. [4])] as $H = H_0 + V$, with $H_0 = H_m^0 + H_r^0$, $V = V^m + V^r$, one can describe the total system by linear combinations of the eigenstates of H_0, that is, $H_0|E_a, n_a\rangle = (E_a + n_a\hbar\omega)\,|E_a, n_a\rangle$. The states $|E_a\rangle$ are the electron–vibrational–rotational molecular states we now define.

Here H_m^0 is the usual electron–nuclear Hamiltonian of quantum chemistry, that is, $H_m^0 = H_{el} + H_N$. The eigenstates of H_{el} furnish adiabatic nuclear potentials $U(R)$ for the nuclear motion, that is, $H_{el}|\psi_{el}\rangle = U(R)|\psi_{el}\rangle$. The eigenstates of the adiabatic molecular Hamiltonian can therefore be written as linear combinations of products of adiabatic nuclear radial functions $F_i(R)$, rotational functions $|\psi_{rot}^i\rangle$, electronic functions $|\psi_{el}^i\rangle$, and photon states $|n_i\rangle$,

$$|\Psi\rangle = \frac{1}{R}\sum_i F_i(R)|\psi_{rot}^i\rangle|\psi_{el}^i\rangle|n_i\rangle. \qquad [30]$$

In this basis, any diatomic system in interaction with radiation can be described by a set of coupled differential equations of second order, corresponding to solutions of $H|\Psi\rangle = E|\Psi\rangle$. We can write this in matrix form as

$$F'' + WF = 0, \qquad [31]$$

where $F'' = \partial^2 F/\partial R^2$. The diagonal matrix elements of W are defined as

$$W_{ii}(R) = \frac{2\mu}{\hbar^2}[E - U_i(R) - n_i\hbar\omega_i] - \frac{J(J+1)}{R^2}, \qquad [32]$$

$$W_{ij}(R) = \frac{2\mu}{\hbar^2}V_{ij}(R). \qquad$$

Here μ is the reduced mass of the molecular system, J is the rotational quantum number, and $V_{ij}(R)$ is the nondiagonal radiative or nonradiative coupling matrix element connecting different electronic states $|i\rangle$ and $|j\rangle$. In the case of radiative couplings we write (28)

$$V_{ij}^r(R) = \gamma\left(\frac{2J+1}{2J'+1}\right)^{1/2}\langle J1M\lambda|J'M'\rangle\sum_m\langle J1\Omega m|J'\Omega'\rangle d_m(R)\delta_{n,n'\pm 1}. \qquad [33]$$

Here Ω is the electronic angular momentum projection on the internuclear axis, λ is the field polarization with respect to laboratory axes, d_m is the $m = 0$, ± 1 component of the electronic transition moment along the molecular axis, and γ is a field-dependent unit conversion factor for a field intensity I in watts per centimeters squared, $\gamma = 1.17 \times 10^{-3}\ [I\,(W/cm^2)]^{1/2}\,(cm^{-1}/a.u.)$. The

Clebsch–Gordan factors in Eq. [33] fix the selection rules at $\Delta J = 0, \pm 1$, $\Delta m = 0, \pm 1$.

Using appropriate transition moments to be described in what follows, one obtains as input for the coupled Eqs. [31] the radiative couplings $V_{ij}^r(R)$. The nondiagonal matrix elements $W_{ij}(R)$ can be generalized to include also nonradiative couplings that give rise to nonadiabatic processes such as predissociation (16–18). The coupled Eqs. [31] can thus treat radiative and nonradiative couplings on an equal footing for any coupling strength. We thus conclude that at the formal level, radiative and nonradiative couplings are analogous, so that radiative bound–continuum transitions discussed in the previous section can be treated as predissociation problems (16–18). All that is needed is to specify the functional forms of the nonradiative couplings (next section) simultaneously with the transition moments as input data for the coupled Eqs. [31]. These equations are then integrated numerically by a Fox–Goodwin algorithm (35–37). The numerical continuum functions are then compared to the analytic asymptotic functions,

$$k_j^{1/2} F_j^{(i)} = \sin(k_j R - J\pi/2)\delta_{ij} + R_{ij}\cos(k_j R - J\pi/2), \qquad [34]$$

where $k_i = \lim_{R \to \infty} [W_{ii}(R)]^{1/2}$. The coefficients R_{ij} define a scattering matrix $S = (1 + iR)/(1 - iR)$ from which the scattering and transition amplitudes follow: $S_{ij} = \delta_{ij} - 2\pi i T_{ij}$ (36).

In Section II we discussed the use of artificial channels to facilitate numerical calculations of bound–continuum transitions via a coupled-equation method for the S or T matrices. Here we summarize the method.

i. One Artificial Continuum $|c_1\rangle$ [Figure 2(a)]. This configuration enables one to calculate direct photodissociation for weak fields only, since from Eq. [12], the transition amplitude $T_{a,c}$ implies a well-defined initial state $|a\rangle$. For this case, the radiative coupling V_{ac}^r must be assymmetric (i.e., $V_{ac}^r \neq 0$, $V_{ca}^r = 0$) so that there are no radiative shifts and widths affecting the state $|a\rangle$. Clearly, $T_{a,c}$ may be extended to multiphoton transitions to a final continuum $|c\rangle$ provided the first radiative transition from the state $|a\rangle$ is weak. As an example, we illustrate a configuration in Figure 3 that was used to calculate the direct photopredissociation cross section in IBr (27). The initial radiative coupling V_{ad}^r is assymmetric (i.e., $V_{ad}^r \neq 0$, $V_{da}^r = 0$), whereas the final radiative transition is symmetric ($V_{dc}^r = V_{cd}^r$). In this case one would obtain, via the coupled-equations calculation for $T_{c_1,c}$, the two-photon amplitude dissociation $T_{a,c}$, where the radiative coupling between states $|d\rangle$ and $|c\rangle$ (i.e., the dissociative step) can be varied continuously from the weak to strong limits without restriction. The bound–bound transition $|a\rangle \to |d\rangle$ must be weak for Eq. [12] to be valid.

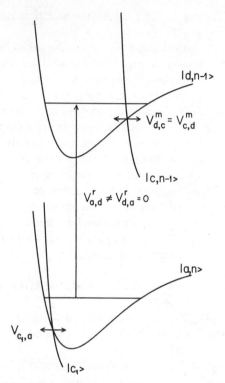

Figure 3. Photopredissociation system with photon occupation numbers and entrance channel $|c_1\rangle$. Here $V_{a,d}^r \neq V_{d,a}^r = 0$ eliminates radiative corrections to $|a\rangle$ (weak-field approximation), and $V_{d,c}^m = V_{c,d}^m$ is nonradiative or radiative coupling of arbitrary strength (weak and strong).

ii. Two Artificial Continua $|c_1\rangle$ and $|c_2\rangle$ (Figure 4). For well-defined initial and final bound states $|a\rangle$ and $|b\rangle$, one needs now two continua, $|c_1\rangle$ as entrance and $|c_2\rangle$ as exit, for the propagation of initial and final waves that allow for the calculation of the S matrix element S_{c_1,c_2}. Clearly, this method allows access to numerical two-photon transition amplitudes for weak fields between bound states $|a\rangle$ and $|b\rangle$ via the intermediate state $|d\rangle$, which can couple to some continuum $|c\rangle$. In this configuration, the assymmetric couplings are used, $V_{a,d}^r \neq 0$, $V_{d,a}^r = 0$, $V_{d,b}^r \neq 0$, and $V_{b,d}^r = 0$, in order to neglect radiative corrections to the initial and final states, in keeping with the weak-field assumption for the transitions $|a\rangle \rightarrow |d\rangle$ and $|d\rangle \rightarrow |b\rangle$. The corresponding expression for T_{c_1,c_2} is arrived at by using Eqs. [7] and [9],

$$T_{c_1,c_2} = V_{c_1,a} G_{ab} V_{b,c_2} \exp i(\eta_1 + \eta_2),$$ [35]

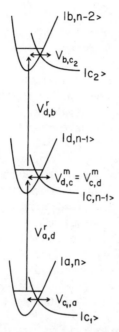

Figure 4. Two-photon transition system with photon occupation numbers. Here $|c_1\rangle$ and $|c_2\rangle$ are entrance and exit channels, $|c\rangle$ is a physical continuum, $V^r_{a,d} \neq V^r_{d,a} = 0$, $V^r_{db} \neq V^r_{bd} = 0$, implies no radiative corrections to $|a\rangle$ and $|b\rangle$ (weak-field approximation), and $V^m_{d,c} = V^m_{c,d}$ is nonradiative or radiative interaction of arbitrary strength.

$$T_{ab} = 2\pi(E - E_a + \tfrac{1}{2}i\Gamma_a)(E - E_b + \tfrac{1}{2}i\Gamma_b)(\Gamma_a\Gamma_b)^{-1/2}$$
$$\times\, T_{c_1,c_2}\exp\left[-i(\eta_1 + \eta_2)\right]. \qquad [36]$$

In this last equation we have used the Fermi golden rule relation $\Gamma_a = 2\pi|V_{c_1,a}|^2$, valid for the isolated resonance $|a\rangle$ due to weak coupling to the continuum $|c_1\rangle$ (and similarly for Γ_b).

Thus, using two artificial continua $|c_1\rangle$ and $|c_2\rangle$, one can calculate rigorously any two-photon transitions at weak fields. It is to be emphasized that there are no restrictions on the number of intermediate states $|d\rangle$ and $|c\rangle$ nor on the couplings between these states. The only restriction is that the levels $|a\rangle$ and $|b\rangle$ do not undergo radiative perturbations for Eq. [36] to be valid. Clearly, the couplings between the $|d\rangle$ and $|c\rangle$ states can be symmetric radiative transitions so that one can treat radiative interactions for any field strengths in that manifold using Eq. [36]. In the next section we will illustrate an example of this configuration as used for the third harmonic generation in Cl_2 (31).

iii. One Artificial Continuum $|c_1\rangle$ Plus One Artificial Bound $|d\rangle$ Channels. This was the case discussed in Section III for direct photodissociation in intense fields. In the previous two methods, (i) and (ii), initial and bound states were assumed to undergo no radiative perturbations, as occurs for weak-field cases. In order to take into account radiative perturbations to the initial state, one projects the initial unperturbed (zero-field) state $|d\rangle$ onto the stationary field–molecule states, that is, the dressed states [see Figure (2b)]. Thus, $|c_1\rangle$ is the entrance artificial channel for the state $|d\rangle$ outside the field. The assymmetric coupling $V_{da}^r = 1$, $V_{ad}^r = 0$, picks out the initial state $|d\rangle = |a\rangle$, the latter being distributed over the complex dressed states $|L\rangle$. It is to be noticed that the back coupling $V_{ad} = 0$ so that there is no effective interaction between states $|d\rangle$ and $|a\rangle$ but rather we obtain only the overlap of the unperturbed $|d\rangle$ state with its equivalent, the $|a\rangle$ state, which is "diluted" over the dressed states. Coupling the artificial set $|c_1\rangle + |d\rangle$ to the state $|a,n\rangle$ in the manifold of complete dressed states, Figure 1(b), provides a means of evaluating resonance fluorescence from the dressed intense-field manifold (10). This method can further be generalized to other strong-field multiphoton transitions such as CARS when the initial state is also strongly perturbed by the incident fields (30). (Coherent Antistokes Raman Spectroscopy)

V. NONADIABATIC EFFECTS IN MULTIPHOTON TRANSITIONS

We shall now present an example of the numerical calculation of transition amplitudes for multiphoton processes in diatomics and the effect of non-adiabatic couplings on these. The necessary input data are the electronic transition moments μ multiplied by the appropriate field strengths E, Eq. [33], the field electronic potentials, Eq. [32], and finally the remaining interstate couplings such as spin–orbit interactions or nonadiabatic interactions. Quantum chemistry is capable of furnishing this information, albeit in what is called the adiabatic representation. In particular, we shall limit ourselves to three-photon processes in the Cl_2 molecule as a test case for our method, the reason being that an extensive series of adiabatic electronic states has been published by Peyerimhoff et al. (38, 39) over the last few years. Such ab initio calculations leave open the question of nonadiabatic corrections to spectroscopic states. Such corrections are known to alter considerably intensities as measured in various spectroscopies (40) and can induce phenomena such as predissociation (16–18).

We illustrate in Figure 5 the adiabatic states of Cl_2 relevant to the present study. We discuss double excitation with $\omega_1 = 33,000\,\text{cm}^{-1}$ incident laser radiation frequency. This corresponds to near resonance with the $^1\Pi_u$ electronic state and allows the second photon to be nearly resonant with the

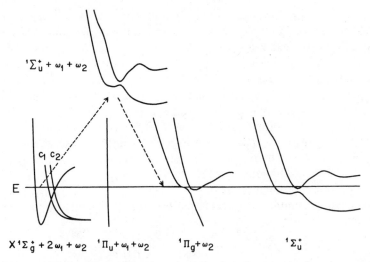

Figure 5. Resonant Cl_2 states for $\omega_1 = 33,000\,cm^{-1}$, $\omega_2 = 17,000\,cm^{-1}$, in a three-photon absorption at total energy E. State $^1\Sigma_u^+ + \omega_1 + \omega_2$ is nonresonant [virtual dressed state allowing for a nonresonant path (ii)].

avoided crossing region between the first two $^1\Pi_g$ states. The third incident photon is tuned at $\omega_2 = 17,000\,cm^{-1}$ in order to give $2\omega_1 + \omega_2 = 83,000\,cm^{-1}$, the total vertical excitation energy. Such excitation allows one to reach the low adiabatic vibrational levels of the $2\Sigma_u^+$ electronic state. In Figure 5 we have included the *virtual* $^1\Sigma_u^+$ states that are known to be responsible for anomalous Raman intensities in Cl_2 (41, 42). Other states representing virtual, nonrotating corrections to the rotating-wave approximation (RWA) would occur at energy $^2\Sigma_u^+ + 3\omega_1 + \omega_2$ and are even less important. Thus, all electron–vibrational–rotational states on the energy line E are resonant, whereas all others are nonresonant or virtual. In particular, the $^1\Sigma_u^+$ states that are two-photon energies $(\omega_1 + \omega_2)$ higher than the total initial energy $E = E_a + (2\omega_1 + \omega_2)$ can contribute to the three-photon process illustrated in Figure 5 via the nonresonant transition $X\ ^1\Sigma_g^+ \rightarrow ^1\Sigma_u^+ \rightarrow ^1\Pi_g^+ \rightarrow ^1\Sigma_u^+$. Thus, the same electronic state $^1\Sigma_u^+$ can act as intermediate virtual and a final resonance state. The six electronic states included in the present numerical study were taken from the work of Peyerimhoff et al. (38). Some of the transition moments were taken from the work of Grein and Peyerimhoff (39). The other transition moments were reconstructed from the known configurations of the electronic states. In particular, the $X\ ^1\Sigma_g^+ \rightarrow 2\ ^1\Sigma_u^+$ and $1\ ^1\Pi_u \rightarrow 1\ ^1\Pi_g$ valence transitions depend on the electronic moments $\langle \sigma_g |z| \sigma_u \rangle$ and $\langle \pi_u |z| \pi_g \rangle$, for which one can readily show that these transition moments are equal to $\frac{1}{2}R$ (31).

The avoided crossings evident in Figure 5 occur as a result of a change in the electronic configuration of the electronic states at the point R_c of closest approach. In particular, for the Cl_2 excited states Π_g and Σ_u^+, one is dealing with changes from *valence* to *Rydberg* configurations (38). Such changes in configurations are a result of the adiabatic approximation, which leads to abrupt changes in the adiabatic potential surfaces at the avoided crossings. It is at such avoided crossings that one expects important nonadiabatic couplings of the derivative type ($\langle i|\partial/\partial R|j\rangle \partial/\partial R$), corresponding to exchange of electronic and nuclear momenta as a result of the conservation of total momentum (14). Such derivative couplings complicate the numerical solutions of the coupled equations (31). It is therefore advantageous to adopt a diabatic representation that can be obtained from two-state adiabatic systems. Thus, defining a unitary transformation U that relates adiabatic functions ϕ_{ad} to diabatic functions ϕ_d by $\phi_{ad} = U\phi_d$, that is,

$$\begin{pmatrix} \phi_{ad}^1 \\ \phi_{ad}^2 \end{pmatrix} = \begin{bmatrix} \cos\theta & \sin\theta \\ -\sin\theta & \cos\theta \end{bmatrix} \begin{pmatrix} \phi_d^1 \\ \phi_d^2 \end{pmatrix}, \qquad [37]$$

the adiabatic electronic eigenvalue equation

$$H_{el} \begin{pmatrix} \phi_{ad}^1 \\ \phi_{ad}^2 \end{pmatrix} = \begin{bmatrix} V_+(R) & 0 \\ 0 & V_-(R) \end{bmatrix} \begin{pmatrix} \phi_{ad}^1 \\ \phi_{ad}^2 \end{pmatrix} \qquad [38]$$

becomes a diabatic electronic Hamiltonian equation

$$H_{el} \begin{pmatrix} \phi_d^1 \\ \phi_d^2 \end{pmatrix} = \begin{bmatrix} V_1(R) & V_{12}(R) \\ V_{21}(R) & V_2(R) \end{bmatrix} \begin{pmatrix} \phi_d^1 \\ \phi_d^2 \end{pmatrix}, \qquad [39]$$

where

$$V_1 = V_+ \cos^2\theta + V_- \sin^2\theta,$$

$$V_2 = V_+ \sin^2\theta + V_- \cos^2\theta,$$

$$V_{12} = (V_+ - V_-)\cos\theta\sin\theta, \qquad [40]$$

$$2\theta = \arctan\frac{2V_{12}}{V_2 - V_1}, \qquad 0 \leqslant \theta \leqslant \pi/2.$$

The corresponding adiabatic transition moments $\mu_{1(2)}^{ad} = \langle \phi_0|\boldsymbol{\mu}|\phi_{1(2)}^{ad}\rangle$ are also transformed to diabatic transition moments:

$$\begin{pmatrix} \mu_1^d \\ \mu_2^d \end{pmatrix} = \begin{bmatrix} \cos\theta & -\sin\theta \\ \sin\theta & \cos\theta \end{bmatrix} \begin{pmatrix} \mu_1^{ad} \\ \mu_2^{ad} \end{pmatrix}. \qquad [41]$$

TABLE 1 Three-Photon Amplitudes[a]

Intensity, $I_1 = I_2$ (W/cm²)	(i) $X^1\Sigma_g^+ \to \Pi_u \leftrightarrow \Pi_g \to 2^1\Sigma_u^+$		(ii) $X^1\Sigma_g^+ \to 2^1\Sigma_u^+ \leftrightarrow \Pi_g \to 2^1\Sigma_u^+$		(iii) $X^1\Sigma_g^+ \to \Pi_u \leftrightarrow \Pi_g \to 1^1\Sigma_u^+$	
	Adiabatic	Exact	Adiabatic	Exact	Adiabatic	Exact
10^8						
Re	3.0(−5)	2.2(−5)	1.9(−6)	7.2(−6)	8.8(−5)	1.3(−6)
Im	−1.3(−5)	−4.5(−6)	−2.4(−8)	−3.5(−6)	−5.9(−5)	−1.5(−5)
10^9						
Re	9.6(−4)	7.0(−4)	6.1(−5)	2.3(−4)	2.8(−3)	4.1(−5)
Im	−4.1(−4)	−1.4(−4)	−7.6(−7)	−1.1(−4)	−1.9(−3)	−4.8(−4)
10^{10}						
Re	3.0(−2)	2.2(−2)	1.9(−3)	7.2(−3)	8.6(−2)	1.2(−3)
Im	−1.3(−2)	−4.4(−3)	−2.4(−5)	−3.5(−3)	−5.8(−2)	−1.5(−2)
10^{11}						
Re	8.9(−1)	6.8(−1)	−6.1(−2)	2.3(−1)	2.3	2.0(−2)
Im	−5.0(−1)	−9.9(−2)	−7.6(−4)	−1.1(−1)	−1.8	−4.5(−1)
10^{12}						
Re	9.9	1.4(+1)	1.9	7.2	1.9	2.2
Im	−2.7(+1)	−3.9	−2.4(−2)	−3.5	−3.4(+1)	6.3

[a] $2\omega_1 = 66,000 \text{ cm}^{-1}$, $\omega_2 = 17,000 \text{ cm}^{-1}$; $T^3 (\text{cm}^{-1}) = \text{Re } T^3 + i \text{Im } T^3$, for Cl_2 (see Figure 5 for potentials).

TABLE 2 Semiclassical (Γ_r^{sc}), Exact coupled-Equation Calculation (Γ_r^{ex}), and Fermi Golden Rule (Γ_r^0) Linewidths of Photodissociating States of Ar_2^+ at Field Intensity $I = 10^{11}$ W/cm^2, $J = 0$

$E_r(cm^{-1})$	u_2	$\Gamma_r^{ex}(cm^{-1})$	$\Gamma_r^{sc}(cm^{-1})$	$\Gamma_r^0(cm^{-1})$
1300	3.48	2	0.3	308
3600	1.46	17	12	113
5500	1.08	3	3.2	70

The further advantage of the use of the diabatic transition moments rather than the quantum chemical adiabatic ones is that now all couplings, the nonradiative couplings V_{12}^d (Eq. [40]) and the radiative couplings $\mu_{12}^d \cdot E$ are all nondiabatic so that the same numerical procedure can be used to solve numerically the coupled Eqs. [31] from which one obtains the appropriate transition matrices as described in Section IV. In the present Cl_2 calculation, the estimated values of the nondiabatic nonradiative couplings were $V_{12}^d(^1\Sigma_u^+) = 950\,cm^{-1}$, $V_{12}^d(^1\Pi_g) = 1370\,cm^{-1}$.

We present in Table 1 various three-photon amplitudes obtained from the coupled-equations calculation of the transition amplitude T_{ab}, Eq. [36]. In these calculations, only the radiative moment $\langle \Pi_u|z|\Pi_g \rangle = \frac{1}{2}R$ is taken to be symmetric as it is the strongest resonant radiative coupling. The $\langle X\Sigma_g^+|z|2\Sigma_u^+ \rangle$ moment is also $\frac{1}{2}R$ but is nonresonant (virtual), Figure 5. This radiative coupling is taken to be assymmetric (as are the others, in order to avoid radiative corrections to the initial and final $|a\rangle$ and $|b\rangle$ states) so as to render the expression for T_{ab}, Eq. [36], valid. We first note that both the real (Re) and imaginary (Im) parts of the three-photon amplitude T^3 scale essentially as $I^{3/2}$. This reflects the validity of the perturbative expression $T^3 = V_0'G_0V'^rG_0V^r$, Eq. [9], where G_0 is the unperturbed resolvent. Pathways (i) and (iii) are resonant whereas pathway (ii) is nonresonant or virtual (Figure 5). [We emphasize that this does not correspond to corrections to RWA but rather is the case where a photon is absorbed virtually first, by the Σ_u^+ state, and then a second photon is adsorbed by the Π_g state.] The results of Table 2 show that at the ω_1 frequency chosen, $\omega_1 = 33,000\,cm^{-1}$, where the photon is nearly resonant at the crossing point of the two Π_g states, the resonant three-photon transition to the $^1\Sigma_u^+$ states predominates, whereas the virtual pathway (ii) contributes an intensity by about one order of magnitude less. In the resonant case, the adiabatic amplitudes generally overestimate the true amplitudes. A large overestimation of resonant transition probabilities by adiabatic calculations has been observed in a resonant four-wave-mixing calculation on Cl_2 using the same potentials as here (31). We conclude therefore that nonadiabatic effects should not be neglected in treating multiphoton transitions in molecules.

VI. SEMICLASSICAL S MATRIX

Intense laser fields lead to situations where treatments going beyond perturbation theory are needed, as has been discussed in the previous sections. Because of the strong radiative coupling, it is not posible to predict interesting features in the scattering cross sections from a knowldege of the molecular levels alone, as is usually done in weak-field, traditional spectroscopy, where the Fermi golden rule and hence Franck–Condon approaches apply. In discussing intense-field multiphoton processes, only the levels of the total system molecule and photons are significant. As first shown by Bandrauk and Sink (6, 7), direct molecular photodissociation in strong fields is the analog of predissociation for which the original semiclassical theory by Bandrauk and Child (16) covers the continuous range of coupling strengths, that is, the weak-coupling diabatic regime to the strong-coupling adiabatic regime. Multistate curve crossings have also been analyzed in the semiclassical approach (18, 19). Such situations occur frequently in multiphoton processes when one takes rotations and many electronic states into consideration (10, 11). The purpose of this section is therefore to derive transition amplitudes between entrance and exit channels by applying the uniform semiclassical description (17) that covers a wide variety of tunneling and curve-crossing problems. The semiclassical scattering amplitudes can be entirely expressed, apart for a phase factor, with the action angles involving only the diabatic and adiabatic phase integrals (16–19).

A general multistate curve-crossing problem that leads to a transition amplitude from channel u to channel v via several intermediate channels (say, M diabatic channels presenting N crossings) is viewed through the S matrix element S_{vu}. The method relies on following changes in the amplitudes W' and W'' of ingoing and outgoing exponentials of WKB wavefunctions around a crossing point R_N. The typical semiclassical wavefunction is

$$\psi_i(R) \propto k_i^{-1/2} \left\{ W_i' \exp\left[i \int_{R_N}^{R} k_i(R')\, dR' \right] \right.$$
$$\left. + W_i'' \exp\left[-i \int_{R_N}^{R} k_i(R')\, dR' \right] \right\}. \qquad [42]$$

The wave numbers $k_i(R)$ are to be calculated using adiabatic potentials. Several adiabatic potentials contribute to $k_i(R)$. A typical situation is depicted in Figure 6 [see Figure 1(b) for the equivalent dressed potentials], where curve crossings between channels i and j occur at R_p. Here $k_i(R)$ is made up from two adiabatic potentials, $V_+(R)$ for $R \leqslant R_p$ and $V_-(R)$ for $R \geqslant R_p$, with the definitions of the adiabatic potentials $V_\pm(R)$ in terms of the diabatic potentials

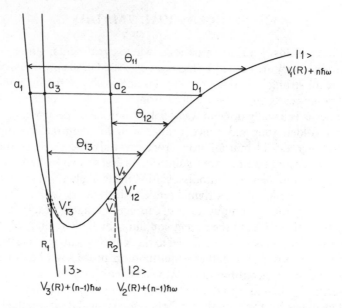

Figure 6. The Ar_2^+ dressed potentials in direct photodissociation from ground state $V_1(R)$ $(^2\Sigma_u^+)$ to first two excited states $V_2(R)$ $(^2\Sigma_g^+)$ and $V_3(R)$ $(^2\Pi_g)$, which are continua. Here V_{12}^r and V_{13}^r are radiative interactions that at high intensities produce avoided crossings between diabatic potentials $V_i(R)$ to give rise to new adiabatic potentials $V_+(R)$ and $V_-(R)$, Θ_{11} is diabatic phase (action) integral for the bound states of potential $V_1(R)$, and Θ_{12} (Θ_{13}) is an adiabatic phase (action) integral for the bound states of $V_+(R)$.

$V_{i(j)}(R)$:

$$V_\pm(R) = \tfrac{1}{2}[V_i(R) + V_j(R)] \pm \tfrac{1}{2}\{[V_i(R) - V_j(R)]^2 + 4V_{ij}^2(R)\}^{1/2}, \quad [43]$$

where $V_{ij}(R)$ is the radiative or nonradiative coupling between the two diabatic potentials.

In order to take into account curve crossings such as occur in the dressed picture, Figure 1(b), Stueckelberg-connecting matrices (16–19) corrected by uniform phase factors $\exp(\pm i\chi)$ permit transitions of the wavefunction [42] across the crossing points. These connecting matrices are symbolized by T_p'' and T_p' in Figure 7, where we display in increasing order a multiple-curve-crossing point system, $R_1 \leqslant \cdots \leqslant R_p \leqslant \cdots \leqslant R_N$. For the isolated crossing point R_p involving the two channels i and j, T_p'' is an $M \times M$ matrix containing unity on its diagonals and zero elsewhere with the exception of four elements situated at the intersection of columns and lines i and j labeled according to the system depicted in Figure 6. More precisely, we have

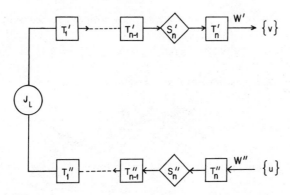

Figure 7. Transition matrix diagram for evaluating the S matrix between continua $\{u\}$ and $\{v\}$. Here W'' and W' are ingoing and outgoing semiclassical amplitudes, T_N is the transition matrix at crossing point R_N, S_N is the free phase propagation matrix between turning point R_N and R_{N-1}, and J_l is reflection matrix at the classical turning point of a particular potential.

$$T_p'' = \begin{bmatrix} \lambda_p & (1 - \lambda_p^2)^{1/2} \exp(i\chi_p) \\ -(1 - \lambda_p^2)^{1/2} \exp(-i\chi_p) & \lambda_p \end{bmatrix} = (T_p')^+. \quad [44]$$

Here T_p' is the transposed matrix of T_p'' and vice versa. The parameters λ_p and χ_p are defined for arbitrary coupling by means of a parameter v_{ij}, which in the weak-interaction (diabatic) limit is the well-known Landau–Zener parameter (16):

$$v_{ij} = V_{ij}^2 / (\hbar^2 v_{ij} \Delta F_{ij}), \quad [45]$$

where v_{ij} is the nuclear velocity at the crossing point and ΔF_{ij} is the difference in diabatic potential slopes at the same point. A more general expression for v_{ij} is (17)

$$v_{ij} = (2\pi)^{-1} \operatorname{Im}\left\{ \int_{R+}^{R-} [k_-^p(R) - k_+^p(R)]\, dR \right\}, \quad [46]$$

where $k_\pm(R)$ is calculated using the adiabatic potentials, Eq. [43], and the integration limits are the so-called complex transition points at which the adiabatic terms $V_\pm^p(R)$ intersect near R_p. Finally, the expressions for λ_p and χ_p are

$$\lambda_p = \exp(-\pi v_{ij}), \qquad \chi_p = \arg \Gamma(i v_{ij}) - v_{ij}\ln v_{ij} + v_{ij} + \tfrac{1}{4}\pi. \quad [47]$$

Here χ_p takes the limits $-\frac{1}{4}\pi$ in the weak-coupling (diabatic) limit and zero in the strong (adiabatic) limit (16).

In connecting neighboring crossing points R_p and R_{p-1}, one has free propagation so that one replaces the transition matrices T by phase accumulation matrices S, of dimension $M \times M$,

$$S_p'' = \begin{bmatrix} \exp(-i\phi_1^p) & 0 \\ 0 & \exp(-i\phi_N^p) \end{bmatrix} = (S_p')^+, \qquad [48]$$

where the phase integrals are defined for channel i by

$$\phi_i^p = \int_{R_p}^{R_{p-1}} k_i(R)\,dR. \qquad [49]$$

The preceding free-propagation matrices S must be modified at turning points for which one has pure reflection. The correct reflection phase behavior is obtained by comparison with the asymptotic form of the regular solution to Airy's equation (16). The linkage conditions at the left turning points a_i $(i = 1, \ldots, n)$ is given by an $M \times M$ matrix J_l (see Figure 7),

$$J_l = -i \begin{bmatrix} \exp(-2i\alpha_1) & 0 \\ 0 & \exp(-2i\alpha_M) \end{bmatrix}. \qquad [50]$$

The factors $-i$ account for the phase change by $-\frac{1}{2}\pi$ due to the left turning points and the α_i's for phase accumulation due to free propagation on the potential $V_i(R)$ between the first crossing point R_1 and the a_i's,

$$\alpha_i = \int_{R_1}^{a_i} k_i(R)\,dR. \qquad [51]$$

The connection between the ingoing and outgoing amplitudes, Eq. [42], before the introduction of the effect of channels closed to the right (bound states) is obtained as a matrix product combining the previous local solutions in the order they appear following the direction of the propagation, as shown by the arrow in Figure 7,

$$W' = AW'',$$

$$A = (T_N'S_N')\cdots(T_2'S_2')(T_1'J_lT_1'')(S_2''T_2'')\cdots(S_N''T_N''). \qquad [52]$$

Following the work of Atabek et al. (19), Eq. [52] can be recast as a product of three effective matrices,

$$A = B\bar{A}B,$$

$$B = S'_N \cdots S'_2 L, \qquad L^2 = J_l,$$

$$\bar{A} = (\bar{T}'_N \cdots \bar{T}'_1)(\bar{T}'_N \cdots \bar{T}'_1)^+, \tag{53}$$

where $\bar{T}'_p$ is modified vis-à-vis T'_p by some phases,

$$\bar{T}'_p = \begin{bmatrix} \lambda_p & -(1-\lambda_p^2)\exp[-i(\psi_{ij}-\psi_{jj})] \\ (1-\lambda_p^2)^{1/2}\exp[i(\psi_{ij}-\psi_{jj})] & \lambda_p \end{bmatrix}, \tag{54}$$

and B is diagonal and expressible as

$$B_{ij} = \delta_{ij}\exp(i\psi_{ii}). \tag{55}$$

The phase factors ψ_{ii} and ψ_{ij} can now be identified as the diabatic scattering angle ψ_{ii},

$$\psi_{ii} = -\alpha_i - \sum_{q=2}^{N} \phi_i^q, \tag{56}$$

associated with the ith diabatic potential and the adiabatic scattering angle ψ_{ij},

$$\psi_{ij} = \chi_p - \alpha_i - \sum_{q=2}^{N} \phi_i^q - \sum_{q=p+1}^{N} \phi_j^q, \tag{57}$$

associated with the adiabatic potential $V_+^p(R)$ of Eq. [43].

The presence of closed channels, which will give rise to *resonances*, requires the stationary state condition (16),

$$W''_{n_i} = -i\exp(2i\beta_{n_i})W'_{n_i}, \tag{58}$$

with

$$\beta_{n_i} = \int_{R_N}^{b_{n_i}} k_{n_i}(R)\,dR. \tag{59}$$

Here the closed channels are labeled n_i ($i = 1, \ldots, K$), where K is the total number of closed channels, and b_{n_i} are the corresponding right turning points.

To define completely the scattering problem, we specify the unit ingoing flux condition for the entrance channel, $W''_u = -1$, whereas for all other channels ingoing fluxes are taken to be zero, $W''_k = 0$, $k \neq \mu$, n_i (18). The scattering matrix is then simply the outgoing amplitudes of the exit channels (18),

$$S_{vu} = W'_v. \qquad [60]$$

The right turning point phase conditions, Eq. [58], introduce extra phases β for the closed channels, which when added to the scattering angles ψ_{ii} and ψ_{ij} lead to action angles defined either as diabatic,

$$\theta_{ii} = \psi_{ii} + \beta_i, \qquad [61]$$

or adiabatic,

$$\theta_{ij} = \psi_{ij} + \beta_j. \qquad [62]$$

The Bohr–Sommerfeld quantization rules applied to these angles give energies associated to diabatic and adiabatic resonances. We apply these semiclassical techniques as a tool for interpreting laser-induced resonances in bound–continuum radiative transitions at high intensities.

VII. PREDISSOCIATION–PHOTODISSOCIATION

As discussed in the previous sections, when a repulsive state crosses and interacts with a bound state of a diatomic molecule, through some perturbation V, the energy levels of the bound state are affected in two ways. First, the bound levels described by a wavefunction $\psi_v(R)$ undergo a transition to the continuum state $\psi_E(R)$ with a probability rate given in second-order perturbation theory by

$$\gamma_{v,E} = \frac{2\pi}{\hbar} |\langle \psi_v | V | \psi_E \rangle|^2. \qquad [63]$$

This determines the width Γ_v that the level at energy E_v acquires, which is given by the golden rule expression as $\Gamma = \hbar\gamma$. The second effect is a shift Δ_v that occurs in the energy levels E_v due to the interaction with the continuum,

$$\Delta_v = \text{PP} \int dE \frac{|\langle \psi_v | V | \psi_E \rangle|^2}{E_v - E}, \qquad [64]$$

where PP means principal part.

Treating the bound states coupled to the continuum as scattering resonances within a scattering (S) matrix formalism, either numerically or semiclassically, allows the energy shifts and widths to be determined for any coupling strength from the complex poles of the S matrix (16–19).

Thus, in a bound-free transition as depicted in Figure 1, one calculates the scattering amplitude in the open channel 2, $(|c\rangle)$, with channel 1, $(|a\rangle)$, closed. This leads for *arbitrary* interaction strength to the following semiclassical expression:

$$S_{22} = \{\cos\theta_{11} + u\cos\theta_{21}\exp[i(\theta_{21} - \theta_{11})]\}/D, \qquad [65]$$

$$D = \cos\theta_{11} + u\cos\theta_{21}\exp[i(\theta_{21} - \theta_{11})], \qquad [66]$$

where

$$u = (1 - \lambda^2)^{1/2}/\lambda, \qquad [67]$$

and λ is defined in Eq. [47]. We note that u measures the effective interaction strength, that is, $u = 0$ $(\lambda = 1, V = 0)$ and $u = \infty$ $(\lambda = 0, V = \infty)$ in the weak and strong interaction limits, respectively. Complex resonance energies are obtained as zero of the denominator D of S_{22}. In the weak interaction limit $(u, V \to 0)$, the Bohr–Sommerfeld quantization is applied to the diabatic angle θ_{11}, which is linearly expanded in the vicinity of E_v as

$$\theta_{11}(E) = (v + \tfrac{1}{2})\pi + \left(\frac{\partial\theta_{11}}{\partial E}\right)_{E_v}(E - E_v). \qquad [68]$$

Here $\partial\theta_{11}/\partial E$ defines the local energy spacing $\pi/\hbar\omega_1$, where ω_1 is the diabatic unperturbed vibrational frequency at $E = E_v$. Similarly in the reverse situation of very strong interaction $(u, V \to \infty)$, the bound levels E_v^+ of the upper adiabatic potential $V_+(R)$ are given by the Bohr–Sommerfeld quantization condition applied to the adiabatic angle θ_{21},

$$\theta_{21}(E) = (v_+ + \tfrac{1}{2})\pi + \frac{\pi}{\hbar\omega_+}(E - E_v^+), \qquad [69]$$

where ω_+ is the adiabatic level frequency.

The respective diabatic and adiabatic energy shifts and widths are:

i. Diabatic Limit: Describes Weak Interaction

$$\Delta_v = \frac{\hbar\omega_1}{\pi}u\sin\theta_{21}\cos\theta_{21}, \qquad \Gamma_v = \frac{\hbar\omega_1}{\pi}u\cos^2\theta_{21}. \qquad [70]$$

The adiabatic angle θ_{21} is essentially an angle obtained from the Franck–Condon factor, so that the preceding results reduce to semiclassical golden rule expressions (16–19).

ii. Adiabatic Limit: Describes Strong Interaction

$$\Delta_{v_+} = \frac{\hbar\omega_+}{\pi} u^{-1} \sin\theta_{11} \cos\theta_{11}, \qquad \Gamma_{v_+} = \frac{\hbar\omega_+}{\pi} u^{-1} \cos^2\theta_{11}. \qquad [71]$$

In this case, we see the role of diabatic angles that define adiabatic resonances.

iii. Intermediate Coupling. When there is proximity between a diabatic and an adiabatic level, sharp resonances will occur at energy E_r, which is a weighted average of E_v and E_v^+ (7, 17),

$$E_r = \frac{E_v + \varepsilon u E_v^+}{1 + \varepsilon u}, \qquad [72]$$

with $\varepsilon = \omega_1/\omega_+$, the ratio of diabatic (unperturbed) and adiabatic level frequencies. The corresponding width is

$$\Gamma_r = \frac{2\pi\varepsilon^2 u(1 + u)}{\hbar\omega_1(1 + \varepsilon u)^3}(E_v^+ - E_v)^2. \qquad [73]$$

VIII. SEMICLASSICAL PHOTODISSOCIATION

Photodissociation presents a perfect analogy with predissociation when the dressed picture, Figure 1(b), of multiphoton processes is considered. One has thus a situation where bound states are embedded in a continuum and interact with a radiative coupling $\boldsymbol{\mu}\cdot\mathbf{E}$. As in the case of predissociation, zero-order bound states undergo energy shifts and are broadened by some width in the presence of the electric field. At weak fields, Fermi golden rule expressions for linewidths may still be valid. However, for strong fields, renormalization of the spectrum via multiphoton processes will invalidate this expression, and new bound states induced by the laser will be of importance. A semiclassical version of the theory of photodissociation valid for any field strength is thus a great help in understanding exact close-coupled channel calculations.

An example of these laser-induced resonances was studied theoretically and numerically in the photofragmentation of Ar_2^+ in an intense laser field by Bandrauk et al. (7, 9–11). As is seen in Figure 6, the photodissociation process in the dressed-molecule picture becomes a curve-crossing problem between the bound-state field potential $V_1(R) + n\hbar\omega$ and the continuum $V_3(R) + (n-1)\hbar\omega$. The interaction is mediated by the field through the radiative coupling $V^r_{13}(R) = \mu_{13}(R) \cdot \mathbf{E}$. Multiphoton absorption would correspond to further crossings that will be neglected, since in this particular case other electronic excited states are situated at about four photons energies from the ground state. Thus, for Ar_2^+, two continua are actually available in the visible and ultraviolet spectra, leading to a second curve crossing with the potential $V_2(R) + (n-1)\hbar\omega$, and the most intense coupling is V^r_{12}. Channel 1 ($^2\Sigma_u^+$ bound states) undergoes a C^+-type predissociation via channel 2 ($^2\Sigma_g^+$ continuum) due to strong radiative interaction V^r_{12} since for $\Sigma_g \to \Sigma_u$ transitions in ionic systems such as Ar_2^+ one obtains a large transition moment, $\mu_{12}(R) \propto \frac{1}{2}R$, which grows with increasing distance. A similar behavior occurs in valence–Rydberg transitions (see Section V). Information concerning the shifts and widths of resonances can be obtained from the poles of the scattering matrix in its full version. In particular, the inelastic matrix element S_{32} corresponds to the physical process of *radiative recombination* on surface 3 with an infrared laser followed by photodissociation to surface 2 via an intense ultraviolet field ($\omega \sim 30,000\,\text{cm}^{-1}$). The resulting semiclassical expression for S_{32} is (7, 19)

$$S_{32} = 2u_1^{1/2}u_2^{1/2}(1+u_2)^{1/2}\exp\left[-i(\alpha_2+\alpha_3)\right](N/D)\exp\left[2i(\theta_{21}-\theta_{11})\right],$$

$$N = \cos\theta_{21}\sin(\theta_{31}-\theta_{11}) - (1+u_2)^{-1}\cos\theta_{21}\sin\theta_{21}\exp\left[-i(\theta_{21}-\theta_{11})\right],$$

$$D = \cos\theta_{11} + u_1\cos\theta_{31}\exp i(\theta_{31}-\theta_{11})$$
$$+ (1+u_1)u_2\cos\theta_{21}\exp\left[i(\theta_{21}-\theta_{11})\right]. \qquad [74]$$

The zeros of the denominator D of Eq. [74],

$$\cos\theta_{11}\exp(i\theta_{11}) + u_1\cos\theta_{31}\exp(i\theta_{31}) + u_2\cos\theta_{21}\exp(i\theta_{21}) = 0, \quad [75]$$

corresponds to the manifestation of laser-induced resonances. [We have replaced $1 + u_1$ by 1 since in the present case $u_1(V_{13}) \to 0$, whereas $u_2(V_{13}) \to \infty$.] Clearly, from Eq. [75] a pole of S_{32} can be associated with a situation where all phases, diabatic (θ_{11}) and adiabatic (θ_{31}, θ_{21}), correspond to bound states, bringing all cosines close to zero. Such a coincidence of diabatic and adiabatic quasi–bound states has been proposed as a general condition for the presence of sharp resonances, that is, small-width resonances (6, 7, 16, 17). Introducing Taylor expansions for the diabatic (θ_{ii}) and adiabatic (θ_{ij}) action angles, as in the previous section, one obtains (from the real and imaginary parts of the general D (Eq. [75]), the resonance energies E_r, and the widths Γ_r)

$$E_r = \frac{E_v + \sum_i \varepsilon_i u_i E_{v_i}^+}{1 + \varepsilon_i u_i} + \Delta E', \qquad [76]$$

$$\Gamma_r = \frac{\pi(E_r - E_v)^2 (\hbar\omega)^{-1} + \sum_i \varepsilon_i u_i (\hbar\omega_i^+)^{-1}(E_r - E_{v_i}^+)^2}{1 + \sum_i \varepsilon_i u_i}. \qquad [77]$$

The sum over i includes all crossings (two for the resonant process pictured in Figure 6). As previously, E_v is the diabatic bound-state energy, whereas $E_{v_i}^+$ corresponds to the adiabatic states newly created by the laser field; $\hbar\omega$ and $\hbar\omega_1^+$ are the corresponding local energy spacings and the ε_i's are their ratios: $\varepsilon_i = \omega/\omega_i^+$. The background shift $\Delta E'$ is not given by a simple semiclassical theory but results from strong nonlinear variation in the diabatic potential or close turning points (43, 44). In the Ar_2^+ example, the coupling at crossing R_2 being much more important than at crossing R_1, we can expect identifiable adiabatic states to occur at R_2. The θ_{11} and θ_{21} being quantized according to Eq. [68 and 69], simplified expressions are obtained for E_r and Γ_r,

$$E_r = \frac{E_v + \varepsilon_2 u_2 E_{v_2}^+ + \Delta E_r^{(1)}}{1 + \varepsilon_2 u_2}, \qquad [78]$$

$$\Gamma_r = \frac{\pi\varepsilon_2 u_2(1 + u_2)}{\hbar\omega_2^+(1 + \varepsilon_2 u_2)^3}(E_v^+ - E_v)^2 + \frac{\Gamma_r^{(1)}}{1 + \varepsilon_2 u_2}. \qquad [79]$$

The first-order corrections to terms readily identified to corresponding expressions obtained for the weak-coupling case in predissociation Eq. [70] are given by

$$\Delta E_r^{(1)} = \frac{\hbar\omega}{\pi} u_1 \sin\theta_{31} \cos\theta_{31}, \qquad [80]$$

$$\Gamma_r^{(1)} = \frac{\hbar\omega}{\pi} u_1 \cos^2\theta_{31}, \qquad [81]$$

which are the shift and width due to the single crossing R_1 in predissociation.

Two remarks are in order: (i) When one coupling is dominant $[u_2(V_{12}) \gg u_1(V_{13})]$, all other perturbations are reduced by a factor $(1 + \varepsilon_2 u_2)$. Thus, all other resonant and nonresonant multiphoton processes are made less important. (ii) Neglecting all these side processes, the semiclassical theory predicts that no matter what the strength of the interaction is, one will obtain sharp states induced by the laser field. Their energies result from a mixing of the energies of the initial diabatic bound states and the new adiabatic bound states supported by a new adiabatic potential formed by radiatively induced avoided crossings, Figures 1(b) and 6. In particular, for a *single* continuum, no photoabsorption will occur at energies where diabatic states are accidentally degenerate with adiabatic states, as in this case one expects the new resonances E_r to be very sharp, their width $\Gamma_v \propto (E_v^+ - E_v)^2$ goes to zero (7). We emphasize that photodissociation under the preceding conditions can be more usefully envisaged as a predissociation induced by the laser field, in the sense that the fragmentation proceeds through sharp resonances.

Numerical comparisons between semiclassical and exact coupled-equation calculations have been presented for Ar_2^+ for various field strengths (7, 9, 10). Very sharp resonances are indeed obtained from these quantum calculations when coincidences occur by increasing the angular momentum J, in conformity with semiclassical predictions. Some of these resonances calculated for the intense field $I = 10^{11}$ W/cm^2 and giving an electronic Rabi frequency, Eq. [10], $\omega_{ac} = 370$ cm^{-1} are displayed in Table 2, together with Fermi's golden rule expression for Γ_r (7), Eq. [70],

$$\Gamma_r^0 = \frac{\hbar\omega}{\pi} u_2 \cos^2 \theta_{21} \simeq \frac{\hbar\omega}{2\pi} u_2. \qquad [82]$$

Since the Rabi frequency ω_{ac} is of the order of diabatic ground-state vibrational frequencies $\omega = 300$ cm^{-1}, we certainly cannot expect the Fermi golden rule expression and hence the Franck–Condon factor approach to photodissociation to be valid for such field intensities. Equations [78 and 79] indicate that increasing u_2 leads to more and more mixing between the diabatic and adiabatic states. [It is to be realized that the adiabatic states are *not* orthogonal to the diabatic states but are rather linear combinations of these, as can be seen from the Fano treatment of interacting resonances (45, 46).] Thus, the adiabatic and diabatic state combination is a useful way of succinctly representing semiclassical resonances. The semiclassical expressions for Γ_r, Eq. [79], agree well with the exact results at energies well above the crossing point (3600 and 5500 cm^{-1}). Better agreement closer to the crossing point (1300 cm^{-1}) can be obtained by including the more elaborate complex semiclassical phases that rely on the crossing point of adiabatic states in the complex R plane, Eqs. [46 and 47].

In actual calculations, one includes the $(2J + 1)M_J$ sublevels of each rotational quantum number J in addition to further J levels excitations that follow from the $\Delta J = 0$, ± 1 selection rules (see Section IV). It is not uncommon to include transitions up to $\Delta J = \pm 10$ in order to obtain convergence for a single process such as direct photodissociation in Ar_2^+ (9–11). The remarkable observation is that the resonances predicted to be sharp in the semiclassical theory of one bound potential coupled to one continuum potential, Eq. [73], remain stable when rotational transitions are included for convergence, thus transforming the problem into a multibound–multicontinua problem. Thus, laser-induced resonances, which occur for particular M_J and J values (i.e., particular orientation) can be quite stable. Applying the extra bound artificial method of Section II to describe preparation of the dressed states from a well-defined initial zero-field state, it has been shown in a calculation of the angular distribution of the photofragmentation of Ar_2^+ that these sharp stable laser-induced states produce *nonstatistical* angular distributions. Broad resonances (i.e., initial states that are very unstable in a strong laser field due to rapid photodissociation) will invariably produce *statistical* angular distributions. Such properties of laser-induced resonances and their stabilities are most conveniently described in the semiclassical framework presented in the preceding. Similar considerations were shown to apply in higher order multiphoton processes in Ar_2^+, such as CARS, which involves a stimulated emission from a resonant continuum to an intermediate bound state (30). The corresponding dressed states were derived by a semiclassical model that included coupling the resonant continuum to another continuum in order to mimic further multiphoton transitions out of the continuum. It was found again that due to the stability of certain laser-induced resonances as predicted by the semiclassical theory, dressed states can be quite stable to further multiphoton transitions to other electronic states in a configuration where an initial-field photodissociation is followed by a strong-field-stimulated emission (30).

We have limited our discussion to diatomics since it is clear from the previous two sections that for these molecules semiclassical techniques are an indispensable tool in understanding strong-field effects. For polyatomics, semiclassical quantization is limited in particular in the high-energy, chaotic regime (47). Furthermore, diabatic and adiabatic approximations in many dimensions seem also to be less useful in the chaotic regime (48). The existence of field-induced resonances in polyatomics is related therefore to the interesting questions of integrability, nonintegrability, and chaos in both classical and quantum regimes. We intend to pursue these questions in future work on the coupled-equations approach to multiphoton transitions in larger molecules.

APPENDIX

For Eq. [19], we first evaluate $\langle c|t|c_1 \rangle$,

$$\langle c|t|c_1 \rangle = \langle c|V|c_1 \rangle + \langle c|VP(E - PHP)^{-1}PV|c_1 \rangle. \qquad [A.1]$$

Now $P = |c\rangle + |c_1\rangle$, and we further have $\langle c|V|c\rangle = \langle c_1|V|c_1\rangle = \langle c|V|c_1\rangle = 0$ [see Figure 2(b)]. Thus, Eq. [A.1] vanishes identically, $\langle c|t|c_1\rangle = 0$. We therefore obtain, for Eq. [19],

$$T_{c,c_1} = \sum_d \langle c|tQ'G|d\rangle\langle d|t|c_1\rangle + \sum_a \langle c|tQ'G|a\rangle\langle a|t|c_1\rangle. \qquad [A.2]$$

Following the same procedure as in evaluating [A.1], one easily obtains $\langle a|t|c_1 \rangle = 0$, so that a further simplification occurs,

$$T_{c,c_1} = \sum_d \langle c|tQ'G|d \rangle \langle d|t|c_1 \rangle. \qquad [A.3]$$

We first use the identity $Q' = |d\rangle\langle d| + Q$ to obtain

$$\langle c|tQ'G|d \rangle = \langle c|tQG|d \rangle + \sum_d \langle c|t|d \rangle \langle d|G|d \rangle. \qquad [A.4]$$

Evaluating $\langle c|t|d \rangle$ as in [A.1], one obtains $\langle c|t|d \rangle = 0$ to give, for T,

$$\langle c|T|c_1 \rangle = \sum_d \langle c|tQG|d \rangle \langle d|t|c_1 \rangle. \qquad [A.5]$$

The matrix element $\langle d|t|c_1 \rangle$ can be evaluated as in [A.1] to give the first-order matrix elements

$$\langle d|t|c_1 \rangle = \langle d|V|c_1 \rangle, \qquad \langle a|t|d \rangle = \langle a|V|d \rangle. \qquad [A.6]$$

Now in analogy with the exact relation $G = G_0 + G_0 VG$ (23), one can obtain, using the operators P and Q (32),

$$QG|d \rangle = \sum_{d'} Q(E - H_0 - QtQ)^{-1} Qt|d' \rangle \langle d'|G|d \rangle. \qquad [A.7]$$

The state $|d\rangle$ is only coupled symmetrically to $|c_1\rangle$, so that it will undergo, for weak coupling, a perturbation $\langle d|t|d \rangle = \Delta E_d + \frac{1}{2} i \Gamma_d$. One can therefore write $\langle d'|G|d \rangle = (E - E_d + i\Gamma_d/2)^{-1} \delta_{d,d'}$. Using this result as well as A.7 and A.6, A.5 reduces to

$$\langle c|T|c_1 \rangle = \langle c|tQ(E - H_0 - QtQ)^{-1} Q|a \rangle$$
$$\times \sum_d \langle d|V|c_1 \rangle (E - E_d + \tfrac{1}{2} i \Gamma_d)^{-1}. \qquad [A.8]$$

We now expand Q in terms of the complex dressed states $|L\rangle$, which are eigenstates of $Q(E - H_0 - QtQ)^{-1}$ (32–34), that is,

$$Q(E - H_0 - QtQ)^{-1} Q = \sum_L |L\rangle (E - E_L + \tfrac{1}{2} i \Gamma_L)^{-1} \langle L|. \qquad [A.9]$$

We therefore obtain, for [A.8],

$$\langle c|T|c_1 \rangle = \sum_L \frac{\langle c|t|L \rangle \langle L|a \rangle}{(E - E_L + \tfrac{1}{2} i \Gamma_L)} \frac{\langle d|V|c_1 \rangle}{(E - E_d + \tfrac{1}{2} i \Gamma_d)}. \qquad [A.10]$$

In this final expression, we have invoked the isolated resonance approximation for the weakly coupled $|d\rangle$ state, which has a width $\Gamma_d = 2\pi |\langle d|V|c_1\rangle|^2$. We can therefore write [A.10] as

$$\langle c|T|c_1\rangle = T_{c,a}\frac{(\Gamma_d/2\pi)^{1/2}}{E - E_d + \frac{1}{2}i\Gamma_d}. \qquad [\text{A.11}]$$

It is this final expression that generates the amplitude $T_{c,a}$ from numerical calculations of the S matrix element S_{c,c_1}, Eq. [24].

Acknowledgments

We thank the National Sciences and Engineering Research Council of Canada for operating and traveling grants supporting this work.

References

1. M. H. Mittleman, *Introduction to the Theory of Laser–Atom Interactions*, Plenum, New York, 1982.

2. N. M. Kroll and K. M. Watson, *Phys. Rev.* **A13**, 1018 (1976).

3. A. I. Voronin and A. A. Samokhin, *Sov. Phys. JETP* **43**, 4 (1976).

4. A. M. Lau and C. K. Rhodes, *Phys. Rev. A* **16**, 2392 (1977).

5. J. M. Yuan and T. F. George, *J. Chem. Phys.* **68**, 3040 (1978).

6. A. D. Bandrauk and M. L. Sink, *Chem. Phys. Lett.* **57**, 569 (1978).

7. A. D. Bandrauk and M. L. Sink, *J. Chem. Phys.* **74**, 1110 (1981).

8. T. F. George, *J. Phys. Chem.* **86**, 10 (1982).

9. A. D. Bandrauk and G. Turcotte, in *Collisions with Lasers*, N. K. Rahman and C. Guidotti, eds., Harwood Academic, Amsterdam, 1984, pp. 351–373.

10. A. D. Bandrauk and G. Turcotte, *J. Chem. Phys.* **77**, 3867 (1982).

11. A. D. Bandruak and G. Turcotte, *J. Phys. Chem.* **87**, 5098 (1983).

12. T. T. Nguyen Dang and A. D. Bandrauk, *J. Chem. Phys.* **79**, 3256 (1983).

13. T. T. Nguyen Dang and A. D. Bandrauk, *J. Chem. Phys.* **80**, 4926 (1984).

14. A. D. Bandrauk and T. T. Nguyen Dang, *J. Chem. Phys.* **83**, 2840 (1985).

15. A. D. Bandrauk, O. Kalman, and T. T. Nguyen Dang, *J. Chem. Phys.* **84**, 6761 (1986).

16. A. D. Bandrauk and M. S. Child, *Mol. Phys.* **19**, 95 (1970).

17. M. S. Child, *J. Mol. Spectrosc.* **53**, 280 (1974).

18. M. L. Sink and A. D. Bandrauk, *J. Chem. Phys.* **66**, 5313 (1977).

19. O. Atabek, R. Lefebvre, and M. Jacon, *J. Chem. Phys.* **81**, 3874 (1984).

20. A. D. Bandrauk and G. Turcotte, *J. Phys. Chem.* **89**, 3039 (1985).

21. C. Cohen-Tannoudji, R. Balian, and S. Haroche, eds., *Frontiers in Laser Spectroscopy*, North-Holland, Amsterdam, 1975.

22. M. L. Goldberger and K. M. Watson, *Collision Theory*, Wiley, New York, 1964, Chapter 8.

23. K. M. Watson and J. Nuttall, *Topics in Several Particle Dynamics*, Holden-Day, San Francisco, 1967, Chapter 1.

24. M. Shapiro, *J. Chem. Phys.* **56**, 2582 (1972).

25. R. Loudon, *The Quantum Theory of Light*, Oxford Press, Oxford, 1973.

26. O. Atabek, J. A. Beswick, and R. Lefebvre, *J. Chem. Phys.* **65**, 4035 (1976).

27. A. D. Bandrauk, G. Turcotte, and R. Lefebvre, *J. Chem. Phys.* **76**, 225 (1982).

28. O. Atabek, R. Lefebvre, and M. Jacon, *J. Chem. Phys.* **72**, 2670, 2683 (1980).

29. K. Kodama and A. D. Bandrauk, *Chem. Phys.* **57**, 461 (1981).

30. A. D. Bandrauk, M. Giroux, and G. Turcotte, *J. Phys. Chem.* **89**, 4473 (1985).

31. A. D. Bandrauk and N. Gélinas, *Chem. Phys. Lett.* **129**, 362 (1986).

32. L. Mower, *Phys. Rev.* **142**, 799 (1966).

33. B. W. Shore, *Rev. Mod. Phys.* **39**, 439 (1967).

34. J. P. Laplante, A. D. Bandrauk, and C. Carlone, *Can. J. Phys.* **55**, 1 (1977).

35. L. Fox, *The Numerical Solution of Two-Point Boundary Value Problem*, Oxford Press, London, 1957.

36. D. W. Norcross and M. J. Seaton, *J. Phys.* **B6**, 614 (1973).

37. O. Atabek and R. Lefebvre, *Multichannel Energy Quantization*, CNRS report, Orsay, 1983.

38. S. D. Peyerimhoff and R. J. Buenker, *Chem. Phys.* **57**, 279 (1981).

39. F. Grein, S. D. Peyerimhoff, and R. J. Buenker, *Can. J. Phys.* **62**, 1928 (1984).

40. G. Orlandi and W. Siebrand, *J. Chem. Phys.* **58**, 4513 (1975).

41. H. Chang and D. M. Hwang, *J. Raman Spectrosc.* **7**, 254 (1978).

42. F. Ghandour, M. Jacon, E. N. Svendsen, and J. Oddershedl, *J. Chem. Phys.* **79**, 2150 (1983).

43. M. S. Child and R. Lefebvre, *Mol. Phys.* **34**, 979 (1977).

44. M. L. Sink and A. D. Bandrauk, *Chcm. Phys.* **33**, 205 (1978).

45. U. Fano, *Phys. Rev. A* **17**, 93 (1978).

46. R. Colle, *J. Chem. Phys.* **74**, 2910 (1981).

47. P. Brumer, *Adv. Chem. Phys.* **47**, 201 (1981).

48. T. T. Nguyen-Dang, *J. Chem. Phys.* **83**, 5019 (1985).

CHAPTER XX

SQUEEZED STATES OF LIGHT

H. J. KIMBLE

Department of Physics, University of Texas at Austin, Austin, Texas 78712

In the past two years a great deal of excitement has been generated in the community of optical physicists by the observation of squeezed states of light in several laboratories (1–7). Although the prospect of "squeezing light" may conjure up a variety of images regarding the "elasticity" of the electromagnetic field, what has in fact been observed are quantum states that directly display the granularity associated with the quantum nature of light. In addition to the intrinsic interest in these nonclassical states of the electromagnetic field, there are as well a number of exciting applications in measurement science and in optical communication associated with sensitivity beyond the "shot noise" limit, which is a limit determined by the zero-point, or vacuum, fluctuations of the field. New phenomena should also be observed in atomic spectroscopy since the interaction of an atom with squeezed radiation modifies radiative decay rates and level shifts.

The schematic representation of the field by the phasor diagrams shown in Figures 1 and 2 attempts to convey the essential qualitative ingredients in the phenomenon of squeezing (8, 9). We write the electric field $\hat{E}(t)$ (where the caret denotes a Hilbert space operator) as the sum of a mean amplitude plus fluctuations,

$$\hat{E}(t) = \hat{E}^{(+)}e^{-i\omega t} + \hat{E}^{(-)}e^{+i\omega t},$$

$$\hat{E}^{(+)} = \mathscr{E}_0 + \delta\hat{\mathscr{E}}, \tag{1}$$

where $\mathscr{E}_0 = \langle \hat{E}^{(+)} \rangle$ and the fluctuating field $\delta\hat{\mathscr{E}}$ is of zero mean amplitude. Of course, for an optical field the amplitude of the electric field varies sinusoidally at a frequency $\omega \sim 10^{15}\,\mathrm{s}^{-1}$, so that the axes of Figure 1 rotate at this frequency when viewed from the laboratory frame. We next introduce the quadrature phase amplitude $\hat{X}_\theta$, where

$$\hat{X}_\theta = \delta\hat{\mathscr{E}}e^{-i\theta} + \delta\hat{\mathscr{E}}^\dagger e^{i\theta}. \tag{2}$$

For the particular choices $\theta = (0, \tfrac{1}{2}\pi)$, the quadrature phase amplitudes are

859

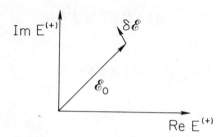

Figure 1. Phasor diagram for the fluctuating field $E^{(+)} = \mathscr{E}_0 + \delta\mathscr{E}$. The field is written in terms of its mean amplitude $\mathscr{E}_0$ plus a fluctuating component $\delta\mathscr{E}$. In a frame rotating at the carrier frequency ω, the field $\mathscr{E}_0$ is stationary with respect to the coordinant axes while the "noise" field $\delta\mathscr{E}$ fluctuates randomly as a result of the quantum nature of the electromagnetic field.

proportional to the components of the fluctuating electric field along the orthogonal axes drawn in Figure 1. If we denote $\hat{X}_\theta$ ($\theta = 0$) by $\hat{X}_+$ and $\hat{X}_\theta$ ($\theta = \pi/2$) by $\hat{X}_-$, the electric field operator $\delta\hat{E}(t) \equiv \hat{E}(t) - \langle \hat{E}(t) \rangle$ can be written as

$$\delta\hat{E}(t) = \hat{X}_+ \cos \omega t + \hat{X}_- \sin \omega t. \tag{3}$$

The two quadrature operators $(\hat{X}_+, \hat{X}_-)$ form a pair of canonically conjugate variables analogous to the position and momentum operators for a harmonic oscillator. For a single-mode field the quadrature operators obey the commutation relation $[\hat{X}_+, \hat{X}_-] = 2i$ with corresponding uncertainty product $\Delta X_+ \Delta X_- \geq 1(9)$. The fluctuations expressed by the uncertainty product

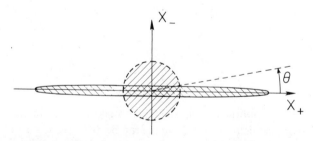

Figure 2. Schematic representation of the fluctuations of the field $\delta\mathscr{E}$. The dashed curve is drawn for the vacuum field or for a field in a coherent state. In this case the fluctuating field $\delta\mathscr{E}$ would uniformly fill a symmetric region around the tip of the vector $\mathscr{E}_0$ in Figure 1. The full curve represents the fluctuations for a squeezed state and clearly displays the asymmetric distribution of fluctuations associated with this state. *Note*: in either a vacuum state or a squeezed state, the fundamental level of fluctuations is constrained by an uncertainty relation, which in qualitative terms fixes a common lower bound for the area of the circle and of the ellipse shown in the figure.

can be graphically represented, as shown in Figure 2, where a symmetric error circle (dashed line) of unit radius is obtained for the vacuum state or for a coherent state (which is approximated by a single-mode laser). For a squeezed state, on the other hand, the symmetric error circle is "squeezed" into an error ellipse (solid line). The variance of the field $\hat{X}_\theta$ is just the projection of the error ellipse along the direction specified by θ (9). As the angle θ is swept, the variance of $\hat{X}_\theta(t)$ will therefore vary both above and below the level of fluctuations set by the vacuum state of the field. The reduction of fluctuations below the vacuum level (either $\Delta X_+ < 1$ or $\Delta X_- < 1$) is the signature of a squeezed state. We note in passing that squeezed states may or may not be minimal uncertainty states, for which not only is either $\Delta X_+ < 1$ or $\Delta X_- < 1$ but also for which the product of uncertainties remains a minimum, $\Delta X_+ \Delta X_- = 1$.

While it is not apparent from the preceding analysis, one can show that squeezed states of light lie outside the realm of any classical theory of the electromagnetic field (8). In qualitative terms it is reasonable to expect that a classical field theory would be adequate for a field with a distribution of fluctuations that is broad with respect to the error circle shown in Figure 2. However, when the fundamental level of fluctuations set by quantum mechanics is reached, and when a repartitioning of these fundamental fluctuations into $\hat{X}_\pm$ is considered, one surely expects that a fully quantum theory is essential. It is precisely the quantum nature of squeezed states that gives rise to a range of intriguing possibilities associated with measurement at the quantum level.

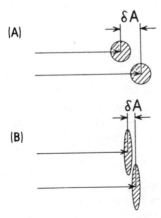

Figure 3. The minimum detectable amplitude change δA of the electromagnetic field is limited by quantum fluctuations. (a) The symmetric distribution of fluctuations of the vacuum state sets a limit on precision known as the shot noise limit. (b) An improvement in precision can in principle be achieved with a properly oriented squeezed state.

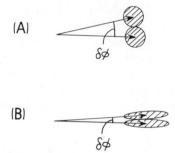

Figure 4. The minimum detectable phase change $\delta\phi$ of the electromagnetic field is limited by quantum fluctuations. (*a*) Symmetric distribution of fluctuations of a coherent state. (*b*) Asymmetric distribution of a squeezed state.

Figures 3 and 4 illustrate how the reductions in fluctuations of a squeezed state can be employed to improve the precision of amplitude and phase measurements of the field. In Figure 3 the amplitude of the electric field is changed by a small amount δA, as might result from absorption by an atomic or molecular sample. The minimum detectable amplitude change is limited by fluctuations of the field that may be either of technical or fundamental origin. In traditional spectroscopy without squeezed states, the best that one can do is to reach the so-called shot noise limit, which is a limit set by the symmetric distribution of the vacuum state [Figure 3(*a*)]. In this case, $(\delta A/\mathscr{E}_0)^2 \sim 1/n$, where n is the number of photons received in the bandwidth of the measurement. With a squeezed state, one can in principle improve this minimum detectable change to a level $(\delta A/\mathscr{E}_0)^2 \sim (\Delta X_+)^2/n$, where $(\Delta X_+)^2 \ll 1$ for a state with a large degree of squeezing [Figure 3(*b*)].

In Figure 4 we illustrate how the sensitivity for phase measurements can be improved with squeezed states. A phase shift $\delta\phi$ is introduced as, for example, by propagation through an atomic sample or through an interferometer. In the best case in traditional interferometry, the minimum detectable phase shift is again set by the level of vacuum fluctuations, with the shot noise limit for phase changes given by $(\delta\phi)^2 \sim 1/n$ [Figure 4(*a*)]. If, however, a squeezed state is employed, an improvement in principle to a level $(\delta\phi)^2 \sim (\Delta X_-)^2/n$ could be achieved, as illustrated in Figure 4(*b*).

Naturally, the realization in practice of reduced noise levels through the use of squeezed light is associated with a reasonable degree of technical complexity. There are also issues of principle associated with system losses and with the actual energy content of the squeezed state that must be addressed. However, it seems clear that the properties of squeezed states offer considerable potential for advances in measurement precision beyond the shot noise limit. We have recently demonstrated improvements in signal to noise relative to the vacuum level in the detection of both amplitude modulation (2.5dB improvement) and phase modulation (3.0dB improvement) by employing the squeezed light produced by an optical parametric oscillator (10).

As for the question of the generation of squeezed states, we mention only briefly the various experiments that have thus far been reported. Each of these experiments is documented in detail in Ref. 1. In simplest terms a squeezed state is generated by mixing a field $\hat{E}^{(+)}$ with its conjugate $\hat{E}^{(-)}$ to produce a new field $\hat{D} = \mu\hat{E}^{(+)} + \nu\hat{E}^{(-)}$, where (μ, ν) are parameters that depend on the particular interaction considered. In light of this simple prescription, much attention has been directed to the process of four-wave mixing in which conjugate signal and idler fields are produced via the interaction with a strong pump field in a nonlinear optical medium. It is this process that first led to the landmark observation of squeezing by Slusher and co-workers (2) at AT&T Bell Laboratories, who reported a 7% reduction in noise level relative to the vacuum state noise limit for squeezed light generated by an optical cavity containing sodium atoms. Recent improvements in the experiment have increased this figure to a 25% reduction. Subsequently, other workers (3–7) have observed deamplification of quantum noise (squeezing) in a variety of physical systems. Shelby, Levenson, and co-workers (3) at IBM San Jose have achieved a 20% reduction in the total noise level below the vacuum state limit for four-wave mixing in an optical fiber cooled below 4.2 K. Maeda, Kumar, and Shapiro (5) at MIT reported a 4% reduction for four-wave mixing in sodium vapor.

In our laboratory at the University of Texas at Austin, two different experiments have produced squeezed states (4, 7). Raizen et al. have observed a 30% noise reduction in a system of two-level sodium atoms in a small (830-μm-length) optical cavity (7). This work exploits the normal-mode structure of a coupled atom–field system to generate squeezed states. The second set of experiments by Wu et al. (4) has investigated the process of degenerate parametric down conversion in which photons at frequency ω_2 bifurcate to produce correlated pairs of photons at frequency $\omega_1 = \frac{1}{2}\omega_2$, with the field at ω_1 in a squeezed state. This process of subharmonic conversion has served as one of the prototypical interactions in theoretical investigations but was thought to be an unsuitable laboratory candidate for a variety of technical reasons. In our experiments on down conversion in an optical cavity, we have observed more than a 60% reduction in noise level as compared to the vacuum level. More importantly, if an accounting is made for the separately measured linear loss mechanisms that degrade the observed squeezing relative to that actually produced in the nonlinear interaction, one infers that the observed noise reduction resulted from a field that was within 10% of an ideal squeezed state. Hence, there is some cause for optimism about the prospects for generating squeezed states of the electromagnetic field that would exhibit a 10-fold reduction in the variance of one field quadrature relative to the vacuum-state limit. Beyond a large degree of squeezing, we have also demonstrated that the field state produced by the down-conversion process is a minimum-uncertainty state. We have given an explicit experimental demonstration of the Heisenberg uncertainty principle for light (Wu et al. in Ref. 1).

All of the experiments described in the preceding rely on optical homodyne detection for the observation of squeezing. With reference to Eq. 3, it is clear that to observe the fluctuations in the quadrature amplitudes $\hat{X}_\pm$ of the field, one must employ a process with sensitivity to the optical phase. In simplest conceptual terms, homodyne detection extracts information about $\hat{X}_\pm$ by forming the product $\hat{I}(t) = \overline{\hat{E}_{LO}(t)\,\delta E(t)}$ at the surface of a square-law detector (e.g., a photodiode). Here $\hat{E}_{LO}(t) = \hat{E}_{LO}\cos(\omega t + \beta)$ is a strong local oscillator field in a coherent state, and the overbar denotes a time average over many optical periods. For $\beta = (0, \frac{1}{2}\pi)$, we have $\hat{I}_\pm \propto \hat{X}_\pm \hat{E}_{LO}$. Hence, the moments of the fluctuations of the photocurrent directly reflect the fluctuations of the squeezed field, which are given by the moments of $\hat{X}_\pm$. In more technical terms, the ratio of the spectral density of photocurrent fluctuations with no squeezing present to that with the squeezed field present is related in a remarkably simple fashion to the "spectrum of the squeezing" of the squeezed field (Ref. 1 contains several detailed treatments of homodyne detection).

As a final note on the experimental realization of squeezed states, we should point to research on reduced fluctuations in photon number. Squeezed states as we have described them refer to a reduction in the fluctuations in one quadrature component of the field. Another type of squeezing attempts to reduce the fluctuations in the modulus of the field (in the photon number). Experiments by several groups (6, 11–14) have demonstrated sub-Poissonian photon statistics, corresponding to a "squeezed photon number." The largest effect reported thus far is by Yamamoto's group at NTT (6), who observe a 7% reduction in detector noise below the vacuum level.

Undoubtedly, this brief overview of the generation and application of squeezed states will be badly outdated before it is published. Such an occurrence will, however, only serve to emphasize the vitality of the field. Quite apart from the research described here, experiments are in progress to observe squeezing in soliton propagation, to produce "twin" photon beams, and to explore further the question of quantum nondemolition measurements. The phenomenon of squeezing will not remain confined to the optical domain; experiments to generate squeezed states of the electromagnetic field at 20 GHz will soon succeed. The next few years should be an exciting time as the diversity of sources of squeezed radiation increases and as the applications to a number of exciting problems come to fruition.

Acknowledgments

This work was supported by the Office of Naval Research, by the Venture Research Unit of British Petroleum of North America, and by The National Science Foundation

References

1. For an overview of the field, see *Journal of Optical Society of America B*, Special Issue on Squeezed States of the Electromagnetic Field, H. J. Kimble and D. F. Walls, eds., October 1987.

2. R. E. Slusher, L. W. Hollberg, B. Yurke, J. C. Mertz, and J. F. Valley, *Phys. Rev. Lett.* **55**, 2409 (1985).

3. R. M. Shelby, M. D. Levenson, S. H. Perlmutter, R. G. DeVoe, and D. F. Walls, *Phys. Rev. Lett.* **57**, 691 (1986); B. L. Schumaker, S. H. Perlmutter, R. M. Shelby, and M. D. Levenson, *Phys. Rev. Lett.* **58**, 357 (1987).

4. L. A. Wu, H. J. Kimble, J. L. Hall, and H. Wu, *Phys. Rev. Lett.* **57**, 2520 (1986).

5. M. W. Maeda, P. Kumar, and J. H. Shapiro, *Opt. Lett.* **12**, 161 (1987).

6. S. Machida, Y. Yamamoto, and Y. Itaya, *Phys. Rev. Lett.* **58**, 1000 (1987).

7. M. G. Raizen, L. A. Orozco, M. Xiao, T. L. Boyd, and H. J. Kimble, *Phys. Rev. Lett.* **59**, 198 (1987).

8. D. F. Walls, *Nature (Lond.)* **306**, 141 (1983).

9. C. M. Caves and B. L. Schumaker, *Phys. Rev.* **A31**, 3068, 3093 (1985); *Quantum Optics IV*, J. H. Harvey and D. F. Walls, eds., Springer-Verlag, New York, 1986, p. 20.

10. H. J. Kimble, M. Xiao, and L.-A. Wu, reported at the Fifteenth International Conference on Quantum Electronics, Baltimore, MD, April 27–May 1, 1987, and *Opt. Lett.* (1988); H. J. Kimble, M. Xiao, and L.-A. Wu, reported at the 1987 Division of Atomic, Molecular, and Optical Physics Meeting of the APS, Cambridge, MA, May 18–20, 1987, and *Phys. Rev. Lett.* **59**, 278 (1987).

11. R. Short and L. Mandel, *Phys. Rev. Lett.* **51**, 384 (1983).

12. M. C. Teich and B. E. A. Saleh, *J. Opt. Soc. Am.* **B2**, 275 (1985).

13. C. K. Hong and L. Mandel, *Phys. Rev. Lett.* **56**, 58 (1986).

14. J. G. Walker and E. Jakeman, *Opt. Acta* **32**, 1303 (1985).

CHAPTER XXI

NOTES ON CLASSICAL AND QUANTUM THEORIES OF DRIVEN NONLINEAR SYSTEMS

P. W. MILONNI, J. R. ACKERHALT, and M. E. GOGGIN

*Theoretical Division (T-12), Los Alamos National Laboratory, Los Alamos,
New Mexico 87545*

CONTENTS

I. INTRODUCTION

Recently the first author asked a very distinguished theoretical physicist for his thoughts on quantum chaos. He replied that "so many otherwise intelligent people have said such foolish things about the subject that I've decided to stay away from it." However, in spite of its difficulty, most people would probably agree that the main question for quantum chaos—In what way, if any, does classical chaos carry over into quantum theory?—is an interesting one.

The hallmark of classical chaos is the property of very sensitive dependence on initial conditions, that is, the property that at least one of the Lyapunov exponents of the system is positive. By numerical computation of Lyapunov exponents, we can therefore determine whether a system is chaotic. If the system is chaotic, it will have a broadband power spectrum and decaying correlation functions, but these properties by themselves do not necessarily imply chaos in the sense of a positive Lyapunov exponent.

The problem is that when we look at model quantum systems that are classically chaotic, we do not find the extreme sensitivity to initial conditions that is possible classically. That is, we can make accurate long-term predictions about the state vector, whereas in a classically chaotic system we lose long-term predictability. Is there anything "left over" from classical chaos when we proceed to a quantum description?

The classical Poincaré recurrence theorem says that an initial state in the phase space of a system of finite volume will be repeated as often as desired if we wait long enough. Similarly, a quantum recurrence theorem may be proved. Bocchieri and Loinger state the theorem this way: "Let us consider a system with discrete energy eigenvalues E_n; if $\psi(t_0)$ is its state vector in the Schrödinger picture at the time t_0 and ε is any positive number, at least one time T will exist such that the norm $\| \psi(T) - \psi(t_0) \|$ of the vector $\psi(T) - \psi(t_0)$ is smaller than ε" (1). This quantum recurrence theorem is more far-reaching than the classical one because, whereas nearby orbits in classical phase space may have quite different recurrence times, there can be many similar quantum states with the same recurrence time. More recently Hogg and Huberman (2), using arguments similar to Bocchieri and Loinger, proved that "under any time-periodic Hamiltonian, a nonresonant, bounded quantum system will reassemble itself infinitely often in the course of time." According to them, "This in turn implies that no strict quantum stochasticity [chaos] is possible, a result which disagrees with recent predictions."

But, as in the classical case, the recurrence time may be exceedingly large. Peres (3) has considered a simple example in which the recurrence time is greater than the age of the universe and concludes that the recurrence argument against quantum chaos may be of no practical concern unless there happens to be only a small number of incommensurate energy levels.

These questions are relevant to the theory of laser–matter interactions (4). One point we emphasize in what follows is simply that *state vectors can have decaying correlations and broadband power spectra*. In other words, some features we associate with classical chaos can have quantum analogs, even though there may be no quantum chaos in the rigorous, classical sense of positive Kolmogorov entropy.

Following Professor Hirschfelder's suggestion for this contribution, we have expanded the topics covered in the presentation at the workshop and included a general review of some of the background material.

II. IS CLASSICAL PHYSICS REALLY DETERMINISTIC?

We are all familiar with the idea that classical physics is deterministic in the sense of Laplace: If we have unlimited computing power at our disposal and know the correct equations of motion of any physical system as well as the initial state of the system at some time $t = 0$, we can predict to any desired accuracy the state of the system at any time t.

But is it really true that the classical world view is deterministic? A strong case can be made that, as a practical matter, it is *not*. Consider the following

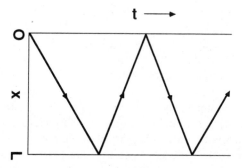

Figure 1. $x(t)$ for a particle bouncing between two walls separated by L.

question: If the initial conditions of a system are known only to within some small error, can we nevertheless make reasonably accurate predictions about the system for any time $t > 0$? As early as 1873, Maxwell seemed to recognize that, in general, we cannot (5). Later Max Born (6) also cautioned against a naive doctrine of classical determinism and used a simple example to illustrate his point. Consider a particle moving in a straight line between two walls at $x = 0$ and $x = L$, where it is elastically reflected. Given the initial values x_0 and v_0 of its position and velocity, we can easily determine $x(t)$ for any t, as illustrated in Figure 1. But if v_0 is changed by Δv, we obtain for any t an $x(t)$ differing from our original $x(t)$ by an amount we denote by $\Delta x(t)$. As illustrated in Figure 2, we have $|\Delta x(t)| = |\Delta v|t$, so that after a time $t = L/|\Delta v|$ the variation $|\Delta x(t)|$ of x is equal to the entire range (L) of possible x values.

In particular, if Δv represents our *uncertainty* in the initial data, our uncertainty Δx in the position of the particle eventually becomes as large as the whole range over which x can vary. If Δv is finite, however small, we cannot make (accurate) long-term predictions. It becomes operationally meaningless

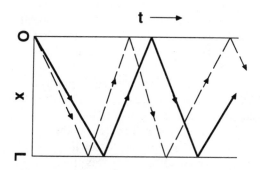

Figure 2. $x(t)$ for two different values of the initial velocity.

to say that the system is deterministic. It is no longer what Born called *determinable*.

In Born's example the sensitivity to initial conditions is linear in the sense that $\Delta x(t)$ is proportional to t. In systems described by nonlinear equations it is possible for uncertainties to grow *exponentially*. Systems with such extreme sensitivity to initial conditions are called *chaotic*. (We exclude from consideration unbounded systems, such as $\dot{x} = x$, which have a trivial sort of very sensitive dependence on initial conditions.) Chaotic systems are non-quasi-periodic, but systems that are not quasi-periodic are not necessarily chaotic.

There are several features of chaos that are new. One is the widespread recognition that chaos can be found even in "simple" systems of three or more coupled first-order differential equations (i.e., a system with phase space dimension greater than or equal to 3). Another is the discovery of certain prevalent, or "universal," routes to chaos.

Consider the following simple example:

$$\dot{x} = p \tag{2.1a}$$

$$\dot{p} = -\beta p + \frac{\beta}{T}(3x - 4x^2) \sum_{-\infty}^{\infty} \delta\left(\frac{t}{T} - n\right) \tag{2.1b}$$

This is a so-called periodically kicked system with kicking period T. Let x_n be the value of x just before the nth kick. Integrating 2.1, using the properties of the delta function, and taking the limit $\beta \to \infty$ of strong dissipation, we obtain $x_{n+1} = 4x_n(1 - x_n)$ or, letting $x_n = \sin^2 \pi\theta_n$, $\theta_{n+1} = 2\theta_n(\text{mod } 1)$. The solution is simply $\theta_n = 2^n\theta_0(\text{mod } 1)$.

This solution obviously has very sensitive (exponential) dependence on initial conditions. If we replace θ_0 by $\theta_0 + \varepsilon_0$, θ_n will change by $\varepsilon_n = 2^n\varepsilon_0 = \varepsilon_0 \exp(n \log 2)$. Thus, there is exponential growth with n of the perturbation of θ_n due to a perturbation ε_0 of the initial θ_0. The number $\log 2 = 0.693\ldots$ is the *Lyapunov exponent* in this example, and the fact that it is positive implies we have chaos. In general, a system has a whole spectrum of Lyapunov exponents, and the system is chaotic if one of them is positive.

The map $\theta_{n+1} = 2\theta_n$ is an example of a Bernoulli shift. Suppose we represent θ_0 in binary notation, as is done on a digital computer. The map is then equivalent to a shift of the "decimal" point to the left. Thus, if $\theta_0 = 0.0110101001\ldots, \theta_1 = 0.110101001\ldots, \quad \theta_2 = 0.10101001\ldots, \quad \theta_3 = 0.0101001\ldots$, and so on, and it is obvious that for large n the value of θ_n depends sensitively on the nth and higher digits in the binary representation of θ_0. In fact, any computer will introduce its own "uncertainty" in the initial conditions due simply to roundoff error. In this sense it may be said that chaos will eventually beat any computer—detailed, long-term predictions on chaotic systems are impossible. It should be emphasized that *the system (2.1) is perfectly deterministic, and yet it is not determinable in the sense of Born.* That is, we cannot make detailed long-term predictions about the system because (a) we can never know initial conditions with perfect precision and (b) any computing machine will introduce roundoff errors.

III. ERGODICITY, CHAOS, AND MIXING

Often a detailed knowledge of the long-term dynamics of a system is not necessary, and we get by with only statistical information. Of course, when the phase space is very large, it is simply impractical to try to specify the dynamics in great detail; instead we resort to techniques that collectively are called statistical mechanics. Here the primary goal is to predict *average* values of things that can be measured. Because it deals in probabilities, statistical mechanics makes use of hypothetical ensembles of identically prepared systems, whereas what is ultimately measured is an average over time in the evolution of one system. This distinction leads to Boltzmann's *ergodic hypothesis*, the assumption that these ensemble and time averages are equal. The ergodic hypothesis may be phrased in a more physically direct way: Over the course of time a trajectory will uniformly cover the $(2N - 1)$-dimensional surface of constant energy in the $2N$-dimensional phase space of the system; averages on this energy surface are then equal to a time average.

The problem is to justify this ergodic hypothesis. For a system with a single degree of freedom $(N = 1)$, it is trivially true, but it is demonstrably not true for any of the other analytically solvable textbook problems (n-dimensional harmonic oscillators, a particle in a Coulomb potential, etc.). For in such problems there are other quantities that are constant on a trajectory, so that any trajectory will explore a region of phase space smaller than the $(2N - 1)$-dimensional energy surface. Obviously, then, only part of the energy surface is covered in the course of time, and so the ergodic hypothesis for these integrable systems is false.

The ergodic problem stimulated the development of a part of mathematics called "ergodic theory," but it seems fair to say that this abstract study has not produced any proofs of ergodicity under assumptions useful to physicists. For about half a century those physicists who worried about the problem felt that small "nonintegrable" perturbations of an integrable system would give rise to ergodicity.** Fermi, for instance, felt this way, and when electronic computers

---

** Recall the definition of an integrable (also known as "completely integrable" or "separable") system: A system with N degrees of freedom (N-dimensional phase space) is integrable if there are N independent functions F_n of the p's and q's, with Poisson brackets $\{F_n, F_m\} = 0$, that are constant on each trajectory. Equivalently, the Hamilton–Jacobi equation is separable into one equation for each of N degrees of freedom. Trajectories of an integrable system are confined to an N-dimensional surface (torus) of phase space.

like the MANIAC at Los Alamos became available in the early fifties, he decided to test this hypothesis "experimentally." With Pasta and Ulam (7) he studied examples like the following:

$$\ddot{x}_j = (x_{j+1} - 2x_j + x_{j-1}) + \alpha[(x_{j+1} - x_j)^2 - (x_j - x_{j-1})^2]$$
$$j = 1, 2, \ldots, N \quad (3.1)$$

that is, coupled oscillators with small nonlinearities. Ergodicity in such an example would mean that energy initially concentrated in one or a few vibrational modes would eventually be shared on average among all possible modes. What was found was just the opposite: The energy stayed within only a few modes. Fermi was surprised by this violation of his renowned intuition and considered the results an interesting "little discovery" (8).

At about the same time Kolmogorov stated a theorem, later proved by Arnol'd and Moser, to the effect that if the perturbation of an integrable system is sufficiently small, most trajectories will remain confined to an N-dimensional surface in phase space. This celebrated KAM theorem (9) takes some of the surprise out of the Fermi–Pasta–Ulam results, but it does not of itself invalidate the ergodic hypothesis. For one thing, "small" really means infinitesimally small for the proof of the theorem.

[We should also emphasize, of course, that not everyone regards the ergodic hypothesis as necessary for the foundations of statistical mechanics. Some argue that statistical mechanics should not be based on long-term averages but should be developed starting from the recognition that we can only make probabilistic statements subject to our uncertainty about the precise Hamiltonian or initial conditions. Jaynes, for instance, formulates statistical mechanics in terms of information theory (10). Furthermore, we usually need only a reduced probability distribution for $n \ll N$ particles, and the N-particle system need not even be equilibrated, let alone ergodic.]

There is still not very much that can be said about the general validity of the ergodic hypothesis. One important result that follows from rigorous mathematical work of Sinai, however, is that a hard-sphere gas of two or more particles in a box is ergodic (and mixing). The proof is very long and involved, and this apparently unavoidable circumstance may be one reason why many people are "turned off" by ergodic theory. But if we accept this result, admittedly a very special case, then at least we have something to anchor to in this sea of uncertainty.

The concept of *mixing* is often introduced following Arnol'd: Rum and cola are poured into a shaker, 20% rum and 80% cola, and the liquid is stirred n times. As $n \to \infty$, every little cell of the liquid contains approximately 20% rum and 80% cola; the drink is mixed. If we think of points in the liquid as points in the phase space of some system and the stirs as iteration of a map or time evolution of a continuous flow, we say that the system is mixing. Intuitively, we would expect a mixing system to be ergodic, and in fact, this is true and not difficult to prove; it is also easy to find examples showing the converse to be false.

An amusing illustration of mixing is provided by the famous Arnol'd cat map, the area-preserving, two-dimensional discrete map,

$$x_{n+1} = x_n + y_n (\text{mod } 1) \tag{3.2a}$$

$$y_{n+1} = x_n + 2y_n (\text{mod } 1) \tag{3.2b}$$

Figure 3, from Arnold and Avez (11), shows what a "crazy-mixed-up cat" we have after just three iterations of the mixing system 3.2 (12)!

The relevance of the concept of mixing for statistical mechanics is often illustrated by consideration of the so-called baker's transformation:

$$x_{n+1} = \begin{cases} 2x_n & \text{for } 0 \leqslant x_n < \tfrac{1}{2} \\ 2x_n - 1 & \text{for } \tfrac{1}{2} \leqslant x_n < 1 \end{cases} \tag{3.3a}$$

$$y_{n+1} = \begin{cases} y_n/2 & \text{for } 0 \leqslant x_n < \tfrac{1}{2} \\ (y_n + 1)/2 & \text{for } \tfrac{1}{2} \leqslant x_n < 1 \end{cases} \tag{3.3b}$$

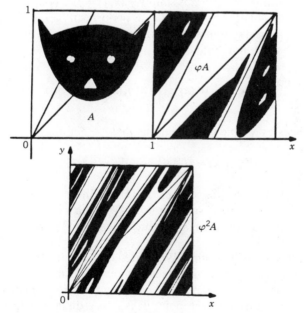

Figure 3. Consequences of mixing behavior for Arnold's cat.

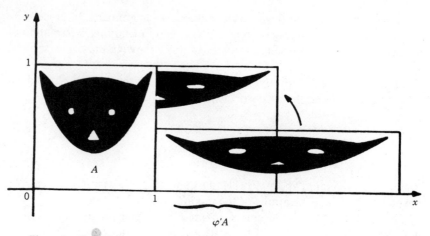

Figure 4. Stretching, cutting, and stacking character of the baker's transformation.

This transformation of the unit square is reminiscent of a baker working dough by stretching, cutting, and stacking (Figure 4). As such, it is no surprise that the mapping is mixing. Now consider some normalized distribution function $f_n(x, y)$ on the unit square defined after n iterations of the map. The "Louiville equation" for $f_n(x, y)$ is simply $f_n(x, y) = f_{n-1}(\frac{1}{2}x, 2y)$ for $0 \leqslant y < \frac{1}{2}$, $f_{n-1}(\frac{1}{2}(x + 1), (2y - 1))$ for $\frac{1}{2} \leqslant y < 1$. Let us perform an average of $f_n(x, y)$ over y, which for a physical system would correspond to averaging over phase space coordinates that are not "interesting":

$$W_n(x) \equiv \int_0^1 dy f_n(x, y) \tag{3.4}$$

Then it follows by simple manipulations that

$$W_{n+1}(x) = \frac{1}{2}\{W_n(\tfrac{1}{2}x) + W_n[\tfrac{1}{2}(x + 1)]\} \tag{3.5}$$

This equation has an obvious solution: $W_n(x) = 1$. In fact, any initial $W_0(x)$ will approach this "equilibrium" solution as $n \to \infty$, which is reasonable because 3.4 corresponds to an averaging procedure that would be expected to smooth out the distribution.

This result is quite interesting. We started out with the baker's transformation, which is an area-preserving (Jacobian of transformation is 1), reversible (x_n, y_n determine x_{n-1}, y_{n-1} uniquely) mapping and allows one and only one trajectory to pass through each point. It is thus a discrete-map version of a Hamiltonian system. And yet the "coarse-grained" average 3.4 of the distribution function satisfies an irreversible rate equation. So what we have is a model for the irreversible approach to equilibrium of a Hamiltonian system (12).

Another interesting thing is that 3.5 looks like the rate equation for a one-dimensional random walk in which from $\frac{1}{2}x$ and $\frac{1}{2}(x + 1)$ we step to x. This suggests that the baker's transformation generates a "random" sequence, and in fact this can be shown rather easily. The situation here is much like that in the case of the Bernoulli shift discussed earlier. Consider again the Bernoulli shift $\theta_{n+1} = 2\theta_n(\text{mod } 1)$, and write the θ's in base 2, so that $\theta_n = d_n d_{n+1} d_{n+2} \cdots$ with each d_i equal to 0 or 1. Call θ_n "heads" if it lies in the left half of the unit interval, in which case $d_n = 0$, and "tails" otherwise, in which case $d_n = 1$. Suppose that in coin tossing we come up with some sequence of heads and tails such as THHTHTTH.... We can produce this same sequence from our Bernoulli shift by simply choosing θ_0 properly, that is, by choosing the d_i properly. Thus, θ_n will be heads if d_n in the base-2 representation of θ_0 is 0. In other words, the Bernoulli shift produces a sequence as "random" as coin tossing. In particular, *any* sequence of heads and tails generated in coin tossing can be generated also by the Bernoulli shift, and we cannot distinguish the results of coin tossing from a Bernoulli shift.

We have noted that mixing implies ergodicity. Next up in the hierarchy of "disorder" are the so-called K systems (K for Kolmogorov). These are characterized by the property we have emphasized for chaos: very sensitive dependence on initial conditions, meaning exponential separation on average of initially close trajectories. The K systems are mixing, as we might expect intuitively because of their extreme sensitivity to initial conditions.

By definition, K systems have a positive Kolmogorov entropy. Suppose we partition phase space at $t = 0$ into a set $\{A_j(0)\}$ of small cells of finite measure. The backward evolution of the system by a unit time step transforms this set to $\{A_j(-1)\}$. The intersection $B(-1)$ of this new set of cells with the first set will usually have a smaller measure than $\{A_j(0)\}$. Backward evolution by another unit time step provides us with a new set $\{A_j(-2)\}$, and we consider the intersection $B(-2)$ of this set with the previous intersection $B(-1)$. Continuing in this manner, we say that the system has positive K entropy if the average measure of each element of $\{B_j(t)\}$ decreases exponentially as $t \to -\infty$. The Kolmogorov entropy may also be defined in such a way that for most systems it is equal to the sum of the positive Lyapunov exponents. Thus, a positive K entropy implies chaos.

From the chain K system $\to$ mixing $\to$ ergodicity we can begin to sense the relevance of chaotic dynamics to fundamental problems of statistical mechanics (e.g.). However, we very often are confronted with "near-integrable" systems (such as the Henon–Heiles system in what follows), which have positive Lyapunov exponents in some regions of phase space but not in others and so cannot be characterized so neatly. In general, we have intermingled regions of chaotic and quasi-periodic behavior.

Near-integrable systems may be regarded as perturbations of integrable systems. Even the simplest nontrivial case, $N = 2$, has remarkably rich behavior, which now seems more or less well understood. The chaotic trajectories are able to "escape" confinement to invariant surfaces (KAM tori) because their initial conditions put them in regions of phase space where there are "overlapping resonances"—resonance in the sense that the frequencies of the unperturbed motion, which for nonlinear systems will generally depend on the energy of oscillation, are commensurate in certain zones of phase space, leading to vanishing frequency denominators in perturbation theory.

IV. SOME EXAMPLES

We will now briefly discuss a few examples of chaotic behavior, gradually making our way to models that are relevant to our understanding of the interaction of light and matter.

The Henon–Heiles model (13) has become one of the paradigms for the study of chaos in Hamiltonian systems. It was considered by Henon and Heiles as a test model for the motion of a star in the gravitational field of the other stars in its galaxy. They assumed a cylindrically symmetric potential, so there are two obvious integrals of motion, the total energy and the z component of angular momentum. Since $N = 3$, the question is whether there is a third integral of motion. The abstract of the Henon–Heiles paper reads: "The problem of the existence of a third isolating integral of motion in an axisymmetric potential is investigated by numerical experiments. It is found that the third integral of motion exists only for a limited range of initial conditions." (We note parenthetically that this "experimental" study was done at about the same time as that of E. N. Lorenz in connection with the Lorenz model. As anticipated early on by people such as Fermi and Ulam, by this time the computer had already become a theorist's "laboratory" for testing ideas.)

Cylindrical symmetry reduces the problem effectively to motion in a plane. Henon and Heiles chose the following specific potential for numerical experiments:

$$V(x, y) = \tfrac{1}{2}(x^2 + y^2) + x^2 y - \tfrac{1}{3}y^3 \qquad (4.1)$$

The particle is trapped in the well for energy $E < \tfrac{1}{6}$. (It is generally believed, based on Sinai's proof for the hard-sphere gas, that a gas with repulsive interparticle forces will be ergodic. In the Henon–Heiles model we have an attractive potential.)

Trajectories of small energy were found to lie on invariant two-dimensional surfaces (tori) in the four-dimensional phase space. The evidence for this is a Poincaré map whose points form closed loops. This suggests a third integral of motion. As E is raised, however, Henon and Heiles found trajectories that appear to give rise to a random sprinkling of points in the Poincaré map, suggesting ergodic behavior. Near the limiting energy $E = \frac{1}{6}$ the system appears to be completely ergodic. In fact, there are positive Lyapunov exponents associated with these ergodic regions of phase space, which Henon and Heiles found to occupy a finite portion of the phase space defined by $E \leqslant \frac{1}{6}$.

We now turn to a second example. In 1949 Fermi proposed that the high energies found in cosmic ray particles might be due to the influence on these particles of magnetic fields in interstellar space. As an analog for such an acceleration mechanism, he considered a ball bouncing between two walls, one fixed and one oscillating. If the phase of oscillation of the latter is taken to be random each time the ball strikes the wall, the ball is accelerated on average, and its energy grows monotonically with time. This is called *Fermi acceleration*.

In 1961 Ulam considered the case where the phase of oscillation has no randomness; that is, he considered (in terms of a discrete mapping) a completely deterministic version of Fermi's example. He found that energy did not increase on average but also that the motion of the ball could be "chaotic." Since then several others have thoroughly investigated the deterministic Fermi problem. Like the Henon–Heiles system, there are orderly and chaotic regions interspersed in phase space.

Monotonic average energy growth is also an interesting aspect of our third example, the kicked rotator described by Casati et al. (14). In this example the equations of motion are

$$\dot{p}_\theta = -(ml^2\omega_0^2)\sin\theta \sum_{-\infty}^{\infty} \delta\left(\frac{t}{T} - n\right) \tag{4.2a}$$

$$\dot{\theta} = p_\theta / ml^2 \tag{4.2b}$$

where l and m are the mass and length, and ω_0 is the natural frequency for small displacements. From 4.2 it follows, upon integration, that

$$p_{n+1} = p_n - K\sin\theta_n \tag{4.3a}$$

$$\theta_{n+1} = \theta_n + p_{n+1} \tag{4.3b}$$

where p_n is proportional to the value of p_θ just before the nth kick, θ_n is the corresponding value of θ, and $K = (\omega_0 T)^2$. The area-preserving map 4.3 is called the *standard map*.

For $K \to 0$ the gravitational potential acting on the pendulum is on continuously, and of course, the system is integrable. For $K \neq 0$ we can summarize the numerical results of Casati et al. as follows. For small K most trajectories lie on invariant curves in the phase space of the near-integrable system. Some of these KAM curves remain as $K \to 1$, but for $K \gg 1$ they are all "broken," and most trajectories are chaotic. In fact, it was found that the average of the square of the angular momentum grows in a random-walk, diffusive manner:

$$\bar{p}^2 \approx \tfrac{1}{2} K^2 t \tag{4.4}$$

where the average is taken over many orbits with $p_0 = 0$ but different θ_0. This result may be understood if, from 4.3, we note that

$$p_n - p_0 = -K \sum_{j=0}^{n-1} \sin \theta_j \tag{4.5}$$

and then calculate the average of $(p_n - p_0)^2$ assuming the θ_j are uniformly distributed random variables. In this way we can attribute the result 4.4 to chaos. In fact, if from 4.5 we regard p_n as a sum of random variables, we might expect, from the central-limit theorem, the Gaussian distribution

$$f(p) = (K\sqrt{\pi t})^{-1} e^{-p^2/K^2 t} \tag{4.6}$$

which in fact was found "experimentally." It is important to note, however, that there are no random forces of any kind in the model. All the "randomness" results from purely deterministic equations of motion.

Our next example is interesting as a model of multiple-photon excitation of molecular vibrations (15). In the field of a relatively intense laser the vibrational excitation of a large number of polyatomic molecules seems to depend mainly on the fluence (time integral of intensity) rather than the intensity (16, 17). A possible explanation has been based on a model consisting of an anharmonic vibrator (representing a pumped vibrational mode) coupled to a "quasi-continuum" of infrared-inactive background modes, the latter being able to exchange vibrational energy with the pumped mode. In this model the vibrational degrees of freedom are treated classically. The laser is assumed to have a constant amplitude, representing a pulse of long duration. For the complex amplitude of the pumped-mode vibration we obtain the nonlinear delay-differential equation (15)

$$\dot{a}(t) = -i\,\Delta a(t) - \frac{\gamma}{2} a(t) + 2i\chi |a(t)|^2 a(t) - i\Omega$$

$$- \gamma \sum_{n=1}^{\infty} a(t - nT)\Theta(t - nT) \tag{4.7}$$

where Δ is the detuning of the laser from the vibrational resonance frequency, χ is the anharmonicity, and Ω is the Rabi frequency, proportional to the electric field strength. Here γ is the golden rule transition rate from the pumped mode to the quasi-continuum of background modes, but it must be emphasized that we do not use the golden rule, which would imply irreversible energy transfer to the quasi-continuum. In particular, the energy density ρ of background modes may not be large enough in a real molecule to justify the assumption of irreversible decay; the possibility of "recurrences" is seen in the appearance of the unit step function $\Theta(t - nT)$, where $T = 2\pi\rho$ is a natural recurrence time that finds its way into the model via the Poisson summation formula (15). Here T is basically just the inverse of the energy spacing of the background modes. Only if $T \rightarrow \infty$ is there effectively irreversible decay into the background modes because then the last term in 4.7 makes no contribution, and the only effect of the background is in the second term.

Using realistic parameter values for a molecule such as SF_6, we find that solutions of 4.7 undergo chaotic time evolution, and moreover, *the energy pumped into the molecule grows approximately linearly with time.* This energy growth, which implies that the absorbed energy in this model is linearly proportional to the laser fluence, may be traced to a decay of correlations typical of chaotic systems, as discussed in the literature (15).

In the model just described we have ignored molecular rotations. Rotational degrees of freedom have been included in a model of a harmonic oscillator in an applied field. (The harmonic vibrator without rotation does not have chaotic behavior, of course, and so we assume here a harmonic vibrator *with* rotations in order to isolate, in a sense, the effects of anharmonicity and rotation.) In this case the equations of motion are (18)

$$\dot{\mathbf{a}} = i\,\Delta\mathbf{a} - i\Omega\mathbf{P} \tag{4.8a}$$

$$\dot{\mathbf{P}} = -2B_0(\mathbf{J} \times \mathbf{P}) \tag{4.8b}$$

$$\dot{\mathbf{J}} = \Omega\dot{\mathbf{P}} \times (\mathbf{a} + \mathbf{a}^*) \tag{4.8c}$$

where $\mathbf{P} = \varepsilon \cdot \mathbf{C}$, ε being the polarization unit vector for the laser field, and $\mathbf{C}$ is the 3×3 orthogonal matrix relating the laboratory and molecular body frames. Here B_0 is the rotational constant characteristic of the molecule. We have assumed a triply degenerate vibrational mode, and again everything is treated classically.

Again we find chaotic time evolution for realistic parameters. However, for large values of the initial vibrational angular momentum $\mathbf{L} = -i(\mathbf{a}^* \times \mathbf{a})$, there is a gyroscopic stabilization such that the evolution of 4.8 is orderly rather than chaotic. These results suggest that it may not be possible in vibrational models to treat molecular rotations simply as a sort of inhomogeneous broadening of the vibrational frequencies.

It is often a good approximation to include only two atomic or molecular states in the description of a resonant interaction of laser light with atoms or molecules. Our next example is based on the Schrödinger equation for a two-state atom (molecule) in an electric field, which takes the well-known Bloch form

$$\dot{x} = -\omega_0 y \tag{4.9a}$$

$$\dot{y} = \omega_0 x + (2d/\hbar)Ez \tag{4.9b}$$

$$\dot{z} = -(2d/\hbar)Ey \tag{4.9c}$$

where ω_0 is the Bohr transition frequency and d is the transition electric dipole moment. If N two-state atoms are contained in a cavity supporting a single field mode, we can write for the electric field E the Maxwell equation

$$\ddot{E} + \omega_0^2 E = 4\pi N\, dx \tag{4.10}$$

The field is assumed to be exactly resonant with the atoms, and the N atoms per unit volume are all lumped together (within a wavelength) so that they all see the same field. (It may be possible to realize such a system experimentally with Rydberg atoms.)

When the parameter $\beta = 8\pi Nd^2/\hbar\omega_0 > 1$, we find that the system 4.9 plus 4.10 shows strongly chaotic behavior. It is interesting that if we make the so-called rotating-wave approximation, which is almost always used in quantum optics, the system is precluded by certain constants of the motion from being chaotic (19).

Let us consider one more classical example. Leopold and Percival (20) reported a classical study of an atom in a monochromatic field, the system with Hamiltonian

$$H = p^2/2m - 1/r + zE_0 \cos \omega t \tag{4.11}$$

in units for which $e = m = 1$. Their calculations were motivated by experiments (21) in which highly excited hydrogen atoms ($n \approx 66$) were further excited and ionized in a microwave field. Leopold and Percival chose initial conditions for the system 4.11 by a Monte Carlo method based on a microcanonical distribution of states, assuming equal populations of degenerate states associated with a fixed principal quantum number n. Some of the classical trajectories were confined to invariant KAM surfaces, others reached high excitation levels with or without subsequent ionization, and finally some trajectories rapidly diffused over the ionization threshold. The agreement of the predictions of this classical model with experiment was surprisingly good for the computations reported. Apparently, the results can be explained in terms of a "stochastic excitation" associated with those (chaotic) trajectories not confined to a KAM surface. Recent experimental and theoretical work on the microwave ionization of hydrogen have shown excellent agreement between classical computations and experiment (22).

V. QUANTUM CHAOS

"Is classical chaos a kind of premonition of quantum uncertainty or something separately?" *Nature*, vol. 300, p. 311 (1982).

It is something separately. If we follow Born's dictum (6) and formulate all classical theory in a statistical way, because things can be indeterminable, we will always be dealing with nonnegative probabilities. The observables of classical physics are "real" things whether we observe them or not; it makes sense to talk about simultaneously real values of x and p, for instance. Quantum mechanics is just different. It is not possible to construct a joint (nonnegative) probability distribution for x's and p's. We cannot build quantum physics on classical premises.

But we can ask a more sophisticated question: If we have a classically chaotic system and treat it quantum mechanically, does the classical chaos manifest itself in any way? Is there some "chaotic" feature of the quantum system? How do we characterize it? At present there is no consensus of opinion on these questions, and we will not try to provide answers here. Instead, we will consider two simple examples of driven quantum systems and suggest that some features we associate with classical chaos are also possible in quantum mechanics even though the state vector seems incapable of evolving in a truly chaotic way (i.e., with very sensitive dependence on initial conditions.)

Consider a quantum system with a purely discrete spectrum of energies. Introduce the perturbation

$$V(x,t) = A(x)F(t) \sum_{-\infty}^{\infty} \delta\left(\frac{t}{T} - n\right) \tag{5.1}$$

Now

$$\sum_{-\infty}^{\infty} \delta\left(\frac{t}{T} - n\right) = \sum_{-\infty}^{\infty} e^{in\omega t}, \qquad \omega = \frac{2\pi}{T} \tag{5.2}$$

and so 5.1 really represents a perturbation consisting of a sum of cosines with frequencies $0, \omega, 2\omega, 3\omega, \ldots$. It may be that one of these frequencies (say, ω_0) is nearly resonant with some allowed transition, whereas the others are way off resonance and therefore of little consequence. Then, if $F(t) = 1$, we have a good approximation to a potential $V = A(x)\cos\omega_0 t$. But the real reason for considering a potential such as 5.1, of course, is that it is much easier to treat numerically than a potential varying continuously in time. In particular, 5.1 leads to a "quantum map" that we can solve by iteration rather than integration.

If we let $|\psi(k)\rangle$ be the state vector just before the kth delta function kick,

$$|\psi(k + 1)\rangle = e^{-iH_0T/\hbar}e^{-iA(x)F(kT)T/\hbar}|\psi(k)\rangle \tag{5.3}$$

where H_0 is the unperturbed Hamiltonian. Equation 5.3 holds because just after the kth kick the state vector is $\exp[-iA(x)F(kT)T/\hbar]|\psi(k)\rangle$, and between kicks we have free evolution governed by $\exp[-iH_0T/\hbar]$. Writing

$$|\psi(k)\rangle = \sum_m a_m(k)|\psi_m\rangle \tag{5.4}$$

where the $|\psi_m\rangle$ are the eigenstates of the unperturbed system $(H_0|\psi_m\rangle = E_m|\psi_m\rangle)$, we obtain from 5.3 the quantum map

$$a_n(k + 1) = \sum_n V_{nm}(k)a_m(k) \tag{5.5a}$$

$$V_{nm}(k) \equiv \langle\psi_n|e^{-iA(x)F(kT)T/\hbar}|\psi_m\rangle e^{-iE_nT/\hbar}. \tag{5.5b}$$

The simplest case of interest is the two-state system. In this case we can take $H_0 = (\hbar\omega_0/2)\sigma_z$, where σ_z is the Pauli spin $\frac{1}{2}$ operator in the conventional notation and ω_0 is the transition frequency $(E_2 - E_1)/\hbar$. For $A(x)$ we can use $-\hbar\Omega\sigma_x$, in which case

$$e^{-iA(x)F(kT)T/\hbar} = e^{i\Omega F(kT)T\sigma_x} = \cos[\Omega(k)T] + i\sigma_x\sin[\Omega(k)T] \tag{5.6}$$

where $\Omega(k) \equiv \Omega F(kT)$. Defining $c_n(k) \equiv a_n(k)e^{ikE_nT/\hbar}$, $n = 1, 2$, we may write the quantum map 5.5 in the form

$$c_1(k + 1) = \cos[\Omega(k)T]c_1(k) + i\sin[\Omega(k)T]e^{-ik\omega_0T}c_2(k) \tag{5.7a}$$

$$c_2(k + 1) = i\sin[\Omega(k)T]e^{ik\omega_0T}c_1(k) + \cos[\Omega(k)T]c_2(k) \tag{5.7b}$$

The system 5.7 is simpler than the Jaynes–Cummings model considered earlier (19) because the field is treated as a *prescribed* quantity whose dynamics are independent of the two-state dynamics. If the applied field is described by a coherent state (as is the field from an ideal laser), it is perfectly permissible to describe it classically (23). In this case the quantum map 5.7 for a kicked two-state system is *exact*, not semiclassical, and we can use it to address questions about quantum chaos.

We introduce the autocorrelation function of the state vector, defined by

$$C(\tau) = \lim_{T \to \infty} \frac{1}{T} \int_0^T dt \langle \psi(t) | \psi(t + \tau) \rangle$$

or

$$|C(k)| = \lim_{N \to \infty} \frac{1}{N} \left| \sum_{n=0}^{N} c_1^*(n) c_1(n + k) + c_2^*(n) c_2(n + k) e^{-ik\omega_0 T} \right| \qquad (5.8)$$

for the two-state system. One of the characteristics of classical chaos is a decay of correlations, and so 5.8 will be a useful tool for the characterization of the quantum map 5.7.

Consider the case of periodic kicking, where $\Omega(k) \to \Omega$ independent of k. In this case we find that $C(\tau)$ is quasi-periodic. When the kicking is quasi-periodic, however, we find that $C(\tau)$ may be a rapidly decaying function of τ. In particular, for $F(t) = \cos \omega' t$ this happens when (a) $x = \omega'/\omega$ is irrational and (b) ΩT is large. Figure 5, for instance, shows $|C(\tau)|$ for $\Omega T = 500$ and $x = 4637/13313$—an "irrational" from the standpoint of machine computation.

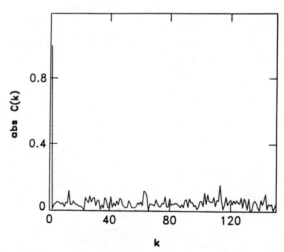

Figure 5. $|C(\tau)|$ for $\Omega T = 500$, $\omega_0 T = 3$, and an irrational frequency ratio x.

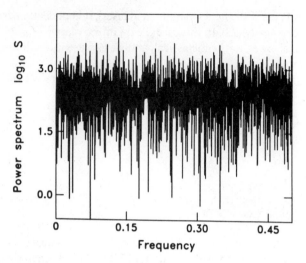

Figure 6.   Power spectrum of the upper state probability amplitude for the case of Figure 5.

We also show in Figure 6 the power spectrum of the "time" series $c_2(k)$.

The decay of correlations (Figure 5) and the broadband power spectrum (Figure 6) for large ΩT and irrational x are reminiscent of classically chaotic behavior. However, the map (5.7) is not chaotic in the sense of a positive Lyapunov exponent (very sensitive dependence on initial conditions). In fact, if we iterate out a long way and then iterate backward in "time," we recover the initial state with which we chose to start. Furthermore, the correlations do not decay to zero; rather, they drop rapidly toward zero but then hover around small values near zero, as in Figure 5. So we cannot conclude from the "decay of correlations" that we have mixing behavior.

These results are consistent with those of Pomeau et al. (24), who integrated the Bloch equations for a two-state system in the case of a bichromatic driving force. For an intuitive understanding of such "chaotic" behavior, consider the factors $\cos[\Omega(k)T] = \cos[\Omega T \cos(2\pi kx)]$ and $\sin[\Omega T \cos(2\pi kx)]$ appearing in the Schrödinger equation 5.7. For large ΩT and irrational x, these functions alone vary erratically with k. We plot $\cos[\Omega T \cos(2\pi kx)]$ in Figure 7, together with its autocorrelation function, for the case of Figures 5 and 6. The correlations are seen to decay rapidly, but this does not occur unless ΩT is large and x is irrational. [Note that the angles $\Theta_k \equiv kx$ satisfy $\Theta_{k+1} = \Theta_k + x$, the circle map with winding number x, and for irrational x the motion on the 2-torus is well known to be ergodic. That is, the map $y_{k+1} = (y_k + x) \bmod 1$ covers the unit interval.] We can attribute the "chaotic" behavior of Figures 5 and 6, therefore, to the fact that the system is being driven by a force that itself varies in an erratic (but not chaotic) way and, in particular, has decaying correlations and a broadband spectrum.

From system 5.7 we can construct an equivalent mapping in terms of the three real Bloch variables and construct Poincaré maps in three orthogonal planes. We infer from these surfaces of section that the motion on the Bloch sphere is ergodic for the case of Figures 5 and 6 (25).

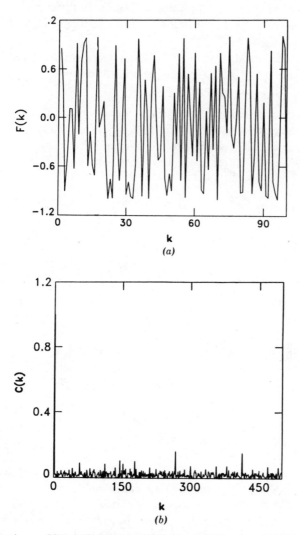

Figure 7. Function $\cos[\Omega T \cos(2\pi kx)]$ (a) and its autocorrelation function (b) for the case of Figure 5.

This is a very simple example, but it illustrates several important points:

1. Ergodicity and chaos are not the same. In this example the dynamics can be ergodic but never chaotic in the rigorous classical sense of a positive Lyapunov exponent.

2. Quantum systems can show features such as broadband power spectra and decaying correlations, which are consequences of chaotic behavior in classical systems, without being chaotic in the classical sense. These features may be the strongest possible manifestations of any sort of "quantum chaos."

3. Quasi-periodically driven quantum systems can display a qualitatively different type of behavior than periodically driven systems. This leads us to our next example.

Earlier we considered the periodically kicked rotator. The kicked rotator is defined by the Hamiltonian 5.1 with $H_0 = p_\Theta^2/2m$ and $A(x) = -(ml^2\omega_0^2)\cos\Theta$. The eigenvalues of H_0 are $E_n = n^2\hbar^2/2ml^2$, and the eigenstates are $\psi_n(\Theta) = (1/2\pi)^{1/2}e^{in\Theta}$. Again taking $F(t) = \cos\omega't$, we obtain, from 5.5, the map

$$c_n(k+1) = \sum_m b_{n-m}[K(k)]e^{-in^2\tau/2}c_m(k) \qquad (5.9)$$

where

$$K(k) = K\cos(2\pi kx) \qquad (5.10a)$$

$$K = ml^2\omega_0^2 T/\hbar \qquad (5.10b)$$

$$b_s(y) = i^s J_s(y) \qquad (5.10c)$$

$$\tau = \hbar T/ml^2 \qquad (5.10d)$$

Here J_s is the Bessel function of order s, and again x is the ratio of the two driving frequencies.

The *classical* kicked pendulum is described by the map

$$P_{k+1} = P_k - \bar{K}(k)\sin\Theta_k \qquad (5.11a)$$

$$\Theta_{k+1} = \Theta_k + P_{k+1} \qquad (5.11b)$$

where $\bar{K}(k) = (K\tau)\cos(2\pi kx)$. In the case of periodic kicking with $F(t) = 1$, $\bar{K}(k) \to \bar{K} \equiv K\tau = (\omega_0 T)^2$, and we have the standard map 4.3.

Let us recall the results for the case of periodic kicking (14). In the classical case the system is chaotic for $\bar{K} > 1$, and the energy growth is diffusive; that is, the energy of the pendulum is on average proportional to the time, as noted earlier. Figure 8, for instance, shows the energy as a function of time (kick

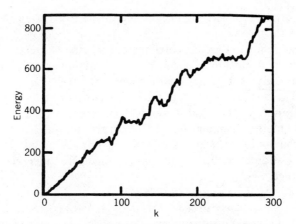

Figure 8. Energy of the classical kicked pendulum for $\bar{K} = 10$.

number) obtained by averaging over a set of 40 initial angles for $\bar{K} = 10$. When the kicked pendulum is treated quantum mechanically, however, this diffusive energy growth persists only up to some "break time," after which the energy growth is greatly suppressed (14). Figure 9 shows the result of computing the following energy expectation value from the quantum map 5.9:

$$\langle E(k) \rangle = \tau K^{-1} \sum_{-\infty}^{\infty} n^2 |c_n(k)|^2 \tag{5.12}$$

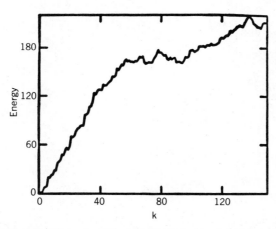

Figure 9. Energy expectation value of the quantum kicked pendulum for case with periodic kicking.

Obviously, there is a very significant suppression of the classical diffusive energy growth.

This suppression of the classical diffusive behavior was related by Grempel et al. (26) to the Anderson localization of a quantum particle in a lattice with random site energies. The random diagonal terms of the tight-binding model correspond in the kicked quantum rotator to a *pseudo-random* sequence, with the lattice points of the tight-binding model corresponding to the integer values n of quantized angular momentum in the rotator. The suppression of diffusive behavior in the kicked rotator is then fully analogous mathematically to the localization in configuration space of an electron in a one-dimensional random lattice.

The localization analogy certainly suggests that the time evolution of driven quantum systems can be more orderly than the corresponding classical evolution. However, it does not necessarily imply that some consequences of classical chaos—such as decaying correlations or diffusive energy growth— are *generally* impossible in quantum mechanics. As in the previous example, let us consider a *quasi*-periodically kicked rotator. Figure 10 shows the energy expectation value for the case $K = 10$, $\tau = 1$, and $x = 4637/13,313$, as in Figures 5 and 6. After 300 iterations there is no evidence of any quantum suppression of energy growth. Shepelyansky (27) has also found that two-frequency kicking can greatly extend the time scale over which diffusive energy growth occurs in the kicked quantum rotator and has suggested that the diffusive time scale increases exponentially with K. In a sense, the addition of

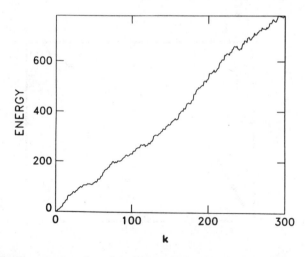

Figure 10.   Energy expectation value of the quantum kicked pendulum for $K = 10$, $\tau = 1$, and x irrational.

more kicking frequencies effectively increases the dimensionality of the system, and since Anderson localization is most effective in one dimension, it might be expected that quasi-periodic kicking acts to weaken the localization effect, as we have found.

This is a very artificial example, of course, but it may provide a useful paradigm for more complex systems. Consider our *classical* model of the multiple-photon excitation of molecular vibrations discussed earlier, wherein the observed fluence dependence is attributed to chaos. At the workshop in Los Alamos, Wyatt (28) presented results of computations indicating that, within the approximations made for the tractability of the quantum computations, the diffusive energy growth found classically is eventually suppressed. This is reminiscent of the situation with the kicked rotator and raises the question of whether the classically predicted diffusion will be found in the case of a *bichromatic* laser field.

There has been a tendency in some of the literature to call ergodic behavior "chaotic" and to study chaos in terms of surfaces of section alone. It should be clear from our first example that chaos and ergodicity are not the same. In particular, a quantum system may cover its available state space uniformly, without being chaotic in the sense of positive Kolmogorov entropy.

There has also been a tendency to define chaos in terms of broadband power spectra. Our first example also shows that the state amplitudes of a system may have broadband power spectra and decaying correlations, but their time evolution is not chaotic in the rigorous sense of the term. In fact, the time evolution of the system is dynamically stable.

If we define "quantum chaos" in terms of features that are *consequences* of classical chaos (e.g., broadband spectra and decaying correlations), quantum chaos is certainly possible. But if we use the classical definition of chaos, that is, positive Kolmogorov entropy, quantum chaos may be impossible. Nevertheless, those aspects (consequences) of classical chaos that are also possible quantum mechanically may lead to experimentally interesting phenomena, such as diffusive energy growth.

Acknowledgment

This work was supported in part by National Science Foundation grants PHY-83-08048 and PHY-84-18070 at the University of Arkansas.

References

1. P. Bocchieri and A. Loinger, *Phys. Rev.* **107**, 337 (1957).

2. T. Hogg and B. A. Huberman, *Phys. Rev. Lett.* **48**, 711 (1982).

3. A. Peres, *Phys. Rev. Lett.* **49**, 1118 (1982).

4. P. W. Milonni, M.-L. Shih, and J. R. Ackerhalt, *Chaos in Laser–Matter Interactions*, World Scientific, Singapore, 1987.

890   P. W. MILONNI, J. R. ACKERHALT, AND M. E. GOGGIN

5. See, e.g., M. V. Berry, *American Institute of Physics Conference Proceedings*, Vol. 46, American Institute of Physics, New York, 1978.

6. See, e.g., M. Born, *Dan. Mat. Fys. Medd.* **30**(7) (1955).

7. *Collected Works* of E. Fermi, paper no. 266, University of Chicago Press, Chicago, IL, 1965.

8. See the commentary of S. Ulam in Ref. 7.

9. A. J. Lichtenberg and M. A. Lieberman, *Regular and Stochastic Motion*, Springer-Verlag, New York, 1983.

10. See, e.g., A. Hobson, *Concepts in Statistical Mechanics*, Gordon and Breach, New York, 1971.

11. V. I. Arnol'd and A. Avez, *Ergodic Problems of Classical Mechanics*, Benjamin, New York, 1968.

12. J. Ford, *Fundamental Problems in Statistical Mechanics*, Vol. 3, North-Holland, Amsterdam, 1975.

13. M. Henon and C. Heiles, *Astron. J.* **69**, 73 (1964).

14. G. Casati, B. V. Chirikov, F. M. Izrailev, and J. Ford, in *Stochastic Behavior in Classical and Quantum Hamiltonian Systems*, G. Casati and J. Ford, eds., Springer Lecture Notes in Physics 93, Springer, New York, 1979.

15. J. R. Ackerhalt and P. W. Milonni, *Phys. Rev. A* **34**, 1211, 5137 (1986).

16. See, e.g., J. L. Lyman, G. P. Quigley, and O. P. Judd, in *Multiple-Photon Excitation and Dissociation of Polyatomic Molecules*, C. D. Cantrell, (ed., Springer-Verlag, New York, 1987).

17. T. B. Simpson, J. G. Black, I. Burak, E. Yablonovich, and N. Bloembergen, *J. Chem. Phys.* **83**, 628 (1985).

18. H. W. Galbraith, J. R. Ackerhalt, and P. W. Milonni, *J. Chem. Phys.* **79**, 5345 (1983).

19. P. W. Milonni, J. R. Ackerhalt, and H. W. Galbraith, *Phys. Rev. Lett.* **50**, 966 (1983).

20. J. G. Leopold and I. C. Percival, *Phys. Rev. Lett.* **41**, 944 (1978).

21. J. E. Bayfield and P. M. Koch, *Phys. Rev. Lett.* **33**, 258 (1974).

22. K. A. H. van Leeuwen, G. v. Oppen, S. Renwick, J. B. Bowlin, P. M. Koch, R. V. Jensen, O. Rath, D. Richards, and J. G. Leopold, *Phys. Rev. Lett.* **55**, 2231 (1985); J. E. Bayfield and L. A. Pinnaduwage, *Phys. Rev. Lett.* **54**, 313 (1985) and references therein.

23. P. W. Milonni, *Phys. Rep.* **25**, 1 (1976).

24. Y. Pomeau, B. Dorizzi, and B. Grammaticos, *Phys. Rev. Lett.* **56**, 681 (1986).

25. P. W. Milonni, J. R. Ackerhalt, and M. E. Goggin, *Phys. Rev. A* **35**, 1714 (1987).

26. D. R. Grempel, R. E. Prange, and S. Fishman, *Phys. Rev. A* **29**, 1639 (1984).

27. D. L. Shepelyansky, *Physica* **8D**, 208 (1983).

28. R. E. Wyatt, presentation at Los Alamos National Laboratory Workshop on Lasers, Molecules, and Methods, July 7–11, 1986.

CHAPTER XXII

CLASSICAL CHAOS VERSUS QUANTUM DYNAMICS: KAM TORI AND CANTORI AS DYNAMICAL BARRIERS

G. RADONS*

Department of Physics and Astronomy, University of Maryland, College Park, Maryland 20742

T. GEISEL**

Institut für Theoretische Physik, Universität Regensburg, 8400 Regensburg, F.R.G.

J. RUBNER

Institut für Theoretische Physik, Technische Universität München, 8046 Garching, F.R.G.

CONTENTS

Note: references in this chapter are enclosed in brackets—Ed.

*Present address: Institut für Theoretische Physik, Universität Kiel, 2300 Kiel, F.R.G.
**Present address: Physikalisches Institut, Der Universitat Wurzburg, Wurzburg, F.R.G.

891

Anharmonic systems such as vibrating molecules often exhibit chaotic motion when they are treated classically. Invariant KAM tori can act as barriers in phase space, confining the probability flow, whereas for a quantum system one expects spreading of wavepackets. We investigate the role of these barriers for the quantum dynamics of a chaotic system. In their vicinity the asymptotic distribution decays exponentially. The penetration depth across a critical KAM torus scales as $\hbar^\sigma$ with a critical exponent $\sigma \simeq 0.66$. The time-dependent transition probability into classically inaccessible regions grows algebraically. The diffusive growth of mean-square displacements can be suppressed by the presence of cantori; they may act as barriers even more severely than in the classical case.

I. INTRODUCTION

In the last decade there has been much progress in the understanding of the classical mechanics of nonintegrable Hamiltonian systems [1, 2]*. In particular, it is now understood that there exist barriers in phase space that can prevent the chaotic flow from exploring energetically allowed regions of phase space. According to the Kolmogorov–Arnol'd–Moser (KAM) theorem [1, 2], these invariant manifolds must exist in generic near-integrable systems. As some parameter is varied, these impenetrable barriers (KAM tori) may break up and give rise to partially penetrable barriers called broken tori or cantori [3]. Although cantori are penetrable, they may provide major obstacles for the flow in phase space and therefore determine many transport properties in Hamiltonian systems [4].

We have asked the question of whether these classical structures in phase space have a counterpart in quantum mechanics. Especially, do dynamical barriers also exist for quantum systems? From the correspondence principle one would expect that such barriers exist, at least in the limit $\hbar \to 0$. On the other hand, quantum mechanics exhibits many features that are in strong contrast to classical mechanics. A simple example is the spreading of wavepackets versus classical confinement by surfaces in phase space. The concept of phase space including surfaces and lines in it does not apply to quantum systems. This difficulty, which of course arises as a consequence of the uncertainty relations, can be circumvented partially by adopting the Wigner representation [5] of quantum states, which provides a quantum analog of classical phase space distributions. There are, however, properties of quantum systems that do not turn into classical behavior ($\hbar = 0$) in the limit $\hbar \to 0$. In particular, the distinction between chaotic behavior [characterized by sensitive dependence on initial conditions or positive Kolmogorov entropy [2] and regular behavior on tori, cantori, or periodic points with zero Lyapunov exponent fails in quantum systems: The quantum analog of the

Kolmogorov entropy is found to be zero for all $\hbar \neq 0$ [6], and sensitivity on initial conditions seems to be absent [7, 8]. As a consequence, one observes suppression of chaotic diffusion in phase space [7–14] and quasi-periodic behavior at long times [15]. Thus, we cannot speak of chaotic flow and its confinement by tori in quantum systems, not even in the limit $\hbar \to 0$. Nevertheless, we will see that the notion of dynamical barriers has also a meaning for quantized, classically chaotic systems in the sense that these barriers determine the accessible states or the available regions of quantum phase space. Our results clearly indicate that these quantum barriers are related to classical KAM tori and cantori. We have investigated the role of quantum barriers in simple model systems such as the periodically kicked rotator, and we will present detailed results for the $\hbar$ dependence and the dependence on the nonlinearity parameter.

It is obvious that dynamical barriers in (quantum) phase space have implications for many fields in physics and chemistry. Applications range from plasma physics [16, 17] to chemistry and physics of molecules and atoms. In investigations of intramolecular energy transfer, it was observed that there exists a "KAM-like" transition in quantized model systems described by, for example, the Henon–Heiles Hamiltonian [18, 19]. These findings have consequences for the possibilities of photoselective chemistry [20], the theory of unimolecular reaction rates [21], and more generally questions of quantum ergodicity [22]. The importance of cantori and their quantum analogs for molecular physics has first been recognized in the context of semiclassical quantization procedures, where they were called "vague" tori [23]. Only recently their role as dynamical barriers for the intramolecular energy transfer was investigated. Classical calculations for OCS and HeI$_2$ molecules [24, 25] demonstrated that cantori act as "intramolecular bottlenecks", and modifications of the usual RKKM theory [21] of unimolecular reactions were suggested to deal with these effects [24–26]. In quantum systems cantori may act as barriers even more drastically, as was observed in model calculations for multiphoton dissociation of HF molecules [27] and for the kicked rotator [28].

Experimental evidence for the role of KAM tori in quantum systems has accumulated in a series of microwave ionization experiments on highly excited hydrogen atoms [29–31]. Although these experiments were successfully analyzed classically [30, 32–34], it is clear that a quantum-mechanical treatment is more appropriate. Considerable progress was made recently [8, 12, 13, 31, 35–37] showing a certain correspondence between classical and quantum stochasticity thresholds (KAM barriers) and also limitations of the classical picture. We expect that the effect of the quantum analog of cantori can also be observed in these experiments.

In Section II we will review some of the properties of cantori and KAM tori in classical Hamiltonian systems. In Section III we introduce the model systems used for our numerical investigations and demonstrate the classical confinement due to tori in these systems. In order to compare classical and quantum-mechanical quantities and to interpret our results, we review some properties of the Wigner function in Section IV. Our results on the quantum manifestation of cantori and KAM tori are presented in Section V and summarized and discussed in Section VI.

II. DYNAMICAL BARRIERS IN CLASSICAL HAMILTONIAN SYSTEMS

In this section we review the important aspects of KAM tori and cantori with respect to their role for the chaotic flow in phase space. For a more comprehensive review of the KAM theorem and classical phase space structures, we recommend an article by Berry [1]. A more technical representation of the subject is given in the book of Lieberman and Lichtenberg [2], and the application of these concepts to intramolecular energy transfer is discussed by, for example, Rice [18] and Brumer [38]. The role of cantori in Hamiltonian systems was investigated only in the last few years. This subject is discussed and reviewed in great detail in Refs. [4] and [39]. We will proceed in discussing classical Hamiltonian dynamics in three steps: First, we deal with the two extreme cases ergodic behavior on the energy surface and completely integrable behavior. The generic case, which is in a sense intermediate because stochastic and regular behavior occurs in the same system, is discussed in a third section. It is also in this section that we review some important aspects of KAM tori and cantori.

A. Ergodic Behavior on the Energy Surface

The dynamics of a Hamiltonian system with N degrees of freedom is governed by the canonical equations. Their solutions describe trajectories in $2N$-dimensional phase space. Let us consider for the moment an autonomous system for which the Hamiltonian contains no explicit time dependence. In the absence of special symmetries, there usually exist only one global invariant, the total energy E. This constraint confines the motion to a $(2N - 1)$-dimensional manifold, the energy surface described by the equation $H(q_1,\ldots,q_N, p_1,\ldots,p_N) = E$. The simplest case in which stochastic behavior can occur is $N = 2$. Here the energy surface is three-dimensional, but even this hypersurface is not easy to visualize because it is not an Euclidean manifold like ordinary position space. For instance, in the case of a two-dimensional harmonic oscillator and other potentials with only one minimum the energy surface is topologically a three sphere S^3. A system is ergodic on the energy surface if almost every trajectory comes arbitrarily close to every point on the energy surface in the course of time. A convenient way to depict such a behavior for $N = 2$ is the Poincaré section, an intersection of the energy surface with some plane such that trajectories repeatedly cross this plane. For example, if one chooses the plane $q_2 = C$ in the space with axes q_1, p_1, and q_2 and marks this plane every time it is crossed with $p_2 > 0$, one gets a sequence of points $X(n) = (q_1(n), p_1(n))$ that form a unique trace of a trajectory. The map describing the transformation from $X(n)$ to $X(n + 1)$ is the Poincaré map, which cannot be given analytically in most cases but can be shown to be area preserving [1, 2]. If a system is ergodic on the energy surface, the intersection of one typical trajectory fills an *area* in the Poincaré surface of section in the course of time. The boundary of this area is determined only by the constraints of energy conservation.

For systems with more than two degrees of freedom the method described in the preceding can also be applied; but, for example, for $N = 3$ the Poincaré surface of section is four-dimensional and therefore is less useful than for $N = 2$. Later we will deal with Hamiltonians with an explicit time dependence. These systems are equivalent to the cases discussed before since they can be mapped onto autonomous systems with an additional degree of freedom: If the Hamiltonian has the form $H = H(q_1, \ldots, q_{N-1}, p_1, \ldots, p_{N-1}, t)$, one can regard t and $-H$ as an additional pair of canonically conjugate variables q_N and p_N of a new Hamiltonian $\bar{H} = H(q_1, \ldots, q_{N-1}, p_1, \ldots, p_{N-1}, t) - H$ with no explicit time dependence (Ref. 2, p. 14). Note that this system is unbounded in the coordinate $q_N = t$, and therefore, many concepts and theorems do not apply. In the case of a periodic time dependence, however, one can identify states after one period and restrict the dynamics in the coordinate q_N to a circle (q_N is regarded as an angle). A convenient choice for a Poincaré section in these systems is then to fix the angle q_N. For time-dependent systems with one degree of freedom (which is equivalent to the case $N = 2$ discussed before) described by $H = H(q, p, t)$, the resulting Poincaré section thus contains points $X(n) = (q_n, p_n)$, where q_n and p_n are coordinates and momentum at integer multiples of the period $T: X(n) = X(t = n \cdot T)$. Ergodicity then again appears as an area-filling sequence of points $X(n)$.

Simple systems that show the described ergodic behavior and where ergodicity was proved include certain billiards (Sinai billiard [40] and Bunimovich stadium [41]) and some time-dependent systems with driving forces discontinuous in some derivative. These systems are not only ergodic but exhibit even stronger degrees of stochasticity such as mixing and sensitive dependence on initial conditions. Mixing results e.g., in decay of correlations and sensitive dependence on initial conditions which means that nearby trajectories diverge in the mean as $\exp(\lambda t)$ with $\lambda > 0$, where λ is the largest Lyapunov exponent [2] characterizing the chaotic flow in phase space. This kind of chaotic behavior is also observed in the stochastic components of generic nonintegrable cases to be discussed in Section II.C. There exists a hierarchy of different degrees of stochasticity: It can be shown that a positive Lyapunov exponent or equivalently positive Kolmogorov–Sinai entropy implies mixing, and mixing implies ergodicity, but the inverse statements are not true. More details on this hierarchy can be found in Ref. 2 or in books on ergodic theory [42, 43]. At this point we remark that ergodicity is assumed in most statistical theories, as in the standard RKKM theory of reaction rates [21], and it is usually expressed in the ergodic hypothesis of statistical mechanics stating that time averages can be replaced by averages over the energy surface.

B.　Completely Integrable Systems

Completely integrable Hamiltonians are characterized by the fact that there exist N independent global invariants for an autonomous system with N degrees of freedom. This means that there exist N functions C_i of the coordinates and momenta that are constant along each trajectory. In the cases of Section II.A we had only one constant, energy. In addition, the C_i must be in involution (i.e., the Poisson brackets $\{C_i, C_j\}$ must vanish) ensuring that these constants of motion can serve as new momenta. The existence of these constants restricts the motion to N-dimensional manifolds in $2N$-dimensional phase space. If the C_i are smooth enough functions and if the motion is bounded, it can be shown [1, 2, 44] that the manifolds are N-tori (i.e., the direct product of N circles). In this case there exists a canonical transformation to action and angle variables $I_1, \ldots, I_N, \theta_1, \ldots, \theta_N$. In these coordinates the Hamiltonian depends only on the N actions, $H = H(I_1, \ldots, I_N)$, and the θ_i are cyclic variables. The Hamilton equations are especially simple in these coordinates,

$$\dot{I}_i = 0 \qquad \dot{\theta}_i = \frac{\partial H}{\partial I_i} \equiv \omega_i(I_1, \ldots, I_N) \tag{1}$$

with the solutions

$$\theta_i(t) = \omega_i \cdot t + \theta_i(0) (\mathrm{mod}\, 2\pi)$$

$$I_i(t) = I_i(0) \tag{2}$$

where the ω_i are the frequencies. From (2) we see again that all trajectories starting on an N-torus labeled by the N numbers I_i remain on this manifold (invariant torus).

The case $N = 2$ is again very illustrative because the motion is confined to a 2-torus, and we have only two frequencies ω_1 and ω_2. If, for example, we make a Poincaré section as discussed in II.A at the angle $\theta_2 = C$ and record θ_1 and I_1

at every intersection of an orbit with the plane $\theta_2 = C$, we get again a sequence of points $X(n) = (\theta_{1n}, I_{1n})$. In this simple case the Poincaré map $T: X(n) \to X(n+1)$ can be obtained explicitly. Dropping the index 1, we have

$$I_{n+1} = I_n$$
$$\theta_{n+1} = \theta_n + 2\pi\alpha(I_n)(\mathrm{mod}\, 2\pi) \tag{3}$$

where $\alpha = \omega_1/\omega_2$ is the rotation or winding number, which in general depends on the actions I through the frequencies ω_i (or on energy and one action). Depending on whether α is rational or irrational, we get different behavior according to (3): If the winding number α is rational, say, $\alpha = r/s$, the frequencies ω_1 and ω_2 are in resonance, and we obtain s points in the Poincaré section that are visited periodically. These points correspond to one closed periodic trajectory on the 2-torus. If we change the initial angle θ_0 and do not change the initial action, the winding number is the same, and we obtain another periodic orbit with the same period laying on the same torus. We see that such a "rational torus" consists of infinitely many periodic orbits resulting in a "rational circle" in the Poincaré section, where every point is a periodic point. In a measure-theoretic sense, however, almost every α is irrational. In this case one orbit fills a whole circle in the Poincaré section and correspondingly covers one 2-torus densely. The motion is then quasi-periodic and ergodic on the 2-torus. The Lyapunov exponent is zero for these trajectories because they do not separate exponentially. Hence, this behavior is called regular. It is characteristic for regular behavior in two degrees of freedom that a typical orbit fills a *line* in the Poincaré surface of section, in contrast to chaotic orbits, which appear to fill an *area* in the course of time. In the typical case we have both regular and chaotic behavior.

C. Typical Hamiltonian Systems

Typical systems can be obtained by perturbing an integrable system. In such a case a Poincaré section of a two-degree-of-freedom system looks as depicted in Figure 1, where several orbits are shown. Some orbits give rise to lines in the Poincaré section (e.g., at large p in Figure 1), whereas others appear to fill an area (e.g., the points around $p = 0$ in Figure 1). The existence of regular orbits in slightly perturbed integrable systems is guaranteed by the KAM theorem [1, 2]. It asserts that a finite measure of tori characterized by winding numbers that are "irrational enough" persist under the perturbation, and they are deformed but not destroyed. For example, in the case $N = 2$ the condition of irrationality can be expressed as the inequality

$$\left| \frac{\omega_1}{\omega_2} - \frac{r}{s} \right| > \frac{K(\varepsilon)}{s^{2.5}} \tag{4}$$

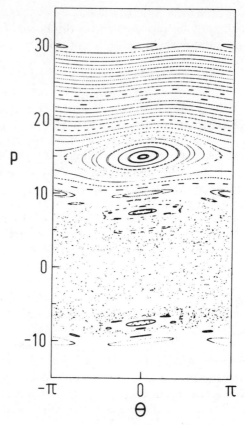

Figure 1. A typical Poincaré section of a system with two degrees of freedom. 80 trajectories consisting of 500 points were generated from equidistant initial values on the $\theta = 0$ axis according to Eq. 7 with $V = \cos \theta$ and $K = 2\pi \ln |p|$. Stochastic orbits around $p = 0$ cannot penetrate into the regular region at large p.

to be fulfilled for all integers r, s. The frequencies ω_i are defined in Eq. 1, and $K(\varepsilon)$ is some constant depending on the strength of the perturbation ε: $H = H_0 + \varepsilon H_1$, where H_0 is a Hamiltonian depending on actions only. The perturbation H_1 must also fulfill certain smoothness conditions, and H_0 must be sufficiently nonlinear [1, 2]. In the Poincaré section the deformation of the 2-tori appear as a deformation of the irrational circles. The rational circles described in Section II.B are destroyed under the perturbation. The fate of the infinitely many periodic points on these circles is governed by the Poincaré–Birkhoff theorem [1, 2]. This theorem states that under small enough perturbations, $2k \cdot s$ $(k = 1, 2, \ldots)$ periodic points of a rational circle with $\alpha = r/s$ survive, half of which are elliptic and half of which hyperbolic.

The elliptic (stable) points are surrounded by new tori (islands of stability) and the hyperbolic (unstable) points give rise to chaotic behavior if their stable and unstable manifolds intersect transversally. The rational circles and their neighborhood thus turn into stochastic layers with islands in it. Some of these islands can be seen in Figure 1 (e.g., a period 1 island at $p = 15$ and two islands visited alternately around $p = 10$). The boundary of the stochastic region consists of undestroyed KAM tori and of new tori bounding the islands. Some of the new tori, however, are destroyed already by the same mechanism as the preceding. This means that inside the islands there are again stochastic layers, and also at the border of the islands one finds new islands chains, and so on. The whole picture repeats itself on a finer and finer scale ad infinitum. This is nicely illustrated for a model system in Ref. 45.

The important aspect of this picture with respect to flow in phase space is that the stochastic regions have boundaries consisting of invariant tori. Due to the uniqueness of the solution of the Hamilton equations, these tori cannot be penetrated by stochastic orbits and act therefore as barriers (most drastically for $N = 2$ but also for more degrees of freedom).

The picture is not yet complete, however. The KAM theorem ensures the existence of tori only for small enough perturbations ε. What happens then to these orbits if the perturbation is enhanced? They break up into invariant sets that have been called cantori, or broken tori. At the moment these are well understood only in the case of two degrees of freedom [3, 4, 39], but they also exist in higher dimensional systems. Their nature is best explained again in the Poincaré section of a system with $N = 2$. Cantori are quasi-periodic orbits characterized by irrational winding numbers similar to tori, but in constrast to tori, they do not fill a line in the Poincaré section but only a fractal object, a Cantor set. This Cantor set can be visualized as a line with infinitely many gaps in it. The gaps lie dense on this line, and the remaining set is of measure zero with respect to the full line. A cantorus is therefore a very "thin" or pointlike object whose fractal dimension even appears to be zero [46], whereas a torus has dimension 1 in the Poincaré section. In spite of their "thin" nature, they can provide major obstacles to the chaotic flow in phase space. The transport across a given cantorus is characterized by an area ΔW in the Poincaré section, which approaches zero as a power of $\Delta\varepsilon = \varepsilon - \varepsilon_c$, where ε_c denotes the critical nonlinearity at which a KAM torus breaks up into the considered cantorus [4, 39]. If the cantorus exists, for example, for $\varepsilon > \varepsilon_c$, then for $\varepsilon < \varepsilon_c$, we have a torus with the same winding number, and ΔW is zero in this case, expressing the fact that tori or invariant circles provide absolute barriers (for $N = 2$). Cantori with $\Delta W \neq 0$ provide partial barriers. From the preceding we see that to every irrational winding number we can attribute at least one quasi-periodic orbit that appears either as a KAM torus or as a cantorus. The most important barriers are, of course, tori and those cantori having the smallest ΔW within a stochastic region. Cantori, at least those with small ΔW, can be regarded as dynamical barriers because they set a new time scale for the chaotic motion. That means that, for example, an ensemble of trajectories started in a chaotic domain relaxes very fast into a region in phase space whose boundaries are provided by the most important cantori and that the transport across these boundaries is slow compared to the fast relaxation within the region [4, 47]. This behavior was demonstrated by MacKay et al. in Ref. 4 (see Figure 2 in that reference). In the same article it is discussed that similar fluxes ΔW can be defined for periodic orbits and minimizing heteroclinic orbits and that the barriers provided by them are much weaker compared to the barriers provided by cantori. Note, however, that the case of flux across minimizing heteroclinic orbits or across broken separatrices can be important for intermolecular energy transfer [25, 26], whereas cantori provide "bottlenecks" for intramolecular energy transfer [24, 27].

We conclude this section about classical dynamics and classical barriers with some remarks on systems with more than two degrees of freedom. For $N = 2$ the effect of tori is most drastic because they divide the energy surface into disjunct regions; stochastic layers are disconnected. This is not true for $N > 2$, where tori similar to lines in three-dimensional space do not divide the phase space. The stochastic layers are connected and constitute the Arnol'd web, a dense web on the energy surface. Therefore, there are in principle only two possibilities: quasi-periodic motion on tori (or cantori) or diffusion through the Arnol'd web where one can reach every point on the energy surface (although the remaining tori are of finite measure). The diffusion process (Arnol'd diffusion), however, can take place on very different time scales: If Arnol'd diffusion is associated with overlapping resonances (as in systems with two degrees of freedom and for modulational diffusion), one obtains a much higher diffusion rate than in the case of nonoverlapping resonances, where the diffusion process can be extremely slow (exponentially small in the perturbation parameter) and is often difficult to observe numerically. For a discussion of these different phenomena, we refer to the book by Lichtenberg and Lieberman [2] and Refs. 48–50. The role of cantori in these systems is not clear at the moment.

III. MODEL HAMILTONIANS: PERIODICALLY KICKED SYSTEMS

In order to investigate the role of KAM tori and cantori in quantum systems, we study the simplest systems that classically exhibit the typical features of nonintegrable systems discussed in Section II.C. These are time-independent systems with two degrees of freedom or equivalently nonautonomous one-degree-of-freedom systems. Among the latter systems are some for which the Poincaré map (introduced in II.A) can be calculated explicitly. The Hamiltonian of these systems is given by

$$H(\theta, p, t) = K(p) + k \cdot V(\theta) \cdot \tilde{\delta}_T \tag{5}$$

where θ is an angle in the interval $(0, 2\pi)$ and p the canonically conjugate momentum. Here $\tilde{\delta}_T(t)$ is the T-periodic δ function

$$\tilde{\delta}_T(t) = \sum_{n=-\infty}^{+\infty} \delta(t - nT) \tag{6}$$

and k is the relevant nonlinearity parameter. For these systems the Poincaré map is easily calculated as

$$p_{n+1} = p_n - kV'(\theta_{n+1})$$
$$\theta_{n+1} = \theta_n + T \cdot K'(p_n) \pmod{2\pi} \tag{7}$$

where θ_n and p_n denote the value of the variable at times $t = nT + 0$, that is, immediately after the nth kicks. For $k = 0$ we obtain the simple twist map of Eq. 3, and the system in integrable. For $k \neq 0$ the typical phase space structures are observed. Figure 1 was obtained by numerically iterating Eq. 7 with $V = \cos\theta$, $K = \pi M \ln p^2$, and parameters $T = 1$, $k = 1$, and $M = 15$. With these functions $V(\theta)$ and $K(p)$ the mapping (7) is the simplified Fermi–Ulam map [2], which served as the model for Fermi acceleration. Not only is the classical behavior of systems with the Hamiltonian (5) governed by simple maps, but also their quantum dynamics is described by so-called quantum maps [7–12, 14–17, 51–53]. These quantum maps connect the state at time $t = nT + t_0$ with the state at $t = (n+1)T + t_0$ by a simple unitary transformation

$$|\psi_{n+1}\rangle = \hat{U}|\psi_n\rangle \tag{8}$$

If one chooses the discrete time steps again immediately after the kicks, the time evolution operator for the systems of Eq. 5 is given by

$$\hat{U} = e^{-(ik/\hbar)V(\hat{\theta})} e^{-(iT/\hbar)K(\hat{p})} \tag{9}$$

where $\hat{\theta}$ and $\hat{p}$ obey the commutation rule $[\hat{\theta}, \hat{p}] = i\hbar$.

Although the Fermi–Ulam model shows the more general phase space structure (Figure 1), there is a special case of Eq. 5 that is more suited for our purpose of investigating the role of tori and cantori in quantum systems. For $V(\theta) = \cos\theta$ and $K(p) = p^2/2J$, Hamiltonian (5) describes a periodically kicked planar rotator with the moment of inertia J. The Poincaré map (7) of this system is known as the Chirikov–Taylor, or standard, map [2, 48], which was thoroughly investigated classically [3, 4, 45, 46–48, 54–60]. With $T = J = 1$, it reads

$$p_{n+1} = p_n + k \sin \theta_{n+1}$$
$$\theta_{n+1} = \theta_n + p_n (\text{mod } 2\pi) \tag{10}$$

It also describes, for example, the local behavior of the Fermi–Ulam map [2] and of many other systems [48]. The attracting feature of this map for our goal is that there exists a critical perturbation strength $k_c = 0.9716\ldots$, where the phase space (which is periodic in p) is divided into cells by *isolated* KAM trajectories. These critical KAM surfaces are the last bounding trajectories for this map as the parameter is varied from $k = 0$ to larger values. This should be understood as follows: For $k = 0$ the phase space consists of irrational and rational tori (see II.B) that would appear as horizontal straight lines in the representation of Figure 1 or 2. As k is enhanced, there appear stochastic layers and island chains (see II.C) where rational lines were located destroying

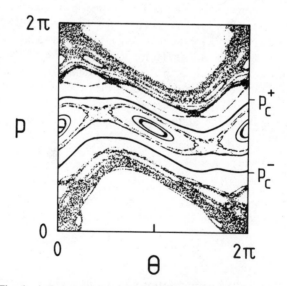

Figure 2. The classical phase space of the kicked rotator for $k = k_c = 0.9716$ showing 5000 iterates of 13 initial points according to Eq. 10. Orbits starting between the two KAM tori remain confined to the momentum interval (p_c^-, p_c^+) as demonstrated in Figure 3.

more and more invariant tori. For $k < k_c$ there still exist many of these layers separated by KAM tori, and the chaotic motion is confined. At $k = k_c$ nearly all layers have merged; we have only two kinds of layers left (apart from secondary layers inside the island chains) separated by the last and most robust KAM trajectories. This situation is depicted in Figure 2. The two kinds of layers or cells are located around integer resonances at $p = 2n\pi$ and half-integer resonances with $p = (2n + 1)\pi$, n integer. Trajectories started in chaotic regions around $p = \pi$ remain in the corresponding cell due to the last-bounding KAM tori. These last KAM trajectories have winding numbers simply related to the golden mean $\gamma = (1 + \sqrt{5})/2$ and show universal scaling properties [54, 57, 58]. The golden mean appears here because γ is the "most irrational" winding number in the sense that it is approximated worst by rationals [61]. As k is enhanced above k_c, these last tori turn into cantori, the stochastic motion is no longer bounded in momentum, and the last stochastic layers have merged into one infinitely large "layer." This large-scale stochastic motion or connected stochasticity [54] is characterized by diffusion in momentum [4, 47, 48],

$$\langle \Delta p_t^2 \rangle = \langle (p - \langle p \rangle)^2 \rangle \sim D \cdot t \tag{11}$$

and is controlled by the remnants of the last tori, at least for parameters near k_c, where $D \sim (k - k_c)^\eta$, $\eta \approx 3$, in accordance [47] with the scaling of the area ΔW transported across these contori.

The confinement of the stochastic motion at $k = k_c$, in contrast to the diffusion for $k > k_c$, can be demonstrated in the following way: Take a normalized initial distribution $\rho_0(\theta, p)$ in the classical phase space of the rotator, for example, $\rho_0(\theta, p) = (1/2\pi)\delta(p - p_0)$ with $p_0 = 3.2$, and determine the asymptotic distribution

$$\rho(\theta, p) = \lim_{N \to \infty} \frac{1}{N} \sum_{t=0}^{N-1} \rho_t(\theta, p) \tag{12}$$

where $\rho_t(\theta, p)$ are the distributions at discrete times t developing according to Eq. 10. If one projects the asymptotic distribution $\rho(\theta, p)$ onto the momentum axis, one obtains the distribution

$$\rho(p|p_0) = \int_0^{2\pi} \rho(\theta, p) \, d\theta \tag{13}$$

For $k = k_c$ this distribution is strictly zero outside the interval (p_c^-, p_c^+), and it is totally confined to this interval. Here $p_c^+ = 4.269$ is the maximum p value of the "golden" torus with winding number $\alpha = v_g^+ = 1/\gamma$, and p_c^- is the minimum of

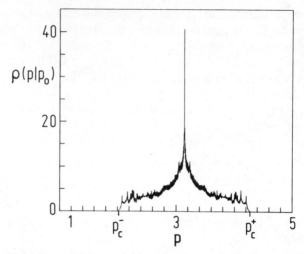

Figure 3.   The projection $\rho(p\,|\,p_0)$ of the asymptotic distribution $\rho(\theta, p)$, Eq. 12, obtained from 70,000 iterates of 1000 initial points equally distributed on line $p = p_0 = 3.2$ for $k = k_c$. Points outside the interval (p_c^-, p_c^+) cannot be reached under these initial conditions due to the phase space structure of Figure 2.

the golden torus with $\alpha = v_g^- = 1 - v_g^+ = 1/\gamma^2$, where α is defined as

$$\alpha = \lim_{n \to \infty} (\theta_n - \theta_0)/2\pi n \tag{14}$$

The projected distribution $\rho(p\,|\,p_0)$ is shown in Figure 3, which was obtained by iterating 1000 points 70,000 times each. The initial values were equally distributed on the line $p = p_0 = 3.2$ and therefore all lay between the above-mentioned tori.

For k infinitesimal larger then k_c and with the same initial conditions, about half of the distribution would flow to $p = \pm \infty$. The rest of the distribution would remain confined because this part of the distribution was placed on stable islands initially. This means that one observes a discontinuous change in the distribution $\rho(p\,|\,p_0)$ as k is varied through k_c. Similar discontinuities occur for values $k < k_c = k_c(v_g)$ because there are other critical values $k_c(\gamma_i)$, where isolated tori (with "noble" winding numbers γ_i also related to the golden mean) break up, but for $k = k_c(v_g)$ the effect is most pronounced.

In Section V we will present results for a quantity analogous to $\rho(p\,|\,p_0)$ for the quantized system as k is varied through k_c. We have seen that the properties of nonintegrable classical systems are mainly properties in phase space. To relate properties of quantum systems with corresponding classical quantities, it is therefore useful to discuss some properties of phase space representations of quantum mechanics, especially of Wigner's distribution function.

IV. CLASSICAL-QUANTUM CORRESPONDENCE VIA WIGNER'S FUNCTION

The Wigner function [5] is a quantum analog of classical phase space distributions. It was therefore of great use in investigation of quantum systems that are classically nonintegrable [19, 51, 62–70]. Here we present only its important properties, providing a better understanding of the relation between our classical and quantum-mechanical calculations.

The Wigner function for one degree of freedom can be defined as

$$W(q,p) = \frac{1}{\pi\hbar} \int \psi(q-x)\,\psi^*(q+x)\,e^{2ixp/\hbar}\,dx \tag{15}$$

where the integration is over the range $(-\infty, +\infty)$ or $(-\pi/2, +\pi/2)$ depending on whether q is a position or an angle coordinate [62, 67]. Its interpretation as an analog of a classical phase space distribution is based on the following facts:

(i) Its projections on the coordinate or momentum axis yield the correct marginal distributions, that is,

$$\int W(q,p)\,dq = |\langle p|\psi\rangle|^2$$

$$\int W(q,p)\,dp = |\langle q|\psi\rangle|^2 \tag{16}$$

where $\int W(q,p)\,dp$ is replaced by $\hbar\sum_n W(q,p_n)$ if q is an angle [62, 67].

(ii) Expectation values of quantum-mechanical operators can be calculated like classical expectation values with the classical distribution function replaced by the Wigner function.

(iii) In the semiclassical limit $\hbar \to 0$, the time evolution of $W(q,p)$ is governed by the classical Liouville operator [5]. In the case of the quantum maps Eqs. 8 and 9, this means that the propagator $K(q,p;q',p')$ of the Wigner function in discrete time,

$$W_{t+1}(q,p) = \int\int K(q,p;q',p')W_t(q',p')\,dq'\,dp' \tag{17}$$

with $W_t(q, p)$ corresponding to $|\psi_n\rangle$ of Eq. 8 via Eq. 15, approaches its classical form for $\hbar \to 0$ [51, 67–69],

$$K(q, p; q', p') \to \delta(q' + T \cdot K'(p') - q) \cdot \delta(p' - kV'(q) - p) \qquad (18)$$

We see that in the semiclassical limit the Wigner function is iterated like a classical distribution obeying Eq. 7 (the first δ function in Eq. 18 is replaced by a 2π-periodic δ function if q is an angle coordinate [67]). It is misleading, however, to conclude that the classical dynamics ($\hbar = 0$) and the quantum dynamics in the limit $\hbar \to 0$ are the same in every respect. The limiting form of $K(q, p; q', p')$ is approached via Airy functions [51, 68, 69] with infinitely rapid oscillations (for $\hbar \to 0$) and therefore in principle contains quantum effects such as tunneling and interference.

Based on Eqs. 17 and 18, it was concluded [51] that for $\hbar \to 0$, Wigner functions representing pure quantum eigenstates must be eigenfunctions of the classical integral kernel (18) (the reverse is not true [51, 70]) and therefore lie on invariant manifolds of the map (7). Before we discuss implications of this relation, we discuss examples of Wigner functions that correspond to simple eigenfunctions demonstrating some of their general features. The Wigner distributions associated with eigenfunctions of position and momentum operators with eigenvalues q_0 and p_0, respectively, are obtained as

$$\langle q | p_0 \rangle = (2\pi\hbar)^{-1/2} e^{ip_0 q/\hbar} \xrightarrow{(15)} W(q, p) = \frac{1}{2\pi\hbar} \delta(p_0 - p) \qquad (19)$$

$$\langle q | q_0 \rangle = \delta(q_0 - q) \longrightarrow W(q, p) = \frac{1}{2\pi\hbar} \delta(q_0 - q) \qquad (20)$$

showing the symmetry of the Wigner representation in the variables q and p. The Wigner distributions (19) and (20) are, however, exceptional since for wavefunctions normalized to unity the corresponding Wigner function remains finite for $\hbar \neq 0$, and it usually also becomes negative (but is always real). Because of this last property, which is a consequence of the uncertainty relations, it is often termed a quasi-probability distribution. For simple one-dimensional potentials (linear or harmonic potentials) the Wigner functions corresponding to the energy eigenstates can be calculated exactly [5b], demonstrating that in one-degree-of-freedom systems the Wigner functions are in general peaked on lines $H(q, p) = E$ in phase space (where E is the eigenvalue of the considered eigenfunction), the peak width is proportional to $\hbar^{2/3}$, and the Wigner function decays exponentially on the convex side of the phase space trajectory and shows oscillations with positive and negative values on the other side. In the limit $\hbar \to 0$ it becomes a δ function on the line $H(q, p) = E$ through infinitely rapid oscillations on the concave side of the trajectory. This last statement means that $W(q, p)$ becomes a δ function on an invariant torus in two-dimensional phase space. The generalization to higher dimensions is expressed in Berry's eigenfunction hypothesis [63]. "Each semiclassical eigenstate has a Wigner function concentrated on the region explored by a typical orbit over infinite times." This gives rise to a distinction between regular and irregular eigenstates, at least for $\hbar \to 0$: The Wigner function of regular eigenstates is supported by the invariant tori of integrable (Section II.B) or near-integrable (II.C) systems, and the Wigner function of irregular eigenstates fills the whole energy surface of ergodic systems (II.A) or in the case of near-integrable systems a stochastic area. From this, one could expect that tori in some sense can act as barriers also in quantum systems, but cantori that are neither tori nor chaotic trajectories do not fit into this picture.

V. QUANTITIES OF INTEREST AND RESULTS

From the foregoing discussion of the Wigner function and its relation to classical phase space, it is clear which calculations are needed for the quantized kicked rotator to compare with the classical calculation described at the end of Section III and resulting in Figure 3. We simply have to replace the classical distributions $\rho_t(\theta, p)$ in Eqs. 12 and 13 by Wigner distributions $W_t(\theta, p)$. As initial condition, we have to choose $W_0(\theta, p = n\hbar) = (1/2\pi\hbar)\delta_{n,m}$ with $m = p_0/\hbar$ corresponding to the classical initial distribution $\rho_0(\theta, p) = (1/2\pi)\delta(p - p_0)$. This initial Wigner function is the discrete p analog of Eq. 19; that is, it corresponds to an eigenstate $|p_0\rangle$ of the angular momentum operator $\hat{p} = -i\hbar(\partial/\partial\theta)$ with eigenvalue $p_0 = m\hbar$. To obtain the analog of $\rho(p|p_0)$ of Eq. 13, however, we do not need to iterate this Wigner function according to Eq. 17 since we are only interested in the projection of the time-averaged Wigner function onto the momentum axis. Using the property (16a) of the Wigner function, this projection is

$$P(p|p_0) = \lim_{N \to \infty} \frac{1}{N} \sum_{t=0}^{N-1} |\langle p|\psi_t\rangle|^2 \qquad (21)$$

Using $|\psi_0\rangle = |p_0\rangle$ and relation (8) one obtains

$$P(p|p_0) = \lim_{N \to \infty} \frac{1}{N} \sum_{t=0}^{N-1} |\langle p|\hat{U}^t|p_0\rangle|^2 \qquad (22)$$

This shows that $P(p|p_0)$ is simply the time-averaged probability of finding the system in the state $|p\rangle$ if it is initially prepared in the state $|p_0\rangle$. Analogous distributions have been discussed elsewhere [18, 19, 22, 64, 71, 72] for the Henon–Heiles potential and in theories of quantum ergodicity. The quantities we will discuss in the following are either related to the time-average Eq. 22 or to the unaveraged quantity

$$P_t(p|p_0) = |\langle p|\hat{U}^t|p_0\rangle|^2 \tag{23}$$

and are therefore determined by the infinite matrix $\langle p'|\hat{U}|p\rangle$, which can be expressed analytically for the kicked rotator $[V(\theta) = \cos\theta, K(p) = p^2/2J$ in Eq. 9] in terms of Bessel functions

$$\langle m'\hbar|\hat{U}|m\hbar\rangle = (-i)^{m'-m} J_{m'-m}\left(\frac{k}{\hbar}\right)\cdot e^{-i(\hbar T/2J)m^2} \tag{24a}$$

We see that the quantum dynamics depends on two dimensionless parameters $k/\hbar$ and $\hbar T/J$. As for the classical map we set $T = J = 1$ and vary k and $\hbar$ in our numerical experiments, where $\hbar$ is now an effective Planck constant; that is, $\hbar$ is measured in units of J/T. Our actual numerical iterations were not performed by matrix multiplication with matrix (24a) but by use of forward and backward fast Fourier transforms [11], which utilizes the fact that one iteration of the wavefunction can be written as

$$\langle m'|\psi_{t+1}\rangle = \frac{1}{2\pi}\int_0^{2\pi} d\theta\, e^{-i\theta m'} e^{-i(k/\hbar)\cos\theta} \sum_{m=-\infty}^{+\infty} e^{i\theta m} e^{-i(\hbar/2)m^2}\langle m|\psi_t\rangle \tag{24b}$$

Depending on the value of $\hbar$ and k, we used basis vectors with up to 16,384 components and verified that the normalization of the iterated vector was conserved. The convergence of time averages such as Eq. 22 was checked by varying the number of iterations. In the following sections we first present our results for the time-averaged quantities and their dependence on $\hbar$ and k and then the results for time-dependent quantities.

A. Asymptotic Distributions

We have investigated the asymptotic distribution $P(p|p_0)$ of Eq. (22) for the kicked rotator near the critical nonlinearity $k = k_c = 0.9716$ where the corresponding classical map Eq. 10 exhibits the transition to global or connected stochasticity, as described in Section III. Let us first describe the results exactly at $k = k_c$ as the effective Planck constant $\hbar$ is varied.

1. $\hbar$ Dependence

In Figure 4 the asymptotic distribution $P(p|p_0)$ for $k = k_c$ is shown on a linear and a logarithmic scale. The eigenvalue $p_0 = 3.2$ of the initial wavefunction

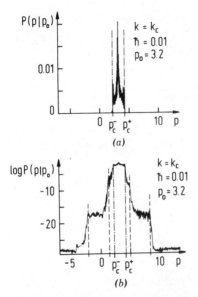

Figure 4. Asymptotic momentum distribution, Eq. 22, (a) on a linear and (b) on a logarithmic scale. Borders of classical confinement due to KAM tori (dot-dashed lines) and the borders of cantori (dashed lines) are indicated. The initial momentum was in the interval (p_c^-, p_c^+) as for the corresponding classical case (Figure 3).

$|p_0\rangle$ is chosen in accordance with the classical situation of Figure 3. On the linear scale one sees that most of the quantum-mechanical distribution $P(p|p_0)$ remains confined to the classically accessible interval (p_c^-, p_c^+), which is determined by the last bounding KAM trajectories (Section III). This demonstrates the role of KAM tori as barriers in quantum systems. The logarithmic display in Figure 4(b) shows that there is also a probability to find the quantum rotator in classically inaccessible states. This probability is clearly a nonclassical effect that can be interpreted as tunneling through classical KAM tori. A closer look at the p dependence of the distribution $P(p|p_0)$ outside the interval (p_c^-, p_c^+) shows that it decays exponentially into the classically forbidden region

$$P(p|p_0) \propto e^{-\lambda|p - p_e|} \tag{25}$$

where $p_e = p_c^\pm$. Moreover, the figure shows also an exponential decay at other values p_e corresponding to the extremal p values of the most important cantori. This relationship will be demonstrated in V.A.2, where the dependence of $P(p|p_0)$ on the nonlinearity is investigated. To understand the exponential decays near tori or cantori, we express $P(p|p_0)$ in terms of quasi-energy eigenstates, which are the eigenfunctions of the time evolution operation $\hat{U}$ of Eq. 9, that is,

$$\hat{U}|j\rangle = e^{-iE_j/\hbar}|j\rangle \tag{26}$$

where E_j is the quasi-energy [73]. Inserting the corresponding spectral decomposition of $\hat{U}$ into Eq. 22 yields

$$P(p|p_0) = \sum_j |\langle p_0|j\rangle|^2 \cdot |\langle p|j\rangle|^2 \tag{27}$$

This expression can be considered as a weighted sum of Wigner functions projected onto the momentum axis

$$P(p|p_0) = \int d\theta \sum_j c_j(p_0) W_j(\theta, p) \tag{28}$$

where the W_j are the eigenfunctions $|j\rangle$ in the Wigner representation Eq. 15, and the c_j are their overlap with the momentum p_0. As discussed in Section IV, Wigner functions, which are located on tori in the semiclassical limit, show an exponential decay on the convex side of a regular trajectory. Attributing the exponential decay to such a regular Wigner function (e.g., on the last KAM torus), one expects the inverse decay length λ to scale as $\hbar^{-1}$, as in the case of tunneling into a potential barrier. In Figure 5 we show the $\hbar$ dependence of λ in the range $0.2 \times 10^{-3} < \hbar < 10^{-1}$. What we observe is an algebraic dependence

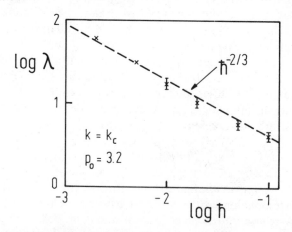

Figure 5. The $\hbar$-dependence of the inverse penetration depth λ, Eq. 25, into classically forbidden regions. The dashed line indicates a slope $-\frac{2}{3}$ for comparison.

of λ as $\hbar^{-\sigma}$ with σ near $\frac{2}{3}$. We attribute this power law to the complexity of phase space near the KAM torus. Especially the classical scaling behavior near the critical KAM trajectory may also be manifest in the quantum system. Indeed, the observed functional dependence on $\hbar$ can be explained in the framework of a renormalization group theory of Grempel, Fishman, and Prange for the quantum system [68, 69]. They have shown that $\hbar$ acts as a relevant variable such that $\hbar$ is rescaled in the course of time. There is a crossover time t^* where the system departs from the classical scaling region into the quantal regime. If one assumes that the exponential dependence (25) has already built up for times $t < t^*$, the Wigner function $W_\infty = \sum_j c_j W_j$ on the right side of Eq. 28 should scale with $\hbar$ as [69] (dropping the θ dependence)

$$W_\infty(|p - p_e|; \hbar) = \hbar^{-\sigma} W^*(\hbar^{-\sigma} \cdot |p - p_e|) \tag{29}$$

The critical exponent σ is determined by the classical scaling factors α and β of angle and momentum [57, 58] near the critical KAM torus

$$\sigma = \left(1 + \frac{\ln|\alpha|}{\ln|\beta|}\right)^{-1} \tag{30}$$

With $\alpha \approx -1.69$ and $\beta \approx -2.56$ (these factors hold in the regions of the tori determining $p_c^\pm$), one obtains $\sigma \approx 0.648$, in good agreement with the critical exponent σ determined numerically in the preceding. The explanation within the scaling theory is further supported by the fact that we find the same exponent at the other points p_e associated with the stablest cantori. These cantori are not far from criticality, and therefore, the scaling theory also applies to them.

We were further interested in the time-averaged probability to find the system in classically inaccessible states, that is, in states outside the interval (p_c^-, p_c^+) and its $\hbar$ dependence. We have thus investigated the probability

$$\bar{W} = \sum_{p < p_c^-} P(p|p_0) + \sum_{p > p_c^+} P(p|p_0) \tag{31}$$

which is the probability outside the dot-dashed lines of Figure 4. As shown in Figure 6, this probability vanishes with a power law $\bar{W} \sim \hbar^\gamma$, with $\gamma \approx 2.5$. This exponent, however, is not independent of the initial wavefunction $|p_0\rangle$. If we choose $|p_0\rangle$, for example, in the larger cell around the integer resonance, we find different exponents, which increase with the distance from the next bounding KAM trajectory. This appears to be due to partial localization of the wavefunction before the bounding torus is reached. Since the probability at the values $p_c^\pm$ depends on p_0 and $\hbar$ in a complicated fashion due to the

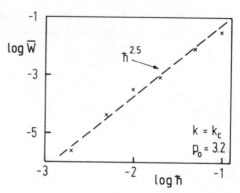

Figure 6. The $\hbar$-dependence of the transition probability $\bar{W}$, Eq. 31, into the classical inaccessible momentum states.

underlying classical structure between p_0 and $p_c^{\pm}$, we do not expect a simple connection between γ and the exponent σ.

The results presented in this section were found for typical values of $\hbar$. There exist values $\hbar$ where a quite different behavior is expected. These special values are, however, of measure zero and therefore atypical. The classical phase space is periodic in p with period 2π in our units. If the ratio of this length to $\hbar$, which is the distance between neighboring momentum eigenstates, is rational we have the case of quantum resonances [9, 14, 52, 53]. In this case the matrix (24) is periodic in $m - m'$. As a result, the quasi-energy eigenstates are extended Bloch states, and the quasi-energy spectrum is continuous. For these values of $\hbar$ there is no confinement due to tori or cantori, and the mean-square displacement in momentum diverges quadratically in time as for ballistic motion. Reference 53 has demonstrated for this case that the (coarse-grained) Wigner function corresponding to certain eigenstates can be associated with the bounding KAM curves. It would be interesting to see whether they show scaling behavior with $\hbar$. Another possibility of anomalous behavior can occur if $\hbar/2\pi$ is a Liouville number [74, 75]. These numbers lie very "close" to rationals and are also exceptional since like rational numbers they have measure zero and were therefore not considered in our investigation.

2. Variation of Nonlinearity

We know from the discussion of Section III that for $k > k_c$ the last bounding tori break up into cantori, resulting in large-scale diffusion. Figure 7 depicts the asymptotic distribution $P(p|p_0)$ (Eq. 22) for two values of the nonlinearity $k > k_c$. The parameters $\hbar$ and p_0 are the same as in Figure 4. A comparison of Figure 7(a) with Figure 4(b) shows that the distribution for $k = 1.1$ is not very different from the one for $k = k_c$, only the weight of the plateaus outside the

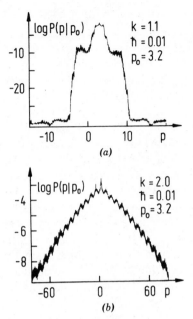

Figure 7. The asymptotic momentum distributions, Eq. 22, for two values $k \geqslant k_c$. Cantori still act as strong barriers for $k = 1.1$ but not for $k = 2$ (note the change in p- and in P-scale).

interval (p_c^-, p_c^+) is enhanced. In light of Eq. 28, this means that the eigenfunctions laying outside (p_c^-, p_c^+) have a slightly larger overlap with p_0. A more important feature of Figure 7(a) is that the exponential decay near $p = p_c^{\pm}$ is still very pronounced and that most of the asymptotic distribution remains in the interval (p_c^-, p_c^+). This demonstrates the role of cantori as dynamical barriers in quantum systems. Apparently, the effect of cantori in quantized systems is even stronger than in classical systems. Thus, the structure and the exponential decay near the values p_e [in Figure 4(b) and in Figure 7(a)] can be explained in terms of the classical quasi-periodic trajectories with the most irrational winding numbers: The values p_e are the minimum p values of (can-) tori whose winding numbers α (Eq. 14) are $v_s^+ - 1$, $1 - v_s^+$, and $1 - \gamma_g^+$ and maximum p values of (can-) tori with $\alpha = v_g^+$, v_s^+ $2 - v_s^+$ [from left to right in Figures 4(a) and 7(b)], where $v_g^+ = 1/\gamma$ and $v_s^+ = (2 + 3\gamma)/(3 + 4\gamma)$ are the two "most irrational" winding numbers. (The cantori with v_g^- and v_s^+ provide also the borders visible in Figure 2 of the classical calculation of MacKay et al. [4].)

In the classical system, the cantori lose their role as barriers with increasing k. An analogous behavior is found in the quantum system, for example, for $k = 2$ in Figure 7(b), where the probability distribution extends to much larger momenta (note the change in the scale of p). The remaining structure has the periodicity of the classical phase space and seems to be related to the large regular islands as it disappears for even larger k. We attribute the exponential decays of Figure 7(b) to quantum-mechanical interference effects in analogy to Anderson localization [11] rather than to classical barriers. It is informative to compare the classical action ΔW transported per iteration across the broken golden torus in the cases of Figures 7(a) and 7(b) with the effective Planck constant $\hbar$. According to Ref. 4a, Table II, the area ΔW is $\simeq 0.002$ in our units for $k = 1.1$. Assuming that the scaling of $\Delta W \simeq A \cdot (\Delta k)^\eta$, and $\eta \simeq 3.0$ is still valid at $k = 2$, one obtains $\Delta W \simeq 0.65$ for $k = 2$. This is to be compared with the area covered by one unperturbed state or a minimum-uncertainty wavepacket, that is, with $\hbar$, which is 0.01 in Figure 7. The difference in $\Delta W/\hbar$ [0.2 in Figure 7(a) versus 65 in 7(b)] explains the large difference in the asymptotic distributions: In Figure 7(a) the cantori still act as barriers, and the probability outside the interval (p_c^-, p_c^+) (which is nearly the same for $k = k_c$ and $k = 1.1$) should be regarded as resulting from tunneling through the cantori, whereas in Figure 7(b) the probability outside (p_c^-, p_c^+) results from flow through the remnants of tori, which stops as a result of destructive interference.

The transition from $k \simeq k_c$ to larger k, which is depicted in Figure 4 and 7, still has another aspect. As discussed at the end of Section III, the classical asymptotic distribution $\rho(p|p_0)$ must change discontinuously at the critical point k_c. Correspondingly, the probability to find the classical system outside the interval (p_c^-, p_c^+) must increase like a step function at k_c, that is, $\bar{W} \sim f(k) \cdot \theta(k - k_c)$, where $\bar{W}$ is defined in Eq. 31 [with $P(p|p_0)$ to be replaced by $\rho(p|p_0)$]. From Figures 4(b), 7(a), and 7(b), one can imagine that quantum mechanically this transition is smoothened due to the role of cantori and the

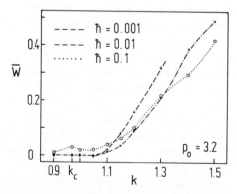

Figure 8. The increase of the transition probability $\bar{W}$, Eq. 31, with increasing nonlinearity k is smoothed due to the role of cantori in the quantum system.

interference effects. This is demonstrated in Figure 8, where the dependence of $\bar{W}$ on the nonlinearity parameter is shown for different values of $\hbar$ [the values of $p_c^{\pm}$ in Eq. 31 remain fixed at $p_c^{\pm}(k = k_c)$]. The main increase of $\bar{W}$ occurs only for values $k \gtrsim 1.1$ due to the inhibiting effect of cantori. The smoothening is stronger for $\hbar$ larger. This smoothening was already observed in the KAM-like transition in the Henon–Heiles system [18, 19]. Our results for the kicked rotator clearly demonstrate the role of cantori in this transition. In Figure 8 one also observes some irregularities in the k dependence, for example, the bump in the vicinity of k_c for $\hbar = 0.1$. These effects are attributed to avoided crossings in the quasi-energy spectrum: Two nearly degenerate eigenstates lying on different sides of the bounding KAM trajectory constitute a tunneling state that enhances the probability in $\bar{W}$. Similar effects have been observed in theories for the ionization of Rydberg atoms in a microwave field leading to "subthreshold ionization" of the atoms [37] and in experiments [29, 76]. We expect that the role of cantori is also observable in these experiments.

B. Time-Dependent Quantities

So far we have discussed only time-averaged quantities. However, for the interpretation of experiments, for example, it is also important to know how and on which time scales the asymptotic distribution is reached.

1. Transition Probability into Classically Inaccessible States

The first quantity we want to investigate is analogous to $\bar{W}$ of Eq. 31, with the difference that the time dependence is retained. We define

$$W_t(p^+|p_0) = \sum_{p > p^+} |\langle p | U^t | p_0 \rangle|^2 \tag{32}$$

which is the probability of finding the system at time t in states with angular momentum larger than p^+ under the condition that it is prepared in p_0 at time $t = 0$. This is therefore the time-dependent transition probability into states with $p > p^+$. The value p^+ is not fixed to p_c^+ as in Eq. 31. We have determined this quantity for k near k_c and for different values of $\hbar$. Figure 9 shows the results for $k = k_c$ and $\hbar = 0.002$ on a log-log scale. The uppermost curve is for $p^+ = 4.0$, which is smaller than $p_c^+ = 4.269$ and therefore contains contributions from the cell where the wavefunction was started ($p_0 = 3.2$). This curve shows a quite different time dependence compared with the other curves, which give the probability above values p^+ that are greater than p_c^+ [i.e., that lie outside the confining cell (p_c^-, p_c^+)]. The probability above $p^+ = 4.0$, which is also several orders of magnitude larger than the probabilities above $p^+ > p_c^+$, oscillates around its mean value in a quasi-periodic manner during the whole time interval. The other curves, which describe the total probability

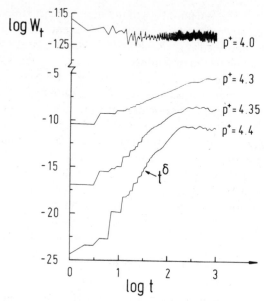

Figure 9. The time dependence of the transition probability from p_0 to states above p^+, Eq. 32 (for $k = k_c$, $\hbar = 0.002$) exhibits characteristic differences for $p^+ < p_c^+$ and for $p^+ > p_c^+$ = 4.269 due to the quantum barrier near p_c^+.

tunneled through the last KAM torus, increase algebraically until a break time t_b is reached and after this time oscillate also quasi-periodically around their mean values $\bar{W}$. The algebraic time dependence is described approximately by

$$W_t(p^+, p_0) \approx W_1(p^+, p_0) \cdot t^\delta \tag{33}$$

with δ given by

$$\delta = \beta \cdot \ln \frac{\bar{W}(p^+, p_0)}{W_1(p^+, p_0)} \tag{34}$$

where $\beta \approx 0.2$ depends only weakly on $\hbar$ in the range $0.2 - 10^{-3} < \hbar < 10^{-1}$. The last statement results from the observation that t_b, which is defined by $W_{t_b} \simeq \bar{W}$, is observed to behave approximately as $t_b \simeq 1/\hbar$, resulting in a logarithmic $\hbar$ dependence of β, namely, $\beta \approx |\ln \hbar|^{-1}$. For the preceding $\hbar$ interval, β varies between 0.16 for $\hbar = 0.2 \cdot 10^{-3}$ and 0.43 for $\hbar = 0.1$. Because of a certain ambiguity in determining t_b, especially for $\hbar = 0.1$, our results are also consistent with $\beta \simeq$ const. The break time $t_b \simeq 1/\hbar$ would have a natural explanation in that the system "recognizes" at this time that it has a discrete quasi-energy spectrum [7, 9] (Eq. 26). We should also mention that in obtaining Figure 9 the data were smoothened by averaging over ± 5 time steps.

The point we wanted to demonstrate in Figure 9 is that correlations between points on the same side of a quantum barrier are different from those between points on different sides of the barrier. This means that quantum barriers are manifest also in dynamical quantities. This point is further demonstrated in Section V.B.2. The validity of the preceding picture is not restricted to $k = k_c$. For instance, for $k = 1.1$, where the remnants of tori act as barriers, we have observed a behavior similar to that of Figure 9. For values $\hbar \gtrsim 0.1$ the algebraic increase is not well defined anymore, and the time dependence of $W_t(p^+, p_0)$ for $p^+ \gtrless p_c^+$ is not characteristically different in constrast to Figure 9.

2. Mean-Square Displacement

The mean-square displacement in (angular) momentum, which behaves diffusively for $k > k_c$ in the classical system (Section III), is bounded in the quantum system. This was first observed for the kicked quantum rotator [9–11, 15] and later in models for the ionization of hydrogen atoms [12, 13]. It is currently the subject of intensive theoretical [9, 35–37] and experimental [29–31] investigations. In this section we will demonstrate that there are different mechanisms leading to limitations of diffusion, namely, dynamical barriers in contrast to limitation due to destructive interference.

The mean-square displacement in momentum $\langle \Delta p_t^2 \rangle$ can be obtained from the iterations of the quantum system Eqs. 8 and 9 as

$$\langle \Delta p_t^2 \rangle = \sum_p p^2 P_t(p|p_0) - \left(\sum_p p\, P_t(p|p_0) \right)^2 \tag{35}$$

where $P_t(p|p_0)$ was defined in Eq. 23. In Figure 10 we show the time dependence of $\langle \Delta p_t^2 \rangle$ on a log-log scale for two values of the nonlinearity k. In both cases initial wavefunction and $\hbar$ are the same. One observes very different behavior in the course of time, and also the scale of the fluctuations differs by two to three orders of magnitude. For $k = 2$ we find an algebraic increase,

$$\langle \Delta p_t^2 \rangle \sim t^\alpha \qquad \alpha = \alpha(\hbar, k) \tag{36}$$

in the time interval shown in Figure 10; for larger times the mean-square displacement saturates and oscillates quasi-periodically. In the case $k = 1.1$ one observes quasi-periodic behavior for all times, and also the fluctuations are much smaller than for $k = 2$. This last observation is in accordance with the width of the corresponding asymptotic distributions shown in Figures 7(a) and 7(b) and the interpretation resulting from different localization mechan-

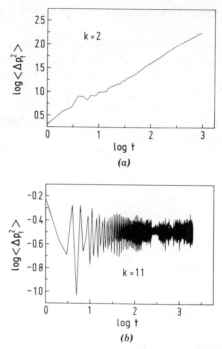

Figure 10.   The mean-square displacement, Eq. 35, for two nonlinearities above k_c and for $\hbar = 0.01$. For $k = 2$ it is limited only at still longer times ($t > 10^3$) by destructive interference. For $k = 1.1$ it remains bounded two to three orders of magnitude below (note the different scales) due to the effect of cantori.

isms: quantum barriers for $k = 1.1$ and destructive interference for $k = 2$. The long-time behavior (quasi-periodicity) was expected [15] to be the same in both cases due to the discrete (pure-point) nature of the quasi-energy spectrum. The characteristic differences for short times must be explained by the differences in the locations and character of the eigenfunctions and in the corresponding quasi-energy spectrum. Note also that for $k = 1.1$ the long-time behavior of the classical and the quantum system are drastically different. In the classical system there is weak but unlimited diffusion, in contrast to the strong localization in quantum system [Figure 7(a)].

We also determined the exponent α for the quantum diffusive behavior at short times, Eq. 36, and its dependence on $\hbar$ and k in the interval $5 \times 10^{-3} < \hbar < 0.1$ and $1.5 < k < 2$. The result is shown in Figure 11. We find in this regime that the diffusive behavior is slowed down compared to the classical case where $\alpha = 1$. For large $\hbar$ and smaller k, $k \simeq 1.5$, the exponent α becomes very small and ceases to be well defined for even smaller k (in this $\hbar$ regime) due to

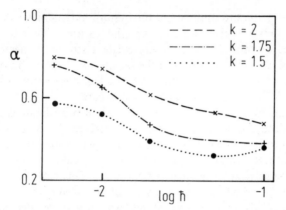

Figure 11. The $\hbar$-dependence of the exponent α, Eq. 36, characterizing the diffusive increase of $\langle \Delta p_t^2 \rangle$ at short times.

the growing importance of cantori. For large k and small $\hbar$, $\alpha \to 1$ in accordance with previous results [7, 69], but the diffusive behavior still persists only for a finite time, in contrast to the classical case.

The results of this section have shown that the quantum analogs of tori and cantori are also observable in certain dynamical quantities. In time-dependent quantities their effect can be even stronger than in the time-averaged quantities (see, e.g., W_t of Figure 9 for short times). The results of V.B.2 showed that they are also essential in the understanding of certain transport properties.

VI. SUMMARY AND DISCUSSION

We have demonstrated that dynamical barriers corresponding to classical KAM tori and cantori exist also for quantum systems that are classically nonintegrable. This is true despite the fact that the distinction between chaos and regular behavior fails in quantum systems. In quantum systems they determine the accessible states to a large extent, similar to the classical case. The intuitive picture that emerges for the situation in the quantum case appears to be more complicated due to typical quantum effects, namely, tunneling through these barriers in quantum phase space and (destructive and constructive) interference effects modifying, for example, the global behavior. It was also demonstrated that cantori can act as even stronger barriers in quantum systems than in classical systems. This is the case if the ratio of the classical flux across a cantorus to Planck's constant is small. The quantum system does not "recognize" the gaps in the cantorus and therefore remains confined. This example shows that a thorough understanding of the classical phase space is helpful to understand the quantum dynamics. We have found that the quantum barriers determine essential features of time-dependent and time-averaged quantities; that is, there exists a strong connection between classical phase space structures and the location and extension of eigenfunctions of the corresponding quantum systems. This connection is still not well understood, especially in the near-integrable case.

We summarize and discuss some of our result in more detail. We have found exponential decays of asymptotic distributions near KAM tori and cantori with the most irrational winding numbers. The penetration depth scales as $\hbar^\sigma$; the exponent $\sigma \approx \frac{2}{3}$ can be explained within a scaling theory [68, 69] for quantum systems near a classical critical point. This is in contrast to usual tunneling depths, which scale as $\hbar$. The total averaged transition probability into the classically inaccessible states was found to decrease like $\hbar^\gamma$, where γ depends on the distance between the initial wavefunction and the nearest bounding KAM torus. Penetration through KAM barriers, was also observed for the accelerator modes of the kicked rotator [17]. The connection with our results is not clear because tunneling out of islands may be different due to a difference in the bounding trajectories. We have investigated KAM tori that are isolated on both sides, whereas for islands the bounding KAM trajectory usually is isolated only on one side (77). An additional complication comes from cantori that surround islands. As we have seen, these act as additional barriers. This effect was recently observed in a model for photo-dissociation of molecules [27]. Furthermore, the role of cantori explains the smoothening of KAM-like transitions observed in other systems [18, 19].

We also found that the dynamical barriers are essential for the understanding of transport properties such as diffusion in these systems. On a short-time scale the effect of the barriers is often more drastical than in the time average, as was demonstrated for the tunneling probability into classically inaccessible states.

Although our investigations were performed for a specific model, the kicked rotator, we expect that the results apply also to other dynamical systems with two degrees of freedom because their phase space is approximated locally by the kicked rotator phase space. The most attractive physical systems in this respect, where accurate experimental verifications of our results can be expected, are probably the microwave ionization experiments of hydrogen atoms and the experiments on multiphoton dissociation of molecules [27]. It is not clear to which extent KAM tori and cantori act as quantum barriers in higher dimensional phase space, but we expect that tunneling through these barriers might be more important than Arnol'd diffusion.

Acknowledgments

We thank S. Fishman and R. E. Prange for useful discussions. One of us (G. R.) acknowledges the hospitality of the Department of Physics and Astronomy at the University of Maryland, where this work was completed. This work was supported by Deutsche Forschungsgemeinschaft.

References

1. For a review see, e.g., M. V. Berry, in *Topics in Nonlinear Dynamics*, AIP Conference Proceedings No. 46, S. Jorna, ed., American Institute of Physics, New York, 1978, p. 16.

2. A. J. Lichtenberg and M. A. Lieberman, *Regular and Stochastic Motion*, Applied Mathematical Sciences, Vol. 38, Springer-Verlag, New York, 1983.

3. I. C. Percival in *Nonlinear Dynamics and the Beam–Beam Interaction*, AIP Conference Proceedings No. 57, M. Month and J. Herrera, eds., American Institute of Physics, New York, 1979, p. 308; S. Aubry, in *Solitons and Condensed Matter Physics*, A. R. Bishop and T. Schneider, eds., Springer-Verlag, Berlin, 1978, p. 264.

4. (a) R. S. MacKay, J. D. Meiss, and I. C. Percival, *Physica* **13D**, 55 (1984); (b) D. Bensimon and L. P. Kadanoff, *Physica* **13D**, 82 (1984).

5. For a review see (a) M. Hillery, R. F. O'Connell, M. O. Scully, and E. P. Wigner, *Phys. Rep.* **106**, 121 (1984); (b) N. L. Balazs and B. K. Jennings, *Phys. Rep.* **104**, 347 (1984).

6. R. Kosloff and S. A. Rice, *J. Chem. Phys.* **74**, 1340 (1981); D. L. Shepelyansky, *Teor. Mat. Fiz.* **49**, 117 (1981).

7. D. L. Shepelyansky, *Physica* **8D**, 208 (1984).

8. G. Casati, B. V. Chirikov, I. Guarneri, and D. L. Shepelyansky, *Phys. Rev. Lett.* **56**, 2437 (1986).

9. G. Casati, B. V. Chirikov, F. M. Izrailev, and J. Ford, in *Stochastic Behaviour in Classical and Quantum Hamiltonian Systems*, G. Casati and J. Ford, eds., Lecture Notes in Physics, Vol. 93, Springer-Verlag, Berlin, 1979, p. 334.

10. B. V. Chirikov, F. M. Izrailev, and D. L. Shepelyansky, *Sov. Sci. Rev.* **C2**, 209 (1981).

11. S. Fishman, D. R. Grampel, and R. E. Prange, *Phys. Rev. Lett.* **49**, 509 (1982); D. R. Grempel, R. E. Prange, and S. Fishman, *Phys. Rev. A* **29**, 1639 (1984).

12. R. Blümel and U. Smilansky, *Phys. Rev. Lett.* **52**, 137 (1984).

13. G. Casati, B. V. Chirikov, and D. L. Shepelyansky, *Phys. Rev. Lett.* **53**, 2525 (1984).

14. B. Dorizzi, B. Grammaticos, and Y. Pomeau, *J. Stat. Phys.* **37**, 93 (1984).

15. T. Hogg and B. A. Huberman, *Phys. Rev. Lett.* **48**, 711 (1982); *Phys. Rev. A* **28**, 22 (1983).

16. E. Ott, in *Long Time Predictions in Dynamics*, C. W. Horton, Jr., L. E. Reichl, and V. B. Szebehely, eds., Wiley, New York, 1982, p. 281.

17. J. D. Hanson, E. Ott, and T. M. Antonson, Jr., *Phys. Rev. A* **29**, 819 (1984).

18. S. A. Rice, in Ref. 20, p. 117, and references therein.

19. J. S. Hutchinson and R. E. Wyatt, *Phys. Rev. A* **23**, 1567 (1981).

20. J. Jortner, R. D. Levine, and S. A. Rice, eds., *Photoselective Chemistry*, Advances in Chemical Physics, Vol. 47, Wiley, New York, 1981.

21. W. Forst, *Theory of Unimolecular Reactions*, Academic, New York, 1973; P. J. Robinson and K. A. Holbrook, *Unimolecular Reactions*, Wiley, New York, 1972.

22. For a review see E. B. Stechel and E. J. Heller, *Ann. Rev. Phys. Chem.* **35**, 563 (1984).

23. R. B. Shirts and W. P. Reinhardt, *J. Chem. Phys.* **77**, 5204 (1982); C. Jaffe and W. P. Reinhardt, *J. Chem. Phys.* **77**, 5191 (1982).

24. M. J. Davis, *J. Chem. Phys.* **83**, 1016 (85).

25. M. J. Davis and S. K. Gray, *J. Chem. Phys.* **84**, 5389 (86).

26. S. K. Gray, S. A. Rice, and M. J. Davis, *J. Phys. Chem.* **90**, 3470 (1986).

27. R. C. Brown and R. E. Wyatt, *Phys. Rev. Lett.* **57**, 1 (1986); *J. Phys. Chem.* **90**, 3590 (1986).

28. T. Geisel, G. Radons, and J. Rubner, *Phys. Rev. Lett.* **57**, 2883 (1986); G. Radons, T. Geisel, and J. Rubner, in *Proceedings of the First Inst. Conference on the Physics of Phase Space*, College Park, MD, May, 20–23 1986, Lecture Notes in Physics, Vol. 278, Springer-Verlag (1987).

29. J. E. Bayfield and P. M. Koch, *Phys. Rev. Lett.* **33**, 258 (1974); J. E. Bayfield, L. D. Gardner, and P. M. Koch, *Phys. Rev. Lett.* **39**, 76 (1977); P. M. Koch, *J. Phys.* **43**(C2), 187 (1982).

30. K. A. H. van Leeuwen, G. v. Oppen, S. Renwick, J. B. Bowlin, P. M. Koch, R. V. Jensen, O. Rath, D. Richards, and J. G. Leopold, *Phys. Rev. Lett.* **55**, 2231 (1985).

31. J. N. Bardsley, B. Sundaram, L. A. Pinnaduwage, and J. E. Bayfield, *Phys. Rev. Lett.* **56**, 1007 (1986).

32. J. G. Leopold and I. C. Percival, *Phys. Rev. Lett.* **41**, 944 (1978); D. A. Jones, J. G. Leopold, and I. C. Percival, *J. Phys.* B **13**, 31 (1980).

33. R. V. Jensen, *Phys. Rev. A* **30**, 386 (1984).

34. J. G. Leopold and D. Richards, *J. Phys.* B **18**, 3369 (1985).

35. N. B. Delone, B. P. Krainov, and D. L. Shepelyansky, *Sov. Phys. Usp.* **26**, 551 (1983).

36. G. Casati, B. V. Chirikov, D. L. Shepelyansky, and I. Guarneri, *Phys. Rev. Lett.* **57**, 823 (1986).

37. R. Blümel and U. Smilansky, *Phys. Rev. Lett.* **58**, 2531 (1987); *Phys. Scr.* **35**, 15 (1987); *Z. Phys.* D **6**, 83 (1987).

38. P. Brumer, in Ref. 20, p. 201.

39. R. S. MacKay, in *Nonlinear Dynamics Aspects of Particle Accelerators*, J. M. Jawett, M. Month, and S. Turner, eds. Proceedings of the Joint US–CERN School on Particle Accelerators, Lecture Notes in Physics, Vol. 247, Springer-Verlag, Berlin, 1986, p. 390.

40. Ya. G. Sinai, *Russ. Math. Surv.* **25**, 137 (1970).

41. L. A. Bunimovich, *Funct. Anal. Appl.* **8**, 254 (1974); *Commun. Math. Phys.* **65**, 295 (1979).

42. V. I. Arnol'd and A. Avez, *Ergodic Problems of Classical Mechanics*, Benjamin, New York, 1968.

43. I. P. Cornfeld, S. V. Fomin, and Ya. G. Sinai, *Ergodic Theory*, Springer-Verlag, New York, 1982.

44. V. I. Arnold, *Mathematical Methods of Classical Mechanics*, Springer-Verlag, New York, 1978.

45. A. J. Lichtenberg, *Physica* **14D**, 387 (1985).

46. W. Li and P. Bak, *Phys. Rev. Lett.* **57**, 655 (1986).

47. I. Dana and S. Fishman, *Physica* **17D**, 63 (1985).

48. B. V. Chirikov, *Phys. Rep.* **52**, 263 (1979).

49. F. Vivaldi, *Rev. Mod. Phys.* **56**, 737 (1984).

50. B. V. Chirikov, M. A. Lieberman, D. L. Shepelyanski, and F. M. Vivaldi, *Physica* **14D**, 289 (1985).

51. M. V. Berry, N. L. Balazs, M. Tabor, and A. Voros, *Ann. Phys. (N.Y.)* **122**, 26 (1979).

52. F. M. Izrailev and D. L. Shepelyansky, *Teor. Mat. Fiz.* **43**, 417 (1980).

53. S. J. Chang and K. J. Shi, *Phys. Rev. Lett.* **55**, 269 (1985); *Phys. Rev. A* **34**, 7 (1986).

54. J. M. Greene, *J. Math. Phys.* **20**, 1183 (1979).

55. S. Aubry, *Physica* **7D**, 240 (1983); *J. Physique* **44**, 147 (1983); S. Aubry and P. Y. Le Daeron, *Physica* **8D**, 381 (1983).

56. J. N. Mather, *Topology* **21**, 457 (1982).
57. L. P. Kadanoff, *Phys. Rev. Lett.* **47**, 1641 (1981); S. J. Shenker and L. P. Kadanoff, *J. Stat. Phys.* **27**, 631 (1982).
58. R. S. MacKay, *Physica* **7D**, 283 (1983).
59. B. V. Chirikov and D. L. Shepelyansky, *Physica* **13D**, 395 (1984).
60. D. K. Umberger and J. D. Farmer, *Phys. Rev. Lett.* **55**, 661 (1985).
61. A. I. Khinchin, *Continued Fractions*, University of Chicago Press, Chicago, 1964, p. 36.
62. M. V. Berry, *Phil. Trans. Roy. Soc. Lond.* **287**, 237 (1977).
63. M. V. Berry, in *Chaotic Behaviour of Deterministic Systems*, G. Iooss, R. H. G. Helleman, and R. Stora, eds., North-Holland, Amsterdam, 1983, p. 172, and references therein.
64. M. J. Davis and E. J. Heller, *J. Chem. Phys.* **80**, 5036 (1984).
65. C. Jaffe and P. Brumer, *J. Chem. Phys.* **82**, 2330 (1984).
66. K. Takahaski and N. Saitô, *Phys. Rev. Lett.* **55**, 645 (1985).
67. G. P. Berman and A. R. Kolovsky, *Physica* **17D**, 183 (1985).
68. D. R. Grempel, S. Fishman, and R. E. Prange, *Phys. Rev. Lett.* **53**, 1212 (1984).
69. S. Fishman, D. R. Grempel, and R. E. Prange, *Phys. Rev. A* **36**, 289 (1987).
70. J. H. Hannay and M. V. Berry, *Physica* **1D**, 267 (1980).
71. E. J. Heller, *J. Chem. Phys.* **72**, 1337 (1980).
72. M. Feingold and A. Peres, *Phys. Rev. A* **31**, 2472 (1985).
73. Ya. B. Zeldovich, *Zh. Eksp. Teor. Fiz.* **51**, 1492 (1966).
74. G. Casati and I. Guarneri, *Commun. Math. Phys.* **95**, 121 (1984).
75. R. E. Prange, D. R. Grempel, and S. Fishman, *Phys. Rev. B* **29**, 6500 (1984).
76. P. Pillet et al., *Phys. Rev. A* **30**, 280 (1984).
77. J. M. Greene, R. S. MacKay, and J. Stark, *Physica* **21D**, 267 (1986).

CHAPTER XXIII

ADIABATIC SWITCHING: A TOOL FOR SEMICLASSICAL QUANTIZATION AND A NEW PROBE OF CLASSICALLY CHAOTIC PHASE SPACE

WILLIAM P. REINHARDT

Department of Chemistry and Laboratory for Research on the Structure of Matter, University of Pennsylvania, Philadelphia, Pennsylvania 19104

CONTENTS

I. INTRODUCTION

During the 1970s there began to be intense interest in development of semiclassical methods of sufficient power to treat the problem of determination of the energies of highly excited states of small and medium polyatomic molecules (1–4). This interest in methods related to those of the "old quantum theory" followed from the many early successes of semiclassical methods in the treatment of simple molecular collisions (5–7) and from the fact that it is almost always possible to carry out numerical solution of the equations of classical Hamiltonian dynamics (8) once a potential surface is chosen, whereas it is quite unusual to be able to make the same statement quantum mechanically, the only exception being the assumption of a quadratic surface, allowing use of path integral or wavepacket methods (9–11). Interest has continued unabated due to the perceived need for methodology to treat many-degree-of-freedom systems with tens of thousands of wavenumbers in vibrational excitation, where recent experiments have revealed the need for new systematics (12–15) and the need to fill, in both a conceptually and computationally useful manner, an obvious gap in our understanding of the correspondence principle (16–23). This latter issue, arising from the extraordinary complexity of classical phase space structure for generic nonseparable systems and related to the existence of nonlinear resonance and chaotic dynamics, has given rise to its own subfield of research, of interest well outside the theoretical molecular physics and chemistry communities (24–26).

It is the purpose of this review to explore recent developments (27–42) in exploiting the method of adiabatic switching (43–55) to carry out primitive Einstein–Brillouin–Keller (EBK) semiclassical quantization (56–58) of nonseparable Hamiltonian systems. Focus will be on determination of bound-state eigenvalues, and thus, quite important and related developments relating to calculation of semiclassical wavefunctions (59), determination of periodic orbit dividing surfaces for use in classical transition state theory (60, 61), and numerical calculation of good angle variables (62) will be omitted from the discussion. However, the fact that the method has been found to succeed (31, 59) in cases where neither the assumptions underlying the adiabatic hypothesis nor those underlying the EBK method are met has led to the extension of the method as a probe of the structure of chaotic phase space, which is, in a way, the complement to the analysis of the destruction of the most irrational invariant tori (63–65).

In Sections II and III the basic methodology is introduced, with several examples from the current literature review in Section IV. Following the hints of Section II.C and III.C.3 that the method is not as simple as at first meets the eye (55, 66–69), Section V summarizes a detailed analysis (70, 71) of the adiabatic method given for a model two-degree-of-freedom system, where the full Hamiltonian dynamics is replaced by the direct simulation of the Poincaré surface of section via a simple area-preserving point map. The close relation of Hamiltonian dynamics to problems in number theory is illustrated through use of the Farey tree (72, 73) decomposition of segments of the real number line. This section ends with a brief discussion of the use of adiabatic switching, well past its original expected regime of validity: Switching into chaotic regions of phase space reveals the structure of unstable manifolds rather than of invariant tori.

II. ADIABATIC SWITCHING IN ONE-DEGREE-OF-FREEDOM SYSTEMS

A. Ehrenfest, Adiabatic Invariants, and Semiclassical Quantization

The idea of the relation of adiabatic invariants to quantization is old. Ehrenfest, in an attempt to understand the fact that classical and quantum treatments of black-body radiation both gave rise to the Wein displacement law (74), pointed out the importance of the concept of adiabatic invariance (43). As discussed by Klein (Ref. 53 contains a discussion on the origins of Ehrenfest's adiabatic principle) and Jammer (54), the concept predates Ehrenfest, and even its relevance to the quantum mechanics of microscopic systems had been implicitly recognized. In response to a question by Lorentz at the 1911 Solvay Congress, Einstein is quoted as saying at once that quantization of a pendulum is retained under adiabatic length changes: "If the length of the pendulum is changed infinitely slowly, its energy remains equal to hv if it was originally hv" (75, 76).

However, Ehrenfest was the first to suggest that adiabatic invariants were the appropriate quantities to "quantize" and that once a fully understood system had been quantized, the adiabatic hypothesis [so named by Einstein, "*Adiabatenhypothese*" (47)] would allow quantization of adiabatically related systems in the sense that "if a system be affected in a reversible adiabatic way, allowed motions are transformed into allowed motions" (45). This last statement is known as Ehrenfest's principle. Assuming the validity of the principle and reading the word "quantizing" for "allowed" in the preceding statement suggests a powerful method for semiclassical quantization.

What is the adiabatic invariant for a one-degree-of-freedom system? For classically bounded motion with two turning points, Ehrenfest (44) derives the adiabatic invariance of the quantity

$$2\bar{T}/v \qquad (2.1)$$

where $\bar{T}$ is the average kinetic energy and v the frequency or inverse period P.

Thus,

$$2\bar{T} = \frac{1}{m} \frac{\oint p^2(t)\, dt}{\oint dt} \tag{2.2}$$

where the path implied by the $\oint$ notation is over one period of the motion. Noting (44, 45) that $(p/m)\, dt = q\, dt$ implies that

$$\frac{2\bar{T}}{v} = \frac{\oint p\dot{q}\, dt}{vP} = \oint p\, dq \tag{2.3}$$

leading to the familiar concepts that it is the action integral that is quantized, that is,

$$\oint p\, dq = \begin{cases} (n + \frac{1}{2})h & \text{for vibrations} \\ nh & \text{for rotations} \end{cases} \tag{2.4a}$$
$$\tag{2.4b}$$

which is the Bohr quantization rule (77), and that it is the classical actions (78, 79) that are the adiabatic invariants (51, 52, 80–82).

Ehrenfest (44, 45) fully realized that the multidimensional generalization of the preceding simple quantization picture was often ambiguous (see Section III) and that, even in the case of one-dimensional motion, his principle did not apply in an equivalent manner on opposite sides of a separatrix. Leaving considerations of these points until later, how does the method work in practice?

B. Examples of Adiabatic Semiclassical Quantization for One-Degree-of-Freedom Systems

Ehrenfest (44, 45) gives several simple examples of the use of the adiabatic principle in obtaining quantization rules for what are essentially one-dimensional systems. Examples of the same type are discussed by Levi-Civita (52). One of the more interesting of these is not quite one dimensional but has aspects that will have parallels in later discussion. Consider central-force motion in three dimensions with $V = V(r, a(t))$, where $a(t)$ is a (adiabatically or otherwise) time-varying parameter. As for all central-force motion the problem is separable, and angular momentum is conserved for any meaningful $a(t)$. The invariant angular momentum is thus also an adiabatic invariant and thus should be quantized. The resulting quantization, $L = nh, n = 0, 1, 2, \ldots$, is equivalent to the original Bohr rule derived from the correspondence principle. In a sense, this is a trivial example, as the "invariant" is determined by an obvious geometric symmetry and thus could be immediately obtained.

What about more complex, and thus more interesting, one-dimensional cases? Johnson (32) has considered, in some detail, several one-dimensional model problems: a linearly forced oscillator and a harmonic oscillator with a time-dependent frequency. The former is analytically soluble, and its analysis illustrates several important points that will arise in the discussion of multidimensional systems. We thus consider it here.

The Hamiltonian for the linearly forced oscillator is

$$H = \tfrac{1}{2}(p^2 + w^2 q^2) + s(t)\mu q \tag{2.5}$$

where we consider $s(t)$ to be a switching function designed to take the system adiabatically from $H^0 = \tfrac{1}{2}(p^2 + w^2 q^2)$ to the fully perturbed system H during the time interval $(0, T)$. Thus, $s(0) = 0$ and $s(T) = 1$. Intuitively, adiabatically follows if $s(t)$ is a smooth function of t, especially at $t = 0$ and $t = T$, and if T is long compared to the fundamental frequency of the system, which in this case is $2\pi/w$. Solov'ev has used the linear switching function

$$
\begin{aligned}
s_1(t) &= 0 & t &\leqslant 0 \\
s_1(t) &= t/T & 0 &\leqslant t \leqslant T \\
s_1(t) &= 1 & t &> T
\end{aligned}
\tag{2.6a}
$$

and Johnson has introduced (29, 32) the function

$$
\begin{aligned}
s_2(t) &= 0 & t &\leqslant 0 \\
s_2(t) &= \frac{t}{T} - \frac{\sin(2\pi t/T)}{2\pi} & 0 &\leqslant t \leqslant T \\
s_2(t) &= 1 & t &> T
\end{aligned}
\tag{2.6b}
$$

which has subsequently been used by many workers. Both functions increase monotonically from 0 to 1 as it runs from 0 to T. Here $s_1(t)$ is continuous, but its derivative is not continuous at either 0 or T; $s_2(t)$ is continuous and has two continuous derivatives at 0 and T, and thus is a smoother function. How does the choice of switching function affect adiabaticity? That is, to what extent does the choice of switching function and the choice of switching time T affect the value of the integral $\oint p\,dq$? How does it affect our ability to calculate a semiclassical eigenvalue by adiabatic switching? As the linear forced oscillator is an analytically soluble problem, whether the switching is adiabatic or not, investigation of these points is straightforward.

Following Johnson (32), if

$$Z(t) = p(t) + iwq(t) \tag{2.7}$$

Hamilton's equations reduce to

$$\frac{dZ(t)}{dt} = iwZ(t) - \mu s(t) \tag{2.8}$$

which has the general solution

$$Z(t) = e^{iwt}\left[Z(0) - \mu \int_0^t s(t)e^{-iwt}\, dt \right] \tag{2.9}$$

The strategy is now to test the conservation of the integral $\oint p\,dq = \oint p^2\,dt$ at the end of the process of switching on the perturbation. Thus, we evaluate the Ehrenfest invariant by integration over one period of the motion, beginning at $t = T$. In this simple case the final period is still $2\pi/w$, and of course, $s(t) = 1$ for all times later than T. Thus, for $t \geqslant 0$, after completion of the switching process,

$$Z(T + t) = e^{iw(T+t)}\left[Z(0) - \mu \int_0^T s(t)e^{-iwt}\, dt - \mu \int_T^{T+t} e^{-iwt}\, dt \right] \tag{2.10}$$

Assuming a specific form for $s(t)$ allows extraction of $p(T + t)$ as $\mathrm{Re}\,\{Z(T + t)\}$. Thus, using $s_1(t)$ of Eq. 2.6a, we find

$$p(T + t) = \cos\,[w(T + t)]p(0) - \sin\,[w(T + t)]wq(0)$$
$$- \mu\{\cos\,(wt) - \cos\,[w(T + t)]\}/(Tw^2) \tag{2.11}$$

which allows evaluation of $\int p^2(t)\,dt$ over a single period following completion of the switching process as

$$\int_0^{2\pi/w} p^2(T + t)\,dt = \{p^2(0)/2 + w^2 q(0)/2\}/w + O(1/T) \tag{2.12a}$$

$$= E^0/w + O(1/T) \tag{2.12b}$$

which is precisely the Ehrenfest result as $T \to \infty$, as for the harmonic oscillator the virail theorem requires $\langle KE \rangle = \langle PE \rangle$ and thus $2\langle KE \rangle = \langle E \rangle$. As similar analysis gives

$$\int_0^{2\pi/w} p^2(T + t)\,dt = \{p^2(0)/2 + w^2 q(0)/2\}/w + O(1/T^3) \tag{2.13}$$

for the switching function $s_2(t)$ of Eq. 2.6b. These results suggest the possibility of a rule: Perhaps the error in the action due to switching for a finite time T goes as $T^{-(\ell+1)}$, where ℓ is the number of continuous derivatives of $s(t)$ at 0 or T. This is indeed the case, as will be seen in Section V.

What is the corresponding error in the semiclassical eigenvalue for the linearly forced oscillator? As the classical energy is not an adiabatic invariant, it changes during the switching process. This change gives the semiclassical estimate of the change in energy of the quantum state of interest as the perturbation is turned on. Here, Johnson has given a complete analysis (32), and we simply quote the result: The uncertainty in energy (after averaging over the initial phase) is of order T^{-1} for the switching function $s_1(t)$ and of order T^{-3} for the function $s_2(t)$.

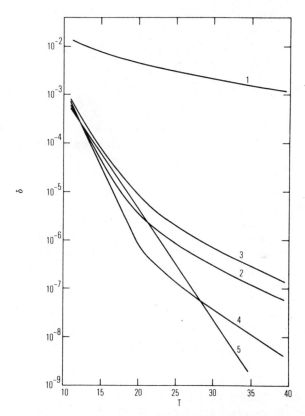

Figure 1. Average error in the semiclassical energy as determined by adiabatic switching of the time-dependent frequency in the Hamiltonian of Eq. 2.14. The effect of switching time and of choice of switching function are shown for the five functions $s_i(t)$ used by Johnson (32) and given in Eqs. 2.6a, 2.6b, 2.15, 2.16, and 2.17. It is seen that "smoothness" of the switching function is generally advantageous. (Figure reproduced from Ref. 32 with permission.)

Johnson has also considered a somewhat more complex one-dimensional problem, namely, an oscillator with a time-dependent frequency. The Hamiltonian is

$$H = \tfrac{1}{2}[p^2 + w^2(t)q^2] \tag{2.14}$$

necessitating numerical evaluation of trajectories. Figure 1 shows the phase-averaged error due to finite switching time T for the switching functions $s_1(t)$ and $s_2(t)$ discussed in the preceding and for the additional functions (shown only for t in the interval $[0, T]$)

$$s_3(t) = \left(\frac{t}{T}\right)^3 \left\{ (6t/T - 15)tT + 10 \right\} \tag{2.15}$$

$$s_4(t) = -\left(\frac{t}{T}\right)^4 \left\{ \frac{[20t/T - 70)t/T + 84]t}{T} - 35 \right\} \tag{2.16}$$

$$s_5(t) = \frac{\mathrm{erf}[\beta(2t/T - 1)] + 1}{2} \tag{2.17}$$

These functions are successively "smoother" and, as might be expected for T large enough, show successively better convergence properties.

C. What Is Invariant: Adiabatic and Geometric Invariants

The integral $\oint p\,dq$, where the $\oint$ denotes integration over one period of the classical trajectory, is Ehrenfest's adiabatic invariant, as illustrated in the previous section. For the special case of one-dimensional motion, there is an alternative interpretation, which may be immediately extended to systems of higher dimensionality: The trajectory for one-dimensional bounded motion with two classical turning points is identical to the isoenergetic curve in phase space defined by the implicit relationship

$$H(p, q) = E \tag{2.18}$$

Generalizing this, Hertz (83) has noted that, for bounded systems of n degrees of freedom, the volume in the $2n$-dimensional phase space enclosed by the implicitly defined $(2n - 1)$-dimensional surface $H(p, q) = \text{const.}$ is an invariant under adiabatic changes of parameters in H. That is, the volume in phase space enclosed by an isoenergetic surface is an adiabatic invariant. This, of course, assumes that during the switching process the phase volume is not divided into disjoint classically allowed regions, nor do two or more such regions merge.

For one-dimensional systems, both the Hertz and Ehrenfest invariants and integration paths are identical. [In higher dimensions the Hertz isoenergetic surface is clearly not a classical trajectory (the dimensionalities are wrong), nor even necessarily a surface, should it exist, defined by all points on a single trajectory.] The standard (e.g., Ref. 82) interpretation of the Ehrenfest adiabatic invariant for one-dimensional motion is in terms of the Hertz invariant: It is the area (i.e., the volume of phase space enclosed by an isoenergetic surface) in the p–q phase plane enclosed by the classical trajectory that is the adiabatic invariant, "and the quantum postulate states that the area of the closed curve described in the pq plane (phase plane), in one period of the motion, is an integral multiple of h" (82).

One is thus led, quite naturally, to view conservation of area in the phase space as being intimately associated with adiabatic invariance. However, this relationship is only correct in one direction. Adiabatic transformations do conserve area, but the converse is false: Phase space areas are invariant to any time-dependent transformation of the Hamiltonian, adiabatic or not! Thus, conservation of phase area does not imply adiabaticity. In fact, it has nothing to do with it. This perhaps surprising and seemingly paradoxical realization follows from arguments equivalent to Stokes's theorem or Gauss's law for magnetic flux lines and thus is completely general (84, 85). Following Arnold (84) and giving the general n-degree-of-freedom result, we note that for a general time-dependent Hamiltonian $H(\bar{p}, \bar{q}, t)$, where

$$p_i = \frac{-\partial H}{\partial q_i} \tag{2.19a}$$

$$q_i = \frac{\partial H}{\partial p_i} \tag{2.19b}$$

it is the case that (Poincaré lemma)

$$\int_{\gamma_1} \bar{p} \cdot d\bar{q} = \int_{\gamma_2} \bar{p} \cdot d\bar{q} \tag{2.20}$$

where γ_1 is any closed curve in the $(\bar{p}, \bar{q})$ phase space at time t_1, and (the essential point) γ_2 is the image of γ_1 propagated to $t = t_2$ under the dynamics of Eqs. 2.19 with Hamiltonian $H(\bar{p}, \bar{q}, t)$. The lemma is illustrated in Figure 2.

Returning to the one-degree-of-freedom case, it is straightforward to state and visualize the Poincaré lemma: Given any closed curve in the pq plane, an area (in the normal sense) is defined. If this boundary curve is propagated via a general time-dependent Hamiltonian, it will (at any fixed time!) encircle the same area. This is illustrated for a time-independent Morse oscillator in Figure 3, where a time-evolving boundary is seen to contain constant area. The resolution of the apparent paradox that area (as defined in the Poincaré

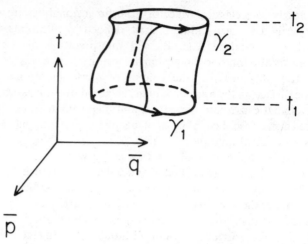

Figure 2. Geometric invariance of the integral $\oint \bar{p} \cdot d\bar{q}$ in extended phase space. The Poincaré lemma discussed in the text states that if the phase space curve γ_1 is taken to define a continuous line of initial conditions, the "tube" of trajectories generated via a time-independent or time-dependent Hamiltonian is a Lagrangian manifold, and thus distortions of γ on the manifold leave the integral invariant. In particular, if γ_1 is at fixed time t_1 and γ_2 is at time t_2, the lemma holds, with the result that for a swarm of trajectories started on a closed loop in phase space and integrated for a fixed time, there is no change in the "action" defined as $\oint \bar{p} \cdot d\bar{q}$ along the path defined by the "tips" of the trajectories. Thus, the phase integral is a geometric invariant, and this result has nothing to do with adiabaticity. Preservation of "area" (defined as is done here) under a time-dependent switching is thus not a criterion for adiabaticity. Note carefully the difference between area as defined here and as defined by the Ehrenfest adiabatic criterion, wherein actual trajectories are implied.

lemma) is always conserved whereas not all time-dependent Hamiltonians have adiabatic invariants is now easily given. Suppose that the curve γ is the phase portrait of a periodic classical orbit evolving according to the dynamics of a time-independent Hamiltonian $H(p, q)$. Then γ will be an invariant curve in the sense that any point on the curve evolves into another point on the curve after an arbitrary time evolution governed by $H(p, q)$. If after γ has been so defined, a general time dependence of the Hamiltonian, $H(\bar{p}, \bar{q}, t)$ is introduced, there is no reason to expect that γ remains an invariant curve. Said differently, there is no reason to expect that γ is a trajectory, and thus, no reason to expect that all points on γ have the same energy. Thus, the exactly conserved "area" of the Poincaré lemma has no relation to the area defined by a periodic orbit, as is implicit in the original Ehrenfest definition of Eq. 2.2. Of course, should the transformation be adiabatic, the curve will be a trajectory, and thus isoenergetic, an invariant curve, and the Ehrenfest and Poincare areas are identical.

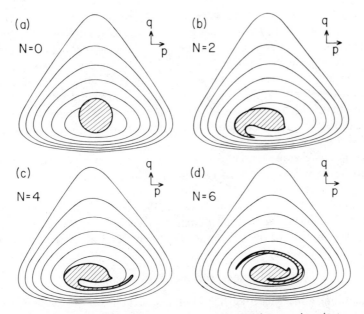

Figure 3. Illustration of the Poincaré lemma (see Figure 2) using a noninvariant curve in the phase plane of a Morse oscillator as an example. (a) Initial curve in the qp phase plane. (b–d) Time-evolved curve at 2, 4, and 6 periods (for small-amplitude motion). As the initial curve is not an invariant curve (i.e., it is not a trajectory), it time evolves, growing tendrils, as the frequency of individual points on the curve is different as the curve is not isoenergetic. However, phase plane area is conserved to numerical accuracy, as expected from the discussion of geometric invariants. The adiabatic hypothesis involves transformation of invariant curves (or surfaces) into invariant curves (or surfaces) as well as conservation of phase plane area.

In what follows we will refer to Poincaré lemma invariants as geometric invariants, to be distinguished from adiabatic invariants. The reason for making this distinction will emerge in Sections III and IV, where it is seen that most practitioners of the adiabatic method carry out actual calculations using geometric rather than adiabatic invariants!

III.   INTEGRABLE SYSTEMS OF HIGHER DIMENSIONALITY

What is the multidimensional analog, for integrable systems, of the Ehrenfest–Hertz adiabatic picture for one-degree-of-freedom systems discussed in Section II, and what is its relation to semiclassical quantization? Does multidimensional semiclassical quantization have a geometric interpretation? Without attempting to survey the development of semiclassical methods for nonseparable systems (3, 4, 18, 23, 54, 56–58), a few selected historical remarks are in order: Epstein (79) and Schwartzchild (86) introduced the use of angle action variables (78) for quantization of the dynamics of separable systems, Burgers (48, 49) related this work to certain aspects of the theory of adiabatic invariants, and Burgers (50) and Einstein (56) found generalizations leading to quantization of nonseparable but integrable systems. Others (56, 79, 86) have used actions for quantization, assuming their existence. Burgers (48–50) showed that, with appropriate assumptions, actions are adiabatic invariants, suggesting their use as a constructive computational tool as well as for quantization.

In the following subsection we review the Einstein (56) picture of quantization on invariant tori. This allows introduction of several important concepts and the notation used in what follows, and a brief review of current techniques for implementation of the EBK methods. This is followed, in Section III.B, by an overview of adiabatic and geometric invariance in relation to the assumption of the existence of such tori and an introduction of the Solov'ev (27) method.

A. Invariant Tori and EBK Quantization

Einstein (56), concerned that line integrals of the form

$$\oint p_i \, dq_i \tag{3.1}$$

are not, in general, canonical invariants, suggested the use of the first Poincaré invariant

$$\sum_i p_i \, dq_i = \bar{p} \cdot d\bar{q} \tag{3.2}$$

whose line integrals

$$\oint_\gamma \bar{p} \cdot d\bar{q} \tag{3.3}$$

are canonical invariants and are also invariant to continuous distortions of γ on Lagrangian manifolds. The expected N quantization conditions for an N-freedom system are then obtained by choosing the N topologically independent 1-cycles on the N-dimensional invariant torus characterizing integrable N-freedom systems. Thus (3, 4, 18, 23, 56–58, 86), the EBK conditions are

$$\frac{1}{2\pi} \oint_{C_i} \bar{p} \cdot d\bar{q} = (n_i + \tfrac{1}{4}\alpha_i)h \quad i = 1, 2, \ldots, N \tag{3.4}$$

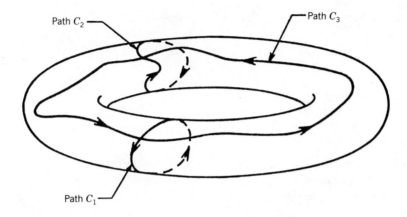

Invariant torus

Figure 4. The EBK (Einstein-Brillouin-Keller) quantization on an invariant torus as illustrated for a two-freedom system. Two topologically independent paths for implementation of the quantization condition of Eq. 3.5 may be taken as C_1 and C_3. As the line integral of $p \cdot dq$ is invariant to distortions of the paths on the Lagrangian manifold (the Poincaré lemma strikes again!) the choice of actual path is arbitrary; thus, path C_2 gives the same result as C_1. The torus is a Lagrangian manifold as it may be defined by all trajectories passing through any 1-cycle.

This now familiar geometric quantization, referred to as EBK quantization, is illustrated in Figure 4 for a two-freedom system. Note at once that the integration paths C_i are not classical trajectories but simply "paths" on the invariant torus. The assumption that all trajectories evolve on invariant tori is equivalent to the assumption of complete integrability of the problem, that is, the existence on a canonical transformation of the type

$$H(\bar{p}, \bar{q}) \rightarrow \tilde{H}(\bar{I}) \qquad (3.5)$$

reducing the N-freedom Hamiltonian that would normally be an explicit function of $2N$ phase space variables to a function of N "actions" with N ignorable coordinates (87) (the angle coordinates), which can be taken to give a coordinate system on the N-torus. The actions are constants of the motion, as the Hamiltonian lacks explicit dependence on the angles θ_i:

$$\dot{I}_i = -\frac{\partial \tilde{H}(\bar{I})}{\partial \theta_i} = 0 \qquad (3.6)$$

The angles evolve linearly in time, as their time derivatives are independent of the θ_i's:

$$\dot{\theta}_i \equiv \omega_i(\bar{I}) = \frac{\partial \tilde{H}(\bar{I})}{\partial I_i} \tag{3.7}$$

Trajectories on a specific torus thus evolve as

$$\theta_i = \omega_i(\bar{I})t + \delta_i \tag{3.8}$$

where the phases δ_i are determined by $t = 0$ initial conditions.

Einstein was well aware of the work of Poincaré (88) showing that existence of such a global transformation cannot be generally assumed and thus that his

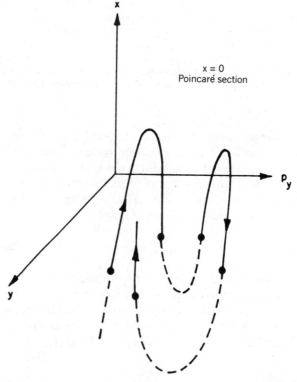

Figure 5. Illustration of the definition of the $x = 0$ Poincaré surface of section for a two-freedom system. Taking x, y, and p_y as independent variables, which is always possible for time-independent Hamiltonian, points are collected on (y, p_y) plane whenever x passes through 0. See also Figure 6.

quantization method was not of general applicability. However, Poincaré's geometric picture of quantization on tori has laid much of the basis for the next 70 years work on integrable and nearly integrable systems. This is in part due to the prevalence of tori and their persistence under perturbation, even for nonintegrable systems, a fact not appreciated by Einstein and Poincaré and only developed much later (18, 89–94).

Since the early 1970s several methods have been developed for implementation of EBK quantization on invariant tori, and reviews have appeared (3, 4, 23, 41). Marcus and co-workers (1, 4) have used direct integration of Hamilton's equations combined with the Poincaré surface of section to find individual tori, as illustrated in Figures 5 and 6. Once such tori are found, and

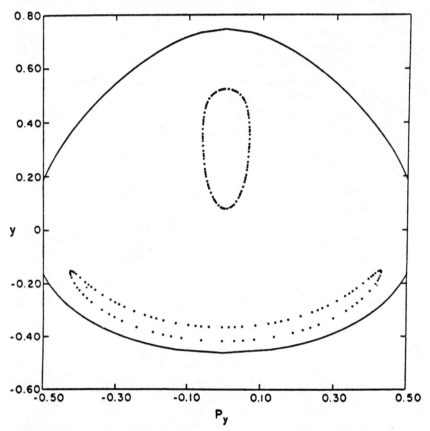

Figure 6. Poincaré section for a high-angular-momentum trajectory on the Henon–Heiles potential surface. The trajectory seems to lie on a phase space surface cut twice by the sectioning plane. The Surface is (to numerical precision) an invariant torus.

their topology elucidated, the appropriate line integrals of the invariant action 1-form, $\bar{p} \cdot d\bar{q}$, may be carried out. Initial conditions and the energy are then adjusted until a "quantizing" torus is found, thereby determining the energy from an iterative search. The major difficulty with this approach, which works well for integrable (and almost integrable) systems of two degrees of freedom, is that no simply calculable equivalent of the Poincaré section method exists for systems of higher dimensionality. The problem of numerical integration of a trajectory in, say, 10-dimensional phase space is simple: Determination of the topology of the 5-dimensional invariant torus and the appropriate integration paths of Eqs. (3.4), given that the trajectory indeed lies on such a torus, is not straightforward. Thus, few examples (an exception being Ref. 95), of the application of direct-trajectory-based determination of tori have been carried out for systems of higher dimensionality. Methods based on determination of the actions for nonseparable systems by Fourier representation of trajectories in initially unknown angle action variables (2, 3) as well as direct calculation of actions via formal "normalization" of the Hamiltonian (23, 96–98) via the Birkhoff (99) and Birkhoff–Gustavson (100) methods have been plagued by convergence problems or simply been too algebraically cumbersome to extend to many-degree-of-freedom systems. Or so it seemed in 1984.

Within the past two years three methods have emerged that circumvent these problems: A new Fourier analysis method, introduced by Heller and DeLeon (101, 102) and refined by Ezra and Martens (103), has served to allow routine quantization of multidimensional systems; additionally, Fried and Ezra (104), using sophisticated Lie transform methods (105, 106), have made substantial advances in the Birkhoff–Gustavson method. These developments are reviewed in Ref. 41 and are not discussed further here, except to note that in finding appropriate zero-order variables for initiation, the adiabatic switching where the work of Farrelly is of importance (106). Finally, the method of adiabatic switching for EBK quantization of multidimensional systems, introduced as a practical computational tool by Solov'ev (27, 28) and independently by Johnson (29), has developed rapidly (30–40) during this same time period and has proved to be a convenient computational tool.

B. Adiabatically Switched Tori

Ehrenfest clearly understood the importance of adiabatic invariance both at the fundamental level, that adiabatic invariants are the quantities to be quantized, and at the practical level, that adiabatic invariance might be used to quantized all motions adiabatically connected to harmonic motion, where it may, under appropriate restrictions, be assumed that quantization is understood (45). Solov'ev is the father of more recent attempts to implement this idea as a practical computational tool for multidimensional nonseparable systems.

Stated as an implementation of the EBK quantization scheme, a possible

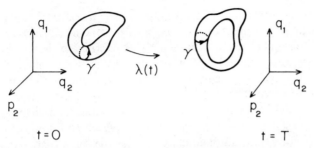

Figure 7. The EKB quantization on tori as implemented by adiabatic switching. As the coupling $\lambda(t)$ is changed, invariant tori are supposed to distort smoothly. It can be imagined that a continuous one-parameter family of invariant tori exist between the initial and final tori; if the initial torus satisfies the EBK quantum conditions, so does the final, and all that need be done is to evaluate H at any point on the final torus. This is the ASTA picture discussed in the text. However, if at any stage of the switching process a periodic orbit, or even an invariant surface of dimension less than N for an N-freedom system, is encountered, the ASTA is false, yet the usually method works anyway!

scenario for adiabatic switching of tori is introduced in Figure 7. Imagine a zero-order Hamiltonian H^0 that is separable in Hamilton–Jacobi co-ordinates-ordinates or integrable with known angle action variables. In either case an unperturbed torus T^0 satisfying (unperturbed) EBK conditions is assumed to be easily analytically found, and such a torus defines the starting point of the Solov'ev method. We now imagine that under adiabatic introduction of a nonseparable perturbation,

$$H(\bar{p}, \bar{q}, t) = H^0(\bar{p}, \bar{q}) + s(t)V(\bar{q}) \tag{3.9}$$

where $s(t)$ is a switching function, as in Eq. 2.6, during the time T the initial quantized torus T^0 is continuously switched into T, an invariant torus of the perturbed Hamiltonian $H(\bar{p}, \bar{q}) = H^0 + V$. In particular, for intermediate times there is a continuous one-parameter family of tori intermediate between T^0 and T. Under restrictions equivalent to these geometrically expressed assumptions, Burgers, a student of Ehrenfest's, showed in a landmark paper (50) that line integrals of the form

$$\int_C \bar{p} \cdot d\bar{q} \tag{3.10}$$

are adiabatic invariants, that is,

$$\oint_C \bar{p} \cdot d\bar{q} = \oint_{C^0} \bar{p} \cdot d\bar{q} \tag{3.11}$$

It is important to note that in the earlier work of Epstein (79) and Schwartzchild (86) and even in Burgers' earlier work (48, 49) it was necessary to assume separability of the system: This restriction is removed, under appropriate restrictions, in Ref. 50. As in the EBK method, C is an independent 1-cycle on the torus $\mathbf{T}$, which is now chosen to be continuously connected to C^0 on $\mathbf{T}^0$ via the continuous family of tori between $\mathbf{T}^0$ and $\mathbf{T}$. Simply stated, the Burgers result implies that if we start on an EBK quantized torus and if adiabatic switching takes us through a continuous family of tori to a new final torus that automatically satisfies the same EBK conditions, the energy of this new torus (which is simply the energy at any point on it) is the semiclassical eigenvalue for the coupled system. We refer to the scenario of Figure 7 and the present discussion as the adiabatically switched torus assumption (ASTA).

C. Critique of the Assumption of Adiabatically Switched Tori

Several questions arise at once:

- Under what conditions does the ASTA actually apply, that is, when does adiabatic switching produce a continuous family of tori as suggested in Figure 7?
- If the assumptions of the ASTA are false, can the method work anyway?
- How can one best implement the ASTA, assuming that it makes sense to do so?

Our strategy will be to discuss these questions from a qualitative point of view here and in the following section and to undertake a more detailed analysis in Section V. This round-about approach follows from the fact that the one-line answer to the first question in the preceding is "almost never," and yet the method works anyway!

In his original work, Ehrenfest was fully aware of the problem of classical degeneracy and its impact on a multidimensional generalization of his results on adiabatic invariants for one-dimensional systems. Classical degeneracy occurs when the frequencies ω_i are rationally related. For two-freedom systems this would imply that $n\omega_1 = m\omega_2$, whereas for the more general case a system has a degeneracy if $\bar{n}\cdot\bar{\omega} = 0$, where $\bar{\omega}$ is a vector whose components are the frequencies ω_i (see Eq. 4.9) and $\bar{n}$ is a nonzero vector whose component are positive and negative integers. When, at any value of $s(t)$, a degeneracy occurs, the system is said to be in resonance, and the orbit will be periodic. Such a periodic orbit does not fully explore an invariant torus. In contrast, an orbit with incommensurate frequencies is ergodic on its underlying torus and thus self-defines the torus.

Suppose now that we wish to implement Solov'ev's quantization by running a single long-time trajectory, that is, by starting a trajectory on T^0 and allowing it to evolve with the Hamiltonian $H(\bar{p}, \bar{q}, t)$ until it arrives on the perturbed torus T, having been instantaneously on one of the intermediate tori at all times during the process. There are then two problems: (a) If the orbit does not fully explore the torus on a time scale short compared to the switching time, the averaging needed to ensure that the result of an adiabatic switching estimate of the EBK energy will be completely independent of initial conditions may not take place and (b) as a periodic orbit does not itself define a torus, existence of degeneracies may well imply that the family of tori of the ASTA does not exist. In fact, Burgers (50), in proving adiabatic invariance for multidimensional systems, explicitly assumes that along the whole switching process $\bar{n} \cdot \bar{\omega} \neq 0$. This is basically equivalent to the assumption that if, on an initial quantizing tours T^0, $\bar{n} \cdot \bar{\omega} \neq 0$ for all $\bar{n}$, then, even though the ω_i, can generally be independent functions of $s(t)$, $\bar{n} \cdot \bar{\omega}$ will also not vanish during the switching process. Thus, at each stage of the switching process an individual trajectory does indeed define a torus, thus assuring the ASTA. As the rationals are dense on the real line, the Burgers assumption is almost always false.

Does any of this matter? It often does. Consider an unperturbed Hamiltonian consisting of two (or more) harmonic oscillators of equal frequencies:

$$H^0 = \tfrac{1}{2}(p_1^2 + p_2^2) + \tfrac{1}{2}\omega^2(q_1^2 + q_2^2) \tag{3.12}$$

All orbits of such a pathological (but important) system are periodic. In such a case the trajectories do not geometrically define any zero-order tori and thus do not allow even a naive trajectory-based implementation of the canonically invariant EKB conditions of Eq. 3.5. Thus, for this very simple case it is not clear what the adiabatic invariants will be, and thus it is not clear what the appropriate quantization conditions are! A similar degeneracy occurs in the hydrogen atom, and Ehrenfest (44, 45) discussed the ensuing difficulties, which were resolved in that case by Epstein (79), who suggested separation of variables for the corresponding Hamilton–Jacobi equation. Addition of a coupling term will generally break the degeneracy, and thus one might think that appropriate zero-order tori could be defined by starting at a finite coupling and (adiabatically) switching back to zero coupling. Ehrenfest (45) actually discussed this possibility and noted its futility as the process would take infinite time to implement. The problem is similar to that of degenerate perturbation theory in quantum mechanics: The correct zero-order variables are not known until the perturbation is diagonalized. We conclude that a degeneracy at $t = 0$ does indeed disturb the implementation of EBK quantization via adiabatic switching. Johnson and Pechukas discuss several illuminating cases of invariance (and noninvariance) of actions under adiabatic switching beginning with such degenerate orbits (68), with the conclusion that even if the system is integrable during the whole switching process, new subsidiary conditions need to be introduced to ensure adiabaticity of actions.

Is this the only difficulty? Suppose that the unperturbed separable problem is nondegenerate, and thus a single trajectory defines a torus, which we assume to satisfy the EKB conditions. Can we switch with impunity? Not at all. For a two-degree-of-freedom system, if the two frequencies *are* independent functions of the coupling, ω_1/ω_2 will pass through rational values infinitely often during the switching process. Each time this happens, an instantaneous torus is not defined, and in fact, each periodic orbit will generally lie at the center of a resonance: If the commensurability is of low order (i.e., $\omega_1/\omega_2 = p/q$ for small relatively prime integers, say, $1:1$ or $2:3$), the extent of the resonance may be substantial, seemingly (but see Section IV.C2) destroying any possibility of adiabatic switching anywhere near the resonance, as illustrated in Figures 8 and 9, which will be of interest later.

It would thus seem that resonances will almost always spoil application of the method. Quoting Johnson and Pechukas (68): "Solov'ev deserves credit not so much for the originality of his suggestion... as for the courage to put it

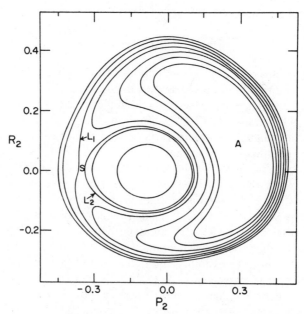

Figure 8. Nonlinear resonance in the dynamics of H_2O. The composite Poincaré section shows distinct regions of phase space corresponding to (a) normal-mode trajectories [see Figure 9(a)] and (b) local-mode trajectories [see Figure 9(b)]. Adiabatic switching will have difficulty in connecting these two types of motion, as crossing a separatrix will give a zero frequency, and no switching is slow compared to infinite time needed to cross a separatrix. (Figure from Ref. 115 with permission.)

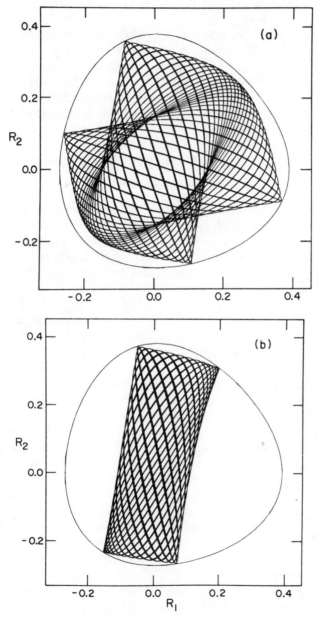

Figure 9. Normal-mode trajectory that (*a*) has the symmetry of the full Hamiltonian, is compared with a local-mode type trajectory and (*b*) appears to break the symmetry. Adiabatic switching has difficulty in connecting these two quite distinct types of motion; see Figure 8 and the numerical results of Table 2. (Reproduced from Ref. 115 with permission.)

into print, because it seems on the face of it that the method cannot work. Resonances will ruin it."

Solov'ev, aware of the potential difficulties due to resonances, based his hopes for success of the method not on the ASTA but on a general faith in the correspondence principle, following an analogy to the quantum adiabatic theory of Born and Fock (107). His numerical results (27, 28) and those of other (29–40) indicate that the method does work despite mathematical objections. We illustrate this in the following section and only provide an analysis of the switching process in Section V.

IV. APPLICATIONS OF THE ADIABATIC EBK METHOD

A.   Implementation

Given that nonlinear variational approaches to determination of invariant tori have proven problematic (see Refs. 2 and 3), Solov'ev chose to implement the adiabatic quantization via integration of classical trajectories. With the exception of Ref. 39, all subsequent workers have followed suit once an appropriate T^0 is known. Implementation is simple in principle: Given a trajectory starting on T^0, integration of Hamilton's equations will yield, assuming that the switching is indeed adiabatic, a trajectory that at every point in time is on one of the intermediate tori (generated by the same dynamics) until, at times later than T, it will continue to evolve on the final fully perturbed torus, and all we need do is evaluate the Hamiltonian at any time $t \geqslant T$ to obtain the semiclassical energy. One would, of course, lengthen the switching time or change the switching function $s(t)$ to test convergence. However, assuming validity of the ASTA, only a single (long-time) trajectory need be run; no searches for nonseparable quantizing tori need be carried out as one need not visualize the torus at any stage of the calculation; and thus, multidimensional quantization is no more difficult in principle than one dimensional. This combination of features is what makes exploration and exploitation of the method so attractive.

However, in practice, as suggested by Solov'ev, most workers have adopted the strategy of integrating not one but a family (or "swarm") of trajectories chosen to give a sampling of the process of transforming the initial into the final torus. An average energy is then calculated after switching is complete, and ΔE^{rms}, the rms deviation from this average, is then a useful figure of merit for the calculation. Solov'ev suggests (27) sampling in proportion to the square of the quasi-classical wavefunction. Others have simply used sampling techniques to average over the initial-angle variables on the unperturbed torus. Johnson (32) gives a detailed analysis of this latter averaging. Skodje et al. (31) have pointed out its relation to classical Hamilton–Jacobi perturbation theory. Jaffe (39), by construction of the tori themselves via sequences of canonical transformations (rather than by integration of Hamilton's equations), calculates an energy averaged over the final torus. There is, of course, an analog with purely quantum calculations: Given an approximate wavefunction, the "local energy" (108)

$$E(\bar{x}) \equiv \frac{H\Psi(\bar{x})}{\Psi(\bar{x})} \tag{4.1}$$

will usually be a strong function of coordinate-ordinate, while $\bar{x}$ the averaged energy

$$\bar{E} = \langle \Psi, H\Psi \rangle / \langle \Psi, \Psi \rangle \tag{4.2}$$

is the usual variational functional and has second-order stability. Given the correspondence between the classical phase space torus and the quantum wavefunction (3), calculation of the semiclassical energy via averaging over the torus is a seemingly appropriate analogy. From a more philosophical perspective, Brillouin (69) has argued in a similar context that it is always a swarm of trajectories that should be considered in relation to the corre-

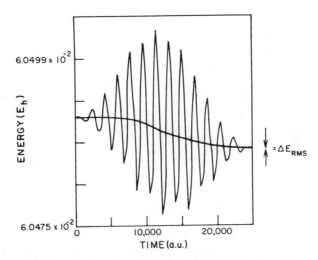

Figure 10. Adiabatic switching for local-mode motion in HOD. The oscillatory curve is the instantaneous energy as a function of time along a single classical trajectory. The smooth curve results from an average of ~ 25 such calculations, providing an average over the angle variables. Averaging smooths the oscillations and provides a result independent of any specific initial condition. The spread of energies about the average also provides a self-diagnosis of the efficacy of the procedure.

spondence principle, although Heller (109, 110) has recently discussed cases where individual periodic orbits may play a role in eigenfunction support.

One practical result of averaging is illustrated in Figure 10, where the energy along the switching process is shown as a function of time for a single trajectory and for a time-dependent energy calculated as the instantaneous average during the turning on of the coupling in a model two-degree-of-freedom system (31). Averaging evidently washes out the phases of local oscillations appearing in the energy of a single trajectory. At the end of the switching process the vanishing of ΔE^{rms} provides a necessary (but not sufficient) condition for the success of the method, and if ΔE^{rms} is nonvanishing, it provides an error estimate for a given semiclassical eigenvalue. Convergence of ΔE^{rms} as a function of switching time T and as a function of the switching function itself provides a useful diagnostic; a typical result is illustrated in Figure 11, where the switching function $s_2(t)$ of Eq. 2.6b suggested by Johnson (29) has been used. A T^{-3} power law was found (31). That this is not an accident is shown in Section V.

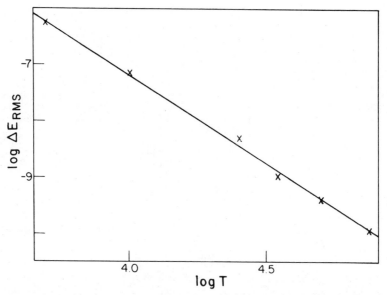

Figure 11. Convergence of the root mean square (rms) deviation of the energy from the average value as a function of total switching time T using switching function $s_2(t)$ of Eq. 2.6b. This switching function has two continuous derivatives at $t = 0$ and $t = T$. As discussed in Section V, the observed T^{-3} inverse-power-law dependence is not a surprise.

B. Geometric Invariance of the EBK Condition

Conservation of line integrals of the form

$$I_\gamma = \oint_\gamma \bar{p} \cdot d\bar{q} \tag{4.3}$$

assuming continuous adiabatic distortion of the underlying torus (following the ASTA) is one statement of adiabatic invariance for multidimensional systems. The implication is that a trajectory starting on T^0 with a corresponding action I will end up on T with the action unchanged. An alternative, and surprisingly inequivalent, interpretation of this invariance involves viewing the invariance of the phase space line integral as it continuously distorts on the time-evolving torus during the switching process. That is, in operational terms, we can imagine direct propagation of γ forward in time via the full dynamics rather than propagation of a single trajectory, which would then be allowed to fully cover the perturbed torus, allowing subsequent construction of γ once the torus was displayed. That distortion of γ on a single invariant torus does not change the value of the integral is a usual invariance of line integrals of 1-forms on Lagrangian manifolds (109).

Reinhardt and Skodje (112) naively imagined that failure of the constancy of I_γ if γ was propagated forward (pointwise by integration of a large number of trajectories) could be used as an alternate figure of merit to ΔE^{rms} during the switching process: If I_γ varied by more than a single unit of h, the system would no longer have retained constancy of action to a degree sufficient to allow enforcement of the EBK conditions. To their initial surprise, propagation of I_γ forward in time gave absolutely constant values whether or not the switching was adiabatic! What had been observed was that if γ is thought to define a line of initial conditions in the phase space, the family of all trajectories passing through said line is a Lagrangian manifold whether the dynamics is governed by a time-independent or a time-dependent Hamiltonian, and thus the line integral of $\bar{p} \cdot d\bar{q}$ is invariant to continuous distortions on the (time-dependent) manifold. This is none other than the geometric invariance discussed in Section II and illustrated in Figure 2. Stated another way, the EBK quantization conditions are identically satisfied in the extended phase space of Figure 2, and thus even though not all points on the forward-propagated γ necessarily even have the same energy, the (classical) state arrived at may be characterized by a unique quantum number. This geometric invariance of quantization condition, obtained by using a family of trajectories on the initial torus, is clearly an underlying reason for the success of the swarm method, again confirming the analysis of Brillouin (69). That the method succeeds in cases where it might well be thought to necessarily fail is illustrated in the following section.

C. Examples of Adiabatic Quantization of Multidimensional Bound Motion

The recent literature contains applications of EBK quantization via adiabatic switching to model coupled oscillator problems in two and higher degrees of freedom (27, 29, 30–32, 34–36, 39), semiclassical quantization of asymmetric rotor problems (33, 37), quantization (40) of three-freedom model dynamics for H_3^+, D_3^+ and T_3^+, the problem of the electronic structure of Rydberg states in external fields (28, 113), and the two-degree-of-freedom model (114) of Jahn–Teller dynamics (38). It will suffice to give a brief discussion of a few representative examples. These are chosen to illustrate the power of the method when it works well and to indicate where failure may be expected.

1. Two Oscillators: Ambiguity in Choice of Zero-Order Variables

If the trajectories of the zero-order dynamics uniquely define a torus, application of the method for small perturbations is straightforward: Skodje et al. (31) have considered a model for the local-mode vibrations of HOD. Taking the zero-order Hamiltonian as uncoupled Morse local oscillators for the OH and OD stretches, the OH and OD frequencies for small oscillations may be taken as (at least approximately) incommensurate, and any trajectory with initial energy in both modes will ergodically fill an invariant torus: Switching on coupling between the modes typically yields results of the quality illustrated in Table 1 and Figures 10 and 11.

The situation is markedly different in the case of low-order degeneracy or near degeneracy. A classical degeneracy occurs if two frequencies are

TABLE 1 Vibrational Energy Levels for HOD in a Two-Degree-of-Freedom Model as Determined by Adiabatic Switching[a]

Local-Mode Quantum Numbers	$E^{\text{adiabatic}}$	$E^{\text{exactquantum}}$	ΔE^{rms}
(0, 0)	1.50896	1.50893	4.8×10^{-7}
(1, 1)	4.438491	4.43852	3.8×10^{-7}
(2, 1)	6.048933	6.04898	1.1×10^{-6}
(4, 0)	7.799228	7.79926	2.7×10^{-5}

[a]The zero-order Hamiltonian was taken to be the local-mode model consisting of two uncoupled Morse oscillators, one for the OH stretch and one for the OD vibrational motion. The coupling $G_{12}p_1p_2$ was then switched on adiabatically using the switching function $s_2(t)$ of Eq. 2.6b. This is an ideal case for the adiabatic method as OD and OH frequencies are incommensurate (at least to low order), and thus no major nonlinear resonances affect the switching. Shown are the adiabatic, exact quantum, and ΔE^{rms} for a switching time of 25,000 atomic units of time, which is about 70 OH vibrational periods. A more extensive summary of these calculations appears in Ref. 31, which includes all parameter values. Energies are in Hartree atomic units.

rationally related. Consider H_2O as two coupled Morse oscillators:

$$H = \sum_i \left\{ \frac{p_i^2}{2\mu_i} + D[1 - \exp(aq_i)] \right\} + \frac{\cos(\Omega)p_1 p_2}{M_0} \qquad (4.4)$$

The classical dynamics of this simple system falls into two types: one with equal or nearly equal energies, and thus frequencies, in each OH oscillator; the other with widely disparate frequencies. It is the possibility of a low-order matching of frequencies (in this case 1:1 degeneracy) that leads to these two dynamical types. Jaffé and Brummer (115) have discussed the classical dynamics of these normal- and local-mode motions; Sibert, Hynes, and Reinhardt (116, 117) have recast and extended this analysis in the language angle action variables and low-order resonances. At its most elementary level, the content of this analysis is illustrated in Figure 8, where a composite Poincaré section is shown for the fully coupled two-freedom dynamics. It is evident that the phase space motion is of two basic types, illustrated by the trajectories of Figure 9(a) and 9(b), giving rise to two families of concentric invariant tori. This behavior is our first example of a nonlinear resonance, and the one illustrated is strong! In the present case the two types of motion correspond to classical normal- and local-mode tori. The boundary separating them is referred to as a separatrix. Stated baldly, adiabatic switching can be expected to fail when the torus crosses such a separatrix. This is illustrated in

TABLE 2 Energy Levels for the Vibrational States of the H_2O Molecule in a Two-Degree-of-Freedom Model[a]

State Type and Quantum Number	$E^{adiabatic}$	$E^{exactquantum}$	ΔE^{rms}
$N(0,0)$	1.749335	1.744533	6.6×10^{-3}
$L(2,0)$	5.038906	5.038517	3.2×10^{-6}
$L(4,0)$	8.030374	8.030257	2.7×10^{-6}
$L(5,0)$	9.410856	9.410772	1.8×10^{-6}

[a] The zero-order Hamiltonian consisted of one of two choices: (1) H^{local}, which is an identical Morse oscillator description for each of the OH stretches; (2) H^{normal}, which is the usual normal-mode Hamiltonian. In either case the difference $(H - H^0)$ was taken as the perturbation and switched on via the adiabatic method, switching using the function $s_2(t)$; depending on the type of classical motion (see Figures 8 and 9) the local or normal Hamiltonian must be chosen. If the incorrect choice were made, the figure of merit ΔE^{rms} would be several orders of magnitude larger than those shown. The method is thus self-diagnosing. In the table, taken from Ref. 31, N denotes a normal-mode state (which may be thought of as a strong 1:1 resonance in the local-mode picture, see Refs. 115–117). L denotes a local-mode state. The switching time is as in Table 1. Tunneling corrections are not made. The exact quantum result is the average of symmetric and antisymmetric states.

Table 2, where it is seen that the local-mode zero-order Hamiltonian

$$H^{\text{local}} = \frac{p_1^2}{2\mu_1} + \frac{p_2^2}{2\mu_2} + V(q_1) + V(q_2) \tag{4.5}$$

allows quantization of the local-mode coupled states, whereas the normal-mode zero-order Hamiltonian

$$H^{\text{Normal}} = \sum_{i=1}^{2} \frac{P_i^2 + Q_i^2}{2} \tag{4.6}$$

allows adiabatic quantization of the normal modes. Using ΔE^{rms} as the figure of merit clearly signals difficulties of interpretation and possible failure of the method when local-mode coupled states are generated from a normal-mode zero-order Hamiltonian, and vice versa. The lesson is clear: When degeneracy leads to a large-scale classical resonance structure, different zero-order Hamiltonians need to be introduced to get into the different resonance zones. Said another way, the zero-order torus must yield a caustic structure of the semiclassical wavefunction that maps smoothly into that of the fully perturbed system.

2. EBK Quantization of Initially Degenerate Problems

Solov'ev (27, 28), Skodje et al. (30, 31), Grozdanov et al. (34–36), Patterson et al. (33, 37), and Sinai and Farrelly (113) have discussed adiabatic quantization of degenerate systems. As discussed in Section III.C.1, if the degeneracy is present in the zero-order separable Hamiltonian, the "correct" zero-order EBK tori that is valid in the neighborhood of zero coupling must be found by a method not involving running trajectories for the zero-order problem. One can use the methods of Marcus et al. (1, 4) to carry out EBK quantization for a small *but finite* value of the coupling constant (assuming that the degeneracy is broken) or proceed analytically. Normal-form (99, 100) methods used by Swimm and Delos (96) and Jaffe and Reinhardt (97) to perturbatively resolve the degeneracy for the initially 1:1 degenerate Henon–Heiles problem have been used by Skodje et al. to carry out adiabatic EBK quantization. The Lie algebraic methods mentioned in Section III.C (104–106) provide a more powerful set of tools to find the correct zero-order variables needed to resolve the classical degeneracy. Or one can proceed more intuitively based on a physical understanding of the dynamics. Prospective zero-order tori can be easily tested: If under adiabatic development the caustic structure of the classical tori retain their topological form, the correct zero-order picture has been found.

An interesting example of a successful application via an intuitive approach that also illustrates the role of low-order resonances in determining the range of validity of the method is the adiabatic quantization of the triply degenerate Hecht Hamiltonian (118) carried out by Patterson (33).

Patterson has treated the Hecht model Hamiltonian (118):

$$H_v = w_3 \underline{n} + (G_{33} + 2T_{33})\underline{l}^2 + T_{33}\{10\underline{m} - 6\underline{n}^2 - 8\underline{n}\} + X_{33}\underline{n}^2 \qquad (4.7)$$

where

$$\underline{n} = \underline{n}_x + \underline{n}_y + \underline{n}_z \qquad n_i = \tfrac{1}{2}(q_i^2 + p_i^2) \qquad (4.8a)$$

$$\underline{l}^2 = \underline{l}_x^2 + \underline{l}_y^2 + \underline{l}_z^2, \underline{l}_x \underline{l}_y, \underline{l}_z \qquad (4.8b)$$

being the usual angular momentum operators, and

$$\underline{m} = \underline{n}_x^2 + \underline{n}_y^2 + \underline{n}_z^2 \qquad (4.8c)$$

as a model of vibrations in molecules such as SF_6 and SiF_4. Of interest in the Patterson study is quantization as a function of the "anharmonicity" T_{33}. As the system is triply degenerate at $T_{33} = 0$, correct zero-order initial conditions must be determines by a method that correctly anticipates the caustic structure of the fully coupled dynamics. Patterson has accomplished this by an averaging procedure (33) yielding the integrable Hamiltonian

$$H^0 = (w_3 - 8T_{33})\underline{n} + G_{33}\underline{l}^2 - \frac{5T_{33}}{2}\left[\frac{3(2n+3)^2}{8} - \frac{l(l+1)}{2}\right]$$

$$\times \frac{3\underline{l}^4/5 - [\underline{l}_x^4 + \underline{l}_y^4 + \underline{l}_z^4]}{\underline{l}^2}(l+1)^2 \qquad (4.9)$$

which allowed a correct zero-order quantization for small T_{33}. Figure 12 shows the adiabatically determined semiclassical energy as a function of switching speeds. What is observed is a classical analog of the difference between adiabatic and diabatic switching in quantum mechanics near a level crossing: Rapid passage forces retention of structure, whereas slow switching samples a larger region of phase space with some trajectories seemingly following the initial torus, and other trajectories, as a function of initial conditions, pass through a resonance separatrix zone, ending on a torus occupying a different region of phase space and thus giving a very different energy dependence. (The fact that a torus can fragment during the switching process is also seen in the Johnson–Pechukas models of Ref. 68.) Comparison with the exact levels (Figure 13) shows an uncanny resemblance, even in the region of coupling, where ΔE^{rms} is large, indicating a strong classical-quantum correspondence and that the adiabatic method contains information not

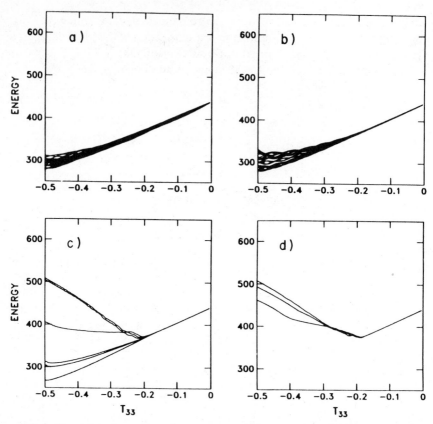

Figure 12. Adiabatic switching through a resonance zone of the Hecht Hamiltonian. In going from (a) through (d) the switching is slowed, revealing a transition from diabatic to adiabatic behavior; the initial quantum numbers were $n = 20$, $C = 20$. Comparison with the actual quantum levels, shown in Figure 13, indicates a striking similarity. (Figure reproduced from Ref. 32 with permission.)

envisaged in the original formulations. Patterson, Smith, and Shirts (37) have shown, in a model problem of the adiabatic quantization of an asymmetric rotor as a function of asymmetry, that quantization across a separatrix may be adiabatically performed using an appropriate reinterpretation of the quantization condition past the separatrix.

3. Adiabatic Quantization of Classical Chaos

The preceding results empirically demonstrate that the adiabatic method can yield useful information when the ASTA is false in the sense that between the initial and final tori, degeneracy may occur, and only if stringent (and unlikely)

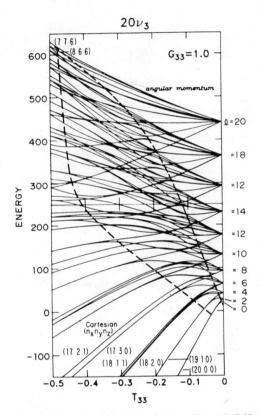

Figure 13. Quantum energy levels of Hecht Hamiltonian. (From Ref. 32, with permission.)

conditions (55) are satisfied will the actions be rigorously adiabatic invariants. Thus, the method can succeed when the continuous family of tori "intermediate" between the initial and final tori does not exist. But what if there is no final torus? Figure 14 shows adiabatically obtained semiclassical eigenvalues for the Henon–Heiles potential as obtained by Skodje et al. (31) as a function of the coupling constant. The parts of the correlation digram that extend into the upper right part of the figure, as indicated by the dashed line, indicate adiabatic switching into regions of separatrix chaos, where final tori do not exist. There is no signature of this in the level diagram, but switching for long times indicates (31) eventual growth of ΔE^{rms}, rather than the T^{-3} decay of Figure 11 obtained using the same switching function. Except in the immediate vicinity of crossings (where a uniform quantization is indicated but not carried out here), agreement with exact quantum mechanics is of the same quality as before the transition to chaos. Thus, at the least, an asymptotically

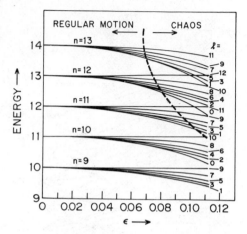

Figure 14. Semiclassical levels as determined by adiabatic switching for the Henon–Heiles Hamiltonian. Levels in this correlation diagram, where the epsilon measures the nonlinear coupling, include a region of classical phase space where no final tori exist, as indicated by the region to the right of the dashed line. The adiabatic method gives accurate semiclassical eigenvalues even in the presence of chaos.

valid method has been found and clear indication given that the method has a validity well beyond the restrictive assumptions leading to the analysis of Ehrenfest, Burgers, and Dirac. What is going on?

V. ANALYSIS OF THE METHOD FOR NONINTEGRABLE SYSTEMS

Adiabatic switching seems to work despite the problem of resonances and even in the presence of at least some types of chaos. This needs to be explained. Solov'ev (27) and Berry (66) have put forward the idea that even though an infinite number of resonances are encountered during the switching process, most have little effect. A simple argument to this end is as follows for two-freedom systems: for most resonances, $w^1/w^2 = p/q$, with p and q reduced so as to contain no common factors, will be of high order, that is, q will be a "large" integer. This implies that the time scale for actually completing one full period of the motion is correspondingly long: $s(t)$ will thus have changed, breaking the resonance, before the period is complete. What difference, then, can it possibly make whether an instantaneous orbit is periodic or nonperiodic?

In what follows this simple physically based idea is made quantitative through consideration of adiabatically switched point maps, which also allow analytic investigation of the role of different switching functions for these analogs of two-freedom Hamiltonian systems. The discussions of Sections V.A, V.B and V.C follow that of Dana and Reinhardt (70, 71).

A. Point Maps and Adiabatic Switching

Analytic treatment of the dynamics of nonintegrable two-degree-of-freedom Hamiltonian systems is complex and visualization of two-dimensional tori in the four-dimensional phase space difficult. The Poincaré section technique of Figures 5 and 6 attempts to circumvent the visualization process by taking the $x = 0$ projection of the dynamics. It has long been realized that analogs of the Poincaré sections may be directly calculated by two-dimensional point maps,

$$\begin{pmatrix} q_{n+1} \\ p_{n+1} \end{pmatrix} = M \begin{pmatrix} q_n \\ p_n \end{pmatrix} \qquad (5.1)$$

where M is a vector-valued function of the input vector (q_n, p_n). The map M recursively generates a series of points in the p–q plane. The analog of energy and phase volume preservation of Hamiltonian dynamics is expressed in the requirement that the mapping function M preserve area.

Two point maps of interest in what follows will be the Chirikov, or "standard," map

$$I_{n+1} = I_n + k \sin \theta_n \qquad (5.2a)$$

$$\theta_{n+1} = \theta_n + I_{n+1} \qquad (5.2b)$$

(which is conventionally written in the angle action form shown, with k and a perturbation parameter, measuring the strength of the nonlinear term) and the Siegel–Henon map

$$p_{n+1} = p_n \cos \alpha - (q_n - k p_n^2) \sin \alpha \qquad (5.3a)$$

$$q_{n+1} = p_n \sin \alpha + (q_n - k p_n^2) \cos \alpha \qquad (5.3b)$$

where k is, again, a strength parameter and α is a rotation angle. Typical composite mappings are shown for these two area-preserving maps in Figures 15 and 16.

Adiabatic switching may be introduced in either of the preceding maps by replacing k by

$$k(n/N) \equiv k s(n/N) \qquad (5.4)$$

Here $s(n/N)$ is the switching function, analogous to those of Eqs. 2.6a and 2.6b, where n is the iteration number (which measures the "time" since the switching

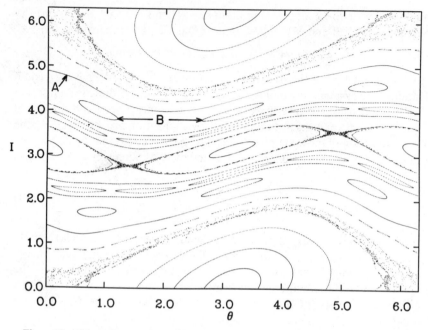

Figure 15. Composite overview of the phase plane for the standard map for an "intermediate" value of the nonlinear parameter k. The symbol A denotes a typical rotational torus, and B denotes a typical resonance island chain.

process began) and N is the number of iterations needed to let the coupling grow from 0 to k. The correspondence to continuous time switching is given by $t/T = n/N$. Typical switching functions are thus (analogs of Eqs. 2.6a, b)

$$s(n/N) = \frac{n}{N} \tag{5.5a}$$

$$s(n/N) = \frac{n}{N} - \frac{\sin\left[2\pi(n/N)\right]}{2\pi} \tag{5.5b}$$

and others are analogous to those of Eqs. 2.15, 2.16, and 2.17.

These allow varying degrees of "smoothness" in the switching process. Because the switching functions need only be defined for integer values of n, it might seem that their smoothness is irrelevant, but we will see that this is not the case.

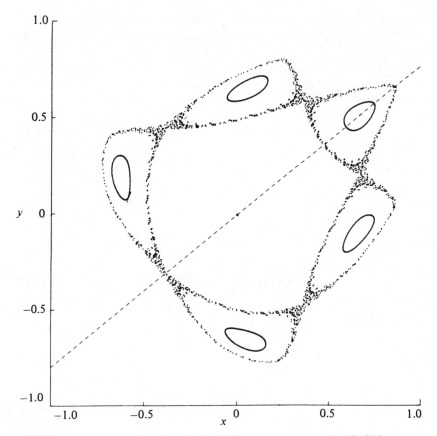

Figure 16. Separatrix chaos for the Siegel–Henon map surrounds a fivefold resonance island chain.

B. Neglect of Resonances: Perturbative Adiabatic Analysis of the Standard Map

For small k most of the I–θ phase plane is filled with continuous invariant curves running from 0 to 2π and is periodic in 2π, as illustrated in Figure 15 by curve A. The action for such *rotational tori* is given by

$$J = \int_0^{2\pi} I(\theta)\, d\theta \qquad (5.6)$$

Such curves are an analog of the invariant tori as seen in surfaces of section for two-freedom Hamiltonian systems. As the rotational tori are single-valued functions of θ, we denote them as $I(\theta)$. In contrast, the tori associated with island chains (see, e.g., curve B, Figure 15) are multivalued and thus are not rotational tori.

Perturbation theory may be used to approximate rotational tori: Assuming existence of a continuous curve $I(\theta)$, where $I(\theta = \theta_n) \equiv I_n$, the standard map is equivalent to the functional equation

$$I(\theta + I(\theta) + k \sin \theta) = I(\theta) + k \sin \theta \tag{5.7}$$

which may be solved perturbatively,

$$I(\theta) = J + \sum_{i=1}^{\infty} k^i h_i(\theta) \tag{5.8}$$

by enforcement of the boundary conditions

$$\int_0^{2\pi} h_i(\theta)\, d\theta = 0 \qquad i = 1, 2, 3, \dots \tag{5.9}$$

Substitution, followed by iteration gives

$$h_1(\theta) = -\frac{\cos(\theta - J/2)}{2 \sin(J/2)} \tag{5.10a}$$

$$h_2(\theta) = -\frac{\cos(2\theta)}{16 \sin(J) \sin^2(J/2)} \tag{5.10b}$$

$$h_3(\theta) = -\frac{\cos(J)}{128 \sin(J) \sin^3(J/2)} \left[\frac{3\cos(3\theta)}{\sin(3J/2)} + \frac{\cos(\theta)}{\sin(J/2)} \right] \tag{5.10c}$$

$$h_4(\theta) = \cdots \tag{5.10d}$$

suggesting, correctly, that denominators contain factors of the form $\sin(pJ/q)$. These perturbative results give a useful reference, assuming their validity at small k, at least as an asymptotic approximation.

Can we regain the results of Eqs. 5.10 via adiabatic turning on of the coupling, thus verifying, at least formally, the use of adiabatic methods for point maps (and thus implicitly for more general two-freedom Hamiltonians dynamics) and perhaps additionally revealing the role of the switching process in the rate of convergence of the method? In carrying out such an analysis, we will be restricted to small k and to values of J such that

$$J/\pi \neq s/r \tag{5.11}$$

for small integers r, s. That is, we will neglect high-order island chains; the effect of these will be discussed in the following section.

We start from the fact that any switching function $s(x)$ defined on a discrete set of points $x_n = n/N$, $n = 0, 1, \ldots, N - 1$, can be expressed as

$$s(x_n) = \sum_{m=1}^{N} a_m \left(\frac{n}{N} \right)^m, \tag{5.12}$$

where

$$\sum_{m=1}^{N} a_m = 1 \tag{5.13}$$

To calculate I_N in 5.2a to the sth order, it is sufficient to have θ_n to the $(s-1)$th order. To order 0 one has

$$\theta_n^{(0)} = \theta + nJ \tag{5.14}$$

By substituting 5.14 into 5.2a and using 5.12 and 5.13, we obtain I_N to first order:

$$I_N^{(1)} = J + \sum_{n=0}^{N-1} ks(x_n) \sin(\theta + nJ)$$

$$= J + \sum_{m=1}^{N} k \frac{a_m}{N^m} \frac{\partial^m}{\partial J^m} \sum_{n=0}^{N-1} \sin\left(\theta + nJ - \frac{m\pi}{2} \right)$$

$$= J + \sum_{n=1}^{N} k \frac{a_m}{N^m} \frac{\partial^m}{\partial J^m} \left[\frac{\cos(\theta + NJ - J/2 - m\pi/2) - \cos(\theta - J/2 - m\pi/2)}{2 \sin(J/2)} \right]$$

giving, finally,

$$I^1(\theta) = J - \frac{k}{2 \sin(J/2)} \cos\left(\theta - \frac{J}{2} \right) + \Delta_N^{(1)}(\theta) \tag{5.15}$$

where the error $\Delta_N^{(1)}(\theta) \to 0$ as $N \to \infty$. Comparison of 5.15 with 5.10a shows formal adiabatic invariance to first order in k.

Going to the second order in k, we shall restrict ourselves, for simplicity, to the case of the linear switching function $\lambda(x) = x$:

$$\theta_N^{(1)} = \theta + \sum_{m=1}^{n} I_m = \theta + nJ + \sum_{m=1}^{n} \sum_{r=0}^{m-1} k \frac{r}{N} \sin(\theta + rJ)$$

$$= \theta + nJ - \frac{kn}{4N \sin^2(J/2)} [\sin(\theta + nJ) + \sin\theta]$$

$$- \frac{k}{4N \sin^3(J/2)} \left[\cos(\theta + nJ) + \frac{J}{2} - \cos\left(\theta + \frac{J}{2}\right) \right] \quad (5.16)$$

Substituting and expressing θ as a function of θ_N, we obtain, after considerable algebra,

$$I^2(\theta) = J - k\frac{\cos(\theta - J/2)}{2\sin(J/2)} - k^2 \frac{\cos(2\theta)}{16 \sin(J) \sin^2(J/2)} + \Delta_N^{(2)}(\theta) \quad (5.17)$$

Here $\Delta_N^{(2)}(\theta)$ is a complicated expression again satisfying $\Delta_N^{(2)}(\theta) \to \theta$ as $N \to \infty$. Comparison with 5.10b again shows formal adiabatic invariance to second order in k.

Perhaps not surprisingly, perturbative analysis of the adiabatic switching process yields results consistent with direct perturbative analysis. However, inspection of the leading order correction $\Delta_N^{(1)}(\theta)$ gives considerable insight into the effect of the choice of the functional form of the switching function on the rate of convergence.

C. Effect of Switching Function on Rate of Convergence: Perturbative Analysis

The switching function $s(x)$ turns the perturbation on as x runs from 0 to 1. As the initial conditions are given in terms of the $s(x) = 0$ Hamiltonian (or map), it is implicitly assumed that $s(x) = 0$ for $X \leqslant 0$. Thus, if $s(x)$ is an analytic function of x (say, a polynomial) on $(0, 1)$, it cannot be analytic at $x = 0$. Similarly, if we assume that $s(x) = 1$ for $x \geqslant 1$, $s(x)$ cannot be analytic at $x = 1$. Intuitively, as we wish to introduce as few transients as possible at the beginning and end of the switching process, $s(x)$ should be as smooth as possible at $x = 0$ and $x = 1$. This is in accord with the results of Johnson (see the discussion of Section II, Figure 1, and the discussion of Ref. 32) for conservation of classical action in one-degree-of-freedom systems. It is also consistent with known results for fully quantum-mechanical treatment of nondegenerate eigenvalues: Sancho (119) has shown that if $s(x)$ has l continuous derivatives at 0 and 1 and is otherwise analytic in $(0, 1)$, the error due to a finite switching time T goes as

$T^{-(l+1)}$. More generally, if $s(x)$ is infinitely differentiable at $x = 0$ and $x = 1$ (although not analytic at these points), asymptotic convergence will be "faster" than any algebraic function of T. This latter behavior will thus be expected for a switching function such as

$$s(x) = \exp\left[-\frac{a}{x}\exp\left(-\frac{b}{x-1} \right) \right] \qquad (5.18)$$

which has essential singularities at $x = 0$ and $x = 1$ and is C^∞ at these points of nonanalyticity.

Dana and Reinhardt (71) have shown perturbatively that precisely the same results apply in the adiabatic switching of area-preserving point maps and thus are expected to hold for the classical adiabatic switching of general Hamiltonian systems, unless, of course, the switching takes the system into a region of strong enough resonant behavior to cause perturbative estimates to fail. Following their analysis, we assume that $s(x)$ has all derivatives continuous in the interval $(0, 1)$ and that the first l derivatives vanish at the endpoints $x = 0, 1$,

$$\left.\frac{\partial^r s(x)}{\partial x^r}\right|_{x=0,1} = 0 \qquad r = 1,\ldots,l \qquad (5.19)$$

and further that

$$\left.\frac{\partial^{l+1} s(x)}{\partial x^{l+1}}\right|_{x=1} = (-1)\left.\frac{\partial^{l+1} s(x)}{\partial x^{l+1}}\right|_{x=0} \equiv a \neq 0 \qquad (5.20)$$

This latter symmetry condition, satisfied by any switching function such that

$$s(1 - x) = 1 - s(x)$$

leads to fairly simple closed-form results but is not essential to the analysis. Focusing the discussion on the standard map, we have, from Eq. 5.15,

$$\Delta_N^1(\theta) - \frac{k\cos(\theta - J/2)}{2\sin(J/2)} = \sum_{n=0}^{N-1} A\left(\frac{n}{N}\right) \qquad (5.21a)$$

where

$$A(x) \equiv s(x)\sin(\theta + NJx) \qquad (5.21b)$$

Using the Euler–Maclaurin expansion formula and noting that $s(0) = 0$, we obtain

$$
\sum_{n=0}^{N-1} A(x) = N \int_0^1 A(x)\,dx + \sum_{r=L}^{\infty} \left(\frac{1}{N}\right)^{2r-1} \frac{1}{(2r)!}
$$

$$
\times B_{2r}\left[\frac{\partial^{2r-1} A(x)}{\partial x^{2r-1}}\bigg|_{x=1} - \frac{\partial^{2r-1} A(x)}{\partial x^{2r-1}}\bigg|_{x=0}\right] + \frac{\sin(\theta)}{2}
$$

$$
-\sum_{r=1}^{\infty} \frac{(-1)^r}{(2r)!} B_{2r} J^{2r-1} \cos(\theta) \tag{5.22}
$$

Here L is the smallest integer satisfying $2L - 1 > l$, and B_{2r} are Bernoulli numbers. The integral in Eq. 5.22 can be evaluated by integrating m times by parts,

$$
N \int_0^1 A(x)\,dx = -k\frac{\cos(\theta)}{J} + \frac{2k\sin(NJ/2 + l\pi/2)}{J} \frac{}{(NJ)^{l+1}}
$$

$$
\times \cos(\theta - NJ/2) + O(N^{-l-2}) \tag{5.23}
$$

Combining these gives the estimate

$$
\Delta J(N) \equiv \left\{\frac{1}{2\pi} \int_0^{2\pi} [\Delta_N(\theta)]^2\,d\theta\right\}^{1/2} \tag{5.24a}
$$

$$
= k \sum_{n>l} \frac{\beta_n(N, J)}{(NJ)^n} \tag{5.24b}
$$

where $\beta_n(N, J)$ are generally proportional to oscillatory functions in N. In particular, the leading order nonzero contribution is

$$
\frac{\beta_{l+1}(N, J)}{N^{l+1}} \propto \frac{|\sin(NJ/2 + l\pi/2)|}{N^{l+1}}
$$

In simple cases $\Delta J(N)$ can be explicity evaluated. For example, the rms error due to the finite switching time for the simple linear ($l = 0$) switching function

$$
s(x) = x \tag{5.25a}
$$

is found to be

$$
\Delta J(N) = \frac{k}{\sqrt{2\pi}\,4\sin^2(J/2)} \frac{|\sin(NJ/2)|}{N} \tag{5.25b}
$$

while, for the switching function

$$s(x) = x - \frac{\sin(2\pi x)}{2\pi} \qquad (5.26a)$$

($l = 2$), we obtain, after a lengthy calculation,

$$\Delta J(N) = \frac{\pi^2 k}{\sqrt{2\pi} \, 16 \sin^4(J/2)} \frac{|\sin(NJ/2)|}{N^3} + O(N^{-5}) \qquad (5.26b)$$

Note the N^{-3} decay law in 5.26 is equivalent to that empirically found in the data of Figure 11, where the continuous version of the same switching function was employed.

The general rule is that l continuous derivatives at $x = 0$ and $x = 1$ give a power law convergence of the form $N^{-(l+1)}$, which is the same as the quantum result (119).

That the oscillatory factors of Eq. 5.26b are in fact observable is easily demonstrated. An interesting example of this concerns switching with an initial action corresponding to Golden mean winding number. In the standard map the winding number

$$\omega = \frac{1}{2\pi} \lim_{n \to \infty} \frac{\theta_n - \theta_0}{n} \qquad (5.27)$$

takes the place of the frequency ratio w_2/w_1 in a general two-freedom Hamiltonian system. The Golden mean is that irrational number with the most slowly convergent continued fraction representation of the form (see, e.g., Ref.)

$$a_0 + \cfrac{1}{a_1 + \cfrac{1}{a_2 + \cfrac{1}{a_3 + \cdots}}} \qquad (5.28)$$

with a_0 a positive or negative integer and $a_1, \ldots, a_m$ positive integers.

All numbers on the real line have a unique representation in this form, and truncations of the continued fraction give successive upper and lower rational approximants p_i/q_i to the real number in question, with the subscript denoting the level at which the continued fraction is truncated. These bounding approximants are the best obtainable in the sense that if in a rational approximation to α

$$\left|\frac{r}{s} - \alpha\right| < \left|\frac{p_i}{q_i} - \alpha\right|$$

then

$$s \geqslant q_{i+1}$$

it being taken for granted that the integers p/q in such ratios have no common factors.

The Golden mean may be defined as

$$\omega^{au} = \cfrac{1}{1 + \cfrac{1}{1 + \cfrac{1}{1 + \cdots}}} \tag{5.29a}$$

and thus,

$$w^{au} = \frac{1}{1 + w^{au}} \tag{5.29b}$$

$$w^{au} = \frac{\sqrt{5} - 1}{2} \tag{5.29c}$$

The Golden mean rational approximants are giving the explicit expression given by ratios of the famous Fibonacci numbers $1, 1, 2, 3, 5, 8, 13, 21, \ldots$, where

$$n_i = n_{i-1} + n_{i-2}$$

as is easily verified by induction, the key step being

$$\frac{p}{q} = \frac{1}{1 + n_{i-2}/n_{i-1}} = \frac{n_{i-1}}{n_{i-1} + n_{i-2}} = \frac{n_{i-1}}{n_i} \tag{5.30}$$

We may now ask, what are the optimal values of N for adiabatic switching via the C^2 function $s(x) = x - \sin(2\pi x)/2\pi$ for $J/2\pi = \omega^{au}$, which is the initial action corresponding to the Golden winding number? Equation 5.26b suggests an overall N^{-3} convergence, but with oscillations occurring for values of N such that for any integral m,

$$NJ/2 = N2\pi\omega^{au}/2 \approx m\pi \qquad (5.31)$$

Rewriting this as

$$\omega^{au} \approx m/N \qquad (5.32)$$

gives the immediate result that if m/N is a good rational approximant to ω^{au}, the factor $|\sin(JN/2)|$ will be minimized. The continued-fraction representation of ω^{au} then implies that the optimal N's are Fibonacci numbers. An illustration is given in Figure 17 for $K = 0.001$, where these numbers-theoretic expectations may be confirmed by inspection.

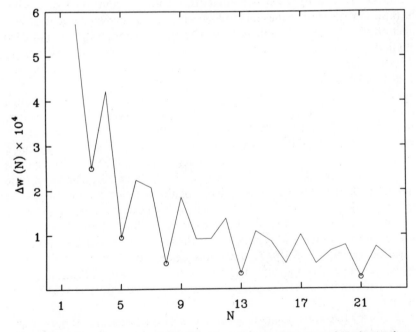

Figure 17. The rms error in the Golden mean torus as a function of N, the number of iterations of the standard map during which the perturbation was turned on. In this case the nonlinearity is quite small, and the minima in the error occur at the Fibonacci numbers 3, 5, 8, 13, 21,... as predicted by the perturbative analysis leading to Eq. 5.26b. (Figure reproduced from Ref. 71 with permission.)

D.   Switching through Resonance Zones: The Farey Analysis

What lies beyond the perturbative analysis of Section V.B? It is clear from the results of Eq. 5.10 that as one proceeds to higher order, the winding number must be kept away from rationals, r/s, of higher and higher order. As k increases and higher order terms are necessary to give an adequate description of the rotational tori, this avoidance of rationals becomes at first difficult and then impossible: The rationals are dense on the real line and cannot be avoided forever as the integers r and s increase. The result is that the rotational tori with rational winding number break up into resonance island chains surrounded by zones of chaos, as illustrated in Figure 15. Eventually, even the rotational tori with irrational winding number break up and do so in the order of the ease with which the winding number is rationally approximated. As shown by Greene (120), beyond a critical value of k ($k_{crit} = 0.97163540631\ldots$) no rotational tori exist, the last such torus being that with Golden mean winding number, that is, the winding number most difficult to approximate by rationals, as discussed in Section V.B.

Suppose we attempt to switch adiabatically in to regions of large k: Since the rationals are dense in the reals and the winding number will, in general, be a function of k, it is not possible to keep away from rational winding numbers and thus from the resonances discussed in Section III. However, the dynamics of the standard map is well enough understood (120, 121) to allow an analysis of the breakdown of adiabaticity: Thus, we can predict the specific resonances that will be most intrusive and determine how quickly we must switch through them to avoid being trapped; see also Ref. 67. The result is a balance of the type familiar in asymptotic expansions: We must switch slowly to avoid transients and take advantage of the decrease in error with increasing switching times; but in order to avoid resonances, we should switch quickly! A detailed analysis has been given by Dana and Reinhardt in Refs. 70 and 71. The key to the analysis is to determine the dependence of the winding number w on the coupling and action. Once $w(k, I)$ is known (it may be determined by adiabatic switching!), an analysis of rational approximations due to Farey (see Refs. 72 and 73) allows systematic generation of those rationals that lie within the range of variation of $w(k, I)$ and thus allows determination of which resonances might be important. The residue analysis of Refs. 120 and 121 then allows determination of the strength and time scale associated with each resonance.

The Farey tree appropriate to w ranging between approximating rationals p/q and p'/q' (common factors always assumed absent in such ratios) is easily understood if these rational ratios are neighbors, that is, if $pq' - p'q = 1$. Assuming such neighbors bracket the range of variation of w, a tree (see Figure 18) of rational approximants is constructed by use of Farey means: $p/q + p'/q' = (p + p')/(q + q')$. These means systematically fill the interval of

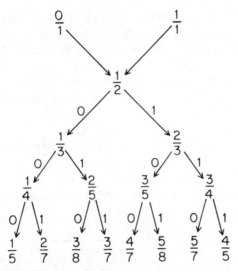

Figure 18. Construction of a Farey tree for the interval $[0, 1]$. The interval is equivalent to that "between" the fractions $0/1$ and $1/1$, allowing construction of Farey mediants via the seemingly naive rule for adding fractions: $\{r/s + p/q\} = (r + p)/(s + q)$. The Farey addition provides mediants (i.e., the result always lies in between the original fractions), provides a systematic and unique decomposition of the real line in that all rationals in the interval occur once and only once in the tree, and of great use for the present purpose, the denominators of the reduced ratios increase slowly, allowing identification of dominant resonances in nonlinear dynamics.

variation of w with rationals in order of increasing denominators, that is, in order of lowest order resonances first. As high-order resonances, which correspond to periodic orbits with long period, will have a smaller effect than low-order ones, the Farey tree provides a systematic listing of just those resonances that are apt to cause the most difficulty.

With the Farey analysis providing the list of possible resonances due to the winding number passing through rational values p/q as the switching proceeds, the optimal switching time may be determined. This optimal time, chosen to be as long as possible but not too long, may be defined as the minimum of N^q for all integers q appearing as denominators in the Farey decomposition. Here N^q measures the influence of an island chain of order q in that for times longer than N^q a trajectory near the island chain will be trapped by it. To leading order $(70, 71)$,

$$N^q = k^{-1}s'(n/N)k^{-q} \qquad (5.33)$$

This result has a simple physical interpretation: If at step n the winding number is at or near a rational of form p/q, the effect of the qth-order resonance must be taken into account. The factor $s'(n/N)$, the derivative of the switching function as the resonance is encountered, is a measure of the time spent at a particular winding number: If the derivative is large, the resonance is passed quickly and does little damage, thus allowing relatively longer switching times. The factor k^{-q} arises from the residue analysis of Greene (120) and MacKay, Meiss, and Percival (121):

$$k^{-q} \approx \frac{\text{diffusion time around island of order } p/q}{\text{width of } p/q \text{ island chain}} \qquad (5.34)$$

Thus, if the diffusion time around an island is large and the width of the island

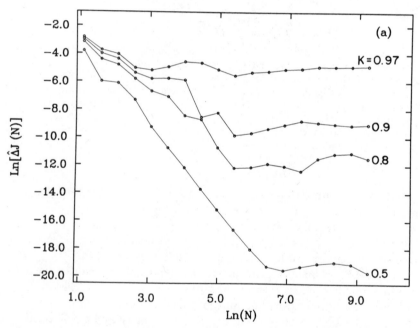

Figure 19. Failure of adiabatic switching for large k for the standard map. Shown is rms deviation of the action (see Eq. 5.24a) for adiabatic switching to $k = 0.5, 0.8, 0.9, 0.97$, the latter being close to the "critical" value beyond which no rotational tori exist. This latter value is certainly beyond the perturbative regime. As adiabaticity fails, the error ceases to decrease as the switching time increases. For the case illustrated here, the switching process was begun with the unperturbed torus with Golden mean winding number. As the winding number changes with increasing k, resonances are encountered, and a combination of the Farey and residue analyses allows specific predictions of the critical switching times, that is, the times beyond which a given resonance dominates the intrinsic nonadiabaticity. (Figure from Ref. 71 with permission.)

is small, a "long" time may be spent with the winding number at or near p/q without the resonance being felt. Mathematically, Eq. 5.33 effectively summarizes the simple physical argument given at the beginning of Section V: For small k and large q (i.e., for high-order resonances), k^{-q} will be enormous and will allow very long switching times. These dependences are illustrated in Figure. 19.

The result of Eq. 5.33 not only justifies the use of adiabatic methods in the presence of resonances but also allows specific predictions to be made. An example is shown in Figure 20, where switching beginning at initial winding number $w = (3 - \sqrt{5})/2$ was carried out to a final k of 0.8, well beyond the perturbative regime. The coupling was then switced (adiabatically) back to zero (while continuing iteration in the positive direction, i.e., n increasing), and the resulting rotational torus was examined and compared to the initial torus. If this on-off switching were adiabatic, the final torus would be the same as the initial: The Farey analysis predicted that an island chain of order $p/q = \frac{11}{29}$

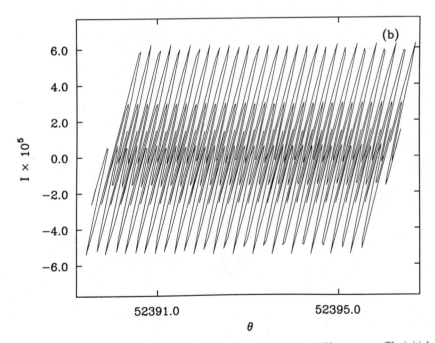

Figure 20. Signature of failure of adiabatic switching due to an 11/29 resonance. The initial winding numbers was $(3 - \sqrt{5})/2$. Switching was carried out to a value of $k = 0.8$ and then back to 0. The Farey analysis of Refs. 70 and 71 predicted that the dominant resonance in the appropriate winding number regime was the 11/29, which is indeed observed (at the part in 10^5 level) at the end of the process.

would dominate the nonadiabaticity, and inspection of Figure 20 shows 29 peaks in the difference $I^{\text{switched}} - I^{\text{initial}}$ as a function of θ.

The analysis of adiabaticity of the local action once a resonance zone is entered appears in the work of Cary et al. (67).

E. Adiabatically Generated Pseudoinvariants in Regions of Large-Scale Chaos

What about switching into regions of chaos, where there are no final tori? The results indicated in Figure 14 seem to show that semiclassical quantization is not strongly affected by the lack of a final torus. But to what final structure, if any, does the method converge?

Preliminary studies (42, 122) indicated that the result of adiabatic switching into regions of bounded separatrix chaos is successive approximations to the unstable manifold and that the homoclinic oscillations associated with that manifold are easily seen. Figure 21 shows a sequence of curves obtained for the Siegel–Henon map by adiabatically switching into the region of separatrix

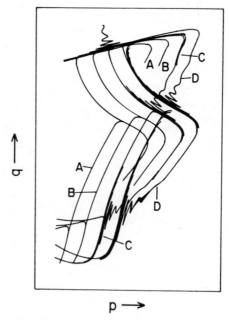

Figure 21. Curves obtained from adiabatic switching into the region of separatrix chaos (see Figure 16) of the Siegel–Henon map of Eq. 5.3. The curves A–D were obtained by increase of the switching time from ~ 600 to ~ 2000 time units. The curves are displaced for clarity, as they overlay almost precisely, all converging to the unstable manifold (see Ref. 122). (Figure from Ref. 42 with permission.)

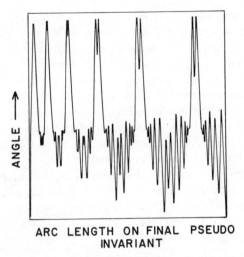

Figure 22.   Fine structure of curve D of Figure 21 displayed as a function of arc length along the final pseudoinvariant curve, which approaches the unstable manifold. It is evident that a quite regular structure has been encountered. So much for the idea that chaos is "irregular"! The Curve is evidently a fat fractal (123) as further time evolution increases the level of complexity on all scales simultaneously.

chaos shown in Figure 16. Curves $A–D$ result from successively longer switching times: 600, 601, 1200, and 2000 iterations of the map. The curves overlay almost exactly, except at the tips of the tendrils, where rapid growth takes place on finer and finer scales. This is illustrated in Figure 22, where curve D is unfolded by plotting the angle (with respect to $p = 0$, $q = 0$) as a function of arc length along the curve. An approximate self-similar structure appears, as oscillations associated with one homoclinic point encounter another. As this curve further time evolves, the structure becomes more complex at all scale levels simultaneously, leading to a fractal dimensionality of 2 and necessitating a generalization of the "fat"-fractal concept of Farmer and co-workers (Ref. 123).

VI.   SUMMARY AND DISCUSSION

Much has been accomplished to date. The efforts of many workers have led to many applications of adiabatic techniques for multidimensional semiclassical quantization. The method appears to be as promising as any currently available as a practical tool for application to multidimensional systems. The simultaneous development of powerful algebraic tools to resolve multidimensional zero-order resonances will allow application to be made to realistic systems. The theoretical understanding of why the method works at all has also improved; however, the problem of semiclassical quantization of strong global chaos (as opposed to bounded separatrix chaos) has not been solved, and it is not known how to apply adiabatic techniques in that case. Perhaps it is not possible! It is also amusing to note that all of the adiabatic switching discussed here relates to use of an artificially introduced switching, simply as a tool for carrying out semiclassical quantization. The whole area of interaction of atoms or molecules with adiabatically time-varying external fields has yet to be approached semiclassically. Jensen (124) has noted that in using classical mechanics to attempt to understand chaotic behavior in the microwave ionization of Rydberg atoms (125), the nature of the state preparation process plays a key role, adiabatic turning on of the microwave field leading to a greater role of classical resonance islands. It would thus seem natural to begin to use the techniques discussed here as an aid to understanding physically as well as mathematically adiabatic processes.

Acknowledgments

The author is most grateful to the many workers who have communicated their unpublished work and thoughts relating to adiabatic switching, in particular, E. A. Solov'ev, B. R. Johnson, C. Jaffé, C. Patterson, R. Shirts, D. Farrelly, G. Ezra, and H. S. Taylor. J. O. Hirschfelder pointed out Ref. 69, and his former student Bruce Johnson pointed out Ref. 119 to the present author. Ms. Carol Carr of the University of Pennsylvania Chemistry Library provided invaluable assistance in obtaining copies of many "early" papers. An even greater debt is owed to my collaborators R. T. Skodje, F. Borondo, I. Dana, and most recently R. Waterland. Without the encouragement and patience of J. O. Hirschfelder the present review would not have seen the light of day. Finally, Grants CHE84-16459 and DMR85-19059 from the National Science Foundation to the University of Pennsylvania have made possible the work reported here and are most gratefully acknowledged. The hospitality of the Telluride Summer Research Center, where the final draft of the manuscript was completed, is also acknowledged.

References

1. W. Eastes and R. A. Marcus, *J. Chem. Phys.* **61**, 4301 (1974).

2. S. L. Chapman, B. R. Garrett, and W. H. Miller *J. Chem. Phys.* **64**, 502 (1976).

3. I. C. Percival, *Adv. Chem. Phys.* **36**, 1 (1977).

4. D. W. Noid, M. L. Koszykowski, and R. A. Marcus, *Ann. Rev. Phys. Chem.* **32**, 267 (1981).

5. W. H. Miller, *Adv. Chem. Phys.* **25**, 69 (1974).

6. W. H. Miller, *Adv. Chem. Phys.* **30**, 77 (1975).

7. M. S. Child, *Molecular Collision Theory*, Academic, New York, 1974.

8. R. N. Porter and L. N. Raff, in *Dynamics of Molecular Collisions*, W. H. Miller, ed., Plenum, New York, 1976, Part B, p. 1.

9. R. P. Feynman and A. R. Hibbs, *Quantum Mechanics and Path Integrals*, McGraw-Hill, New York, 1965.

10. L. S. Schulman, *Techniques and Applications of Path Integration*, Wiley-Interscience, New York, 1981.

11. E. J. Heller, *J. Chem. Phys.* **62**, 1544 (1975).

12. H. L. Dai, R. W. Field, and J. L. Kinsey, *J. Chem. Phys.* **82**, 1606, 2161 (1985).

13. K. V. Reddy, D. F. Heller, and M. J. Berry, *J. Chem. Phys.* **76**, 2814 (1982).

14. H. Petek, D. J. Nesbit, D. C. Darwin, and C. B. Moore, *J. Chem. Phys.* **86**, 1172, 1189 (1987).

15. J. E. Baggott, M.-C. Chuang, R. N. Zare, H. R. Dubal, and M. Quack, *J. Chem. Phys.* **82**, 1186 (1985).

16. I. C. Percival, *J. Phys. B* **6**, L229 (1973).

17. M. V. Berry, *J. Phys. A* **10**, 2083 (1977).

18. M. V. Berry, in *Topics in Nonlinear Dynamics, a Tribute to Sir Edward Bullard*, S. Jorna, ed., AIP Conference Proceedings, Vol 46, American Institute of Physics, New York, 1978, p. 16.

19. M. V. Berry and M. Robnik, *J. Phys. A* **17**, 2413 (1984).

20. M. Gutzwiller, *J. Math. Phys.* **12**, 343 (1971).

21. M. Gutzwiller, *Physica D* **5**, 316 (1982).

22. E. J. Heller, *Phys. Rev. Lett.* **53**, 1515 (1984).

23. W. P. Reinhardt, in *Mathematical Analysis of Physical Systems*, R. E. Mickens, ed., Van Nostrand Reinhold, New York, 1985, p. 169.

24. R. V. Jensen and S. M. Susskind, in *Proceedings of the Workshop on Photons and Continuum States of Atoms and Molecules, Cortnona*, 1986, Springer, New York, 1987.

25. J. N. Bardsley, B. Sundaram, L. A. Pinnaduwage, and J. E. Bayfield, *Phys. Rev. Lett.* **56**, 1007 (1986).

26. R. E. Prange, D. R. Gempel, and S. Fishman, in *Chaotic Behavior in Quantum Systems*, G. Casati, ed., Plenum, New York, 1985, p. 205.

27. E. A. Solov'ev *Sov. Phys. JETP* **48**, 635 (1978).

28. T. P. Grozdanov and E. A. Solov'ev, *J. Phys. B* **15**, 1195 (1982).

29. B. R. Johnson, aerospace report, unpublished, 1983.

30. R. M. Hedges, Jr., R. T. Skodje, F. Borondo, and W. P. Reinhardt, *ACS Symp. Ser.* **263**, 323 (1984).

31. R. T. Skodje, F. Borondo, and W. P. Reinhardt, *J. Chem. Phys.* **82**, 4611 (1985).

32. B. R. Johnson, *J. Chem. Phys.* **83**, 1204 (1985).

33. C. W. Patterson, *J. Chem. Phys.* **83**, 4618 (1985).

34. T. P. Grozdanov, S. Saini, and H. S. Taylor, *Phys. Rev. A* **33**, 55 (1986).

35. T. P. Grozdanov, S. Saini, and H. S. Taylor, *J. Chem. Phys.* **84**, 3243 (1986).

36. T. P. Grozdanov, S. Saini, and H. S. Taylor, *J. Phys. A* **19**, 691 (1986).

37. C. W. Patterson, R. S. Smith, and R. B. Shirts, *J. Chem. Phys.* **85**, 7241 (1986).

38. J. W. Zwanziger, E. R. Grant, and G. Ezra, *J. Chem. Phys.* **85**, 2089 (1986).

39. C. Jaffe, *J. Chem. Phys.* **85**, 2885 (1986).

40. B. R. Johnson, *J. Chem. Phys.* **86**, 1445 (1987).

41. G. Ezra, C. C. Martens, and L. E. Fried, *J. Phys. Chem.* **91**, 3721 (1987).

42. W. P. Reinhardt and I. Dana, *Proc. Roy. Soc. (Lond.) A* **413**, 157 (1987).

43. P. Ehrenfest, *Proc. Amsterdam Acad.* **16**, 591 (1914).

44. P. Ehrenfest, *Proc. Amsterdam Acad.* **19**, 576 (1917).

45. P. Ehrenfest, *Philos. Mag.* **33**, 500 (1917); reprinted in *Sources of Quantum Mechanics*, B. L. Van Der Waerden, ed., Dover, New York, 1967, p. 79.

46. P. Ehrenfest, *Naturwiss.* **11**, 543 (1923).

47. A. Einstein, *Verh. d. D. Phys. Ges.* **16**, 826 (1914).

48. J. M. Burgers, *Verslag van de Gewone Vergaderingen de Wis-en Natuurkundige Afdeeling Konieklijke Akedemie va Wetenschappente Amsterdam* **25**, 848 (1916).

49. J. M. Burgers, *Verslag van de Gewone Vergaderingen de Wis-en Natuurkundige Afdeeling Konieklijke Akedemie va Wetenschappente Amsterdam* **25**, 918 (1916).

50. J. M. Burgers, *Verslag van de Gewone Vergaderingen de Wis-en Natuurkundige Afdeeling Konieklijke Akedemie va Wetenschappente Amsterdam* **25**, 1055 (1916).

51. T. Levi-Civita, *Hamburger Ab. Math. Sem. Univ.* **6**, 323 (1928); reprinted in Vol. 4 of T. Levi-Civita, *Collected Mathematical Works*, Vol. 4, N. Zanichelli, Bologna, 1960, p. 499.

52. T. Levi-Civita, *J. Math. Phys.* **13**, 18 (1934).

53. M. Klein, *Proc. Xth Int. Cong. Hist. Sci. Cornell, 1962*, Hermann, Paris, 1964, p. 801.

54. M. Jammer, *The Conceptual Development of Quantum Mechanics*, McGraw-Hill, New York, 1966, pp. 89–109.

55. P. A. M. Dirac, *Proc. Roy. Soc.* **107**, 725 (1925).

56. A. Einstein, *Verh. Phys. Ges.* **19**, 82 (1917).

57. L. Brillouin, *J. Phys. Radium* **7**, 353 (1926).

58. J. B. Keller, *Ann. Phys. (N.Y.)* **4**, 180 (1958).

59. R. T. Skodje and F. Borondo, *J. Chem. Phys.* **85**, 2760 (1986).

60. R. T. Skodje and F. Borondo, *J. Chem. Phys.* **84**, 1533 (1986).

61. C. Jaffe, private communication, 1987.

62. R. T. Skodje and F. Borondo, *Chem. Phys. Lett.* **118**, 409 (1985).

63. D. Bensimon and L. P. Kadanoff, *Physica D* **13**, 82 (1984).

64. R. S. MacKay, J. D. Meiss, and I. C. Percival, *Physica D* **13**, 55 (1984).

65. M. J. Davis, *J. Chem. Phys.* **83**, 1016 (1985).

66. M. V. Berry, *J. Phys. A* **17**, 1225 (1984).

67. J. R. Cary, D. F. Escande, and J. L. Tennyson, *Phys. Rev. A* **34**, 4256 (1986).

68. F. R. Johnson and P. Pechukas, *Proceedings of the First International Conference on the Physics of Phase Space*, Lecture Notes in Physics, Springer, New York, Vol 278, p 140, 1987.

69. L. Brillouin, *Arch. Rat. Mach. Anal.* **5**, 76 (1960).

70. I. Dana and W. P. Reinhardt, *Proceedings of the First International Conference on the Physics of Phase Space*, Lecture Notes in Physics, Springer, New York, Vol. 278, p. 146, 1987.

71. I. Dana and W. P. Reinhardt, *Physica D*, **28**, 115 (1987).

72. G. H. Hardy and E. M. Wright, *An Introduction to Number Theory*, 4th ed., Clarendon, Oxford, 1954.

73. S. Kim and S. Ostlund, *Phys. Rev. A* **34**, 3426 (1986).

74. W. Wien, *Wiedimannsche Ann. Phys.* **52**, 132 (1894).

75. Jammer, in Ref. 54, p. 98.

76. A. Einstein, in *La Theorie du Rayonnement et les Quanta–Rapports du Discussions de la Reunion Tenue a Bruxelles, 1911*, P. Langevin and M. de Broglie, eds., Gauthier-Villars, Paris 1912, p. 450.

77. N. Bohr, *Phil. Mag.* **26**, 1 (1913).

78. C. E. Delauney, *Theorie du Mouvement de la Lune*, Paris, 1860.

79. P. S. Epstein, *Ann. Phys.* **51**, 168 (1916).

80. I. C. Percival and D. Richards, *Introduction to Dynamics*, Cambridge University Press, Cambridge, 1982, p. 141 ff.

81. V. I. Arnold, *Mathematical Methods of Classical Mechanics*, Springer, New York, 1978, pp. 297–300.

82. M. Born, *Atomic Physics*, 7th ed. Hafner, New York, 1961, p. 115 ff.

83. P. Hertz, *Ann. Phys.* **33**, 225, 537 (1910).

84. See Ref. 81, pp. 233–238.

85. E. C. G. Sudarshan and N. Mukunda, *Classical Dynamics, a Modern Perspective*, Wiley-Interscience, New York, 1974, Chapter 6.

86. K. Schwarzchild, *Berliner Berichte*, 1916, p. 548.

87. H. Goldstein, *Classical Mechanics*, Addison-Wesley, Reading, MA, 1951.

88. H. Poincaré, Les Methods Nouvelles de la Mechanique Celeste, Gauthier-Villars, Paris, p 233 (1892).

89. V. I. Arnold and A. Avez, *Ergodic Problems in Classical Mechanics*, Benjamin, New York, 1968.

90. J. Moser, *Stable and Random Motions in Dynamical Systems*, Princeton University Press, Princeton, NJ, 1973.

91. J. Ford, in *Fundamental Problems in Statistical Mechanics*, Vol. 3, E. G. D. Cohen, ed., North-Holland, New York, 1975, p. 215.

92. R. Abraham and J. E. Marsden, *Foundations of Mechanics*, 2nd ed., Benjamin/Cummings, Reading MA, 1978.

93. A. N. Kolmogorov, 1954 Address, in Ref. 92, pp. 741–757.

94. R. H. G. Helleman, in *Fundamental Problems in Statistical Mechanics*, Vol. 5, E. G. D. Cohen, ed., North-Holland, New York, 1980, p. 165.

95. D. W. Noid, M. L. Koszykowski, and R. A. Marcus, *J. Chem. Phys.* **73**, 391 (1980).

96. R. T. Swimm and J. B. Delos, *J. Chem. Phys.* **71**, 1700 (1979).

97. C. Jaffe and W. P. Reinhardt, *J. Chem. Phys.* **77**, 5191 (1982).

98. M. Robnik, *J. Phys. A* **17**, 109 (1984).

99. G. D. Birkhoff, *Dynamical Systems*, American Mathematical Society, Providence, RI, 1979.

100. F. G. Gustavson, *Astron. J.* **71**, 670 (1966).

101. N. DeLeon and E. J. Heller, *J. Chem. Phys.* **78**, 4005 (1983).

102. C. W. Eaker, G. C. Schatz, N. DeLeon, and E. J. Heller, *J. Chem. Phys.* **81**, 5913 (1984).

103. C. C. Martens and G. S. Ezra, *J. Chem. Phys.* **86**, 279 (1987).

104. L. E. Fried and G. S. Ezra, *J. Comp. Chem.* (in press).

105. A. J. Dragt and J. M. Finn, *J. Math. Phys.* **20**, 2649 (1979).

106. D. Farrelly, *J. Chem. Phys.* **85**, 2119 (1986).

107. M. Born, V. Fock, *Zeit. f. Phys.* **51**, 165 (1928).

108. D. K. Harriss and A. A. Frost, *J. Chem. Phys.* **40**, 204 (1964).

109. E. J. Heller, *Phys. Rev. Lett.* **53**, 1515 (1984).

110. P. O'Connor, J. Gehlen, and E. J. Heller, *Phys. Rev. Lett.* **58**, 1296 (1987).

111. See, e.g., M. Spivak, *Calculus on Manifolds*, Benjamin, New York, 1965.

112. W. P. Reinhardt and R. J. Skodje, unpublished.

113. S. Saini and D. Farrelly, *Phys. Rev. A* **36**, 3556 (1987).

114. H.-D. Meyer and W. H. Miller, *J. Chem. Phys.* **70**, 3214 (1979).

115. C. Jaffe and P. Brumer, *J. Chem. Phys.* **73**, 5546 (1980).

116. E. L. Sibert III, W. P. Reinhardt, and J. T. Hynes, *J. Chem. Phys.* **77**, 3583 (1982).

117. E. L. Sibert III, J. T. Hynes, and W. P. Reinhardt, *J. Chem. Phys.* **77**, 3595 (1982).

118. K. T. Hecht, *J. Mol. Spec.* **5**, 355 (1960).

119. F. J. Sancho, *Proc. Roy. Soc.* **89**, 1 (1966).

120. J. M. Greene, *J. Math. Phys.* **20**, 1183 (1979).

121. R. S. MacKay, J. D. Meiss, and I. C. Percival, Physica D **27**, 1 (1987).

122. R. Waterland and W. P. Reinhardt, unpublished work in progress.

123. D. K. Umberger and D. Farmer, *Phys. Rev. Lett.* **56**, 661 (1985).

124. R. V. Jensen, Effects of Classical Resonances on the Chaotic Microwave Ionization of Highly Excited Hydrogen Atoms, preprint, 1986.

125. For example, K. A. H. vanLeeuwen, G. V. Oppen, S. Renwick, J. B. Bowlin, P. M. Koch, R. V. Jensen, O. Rath, D. Richards, and J. G. Leopold, *Phys. Rev. Lett.* **55**, 2231 (1985).

AUTHOR INDEX

NOTE: Page numbers in **boldface** type refer to reference pages in text.

SUBJECT INDEX